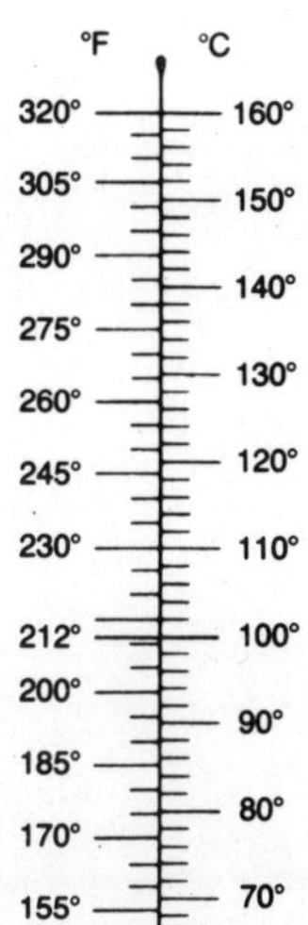

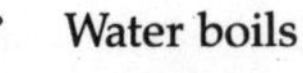

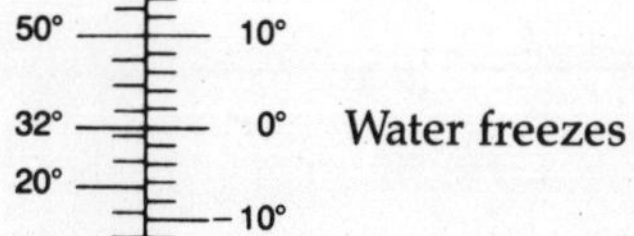

To convert Fahrenheit to centigrade, use this formula:

$$°C = \frac{5}{9}(°F - 32)$$

To convert centigrade to Fahrenheit, use this formula:

$$°F = \frac{9}{5}(°C + 32)$$

ESSENTIALS OF MICROBIOLOGY

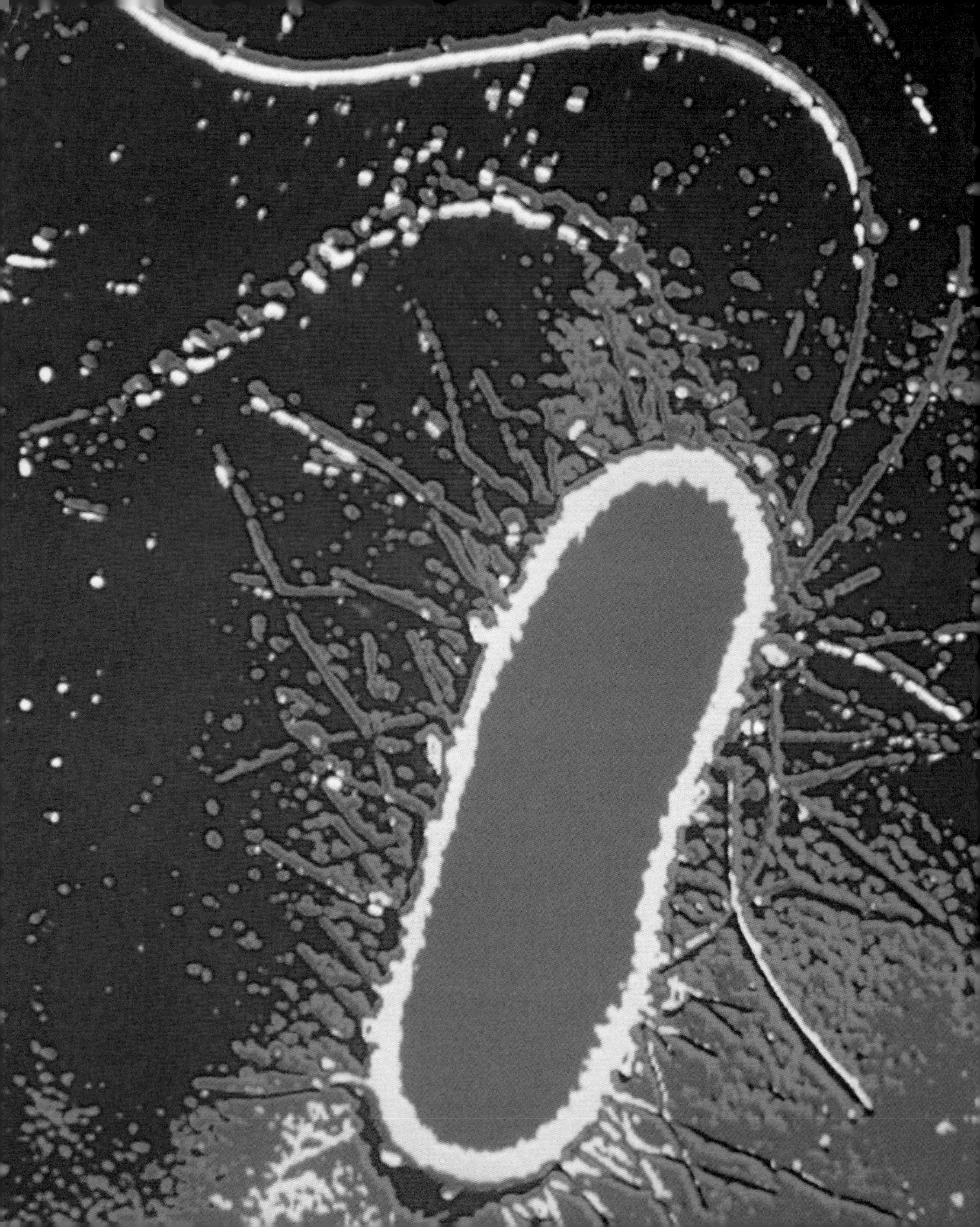

ESSENTIALS OF MICROBIOLOGY

RAÚL J. CANO
JAIME S. COLOMÉ
California Polytechnic
State University
San Luis Obispo

WEST PUBLISHING COMPANY
St. Paul New York Los Angeles San Francisco

We would like to dedicate this book to our families.

My parents Raoul and Josefina Cano, my wife Pat, and my children Monica, Sara, and Anthony.

My parents Jaime and Rosamond Colomé.

PRODUCTION CREDITS

Composition: Graphic Typesetting Service, Inc.
Interior design: Judith Fletcher Getman
Cover design: David Farr/Imagesmythe, Inc.
Cover and title page photo: © Howard Sochurek 1980/Woodfin Camp & Associates
Copyediting: Virginia Dunn
Indexing: Sonsie Carbonara Conroy
Illustrators: Cyndie Clark-Huegel, Carla Simmons, Marilyn Hill, Wayne S. Clark, Parry Clark, John Waller, Judy Waller, Darwen Hennings, Vally Hennings, Ginny Mickelson, Tom Mallon

Figure credits follow index.

A study guide has been developed to assist you in mastering concepts presented in this text. The study guide reinforces concepts by presenting them in condensed concise form. Additional illustrations and examples are also included. The study guide is available from your local bookstore under the title, *Study Guide to Accompany Essentials of Microbiology*, prepared by Michael Womack with the assistance of Dr. Geraldine Ross.

Printed in the United States of America

96 95 94 93 92 91 90 8 7 6 5 4 3 2

Library of Congress Cataloging-in-Publication Data

Cano, Raúl J.
Essentials of microbiology.

Includes index.
1. Microbiology. I. Colomé, Jaime S.
II. Title.
QR41.2.C34 1988 576 87-34601
ISBN 0-314-60534-7

CONTENTS

CHAPTER 13
THE IMMUNE RESPONSE

CHAPTER 14
DISORDERS ASSOCIATED WITH THE IMMUNE RESPONSE

CHAPTER 15
EPIDEMIOLOGY OF INFECTIOUS DISEASES

PREFACE

Since its origins a century ago, microbiology has developed from a mere curiosity into a mature and challenging science. Much has been discovered about microbial life and the tremendous impact that microorganisms have on human life. This presents a challenge for the teacher, who must present a vast amount of pertinent information to the student, make it logical, understandable, and appetizing.

Because microbiology encompasses such a wide variety of topics, it can be taught from many different points of view and to a variety of different audiences. We have designed *Essentials of Microbiology* for freshman or sophomore college students who are taking their first course in microbiology and have little or no background in the biological sciences and chemistry in their college careers. Students in majors such as nursing, allied health sciences, dietetics, food sciences, liberal arts, and health education, who only take a one-semester course in microbiology, will find *Essentials of Microbiology* admirably suited for their needs.

We have written *Essentials of Microbiology* with students in mind. Throughout the text we have strived to convey basic biological principles in a logical fashion and have emphasized the enormous impact that microorganisms have on every facet of human life. *Essentials of Microbiology* provides a balanced discussion of fundamental principles and applications and liberally uses illustrations, figures, tables, and real events to illustrate our points. First-time microbiology students will come to appreciate and be awed by the large variety of microorganisms that share our biosphere, their role in health and disease, and their impact on everyday life.

ORGANIZATION OF THE TEXTBOOK

In order to provide students with a logical and flexible approach to studying general microbiology, we have organized the textbook into four parts: "The Basic Concepts"; "A Survey of Microorganisms"; "Medical Microbiology"; and "Applied and Environmental Microbiology." Chapters within each part can be selected and organized so that the instructor can create a cohesive unit that is best suited for the intended curriculum.

PART 1: THE BASIC CONCEPTS constitutes a core of topics that should be common to all curricula. The six chapters in Part 1 cover important concepts of microbial biology and serve as a theoretical basis for the other chapters in the textbook.

PART 2: A SURVEY OF MICROORGANISMS consists of five chapters in which we discuss the various groups of microorganisms, their biology, and importance to humans. A chapter discussing the classification of microorganisms is included in this part to emphasize the relationship among the various microbial groups and to unify the chapters in Part 2. We also include in this chapter a section on methods of diagnosing infectious diseases in order to give students the basics of specimen collection and processing. This section will be particularly useful for nursing and allied health students who will be routinely collecting and processing specimens for laboratory analysis.

PART 3: MEDICAL MICROBIOLOGY consists of 11 chapters in which we discuss various aspects of medical and clinical microbiology. The first three chapters discuss host-parasite relationships and include: Determinants of Health and Disease, The Immune Response, and Disorders Associated with the Immune System. These are followed by a chapter on Epidemiology of Infectious Diseases, six chapters discussing important infectious diseases by the organ system approach, and a chapter on Chemotherapy of Infectious Diseases. The chapters dealing with the infectious diseases emphasize, not only the clinical and epidemiological features of the diseases, but also the intricate interactions that take place between the infecting microorganism and its host and the consequences of these interactions.

PART 4: APPLIED AND ENVIRONMENTAL MICROBIOLOGY consists of two chapters in which we discuss some of the important activities that microorganisms carry out in nature and how these activities can be employed to improve our quality of life. In this part we include a chapter in Biotechnology to emphasize the importance that microorganisms have in our economy and how they positively affect our lives. In this chapter we discuss, not only the more traditional methods of obtaining consumer goods, but also the exciting new field of genetic engineering.

LEARNING FEATURES

In order to be an effective learning tool, a textbook must make the subject matter understandable and motivate the student to learn the subject. This is sometimes difficult, especially when dealing with a subject like microbiology, which many first-time students think is "too hard." We have incorporated several features in *Essentials of Microbiology* that will aid students in understanding the concepts and become aware that microbiology is a pertinent and timely science.

CHAPTER PREVIEWS introduce the student to practical applications of the subject matter discussed in the chapter or significant events in microbiology. A few examples are how to control the spread of infections in hospitals, how to trace the source of an epidemic, or what scientists are doing to make human life on Mars possible. Chapter previews are aimed at illustrating that microbiology is **not** a sterile laboratory science of interest only to scientists, but a dynamic field of endeavor with practical applications and of vital importance to humans.

CLINICAL PERSPECTIVES are summaries of actual clinical cases used to illustrate the interactions that exist between the patient and his or her health provider. We have included a Clinical Perspective in each chapter of Part 3 to bring together the various aspects of medical microbiology as it applies to patients.

FOCUS BOXES introduce practical, interesting, or supplementary information to expand upon the topics discussed in the chapter. Several focus boxes are historical vignettes that serve to remind students that the information covered in class and in the text is based upon data provided by scientists and teachers like their instructors. Most deal with topics of current interest including: The Ames Test Is Used to Detect Cancer-Causing Chemicals; *Neisseria gonorrhoeae* Received Genes from Other Bacteria; Did Viruses and Bacteria Originate from Outer Space?; and An Allergic Reaction to a Vaccine. Through these focus boxes, we attempt to capture the interest of students and encourage them to read further into the chapter.

KEY TERMS in **bold face type** will help students recognize the important terms in the chapter. Some terms are printed in boldface in more than one chapter so that students will come across important terms, even if their reading assignments skip some chapters. Many of these key terms are defined also in the GLOSSARY, found at the back of the text.

TABLES in most chapters compile pertinent data to aid the student in quickly grasping large quantities of related information. In addition, the tables serve as easily accessible reference material.

LINE DRAWINGS and PHOTOGRAPHS complement the textual material. All chapters in *Essentials of Microbiology* have many line drawings rendered in two colors to aid students in understanding the subject matter. In addition, we have made extensive use of photographs and electron micrographs to illustrate important structures, diseases, or concepts. Full-color photographs have been included in chapters where they will aid students' comprehension of a concept or will significantly enhance the discussion of a topic.

The SUMMARIES are extensive and written in outline form to aid students in studying the subject matter. Each summary recaps the major sections in the chapter and provides students with an overview of most of the important concepts discussed in the chapter.

STUDY QUESTIONS and SELF-TESTS are included at the end of each chapter to help students assess their grasp of the main ideas discussed in the chapter. The study questions and self-tests frequently require students to analyze and gather information from various parts of the chapter before answering the question.

SUPPLEMENTAL READINGS provide sources of information outside the textbook to expand upon some of the key concepts discussed in the chapter. Many of the readings are recent articles published in *Scientific*

American, Hospital Practice, Discover, and other publications that write scientific topics for the layperson.

An extensive GLOSSARY AND PRONUNCIATION GUIDE has been prepared to provide easy access to the definitions of many of the important terms used throughout the text and the accepted way of pronouncing them. Pronunciations have also been provided for many of the microorganisms discussed in the text.

Essentials of Microbiology is accompanied by an instructor's manual, written by the authors; a Study Guide, written by Michael Womack at Macon Jr. College; and a laboratory manual (also accompanied by an instructor's manual), *Laboratory Exercises in Microbiology,* written by J. S. Colomé, A. M. Kubinski, R. J. Cano, and D. V. Grady at California Polytechnic State University, San Luis Obispo. These study aids should further enhance the student's understanding and appreciation of microbial life and activities.

We would like to express our appreciation to all those who helped us complete this text; without their help, our tasks would have been extremely difficult or impossible. We wish to thank our students for helping us test our ideas and approaches to teaching, and for challenging us to the limit. We also would like to express our deep gratitude to the following reviewers of our text for their insightful, although sometimes painful, comments, suggestions, and ideas.

Jerry Sipe, Anderson College
William Finnerty, Univ. of GA-Athens
Les Albin, Austin Comm. Col-Rio Grande Campus
Michael Womack, Macon Jr. College
Ronald Trudel, Southeastern Mass. Univ.
David Balkwill, Florida St. Univ.
Glendon Miller, Wichita St. Univ.
Jonathan Brosin, Sacramento City Col.
Joan Burns, Fairmont St. Co.
Timonthy Kral, Univ of Arkansas
Jay Kim, CA St. Univ-Long Beach
Ronald Siebeling, Louisiana St. Univ.
Hugh Potter, Union County Col.-Scotch Plains
Peter Gaffney, GA State Univ.
Don Niederpruem, Indiana Univ.-Medical Ctr.
Harold E. Laubach, Southeastern Col. of Osteopathic Medicine
Richard Fox, American Univ.
Allen Schark, Univ of Maine

I (rjc) would like to express my eternal gratitude to Drs. Evelio J. Perea, José Carlos Palomares, Javier Aznar, Maria Victoria Borobio, and Maria del Carmen Lozano at the University Hospital, University of Seville, Spain, for their friendship, moral support, and valuable comments.

We would like to thank our families for "putting up" with us during the preparation of this book. Those hours that we spent working on the text were hours during which we neglected them.

Finally, we would like to express our appreciation to Peter Marshall, Denise Simon, and Deanna Quinn, our editors, who provided encouragement and many excellent ideas. Their attention to detail, hard work, good taste, and organization made this beautiful book possible.

Raúl J. Cano
and
Jaime S. Colomé

PART 1

THE BASIC CONCEPTS

CHAPTER 1

AN INTRODUCTION TO MICROBIOLOGY

CHAPTER PREVIEW MICROORGANISMS AND HUMAN DISEASE

A change in behavior or living conditions may cause the incidence of common infectious diseases to increase or new diseases to appear. There are many examples throughout history. Whenever rural people congregated in cities, the incidence of diseases such as typhoid fever, hepatitis, typhus, plague, and influenza often increased because of poor sanitation, filth, malnutrition, and overcrowding. Conversely, when people improved their living conditions, the incidence of many infectious diseases decreased significantly.

When humans traveled and made contact with isolated populations, new diseases were often acquired by the travelers or given to the isolated populations. These diseases have often destroyed whole cultures. For example, in the early 1500s, when Cortez was conquering Mexico, smallpox spread from the invading army to the Indian population. It is estimated that smallpox killed 13 million Indians, of a population of 25 million, in the first year of the epidemic. Before 1490, syphilis did not exist in Europe. It was introduced by explorers or soldiers sometime in the 1490s. It is not clear exactly where the disease originated, but it is hypothesized that it may have come from Africa or from America as a mutated form of a disease that existed on these continents. Syphilis, when first introduced, was a deadly disease that ravaged Europe for many years. It rapidly spread from Europe to India, China, and Japan, and then to Africa and the American continent.

In the early 1800s, when Americans settled in the Hawaiian Islands, they introduced measles. This disease almost completely eliminated the native Hawaiians. A consequence was the immigration of Chinese and Japanese to the Hawaiian Islands.

Recently, leprosy has been on the increase in the United States because of the recent wave of immigration from Southeast Asia. There are now about 4000 lepers in the United States who have to take drugs for the rest of their lives to control the disease. Leprosy is a major problem in certain parts of the world, despite the development of antimicrobials. There are approximately 12 million lepers worldwide and 5 million of them are not being treated. The untreated lepers are crippled by the disease and generally abandoned by their families and friends.

The viruses that cause influenza are responsible for many millions of cases of the disease and over a million deaths per year worldwide. During the 1918 flu epidemic, 20 million died worldwide while 500,000 died in the United States alone. More recently, the 1981–82 influenza epidemic in the United States was responsible for over 60,000 deaths.

Until the last century, people did not understand what was responsible for infectious diseases. Many believed that angry gods or immoral acts were responsible for these afflictions. It is now known that various types of microorganisms are responsible for infectious diseases of plants, animals, and humans. Microorganisms are also involved in many natural beneficial processes and are being manipulated to benefit humankind. In this chapter you will be introduced to some of the discoveries that led us to our understanding of the roles microorganisms play in nature.

CHAPTER OUTLINE

As PEOPLE HAVE BECOME MORE AWARE OF their environment and themselves, they have discovered a world of microscopic biological forms called **microorganisms** (fig. 1.1). Microorganisms have helped to shape the world we take for granted, and they have also influenced the evolution of all plants and animals. In fact, life is completely dependent upon the activities of numerous different microorganisms. Many of these activities result in the production of life-saving antibiotics and other drugs. Scientists can now prepare vaccines and drugs that can be used to prevent and treat fatal human and animal diseases using genetically engineered microorganisms.

Although microorganisms carry out numerous chemical transformations that are essential to human existence, many people not acquainted with microorganisms believe that all are harmful and undesirable. This erroneous view is due to the fact that some microorganisms spoil foods or cause serious diseases. The notion that microorganisms are for the most part harmful is quickly dispelled once their roles in nature are understood and appreciated. In fact, the beneficial activities of microorganisms far outweigh the harmful effects of a few disease-causing species.

WHAT ARE MICROORGANISMS?

Microorganisms are a heterogeneous group of organisms (fig. 1.1) that are too small to be seen with an unaided eye. For this reason, an optical device that magnifies objects called the **microscope** is normally used to view them (see Chapter 2). Microorganisms are generally single-celled or a simple aggregations of cells. Although their sizes vary considerably (fig. 1.2), microorganisms measure less than 0.1 mm (0.1 mm = 1/10,000 of a meter or approximately 0.0039 inches) in diameter. It was because of their small size that they were not seen until the 1600s (when the first microscopes were made), although their activities have been felt since the beginning of time. Traditionally, microorganisms include bacteria, protozoa, some algae, fungi, and the viruses.

MICROBIOLOGY: A MULTIFACETED SCIENCE

Microbiology is much more than just a taxonomic study of microorganisms. It is also concerned with the roles microorganisms play in causing disease, the changes they make in the environment, and the products they generate. Microbiology encompasses such a vast area of knowledge that it is broken down into a number of specialty areas. Scientists often divide microorganisms into groups based upon shape and physiology, and restrict their study to one or more groups (fig. 1.3a). This approach to studying microorganisms, sometimes called the taxonomic approach to microbiology, divides microorganisms into viruses, bacteria, protozoa, algae, fungi, and microscopic animals. **Virology** is the study of the noncellular organisms called viruses. It is concerned with the structure and reproduction of viruses, how they cause disease, and how they can be controlled. **Bacteriology** is the study of the nonnucleated cellular organisms known as bacteria. Bacteriology deals with the uses of bacteria in industry; the role of bacteria in the ecology of the world; bacterial genetics, structure, and multiplication; the mechanisms by which bacteria cause disease; and the means by which bacteria can be controlled. **Protozoology, phycology** (algology), and **mycology** are the studies of protozoa, algae, and fungi, respectively. Although some of the organisms in these groups are not microscopic, the vast majority of them are; thus, their study is considered part of microbiology. Protozoans, parasitic animals, and arthropods of medical importance are the subject matter of **parasitology**.

Scientists often study microorganisms from a functional rather than a taxonomic point of view (fig. 1.3b). For example, the science of **microbial ecology** is concerned with the interactions that take place between microorganisms and their environment.

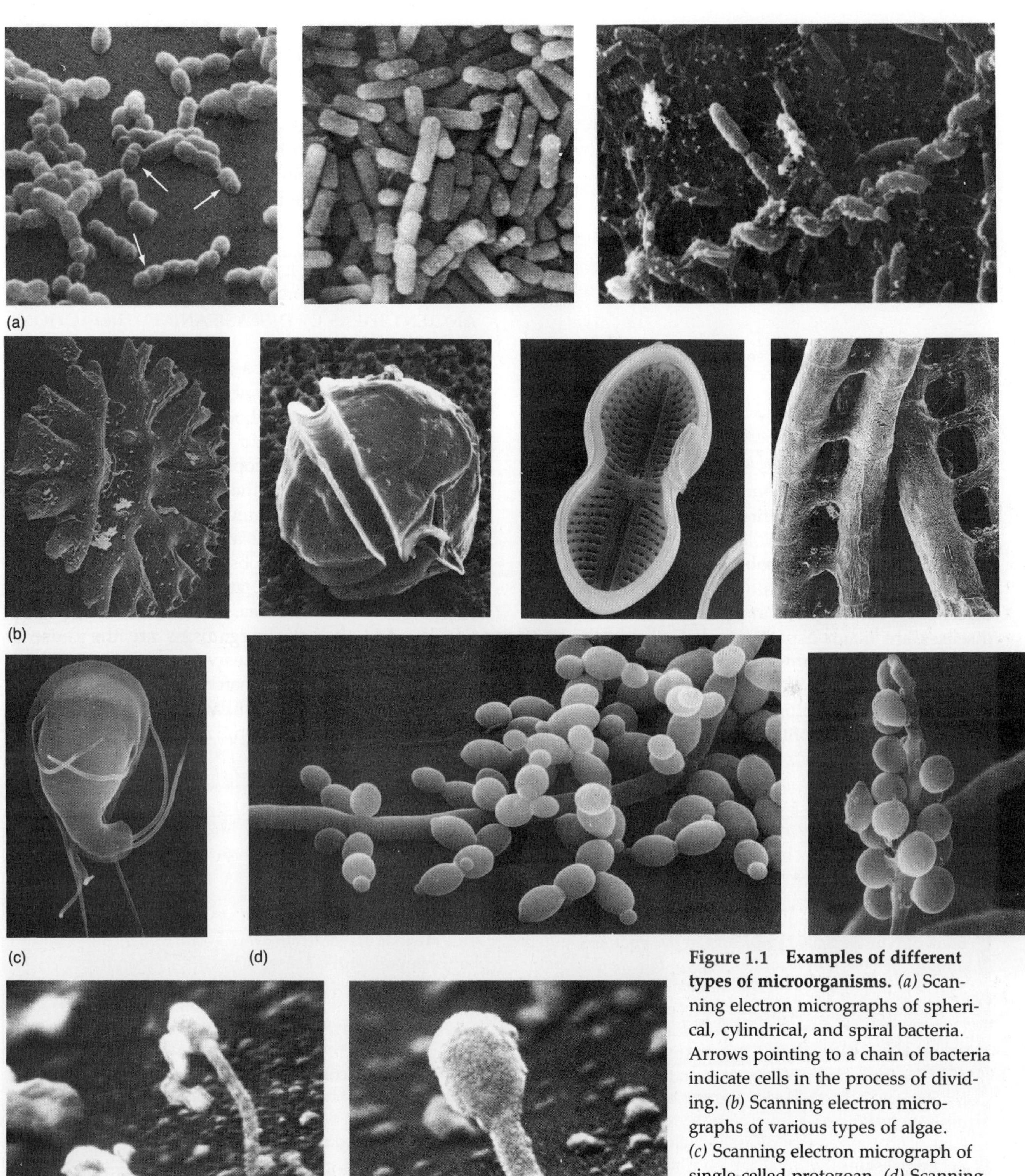

Figure 1.1 Examples of different types of microorganisms. *(a)* Scanning electron micrographs of spherical, cylindrical, and spiral bacteria. Arrows pointing to a chain of bacteria indicate cells in the process of dividing. *(b)* Scanning electron micrographs of various types of algae. *(c)* Scanning electron micrograph of single-celled protozoan. *(d)* Scanning electron micrographs of a yeast and a filamentous fungus. *(e)* Scanning electron micrographs of bacterial viruses.

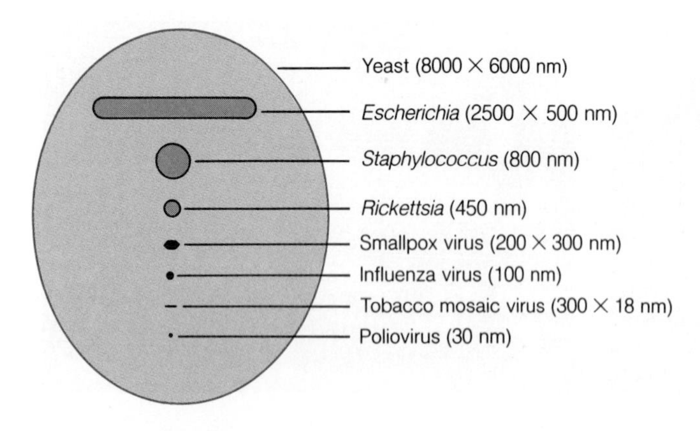

Figure 1.2 Relative size of some important microorganisms

Industrial microbiology is a study of microbial activities that are valuable to humans. The production of antibiotics, alcoholic beverages, and cheeses is carried out by industrial microbiologists. **Medical microbiology** deals with the microbial activities that focus on health and disease. The study of how organisms cause disease and the diagnosis and prevention of infectious diseases are major aspects of medical microbiology. Studies of microorganisms involved in soil fertility, livestock health, and plant disease are in the realm of **agricultural microbiology**. Since **immunology** is concerned with aspects of the immune systems that protect animals from microorganisms, it is generally considered to be part of microbiology.

Microbiology is an integral part of many fields of study, such as immunology, genetics, molecular biology, physiology, and ecology. Its central role in medicine, agriculture, industry, and genetic engineering indicates that microbiology is a major field of study and that an understanding of this field is essential to our welfare.

ANCIENT IDEAS OF DISEASE AND PUTREFACTION

Microorganisms include a myriad of free-living and parasitic forms. They are ubiquitous (everywhere): in the environment, on our bodies, in the air we breathe, in the food we eat, and in the water we drink. Throughout the ages, people have been aware of the effects of microbial activities even though they were not aware of the microorganisms as biological entities. Most people have made and consumed fermented foods and beverages, observed the spoilage and decomposition of matter, and suffered the misery of various infectious diseases.

The fact that microorganisms are the cause of spoilage and many diseases was unknown until relatively recently. Nonsupernatural explanations of these phenomena generally postulated that "poisoned air" or "seeds of decomposition," rather than unseen bio-

FOCUS A PRIMITIVE HEALTH CODE

Many of the ideas and laws of Moses found in the Book of Numbers (1000 B.C.) and in the Book of Deuteronomy (650 B.C.) represent an early public health code. The Israelites clearly realized that many diseases could be contracted from the people they subjugated. For instance, venereal diseases could be acquired from infected men and women. Moses' solution was to spare only the lives of virgin females from the conquered Medianites, hence reducing the risk of Israelites acquiring venereal diseases from these people. It must also have been realized that diseases could be transmitted on animal skins (anthrax) and on clothing (louse-infected materials can transmit typhus fever), since the codes recommended that these items be purified by washing.

The laws attributed to Moses in Deuteronomy indicate that the Israelites were aware of the connection between pork and disease. Swine are known to be infected by parasites (in particular, roundworms that burrow into the muscles) that cause very serious diseases (trichinosis). Eating undercooked pork spreads the parasites to humans. In addition, it may have been noticed that the consumption of shellfish often resulted in diseases (typhoid fever, cholera, hepatitis, polio, etc.). The easiest way to deal with the problem was to ban the food. Today we know that shellfish in sewage-polluted waters frequently accumulate bacteria and viruses that can cause disease.

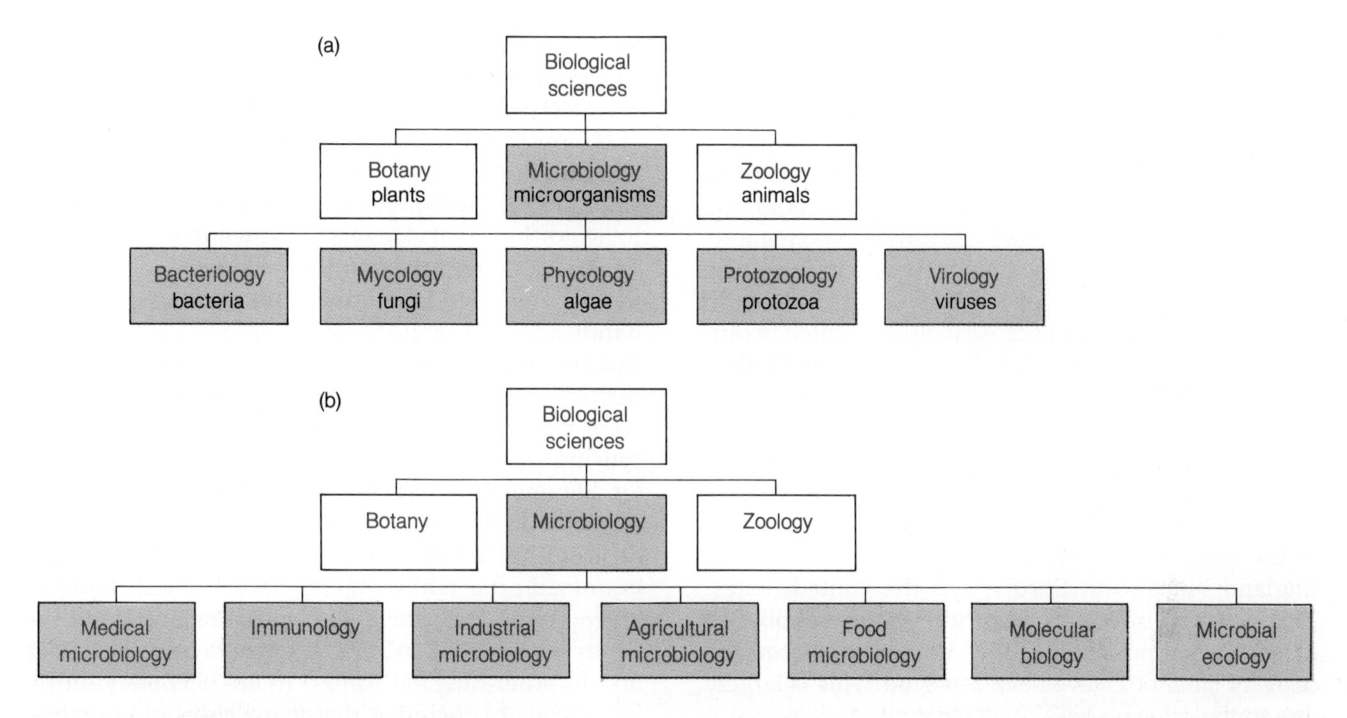

Figure 1.3 Disciplines within the field of microbiology. *(a)* Disciplines within the field of microbiology based on a taxonomic approach. *(b)* Microbiology may also be divided into fields of study based on a functional approach.

logical forms, were responsible. Diseases were also known to be contagious.

One of the oldest manuscripts on diseases was written by an Italian scientist, Girolamo Fracastoro (1478–1553). In his 1546 manuscript *About Contagion and Contagious Diseases,* Fracastoro proposed the idea that diseases were due to a *contagion* that could be transferred from one person to another. Contagions, according to Fracastoro, were transferred by direct contact; by touching contaminated inanimate objects called **fomites**, such as clothing, cups, and eating utensils; or through the air. He believed (incorrectly) that contagions were not responsible for the putrefaction of meat or the spoilage of milk, but he understood that fruit decay, as well as plant and animal infections, were due to contagions. Fracastoro's writings indicate that he considered contagions to be destructive particles whose heat, moisture, or other characteristics caused the destructive changes. He believed that foot and mouth disease was spread through the air (incorrect); that tuberculosis was transmitted through the air or on contaminated clothing (correct); that rabies was transmitted by the bite of a rabid dog (correct); and that syphilis was transmitted to children in the milk of infected mothers (possible, but it is usually transmitted through the placental barrier before the child is born) and through sexual intercourse (correct). In a manuscript published in 1530 *(Syphilus Suffering from the French Disease),* Fracastoro described the symptoms of venereal disease in a shepherd named Syphilus. Eventually, this disease—which was known as the French Disease in Italy and as the English Disease in France—became known as syphilis. Fracastoro also pointed out that many contagions attacked only certain crops or trees, while others attacked only specific animals or humans. He observed that diseases often attacked specific organs of the body. For example, tuberculosis generally affected the lungs, while trachoma damaged the eyes.

Even though most scientists were ignorant as to the actual cause of disease until the 1880s, they discovered how to deal with one of the most feared contagious diseases: smallpox. This disease killed and disfigured great numbers of individuals each year worldwide. Indian Vedic writings of about 1100 B.C. explained how to protect against smallpox. It was suggested that pus from the pox blisters should be injected into individuals wishing to be protected. A fever and a mild disease would result, but the person would become resistant to smallpox. This advice clearly indicated that smallpox was due to something in the pox blisters that could be transferred to other persons. By the 1700s, Europeans who had learned about the ancient Eastern practice of smallpox "vaccination" (known as **variolation**) were attempting to introduce this procedure into Europe. There was resistance to this practice, however, because it often resulted in outbreaks of smallpox. In 1798 Edward Jenner, an English physician, reported that material from cowpox lesions could be used to vaccinate humans without causing smallpox and that vaccinated individuals became resistant to smallpox (fig. 1.4). Even though Jenner's observations were resisted by the medical profession of the time, vaccination soon caught on, not only in England but also in Europe and the United States. This resulted in a large reduction in the number of cases of smallpox. Since 1976, no naturally acquired cases of smallpox have been reported. This is largely due to the effectiveness of vaccination.

In the 1800s, **puerperal sepsis**, better known as childbed fever, often killed as many as 15% of the women giving birth in some hospital wards. Childbed fever is due to a blood infection that can occur during childbirth, when many blood vessels are ruptured and microorganisms can enter the mother's bloodstream. The frequency of infection increases if improperly sterilized instruments are brought into contact with the damaged tissue or if midwives, nurses, or doctors introduce microorganisms into the wound.

An Austrian physician, Ignaz Semmelweis, noticed that one of the wards in the hospital where he worked had four times as many deaths due to puerperal sepsis as did other wards, and that expectant mothers tried to avoid this ward. He was also aware of the fact that doctors attending the ward often examined expectant mothers or aided in childbirth without washing their hands after doing autopsies. Semmelweis concluded that the doctors were picking up "putrefying organic matter" on their hands from corpses and that they were transmitting this matter to the mothers, thus causing puerperal sepsis. To alleviate this situation, Dr. Semmelweis required that all hospital aides wash their hands carefully in a solution of lime chloride to destroy the "putrefying organic matter" before they examined expectant mothers or women who had just given birth. This preventive measure decreased the death rate from 8.3 to 2.3%, the rate in private practice and in other hospital wards. In an 1850 manuscript, Semmelweis concluded that many cases of puerperal fever were due to something on instruments or on the hands of student obstetricians and physicians, and that the medical staff was responsible for infecting the patients. Most physicians refused to accept this idea and did not bother to test its validity. Instead, the physicians reviled Semmelweis and were instrumental in having him dismissed from his position at the hospital where he worked. It was another thirty years

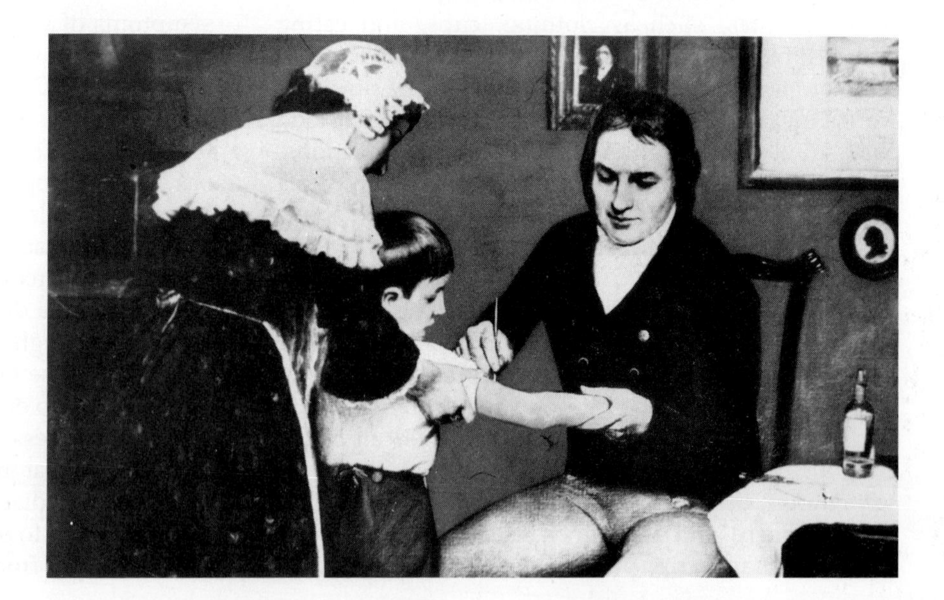

Figure 1.4 Edward Jenner. A painting illustrating Edward Jenner (1749–1823) administering the cowpox vaccine.

before physicians realized that Dr. Semmelweis was indeed correct in his assessment of the source of infection in childbed fever.

THE DISCOVERY OF MICROORGANISMS

Our present understanding of the microbial world and the causes of infectious diseases has depended upon the development of the microscope. This tool allows scientists to view many microorganisms that cannot be seen by the unaided eye. It would have been very difficult to go beyond Girolamo Fracastoro's observations on contagions if microorganisms could not have been visualized in some manner. Although a number of scientists were using primitive microscopes in the 1600s, microorganisms were first described by Antony van Leeuwenhoek (1632–1723), a Dutch dry goods merchant and minor city official who built simple microscopes as a hobby (fig. 1.5a). His microscopes resembled magnifying glasses (fig. 1.5b). They consisted of a single, tiny lens sandwiched between two thin metal plates. The object of interest was placed on one side of the lens, on the point of a needle, or inside a fine glass tube. The object was brought into focus by turning a screw that moved the object farther from or closer to the lens. The eye was positioned on the other side of the lens. Van Leeuwenhoek's simple microscope was capable of magnifying objects approximately 275 times, enough to see protozoans, algae, fungi, and many of the larger bacteria clearly. For many years he reported his microscopic observations of water, pepper infusions, dental plaque, blood, and semen in papers in the *Philosophical Transactions of the Royal Society of London*. In these papers he meticulously described many different types of bacteria, fungi, protozoans, algae, blood cells, and sperm (fig. 1.5c). Van Leeuwenhoek called the microorganisms he discovered "**animalcules**," because many of them were motile and darted around rapidly like tiny animals in the drops of liquid where they were found. Van Leeuwenhoek provided the groundwork for future microbiologists, for he introduced the scientific com-

(a)

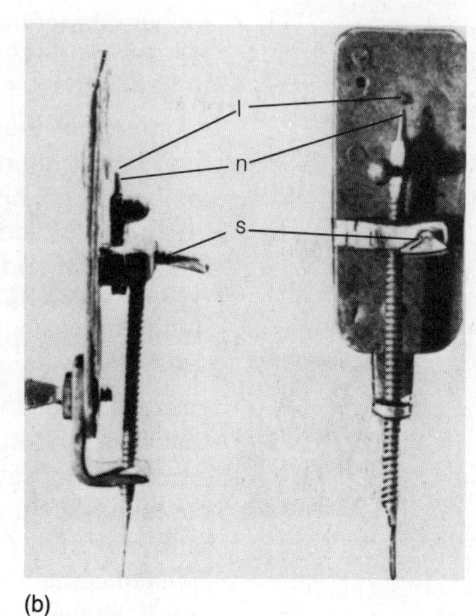

(b)

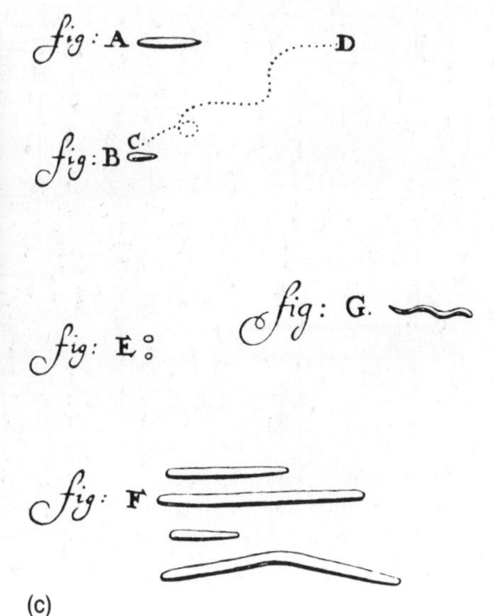

(c)

Figure 1.5 Antony van Leeuwenhoek and his microscope. *(a)* A portrait of Antony van Leeuwenhoek (1632–1723). *(b)* Front and side views of a hand-held microscope similar to that used by van Leeuwenhoek. The sample was placed at the tip of the needle (n) and brought into focus by turning the screw (s). The sample was viewed through the lens (l). The light from a candle or from a bright window was used to illuminate the specimen. *(c)* Antony van Leeuwenhoek's drawings of "animalcules" from the human mouth. The "animalcules" in these drawings are large bacteria. In B, the movement of a motile organism is illustrated.

munity to a previously unseen world and showed that animalcules were found nearly everywhere they were sought. The idea that microbes are intimately involved in our everyday lives, and that they are the cause of putrefaction and disease, was not seriously considered for another 150 years.

Figure 1.6 Francesco Redi's experiment disproving spontaneous generation of maggots. Francesco Redi (1626–1697), in an experiment similar to the one illustrated, demonstrated that maggots (fly larvae) do not arise spontaneously from meat or fish but from eggs laid by flies. The meat that is open to the flies develops maggots in 2–3 days *(c)*, but the meat that is covered with cheesecloth *(b)* or glass *(a)* does not develop maggots. Maggots appear on the cheesecloth covering the meat because the flies lay their eggs in response to the smell of the meat.

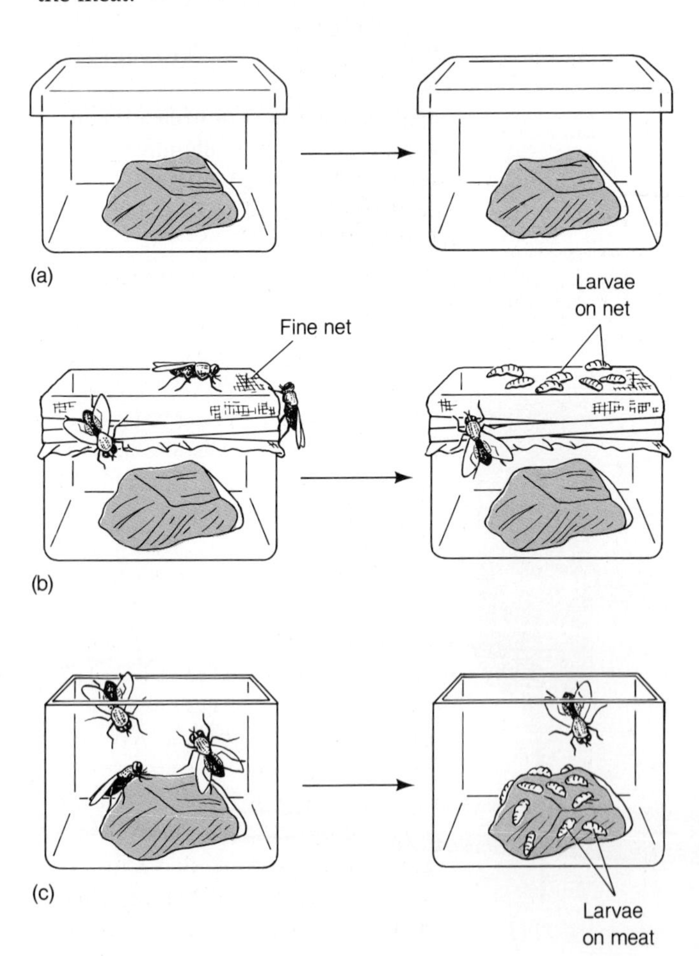

THE SPONTANEOUS GENERATION CONTROVERSY

The idea that life forms could arise from nonliving materials was held by many scientists and philosophers until the late 1860s. Many people believed that horse hairs in stagnant water could give rise to snakes and that rotting flesh would breed maggots. One of the most outlandish ideas was a recipe for making mice from soiled undergarments and wheat, written by J. B. van Helmont (1577–1644), a prominent chemist and philosopher:

> *If a dirty undergarment is squeezed into the mouth of a vessel containing wheat, within a few days (say 21) a ferment drained from the garments and transformed by the smell of the grain, encrusts the wheat itself with its own skin and turns it into mice. And what is more remarkable, the mice from the grain and undergarments are neither weanlings or sucklings nor premature but they jump out fully formed.*

Van Helmont's observations on how to make mice is an indication of the state of biology in the 1600s. Much opinion and pure fabrication substituted for experimentation and basic facts. During the next 250 years, a number of scientists showed convincingly that the spontaneous generation hypothesis was untenable. Francesco Redi (1626–1697), a physician and poet, was one of the first to challenge seriously the idea of spontaneous generation of large organisms by showing that rotting meat and fish did not breed maggots. In about 1665, he carried out an experiment that demonstrated the fact that maggots develop from fly eggs laid on meat and do not arise spontaneously from the meat (fig. 1.6). In his experiment, he placed meat inside each of three containers. One of the containers was covered with a glass top, another was covered with gauze, and the third was left uncovered. Redi observed that flies landed on the uncovered meat and on the gauze. His observations showed that the meat in the container left uncovered eventually became infested with maggots, while the meat in the glass- and gauze-covered containers remained free of maggots, although maggots were found on the gauze. From these observations, Redi correctly reasoned that the odor of rotting meat attracts flies, which then lay their eggs on the meat or as close to it as possible. The eggs hatch within a few days, giving rise to maggots that consume the meat or crawl around on the gauze.

The idea that microorganisms could arise spontaneously from nonliving matter was challenged by a number of scientists, including Lazzaro Spallanzani (1729–1799) and Theodor Schwann (1810–1882). In their experiments they demonstrated that vegetable and meat juices, called infusions, did not give rise to microorganisms if the infusions were heated to kill all organisms and were not subsequently contaminated by organisms in the air. Spallanzani prevented contamination by sealing the infusions in flasks, while Schwann heated a bent or coiled tube that opened the flasks to the air (fig. 1.7). Their experiments showed that air was laden with microbes and capable of contaminating infusions. If microorganisms present in the air could not come into contact with the infusions, or were killed by heating the air before it entered the infusions, the infusions remained free of life. Although these experiments provided evidence that even microorganisms do not arise spontaneously, proponents of the theory of spontaneous generation (of whom John Needham was one of the most active) still found fault with them. When spontaneous generation did not occur in heated infusions, proponents argued that the heating process necessary to sterilize the air also destroyed a "vegetative force" essential for the spontaneous development of microorganisms. Support for the spontaneous generation of microorganisms came from many eminent scientists in the early 1800s, because they often found microorganisms within infusions when they repeated Spallanzani's and Schwann's experiments. In retrospect, it is possible to under-

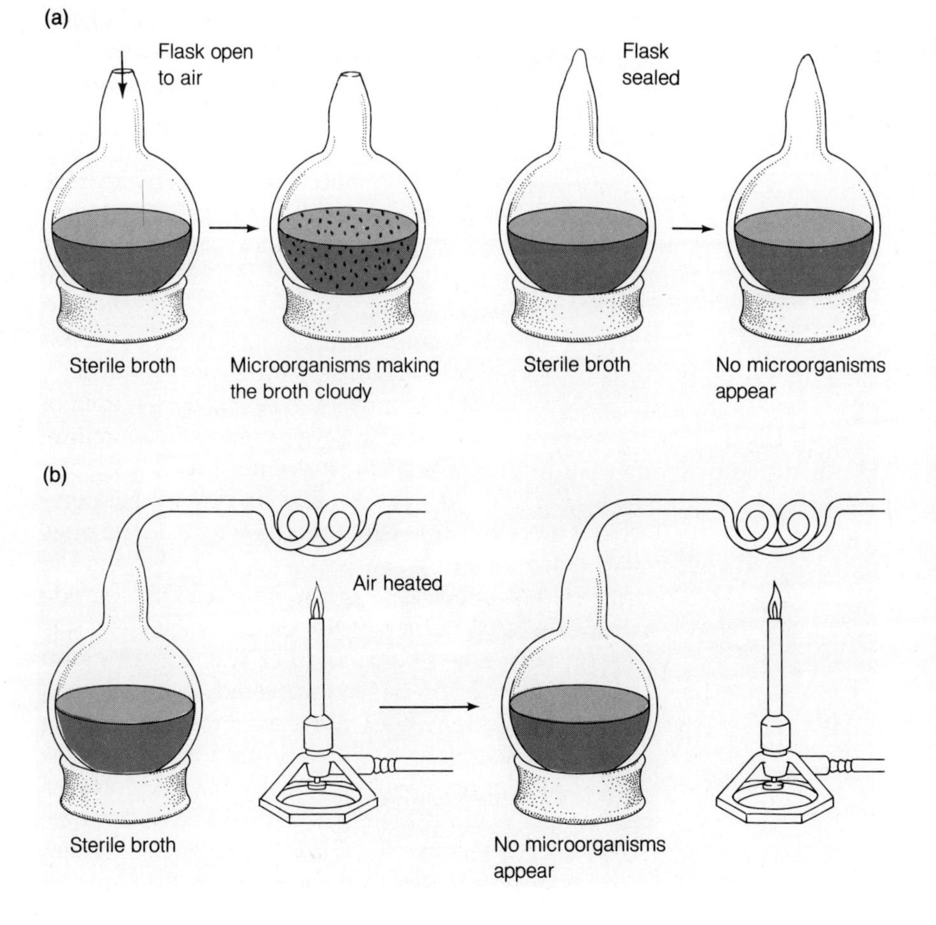

Figure 1.7 Experiments by Spallanzani and Schwann, aimed at disproving that microorganisms arise by spontaneous generation. *(a)* Spallanzani (1729–1799) heated flasks containing nutrient infusions and sealed the hot flasks. These flasks showed no evidence of growth after an extended incubation period. Once the seal was broken and air entered into the flasks, microbial growth was evident within a day. Spallanzani concluded that microorganisms did not arise spontaneously but came from contaminating air. Proponents of the spontaneous generation theory argued that the spontaneous generation of microorganisms required fresh air. *(b)* Like Spallanzani, Schwann (1810–1882) heated flasks containing nutrient infusions. Schwann, however, allowed heated air to reenter the flasks. The unsealed flasks showed no evidence of growth no matter how long he allowed heated air to enter the flask. Schwann concluded that microorganisms in the air were killed by the heating and this was why there was no growth in the infusion. Proponents of the spontaneous generation theory argued that Schwann was destroying a "vital force" when he heated the air entering the flask.

stand why heated and sealed infusions sometimes showed growth: some microorganisms are extremely resistant to heat and are not always killed by boiling for short periods of time. The idea that microorganisms could develop by spontaneous generation died slowly and was accepted on faith by many scientists, possibly because they wanted a nonsupernatural explanation for the origin of life.

Louis Pasteur (1822–1895) showed convincingly that certain groups of microorganisms are responsible for specific types of metabolism (fermentations) and that microorganisms do not arise spontaneously within infusions (fig. 1.8). Pasteur hypothesized that the appearance of microbes within infusions was due to contaminating organisms from the air. To test his hypothesis, Pasteur filtered large volumes of air through "clean" guncotton to catch any microorganisms that might be in the air. He then dissolved the guncotton in a mixture of alcohol and ether and collected the freed microorganisms that precipitated in the solvent. He showed that microorganisms were in the air by comparing the number of organisms on "clean" guncotton to that on "used" guncotton under a microscope.

The most convincing evidence against the idea of spontaneous generation was provided by Pasteur in 1861 when he showed that heated infusions open to the air did not become contaminated. Pasteur placed infusions such as sugared yeast water, sugar beet juice, or pepper water into round-bottomed flasks. He heated the openings of the flasks and then drew them out so that the flasks had thin, curved necks (fig. 1.9). Pasteur then boiled the infusions to kill all the microorganisms already present. The flasks were then allowed to stand at room temperature for many days, and although the infusions were exposed directly to unheated air—which, according to many who believed in spontaneous generation, contained the "vegetative force"—no growth occurred in the flasks. Pasteur reasoned that the flasks remained sterile because bacteria in the air settled out along the curved necks of the flasks before they could reach the infusions. To see if microorganisms were really present in the necks, he tipped several of the flasks so that the infusions flowed into their necks and then back into the flasks. When he did this, the infusions became turbid within a few days. Pasteur's experiments left little doubt that the spontaneous generation hypothesis was untenable and that living organisms that appeared in infusions originated from contaminating organisms in the infusion or in the air.

By 1857, Pasteur had established that alcoholic fermentations were carried out by yeasts and that lac-

Figure 1.8 Louis Pasteur. Portrait of Louis Pasteur (1822–1895) in his laboratory, checking infusions for evidence of microbial growth.

tic acid fermentations were due to much smaller, spherical organisms (now known as the lactic acid bacteria). In 1861, one of Pasteur's manuscripts indicated that butyric acid fermentations were due to small rod-shaped organisms that required an oxygen-free environment. Pasteur's experiments disproving the spontaneous generation hypothesis, and those showing that particular microorganisms were responsible for specific fermentations, finally convinced many scientists that microorganisms did not arise spontaneously and that they could be the cause of fermentations, putrefaction, and disease.

THE GERM THEORY OF DISEASE

Until about 1880 there were many erroneous ideas regarding the cause of infectious diseases. Most people believe that infectious diseases were not caused by organisms, but were a punishment sent by the gods. Even today, there are many people who believe that sexually transmitted diseases are God's punishment for sexual transgressions. Those who had nonsupernatural explanations to account for infectious diseases invariably blamed nonliving entities, such as "contagions" (Fracastoro, 1546), "morbid matter" (Jenner, 1798), or "putrefying organic materials" (Semmelweis, 1850).

The first disease-causing microorganisms discovered were fungi. In 1807, Bénédict Prévost observed the germination of wheat-bunt spores and proposed that the fungus caused the plant disease. Agostino Bassi in 1835 showed that a fungus was causing a disease (muscardine) of silkworms, and in 1839 J. Schoenlein and David Gruby demonstrated that fungi were responsible for a chronic ringworm (such as athlete's foot). At about the same time, B. Lagenbeck showed that yeasts were the cause of thrush, an infection of the mouth. In 1845 M. J. Berkeley characterized the fungus that causes late blight of potato. Anton De Bary's 1853 manuscripts described fungi as the cause of plant smuts and rusts. In addition, De Bary described the life cycle of the fungus that causes late blight of potato. Despite this evidence, however, it was not until Louis Pasteur and Robert Koch published their observations that the role of microorganisms in causing disease became generally accepted. These two men could be called the fathers of the **germ theory of disease**, which postulates that infectious diseases are caused by microorganisms.

Louis Pasteur contributed considerable evidence to support the idea that microorganisms were responsible for disease. For instance, he developed vaccines that protected humans and domestic animals from diseases such as swine erysipelas, chicken cholera, anthrax, and rabies. He made the vaccines by selecting organisms that were no longer able to cause disease. The nonvirulent organisms, when injected into

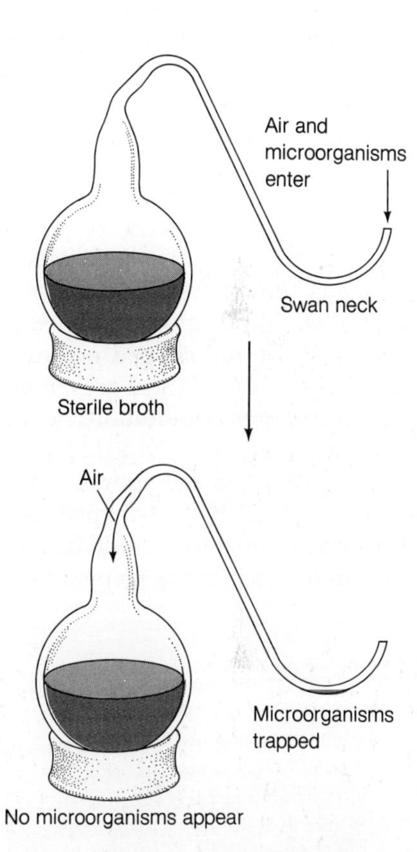

Figure 1.9 Louis Pasteur's "swan-necked" flasks. Pasteur heated flasks containing nutrient infusions and then drew out the necks of the flasks so that microorganisms would have difficulty reaching the infusions. Many of the flasks were not sealed, so as to allow fresh, cool air to enter the flasks. The open flasks showed no evidence of growth unless the infusion was tipped into the neck area, where microorganisms were trapped. These experiments showed that contaminating microorganisms not spontaneous generation, were responsible for growth in the flasks.

an animal or human, did not cause the disease but made the animal or human immune or resistant to it. The successful development of these vaccines was very strong circumstantial evidence that these microorganisms caused the diseases.

Robert Koch (1843–1910) provided irrefutable evidence linking specific microorganisms with infectious diseases (fig. 1.10). His first series of experiments demonstrated that anthrax was caused by a bacterium. Anthrax is a devastating disease that killed large numbers of livestock and inflicted severe financial losses on farmers in Europe during Koch's time. In his initial studies (1876), Koch noticed the presence of rod-shaped structures in the spleen and blood of animals killed by anthrax but not in the spleen or blood of normal animals. Consequently, he guessed that these rods might be the cause of anthrax. He transferred some of the rods to a drop of sterile beef serum or aqueous humor from the eye of a cow and noticed that they grew and multiplied in the media. He observed the rod-shaped organisms with his microscope and watched them grow into long filaments and produce spores. In addition, he found that the spores germinated into rod-shaped bacteria similar to the ones that gave rise to the spores. This observation demonstrated that the rod-shaped structures were living bacteria and that they could be grown outside the living animal for many generations. By growing the microorganisms in fresh media a number of times, he was diluting out any "contagion" or "putrefying organic materials" that might be the cause of the disease. Koch then injected the bacteria into mice, which subsequently died of anthrax. When the diseased animals were examined, the bacteria that had been injected were found in large numbers in the animals' blood. These studies provided the first experimental evidence that a specific bacterium caused an infectious disease.

Koch is also credited with proving, in 1882–1884, that tuberculosis is caused by a bacterium. He removed infected mouse spleen or tubercles (the primary lung lesion of tuberculosis) from the lungs of guinea pigs and spread the material over the surface of jelled cow or sheep blood in a tube. It took approximately 10 days at 37°C for very small colonies of bacteria to appear on the denatured blood. Koch was able to show by a special staining procedure that the colonies were made up of one type of bacterium. Thus, he was dealing with a pure culture of bacteria. When the pure culture was injected into mice, the animals developed tuberculosis, and the bacteria could be isolated once more in pure culture.

To prove that an organism was the cause or **etiological agent** of a disease, Koch applied a specific series of objective criteria that eventually became known as **Koch's postulates** (fig. 1.11). This series of criteria generally must be satisfied if an etiological agent is to be accepted as the cause of an infectious disease.

Figure 1.10 Robert Koch (1843–1910)

Koch's Postulates

1. The specific microorganism thought to be the cause of a disease must be consistently isolated from individuals suffering from the disease, but not from healthy individuals.
2. The suspected etiologic agent of the disease must be cultivated in pure form outside the host, *in vitro*.
3. Pure cultures of the suspected pathogen, when introduced into a suitable and susceptible host, must produce the signs and symptoms characteristic of the disease.
4. The same organisms must be consistently isolated in pure culture from the afflicted experimental host and be cultivated, again, *in vitro*.

Koch's postulates cannot always be followed in establishing the cause of a disease, because in some

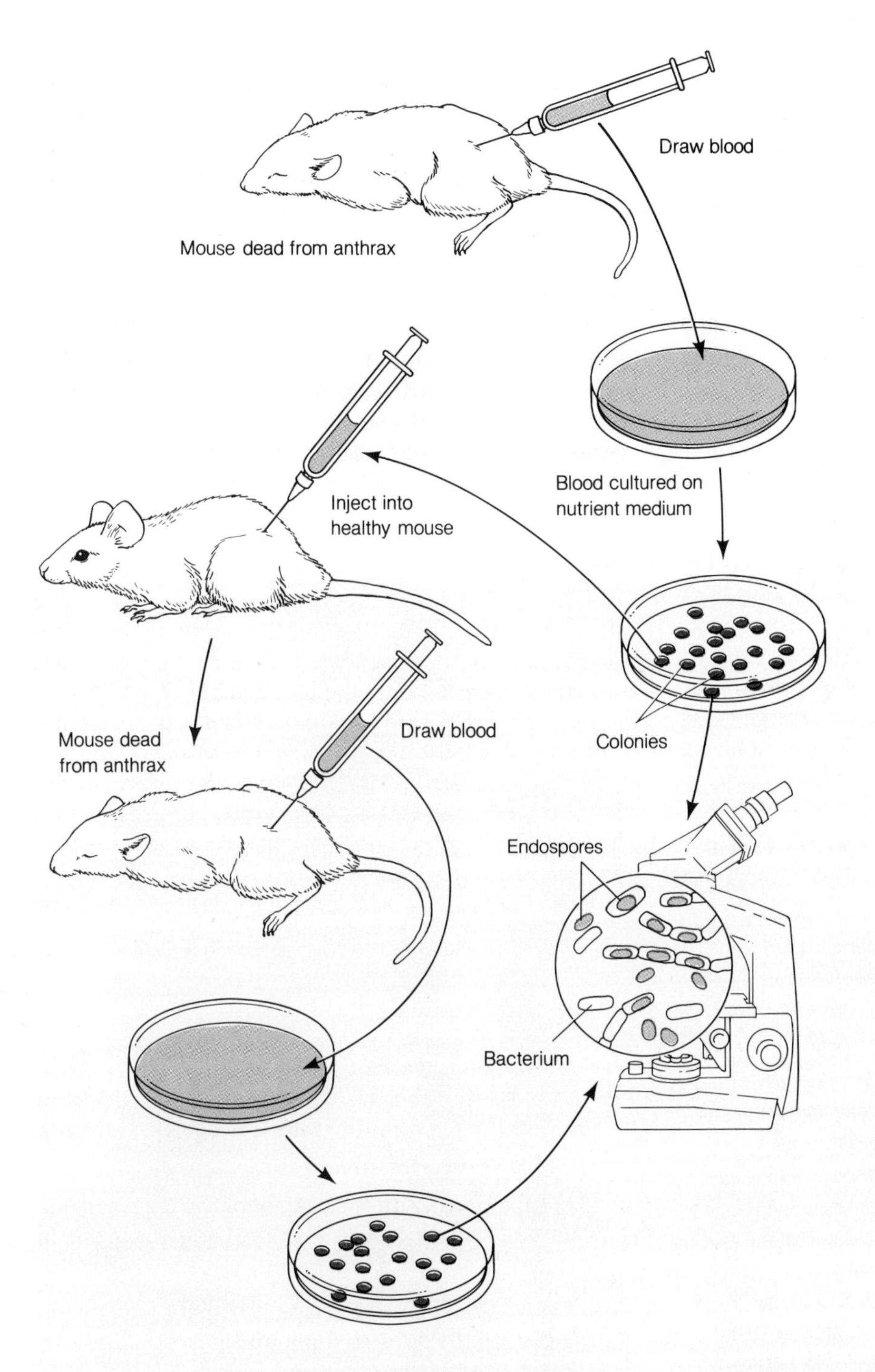

Figure 1.11 Illustration of Koch's postulates. Koch's postulates are a set of rules that are followed to prove that an organism is responsible for a disease. The organism must be isolated and observed in each case of the disease. In this illustration, a bacterium is isolated on a solid nutrient from the blood of a dead animal. The organism can be observed with the aid of the microscope and characterized. A pure culture of the organism is injected into healthy animals and then isolated and observed in each case of the disease. The organism isolated initially should be identical to the organisms isolated from the inoculated animals.

cases it is impossible to find the etiological agent in the diseased organism. For example, some viruses that cause cancer do not behave as typical viruses, but simply exist as a piece of genetic material (DNA) integrated into the host's genetic information. Also, microorganisms that cause diseases such as food poisoning by producing a toxin are not present during the disease. In many cases it is not possible to cultivate the organism outside the host. For instance, the bacteria that cause syphilis and leprosy and certain viruses that cause disease cannot be grown *in vitro* and so pure cultures are difficult to prove. Finally, many diseases are caused by organisms that disappear from the host, sometimes long before the onset of the disease; for example, the viruses that cause an animal's immune system to turn upon itself.

SIGNIFICANT DISCOVERIES IN MICROBIOLOGY

Louis Pasteur and Robert Koch were instrumental in establishing some of the roles microorganisms play in nature, but more importantly, they stimulated a systematic study of microbes by many other scientists. By showing that vaccines could be made that afforded protection against a number of diseases, and that some infectious diseases were caused by specific microorganisms, Pasteur and Koch promoted the discovery of new vaccines and of other pathogens. Table 1.1 illustrates the rapid discovery of over 25 pathogens by numerous scientists over a period of about 40 years (1875–1915), sometimes called the **golden age of microbiology**.

Technology

Robert Koch, his students, and his colleagues, in particular, are credited with the development of techniques that made the study of microbiology relatively simple. For example, stained smears of bacteria on

TABLE 1.1
INFECTIOUS AGENTS DISCOVERED IN THE GOLDEN AGE OF MICROBIOLOGY

Disease	Infectious agent*	Year	Discoverer(s)
Pear fire blight	*Erwinia amylovora*	1877	Burrill
Anthrax	*Bacillus anthracis*	1877	Koch
Gonorrhea	*Neisseria gonorrhoeae*	1879	Neisser
Malaria	*Plasmodium malariae*	1880	Laverans
Wound infections	*Staphylococcus aureus*	1881	Ogston
Tuberculosis	*Mycobacterium tuberculosis*	1882	Koch
Erysipelas	*Streptococcus pyogenes*	1882	Fehleisen
Diphtheria	*Corynebacterium diphtheriae*	1883	Klebs & Loeffler
Cholera	*Vibrio cholerae*	1883	Koch
Typhoid fever	*Salmonella typhi*	1884	Eberth & Gaffky
Bladder infections	*Escherichia coli*	1885	Escherich
Smallpox	Smallpox Virus	1887	Buist
Food poisoning	*Salmonella enteritidis*	1888	Gaertner
Tetanus	*Clostridium tetani*	1889	Kitasato
Gas gangrene	*Clostridium perfringens*	1892	Welch & Nuttall
Plague	*Yersinia pestis*	1894	Yersin & Kitasato
Botulism	*Clostridium botulinum*	1897	Van Ermengem
Bacillary dysentery	*Shigella dysenteriae*	1898	Shiga
Foot and mouth disease	Foot and Mouth Disease Virus	1898	Loeffler & Frosch
Tobacco mosaic disease	Tobacco Mosaic Virus	1898	Beijerinck
Yellow fever	Yellow Fever Virus	1900	Reed
Syphilis	*Treponema pallidum*	1905	Schaudinn & Hoffman
Whooping cough	*Bordetella pertussis*	1906	Bordet & Gengou
Rocky mountain spotted fever	*Rickettsia rickettsii*	1909	Ricketts
Typhus	*Rickettsia prowazekii*	1910	Ricketts & Wilder
		1916	Lima
Tularemia	*Franciscella tularensis*	1912	McCoy & Chapin

*The names for the infectious agents are those in current use rather than those given at the time of their discovery.

glass slides not only made bacteria easier to see using the microscope but also helped in the identification of bacteria, such as the tubercle bacillus. Paul Ehrlich (1854–1915), one of Koch's colleagues, is credited with the development of a staining procedure similar to the one used today to identify the tubercle bacillus. Ehrlich's staining technique, developed in 1882, was used to test for the tubercle bacillus in the sputum of suspected tuberculosis patients at medical centers all over Europe, including the hospital where Koch worked.

Walther Hesse (1846–1911), one of Koch's students, initiated the use of agar to solidify or jell nutrients upon which microorganisms would grow. Unlike gelatin, which was used by Koch, agar remains solid at temperatures at which many microorganisms grow best (35 – 40°C) and it is not degraded by most microorganisms. Solidified growth media are extremely important in microbiology because microorganisms can be separated and purified on these media. Richard Petri (1852–1921), another of Koch's students, inspired the use of shallow flat-bottomed dishes with flat, overhanging covers to grow microorganisms (1887). The nutrient-agar dishes provided a large area over which microorganisms could be spread, and the lids helped to reduce contamination. These dishes are currently used in all microbiology laboratories and are known as **petri dishes** or petri plates.

The Danish scientist Christian Gram (1853–1935) developed a staining method that demonstrated bacteria in infected animal tissues (1884) and was used years later to distinguish between two basic types of bacteria. The staining procedure became known as the **Gram stain** and is still used universally today.

From about 1890 to 1910, Raymond Sabouraud intensively studied the fungi that cause ringworm, and in 1910 he published a classic manuscript called *Les Teignes* (the skin diseases). He also developed a culture medium that is still used to grow fungi.

Pasteur's studies of microorganisms demonstrated that they are responsible for many of the chemical changes that take place in foods and beverages. For instance, in the early 1860s Pasteur noticed that the presence or absence of oxygen had an effect on the growth rate and yield of yeasts in a juice. The presence of oxygen provoked a rapid growth rate and a high yield of yeast, but no alcohol production. The absence of oxygen resulted in a slow growth rate and a low yield, but ethanol was produced. Pasteur noticed that organisms grown in the presence of oxygen used glucose less rapidly than organisms grown in the absence of oxygen. The slower rate of glucose utilization and the lack of alcohol or acid production when an organism begins to grow aerobically (with O_2) is known as the **Pasteur effect**. Pasteur also demonstrated that many microorganisms produce specific waste materials (or combinations of waste materials), such as ethanol, butyric acid, and lactic acid. These discoveries have led to the manipulation of environmental conditions and to the selection of specific microorganisms to produce valuable products. For instance, certain microorganisms are used to make ethanol, vinegar, lysine, monosodium glutamate, and antibiotics. The use of microorganisms to make various products has resulted in the development of the very profitable branch of microbiology called **industrial microbiology**.

Pasteur reasoned that he could prevent spoilage of wines and beers by inhibiting or killing the microorganisms that chemically alter them. He found that the simplest way of destroying microorganisms in wines and beers was to heat the alcoholic beverages, after they were bottled, to temperatures just below their boiling point. This treatment generally prevented spoilage because it killed the spoilage organisms. Heating at temperatures less than boiling to kill spoilage or disease-causing microorganisms is now referred to as **pasteurization**. It is used not only on beer, wine, juices, and nonalcoholic beverages to prevent spoilage, but also on milk to kill pathogenic organisms, such as those that cause tuberculosis, undulant fever (brucellosis), and Q fever.

Immunology

The development of solidified media, staining techniques, and pure culture methods was of great importance to the study of microbiology. These developments promoted the widespread study of microorganisms and the discovery and characterization of pathogens. In addition, Pasteur's development of vaccines that afforded protection against anthrax, rabies, chicken cholera, and swine erysipelas stimulated scientists to attempt to find vaccines against other diseases. Such discoveries were possible for an increasing number of pathogens once pure culture techniques had been developed.

Paul Ehrlich and Emil von Behring worked together on the development of a diphtheria antitoxin from blood serum during the 1890s (fig. 1.12). Von Behring's studies showing that diphtheria toxins could be neutralized by antitoxin won him the first Nobel Prize in Medicine or Physiology in 1901, and the diphtheria antitoxin made von Behring a rich man. Ehrlich, on the other hand, signed away his interest in the diphtheria antitoxin in order to obtain a government post.

(a)

(b)

Figure 1.12 Paul Ehrlich and Elie Metchnikoff. *(a)* Paul Ehrlich (1854–1915) proposed that the immune response consists of circulating molecules, now known as antibodies, that destroy foreign molecules. His idea is known as the humoral theory. In addition, Ehrlich worked with Emile von Behring to develop a diphtheria antitoxin. *(b)* Elie Metchnikoff (1845–1916) proposed that the immune response consists of cells, now called phagocytes, that ingest and destroy foreign materials. His idea is known as the cellular theory of immunity.

In the early 1890s, Ehrlich observed that immunity to toxins was transferred from immune mouse mothers to nursing baby mice. In addition, he noticed that immunity could be transferred from one mouse to another by injection of blood serum from immune mice. Based upon these observations, Ehrlich proposed one of the first good explanations for how vaccines protect an animal. He hypothesized that a foreign material or toxin entering the body would bind to receptors on cells and stimulate the multiplication of the receptors, which were then released into the blood, where they neutralized the toxins. On the other hand, Elie Metchnikoff, a student of Koch's, thought that immunity was mediated by cells that phagocytize (ingest) foreign materials and microorganisms. Today, we know that the immune system is much more complicated than envisioned by Ehrlich (**humoral hypothesis**) and Metchnikoff (**cellular hypothesis**). There are specific cells that produce antibodies, others that synthesize toxic inhibitory substances, some that regulate immune responses, and still others that are involved in phagocytosis. Ehrlich shared the 1908 Nobel Prize with Elie Metchnikoff for their experiments and ideas about the immune system (fig. 1.12). The contributions made by Pasteur, Ehrlich, Metchnikoff, von Behring, and other scientists of that period provided the foundation for the dynamic branch of microbiology known as immunology.

Industrial Microbiology

Paul Ehrlich was also intimately involved in the search for chemicals that might inhibit or kill pathogenic microorganisms. In 1910 he reported the discovery of an arsenic compound he called **Salvarsan**, which inhibited and killed the bacteria that cause yaws, syphilis, and relapsing fever. Although the drug was very toxic to humans, it proved to be most effective against syphilis. Not until 1935 was another drug discovered that would cure diseases. Gerhard Domagk, in Germany, reported that a **sulfonamide** called **prontosil** was effective in controlling a number of bacteria.

Drugs that could be used to cure diseases were

discovered and produced in the 1940s with increasing frequency. In 1941, Howard Florey and Ernst Chain were able to isolate from the fungus *Penicillium* the antibiotic **penicillin**. Antibiotics are organic chemicals produced by various organisms which inhibit or kill other organisms. Alexander Fleming had reported the discovery of penicillin in 1928 but had been unable to purify it. By the mid 1940s, Selman Waksman and his colleagues had discovered a number of antibiotics produced by various species of the bacterial genus *Streptomyces*. The most important of the antibiotics were streptomycin and tetracycline, which were effective against a large spectrum of microorganisms. Since the 1940s, thousands of antibiotics have been isolated from microorganisms and used to treat infectious diseases. Numerous multi-billion-dollar pharmaceutical industries exist in many of the industrialized countries and capitalize on the antibiotic-producing capabilities of microorganisms. New antibiotics and drugs continue to be developed and tested in order to find chemical agents that are more effective against specific pathogens and cancers and that have less severe side effects on the host.

Virology

During the golden age of microbiology, numerous scientists discovered that certain diseases were due not to typical bacteria but to something that was much smaller. In 1892, Dmitri Iwanowski reported that mosaic disease of tobacco could be transmitted among plants by something that passed through filters fine enough to stop all known bacteria. Beijerinck in 1899 reported that this infectious agent could be dried and heated to temperatures as high as 90°C and still remain infectious. These filterable agents (viruses) were much smaller than endospores and therefore invisible in the light microscope; nor could they be grown on artificial media. In 1898, Friedrich Loeffler and Paul Frosch showed that foot and mouth disease was also caused by a filterable agent that could not be observed with the microscope. William Twort in 1915 and Felix-Hubert d'Herelle in 1917 reported a filterable agent that destroyed bacterial lawns (confluent growth on solid media) or cleared turbid broth cultures. The filterable agent was believed to be a new microscopic life form, and was named **bacteriophagium** (bacterial eater). Eventually, all the filterable agents were regarded as one type of organism and were given the name virus.

With the development of the electron microscope during the late 1930s and early 1940s, it soon became possible to "see" that viruses and bacteria were very different from each other. The bacteria and other microorganisms were found to be essentially typical cells, while the viruses had no cellular organization and consisted of a protein coat enclosing the genetic material. Because its resolution is much greater than that of the light microscope (1 nm vs. 200 nm), the electron microscope has been extremely useful, not only in characterizing viruses but also in establishing the fundamental differences between bacteria and cells of higher organisms.

Medicine

During the golden age of microbiology (1875–1915), the scientific community became convinced that infectious diseases were caused by microorganisms. In addition, the microorganisms responsible for many diseases were being isolated and characterized. Scientists, for the first time, had specific targets to attack and they did so by developing vaccines (Pasteur and von Behring) and chemotherapeutic agents (Ehrlich).

The knowledge that microorganisms cause infectious diseases prompted Joseph Lister to employ carbolic acid as a disinfectant during surgery (fig. 1.13). Lister based his technique on the assumption that airborne microbes falling on the wound were the cause of postoperative infections. Lister was successful in reducing the number of postoperative infections by using dressings soaked in carbolic acid. Pasteur's study of microorganisms led to an understanding of how pathogens might spread through the air; Lister used this information and his knowledge about carbolic acid to stop the spread of pathogens in wounds.

The understanding that microorganisms were responsible for spoilage and disease led to food pasteurization, development of sewage and garbage systems that improved sanitation, and the filtering and eventually the chlorination of drinking water supplies. During the golden age of microbiology there were major new discoveries almost every year (table 1.2).

Today, medical microbiology remains just as fertile. Research continues to explore the fundamental nature of pathogenic microorganisms, how they cause disease, and ways of combating them. As human populations increase in density and their patterns of behavior change, unusual diseases and pathogens become important while others almost disappear. For example, outbreaks of Legionnaires' disease are now occurring at numerous locations where individuals congregate; genital herpes has become epidemic in industrialized nations; and an immunodeficiency disease called AIDS (acquired immune deficiency syndrome) is now prevalent among homosexuals, hemophiliacs, and intravenous drug abusers.

(a)

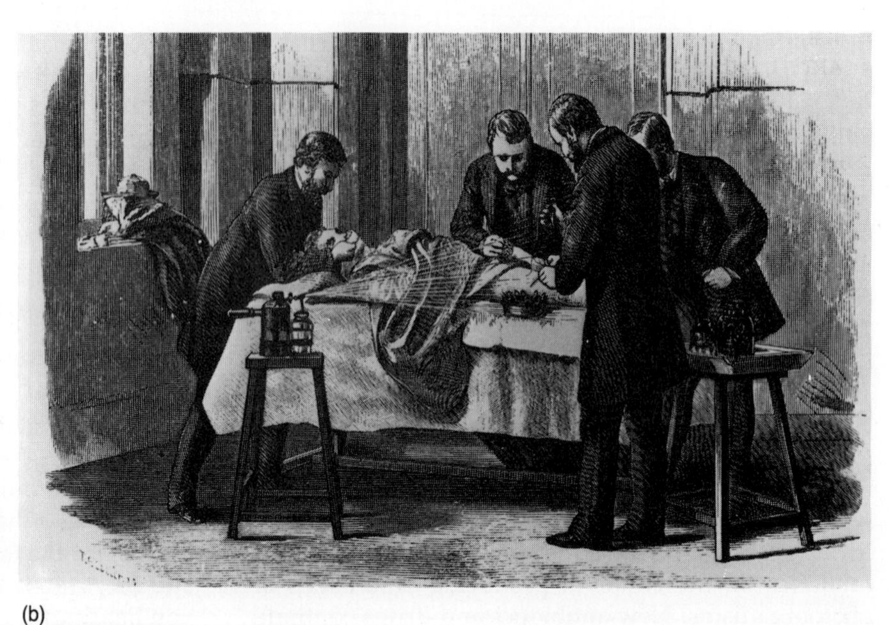

(b)

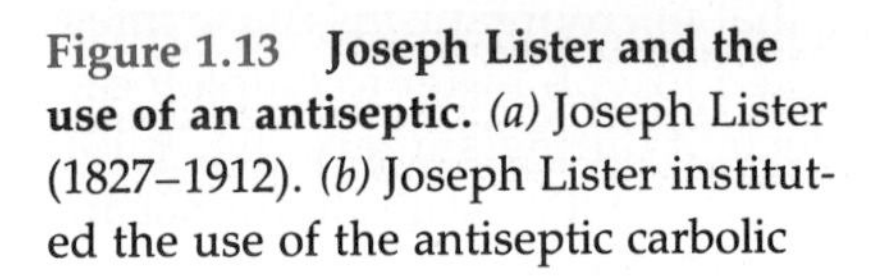

Figure 1.13 Joseph Lister and the use of an antiseptic. *(a)* Joseph Lister (1827–1912). *(b)* Joseph Lister instituted the use of the antiseptic carbolic acid (phenol) to kill microorganisms and prevent infection during surgery. The carbolic acid was sprayed over the area of the incision and used on dressings. Lister's use of carbolic acid during and after surgery reduced the death rate caused by infection from 45% to about 15%.

The identification of pathogens is still an important part of microbiology. Since each pathogen is sensitive to particular treatments, it must be identified promptly to combat it effectively. For this reason, constant research is directed toward the development of faster and more reliable methods of isolating and identifying bacteria.

Molecular Biology and Biotechnology

During the last 30–35 years, the genetics of microorganisms has provided fundamental insights into all areas of genetics and physiology. Two important scientific papers initiated the present era of microbiology, which is dominated by molecular biology: the 1944 report by Oswald Avery, Colin MacLeod, and Maclyn McCarty, describing bacterial transformation by DNA, and the 1953 paper by James Watson and Francis Crick which described the structure and possible functioning of DNA. The work of Francois Jacob and Jacques Monod, published in 1960, initiated numerous studies that clarified how the information contained in DNA directed the synthesis of proteins and how the synthesis of proteins could be regulated. Their work, and the research of many scientists all over the world, showed how various sugars and amino acids could turn specific gene systems on or off and drastically alter the physiology of a cell. The nature of the DNA's genetic code was worked out in the early 1960s by Marshall Nirenberg, Heinrich Matthaei, and Har Gobind Khorana. Also, during the late 1960s, a large number of researchers were determining how the genetic information (DNA) directed the synthesis of proteins and how the antibiotics blocked cellular processes.

During the 1960s and 1970s, our idea of how cells obtain their energy was revolutionized almost singlehandedly by Peter Mitchell. He proposed that some bacteria generate electrical potentials across their cytoplasmic membranes and then use the electrical potential to concentrate nutrients, propel the cell, or synthesize energy-rich compounds.

In the 1970s and early 1980s, microbiologists learned how to cut DNA with specific enzymes and integrate and exchange pieces of DNA in the test tube. Microbiologists are now able to put almost any gene into a bacterial cell and produce billions of copies of it. This new science, known as genetic engineering, has already made possible the synthesis of lifesaving products such as insulin, interferon, a foot and mouth

TABLE 1.2
A PARTIAL LIST OF PIONEERS IN MICROBIOLOGY

Investigator	Year	Contribution
Antony Van Leeuwenhoek	1677	Reported the discovery of microorganisms in a letter to the Royal Society of London. Built simple microscopes that could magnify more than 100×.
Lazzaro Spallanzani	1767	Conducted the first experiments that challenged the idea that microorganisms arise by spontaneous generation.
Edward Jenner	1796	Carried out the first successful vaccination with cowpox against smallpox.
Benedict Prevost	1807	Proposed that a fungus was responsible for wheat-bunt disease, and promoted the germ theory of disease.
Theodor Schwann	1837	Reported that yeasts are required for an alcoholic fermentation to occur.
Anton De Bary	1853	Reported that fungi were responsible for plant smuts and rusts. Challenged the theory of spontaneous generation and promoted the germ theory of disease.
Louis Pasteur	1857	Demonstrated that each type of fermentation is dependent upon a specific type of organism. Yeast carry out an alcoholic ferment, specific bacteria carry out a lactic acid ferment.
	1861	Showed that oxygen inhibits butyric acid ferments and kills the bacteria responsible. The theory that microorganisms can evolve from nutrients in a sterile medium is disproved. "Spontaneous generation" is shown to be due to contaminating microorganisms.
	1864	Developed a technique, now called pasteurization, to destroy spoilage organisms in wine.
	1880	Developed vaccines against chicken cholera, anthrax, swine erysipelas, and rabies.
Joseph Lister	1867	Published a study on the use of antiseptics during surgery.
Robert Koch	1876	Observed the growth of *Bacillus anthracis* and spore formation. Showed that this bacterium caused anthrax in animals.
	1881	Obtained the first pure culture of a bacterium on solid medium. Developed techniques for staining microorganisms. Developed various complex media to isolate different bacteria in pure culture. Koch's postulates were made public.
Ferdinand Cohn	1877	Discovered bacterial endospores and their extreme resistance to heat.
Joseph Lister	1878	Obtained pure cultures of a lactic acid bacterium by a dilution method.
Elie Metchnikoff	1884	Proposed a cellular theory of immunity.
Hans Christian Gram	1884	Developed a differential staining technique, now called the Gram stain, for bacteria.
Richard J. Petri	1887	Introduced a covered dish for growing microorganisms on a solidified medium.
Dmitri Iwanowski	1892	Discovered a filterable organism that caused tobacco mosaic disease.
Martinus Beijerinck	1898	Further characterized the organism that caused tobacco mosaic disease.
Erwin F. Smith	1890s	Showed that bacteria were the cause of a number of plant diseases.
Emil Von Behring	1890s	Developed a diphtheria antitoxin.
Paul Ehrlich	1890s	Proposed a humoral theory of immunity which proposed that antibodies were responsible for the immunity.
	1910	Announced the discovery of an arsenic compound that could be used to treat syphilis.

disease vaccine, and a hepatitis vaccine. In the future, new products made by genetic engineering will save countless human lives. For example, vaccines for malaria, AIDS, and sleeping sickness are presently in the making.

The tremendous amount of knowledge gained by studying microbial genetics and physiology and the powerful tool of genetic engineering are providing a solid foundation for an understanding of the genetics, physiology, and evolution of higher organisms.

Prominent scientists who have contributed significantly to our knowledge of life are sometimes awarded the Nobel Prize. Nobel laureates who have been honored for their work in the fields of chemistry and medicine or physiology are listed in table 1.3, along with a brief summary of their accomplishments.

TABLE 1.3
NOBEL LAUREATES WHO CONTRIBUTED TO MICROBIOLOGY

Nobel laureates	Year	Contribution
Emil von Behring	1901	Developed a diphtheria antitoxin.
Ronald Ross	1902	Discovered that the malaria parasite entered the human host from the bite of certain mosquitos.
Robert Koch	1905	Proved that a bacterium was the cause of tuberculosis and developed techniques for diagnosing the disease.
Charles Laveran	1907	Discovered that protozoans can cause human disease.
Paul Ehrlich and Eli Metchnikoff	1908	Hypotheses and studies on the immune response.
Charles Richet	1913	Described one type of allergic response, called anaphylaxis.
Jules Bordet	1919	Discovered and studied the complement system.
Charles Nicolle	1928	Studied the cause of typhus.
Karl Landsteiner	1930	Described the ABO scheme for grouping human blood.
Gerhard Domagk	1939	Discovered the antimicrobial sulfa drug prontosil.
Alexander Fleming, Ernst Chain, and Howard Flory	1945	Discovered penicillin and how to isolate it.
James Sumner, John Northrup, and Wendell Stanley	1945	Crystallized enzymes and viruses.
Max Theiler	1951	Developed a yellow fever vaccine.
Selman Waksman	1952	Discovered and isolated the antibiotic streptomycin.
Frits Zernicke	1953	Developed the phase contrast microscope.
John Enders, Thomas Weller, and Frederick Robbins	1954	Developed culture techniques for the polio virus.
Edward Tatum, George Beadle, and Joshua Lederberg	1958	Studied the relationship between genes and biochemical defects in microorganisms.
Severo Ochoa and Arthur Kornberg	1959	Isolated and characterized enzymes that synthesized nucleic acids.
Macfarlane Burnet and Peter Medawar	1960	Studied immunological tolerance.
Francis Crick, James Watson, and Maurice Wilkins	1962	Proposed a structure for DNA.

SUMMARY

WHAT ARE MICROORGANISMS?

1. Protozoa, some algae, fungi, bacteria, and the noncellular viruses are referred to as microorganisms.
2. Generally, microorganisms are too small to be seen without the aid of a magnifying glass or microscope.

MICROBIOLOGY: A MULTIFACETED SCIENCE

1. Microbiology is often subdivided into fields of study based upon various types of microorganisms. This subdivision is sometimes known as the taxonomic approach to microbiology. Virology, bacteriology, mycology, phycology, and protozoology are studies that are concerned with viruses, bacteria, fungi, algae, and protozoa, respectively.
2. Microbiology is often subdivided into fields of study based upon microbial activity. This subdivision may be referred to as the functional approach to microbiology. Microbial ecology, industrial microbiology, agricultural microbiology, medical microbiology, physiology, genetics, and biochemistry are subdivisions of microbiology.

ANCIENT IDEAS OF DISEASE AND PUTREFACTION

1. For most of recorded human history, disease has generally been regarded as a punishment from

TABLE 1.3 (continued)

Nobel laureates	Year	Contribution
Francois Jacob, Andre Lwoff, and Jacques Monod	1965	Proposed a model for the control of enzyme synthesis.
Peyton Rous and Charles Huggins	1966	Discovered a cancer-producing virus and discovered that hormones could be used to treat some cancers.
Robert Holley, Har Gobind Khorana, and Marshall Nirenberg	1968	Determined how tRNA was involved in protein synthesis and what the genetic code was.
Max Delbruck, Alfred Hershey, and Salvador Luria	1969	Studied the genetics of bacterial viruses.
Earl Sutherland	1971	Discovered the effects of cAMP in bacteria and in eukaryotes.
Gerald Edelman and Rodney Porter	1972	Studied the structure of antibodies.
Albert Claude, George Palade, and Christian De Duve	1974	Elucidated the structure and function of eukaryotic organelles.
Renato Dulbecco	1975	Studied DNA tumor viruses.
Howard Temin and David Baltimore	1975	Studied RNA tumor viruses and discovered reverse transcriptase.
Carlton Gajdusek and Baruch Blumberg	1976	Studied the epidemiology of slow virus diseases. Discovered the relationship between Australian antigen and hepatitis B virus.
Werner Arber, Hamilton Smith, and Daniel Nathans	1978	Discovered restriction enzymes and used them to map genes on DNA molecules.
Peter Mitchell	1978	Proposed the chemiosmotic hypothesis to explain oxidative phosphorylation.
Baruj Benacerraf, George Snell, and Jean Dausset	1980	Studied the genetic regulation of the immune response.
Paul Berg, Walter Gilbert, and Fredrick Sanger	1980	Developed various techniques to sequence nucleic acids.
Barbara McClintock	1983	Discovered mobile genetic elements.
Niels Jerne, George Koehler, and Cesar Milstein	1984	Studied the development and control of the immune system. Developed techniques for production of monoclonal antibodies.
Susumu Tonegawa	1987	Studied the regulation of the immune response.

gods or demons. People have also known, since very ancient times, that some diseases were spread from person to person, on fomites, and through the air.

2. Girolamo Fracastoro proposed in a manuscript (1546) that disease and some types of putrefaction were due to things he called "contagions," which were transferred by direct contact, by touching fomites, or through the air.
3. Edward Jenner (1798) developed a smallpox vaccine from material taken from cowpox lesions. Soon after his discovery, smallpox vaccination was commonly practiced in England, Europe, and the United States.
4. Ignaz Semmelweis (1850) proposed that childbed fever was due to "putrefying organic matter." He believed that doctors were picking up this material from corpses they were studying and transmitting it to new or expectant mothers.

THE DISCOVERY OF MICROORGANISMS

1. The person credited with first observing bacteria and other microorganisms through a microscope is Antony van Leeuwenhoek. In the late 1600s and early 1700s van Leeuwenhoek published a number of papers in which he described bacteria, fungi, protozoans, algae, blood cells, and sperm.
2. Van Leeuwenhoek used a simple microscope that resembled a magnifying lens to observe his "animalcules."

THE SPONTANEOUS GENERATION CONTROVERSY

1. Franceso Redi showed with a simple experiment that maggots do not arise spontaneously from rotting meat, but instead arise from eggs laid by flies.
2. Spallanzani and Schwann demonstrated that boiled infusions remain sterile until contaminated with air.
3. Louis Pasteur showed that boiled infusions remain sterile as long as airborne microorganisms do not contaminate the infusions. His experiments demonstrated that microorganisms do not arise by spontaneous generation in sterile infusions.

THE GERM THEORY OF DISEASE

1. In the 1800s, scientists believed that putrefaction and infectious diseases were due to either (a) decomposing substances that imparted their characteristics to living or nonliving things, or (b) microorganisms carrying out normal metabolic processes on living or nonliving things.
2. Not until the 1880s was it generally accepted that microorganisms were responsible for infectious diseases and putrefaction. The French microbiologist Louis Pasteur and the German physician Robert Koch were largely responsible for convincing the scientists of the 1880s that microorganisms have a central role in fermentations, putrefactions, and infectious diseases.
3. Louis Pasteur demonstrated that specific organisms were responsible for a particular type of fermentation, and that microorganisms that frequently appeared in infusions were not arising spontaneously but were contaminants from the air. Pasteur's experiments and conclusions stimulated scientists all over the world to realize the importance of microorganisms in fermentations, putrefactions, and infectious diseases.
4. Robert Koch, together with some of his students and colleagues, developed many of the basic techniques, media, and equipment necessary for handling and studying microorganisms. His laboratory is credited with the development of agar-solidified media (Hesse), covered culture dishes (Petri), and staining techniques (Koch and Ehrlich). In addition, Koch demonstrated convincingly that the bacteria *Bacillus anthracis* and *Mycobacterium tuberculosis* were responsible for anthrax and tuberculosis, respectively. The technical developments credited to Koch's laboratory allowed many other scientists to purify and begin studying microorganisms in a sensible way.
5. The rules for establishing whether or not an organism is responsible for a disease, now known as Koch's postulates, helped develop microbiology into a systematic and believable science.

SIGNIFICANT DISCOVERIES IN MICROBIOLOGY

1. Louis Pasteur is responsible for the development of vaccines against rabies, anthrax, chicken cholera, and swine erysipelas.
2. Paul Ehrlich is credited with one of the first good models to explain immunity. He is also responsible for the development of an arsenic compound, called Salvarsan, that was useful in treating syphilis.
3. Elie Metchnikoff is credited with the idea that phagocytosis was an important part of the immune system.
4. Emil von Behring developed an antitoxin vaccine against the diphtheria toxin.
5. Joseph Lister obtained the first pure culture of a bacterium in 1878, but is most famous for his use of carbolic acid as an antiseptic during surgery.
6. In 1944 Oswald Avery, Colin MacLeod, and Maclyn McCarty demonstrated that the transforming material was DNA.
7. James Watson and Francis Crick in 1953 published a model for the structure of DNA and how its replication might occur.
8. In 1960 Francois Jacob and Jacques Monod presented a model for how genes might be regulated in bacteria and in viruses.
9. Marshall Nirenberg, Heinrich Matthaei, and Har Gobind Khorana are credited with working out the genetic code during the early 1960s.
10. During the 1960s and 1970s, Peter Mitchell demonstrated that some bacteria generate electrical potentials across the cytoplasmic membrane. Bacteria use this potential to concentrate nutrients, rotate flagella, and synthesize ATP.

STUDY QUESTIONS

1. Discuss two nonsupernatural explanations for infectious diseases. Who were the early supporters of these ideas?

2. What is the current scientific explanation for infectious diseases?
3. Explain the idea of spontaneous generation and why so many people believed in it. Consider the role that Redi and Pasteur played in disproving spontaneous generation.
4. Did Pasteur prove that life could not have developed naturally on earth? Explain.
5. Explain the germ theory of disease and putrefaction. Did Fracastoro, Jenner, or Semmelweis believe that microorganisms were responsible for infectious diseases and putrefaction?
6. Describe the contributions of Pasteur and Koch to the advancement of microbiology.
7. List and explain Koch's postulates. How do you prove that an organism is responsible for a disease if you cannot fulfill one or more of Koch's postulates?
8. List the contributions of Ehrlich and Metchnikoff to the science of immunology.

SELF-TEST

1. Microorganisms, for the most part, are responsible for many beneficial processes in our planet and should not be considered to be harmful. a. true; b. false.
2. The presence of microorganisms has been felt by all human populations but it was not until the 17th century that they were actually seen and recognized as biological entities. a. true; b. false.
3. Edward Jenner was an English physician who introduced a technique that is presently used and that has saved countless lives. The technique introduced by Jenner was:
   a. sterilization;
   b. the use of antibiotics;
   c. vaccination;
   d. genetic engineering;
   e. disinfection.
4. The spontaneous generation controversy raged on for almost 200 years before pioneer microbiologists showed by experiment that living organisms can arise only from organisms like themselves and not from inanimate objects. Which of the following scientists was responsible for offering conclusive evidence against the spontaneous generation theory?
   a. Van Helmont;
   b. Koch;
   c. Metchnikoff;
   d. Pasteur;
   e. Needham.
5. Pasteur was a very prominent scientist during the golden age of microbiology. Which of the following contributions to the field of microbiology are attributed to Pasteur?
   a. alcoholic fermentations are carried out by yeasts;
   b. vaccination against chicken cholera, anthrax, and rabies;
   c. antibiotics are produced by fungi;
   d. a and c;
   e. a and b.
6. The germ theory of disease was significant because scientists became aware that many diseases were caused by specific infectious agents. Which of the following diseases was the first shown to be caused by a bacterium?
   a. anthrax;
   b. tuberculosis;
   c. cholera;
   d. whooping cough;
   e. plague.
7. Which of the following **is not** one of Koch's postulates?
   a. the suspected pathogen must always be isolated from afflicted individuals;
   b. the suspected pathogen must be cultured in pure form outside the host;
   c. when the suspected pathogen is injected into animals, the animals become diseased;
   d. the same organism must be consistently isolated in pure form from afflicted experimental hosts;
   e. when animals are subjected to the suspected pathogen, the animals must develop the disease associated with the pathogen.
8. Which of the following scientists is responsible for the use of agar as a solidifying agent for culture media?
   a. Pasteur;
   b. Petri;
   c. Ehrlich;
   d. Hesse;
   e. Spallanzani.
9. Which of the following scientists discovered that the cause of malaria was a parasite called *Plasmodium malariae*?
   a. Kitasato;
   b. Laverans;
   c. Yersin;
   d. Ricketts;
   e. Koch.
10. The discovery of the chemical nature of the genetic material has allowed scientists to develop the very dynamic and exciting field of genetic engineering. The discovery, using transformation studies, that DNA was the genetic material was made by the following scientists:
    a. Watson and Crick;
    b. Linus Pauling;
    c. Beadle and Tatum;
    d. Avery, MacLeod, and McCarty;
    e. Jacob and Monod.

SUPPLEMENTAL READINGS

Ada, G. L. and Nossal, G. 1987. The clonal-selection theory. *Scientific American* 257(2):62–69.

Bendiner, E. 1980. Ehrlich: Immunologist, chemotherapist, prophet. *Hospital Practice* 15(11):129–157.

Bendiner, E. 1986. Liberator of surgery from shackles of sepsis. *Hospital Practice* 21(6):126C–126SS.

Bendiner, E. 1987. Semmelweiss: lone ranger against puerperal fever. *Hospital Practice* 22(2): 194–225.

Brock, T. D. 1961. *Milestones in microbiology*. Englewood Cliffs, NJ: Prentice-Hall Biological Sciences Series.

Bulloch, W. 1938. *The history of bacteriology*. London: Oxford Univ. Press.

Cartwright, F. and Biddiss, M. 1972. *Disease and history.* New York: T. Y. Crowell.

De Kruif, P. 1926. *Microbe hunters*. New York: Harcourt, Brace & World.

Dobell, C. 1932. *Antony van Leeuwenhoek and his little animals*. London: Constable.

Dubos, R. 1976. *Louis Pasteur: free lance of science*. New York: Charles Scribner's Sons.

Groschel, D. 1982. The etiology of tuberculosis: A tribute to Robert Koch on the occasion of the centenary of his discovery of the tubercle bacillus. *American Society for Microbiology News* 48(6):248–250.

Hoyle, F. and Wickramasinghe, N. C. 1979. *Diseases from space*. New York: Harper & Row.

Lechevalier, H. and Solotorovsky, M. 1965. *Three centuries of microbiology.* New York: McGraw-Hill.

Porter, J. R. 1972. Louis Pasteur sesquicentennial (1822–1972). *Science* 178:1249–1254.

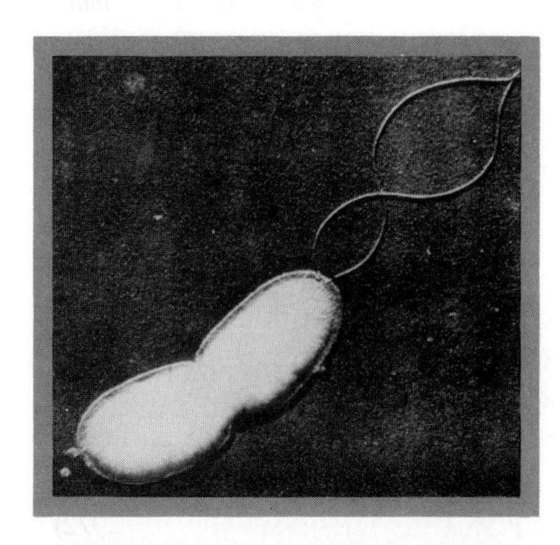

CHAPTER 2

STRUCTURE OF PROKARYOTIC AND EUKARYOTIC CELLS

CHAPTER PREVIEW THE BACTERIAL CELL WALL IS IMPORTANT TO HUMANS TOO

Most bacteria have a tough, rigid outer layer called the cell wall, which gives the cell its shape and protects it from environmental damage. In addition, the bacterial cell wall may control the passage of certain chemicals into and out of the cell. Clearly, the bacterial cell wall is an asset to bacteria because it improves their chances for survival in hostile environments.

Research has shown that most bacteria have one of two basic types of cell walls, which can readily and easily be differentiated by performing the **Gram stain**. Using the Gram stain, bacteria can be classified as gram-positive or gram-negative. This staining procedure is used routinely in microbiology laboratories and constitutes a very important criterion for the identification of bacteria.

The type of cell wall a bacterium has is extremely important to physicians who are interested in how to treat a patient suffering from a bacterial infection. For example, the use of the antibiotic penicillin is recommended primarily for the control of infections caused by gram-positive bacteria, since these bacteria, not the gram-negative ones, are more susceptible to the antibiotic. Similarly, there are numerous antibiotics that are used against gram-negative bacteria because they usually work better than penicillin.

The cell wall of gram-negative bacteria is important to humans in another way. It is now known that gram-negative cell walls have a component called **lipopolysaccharide**, or simply "LPS," that has toxic properties. LPS is also commonly known as "endotoxin." Persons whose blood becomes infected by gram-negative bacteria may suffer severe ill effects, including high fever, extremely low blood pressure, and sometimes death. It is important to insure that medicines and pharmaceuticals administered by injection are free of gram-negative bacteria or cell wall lipopolysaccharide, because injection of bacteria or LPS can have grave consequences.

Scientists have also learned that LPS has a significant effect on the immune system, and it is now used as a tool to study the immune system.

As can be seen from these examples, the bacterial cell wall has characteristics that are important to humans. The study of biological structures such as bacterial cell walls can lead to our understanding of how microorganisms affect us. In this chapter we discuss the major structural characteristics of microorganisms, emphasizing their biological importance and impact on humans.

CHAPTER OUTLINE

THE STRUCTURE OF MICROORGANISMS HAS been studied in detail because scientists are interested in understanding how organisms are constructed and how their various parts function. In addition, a knowledge of what microorganisms are and the various forms that they can have is often helpful in their identification, control, and industrial exploitation.

Much of our knowledge about the structure of microoganisms has been linked to technological advances in the field of microscopy. Before the advent of the electron microscope, our understanding of the structure of microorganisms was restricted to those structures that were visible using a light microscope. As a consequence, a great many unique features of microbial anatomy were unknown. With the development of the electron microscope in the 1930s, much has been learned about the anatomy of microorganisms. Even today, the electron microscope is being used to learn new things about the anatomy of microorganisms and cells from higher organisms.

All microorganisms (except for the viruses, viroids, and prions) are cells or are composed of cells. **Cells** are chambers defined by a lipid–protein membrane. The chambers are filled with an aqueous–gelatinous material called the **cytoplasm**. The cytoplasm contains numerous small and large molecules that are involved in the many cellular processes such as metabolism and reproduction. All cells, except for the highly differentiated red-blood cells of primates, have DNA as their hereditary information. Cells are considered by most biologists to be the smallest unit of life.

The term "cell" was first introduced by Robert Hooke in 1665 to describe the chambers in thin sections of cork and wood. Although we now know that these chambers represented the remains of dead plant cells, Hooke's observations were the first of many that led to the **cell theory of life**. Two important ideas in the cell theory of life are that all living organisms are made up of cells, and that all cells come from preexisting cells. They are not capable of arising by spontaneous generation.

In this chapter we discuss the most important structures found in microbial cells, their biological role, and their importance to humans.

THE LIGHT MICROSCOPE

For observations with microscopes to have any meaning, the observer must have a standard means of measurement and a way to determine the size of the specimens under investigation. Many of the early investigators used no standard system of measurement, but instead compared their specimens to common objects. For example, in his letters to the Royal Academy, Van Leeuwenhoek described his "animalcules" and compared them to the size of a grain of sand or to a louse's hair. Since not all sand grains nor all louse hairs are of the same size, these types of standards are not very useful when precise measurements are required. A system of measurements called the **metric system** has been adopted by the scientific community in order to establish a uniform standard of measurement. The metric system is now used by most scientists for all quantitative determinations (table 2.1).

Simple and Compound Microscopes

Microscopes are optical devices that are used to view objects too small to be seen with an unaided eye. Most microscopes used in the laboratory use visible light to illuminate the object and are called **light microscopes**.

Light microscopes that use two sets of lenses in series to view an object are called **compound micro-**

TABLE 2.1
THE METRIC SYSTEM

Unit	Measure		Symbol	English equivalent
Linear measure				
1 kilometer	= 1000 meters	10^3 m	km	0.62137 mile
1 meter		10^0 m	m	39.37 inches
1 decimeter	= 1/10 meter	10^{-1} m	dm	3.937 inches
1 centimeter	= 1/100 meter	10^{-2} m	cm	0.3937 inch
1 millimeter	= 1/1000 meter	10^{-3} m	mm	
1 micrometer	= 1/1,000,000 meter	10^{-6} m	μm	English equivalents
1 nanometer	= 1/1,000,000,000 meter	10^{-9} m	nm	infrequently used
1 angstrom*	= 1/10,000,000,000 meter	10^{-10} m	Å	
Measures of capacity (for fluids and gases)				
1 liter			l	1.0567 U.S. liquid quarts
1 milliliter	= 1/1000 liter = Volume of 1 g of water at standard temperature and pressure (STP)		ml	
Measures of volume				
1 cubic meter			m^3	
1 cubic decimeter	= 1/1000 cubic meter = 1 liter (l)		dm^3	
1 cubic centimeter	= 1/1,000,000 cubic meter = 1 milliliter (ml)		cm^3 = ml	
1 cubic millimeter	= 1/100,000,000 cubic meter		mm^3	
Measures of mass				
1 kilogram	= 1000 grams		kg	2.2046 pounds
1 gram			g	15.432 grains
1 milligram	= 1/1000 gram		mg	0.01 grain (about)
1 microgram	= 1/1,000,000 gram		μg (or mcg)	

*The angstrom is not part of the metric system but is so frequently encountered in the literature that it is included.

scopes. The **objective** lenses are close to the object and produce a magnified image of it. The **ocular** lenses are near the eye and magnify the image. Van Leeuwenhoek's microscopes were **simple** microscopes because only one lens was used between the object and the eye. Many of the modern compound microscopes have two eyepieces to reduce eyestrain. Because of the two eyepieces, they are called **binocular microscopes**. Microscopes that have only one eyepiece are called **monocular microscopes**.

Magnification and Resolution

An important property of microscopes is **magnification**, the enlargement of the apparent size of an object. Magnification is produced by the combined effects of the objective and ocular lens systems (fig. 2.1). The objective lens provides an enlarged image of an object, usually 4–100 times, and the ocular lens magnifies the image projected by the objective lens. Ocular lenses generally magnify the image 10–15 times. The **total magnification** of the lens system is the product of the objective's magnification and the ocular's magnification. For example, if the object is viewed with a 100× objective lens, the image in the microscope tube will be 100 times larger than the object. If this image is then viewed with a 10× ocular lens, it will be magnified 10 times. The total magnification of the object will be 100 × 10 = 1000 times.

Light microscopes should also provide clear images of the object. The clarity or **resolution** of a light microscope is defined as the ability of a lens system to allow the viewer to distinguish between two structures near each other in a specimen.

Although it might appear that magnifications much greater than 1000× could be obtained by using objective and ocular lenses above 100×, such magnification generally is not useful. If the total magnification of microorganisms exceeds about 1500×, the resolution

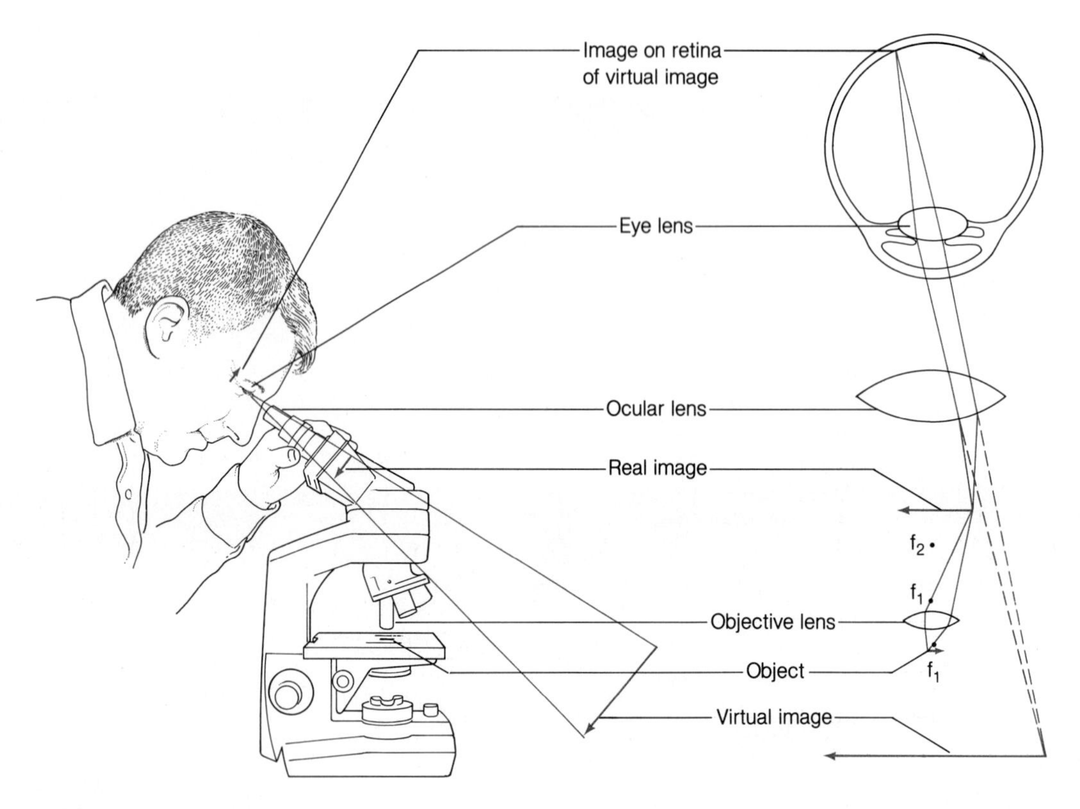

Figure 2.1 How the bright field light microscope works. Light from the specimen is bent by the objective lens so that an inverted, magnified real image is created between the objective lens and the ocular lens. The image is magnified further by the ocular lens, so that an inverted, magnified virtual image is created. The lens in the eye focuses the inverted magnified virtual image on the retina so that the specimen can be seen.

and clarity of the image decrease because the light becomes **diffracted** (spread out) as it passes through very small lenses. The diffraction becomes worse as the magnification increases because the lens becomes smaller. Consequently, the image becomes more and more blurry (resolution decreases) and the magnification is useless.

The basic light microscope can be modified by altering the lens systems or the light source so that special effects can be obtained. Examples of such modifications include dark field, phase contrast, and fluorescence microscopes. Images of the same object viewed with some of these microscopes can be seen in figures 2.3 and 2.4.

LIGHT MICROSCOPY TECHNIQUES

Wet and Hanging Drop Mounts

Microorganisms are studied using the light microscope to determine their size and shape, whether they are motile, how they divide, and their cellular structures. **Wet mount** (fig. 2.5) preparations are generally used to study living microorganisms. Wet mounts are made by placing an aqueous suspension of microorganisms on a glass slide and covering it with a cover glass (fig. 2.5a). The slide can then be viewed with a

FOCUS RESOLUTION AND THE WAVE NATURE OF LIGHT

Because of its wave nature, light behaves in an unexpected manner when it passes through small apertures. Instead of producing a sharp image, light under these conditions results in a large, diffuse image. The smaller the hole, the greater is the diffraction or spreading out of the light.

Diffraction is a problem in microscopes because high-magnification lenses are quite small. Lenses that magnify 50× have diameters of about 5 mm, while those that magnify 100× have diameters of about 2–3 mm. If you check your microscope, you will discover that the diameter of the lens decreases as the magnification increases. The smaller the lens, the more the light spreads out (diffracts) and the more the image is diffused and blurred (fig. 2.2).

A lens that magnifies 50× results in very little visible diffraction because of its large diameter, but a 100× objective lens spreads out the light and makes the images more diffuse than expected. The images may blur into each other, yet still be distinguished. A 150× lens severely diffracts the light, however, so that the images may overlap completely and two objects would appear to be one. Thus, an increase in the magnification, which would be expected to make the images larger and clearer, actually results in a decrease in the resolution of enlarged objects.

Since the diffraction is due to the waves of light interacting with the edge of the lens, it can be minimized by using a large lens and light with a short wavelength.

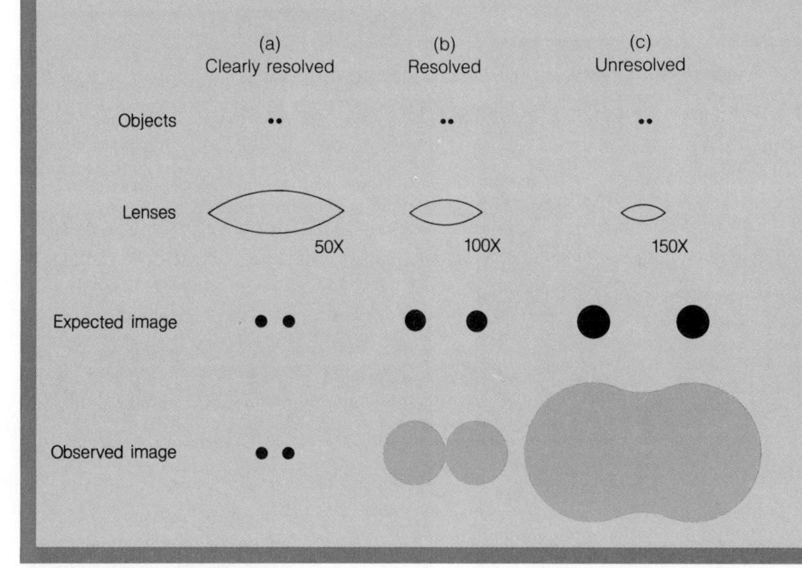

Figure 2.2 **Resolution and the wave nature of light.** Very little diffraction of the light is produced by the 50× objective lens. The 100× objective lens causes some diffraction because of its size. Consequently, the image is more diffuse and blurred than expected. Nevertheless, the objects can be resolved. A 150× objective lens results in a severe diffraction of the light, so that the objects cannot be resolved. The light from the objects is so diffuse that it overlaps, producing a single, blurred image.

light microscope. A variation of the wet mount is the **hanging drop** preparation (fig. 2.5b), which requires the use of a **depression slide**, a cover glass, and some petroleum jelly. The advantage of this technique is that it allows extended observations of the microbial population in a fluid environment. The petroleum jelly is used to prevent evaporation of the drop.

Examination of Stained Preparations

Although much can be learned from observing unstained microorganisms, the addition of colored substances can reveal structures that are invisible in unstained preparations. Colored organic compounds called **dyes** are often used to stain microbiological specimens on slides to enhance the contrast between the microorganisms and their background. Staining techniques are used routinely to differentiate among major groups of microorganisms, to detect chemical and structural differences in bacterial cell walls (Gram stain and acid fast stain), and to observe specific cellular components.

Dyes are made up of a charged colored portion, called the **chromophore**, and a complementary **anion** (negatively charged ion) or **cation** (positively charged

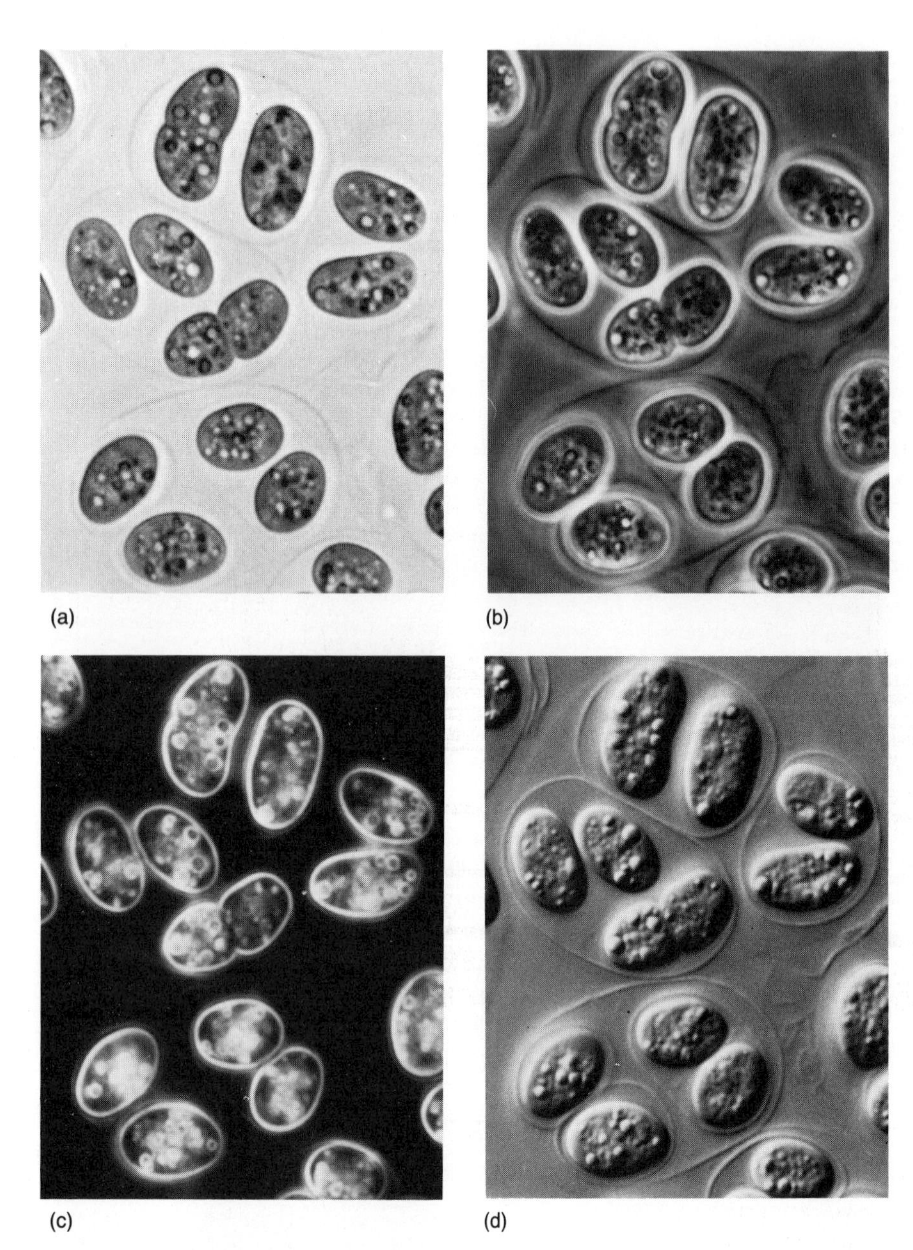

Figure 2.3 Images produced with bright field, phase contrast, dark field, and interference microscopes. Shown are the cells of a blue-green alga as seen using different microscope techniques. *(a)* Bright field. *(b)* Phase contrast. *(c)* Dark field. *(d)* Interference.

Figure 2.4 Fluorescent staining. This photograph illustrates the fluorescent staining of *Treponema pallidum*, the cause of syphilis. When cells "stained" with appropriate fluorescent dyes or reagents are subjected to ultraviolet light, they give off visible light and therefore appear bright. The background is dark because the ultraviolet light is invisible and filtered out so that it cannot reach the viewer.

ion). Basic dyes, such as methylene blue and crystal violet, have positively charged chromophores, while acidic dyes, such as eosin and picric acid, have negatively charged chromophores. Since cells generally have many more negatively charged groups than positively charged groups, they will be stained readily by basic dyes because the opposite charges attract each other. Acidic dyes are almost always used to stain the background.

Before microorganisms can be stained, usually they must be attached or fixed to a glass slide. To accomplish this fixation, a thin **smear** of the microorganism is prepared by spreading a small portion of an aqueous suspension of microorganisms on a clean glass slide. The smear is allowed to air-dry and is then heated for about 3 seconds over an open flame. The heating process is called **heat fixation**. It serves to attach the microorganisms to the slide so that they will not wash away during the staining procedures.

Simple staining techniques are used to color microorganisms or cellular inclusions using a single dye so that they can be seen readily. Basic dyes are generally used in simple staining procedures, because they attach efficiently to the many negatively charged groups in the cytoplasm, cell membrane, and cell wall. Sometimes, dyes have a higher affinity for some components within the cell than others and an unevenly stained cell results.

Differential staining techniques are commonly used in microbiology because they provide a great deal of information about the structure and chemical composition of the organisms being stained. There are two differential staining procedures in common use: the **Gram stain** and the **acid-fast stain**.

The Gram stain, developed by Hans Christian Gram in 1884, was originally used to reveal bacteria in diseased animal tissue. Soon after its development, however, microbiologists realized how valuable the Gram stain could be in distinguishing among bacteria. The Gram stain is undoubtedly the most widely utilized differential staining procedure in the bacteriology laboratory, because it allows microbiologists to divide bacteria into two types: the **gram-positive** and the **gram-negative** bacteria (table 2.2). The separation of bacteria into these two types is one of the first steps in the identification of a bacterium. After being Gram stained, bacteria are colored either by **crystal violet** (blue-violet) or by **safranin** (light red). The blue-violet bacteria are called gram-positive, while the light red bacteria are gram-negative (table 2.2). It is believed that differences in the thickness, charge concentration, and chemical composition of the cell wall are

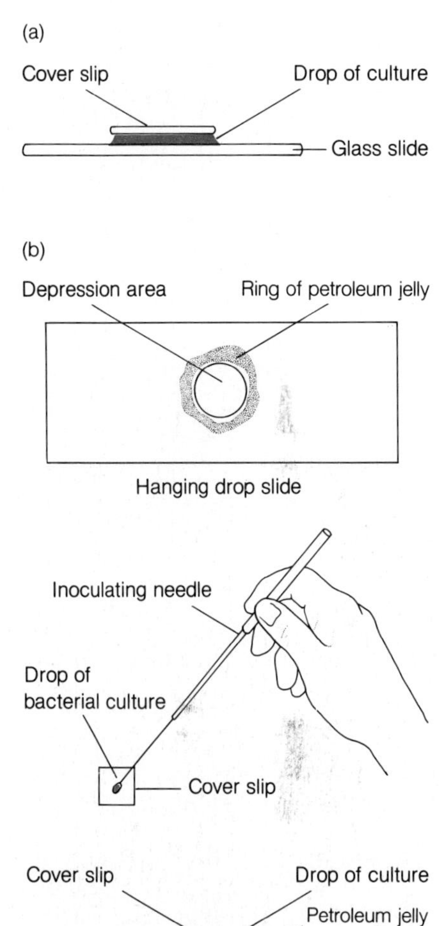

Figure 2.5 Wet mount and hanging drop mount. *(a)* The wet mount is prepared by placing a small drop of culture onto a clean glass slide and then covering the drop with a coverslip. *(b)* The hanging drop mount is prepared by placing a small drop of culture onto a coverslip and then placing the inverted coverslip onto a ring of petroleum jelly that surrounds the depression in a depression slide.

TABLE 2.2
THE GRAM STAIN PROCEDURE

Reagents	Time applied	Reactions	Appearance
Unstained smear	—	—	Cells are colorless and difficult to see.
Crystal violet	1 minute, then rinse with water.	Basic dye attaches to negatively charged groups in the cell wall, membrane, and cytoplasm.	Both gram-negative and gram-positive cells are dark blue-violet.
Gram's iodine (mordant)	1 minute, then rinse with water.	Iodine strengthens the attachment of crystal violet to the negatively charged groups.	Both gram-negative and gram-positive cells remain dark blue-violet.
Ethanol or acetone–ethanol mix (decolorizer or leaching agent)	10–15 seconds, then rinse with water.	Decolorizer leaches the crystal violet and iodine from the cells. The color diffuses out of gram-positive cells more slowly than out of gram-negative cells because of the chemical composition and thickness of the cell walls.	Gram-positive cells remain dark blue-violet, but the gram-negative cells become colorless and difficult to see.
Safranin (counterstain)	1 minute, then rinse thoroughly, blot dry, observe under oil immersion.	Basic dye attaches to negatively charged groups in both cell types. Few negative groups are free in gram-positive cells, while most negative groups are free of crystal violet in gram-negative bacteria. Consequently, gram-positive bacteria remain dark blue-violet, while gram-negative bacteria become light red.	Gram-positive cells remain dark blue-violet, while gram-negative cells are stained a light red.

responsible for the differential staining of bacteria. The gram-positive bacteria retain the purple dye better than the gram-negative bacteria because the gram-positive bacteria have thicker walls and more positive charges.

The acid-fast stain measures the resistance of a bacterial cell to decolorization by acid. It was originally developed by Paul Ehrlich and later modified by Franz Ziehl and Friedrich Neelsen to demonstrate the presence of *Mycobacterium tuberculosis* in patients afflicted with tuberculosis. The acid-fast characteristic is a distinctive property of the genus *Mycobacterium* and some species of *Nocardia*. Pathogenic bacteria in these genera include the agents of tuberculosis, leprosy, and lumpy jaw (nocardiosis). Acid-fast organisms, stained with a hot phenolic solution of carbol fuchsin, retain the red dye even after treatment with dilute sulfuric or hydrochloric acid-alcohol solutions. On the other hand, non-acid-fast organisms are decolorized when treated with acid–alcohol solutions. Non-acid-fast organisms appear blue after the acid-fast staining procedure because they are counterstained with methylene blue. The acid-fast characteristic is generally associated with the very high content of complex waxes in the cell wall.

ELECTRON MICROSCOPES

The **electron microscope** has enabled microbiologists to "see" and study structures that are too small to be resolved with the light microscope. The electron microscope uses electrons rather than light to produce images. This is possible because electrons have wavelike properties similar to those of light. Since electron beams have extremely short wavelengths (0.0055 nm for an accelerating voltage of 50,000 volts), the theoretical limit of resolution is much lower than for the light microscope.

There are two types of electron microscopes: the **transmission electron microscope** (TEM) and the **scanning electron microscope** (SEM). Both the TEM and SEM are used extensively by microbiologists. The TEM is used when information regarding subcellular details is desired, while the SEM is used for visualization of a surface structure (fig. 2.6). In practice, the transmission electron microscope can resolve objects as close as 2.5 nm. Objects are generally magnified between 10,000 and 100,000×. In contrast, the scanning electron microscope can resolve objects as close as 20 nm. With this microscope, objects are usually magnified between 1000 and 10,000×.

(a)

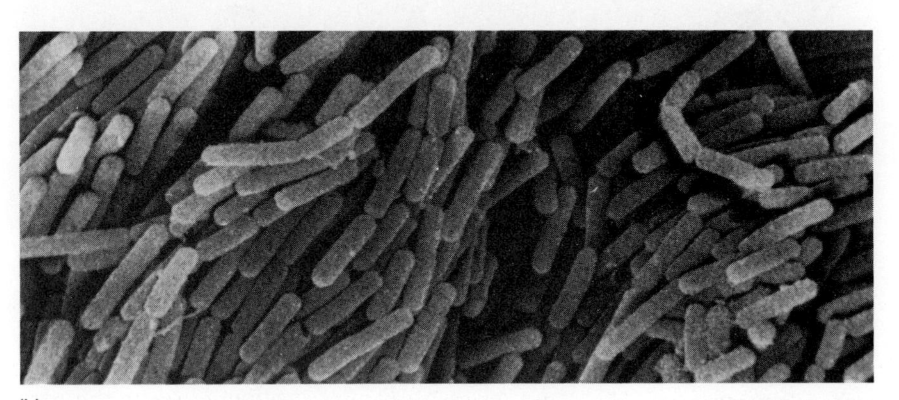

(b)

Figure 2.6 Images produced by transmission electron microscope and scanning electron microscope. *(a)* Transmission and *(b)* scanning electron micrographs of *Bacillus*. Notice that internal details are best seen using the TEM while surface features are readily viewed with the SEM.

PROKARYOTIC AND EUKARYOTIC CELLS: A DEFINITION

Examination of a variety of cells with light and electron microscopes has shown that there are two different types of cells: **prokaryotic** and **eukaryotic**. Prokaryotic cells (prokaryotes) get their name from the fact that they do not contain a nucleus: *pro* means "primitive," while *karyote* (from the Greek *karyon*) refers to the nucleus. Thus, prokaryote denotes a cell with a primitive or nonexistent nucleus. Eukaryotic cells (eukaryotes) get their name because they have a nucleus (*eu* means "normal" or "true"). Prokaryotic cells are generally much simpler than eukaryotic cells (fig. 2.7). Electron photomicrographs of typical prokaryotes reveal a cell wall, a plasma membrane, a homogeneous granular cytoplasm, and a compact mass of chromatin (DNA and protein), which will be called the **bacterial chromosome**, in the cytoplasm. In some prokaryotic cells, membranes such as mesosomes, chromatophores, and thylakoids are visible within the cytoplasm. Granules of starch, glycogen, polyphosphate, sulfur, and poly-β-hydroxybutyric acid, as well as gas vesicles, are also found in some prokaryotic

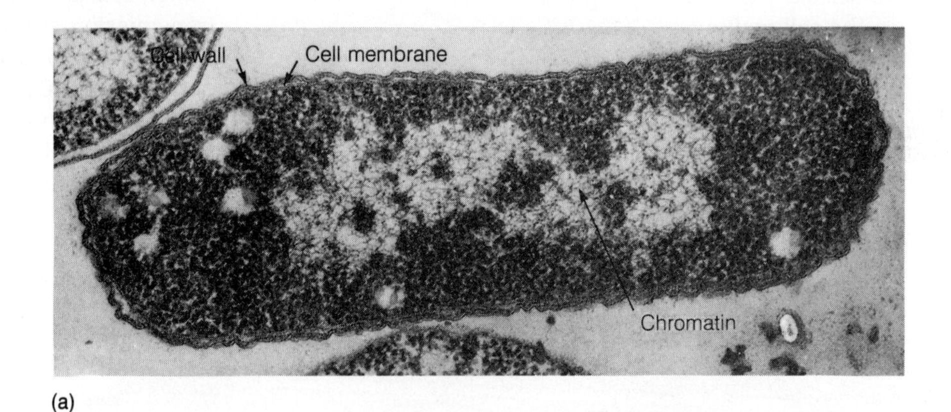

(a)

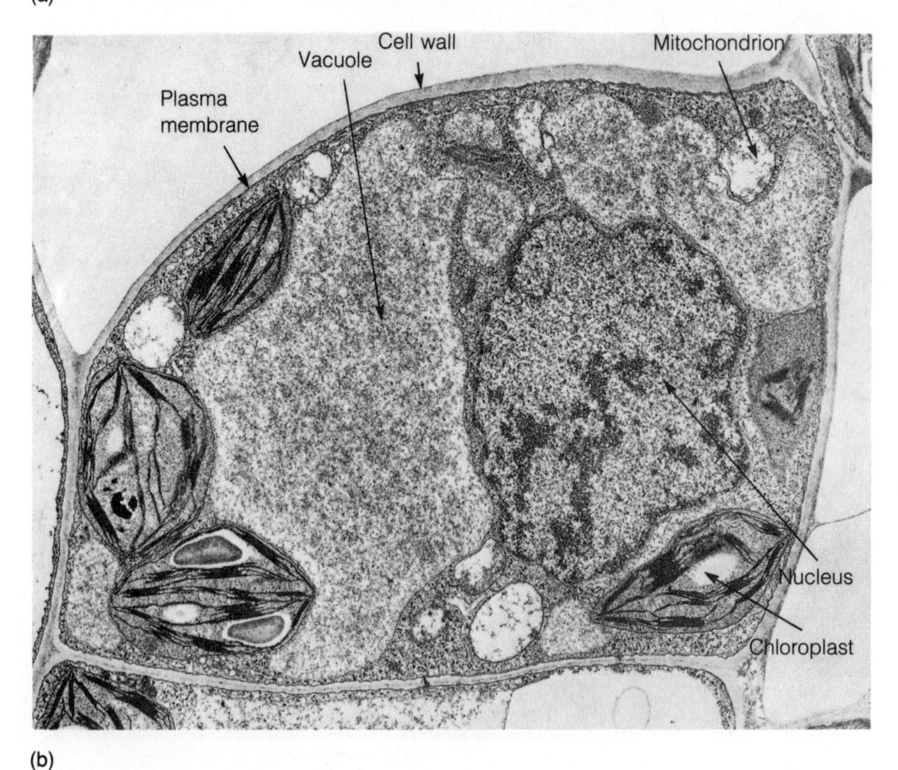

(b)

Figure 2.7 Prokaryotic and eukaryotic cells. *(a)* A transmission electron micrograph shows a cross section along the length of a prokaryotic (bacterial) cell, *Bacillus cereus*. The cell wall, cell membrane, and a dispersed chromosome, referred to as chromatin, are regularly seen in prokaryotes. *(b)* A transmission electron micrograph shows a cross section through a eukaryotic (plant) cell. The plant cell is from the bean plant, *Phaseolus vulgaris*. The cell wall, plasma membrane, nucleus with dispersed chromosomes called chromatin, chloroplasts, mitochondria, vacuoles, Golgi membranes, endoplasmic reticulum, and ribosomes are generally seen in plant cells.

cells. All the different types of bacteria (**eubacteria, archaebacteria, cyanobacteria, chlamydiae, rickettsiae,** and **mycoplasmas**) are prokaryotic cells. In fact, the terms prokaryotic and bacterial may be used interchangeably.

Eukaryotic cells are generally much larger than prokaryotic cells, and have organelles in their cytoplasm (fig. 2.7). **Organelles** are membranous structures inside cells that carry out special functions. Examples of organelles are the nucleus, the mitochondrion (plural, mitochondria), the chloroplast, the endoplasmic reticulum, and the Golgi system. Some of the differences between prokaryotic cells and eukaryotic cells are outlined in table 2.3.

GROSS CELLULAR MORPHOLOGY OF BACTERIA

Bacteria exhibit a wide range of sizes. Some, like the agents of typhus fever and Rocky Mountain spotted fever (rickettsia), are quite small and are barely resolved with the light microscope. Others are very large and can be seen with a magnifying lens or simple microscope. Most commonly encountered bacteria, however, have diameters between 0.2 and 3.0 μm. Rod-shaped bacteria usually have lengths between 0.5 and 15μm. The average rod-shaped bacterium is about the size of an average mitochondrion in a eukaryotic cell. Eukaryotic cells generally have diameters between 7 and 100 μm. A mouse cell infected with an intracellular bacterium, *Streptococcus pyogenes*, illustrates the difference in size between a prokaryote and a eukaryote (fig. 2.8). Because bacteria are so small, they have a large surface-to-volume ratio. Thus, they are able to concentrate nutrients rapidly and the nutrients are able to diffuse quickly to all sites inside the cell. Their small size is one of the factors that enables some bacteria to multiply rapidly. Bacteria are the fastest-growing life forms on earth. For example, the bacterium *Escherichia coli* can duplicate itself in about 20 minutes in a rich medium at 37°C. In contrast, a mammalian cell in culture takes 13–24 hours to divide.

The shape of a bacterial cell is a characteristic that is often helpful in identifying it. Most of the commonly encountered bacteria are spheres, rods, curved rods, spiral rods, or helical rods (fig. 2.9). The spherical bacteria are commonly called **cocci** (singular, **coccus**); the rods are known as **bacilli** (singular, **bacillus**); the curved rods are designated **vibrios**, (singular, **vibrio**); the spiral rods are referred to as **spirilla** (singular, **spirillum**); and the helical rods are identified as **spirochetes** (singular, **spirochete**). The shape of a

TABLE 2.3
SUMMARY OF CHARACTERISTICS OF PROKARYOTIC AND EUKARYOTIC CELLS

Characteristics	Prokaryotic cell	Eukaryotic cell
Genetic material	Single chromosome in cytoplasm. No discrete nucleus.	Many chromosomes inside nucleus.
Double-membraned organelles	Absent.	Mitochondria, chloroplasts.
Single-membraned organelles	Chlorobium vesicles, chromatophores, thylakoids.	Golgi membranes, lysosomes, endoplasmic reticulum, vacuoles.
Ribosomes	70S. Free in cytoplasm.	80S. On endoplasmic reticulum and free in cytoplasm.
Locomotion	Flagella that rotate, gliding motion.	Flagella and cilia that undulate. Ameboid motion.
Plasma membrane	Sterols usually absent.	Sterols usually present.
Size of cell	Usually <5 μm in diameter.	Usually >10 μm in diameter.
Cell wall	Peptidoglycan or protein.	Cellulose or chitin.
Mitosis and meiosis	Absent.	Present.
Site of respiration	Plasma membrane.	Mitochondria.
Site of photosynthesis	Chromatophores, chlorobium vesicles, thylakoids.	Chloroplasts.

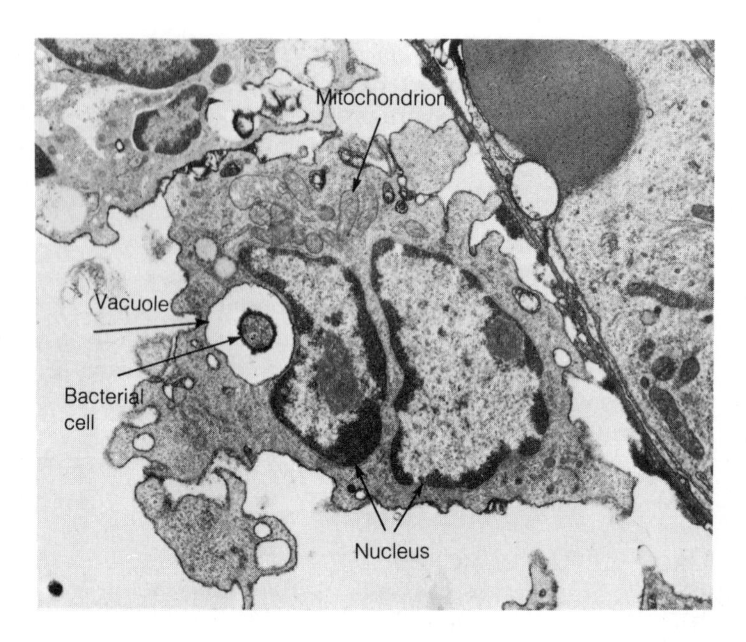

Figure 2.8 **Comparison between a eukaryotic cell and a prokaryotic cell.** A lung mononuclear phagocyte (a eukaryotic cell), which has ingested a cell of *Streptococcus pyogenes* (a prokaryotic cell) is shown here. The bacterial cell is contained within a vacuole. The phagocyte's nucleus and mitochondrion are also indicated.

bacterium is determined by the shape of a rigid layer called the **cell wall**. If bacteria are treated with chemicals that disrupt or remove their cell walls and are placed in an isotonic environment, they assume a spherical shape, regardless of their original shape. A cell wall determines the shape of some eukaryotic cells, but many eukaryotic cells maintain a rigid shape without one. In these cells, the cytoplasmic membrane is strengthened by a layer of **stress fibers**, called the **cortex**, along the inside surface of the cytoplasmic mem-

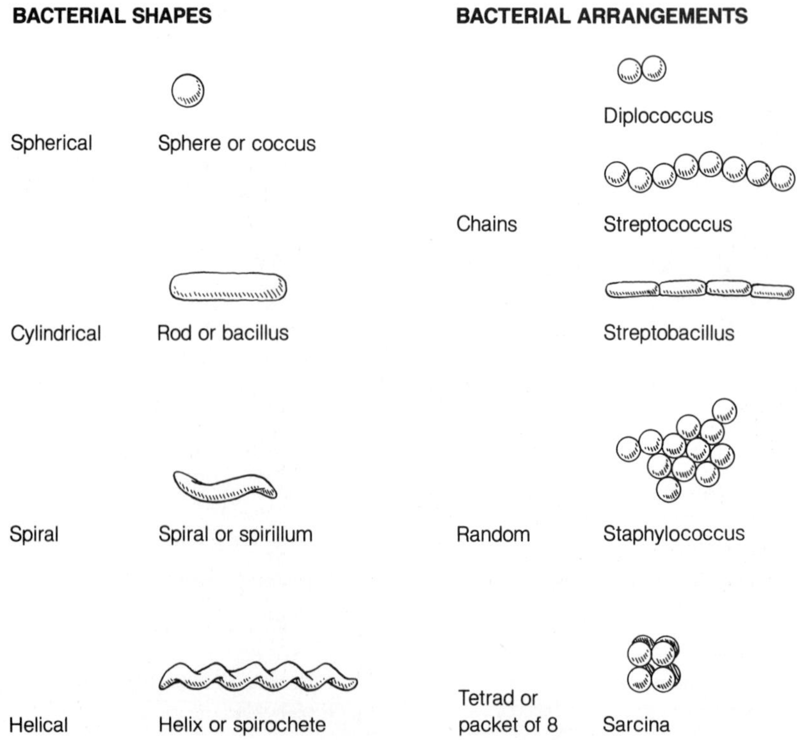

BACTERIAL SHAPES
Spherical — Sphere or coccus
Cylindrical — Rod or bacillus
Spiral — Spiral or spirillum
Helical — Helix or spirochete

BACTERIAL ARRANGEMENTS
Diplococcus
Chains — Streptococcus
Streptobacillus
Random — Staphylococcus
Tetrad or packet of 8 — Sarcina

Figure 2.9 **Bacterial shapes and arrangements.** Bacterial shapes are useful in identifying organisms. Often cocci remain attached to each other and form typical arrangements such as diplococci, streptococci, and staphylococci. Rod-shaped bacteria in chains are sometimes called streptobacilli.

brane. The stress fibers consist of proteins, such as microtubules, actin, and myosin.

Bacteria primarily multiply asexually and amitotically by a process called transverse **binary fission.** During this process, a mature bacterial cell divides in half to produce two identical "daughter" cells. Following cell division, each daughter cell grows until it is capable of dividing again. Among certain groups of bacteria, the daughter cells remain attached after cell division and characteristic arrangements of the cells occur. For example, bacteria in the genus *Streptococcus* often form long chains, while those in the genus *Staphylococcus* frequently form grapelike clusters. Some of the basic cell arrangements that bacteria may assume are illustrated in figure 2.9.

FINE STRUCTURE OF BACTERIAL CELLS

Although the light microscope is very useful for identifying microorganisms, the electron microscope has been an invaluable tool in the study of subcellular structures and has allowed scientists to uncover the detailed structure of the prokaryotes. All bacteria have a plasma membrane, a genome (hereditary material) that is double-stranded DNA, and ribosomes that carry out protein synthesis. All bacteria have some kind of cell wall, except for the mycoplasmas. There are many structures, such as endospores, capsules, thylakoids, gas vesicles, and reserve materials, that are associated only with certain bacteria. In the sections that follow, some of the major bacterial structures are discussed.

Cell Envelopes

Plasma Membrane The **plasma membrane** (or cytoplasmic membrane) is the thin, pliable, lipid and protein envelope that defines a cell and controls the movement of small molecules in and out of the cell. Plasma membranes are approximately 7 nm in thickness and are therefore below the limit of resolution of the light microscope (approximately 280 nm), but they can be seen with the TEM (fig. 2.7). The membrane consists of two layers of phospholipids (lipid bilayer) in which proteins and glycoproteins are embedded (fig. 2.10). According to the **fluid mosaic model**, the proteins in the lipid bilayer are free to move like "icebergs" in the phospholipid "sea." This model suggests that the membrane is a dynamic structure, capable of rapidly changing its protein and lipid composition.

The phospholipid bilayer is the barrier that limits the movement of material back and forth between the cytoplasm and the outside environment. Many of the proteins in the membrane specifically facilitate the movement of materials through the membrane. For

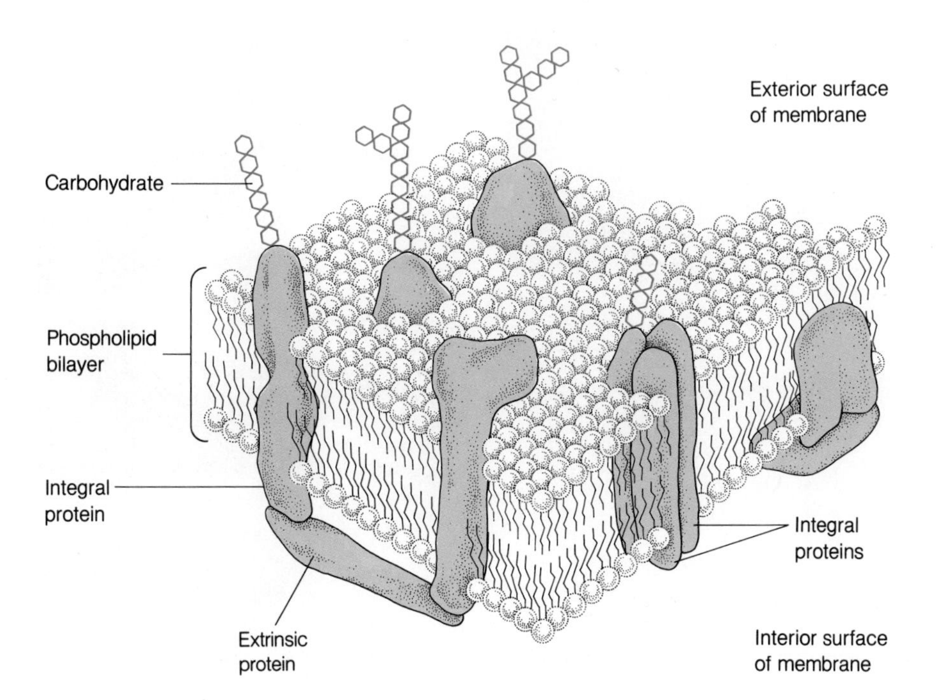

Figure 2.10 **Bacterial plasma membrane.** The bacterial plasma membrane consists of a lipid bilayer and various proteins that are associated with the lipid bilayer. Proteins that penetrate into the interior of the lipid bilayer are called integral proteins. Some integral proteins form channels, function in the transport of materials, and strengthen the membrane. Proteins that lie on the surface and do not penetrate the lipid bilayer are called peripheral or extrinsic proteins. Extrinsic proteins generally strengthen the membrane, bind nutrients, and carry out chemical reactions. Polysaccharides are frequently attached to membrane proteins and membrane lipids on the exterior side of the membrane.

FOCUS PHOTOSYNTHETIC MEMBRANES: CHLOROBIUM VESICLES, LAMELLAR MEMBRANES, AND CHROMATOPHORES

A few groups of bacteria have developed membranes with specialized functions. Some of the membranes contain enzymes, chlorophylls, and accessory pigments that allow photosynthesis (fig. 2.11). Some nonphotosynthetic bacteria, such as the nitrogen-fixers (*Azotobacter*) and the nitrifying bacteria (*Nitrobacter*), have invaginating membranes that are not involved in photosynthesis. The function of these invaginating membranes is not known. It is believed that they may contain the enzymes necessary for the fixation of CO_2 into organic molecules or for the generation of extra ATP used in fixing CO_2.

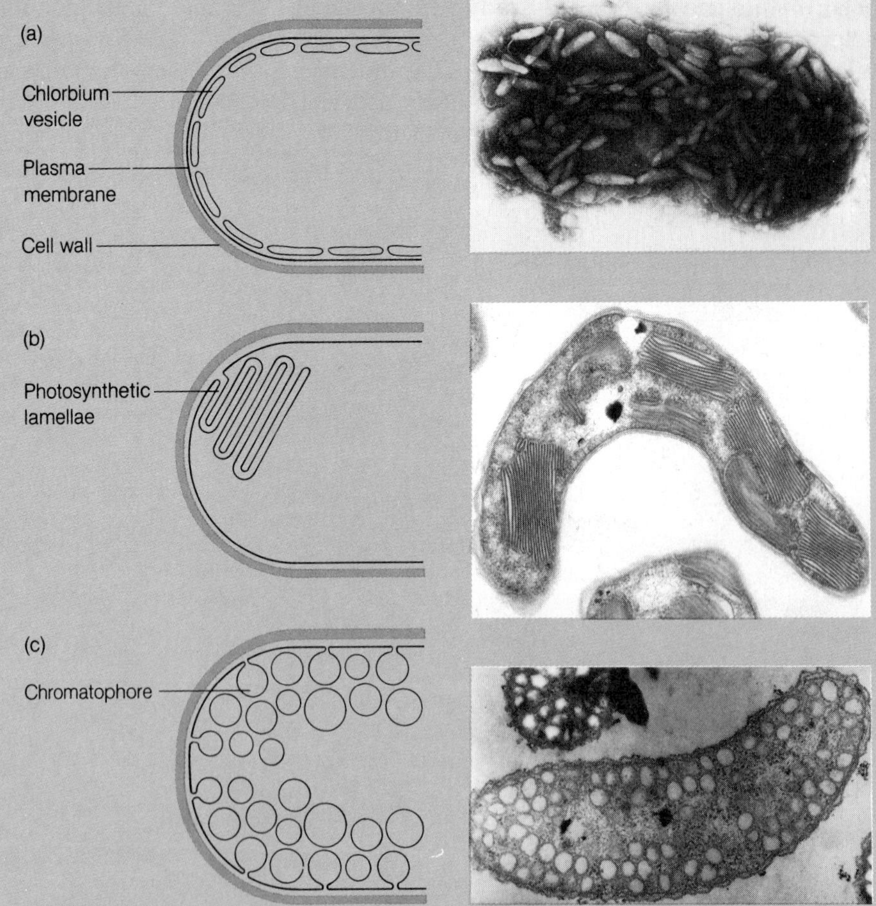

Figure 2.11 Photosynthetic membranes. *(a)* The chlorobium vesicles or chlorosomes of the green sulfur bacterium *Pelodictyon* are cigar-shaped vesicles and lie just under the plasma membrane. They are involved in the synthesis of ATP, which is required in large quantities to fix CO_2. The chlorobium vesicles are not continuous with the plasma membrane. *(b)* The lamellar membranes of the purple sulfur bacterium *Ectothiorhodospira* resemble stacked, folded sheets. The membranes are continuous with the plasma membrane. *(c)* Chromatophores of *Rhodospirillum rubrum*. The photosynthetic membranes are continuous with the plasma membrane.

example, a number of membrane proteins form **pores** that allow water and a few other materials to rapidly move back and forth across the membrane. In addition, membrane proteins called **permeases** transport nutrients into the cell.

In some bacteria, respiratory enzymes are associated with the bacterial plasma membrane. Thus, the plasma membrane in some prokaryotes is involved in **respiration**. Respiration is a type of oxidative metabolism whereby organisms generate ATP.

The bacterial cytoplasmic membrane also plays a role in the distribution of the genetic information to daughter cells. The genetic information is contained in a single chromosome that is bound to the plasma membrane. After the chromosome replicates, both copies are attached to the plasma membrane. As the membrane grows, a copy of the bacterial chromosome ends up near each end of the cell; therefore, daughter cells have a complete copy of the genetic material after cell division.

Bacterial flagella are anchored in the plasma membrane and are powered by the energy derived from energy-generating systems found in the membrane.

Cell Wall Almost all bacteria have a rigid layer of material that surrounds the plasma membrane. This rigid layer is called the **cell wall** and is analogous to the cell walls found in plants and fungi. The cell wall protects the plasma membrane and maintains the shape of the cell (fig. 2.7). All bacteria (except the **mycoplasmas**, which lack a cell wall, and the **archaebacteria**, which have unusual polysaccharide or proteinaceous walls) have a wall that is constructed of a polysaccharide made up of two alternating sugars, *N*-acetylglucosamine and *N*-acetylmuramic acid, joined together to form long strands. The strands are joined together by short peptides into a three-dimensional framework called **peptidoglycan** or **murein** (fig. 2.12).

The cell walls of gram-positive bacteria are 30 to 60 nm thick and consist almost exclusively of the peptidoglycan layer. The cell walls of gram-negative bacteria contain a thin peptidoglycan layer that is about

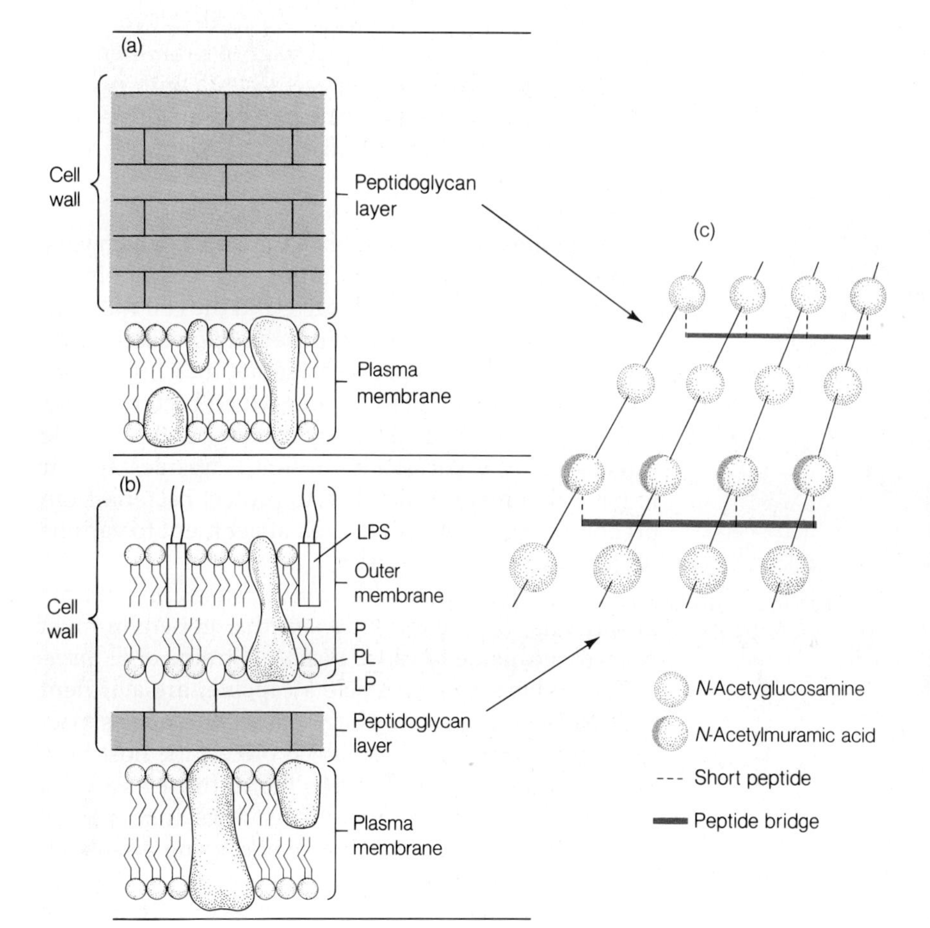

Figure 2.12 Cell wall of gram-positive and gram-negative bacteria. *(a)* The cell wall and cell membrane of a gram-positive bacterium are illustrated. The cell wall consists primarily of peptidoglycan. *(b)* The cell wall and cell membrane of a gram-negative bacterium are illustrated. The cell wall consists of a layer of peptidoglycan and a lipid bilayer called the outer membrane. The outer membrane contains lipopolysaccharides on the outer surface, which are called endotoxins because they cause allergic reactions in animals. The cell membrane of gram-negative bacteria is similar to the cell membrane in gram-positive bacteria. *(c)* Most bacteria have a peptidoglycan layer in their cell wall. The peptidoglycan consists of a polysaccharide made up of alternating *N*-acetylglucosamine and *N*-acetylmuramic acid and short peptides attached to the *N*-acetylmuramic acid. LPS, lipopolysaccharide; P, protein; PL, phospholipid; LP, lipoprotein.

2–3 nm thick. This layer contains no teichoic acids. An **outer membrane** 7 nm thick surrounds the peptidoglycan layer in the gram-negative bacterium and is considered to be part of the cell wall. The outer membrane is similar to the plasma membrane in that it consists of a lipid bilayer. Many of the lipids making up the external side of the outer membrane have polysaccharide attached to them. Consequently, these lipids are referred to as **lipopolysaccharides**. The outer membrane is a selective permeability barrier that excludes compounds with molecular weights above 800. Many small molecules pass through the outer membrane, however, because it contains various types of pores. The peptidoglycan layer and the outer membrane form a wall that is approximately 10 nm thick. In some bacteria, the lipopolysaccharide fraction of the outer membrane is toxic to mammals and is known as **endotoxin**.

The concentration of solutes (sugars, amino acids, and salts) in an environment have an effect on the flow of water into and out of cells. A high concentration of small molecules attracts water and is a region with a high **osmotic pressure**. With regard to cells, an **isotonic** environment is one in which the osmotic pressure is the same inside and outside the cell. A **hypotonic** environment is one that has a lower concentration of solutes than the cell's cytoplasm, while a **hypertonic** environment is one that has a higher concentration of solutes than the cell's cytoplasm. Bacteria in their natural habitat often grow in aqueous environments where the concentration of salts and sugars is much lower than inside the cell. In such environments, water tends to diffuse into the cell. Without a cell wall, the influx of water would cause the cell to swell and ultimately burst. The cell wall prevents this from happening by offering resistance to the excessive influx of water and the internal water pressure (osmotic pressure). Water flows into the cell until the osmotic pressure is balanced by the resistance of the cell wall. Prokaryotes without walls assume a spherical shape in an isotonic environment. When cells lacking walls are placed in a hypotonic environment, however, they fill with water and eventually burst, while cells introduced into a hypertonic medium lose water and shrink. A stable **protoplast** is said to form when the cell wall is completely removed from gram-positive bacteria in an isotonic environment. **Spheroplasts** look very much like protoplasts, but these structures result from the partial destruction of the cell wall of gram-negative bacteria. The treatment of these bacteria with penicillin in isotonic media for a number of generations results in daughter cells that are missing the peptidoglycan layer of the cell wall.

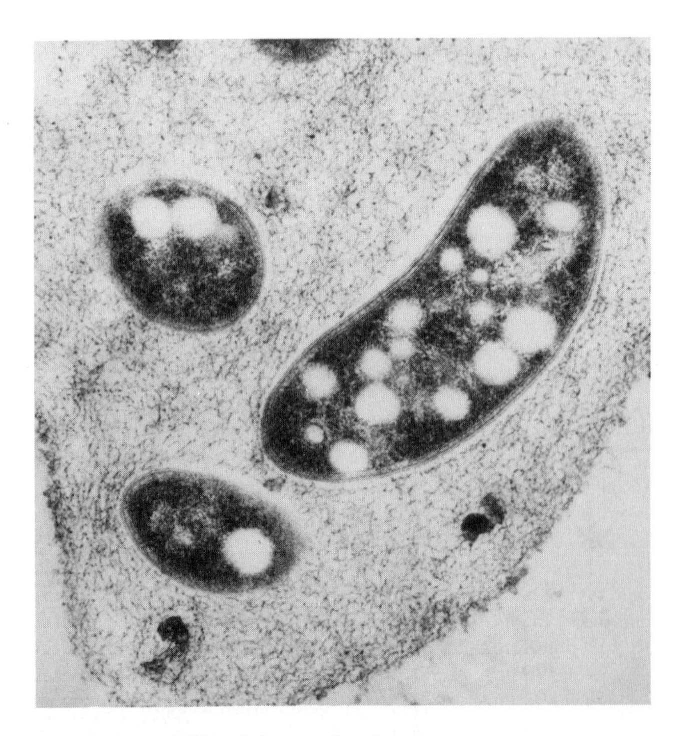

Figure 2.13 Bacterial capsule. A transmission electron micrograph of a thin section of *Rhizobium trifolii* shows the capsule around these bacteria.

Capsules and Glycocalyx **Capsules**, which vary in thickness, are slimy or gummy layers of polysaccharide or polypeptide that surround the cell wall (fig. 2.13). Most capsules are composed of polysaccharide. The production of capsules is sometimes influenced by the cultural conditions under which bacteria grow. For example, some organisms produce polysaccharide capsules only if sucrose is present. Capsules appear to have two major functions: to protect bacteria from predators, and to promote their attachment to various objects and to each other.

The bacterium that frequently causes pneumonia in humans, *Streptococcus pneumoniae*, generally must possess a capsule to cause disease. Strains of *S. pneumoniae*, unable to synthesize a capsule, are avirulent and readily ingested by phagocytic cells, such as macrophages and neutrophils, which protect the host from microorganisms. On the other hand, those strains of *S. pneumoniae* that are encapsulated resist ingestion by phagocytic cells and are capable of causing disease. It is believed that similar functions can be ascribed to capsules of microorganisms that normally inhabit

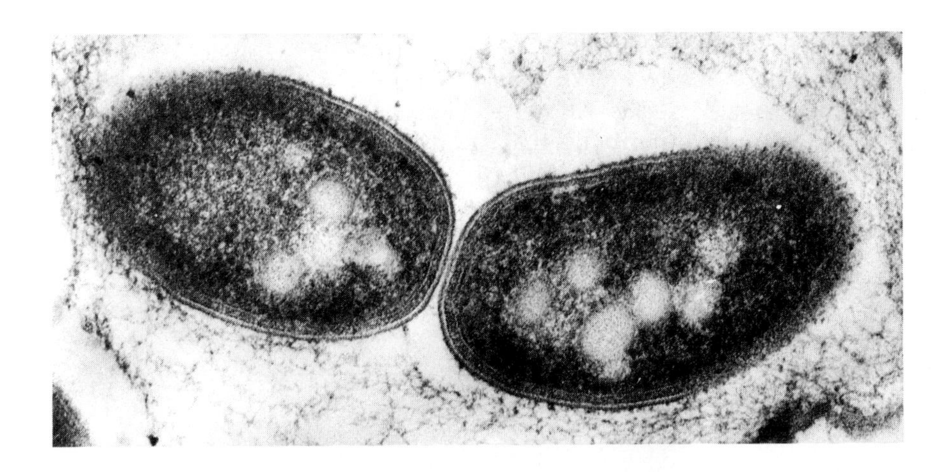

Figure 2.14 Bacterial glycocalyx. A transmission electron micrograph of a bacterium with glycocalyx. Bacterial glycocalyx is a network of polysaccharide that extends from the surface of many bacteria. A thick, dense glycocalyx is equivalent to a capsule. In this transmission electron micrograph, bacteria are attached by their glycocalyx to each other.

extrahuman habitats. The capsule may prevent predatory microorganisms from ingesting the encapsulated bacterium or interfere with its destruction once ingested. The presence of a capsule does not mean that an organism is pathogenic. There are many capsule-forming organisms, such as *Azotobacter* and *Leuconostoc,* that do not cause disease.

The **glycocalyx** consists of polysaccharide fibers that extend from the bacterial surface (fig. 2.14) and that allow bacteria to attach to inanimate objects, other bacteria, plant cells, or animal cells. The glycocalyx may become so extensive that it forms a capsule. Adhesion to inanimate objects can be advantageous to microorganisms in a rapidly moving stream, while attachment to plant or animal tissue provides bacteria with nutrients and a physical environment that is conducive to growth. Many diseases are caused by bacteria partly because they are able to synthesize a glycocalyx or a sticky capsule. For example, the bacteria that cause dental caries are able to do so because they can attach to tooth surfaces and reproduce there. The network of glycocalyx and bacteria on tooth surfaces, called **dental plaque**, often grows quite thick. This means that bacterial acids, which degrade the dental enamel, are highly concentrated at some locations and consequently cause cavities. In addition, it is believed that the bacteria that cause cholera and urinary tract infections are able to do so partly because they are able to attach to specific tissues, because of either their glycocalyx or the glycocalyx of the host tissue.

A number of other reasons for capsules have been proposed. For instance, capsules may be a way of storing large amounts of organic materials, protecting against desiccation, and protecting against virus infections. Capsules are best seen with negative stains, although simple stains have been developed to demonstrate their presence in biological materials.

Sheath **Sheaths** are stiff polysaccharide coverings that some bacteria secrete. Polysaccharides form a stiff sheath around a chain or group of bacteria (fig 2.15). The sheath provides an extra layer of material that protects the bacteria.

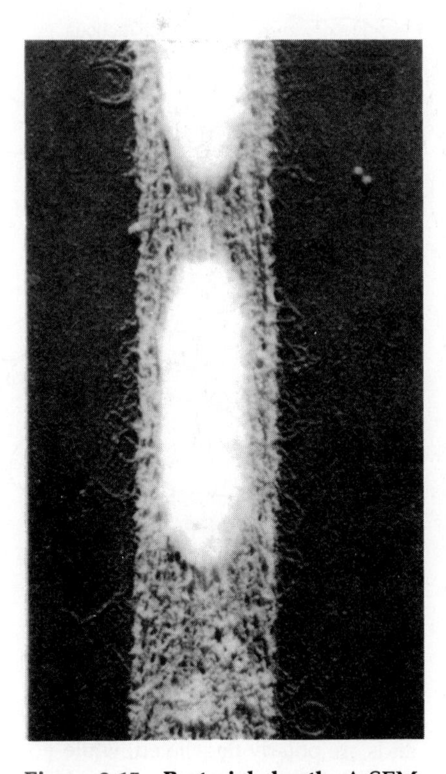

Figure 2.15 Bacterial sheath. A SEM of bacteria covered by a sheath.

Appendages and Structures Used for Locomotion

Flagella Bacterial **flagella** are hairlike appendages 10–20 μm long and about 0.02 μm in diameter that originate in the plasma membrane (fig. 2.16). Because they are so thin, bacterial flagella cannot be seen with the light microscope and special staining techniques must be used to demonstrate their presence. These flagellar stains employ compounds called **mordants** that coat the flagella and make them thicker. The mordant is stained with a dye, thus coloring it

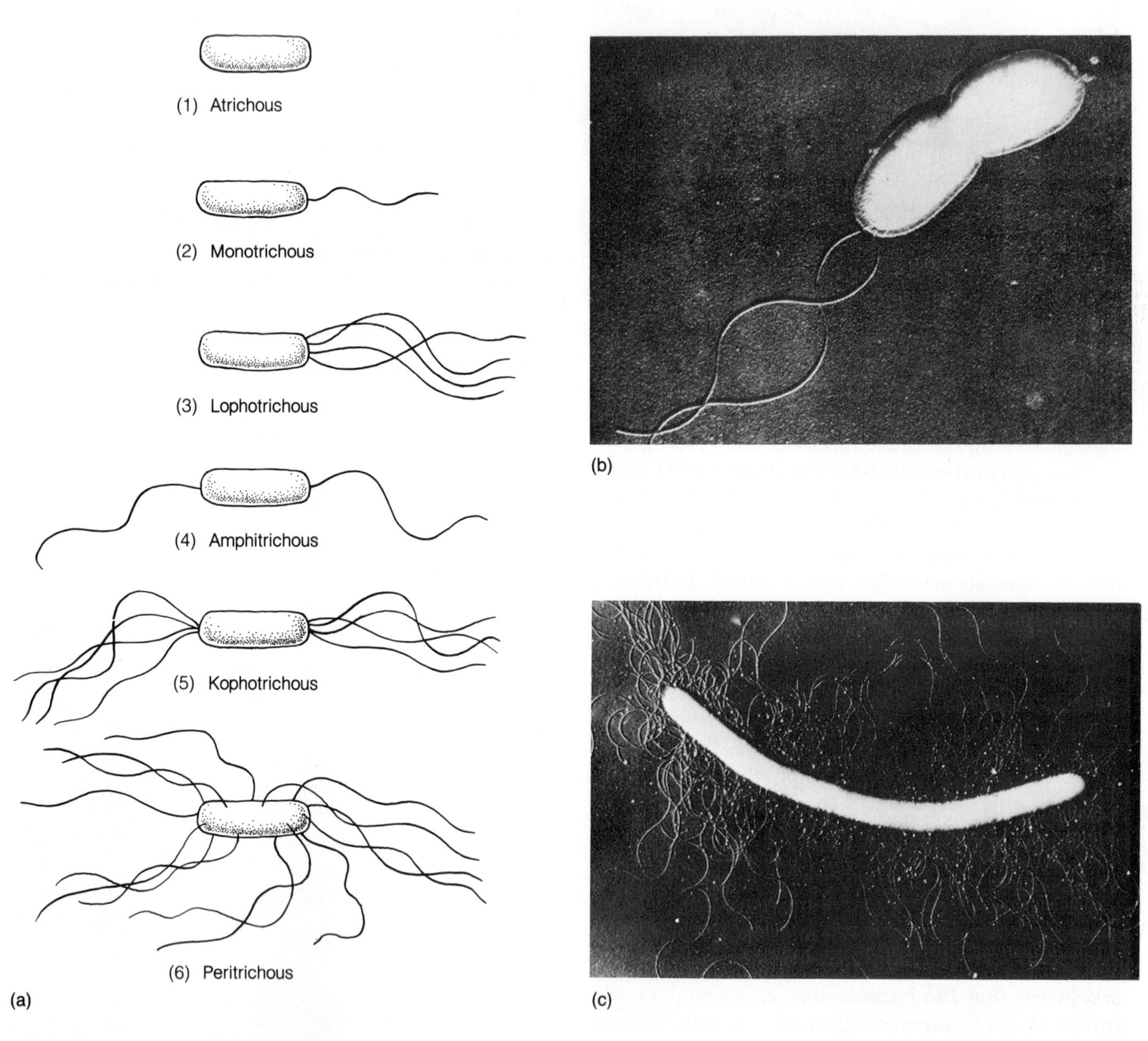

Figure 2.16 Arrangement of bacterial flagella. *(a)* This drawing illustrates the arrangement of bacterial flagella. Bacteria that have flagella only at their ends are polarly flagellated, while those that have flagella over much of their surface are peritrichously flagellated. The following nomenclature is sometimes used to describe the types of flagellation:
(1) atrichous; (2) monotrichous;
(3) lophotrichous; (4) amphitrichous;
(5) kophotrichous; (6) peritrichous.
(b) This transmission electron micrograph of *Pseudomonas fluorescens*, shadowed with a heavy metal, shows that it has two flagella at one pole. *(c)* This transmission electron micrograph of *Proteus vulgaris*, shadowed with a heavy metal, shows the flagella over the entire surface.

and making the flagella easier to see. The arrangement of flagella as seen in stained preparations (fig. 2.16) is sometimes used in identifying an unknown bacterium.

When a flagellated bacterium moves in a liquid, it generally moves in one direction for a while and then **tumbles** before moving off in another direction (fig. 2.17). Some bacteria respond to chemicals or to light by moving toward or away from them. For example, when bacteria move toward a source of nutrients, they tumble less frequently than usual and therefore have long runs. When they are moving away from a nutrient, however, they tumble more frequently than usual. Bacteria eventually end up where the concentration of the nutrient is greatest.

Axial Filaments The spirochetes have modified flagella, called **fibrils**, which form **axial filaments** (fig. 2.18). Axial filaments coil around the body of the cell between the peptidoglycan layer and the outer membrane (fig. 2.18). The rotating axial filaments cause the spirochete to corkscrew through an aqueous environment.

Gliding Microorganisms The **gliding bacteria** have neither flagella nor axial filaments, yet they are capable of a slow, steady movement over surfaces. There are a number of explanations for **gliding motility**, but none satisfactorily explain the phenomenon.

Pili **Pili** (singular, **pilus**), or **fimbriae** (singular, **fimbria**), are hollow tubes that protrude peritrichously from some bacteria and are made of a protein called **pilin** (fig. 2.19). Pili are about 0.02 μm in diameter and 0.2–20 μm in length. There are at least six different groups of pili, classified on the basis of their

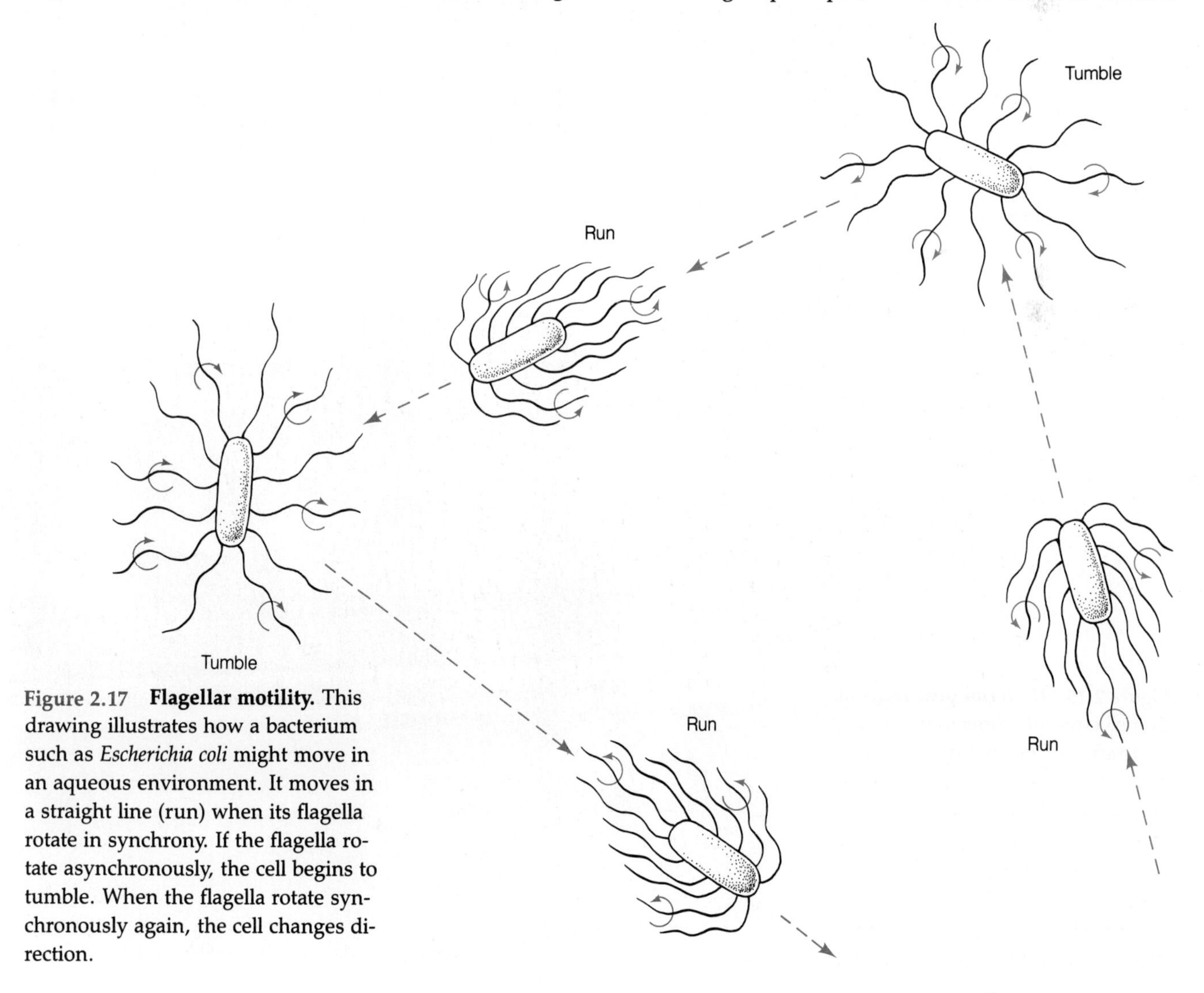

Figure 2.17 **Flagellar motility.** This drawing illustrates how a bacterium such as *Escherichia coli* might move in an aqueous environment. It moves in a straight line (run) when its flagella rotate in synchrony. If the flagella rotate asynchronously, the cell begins to tumble. When the flagella rotate synchronously again, the cell changes direction.

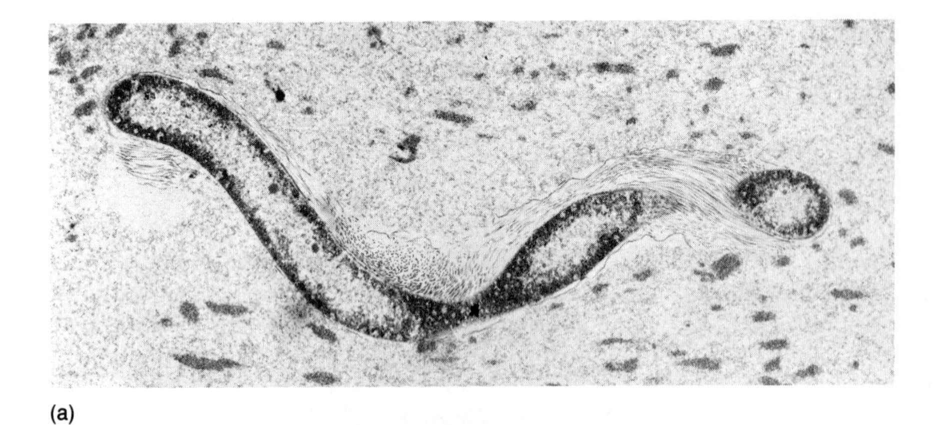

(a)

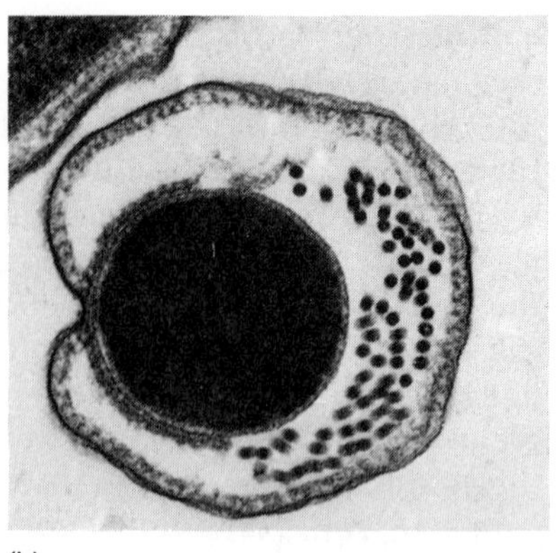

(b)

Figure 2.18 Axial filaments. These are transmission electron micrographs of a spirochete from the human mouth. *(a)* Numerous fibrils that make up the axial filament can be seen wound along the length of this helical cell. Fibrils originate near each end of the cell and wind back along more than half a cell length, where they overlap. *(b)* A cross section of a spirochete shows many fibrils between the peptidoglycan layer and the outer envelope (or membrane) of the cell wall.

adhesive and morphological properties. Pili are for the most part found in gram-negative bacteria, such as the **enterics** and the **psuedomonads**, although at least one gram-positive bacterium, *Corynebacterium renale*, has them.

Pili allow bacteria to stick to one another, to other organisms, and to inanimate objects. Thus, like the glycocalyx, pili contribute to the adhesiveness of bacteria. There are many examples indicating that the adhesiveness of piliated cells is one factor that determines the **virulence** (disease-causing ability) of some bacteria. For example, *Neisseria gonorrhoeae*, which causes gonorrhea, and *Escherichia coli*, a common cause of urinary tract infections, must both be piliated to

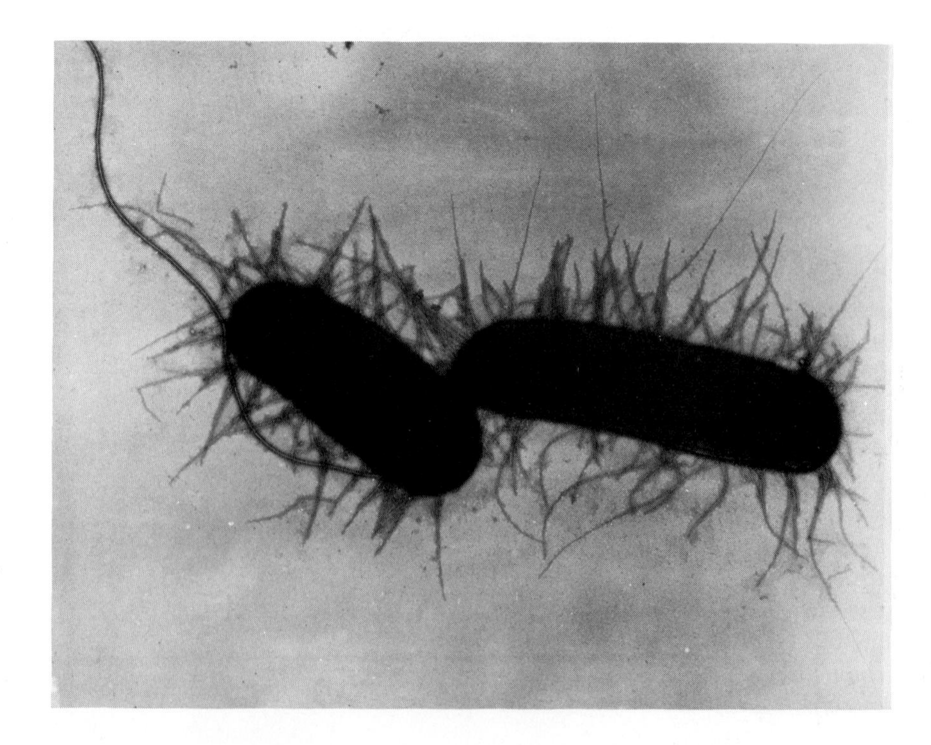

Figure 2.19 Bacterial pili. Transmission electron photomicrograph of *Escherichia coli* exhibiting pili and flagella. Pili are the short appendages that protrude all around the cell and may help bacteria attach to each other or to solid surfaces. One type of pilus, the F pilus is involved in transferring genetic information from one bacterium to another.

cause disease. This requirement is the exception rather than the rule, however, since pili are not required by most pathogenic bacteria.

Pili and the glycocalyx allow bacteria on the surface of a liquid medium to form thin films, or **pellicles**. The formation of pellicles that do not sediment is advantageous because the bacteria float on the surface, where the oxygen concentration is the greatest.

One group of pili (**sex pili**) are involved in the unidirectional transfer of DNA from donor or "male" bacteria to recipient or "female" bacteria. The sex pili bind a male and female bacterium together and facilitate the transfer of DNA from the donor to the recipient. The pilus functions by drawing the cells together so that a more intimate connection forms between the cells.

Intracellular Structures

The Chromosome The genetic material in bacteria is a double-stranded DNA molecule called the **chromosome** (fig. 2.20). It contains all the information for controlling the development and metabolic activities characteristic of the cell. The sum total of an organism's hereditary information is referred to as its **genome**. The typical bacterial chromosome consists of a single, circular molecule of DNA with a contour length (circumference) of about 1100 μm and containing about 4000 genes. The chromosome is folded into a tight mass, often less than 0.2 μm in diameter, and is complexed with small amounts of protein and RNA (fig. 2.20). Thin sections of bacteria examined with a transmission electron microscope show a tightly bundled chromosome occupying 15–25% of the cell's cytoplasm (fig. 2.6). Some bacterial chromosomes such as those found in many of the mycoplasmas, are very small and are estimated to contain fewer than 1000 genes, while others, such as those found in some cyanobacteria, are larger than average and may have more than 5000 genes.

Plasmids Many bacteria possess multiple copies of small self-duplicating pieces of circular DNA called **plasmids** in addition to their chromosome (fig. 2.21). Small plasmids contain only a few genes, while the larger plasmids may consist of hundreds of genes. In general, the plasmid's genetic information is not necessary for the survival of the cell and therefore the cell can loose them without dying. Plasmids often carry genetic information that makes the bacterium resistant to certain antibiotics and heavy metals or able to synthesize or break down unusual compounds. Some plasmids can be transferred from one bacterium to another, and some can mobilize the transfer of the main chromosome by a process called **conjugation**.

Ribosomes **Ribosomes** are cellular structures composed of proteins and RNA. They are involved in

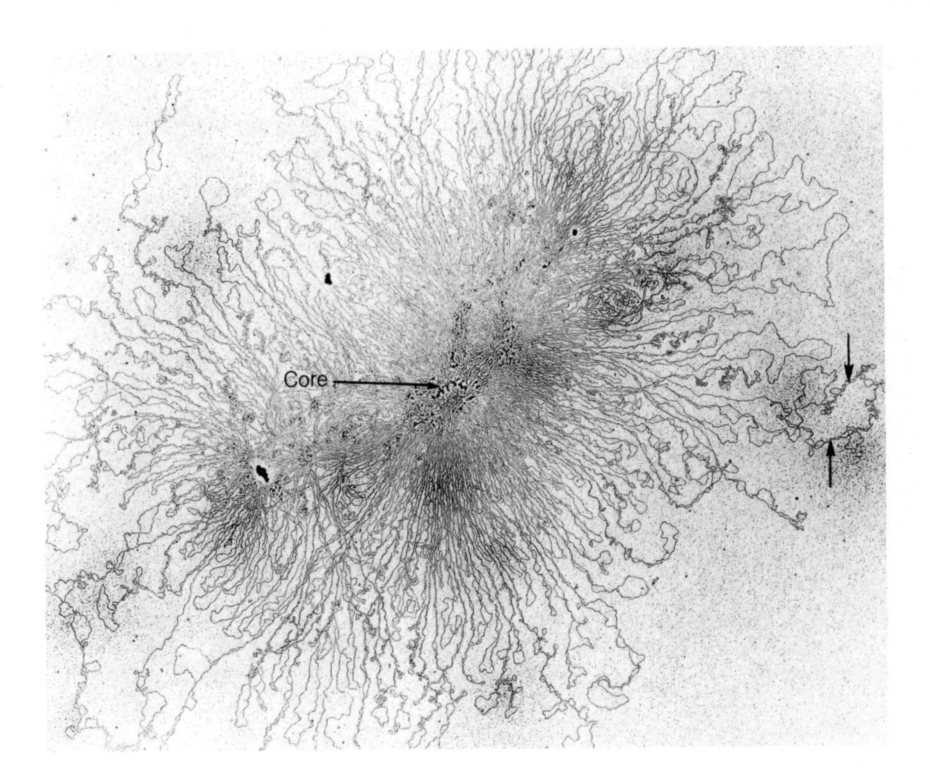

Figure 2.20 **A bacterial chromosome.** The bacterial chromosome from *Escherichia coli* consists of a circular, double-stranded DNA molecule. The chromosome is supercoiled and folded so that it can fit into the cell. The protein–RNA core that keeps the DNA folded and loops of the supercoiled DNA (arrows) are indicated.

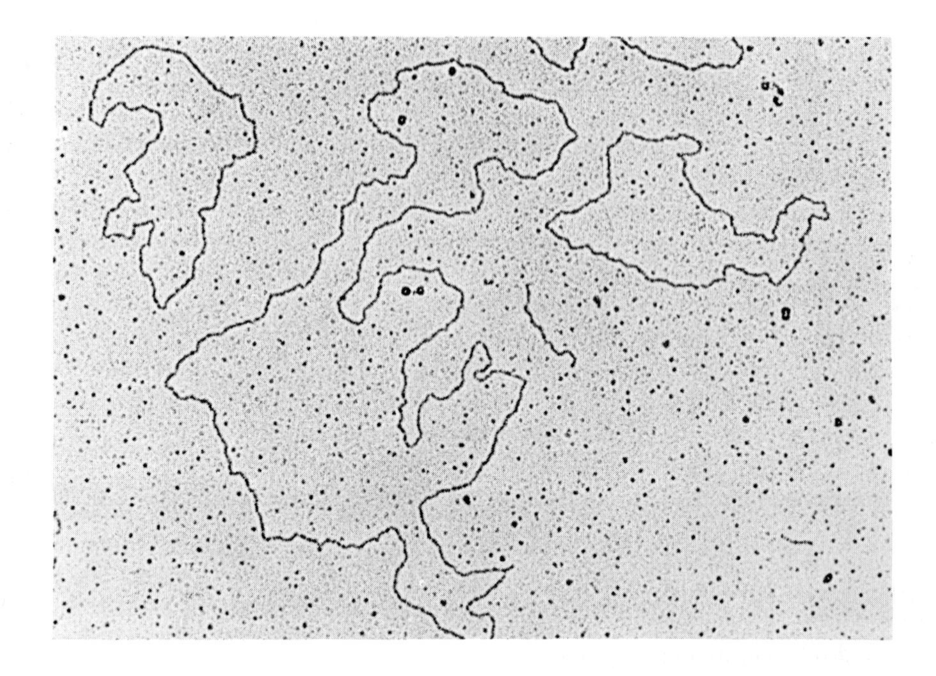

Figure 2.21 Bacterial plasmids. The larger molecule (plasmid) is pBF4 isolated from *Bacteroides fragilis* and codes resistance to clindamycin and erythromycin. The smaller plasmids are pSC101 isolated from *Escherichia coli* and code resistance to tetracycline.

protein synthesis, an essential process for all cells. Protein synthesis in bacteria is carried out by **70S ribosomes** (fig. 2.22). 70S refers to the sedimentation velocity, in Svedberg units, of the ribosome in gradients. Since Svedberg units reflect the size and density of objects, a 50S structure is larger and denser than a 30S structure. When bacterial ribosomes are not engaged in protein synthesis, they split into the 50S and 30S ribosomal subunits. The 50S and 30S subunits are scattered throughout the cytoplasm and join together to form a 70S ribosome when protein synthesis begins.

Endospores A small number of soil bacteria (*Bacillus, Clostridium, Desulfotomaculum, Sporosarcina, Sporolactobacillus, Thermoactinomyces,* and possibly *Metabacterium*) produce a special type of spore that forms within the bacterial cell, called an **endospore** (fig. 2.23). Endospores are extremely resistant to high temperatures and desiccation. For example, *Thermoactinomyces* produces heat-resistant endospores that can survive boiling at 100°C for as long as an hour. *Thermoactinomyces*, in its cellular form, is killed at temperature above 75°C. Endospores are also very resistant to chemicals such as pesticides, antibiotics, and dyes.

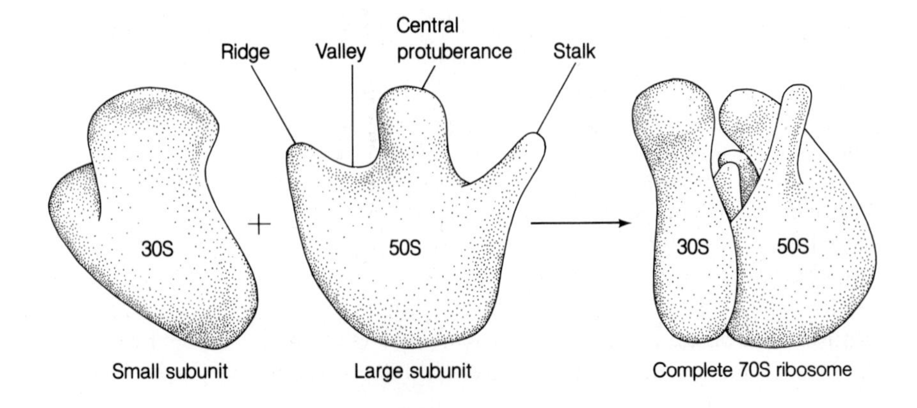

Figure 2.22 70S ribosomes. The bacterial 70S ribosome is approximately 25 nm in diameter and is made up of rRNA and proteins. It consists of two subunits, the 30S subunit and the 50S subunit.

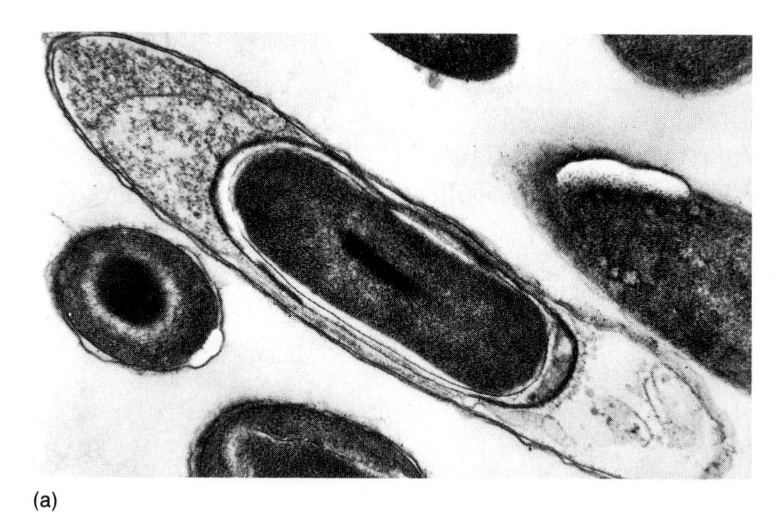

(a)

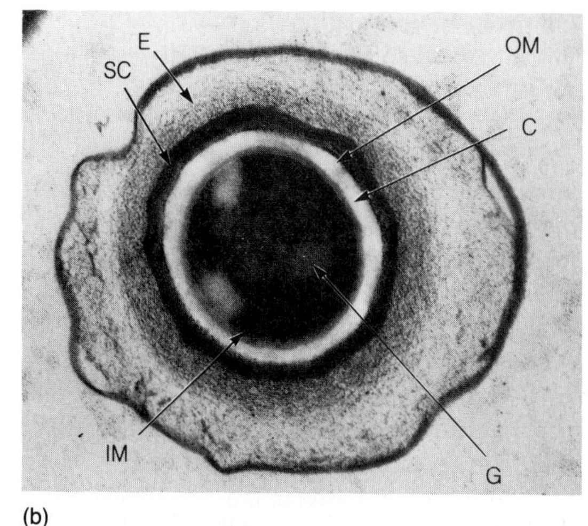

(b)

Figure 2.23 **Endospores.** Endospores may be spherical or oval. Each cell produces only one endospore. *(a)* An endospore produced by *Bacillus fastidiosus* is visible within the cell. *(b)* The endospore consists of a dehydrated core where the genome (G) is found, an inner membrane (IM) that surrounds the dehydrated core, a surrounding cortex (C), an outer membrane (OM), a spore coat (SC), and an exosporium (E).

Endospores contain only about 15% water, in comparison to vegetative cells, which are more than 75% water. Because of their low water content, no metabolism occurs within endospores. As a consequence, they do not assimilate chemicals. Since endospores are resistant to all but extreme conditions, they protect the bacterial chromosome.

Under conditions of limiting carbon, nitrogen, or phosphorus, the endospore-forming bacteria undergo a developmental process called **sporulation**, which results in the formation of an endospore (fig. 2.24). Each cell produces only one endospore, whose size and location within the cell vary depending upon the species. Because only one endospore is produced from each cell and the cell degenerates upon germination, sporulation is not a form of reproduction. Since endospores often survive temperatures as high as 100°C and the lack of nutrients and water, they persist for long periods of time in foods and soils. Even after boiling for as long as 5 minutes, most endospores commonly encountered survive and are capable of germinating when favorable environmental conditions return. **Germination** involves the emergence of a single vegetative cell from the endospore.

The failure to sterilize canned vegetables or meats that are contaminated with *Clostridium botulinum* endospores can result in a deadly situation if these canned goods are subsequently consumed without cooking. Under the anaerobic conditions within the jar, the endospores germinate and the resulting vegetative cells produce one of the most potent poisons known, the **botulism toxin**. This toxin, when ingested even in small amounts, causes severe illness or death.

FINE STRUCTURE OF EUKARYOTIC CELLS

Cell Envelopes

Plasma Membrane The plasma membrane of eukaryotic cells consists of a lipid bilayer and associated proteins, and so resembles that of the prokaryotes. The eukaryotic plasma membrane, however, generally contains high concentrations of lipids such as cholesterol, while the prokaryotic membrane generally does not. One notable exception is the plasma membranes of some of the mycoplasmas, which contain cholesterol. There are numerous proteins in the eukaryotic plasma membrane that are involved in controlling the movement of materials back and forth, but there are no respiratory enzymes located in the eukaryotic plasma membrane. All respiratory enzymes in the eukaryotic cell are located in the inner membranes of **mitochondria**.

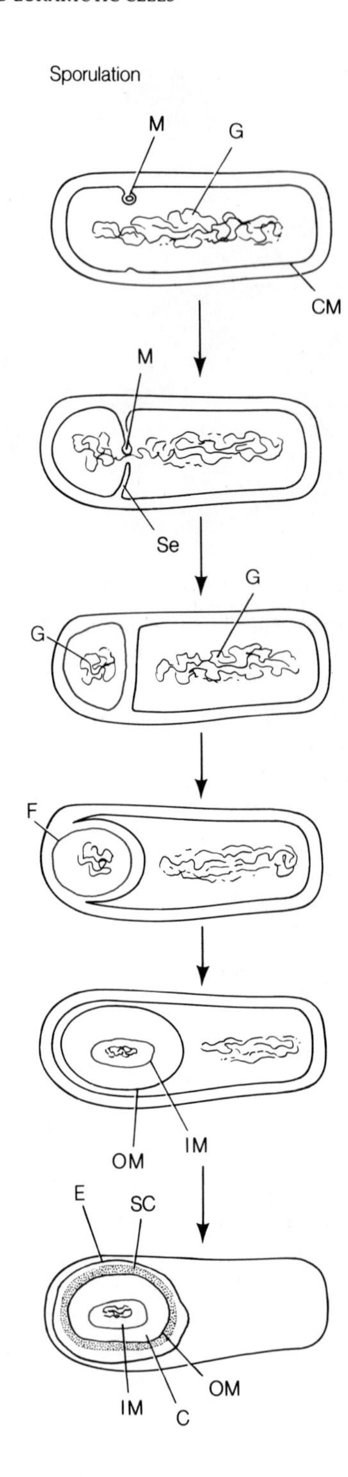

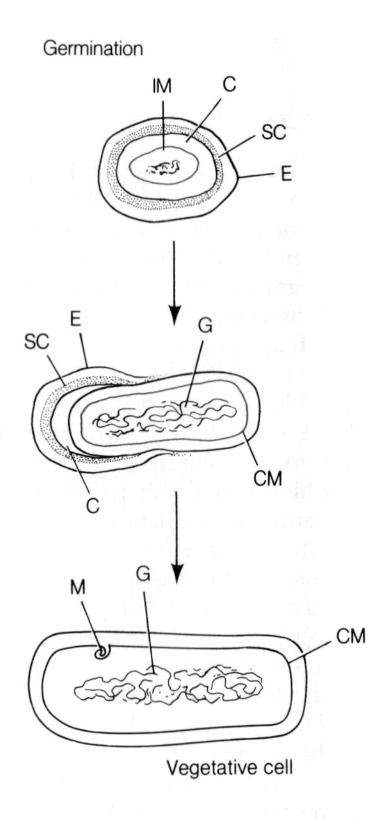

Figure 2.24 Sporulation. Sporulation begins when a committed cell divides into a minicell and a large cell. The sporulation process involves the engulfment of the minicell by the large cell, the loss of water by the minicell, and the development of various layers of material within and around the developing endospore. Eventually the vegetative cell decomposes and the endospore is released. At 37°C, the sporulation process takes approximately 15 hours. **Germination.** The germination process involves the reconstitution of the endospore, the breakdown of the cortex and the rupture of the endospore coat. From the endospore emerges a vegetative cell. Germination takes approximately 1 hour. Mesosome (M), septum (Se), cell membrane (CM), forespore (F); for other labels see figure 2.23.

Cell Wall Most of the photosynthetic microorganisms, the fungi, and plant cells have cell walls that are chemically and structurally very different from one another and from the bacterial cell walls. For example, the cell walls in fungi may be constructed from the polysaccharides cellulose and chitin. The cell walls of the photosynthetic microorganisms can consist of cellulose, silicon, or calcium carbonate, while

FOCUS THE MAGNETIC BACTERIA

Various aquatic bacteria have tiny deposits of magnetic iron in their cytoplasm that orient them along the lines of force of the earth's magnetic field (fig. 2.25). Consequently, when these bacteria swim, they move along magnetic field lines toward the bottom of the water. Bacteria that respond to magnetic fields are called **magnetotactic bacteria**. The magnetic field lines of the earth are vertical at the poles and tangential at the equator, but between the poles and the equator the field lines are inclined at angles that increase with latitude. In the Northern Hemisphere the field is inclined downward, while in the Southern Hemisphere it is inclined upward. The magnetic bacteria in the Northern Hemisphere swim downward and toward the north pole unless a strong magnet is brought near them. These bacteria also swim downward in the Southern Hemisphere, because their magnets are reversed. The north and south poles of a strong magnet determine the direction in which the magnetotactic bacteria swim. If the direction of the magnet changes, the bacteria change their direction.

The electron microscope has helped scientists see the magnets within the magnetotactic bacteria. The magnets consist of small crystals of magnetite (Fe_3O_4) that are usually arranged in a thin line along the length of the bacterial cells (fig. 2.25). The particles of magnetite, called magnetosomes, function like a compass, orienting the bacteria along the lines of force of the earth's magnetic field. The magnetosomes appear to be individually enclosed in membranes.

The bacteria synthesize magnetosomes from iron in the environment. If magnetotactic bacteria are grown in an iron-poor medium, their descendents lack magnetosomes and are not magnetotactic. How the bacteria synthesize a magnetic crystal rather than a nonmagnetic crystal of iron is not understood. The bacteria in the northern and southern hemispheres must orient their magnets in opposite directions with respect to their flagella, so that they swim northward and southward, respectively. This orientation insures that they always swim down toward the sediment. The earth's magnetic field appears to be partially responsible for the orientation of the bacterial magnets.

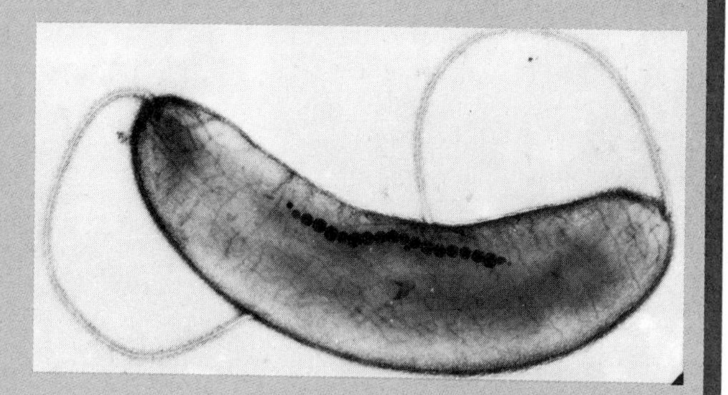

Figure 2.25 **Magnetite crystals within magnetic bacteria.** A transmission electron micrograph of a magnetotactic bacterium shows a chain of magnetite crystals within the bacterium aligned along the length of the cell. The magnetite crystals, called magnetosomes, act like a magnetic compass needle and align the bacterium along the earth's magnetic field in the same direction as the lines of force. This particular bacterium has flagella at both ends so that it is able to swim in either direction.

Magnetotactic bacteria use their magnetosomes to navigate toward the bottom of aqueous environments, where anaerobic conditions prevail. Since most of the magnetotactic bacteria are anaerobic (survive only in the absence of oxygen) or microaerophilic (grow best in low concentrations of oxygen), it is to their advantage to be forced consistently toward the bottom, where conditions are anaerobic and where sedimented nutrients might be plentiful. Although magnetosomes are not absolutely essential for the survival of magnetotactic bacteria, those populations with magnetosomes are more successful.

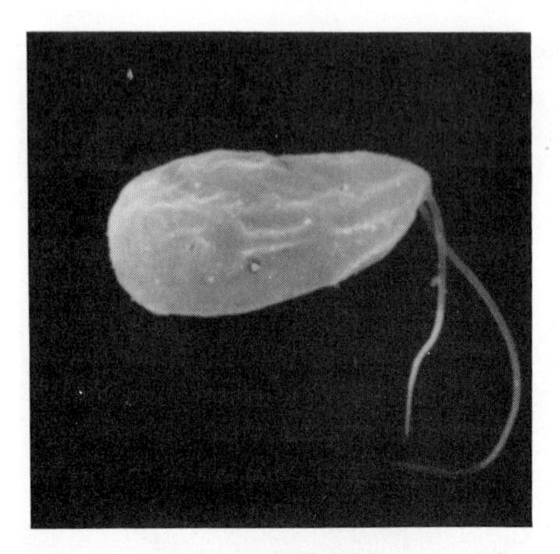

Figure 2.26 **Flagella.** Scanning electron micrograph of *Naegleria* showing its two flagella.

those of plant cells are generally constructed from cellulose and other polysaccharides. A great many eukaryotes do not have cell walls. For example, most of the protozoans (amebas and paramecia) lack walls. Their plasma membranes are reinforced by **stress fibers** that consist of proteins such as microtubules, actin, and myosin.

Appendages and Structures Used for Locomotion

Flagella Eukaryotic flagella (singular, flagellum) (fig. 2.26) and prokaryotic flagella are very different from each other structurally and are powered by completely different mechanisms. The eukaryotic flagellum develops from a **basal body** in the cytoplasm, sometimes called a **centriole** (fig. 2.27), in contrast to the prokaryotic flagellum, which originates from the cytoplasmic membrane. Microtubules and other proteins that make up the skeleton of the eukaryotic flagellum mold the membrane into a thin, whiplike extension of the cell. In contrast, the proteins that make up the prokaryotic flagellum are entirely outside the cell. The eukaryotic flagellum propels the cell by bending and twisting against its watery environment. The bending and twisting of the flagella are due to microtubules in the flagellar skeleton that are forced to slide past each other. Eukaryotic flagella are powered by the hydrolysis of ATP. In contrast, prokaryotic flagella rotate like propeller shafts and are powered by the movement of ions across the cytoplasmic membrane.

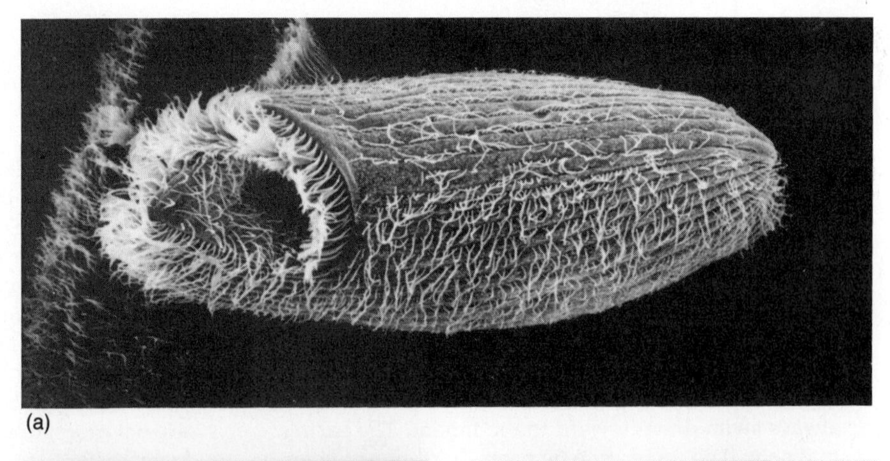

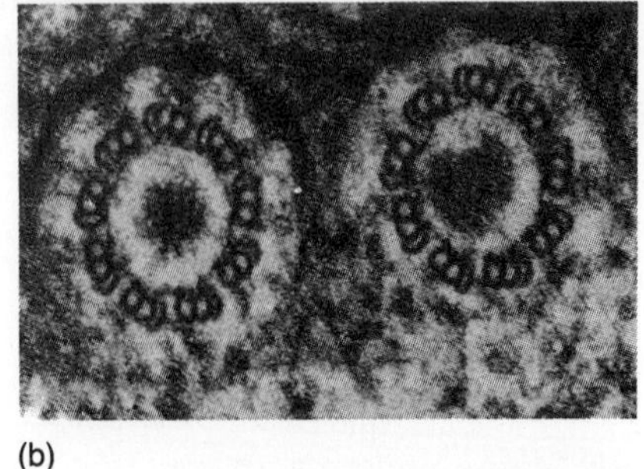

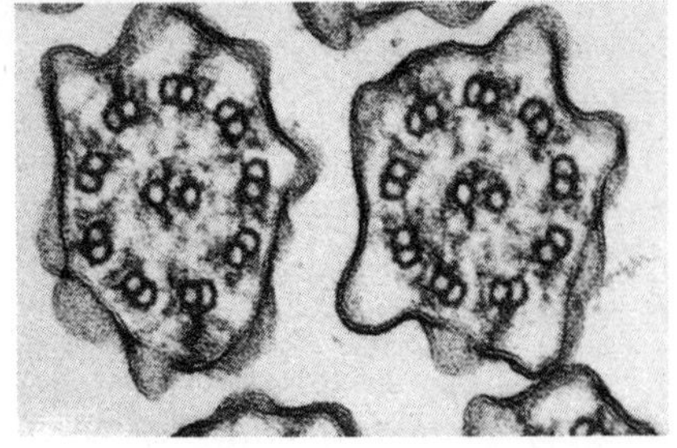

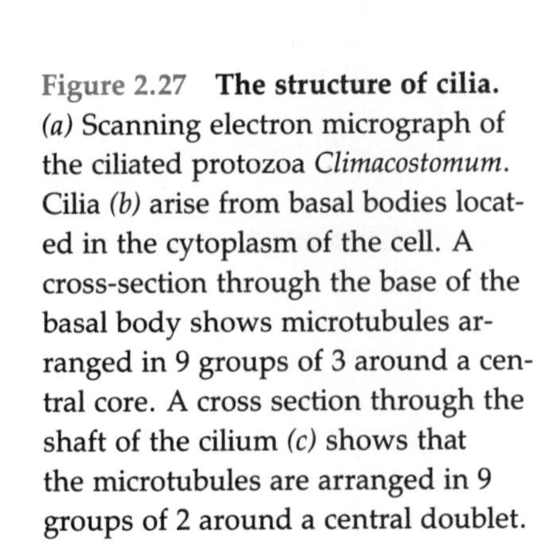

Figure 2.27 **The structure of cilia.** *(a)* Scanning electron micrograph of the ciliated protozoa *Climacostomum.* Cilia *(b)* arise from basal bodies located in the cytoplasm of the cell. A cross-section through the base of the basal body shows microtubules arranged in 9 groups of 3 around a central core. A cross section through the shaft of the cilium *(c)* shows that the microtubules are arranged in 9 groups of 2 around a central doublet.

Cilia Many eukaryotic cells have **cilia** rather than flagella. Cilia are very similar to flagella, both structurally and functionally, but they are much shorter (fig. 2.27).

Intracellular Structures

Nucleus The **nucleus** is a double-membraned organelle that contains the eukaryotic cell's genetic information (fig. 2.7b). The nuclear membranes contain numerous large pores through which proteins and RNA can move. The outer nuclear membrane often gives rise to the **endoplasmic reticulum**, a network of cytoplasmic membranes where proteins are sometimes synthesized and modified. The fluid and dissolved materials within the nucleus constitute the **nucleoplasm**. The nucleoplasm may contain one or more **nucleoli** (singular, **nucleolus**), where the synthesis of eukaryotic ribosomal subunits begins.

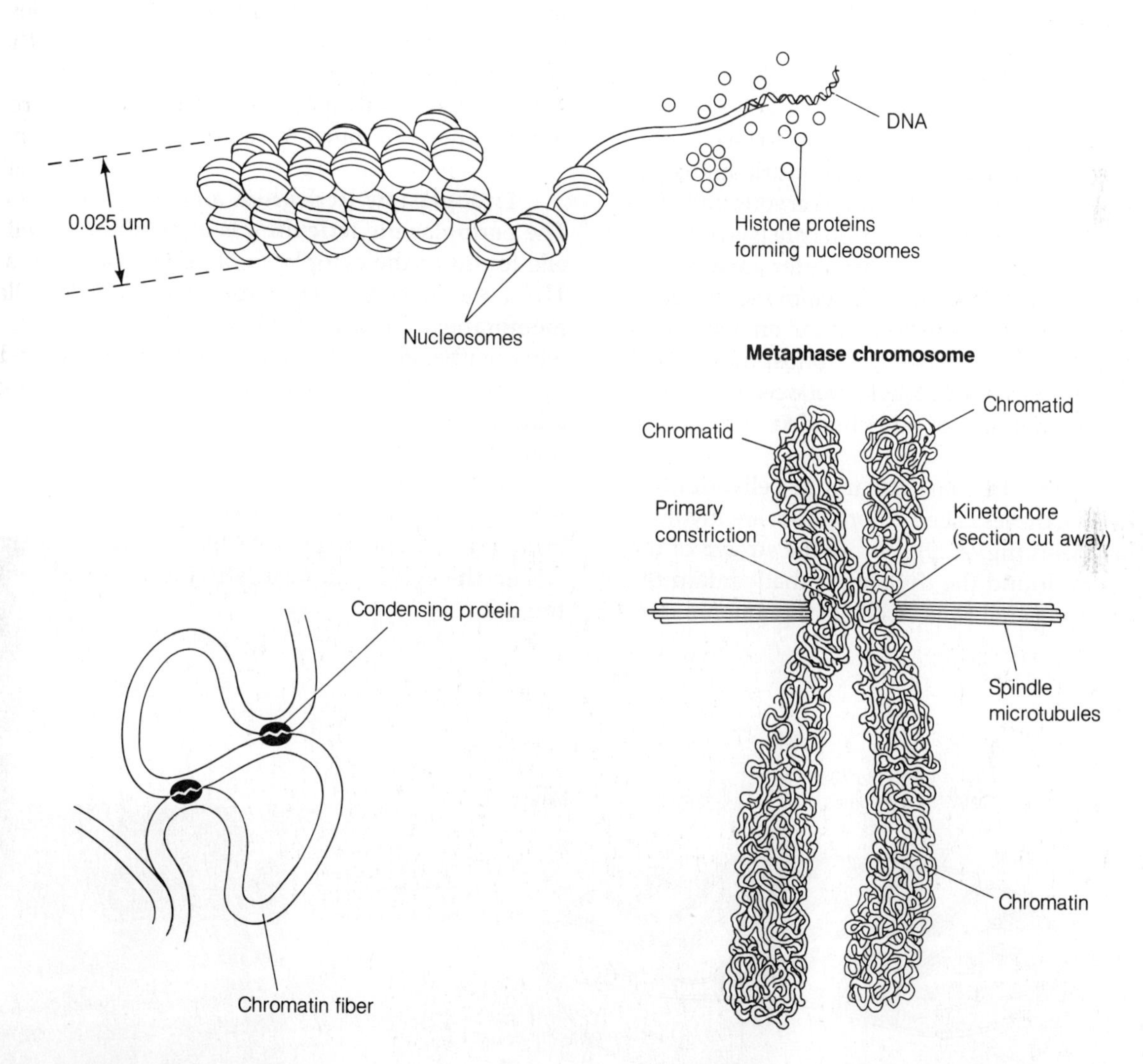

Figure 2.28 Structure of the eukaryotic chromosome. The interphase chromosomes are not visible using the light microscope because they are only 0.025 μm in diameter. The interphase chromosomes consist of DNA wound around nucleosomes. The nucleosomes are complexed so that a 0.025-μm strand is formed. The metaphase chromosomes are visible using the light microscope because they reach diameters of 0.5 μm. The metaphase chromosomes form when the interphase chromosomes condense. The metaphase chromosome illustrated actually consists of two daughter chromosomes that have not separated after DNA replication. The daughter chromosomes are called chromatids.

Eukaryotic Chromosomes The genetic material in eukaryotes is found in the nucleus and looks like a mass of threads when examined with the electron microscope. The threads consist of DNA and protein and are called **chromatin**. During cell division, the chromatin condenses into chromosomes that are visible under the light microscope (fig. 2.28). The eukaryotic DNA is coiled around basic proteins called **histones**. The histones and coiled DNA structures are known as **nucleosomes**. The chromatin in some cells is attached to the inner nuclear membrane; the significance of this attachment is unknown.

Mitochondria **Mitochondria** (singular, **mitochondrion**) are cytoplasmic organelles involved in the production of chemical energy in the form of ATP. Because of this function, the mitochondria are often called the "powerhouses" of the eukaryotic cell (fig. 2.29). Quite simply, a mitochondrion consists of a convoluted inner membrane and an outer membrane. Invaginations formed by the inner membrane are called **cristae** (singular, **crista**) and contain the enzymes that are used to synthesize adenosine triphosphate (ATP). All respiratory enzymes in the eukaryotic cell are located in the inner membranes of mitochondria.

Chloroplasts In some eukaryotic cells, double-membraned organelles called **chloroplasts** are involved in photosynthesis (fig. 2.30). Within the **stroma** of the chloroplast are found the thylakoids that contain the enzymes, chlorophyll, and accessory pigments necessary for photosynthesis. In some chloroplasts, pancakelike thylakoids are stacked like coins. The stacked thylakoids are called **grana**. In contrast, thylakoids in some algal chloroplasts are layered on top of one another like blankets on a bed.

Ribosomes Eukaryotic **ribosomes** are 80S ribosomes and are found in the cytoplasm. Some are attached to the endoplasmic reticulum, while others are attached to the cytoskeleton and appear to be "free" in the cytoplasm (fig. 2.31). The 80S ribosomes found in eukaryotes are larger than the 70S ribosomes found in bacteria. In addition, the two types of ribosomes are sensitive to different sets of antibiotics and drugs. Nevertheless, both types of ribosomes carry out protein synthesis in much the same way.

Endoplasmic Reticulum and Golgi Membranes The **endoplasmic reticulum** is a membranous organelle found in the cytoplasm of most eukaryotic cells. The outer nuclear membrane and the endoplasmic membrane are studded with 80S ribosomes. The proteins synthesized by the 80S ribosomes enter the lumen (interior) of the endoplasmic reticulum. These proteins are often modified in the endoplasmic reticulum (fig. 2.31) and then transported within vesicles to other parts of the cell, or to the cytoplasmic membrane, where they may be excreted. The 80S ribosomes "free" in the cytoplasm synthesize proteins that function as enzymes within the cytoplasm or as structural components of the cell.

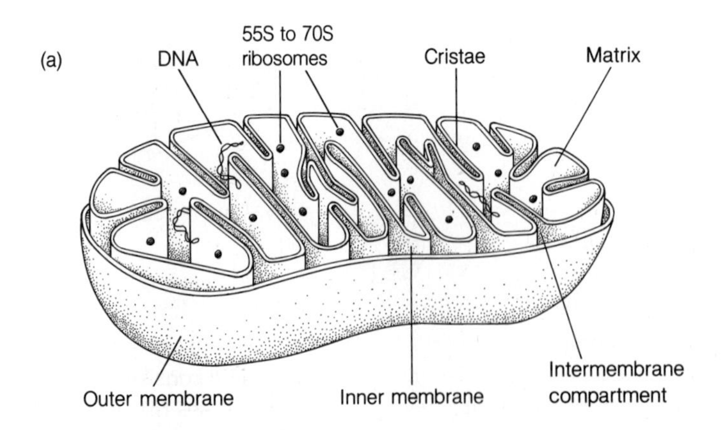

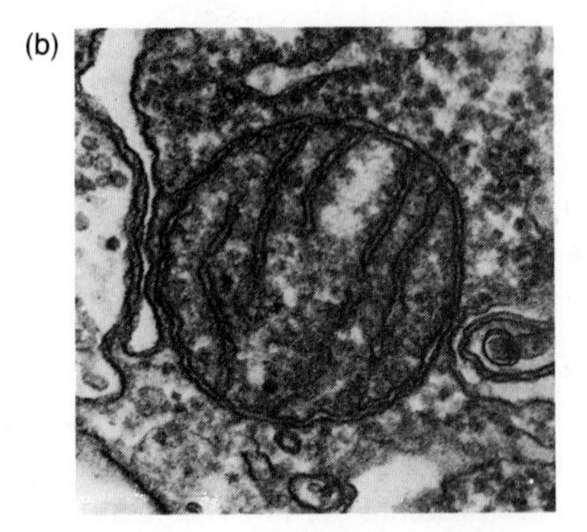

Figure 2.29 Mitochondria. *(a)* Mitochondria are double-membraned organelles that are involved in respiration. The inner membrane invaginates to form cristae. Most of the enzymes involved in respiration are located in the cristae and in the matrix. *(b)* Transmission electron photomicrograph of a mitochondrion.

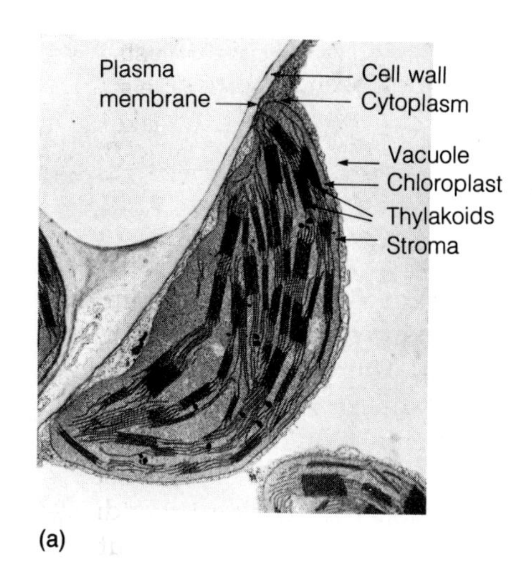

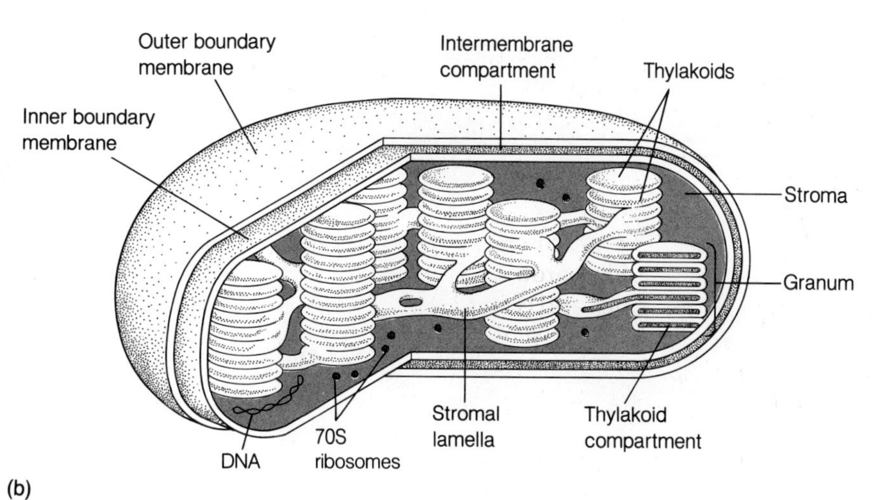

Figure 2.30 Structure of chloroplasts. Chloroplasts are organelles involved in photosynthesis. They consist of two outer membranes and numerous stacked membranes within the stroma. Most of the enzymes involved in photosynthesis are located in the stacked membranes and in the stroma. *(a)* Transmission electron micrograph of a cross section of a corn chloroplast. *(b)* Diagram illustrating the various components of chloroplasts.

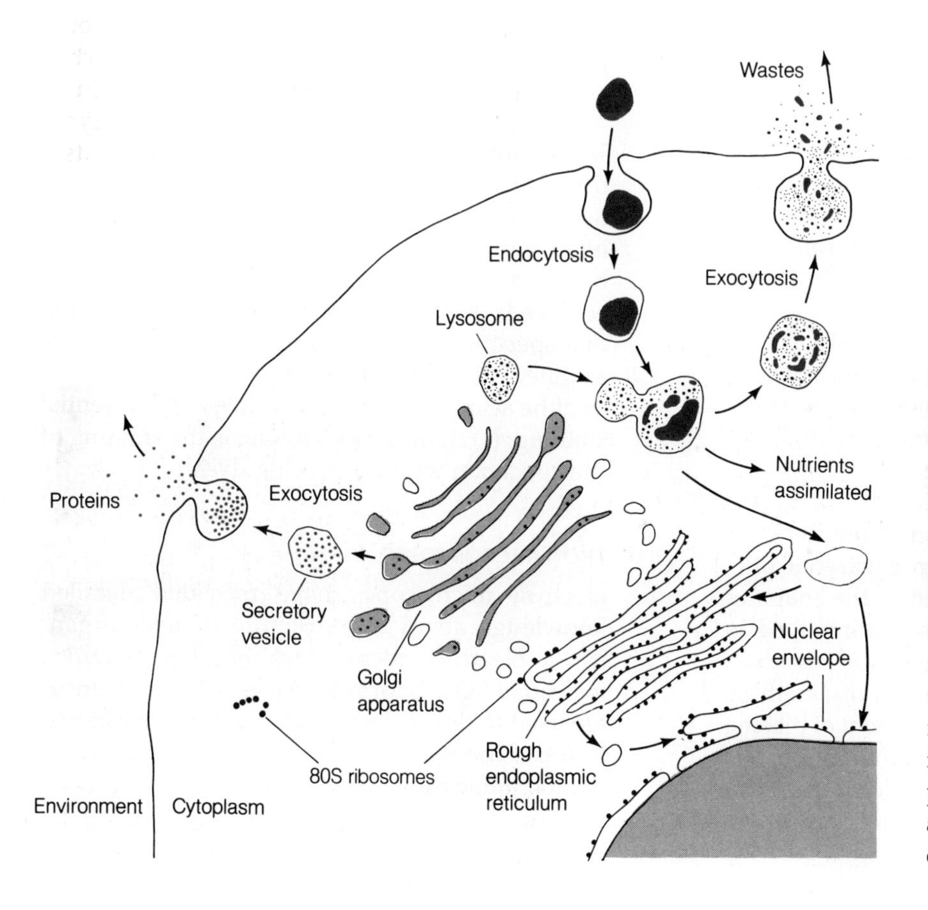

Figure 2.31 Endoplasmic reticulum and golgi membranes. Extensive endoplasmic reticulum and numerous Golgi bodies are often visible in the cytoplasm of eukaryotic cells. The endoplasmic reticulum is a site where protein synthesis occurs and where the proteins are modified. Vesicles that pinch off from the endoplasmic reticulum carry the modified proteins to the Golgi body, where they may be modified further. Membrane material in the nuclear envelope, the endoplasmic reticulum, the Golgi body, and the plasma membrane are cycled continuously, as illustrated.

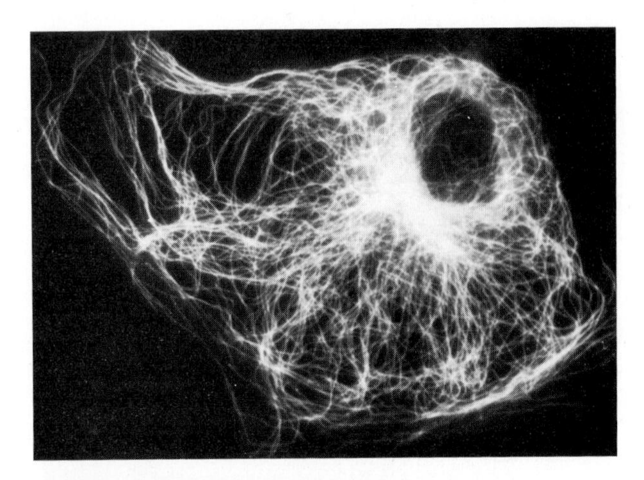

Figure 2.32 **The cytoskeleton of eukaryotic cells.** The cytoskeleton in this photograph is revealed using immunofluorescence microscopy. The cytoskeleton strengthens the cytoplasmic membrane, confers shape to the cell, and is involved in cellular movements.

Golgi membranes or **dictyosomes** are stacked membranes that can be distinguished from the continuous endoplasmic reticulum (fig. 2.31). In the Golgi membranes, proteins are modified and packaged into membranous vesicles for transport to the plasma membrane or to food vacuoles. The Golgi system produces membranous vesicles that carry hydrolyzing enzymes and are referred to as **lysosomes.**

Bacteria do not have any type of endoplasmic reticulum or Golgi system. Proteins that are released into the environment by prokaryotes are generally synthesized on the plasma membrane in such a way that the protein appears outside the cell.

Cytoskeleton The cytoskeleton is the network of proteins that fills the cytoplasm in eukaryotic cells (fig. 2.32). The cytoskeleton determines the shape of the eukaryotic cell and in some cases is involved in cell motility. Ameboid motion displayed by amoebas, for example, is due to the assembly and disassembly of the cytoskeleton as well as the movement of plasma membrane from the rear of the cell to the front of the cell. No bacteria, except for some spirochetes, have anything that approximates a cytoskeleton. Some spirochetes have microtubulelike proteins that run the length of the cell and appear to be responsible for the bending and flexing motion seen in spirochetes.

SUMMARY

THE LIGHT MICROSCOPE

1. A compound microscope is a light microscope that has two lens systems in series: the objective lens and the ocular lens. There are several different types of light microscopes: bright field, dark field, phase contrast, differential interference contrast, and fluorescence microscopes.
2. Objects closer together than the limit of resolution cannot be distinguished as separate entities. In addition, objects that have a dimension much smaller than the limit of resolution generally cannot be seen.
3. The bright field microscope is the most widely used microscope in biology. It generally is used to determine the shape, size, characteristic arrangement, and motility of bacteria and other microorganisms in their natural environments.

LIGHT MICROSCOPY TECHNIQUES

1. Living microorganisms are generally observed by using wet and hanging drop mounts.
2. To see microorganisms more clearly and study their anatomy, various staining techniques are used: (a) simple staining is the staining of a specimen with a basic dye so that a good contrast between the specimen and the background is achieved; (b) differential staining is the staining of a specimen with two or more basic dyes so as to differentiate cellular structures. The Gram stain and the acid fast stain are examples of differential staining; (c) fluorescent staining is the staining of the specimen with fluorescing dyes.

ELECTRON MICROSCOPES

1. Electron microscopes have provided detailed knowledge about the structure of microorganisms. Electron microscopes employ electrons instead of light to form an image. Magnetic lenses are used to direct electrons against the object and to focus the resulting electrons onto a screen or photographic plate. Because electrons have a much shorter wavelength than visible light or ultraviolet radiation, the theoretical limit of resolution

for an electron microscope is about 100× less than the limit of resolution for a light microscope.

2. There are two types of electron microscopes: the transmission electron microscope (TEM) and the scanning electron microscope (SEM). The TEM is generally used when subcellular structures and their relationships to each other are to be studied. In the TEM, the electrons that penetrate the specimen are used to make an image of the object. The SEM is most often employed when information about the surface structure of a microorganism is desired.
3. The electron microscope has been used to visualize viruses that are too small to be seen with any light microscope.

PROKARYOTIC AND EUKARYOTIC CELLS: A DEFINITION

1. The electron microscope has been used to demonstrate that there are two fundamental types of cells: the prokaryotic cell and the eukaryotic cell. All bacteria (eubacteria, cyanobacteria, archaebacteria, rickettsiae, mycoplasmas) have a prokaryotic cell structure. All other types of cells have a nucleus and other organelles, such as mitochondria and chloroplasts, and consequently are considered eukaryotic cells.
2. In general, prokaryotic cells are much smaller than eukaryotic cells and their internal structure is much simpler. Even though prokaryotes lack the organelles commonly found in the eukaryotes (nuclei, mitochondria, chloroplasts, endoplasmic reticulum, Golgi membranes, vacuoles, etc.), prokaryotes carry out many of the same functions associated with eukaryotes. For example, many bacteria are capable of respiration and photosynthesis.

GROSS CELLULAR MORPHOLOGY OF BACTERIA

1. Most bacteria studied in the laboratory can be classified as spheres (cocci), rods (bacilli), spirals (spirilla), or helices (spirochetes).
2. Some bacteria remain attached to each other after cell division and form long chains or clusters. Rod-shaped bacteria in chains are called streptobacilli. On the other hand, cocci in chains are referred to as streptococci, while cocci in random clumps are called staphylococci. Cocci that are found mostly in pairs are called diplococci.
3. Most bacteria studied in the laboratory divide by binary fission.

FINE STRUCTURE OF BACTERIAL CELLS

1. Bacteria may have envelopes that surround their cell membrane. Most bacteria have a cell wall that protects them from osmotic lysis. Many bacteria have a capsule that protects them from predators and mechanical damage. A few bacteria have sheaths that protect them from predators and mechanical damage. Many, if not most, bacteria are believed to have a glycocalyx that helps them attach to each other and to solid substrates.
2. Some bacteria have pili. Some of these pili are believed to help the bacteria stick to each other and to solid substrates, while others are known to be involved in the transfer of genetic information.
3. Many bacteria are motile because they have flagella. One group of bacteria has modified flagella, called axial filaments, that are used to propel them through aqueous environments.
4. Bacteria have many intracellular structures. All bacteria are believed to have only one chromosome, although it may be present in multiple copies when they reproduce rapidly. Some bacteria have small pieces of circular DNA called plasmids.
5. A few bacteria have membranes that invaginate into the cell's cytoplasm. Photosynthetic bacteria have thylakoids and photosynthetic membranes that are involved in photosynthesis.
6. Often, electron micrographs will show numerous very small granules in the cytoplasm. These very small granules are ribosomes, where protein synthesis occurs.
7. Endospores are extremely resistant to heat, desiccation, chemicals, and the lack of nutrients.

FINE STRUCTURE OF EUKARYOTIC CELLS

1. Eukaryotes may have envelopes that surround their plasma membrane. For example, most of the photosynthetic microorganisms and plant cells have cell walls that protect them from osmotic lysis and mechanical damage. Eukaryotic cells may be covered by a glycocalyx, which allows them to bind together.
2. The most obvious appendages of eukaryotic cells are their flagella and cilia.

3. Almost all eukaryotic cells have one or more nuclei. The nucleus contains the cell's genetic information, which is packaged in a number of chromosomes.
4. Almost all eukaryotic cells have one or more mitochondria, which are involved with respiration.
5. All photosynthetic eukaryotes have one or more chloroplasts, which carry out photosynthesis.
6. Most eukaryotic cells have some endoplasmic reticulum. Some of the cell's protein synthesis occurs on the endoplasmic reticulum. In addition, proteins may be modified in the endoplasmic reticulum and packaged in membranous vesicles. These vesicles may form lysosomes or fuse with food vacuoles.
7. Golgi membranes can also be found in most eukaryotic cells; they are another site where proteins are modified and packaged into membranous vesicles.
8. Eukaryotic cells contain a protein skeleton called the cytoskeleton. Actin, myosin, and tubulin are examples of some of the proteins found in the cytoskeleton. The cytoskeleton strengthens the plasma membrane, gives the cell its shape, and is involved in ameboid and gliding motion.

STUDY QUESTIONS

1. What is a compound microscope? What is a monocular microscope?
2. What is the limit of resolution of:
   a. the human eye;
   b. a good bright field microscope;
   c. an electron microscope.
3. Which lens system should be used to see:
   a. a flat human cheek cell 100 μm × 100 μm;
   b. a rod-shaped bacterium 5 μm × 1 μm;
   c. a virus 0.1 μm in diameter.
4. Discuss two ways in which a specimen might be prepared for the bright field microscope.
5. Compare and contrast the SEM and TEM.
6. Compare a prokaryotic cell with a eukaryotic cell.
7. Discuss the structure and chemical composition of bacterial cell walls. Compare the wall of a gram positive bacterium with that of a gram negative bacterium. What is the importance of a cell wall to a bacterium?

SELF-TEST

In items 1–5, match the characteristics of the cell with: a. prokaryote; b. eukaryote.

1. This type of cell has a single chromosome dispersed in the cytoplasm.
2. These cells have single- and double-membraned organelles such as endoplasmic reticula and mitochondria.
3. The flagella originate in the plasma membrane and rotate.
4. These cells are generally less than 5 μm in diameter.
5. Peptidoglycan is usually found in the cell walls of this type of cell.
6. What generally protects bacteria from predators?
   a. cell walls;
   b. pili;
   c. flagella;
   d. capsules;
   e. axial filaments.
7. Bacteria that grow in grapelike clumps are referred to as:
   a. diplococci;
   b. streptococci;
   c. staphylococci;
   d. streptobacilli;
   e. sarcinae.
8. Pili have two major functions:
   a. nutrient uptake and nutrient storage;
   b. locomotion and nutrient uptake;
   c. protection from predators and binding cells together;
   d. conjugation and binding bacteria together and to other materials;
   e. protection from predators and nutrient uptake.
9. Bacteria that have flagella over their entire surface have what type of flagella?
   a. montrichous;
   b. monotrichous;
   c. lophotrichous;
   d. peritrichous;
   e. kophotrichous.
10. Which group of bacteria have axial filaments?
    a. archaebacteria;
    b. mycoplasmas;
    c. spirochetes;
    d. rickettsias;
    e. chlamydiae.
11. What is the term used to describe the small self-duplicating circular DNA molecules in bacteria that frequently carry drug resistance genes?
    a. chromosomes;
    b. genomes;

c. plastids;
d. plasmids;
e. pili.

12. Which are characteristic of endospores?
a. form inside the bacteria cell;
b. highly resistant to boiling;
c. resistant to antibiotics;
d. do not require an aqueous environment or nutrients;
e. all of the above.

13. Flagella in the eukaryote are powered by the:
a. hydrolysis of ATP;
b. electrochemical potential across the plasma membrane;
c. hydrolysis of any phosphorylated substrate;
d. transport systems in the plasma membrane;
e. none of the above.

14. Eukaryotic chromosomes attach to the:
a. plasma membrane;
b. endoplasmic reticulum;
c. inner membrane of mitochondria;
d. Golgi membranes;
e. inner nuclear membrane.

15. Eukaryotic protein synthesis takes place at ribosomes associated with:
a. cytoplasm;
b. endoplasmic reticulum;
c. Golgi membranes;
d. plasma membrane;
e. a and b.

SUPPLEMENTAL READINGS

Allen, R. D. 1987. The microtubule as an intracellular engine. *Scientific American* 256(2): 42–49.

Porter, K. and Tucker, J. 1981. The ground substance of the living cell. *Scientific American* 244(3): 56–67.

Spencer, M. 1982. *Fundamentals of Light Microscopy*. Cambridge Univ. Press, London.

Valkenburg, J. A. C., Woldringh, C. L., Brakenhoff, G. J., van der Voort, H. T. M., and Nanninga, N. 1985. Confocal scanning light microscopy of the *Escherichia coli* nucleoid: Comparison with phase-contrast and electron microscope images. *Journal of Bacteriology* 161(2): 478–483.

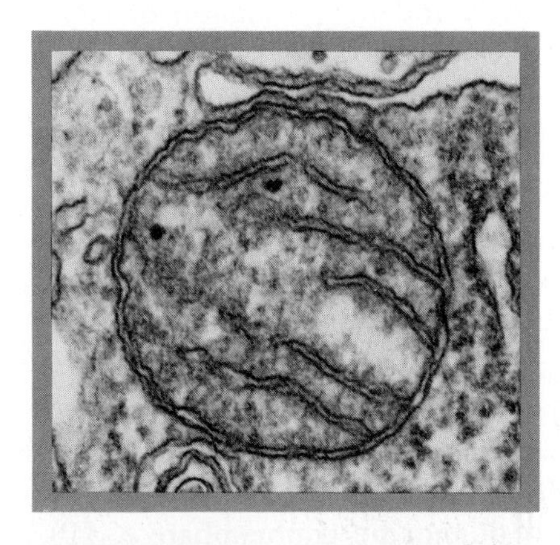

CHAPTER 3

MICROBIAL METABOLISM

CHAPTER PREVIEW GAS GANGRENE RESULTS FROM BACTERIAL METABOLISM

Clostridial myonecrosis, more commonly known as "gas gangrene," is a syndrome that is characterized by rapid death of muscle and soft tissue, inflammation of the affected area, pain, and the production of gas in tissue. The disease is a frequent complication of traumatic lacerations such as those resulting from gunshot wounds, bone fractures, or automobile accidents. Gas gangrene is a characteristic complication of war wounds, and much of what we know about the disease was learned during World War I.

Gas gangrene is most frequently caused by a bacterium called *Clostridium perfringens*. This bacterium is **anaerobic**; that is, it must metabolize and reproduce in environments devoid of oxygen—much like those found in traumatic wounds.

In the anaerobic environment of dirty, lacerated wounds, *C. perfringens* multiplies rapidly, using sugars as its source of energy. During this metabolic process, known as a **fermentation**, *C. perfringens* produces acids and large quantities of gas that are detected by physicians as a wound frothing. Also, during metabolic activities, *C. perfringens* produces a variety of toxic substances that digest muscle and soft tissue and allow the microorganism to spread in the wounded area. The progressive, insidious destruction of tissue and frothing of the wound are a direct consequence of the metabolic activities of the bacterium.

Understanding the metabolism of *C. perfringens* permits physicians to treat the disease. Patients are usually treated with antibiotics and the wounds are cleaned surgically. Sometimes the affected gangrenous areas are also placed in "hyperbaric chambers," where the oxygen concentration is very high. The high concentrations of oxygen in the affected area inhibit clostridial growth and promote the cure of the disease.

In gas gangrene, as in many other infectious diseases, it is essential to know about the metabolism of the causative agent to understand the disease process and implement effective treatment and control measures. In this chapter we learn how microorganisms obtain energy from nutrients in their environment and how this energy is used to drive biological activities.

CHAPTER OUTLINE

METABOLISM IS A TERM USED TO DESCRIBE ALL of the chemical reactions that take place in the cell. Metabolism that involves the assimilation of nutrients and the construction of new cell material is called **anabolism**. During anabolism, or **biosynthesis**, the cell assembles small molecules into more complex ones (fig. 3.1). The formation of proteins from amino acids and the formation of DNA from nucleotides are examples of anabolic processes. As a rule, the assemblage of small molecules into biological polymers during anabolism requires that energy be provided. The energy needed to power anabolic reactions is generated during **catabolism**. Catabolism includes all the chemical reactions of the cell that result in the breakdown of organic molecules into simpler ones. The breakdown of the sugar glucose into lactic acid or carbon dioxide is an example of catabolism. During the breakdown of many organic molecules (e.g., glucose), energy is released. Some of this energy can be conserved in nucleotides, such as **adenosine triphosphate** (ATP) or **guanosine triphosphate** (GTP), and can subsequently be used to power anabolic reactions. In this chapter we study the various ways in which microorganisms conserve energy in ATP molecules, and how these molecules are used to power biosynthetic reactions.

METABOLISM: AN OVERVIEW

Microorganisms can extract energy from foods and conserve it in molecules of nucleotides such as **adenosine triphosphate** (ATP). This extraction requires modification of the food so that it can be brought into

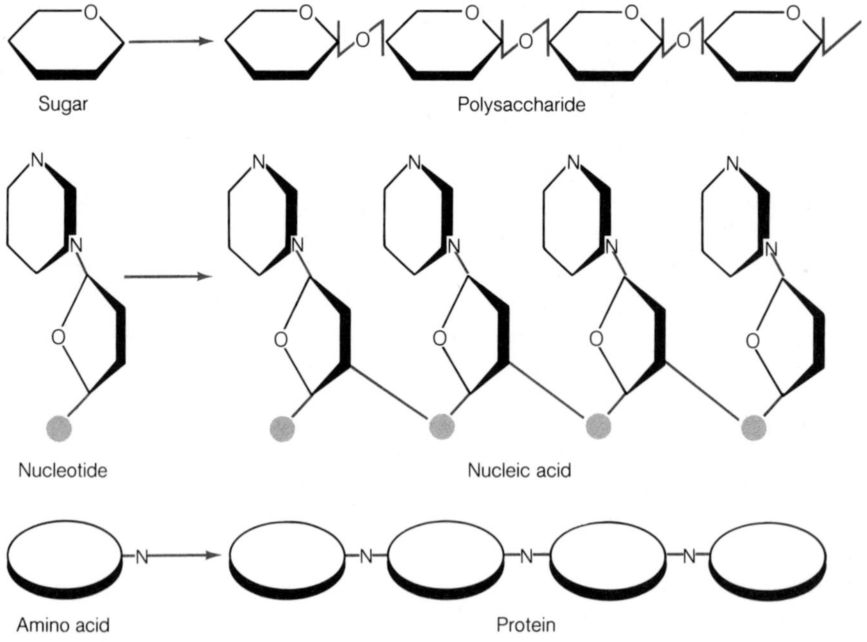

Figure 3.1 Chemical composition of important biological molecules. Biological polymers are composed of building blocks that are joined together by chemical bonds to form the polymers. Polysaccharides are built from simple sugars. Nucleic acids are constructed from nucleotides. Proteins are formed from amino acids.

the cell and completely oxidized to maximize its energy yield (fig. 3.2).

The first stage toward obtaining energy from foods is sometimes called **digestion**. It involves the breakdown of large molecules (macromolecules) into simpler ones, usually by a chemical process called **hydrolysis**. For example, the digestion of proteins yields amino acids, while the digestion of starch yields simple sugars. Bacteria and many eukaryotic microorganisms cannot bring large molecules into the cell; hence, they must digest them extracellularly (outside the cell). This is accomplished by the secretion of **exoenzymes** that catalyze the breakdown of materials outside the cell. Examples of exoenzymes are **proteases** (protein-degrading enzymes), **amylases** (starch-digesting enzymes), and **lipases** (lipid-digesting enzymes). All of these work by using water to break the chemical bonds between the subunits of the macromolecule (i.e., hydrolysis). Sometimes the exoenzymes released by microorganisms in the human body digest host tissue and cause severe damage and disease. The exoenzyme **hyaluronidase**, an enzyme that digests the hyaluronic acid that is part of the connective tissue of humans, is secreted by a bacterium called *Staphylococcus aureus,* which causes diseases such as impetigo and wound infections.

The second stage toward obtaining energy from nutrients consists of the transport of small molecules into the cell. Some transport mechanisms require cellular energy, and consequently are called **active transport** systems. Others, those referred to as **passive transport** systems, do not require an input of energy. We discuss these transport mechanisms later in the chapter.

The third stage in the extraction of energy from foods involves the degradation of small organic molecules into simpler, more oxidized ones. For example, the degradation of glucose gives rise to molecules such as pyruvate or acetate. During this stage, organic molecules are oxidized and the protons and electrons resulting from the oxidation are donated to carrier molecules known as **coenzymes** such as nicotinamide adenine dinucleotide (NAD) or flavin adenine dinucleotide (FAD). Additionally, during this stage, some of the energy that resulted from the oxidation of glucose is conserved in molecules of ATP.

The fourth stage in the catabolism of nutrients involves the complete oxidation of pyruvate to carbon dioxide. The complete oxidation of this molecule results in the formation of additional ATP molecules.

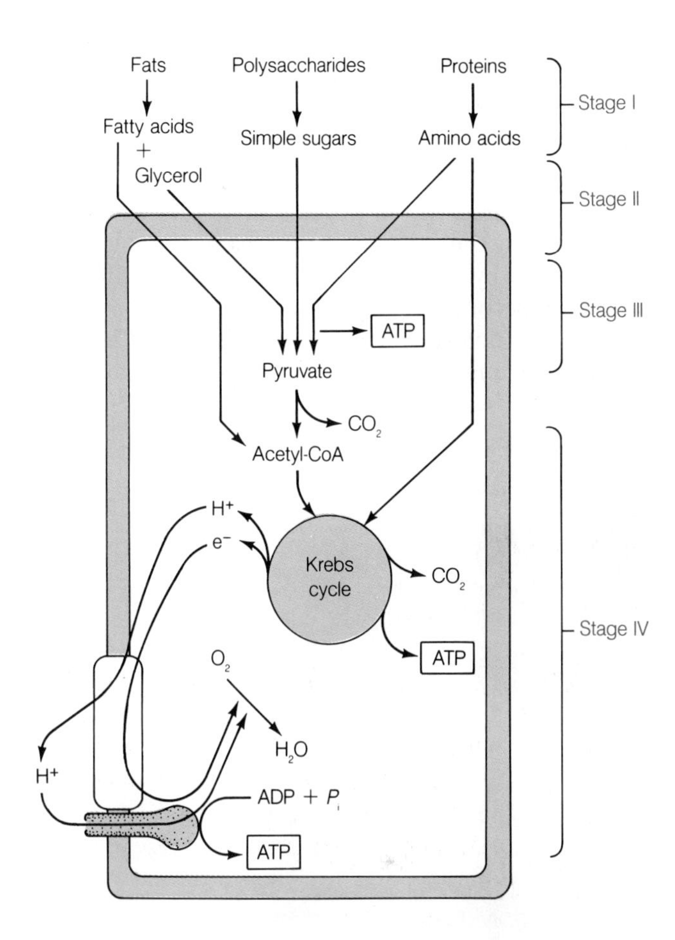

Figure 3.2 Stages of metabolism of foods. Stage I is concerned with the breakdown of large molecules into small ones. Stage II involves the transport of small molecules into the cell. Stage III involves the partial oxidation of the small molecules, with the production of chemical energy (ATP). Some microorganisms cannot catabolize foods further. Stage IV involves the complete oxidation of food molecules and the subsequent production of large quantities of ATP.

WHAT ARE ENZYMES?

Enzymes are proteins that participate in most cellular reactions as catalysts. In the absence of enzymes, cellular reactions would be very slow and uncoordinated. Enzymes speed up chemical reactions without becoming altered in the process (i.e., they are catalysts). Because they are not altered, enzyme molecules can be reused by the cell, sometimes as often as 100,000 times each second.

The function of an enzyme is determined largely by its three-dimensional shape (fig. 3.3). Substrate molecules bind to specific sites on the enzyme molecule called **active sites**. It has been suggested that the active site may have a shape that is complementary to the shape of its substrate. A popular way of expressing this concept is by using a "lock and key" model of enzyme activity, where the substrate (key) fits into the active site of the enzyme (lock) (fig. 3.3). Since it is the shape of the active site that determines with which substrates the enzyme will react, any changes that occur in the three-dimensional shape of the active site will affect the catalytic ability of the enzyme. When enzymes are subjected to extreme environmental conditions such as heat or low pH, the active site of the enzyme becomes so altered that it no longer functions. If this enzyme catalyzes a vital function of the cell, the cell will cease to reproduce and eventually die.

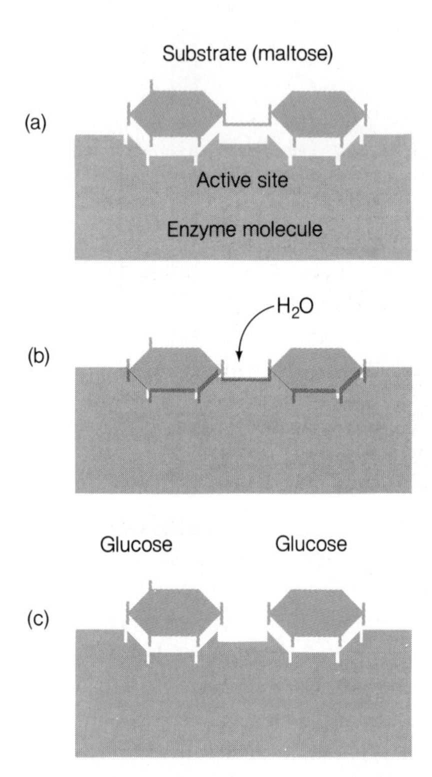

Figure 3.3 Lock and key explanation of enzyme activity. *(a)* A molecule of maltose (two glucose molecules covalently bonded) coming into the active site of the enzyme. Notice that the active site has a shape that is complementary to the shape of the substrate. *(b)* Enzyme–substrate complex. *(c)* Enzyme catalyzes reaction and the products (two glucose molecules) are formed and released from the active site. Enzyme is now available to catalyze the hydrolysis of another molecule of maltose.

Cells can regulate their metabolic activities by modifying the active site of some of their enzymes. For example, an enzyme participating in the first step in the formation of the amino acid tryptophan can be regulated by the concentration of tryptophan so that only the needed amount of tryptophan is made. Tryptophan, binding to the enzyme, changes the shape of the active site of the enzyme so that the rate of substrate conversion to product is limited.

Enzymes speed up chemical reactions by reducing the amount of energy required to start them. This energy, called **activation energy** (fig. 3.4) is required for all chemical reactions. When a chemical reaction requires a large amount of activation energy, the reaction does not take place rapidly enough for a cell's needs. The function of enzymes is to lower the activation energy required so that a chemical reaction can take place rapidly.

NUTRIENT TRANSPORT

Nutrients found in the environment are used by microorganisms as a source of energy and building blocks for growth and reproduction. Before these nutrients can be of use to the cell, however, they must be brought into the cytoplasm of the cell. A few small molecules such as water and glycerol can pass readily into the cell, but the plasma membrane is usually impermeable to ions and molecules to insure that the cell's contents do not leak out. The mechanisms used by the cell to bring nutrients into the cell are collectively called **permeases**.

Living organisms accumulate nutrients and discharge cellular wastes in a number of ways. Figure 3.5 summarizes the most important processes involved

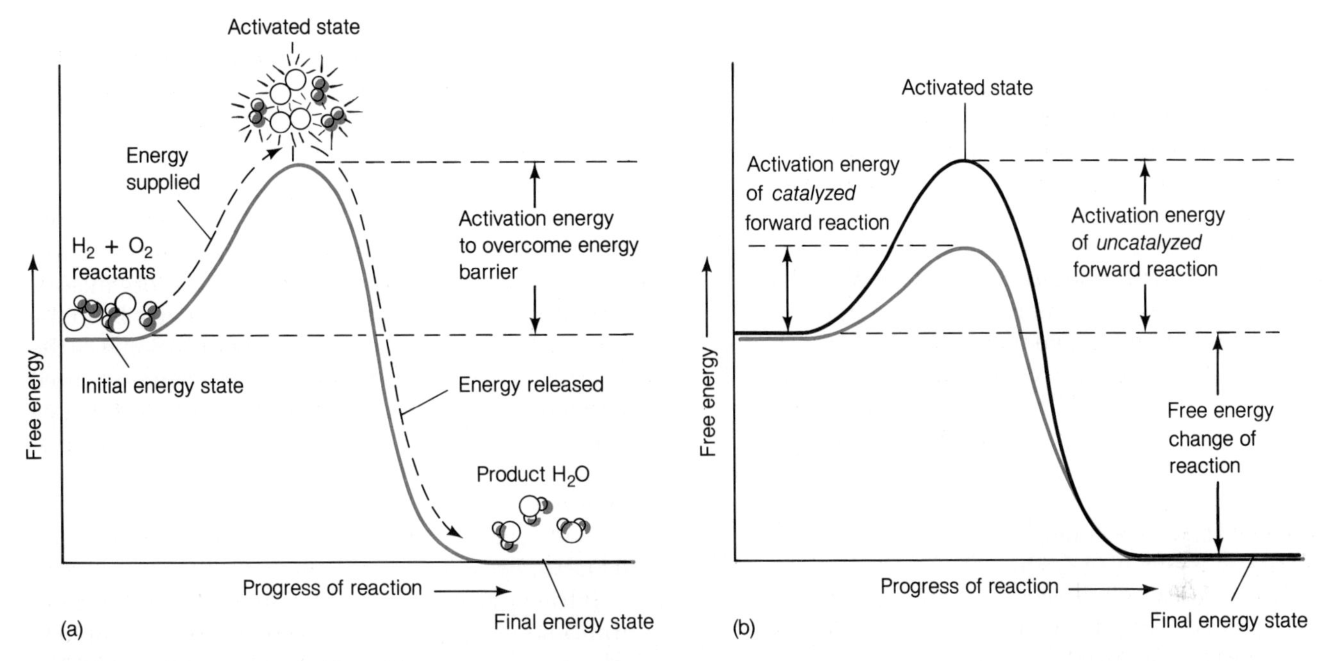

Figure 3.4 Energy of activation. *(a)* Diagram illustrating the amount of energy required to initiate a chemical reaction using water formation as an example. The reactants must acquire additional energy so that they can react with each other and form water. This extra energy required is the energy of activation and is used in the reaction to break some bonds and form others. *(b)* Enzymes reduce the energy of activation required for a chemical reaction.

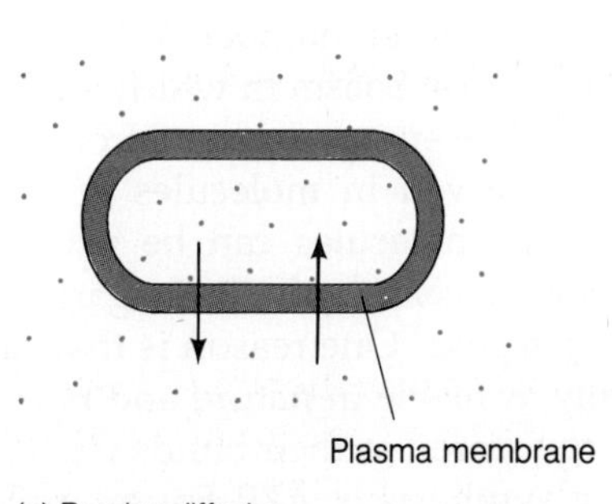

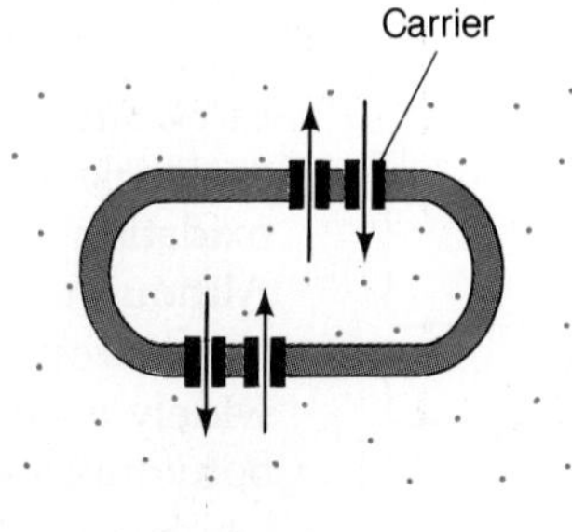

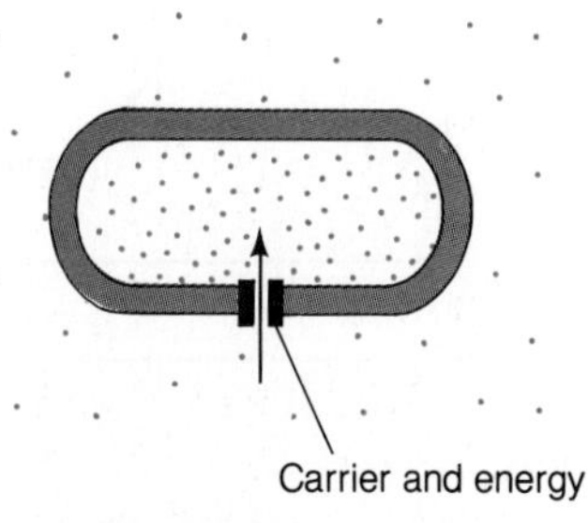

Figure 3.5 Transport of nutrients across biological membrane. *(a)* Passive diffusion. No energy expenditure by the cell is required. The rate at which nutrients pass into the cell is directly proportional to the difference in concentration between the environment and the interior of the cell. The concentration of nutrients inside the cell never exceeds that of the environment. *(b)* Facilitated diffusion. No energy expenditure by the cell is required. Carrier proteins transfer nutrient molecules from the environment into the cell and increase the rate of transport into the cell. The concentration of nutrients inside the cell never exceeds that of the environment. *(c)* Active transport. Requires that the cell use up energy. Proteins called permeases transfer nutrients from the environment into the cell. The rate of transfer is increased (as in facilitated diffusion). Because of the expenditure of energy (ATP → ADP + P_i), it is possible for the cell to concentrate nutrients, even from areas of lower solute concentration to areas of higher concentration.

in moving chemicals across the plasma membrane: **passive diffusion**, **facilitated diffusion**, and **active transport**. In passive and facilitated diffusion (figs. 3.5a and b), molecules pass through the plasma membrane from areas of high concentration to areas of low concentration. Neither of these two mechanisms can account for the transport of molecules against a concentration gradient (i.e., from areas of low concentration to areas of low concentration). Cells do not expend energy in transporting molecules in and out of the cell by passive or by facilitated diffusion. These transport mechanisms, consequently, do not allow the concentration of nutrients inside the cell to exceed that outside the cell. Most microorganisms make only limited use of diffusion as a means of transporting nutrients.

Most of the nutrients that enter the cell are concentrated by active transport (fig. 3.5c). During active transport mechanisms, the cells expends energy to bring nutrients into the cell. The energy derived from the hydrolysis of ATP (ATP $\rightarrow$ ADP + P_i + energy), for example, can be used to bring molecules into the cell against a concentration gradient. In this way, the cell can accumulate nutrients inside the cell at a concentration far exceeding that outside the cell.

(a)

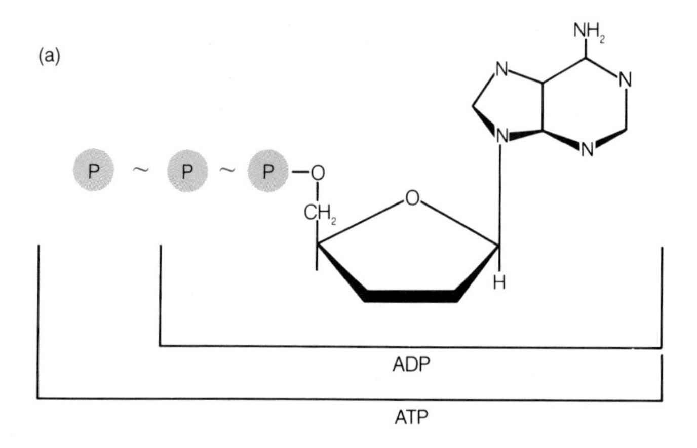

(b)

Figure 3.6 Chemical structure of the nucleotide ATP. *(a)* A ~ indicates a high-energy bond. A Ⓟ indicates a phosphate group. *(b)* The chemical structure of a phosphate group is illustrated.

BIOENERGETICS

Bioenergetics is a field of study concerned with the production and use of energy by living organisms. The energy available to these organisms may be in the form of light or stored in chemical bonds. Some of the energy extracted from light and chemical bonds is stored in nucleotides such as ATP (fig. 3.6) so that it can be used at a later time to power energy-requiring cellular activities. Those organisms that use light as their source of energy are called **phototrophs**, while those that use chemicals as their energy source are called **chemotrophs**.

If the hydrolysis of ATP by the cell results in a net output of energy, then the synthesis of ATP requires that energy be furnished. The energy required by the cell for the synthesis of ATP comes from the metabolism and oxidation of fuel molecules. In the following sections we study some of the ways microorganisms make ATP from energy sources available in their environment.

CHEMOTROPHIC METABOLISM

All eukaryotic cells, and a large number of bacteria, carry out a chemotrophic metabolism in which organic molecules are oxidized. The energy released from this oxidation is partly conserved in molecules of ATP. Although many organic molecules can be used in chemotrophic metabolism, carbohydrates (sugars) are widely used for this purpose. One reason is that carbohydrates are readily available in nature and have a large amount of energy stored in their bonds that can be used to power the synthesis of ATP. For example, the oxidation of 1 mole of glucose to carbon dioxide and water releases approximately 688 kilocalories (1 kcal = 1000 calories). Some of this energy is conserved and can yield as many as 38 moles of ATP.

The oxidation of glucose in chemotrophic metabolism can be subdivided into three stages. The first stage, sometimes called **glycolysis**, involves the oxidation of glucose to 2 pyruvates, resulting in the net production of 2ATP and 2 reduced NADs (2NADH + $2H^+$). This first stage is also called the **Embden-**

Meyerhof-Parnas pathway. The second stage involves the oxidation of the pyruvates to carbon dioxide and the production of $8NADH + 8H^+ + 2FADH_2 + 2GTP$ (another nucleotide in which the cell stores energy). This stage is called the **Krebs cycle**, or the citric acid cycle. The third stage involves the oxidation of the reduced NADs and FADs and the synthesis of as many as 34 ATP. In this oxidative process electrons and hydrogen ions (protons) are transferred from NADH $+ H^+$ and $FADH_2$ to a series of nonprotein and protein electron and hydrogen ion carriers. Some of the protein electron carriers are **cytochromes**. The enzyme, ATPase, is both a hydrogen ion carrier and the enzyme that synthesizes ATP in the third stage. This system of carriers is called the **electron transport system** (ETS). The electron transport system passes the electrons to an inorganic electron acceptor. One of the most common electron acceptors is oxygen (O_2). This process is called **aerobic respiration**. In some organisms, the electrons are instead transferred to nitrate, iron, or sulfate. When these electron acceptors participate in respiration rather than oxygen, the process is called **anaerobic respiration**.

Some microorganisms can carry out only the oxidation of glucose to pyruvate and then a subsequent reduction of the pyruvate. This process, called **fermentation**, results in the production of 2 ATP molecules per glucose oxidized and the release of various types of waste products such as CO_2, ethanol, or lactic acid.

Glycolysis

The Embden-Meyerhof-Parnas pathway (fig. 3.7), also called glycolysis, consists of 10 reactions, each catalyzed by a different enzyme, that oxidize glucose to pyruvate. These enzymes are found in the cytoplasm of both prokaryotic and eukaryotic cells. The first three reactions in glycolysis do not involve the removal of electrons or protons (an oxidation) and they are preparatory steps for a subsequent oxidation. During these three steps, 2 ATP molecules are used. The rest of the reactions in the pathway lead to the synthesis of 4 ATP molecules and 2 reduced NADs ($2NADH + 2H^+$). Hence, after the completion of the Embden-Meyerhof-Parnas pathway, the cell has produced a net total of 2 ATP molecules and $2NADH + 2H^+$.

Adding together all of the reactions of the Embden-Meyerhof-Parnas pathway leading to the formation of pyruvate, we can get an overall view of glycolysis:

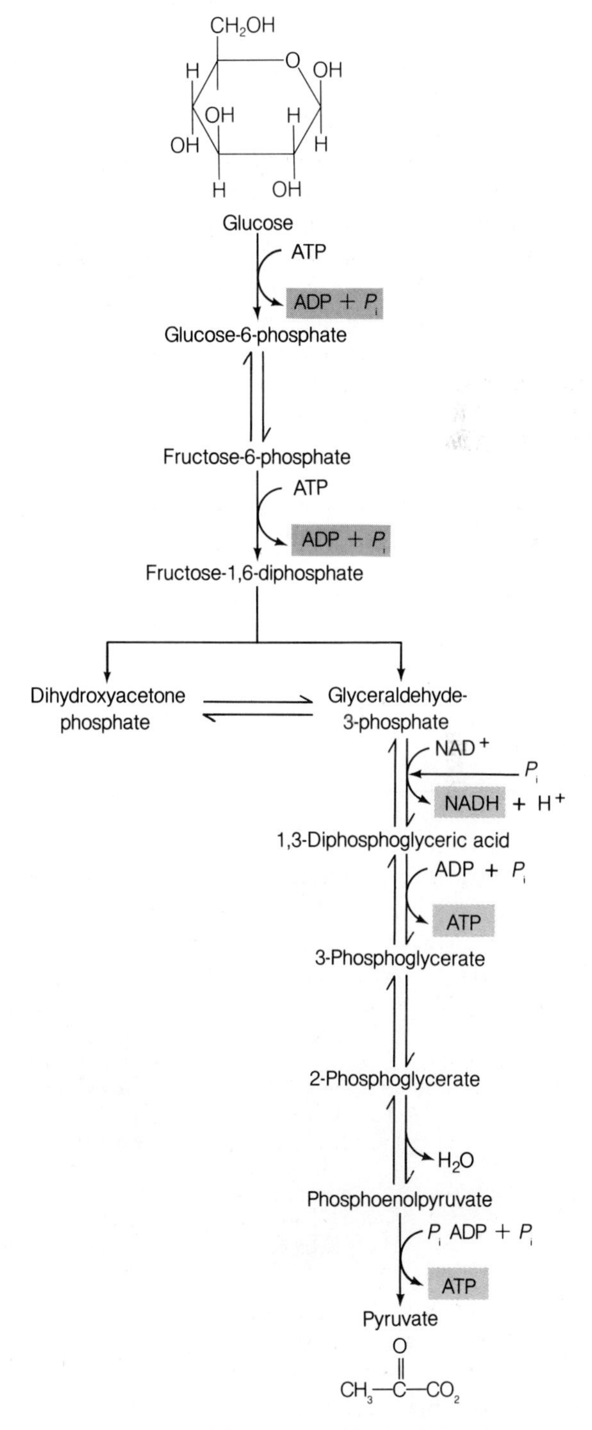

Figure 3.7 The Embden-Meyerhof-Parnas pathway of glycolysis

$$\text{Glucose} + 2\text{ATP} + 2\text{NAD}^+ + 2\text{ADP} + 2P_i \rightarrow$$
$$\rightarrow 2\text{Pyruvate} + 2\text{NADH} + 2\text{H}^+ + 4\text{ATP} + 2\text{H}_2\text{O}$$

If the 2 ATP molecules required to initiate the Embden-Meyerhof-Parnas reactions are subtracted from the total yield, the equation can be summarized as follows:

$$\text{Glucose} + 2\text{NAD}^+ + 2\text{ADP} + 2P_i \rightarrow$$
$$\rightarrow 2\text{Pyruvate} + 2\text{NADH} + 2\text{H}^+ + 2\text{ATP} + 2\text{H}_2\text{O}$$

The fate of pyruvate depends upon a number of factors. In the absence of an external electron acceptor such as oxygen, or if the cell lacks some of the enzymes in the Krebs cycle, the energy-yielding metabolism is restricted to fermentation. If glycolysis is to continue taking place, there must be a readily available supply of oxidized NAD (NAD^+) to accept the electrons that are produced during the oxidation of glyceraldehyde-3-phosphate to 1,3,diphosphoglycerate (fig. 3.7). Since there is a limited supply of NAD^+ in the cell, there must be a mechanism to regenerate it from reduced NAD (NADH + H^+). One such mechanism involves donating the electrons in NADH to organic molecules. Figure 3.8 shows how various chemotrophic bacteria dispose of their electrons. When an organism synthesizes ATP during glycolysis and disposes its electrons by giving them to organic molecules, the organism has carried out a fermentation.

No single organism has all the enzymes required for the fermentation pathways shown in figure 3.8. In

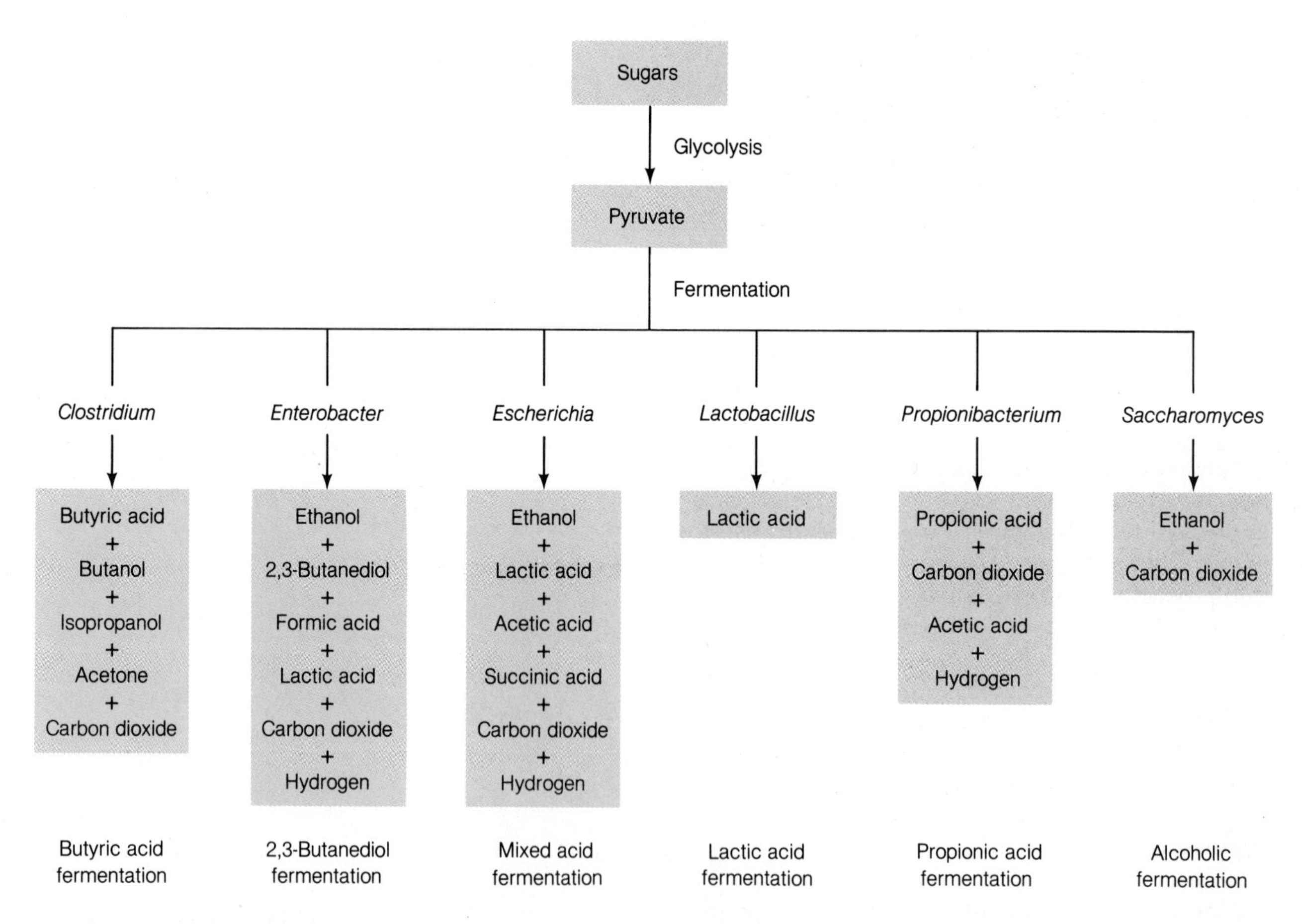

Figure 3.8 **Outline of major fermentation pathways.** Microorganisms produce various waste products when they ferment in order to dispose of their excess electrons. The by-products released are often characteristic of the microorganism and can be used as an aid in their identification.

fact, many microorganisms use only one pathway preferentially, even though they might be able to use a number of pathways. Thus, organisms can often be characterized by the kinds of waste products they release into the environment when they ferment.

Those chemotrophs that can ferment and respire (depending on conditions) are known as **facultative anaerobes**. The term indicates that the organism can generate its energy either by fermentation or by respiration. A facultative anaerobe consumes a carbohydrate such as glucose more rapidly when it ferments than when it respires. Their growth rate is greater when they respire, however, because respiration utilizes carbohydrates more efficiently to produce ATP. This leaves much of the carbohydrate available for the synthesis of cell material. In fermentation, most of the carbohydrate is consumed to generate ATP (energy) rather than new cell material.

Respiration

The oxidation of glucose to pyruvate results in a net release of about 45 kcal of energy, but only about 15 kcal are conserved as ATP. Therefore, the potential energy in the molecule of pyruvate (688 − 45 = 643 kcal/mole) remains largely untapped by the cell. During respiration, nearly half of this energy is conserved in ATP through the oxidation of pyruvate to carbon dioxide and water.

Organisms that are capable of respiration oxidize pyruvate (a three-carbon compound) to acetate (a two-carbon compound) (fig. 3.9). During this conversion,

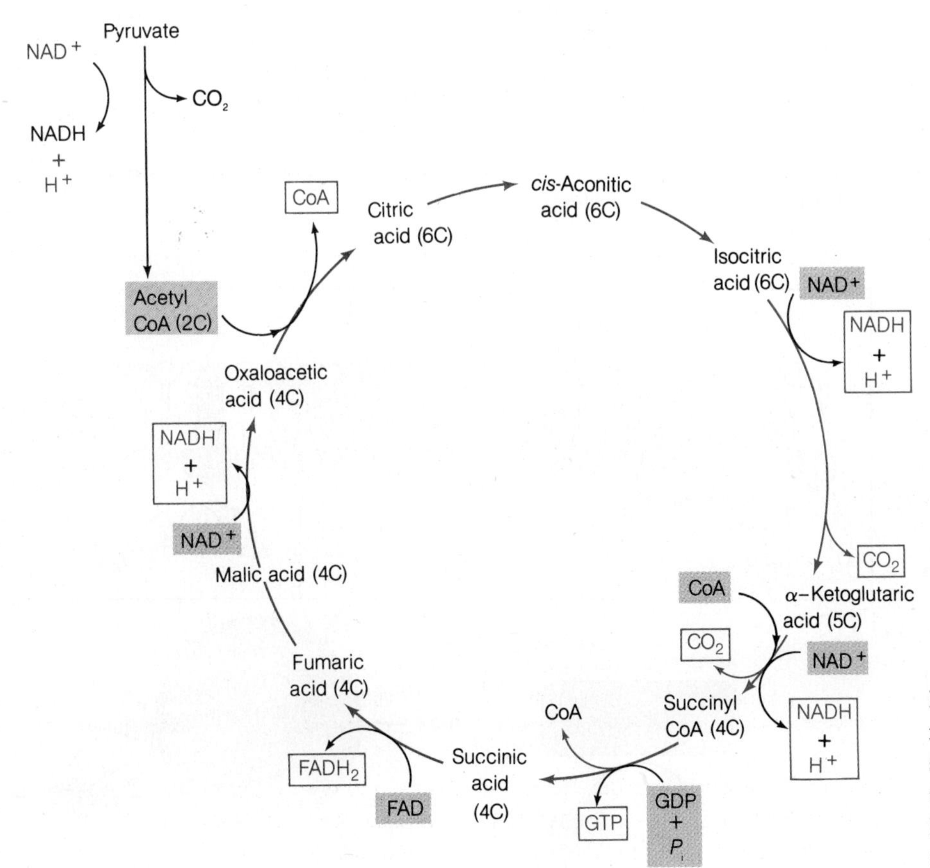

Figure 3.9 The Krebs cycle. Materials entering the Krebs cycle are in shaded boxes. Important products of the Krebs cycle are in outlined boxes. Reduced NAD and FAD donate electrons to an electron transport system.

TABLE 3.1
BALANCE SHEET OF BACTERIAL CHEMOTROPHIC METABOLISM

Reaction	ATP/glucose	Energy (kcal) conserved in ATP	Efficiency of energy conservation
Fermentation			
Glucose → Pyruvate → Lactate	2	15	2.2%
Respiration			
Glucose → 2Pyruvate + 2NADH + 2ATP	8 (6 + 2)		
2Pyruvate → 2AcCoA + 2NADH	6		
2AcCoA → $4CO_2$ + 6NADH + $2FADH_2$ + 2GTP	24 (18 + 4 + 2)		
Glucose + $6O_2$ → $6CO_2$ + $4H_2O$	38	277	40.3%

acetate combines with coenzyme A (CoA), forming acetyl-CoA, a molecule of carbon dioxide, and one NADH + H^+. This is a necessary first step for the complete oxidation of pyruvate. The acetate enters the Krebs cycle (fig. 3.9), where it is further oxidized. In the Krebs cycle, a series of oxidation–reduction reactions takes place in which the acetate becomes oxidized. Notice in figure 3.9 that several molecules of NADH + H^+ are produced as a result of acetate oxidation. Hence, each turn of the Krebs cycle will produce 3NADH + $3H^+$. In addition, during the Krebs cycle there is also the production of 1 $FADH_2$ and 1 GTP. The NADH + H^+ produced during glycolysis and the Krebs cycle and the $FADH_2$ will enter an electron transport system where ATP will be synthesized. The main catabolic function of the Krebs cycle is to completely oxidize the pyruvate. In prokaryotic cells, the enzymes involved in the Krebs cycle are located in the cytoplasm of the cell. In eukaryotic cells, these enzymes are located in the matrix of the mitochondrion.

The electrons and protons released during the oxidation of glucose in the Embden-Meyerhof-Parnas pathway and the Krebs cycle are carried by electron carriers like nicotinamide adenine dinucleotide (NAD) or flavin adenine dinucleotide (FAD) to certain sites on the plasma membrane of prokaryotes, or to the inner mitochondrial membrane of eukaryotes, where an electron transport system is located.

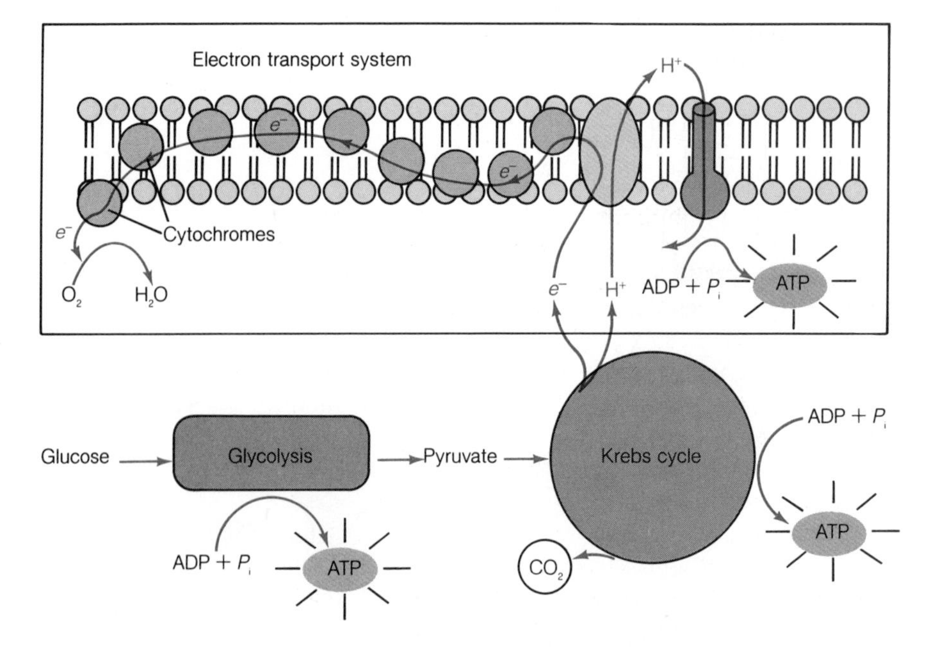

Figure 3.10 Summary of chemotrophic metabolism. Glucose is oxidized to pyruvate with the net production of 2 ATP molecules/glucose oxidized. Pyruvate is then oxidized to acetate, which then enters the Krebs cycle. During the Krebs cycle, the acetate is oxidized to carbon dioxide. The electrons and protons released from the oxidations enter an electron transport system. As the electrons and protons pass through the electron transport system, they release sufficient energy to synthesize several molecules of ATP. In bacteria, ATP yield during respiration can be as high as 38 molecules of ATP/molecule of glucose oxidized.

The electron transport system (fig. 3.9) consists of a sequence of various types of molecules that are capable of becoming reduced and then reoxidized. The transport of electrons and protons across the plasma membrane (in prokaryotes) or the inner mitochondrial membrane (in eukaryotes) can release sufficient energy to power the synthesis of ATP. The electrons, once they have passed through the electron transport system, are used to reduce molecules such as oxygen or inorganic ions called external electron acceptors, which are found in the environment. If oxygen is the final electron acceptor, the process is called **aerobic respiration**. If other inorganic molecules serve as the final electron acceptor then the process is called **anaerobic respiration.**

The overall chemical reaction for an aerobic respiration of glucose can be written as follows:

$$\text{Glucose } (C_6H_{12}O_6) + 38\text{ADP} + 38P_i + 6O_2 \rightarrow 38\text{ATP} + 6CO_2 + 6H_2O$$

Of the 688 kcal/mol of glucose oxidized that are available for ATP synthesis, 277 kcal are conserved in 38 mol of ATP synthesized by microorganisms (table 3.1). This value represents a conservation of about 40% of the potential energy of the glucose. In higher organisms, the remaining 60% is not completely lost because some of those calories are used to provide the heat necessary to maintain the cellular temperature at levels conducive to biochemical reactions. Figure 3.10 summarizes chemotrophic metabolism in bacteria.

The Chemiosmotic Hypothesis of Energy Conservation

The **chemiosmotic hypothesis** explains how electrons and protons (hydrogen ions or H^+) may be used by the cell to make ATP in an electron transport system. According to this hypothesis, H^+ and electrons from glycolysis and the Krebs cycle are separated from each other across a membrane (fig. 3.11) creating a potential (voltage) across the membrane that can be used to synthesize ATP. In this process, the electrons and H^+ are given to membrane-associated proteins. The H^+ travel along the proteins to the outside of the membrane while the electrons travel back across the membrane. The increased concentration of H^+ outside the cell results in both a pH difference and a voltage across the membrane. This energized state of the membrane is called a **proton motive force** or pmf. The pmf provides the energy necessary for enzymes called **ATP synthetases** (ATPases) to catalyze the synthesis of ATP

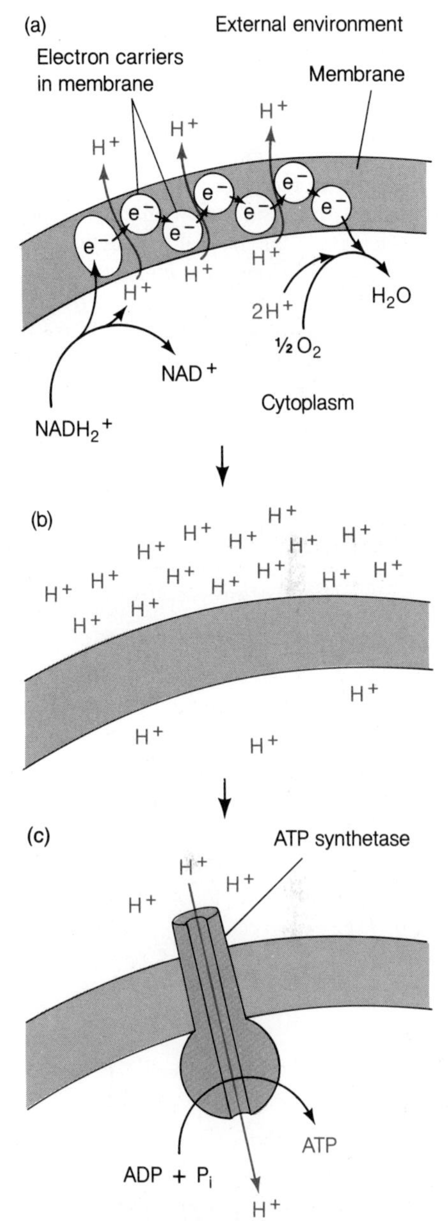

Figure 3.11 The chemiosmotic hypothesis. *(a)* Electron carriers in the electron transport system in the plasma membrane pump protons (H^+) to the outside of the cell. Electrons remain inside the cell. *(b)* The separation of protons and electrons across the membrane creates a potential difference (voltage). *(c)* Protons pass through a membrane-associated ATP synthetase (ATPase). The energy released when the potential is discharged is coupled to the synthesis of ATP.

FOCUS METABOLIC ACTIVITIES BY MICROORGANISMS AND THE LEAVENING OF BREAD

Chemical reactions are a part of all life processes. Some chemical reactions are involved in transporting nutrients into the cell, while others produce the energy necessary for the cell to function. As a consequence of these chemical reactions, microorganisms break down and synthesize numerous chemicals. Generally, waste products are discharged into the environment. Some of these chemical reactions drastically change the environment.

The leavening of bread occurs because a waste product, carbon dioxide, is released by a microorganism, such as a yeast. Bread rises because the leavening agent produces gas that makes the dough light and fluffy.

Bakers' yeast is used to leaven most breads (one notable exception is sourdough bread), coffee cakes, Danish pastries, and a variety of dinner rolls. This yeast produces the gas carbon dioxide and grain alcohol (ethanol), using the sugar in the dough as its source of energy and carbon. The ethanol may contribute to the aroma and flavor of the baked bread.

Sourdough breads are made in much the same way as other leavened breads, except that the microorganisms contributing to the flavor also include **lactic acid bacteria** (such as *Lactobacillus sanfrancisco*). These bacteria produce lactic acid, acetic acid, ethanol, and carbon dioxide from their energy-producing chemical activities. The carbon dioxide causes the dough to rise, while the acids (in particular acetic acid) give the sourdough bread its characteristic tart flavor.

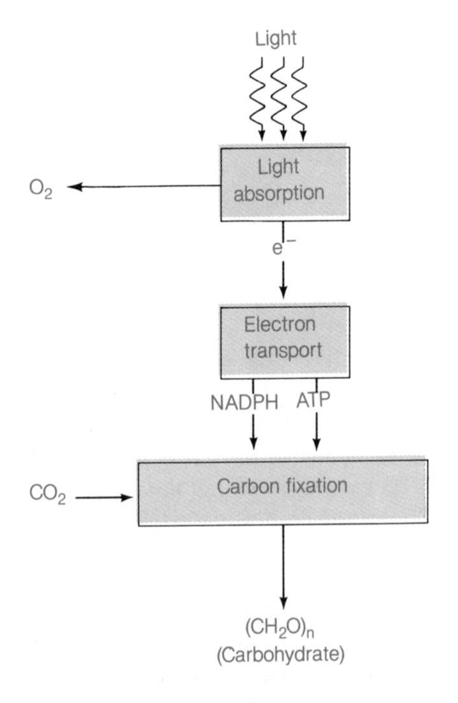

Figure 3.12 An overview of phototrophic metabolism

from ADP and P_i. Thus, the membrane serves as a tiny battery that can power the synthesis of ATP.

PHOTOTROPHIC METABOLISM

Phototrophs are those organisms that absorb light and use it to power the synthesis of ATP. These organisms utilize light-sensitive pigments like chlorophyll to trap light energy and convert it into chemical energy. The process of phototrophic metabolism is known as **photosynthesis** (fig. 3.12) and it takes place in two separate, although coordinated sets of chemical reactions.

Light Reactions of Photosynthesis

The first set of reactions, called **light reactions**, involve the conversion of light energy into ATP and reduced nicotinamide adenine dinucleotide phosphate (NADPH + H^+). During light reactions, light is absorbed by chlorophyll, causing this light-sensitive pigment to release electrons (become oxidized). The electrons are passed through an electron transport system similar to that in respiration to produce ATP and NADPH + H^+. Reduced NADP is also known as reducing power because it can readily donate electrons to other chemicals, thereby reducing them.

FOCUS HALOPHILIC BACTERIA USE BACTERIORHODOPSIN INSTEAD OF CHLOROPHYLL TO CONVERT LIGHT ENERGY INTO CHEMICAL ENERGY

Sometimes dried, salted fish or other consumer goods preserved by the addition of salt display reddish areas that result from bacterial growth. The bacteria that cause this red discoloration are not ordinary bacteria. They are **extreme halophiles** that require very high concentrations of NaCl and magnesium in their environment. Extreme halophiles are found in marine environments such as the Dead Sea where the salt concentration is very high. The red color is due to the fact that these bacteria have red and purple pigments in their plasma membrane. The formation of the purple pigment occurs only under anaerobic conditions.

The purple pigment is of interest because it is involved in an unusual form of energy generation. The purple pigment is **bacteriorhodopsin**. This photosensitive pigment bears a remarkable resemblance to the retinal pigment **rhodopsin**, found in the vertebrate eye. Although both pigments are involved in light-dependent reactions, rhodopsin is intimately involved in vision, while bacteriorhodopsin is involved in energy-yielding metabolism.

Bacteriorhodopsin absorbs light, which drives protons (H^+) out of the cell, creating a proton motive force similar to that created during respiration. The proton motive force is then used to provide the energy for the synthesis of ATP.

The discovery of these photosynthetic bacteria in 1971 provided yet another example of the remarkable versatility that exists in the world of microorganisms.

Each cell has a finite number of chlorophyll molecules that are constantly being oxidized during periods of abundant light. If the process of photosynthesis is to continue for extended periods of time, the light-stimulated chlorophyll molecules must be reduced. The source of electrons for the reduction of chlorophyll may come from water or other inorganic molecules. Cyanobacteria and photosynthetic eukaryotes split water to provide the necessary electrons for chlorophyll and release oxygen as a waste product. Bacteria other than the cyanobacteria use hydrogen sulfide or molecular hydrogen as a source of electrons.

The Calvin Cycle

The second set of reactions in photosynthesis are light-independent or **dark** reactions known as the **Calvin cycle**, in honor of the scientist who described it (fig. 3.13). The Calvin cycle involves the incorporation of carbon dioxide from the atmosphere into a three-carbon molecule called 3-phosphoglycerate to form various sugars that may be used to build cellular components, such as nucleotides and polysaccharides. This process is called **CO_2 fixation** and requires the expenditure of energy. The energy is provided by the hydrolysis of ATP and by reduced NADP. As we have seen, both ATP and NADPH + H^+ (reduced NAD) are produced in the light reactions of photosynthesis.

BIOSYNTHESIS

Microorganisms conserve energy in the form of nucleotides such as ATP and GTP so that they can use them to power cellular movement, nutrient transport, and the synthesis of the various cell components. The synthesis of cellular components, or **biosynthesis**, encompasses all cellular activities in which biological polymers (e.g., proteins and nucleic acids) and their precursors (e.g., amino acids and nucleotides) are made. As a rule, all of these reactions require the net input of energy and much of it comes from the hydrolysis of nucleotides.

Biosynthetic reactions usually take place in a stepwise fashion, where each step is catalyzed by a chemical reaction. The sum of all the steps leading to the synthesis of the product is called a **biosynthetic pathway** (fig. 3.14). The cell can regulate the activity of the enzymes in the pathway, and thus regulate the amount of final product produced. An example of a biosynthetic pathway for an amino acid is illustrated in figure 3.14.

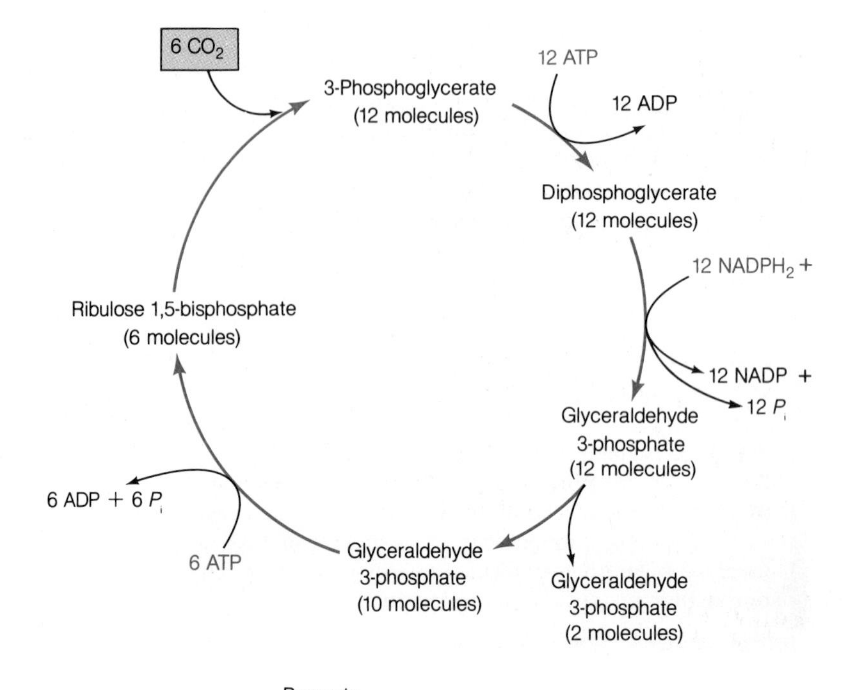

Figure 3.13 A summary of the Calvin cycle. The Calvin cycle, also known as the dark reaction of photosynthesis, is involved in the fixation of carbon dioxide to produce organic molecules. Carbon dioxide molecules enter the cycle one at a time. Six turns of the cycle, summarized in the figure, produce two gylceraldehyde-3-phosphate molecules. The energy required to drive the Calvin cycle is derived from ATP and NADPH, which come from photophosphorylation.

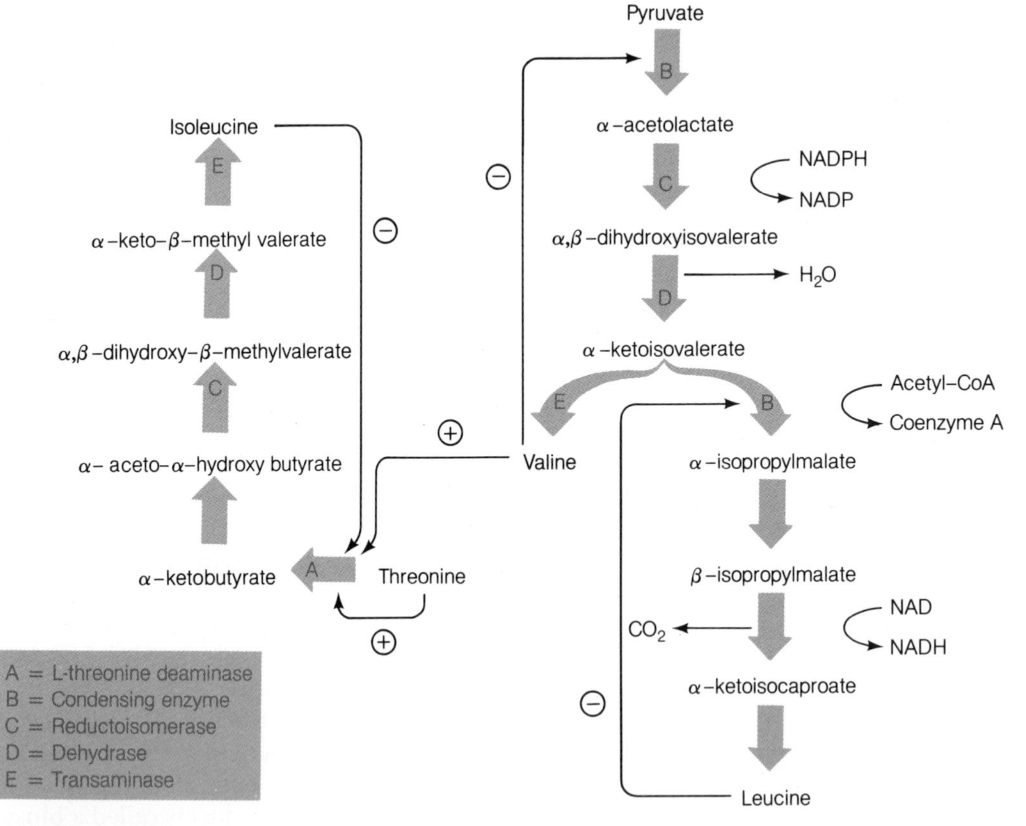

Figure 3.14 Pathways for the synthesis of three amino acids. Simplified pathway for the synthesis of leucine, isoleucine, and valine. Notice that all three pathways are interrelated. A (+) indicates that the product enhances the activity of the allosteric enzyme. A (−) indicates feedback inhibition of the allosteric enzyme. Enzyme stimulation and feedback inhibition work synergystically to produce the needed amounts of the three amino acids.

Carbon Assimilation

Carbon assimilation is an important aspect of microbial metabolism because this element is required for the synthesis of all organic molecules that make up the cell. Some organisms assimilate carbon as carbon dioxide. This type of carbon metabolism, called **autotrophic** (self-feeding) metabolism, takes place in the Calvin cycle. We saw briefly how this cycle functions in the fixation of CO_2 by autotrophic microorganisms. Most nonphotosynthetic microorganisms must assimilate their needed carbon as organic molecules. This type of carbon metabolism is called **heterotrophic** (mix-feeding) metabolism.

Heterotrophic microorganisms must have a ready source of organic molecules for their biosynthesis. These molecules are used by the cell to supply its carbon and as a source of energy. The same metabolic pathways that generate electrons (e.g., the Krebs cycle) can also be used to provide essential metabolic intermediates for biosynthesis (fig. 3.15). For this reason, the Krebs cycle is said to be an **amphibolic** pathway that is used by the cell in both anabolism and catabolism.

For example, the intermediate product of the Krebs cycle, called α-ketoglutaric acid, is also used by the cell as a precursor in the synthesis of the amino acid called glutamic acid, which in turn is used by the cell to build proteins. Similarly, oxaloacetic acid is a Krebs cycle component that also functions as a precursor in the biosynthesis of the amino acid, aspartic acid. This amino acid is a precursor of a component of nucleic acids such as DNA and RNA (see Chapter 6). Even some anaerobes that are unable to respire have most of the Krebs cycle enzymes, mainly for the generation of biosynthetic purposes.

REGULATION OF METABOLISM

The cell "fine tunes" its use and synthesis of materials so as to utilize its resources efficiently. For example, if

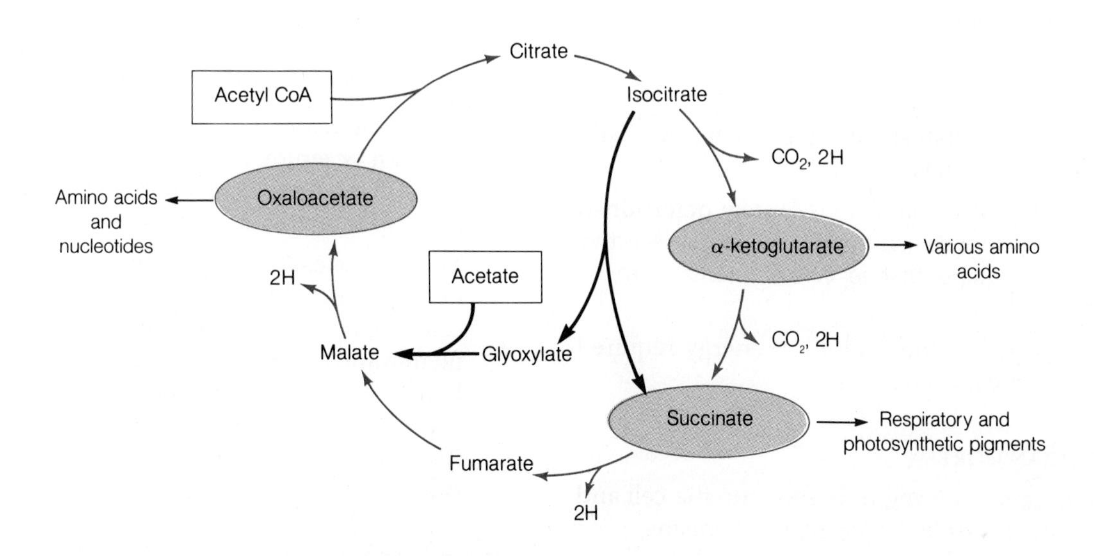

Figure 3.15 An amphibolic pathway. Amphibolic pathways are involved in both anabolism and catabolism. The Krebs cycle (colored lines) and the glyoxylate bypass (black lines) are involved in the production of important anabolic intermediates for the synthesis of various amino acids.

the cell oxidized glucose only to produce ATP, eventually there would be a surplus of ATP and no organic compounds with which to make new cell materials. This would be a waste of glucose. Instead, when a cell "senses" that there is sufficient ATP, it shifts its metabolism toward biosynthesis. In this fashion, some of the intermediates in energy metabolism can be used instead for biosynthesis (fig. 3.15). Additionally, the number of molecules oxidized to obtain energy or synthesized to make cellular material are carefully regulated by the cell so as to conserve energy and nutrients, thus enhancing its chances for survival (fig. 3.14).

SUMMARY

METABOLISM: AN OVERVIEW

1. Metabolism is the term used to describe all the chemical reactions that take place in a cell. Anabolism refers to those reactions that result in the building of cell material, while catabolism refers to processes that break down cell material and nutrients.
2. Microorganisms can extract energy from foods and conserve it in molecules of phosphorylated nucleotides.
3. The steps involved in extracting energy from foods may include digestion, transport, oxidation, and phosphorylation.

WHAT ARE ENZYMES?

1. Enzymes are catalysts that participate in nearly all cellular activities.
2. The activity of an enzyme is largely determined by its three-dimensional shape. An enzyme's active site has a shape that is complementary to its substrate.
3. Enzymes reduce the amount of energy required to start a chemical reaction.

NUTRIENT TRANSPORT

1. Microorganisms bring nutrients into the cell and get rid of wastes by transport mechanisms.
2. Transport mechanisms can be classified as passive or active. Passive transport mechanisms do not require the expenditure of energy, but cannot transport nutrients against a concentration gradient. Active transport mechanisms require the expenditure of energy, but can be used to bring nutrients into the cell even against a concentration gradient.

BIOENERGETICS

1. Bioenergetics is the study of energy production and utilization by the cell.
2. Reactions that hydrolyze ATP or other nucleotides are energy-yielding reactions that are commonly coupled to biosynthetic reactions in the cell.
3. Energy from food molecules and light is conserved by microorganisms in molecules of ATP.

CHEMOTROPHIC METABOLISM

1. Chemotrophs obtain their energy by oxidizing organic molecules. They may ferment or respire the molecules.
2. In fermentation, ATP is synthesized and organic molecules act as the final electron acceptors. The net yield of ATP during a fermentation is generally 2 molecules of ATP/glucose molecule fermented.
3. Respiring microorganisms oxidize substrates and use the resulting electrons and H^+ to synthesize ATP. The electrons and H^+ come from the Embden-Meyerhof-Parnas pathway and the Krebs cycle. The total amount of energy conserved during respiration may be as high as 38 molecules of ATP/glucose molecule respired.
4. In aerobic respiration, oxygen serves as the final electron acceptor in the electron transport system. In anaerobic respiration, other inorganic molecules are the acceptors.
5. The chemiosmotic hypothesis suggests that nutrient and light energy are conserved by the cell when electrons and H^+ are separated by membrane-associated electron transport systems. The concentration of hydrogen ions on one side of a membrane creates a proton motive force (pmf) that can be used to drive the synthesis of ATP by ATPases.

PHOTOTROPHIC METABOLISM

1. Phototrophs absorb light energy and conserve it in ATP. Electrons are obtained when light oxidizes chlorophyll. The electrons are then passed through an electron transport system, where ATP is synthesized.

BIOSYNTHESIS

1. Biosynthesis includes all cellular reactions in which biological polymers and their precursor molecules are made.
2. Some bacteria (autotrophs) can satisfy all of their carbon needs by fixing carbon dioxide in a pathway called the Calvin cycle.
3. Heterotrophs use organic molecules as their source of carbon. They use the Embden-Meyerhof-Parnas pathway and the Krebs cycle intermediates to provide the carbon skeletons for certain steps in biosynthesis.
4. The cell coordinates its metabolism so as to optimize the utilization of nutrients available in the environment.

STUDY QUESTIONS

1. Compare anabolism with catabolism.
2. Discuss how a molecule of starch found in the environment might be used by chemotrophs.
3. Discuss three ways in which ATP can be synthesized by bacteria.
4. How are ATP and NADH (or NADPH) used by the cell in metabolism?
5. Discuss how the following pairs of cellular processes differ from each other:
   a. fermentation and respiration;
   b. aerobic and anaerobic respiration;
   c. active and passive transport.
6. Discuss the nature and function of the Calvin cycle. Why is it also called the dark reaction of photosynthesis?
7. Discuss the function(s) of the Krebs cycle in: a. catabolism; b. anabolism.
8. How does the chemiosmotic hypothesis explain the synthesis of ATP?
9. What are amphibolic pathways?

SELF-TEST

1. A species of bacteria utilizes glucose in an oxygen-free environment to produce 2 moles of ATP per mole of glucose oxidized and an assortment of organic acids. In an atmosphere containing oxygen, the net output of ATP increases 18-fold. In view of this information, this species of bacteria is:
   a. strictly aerobic;
   b. facultatively anaerobic;
   c. aerotolerant anaerobic;
   d. anaerobic.
2. An important function of the Krebs cycle in biosynthesis is:
   a. the oxidation of glucose molecules for ATP synthesis;
   b. the production of reducing power as NADH for oxidative phosphorylation;
   c. the production of amino acid precursor molecules;
   d. the production of GTP;
   e. the production of NADPH.
3. The chemiosmotic hypothesis of energy conservation states that an electrical potential is created across the plasma membrane as a result of the separation of protons and electrons. This electrical potential can provide the necessary energy to power the synthesis of ATP by ATPases. a. true; b. false.
4. The amount of ATP produced from aerobic respiration of glucose is:
   a. much less than that from fermentation;
   b. greater than that from fermentation;
   c. less than that from photophosphorylation;
   d. about the same as in fermentation;
   e. about the same as in photosynthesis.
5. If cultures of *Escherichia coli* are cultivated in nutrient broth supplemented with glucose under aerobic and anaerobic conditions for an identical period of time:
   a. the culture grown aerobically will utilize less glucose than the anaerobic culture;
   b. the anaerobic culture will produce more cells than the aerobic culture;
   c. the anaerobic culture will utilize less glucose than the aerobic culture;
   d. both cultures will use the same amount of glucose;
   e. none of the above.
6. Bacteria carry out energy-yielding metabolism for all of the following *except*:
   a. powering cellular movement;
   b. breaking some chemical bonds and forming others;
   c. synthesizing cellular components;
   d. binary fission;
   e. storing energy in the form of CO_2.
7. The difference between facilitated and passive diffusion is that in passive diffusion:
   a. there is no need for a carrier molecule;
   b. there is a requirement for carrier molecules;
   c. cells can transport nutrients in against a concentration gradient;
   d. involves the expenditure of energy.
8. Most of the nutrients that enter the cell are concentrated by:
   a. active transport;
   b. passive transport;
   c. facilitated transport;

d. osmosis;
e. none of the above.

9. The first step in the assimilation of macromolecules by microorganisms to be used as foods is:
a. nutrient transport by permeases;
b. oxidation by enzymes;
c. reduction by reductases;
d. digestion by exoenzymes;
e. none of the above.

10. Enzyme activity is largely determined by:
a. the type of substrate;
b. the presence of effector molecules;
c. the presence of ATP in the cell;
d. its primary structure;
e. its three-dimensional shape.

11. The amount of energy required for the initiation of biological reactions is reduced by the action of enzymes. a. true; b. false.

12. For the transport of molecules such as water or glycerol into the cytoplasm of the cell, the hydrolysis of ATP (ATP $\rightarrow$ ADP + P_i + energy) must occur. a. true; b. false.

13. During the first three steps in the oxidation of glucose to pyruvate via the Embden-Meyerhof-Parnas pathway the following occurs:
a. 2 ATP molecules are synthesized;
b. glucose is oxidized to carbon dioxide;
c. 4 molecules of NAD are reduced;
d. 2 ATP molecules are used up;
e. 2 NADH + H^+ are used up.

14. The principal catabolic function of the Krebs cycle is:
a. to synthesize GTP;
b. to produce electrons and protons for the electron transport system;
c. to provide biosynthetic intermediates;
d. to reduce pyruvate;
e. to carry out fermentations.

15. Respiration is a process whereby pyruvate is oxidized to CO_2 in the Krebs cycle and the electrons and protons donated by pyruvate are used to produce ATP in the electron transport system. a. true; b. false.

16. What is the maximum number of ATP molecules that can be produced from the complete oxidation of glucose by chemotrophic organisms?
a. 2;
b. 18;
c. 38;
d. 32;
e. 4.

17. During anaerobic respiration, the electrons obtained from the oxidation of glucose are donated to organic molecules to form substances such as ethanol and acetic acid. a. true; b. false.

18. The electron transport system in bacteria consist of a group of proteins called cytochromes that are embedded in the plasma membrane of the cell. a. true; b. false.

19. The light reactions of photosynthesis involve the incorporation of CO_2 from the atmosphere into three-carbon compounds. a. true; b. false.

20. Biosynthetic reactions usually take place in biosynthetic pathways to maximize the quantity of product made during biosynthesis: a. true; b. false.

SUPPLEMENTAL READINGS

Dickerson, R. E. 1980. Cytochrome C and the evolution of energy metabolism. *Scientific American* 242(3):136–153.

Drews, G. 1985. Structure and functional organization of light-harvesting complexes and photochemical reaction centers in membranes of phototrophic bacteria. *Microbiological Reviews* 49:59–70.

Hinckle, P. C. and R. E. McCarthy. 1978. How cells make ATP. *Scientific American* 238(3):104–123.

Ingledew, W. J. and Poole, R. K. 1984. The respiratory chains of *Escherichia coli*. *Microbiological Reviews* 48(3):222–271.

Michels, M. and Bakker, E. P. 1985. Generation of a large, protonophore-sensitive proton motive force and pH difference in the acidophilic bacteria *Thermoplasma acidophilum* and *Bacillus acidocaldarius*. *Journal of Bacteriology* 161(1):231–237.

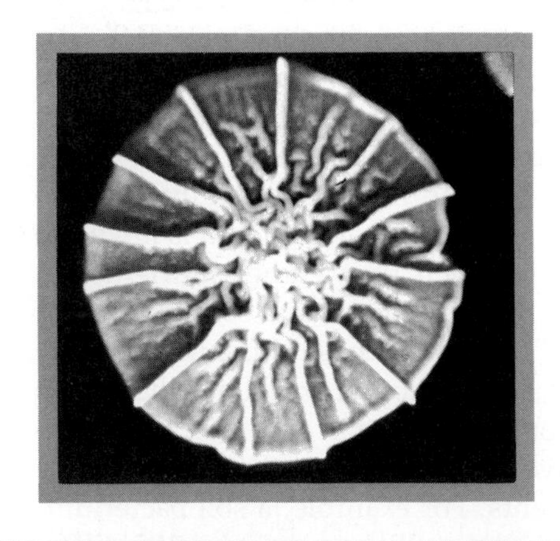

CHAPTER 4

NUTRITION AND GROWTH OF MICROORGANISMS

CHAPTER PREVIEW ANTIBIOTICS: A CONSEQUENCE OF MICROBIAL GROWTH

When certain spore-forming microorganisms reproduce, they release chemicals that inhibit the reproduction of other microorganisms or even kill them. These toxic microbial products are called **antibiotics**.

Microbiology made its debut in the pharmaceutical industry in the 1940s with the isolation of two antibiotics: penicillin by Ernst Chain and Howard Florey, and streptomycin by Selman Waksman. These discoveries revolutionized the pharmaceutical industry. Until then, physicians were virtually helpless in combating infectious diseases. Today, antibiotics save countless lives and the pharmaceutical industry nets billions of dollars annually from the exploitation of antibiotics produced by microorganisms.

Much research has been directed toward the study of the growth of antibiotic-producing organisms, so that optimal conditions for antibiotic production can be achieved. This information is important because, even though a certain microorganism can produce an antibiotic, it may not do so under all growth conditions. Furthermore, a microorganism may not produce the antibiotic at a constant rate throughout its growth cycle. Some may produce antibiotics when they are dividing actively, while others may produce them only when their rate of reproduction is slowing down. For example, many of the bacteria in the genus *Bacillus*, which is characterized by the production of endospores, synthesize antibiotics at about the time they are forming their endospores. Since the production of endospores is influenced by the physiologic state of the cells, antibiotic production can be optimized by establishing the growth conditions that are conducive to endospore formation. Hence, knowing the growth conditions that each microorganism must have to produce antibiotics is essential for the optimal production of these life-saving substances.

The pharmaceutical industry relies heavily on available knowledge of microbial growth and reproduction to maximize the production of antibiotics. Some of the characteristics of microbial growth and the factors that affect it are discussed in this chapter.

ALL LIVING ORGANISMS REQUIRE NUTRIENTS for maintenance, growth, and reproduction. Nutrients provide the raw materials and energy used to build new cellular components and the energy to power cellular processes (fig. 4.1).

The ability to grow and multiply rapidly is essential to the survival of microorganisms and is linked to the efficiency with which they gather nutrients. In nature, nutrients are often limited and competition for them is great. To overcome these difficulties, many microorganisms have developed great flexibility in the utilization of various substrates available in their environment and have evolved forms that aid them in assimilating nutrients. For example, a soil bacterium, *Pseudomonas cepacia*, can utilize as many as 105 different organic compounds as sources of carbon and energy. *Caulobacter*, a bacterium that is commonly found in aquatic environments, has an appendage (prostheca) that allows the bacterium to absorb nutrients more efficiently from highly diluted environments.

To understand microorganisms, we must be able to cultivate and maintain them in pure form in the laboratory. The techniques for growing microorganisms have numerous applications. For example, in the hospital laboratory, these techniques are used routinely for isolating and identifying microorganisms suspected of causing disease. Also, microorganisms are isolated from nature and studied in the laboratory so that we can better understand their roles in natural environments. Many industries, such as the petroleum, agriculture, and food industries, capitalize on

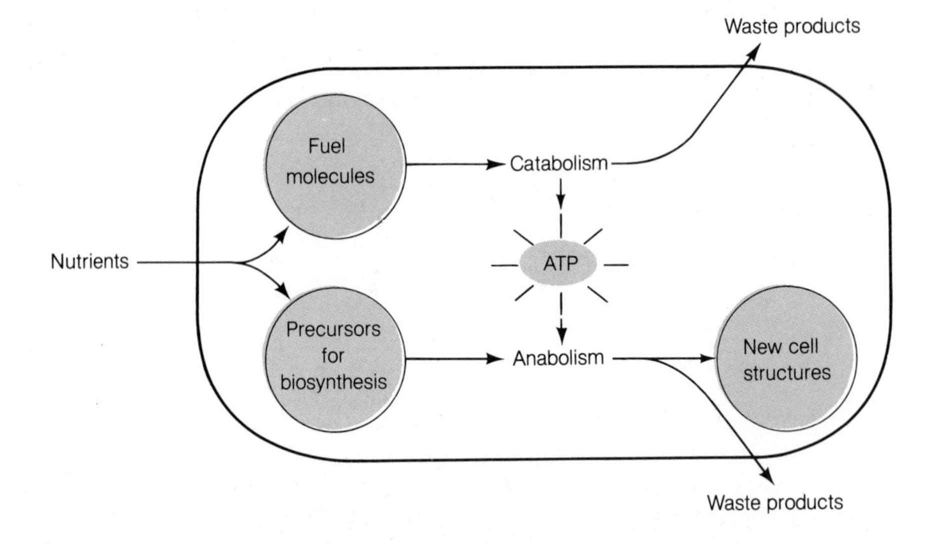

Figure 4.1 Relationship between nutrient intake and cellular activities. Nutrients in the environment are taken up by the cell. These nutrients are then oxidized to produce energy-rich compounds such as adenosine triphosphate (ATP). Nutrients are also used by the cell to synthesize cellular structures. The synthesis of cell structures requires that ATP be present as the source of energy.

the knowledge obtained from laboratory studies of microorganisms to improve their products or to acquire materials from nature.

In this chapter we discuss some of the basic concepts of microbial nutrition and describe the various ways in which microorganisms can satisfy their need for nutrients. We also discuss the basic techniques employed for cultivating microorganisms in the laboratory and the various types of culture media used to grow them.

PRINCIPLES OF MICROBIAL NUTRITION

Microorganisms obtain nutrients from their environment. Although they have the same basic nutritional requirements, they differ in the ways they take up these nutrients. Bacteria and fungi are called **osmotrophs** because they obtain their nutrients in solutions that pass through the plasma membranes. Osmotrophic microorganisms are unable to transport large molecules (such as proteins or polysaccharides) through their membranes, and consequently must digest them outside the cell. Large molecules are broken down by **exoenzymes** that are secreted by the cell to the external environment (fig. 4.2). For example, exoenzymes called **proteases** degrade proteins to amino acids. The small breakdown products are then transported into the cell. Other microorganisms, such as the protozoa, ingest food particles by a process called **phagocytosis** and are called **phagotrophs** (fig. 4.3). During the process of phagocytosis, the organism surrounds the food particle with its plasma membrane and then brings the particle inside the cell surrounded by a piece of plasma membrane (vacuole). Inside the vacuole, enzymes derived from lysosomes digest the food particle into simpler molecules, such as amino acids and sugars. This digestive process resembles the digestion of molecules by exoenzymes. The resulting small molecules are transported into the cytoplasm where they are used by the cell as nutrients.

Once the nutrients are inside the cell, they are chemically modified to meet the specific nutritional needs of the cell. For example, sugars may be converted to amino acids so that proteins can be made. On the other hand, amino acids are transformed into sugars, nucleotides, or other compounds that are needed to build cellular components, or they may be used as energy sources.

NUTRITIONAL REQUIREMENTS OF MICROORGANISMS

Nutrients are used as a source of energy to make cellular components (table 4.1) such as chromosomes, the plasma membrane, and the cell wall. Nutrient requirements are often satisfied in very different ways, depending upon the species of microorganism and the availability of specific nutrients. For example, although all microorganisms require carbon, some can

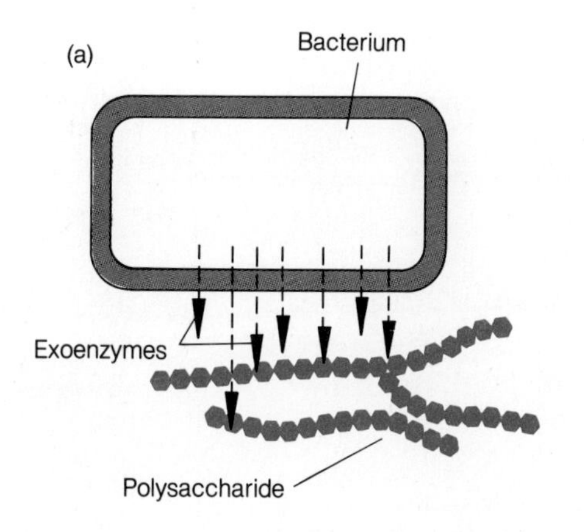

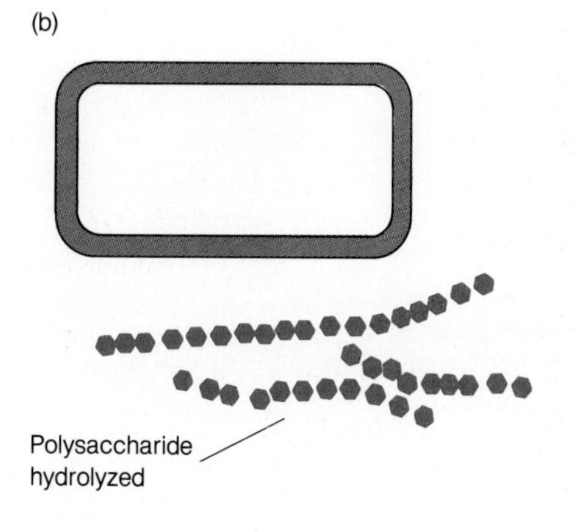

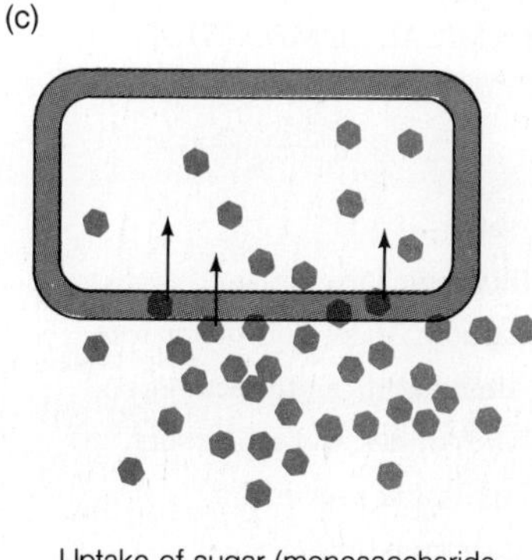

Figure 4.2 **Function of exoenzymes in nutrient digestion.** *(a)* Exoenzymes are secreted by the cell. They act on large, insoluble food molecules (carbohydrates, proteins, fats, etc.) present in the immediate environment. *(b)* The exoenzymes break these large molecules into constituent molecules (simple sugars, amino acids, etc.), which are then transported into the cell to be used as nutrients *(c)*.

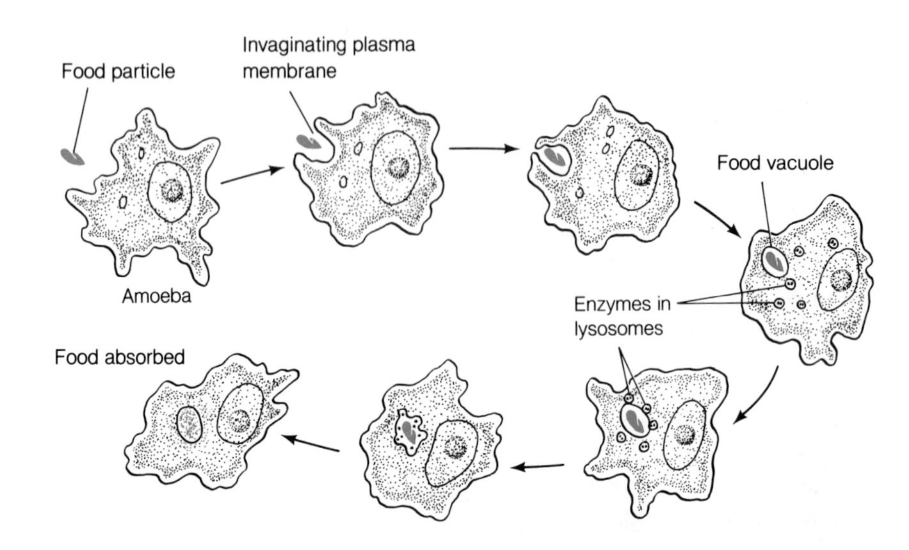

Figure 4.3 Phagocytosis. Phagotrophic organisms obtain their nutrients by engulfing food particles with their plasma membrane. The particle is surrounded by an invaginating piece of plasma membrane, which then forms a food vacuole. The food particles within the food vacuole are acted upon by digestive enzymes, which then break down the food into small molecules (sugars, amino acids, nucleosides, etc.). These simple molecules are then used by the cell as nutrients.

take it up only in an organic form, while others may assimilate it as carbon dioxide. Similarly, nitrogen is universally required by microorganisms, some obtain their nitrogen from amino acids while others can use ammonium- or nitrate-nitrogen. In the following paragraphs, we discuss some of the basic nutritional requirements of microorganisms and indicate how these nutrients are used.

Energy Requirements

All cellular processes, such as the synthesis of new cell material, concentration of nutrients within the cell, and cell motility, require energy. The energy is obtained from sunlight or from the oxidation of chemical compounds. Microorganisms convert this energy into energy-rich compounds such as the nucleotide adenosine triphosphate (ATP), which can be used as "energy

TABLE 4.1
CHEMICAL COMPOSITION OF A TYPICAL BACTERIAL CELL*

Chemical component	% of cell weight	Number of molecules per cell ($\times 10^6$)	Type of each molecule
Water	70	4000	1
Inorganic ions	0.3	200	20
Carbohydrates and precursors	0.1	210	180
Amino acids and precursors	0.4	200	100
Nucleotides and precursors	0.2	100	100
Lipids and precursors	3.0	245	70
Macromolecules			
DNA	1.0	0.000002	1
RNA	6.0	0.3	460
Proteins	16.5	2	1100
Polysaccharides	2.5	1	10

*Values are approximate values of wet weight cells (*Escherichia coli*) growing aerobically at 35°C.

currency" for cellular activities. Those organisms that obtain their energy from sunlight are called **phototrophs**, while those that derive their energy from the oxidation of chemicals are called **chemotrophs**.

Most phototrophs obtain their energy by a process called **photosynthesis** (fig. 4.4). In this process, the radiant energy from the sun is "trapped" by light-sensitive pigments such as chlorophyll. The ultimate result is the production of chemical energy in the form of molecules of adenosine triphosphate, or ATP.

There are two major groups of phototrophic bacteria: the **cyanobacteria** and the **photosynthetic bacteria**. One important difference between these two groups is that the cyanobacteria produce O_2 as a by-product of photosynthesis, while the photosynthetic bacteria produce other compounds.

Chemotrophic microorganisms may obtain their energy from the oxidation of organic and inorganic compounds such as glucose or ammonium. These biological oxidations are known as fermentations and respirations (fig. 4.5) and result in energy being stored in energy-rich compounds such as ATP.

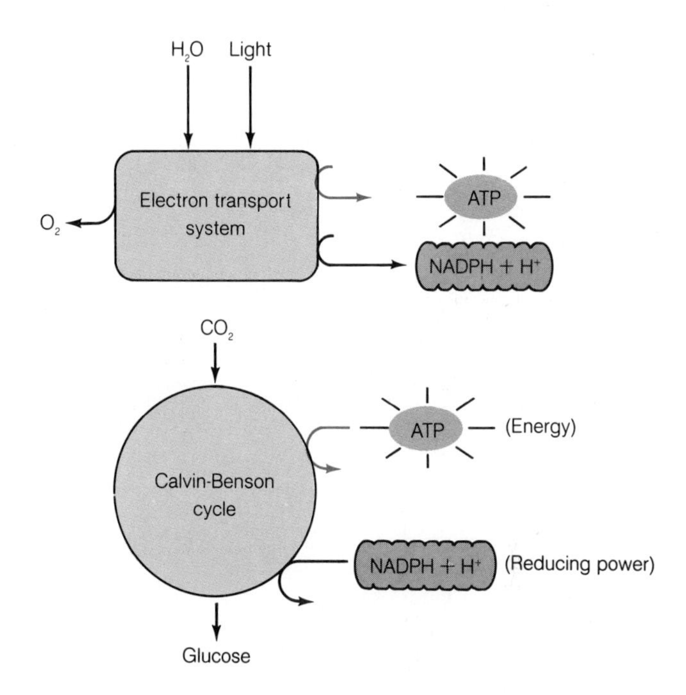

Figure 4.4 Simplified illustration of phototrophic metabolism. Phototrophic organisms obtain their energy from the sun. Radiant energy in light stimulates light-sensitive pigments, initiating a series of light-dependent chemical reactions that result in the production of usable energy in the form of ATP and reducing power in the form of NADPH. The energy and reducing power are used to convert carbon dioxide into carbohydrates.

Carbon Requirements

Carbon (C) atoms are present in all cellular components. Macromolecules such as proteins, lipids, polysaccharides, and nucleic acids, which collectively make up the bulk of the cell's organic material, are mostly made of carbon. Carbon-containing molecules participate in all phases of cellular metabolism. They participate in **catabolic** (biodegrading) activities leading to the production of energy-rich molecules such as ATP, as well as in **anabolic** (biosynthetic) activities (fig. 4.1), which lead to the synthesis of cellular components.

Some microorganisms can satisfy all of their carbon requirements from carbon dioxide in the atmosphere. These organisms are called **autotrophs** (self-feeders) and are able to fix or incorporate CO_2 into complex organic compounds. Other microorganisms must have organic molecules as carbon sources and are called **heterotrophs** (table 4.2). Microorganisms that obtain their energy from the sun, and their carbon in the form of carbon dioxide, are called **photoautotrophs**. Those phototrophic organisms that use organic molecules to satisfy their carbon needs are called **photoheterotrophs**. Similarly, microorganisms that oxidize chemical compounds for their energy and use carbon dioxide as their carbon source are called **chemoautotrophs**, and those that oxidize chemical compounds for their energy and require organic forms of carbon are called **chemoheterotrophs**.

Nitrogen Requirements

Nitrogen is found in almost all structural components of the cell. It is necessary for the synthesis of important building blocks such as amino sugars, nucleotides, and vitamins. Microorganisms exhibit remarkable versatility in their ability to use various nitrogenous compounds to satisfy their nitrogen requirements. Some bacteria use proteins or polypeptides to obtain their nitrogen. These large molecules are digested to amino acids, which in turn are chemically altered to make other nitrogen-containing molecules or are incorporated into cellular proteins. Many microorganisms, however, can obtain all of their needed nitrogen from inorganic salts of nitrate or ammonium.

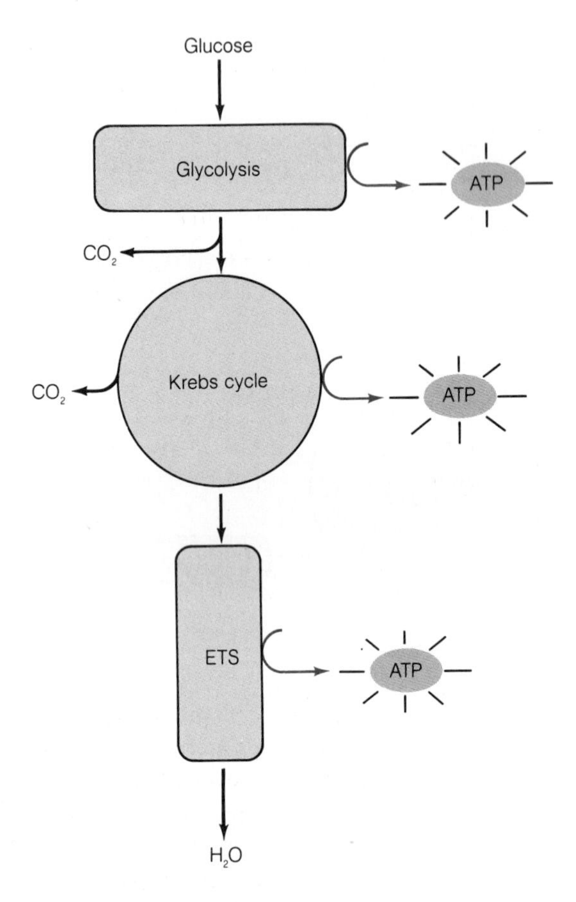

Figure 4.5 Simplified illustration of chemotrophic metabolism. Chemotrophic organisms oxidize fuel molecules, such as glucose, in a series of enzyme-catalyzed reactions such as glycolysis, the Krebs cycle, and the electron transport system (ETS). The end result of these oxidations is the extraction of energy from fuel molecules plus its conservation in ATP molecules. If the oxidation of fuel molecules includes the Krebs cycle and the electron transport system, the process is called respiration. If the oxidation of fuel molecules is only to pyruvate, the process is called a fermentation.

A unique characteristic of a few bacteria is their ability to **fix** or incorporate molecular nitrogen into organic compounds. The process of **nitrogen fixation** involves the reduction of molecular nitrogen (N_2) to ammonium (NH_4^+) and its subsequent incorporation into organic molecules to form amino acids.

Oxygen Requirements

Oxygen is a common atom found in many biological molecules. It is an integral part of amino acids, nucleotides, glycerides, and other molecules and is generally taken into the cell as part of nutrients, such as proteins and lipids. In addition, this element, in the form of molecular oxygen (O_2), is required by most eukaryotes and many prokaryotes to generate energy by respiration.

Sulfur and Phosphorus Requirements

Most of the sulfur in cells is found in the sulfur-containing amino acids cysteine and methionine. Sulfur is also found in some polysaccharides, such as agar, and in certain coenzymes. Microorganisms can obtain their sulfur as inorganic salts of sulfate, hydrogen sulfide, sulfur granules, thiosulfate, or as organic compounds (cysteine or methionine). When microorganisms use sulfate as their source of sulfur, they reduce it to hydrogen sulfide, which is then incorporated into existing organic molecules. If the microorganism cannot reduce sulfate, then the sulfur must be obtained in a reduced form, such as that found in amino acids or in sulfides.

Phosphorus is found in the cell primarily in nucleic acids, phospholipids, and coenzymes. Phosphate salts of sodium or potassium are the most common sources of this element for microorganisms, although some phosphorus is made available when phosphorus-containing organic molecules (e.g., nucleotides) are taken up.

Trace Element Requirements

Numerous minerals such as cobalt (Co), potassium (K), molybdenum (Mo), magnesium (Mg), manganese (Mn), calcium (Ca), and iron (Fe) are required by all cells. Minerals are required for the activity of a number of enzymes and other organic molecules. For example, magnesium is an integral component of the light-sensitive pigment chlorophyll, and cobalt is required for the enzyme **nitrogenase** to catalyze the fixation of nitrogen. Iron is an essential component of cytochromes (proteins involved in cellular respiration), catalase (an enzyme that degrades H_2O_2), and hemoglobin.

TABLE 4.2
SUMMARY OF THE NUTRITIONAL TYPES OF MICROORGANISMS BASED ON THEIR ENERGY AND CARBON SOURCES

Nutritional type	Energy source	Reducing power	Carbon source
a. Photo*autotrophs*	Light	Inorganic molecules	Carbon dioxide
b. Photo*heterotrophs*	Light	Organic molecules	Organic carbon
c. Chemo*autotrophs* (Chemolithotroph)	Inorganic molecules	Inorganic molecules	Carbon dioxide
d. Chemo*heterotrophs*	Organic molecules	Organic molecules	Organic carbon

Minerals are needed only in extremely small amounts, so they are often referred to as **trace elements**. Although trace elements are sometimes supplied to microorganisms in the form of mineral salts, it is not usually necessary to do so. Trace elements are required in such low concentrations by the cell that sufficient quantities usually are found as contaminants in the other nutrients supplied to the cell or dissolved in water.

Growth Factor Requirements

Some microorganisms can grow in an environment consisting of water, inorganic salts, and a single organic compound such as glucose (table 4.3). For example, *Escherichia coli* can synthesize all of its required cellular components (e.g., amino acids, vitamins, and nucleotides) from inorganic chemicals and glucose in its environment. Many microorganisms, however, require additional preformed substances in order to grow and multiply. Lactic acid bacteria such as *Lactobacillus*, for example, lack the ability to synthesize many of their required amino acids and vitamins and can grow only in environments containing these growth factors (table 4.4).

Most organisms cannot grow on a medium that consists of only salts and glucose because they lack the enzymes to convert these compounds to cellular building blocks. Those components that cannot be synthesized by cells and that must be obtained from the environment are called **growth factors**. Compounds such as vitamins, amino acids, nucleic acid bases, and nucleosides that organisms are unable to synthesize from inorganic salts and the carbon and energy source are frequently-encountered growth factors (table 4.3 and table 4.4).

Water Requirements

All metabolic activities of a cell are carried out in aqueous environments. A dry environment generally results in the loss of the cell water to its surroundings. This disrupts cellular activities.

CULTURE MEDIA AND THEIR USE IN THE LABORATORY

A **culture medium** is an aqueous solution of the various nutrients required by a microorganism (tables 4.3 and 4.5). It generally contains a source of energy, carbon, nitrogen, sulfur, phosphorus, hydrogen and oxygen, trace elements, and water. Various growth factors and other ingredients may be added to this basic medium to grow the desired microorganism.

The culture medium may be in a liquid or a gel form. Liquid culture media are usually referred to as **broths**, while gel or semisolid media are often called **agars** because they are solidified with a red algal polysaccharide called agar. This polysaccharide is useful as a jelling agent for several reasons: (a) agar is rarely used as a nutrient by microorganisms and therefore is not readily degraded when microbes grow on culture media solidified with agar; (b) unlike gelatin, which originally was used as a solidifying agent, agar remains solid over a wide range of incubating temperatures (0–80°C); (c) agar liquefies at about the boiling temperature of water, but does not jell until it cools to approximately 42°C.

Categories of Culture Media

The composition of culture media largely determines which types of microorganisms can grow in them. The judicious preparation and use of culture media can afford a microbiologist a powerful tool with which to study the biology of microorganisms and isolate them in pure form. There are three different categories of culture media: (a) chemically defined or simple; (b) chemically undefined or complex; and (c) living.

TABLE 4.3
INGREDIENTS FOR A CHEMICALLY DEFINED MEDIUM AND A COMPLEX MEDIUM

A. Chemically-defined medium for cultivation of *E. coli*	
$NH_4H_2PO_4$	1 g
Glucose (energy & carbon source)	5 g
NaCl	5 g
$MgSO_4 \cdot 7H_2O$	0.2 g
K_2HPO_4	1 g
H_2O	1000 ml
B. Complex medium for cultivation of lactobacilli*	
Casein hydrolyzate	5 g
Glucose	10 g
Solution A	10 ml
Solution B	5 ml
L-Asparagine*	250 ml
L-Tryptophan*	50 mg
L-Cystine*	100 mg
Methionine*	100 mg
Cysteine*	100 mg
Ammonium citrate	2 g
Sodium acetate (anhydrous)	6 g
Adenine; guanine; xanthine; uracil; each*	10 mg
Riboflavin; thiamine; pantothenate; niacin; each*	500 μg
Pyridoxamine*	200 μg
Pyridoxal*	100 μg
Pyridoxin*	200 μg
Inositol and choline, each*	10 mg
p-Aminobenzoic acid*	200 μg
Biotin*	5 μg
Folic acid (synthetic)*	3 μg

Make up to 1l with distilled water.
Solution A. K_2HPO_4 and KH_2PO_4, each 25 g, into distilled water to a volume of 250 ml.
Solution B. $FeSO_4 \cdot 7H_2O$, 0.5 g; $MnSO_4 \cdot 2H_2O$,2.0 g; NaCl, 0.5 g; and $MgSO_4 \cdot 7H_2O$, 10 g. Dissolve in distilled water to a volume of 250 ml.

*Growth factors.
From *Microbiology*, 4th ed., by Pelczar, Reid, and Chan. Copyright © 1977 by McGraw-Hill Book Company. Reprinted by permission. Data from M. Rogosa, et al., *J Bacteriol*, 54:13, 1947.

TABLE 4.4
GROWTH FACTOR REQUIREMENTS OF SOME BACTERIA

Species of bacteria	Growth factor required
Bacillus anthracis	Thiamine (B_1)
Bacteroides melaninogenicus	Vitamin K
Brucella abortus	Niacin
Clostridium tetani	Riboflavin
Lactobacillus sp.	Pyridoxine (B_6)/Cobalamin (B_{12})
Leuconostoc dextranicum	Folic acid
Leuconostoc mesenteroides	Biotin
Proteus morganii	Pantothenic acid
Purple nonsulfur bacteria	Thiamine (B_1)
Purple sulfur bacteria	Cobalamin (B_{12})

TABLE 4.5
SELECTED CULTURE MEDIA FOR THE ISOLATION AND DIFFERENTIATION OF REPRESENTATIVE GROUPS OF MICROORGANISMS

Microorganism or group desired	Selective media (for isolation)	Differential media (for differentation)
Streptococccus	Azide-blood agar base	Mitis-Salivarius agar
Staphylococci	*Staphylococcus* 110 agar Chapman-Stone medium	Mannitol-Salts agar Chapman-Stone medium
Neisseria	Thayer-Martin medium	Phenol red-carbohydrate media
Mycobacterium	Lowenstein-Jensen medium 7H10 and 7H11	7H10 and 7H11
Coliforms	Violet red bile agar Levine EMB MacConkey agar	Violet red bile agar Levine EMB MacConkey agar
Salmonella and *Shigella*	SS agar XLD agar Hektoen-Enteric agar	SS agar XLD agar Hektoen-Enteric agar
Fungi	Sabouraud glucose agar	Corn meal agar Chlamydospore agar

Chemically defined media are made so that the concentration of each ingredient is known. The ingredients used in chemically defined media are usually highly purified inorganic salts and simple organic compounds, such as glucose or purified amino acids. By knowing the exact composition of the culture medium, it is possible to maintain a high degree of consistency from batch to batch. One of the drawbacks of a defined medium is that it can be very expensive to prepare if purified nutrients such as amino acids, vitamins, and nucleotides have to be added. Since a particular defined medium may support the growth of only a few species of bacteria, it cannot be used to grow a large variety of microorganisms; in this case complex media are used.

Complex media are prepared using natural products, such as meat extracts or vegetable infusions. Although natural products contain many of the essen-

FOCUS THE FIRST USE OF AGAR AS A SOLIDIFYING AGENT

The first use of agar as a solidifying agent is credited to Walther Hesse, who worked in Dr. Robert Koch's laboratory. Koch experienced difficulties isolating disease-causing bacteria on culture media solidified with gelatin. At room temperature (18–22°C) the gelatin remained solid, but when the bacteria grew, many digested the gelatin and the medium became liquid. This problem was compounded when disease-causing bacteria were cultured at human body temperature (37°C). At this temperature the gelatin liquefied, regardless of whether the bacteria digested the gelatin. With the gelatin turning liquid, it became impossible for Koch to culture bacteria in pure form.

This problem was solved by Hesse's resourceful wife. Apparently, Koch had been discussing with Hesse the difficulties of preparing a stable solid culture medium, and Hesse mentioned this discussion to his wife. Frau Hesse suggested—to the benefit of countless microbiologists—that Koch try an ingredient her family in the West Indies used to thicken sauces, jams, and the like. This ingredient was called **agar-agar**. It is reported that Hesse told Dr. Koch about his wife's suggestion, Koch tried it, and it worked! Agar is now used by microbiologists throughout the world as *the* hardening agent for nearly all culture media.

tial nutrients necessary for microbial reproduction, the exact concentration of each nutrient is unknown. Complex culture media are used extensively in the microbiology laboratory because they are easy to prepare, inexpensive, and able to support the growth of many different types of microorganisms. The major shortcoming of complex media is the variability of their composition from batch to batch. Variability in the chemical composition of the medium may alter the growth characteristics of the microorganisms. Thus, precise physiological studies may be difficult to conduct using complex media.

Some microorganisms will proliferate only when growing in a living host. For example, viruses require living cells to reproduce. They are grown in the laboratory in living microorganisms and in cells from plants and animals (e.g., a chick embryo or a mouse) called **tissue cultures**. Tissue cultures consist of a nutrient solution that can support the growth of the host cells derived from the tissues of various animals or plants. Viruses, chlamydias, rickettsias, and some spirochetes are cultured in living organisms or in tissue cultures.

Selective and Differential Media

Culture media are sometimes prepared to select for a particular microorganism or group of microorganisms, such as *Staphylococcus* or coliforms. These types of culture media are called selective media (table 4.5). Other media can be used to reveal differences among groups or species of microorganisms and are referred to as differential media.

Selective media are culture media containing at least one ingredient that inhibits the reproduction of unwanted organisms, but permits the reproduction of specific microorganisms. The ingredients include one or more **selective agents**. For example, the isolation of *Staphylococcus* from human skin can be accomplished using nutrient agar supplemented with 7.5% NaCl. Sodium chloride constitutes the selective agent in the culture medium, because it will inhibit the reproduction of skin microorganisms other than *Staphylococcus*. Selective agents often employed in culture media include dyes, such as crystal violet and malachite green; antibiotics, such as penicillin and streptomycin; and chemicals, such as sodium azide or sodium chloride.

Differential media are special formulations designed to reveal differences among microorganisms or groups of microorganisms. Differential media usually contain a chemical that is utilized or altered by some microorganisms but not by others. When different microorganisms grow on this medium, they change its appearance. These changes can be used to differentiate among microorganisms. For example, the selective medium we described previously for the isolation of *Staphylococcus* can be modified so that it is possible to differentiate among various species in the genus. The modification involves the addition of 0.5% mannitol and a pH indicator, such as phenol red. On this medium, both *S. aureus* and *S. epidermidis* will grow because they can multiply in the presence of 7.5% NaCl. *Staphylococcus aureus* will ferment the mannitol, however, releasing acidic by-products, whereas *S. epidermidis* will not. The acidic products released by *S. aureus* will cause the pH indicator surrounding its colonies to change from red to yellow. Since *S. epidermidis* does not ferment mannitol, the area around its colonies remains red. By using this differential medium, it is possible to differentiate between these species of staphylococci. Differential media are widely used in microbiology because they aid in the isolation and identification of microorganisms isolated from nature, foods, or diseased tissues. Some of the most common differential media in current use are listed in table 4.5.

METHODS FOR CULTIVATING MICROORGANISMS

To study and characterize a microorganism, it is first necessary to isolate the microbe from all others and to maintain it in pure form throughout the study. Cultures consisting of only one type of microorganism are called **pure cultures**. The isolation of a microorganism and maintenance of pure cultures are achieved by procedures known as **aseptic techniques**. Aseptic techniques include methods for the isolation of microorganisms from contaminated sources and for the transfer and maintenance of pure cultures.

A common method for isolating pure cultures of microorganisms is the **streak technique** (fig. 4.6). It is performed by spreading a small amount of culture over the surface of an agar-solidified medium with an **inoculating loop**. As the microorganisms are spread over the agar surface, their concentration on the inoculating loop decreases. Eventually, a single microbial cell lands on an isolated area of the agar-solidified medium. As the isolated microorganism multiplies, a population develops, forming a clone of cells called a **colony**. If the original specimen possessed several different types of microorganisms, each type may form

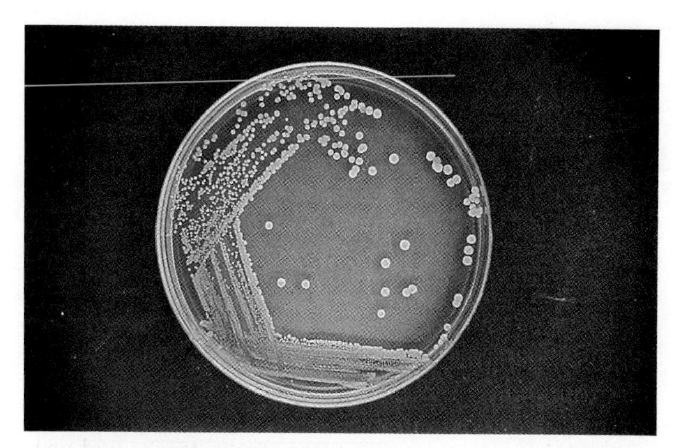

Figure 4.6 The streak plate. Blood agar streaked with *Staphylococcus epidermis*. The streak plate is commonly used to obtain pure cultures of microorganisms. The streak plate is performed by spreading a portion of a bacterial sample over the entire surface of the plate with an inoculating loop. This separates the bacterial cells forming distinct colonies when they multiply. Each colony, theoretically, originates from a single cell; hence it represents a population of genetically identical bacteria. The colonies can then be picked with an inoculating needle and subcultured to obtain a pure culture.

a distinct colony that can generally be distinguished from the others (fig. 4.7). The different organisms can then be purified by picking cells from a colony with an inoculating loop and streaking it onto fresh medium.

GROWTH OF CELLS AND MICROBIAL POPULATIONS

All living organisms are characterized by their ability to grow and reproduce in their environment. Microorganisms grow by increasing their size (volume), a vital process that normally culminates in cell division and, thus, in an increase in the number of cells. To understand how microorganisms carry out decomposition, cause disease, and participate in useful industrial processes, it is necessary that we know how they grow, how they multiply, and what effect the environment has on their rate of reproduction.

Cellular growth can be defined as the sum of all metabolic activities of a cell leading to the orderly increase of all its constituents. It involves the synthesis of the vital cell components, such as the ribosomes, plasma membrane, and genetic material. Cellular growth results in an increase in the size and mass of the cell. An increase in cell mass, however, does not always reflect cellular growth. It is possible for microorganisms to synthesize reserve materials, such as starch or volutin, yet not increase their size.

Although it is relatively simple to study the growth of a single plant or animal, microbial cells are much too small to be studied individually. Instead, microbiologists study microbial populations to gain an understanding of the biology of the microorganism. **Population growth** refers to the increase in the number of cells in a population, whereas cellular growth refers to the increase in the size of the cell. The rate at which populations grow is determined largely by the environment they inhabit and by factors such as moisture, heat, and nutrient availability.

If a flask of nutrient broth is seeded with a bacterium such as *Escherichia coli* and allowed to stand in a warm place for a few hours, it will become turbid. The increase in turbidity of the nutrient broth roughly reflects the growth of the bacterial population inside the flask and can be used as an indication of how fast the population is growing. In general, the dynamics of a growing microbial population can be studied by measuring (a) the number of cells, (b) cell mass, and (c) metabolic activities.

Measuring the Number of Cells

There are two basic procedures commonly employed in the laboratory to measure the increase in the number of individual cells in a microbial population: the direct microscopic count and the viable count.

Direct Microscopic Count The **direct microscopic count** consists of counting the number of cells in the population with the aid of a counting chamber. Counting chambers, such as the **Petroff-Hausser chamber** and the **hemocytometer**, are constructed especially for the direct enumeration of cells in suspensions. These counting chambers are special slides with chambers of known depth that have been ruled into squares of known area (fig. 4.8) so that they can be used to determine the volume of suspension being counted. The specimens may be stained before they

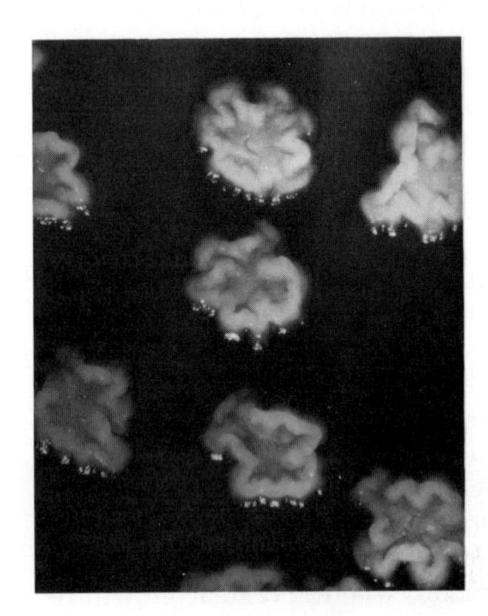

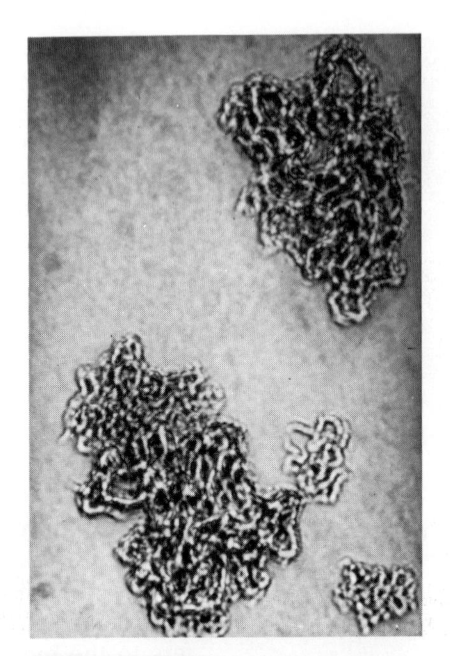

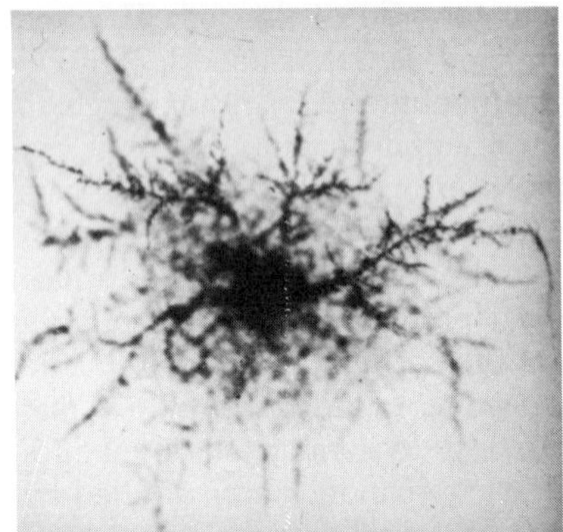

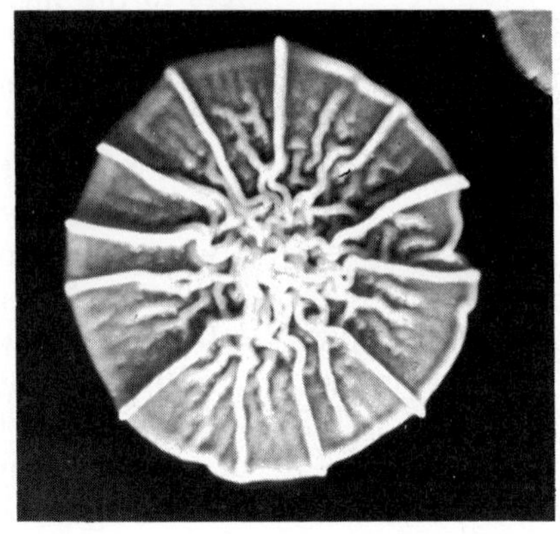

Figure 4.7 Various types of bacterial colonies

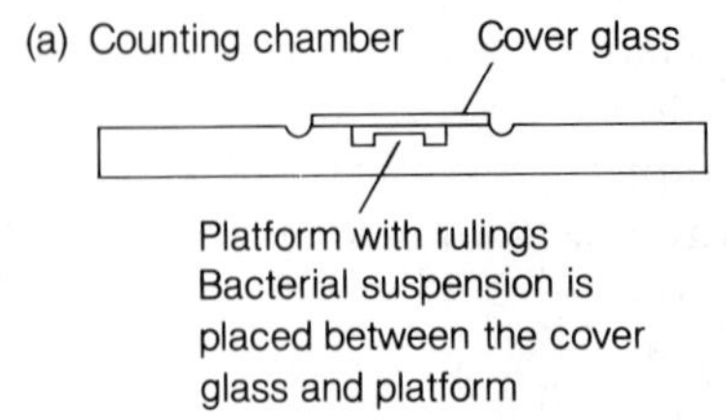

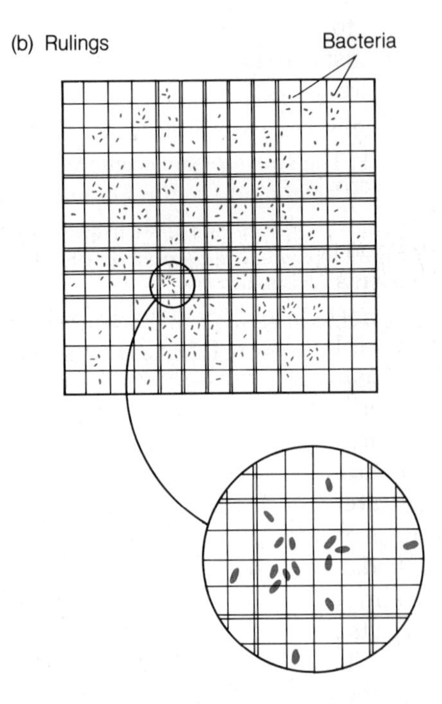

Figure 4.8 Direct enumeration of bacterial populations using the Petroff-Hausser counting chamber. *(a)* The Petroff-Hausser bacterial counter consists of a platform with rulings. The bacterial suspension is placed between the cover glass and the ruled chambers. The suspension enters the chamber by capillary action and the bacteria are distributed throughout the ruled chamber. The bacteria are then counted using a compound microscope. *(b)* The central chamber of the Petroff-Hausser bacterial counter consists of 25 large squares. The bacteria in all the squares are counted and the number is multiplied by 50,000 to give the value of bacteria/ml of suspension.

are counted with these chambers in order to increase the ease with which the microorganisms are viewed.

Direct microscopic counts using the Petroff-Hausser chamber, although fast and easy to perform, are somewhat inaccurate because all microorganisms present, both living and dead, are counted. In addition, since extremely small volumes are used, each microorganism seen in the slide represents a very large number of microorganisms per milliliter of suspension; hence large numbers of microorganisms must be present in the population before they can be counted accurately.

Electronic counters, such as the **Coulter counter** (fig. 4.9), are routinely used in the clinical laboratory to count red and white blood cells. They are less frequently used to count microbial populations, although the results obtained are accurate and reproducible.

Viable Counts The methods most commonly used to determine **viable counts** are based on the assumption that each cell in a solid medium can multiply repeatedly and eventually form a distinct colony. The spread plate and the pour plate are used extensively to obtain these counts.

In the **spread plate** method, a bent glass rod is used to spread 0.1–0.5 ml of a suspension of microorganisms evenly over the surface of a suitable solid culture medium in a petri dish (fig. 4.10a). The glass rod (shaped like a hockey stick) is used to spread the microbial cells over the agar surface so that they are not crowded and can multiply freely, forming colonies. The colonies that develop after an incubation period can be counted to obtain the number of microorganisms originally present in the suspension.

To perform a viable count by the **pour plate** method (fig. 4.10b)(, a sample (0.1–1.0 ml) of a liquid culture or suspension is mixed with a melted agar medium in a sterile petri dish and incubated for a suitable length of time. After incubation, the colonies in (and on) the medium are counted, usually with the aid of a **colony counter** (fig. 4.11). This device may be a magnifying glass or a sophisticated electronic colony counter. The **standard plate count** (SPC) is a standardized pour plate count for determining the number of aerobic and facultative anaerobic chemoheterotrophs in a suspension. The SPC method is used routinely to count the number of microorganisms in water, milk, and foods, because it affords microbiologists a means of comparing results with those of other laboratories.

Some specimens (or cultures) may contain so many microorganisms that, after plating, the developing colonies grow together and cannot be counted. When **confluent** (grown together) growth is expected, the specimens must be diluted before they are plated so that most colonies appearing on the plate will be isolated. The degree of dilution of the specimen before plating is extremely important. Excessive dilution will usually result in too few colonies, while insufficient dilution will result in crowded and confluent growth. Either situation will lead to inaccurate counts. It is best to dilute the sample so that it will yield between 30 and 300 colonies per plate. Insufficient dilution results in microorganisms landing very near one another in the agar, where their populations merge and give rise to a single colony. When this happens, the number of microorganisms estimated to be in the sample is lower

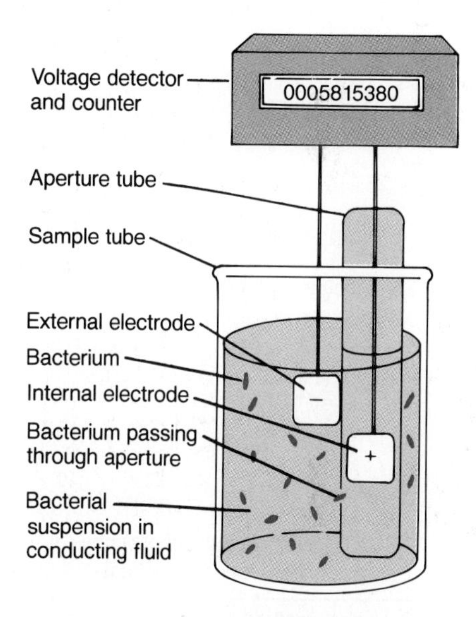

Figure 4.9 Counting bacteria using the Coulter counter. As each bacterial cell passes through a slit in the Coulter counter, it causes changes in conductivity. These changes are recorded and tallied automatically. The results are displayed. The total count represents all of the conductivity changes recorded by the Coulter counter.

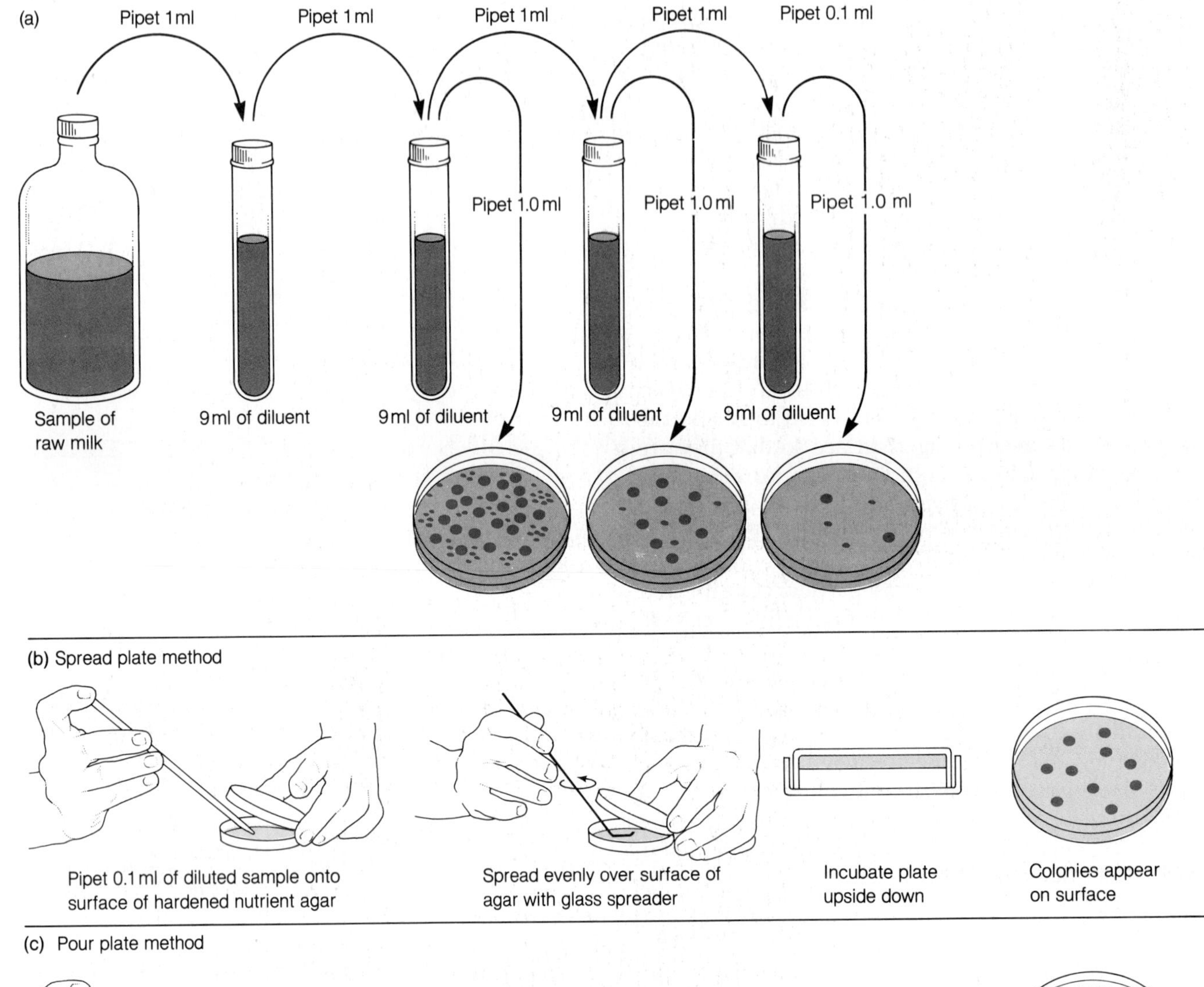

(b) Spread plate method

Pipet 0.1 ml of diluted sample onto surface of hardened nutrient agar

Spread evenly over surface of agar with glass spreader

Incubate plate upside down

Colonies appear on surface

(c) Pour plate method

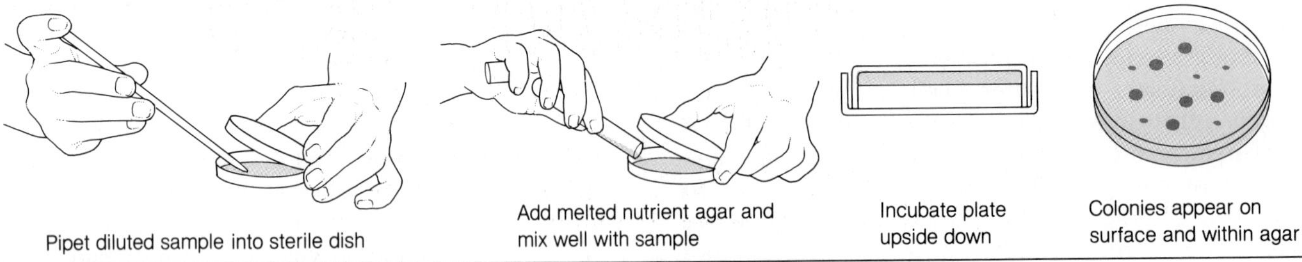

Figure 4.10 **Viable count methods.** *(a)* Viable counts are commonly carried out by dispersing a known volume of microbial suspension on an agar-solidified medium in a petri dish. The petri dish is then incubated for a specified amount of time and the colonies are counted. If the samples are suspected to contain too many bacteria, they must be diluted before plating. *(b)* The **spread-plate method** involves spreading a known volume (usually 0.1 ml) of suspension over the surface of an agar-solidified medium. *(c)* The **pour-plate method** involves mixing a known volume (0.1–1.0 ml) of suspension in 15–20 ml of liquefied, cooled agar medium. After an incubation period, colonies that appear throughout the medium are counted. The product of the number of colonies counted and the reciprocal of dilution plated is equal to the number of viable cells in the undiluted population.

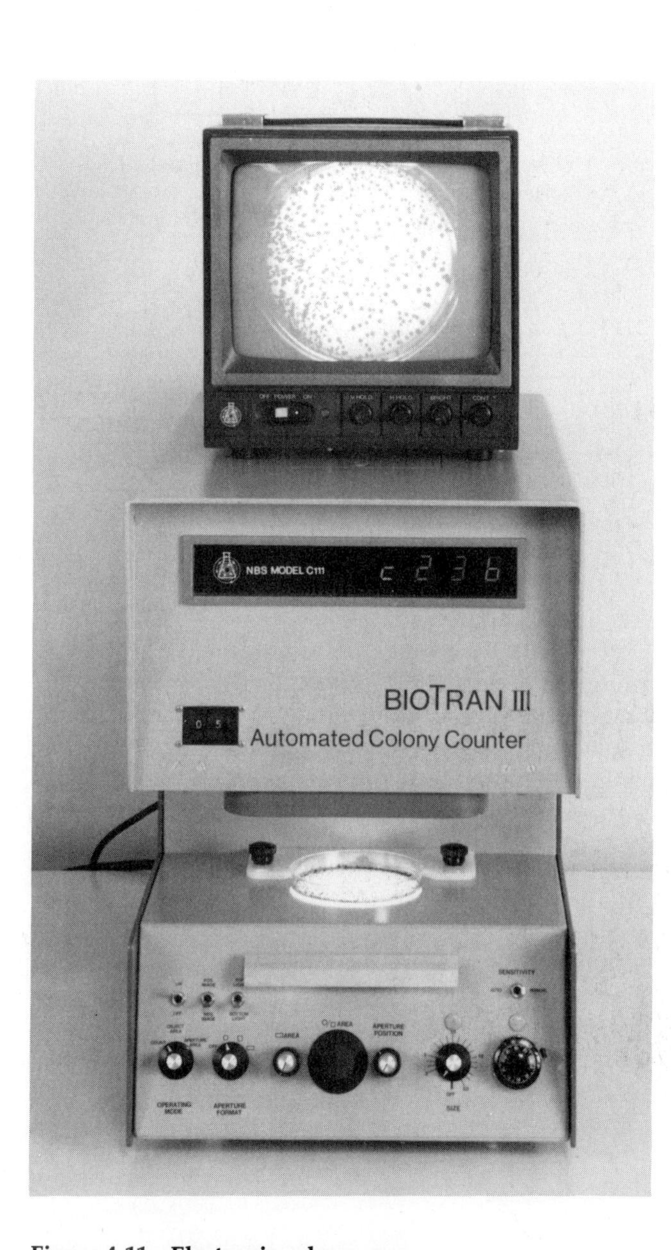

Figure 4.11 Electronic colony counter. The plate to be counted is inserted in the counting chamber. The colonies appearing on the plate are counted electronically and the total count is displayed automatically.

than the actual number. Excessive dilutions will also lead to inaccurate counts.

Viable counts, unlike direct microscopic methods, are extremely sensitive. In theory, even a single microorganism present in a sample can be detected by viable counts. Like all other methods for enumerating populations of microorganisms, however, viable counts have shortcomings. Cultural conditions can affect the viable counts, and the procedure may therefore underestimate the actual number of microbes in the sample. It is obviously impossible to satisfy the growth requirements of all microorganisms with one set of cultural conditions; therefore, viable count methods detect only those microorganisms capable of reproducing under the cultural conditions established. Such methods indicate the *minimum* number of microorganisms present in the sample, because some microorganisms present may not have been able to reproduce due to inappropriate cultural conditions. For instance, a standard plate count of a dairy product that has been incubated aerobically for 48 hours will detect only the number of aerobic and facultative anaerobic microorganisms present. Any anaerobic bacteria present in the dairy product will not form a colony, because they are inhibited by oxygen. In addition, certain groups of microorganisms have a tendency to form aggregates of two or more cells that will give rise to a single colony. This process will also result in viable counts that are lower than the actual number of cells present in the sample. In spite of these shortcomings, viable counts are widely used in microbiology and they yield accurate and acceptable results.

Turbidimetric Measurements **Turbidimetric measurements** are based on the principle that the number of tiny particles (e.g., bacteria) in a suspension is inversely proportional to the amount of light transmitted. Turbidity (the cloudiness of a liquid medium) is measured in **optical density** (O.D.) units that are inversely proportional to the percentage of transmittance (%T).

Turbidimetric measurements of bacterial population size are usually made with a photometer (fig. 4.12), an apparatus that measures the amount of light absorbed or scattered by a suspension. The photometer consists of a light source, a chamber in which the specimen is placed, and a photoreceptor to measure the amount of transmitted light.

Turbidimetric methods provide rapid and reproducible results when properly performed, but they only approximate population size, because they measure the amount of light scattered by the population

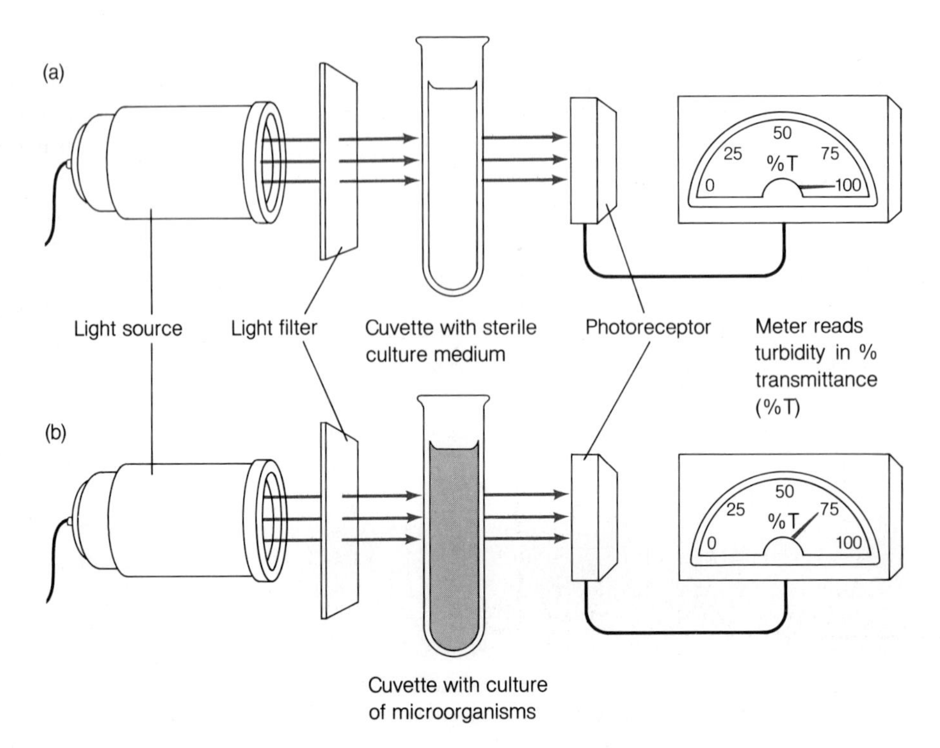

Figure 4.12 Turbidimetric measurements of bacterial populations. Measuring the turbidity of a bacterial population. *(a)* The turbidity of a cell-free medium is measured and the photometer adjusted to read 100%T. *(b)* The turbidity of the culture is measured and the %T recorded. The %T is inversely proportional to the number of bacteria in the sample and can be used to estimate the population size.

rather than counting the cells themselves. For this reason, the technique may yield inaccurate results when the concentration of cells is high.

Measurements of O.D. do not directly yield the number of cells in the culture; the O.D. values are converted to cell numbers or cell mass by using a calibration curve (fig. 4.13). A calibration curve is constructed by plotting the optical density against some parameter that reflects population size, such as viable count or dry weight. Since the nutritional and envi-

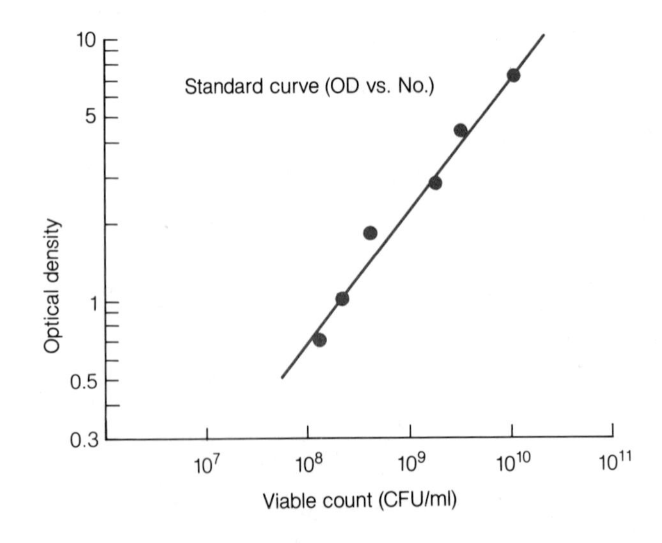

Figure 4.13 Optical density versus viable count calibration curve. The calibration curve is used to estimate the size of a population using optical density measurements. To construct the curve, the optical density and viable counts of various dilutions of a population are made. The results are plotted on a graph as illustrated. A straight line can be obtained if the O.D. values are plotted on a logarithmic scale and the viable counts on an arithmetic scale. CFU, colony-forming units.

ronmental conditions of the culture affect the mass and size of the population, they also affect the O.D. For this reason, it is essential that well-defined conditions be employed when estimating population size by optical density methods.

Bacterial Division **Transverse binary fission** is the method by which most bacteria multiply (fig. 4.14). During **vegetative growth**, a newly formed cell undergoes a gradual increase in volume, reflecting an increase in cellular constituents in preparation for cell division. After the cell reaches a certain size, it begins to delimit (form) a septum that ultimately divides the enlarged cell into two identical daughter cells. During this time, vital cellular components are divided equally between the two developing cells. Each of the daughter cells then begins to increase in volume in preparation for the next division cycle. Thus, each time the cells divide, the population doubles in size. If the starting population consists of 1 cell, the next generation will contain 2 cells, the next 4 cells, then 8, 16, 32, 64, etc. The size of the population can be determined by the following formula:

$$N_f = (N_i)2^n$$

where N_f is the final size of the population, N_i is the initial size of the population, and n is the number of generations (doublings). The speed at which a population increases in size is usually expressed as its **generation time**, defined as the time it takes for a population to double in size. The generation time of microbial populations varies from about 20 minutes to several days, depending on the species of organism and the cultural conditions in which it grows.

The Bacterial Growth Curve

Bacterial populations, like all other populations, increase or decrease in numbers in response to environmental changes. When a pure culture of bacteria multiplies in a liquid culture medium, the population undergoes a predictable sequence of changes in its growth rate. When these changes are plotted on graph paper, they give rise to a curve known as the **growth curve** (fig. 4.15).

The size of the population is expressed on a logarithmic scale rather than on an arithmetic scale in order to observe changes in growth rate. On a logarithmic scale, a constant growth rate is indicated by a straight line, a changing growth rate by a curved line. An arithmetic plot would not clearly delineate changes in growth rate or the characteristic phases of growth displayed by populations. If the number of cells is plotted on a logarithmic scale versus time (fig. 4.16), the growth curve clearly demonstrates the various phases of growth (fig. 4.15). In addition, the slope of the straight line can be used to obtain the growth rate of the population because it graphically reflects the rate of growth. Hence, the steeper the slope, the higher the growth rate and vice versa.

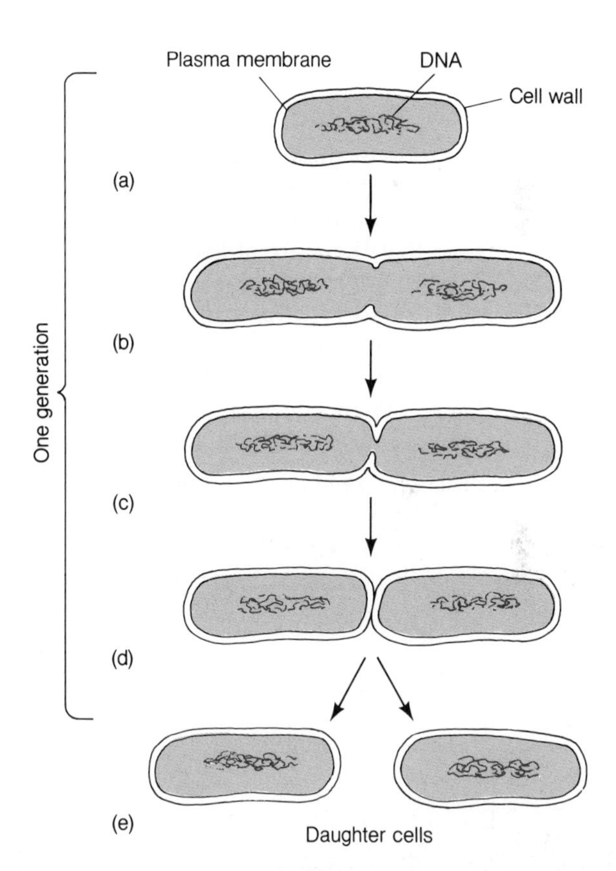

Figure 4.14 Binary fission by bacteria. The process of bacterial cell division is an orderly sequence of steps that results in the formation of two bacteria from a parental cell with an identical genetic makeup. *(a)* Newly formed parental cell. *(b)* Cell elongation and septum formation. *(c)* Invagination of cell wall and distribution of genetic material. *(d)* Separation of two new cells. *(e)* Each new cell then begins a new cell division cycle.

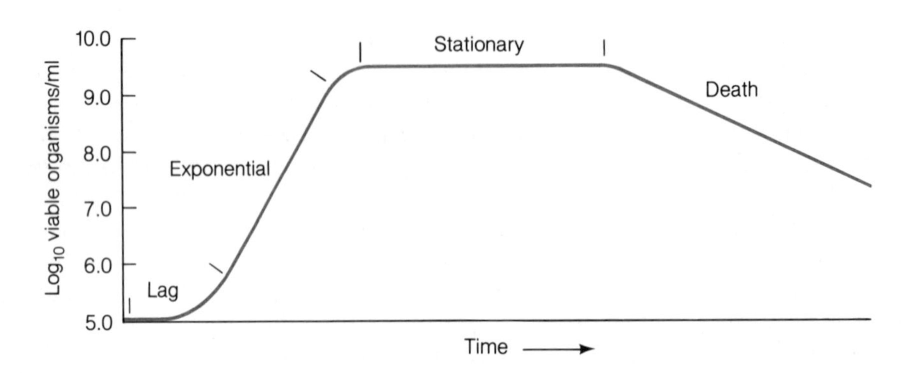

Figure 4.15 The bacterial growth curve. A population of bacteria growing in a broth medium will undergo a series of changes in its rate of growth that reflect changes in its environment and physiology. These phases are the lag phase, the exponential growth phase, the stationary phase, and the death phase.

The different **growth phases** of the culture (fig. 4.15) reflect the physiological states of the cells and the condition of the culture medium. From figure 4.15 it can be seen that there are at least four recognizable phases of growth: (a) the lag phase; (b) the exponential growth phase; (c) the stationary growth phase; and (d) the death phase.

The Lag Phase When a population of bacteria is transfered into a new culture medium its growth may be delayed until it adjusts to the new growth conditions. This period of adjustment is called the **lag phase.** During much of the lag phase, the cells are increasing in size (mass) in preparation for cell division and actively synthesizing the enzymes necessary to reproduce. During the latter part of the lag phase, the cells begin to multiply more quickly until the maximum growth rate is reached.

The duration of the lag phase depends on factors such as the age of the culture and the nature of the

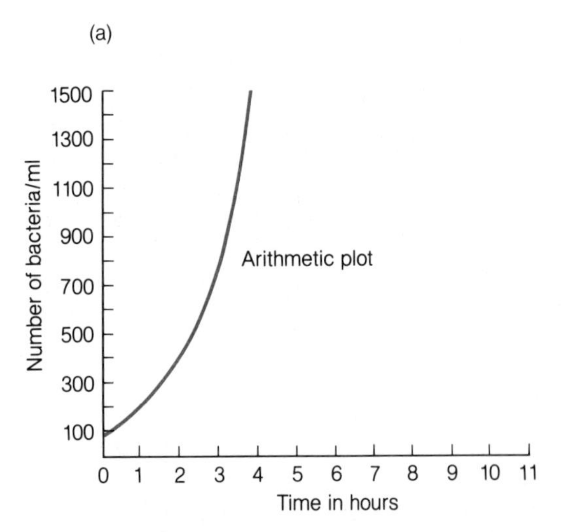

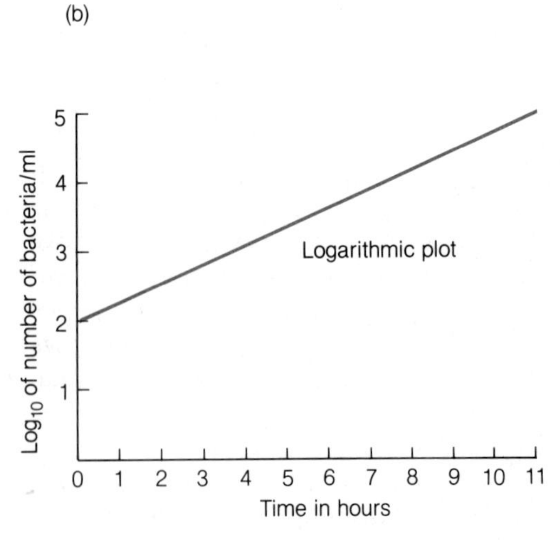

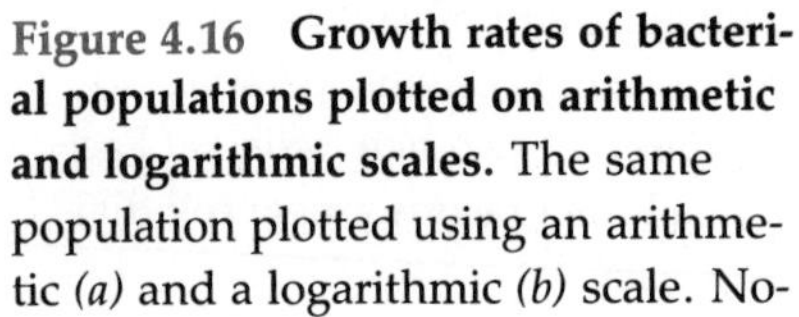

Figure 4.16 Growth rates of bacterial populations plotted on arithmetic and logarithmic scales. The same population plotted using an arithmetic *(a)* and a logarithmic *(b)* scale. Notice the sharp rise of the population growth curve when the population size is plotted arithmetically. At this rate of climb, the curve will shoot off the graph paper in a few generations. The logarithmic plot gives a gradually rising straight line. The slope of the line can be used to visualize the rate of population growth: the steeper the slope, the faster the rate of growth.

medium. If an old culture is inoculated into fresh medium, there is a lag phase. If a young, actively dividing culture is inoculated into a similar medium, however, the lag phase is short or even absent, because young cultures are already dividing and can resume their reproduction in the new medium. Older cultures that have stopped dividing must synthesize the enzymes necessary for growth and division before they can start reproducing. Even actively dividing cells, when introduced into a new environment (e.g., a different type of culture medium), typically exhibit a lag phase.

The Exponential Phase During the **exponential phase** all cells are dividing at a constant and maximum rate. This rate is characteristic of the species of microorganism and is also influenced by cultural conditions. During the exponential phase, the number of growing, dividing cells far exceeds the number of dying cells. The rate of cell reproduction is controlled by a variety of factors, such as temperature, hydrogen ion concentration, and nutrients present in the culture medium. For example, when a bacterium such as *Escherichia coli* is growing in a complex medium, many of the essential amino acids, nucleosides, and vitamins are already present and need not be synthesized by the cell. As a consequence, the bacterial generation time is short. When the same bacterium is growing in a culture medium consisting of mineral salts and glucose, however, it must synthesize all the amino acids, nucleosides, and vitamins required for growth. This synthesis takes time and is reflected in a longer generation time.

The Stationary Phase The exponential growth phase lasts as long as ample nutrients are available in the environment, or until the population pollutes its environment and makes it inhospitable. During exponential growth, large amounts of organic acids and other toxic substances may accumulate in the culture medium and inhibit the growth of the culture. The accumulation of toxic materials and the depletion of essential metabolites are largely responsible for the onset of the **stationary phase** (fig. 4.15). There is little growth or reproduction during the stationary phase. The number of growing, dividing bacteria is offset by an equivalent number of dying cells. The growth rate of the population is zero.

A population of bacteria growing exponentially will enter the stationary phase of growth before all of the available nutrients are depleted. For example, if a sample of broth culture that just entered the stationary phase is filtered to remove all bacteria and is then inoculated with a fresh culture, further growth takes place in the culture medium. This growth indicates that the accumulation of toxic materials, and/or the depletion of essential nutrients, cannot be exclusively responsible for the cessation of growth.

The Death Phase The **death phase** usually follows the stationary phase and it is characterized by a progressive decline in the number of viable cells in the population. During the death phase, the dying cells far outnumber those still capable of reproducing. The rate of population decline during the death phase is a characteristic of the species, and is also influenced by cultural conditions and by the presence of toxic materials. It is important to note that, although the population as a whole is declining, many individual cells may still be viable. For this reason, when a dying population is placed in fresh culture medium, it begins to grow again.

Microbial Growth on Solid Culture Media

Microorganisms growing on solid culture media generally remain together and consequently form colonies. A colony often arises from a single cell (fig. 4.17) and consists of a population of microorganisms at various stages in their growth cycle. Some of the microorganisms are entering the exponential growth phase, while others are already in their stationary phase or even in their death phase.

Since some microorganisms in a colony are not in contact with the culture medium, or have used up the nutrients near them, the rate of nutrient uptake varies from cell to cell. The response of individual microorganisms to nutrient availability can affect colony size. Crowding also influences the size of the colony. This is a result of two adjacent colonies competing for nutrients in their immediate area. Metabolic wastes released by the growing populations may also reduce their growth rate. Both these factors result in a reduction in the sizes of the two adjacent colonies.

FACTORS AFFECTING MICROBIAL GROWTH

The chemical and physical environment can alter the rate at which a population grows. The concentration of oxygen, nutrients, heavy metals, antibiotics, and water, as well as the temperature and electromagnetic radiation, can all dramatically affect the growth of microbial populations.

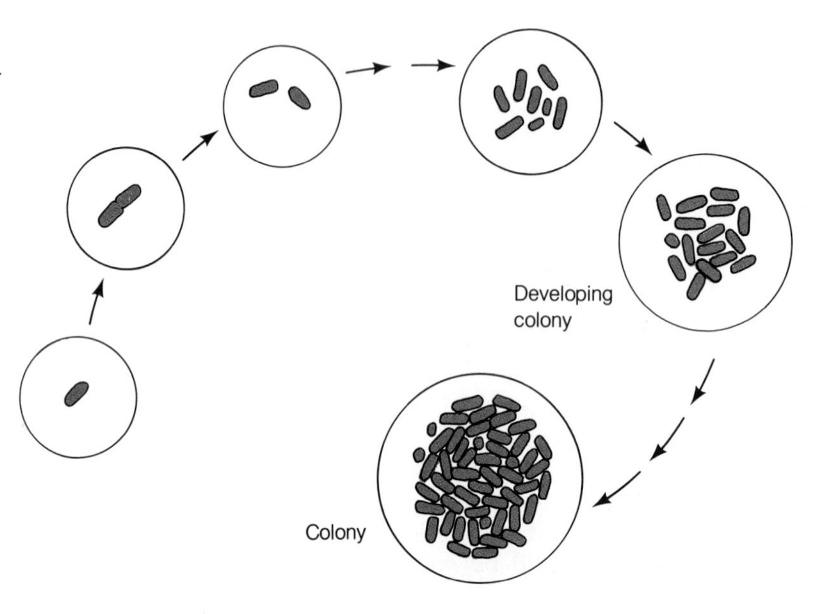

Figure 4.17 Bacterial colony formation on an agar medium. A colony results from sequential cell divisions of a single bacterium. Since they are growing in an agar-solidified medium, the bacterial cells remain together, forming a discrete mass (colony)

Nutrient Concentration

The concentration of nutrients in a culture medium can affect either the microbial population growth rate or the total cell yield of the culture. At low nutrient concentrations, the rate at which populations grow is directly proportional to the concentration of nutrients in the culture medium. This phenomenon may be due to the efficiency with which the cell can transport nutrients. Cells accumulate nutrients by means of proteins called **permeases**, which are part of the plasma membrane. At low nutrient concentrations these permeases are not fully utilized, and cells do not have sufficient nutrients to reproduce at their maximum rate. As the nutrient concentration increases, more and more permeases are utilized, until a point is reached at which all the permeases are transporting nutrients at their maximum rate into the cell. Beyond this point, nutrient concentration no longer affects growth rate and the growth rate reaches a maximum and remains constant.

As previously noted, the concentration of nutrients affects the total cell yield, or **maximum crop**, of a culture. As the concentration of nutrients in a culture medium increases, more of these nutrients can be converted to cellular material, and the result is a net increase in the number of cells in the culture. Studies of maximum crop yield are important when it is necessary to obtain large quantities of cells for physiological studies, or for the preparation of single-cell protein or other industrially valuable products.

Temperature

Microorganisms have evolved enzymes and cell membranes that function best within the temperature range they normally encounter. This temperature range may be as wide as 45°C, such as that exhibited by *Bacillus subtilis* (8 to 53°C), or as narrow as 10°C, as that of *Neisseria gonorrhoeae* (30 to 40°C).

The range of temperatures within which various microorganisms reproduce is largely determined by whether or not cellular enzymes can function. If the temperature is too high, the enzymes denature permanently and the membrane becomes excessively fluid, resulting in autolysis. On the other hand, if the temperature is too low, chemical reactions become exceedingly slow or stop altogether. The lowest temperature at which a microbial population can grow is referred to as the **minimum growth temperature** (fig. 4.18). As the temperature increases from the minimum growth temperature, the population multiplies at an increasing rate until it reaches a point at which it multiplies as a maximum rate. The temperature at which a population grows at its maximum rate is called the **optimum growth temperature** (fig. 4.18). Above the optimum growth temperature, the rate of growth slows down, because higher temperatures cause some of the cellular enzymes to denature. At some temperature above the optimum, one or more essential enzymes become nonfunctional and the population ceases to reproduce. The elevated temperature at which growth

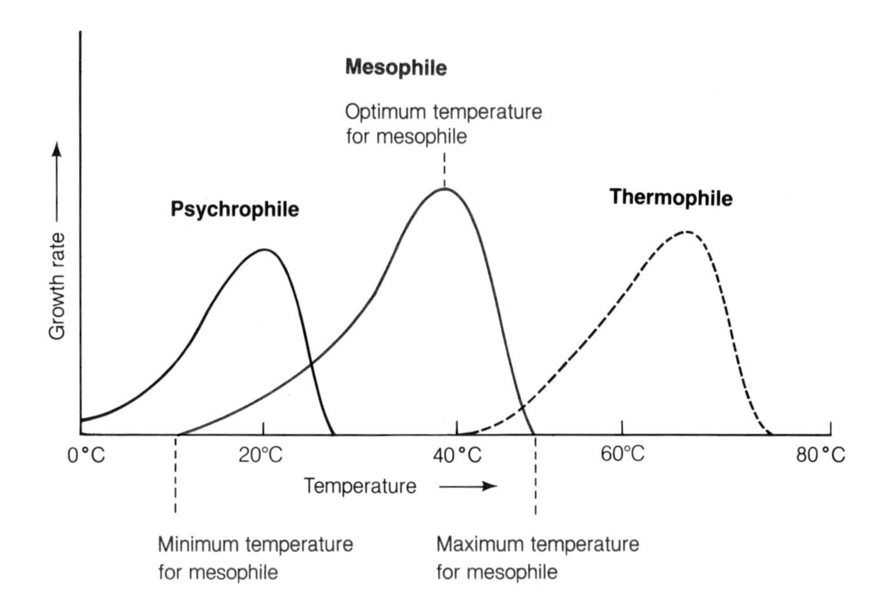

Figure 4.18 Temperature growth range of microorganisms. *Thermophilic* microorganisms grow best at temperatures between 50 and 100°C but grow poorly or not at all at temperatures below 45°C. *Mesophilic* microorganisms have optimum growth temperatures between 20 and 50°C. *Psychrophilic* microorganisms have an optimum growth temperature below 20°C and show significant growth at 0°C within 2 weeks. Some can grow at temperatures as low as −5°C.

ceases is called the **maximum growth temperature**. This response to increasing temperature gives rise to the skewed growth curve.

Microorganisms that grow most rapidly in cold environments, at temperatures between about −5°C (still in liquid state) and +20°C, are called **psychrophilic** (cold-loving) organisms (fig. 4.18). Psychrophiles grow well in refrigerators at temperatures near 4°C and are responsible for the spoilage of refrigerated foods. In nature, these organisms are common in glacier-fed lakes, cold ocean bottoms, and arctic and subarctic environments. Those microorganisms that grow best at temperatures between 20 and 50°C are called **mesophilic** (middle-loving) organisms. Organisms growing in the human body, and in most topsoils in the United States, are mesophiles. **Thermophilic** (heat-loving) organisms grow best at temperatures between 50 and 100°C. These are found predominantly in hot springs and in compost piles where the temperature is above 50°C.

Hydrogen Ion Concentration (pH)

The enzymes, electron transport systems, and nutrient transport systems found in the cell membrane are sensitive to the concentration of hydrogen ions (H^+). The concentration affects the three-dimensional structures of most proteins, including enzymes necessary for growth. Every organism grows most rapidly at its optimum level of H^+. Each organism also has a range of H^+ concentrations outside which it fails to reproduce (fig. 4.19).

Hydrogen ion concentrations are measured in pH units. A high concentration of H^+ makes an environment acidic and gives it a low pH. A low concentration of H^+ results in an alkaline environment and a high pH. A pH below 7 is considered acidic, while a pH above 7 is alkaline or basic.

Most bacteria reproduce only at pH levels between 6 and 8. Very few proliferate at pH levels much below 4. Many fungi, however, grow well at low pH values (5–6).

When microorganisms reproduce, they release waste products that may change the pH of their environment. If the pH change is extreme, their environment becomes inhospitable. Therefore, if extensive microbial growth is desirable, changes in the pH of the environment must be avoided. Buffers (chemicals that resist changes in pH) can be added to the growth environment to eliminate wide changes in pH.

Foods can be preserved by lowering their pH. For example, sauerkraut and pickles are protected from microbial spoilage because their low pH values inhibit microbial growth.

The Osmotic Pressure

Osmotic pressure is defined as the minimum amount of pressure that must be applied to a solution to pre-

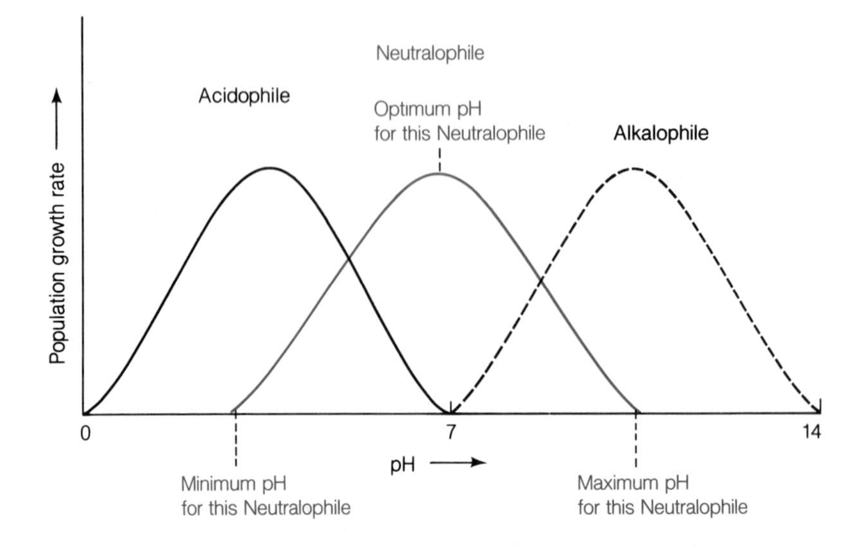

Figure 4.19 Relationship between pH and population growth rate. Each microbial population responds to changes in environmental pH with changes in its growth rate. The response of the growth rate to changes in pH is expressed as a bell-shaped curve with a maximum, a minimum, and an optimum pH.

vent the flow of water across a membrane within the solution. For example, if a dialysis bag (a membrane-like material that allows the free passage of water molecules) is filled with 10% sucrose (table sugar) inside a beaker filled with water, the water in the beaker will tend to flow into the bag to dilute the sucrose solution. The amount of pressure that must be exerted from inside the bag to counteract the flow of water inward is the osmotic pressure.

The osmotic pressure of the environment influences whether or not a particular microorganism can reproduce. Microorganisms usually inhabit slightly **hypotonic** environments, or those in which the concentration of solutes (dissolved nutrients) is lower than that of the cytoplasm (fig. 4.20). In these environments, water tends to move into the cell and make it turgid. The rigid cell wall of bacteria, fungi, algae, and some protozoa eventually limits the amount of water

FOCUS BACTERIA THAT GROW DEEP IN THE OCEAN

In the past few years, bacteria from deep oceanic trenches have been discovered that can grow only at pressures considerably above atmospheric pressure. These bacteria inhabit environments 2500 m below sea level, where the pressure is more than 250 bars (250 times atmospheric pressure). Pressure-dependent bacteria, called **barophiles**, have been discovered that can grow at pressures between 500 and 1000 bars but die at atmospheric pressure. Their death has been attributed to the fact that their gas vesicles expand on decompression, pushing the cytoplasm against the membrane and cell wall with such force that the cell ruptures.

Equally noteworthy are bacteria that reproduce near hot oceanic vents. These vents fill the water with hydrogen sulfide and other minerals and heat the water to temperatures in excess of 350°C. Experiments have indicated that some bacteria may be able to grow at temperatures as high as 120°C and at pressures of more than 250 atmospheres. Even under these environmental conditions, the population doubles approximately every 40 min.

The discovery of barophiles and thermophiles that live thousands of meters below sea level indicates that bacteria may also exist within the earth's crust, where sufficient nutrients exist.

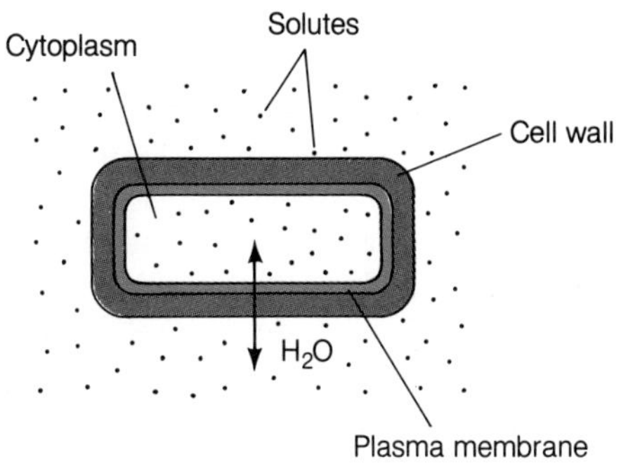

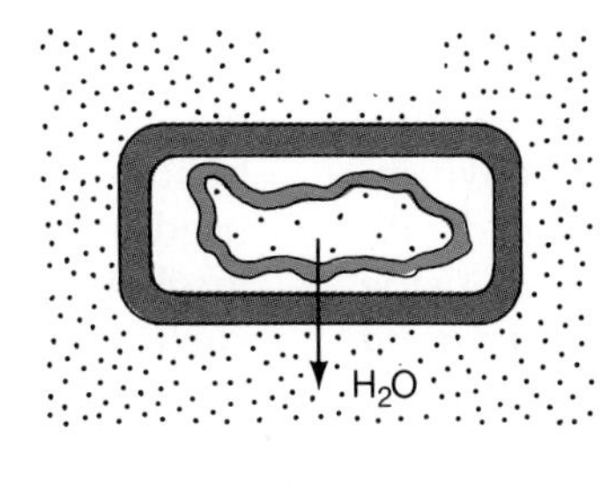

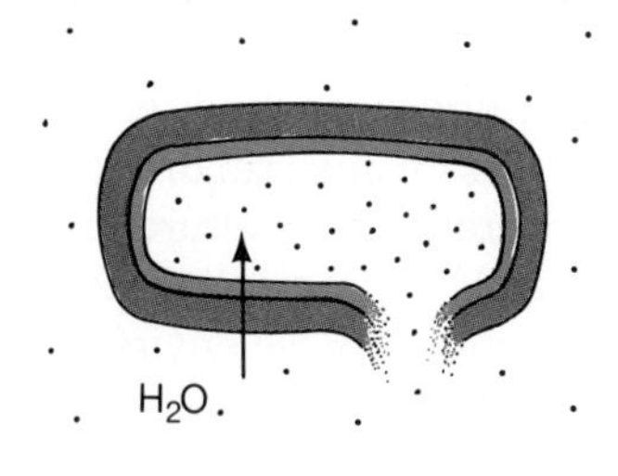

Figure 4.20 **Flow of water in isotonic, hypertonic, and hypotonic environments.** *(a)* Cell in an isotonic environment. In isotonic environments, the concentration of environmental solutes is equal to that of the cell. In this environment, cellular water shows no net movement in or out of the cell. *(b)* Cell in a hypertonic environment. In hypertonic environments the concentration of solutes is greater than that of the cell. Hence, there will be a tendency for cellular water to flow out into the environment. This results in the dehydration and shrinking of the cell. *(c)* Cell in hypotonic environment. In hypotonic environments the concentration of solutes is lower than that of the cell. Hence, environmental water will tend to flow into the cell. In the process, the cell swells from an excess of water. If the flow of water is too great, the cell may burst. Although cells with cell walls may simply swell and not burst, cells with weak cell walls (gram-negative bacteria) may burst due to excessive water intake.

coming into the cell and also prevents the cell from swelling and bursting. On the other hand, if the solute concentration of the environment is greater than that of the cell (**hypertonic** environment), the cell becomes dehydrated and ceases to function because cellular water tends to flow out. The osmotic pressure of the culture environment is particularly important when culturing bacteria such as the mycoplasmas that lack a rigid wall and hence are susceptible to osmotic lysis. The cultivation of pure cultures of mycoplasmas and bacteria with weak cell walls requires that the proper concentration of nutrients be used, so that the osmotic pressure of the environment does not cause the lysis of cells.

Foods such as fruit preserves and salted fish are resistant to spoilage by microorganisms because of their high osmotic pressure. Microorganisms present in these foods are unable to grow because the hypertonic environment draws out the cell's water, thus inhibiting microbial metabolism.

Bacteria that normally inhabit salty environments are called **halophiles** (salt-lovers). Some halophiles require more than 15% salt to reproduce and are called **extreme halophiles.** Other microorganisms can grow over a wide range of salinity (salt concentration) and are called **moderate halophiles** or facultative halophiles. Most of these organisms are marine bacteria.

The Atmospheric Environment

Many microorganisms require free molecular oxygen (O_2) to reproduce.

Organisms requiring O_2 for cellular respiration are called **aerobes**. Aerobes constitute a very large group of organisms that includes most of the algae, the fungi, and many of the protozoa and bacteria. There are some microorganisms, mostly bacteria and protozoa, that have oxygen-sensitive enzymes and cannot function in the presence of molecular oxygen. As a consequence, these organisms are unable to multiply in any environment that contains O_2 and are called **anaerobes**. Anaerobes obtain their energy from oxygen-independent metabolism. **Microaerophilic** organisms are aerobes that require low O_2 tensions. Other microorganisms, called **facultative anaerobes,** can multiply in either the presence or the absence of O_2.

There are groups of microorganisms that require an atmosphere composed of O_2 and elevated levels of carbon dioxide. For example, *Neisseria gonorrhoeae*, the bacterium that causes gonorrhea, grows best in an atmosphere enriched with 5–10% carbon dioxide. Microorganisms requiring elevated CO_2 environments are usually cultured in incubators equipped with devices that permit the regulation of the gases inside the chamber. These environments can also be achieved with a device called a **candle jar**, which consists of a large jar (a one-gallon jar will do) with a candle in it. After the cultures are placed inside the jar, the candle is lit and then the jar is sealed. The candle inside the jar burns until there is not enough oxygen left inside the jar to maintain combustion. The jar then contains a reduced concentration of free oxygen and about 3.5% CO_2.

Organisms that require oxygen-free environments must be cultured in special incubators or chambers that exclude molecular oxygen. **Anaerobic incubators** are specially designed to culture anaerobes. These incubators are equipped with ports and valves so that the air can be extracted from the growth chamber and replaced with inert gases such as nitrogen, helium, or carbon dioxide. Hydrogen gas is seldom used because of its explosive nature. Sometimes the incubator chamber is filled with a mixture of inert gases, such as nitrogen and carbon dioxide, in a ratio of 95:5. Carbon dioxide is important to the reproduction of certain anaerobes, hence it must be made available in the growth environment.

In many clinical and industrial laboratories, the **anaerobic jar** (fig. 4.21) is routinely used as an economical alternative to the anaerobic incubator. There are several versions of the anaerobic jar in current use

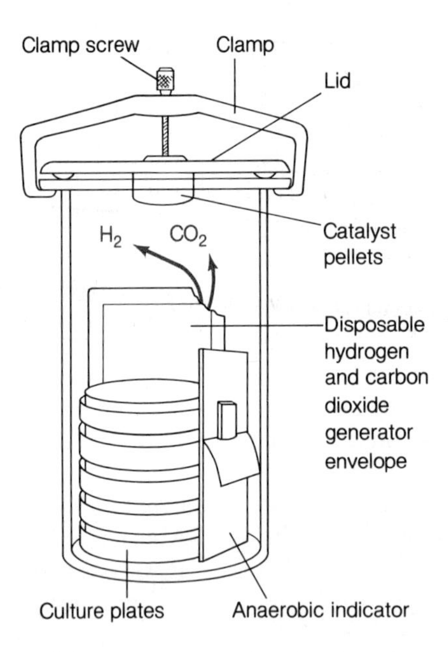

Figure 4.21 An anaerobic system. The anerobic system illustrated here is commonly used in microbiology laboratories to cultivate anaerobic microorganisms. It consists of a jar with a screw-on lid, a hydrogen–carbon dioxide generator envelope, and a palladium catalyst. Water is added to the hydrogen–carbon dioxide generating envelope to activate it, and the envelope is placed inside the jar along with the cultures. An anaerobic indicator strip may also be included. The strip consists of a pad saturated with methylene blue solution, and it changes from blue to colorless when an anaerobic atmosphere inside the jar is developed. The hydrogen gas generated by the envelope reacts with any oxygen that may be present in the jar to form water. The reaction is catalyzed by the palladium catalyst, which permits the formation of water from hydrogen and oxygen to take place at room temperature. This reaction removes any free oxygen from inside the jar, creating an anaerobic environment. The CO_2 is needed to stimulate the growth of certain anaerobes.

d. in mitochondria;
e. in chloroplasts.

1. Soil bacteria such as *Bacillus* and *Clostridium*.
2. Phagocytizing cells such as *Amoeba*.
3. Photosynthetic eukaryotic organisms such as Euglena.

For items 4–8, match the choices with the statements below:
a. chemoheterotroph;
b. chemoautotroph;
c. photoheterotroph;
d. photoautotroph;
e. none of the above.

4. All plants.
5. All fungi.
6. All animals.
7. *Amoeba* (phagocytic eukaryotic cell).
8. Bacteria that use ammonium as a source of energy and carbon dioxide as a source of carbon for making organic molecules.
9. Which of the following is not part of a defined medium?
a. trace elements;
b. yeast extract;
c. K_2HPO_4;
d. $(NH_4)_2 SO_4$;
e. glucose.
10. What is produced by the Calvin-Benson cycle?
a. ATP;
b. carbohydrates;
c. CO_2 ;
d. reducing electrons;
e. amino acids.
11. What important molecule is derived from glycolysis and the Krebs cycle by substrate level phosphorylation?
a. ATP;
b. glucose;
c. CO_2;
d. reducing electrons;
e. water.
12. What important molecule is derived from oxidative phosphorylation?
a. ATP;
b. glucose;
c. CO_2;
d. reducing electrons;
e. water.
13. An organism, which normally grows on a defined medium, sustains a mutation that blocks its ability to synthesize an amino acid. The mutant will grow if what missing component is added to the defined medium?
a. trace element;
b. defined compound;
c. growth factor;
d. cofactor;
e. coenzyme.
14. Most bacteria use agar as their carbon and energy source.
a. true; b. false.

For items 15–17, match the medium with the statements below:
a. complex;
b. defined;
c. selective;
d. differential;
e. living cell medium or tissue culture.

15. Media that contain additions such as yeast extract, beef extract, and red blood cells.
16. Media that allow only a small group of organisms to grow.
17. A medium used to grow viruses.

For items 18–20, suppose 0.1 ml of a culture is plated onto a complex medium and 25 colonies (10 different ones) develop after incubation for 48 hours.

18. What is the total number of bacteria in each ml of the culture?
a. 2.5;
b. 25;
c. 250;
d. 10;
e. 100.
19. Is the number calculated in the previous question
a. an overestimate;
b. an underestimate;
c. exactly the number in the sample plated.
20. It would be safe to assume from the results described above that there are more than 10 different types of organisms in the culture. a. true; b. false.
21. If a cell divided into 25 new cells rather than just 2 (as in binary fission), the formula $N_f = N_i\ 25^n$ could be used to describe the growth of a population of these cells. a. true; b. false.

For items 22–26, match the phase of the growth curve with the statements below:
a. depression;
b. stationary;
c. exponential;
d. death;
e. lag.

22. During this phase, cells produce antibiotics.
23. During this phase, cells are growing at a constant rate.
24. At the end of this phase, cells are growing at an ever increasing rate.

25. At the end of this phase, cells are running low on external nutrients or being inhibited by toxic materials.
26. At the end of this phase, cells are running out of stored nutrient and cannot generate enough ATP to make cell repairs.

For items 27–28, match the statements with the term below:
   a. neutralophile;
   b. mesophile;
   c. psychrophile;
   d. acidophile;
   e. alkalinophile.

27. An organism that grows best when the pH is below 5.
28. An organism that grows best when the temperature is between 20 and 50°C.

For items 29–30, match the statements with the term below:
   a. aerotolerant anaerobe;
   b. anaerobe;
   c. microaerophile;
   d. obligate anaerobe;
   e. facultative anaerobe.

29. An organism that only ferments and is not inhibited by oxygen.
30. An organism that is capable of fermenting and respiring, depending upon the oxygen concentration.

SUPPLEMENTAL READINGS

American Society for Microbiology. 1981. *Methods in general microbiology*. Washington: American Society for Microbiology.

Difco Laboratories. 1984. *Difco manual of dehydrated culture media and reagents for microbiological and clinical laboratory procedures*. 10th ed. Detroit: Difco Laboratories.

Ingraham, J., Maaløe, O., and Neidhardt, F. 1983. *Growth of the Bacterial Cell*. Sinauer Associates Inc., New York.

Lenette, E. H., Balows, A., Hausler, W. J., Jr., and Shadomy, J. P. 1985. *Manual of clinical microbiology*. 4th ed. Washington: American Society for Microbiology.

Payne, W. J. and Wiebe, W. J. 1978. Growth yield and efficiency in chemosynthetic microorganisms. *Annual Review of Microbiology* 32:155–183.

Randall, L. L. and Hardy, S. 1984. Export of protein in bacteria. *Microbiological Reviews* 48(4):290–298.

Shapiro, J. A., 1985. Photographing bacterial colonies. *ASM News* 51(2):62–69.

American Public Health Association. 1985. *Standard methods for the analysis of water and wastewater*. 16th ed. Washington: American Public Health Association.

Koch, A. L. 1984. Turbidity measurements in microbiology. *ASM News* 50(10:473–477.

Kushner, D. J. 1978. *Microbial life in extreme environments*. New York: Academic Press.

Monod, J. 1949. The growth of bacterial cultures. *Annual Review of Microbiology* 3:371–394.

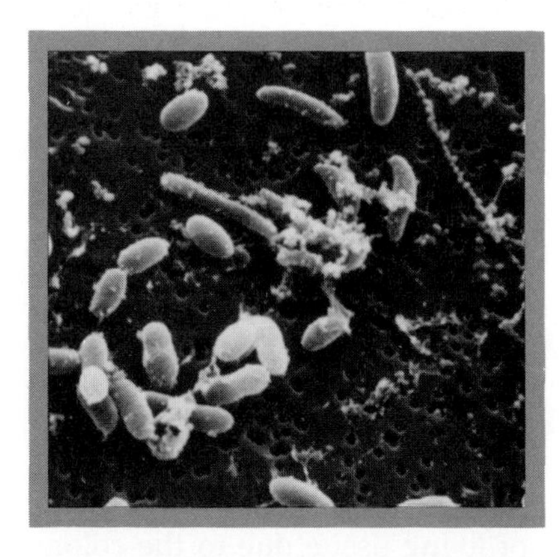

CHAPTER 5

CONTROL OF MICROORGANISMS

CHAPTER PREVIEW CONTROL OF INFECTIONS ACQUIRED IN A HOSPITAL

Between 5 and 15% of hospital patients develop infections during the course of their hospitalization. Sometimes as many as 5% of these infected patients die as a direct result of the infections. Since these statistics amount to more than two million infections and as many as a hundred thousand deaths each year, hospitals constantly study outbreaks of infections to discover their source and to develop procedures to control the microorganisms responsible. Infections acquired in hospitals are referred to as **nosocomial infections**.

In one outbreak of hospital infections, laboratory technicians observed an increase in isolates of the bacterium *Acinetobacter calcoaceticus*, which can cause lung and blood infections. To determine which type of patients were at risk and why, the hospital's bacteriology laboratory technicians checked personnel, patients, air, floors, walls, sinks, hospital equipment, and furniture for microorganisms. It was discovered that the ventilators used to help patients breathe were contaminated. Seventy-five percent of the spirometers (used to measure air volume) on the ventilators were contaminated with *Acinetobacter*. Even after the ventilator parts were sterilized, approximately 30% of the spirometers were found to be contaminated with *Acinetobacter*. A review of records indicated that *Acinetobacter* had been cultured from the respiratory tracts of 45 patients during a six-month period. All the patients had been mechanically ventilated in intensive care units. In addition, *Acinetobacter* was found on the hands of approximately 10% of the respiratory therapists and nurses working in the intensive care wards.

After it was discovered that condensate was being wiped from the inside of the spirometers with paper towels six to nine times a day, investigators suspected that respiratory technologists and nurses were contaminating the spirometers. The spirometers served as a growth chamber and reservoir for *Acinetobacter* because of the condensate. The investigators also assumed that the hospital personnel were responsible for transmitting the bacteria among the patients and the ventilators, but they did not rule out the possibility that some patient infections were caused by contaminated ventilators.

To control the outbreak of *Acinetobacter* infections in the intensive care units, all spirometers were removed from the ventilators, hands were washed religiously, and personnel used sterile gloves. Approximately six weeks after the control measures were instituted, the infection rate for *Acinetobacter* in the intensive care units decreased steadily from 9% to about 0.5%.

Acinetobacter infections in hospitals can be serious. Studies indicate that as many as 35% of patients with *Acinetobacter* pneumonia die as a result of the pneumonia. In one outbreak where 53 patients were infected with *Acinetobacter*, 25 had pneumonia. Of these 25 patients, 11 died. Nine of the deaths were caused directly by the pneumonia.

Hospital programs that monitor the incidence of infections are extremely important because they allow hospital personnel to discover and eliminate outbreaks of infection. Microorganisms are controlled in a number of ways that depend upon the situation. For example, the control of *Acinetobacter* that was growing in ventilators required that the equipment be modified and that personnel wash their hands and wear sterile gloves. Because control is such an important part of microbiology, the methods and principles of control are considered in the following chapter.

CHAPTER OUTLINE

MICROORGANISMS ARE RESPONSIBLE FOR THE death and suffering of millions of people each year. These organisms also cause disease in farm animals and crops and spoil food and other manufactured goods, thus inflicting serious financial losses on farmers and on many industries. To reduce the death and suffering caused by microorganisms and to alleviate financial losses, much research in microbiology and in industry is devoted to discovering new and more efficient ways of controlling microorganisms. In this chapter we discuss some of the materials and methods commonly used to control populations of microorganisms.

AGENTS USED TO CONTROL MICROORGANISMS

Microbiostatic and Microbiocidal Agents

Chemical and physical agents (such as heat and radiation) play an important role in the control of microorganisms and are classified on the basis of their effect on microbes. If the control agent inhibits or completely stops the growth of a population, it is called a **microbiostatic** agent (fig. 5.1). If, on the other hand, the agent actually kills and therefore reduces the number of viable organisms in the population, it is called a **microbiocidal** agent.

There are various related terms commonly used in place of microbiostatic and microbiocidal. **Bactericidal** agents kill bacteria, while **bacteriostatic** agents inhibit the growth of bacteria. Agents that inactivate viruses and kill fungi or algae are called **virucidal, fungicidal,** or **algicidal**, respectively. Antimicrobial agents that destroy a wide variety of different microorganisms are called **germicides**.

Categories of Chemical Control Agents

The diseased state of living tissue due to the colonization and growth of pathogenic microorganisms is called **sepsis**. This is a condition that must be prevented or eliminated to insure the health and well-being of the individual. Sepsis is often prevented by using selectively toxic chemicals known as **antimicrobial agents**.

Antimicrobial agents are somewhat artificially divided into the following categories: antiseptics, disinfectants, sterilants, chemotherapeutic agents, pesticides, and preservatives. The distinction is sometimes blurred because some of the chemicals can fall into a number of categories depending upon their concentration. For instance, at 20% formaldehyde is a sterilant, at 4% it may be a disinfectant, and at 0.1% it may be used as an antiseptic. Phenol at 5% is a toxic disinfectant but at 0.05% can be used as an antiseptic.

Antiseptics are chemicals that inhibit or kill microorganisms but are relatively safe to use as topical agents on humans and animals. Antiseptics include 70% isopropyl alcohol, 3% hydrogen peroxide, 1% tincture of iodine, 1% silver nitrate, and 1% mercurochrome (a mercury-containing compound).

Disinfectants, chemicals that are usually more toxic than antiseptics, are used to kill microorganisms on floors, toilets, showers, bench tops, and equipment. Disinfectants include 0.5% chlorine bleach (Clorox), 3% phenol (carbolic acid), 5% quarternary ammonium compounds (Roccal and Zephiran), 80% ethyl alcohol, and 80% isopropyl alcohol. The alcohols are classified as disinfectants rather than antiseptics because at these concentrations they are not safe to use on mucous membranes. Many disinfectants are composed of a mixture of chemicals: Lysol disinfectant consists of 0.1% *o*-phenylphenol and 79% ethyl alcohol while Lysol cleaner is made up of 2.7% alkyldimethylbenzyl ammonium chlorides, 0.13% tetrasodium ethylenediamine tetraacetate, and 0.34% ethyl alcohol.

Chemical sterilants are compounds that are very toxic to life and must be used with great caution. Chemical sterilants such as the gases ethylene oxide (12%) and propiolactone are used to kill organisms in heat-labile materials such as plastic petri dishes, plastic pipets, plastic filtering systems, and surgical implants. Liquid gluteraldehyde (20%) has been used to sterilize anesthesia tubing and surgical instruments since it rinses off easily with water.

Pesticides are chemicals that are used against insects, arachnids, nematodes, algae, fungi, and bacteria. For example, solutions of copper sulfate (Bordeaux mixtures) kill algae and fungi, 1% copper-8-hydroxy quinolate kills fungi and bacteria, and 400 mg/l ethylene bromide kills insects and arachnids in grains and on fruits as well as nematodes in soils.

Food preservatives are chemicals that are added to foods to inhibit the growth of microorganisms. Some of the preservatives are natural compounds such as vitamin C and propionic acid and probably are not dangerous to the consumer at the added concentrations. Many of the preservatives such as sodium nitrite, ethyl formate, sulfur dioxide, and polymyxin B may be dangerous if consumed in large amounts. Common food preservatives include sodium chloride, sugars, and acetic (vinegar), ascorbic (vitamin C), benzoic, lactic, propionic, sorbic, and sufurous acids. Fungi are inhibited by preservatives such as benzoic acid (benzoates), sorbic acid (sorbates), propionic acid (propionates), sulfur dioxide (sulfites), sodium diacetate, and ethyl formate while bacteria are commonly inhibited by vitamin C, sodium nitrite, sodium chloride, sugars, and polymyxin B.

Chemotherapeutic agents are chemical substances possessing a high degree of antimicrobial activity and can safely be used internally. Many of the chemotherapeutic agents used to inhibit or kill microorganisms are called **antibiotics** and are produced by microorganisms. Most of the useful antibiotics are produced by bacteria such as *Streptomyces* and *Bacillus* and by fungi such as *Penicillium* and *Cephalosporium*. The antibiotics synthesized by bacteria and fungi are usually produced during late exponential or early stationary phase, when these organisms begin to sporulate. The chemotherapeutic agents that are synthesized by scientists are called **synthetic drugs**. Antibiotics and synthetic drugs affect the growth of microorganisms by interfering with specific cellular functions.

The best chemotherapeutic agents show **high selective toxicity**, that is, they are very effective against the microbe but have little effect on the patient. The selective toxicity of many drugs relies on morphological or chemical differences between the host and the infecting microbe. For example, penicillin exhibits high selective toxicity against bacteria because it acts by inhibiting the synthesis of peptidoglycan and consequently affects only multiplying bacteria. Animal cells do not have cell walls, let alone any peptidoglycan, so they are not adversely affected by penicillin.

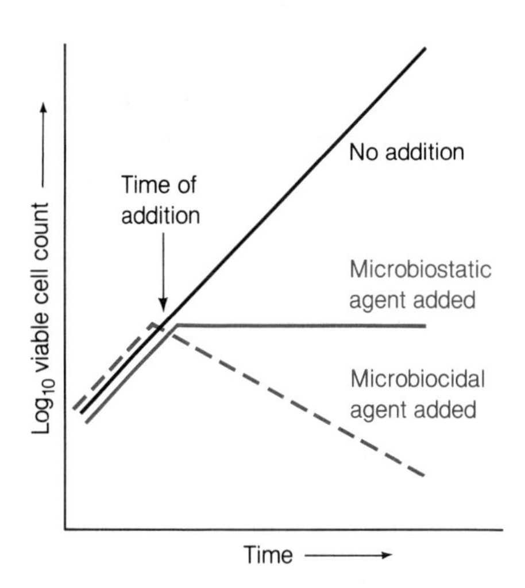

Figure 5.1 Microbiostatic and microbiocidal agents. Agents that kill microorganisms (microbiocidal) stop the growth of a population and decrease the number of viable cells, while those that simply inhibit them (microbiostatic) stop growth without an immediate decrease in the number of viable cells.

Some drugs show little selectivity to their target. This is particularly true of those drugs developed to combat eukaryotic pathogens. For example, **Amphotericin B**, a very effective antibiotic against a number of pathogenic fungi, often causes serious damage to human cells.

PROCEDURES USED TO CONTROL MICROORGANISMS

General Considerations

Disinfection is the treatment of materials with disinfectants in order to eliminate or reduce the number of disease-causing microorganisms. This definition implies that there may be residual living microorganisms present in or on the material after disinfection. Disinfectants are usually employed to reduce the possibility of infection when handling contaminated materials. Often the term **decontamination** is used in place of disinfection, but decontamination is also used to describe the reduction or elimination of harmful chemicals or radioactive substances.

Sanitization is a procedure that reduces the number of microorganisms to a low level so that they will not be a problem. The reduction in microorganisms is achieved through the use of chemical or physical agents.

Sterilization is a process that removes or *kills all* living microorganisms. In addition, sterilization generally implies that infectious agents such as viruses, viroids, and prions are inactivated. A substance is said to be **sterile** if it is free of all cellular microorganisms and infectious agents. Sterilization can involve the use of heat, radiation, or chemicals. Each of these has advantages and disadvantages that should be considered before one attempts to sterilize any substance. Sterilization procedures are routinely employed in the laboratory to remove contaminating microbes from culture media and the other materials employed in the isolation and cultivation of microorganisms. Sterilization procedures are also used in hospitals to remove all microorganisms from surgical supplies and equipment to prevent postsurgical sepsis.

Aseptic techniques are procedures used to prevent contamination of previously uncontaminated materials and to obtain and perpetuate pure cultures of microorganisms. A surgical team performing even the most insignificant of procedures employs aseptic techniques to prevent the contamination of the surgical incision or internal organs with unwanted microorganisms. Rudimentary aseptic techniques include personnel washing their hands with antiseptics and mild disinfectants; treating bench tops and tables with disinfectants; sterilizing all materials and equipment used in an operation, such as instruments, gloves, surgical drapes, masks and gowns; and maintaining the sterility of materials by appropriate wrapping and handling. This control of microorganisms is essential to prevent infection or sepsis. **Asepsis** is a nondiseased state of living tissue that is free of proliferating pathogenic microorganisms. In the case of aseptic meningitis, aseptic means that cellular organisms cannot be cultured from the meninges.

The Use of Chemical Agents

Many chemicals disrupt cellular metabolism or vital structures and consequently inactivate or kill microbes. Some of these substances are used in laboratories and in hospitals to decontaminate or disinfect work areas, such as surgical suites and media preparation rooms. Other chemicals are used as antiseptics to prevent infections.

Any one chemical may not be effective under all conditions or against all microorganisms. Consequently, to choose an effective sterilant or disinfectant it is important to understand which compounds are best to use in specific situations.

The type of microbial contamination determines which chemical control agent to use. Chemical control agents are generally effective against most vegetative bacteria, viruses, and fungi, but few are effective against endospores. Glutaraldehyde, formaldehyde, ethylene dibromide, and ethylene oxide are four of the most effective sporicides in common use. They are alkylating agents capable of inducing mutations, not only in microorganisms but also in humans. Consequently, these mutagens and carcinogens (cancer-causing agents) should be used carefully. The control of spore-forming bacteria is important because they are common airborne organisms that often contaminate pharmaceuticals and culture media. For these reasons, it is important to select a chemical control agent with sporicidal properties when attempting to decontaminate areas that are likely to contain these microorganisms.

The tubercle bacillus, *Mycobacterium tuberculosis*, is another bacterium that is highly resistant to chemical control agents and thus poses a considerable health hazard. Viruses exhibit varying degrees of resistance to chemical control agents. The hepatitis virus, the rhinoviruses (which cause the common cold), and the enteroviruses (which include the poliovirus) are among the more resistant ones.

The strength of chemical control agents is a factor in determining which should be used. Chemical germicides are classified on the basis of how effective they are (table 5.1). Germicides classified in the **high level** category are effective against all forms of life, including bacterial endospores. High level germicides include ethylene oxide and 2% buffered glutaraldehyde. Even though these agents have a high level of activity, they may require 10 hours before they kill an entire population of endospore-forming bacteria.

TABLE 5.1
ANTIMICROBIAL ACTIVITY OF COMMONLY USED COLD "STERILANTS"

Microbicidal agent	Range of effectiveness: Bacteria	Mycobacterium tuberculosis	Endospores	Fungal spores	Viruses	
Ethylene oxide (450 to 800 mg/L)	+ + +	+ + +	+ + +	+ + +	+ + +	High
β-Propiolactone (1.6 mg/L)	+ +	+ +	+ +	+ +	+ +	High
Glutaraldehyde (buffered, 2%)	+	+	+ +	+	+	High
Formalin (37% aqueous formaldehyde)	+	+	+	+	+	High
Alcohol: formaldehyde (70%:8%)	+ +	+	+	+	+	Intermediate
Alcohol: iodine (70%:2%)	+ +	+	±	+	+	Intermediate
Iodine (2 to 5% aqueous)	+ +	+	±	+	+	Intermediate
Iodophors (1%)	+	+	±	±	+	Intermediate
Sodium hypochlorite (5%)	+ +	+	0	+	+	Intermediate
Ethyl alcohol (70 to 90%)	+	+	0	+	±	Intermediate
Isopropyl alcohol (70 to 90%)	+ +	+	0	+	±	Intermediate
Phenolic derivatives (1 to 3%)	+	+	0	+	±	Intermediate
Quaternary ammonium compounds (3%)	+	0	0	+	±	Low
Hexachlorophene (3%)	+	0	0	+	±	Low
Benzalkonium chloride: H_2O (1:750)	+ +	0	0	+	0	Low
Merthiolate: H_2O (1:1000)	±	0	0	+	±	Low

+ + +, Superior; + +, very good; +, good; ±, fair (greater concentration or more time needed); 0, no activity.
**Based on data from DiPalma, J. R., ed. 1971. Drill's pharmacology in medicine. 4th ed. New York: McGraw-Hill. Reprinted by permission.*

Intermediate level control agents are defined as **tuberculocidal**; that is, they are capable of killing *M. tuberculosis*. In addition, these chemical control agents are usually effective against the more resistant viruses, such as the hepatitis viruses and the rhinoviruses. Intermediate level control agents are not very effective against endospores. **Low level** chemical control agents are not effective against *M. tuberculosis*, bacterial endospores, many fungal spores, or viruses that lack a membrane. They are effective against most vegetative bacteria and fungi, however, and are used extensively in routine decontaminations because they are economical and not excessively toxic to humans. Many of these chemicals are available in supermarkets and are sold as mouthwashes, disinfectants, and room deodorizers. Detergents such as quaternary ammonium compounds, Lysol (a phenol derivative), and mercurials (such as Merthiolate) fall in this category.

Evaluating the Germicidal Activity of Chemical Agents The need to control infectious organisms in hospitals and in other institutions has been an incentive for chemical industries to develop new and more effective germicides. To test the effectiveness of a new germicide, it is necessary to compare its activity to known, proven standards, such as the disinfectant phenol (table 5.2). There are several methods for comparing the activity of germicides, but the best known is the **phenol coefficient test**. The **phenol coefficient** is a measure of the effectiveness of a germicide expressed as the ratio of the effectiveness of the test germicide to that of phenol against a test organism (fig. 5.2). For example, if a 1:250 dilution of a test germicide kills a standard population of *Staphylococcus aureus*, while the highest dilution of phenol showing the same results is 1:60, the phenol coefficient of the test germicide is 250/60 or 4.2. This means that the test germicide is 4.2 times more effective than phenol in killing *S. aureus* at least *in vitro*.

Chemical control agents can usually be placed in one of the following categories: alcohols, aldehydes, detergents, halogens, heavy metals, phenols, or alkylating gases. Some of these chemicals are extremely toxic to humans and should be used with care in appropriate facilities. Some of the more important chemicals are discussed here as an aid to choosing the best possible chemical control agent and to understand how they work and why they may be dangerous.

Phenolic Compounds **Phenols** are aromatic compounds consisting of a benzene ring with a hydroxyl group attached (fig. 5.2). Phenol, or carbolic acid, has long been recognized for its germicidal properties. In the late 1800s, Joseph Lister used phenol extensively as an antiseptic to treat surgical incisions and as a disinfectant to decontaminate surgical areas. Using this germicide, Lister was successful in reducing the number of postsurgical sepsis cases in the hospital where he worked. Many products currently available are derivatives of phenol (table 5.2). Most of these phenol derivatives are more germicidal than phenol, with phenol coefficients ranging from 2.3 to about 300 when *Staphylococcus aureus* is used as the test organism. Most of the phenol derivatives are used as disinfectants rather than as antiseptics. A noteworthy exception is **hexachlorophene**, which is mixed with some soaps. These soaps were once used in hospitals to wash newborn babies to prevent staphylococcal infections. It is now known, however, that prolonged use of 3% hexachlorophene results in brain damage in rats, so it is no longer used routinely to wash babies. Nevertheless, hexachlorophene is very effective against staphylococci and its judicious use is still recommended to control staphylococcal infections in nurseries.

TABLE 5.2
PHENOL COEFFICIENTS OF SOME COMMON PHENOLICS

Name	Phenol coefficient*
Phenol	1.0
o-Cresol (Lysol)	2.3
m-Cresol	2.3
p-Cresol	2.3
2,4-dimethylphenol	5.0
Butylphenol	43.7
Hexachlorophene	125.0
Hexylresorcinol	313.2

*Phenol coefficients were determined at 37°C against *Staphylococcus aureus*.

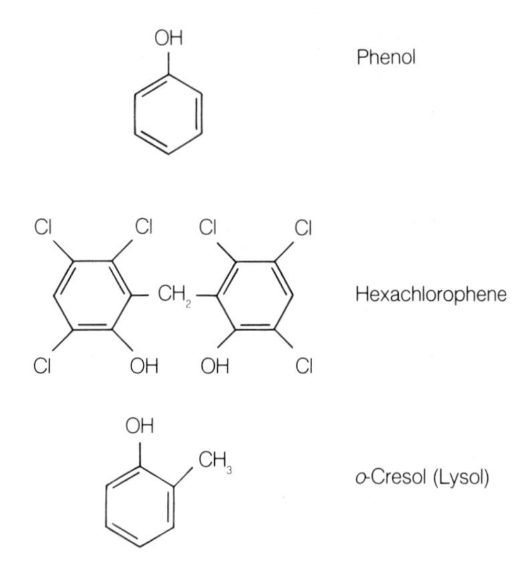

Figure 5.2 Chemical structures of some common phenolics

Phenolics can be either germistatic or germicidal, depending upon the concentration used. If the concentration is high, phenolics are bactericidal, tuberculocidal, fungicidal, and virucidal, but they are not effective against endospores. The phenolics kill cells by disrupting the plasma membrane, resulting in the cessation of vital membrane-associated functions of the cell, and by inactivating intracellular components such as proteins and nucleic acids. **Amphyl**, a preparation used in many laboratories to disinfect work areas, contains phenolic compounds, alcohol, and soap.

Alcohols Ethanol (CH_3CH_2OH) and isopropanol [$(CH_3)_2CH_2OH$] are very effective germicides. At concentrations of 70 to 80%, these alcohols kill fungi, the vegetative forms of most bacteria, and inactivate most viruses. Ethanol and isopropanol are widely used, primarily as antiseptics for cleansing wounds and for disinfecting the skin before injections. They are also used to reduce the microbial flora on thermometers.

Alcohols are believed to kill cells by denaturing vital proteins and by solubilizing and disrupting the

TABLE 5.3
EFFECTIVENESS OF ETHANOL AS A BACTERICIDAL AGENT

Concentration* (%)	Time needed to sterilize† (s)
100	>60
95	20
90	<10
80	<10
70	<10
60	<10
50	30
40	>60
30	>60

Source. Morton, E. H. 1950. *Annals of the New York Academy of Sciences* 51:191–196

*Percentages refer to the volume of 100% ethanol mixed in distilled water.

†Time of sterilization is an approximation of the time required to sterilize (no colonies formed) a standard suspension of *Streptococcus pyogenes*.

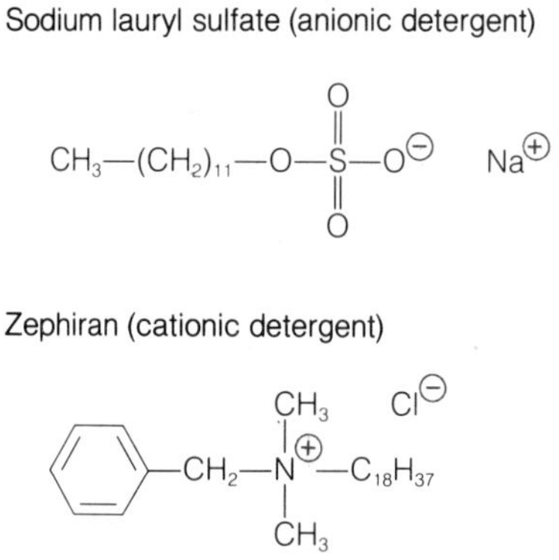

Sodium lauryl sulfate (anionic detergent)

$CH_3—(CH_2)_{11}—O—S(=O)_2—O^{\ominus}$ $Na^{\oplus}$

Zephiran (cationic detergent)

$C_6H_5—CH_2—N^{\oplus}(CH_3)_2—C_{18}H_{37}$ $Cl^{\ominus}$

Figure 5.3 Detergents. Anionic detergents are those that have a negative charge, while cationic detergents have a positive charge.

integrity of the plasma membrane. Since 100% alcohol is a dehydrating agent, it can extract intracellular water and thus enhance the survival of treated cells. This fact may explain why 70% ethanol is more effective than 100% ethanol in killing cells (table 5.3).

Detergents Detergents are organic molecules that consist of hydrophobic (water-repelling) and hydrophilic (water-loving) chemical groups on the same molecule (fig. 5.3). Two examples of common detergents are **sodium lauryl sulfate** (SLS) and **Zephiran** (benzalkonium chloride). The organic portions of these molecules have detergent properties because they bind tightly to organic materials, such as greases and oils. Detergents such as SLS with negative charges on the organic portions of the molecule are called **anionic detergents**. Detergents such as Zephiran with positive charges on the organic portion of the molecule are called **cationic detergents**. Some detergents do not ionize and are therefore called **nonionic detergents**. The nonionic detergents do not possess germicidal activity.

Detergents are used primarily to cleanse surfaces by removing microorganisms and organic matter. Detergents are microbiocidal because they emulsify and disrupt lipids that make up the plasma membrane, causing leakage of intracellular materials and ultimately cell death.

Quaternary ammonium compounds (fig. 5.3) are cationic (positively charged) detergents that have considerable bactericidal and fungicidal and some virucidal activity. In addition to disrupting the plasma membrane, quaternary ammonium compounds have been reported to inactivate enzymes and other cellular proteins. Zephiran and Roccal are two brand names for preparations that contain quaternary ammonium compounds and are used in many laboratories. These detergents are inactive, however, against endospores and the tubercle bacillus. Another, very important exception to the bactericidal activity of quaternary ammonium compounds is their inability to kill *Pseudomonas aeruginosa*. This bacterium is characteristically resistant to the action of quaternary ammonium compounds. Failure to recognize this fact has led to serious outbreaks of pseudomonal infections in hospitals and nurseries treated only with such disinfectants.

Halogens The halogens, fluorine (F), bromine (Br), chlorine (Cl), and iodine (I), constitute a family of chemical elements with a high affinity for electrons. Because of this affinity, the halogens are extremely reactive with and toxic to most forms of life. The halogens often react with cellular components or produce powerful oxidizing agents that react with various parts of the cell.

Chlorine compounds are not used as antiseptics, but are routinely used as disinfectants in water supplies, hot tubs, and swimming pools. Household bleach is a 5% aqueous solution of sodium hypochlorite (NaClO). Bleach or chlorine should not be used with compounds that contain ammonium ions or with acids because explosive and poisonous gases (nitrogen trichloride and chlorine, respectively) are produced.

Iodine has been used for over a century as an antiseptic for treating superficial wounds. Dissolved in 70% ethanol, it is known as **tincture of iodine** and is one of the most effective antiseptics in use today. **Betadine** is also an effective iodine-containing antiseptic used to treat superficial wounds and for preparing areas of skin for surgery. Betadine consists of an iodine-containing compound mixed with a detergent. This mixture is referred to as an **iodophore**. Since betadine is not painful and does not irritate the skin, it is favored over tincture of iodine as a disinfectant, especially for children.

Figure 5.4 **Heavy metals.** Many heavy metal compounds containing mercury, lead, copper, silver, and arsenic are extremely toxic to living organisms because heavy metals react with proteins and inactivate them. The inhibition and killing of microorganisms by small amounts of a heavy metal is referred to as the **oligodynamic action** of the metal. The dime that contains silver and the heart with gold inhibit the growth of the bacterial lawn near them.

Heavy Metal Ions Heavy metal ions of arsenic, copper, mercury, and silver are toxic to most forms of life because they combine with cellular proteins and denature them. Some of the heavy metals (e.g., mercury and silver) are toxic even in minute quantities. This toxic property is sometimes called the **oligodynamic action** of metals (fig. 5.4).

Mercury in the form of inorganic salts, such as mercuric chloride, is germicidal and has been used as the active ingredient in antiseptic ointments. Its activity as a germicide is reduced by extraneous organic material: therefore, skin surfaces should be scrupulously clean before application. Merthiolate and Mercurochrome are organic mercurials that are often used as antiseptics in the home to treat minor skin abrasions and wounds, although they are not very effective antiseptics (fig. 5.5).

Silver nitrate ($AgNO_3$) is an antiseptic that is sometimes used on newborn babies. Eye drops containing 1% $AgNO_3$ have been used to prevent a disease called **opthalmia neonatorum**, which results from an infection of the eye by *Neisseria gonorrhoeae*. The bacterium is acquired during delivery as the baby passes through an infected birth canal. $AgNO_3$ is now infrequently used, however, because it is irritating and better agents are available such as the antibiotic erythromycin.

Alkylating Agents **Alkylating gases** are chemicals that attach methyl or ethyl groups to cellular molecules. The alkylating gases generally cause the death of microorganisms by alkylating proteins and DNA. The degree of alkylation can be so great that proteins and DNA become totally nonfunctional. The two most commonly used alkylating gases are **ethylene oxide** and **β-propiolactone**. These chemosterilants can be used to sterilize heat-sensitive and packaged materials because they are capable of penetrating such materials. Ethylene oxide is extremely flammable, however, and β-propiolactone causes blisters when it comes in contact with the skin: thus, special precautions must be taken to reduce the hazards they present. Because these gases are so toxic to humans, materials that have been sterilized with ethylene oxide or β-propiolactone must be set aside in detoxification chambers for days to allow the gases to dissipate. Nevertheless, materials sterilized with ethylene

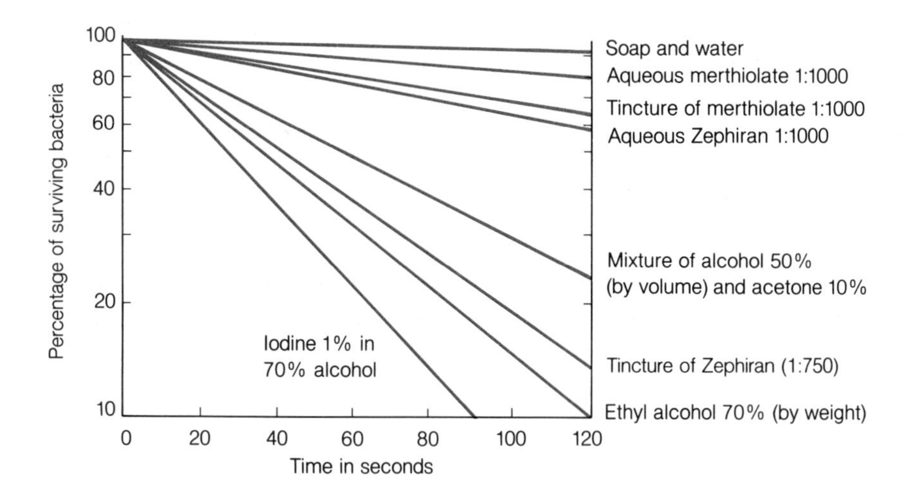

Figure 5.5 Effectiveness of antiseptics. The effectiveness of various antiseptic solutions on skin bacteria indicates that alcoholic solutions are more effective antiseptics than aqueous solutions. (Based on P. B. Price. 1957. Skin antisepsis. *Lectures on sterilization*, J. H. Brewer, ed. Durham, NC: Duke Univ. Press.)

oxide and ethylene dibromide still retain residues of the gases.

Alkylating solutions such as a 37% aqueous solution of formaldehyde gas (formalin) and a 2% aqueous solution of glutaraldehyde (buffered at pH 7.5–8.5) are sporicidal, tuberculocidal, virucidal, bactericidal, and fungicidal. Aldehydes inactivate cells primarily by alkylating proteins and cross-linking them, but nucleic acids and some membrane lipids apparently are also affected. Since aldehydes are such effective chemosterilants, they can be used to sterilize heat-sensitive materials such as anesthesia tubing and surgical instruments (table 5.4). Glutaraldehyde is more effective and less irritating to the skin than formaldehyde. In addition, glutaraldehyde rinses off easily

FOCUS DANGERS OF GASEOUS STERILANTS

In 1981 consumer groups presented evidence that ethylene oxide caused an increase in certain types of cancer and genetic damage. One consumer group sued the Occupational Safety and Health Administration (OSHA) to force the government agency to establish a limit of one part per million (1 ppm) per exposure and a limit of five parts per million (5 ppm) for repeated exposures in the work place. OSHA refused to issue these standards and stalled in setting any standards.

In 1982 a report in the *British Medical Journal* indicated that thousands of female hospital workers in the United States, at exposure levels of only 0.01 ppm, had a miscarriage rate three times that of the general population. This finding indicated that a limit of 1 ppm may not be sufficient. About 144,000 American workers are consistently exposed to ethylene oxide. 100,000 are affected when it is used as a sterilant in the workplace and as a fumigant in libraries, in museums, and on crops. The Public Citizen Health Research Group associated with Ralph Nader has urged the House Science and Technology Subcommittee to look into the matter and to force OSHA to set a limit for ethylene oxide exposure in the workplace.

In 1984, a number of muffin mixes were withdrawn from California stores after state tests showed that they had concentrations of ethylene dibromide ranging from 0.372 to 5.4 ppm. These muffin mixes were not withdrawn from stores in any other state. Ethylene dibromide is a known cause of cancer in laboratory animals and is a suspected cause of cancer in humans. It gets into wheat and corn products when the grains are treated with ethylene dibromide to kill insects and insect larvae. The California State Department of Health Services reported finding traces of ethylene dibromide, between 0.001 and 0.189 ppm, in 30 other baking products, but these were not removed from stores. Safe levels have not been established; nevertheless, government agencies have established that 0.030 ppm will be accepted in foods.

TABLE 5.4
CONTROL OF MICROORGANISMS IN HOSPITALS AND LABORATORIES

Chemical control	Time, temperature	Material
Ethylene oxide	2–12 h, 70°C	Artificial implants (breast, heart, penis)[+] Plastic petri dishes[+] Instruments with optical lenses (endoscopes, fiberoptic duodenoscope)[+] Polyethylene tubing[°] Rubber tubing[+] Silk catheters[+]
2% Aqueous glutaraldehyde	10 h	Polyethylene tubing[+] Thermometers[+] Instruments with optical lenses[−]
2% Amphyl (*o*-phenylphenol, paratertiary amylphenol, potassium ricinoleate, propylene glycol, ethanol)	20 min	Floors and walls[°] Toilets[°] Furniture[+] Surgical instruments[°] Rubber tubing[−] Silk catheters[−]
37% Aqueous formalin (formaldehyde)	12 h	Instruments with optical lenses[°]
70–90% Isopropyl alcohol plus 2% iodine	20 min	Thermometers[−]
Iodophor (500 ppm available iodine + detergent)	20 min	Rubber tubing[−]
Physical control	**Time, temperature**	**Material**
Autoclaving	15–30 min, 121°C	Heat stable media[+] Water and heat-resistant equipment (surgical tools, hypodermic syringes and needles)
Boiling	30 min, 100°C	Water and heat-resistant surgical instruments[+] Baby bottles and rubber nipples[+]
Pasteurizing	30 min, 64°C	Foods (milk, wine)[+]
Incinerating	Complete oxidation in flame	Inoculating loops[+] Bodies
Heating in hot air ovens	1–2 h, 180°C	Pipets[°]
Refrigerating	Constant 4°C	Foods[+] Chemicals[+]
Freezing	Constant −196°C	Foods[+] Chemicals[+]
Filtering	0.2-μm pore	Heat-sensitive liquid media and solutions[+]
Radiating	Ultraviolet	Air[+] Table tops[+]
	Radioactive elements $2-4 \times 10^6$ rads	Foods (bacon, canned meats, chicken, and vegetables)[−] Artificial implants[°]

[+] Best method for sterilization or disinfection.
[°] Alternative method for sterilization or disinfection.
[−] Least acceptable method for sterilization or disinfection.

with running water, so sterilized materials can be washed with sterile water to eliminate toxic residues. Low concentrations of glutaraldehyde are used as antiseptics, but long-term use damages the skin.

The Use of Physical Agents

Heat as a Physical Control Agent Of all the control agents, high heat is the most efficient and cost-effective sterilant. If a material is not likely to be damaged at high temperatures, its sterilization with heat is recommended. Generally, when psychrophilic and mesophilic microorganisms are exposed to temperatures above 55°C, their proteins denature irreversibly and they die. As the temperature is increased, they die at a faster rate (fig. 5.6). The loss of viability in a population exposed to high heat correlates well with the denaturation of proteins essential for microbial activities, particularly when the heat is applied in moist environments (moist heat). It also appears, however, that membranes become much more fluid and extremely permeable at high temperatures. The increased permeability results in the leakage of intracellular materials and the loss of membrane potentials. This blocks the transport of nutrients into the cell and, in some cases, the generation of ATP. Thus, denaturation of cellular proteins and disruption of membrane integrity, whether operating alone or together, contribute to efficient sterilization.

Heat affects microorganisms in dry environments differently than those in moist environments. Dry heat tends to dehydrate the cell, causing reversible protein denaturation. Desiccation tends to "preserve" some cells in an inactive state; consequently, higher temperatures are required to kill cells in dry environments than in moist ones. The physiological state of desiccated cells is such that their cellular contents become much more resistant to irreversible damage. In general, it takes considerably more time to kill endospores using dry heat at 180°C than it does using moist heat at 121°C. Bacteria and other microorganisms eventually are killed in dry heat environments when their proteins denature and they become charred (oxidized).

Several terms are used to express the susceptibility of microorganisms to high heat: (a) thermal death point (TDP); (b) thermal death time (TDT); and (c) decimal reduction time (*D*). The **thermal death point** for a particular population of microorganisms in an aqueous suspension is defined as the lowest temperature that sterilizes the culture within 10 minutes. The **thermal death time** is the shortest period of time necessary to achieve sterility at a given temperature under standardized conditions. **Decimal reduction time,** or

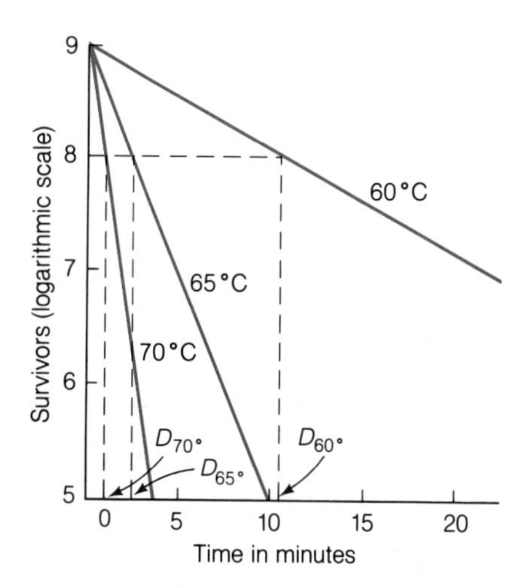

Figure 5.6 Decimal reduction times. The decimal reduction time is the time required to reduce a population by 90% at a particular temperature. The decimal reduction time decreases as the temperature increases. For example, in the graph above, the D_{60} is about 10.5 min, the D_{65} is 3 min, and the D_{70} is 1 min. The decimal reduction time is calculated by finding the points on the death curve between which the population changed by one log unit (e.g., from 10^4 to 10^3), and then determining the time it took for this change.

***D* value**, is another way of expressing the susceptibility of microbes to the killing effect of heat. The *D* value is defined as the time required for a pure culture of microorganisms to decrease by 90% when exposed to a specific temperature (fig. 5.6). The *D* value expresses the rate at which a population dies when exposed to heat: the smaller the *D* value, the more rapidly the organisms die. This measure is used mostly in the food industry, particularly in the canning industry.

Steam Sterilization Steam is an ideal sterilizing agent because, when under pressure, it can be used to heat materials to temperatures above 100°C, and because it heats porous substances evenly and quickly. To increase the efficiency of sterilization using steam, the materials are usually sterilized in a chamber in which the steam pressure is above that of the atmosphere. At elevated pressures, steam can reach temperatures above 100°C. The increase in steam tem-

TABLE 5.5
RELATIONSHIPS BETWEEN TIME OF AUTOCLAVING, PRESSURE, AND TEMPERATURE

Steam pressure (PSI)*	Temperature (°C)	Temperature (°F)	Time of exposure (min)†
0	100	212	>60
5	110	230	>60
10	115	239	60
15	121	250	15
20	126	259	10
25	134	273	5

*PSI refers to pounds per square inch above atmospheric pressure.
†Approximate time required to kill a standard suspension of endospores.

perature is proportional to the pressure (table 5.5). Routine laboratory sterilizations are carried out in **autoclaves** at 121°C for 15 to 30 minutes, depending upon the volume of material to be sterilized and the amount of contamination expected (fig. 5.7). It is usually recommended that large volumes of materials, such as culture media or surgical dressings, be dispensed into small volumes or portions to reduce the time required for sterilization and, thus, decrease the chance of overcooking or damaging the material (table 5.6). Although autoclaving is a highly recommended sterilization procedure, its major drawback is that it may cause damage to heat-sensitive materials. Therefore, steam sterilization cannot be used to sterilize all materials.

Boiling Boiling is an inexpensive and relatively effective means of disinfecting substances. Many contaminating microorganisms and pathogens are readily destroyed by boiling for a few minutes. Boiling is often used to disinfect and sterilize materials at home and in places where autoclaves are not readily available. This method of moist heat sterilization has one major shortcoming, however: it does not destroy many of the bacterial endospores and some of the viruses. As a consequence, boiled materials are often not sterile. Nevertheless, boiling has been used extensively for centuries and has been found quite satisfactory for sanitizing certain materials, such as baby bottles and rubber nipples.

Pasteurization Pasteurization is not a method of sterilization but rather a process whereby disease or

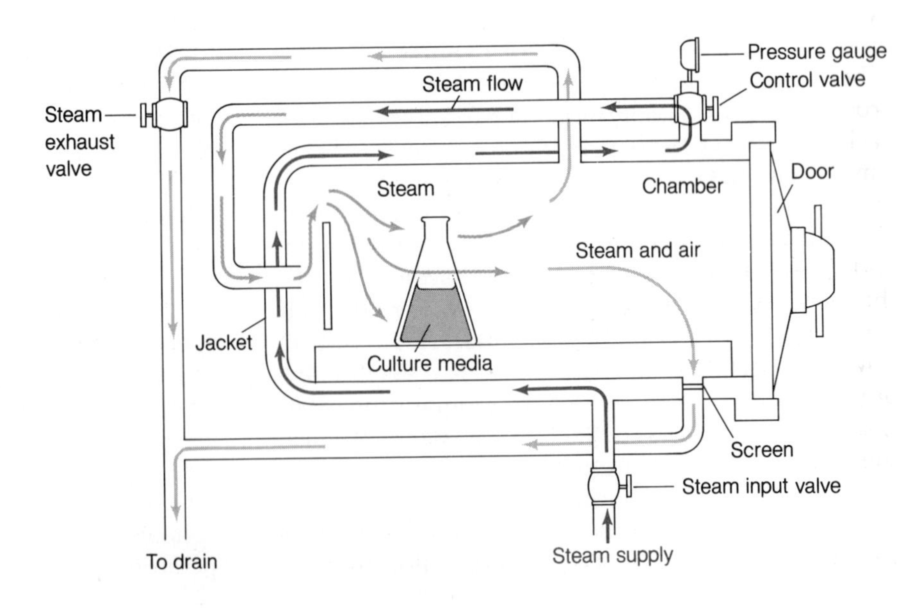

Figure 5.7 **Autoclave.** Autoclaves are used to sterilize heat-stable media and equipment. Steam initially enters the steam jacket around the autoclave chamber and heats up the autoclave. An object is placed in the chamber, the door is shut securely, and steam from the jacket is allowed to enter the chamber. Air is forced out of the chamber by the incoming steam until only pure steam is being forced out. Then the air exit is closed by the high temperature of the steam, and the steam pressure builds to 15 pounds per square inch. Eventually, the temperature rises to 121°C. When the sterilization is completed, the steam exhaust valve opens and the steam flows out of the chamber. When liquid media are being sterilized, the steam must be cooled and released slowly to avoid evaporation and boiling over of the liquid media.

TABLE 5.6
STERILIZATION TIMES AND METHODS FOR SELECTED INSTRUMENTS, SUPPLIES, AND CULTURE MEDIA

Items to be sterilized	Time required (min)	Method of sterilization		
		Autoclaving (121°C)	Dry heat (160°C)	Filtration
Gloves and gauze pads	30	+		
Surgical drapes and dressings	30	+		
Surgical instruments	30	+		
Glassware and metalware	30	+		
Culture media in tubes	15	+		
Culture media in flasks	15	+		
Culture media in large volumes (more than 10 L)	70	+		
Glassware	60		+	
Surgical instruments	120		+	
Oils, petroleum jelly, gauze	120		+	
Sharp instruments	270		+	
Heat-sensitive solutions	—			+
Serum-containing media	—			+

spoilage-causing microorganisms in liquids are destroyed by heating the liquids to temperatures below 100°C. For example, some liquids can be **batch pasteurized** in large tanks by subjecting them to temperatures of 63°C for 30 min, or **flash pasteurized** by heating them to 72°C for 15 seconds as they flow through a heater. Pasteurization was developed by Louis Pasteur in the mid 1800s to kill microorganisms that were spoiling French wines. Unlike boiling, the pasteurization process did not alter the flavor or quality of the wines. The procedure developed by Pasteur is now used to kill pathogenic organisms in a number of beverages and foods. For instance, most of the milk products on supermarket shelves are pasteurized before packaging to kill all the disease-causing microorganisms that may be present. Of particular importance are the bacteria that cause tuberculosis *(Mycobacterium)*, listeriosis (*Listeria*), undulant fever *(Brucella)*, and Q fever *(Coxiella)*. Since pasteurization kills all of the heat-sensitive microorganisms, it prolongs the products' shelf life. Pasteurized milk is not sterile, however, because many **thermodurics** (heat-resistant organisms) survive. Foods and beverages usually are not boiled or sterilized, because these procedures alter their flavor, texture, and quality.

Incineration Microorganisms exposed to open flames burn and consequently are removed from materials that can stand the heat. This form of sterilization is used to sterilize inoculating loops and some other equipment used in the cultivation of microorganisms. In addition, many contaminated materials and carcasses of infected animals are **incinerated** to kill potentially dangerous organisms.

Hot Air Ovens Dry heat is routinely used in laboratories and hospitals to kill microorganisms. Glassware such as pipets, petri dishes, and other heat-resistant materials are often sterilized in hot air ovens, usually at 160 to 180°C for 1 to 2 hours. These ovens are equipped with a heating unit to raise the temperature inside the chamber, and a fan to circulate the hot air so that the materials are heated evenly. Successful sterilizations can be carried out using a gas or electric stove.

Dry heat sterilization has at least one advantage over steam sterilization in that it does not require water, which may damage many materials or alter their properties. In addition, certain materials, such as glass pipets and glass petri dishes, need to be completely dry before they can be used. With moist heat, some of the water vapor condenses on the glass.

Refrigeration Refrigeration is employed to control microbial growth in order to preserve easily spoiled materials. In general, the lack of heat (low temperature) does not kill microorganisms; rather, it inhibits their growth by slowing enzymatic reactions. It is only at temperatures much below 0°C that cellular water begins to freeze and to create an environment that is not conducive to metabolic activities. A certain amount of cell damage occurs during freezing because of the formation of ice crystals within the cell. Leakage of cell materials may occur as a result of damage to the plasma membrane. In addition, ice forming around the exterior of the cell causes water to be lost from within the cell, thus causing some proteins to be denatured. Although normal metabolism and cell repair may cease, some enzymes may remain active and degrade parts of the cell. The damage these enzymes cause may in some cases be so extensive that the cells are killed. If cells are frozen rapidly at a low temperature of −80°C in an appropriate medium, however, they can be preserved indefinitely in a viable form. The low temperature and the frozen state of the cytoplasm protect these cells from damage.

Freezing is commonly used to preserve foods and other rapidly spoiled materials. At temperatures of −20°C (the approximate temperature inside the freezer compartment of most refrigerators) there is no spoilage due to microbial growth.

Radiation as a Physical Control Agent Ultraviolet light, x rays, and gamma rays cause cell death by inducing extensive changes in the cell's DNA or by ionizing cellular components (fig. 5.8). X rays, gamma rays, high-energy alpha and beta particles, and neutrons are known as **ionizing radiation**. When a cell is in the path of any of these types of radiation, considerable ionization occurs. Cellular constitutents hit by the radiation lose electrons and protons. This process results in the formation of ionized compounds and free radicals, chemicals that are highly reactive and can cause severe damage to the cell by reacting with proteins and nucleic acids. One of the more permanent changes believed to occur as a result of damage caused by ionizing radiation is the formation of large numbers of single- and double-strand breaks in the DNA molecule. These breaks can alter the DNA so

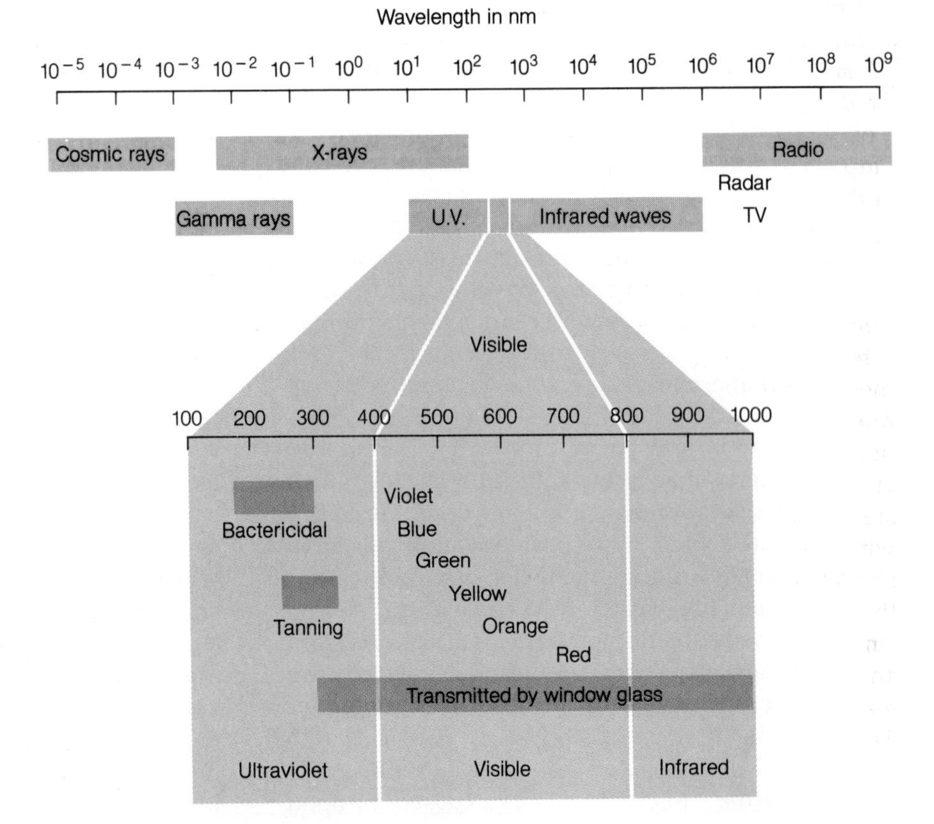

Figure 5.8 Electromagnetic radiation. Ultraviolet light, x rays, gamma rays, and cosmic rays all destroy cells and viruses. Ultraviolet light is often used to sanitize the air and the tops of work benches in laboratories and manufacturing plants where microbial contamination must be minimal. Gamma rays from radioactive elements are often used to sterilize packaged materials, such as plastic petri dishes and filtering equipment. X rays and cosmic rays are not used to sterilize.

that it can no longer function as a template for its own replication or for the synthesis of RNA.

Ultraviolet Light Electromagnetic radiation with wavelengths between 100 and 400 nm is called **ultraviolet (UV) light** (fig. 5.8). These wavelengths, especially those near 260 nm, are particularly important from a biological point of view because they are absorbed by nucleic acids. This energy absorption induces chemical changes that alter the structure of the DNA. During the repair process, the sequence of nucleic acids may be altered, thus leading to the death of the organism. The amount of damage to the DNA molecule is directly proportional to the amount of radiation absorbed by the cell.

Most ultraviolet radiation does not readily penetrate glass, opaque solids, or liquids, so it is generally used to sterilize surfaces in work areas or the air in rooms and surgical suites.

Radioactivity Unlike UV radiation, ionizing radiation penetrates nonliving and living organisms readily and hence is used to sterilize heat-sensitive medical supplies, such as plastic syringes, drugs, and single-service surgical equipment. Cobalt-60 is one of the most common sources of gamma rays employed in industry. It is useful because it produces high-energy radiation capable of penetrating materials so that the products can be prepackaged before sterilization. Aside from medical supplies, ionizing radiation is seldom employed as a routine sterilizer because it is expensive and dangerous to use.

Filtration as a Physical Control Agent Filtration is an effective and reasonably economical way to "sterilize" liquids and gases (fig. 5.9). It is recommended for the elimination of microorganisms in **thermolabile** (heat-sensitive) fluids and air. Sugar solutions used for the cultivation of microorganisms tend to caramelize during autoclaving, so they are often filter-sterilized. Beer and wine, which some breweries and wineries have traditionally pasteurized after the fermentation process is completed, are now being passed through special filters to remove spoilage microorganisms. An advantage of this process is that it sterilizes and clarifies the beverage in one step. Many pharmaceuticals, media, and chemical solutions are sterilized by filtration. Currently, filters are manufactured that can remove viruses.

Materials such as asbestos, sintered glass, porcelain, and diatomaceous earth have been used successfully to filter liquids and gases. Because these materials usually exclude only bacteria and large microorganisms, and because they are expensive, they have been replaced by economical **membrane filters** made of cellulose acetate and similar materials. These filters are manufactured with pores of different sizes. The diameter of the pore and the electrical charge on the filter determine the kinds of microorganisms that will be filtered out. For example, filters with pore size of 0.1–0.2 μm are routinely used to filter out contaminating bacteria. Many membrane filters have the advantage of being semitransparent when moistened, so that direct microscopic examinations can be made of organisms trapped on the filter. In addition, nutrients can pass virtually unimpeded through these filters; thus, microorganisms can be grown on the surface of a filter by placing the filter on an agar medium. This procedure allows microbiologists to count the number of microorganisms captured on the filter and consequently to estimate the number of organisms per milliliter of fluid or gas.

All filters work in much the same way, whether the material to be filtered is a fluid or a gas. When a substance is filtered, the fluid or gas passes through the pores, but microorganisms and inert particles that are too large to fit into the pores remain trapped (fig. 5.9). In addition, many particles and microorganisms that are small enough to pass through the pores are trapped by the filter due to electrostatic charges. Because suction devices are generally used on filtering apparatuses to speed up the process, certain bacteria, such as the **mycoplasmas** (which lack cell walls and are therefore pliable), can squeeze through pores as small as 0.2 μm. It must also be recognized that materials filtered through pores as small as 0.1 μm in diameter may not really be sterile, because many viruses can pass through these pores.

PRINCIPLES OF MICROBIAL KILLING

To determine the most efficient way of dealing with contaminating microbes, scientists have studied the behavior of microbial populations exposed to various killing agents. Some chemical and physical control agents are effective over a wide range of microorganisms, while others are more restricted in their scope. Regardless of the nature of the control agent, its effec-

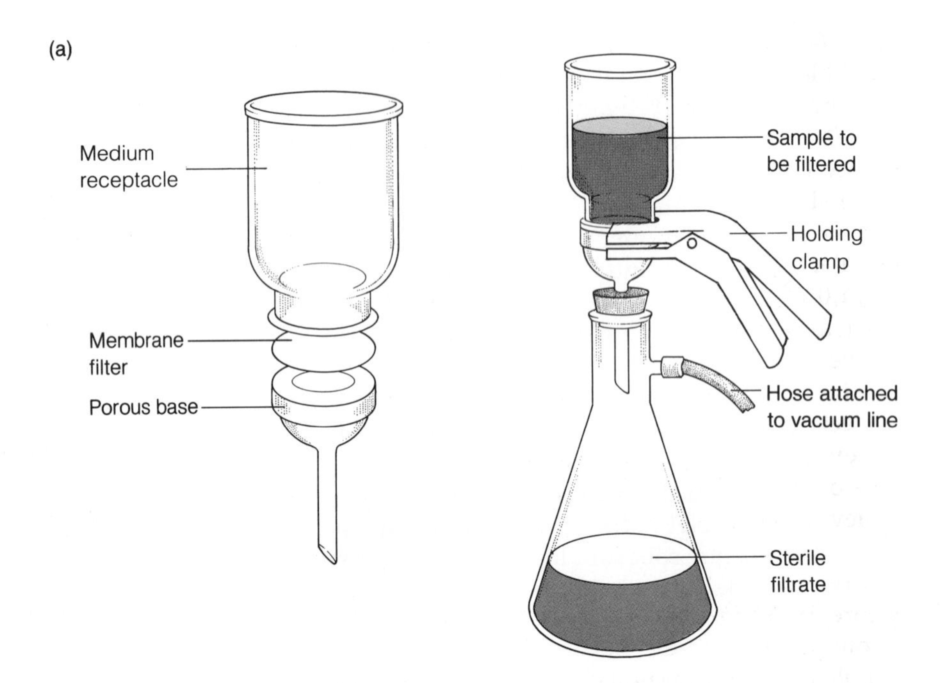

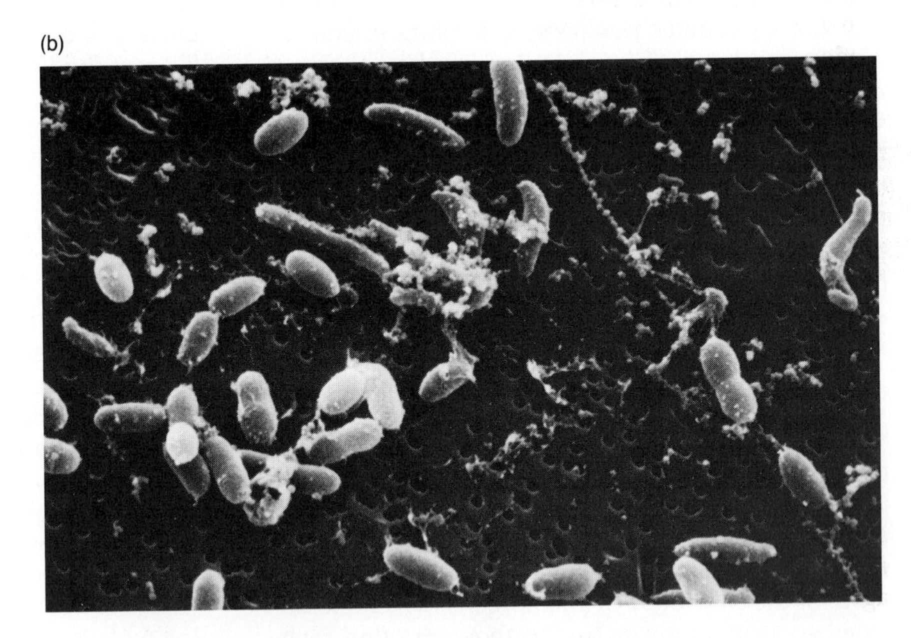

Figure 5.9 Filtration. *(a)* Heat-labile (sensitive) media are generally "sterilized" by filtration through membrane filters. In fact, the medium is not really sterilized because viruses generally pass through the filters. Water samples are often checked for microorganisms by filtering. The microorganisms adhering to the filter can be detected by placing the filter on a selective and/or differential agar medium. The resulting colonies can then be counted and identified. *(b)* Bacteria are shown caught on a Millipore membrane filter, in which the pores are too small to allow the organisms to pass through.

tiveness is influenced by the conditions under which it is applied. The following discussion covers some of the most important factors that influence the effectiveness of control agents.

Population Size and Volume

When a population of microorganisms is exposed to a killing agent, only a fraction of the microbes present dies during a given time interval (fig. 5.10). The longer the population is exposed to a killing agent, the more individuals are killed. Generally, the decline in viable count of a population occurs exponentially, and the time it takes to achieve sterility is directly proportional to the number of organisms initially present. For example, if the initial population contains 100,000 viable cells/ml of culture medium, and it declines by 90%

every 60 seconds, the death curve (or survivor curve) would look like curve C in figure 5.10. It can be seen that the initial size of the population (at time 0) was 100,000 viable cell/ml and that it took 5 minutes to reduce the population to 1 cell/ml. Curve B has the same slope—that is, the population is dying at the same rate—but it takes 1 min longer to reduce the population to 1 cell/ml, since the initial population was 1,000,000 viable cells/ml instead of 100,000. It takes even longer to reduce a population of 10^7 cells/ml (curve A).

In general, the larger the size of the population subjected to a sterilizing agent, the longer it takes to achieve sterility. It follows also that the larger the volume of material to be sterilized, the longer it takes to achieve sterility.

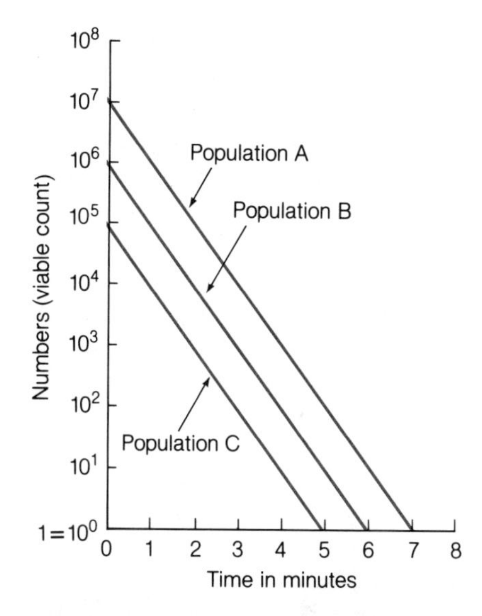

Figure 5.10 Death curves. The time required to kill a population of microorganisms is directly proportional to the size of the population. The graph shows three hypothetical populations of the same organism exposed to the same killing agent (e.g., heat). Notice that populations A, B, and C die at the same rate (all three lines are parallel) although population C dies sooner than B and B dies sooner than A because the initial sizes (microorganisms/ml) are smaller.

Exposure Time

Figure 5.10 shows how long it takes different populations to be reduced to one organism. By treating the populations for another minute, the chances of having one viable organism is reduced one-tenth. If the population were treated for longer periods of time, the chances of having one viable organism in the population is reduced even further. The question is: How much longer should the microbial population be treated with a sterilizing agent before it can be considered sterile? Many microbiologists consider a culture medium sterile if there is only one chance in a million of having a living microorganism left in the culture. This would require 6 extra minutes of sterilization (fig. 5.10). This time was calculated by observing that the population was decreased by one order of magnitude every minute (i.e., 7 min/7 orders). To reduce it 6 orders of magnitude further, (one chance in a million) 6 min are required {(1 min/1 order) $\times$ (6 orders) $=$ 6 min.} The actual number of microorganisms in a contaminated solution is seldom known. Consequently, the length of treatment of the solution with a sterilizing agent must be determined by experimentation.

Generally, sterilization times are determined using microorganisms that are highly resistant to the sterilizing agent. For example, the endospores of *Bacillus stearothermophilus* are highly resistant to heat and are therefore used to establish routine times for heat sterilization. Routine laboratory sterilizations using steam under pressure are carried out at 121°C for 15 minutes. During this time and at this temperature, there is only an infinitesimal chance that even one endospore of *B. stearothermophilus* remains viable. Standard times for achieving sterilization of various materials with different killing agents are determined using highly resistant organisms to insure the sterility of the material, even when it is heavily contaminated with other types of microorganisms.

Population Susceptibility

Another characteristic of microbial populations to keep in mind when developing an appropriate control method is that different microorganisms die at different rates when exposed to a given sterilizing agent. In other words, not all microorganisms have the same susceptibility to a given sterilizing agent.

The rate at which a given population of microorganisms dies is also a function of its age. Generally, young, actively growing cultures are more susceptible to antimicrobial agents than are older ones in the stationary phase. This tendency appears to be true of populations exposed to heat and to disinfectants. It

FOCUS VISITS TO THE DOCTOR RESULT IN A RASH OF MEASLES CASES

The waiting rooms at pediatricians' offices and clinics are becoming dangerous settings for babies. During the years 1980–1982, seven out of every 1000 cases of measles were acquired at a doctor's office or family practice clinic. In 1985, the number of cases had increased to 47 out of 1000. Since there are now about 6000 cases of measles each year, this means that approximately 300 cases of measles are acquired at medical offices and clinics.

The transmission of measles at doctors' offices is promoted in babies between 5 and 15 months old because most have not been vaccinated yet and consequently are susceptible. In addition, since many of the offices are small, sick babies are not separated from healthy ones, and many babies are brought into the office each day, airborne transmission from one infected baby to many can occur over a period of several hours.

A well-documented outbreak of measles involving 21 patients, showed that nine babies were exposed to measles at a pediatricians' office and at an adjoining family practice clinic during visits at different times between 8:30 a.m. and 2 p.m.

Since measles transmissions in a small office can occur over a period of 5 hours or more, it is essential that pediatricians separate diseased children from healthy children by having them enter separate waiting and examination rooms. The rooms used for diseased children should be ventilated so as to remove and destroy the measles virus in the air. Removal of viruses can be achieved using special air filters. In addition, before the next patient enters the rooms, areas where saliva or nasal secretions may have been deposited should be disinfected.

may be explained, at least in part, by the fact that older cells are biochemically less active than younger ones. Because they are biochemically less active, they also have a diminished capacity to take up chemical control agents that damage them. Since many antimicrobials must penetrate the cell to be active, they are not as effective on stationary phase cells as they are on actively metabolizing cells.

Hydration

The presence of water in an environment affects the rate at which a particular agent kills a microbial population. For example, a high water content expedites the coagulation or jelling of cellular proteins when heat is the killing agent. A 50% solution of egg albumin in water is coagulated when the solution is heated to 55°C, whereas a 75% solution is not coagulated until it reaches 75°C. Hot, moist air kills more efficiently than hot, dry air, because the moist air has a greater heat content than does dry air. In addition, many chemicals must be in solution to be effective as control agents. Water affects the activity of most chemical control agents by promoting chemical reactions and the ionization of the agents.

Temperature

Elevated temperatures usually enhance the activity of disinfectants because they promote chemical reactions, metabolic activities, and in many cases a rapid uptake of disinfectants by the cell. As the concentration of disinfectant increases inside the cell, so does the rate of death. The effect temperature has on antimicrobial agents is illustrated by the following example. A 1.1% aqueous solution of phenol, maintained at 10°C, sterilizes a standard bacterial culture in approximately 160 minutes. The same solution of phenol at 20°C sterilizes a similar culture in less than 60 minutes.

Antimicrobial Concentration

Generally, the rate of killing is directly proportional to the concentration of the antimicrobial. That is, the higher the concentration of the antimicrobial applied, the faster it kills microorganisms. Optimal concentrations for each routinely used antimicrobial should be determined empirically. A compromise should be reached between rate of killing and economy. A highly concentrated solution of antimicrobial undoubtedly is more effective than a more dilute one, but it is also more expensive to use. Thus, the use of optimal concentrations insures that the antimicrobial will be effective for its desired purpose but not overly expensive.

QUALITY ASSURANCE

The use of contaminated medical supplies, drugs, vaccines, or food can result in infections and death. Consequently, it is important to insure that products are free of dangerous microorganisms by having a **quality assurance** check as part of the production procedure. Quality assurance is simply a set of controls to insure that the product does what it is supposed to do and that it contains few microorganisms or is sterile. If a drug used for the treatment of a particular disease is administered by injection, not only should it be effective against the disease being treated, it should also be sterile, so that it does not cause an infection when injected. In any quality assurance program, a representative portion of the drug should be tested for sterility before it is marketed. In addition to testing the final product for sterility, it is customary to test the effectiveness of the sterilizing agents and the sterilization protocol on a periodic basis. For example, routine laboratory sterilizations using steam are carried out at 121°C for 15 to 30 minutes. Under these conditions even the most resistant of life forms should be killed. To insure that this assumption is indeed valid, a standard suspension of *Bacillus stearothermophilus* (usually 1,000,000 endospores/ml) is subjected to a routine sterilizing procedure, together with other materials such as culture media and dressings. If the standard suspension of test bacteria is sterile, it can be assumed that the sterilization procedure is working correctly. If the suspension is not sterile, then it is necessary to reevaluate the sterilization procedure. Quality assurance should be as routine in microbiology laboratories as lighting Bunsen burners or inoculating culture media.

SUMMARY

AGENTS USED TO CONTROL MICROORGANISMS

1. Controlling the growth of microbial populations is important because it allows us to prevent the transmission of disease, spoilage of foods, and contamination of materials.
2. Microbiostatic drugs *inhibit* the growth of microorganisms, while microbiocidal drugs *kill* microorganisms.
3. Antiseptics are chemical compounds that can be used on the surfaces of animals and plants to kill or reduce the number of microorganisms.
4. Disinfectants are chemical compounds that generally cannot be used on the surface of animals and plants, because they are too toxic. They are usually used to kill microorganisms on the surface of inanimate objects.
5. Chemotherapeutic agents are drugs that show selective toxicity for the pathogen. These drugs are used to control infections of microorganisms and can be injected into the human body.
6. Antibiotics are microbial products that are active against other microorganisms.

PROCEDURES USED TO CONTROL MICROORGANISMS

1. Disinfection is the treatment of materials with chemicals that eliminate or reduce the number of disease-causing microorganisms.
2. Sanitization is a procedure that reduces the number of microorganisms to a low level so that they will not be a problem.
3. A material is said to be sterile if it is free of all microorganisms.
4. Aseptic techniques are procedures designed to prevent the contamination of sterile materials.
5. Microorganisms can be controlled by exposing them to phenols, alcohols, detergents, halogens, heavy metals, alkylating gases, and aldehydes.
6. Alcohols, such as ethanol and isopropanol, are good antiseptics and disinfectants.
7. Aldehydes, such as formaldehyde and glutaraldehyde, are infrequently used as disinfectants.
8. Detergents, such as Roccal and Zephiran, are frequently used as disinfectants to clean work spaces.
9. The halogen chlorine is used as a disinfectant in swimming pools, while the halogen-containing compound known as bleach is used as a disinfectant on floors and on work benches. Iodine-containing compounds, such as tincture of iodine and betadine, are used as antiseptics.
10. Compounds containing heavy metals, such as

Mercurochrome, silver nitrate, and copper sulfate, are used as antiseptics.

11. Phenols are generally used as disinfectants, but some phenols, such as carbolic acid and hexachlorophene, can also be used as antiseptics.
12. Alkylating gases, such as ethylene oxide and propiolactone, are used to sterilize heat-labile materials.
13. Microorganisms can be controlled by steaming under pressure (autoclaving), boiling, pasteurizing, incinerating, heating in ovens, refrigerating, irradiating, and filtering.
14. Steaming under pressure is the most common method of sterilizing heat-stable materials, such as glassware and most media.
15. Boiling generally is not used to sterilize materials, because some endospores are quite resistant to boiling.
16. Pasteurization is used to destroy spoilage- and disease-causing microorganisms in beverages, because generally it does not alter the taste of the beverage.
17. Incineration is used to destroy materials that are no longer wanted, but which are contaminated with dangerous microorganisms.
18. Heating in ovens is used not only to kill microorganisms but also to dry heat-stable equipment.
19. Refrigeration is used to inhibit the growth of microorganisms, but even at very low temperatures, it does not kill all microorganisms.
20. Ultraviolet light is often used to sterilize bench tops and the air in rooms.
21. Filtration is used to sterilize heat-labile liquid media.

PRINCIPLES OF MICROBIAL KILLING

1. Populations of microorganisms, when exposed to the influence of a killing agent, exhibit a progressive decline in numbers until the culture becomes sterile.
2. Many factors affect the killing activity of control agents. Some of these factors are exposure time, age of the culture, pH, amount of organic matter present, concentration of microorganisms, temperature, and hydration.

QUALITY ASSURANCE

1. Quality assurance programs are essential in industrial and other processes involving the control of microorganisms to insure that the products are safe to use.
2. To insure that sterilization procedures are adequate, a standard suspension of *Bacillus stearothermophilus* is used as a control.

STUDY QUESTIONS

1. Define the following terms:
   a. sepsis;
   b. asepsis;
   c. aseptic technique;
   d. antiseptic;
   e. disinfectant;
   f. sterilant;
   g. antimicrobial.
2. What is the difference between a microbiostatic agent and a microbiocidal agent? Give an example of each.
3. What evidence demonstrates that the time it takes to achieve sterility is directly proportional to the number of organisms present?
4. Initially, a population consists of 10^5 organisms. After exposure to a killing agent for 5 min, the population is reduced to 1 organism. How much longer should the population be sterilized to be sure there is only a one in a million chance that an organism remains alive?
5. Explain how the hydration of the environment affects the rate of sterilization of microbial populations.
6. Explain how the concentration of an antimicrobial affects the rate of killing.
7. Explain how temperature affects the rate of killing by antimicrobials.
8. Explain why moist heat kills microorganisms more efficiently than dry heat.
9. Why does freezing at low temperatures preserve many microorganisms?
10. List the methods used to sterilize materials, and discuss any problems associated with the following methods:
    a. steam under pressure;
    b. boiling;
    c. incineration;
    d. hot air ovens;
    e. radiation;
    f. ethylene oxide.

11. What is pasteurization, and how does it differ from sterilization?
12. List the methods used to control the growth and multiplication of microorganisms in foods, and discuss problems associated with the following methods:
    a. pasteurization;
    b. refrigeration;
    c. desiccation (drying).
13. Explain the difference between high-level, intermediate-level, and low-level chemical control agents.
14. Under which categories (antiseptic, disinfectant, sterilant, chemotherapeutic agent) would the following be placed?
    a. isopropanol and ethanol;
    b. formaldehyde and glutaraldehyde;
    c. sodium lauryl sulfate;
    d. Roccal and Zephiran;
    e. tincture of iodine;
    f. sodium hypochlorite;
    g. arsenic;
    h. mercuric chloride;
    i. silver nitrate;
    j. copper sulfate;
    k. hexachlorophene;
    l. ethylene oxide.
15. Discuss the importance of selective toxicity.
16. Describe a typical quality assurance check.

SELF-TEST

1. Approximately how many nosocomial infections each year result in death in the United States?
    a. 100;
    b. 1000;
    c. 10,000;
    d. 100,000;
    e. one million.

For items 2–6, match the statements below with the best term that follows:
    a. antiseptic;
    b. disinfectant;
    c. sterilant;
    d. chemotherapeutic agent;
    e. pesticide.

2. 20% formaldehyde
3. 4% formaldehyde
4. 0.1% formaldehyde
5. Bordeaux mixture (1% copper sulfate)
6. Penicillin
7. Asepsis is the disease state of living tissue due to the presence of reproducing microorganisms. a. true; b. false.

For items 8–12, list in order, starting with the most effective sterilant, the following chemicals:
    a. 1% hexachlorophene;
    b. 70% ethanol;
    c. 2% glutaraldehyde;
    d. 0.5% ethylene oxide;
    e. Zephiran (1/750 benzalkonium chloride).

8. 1st
9. 2nd
10. 3rd
11. 4th
12. 5th
13. Zephiran (benzalkonium chloride) is used as a disinfectant. What catagory of compound is Zephiran?
    a. heavy metal;
    b. alcohol;
    c. quaternary ammonium detergent;
    d. alkylating agent;
    e. oxidant.
14. Detergents kill microorganisms by:
    a. breaking down cell walls;
    b. solubilizing cell membranes;
    c. inhibiting ribosomes;
    d. blocking the replication of DNA;
    e. inhibiting exoenzymes.
15. A tincture is a:
    a. heavy metal;
    b. alcohol;
    c. quaternary ammonium detergent;
    d. alkylating agent;
    e. chemical, usually a heavy metal that imparts color, dissolved in 70% ethanol.
16. A iodopore is a:
    a. heavy metal;
    b. alcohol;
    c. quaternary ammonium detergent;
    d. alkylating agent;
    e. iodine-containing compound mixed with a detergent.

For items 17–21, list in order, beginning with the most effective, the mixtures that are the best antiseptics:
    a. soap and water;
    b. 1/1000 tincture of Merthiolate;
    c. 1/750 tincture of Zephiran;

d. 70% ethyl alcohol;
e. 1% iodine in 70% ethanol.

17. 1st
18. 2nd
19. 3rd
20. 4th
21. 5th
22. Pasteurization is a special type of sterilization process. a. true; b. false.
23. Freezing is an effective method of sterilization. a. true; b. false
24. Filtration is an effective procedure for removing bacteria and organisms larger than bacteria from media that might be damaged by autoclaving. a. true; b. false.

SUPPLEMENTAL READINGS

Baldry, P. 1977. *The battle against bacteria*. Cambridge, England: Cambridge University Press.

Ball, A. P., Gray, J. A., and Murdoch, J. M. 1978. *Antibacterial drugs today*. Baltimore: University Park Press.

Block, S. S., ed. 1977. *Disinfection, sterilization, and preservation*. 2d ed. Philadelphia: Lea & Febriger.

Castle, M. 1980. *Hospital infection control*. New York: Wiley.

Laskin, A. I. and Lechevalier, H. 1984. *CRC handbook of microbiology*. Vol. 6. *Growth and metabolism*. 2d ed. Boca Raton, FL: CRC Press.

McInnes, B. 1977. *Controlling the spread of infection*. 2d ed. St. Louis: Mosby.

Spaulding, E. H. and Gröschel, D. H. M. 1985. Hospital disinfectants and antiseptics. *Manual of clinical microbiology*. 4th ed. Lennette et al., eds. Washington, DC: American Society for Microbiology.

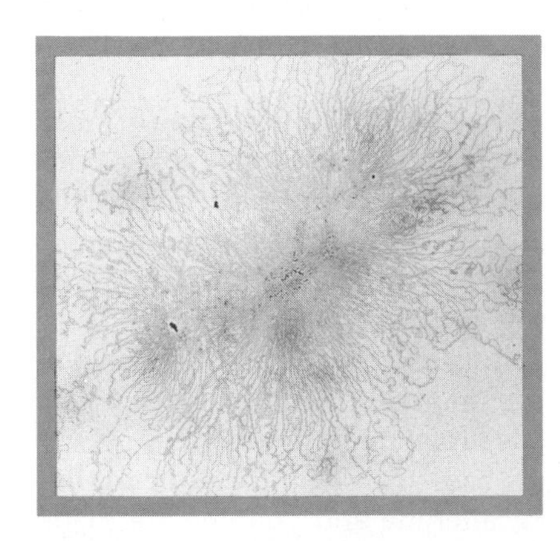

CHAPTER 6

MICROBIAL GENETICS

CHAPTER PREVIEW GENETICS IS USED TO DETECT EPIDEMICS AND TRACE THE SOURCE OF THE PATHOGEN

Each year in the United States there are more than 40,000 reported cases and 500 deaths as a result of salmonellosis. Medical cost due to *Salmonella* infections are estimated to run in excess of $50 million. Since only 1 to 10% of the actual cases of salmonellosis are diagnosed and reported, the actual number of cases exceeds 400,000 each year.

Analyses of the hereditary information of certain strains of *Salmonella* isolated during epidemics of salmonellosis have indicated that many of the bacteria tested are resistant to antimicrobial drugs and that these strains originated from animals that had been treated with antibiotics to increase food production. The resistance to antimicrobials was due to genes that were carried in the bacterial chromosomes or in pieces of circular, extrachromosomal DNA known as **plasmids**.

A number of studies have demonstrated that antimicrobial drugs given to animals to increase meat production select for bacteria that are resistant to those antimicrobials. From 1979 to 1984, the frequency at which antimicrobial-resistant strains of *Salmonella* were isolated increased from 16 to 24%. In those outbreaks of food poisoning where resistant *Salmonella* were isolated, 69% of the cases were due to contaminated foods prepared with meat, eggs, or dairy products. The remaining 31% resulted either from person-to-person contact or from undetermined causes.

Most of the antimicrobial-resistant *Salmonella* contain genes conferring resistance to ampicillin, chloramphenicol, streptomycin, and/or tetracycline. Thus, *Salmonella* carrying antimicrobial resistance genes are a serious threat to human and other animals. The use of antimicrobials by poultry breeders, ranchers, and dairies represents a potential health hazard to the general public plus an even greater danger to persons undergoing antimicrobial treatment for other infections.

An outbreak of salmonellosis demonstrating a link between antimicrobial treatment of animals, antimicrobial-resistant bacteria, and human disease occurred in California in 1985. A strain of *Salmonella newport* that was resistant to a number of antimicrobials was isolated from approximately 1000 persons afflicted with food posioning. The drug-resistant genes that the bacteria carried (chloramphenicol resistance) indicated that all the persons were part of one outbreak. A check on a common food source indicated that all of them had consumed hamburgers prepared from a single batch of ground beef. The ground beef was contaminated with the same strain of *Salmonella newport* responsible for the outbreak of food poisoning and the strain carried chloramphenicol-resistance genes. Similarly, meat packers associated with the meat slaughter house which processed the ground beef carried the same strain of *Salmonella newport*, as did the cows and calves on the ranches where the chloramphenicol was used illegally.

Analysis of genes from *Salmonella* isolated from supposedly sporadic or unrelated outbreaks indicates that many cases are related and part of an epidemic. Thus, a study of the types of genes that bacteria carry can be extremely useful in detecting an epidemic and its cause.

Because of the growing importance of genetics in medical microbiology, it is necessary to know and understand how hereditary information is processed by microorganisms and transferred to other cells. In this chapter you will learn how the hereditary information is expressed, duplicated, and transferred to the same or to other organisms. In addition, you will learn about special hereditary information that is carried by the genes for drug resistance.

GENETICS IS THE STUDY OF HEREDITY. IT IS concerned with how the information contained in nucleic acids is expressed; how this type of molecule is duplicated and transmitted to progeny; and how these processes account for the characteristics of microorganisms and noncellular biological entities (e.g., viruses and viroids). In this chapter you will be introduced to the basic concepts of microbial genetics, including some of the ways in which the genetic traits of microorganisms are expressed and passed from one organism to another.

NUCLEIC ACIDS ARE THE HEREDITARY MATERIAL OF ALL ORGANISMS

Nucleic acids are large organic molecules that are found in all cells. Their name is derived from the fact that they are acidic molecules found predominantly around the nucleus of the cell.

We are aware that certain characteristics are passed on from parents to offspring. The information for creating these characteristics is contained in a type of nucleic acid called **deoxyribonucleic acid**, abbreviated **DNA**. Within the cell, it directs the production of proteins, which are responsible for all cellular characteristics. These proteins are made with the intervention of another type of nucleic acid, called **ribonucleic acid** or **RNA**.

Nucleic acids are constructed from a string of small molecules called **nucleotides**. Nucleotides consist of a five-carbon sugar (a pentose), one or more phosphate groups, and bases with nitrogen-containing rings (nitrogenous bases) arranged as illustrated in figure 6.1. The pentose sugar and the phosphate groups make up the backbone of the nucleic acid and the nitrogenous bases are the side branches. A simple nucleic acid (an RNA molecule) is illustrate in figure 6.2.

The nitrogenous bases can be **purines** or **pyrimidines**. The purines, adenine and guanine are made up of two nitrogen-containing rings, while the pyrimidines, cytosine, thymine, and uracil contain a single nitrogen-containing ring. The nitrogenous bases are usually referred to by the first letter of their name. For example, adenine is simply referred to as A, cytosine as C, guanine as G, uracil as U, and thymine as T. The bases A, C, G, and T are normally found in DNA

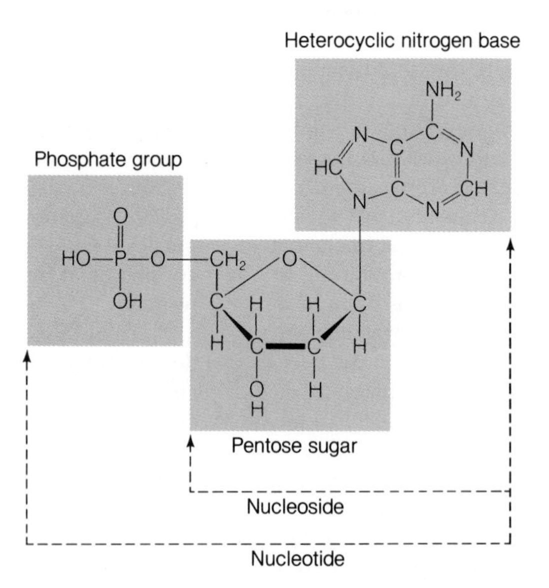

Figure 6.1 Components of nucleotides. Nucleotides are the building units of nucleic acids. Each is composed of a pentose, a nitrogen base, and a phosphate group. A nitrogen base bonded to the pentose forms a nucleoside. A phosphate group bonded to a nucleoside forms a nucleotide.

molecules, while T is replaced by U in RNA. The 5-carbon sugar in DNA is deoxyribose and that in RNA is ribose.

The backbone of a nucleic acid such as RNA consists of a long string of five-carbon sugars joined together by phosphate groups (fig. 6.2). The nitrogenous bases are joined to the sugars by chemical (covalent) bonds. RNA molecules are normally single-stranded molecules and can be of three types: messenger RNA (**mRNA**), transfer RNA (**tRNA**), or ribosomal RNA (**rRNA**). Each of these three types of RNA molecules are important in the expression of the genetic information and are discussed in other sections of this chapter.

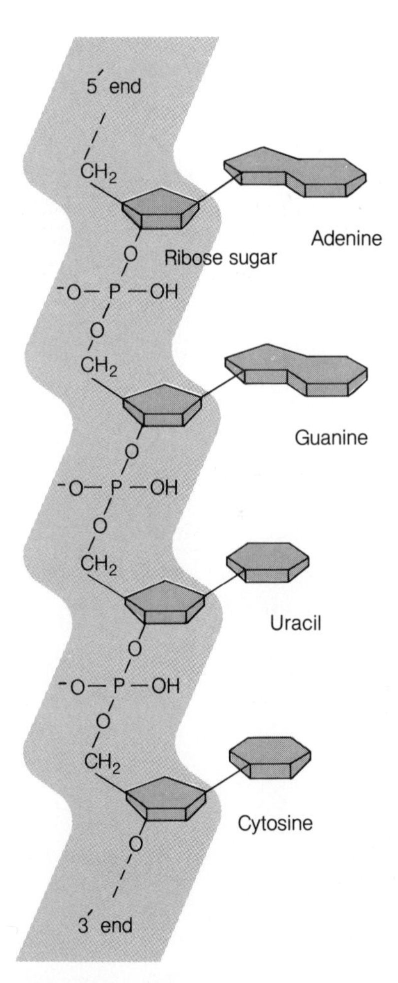

Figure 6.2 Primary structure of a ribonucleic acid (RNA) molecule

DNA molecules are double-stranded, with each strand wrapped around the other in a helical fashion forming what is called a **double helix** (fig. 6.3). The double helix is maintained by hydrogen bonds between the nitrogenous bases on both strands. The hydrogen bonding is quite specific since A binds only with T (or with U in RNA) and G binds only with C. If the nucleotide sequence of one strand of the DNA molecule is known, it is possible to construct the other strand because the bases that it contains are complementary to those in the other strand. For example, if one strand of the DNA molecule is ATCGATCG, then the complementary strand will be TAGCTAGC. The double stranded nature of the DNA molecule makes it very stable. For example, the helical structure of DNA can survive temperatures as high as 70°C, high salt concentrations and even relatively high concentrations of acids (3-6N) for extended periods of time. Moreover, the more G-C base pairs the stronger the two strands of DNA are held together.

DNA determines the characteristics of an organism and maintains and controls the vital processes of all cells. The genetic information in DNA is normally expressed by the cell in two steps. The first step, called **transcription**, involves the formation of an RNA molecule using DNA as a template. The second step, termed **translation**, consists of the synthesis of a protein using the genetic information in the RNA.

The Central Dogma

The unit of genetic information or hereditary material contained in the DNA molecule is called a **gene**. A gene is a sequence of nucleotides in the DNA molecule that codes for an RNA molecule, and ultimately for the synthesis of a protein. The **central dogma** is the theory stating that genes guide the synthesis of mRNA and that mRNA, in turn, directs the order in which amino acids are assembled to form proteins (fig. 6.4). The central dogma also postulates that a DNA molecule can direct its own replication by giving rise to two indentical DNA molecules. This replication is necessary when a cell is reproducing because each daughter cell must inherit a complete set of all the genes from the parental cell in order to carry out life processes. Certain cancer-causing viruses (retrovi-

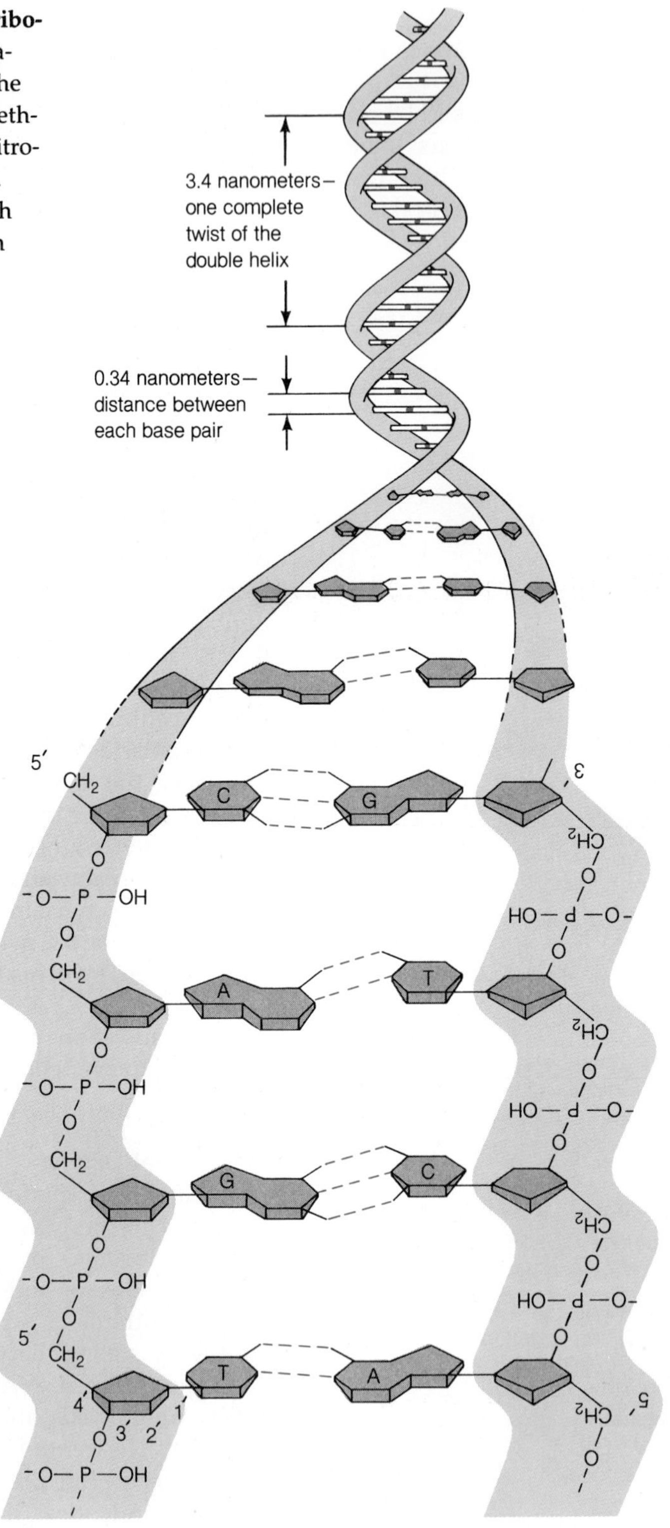

Figure 6.3 **Structure of a deoxyribonucleic acid (DNA) molecule.** Diagram of the DNA double helix. The two helical strands are joined together by hydrogen bonds between nitrogenous bases in opposite strands. Adenine (A) hydrogen bonds with thymine (T) and cytosine (C) with guanine (G).

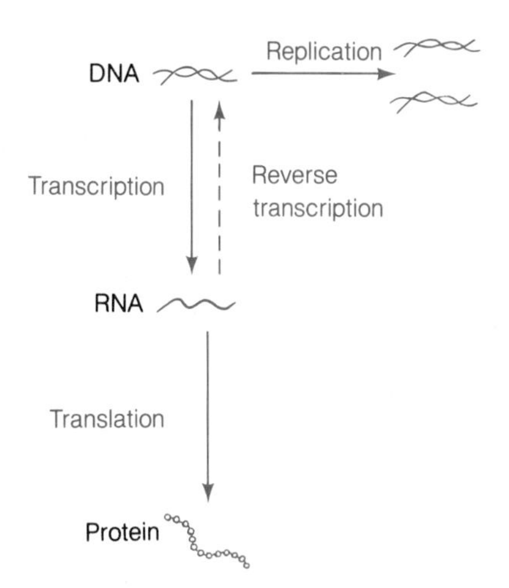

Figure 6.4 Central dogma. The central dogma illustrates how DNA controls its own replication and the synthesis of RNA. The relationship between RNA and the synthesis of protein is also indicated.

ruses) are able to synthesize DNA using RNA as a template. This process is called **reverse transcription**.

DNA REPLICATION IN BACTERIA

The total genetic information in bacteria, also called the **genome**, consists of circular DNA molecules found within the cell. Most of the genome is contained in the single **bacterial chromosome**, although smaller pieces of circular DNA called **plasmids** may also carry a few important genes such as those coding for resistance to antimicrobial drugs. The bacterial chromosome contains most of the genetic information of bacteria and is attached to the plasma membrane. The size of the chromosome varies from species to species. For example, a group of bacteria called mycoplasmas have very small chromosomes with fewer than 1 million nucleotide base pairs, a genome large enough to code for about 1000 proteins. In contrast, the common intestinal bacterium *Escherichia coli* has a chromosome with about 4.5 million nucleotide base pairs that can code for approximately 4500 proteins.

When a bacterial chromosome replicates, both of its strands are duplicated (fig. 6.5). Each strand of DNA functions as a **template** or pattern that specifies the sequence of bases in the newly-formed complementary strand because of the specific hydrogen bonding that occurs between bases (A-T and G-C). Figure 6.6 illustrates how a complementary strand is synthesized from a template. During DNA replication, enzymes called **DNA polymerases** process nucleotides from the cytoplasm that are complementary to the template and fit them into place. Hence, after one round of DNA replication, the resulting two DNA molecules will consist of a parental strand and a new strand. For this reason DNA replication is said to be **semiconservative**; that is, one of the parental strands is kept (conserved) in each new molecule made.

DNA replication in bacteria begins at a specific site on the DNA strand called the **origin**. The replication process proceeds in both directions from the origin; hence, DNA replication ends approximately halfway around the circular chromosome. The time it takes for a bacterial chromosome to replicate depends upon the growth conditions, but a bacterium like *Escherichia coli* growing in nutrient broth may replicate its chromosome in approximately 20 minutes.

RNA SYNTHESIS IN BACTERIA

The synthesis of RNA (or **transcription**), involves the assembly of nucleotides by an enzyme called RNA polymerase that uses a strand of DNA as a template. The process of transcription begins when the RNA polymerase binds to the DNA at a **promoter site** near the gene to be transcribed. After attaching to the promoter site, the RNA polymerase travels along the length of the strand of DNA until it reaches a **termination site** (fig. 6.7). Notice in figure 6.7 that as the RNA is made, it is displaced from the DNA template, even before it is completed because hydrogen bonding between two strands of DNA is more stable than hydrogen bonding between DNA and RNA.

The process of transcription is essentially the same, regardless of which type of RNA is synthesized. After messenger RNA is made, it is used by the cell as a guide to make proteins. Ribosomal RNA, after its synthesis, becomes associated with certain proteins to form ribosomes. Transfer RNA is made as a linear molecule, which then becomes folded to assume the characteristic shape of tRNA (fig. 6.8).

Transfer RNAs are small RNA molecules that are involved in translating the information in the mRNA

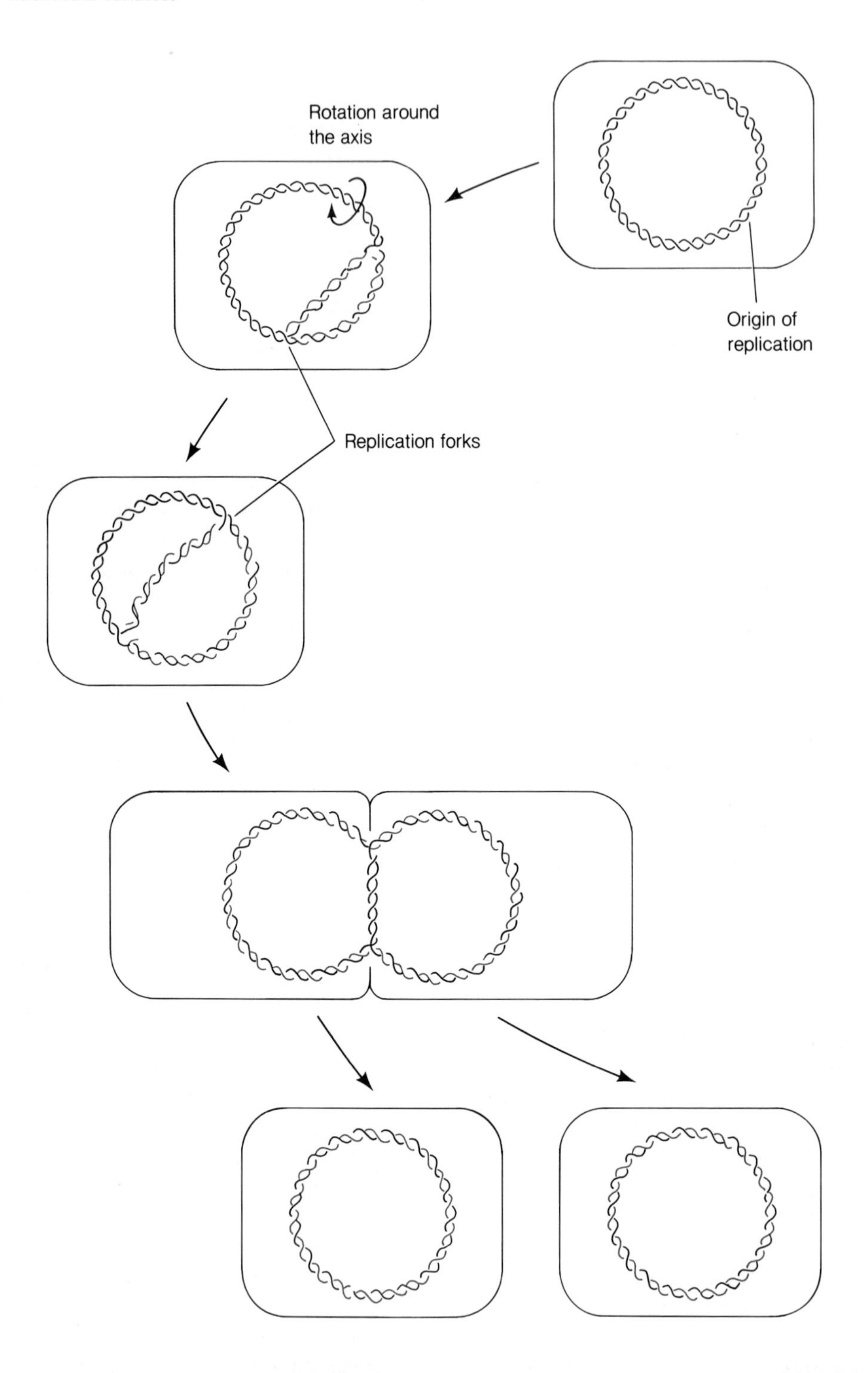

Figure 6.5 DNA replication and its segregation in the prokaryote. The chromosome in bacteria is a circular DNA molecule that is replicated bidirectionally. The replication forks are indicated. The figure illustrates how daughter chromosomes are believed to be segregated from each other in bacteria. The two original strands of the chromosomes are apparently attached to different sites on the plasma membrane. As the membrane grows between the attachment points, the daughter chromosomes are moved to the poles of the cell. When the cell membrane invaginates between the daughter chromosomes, two new cells are formed, containing identical chromosomes.

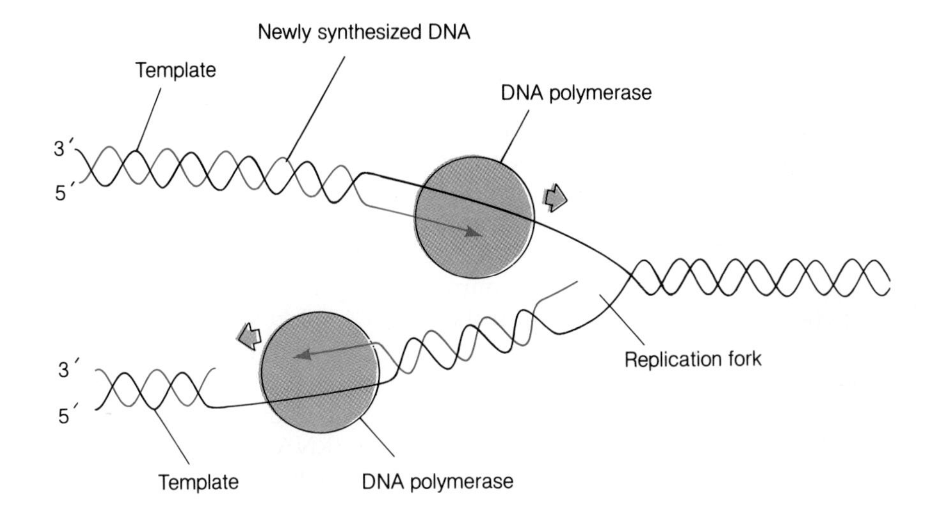

Figure 6.6 The replication fork. A detailed view of the replication fork illustrates how DNA polymerases synthesize new strands. Because the templates are oriented in opposite directions, the DNA polymerases are moving in opposite directions in the replication fork.

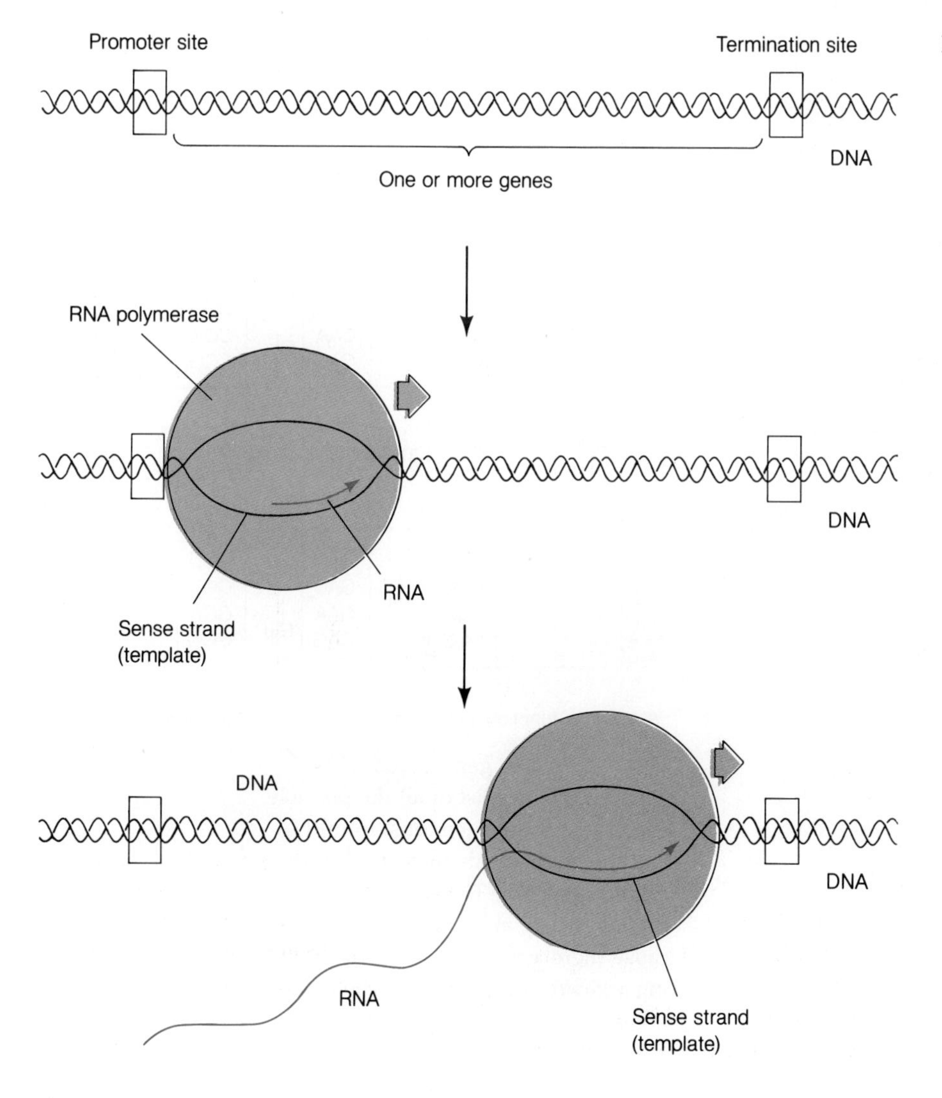

Figure 6.7 Transcription. Transcription begins at promoter sites and ends at termination sites on the DNA. An RNA polymerase transcribes one strand of the DNA, sometimes called the sense strand, into RNA.

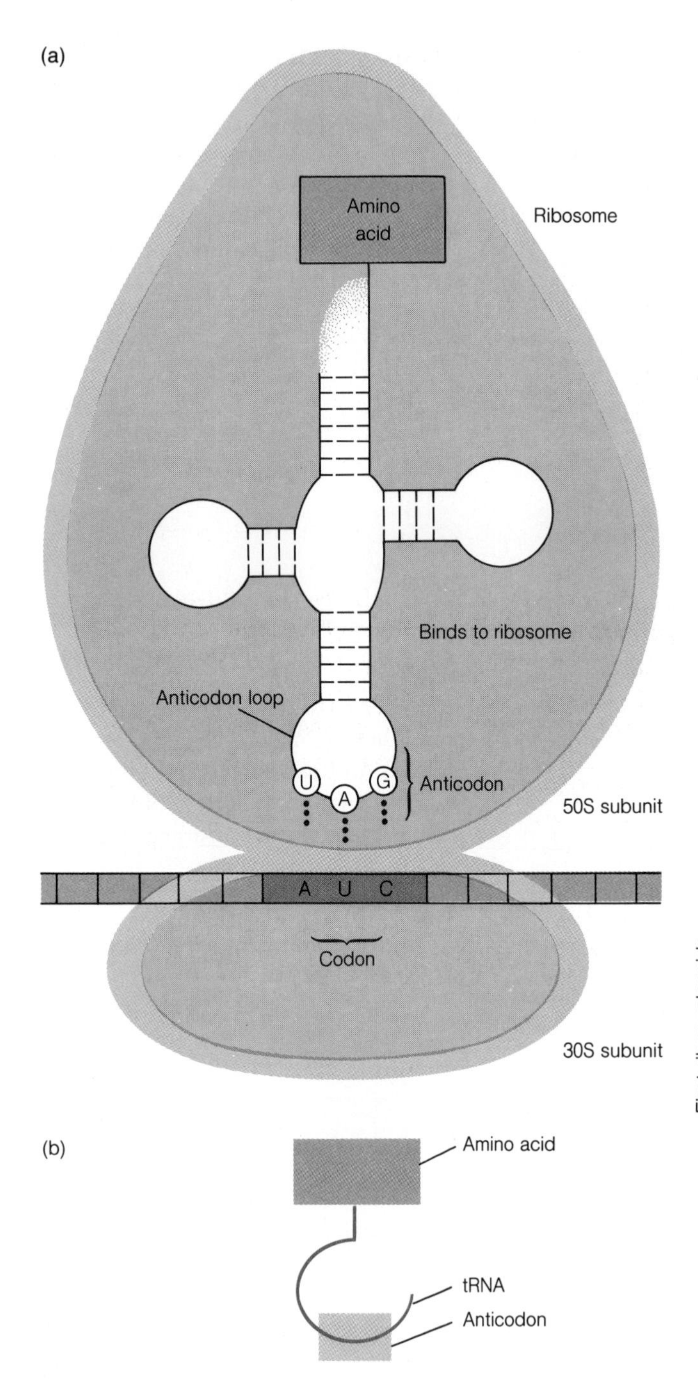

Figure 6.8 The structure of transfer RNA. *(a)* Transfer RNA (tRNA) is processed from various large RNA transcripts. The final tRNAs are all about the same size. *(b)* Simplified notation used to illustrate a tRNA molecule.

into proteins. A typical folded tRNA molecule has a clover leaf shape with three loops. At one end of the molecule is an amino acid binding site for attachment of specific amino acids. Each of the 20 amino acids that form proteins has at least one specific tRNA to which it attaches. Opposite the amino acid binding site in the tRNA is the **anticodon loop** (fig. 6.8). This loop is a region of the tRNA molecule with three nitrogenous bases that are complementary to a set of three nucleotides present in the mRNA molecule called a **codon**. Most codons in the mRNA molecule code for a specific amino acid (fig. 6.9); the few that don't are known as **termination codons**. The codons that code for amino acids are recognized by the tRNA molecule with the complementary anticodon. In the following section we see how the mRNA molecule is translated by specific tRNAs.

Second ribonucleoside

First ribonucleoside	U		C		A		G		Third ribonucleoside
U	UUU	Phe	UCU	Ser	UAU	Tyr	UGU	Cys	U
	UUC		UCC		UAC		UGC		C
	UUA	Leu	UCA		UAA	non	UGA	non	A
	UUG		UCG		UAG	non	UGG	Trp	G
C	CUU	Leu	CCU	Pro	CAU	His	CGU	Arg	U
	CUC		CCC		CAC		CGC		C
	CUA		CCA		CAA	Gln	CGA		A
	CUG		CCG		CAG		CGG		G
A	AUU	Ile	ACU	Thr	AAU	Asn	AGU	Ser	U
	AUC		ACC		AAC		AGC		C
	AUA		ACA		AAA	Lys	AGA	Arg	A
	AUG	Met	ACG		AAG		AGG		G
G	GUU	Val	GCU	Ala	GAU	Asp	GGU	Gly	U
	GUC		GCC		GAC		GGC		C
	GUA		GCA		GAA	Glu	GGA		A
	GUG		GCG		GAG		GGG		G

AUG when functioning as a start codon codes for N-formylmethionine

Figure 6.9 The genetic code. The genetic code is a list of all the possible triplet nucleotides (codons) in mRNA and the amino acids these triplet nucleotides specify. Three of the codons do not specify any amino acid because there are no tRNAs with matching anticodons. These are called nonsense or termination codons.

PROTEIN SYNTHESIS IN BACTERIA

Protein synthesis is carried out in the cytoplasm of the cell in the presence of ribosomes, mRNA, and tRNA molecules with their corresponding amino acid attached (fig. 6.10). In addition to these essential cellular components, an energy source in the form of GTP (similar to ATP) and some enzymes are necessary (fig. 6.11).

The amino acid sequence in proteins is determined by the genetic code (fig. 6.9) in the nucleotide sequence in the mRNA. The genetic code is "read" (translated) by tRNA molecules when the anti-codon on the tRNA forms complementary base pairs with the codon of the mRNA. Ribosomes promote the codon-anticodon interaction and catalyze the formation of bonds between adjacent amino acids (called peptide bonds). The synthesis of proteins is a continuous process but to study and understand the process scientists have divided it in three stages: initiation, elongation, and termination.

Initiation of Translation

Translation in bacteria begins when the 30S subunit of the ribosome binds to a **start codon** (AUG) in the messenger RNA (fig. 6-11a). The start codon is recognized by a tRNA (with a UAC anticodon) carrying an amino acid called formylmethionine. The joining together of the 30S ribosomal subunit, the mRNA, and the tRNA requires energy from the hydrolysis of GTP. This energy also promotes the binding of the 50S subunit of the ribosome to form the **initiation complex**, consisting of the complete ribosome (70S), the mRNA, and the tRNA with the formylmethionine attached.

Elongation of Translation

Once the initiation complex is formed, subsequent tRNAs with their corresponding amino acid attach to the mRNA on the 70S ribosome (fig. 6.11b). At this point, the tRNA with the formylmethionine (fMet) is attached to a site in the ribosome called the **P** (or peptidyl) site. The codon following the start codon determines which of the tRNAs with the appropriate amino acid (aminoacyl-tRNA) will attach next. The incoming amino acyl-tRNA will attach to the ribosome and the mRNA at another site called the **A** (aminoacyl) site. Energy in the form of GTP is also required for the attachment of the aminoacyl-tRNA to the A site. At this time, there is a fMet-tRNA on the P site and the new amino acyl-tRNA on the A site. An enzyme called **peptidyl transferase** located on the 50S subunit catalyzes the formation of a chemical bond between the two adjacent amino acids so that now, both amino acids are attached to the tRNA (**peptidyl-tRNA**) on the A site and the tRNA on the P site is free. Another enzyme called **translocase** provides the stimulus, along with GTP, for ejecting the free tRNA from the P site and moving the peptidyl-tRNA to the P site, freeing the A site for the incoming amino acyl-tRNA (fig. 6.11b). Since the peptidyl-tRNA is still attached to the codon next to the start codon, the mRNA is pulled past the ribosome. Amino acids are joined together into proteins by repeating these elongation steps.

Termination of Translation

Protein synthesis ends (fig. 6.11c) when the ribosome come to a **termination codon**: UAA, UAG, or UGA. There are no normal tRNAs with anticodons that recognize these 3 codons. The termination codons are commonly called **nonsense codons** because they provide no sense of information for making a protein.

Proteins called **release factors** bind to the ribosome when it is stalled at termination codon and stimulate the peptidyl transferase to cut the newly formed protein from the tRNA. When this happens, the ribosome comes apart and the mRNA falls off the ribosome. The proteins are synthesized as a linear arrangement of amino acids and assume their functional three-dimensional shape by folding after they are released from the ribosome.

CHANGES IN THE DNA MOLECULE CAN CAUSE MUTATIONS

The Nature of Mutations

The sequence of nucleotides in DNA determines the amino acid sequence of the multitude of proteins required for the life processes of all cells. The genetic information must be accurately replicated and passed on intact to daughter cells if a species is to survive. If the DNA is not faithfully replicated, or if only part of it is replicated, one or both of the daughter cells may not survive.

An inheritable change in an organism's genetic information is known as a **mutation**. A mutation can occur in a number of ways. For example, a mutation may occur when: (a) one base pair is exchanged for another in the DNA molecule; (b) one or more bases are inserted in the DNA molecule; (c) one or more bases are deleted from the DNA molecule (fig. 6.12).

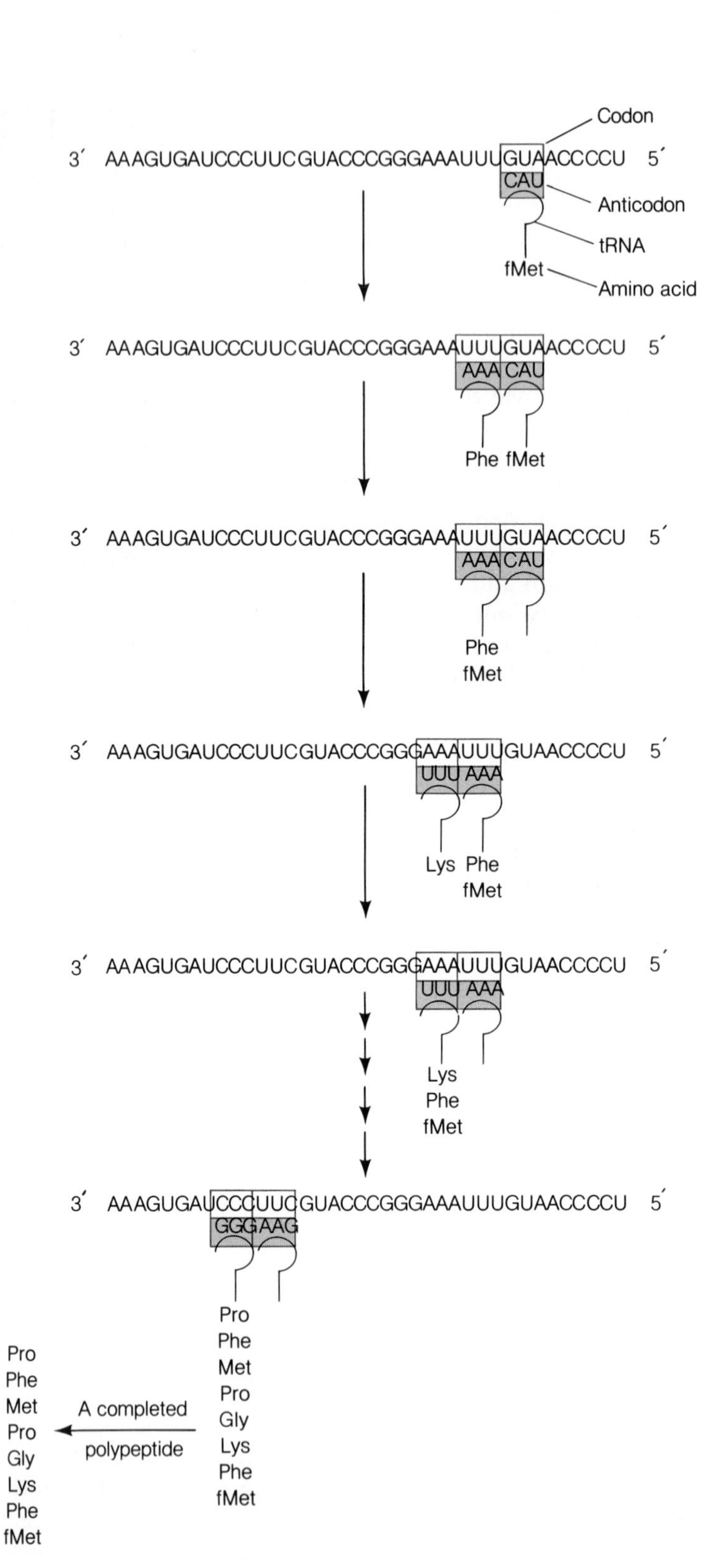

Figure 6.10 Relationship between messenger RNA and protein. The nucleotide sequence in mRNA determines the amino acid sequence in a protein. Protein synthesis in bacteria begins near the 5′ end of a mRNA with the first start codon 5′AUG 3′, and ends at the first nonsense (termination) codon: 5′UAA 3′, 5′UAG 3′, or 5′UGA 3′. The anticodon of the appropriate tRNA hydrogen bonds to triplet codons starting with 5′AUG 3′. The ribosome promotes the hydrogen bonding between the codons and anticodons and catalyzes the formation of peptide bonds between adjacent amino acids attached to the tRNAs.

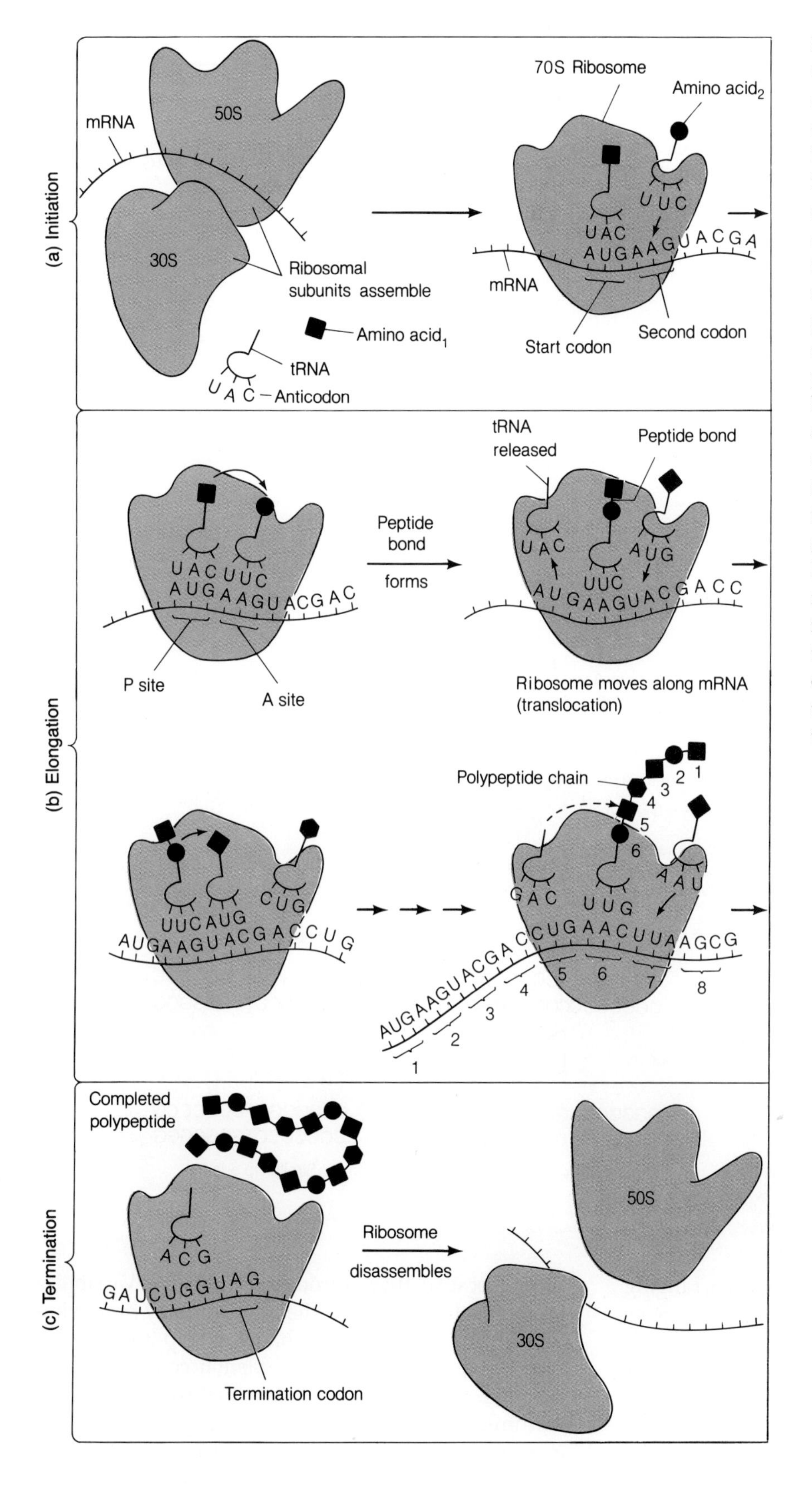

Figure 6.11 Protein synthesis. Protein synthesis, or translation, is divided into three stages: initiation, elongation, and termination. *(a)* During the initiation of translation, the 30S and 50S subunits of the 70S ribosome bind to the mRNA at a start codon. In addition, the first aminoacyl-tRNA (*N*-formylmethionyl-tRNA) is attached to the start codon and to the peptidyl-tRNA site (P-site) on the 50S subunit. *(b)* During the elongation process, aminoacyl-tRNAs bind one by one to the mRNA and the aminoacyl-tRNA site (A-site) on the 50S subunit. The ribosome catalyzes the formation of a peptide bond between the amino acid in the A-site and the growing peptide in the P-site. *(c)* Eventually, nonsense codon are reached and the termination process is initiated. A release factor binds to the ribosome and causes the release of the growing peptide from the tRNA and a dissociation factor causes the dissociation of the 70S ribosome.

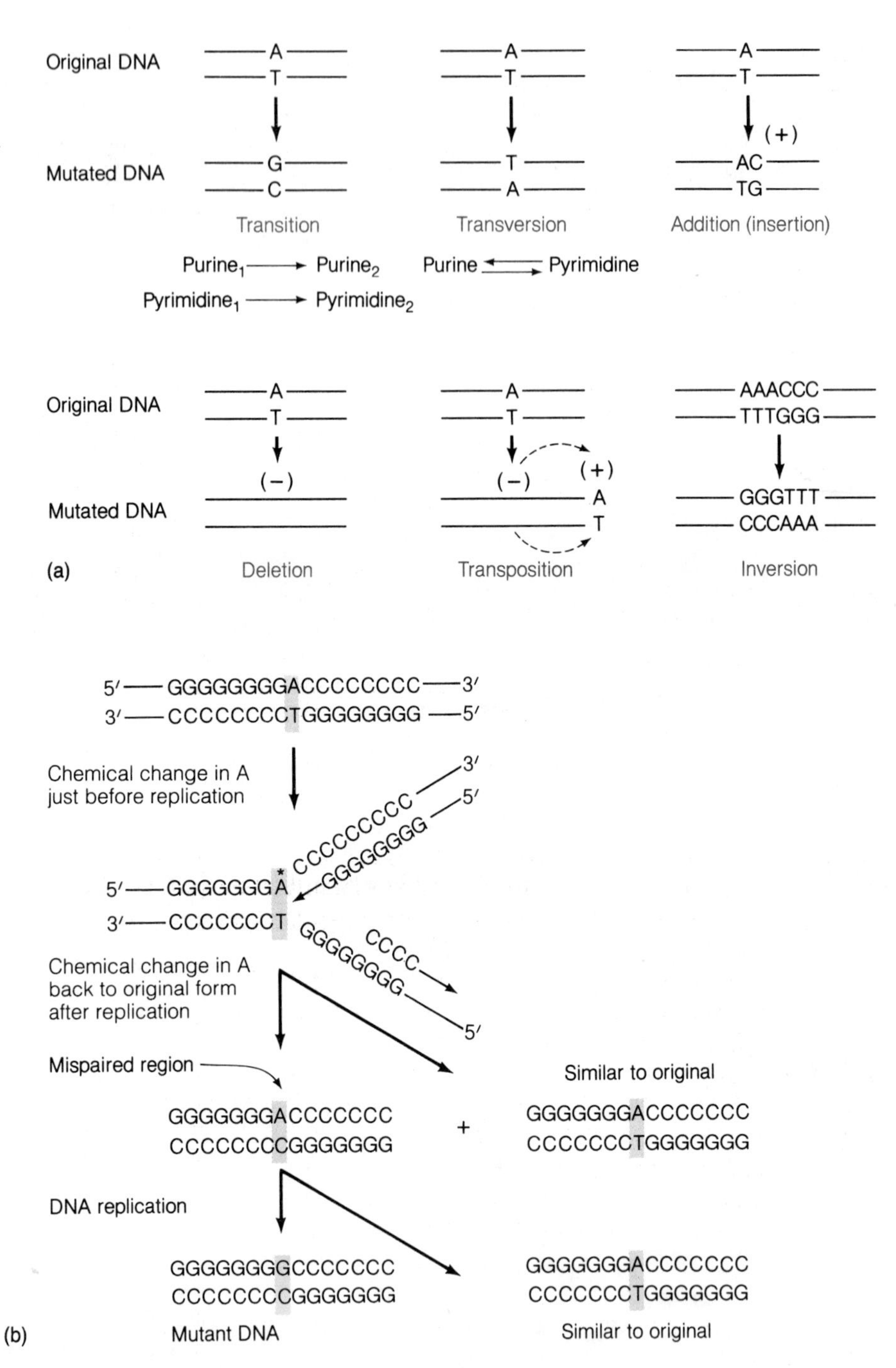

Figure 6.12 Types of mutations and the consequences. *(a)* Common types of mutations. *(b)* Consequence of abnormal base pairing during DNA synthesis. Chemical changes in the structure of bases during DNA synthesis cause the wrong nucleotides to be incorporated into the new strands of DNA. When the DNA is replicated again, one of the daughter DNAs contains a mutation.

A DNA molecule can also be altered by rearranging sections of the molecule or by exchanging DNA regions with other DNA molecules. This latter process is called **recombination** and is discussed later in this section.

Although most mutations are harmful because they drastically alter the function of essential proteins, some mutations are beneficial to the cell and to the population because they introduce variability. Variability promotes the survival of the population by producing a variety of characteristics that may help it to survive a broad spectrum of environmental conditions. For example, the mutation of a gene that codes for a protein involved in cell division may make this protein more tolerant to heat than the normal protein. Bac-

TABLE 6.1
MUTAGENS AND THEIR MODE OF ACTION

Mutagen	Mode of action
Heat	Causes chemical changes in the structure of bases.
Chemicals	Interact directly with the DNA and cause disruptive changes.
Ultraviolet light	Chemically alters the DNA and stimulates excessive repair.
X-rays, Gamma rays (electromagnetic waves)	Chemically alter the DNA by ionization and cause disruptive changes and excessive repair.
Radioactive atoms beta particles alpha particles (subatomic particles)	Chemically alter the DNA by ionization and cause disruptive changes and excessive repair.

terial cells possessing this mutant protein will be able to survive and perpetuate the species if the normal environment in which the population lives becomes hotter.

Mutations are caused by many chemical and physical agents, which are known as **mutagens** (table 6.1). Ultraviolet light is a powerful mutagen because it is absorbed by pyrimidines (C and T). The absorption of UV by adjacent thymines in the same strand of DNA will cause them to react with each other (fig. 6.13). The altered thymines, known as thymine dimers, distort the DNA, causing errors in replication and transcription. If not corrected, this can lead to cell death. Cells, however, have a number of repair systems that can correct the distortion. One such repair system is illustrated in figure 6.13.

Chemicals are frequently used to induce mutations in the laboratory. For example, the mutagen nitrous acid (HNO_2) alters the chemical structure of A, C, and G so that they change their base pairing. This introduces mutations during DNA replication.

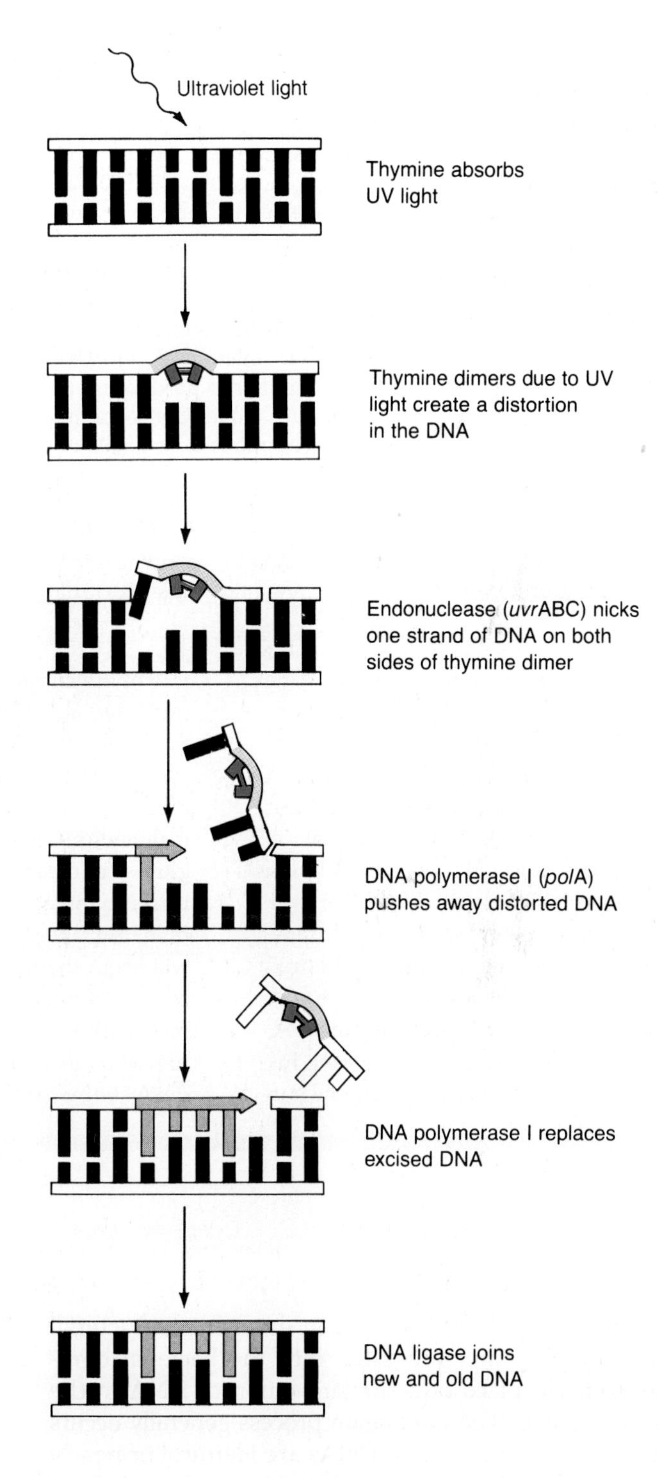

Figure 6.13 **The formation of thymine dimers.** When two thymine nucleotides lie next to each other in the same DNA strand, they can react together if they are destabilized by the absorption of ultraviolet light with a wavelength of 280 nm. The thymine form thymine dimers. Thymine dimers distort the DNA molecule and prevent DNA replication and transcription. Most bacteria have a repair system that removes thymine dimers and replaces them with the appropriate nucleotides. Some of the steps involved in the repair of DNA distorted by thymine dimers are outlined.

FOCUS THE AMES TEST IS USED TO DETECT CANCER-CAUSING CHEMICALS

Each year in the United States, more than 700,000 people are diagnosed as having cancer and approximately 400,000 die of the disease. Available evidence indicates that 90% of the cases are due to chemicals and radiation that humans could avoid.

Thousands of chemical carcinogens (cancer-causing substances) are found in our water, food, and air. These chemicals often work together to produce more cancers than each would produce alone. Consequently, it is essential to detect which chemicals are capable of causing mutations that often lead to cancer and to limit the release of these chemicals into our environment.

The **Ames** test is a relatively inexpensive test for determining whether or not a chemical is mutagenic and therefore potentially carcinogenic. It uses bacteria, instead of expensive laboratory animals to test the potential cancer-causing agents. The Ames test (fig. 6.14) indicates whether or not a chemical can induce a mutation in *Salmonella typhimurium* that would transform an amino acid-(histidine) requiring strain into a strain that does not require the growth factor.

Because many chemicals are not carcinogenic until they have been altered by animal cells, chemicals are first treated by extracts from mammalian liver cells before they are tested on bacteria. If a chemical is found to be mutagenic for *Salmonella typhimurium*, there is a good chance that it is carcinogenic for humans. The chemical can then be tested more thoroughly on laboratory animals.

Figure 6.14 **Ames test.** The Ames test is a simple method to determine whether or not a chemical is a mutagen. Since mutagens are potential carcinogens, the Ames test indicates which chemicals may cause cancer. To test a chemical, a sample is mixed with mouse liver homogenate and various concentrations of the mixture are poured into petri dishes containing a culture medium lacking histidine. A mutant of *Salmonella* requiring histidine is then spread over the surface of the culture medium and incubated at 35°C for 24–48 hours. Since the bacteria are unable to synthesize the amino acid histidine, they will not grow on a minimal medium that contains only glucose and salts. *Salmonella* that undergo mutations that allow them to make their own histidine will begin to grow and form colonies: The potency of a mutagen is indicated by the number of colonies: the more colonies forming on the histidine-free medium, the more efficient is the mutagen. ▶ Some chemicals must be incubated with homogenized liver before they are tested, because it is not the original chemical that is mutagenic and carcinogenic but rather a breakdown product. Since bacteria lack liver enzymes that alter compounds, the effect of these compounds in animals would be missed if the compounds were not mixed with liver extract.

Recombination

Sometimes pieces of DNA can enter into a bacterial cell and become integrated into the bacterial chromosome in place of a similar section of DNA in the chromosome. The integration process generally occurs at points where the two DNAs are identical or nearly so. Alteration of DNA by exchanging regions is known as **recombination** (fig. 6.15). Bacteria frequently acquire pieces of DNA from the environment or from other bacteria and incorporate them into their chromosomes by recombination. Recombination is a desirable process because it introduces new genetic information into a population of bacteria. This may allow the population to survive in adverse or changing environmental conditions. The processes by which bacteria take up foreign DNA and incorporate it into their genomes is discussed in the next section.

Transposons

Transposons are small pieces of DNA (containing 2000 to 20,000 base pairs) found in chromosomes and plasmids. They are able to direct the synthesis of copies of themselves and become incorporated in various places in the bacterial chromosome (fig. 6.16). Because

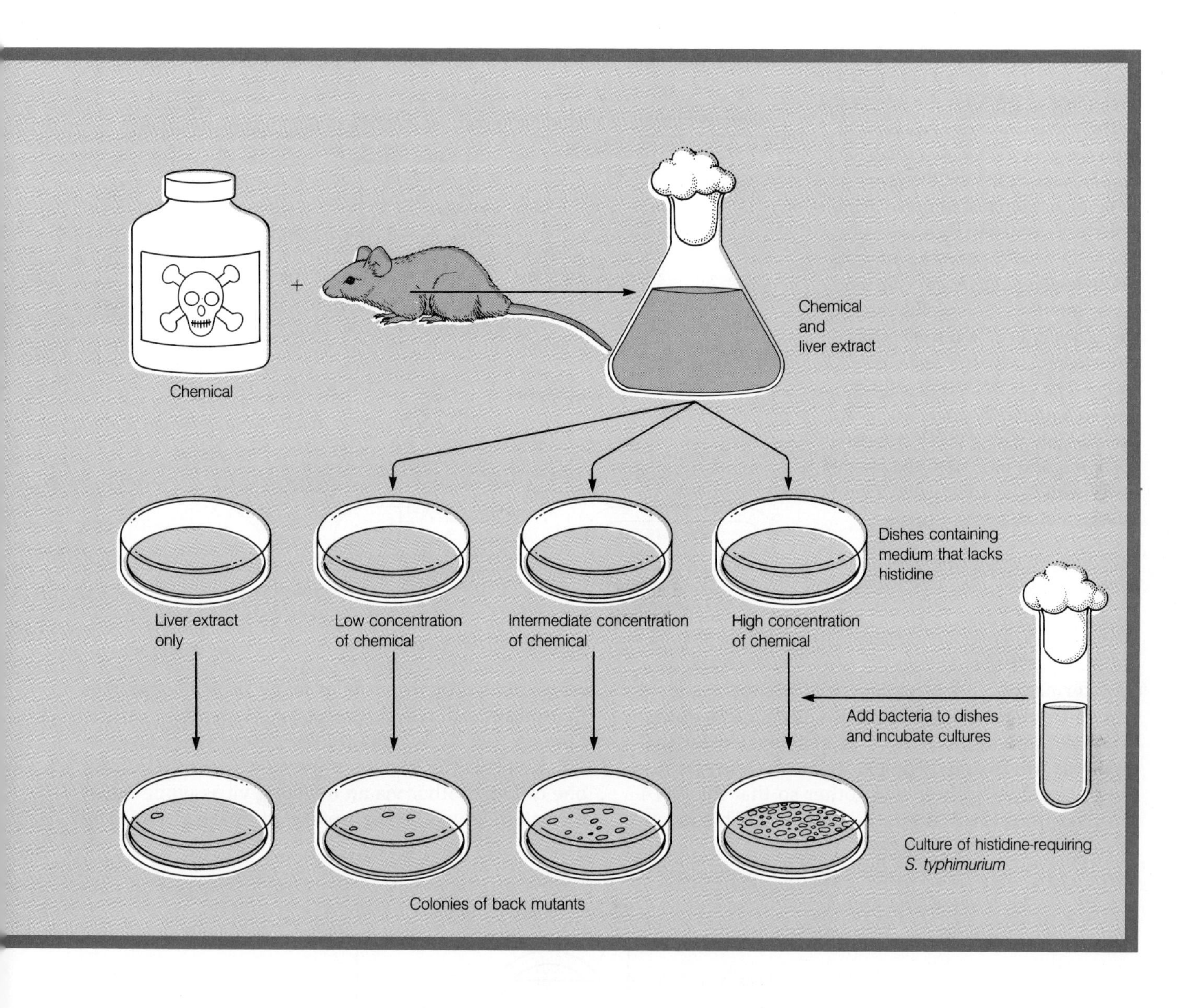

of this ability to insert copies of themselves in a chromosome or change their location within a chromosome, transposons are sometimes called **jumping genes**.

Transposons often contain genes that confer antibiotic resistance to the bacteria. For instance, the transposon Tn4 carries the genes that make some bacteria resistant to ampicillin, streptomycin, and sulfonamides. Other transposons may contain the genes for resistance to kanamycin, trimethoprim, chloramphenicol, tetracycline, and erythromycin. Because transposons can become inserted into various regions of the bacterial chromosome, they induce mutations when they "jump" from one part of the chromosome to another.

TRANSFER OF GENETIC INFORMATION IN BACTERIA

Most bacteria have evolved mechanisms for the uptake of foreign DNA to create more genetic variability than could be possible from spontaneous mutations alone. Three mechanisms, **conjugation**, **transduction**, and

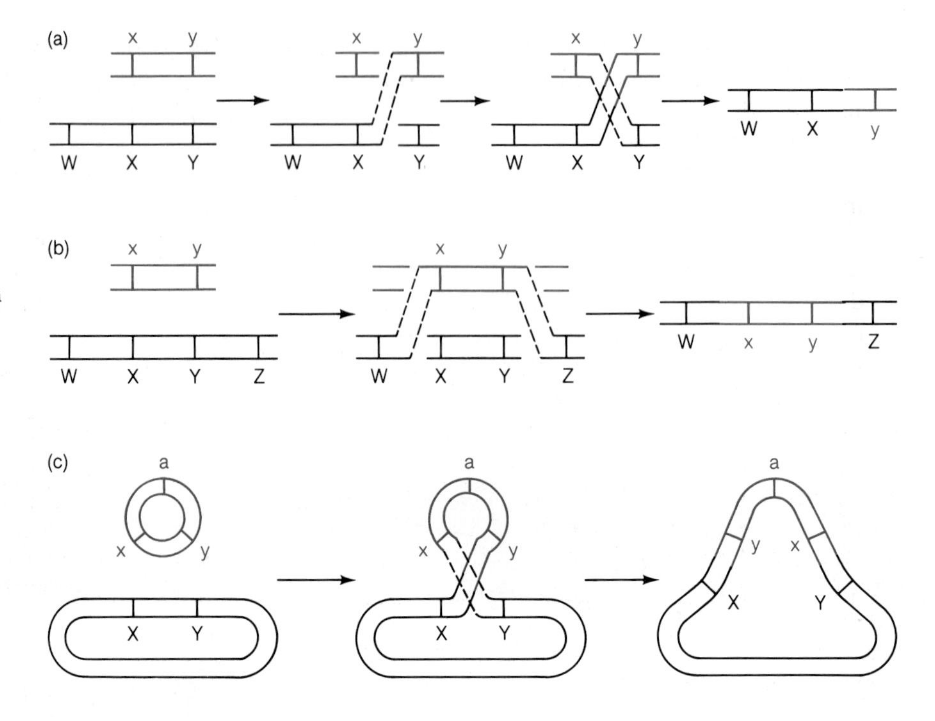

Figure 6.15 The process of recombination. *(a)* Only one cut is required in each piece of DNA for the integration of the y gene into the chromosome with the genes WXY. A recombinant chromosome, carrying the genes WXy, is the result of this recombination. *(b)* Two cuts in each piece of DNA are necessary for the integration of the region of DNA carrying the xy genes into the chromosome carrying the genes WXYZ. A recombinant chromosome with the genes WxyZ forms. The cut DNA is usually degraded by the cell's enzymes. *(c)* A plasmid integrating into a chromosome requires one cut in the plasmid and one in the chromosome. The resulting molecule is still circular.

transformation, allow a beneficial characteristic to spread throughout a population within a few hours. Conjugation is a mechanism of genetic transfer that requires donor and recipient cells to form a cytoplasmic bridge between each other so that the DNA can pass from the donor to the recipient. This mechanism allows the passage of many genes, sometimes the entire bacterial chromosome, depending on the time the two cells remain joined together. Transduction involves the transfer of genetic information from one cell to another via an infecting virus while transformation involves the uptake of "naked" DNA by

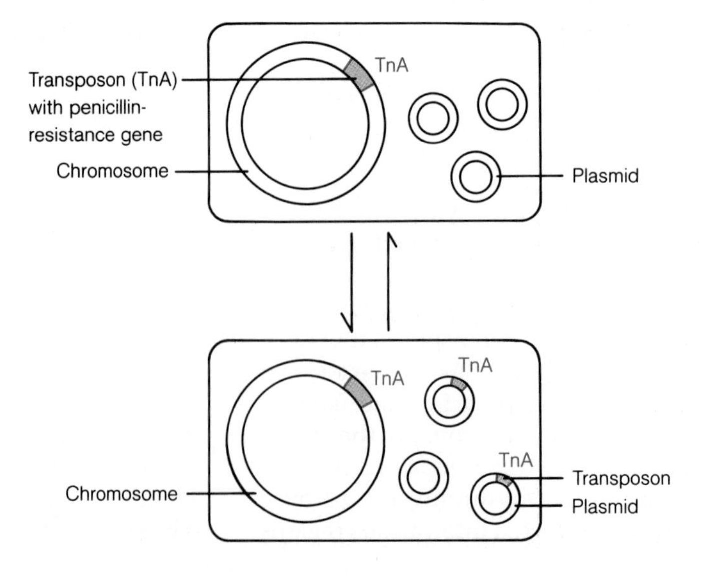

Figure 6.16 Movement of a transposon. Transposons can make copies of themselves within chromosomes or plasmids. Thus, genes on the chromosome that might make an organism resistant to a drug can be transferred to a plasmid, and this plasmid, in turn, can be transferred to another bacterium, making the recipient resistant to the drug.

recipient cells. As a general rule, only a few genes can be transferred by bacteria through transduction or transformation.

Conjugation

Numerous bacteria have, in addition to their chromosome, small, circular DNA molecules called plasmids. Plasmids usually have between 20 and 100 genes and are able to replicate themselves. Plasmids frequently carry genes that confer antimicrobial resistance to microorganisms. In addition, certain groups of plasmids are required for bacterial conjugation.

Conjugative plasmids have genes that give the cell the ability to transfer genes from one cell to another. One group of such plasmids are the **fertility factors** (F plasmids) of certain gram-negative bacteria. These fertility factors code for a protein called **pilin** that self-assembles to form sex pili. The hollow sex pili bind bacteria and draw them together so that the outer membranes fuse and a passageway is created for the transfer of the DNA (fig. 6.17). One strand of the plasmid DNA is able to pass from a donor bacterium (i.e., the one with the F plasmid, F^+) to a recipient bacterium (i.e., the one lacking the F plasmid, F^-) through the cytoplasmic bridge that is created between the bacteria.

One model for the transfer of DNA suggests that the plasmid DNA is cut at a specific site. One of the DNA strands enters the recipient, which makes a complementary strand. The end result is a double-stranded molecule. Even though the plasmid in the donor cell is transferring one of its strands to the recipient, it remains a double-stranded molecule because the donated strand is being replaced as fast as it leaves the cell (fig. 6.18).

When a recipient bacterium obtains an F plasmid, it is converted into a donor and can make sex pili. Not all cells in a population become donors because some donors lose their plasmids during cell division.

The plasmid usually transmits genes that are not found on the chromosome. Genes that make the bacterium resistant to a variety of antimicrobials are common and they protect bacteria from antimicrobials produced by other microorganisms in their environment. These genes are found in **resistance plasmids** or **R plasmids**. Some R plasmids also carry genes that allow their transfer by conjugation, while others are passed on to bacteria only when an F plasmid is present. For example, the bacterium that causes gonorrhea, *Neisseria gonorrhoeae*, has an F plasmid that carries a gene for tetracycline resistance as well as other small R plasmids that confer resistance to penicillin, ampicillin, and trimethoprim. The small R plasmids can only be transferred to other cells by conjugation when the large F plasmid is present. The large F plasmid, however, can transfer itself to recipient bacteria, along with its gene for antimicrobial resistance.

In addition to resistance to antimicrobials, other functions that plasmids encode include the following: (a) resistance to heavy metals, (b) virulence factors (which impart the ability to cause disease), (c) cell adhesion and attachment to epithelial tissues, and (d) resistance to bacteriophage (bacterial virus) invasion.

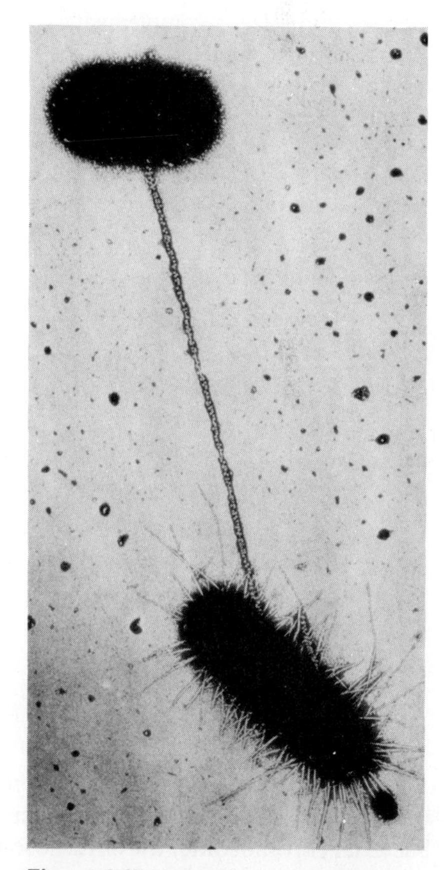

Figure 6.17 Bacteria conjugating. A male bacterium (bottom) has produced a long (about 4μm) sex pilus that has attached to a female bacterium (top). Other types of pili are visible on the male bacterium, but the female bacterium has no pili on its surface.

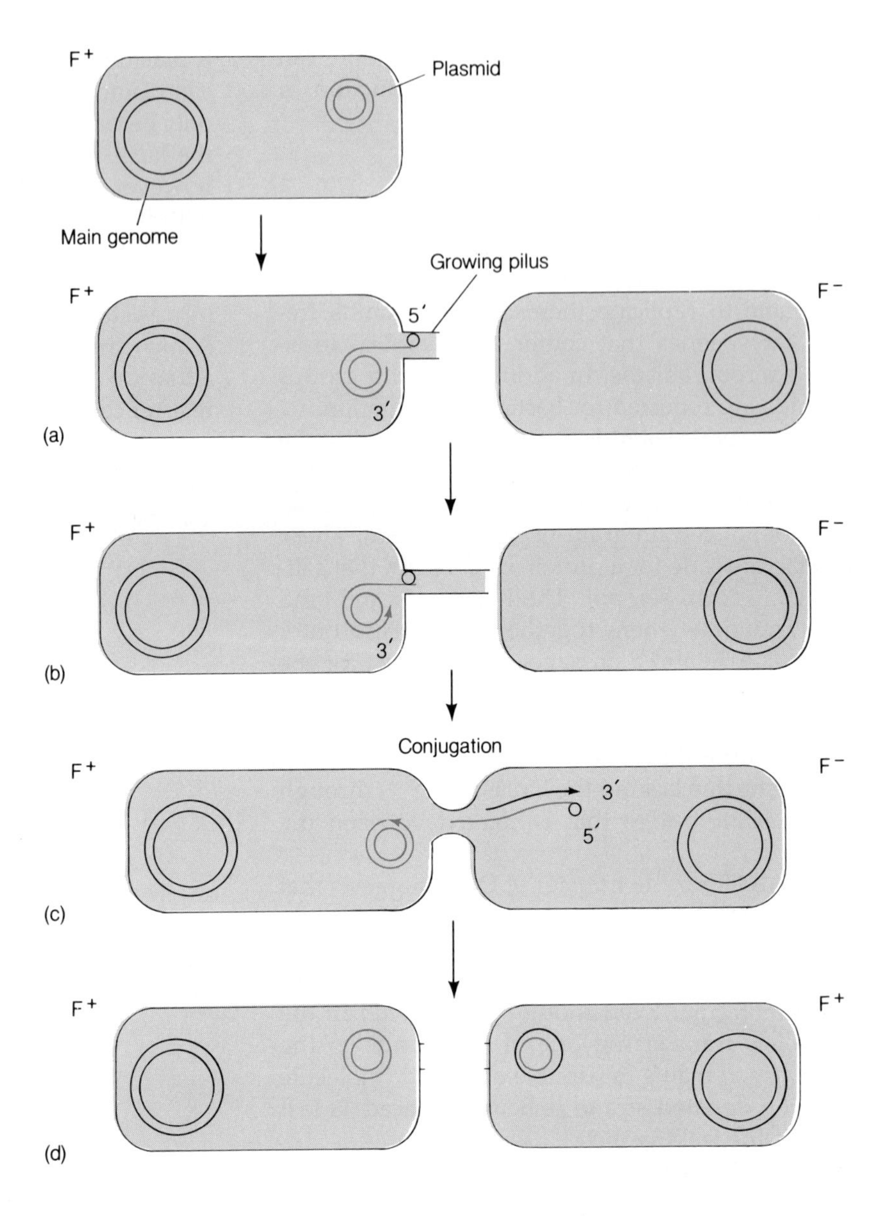

Figure 6.18 Transfer of DNA during conjugation in bacteria. *(a)* One strand of the plasmid in an F^+ bacterium is cut and the 5′ end of the strand attaches to the inside of a growing pilus. *(b)* As the pilus grows the DNA strand unwinds from the plasmid. The 3′ end of the strand is elongated by a DNA polymerase so that the plasmid continues to be double-stranded. *(c)* Eventually, the pilus makes contact with a F bacterium that absorbs the pilus. When the pilus is absorbed by the recipient bacterium, the unwound strand of DNA enters the recipient, in which a DNA polymerase synthesizes a complementary strand. *(d)* If a complete plasmid enters the recipient it forms a circular plasmid and the cell becomes a donor.

Some plasmids are able to integrate themselves into the chromosome. When a plasmid is part of the bacterial chromosome, the chromosomal genes can be transferred to recipient cells by conjugation in much the same way plasmids are transferred. Since the chromosome (approximately 4500 genes) is so much larger than the plasmid (20–100 genes), the entire chromosome is normally not transferred (although such a transfer is possible). The fragment of chromosomal DNA that enters the recipient cell does not form a plasmid; instead it recombines with the recipient's chromosome. This way, the recipient may acquire a number of different traits that are present in the donor's DNA.

Transduction

The transfer of bacterial genes from a donor to a recipient by a bacteriophage is called transduction. Transducing bacteriophages are categorized as either **generalized** or **specialized**. Generalized transducing phages reproduce by injecting their DNA into a bacterial host forming viral progeny in the cytoplasm and eventually rupturing the host cell (fig. 6.19). During the reproduction of these viruses, the bacterial chro-

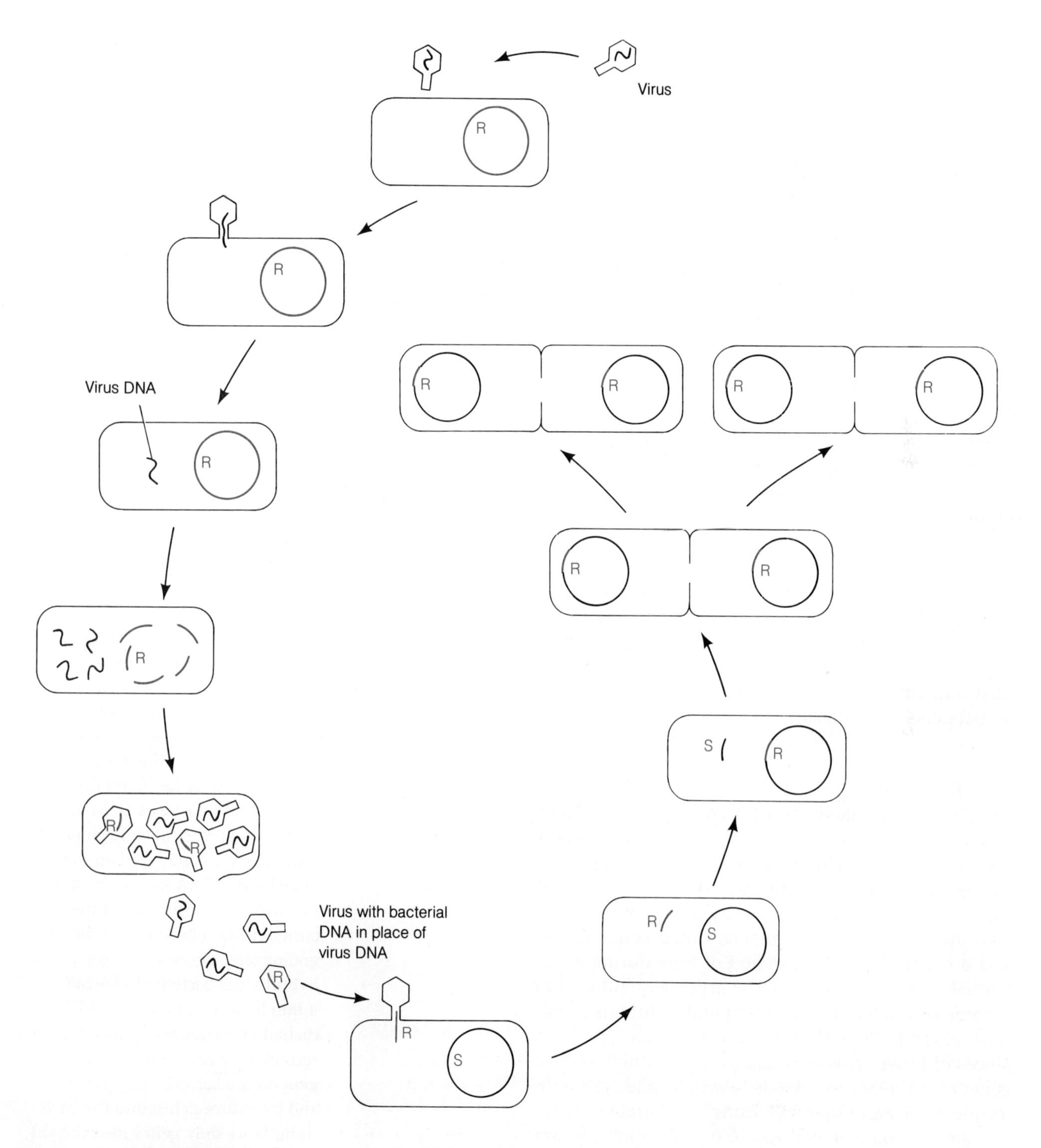

Figure 6.19 Generalized transduction. Generalized transducing bacterial viruses invade bacteria by injecting their hereditary material into the bacteria. The virus DNA directs the synthesis of new virus DNA and proteins. As the viruses develop within the bacterial cytoplasm, some of the fragmented bacterial DNA is incorporated within virus coats instead of virus DNA. The "pseudoviruses" with the bacterial DNA are able to inject the bacterial DNA into other bacteria. The bacterial DNA injected by the pseudoviruses may recombine with the bacterial chromosome giving the bacterium a new trait. R, resistance to an antibiotic; S, sensitivity to an antibiotic.

mosome breaks into fragments and some of the newly formed viruses pick up (package) the bacterial DNA fragments instead of viral DNA. Thus, a small fraction of the virus population functions as carriers of bacterial DNA that can infect recipient bacteria and transfer the donor's DNA. The bacteria that are infected by the viruses carrying bacterial genes will acquire new genetic information. The new information may be beneficial in that it may help the bacteria to survive in the presence of antimicrobials or to use nutrients that could not be used before.

In contrast, a specialized transducing phage frequently integrates its DNA into a bacterial chromosome at one specific site without destroying its host (fig. 6.20). When the virus excises itself from the bacterial chromosome, it sometimes carries a region of the bacterial DNA near its site of integration. Since these viruses can transduce only genes near their integration site, they are called specialized transducing phages.

Both types of transduction promote the acquisition of new genetic information by bacteria. For example, *Staphylococcus aureus*, a bacterium that is unable to conjugate, has its plasmids transferred by a generalized transducing phage. Thus, *S. aureus* may acquire R plasmids by transduction. Generalized and specialized transducing phages also transfer many chromosomal genes of *S. aureus*.

Transformation

Small pieces of naked DNA (DNA isolated from cells and viruses) can be taken up by bacteria and incorporated into their chromosomes. This phenomenon occurs frequently in nature and may be an important means of creating variability in bacteria. The ability to take up DNA in this manner is called **competency**, and it occurs for a short period of time during exponential growth. Bacteria can be artificially induced to become competent by treatment with solutions of calcium chloride. The bacteria can then take up larger pieces of DNA. This technique is commonly used by genetic engineers to introduce whole plasmids into recipient cells and thereby "clone" a desirable gene.

One well-known example of transformation occurs when heat-killed, encapsulated *Streptococcus pneumoniae* transfer the ability to produce capsules to pneumococci that lack this ability. Encapsulated pneumococci cause death in mice within 24 hours while unencapsulated bacteria are harmless. However, if heat-killed, encapsulated cells are mixed with live, unencapsu-

Figure 6.20 **Specialized transduction.** Specialized transducing bacterial viruses invade bacteria by injecting their hereditary material into the bacteria. The virus DNA generally directs the synthesis of new virus DNA and proteins, but occasionally it recombines with the bacterial chromosome and becomes integrated. Under certain conditions, the integrated virus DNA excises itself and directs the synthesis of new virus DNA and proteins, which eventually form new viruses. Occasionally, when the integrated virus DNA excises itself, it carries with it a short piece of the bacterial chromosome near the integration site. Viruses that contain the hybrid virus-bacterial DNA can inject it into bacteria. The bacterial DNA carried on the viral chromosome can recombine with the homologous region on the bacterial chromosome and transduce genes into the bacterium. Since only genes near the single integration site are picked up by the viral chromosome, all the viruses pick up similar genes. Consequently, the virus is a specialized transducing phage rather than a generalized transducing phage.

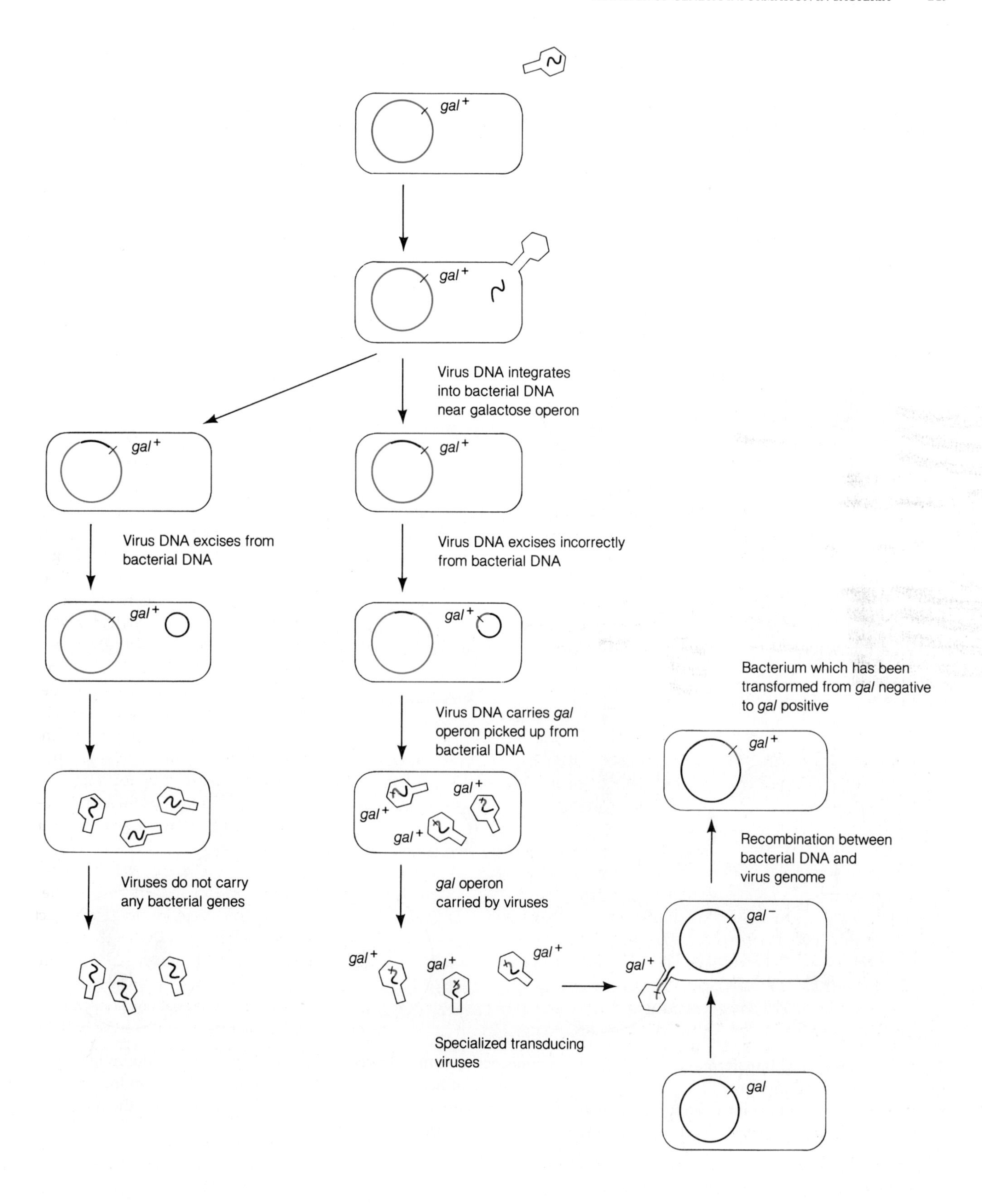

gal+
gal+
Virus DNA integrates into bacterial DNA near galactose operon
gal+
gal+
Virus DNA excises from bacterial DNA
Virus DNA excises incorrectly from bacterial DNA
gal+
gal+
Virus DNA carries gal operon picked up from bacterial DNA
Bacterium which has been transformed from gal negative to gal positive
gal+
gal+
gal+
gal+
Recombination between bacterial DNA and virus genome
Viruses do not carry any bacterial genes
gal operon carried by viruses
gal−
gal+
gal+
gal+
gal+
Specialized transducing viruses
gal

lated cells and then injected into mice, the mice will die as though they were injected with live, encapsulated cells. In fact, when the mice are autopsied, only encapsulated bacteria are isolated. This phenomenon occurs because DNA of the dead, encapsulated bacteria, which possesses the gene for capsule production, is taken up by competent, unencapsulated cells. By acquiring the gene for capsule production and subsequently incorporating it into their chromosomes, the recipient cells acquire the trait to form capsules and thus to cause fatal infections in mice. This phenomenon is illustrated in figure 6.21.

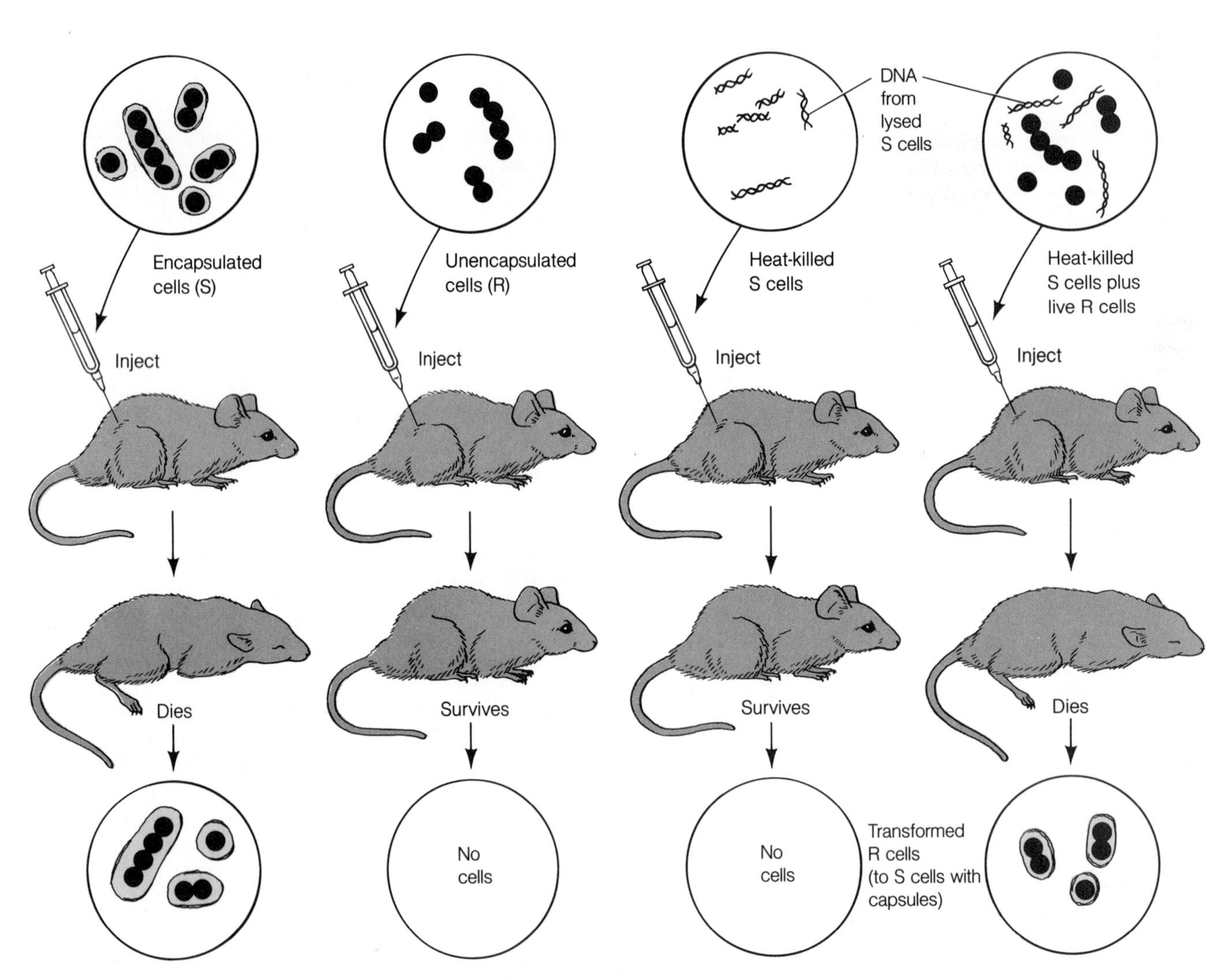

Figure 6.21 **Bacterial transformation.** Encapsulated (smooth or S) pneumococci cause a fatal infection in mice, while nonencapsulated (rough or R) pneumococci do not. Heat-killed smooth pneumococci do not kill mice; however, if a mixture of heat-killed smooth strains and living rough strains are injected into mice, the mice are killed. DNA from the heat-killed bacteria enters the rough bacteria and transforms them into smooth bacteria, which can kill the mice.

FOCUS *NEISSERIA GONORRHOEAE* **RECEIVED GENES FROM OTHER BACTERIA**

Neisseria gonorrhoeae is the bacterium that causes gonorrhea. This organism enters the genital system of humans during sexual intercourse and causes an infection. The most common course of the infection in males is the production of pus and pain in the affected area. This ancient disease occurs in human populations in epidemic proportions and it is very difficult to control. In the United States alone, more than 2 million cases of gonorrhea occur each year.

The antibiotic penicillin has traditionally been the drug of choice for the treatment of this disease and a single injection of 1–2 million units was normally sufficient to cure the patient of the disease. In 1976, however, there were reports from health officials in the United States, the United Kingdom, and Southeast Asia of cases of gonorrhea that were not treatable with penicillin. Since then, there has been a constant increase in the number of cases of gonorrhea that do not respond to penicillin treatment. In 1986, the proportion of cases of gonorrhea that did not respond well to penicillin treatment was about 10% in the United States, 18% in The Netherlands, and 9% in Great Britain.

The appearance of penicillin-resistant strains of *Neisseria gonorrhoeae* occurred because the bacteria acquired a gene which coded for a protein that inactivates penicillins. This protein is an enzyme called **β-lactamase** and the gene that codes for it is named *bla*. Genetic analysis of penicillin-resistant *Neisseria gonorrhoeae* revealed that the gene *bla* was located in a small plasmid that was obtained from another bacterium called *Haemophilus*. Apparently, at some time prior to 1976, *Neisseria gonorrhoeae* came in contact with DNA from *Haemophilus* (either in the human throat or the genital area and by the genetic process called **transformation**, brought the DNA inside the cell and incorporated it as its own. From this point on, daughter cells from this original *Neisseria gonorrhoeae* acquired the genetic capacity of penicillin resistance.

This example illustrates the remarkable genetic flexibility of bacteria and the impact that a single genetic event taking place in a single bacterial cell can have in human populations.

SUMMARY

NUCLEIC ACIDS ARE THE HEREDITARY MATERIAL OF ALL ORGANISMS

1. The genetic information in all cellular organisms is DNA.
2. DNA is a double-stranded molecule consisting of a sugar-phosphate backbone and nitrogenous bases on side chains.
3. The nitrogenous bases in DNA are adenine (A), guanine (G), cytosine (C), and thymine (T). In RNA, T is replaced by uracil (U).

DNA REPLICATION IN BACTERIA

1. Bacteria generally contain a single circular DNA chromosome attached to the plasma membrane.
2. The bacterial chromosome is replicated bidirectionally from a single start site. Some chromosomes can be replicated in less than one half hour. Numerous enzymes and proteins are involved in DNA replication. The most noteworthy is DNA polymerase.
3. In bacteria, daughter chromosomes are separated from each other by the growth of the plasma membrane between their points of attachment. Daughter cells form when the plasma membrane and the cell wall invaginate and create a partition between the chromosomes.

RNA SYNTHESIS IN BACTERIA

1. Prokaryotes (as well as eukaryotes) synthesize three types of ribonucleic acid (RNA): ribosomal RNA (rRNA), transfer RNA (tRNA) and messenger RNA (mRNA).
2. Ribosomal RNAs are structural components of the ribosomes.

3. There are 1 to 6 different transfer RNAs for each of the 20 amino acids found in protein. Because of their anticodons, tRNA molecules hydrogen-bond to specific triplet codons in the mRNA. The mRNA determines the order in which charged tRNAs (tRNA molecules with covalently attached amino acids) bind to an mRNA-ribosome complex.
4. Ribosomes catalyze the formation of peptide bonds between adjacent amino acids.

PROTEIN SYNTHESIS IN BACTERIA

1. Protein synthesis consists of three stages: initiation, elongation, and termination. During initiation, a ribosome forms from subunits at a start codon (AUG) on the mRNA. During elongation, amino acids are linked in a sequential manner as the mRNA moves along the ribosome. Termination of protein synthesis occurs when the ribosome comes to a nonsense codon (UAA, UGA, or UAG). The peptide is cleaved from the tRNA holding it to the last codon of the mRNA and the ribosome dissociates into its subunits.
2. The genetic code is a triplet code because it is based upon groups of three nucleotides, called codons. The codon consists of any three of the four nucleotides (A, U, G, C) in any order. All except the three nonsense codons are recognized by tRNAs.

CHANGES IN THE DNA MOLECULE CAN CAUSE MUTATION

1. In bacteria, spontaneous mutations in a particular gene occur once in about every 10^7 to 10^9 bacteria. Most mutations are caused by heat, radiation, and chemicals.
2. Scientists commonly induce mutations with ultraviolet light, aklylating chemicals, nitrous acid, and base analogues.
3. The cutting and rejoining of DNA are necessary for recombination.

TRANSFER OF GENETIC INFORMATION

1. There are three types of donor (male) bacteria: F^+, Hfr, and F'. Male bacteria are capable of transmitting their genetic information to recipient females (F^-) by conjugation.
2. Gram-negative donor bacteria generally attach to recipient bacteria by long, hollow sex pili. Transfer of DNA takes place through a cytoplasmic bridge.
3. Transduction occurs when a virus introduces DNA into a cell and recombination occurs. Transductions are carried out by generalized and specialized transducing bacteriophages.
4. Transformation occurs when naked DNA enters a cell and recombines with the cell's genetic information.

STUDY QUESTIONS

1. What is the central dogma theory?
2. Explain:
   a. DNA replication;
   b. reverse transcription;
   c. RNA replication;
   d. transcription;
   e. translation.
3. Describe how DNA is replicated.
4. Define semiconservative DNA replication.
5. Explain what a gene is, and the relationship between a gene's structure and an organism's phenotype.
6. Describe how each of the different types of RNA (rRNA, tRNA, mRNA) functions in a cell.
7. Outline the details of protein synthesis in the prokaryote.
8. Compare and contrast:
   a. codons and anticodons;
   b. template and primer;
   c. start codon and nonsense codon;
   d. promoter site and termination site.
9. Explain what the genetic code is.
10. Explain how spontaneous mutations occur.
11. Explain how ultraviolet light and nitrous acid induce mutations.
12. Explain the difference between F^- and F^+ bacteria.
13. Contrast conjugation, transduction, and transformation.
14. What is the difference between generalized and specialized transducing bacteriophages?
15. Draw a series of pictures illustrating recombination between:
    a. a plasmid and a circular chromosome;
    b. a linear piece of DNA and a similar region in the middle of the chromosome.

SELF-TEST

1. What is the major source of drug-resistant *Salmonella* isolated in outbreaks of salmonellosis in the U.S.?
   a. person-to-person contact;
   b. contaminated soil;
   c. contaminated meats, eggs, or dairy products;
   d. contaminated water;
   e. animal-to-human contact.

2. Analysis of genes from *Salmonella* isolated from supposedly sporadic outbreaks can be used to show whether or not the bacteria are all part of the same epidemic. a. true; b. false.

For items 3–7, match the function or characteristic with the RNA below:
   a. mRNA;
   b. tRNA;
   c. rRNA.

3. These nucleic acids are structural components of ribosomes.

4. This nucleic acid contains anticodons.

5. Amino acids are carried to the ribosome and messenger RNA by this small nucleic acid.

6. This nucleic acid contains codons.

7. This nucleic acid codes for proteins.

8. A genome is the total complement of hereditary information found in a cell. a. true; b. false.

9. What is the codon called where protein synthesis begins?
   a. nonsense;
   b. initiation;
   c. termination;
   d. start;
   e. beginning.

10. What are codons called where protein synthesis ends?
   a. nonsense;
   b. final;
   c. end;
   d. stop;
   e. missense.

11. Which nucleotide powers protein synthesis?
   a. ATP;
   b. GTP;
   c. UTP;
   d. TTP;
   e. CTP.

12. What frequently causes pairing nucleotides to no longer pair?
   a. hydration;
   b. ultraviolet light;
   c. nuclear radiation;
   d. oncogenic viruses;
   e. nitrous acid.

13. What is responsible for the formation of thymidine dimers?
   a. base analogues;
   b. alkylating agents;
   c. nitrous acid;
   d. ultraviolet light;
   e. nuclear radiation.

For items 14–16, match the statements with the items below.
   a. conjugation;
   b. transduction;
   c. transformation;
   d. recombination.

14. The unidirectional transfer of genetic information from one bacterium to another.

15. The acquisition of naked DNA by a bacterium.

16. The transfer of genetic information from one bacterium to another by a virus.

17. Viruses that pick up any gene in a cell and transmit the gene to other cells are known as:
   a. oncoviruses;
   b. specialized transducing viruses;
   c. generalized transducing viruses;
   d. retroviruses;
   e. arboviruses.

18. When heat-killed encapsulated pneumococci are mixed with unencapsulated pneumococci and the mixture is injected into mice, how do the unencapsulated bacteria become pathogenic?
   a. conjugation;
   b. transduction;
   c. transformation;
   d. translation
   e. transcription.

19. If a linear piece of DNA integrates in the middle of a chromosome, how many cuts have to be made in each DNA for recombination to occur? a. 1, b. 2.

20. If a plasmid integrates into a circular chromosome, the resultant molecule is: a. linear; b. circular.

21. Transposons are frequently found in plasmids. a. true; b. false.

22. Transposons are often found in chromosomes. a. true; b. false.

23. If a bacterium has a plasmid, it is able to conjugate with recipient bacteria. a. true; b. false.

SUPPLEMENTAL READINGS

Barany, F. and Kahn, M. 1985. Comparison of transformation mechanisms of *Haemophilus parainfluenzae* and *Haemophilus influenzae*. *Journal of Bacteriology* 161(1):72–79.

Croce, C. M. 1985. Chromosomal translocations, oncogenes, and B-cell tumors. *Hospital Practice* 20:41–48.

Ehrenfeld, E. E. and Clewell, D. B. 1987. Transfer functions of the *Streptococcus faecalis* plasmid pADI: organization of plasmid DNA encoding response to sex pheremone. *Journal of Bacteriology* 169(8):3473–3481.

Hartstein, A., Valvano, M., Morthland, V. Fuchs, P., Potter, S., & J. Crosa. 1987. Antimicrobic susceptibility and plasmid profile analysis as identity tests for multiple blood isolates of Coagulase negative Stapylococi. J. *Clinical Microbiology* 25(4):589–593.

Howard-Flanders, P. 1981. Inductible repair of DNA. *Scientific American* 245(5):72–80.

Judson, H. F. 1979. *The eighth day of creation*. New York: Touchstone Books, Simon & Schuster.

Kazazian, H. H., Jr. 1985. The nature of mutation. *Hospital Practice* 20:55–69.

Kelly, R., Atkison, M., Huberman, J., and Kornberg, A. 1969. Excision of thymine dimers and other mismatched sequences by DNA polymerase of *Escherichia coli*. *Nature* 224:495–501.

Kornberg, A. 1969. Active center of DNA polymerase. *Science* 163:1410–1418.

Lake, J. 1981. The ribosome. *Scientific American* 245(2):84–97.

Nomura, M. 1984. The control of ribosome synthesis. *Scientific American* 250(1):102–114.

Varmus, H. 1987. Reverse transcription. *Scientific American* 257(3):56–64.

Walker, G. 1984. Mutagenesis and inducible responses to deoxyribonucleic acid damage in *Escherichia coli*. *Microbiological Reviews* 48(1):60–93.

PART 2

A SURVEY OF MICROORGANISMS

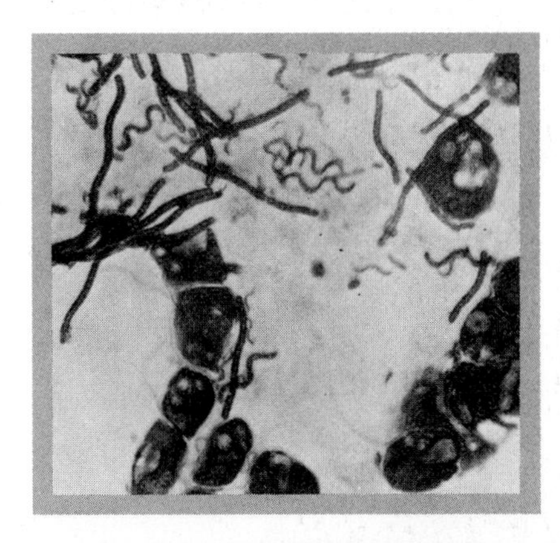

CHAPTER 7

CLASSIFICATION AND IDENTIFICATION OF MICROORGANISMS

CHAPTER PREVIEW TRACKING DOWN THE AGENT OF LEGIONNAIRES' DISEASE

The Pennsylvania Department of the American Legion held its 58th annual convention in 1976 at the Bellevue-Stratford Hotel in Philadelphia from July 21 to July 24. Of the more than 4000 people attending the convention, about 221 contracted a previously unknown type of pneumonia that killed 34 individuals. This serious illness was promptly labeled by the press as "Legionnaires' disease" although not all those who became ill were Legionnaires; some were hotel or convention employees and visitors. This outbreak of a previously unknown disease represents an important landmark in the history of infectious disease diagnosis.

The outbreak of Legionnaires' disease prompted an extensive and ultimately productive campaign to find the cause of the disease. It took about five frustrating months of research until the organism was found.

The first step in establishing the cause of Legionnaires' disease was to define the characteristics of the disease so that the population at risk could be established. The disease included a high fever, coughing, and pneumonia. Unfortunately, the clinical symptoms of the disease did not narrow down the cause sufficiently; it could have been caused by toxic organic substances, heavy metals, bacteria, fungi, or viruses.

In view of the clinical findings, the decision was made to collect specimens from diseased and normal individuals in an attempt to isolate the causative agent. The samples included bits of tissue (biopsy materials), sputum, and blood. They were cultured in a variety of media and tissue cultures and incubated under various environmental conditions.

Finally, in January 1977, the agent of Legionnaires' disease was discovered. Joseph E. McDade, Charles C. Shepard, and other scientists at the National Centers for Disease Control (CDC) made the discovery when they examined yolk sacs of eggs that had been injected with infected material. The cause of Legionnaires' disease was a rod-shaped, gram-negative bacterium, which they named *Legionella pneumophila*. This discovery represented the culmination of one of the most extensive searches for a pathogen recorded in recent history.

Once the organism of **legionellosis** was discovered, techniques were developed to culture the organism in laboratory media, and a fluorescent antibody test was introduced as a means of identifying the pathogen.

The scientists at CDC were successful in their goal because they had previously mastered techniques employed for the isolation, cultivation, and identification of infectious disease agents. They were careful to collect the proper specimens, using aseptic techniques; to transport them to the laboratory in a safe and suitable manner; and to employ cultural techniques that had proved successful in isolating other pathogens. This chapter introduces the field of diagnostic microbiology. It covers the basic principles of specimen collection and processing and discusses some of the more common techniques presently used to identify microorganisms.

CHAPTER OUTLINE

CLASSIFYING AND IDENTIFYING MICROORganisms are important areas of study in microbiology. Due to microbiologists' work in these fields, medical specialists have been able to improve the quality of human life by isolating and identifying pathogens more rapidly and accurately. In this chapter we discuss classification and identification of microorganisms and their application in diagnosing infectious diseases.

TAXONOMY AND THE KINGDOM OF ORGANISMS

Taxonomy is the science that identifies, names, and arranges organisms into categories. It can be subdivided into three disciplines: **classification** (ordering); **nomenclature** (naming); and **identification** (distinguishing). Microbiologists are concerned with all aspects of taxonomy; in the clinical laboratory, however, they are mainly concerned with identification. It is through this aspect of taxonomy that infectious diseases are diagnosed after isolation and identification of disease-causing organisms present in clinical specimens.

Microorganisms Are Classified into Kingdoms

In the mid 1700s, Carolus Linnaeus began a system of classification of living organisms that is still used today. He divided all living organisms into two obvious kingdoms, the **Plantae** and the **Animalia** (table 7.1). In Linnaeus's system of classification, plants and animals are grouped to indicate their degree of similarity. A frequently used system of classification places organisms into groups on the basis of similarities in their morphology and physiology. Organisms are first assigned to a **kingdom**, then to a division or **phylum**,

TABLE 7.1
THE CLASSIFICATION OF MICROORGANISMS INTO KINGDOMS

Number of kingdoms	Scientist credited	Kingdoms and organisms
2	Linnaeus (1753)	Plantae Plants, Algae, Fungi, Bacteria Animalia Animals, Protozoans
3	Haeckel (1865)	Plantae Plants, Multicellular algae Animalia Animals Protista Protozoans, Single-celled Algae, Fungi, Bacteria, and Sponges (which are animals).
5	Whittaker (1969)	Plantae Plants, Multicellular algae Animalia Animals Protista Protozoans, Single-celled algae Fungi Molds and Yeasts Monera Cyanobacteria (Blue-green Algae), Eubacteria (Photosynthetic and Nonphotosynthetic Bacteria), Archaebacteria.
3	Woese (1980)	Archaebacteria Halobacteria, methanogens, extreme thermophiles Eubacteria Gram-positive bacteria, purple bacteria, cyanobacteria, green sulfur bacteria, spirochetes, flavobacteria, planctomyces, chlamydia, green nonsulfur bacteria. Eukaryotes Microsporidia, flagellates, slime molds, ciliates, fungi, plants, animals

a **class**, an **order**, a **family**, a **genus**, and finally a **species**. Organisms in a kingdom may be very different from one another, but they share some major characteristics. Organisms within a family, genus, or species are more similar to one another and share major and minor characteristics. Sometimes species are subdivided into **races**, **subspecies**, or **strains**. Linnaeus is also credited with establishing a **binomial system of nomenclature** that gives each organism a generic (genus) name and a specific (species) name that, in many cases, indicate something about the organism. For example, *Streptococcus pyogenes* indicates that it is a spherical (coccus) bacterium arranged in chains (strepto) that causes pus-containing (pyogenic) lesions. Very closely related organisms (species) that do not normally interbreed may be in the same genus (plural, genera). Many genera of bacteria are not placed in categories higher than families. Instead they are placed in parts.

After Linnaeus established a two-kingdom system of classification, it became obvious that the kingdoms Plantae and Animalia were not appropriate for microorganisms such as the fungi, protozoans, and bacteria. Ernst Haeckel, one of Charles Darwin's students, proposed in 1866 that microorganisms be placed in a third kingdom called the **Protista** (primitive or first organisms) (table 7.1). Because bacteria were found to be fundamentally different from other microorganisms, scientists later proposed that they be placed in their own kingdom. In 1969 R. H. Whittaker proposed that all cellular organisms be placed in one of five kingdoms: Animalia, Plantae, Fungi, Protista, and Monera. The prokaryotic bacteria and cyanobacteria are single-celled organisms, so different from the eukaryotic, multicellular animals and plants that the kingdom **Monera** (or **Prokaryotae**) was established for them. In addition, because the single-celled algae and protozoans have very complex cell structures and long evolutionary histories that make them very different from bacteria, plants, and animals, a kingdom called the **Protista** was established for them. The fungi, which are quite different physiologically and morphologically from bacteria, protista, plants, and animals, are placed in a kingdom of their own, called the **Fungi**.

Numerous scientists studying ribosomal RNAs (rRNAs) and how they differ have come to the conclusion that there are two groups of distantly related bacteria that should each have kingdom status. The two groups of bacteria are the eubacteria (true bacteria) and the archaebacteria (ancient bacteria). Plants, animals, fungi, and protista, which are very different structurally, have similar rRNAs. A comparison of rRNAs indicates that the evolutionary distance between two eukaryotes such as a paramecium (ciliate) and a human (animal) is much smaller than the distance between an eubacterium and an archaebacterium or the distance between a bacterium and an eukaryote (fig. 7.1).

On the basis of rRNA studies, C. R. Woese and other scientists have proposed that the five-kingdom system of classification does not accurately reflect the evolutionary relationship among organisms and that it be replaced with a three-kingdom system of classification: eubacteria, archaebacteria, and eukaryotes. Future study will indicate whether any of the currently accepted eukaryotic kingdoms (Animalia, Plantae, Fungi, Protista) should be preserved as subkingdoms. It appears that animals, plants, fungi, and ciliates are more closely related than the five-kingdom system of classification implies and that slime molds, flagellates, and microsporidia may represent three new subkingdoms (fig. 7.1). The archaebacteria may be divided into two subkingdoms or phyla: the methanogens and the sulfur-utilizing extreme thermophiles. The halobacteria appear to be closely related to the methanogens and may not warrant a subkingdom of their own. The ribosomal RNAs from the eubacteria indicate that they have diverged into as many as 11 groups that may be given subkingdom or phylum status.

Presently, there is insufficient information to decide what should constitute a subkingdom or phylum. Because these RNA studies and the derived phylogenetic relationships so drastically change all previously used classification schemes, change will be slow.

The Whittaker and Woese systems of classification are concerned only with cellular microorganisms, and ignore the noncellular microorganisms like viruses (table 7.1). Some writers do not consider the viruses, viroids, and prions to be organisms, because they are noncellular in nature. Because these particles are "organized" structures (some of the viruses are extremely complex) that engage in many processes associated with life, we will refer to them as microorganisms as well as noncellular infectious agents.

METHODS USED TO CLASSIFY AND IDENTIFY MICROORGANISMS

As we have seen in Chapter 6, all of the information for the characteristics of a cell is stored as genes in the

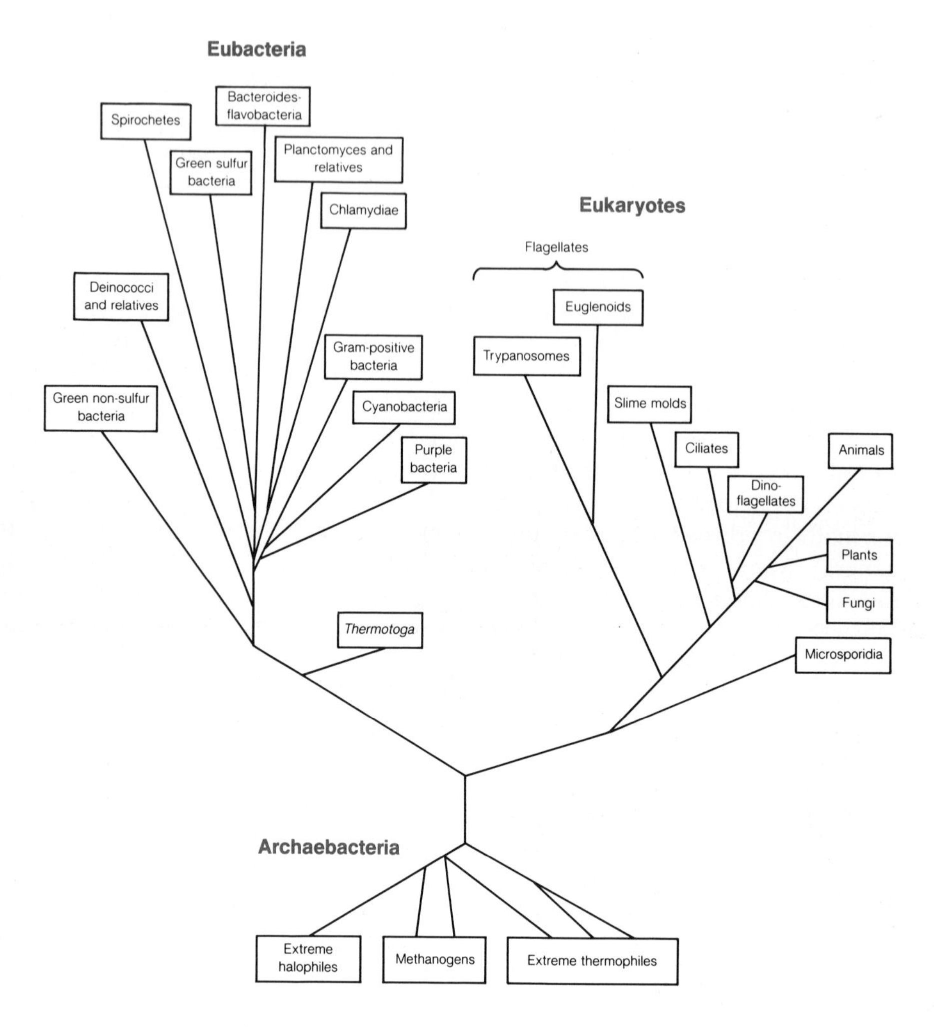

Figure 7.1 Three-kingdom system of classification. A study of the nucleotide sequences of ribosomal RNA (rRNA) from numerous organisms indicates that their rRNAs fall into three distinct groups: archaebacteria, eubacteria, and eukaryotes. Based on rRNA sequences, the bacterial groups are no more related to each other than they are related to the eukaryotes. The evolutionary distance between organisms is approximated by the length of the line that separates them.

DNA. The chemical composition of the cell's nucleic acids (both DNA and RNA), the amino acid sequence of its proteins, its physiological activities, and its morphological characteristics reflect the uniqueness of the genetic information of that organism and can be used to classify and identify the microorganism. Traditionally, microbiologists have relied heavily on morphological, physiological, and staining characteristics of microorganisms to classify and identify them (table 7.2). More recently, with the development of molecular biology techniques, microbiologists have included more sophisticated methods, such as DNA hybridizations and nucleotide sequencing of nucleic acids, to classify and identify microorganisms. In this section we discuss some of the more common methods used by microbiologists to classify and identify microorganisms.

Morphology and Physiology

Microorganisms are generally classified on the basis of readily determined characteristics such as shape, staining properties, and physiology, but for closely related species it is sometimes necessary to employ other methods such as growth and genetic characteristics (table 7.2). For instance, the ninth edition of *Bergey's Manual of Systematic Bacteriology* groups bacteria into 30 sections (see appendix A) on the basis of such criteria as cell shape, Gram-staining reaction, motility, and characteristic structures like flagella and endospores (fig. 7.2). Those bacteria that cannot be separated using these criteria, can be classified according to their biochemical reactions, such as carbohydrate fermentation and amino acid utilization, and genetic characteristics, such as the nucleotide base composition of their DNA.

TABLE 7.2
CRITERIA FOR CLASSIFYING MEDICALLY IMPORTANT BACTERIA

Structural characteristics of bacteria			
Cellular	**Staining**	**Colony**	**Growth in broth**
Shape Size Arrangement Flagella and their arrangement Endospores Capsules	Gram Acid fast	Color Texture Shape Size Diffusible pigment	Surface (pellicle formation) Sediment formation Diffusible pigment

Physiological and molecular characteristics of bacteria			
Nutritional and environmental requirements	**Physiological**	**Genetic**	**Biochemical**
Carbon Energy Nitrogen Amino acid Vitamin Complex Atmospheric: Aerobic Microaerophilic Anaerobic	Range of carbohydrates that can serve as carbon and energy sources Optimum temperature Range of temperature Optimum atmospheric conditions (aerobic, microaerophilic, anaerobic) Optimum osmotic pressure and range Optimum pH Range of pH Optimum salt concentration and range Sensitivity to antibiotics and drugs Products of respirations and fermentations	Tm and density of DNA indicates % G + C Amount of DNA hybridizations indicates DNA similarities Ribosomal RNA	Types of storage granules Antigenic determinants Chemical structure of cell wall Serotyping

Morphology and Staining Characteristics Two of the primary criteria used to classify bacteria are the staining characteristics and the shape of the cells (fig. 7.2). Most bacteria can be characterized as spheres (cocci), rods (bacilli), spirals (spirilla), coils (spirochetes), or branches. The cells may exist singly, in chains, in clusters, or in other arrangements. Hence, when a bacterium is referred to as a *Streptococcus*, it means that the organism forms chains (strepto) of cocci.

Cell shape and arrangement are usually determined on Gram-stained smears (see Chapter 2) so that the Gram reaction of the bacteria is noted simultaneously with the cell shape and arrangement. Using this staining technique, bacteria can be classified as either gram-positive or gram-negative. This criterion is very important, not only in the classification of bacteria, but also in the judicious selection of antimicrobials since gram-positive and gram-negative bacterial infections are sometimes treated with different types of antibiotics.

Other staining methods may be necessary to determine different characteristics. For example, bacteria in the genus *Mycobacterium*, which include the agents of tuberculosis and Hansen's disease (leprosy), and some bacteria in the genus *Nocardia*, a causative agent of pneumonia and lung abscesses, are identified by the **acid-fast** stain. The acid-fast stain reflects the fact that acid-fast bacteria have a high lipid content in their cell walls, a substance that is absent in most bacterial walls.

Staining procedures for the detection of capsules or endospores are sometimes done to detect the presence of these structures in the bacterial cell. Bacteria in the genera *Bacillus* and *Clostridium*, both of which

Figure 7.2 A key for the identification of common bacteria. *(a)* Key to selected gram-positive bacteria. *(b)* Key to selected gram-negative bacteria.

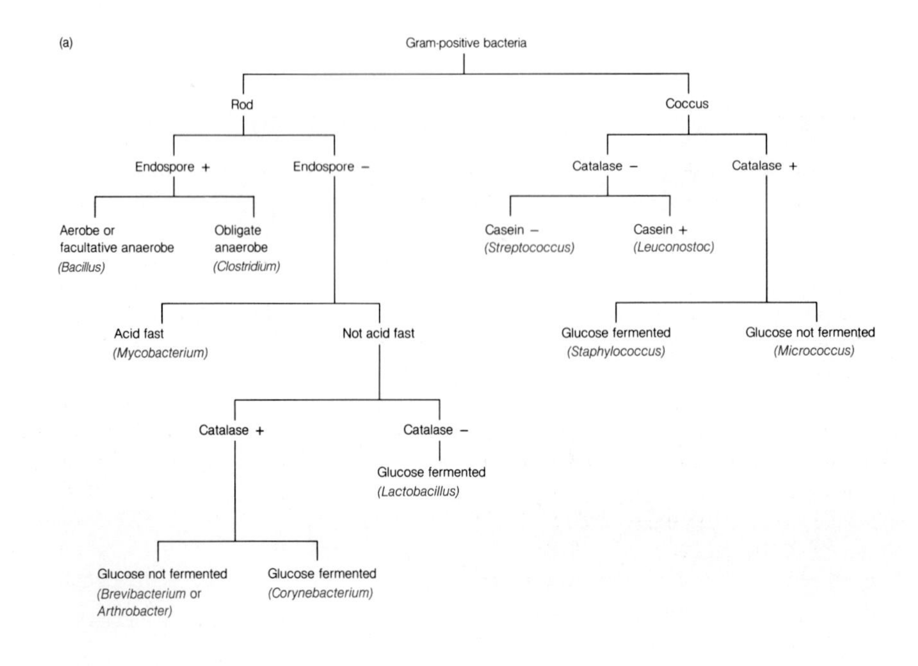

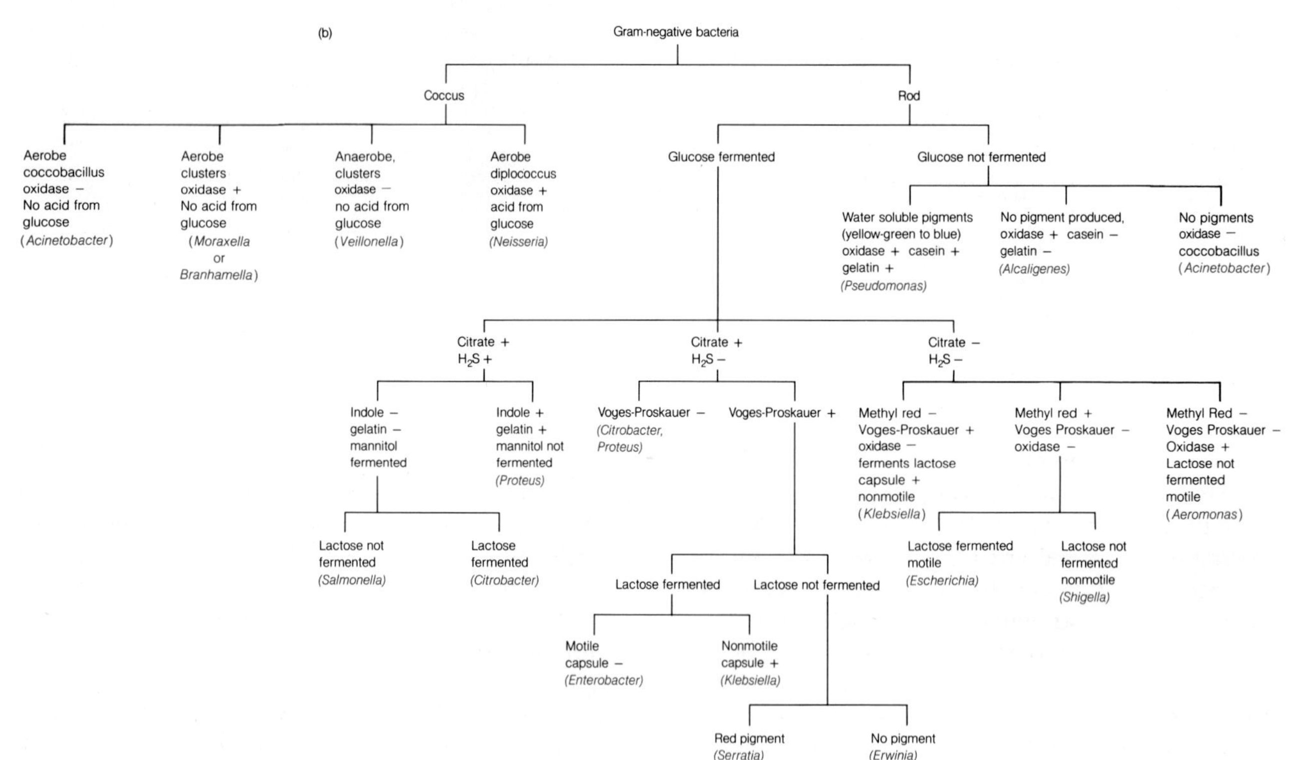

FOCUS DID VIRUSES AND BACTERIA ORIGINATE IN OUTER SPACE?

Some astronomers have hypothesized that viruses and bacteria evolved in outer space and became part of the earth during its formation or soon after it had cooled. In addition, some astronomers have proposed that microorganisms from outer space are constantly raining down upon the earth and are responsible for the epidemics that occur from time to time. Few biologists believe that there is any evidence to support the hypothesis that microorganisms first populated the earth from space (**pangenesis**) or the idea that epidemics are due to alien microorganisms presently hitting the earth.

The idea that viruses could have evolved in outer space, or in the icy heads of comets, does not fit in with what is known about viruses and chemical evolution. The concentration of organic molecules and the temperatures in space are so low that chemical reactions would not lead to the formation of macromolecules, such as proteins, nucleic acids, polysaccharides, or lipids. In fact, astronomers have found no evidence for these macromolecules anywhere in space or on numerous types of rock samples from space (meteorites and moon rocks). If viruses or cellular organisms have been raining down upon the earth and moon from outer space, their remains should have been found in the moon rocks, which have been studied extensively. On the other hand, a myriad of small molecules such as amino acids and macromolecules such as polypeptides have been easily produced in the Miller-Urey experiments, evidence that the earth is a much better environment than outer space for the production of such molecules.

Chemical reactions in living organisms have evolved in an aqueous environment and would not occur in the nonaqueous environment of outer space or in the icy heads of comets. Only when comets are near a sun might the temperature be favorable for chemical reactions to build organic molecules. Energetic radiations (such as ultraviolet light) found near a sun would be expected to ionize and rapidly destroy any guiding templates, such as nucleic acids or proteins, that might form in space. All forms of life on earth, from the viruses and bacteria to animals and plants, are extremely sensitive to radiation and are rapidly inactivated by it. Even if the concentration of molecules, the temperature, the nonaqueous environment, and the radiation in space were not sufficient constraints for macromolecules to evolve by chemical evolution, it would not be expected that viruses could evolve in space in the absence of specific types of cells. Virus replication is a cell-dependent process and would not occur in space or on the prebiotic earth.

include human pathogens, have endospores that can be detected by staining procedures.

Fungi, algae, and protozoa can also be classified using morphological criteria, for example, the color and shape of the fungal colonies, and the characteristic shape and arrangement of their spores (see Chapter 9). Similarly, the protozoa and larger multicellular parasites (e.g., flukes, tapeworms, and nematodes) are classified based on their morphology and the presence of characteristic structures (see Chapter 10).

Physiology Many closely related organisms cannot be classified by shape, arrangement of cells, or staining properties because they share the same characteristics. In these cases, physiological and biochemical characteristics are used. The use of physiological characteristics is based on the concept that the genetic makeup of the cell is reflected in its ability to utilize and metabolize nutrients in its environment. Whether or not a given bacterium can break down starch or ferment glucose and produce acid and gas is a direct result of the bacterial genetic information and can be used to identify or classify the bacterium. Figure 7.2 illustrates how a combination of morphological and physiological characteristics can be used to identify bacteria.

Some molecular biology techniques are used to study the nature of nucleic acids of different microorganisms. Two of these techniques are **nucleic acid sequencing** and **DNA hybridization**.

In nucleic acid sequencing, the nucleotide sequence in DNA or RNA isolated from bacteria is determined biochemically and then compared with that of other bacteria to determine the degree of relatedness that exists among them. By sequencing the RNA in bacterial ribosomes, scientists have been able to prove that there exists a very ancient and unique group of bacteria, called the archaebacteria. In addition, the eukaryotic line has been shown to be much older than previously suspected.

DNA Hybridization

When complementary DNA strands from closely related organisms hydrogen bond with each other, there is an almost perfect match and the two strands hybridize (fig. 7.3). The nucleotide sequences from two unrelated organisms, however, are quite different. The more unrelated the organisms are, the more poorly their DNAs hydrogen bond with each other and they will hybridize poorly or not at all. Thus, the degree of relatedness of organisms can be determined by the extent of hybridization.

DIAGNOSIS OF INFECTIOUS DISEASES

Infectious diseases generally result from microbial growth, or toxin production by microorganisms, in or on our body tissues. While certain infectious diseases produce unique signs and symptoms, most diseases result in signs and symptoms that are not characteristic. In these cases, further steps must be taken to ascertain the cause. In these cases, it is imperative that we isolate and identify the pathogen so that the patient can be treated effectively.

To isolate and identify infectious agents from **clinical specimens**, it is necessary to collect and handle the specimens properly. The importance of proper specimen collection and handling cannot be overemphasized. It is at this stage that successful pathogen

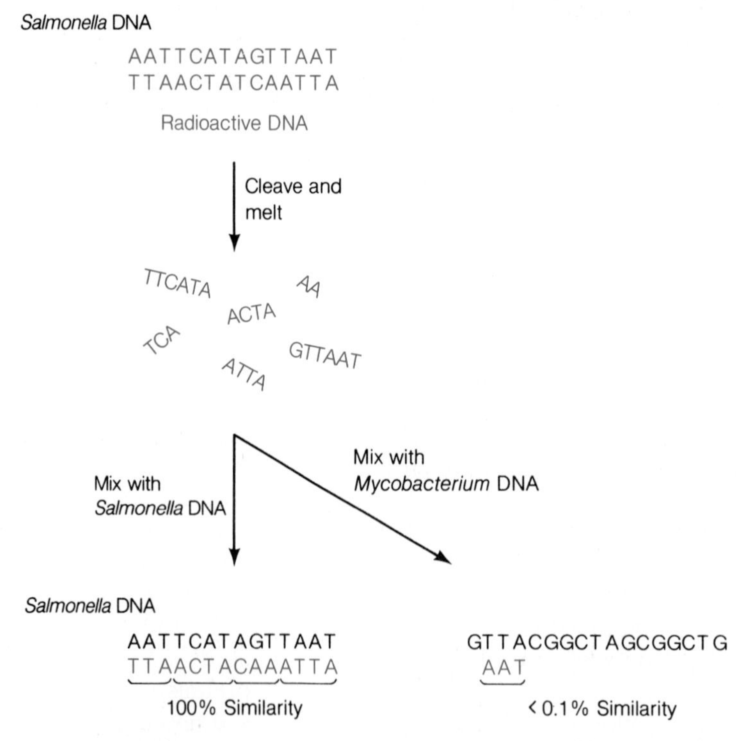

Figure 7.3 Hybridization of DNA. Strands of DNA from very closely related organisms are, for the most part, complementary and consequently will hydrogen-bond to each other and form hybrid DNA. On the other hand, strands of DNA from unrelated organisms are usually not very complementary and so there will be little or no hybrid DNA formed. The degree of hybridization between two DNAs is directly proportional to how similar the DNAs are.

isolation begins. Once the specimen is safely in the laboratory, it is then cultured for the isolation and identification of the pathogen. Identification of microorganisms generally involves (a) direct microscopic observations, (b) cultivation and biochemical tests, and (c) serological tests. In the remainder of this chapter we discuss techniques used in the collection of specimens and in the isolation and identification of microorganisms suspected of causing disease.

Collection and Transport of Specimens

The diagnosis of an infectious disease is based not only on its signs and symptoms but also on pathogenic microorganisms found in diseased tissues or exudates. The clinical microbiologist's task is to culture these specimens in order to isolate and identify pathogens. In addition, the microbiologist must ascertain that the results are accurate by collecting and handling the specimens properly. Often, the failure to isolate a pathogen from a clinical specimen is due not to faulty testing but to faulty collection. The following are important considerations for the successful isolation of pathogens in clinical specimens. The specimens should be:

1. Collected before antimicrobials are administered;
2. Collected at the time the pathogen is likely to be present;
3. Collected from the active site of the infection;
4. Collected in sufficient amounts;
5. Collected with the proper tools and placed in suitable containers;
6. Delivered promptly to the laboratory for analysis;
7. Stored under appropriate conditions until processed;
8. Processed as soon as possible to increase the chances of isolation.

Specimens should be collected before the administration of antimicrobials. Sometimes physicians diagnose infectious diseases based on signs and symptoms and then initiate antimicrobial therapy without the identification of the causative agent. Whether or not this therapy is successful, it may be desirable at a later date to identify the agent; however, inhibitory levels of the antimicrobials may be present in the clinical specimens, thus making the isolation of the pathogen more difficult (or impossible). For this reason, attempts to isolate the pathogen should be made before antimi-

FOCUS USING DNA PROBES TO IDENTIFY INFECTIOUS AGENTS

Frequently life or death situations are encountered where the rapid and accurate diagnosis of a disease is essential for the initiation of life-saving therapy. Countless lives have been saved because of the many advances that have been made in this area of taxonomy.

Recently, a new technique that originated in genetic engineering research has been introduced in the field of diagnostic microbiology. This technique, known as DNA–DNA hybridization, involves mixing an unknown bacterial DNA with DNA of known origin. The extent to which the unknown DNA attaches to the known DNA determines the identity of the unknown organism. For example, there is a commercially available kit for the detection of *Salmonella* in foods. The kit contains a fraction of *Salmonella* DNA, called a **probe**, as well as a variety of reagents used to detect this DNA. Samples of foods that are suspected of being contaminated with *Salmonella* are made to react with the DNA probe. If the sample is contaminated with *Salmonella*, the probe will react with the DNA in the sample and give visible results after the addition of the special reagents. If there is no *Salmonella* in the sample, no visible results will be detected. Similar kits are now available for the detection of important pathogens in clinical specimens, such as gonorrhea, brucellosis, listeriosis, and tuberculosis (to name but a few).

The field of diagnostic microbiology is a dynamic one where new techniques are constantly being evaluated to find faster and more accurate methods for the identification of pathogenic microorganisms and the diagnosis of infectious diseases.

crobial therapy has been initiated, or after treatment has been discontinued for a time, so that the concentration of antimicrobials in the specimens is no longer inhibitory.

Specimens should be collected when the parasites are likely to be present. Some parasites, including many protozoa and helminths, appear in a particular anatomical site at regular intervals and then disappear. An example is the nematode (roundworm) *Wuchereria bancrofti*. The larvae of this parasite, called **microfilariae**, can be observed in blood specimens collected during the night. Attempts to locate the microfilariae in specimens collected during the day would almost certainly fail. The protozoa that cause malaria and African sleeping sickness also make cyclic appearances in the blood of infected hosts. Many laboratories take three samples on three consecutive days to improve their chances of detecting the parasites in the specimens.

The clinical specimen should be sufficiently large to allow the microbiologists to conduct all the necessary tests. Frequently, a clinical specimen is split into two or more aliquots (volumes) that are treated in various ways. For these procedures to be carried out, there must be an ample supply of specimen. For example, a sputum sample may be split into two or three 5-ml aliquots to be cultured for bacteria and fungi. Hence, at least 10 ml of sputum must be collected so that all these procedures can be carried out.

Care must be exercised to maintain the viability of the pathogens in the specimen. If the specimen is taken with a tool such as a swab or a syringe containing toxic chemicals, the chances of isolating the pathogen diminish. For example, cotton swabs may contain materials such as lipoproteins that are inhibitory for certain hard-to-grow (fastidious) microorganisms. Although cotton swabs can frequently be used to collect specimens, it is sometimes preferable to use Dacron® swabs. Swabs made with this material generally are less inhibitory than cotton and are recommended for isolating β-hemolytic streptococci from throats and *Neisseria gonorrhoeae* in vaginal exudates.

The need for sterile containers for collecting clinical specimens cannot be overemphasized. If the specimen is collected in a container contaminated with other microorganisms, the contaminants may overgrow the pathogen and prevent its isolation, particularly if there are few pathogens in the specimen. Contaminating microorganisms may also release toxic substances that are inhibitory to the pathogen. Contaminated specimens also make it difficult to establish the significance of the isolated microorganism.

Specimens should be collected from the active site of the lesion. The reason for this is that some infectious agents are present only in certain portions of the lesions they cause, and attempts to culture the pathogen from other portions of the lesion will fail to yield the pathogen. For example, the fungi that cause ringworm spread radially from the site of infection, forming a circular lesion. The leading edge of the lesion contains the fungi, whereas the central portion of the lesion has very few fungal elements. Therefore, it is necessary to sample the leading edge of the lesion.

The proper collection of specimens requires the cooperation of the physician, the nursing staff, and the microbiologist. If proper precautions are taken and judicious choices are made as to how and when to collect a specimen, the chances for a quick diagnosis are greatly enhanced. A rapid diagnosis will not only result in the best treatment possible for the patient, but also will reduce the cost of hospitalization.

After collection, the specimen should be delivered to the laboratory as soon as possible. In rural communities, where the closest laboratory may be miles away, the specimens often are not delivered immediately to the laboratory for processing. In such cases, it is recommended that the samples be refrigerated shortly after collection until delivery to the laboratory.

There are also problems with stored specimens. For example, toxic materials might be present in the specimen that would inhibit or kill pathogens. Specimens stored improperly even for short periods of time will allow contaminating microorganisms to multiply and outnumber the pathogen. For this reason, several transport media have been developed to preserve the viability of the pathogen while inhibiting microbial growth.

Microscopic Methods in Diagnosis

Microscopic examination of fresh clinical specimens is one quick way of determining the cause of some infectious diseases. Diagnoses of bacterial diseases, however, are generally not made solely by microscopic examination, because bacteria, for the most part, lack distinct morphological characteristics that set them apart from all others. Fungal infections can sometimes be diagnosed by microscopic examination alone, although other diagnostic procedures usually are carried out concurrently. Infectious diseases caused by protozoa or helminths (worms) are almost always diagnosed using microscopic methods such as wet mounts or stained smears.

Vaginitis caused by *Trichomonas vaginalis* is generally diagnosed by examining a wet mount of the vaginal exudate and demonstrating the presence of motile, flagellated protozoa (fig. 7.4). A diagnosis of syphilis can sometimes be made by examining a suspension of material obtained from the chancre, using a dark field microscope. In such preparations, the spirochetes can be seen as bright, motile, helically shaped bacteria against a dark background.

Stained smears are commonly used for examining clinical specimens. The smears should be made with fresh specimens, preferably at bedside. Most clinical specimens are examined using a variety of staining techniques (table 7.3). For specimens suspected of containing bacterial pathogens, the Gram stain is routinely used. The Gram stain allows microbiologists to view the specimens and decide what further steps are needed to identify the pathogen. The Gram stain provides a necessary first step in the identification of many microorganisms.

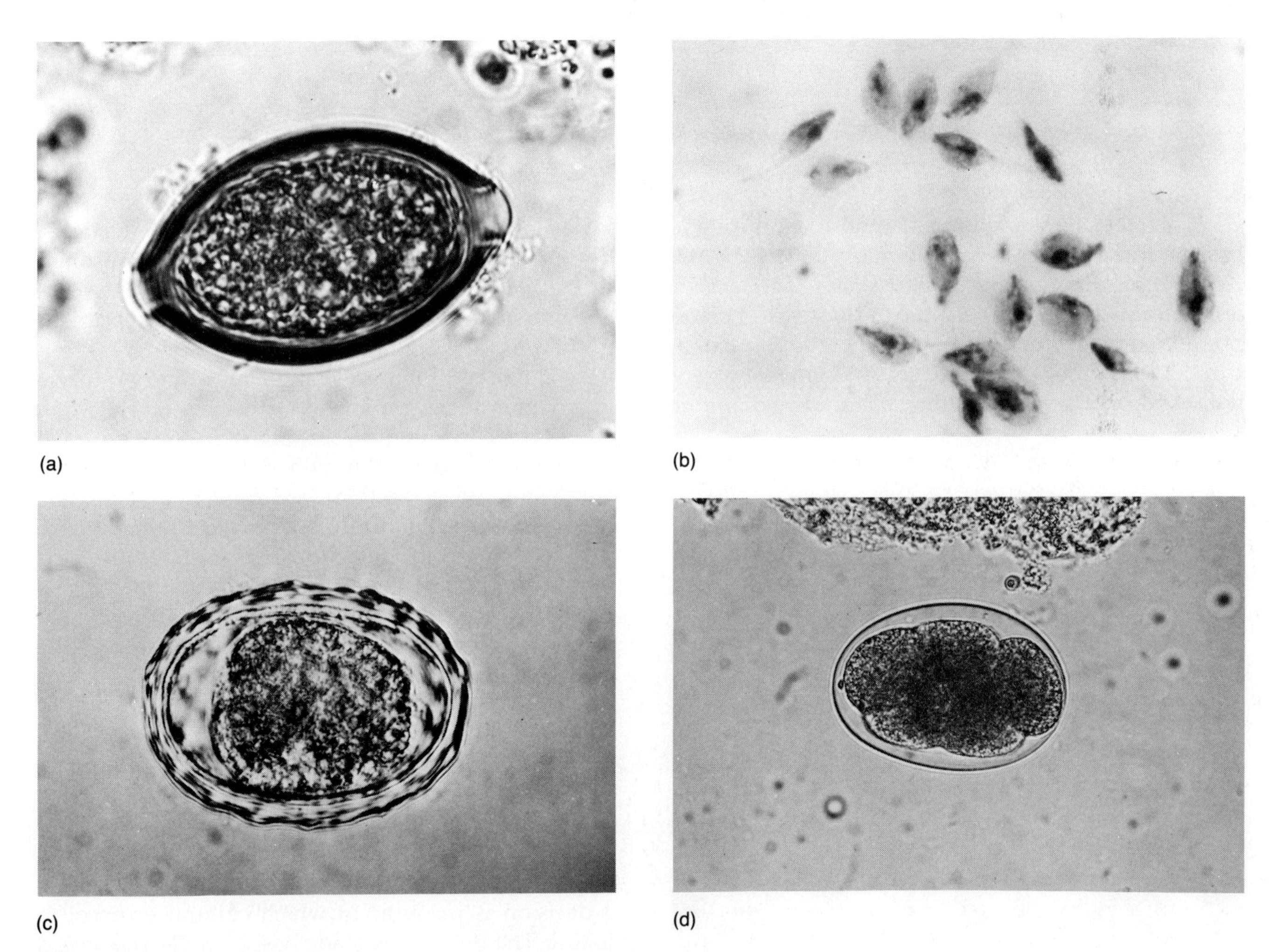

Figure 7.4 Wet mounts of selected clinical specimens. Microscopic appearance of helminths and protozoa in wet mounts of clinical specimens. *(a) Trichuris trichura* (whipworm) egg in human feces. *(b) Trichomonas vaginalis* in vaginal exudate. *(c) Ascaris lumbricoides* egg in human feces. *(d)* Hookworm egg in human feces.

TABLE 7.3
SELECTED STAINING PROCEDURES EMPLOYED IN THE MICROBIOLOGY LABORATORY TO EXAMINE CLINICAL SPECIMENS

Staining procedure	Use in laboratory	Cells or organisms detected
Acid-fast stain	Examine sputum, tissue, skin	*Mycobacterium, Nocardia, Cryptosporidium*
Giemsa stain	Examine blood smears	Bacteria, protozoa, fungi, blood cells
Gram stain	Examine various specimens	Bacteria
Gridley fungus stain	Examine tissue sections	Fungi
Hematoxylin and eosin stain (Wright stain)	Examine tissue sections, blood smears	Bacteria, protozoa, fungi, cancer cells
Papanicolau stain	Examine vaginal smears	Cancer cells
Periodic acid-Schiff stain	Examine sputum, tissue, skin	Fungi
Trichrome stain	Examine stool smears	Protozoa, worms

Sometimes a preliminary diagnosis can be made based on microscopic examination of stained smears. For example, the presumptive diagnosis of gonorrhea is usually made by demonstrating the presence of gram-negative diplococci inside polymorphonuclear leukocytes (PMNs) in Gram-stained smears made from genital exudates (fig 7.5). Protozoal and helminthic infections are usually diagnosed using stained smears. The presence of malarial parasites or trypanosomes in the blood can be detected simply on the basis of Giemsa-stained thin blood smears. Fecal specimens stained using a Trichrome stain can be used to diagnose amebiasis, giardiasis, and many other protozoal and helminthic infections.

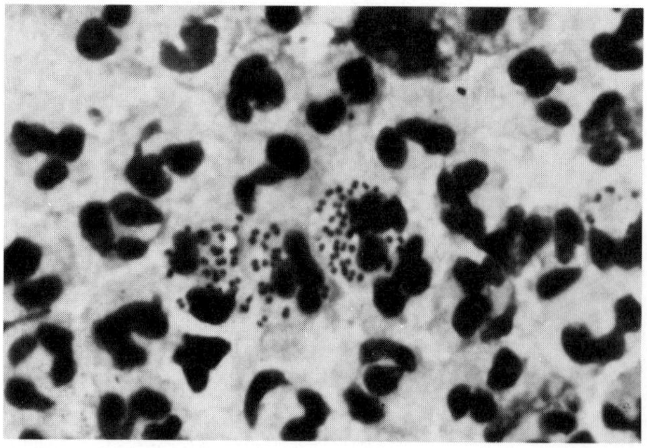

Figure 7.5 ***Neisseria gonorrhoeae* in pus cells**. Gram stain of pus from a patient suffering from gonorrhea. *Neisseria gonorrhoeae* is seen within white blood cells.

Stained histological sections (fig. 7.6) of diseased tissue are sometimes examined for the presence of pathogens. Most sections are examined using the Wright stain, which has the advantage of being able to demonstrate most infectious agents and at the same time give the pathologist an idea of the type of host reaction the pathogen is eliciting. Table 7.3 summarizes some common staining procedures used in the clinical laboratory to aid in the diagnosis of infectious diseases.

Cultural Methods in Diagnosis

The isolation and identification of pathogens from clinical specimens in modern microbiology laboratories proceeds along very logical pathways. Depending upon the source and origin of the clinical specimen, a decision is made as to which cultural protocol to follow. The decision is made based on the anatomical site affected, as well as on the evidence obtained from the microscopic examination of stained specimen preparations.

The first step in the isolation and identification of a pathogen involves culturing the specimen on special

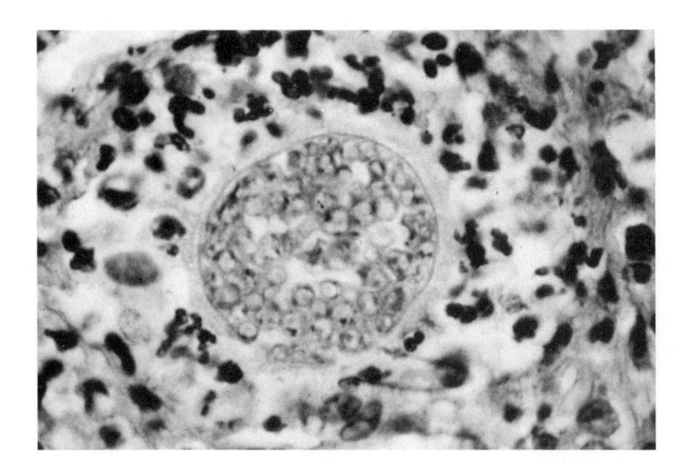

Figure 7.6 Histological section showing the spherule of *Coccidioides immitis*. This structure measures approximately 100 μm in diameter.

media (table 7.4) that allow the growth of a selected group of pathogens. The primary isolation procedure may involve treating the specimen with selective agents (e.g., NaOH, antibiotics) before culturing it on the media to increase the chances of isolating the pathogen.

The choice of appropriate media is essential for the isolation of pathogens from clinical specimens, particularly in the case of specimens such as sputum or feces, which are likely to contain contaminating microorganisms in addition to the pathogen. For example, the isolation of enteric pathogens such as *Salmonella* or *Shigella* is greatly facilitated with the use of selective and differential media. Table 7.5 summarizes various specimens commonly encountered in the

TABLE 7.4
SELECTIVE MEDIA USED IN ISOLATING PATHOGENS

Culture meduim	Specimen	Group of microorganism cultured
Baird-Parker agar	Exudates	*Staphylococcus*
Brucella agar	Blood	*Brucella* sp.
EMB agar	Feces	Enteric bacteria
Hektoen-Enteric agar	Feces	Enteric bacteria
Loeffler agar	Throat swab	*Corynebacterium diphtheriae*
Lowenstein-Jensen agar	Sputum	*Mycobacterium, Nocardia*
MacConkey's agar	Feces	Enteric bacteria
Sabouraud glucose agar	Biopsies and body fluid	Fungi
7H10 agar	Sputum	*Mycobacterium, Nocardia*
SS agar	Feces	Enteric bacteria
Staphylococcus-100 agar	Exudates	*Staphylococcus aureus*
TCBS agar	Feces	*Vibrio cholerae*
Thayer-Martin agar	Exudates	*Neisseria gonorrhoeae*
XLD agar	Feces	Enteric bacteria

TABLE 7.5
REPRESENTATIVE PATHOGENIC MICROORGANISMS AND THE SPECIMENS FROM WHICH THEY ARE ISOLATED

Clinical specimen	Possible pathogens isolated
Blood	*Salmonella typhi, Brucella, Plasmodium, Leptospira, Streptococcus,* yeasts, *Staphylococcus, Neisseria* spp., *Streptococcus pneumoniae,* enteric bacteria, *Rickettsia* spp., *Yersinia, Franciscella,* various viruses
Feces	*Salmonella, Shigella, Escherichia, Campylobacter, Vibrio,* hepatitis virus, mumps virus, polio virus, tapeworms, flukes, nematodes
Urine	*Leptospira, Salmonella, Escherichia, Proteus, Schistosoma* (fluke), yeasts
Pus	*Neisseria gonorrhoeae, Treponema pallidum, Staphylococcus, Brucella, Streptococcus, Borrelia, Yersinia, Franciscella,* herpes virus, yeasts, smallpox viruses
Sputum, throat, and nasal discharges	*Myobacterium tuberculosis, Streptococcus pneumoniae, Klebsiella pneumoniae, Mycoplasma, Haemophilus, Corynebacterium diphtheriae, Neisseria meningitidis, Bordetella pertussis, Streptococcus, Staphylococcus, Treponema,* rhinoviruses, influenza viruses, adenoviruses, pathogenic fungi, yeasts, varicella-zoster (chickenpox) virus, mumps virus. *Ascaris* (nematode), *Paragonimus* (fluke)

clinical laboratory and the possible pathogen(s) that may be isolated.

One additional procedure commonly employed in the clinical laboratory to increase the chances of isolating a pathogen is the **enrichment** procedure. This technique is particularly useful when isolating fastidious microorganisms, or microorganisms that are present in low numbers in contaminated specimens. For example, selenite-cystine broth is used to enrich for *Shigella* in fecal specimens. A portion of fresh, untreated stool is inoculated into a tube containing sterile selenite-cystine broth. This procedure promotes the growth of *Shigella* and inhibits the growth of other bacteria, thus increasing the relative numbers of the pathogen and the chances of isolating it from the stool specimen.

Fungi are routinely cultivated in the clinical laboratory. These organisms, which are important pathogens, are generally cultivated on selective media such as Sabouraud glucose agar. Sabouraud glucose agar selects against bacteria because it has a pH of about 5.7, which is too low for the growth of most bacteria. This culture medium may be supplemented with chloramphenicol and cyclohexamide to inhibit contaminating bacteria and many saprophytic fungi, respectively.

Viruses are usually grown in *in vitro* (in test tubes) tissue cultures of human or animal cells. Many laboratories do not culture viruses routinely; instead they send their specimens for virus culture to specialized laboratories. In view of the importance of viruses as human pathogens, however, and the need to diagnose viral diseases rapidly, many laboratories are growing some of the more commonly-encountered viruses, such as herpes and cytomegaloviruses.

Identification of Microorganisms Isolated from Clinical Specimens Before any attempt is made to identify a microbial isolate, the microbiologist must make sure that the culture is pure. Microscopic examination of Gram-stained smears of suspect colonies can serve as a good indication of the purity of the culture, but there is no substitute for streaking the colony on a suitable culture medium.

The choice of identification protocol is usually made after the morphology and Gram-staining properties of the isolate have been determined. Following the microscopic examination of a Gram stain, various morphological and biochemical properties of the isolate are determined. The proper choice of tests depends upon the isolate. Some microorganisms are simple to identify, while others require extensive testing. Figure 7.7 illustrates a method used to identify organisms in a vaginal sample.

The first step in identifying microorganisms is to separate them into groups that can be differentiated on the basis of their staining and morphological properties. Figure 7.8 illustrates a flowchart designed for the preliminary identification of a selected group of organisms (gram-positive cocci). The second step in identifying microorganisms involves the use of characteristics such as the isolate's ability to grow on selective media, alter differential media, produce catalase, hemolyze (break down) blood, and ferment sugars.

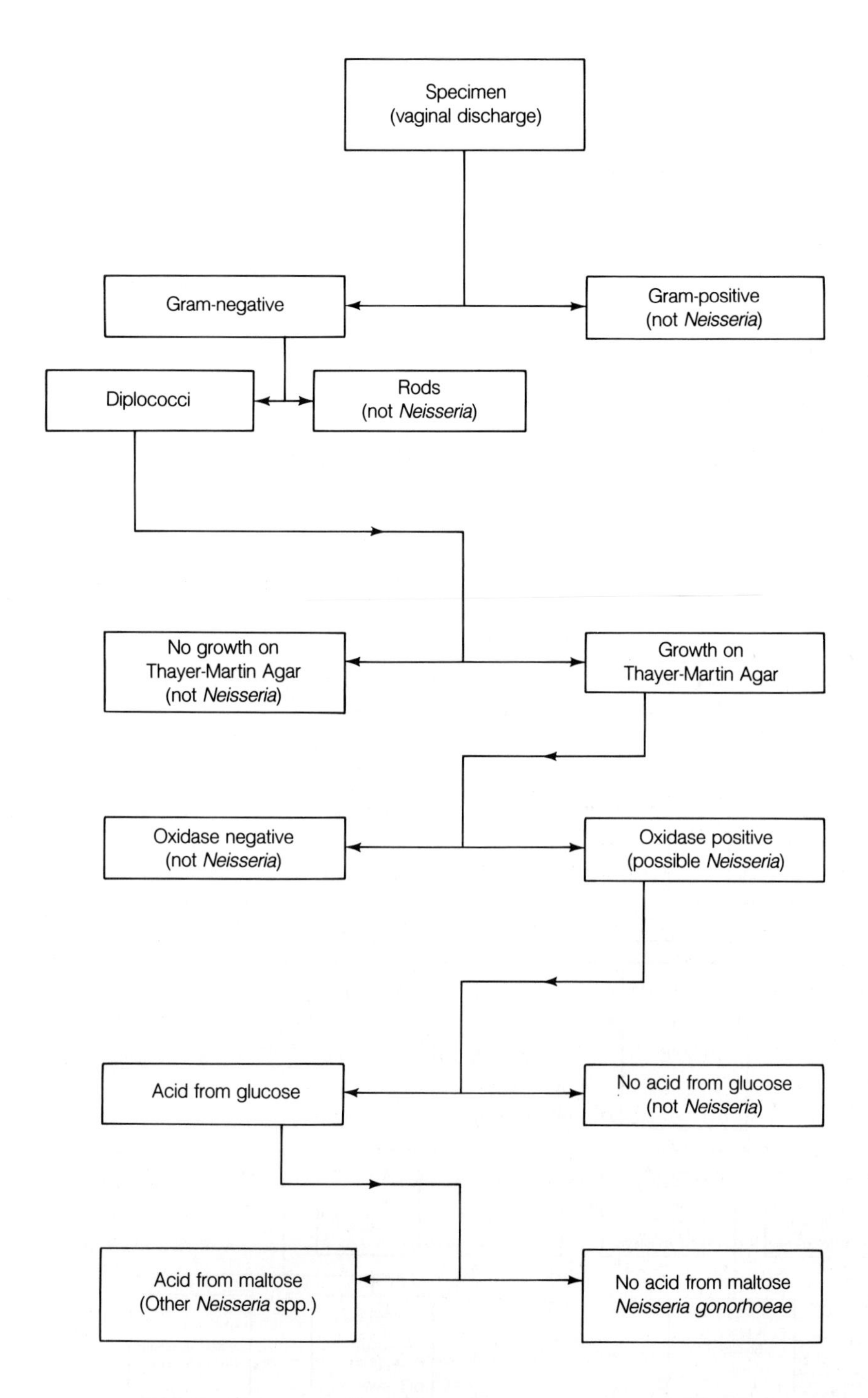

Figure 7.7 Procedure for the presumptive identification of *Neisseria gonorrhoeae*

Not all bacteria are identifiable by these simple procedures, and additional testing may be required for identification. For example, the gram-negative glucose fermenting and gram-negative glucose respiring bacteria are identified on the basis of their ability to utilize amino acids, carbohydrates, and inorganic ions to ferment or respire various carbohydrates or to hydrolyze various polymers.

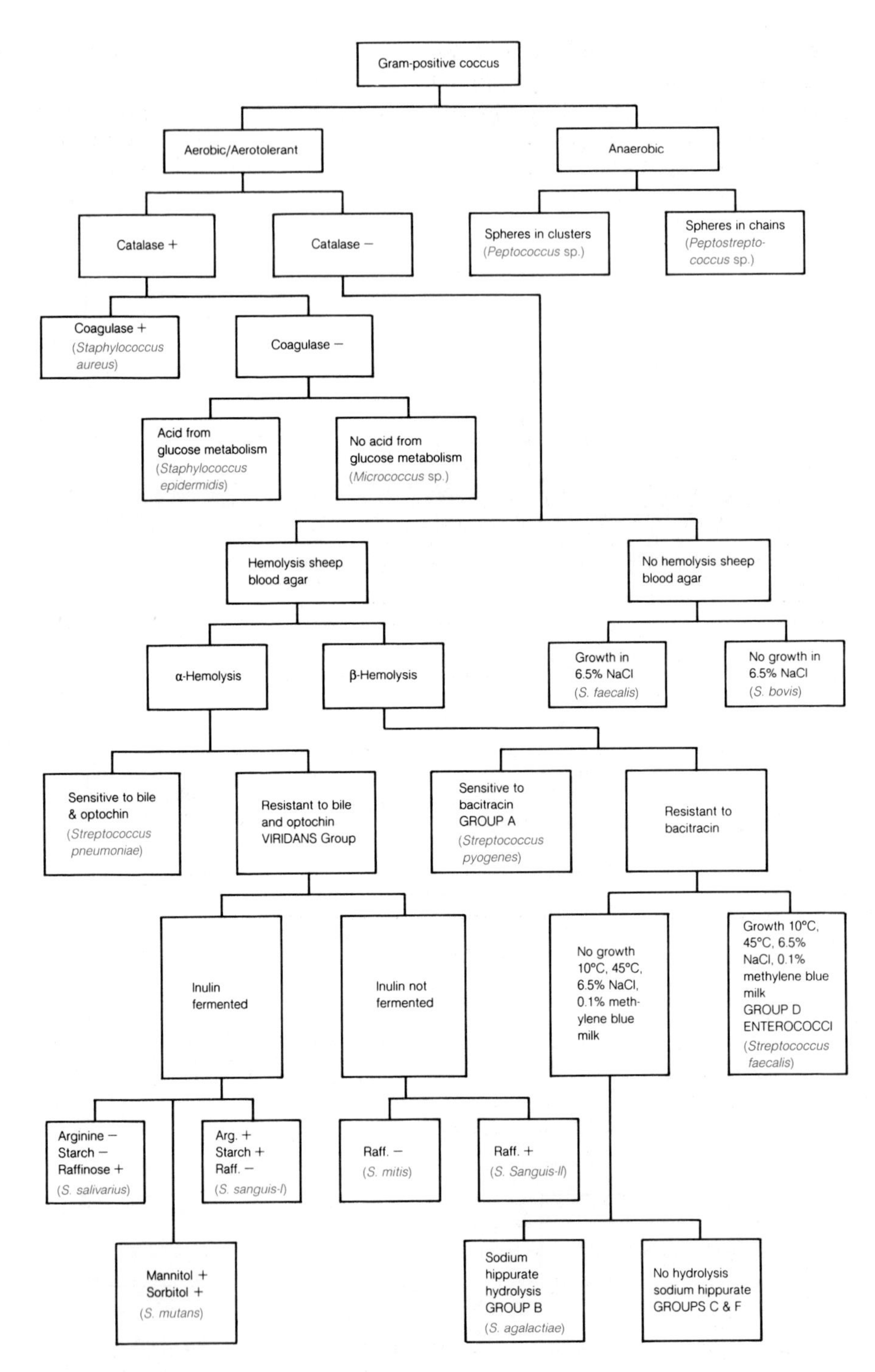

Figure 7.8 Flowchart for the preliminary identification of a selected group of organisms

Rapid Methods for the Identification of Clinical Isolates While isolation and identification of microorganisms is essential for the control of infections, it is also very expensive and time-consuming. As we can see from figure 7.8, a large number of tests may have to be carried out to identify a pathogen properly. Many of these tests involve the cultivation of microorganisms and therefore require 6–96 h of incubation before results can be obtained. In addition, many types of culture media must be prepared, stored, and checked periodically for deterioration. Most of these media must be inoculated and read individually.

Frequently, it is desirable, either for economy or expediency (or both), to use rapid methods or "kits" to identify common pathogens. There are several kits on the market (fig. 7.9) that are extremely well suited for the identification of bacteria and yeasts. Most are made up of strips or tubes containing reagents for a large number of different tests. The kits can be stored for relatively long periods of time and occupy little

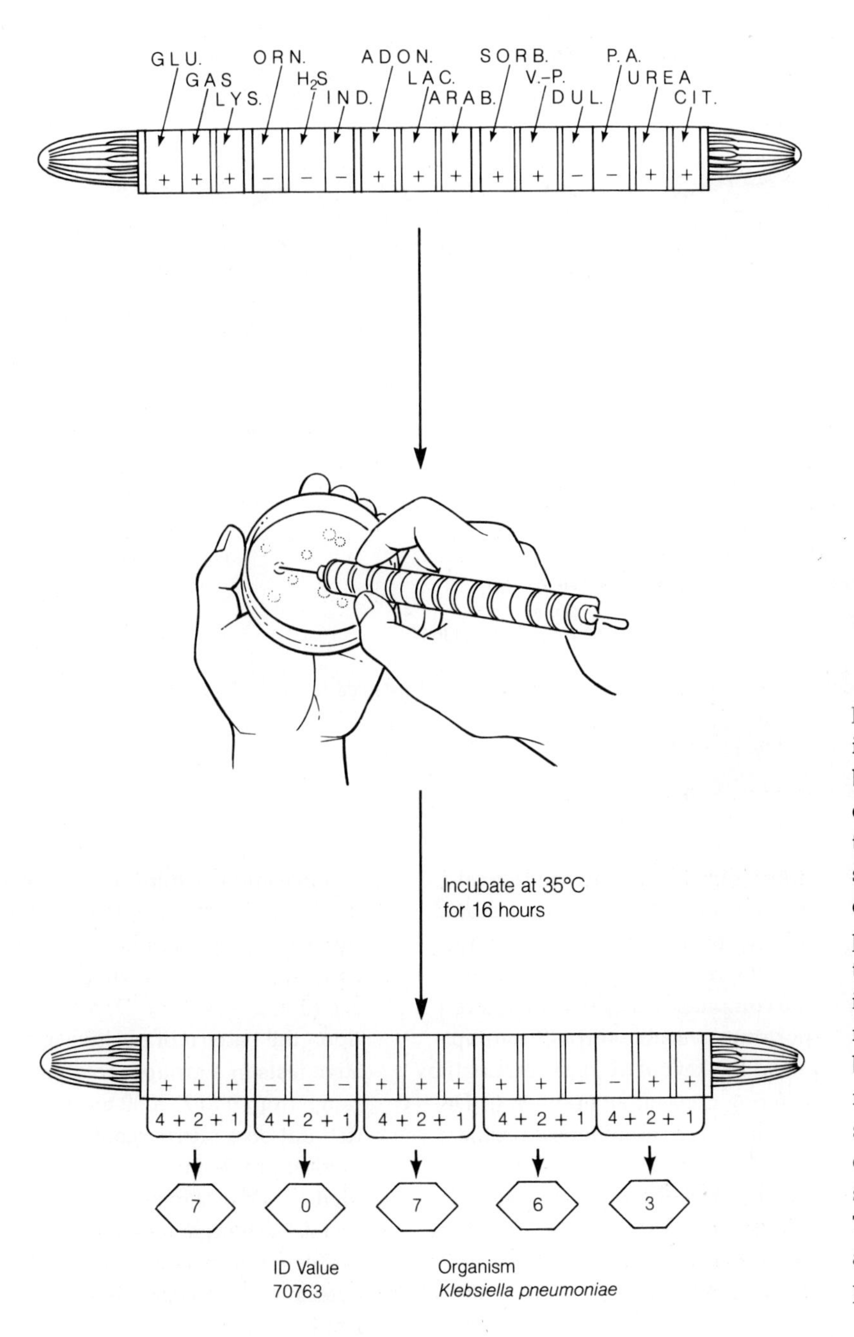

Figure 7.9 Kit used for the rapid identification of medically important bacteria. The Enterotube chambers each contain a different type of culture medium, within a plastic enclosure. All chambers are inoculated at once, using the inoculating needle provided by pulling the needle through the chambers. The source of inoculum is an isolated colony. The results are assessed after 24 h of incubation at 35°C and converted to a numerical code that corresponds to a specific organism. Tests are considered in groups of three. A positive result corresponds to a value 4, 2, or 1. The group number is determined by adding together the value of each positive result.

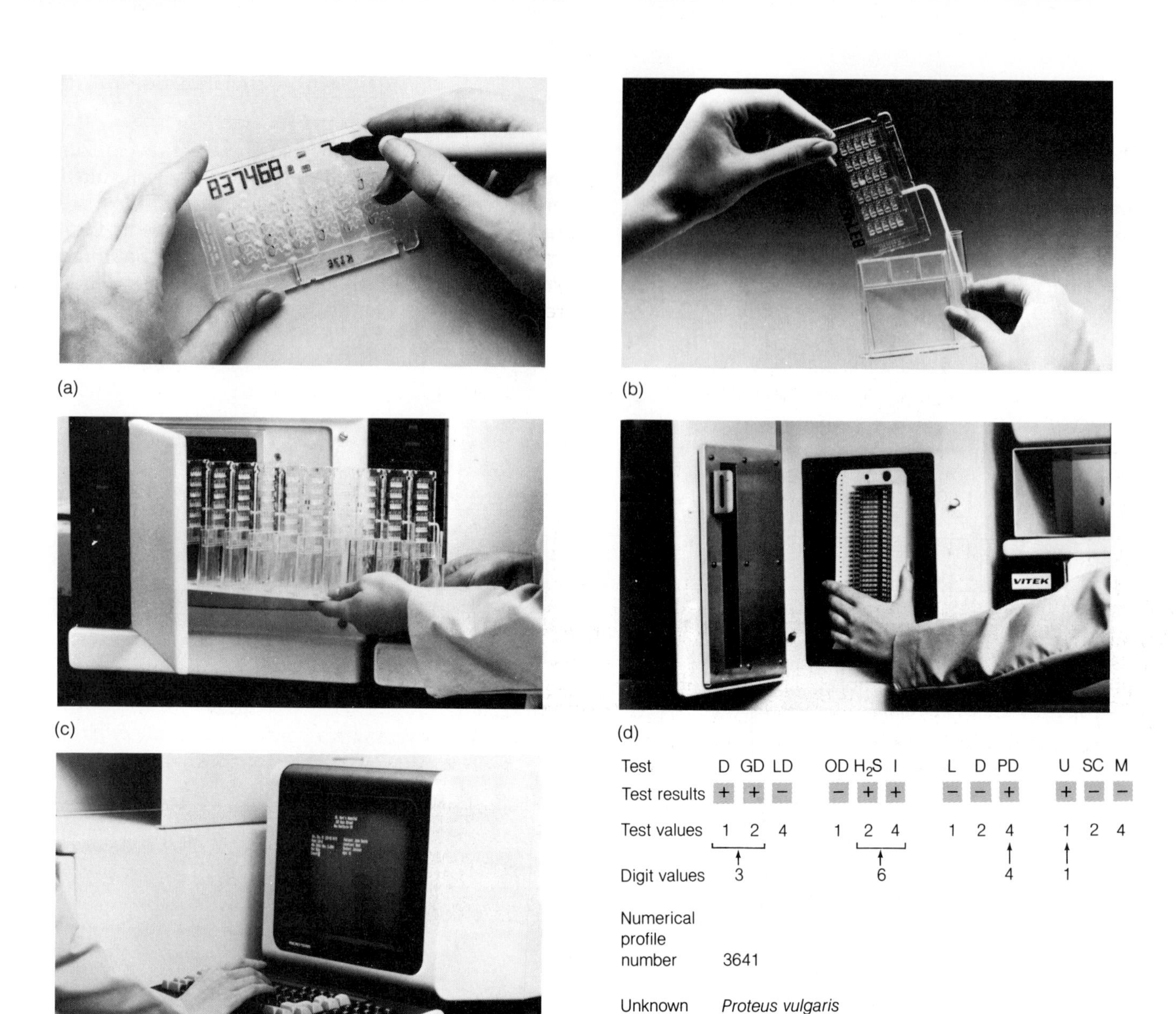

Figure 7.10 A computer used to identify and determine antimicrobial susceptibilities. *(a)* A test card contains as many as 30 wells for determining biochemical characteristics or antimicrobial susceptibilities of a microorganism. The test card is marked with an instrument-readable sample number. *(b)* A test card is placed in a holder, with its capillary straw in an aqueous sample of the microorganism to be tested. *(c)* The wells of the test card are inoculated by the filler/sealer module of the computer. A vacuum is created which draws the sample into the test card. Subsequently, the test card is sealed. *(d)* The inoculated cards are loaded into the reader/incubator module of the computer system. *(e)* Status reports are available through prompting at the keyboard (data terminal) by entering the sample number. *(f)* Determination of a numerical profile number. Tests are divided up into groups of three. In each group, the tests are given the values 1, 2, and 4, respectively. When all the tests are negative, the digit is zero (0 + 0 + 0). If the first and second tests are positive and the third test is negative, the digit is 3 (1 + 2 + 0 = 3). If the first test is negative and the second and third tests are positive, the digit is 6 (0 + 2 + 4 = 6). Depending upon the pattern of positive and negative tests in a group, the digit may range from 0 to 7. If 30 tests are carried out on a microorganism, the numerical profile number will have 10 digits, each of which may range from 0 to 7. Each organism has a numerical profile number. Once the number is determined from the tests, the organism is known.

laboratory space, and the cost per isolate identified usually is much lower than the cost of using conventional methods. Many of the manufacturers of these kits provide computerized identification services at little or no extra cost. The kits are well suited for laboratories with relatively light work loads that would prohibit the maintenance of a large number of conventional media. For clinical laboratories with large work loads, there are several automated microbial identification systems. Most are based on the ability of microorganisms to grow on media with various antibiotics or special nutrients. Microbial growth is generally measured by turbidimetric methods or color changes and identification of the isolate is based on a computer-assisted analysis of the growth pattern of the isolate on the various culture media.

Serological Methods in Diagnosis

Serological methods are those that involve reactions between certain serum proteins called **antibodies** and substances or cells known as **antigens**. Antibodies are produced by the body in response to infections by microorganisms or the presence of foreign substances or particles. Since antibodies react only with those substances that induce their formation, they can be used to determine the cause of an infectious disease. Let us assume that an expectant mother has had contact with a friend who had rubella and her physician is afraid that if she comes down with rubella, it may affect the fetus. The physician orders that a **serological** rubella test be performed in the laboratory. This is done with the assumption that if the expectant mother has been infected by the rubella virus, her immune system will respond by producing antibodies against the rubella virus. If the test is positive, it means that she has antibodies against the rubella virus and was likely infected by the virus. If the test is negative, no such infection has occurred.

Serological methods have the advantage that they are relatively easy to perform and give accurate results, and the results can be obtained in a few hours. Most infectious diseases are presently diagnosed using a serological method as part of the diagnosis protocol. In instances where the causative agent cannot be cultured in the laboratory or the cultural procedure is very time-consuming, serological methods are indicated. This topic of serology is taken up in greater detail in Chapter 13.

Computer-Assisted Identification

Computers are being used to identify organisms in half the time this process once took (fig. 7.10). Computerized systems such as the Vitek system illustrated in figure 7.10 can easily analyze 30 biochemical tests in 10 to 12 hours and then determine which organisms are most likely to fit the results. The computer program compares the test results with a large list of isolates that have previously been characterized. In addition, the computer can indicate which antibiotics are most effective.

Computers are now being used to determine, in one trial, the susceptibility of a microorganism to as many as 13 antimicrobials in multiple concentration. Inoculum for the trial is made from three or four colonies emulsified in 1.8 ml of sterile saline solution. After 4 hours of incubation, the computer checks the control sample to determine whether there has been sufficient growth. Susceptibilities for fast-growing organisms, such as the gram-negative rods, generally can be determined after 4 hours of incubation. If the growth is not adequate, however, the computer automatically expands the incubation period and records results when there is sufficient growth of the control. The computer printout indicates whether the organism is sensitive, intermediate in sensitivity, or resistant to the antimicrobial. The computer system eliminates the time-consuming setup of other procedures, and removes subjective error in reading results. Best of all, the computer system can give susceptibilities within 4 to 10 hours, while traditional tests require 24 hours.

SUMMARY

TAXONOMY AND THE KINGDOM OF ORGANISMS

1. Taxonomy is the science that arranges organisms into categories called taxa, names them, and identifies them. Thus, classification, nomenclature, and identification are important aspects of taxonomy.
2. Organisms are generally classified (grouped) on the basis of similarities in morphology and physiology, and are placed in kingdoms, divisions, classes, orders, families, genera, and species.
3. Every cellular organism is given a genus name and a species name. The system of naming is called the binomial system of nomenclature and was developed by the Swedish botanist Carolus Linnaeus in the mid 1700s.
4. R. H. Whittaker, in 1969, proposed that all cellular organisms be placed in one of five kingdoms: Animalia, Plantae, Fungi, Protista, and Monera. The eubacteria, archaebacteria, and cyanobacteria are in the kingdom Monera. All bacteria are placed in a genus and given a species name. Depending

upon the classification system used, bacteria may also be placed in a family, an order, and a class.

5. Woese and other scientists proposed a three-kingdom system that includes the eubacteria, the archaebacteria, and the eukaryotes.

METHODS USED TO CLASSIFY AND IDENTIFY MICROORGANISMS

1. Since an organism's DNA specifies its RNAs and proteins and since these in turn determine the molecules that make up the cellular structure, an organism can be studied on a number of levels. For instance, cells can be identified and classified on the basis of their nucleic acids, proteins, cell components, or morphology.
2. The degree to which DNAs from different organisms hybridize indicates their similarity.
3. Usually organisms are identified and classified on the basis of their gross morphology and physiology. Many bacteria can be identified and classified by a few characteristics, such as shape (rod, sphere, coil), staining characteristics (gram-positive, gram-negative, etc.), type of metabolism (aerobic, anaerobic, facultative), special structures (endospores, appendages), and type of physiology (growth on selective and differential media).

DIAGNOSIS OF INFECTIOUS DISEASES

1. Isolation and identification of infectious agents essential for the proper management of infectious diseases involves the proper collection and transport of clinical specimens and proper laboratory procedures.
2. Identification of microorganisms may involve microscopic methods, cultural methods, and/or serological methods.
3. Collection and transport are essential to the proper isolation of pathogens.
4. The proper collection of specimens requires knowing how and when to collect the specimen. It also requires that every effort be made to prevent the introduction or removal of any source of toxic materials and to prevent contamination. Specimens should be delivered promptly to the laboratory and maintained refrigerated until they can be processed.
5. Microscopic methods can help in establishing a quick diagnosis of an infectious disease, or at least give a good indication as to the best course to follow to make an accurate diagnosis.
6. Wet mounts, stained smears, and stained histological sections are all useful in viewing clinical specimens for the presence of infectious agents.
7. Clinical specimens are inoculated on selective and/or differential media to increase the chances of isolating the pathogen from the specimen. Sometimes the clinical specimen is treated with selective agents, such as alkali or antibiotics, before inoculating onto culture media, in order to increase further the chances of isolating the pathogen.
8. Once a suspected pathogen has been isolated, it is then identified using logical sequences of testing that include staining procedures, cultural and biochemical tests, and serological tests.
9. Sometimes it is desirable, for economy or expediency, to use rapid methods or kits to identify pathogens. These kits have several advantages over the conventional methods: low cost per isolate, high reliability, and long shelf life.
10. Serological methods for diagnosing infectious diseases are desirable when the suspected agent is difficult or expensive to grow in the laboratory, or when the results are needed quickly.
11. Several different serological tests are commonly used in the laboratory, but all are based on the same principle: an individual infected by a microorganism will respond immunologically to the infectious agent, and the response can be detected with these serological methods.

STUDY QUESTIONS

1. Discuss three important characterictics of a properly collected specimen.
2. Why are transport media used in the clinical laboratory?
3. Outline a procedure that can be used to isolate a pathogen from a urine specimen.
4. Given that you have isolated a bacterium from a stool specimen, outline the steps that you will take to identify this organism.
5. What is the importance of selective and differential media in the identification of pathogens?
6. Why are serological techniques so popular in the clinical laboratory?
7. What is the principle on which serological reactions are based?

8. What are the three areas of taxonomy?
9. Define classification, identification, and nomenclature.
10. Explain how DNA–DNA hybridization is used to identify and classify bacteria.

SELF-TEST

1. The kingdoms Protista and Monera were introduced after Linnaeus' two-kingdom system of classification because microorganisms did not fit well into either the plant or the animal kingdom. a. true; b. false.
2. All of the following are criteria used to classify bacteria *except*:
   a. Gram-staining properties;
   b. presence of endospores;
   c. length of incubation;
   d. byproducts they release during fermentation;
   e. colony morphology.
3. Which of the following methods for identifying and classifying bacteria is *not* routinely used in the clinical laboratory?
   a. Gram-staining properties;
   b. DNA–DNA hybidization;
   c. colony characteristics;
   d. biochemical reactions;
   e. the production of endospores.
4. Which of the following criteria listed below is *not* considered adequate for the proper collection of specimens?
   a. specimens should be collected before antimicrobials are given;
   b. specimens should be collected from the active site of infection;
   c. specimens should be collected when pathogens are likely to be present;
   d. specimens should be stored in the refrigerator for at least 1 hour before processing.
5. The diagnosis of vaginitis (*Trichomonas vaginalis*) and syphilis (*Treponema pallidum*) is normally accomplished after culturing the organisms in suitable culture media. a. true; b. false.
6. Which of the following disease-causing bacteria is *not* normally found in clinical specimens consisting of throat swabs?
   a. *Mycobacterium tuberculosis;*
   b. *Neisseria;*
   c. *Streptococcus;*
   d. *Bordetella;*
   e. *Salmonella.*
7. The first step in the identification of microorganisms isolated from clinical specimens is to make sure that the culture is in pure form (i.e., uncontaminated). a. true; b. false.
8. Which of the following tests is *not* normally used to identify *Neisseria gonorrhoeae?*
   a. the Gram stain;
   b. growth on Thayer-Martin medium;
   c. the catalase test;
   d. the oxidase test;
   e. the production of acid from glucose.
9. Serological methods of diagnosis are based on the principle that infected individuals produce antibodies against the infectious agent. Therefore, the presence of antibodies in the patient's serum indicates present or prior infection. a. true; b. false.
10. Rapid methods of diagnosis of infectious diseases include kits in which several different culture media are packaged in one container and are inoculated at once. a. true; b. false.

SUPPLEMENTAL READINGS

Centers for Disease Control. 1985. Provisional public health service inter-agency recommendations for screening donated blood and plasma for antibody to the virus causing acquired immunodeficiency syndrome. *Morbidity and Mortality Weekly Report* 34(1):5–7.

Delmee, M., Homel, M., and Wauters, G. 1985 Serogrouping of *Clostridium* difficile strains by slide agglutination. *Journal of Clinical Microbiology* 21(3):323–327.

Forney, J. E., ed. 1983. *Collection, handling, and shipment of microbiological specimens.* Public Health Service Publication No. 976. Washington, DC: U.S. Public Health Service.

Grimont, P. A. D., Grimont, F., Desplaces, N., and Tchen, P. 1985. DNA probe specific for *Legionella pneumophila. Journal of Clinical Microbiology* 21(3):431–437.

Jones, W. T., Nagle, D. P., and Whitman, W. B. 1987. Methanogens and the diversity of archaebacteria. *Microbiological Reviews* 51(1):135–177.

Karachewski, N. O., Busch, E. L., and Wells, C. L. 1985. Comparison of PRAS II, RapID ANA, and API 20A systems for identification of anaerobic bacteria. *Journal of Clinical Microbiology* 21:122–126.

Lennette, E. H., Balows, A., Hausler, W. J. Jr., and Shadomy, H. T., eds. 1985. *Manual of clinical microbiology.* 4th ed. Washington, DC: American Society for Microbiology.

Pace, N. R., Stahl, D. A., Lane, D. J., and Olsen, G. J. 1985. Analyzing natural microbial populations by RNA sequences. *ASM News* 51(1):4–12.

Woese, C. R. 1987. Bacterial evolution. *Microbiological Reviews* 51(2):221–271.

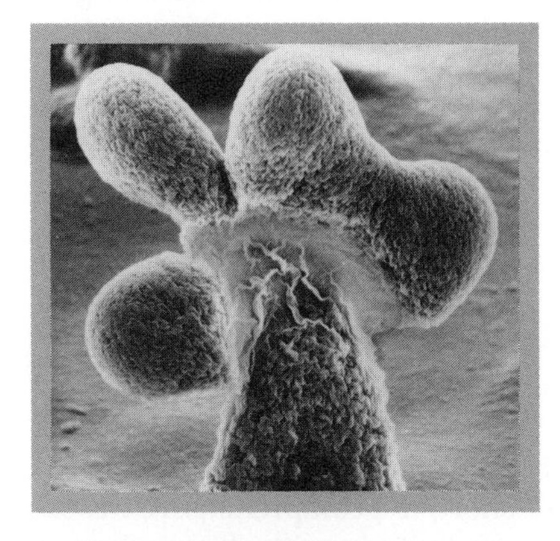

CHAPTER 8

THE BACTERIA

CHAPTER PREVIEW MAKING PROTEIN FROM AIR POLLUTANTS USING BACTERIA

Single-cell protein is a term frequently used to describe protein-rich microorganisms that can be used as human or domestic animal food. The ever-increasing world population and ever-decreasing natural resources have stimulated research in the area of single-cell protein. The idea is to grow large quantities of protein-rich microorganisms using cheap, readily available materials or industrial waste products. Microorganisms such as yeasts, bacteria, fungi, and algae have been grown on a variety of inexpensive substrates, including whey, potato starch, paper pulp waste, coffee, and molasses. Studies on the growth habits of these microorganisms and their nutritional requirements have proved fruitful, and much single-cell protein is being produced using modern technology.

These studies have also revealed that a group of bacteria called the **carboxydobacteria** can use carbon monoxide (CO) from automobile exhausts as their sole source of carbon atoms, as well as their source of energy for life processes.

Carbon monoxide is a principal pollutant from the internal combustion engine, and in sufficiently high quantities can cause death by poisoning. The discovery of the carboxydobacteria served as an important first step in the bioconversion of the poisonous carbon monoxide into a life-giving product.

The bacterium *Pseudomonas carboxydovorans*, one of several CO-consuming bacteria, has been shown to reproduce well in an environment containing automobile exhaust and air in a 50:50 mixture. From this mixture, *P. carboxydovorans* produces a population of cells containing approximately 65% crude protein. It has been determined that the CO produced by the estimated 350,000,000 cars in the world could be used to make about 500,000 tons of dry cell protein/year, or about 2.5 g dry weight protein/car/year, a significant contribution to the nutrition of countless humans who are presently living under conditions of near starvation.

The discovery of the carboxydobacteria represents one of the many discoveries of useful microorganisms by microbiologists. The discoveries were made only after scientists had a clear understanding of the nutritional requirements of the desired microorganisms and how to grow them in the laboratory.

The study of bacteria often results in the discovery of new and exciting life forms that perform important functions in nature. Some of these functions are essential for the survival of the human race, while others are harmful. In this chapter we study some of the prominent microbial groups and emphasize the role that they play in nature or in human affairs.

THE BACTERIA ARE AN EXTREMELY HETEROgeneous group of prokaryotic microorganisms. They inhabit very diverse environments in which they perform numerous functions, many of which are essential for life on earth. For example, bacteria recycle nutrients in the biosphere, making them available to all living organisms. Bacteria also perform activities that are economically and medically valuable to humans such as the production of antibiotics and the fermentation of foods and beverages. Not all bacteria carry out useful activites in nature. There are various groups of bacteria that are best known by their ability to cause disease in plants and animals.

To appreciate the impact that these microorganisms have on human affairs, it is necessary that we know a little about their biology. This chapter summarizes the characteristics of the major groups of bacteria, emphasizing those of medical or industrial importance.

BERGEY'S MANUAL OF SYSTEMATIC BACTERIOLOGY

Bergey's Manual of Systematic Bacteriology also called ***Bergey's Manual*** for short, is a comprehensive treatise on the classification of bacteria with descriptions of bacterial species, taxonomic keys to the various genera in families, and practical tables to identify species. It is used extensively by bacteriologists as a reference source because it classifies bacteria into logical groups. The 8th edition is a scheme of classification based on a few readily determined characteristics (table 8.1). The bacteria are classified into 19 parts, based on features such as gram reaction, cell shape and arrangement, metabolic and nutritional capabilities, and oxygen requirements. The recently published 9th edition (1984) subdivides the bacteria into four groups, each discussed in a separate volume: (1) the gram-negatives of general, medical, or industrial importance; (2) the gram-positives other than actinomycetes; (3) the archaebacteria, cyanobacteria, and remaining gram-negatives; and (4) the actinomycetes.

GRAM-NEGATIVE BACTERIA OF MEDICAL OR APPLIED IMPORTANCE

This is a broad group that includes primarily human pathogens (disease-causing bacteria), for example, the spirochetes, helical and curved bacteria, aerobes, anaerobes and facultative anaerobes, rickettsias, and mycoplasmas.

Spirochetes

The **spirochetes** (fig. 8.1) are slender, flexible bacteria that are helically coiled. Different genera of spirochetes show considerable variability in length, width, and amount of coiling. The human pathogenic species *Treponema pallidum* and *Leptospira interrogans* measure approximately 0.1 μm in diameter and 5–18 μm in length (table 8.2). These organisms are usually invisible using routine light microscopic techniques and special procedures are required to view them. In dark

TABLE 8.1
SUMMARY OF THE VARIOUS GROUPS OF BACTERIA ACCORDING TO THE 8TH EDITION OF *BERGEY'S MANUAL OF DETERMINATIVE BACTERIOLOGY*

Part	Name of group	Brief description of group	Sample genera
1	Phototrophic bacteria	Gram-negative spheres or rods. Carry out anoxygenic photosynthesis. Colonies pigmented red, orange, green, purple.	*Chromatium, Chlorobium, Rhodospirillum*
2	Gliding bacteria	Gram-negative rods surrounded by a slime coat. May aggregate to form fruiting bodies. Move on solid surfaces by gliding.	*Beggiatoa, Cytophaga, Stigmatella*
3	Sheathed bacteria	Gram-negative rods aranged in chains enclosed within a sheath.	*Sphaerotilus, Leptothrix*
4	Budding and/or appendaged bacteria	Gram-negative rods, cocci or vibrios that divide by budding or binary fission. Division products are dissimilar.	*Caulobacter, Hyphomicrobium Gallionella*
5	Spirochetes	Gram-negative, slender, coiled cells with flexible walls. Divide by transverse fission.	*Treponema, Borrelia, Leptospira*
6	Spiral and curved bacteria	Gram-negative, helically curved rods with rigid cell walls.	*Bdellovibrio, Campylobacter*
7	Gram-negative, aerobic rods and cocci	Generally motile organisms that carry out aerobic respiration. Many species are pathogens.	*Pseudomonas, Brucella, Azotobacter*
8	Gram-negative facultatively anaerobic rods	Straight and curved rods and vibrios that carry out various types of fermentation. Many species are motile by flagella.	*Escherichia, Vibrio, Salmonella, Shigella*
9	Gram-negative anaerobic rods	Obligate anaerobes that tend to be pleomorphic. No endospores formed.	*Bacteroides, Fusobacterium, Desulfovibrio*
10	Gram-negative cocci and coccobacilli	Spherical organisms that tend to cluster into parts and chains. Facultative anaerobes or aerobes.	*Neisseria, Acinetobacter*
11	Gram-negative anaerobic cocci	Spherical cells with anaerobic type of metabolism.	*Veillonella*
12	Chemolithotrophic bacteria	Gram-negative rods and cocci. Use inorganic substances as energy sources.	*Nitrobacter, Nitrosomonas, Thiobacillus*
13	Methanogenic bacteria	Gram-positive and gram-negative anaerobes that produce methane during metabolism.	*Methanosarcina, Methanococcus*
14	Gram-positive cocci	Spherical bacteria arranged in clusters and chains. Aerobes, anaerobes, or facultative anaerobes.	*Staphylococcus, Streptococcus, Micrococcus*
15	Endospore-forming rods and cocci	Gram-positive bacteria that form endospores. Aerobes, anaerobes, and facultative anaerobes.	*Bacillus, Clostridium, Sporosarcina*
16	Gram-positive asporogenous rods	Aerobes, anaerobes, and facultative anaerobes.	*Listeria, Lactobacillus*
17	Actinomycetes and related organisms	Gram-positive rods that tend to form filaments. Some species are acid-fast.	*Mycobacterium, Nocardia, Corynebacterium*
18	Rickettsias	Gram-negative bacteria with typical cell walls. Most species are obligate intracellular parasites. Chlamydiae lack peptidoglycan in their cell walls.	*Chlamydia, Rickettsia, Coxiella*
19	Mycoplasmas	Bacteria without cell walls. Highly pleomorphic. Aerobes and anaerobes.	*Mycoplasma, Spiroplasma*

SOURCE: From *Bergey's Manual of Determinative Bacteriology*, 8th ed. Copyright © 1974 by Williams & Wilkins Co., Baltimore. Reprinted by permission.

Figure 8.1 Structural features of the spirochetes. *(a)* Scanning electron micrograph of a spirochete showing the axial filament wound around the body of the cell. *(b)* Transmission electron micrograph of a palladium-platinum shadowed spirochete exhibiting a fibril and the outer sheath. *(c)* Transmission electron micrograph of *Cristispira*. Note the numerous axial fibrils running lengthwise along the cell.

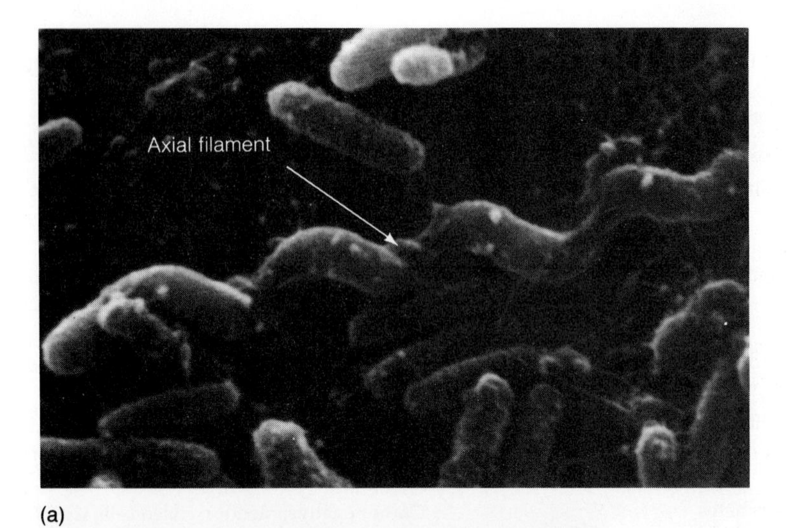

(a)

(b)

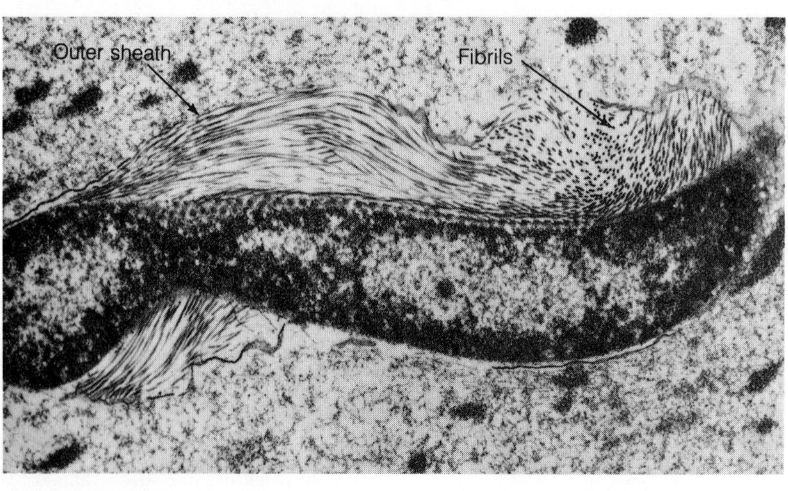

(c)

TABLE 8.2
SUMMARY OF CHARACTERISTICS OF MEDICALLY IMPORTANT SPIROCHETES

Characteristic	*Borrelia*	*Leptospira*	*Treponema*
Length of cell (μm)	5–30	5–18	6–15
Width of cell (μm)	0.4	0.1	0.2
Coils			
Number	<10	30–40	5–20
Arrangement	Irregular	Tightly wound	Regular
Number of fibrils	15–20	1	3
O_2 requirement	Anaerobic	Aerobic	Anaerobic
Energy metabolism	Fermentative	Respiratory	Fermentative
Cultivated *in vitro*	Yes	Yes	No
Diseases caused	Vincent's angina, Lyme's disease	Leptospirosis	Syphilis, Yaws

field preparations, the spirochetes can be seen in tight coils, flexing and bending in a characteristic fashion.

Modified flagella called **fibrils** allow these bacteria to be motile. The fibrils may be arranged in a bundle known as the **axial filament** (fig. 8.1a). The fibrils originate near the two ends of the cell and fold back along the cell, between the peptidoglycan layer and an **outer sheath** (fig. 8.1b). The fibrils are not free to whirl about like flagella of other motile bacteria because they are contained within the outer sheath and wound around the cell. As the fibrils try to rotate, they make the cell spin like a corkscrew, imparting a characteristic form of motility to the cell.

Some species of spirochetes can cause disease in humans. For example, *Treponema pallidum* causes the disease syphilis; *Borrelia recurrentis* causes relapsing fever; and *Leptospira interrogans* causes leptospirosis or Weil's disease. These diseases are of great public health importance and will be discussed in Part 3.

Spiral and Curved Bacteria

The **spiral and curved bacteria** are rigid, helically curved or S-shaped rods (fig. 8.2). Many species of these gram-negative bacteria are motile by means of polar flagella. Some members of this broad group parasitize mammals and even other bacteria. They are readily distinguished from the spirochetes because they possess a rigid cell wall and their flagella are not enclosed by a sheath.

Some species of *Campylobacter* are pathogenic to humans. Some other members of the genus, however, like *C. sputorum*, are found in the human oral cavity in the absence of disease. *Campylobacter jejuni* has been found to infect humans and to cause a variety of foodborne diseases, such as enteritis (inflammation of the intestines), subacute bacterial endocarditis (inflammation of the heart), meningitis (inflammation of the membranes surrounding the brain and spinal cord), and septicemia (invasion of the blood or tissues).

Recent studies have revealed that *C. jejuni* is one of the most common causes of gastroenteritis in human populations. Its role in the disease became apparent when microbiologists started culturing fecal specimens on a routine basis at 40–42°C in an elevated CO_2 atmosphere. These cultural conditions are ideal for the cultivation of *Campylobacter* and until this fact was broadly recognized the organism was infrequently isolated and thus, its role in human disease was not appreciated.

Gram-Negative Aerobic Rods

The **gram-negative aerobic rods** are all chemoheterotrophs. These bacteria are remarkably versatile in their ability to assimilate organic materials and to perform biochemical activities.

Pseudomonas aeruginosa is an important member of this group (fig. 8.3). Because it causes disease only when hosts are weakened or compromised, *P. aeruginosa* is considered to be an **opportunistic** pathogen. This bacterium is commonly found in moist environments and in many hospitals. *Pseudomonas aeruginosa* is of major concern in hospitals because it is capable of aggravating surgical wounds and burns or initiating pulmonary (lung) infections in debilitated patients. This organism grows well on water faucet aerators, so

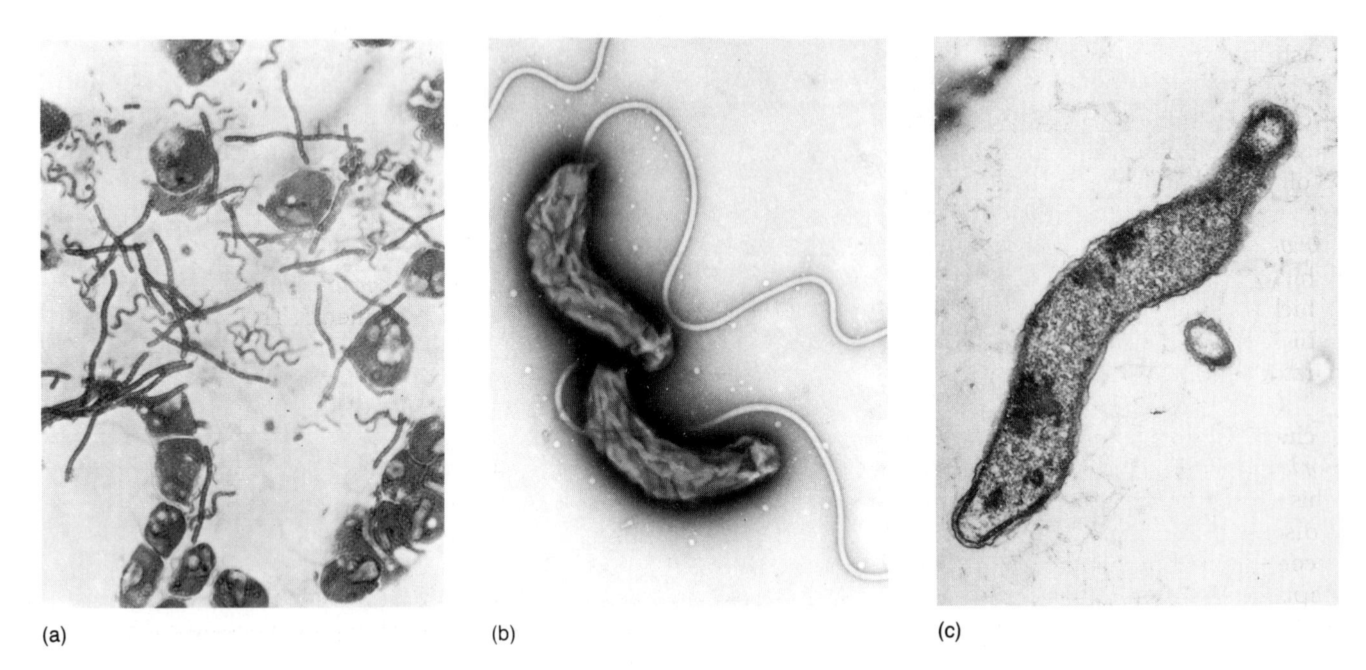

Figure 8.2 **Various species of spiral and curved bacteria.** *(a) Spirillum. (b) Bdellovibrio bacteriovorus. (c) Campylobacter.*

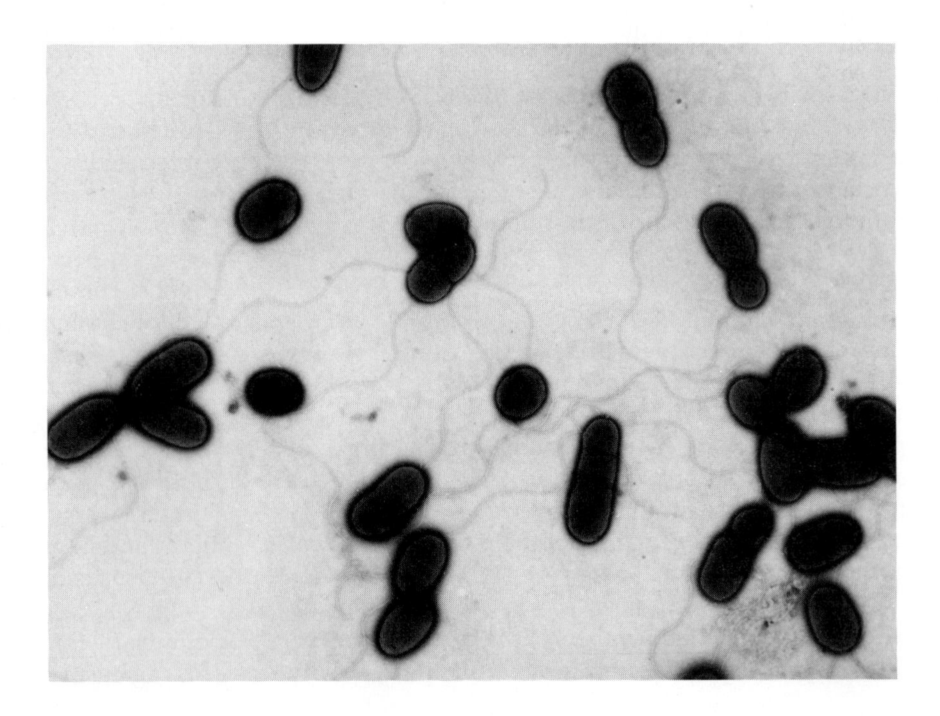

Figure 8.3 ***Pseudomonas,* a typical gram-negative, aerobic bacterium.** A transmission electron photomicrograph of *Pseudomonas aeruginosa* stained with phosphotungstic acid.

washing may contaminate materials such as oxygen tent air hoses, ventilators, nebulizers, and nursery incubators. If these pieces of equipment are not thoroughly dried after washing, contaminating bacteria may proliferate and increase the likelihood of an infection.

Other pathogenic pseudomonads include *Pseudomonas cepacia*, which causes **onion bulb rot**; *P. mallei*, which causes the respiratory tract disease **glanders**, which affects horses and humans; and *P. pseudomallei*, which causes **melioidosis**, another disease of the respiratory tract.

Other medically important bacteria in this group include *Bordetella*, *Francisella*, *Brucella*, and *Legionella*. *Bordetella pertussis* is an important pathogen of humans. This short, gram-negative rod causes whooping cough, a disease afflicting primarily children. Until relatively recently, whooping cough was prevalent in human populations. With the advent of a vaccine, however, the incidence of this disease in children has been significantly reduced.

Legionella pneumophila (see Chapter 7 preview) is an aerobic, gram-negative bacterium that causes a pneumonia (lung inflammation with fluid accumulation) better known as Legionnaire's disease. The bacterium responsible for the 1976 outbreak of Legionnaire's disease in Philadelphia has recently been recognized as a common cause of pneumonia in human populations. *Legionella pneumophila* can exist in many different aquatic environments. The source of the bacterium that infected the legionnaires in Philadelphia was contaminated water used to cool the air in the hotel's air conditioning system.

Bacteria in the genera *Azotobacter* and *Rhizobium* are of applied importance because they carry out a process called nitrogen fixation. These gram-negative, aerobic rods are soil inhabitants that contribute to soil fertility by converting molecular nitrogen, a form of nitrogen largely unusable to living organisms, into organic nitrogen, which can readily be used by most living organisms.

FOCUS *PSEUDOMONAS* HELPS PREVENT PLANT DISEASE

Pseudomonas is a gram-negative rod that is widely distributed in nature. These bacteria have a remarkable ability to assimilate a wide variety of organic substrates, and hence can reproduce in many different environments. Certain species of *Pseudomonas* can extensively colonize plant root surfaces. It has been noted that plants that have these bacteria colonizing their roots (fig. 8.4) are less likely to acquire certain diseases than plants devoid of *Pseudomonas*. Plant pathogens like the fungi *Fusarium*, *Pythium*, and *Phytophthora* are inhibited, or prevented, from causing disease in these plants by the growth of *Pseudomonas*.

The reason that *Pseudomonas* suppresses the growth of plant pathogens is that it produces iron-binding compounds known as **siderophores**. The siderophores bind iron and sequester it from the area around the root. In the absence of available iron, plant pathogens cannot reproduce.

The discovery of a species of *Pseudomonas* that protects plants from disease is one of the many examples supporting the fact that bacteria perform useful functions in nature.

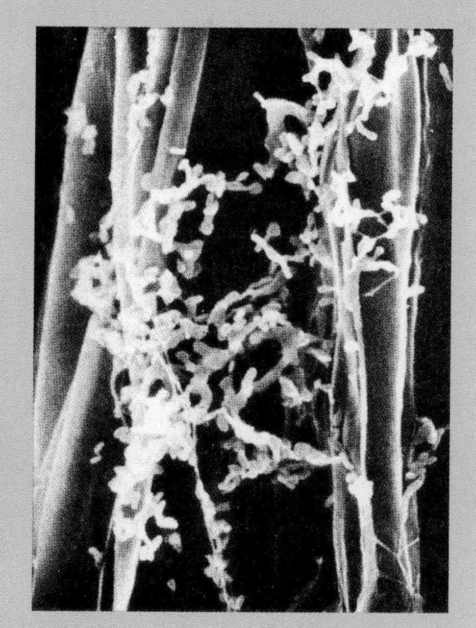

Figure 8.4 ***Pseudomonas* attached to the roots of a plant host**

Gram-Negative, Facultatively Anaerobic Rods

The **gram-negative, facultatively anaerobic rods** are a large group of chemoheterotrophic bacteria that are widely distributed in nature (table 8.3) and includes many important human pathogens. These bacteria are currently grouped into two families: the Enterobacteriaceae and the Vibrionaceae. The two families are differentiated largely on the basis of biochemical and morphological features.

The **enterics**, or the Enterobacteriaceae family, are straight rods that characteristically inhabit the intestines of humans and other animals. Some genera, however, can also be isolated from soils or from aquatic environments. The enterics include some very important pathogens such as the agents of typhoid (*Salmonella typhi*), bacillary dysentery (*Shigella dysenteriae*), and gastroenteritis (*Salmonella* and *Shigella*). Although the enterics cause gastrointestinal disease predominantly, they are also capable of causing urinary tract infections, wound infections, pneumonia, meningitis, and septicemia. *Escherichia coli* has been implicated in many of these diseases.

The **Vibrios**, or the Vibrionaceae family, are gram-negative, curved rods. They are motile by means of polar flagella and obtain their energy for metabolism by respiring and fermenting organic substrates. This group contains several important human pathogens. For example, *Vibrio cholerae* causes the severe and often fatal disease, cholera. The bacterium produces an enterotoxin that induces an excessive fluid loss from

TABLE 8.3
CHARACTERISTICS OF SOME GRAM-NEGATIVE, FACULTATIVELY ANAEROBIC BACTERIA

Organism	Salient characteristics
Family: Enterobacteriaceae (enterics)	Usually straight rods.
Escherichia coli	Inhabit the gastrointestinal tract of mammals. May cause enteric (intestinal) diseases. Used as an indicator of fecal contamination.
Salmonella *S. typhi* *S. enteritidis* *S. arizona* *S. paratyphi*	Pathogenic bacteria. Cause typhoid fever (*S. typhi*) and gastroenteritis. Common agents of food poisoning.
Shigella *S. dysenteriae* *S. sonnei* *S. flexneri* *S. boydii*	Agents of shigellosis or bacillary dysentery. Some species produce exotoxins. Inhabit the gastrointestinal tract of mammals. Transmitted in food and water.
Citrobacter freundii	Saprophyte in water, foods, and body wastes. May be associated with a variety of infections.
Klebsiella pneumoniae	Widely distributed in nature. Commensal in the intestinal tract of mammals. Can cause a variety of diseases including gastroenteritis, pneumonia, and urinary tract infections.
Enterobacter *E. aerogenes* *E. cloacae*	Inhabit the gastrointestinal tract of mammals. Can cause enteric and urinary tract infections.
Serratia marcescens	Opportunistic pathogens. Form red-colored colonies. Widely distributed in nature.
Proteus *P. vulgaris* *P. mirabilis* *P. inconstans*	Some species cause urinary tract infections, others cause diarrheas. All found in intestines of mammals.
Yersinia *Y. pestis*	The agents of bubonic plague. Found in nature infecting small feral rodents. Transmitted to humans by fleas.
Y. enterocolitica	Cause an enteric disease called enterocolitis.
Family: Vibrionaceae	Usually curved rods.
Vibrio *V. cholerae*	Cause cholera. Organisms produce choleragen, a toxin that causes the disease. Found in water.
V. parahaemolyticus	Cause food poisoning. Shellfish are a common source of infection. Found in coastal marine environments.
Aeromonas hydrophila	Associated with wound infections, meningitis and septicemia. Normally found in aquatic environments and marine life.
Plesiomonas shigelloides	May cause gastroenteritis. Found in aquatic environments.

TABLE 8.4
CHARACTERISTICS OF SOME GRAM-NEGATIVE ANAEROBIC BACTERIA

Organism	Characteristics
Bacteroides	Irregularly staining rods with rounded ends. Frequently isolated from clinical specimens. Normally colonizes the human gut.
Fusobacterium	Long, slender rods with tapering ends. Some pleomorphism evident. Colonizes the mouth, urinary tract, genital tract, and gastrointestinal tract. May cause abscesses.
Leptotricha	Long, plump rods forming chains. Normal inhabitant of the mouth. Not considered a pathogen.

intestinal cells. This cellular fluid is discharged into the lumen of the intestine, giving rise to the characteristic "rice water stool" of cholera victims. *Vibrio parahaemolyticus* is another important human pathogen. It is one of the pathogens responsible for outbreaks of food poisoning associated with ingestion of contaminated shellfish. Species of *Plesiomonas* and *Aeromonas* can also infect human tissue and have been isolated from blood, urine, and spinal fluids of patients.

Gram-Negative Anaerobic Bacteria

The **gram-negative, strictly anaerobic bacteria** are typically rod-shaped or curved (table 8.4). One characteristic feature of these bacteria is their ability to produce a variety of organic acids from the fermentation of carbohydrates, such as succinic acid, acetic acid, and lactic acid. The bacteria are identified on the basis of the principal acid and the number of different acids they produce when they ferment. Some species can be isolated from the oral or intestinal cavities of humans and have been implicated in human disease. The genus *Bacteroides* is notably important. Bacteria in this genus are strict anaerobes that normally colonize human cavities (oral, genital, or intestinal). Certain species of *Bacteroides* are capable of causing severe gangrenous infections. *Fusobacterium* is a long, slender rod with pointed ends. These bacteria normally colonize the mouth and occasionally may cause periodontal disease, abscesses, and septicemia.

Gram-Negative Cocci and Coccobacilli

The **gram-negative cocci** (table 8.5) and coccobacilli are chemoheterotrophs whose energy metabolism is generally aerobic, using sugars as the source of energy and carbon. The bacteria in this group may grow as irregular clusters of cells, diplococci, or paired rods (fig. 8.5). The genera *Neisseria*, and *Moraxella (Branhamella)* contain pathogenic species, and the genus *Acinetobacter* contains at least one opportunistic pathogen. Most bacteria in this group, except for *Acinetobacter*, are extremely sensitive to penicillin. Penicillin sensitivity is noteworthy in this group of bacteria because gram-negative bacteria generally are not very sensitive to this antibiotic. The diseases **gonorrhea** and **epidemic cerebrospinal meningitis** are caused by two species of *Neisseria: N. gonorrhoeae* (the gonococcus)

TABLE 8.5
CHARACTERISTICS OF SOME GRAM-NEGATIVE COCCI

Organism	O_2 requirement	Acid from utilization of carbohydrates					Disease(s)
		Glucose	Maltose	Sucrose	Lactose	Fructose	
Moraxella (Branhamella) catarrhalis	Aerobe	–	–	–	–	–	Urinary tract and respiratory tract infections
Neisseria gonorrhoeae	Aerobe	+	–	–	–	–	Gonorrhea
Neisseria meningitidis	Aerobe	+	+	–	–	–	Meningitis
Neisseria sicca	Aerobe	+	+	+	–	+	Nonpathogenic
Veillonella	Anaerobe	–	–	–	–	–	Nonpathogenic

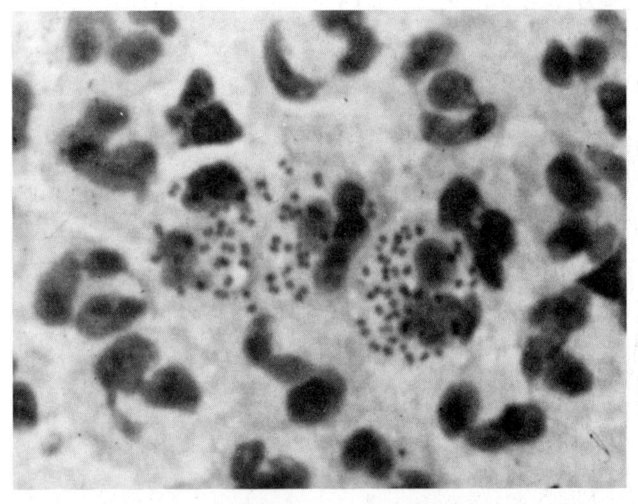

(a)

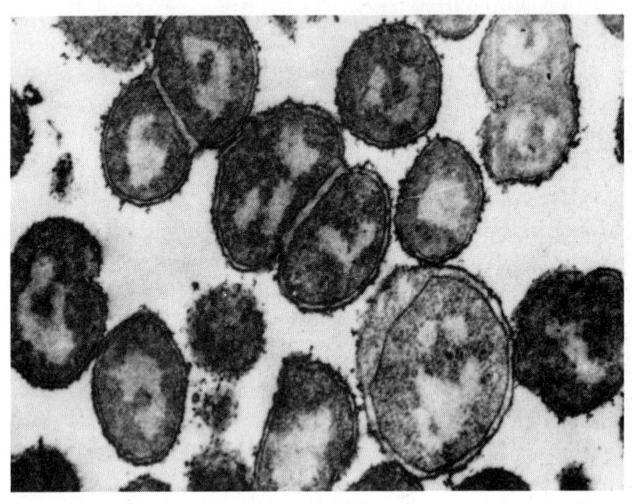

(b)

Figure 8.5 ***Neisseria gonorrhoeae.*** *(a)* Photomicrograph of *Neisseria gonorrhoeae* seen inside neutrophils in vaginal exudate. *(b)* Transmission electron micrograph of *N. gonorrhoeae* illustrating the details of the gonococcus.

and *N. meningitidis* (the meningococcus), respectively. Other neisseriae can frequently be isolated from the nose and throat of humans without apparent disease. Species of *Moraxella* have been isolated from the mucous membranes of humans, and at least one species causes a disease of cattle (and sometimes humans) called pink eye (conjunctivitis).

Rickettsias

The **rickettsias** (fig. 8.6) constitute a group of pleomorphic (variably shaped) rods or coccobacilli that generally give a gram-negative reaction and have cell walls that are similar in composition to those of other gram-negative bacteria. They are obligate intracellular parasites of eukaryotes, with life cycles that usually include an arthropod (joint-legged invertebrate) vector. One notable exception is *Coxiella burnetii*, which is not transmitted by vectors but rather is carried in air currents. The rickettsias live in a state of mutual tolerance with their arthropod hosts and are transmitted to warm-blooded animals by the arthropod's bite.

The rickettsias have their own metabolic enzymes and possess a complete Krebs cycle and electron transport system. These bacteria are obligate parasites because they cannot synthesize NAD or coenzyme A and must obtain these compounds from their host. One important feature of the rickettsias is that their membranes are "leaky;" that is, nutrients and metabolic products pass into and out of the cell through the plasma membrane much more freely than in other bacteria. The increased permeability of their membranes appears to be a necessary adaptation to intracellular parasitism, since they must obtain some of their growth factors from the host.

The rickettsias have an affinity for the cells that constitute the endothelial lining (the inner lining of blood vessels) of capillaries. It is because of this affinity that all diseases caused by rickettsias (table 8.6) are characterized by a rash. The rash results when the rickettsial infection obstructs the small blood vessels of the skin causing lesions (the rash).

The rickettsias can cause severe epidemics in human populations, and have influenced the outcome of wars by causing widespread disease and debilitating entire armies. Epidemic typhus caused by *Rickettsia prowazekii* and Rocky Mountain spotted fever caused by *R. rickettsii* have been responsible for the deaths of many thousands of humans.

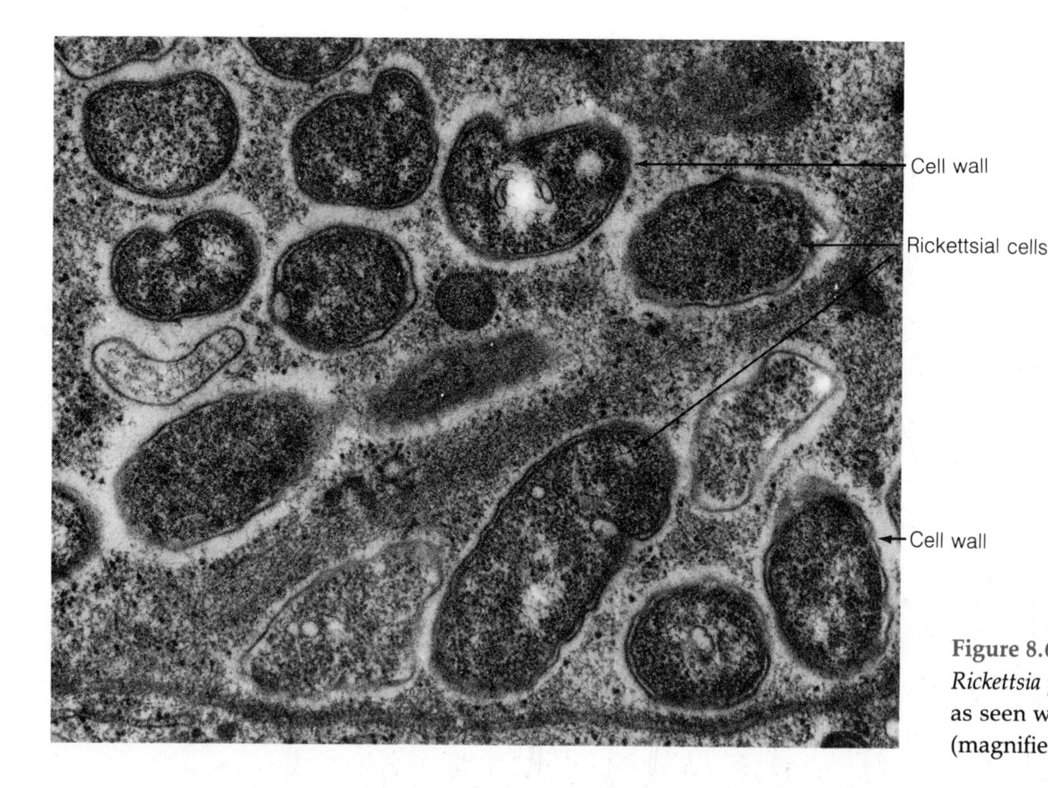

Figure 8.6 ***Rickettsia prowazekii.*** *Rickettsia prowazekii* in a chick embryo as seen with an electron microscope (magnified 58,900 ×).

Chlamydias

The **chlamydias** (fig. 8.7) are very small gram-negative bacteria that, like the rickettsias, are obligate (intracellular) parasites. These microorganisms are much smaller than the rickettsias, however, and have a life cycle that involves two different morphological types: the infectious stage or **elementary body** and the vegetative stage or **reticulate body** (fig. 8.7). The chlamydia cannot synthesize their ATP and hence are sometimes called "energy parasites."

Diseases such as trachoma, psittacosis, and lymphogranuloma venereum are caused by chlamydia (table 8.7). *Chlamydia trachomatis* has been known to cause various diseases, notably trachoma, a highly infectious disease of the eye that often leads to blindness. More recently, it has been noted that this chlamydia is the most common agent of a sexually transmitted disease called nongonococcal urethritis

TABLE 8.6
RICKETTSIAL DISEASES AND THEIR CAUSATIVE AGENTS

Organism	Disease it causes	Vehicle of transmission or vector*
Rickettsia rickettsii	Rocky Mountain spotted fever	Hard (ixodid) ticks
Rickettsia prowazekii	Epidemic typhus	Lice
Rickettsia typhi	Endemic typhus	Fleas
Rickettsia tsutsugamushi	Scrub typhus	Mites (trombiculid)
Rochalimaea quintana	Trench fever	Body louse
Coxiella burnetii	Q-fever	Arthropod? aerosol, food consumption

*A vector is an arthropod that is involved in the transmission of an infectious agent from one individual to another.

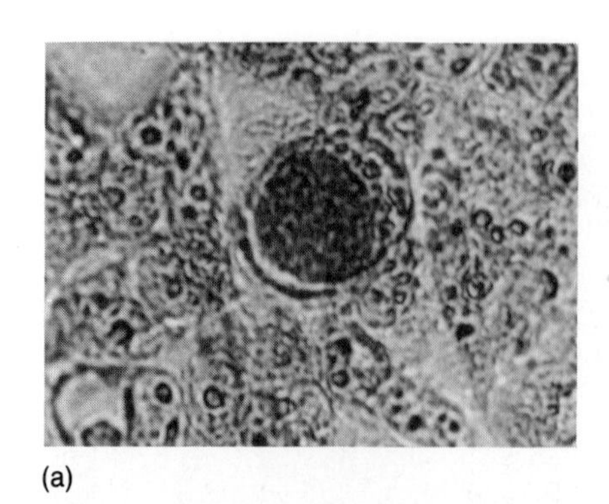

(a)

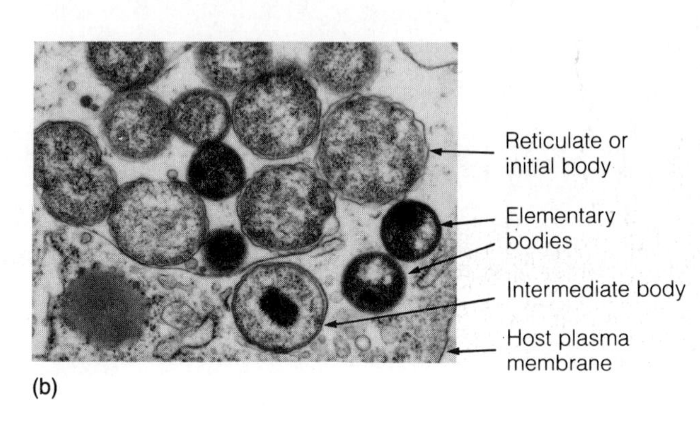

(b)

Figure 8.7 The Chlamydia. *(a)* Inclusion body of *Chlamydia trachomatis*, the cause of some eye infections and sexually-transmitted diseases. *(b)* Transmission electron micrograph of the various stages in the life cycle of *Chlamydia psittaci*.

because the symptoms may resemble those caused by the gonococcus. Nongonococcal urethritis is possibly the most common sexually transmitted disease in the world.

Mycoplasmas

The **mycoplasmas** are a highly variable group of bacteria lacking a cell wall (fig. 8.8). These bacteria are quite small: their average size is just below the limit of resolution of the light microscope (0.1–0.2 μm).

TABLE 8.7
COMMON CHLAMYDIAL DISEASES

Organism	Disease it causes	Mode of infection
Chlamydia trachomatis	Trachoma (an eye infection)	Aerosols, personal contact
	Lymphogranuloma venereum	Sexual intercourse
	Nongonococcal urethritis	Sexual intercourse
Chlamydia psittaci	Psittacosis (a disease of the lungs)	Aerosols, contact with infected birds (psittaccine birds)

The mycoplasmas are the simplest prokaryotic microorganisms that are still capable of a free-living existence. In general, each mycoplasma is a highly pleomorphic, forming spheres, filaments, and irregularly-shaped structures (fig. 8.8a). This characteristic may be due to the absence of the rigid cell wall that maintains the shape of other bacteria. Most mycoplasmas require cholesterol in their diets. Apparently, this lipid is needed to strengthen their plasma membrane in order to prevent osmotic lysis of the cell.

Some species of *Mycoplasma* and *Ureaplasma* are pathogenic for humans and have been described as **membrane parasites**, because they are always found in close association with the host's plasma membrane. *Mycoplasma pneumoniae* is the cause of an inflammation of the lung called primary atypical pneumonia. *Ureaplasma urealyticum* is transmitted from one individual to another by sexual intercourse, causing urethritis (inflammation of the urethra). *Spiroplasma mirum* is the agent responsible for cataracts and chronic brain infections in newborn mice. Spiroplasmas have recently been discovered to be the cause of more than 10 different plant diseases previously thought to be caused by viruses (fig. 8.8b). These helically shaped, motile mycoplasmas have been found in the phloem of infected plants.

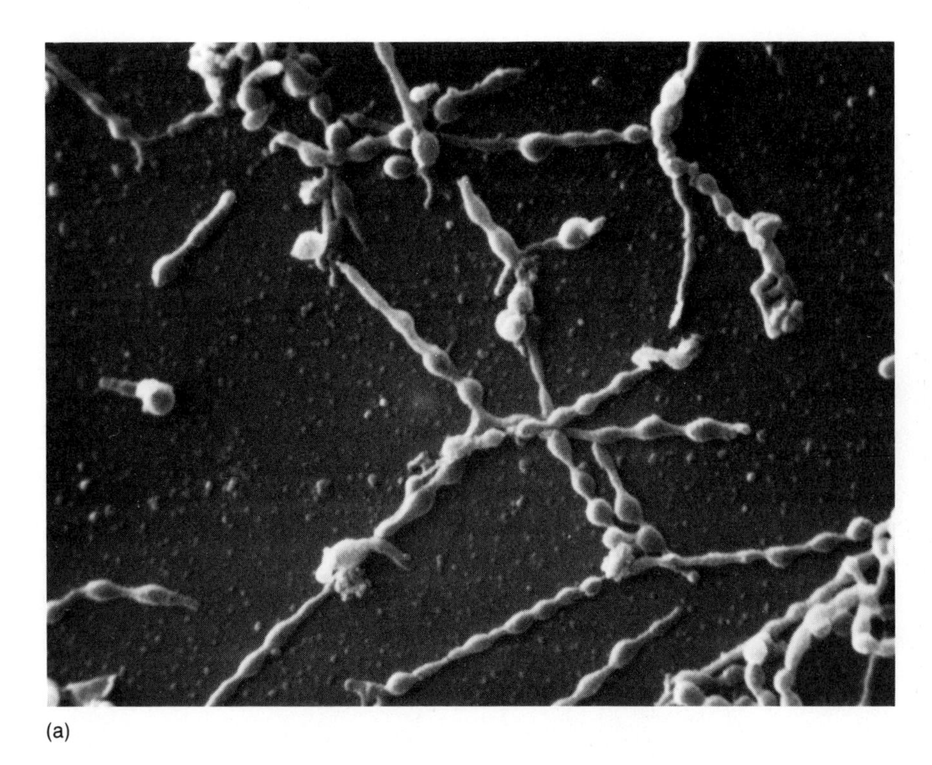

(a)

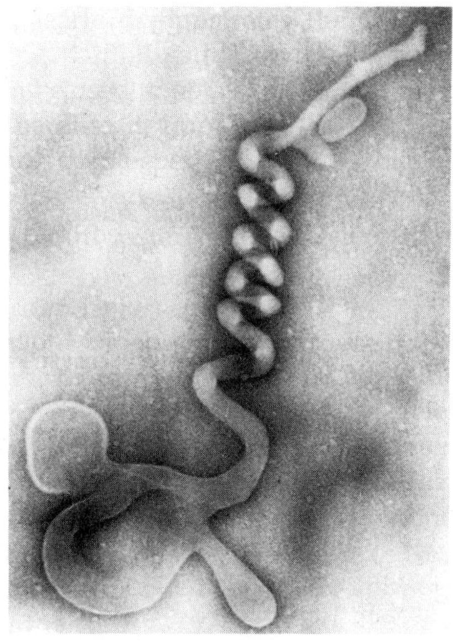

(b)

Figure 8.8 The Mycoplasmas. *(a)* Scanning electron photomicrograph of *Mycoplasma pneumoniae.* *(b)* Transmission electron photomicrograph of a spiroplasma.

GRAM-POSITIVE BACTERIA OF MEDICAL OR APPLIED IMPORTANCE

The gram-positive bacteria of medical or applied importance are predominantly pathogens of humans and mammals. These include the agents of tuberculosis, diphtheria, and streptococcal sore throat. All of these diseases can have serious consequences if not treated promptly. Also included in this group are bacteria that are important in the preparation of fermented foods or in the synthesis of antibiotics and certain insecticides.

Gram-Positive Cocci

The **gram-positive cocci** constitute a large group of spherical bacteria (fig. 8.9) that are extremely important from both a medical and an industrial point of view. Although all the gram-positive cocci are chemoheterotrophic, some obtain their energy exclusively by aerobic respiration while others obtain it by fermentation (table 8.8). The individual cells of these gram-positive bacteria are often arranged in characteristic patterns. They may exist as individual cells, pairs (diplococci), packets of eight, chains, or grapelike clusters. These arrangements are often characteristic of a particular genus and can be used in identifying the organisms. For example, *Streptococcus* (fig. 8.9b) typically forms chains of cocci, while *Staphylococcus* forms an irregular cluster of spherical cells resembling a bunch of grapes.

Some gram-positive cocci are part of the resident microbiota of many animals (including humans), while others are normal inhabitants of soils. Several species in this group cause severe diseases of humans and livestock, most of them characterized by the formation of lesions containing pus. For this reason, these pathogenic bacteria are sometimes called **pyogenic** (pus-forming) cocci. The disease commonly called **strep throat** (because its most obvious manifestation is a sore throat) is usually caused by *Streptococcus pyogenes*. Many wound infections and **folliculites** (inflammation of hair follicles), such as carbuncles, furuncles, and pimples, are caused by another pyogenic bacterium called *Staphylococcus aureus*. More recently, *S. aureus* has been identified as the cause of **toxic shock syndrome**, which

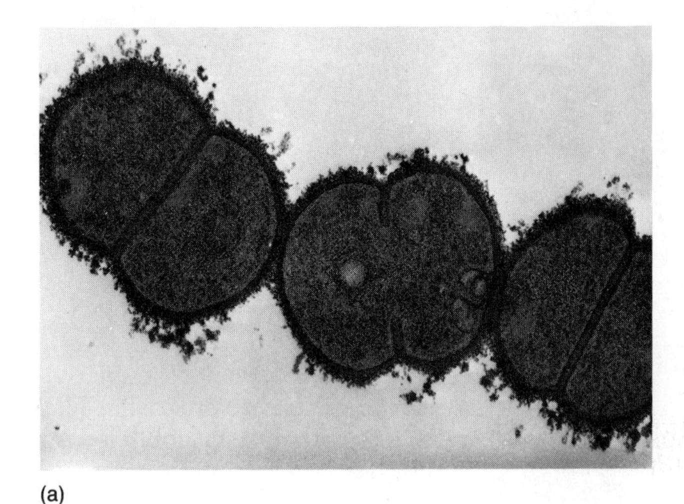

(a)

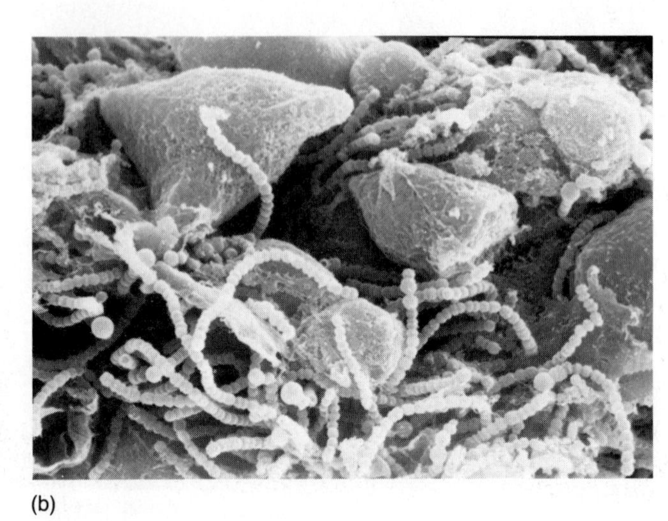

(b)

Figure 8.9 **Electron photomicrographs of representative gram-positive cocci.** *(a)* Transmission electron photomicrograph of *Streptococcus salivarius* showing various stages of cell division. Notice that the chain of cells is fragmenting. *(b)* Scanning electron micrograph of *Streptococcus pneumoniae* adhering to human conjunctival epithelial cells. Notice the long chains of cells.

affects a small but significant proportion of menstruating women and others suffering from certain staphylococcal infections. Mastitis (inflammation of the udder) in cattle, a very difficult disease to control, is caused by *Streptococcus agalactiae* and *Staphylococcus aureus*.

Other species, collectively called **lactic acid bacteria**, participate in the production of fermented goods. The lactic acid bacteria, which include those in the genera *Streptococcus* and *Leuconostoc*, are catalase negative and produce lactic acid as their major fermentation end product. These bacteria are very important economically because they are used to prepare yogurt, sauerkraut, kefir, cheese, buttermilk, and other fermented foods.

Endospore-Forming Rods

This group of prokaryotes includes several genera of gram-positive heterotrophic bacteria that undergo a process of cellular differentiation that results in the formation of an endospore (fig. 8.10). Some are strict anaerobes (e.g., *Clostridium*), while others are aerobes

TABLE 8.8
CHARACTERISTICS OF SOME GRAM-POSITIVE COCCI

Genus	O_2 requirements	Arrangement of cells	Catalase activity*
Staphylococcus	Facultative anaerobe	Clusters (grapelike)	+
Micrococcus	Aerobes	Single, pairs, packets	+
Streptococcus	Aerotolerant anaerobe	Chains, pairs	−
Aerococcus	Microaerophilic	Single, pairs, tetrads	−/+

*Catalase activity refers to the ability to synthesize the enzyme catalse. Activity is detected by the production of gas bubbles (O_2) from hydrogen proxide (H_2O_2).

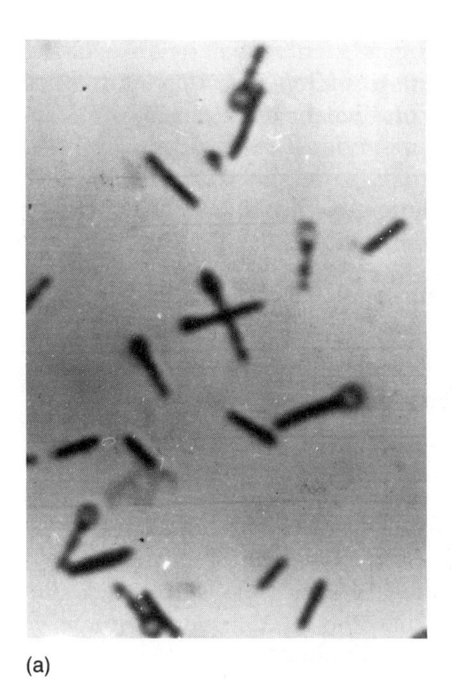

(a)

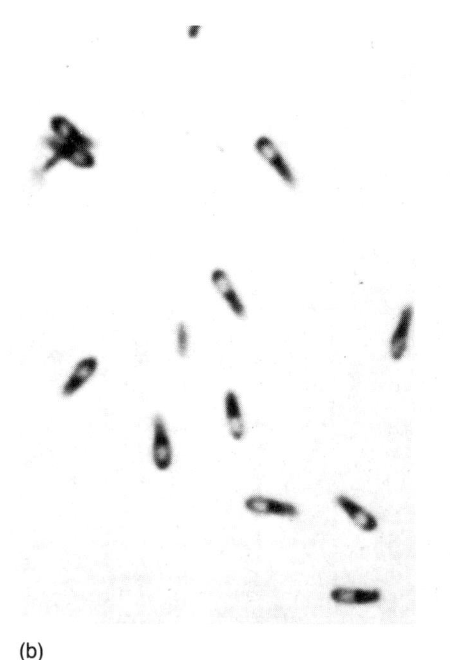

(b)

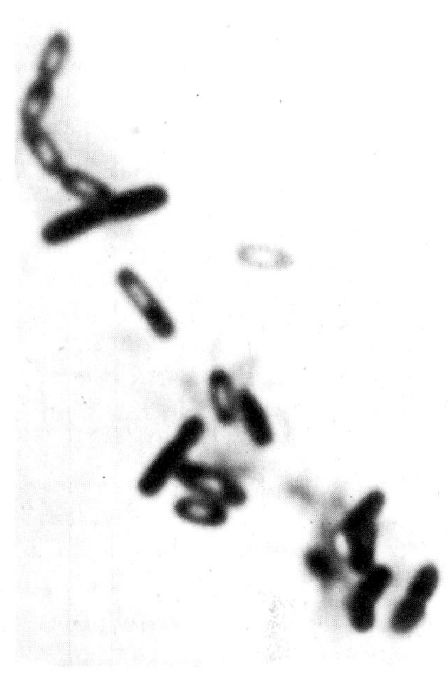

(c)

Figure 8.10 Endospore-forming bacteria. *(a)* Photomicrograph of *Clostridium tetani* showing the round, terminal spores. *(b)* Photomicrograph of *Clostridium subterminale* showing oval, subterminal endospores. *(c)* Photomicrograph of *Clostridium bifermentans* showing oval, central to subterminal endospores.

or facultative anaerobes (e.g., *Bacillus*). The endospore-forming rods are widely distributed in soils and occasionally infect animals (including humans), causing various diseases.

Many of the bacteria in this group are capable of producing antibiotics such as bacitracin and gramicidin. For this reason, some of them, particularly those in the genus *Bacillus*, are studied by the pharmaceutical industry. Some species of this genus, notably *B. thuringiensis*, produce a crystal of protein called the **parasporal body**. This protein, which can be seen adjacent to the endospore, kills the larvae of some insects. Thus the bacterium is grown in large quantities by industrial concerns and sold as an insecticide under various trade names (e.g., Dipel®). Some species of spore-forming rods are capable of causing severe diseases such as anthrax, botulism, gas gangrene, and tetanus.

Coryneform Bacteria

The **coryneform bacteria** constitute an aggregation of ubiquitous, gram-positive, chemoheterotrophic bacteria. These bacteria are generally pleomorphic rods with a tendency to form angular arrangements resembling Chinese characters (fig. 8.11). They do not form endospores and frequently have polyphosphate (volutin) granules.

Some bacteria in this group can cause disease in humans. For example, *Corynebacterium diphtheriae* causes the disease **diphtheria**, which afflicts predominantly children. Vaccines against diphtheria have reduced the incidence of the disease to a few cases each year in the United States.

The **propionic acid bacteria** are coryneform bacteria of economic importance. They are found in dairy products, and some species participate in the ripening of Swiss cheese. The characteristic flavor and appearance of this cheese is partially the result of the reproduction of the propionic acid bacteria on the cheese curd while the "holes" in the cheese are due to the CO_2 produced by the bacterium.

Mycobacteria

The **mycobacteria** are gram-positive, acid-fast rods with a tendency to form filaments. All of these bacteria are aerobic and generally grow slowly. The mycobacteria have a high concentration of lipids associated with their cell wall (as much as 60% of their dry weight).

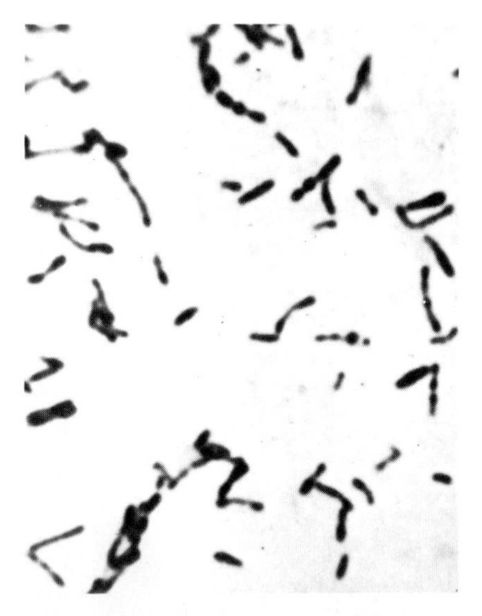

Figure 8.11 ***Corynebacterium diphtheriae.*** Photomicrograph of Gram-stained smear of *Corynebacterium diphtheriae.* Notice the strongly stained, swollen areas that are typical of this organism.

The lipids are responsible for the waxy nature of the mycobacteria, their **serpentine** growth pattern (fig. 8.12), and their acid-fast characteristics.

A cell wall component called **cord factor** is toxic to eukaryotic cells. This toxicity is partly responsible for the disease-causing capabilities of some species of *Mycobacterium.* There are two very important pathogens in this group, *Mycobacterium tuberculosis* and *M. leprae,* which cause tuberculosis and leprosy, respectively. Both of these diseases are of considerable public health importance and afflict millions of people throughout the world.

Nocardias

The **nocardias** are aerobic, gram-positive rods resembling the mycobacteria. Their colonies are often orange- or red-pigmented. Some species, such as *Nocardia asteroides,* are acid-fast. The genus *Nocardia* includes very important human pathogens that cause mycetoma (subcutaneous) and pulmonary (lung) infections.

Streptomycetes and Related Organisms

The **streptomycetes** are all gram-positive bacteria that form well-developed, branching filaments (mycelium), and some species form very elaborate arrays of aeral spores or **conidia** (fig. 8.13). The streptomycetes are primarily soil inhabitants; in fact, the "earthy" odor of moist soils and compost heaps is due to the formation of **geosmin,** a chemical synthesized by various species of streptomycetes.

Some species of streptomycetes are important human pathogens, but the most notable characteristic of these bacteria is that they produce a large array of antibiotics. Many of the antibiotics in common use today are synthesized by these bacteria. **Streptomycin** is one such antibiotic that is synthesized by *Strepto-*

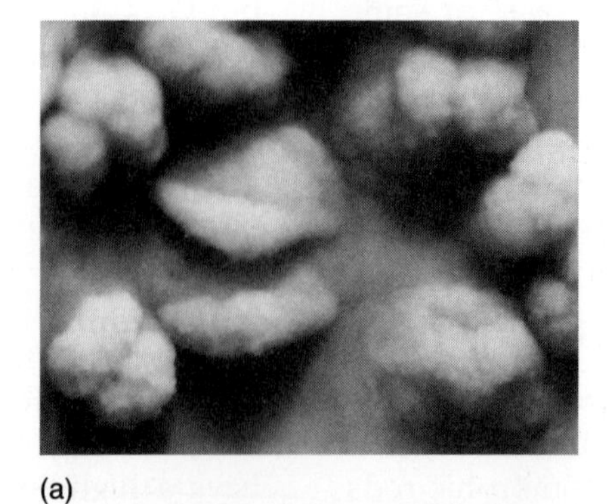

(a)

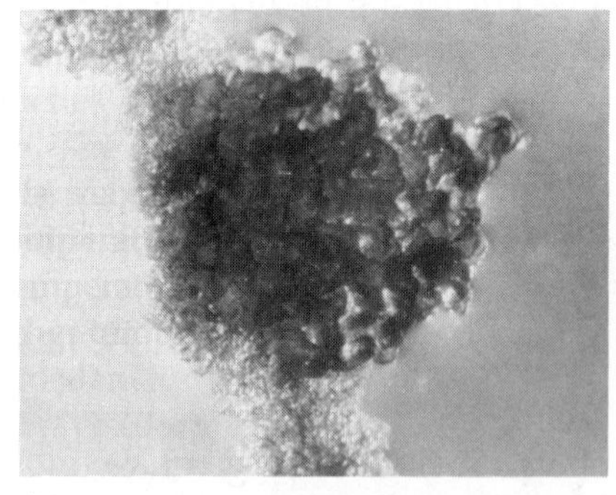

(b)

Figure 8.12 Cultural characteristics of *Mycobacterium tuberculosis.* *(a)* Colonial characteristics of *Mycobacterium tuberculosis* growing on Lowenstein-Jensen agar. *(b)* Colonial characteristics of *Mycobacterium phlei* growing on Lowenstein-Jensen agar.

(a)

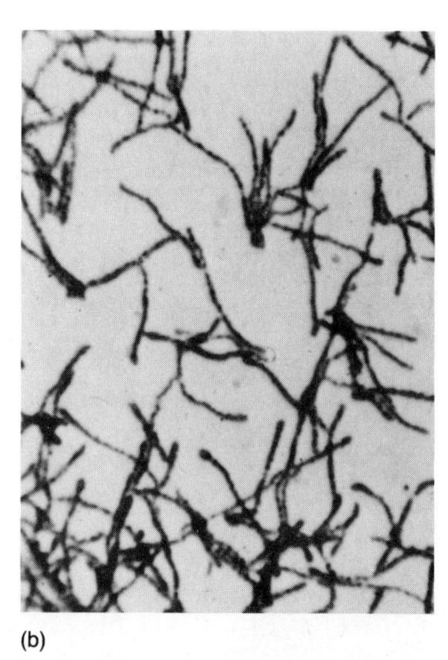
(b)

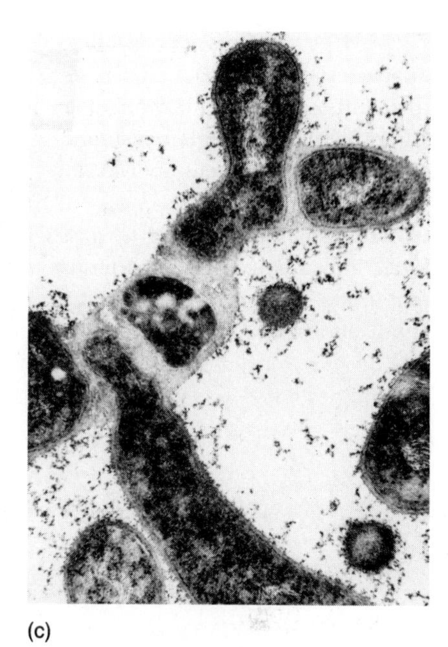
(c)

Figure 8.13 Growth characteristics of some actinomycetes. *(a)* Hyphal filaments of *Streptomyces*. *(b) Actinomyces israelii* stained with the Gram stain. Notice the characteristic branching exhibited by the bacteria. *(c)* Transmission electron micrograph of *Actinomyces* showing a central cell producing branching filaments.

myces griseus. Discovered by Selman Waksman in 1947, this was the first bacterial antibiotic produced in large quantities. Since the discovery of streptomycin, many other antibiotics have been isolated from streptomycetes, and some of these are presently in common use to treat infectious diseases.

OTHER GRAM-NEGATIVE BACTERIA

There are a few groups of bacteria that neither cause diseases nor perform industrially useful functions. Nevertheless, these bacteria are an integral part of our environment, and without their presence, life as we know it could not exist. Bacteria play a central role in recycling of nutrients in the biosphere. Some species convert organic materials into inorganic substances, while others transform these inorganic substances into other inorganic or organic materials. By the concerted effort of all these organisms, nutrients are constantly released from dead organisms as waste products that can be used as nutrients by other organisms.

The Cyanobacteria (Blue-Green Algae)

The **cyanobacteria** are photoautotrophs that share the ability to release molecular oxygen during photosynthesis. These microorganisms are a highly diversified group, consisting of various morphological types (fig. 8.14). The cyanobacteria represent the oldest living ancestors of the eukaryotic algae. Their appearance on the early earth about 3 billion years ago and their subsequent success altered the earth significantly. These cyanobacteria produced most of the oxygen found in the developing atmosphere. The increase in atmospheric oxygen to about 0.6%, approximately 1.5 billion years ago, served as a stimulus for the evolution of aerobic organisms.

All cyanobacteria have intracytoplasmic networks of membranes called **thylakoids** (fig. 8.15). These membranes are distributed throughout the cytoplasm and contain many of the enzyme systems involved in photosynthesis.

The cyanobacteria inhabit a wide range of aquatic and terrestrial environments. Many species of cyanobacteria are tolerant of extreme environments, such as

Figure 8.14 Various species of cyanobacteria. *(a) Calothrix* trichomes (filaments), exhibiting hormogonia (short filaments), a heterocyst, and an akinete (resting cell). *(b) Gloeocapsa*. The spherical cells are enclosed within a gelatinous sheath. *(c) Anabaena*, a filamentous cyanobacterium that forms heterocysts (where nitrogen fixation takes place).

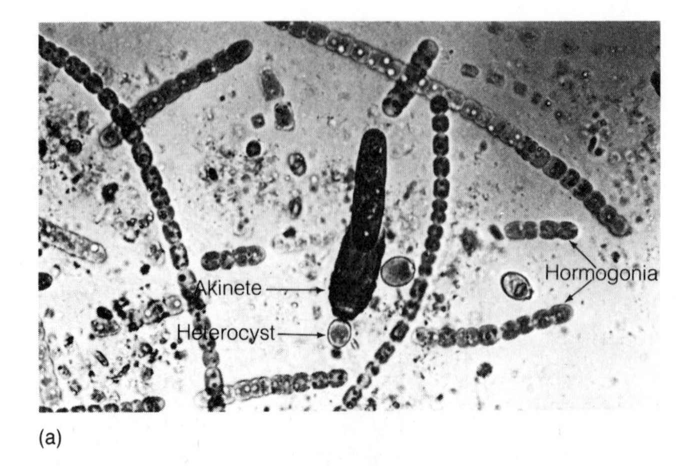

(a)

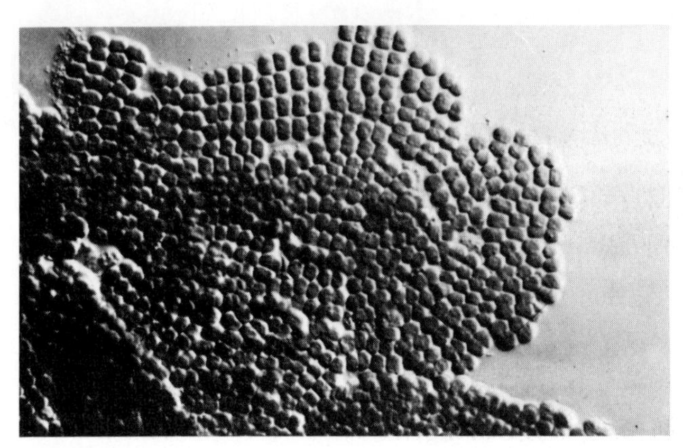
(b)

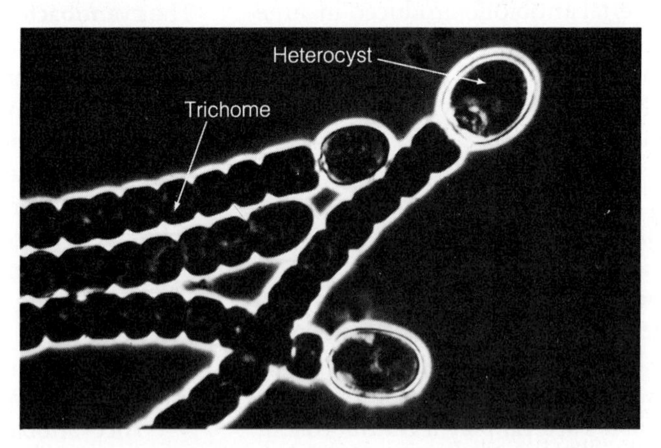

(c)

FOCUS THE ARCHAEBACTERIA, A SEPARATE KINGDOM?

Microbiologists are constantly searching for—and sometimes discovering—new microorganisms inhabiting our biosphere. The search is often directed toward the discovery of new pathogens or useful microorganisms such as antibiotic producers. Much of the research is financially supported by industrial concerns or government agencies. Sometimes microorganisms are discovered that are extremely important in the understanding of biology and evolution. Such is the case with the **archaebacteria**. Some scientists claim that these bacteria represent one of the oldest forms of living organisms on the earth, and that they are different enough from other bacteria to warrant their own kingdom.

The archaebacteria include the methanogens (which produce methane from their metabolism), the extreme halophiles (which live in environments with high salt concentrations) and the thermoacidophiles (which live in acidic, hot environments). These bacteria have in common membranes that are composed of lipids containing ether bonds rather than the usual ester bonds. In addition, none of the archaebacteria have cell walls containing the peptidoglycan murein. Instead, they have other peptidoglycans devoid of murein, or they have proteinaceous materials. The DNAs and RNAs of the archaebacteria are also different from those of other bacteria. This evidence has suggested to some scientists that the archaebacteria were once widely distributed in the primitive earth and for the most part have remained in those anaerobic environments conducive to their growth.

The archaebacteria are considered to be a third form of life, sharing the biosphere with the other bacteria and the eukaryotes. They have participated actively in shaping the earth's biosphere.

those found in alkaline hot springs (where the temperature reaches 74°C) and desert soils.

The cyanobacteria exert a significant influence on human affairs. Their growth in drinking water can impart "earthy" odors and undesirable flavors to the water. Some species of cyanobacteria, such as *Anabaena flos-aquae* and *Lyngbya majuscula*, produce metabolic products that are toxic to humans and animals. For example, the toxin produced by *L. majuscula* can cause skin irritations and dermatitis in humans and other primates.

The cyanobacteria are often the primary colonizers of bare soils, and in this way contribute to soil fertility. Sometimes they form large mats on certain soils, releasing large amounts of organic nutrients into the soil and consequently increasing the soil's fertility. The cyanobacteria have also been used to fertilize rice paddies. Species like *Anabaena*, which can fix nitrogen from the atmosphere into organic nitrogen, are allowed to grow in rice paddies where they fix nitrogen and hence increase the nitrogen content of the paddy. These cyanobacteria represent an inexpensive and natural source of fertilizer that significantly enhances rice yield. Some cyanobacteria, particularly *Spirulina*, have been

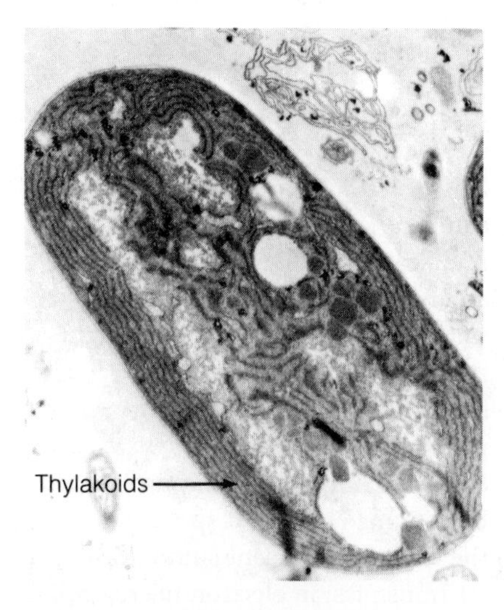

Figure 8.15 **The thylakoids of cyanobacteria.** Thylakoids are intracytoplasmic membranes where photosynthesis takes place. Transmission electron photomicrograph of *Anabaena*. The membranous structures layered around the cell are the thylakoids.

grown on a large scale and are sold as a food supplement.

Phototrophic Bacteria

The **phototrophic bacteria** include all the bacteria (other than the cyanobacteria) that obtain their energy from the sun by photosynthesis (fig. 8.16). Unlike the cyanobacteria, the phototrophic bacteria do not produce O_2 during photosynthesis. The phototrophic bacteria play a very important role in the recycling of nutrients in nature. They are anaerobic organisms that live predominantly in aquatic habitats rich in sulfides or organic matter. Some phototrophic bacteria obtain their carbon from the atmosphere as carbon dioxide, but others obtain it from organic acids. Those that use reduced inorganic compounds such as hydrogen sulfide, sulfur, or molecular hydrogen to reduce carbon dioxide remove these toxic substances from the water. Thus, some phototrophic bacteria convert carbon dioxide and inorganic sulfur compounds into chemical compounds that other organisms can use.

Gliding Bacteria

The **gliding bacteria** constitute a complex group of microorganisms exhibiting a variety of morphological and functional types (fig. 8.17). They have in common, however, the ability to move by gliding when associated with solid surfaces.

The fruiting myxobacteria (fig. 8.17) are gram-negative chemoheterotrophic rods that move by gliding. Most of the myxobacteria are found in terrestrial habitats, where they are associated with decaying organic matter in the form of animal feces or vegetation. These myxobacteria are unique among the pro-

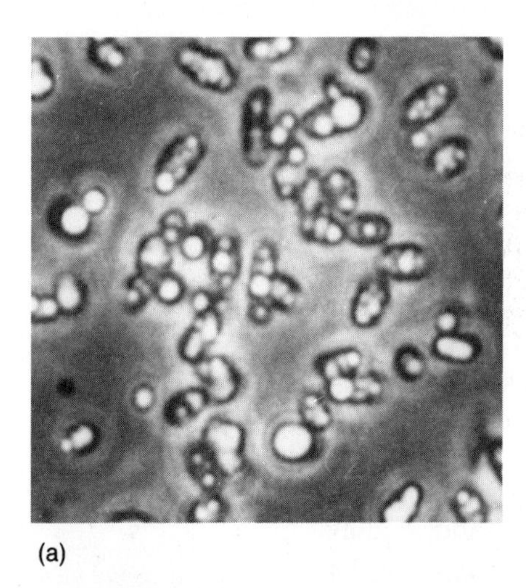

(a)

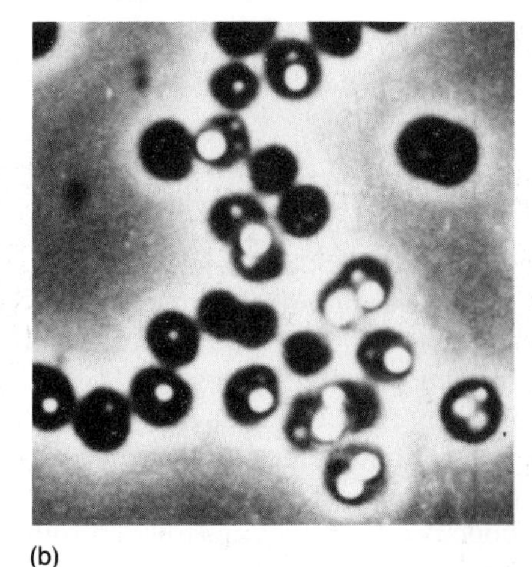

(b)

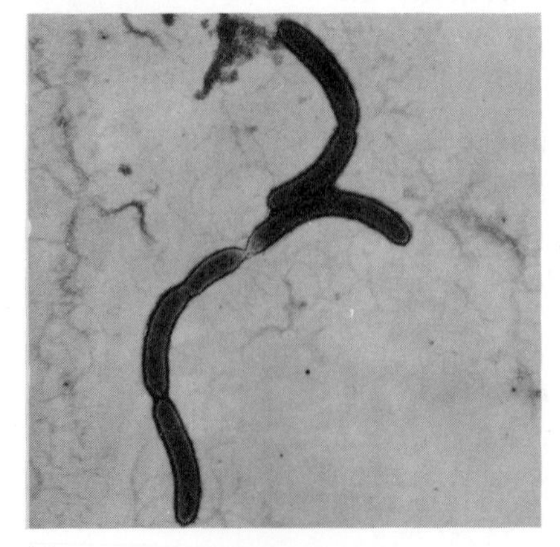

(c)

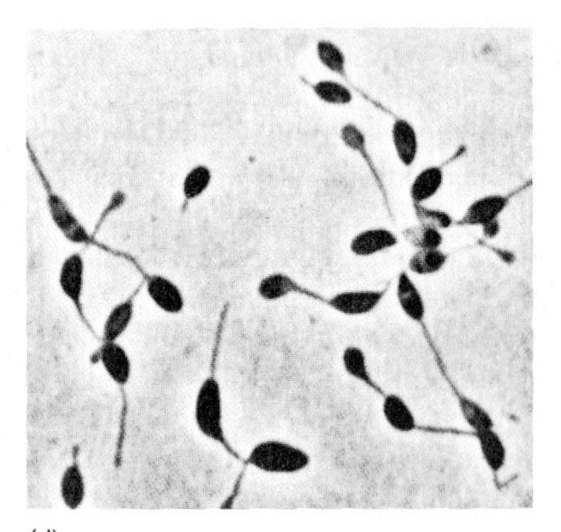

(d)

Figure 8.16 Representative phototrophic bacteria. *(a)* Phase contrast photomicrograph of *Chromatium* sp., an anaerobic, marine photosynthetic, purple-sulfur bacterium. The bright spots within the bacteria represent sulfur granules. *(b)* Phase contrast photomicrograph of *Thiocapsa*, an anaerobic, photosynthetic, purple-sulfur bacterium. The bright spots within the bacteria represent sulfur granules. *(c)* Transmission electron micrograph of negatively stained *Rhodospirillum rubrum*, an anaerobic photosynthetic, purple-nonsulfur bacterium. *(d)* Phase contrast photomicrograph of *Rhodomicrobium vannielii*, an anaerobic, photosynthetic, appendaged, purple-nonsulfur bacterium.

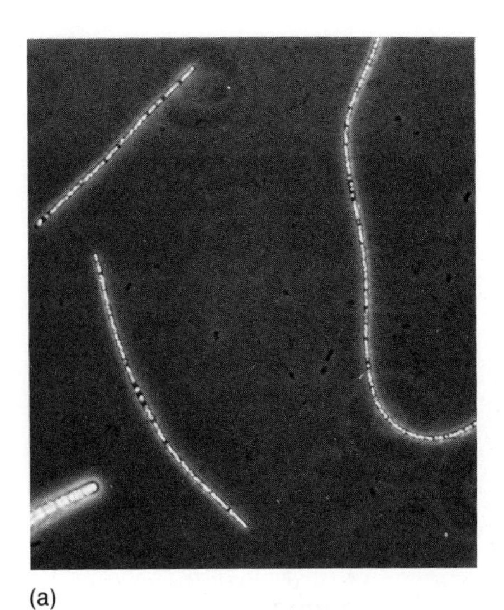

(a)

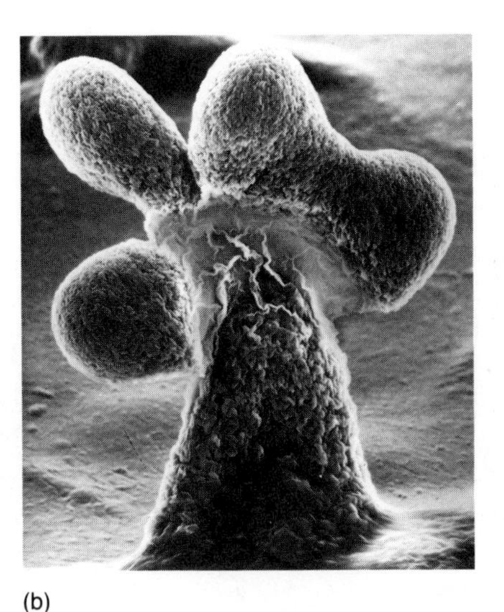

(b)

Figure 8.17 Gliding bacteria. *(a)* Phase contrast micrograph of *Thiothrix* sp., a marine gliding bacterium that uses H_2S as an electron source. Notice the numerous sulfur granules within the cytoplasm. *(b) Stigmatella,* a fruiting myxobacterium. The fruiting body is made up of numerous individual cells aggregated in an organized fashion.

karyotes in that they aggregate and subsequently form **fruiting bodies**. The fruiting bodies are filled with **myxospores**, which are capable of germinating into vegetative cells.

The fruiting myxobacteria form red, yellow, or orange colonies and are found speckling pieces of rotting wood in shady, damp areas of forests. These bacteria are capable of digesting a variety of large, insoluble molecules, such as chitin, cellulose, and peptidoglycan, breaking them down in to smaller units that can be used as nutrients by other organisms.

Sheathed Bacteria

The **sheathed bacteria** are filamentous, gram-negative bacteria that divide by binary fission within a mucilaginous sheath (fig. 8.18). The sheath is composed of proteins, polysaccharides, and lipids. The sheathed bacteria are common in aquatic habitats, where they sometimes form **blooms** (large masses of aquatic microorganisms) in habitats that are rich in nutrients. For example, *Sphaerotilus* can become a nuisance in sewage treatment plants because, in the presence of large quantities of organic matter commonly found there, these bacteria proliferate rapidly and form blooms that foul up filters.

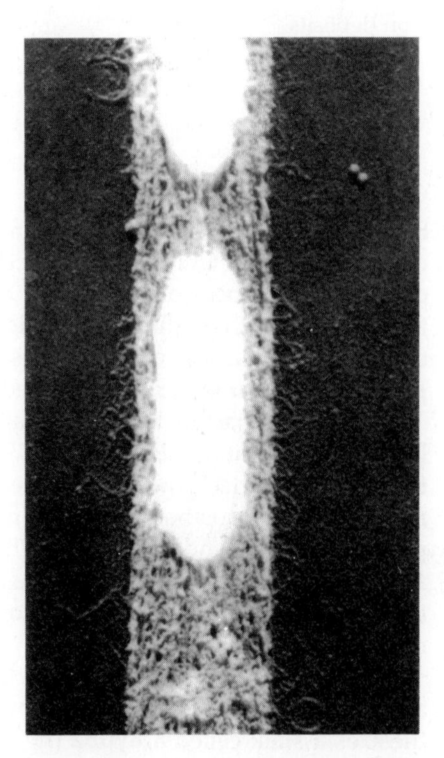

Figure 8.18 The sheathed bacterium *Leptothrix*

Budding and/or Appendaged Bacteria

The **budding** and/or **appendaged bacteria** develop various forms of appendages other than flagella (fig.

(a)

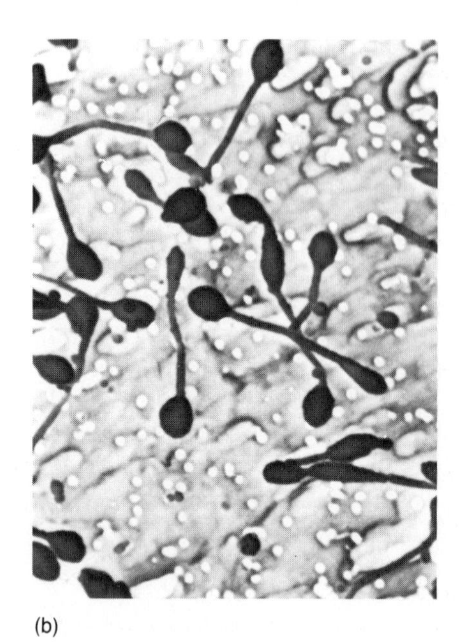

(b)

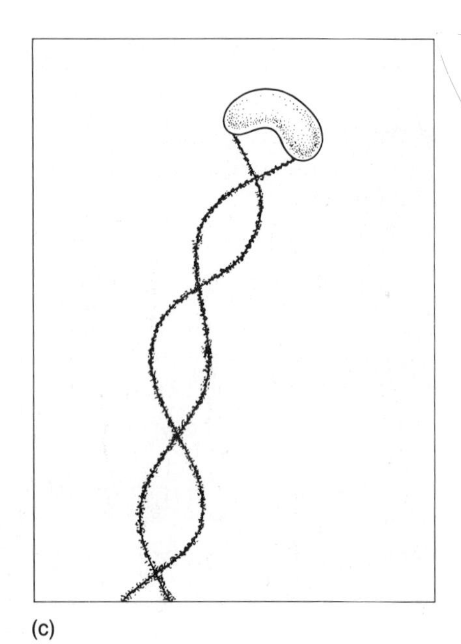

(c)

Figure 8.19 Appendaged bacteria. *(a) Caulobacter.* The appendage is an extension of the cytoplasm *(b) Hyphomicrobium,* a budding bacterim. *(c) Gallionella.* Note the twisted stalk. The stalk is acellular and consists of iron deposits.

8.19). Generally, these microorganisms undergo a form of cell division in which the products are asymmetrical. Some members of this group divide by binary fission, while others divide by budding. The appendages may be either cellular extensions of the cytoplasm or acellular materials deposited outside the cell.

The **Caulobacters** are appendaged bacteria that have a prominent prostheca (fig. 8.19) with a **holdfast** that serves to attach the cell to other cells or to solid materials. The prostheca is an extension of the cytoplasmic material and provides additional membrane area for absorption of nutrients, thus allowing the bacteria to concentrate nutrients more efficiently in nutrient-poor environments.

Hyphomicrobium, another important inhabitant of aquatic environments, reproduces by budding. It synthesizes a stalk called a **hypha** (fig. 8.19b) that has a primarily reproductive function because newly formed cells bud out from the tips of these hyphae.

The appendaged budding bacteria play an important role in the recycling of nutrients. They prefer one-carbon (C-1) organic compounds, such as formate, methanol, and formaldehyde, as carbon sources. Many of the C-1 compounds are toxic to aquatic organisms; hence, these bacteria detoxify their environment as they reproduce by using these C-1 compounds as energy and carbon sources.

Chemolithotrophic Bacteria

The **chemolithotrophic bacteria** are gram-negative microorganisms that oxidize a variety of reduced inorganic ions for their energy. These bacteria, which may be rods or cocci, play a major role in nutrient recycling in both aquatic and terrestrial environments. For example, the **nitrifying bacteria** obtain their energy from the oxidation of ammonium or nitrite. *Nitrosomonas* oxidizes ammonium to nitrite, while *Nitrobacter* oxidizes nitrite to nitrate. Both of these bacteria are commonly found in soils, where they work to convert the ammonium resulting from the decomposition of organic matter into nitrate. This nitrate can be used by plants and other organisms (such as bacteria, algae, and fungi) as their source of nitrogen to build cellular components. The **sulfur bacteria** are the counterparts of the nitrifying bacteria, except that they oxidize inorganic sulfur compounds instead of nitrogen.

Methane-Producing Bacteria

The **methane-producing** (methanogenic) **bacteria** are a heterogeneous group consisting of gram-variable, anaerobic bacteria, all of which can convert carbon dioxide into methane. These bacteria have proteinaceous cell walls that lack peptidoglycan. This characteristic, coupled with other genetic differences, set these bacteria apart from most other prokaryotes. Some scientists believe that these bacteria are sufficiently different to be placed in their own kingdom; the archaebacteria.

The methanogenic bacteria are widely distributed in anaerobic environments that are rich in decayed organic matter. These environments include the intestinal tracts of humans and other animals, the rumen of livestock, stagnant water, swamps (bogs), garbage dumps, and sewage treatment plants. Marsh gas (methane) is produced in profusion by these bacteria in areas that are anaerobic and contain organic materials. The methane produced by these bacteria is sometimes used by sewage treatment plant operators to produce heat or to power some of the mechanical operations of the plant.

SUMMARY

1. Bacteria constitute an extremely heterogeneous group of microorganisms that share a prokaryotic cell type.
2. As a result of their prolific metabolic activities, bacteria perform numerous beneficial functions for humans and their environment.
3. An important role of bacteria in nature is decomposing organic matter so that it can be used by other life forms.

BERGEY'S MANUAL OF SYSTEMATIC BACTERIOLOGY

1. *Bergey's Manual* is a comprehensive treatise on the classification of bacteria. It includes descriptions of all the bacteria, keys to the identification of the various genera, and tables containing useful information on the identification of the various species.
2. In *Bergey's Manual*, all bacteria are classified in the kingdom Monera (Prokaryotae). This kingdom is subdivided into two divisions; the cyanobacteria and the bacteria.

GRAM-NEGATIVE BACTERIA OF MEDICAL OR APPLIED IMPORTANCE

1. The spirochete *Treponema pallidum* is the bacterium responsible for the venereal disease syphilis.
2. The aerobic rod *Pseudomonas* is the cause of numerous infections, especially ear and burn infections. *Legionella* is responsible for Legionnaire's disease.
3. Numerous pathogens are facultatively anaerobic bacilli: *Escherichia* that is responsible for the majority of urinary tract infections, *Salmonella* that causes gastroenteritis, *Shigella* that is associated with dysentery, and *Klebsiella* that is responsible for pneumonia and urinary tract infections.
4. *Vibrio* is a facultatively anaerobic curved bacterium that causes cholera.
5. The gram-negative aerobic diplococcus *Neisseria* is the cause of two serious diseases. *N. gonorrhoeae* is associated with the venereal disease gonorrhea while *N. meningitidis* is responsible for epidemic meningitis.
6. *Rickettsia* and *Coxiella* are coccobacilli and intracellular parasites. Species of *Rickettsia* are responsible for diseases such as Rocky Mountain spotted fever and typhus while *Coxiella* is responsible for Q-fever.
7. *Chlamydia* is the cause of two venereal diseases, nongonococcal urethritis and lymphogranuloma venereum. It is also responsible for an eye infection known as trachoma.
8. *Mycoplasma* and *Ureaplasma* are referred to as mycoplasmas. They lack a cell wall and consequently have no characteristic shape. *Mycoplasma* is responsible for primary atypical pneumonia while *Ureaplasma* is the cause of urethritis and sterility.

GRAM-POSITIVE BACTERIA OF MEDICAL OR APPLIED IMPORTANCE

1. Many gram-positive bacteria such as *Mycobacterium tuberculosis*, *Corynebacterium diphtheriae*, *Streptococcus pyogenes* and *Staphylococcus aureus* are capable of causing serious human diseases. Consequently, they are of considerable public health significance.
2. *Staphylococcus aureus* is a facultative anaerobe that is catalase positive. Strains of this organism are

the cause of toxic shock syndrome and numerous types of skin infections.

3. *Streptococcus pyogenes* is an aerotolerant anaerobe that is catalase negative. Strains of this organism are responsible for strep throat, rheumatic fever, and scarlet fever.
4. Numerous species of *Bacillus* and *Streptomyces* produce antibiotics that are used in medicine.
5. Species of the anaerobic endospore former *Clostridium* are responsible for disease: *C. botulinum* is the cause of the food poisoning called botulism and *C. tetani* is responsible for tetanus.
6. *Corynebacterium diphtheriae* is the cause of diphtheria.
7. *Mycobacterium* is responsible for two serious diseases: tuberculosis and leprosy.

OTHER GRAM-NEGATIVE BACTERIA

1. The cyanobacteria are all those prokaryotic microorganisms that carry out a process of photosynthesis similar to that of the green plants, evolving oxygen from the process.
2. Many cyanobacteria have the ability to convert atmospheric nitrogen into organic nitrogen. Nitrogen-fixing cyanobacteria have been used as "fertilizers" of rice paddies, where they increase the nitrogen content of the paddy.
3. The cyanobacteria are widely distributed in aquatic and terrestrial habitats. Some of them cause undesirable odor and flavor changes in drinking water when they grow in profusion in reservoirs and lakes. Some species of cyanobacteria produce toxins that cause diseases in humans and domestic animals.
4. There are many gram-negative bacteria that neither cause disease nor produce industrially important substances. These bacteria, which include the phototrophic, gliding, appendaged, and chemolithotrophic bacteria, play important roles in the recycling of matter in nature. Their activities make life on earth possible for all other organisms.

STUDY QUESTIONS

1. Describe four characteristics that are common to all bacteria.
2. Explain how the cyanobacteria are different from the phototrophic bacteria.
3. Cite the criteria that we used to differentiate the vibrios from the enterics.
4. What difficulty is encountered by anyone who attempts to classify the bacteria along phylogenetic lines?
5. Construct a table outlining the salient characteristics of the following groups of bacteria. Also include an important function they perform.
   a. Gram-positive bacteria
   b. Gram-negative bacteria
   c. Prosthecate or appendaged bacteria
   d. Chemolithotrophic bacteria
6. How do cyanobacteria differ from other bacteria?

SELF-TEST

1. The 9th edition of *Bergey's Manual* groups the bacteria:
   a. into 19 major groups;
   b. along phylogenetic lines;
   c. into 4 major groups based on cell wall characteristics;
   d. based on mechanism of energy generation;
   e. none of the above.
2. Spirochetes can be differentiated from other bacteria because they:
   a. have a gram-negative cell wall;
   b. are flexible and spiral-shaped;
   c. have a rigid, coiled cell wall;
   d. are gram-positive;
   e. are strict anaerobes.
3. The bacterium *Pseudomonas aeruginosa* is a gram-negative, facultatively anaerobic rod that causes severe gastrointestinal disorders. a. true; b. false.
4. *Campylobacter jejuni* is a curved, rigid, gram-negative, facultatively anaerobic bacillus that causes severe gastrointestinal disorders. a. true; b. false.
5. Enteric bacteria such as *Escherichia coli* differ from the vibrios because the vibrios are:
   a. straight, gram-negative rods;
   b. pleomorphic, gram-positive cocci;
   c. nonmotile;
   d. polarly flagellated, gram-negative curved rods;
   e. not able to ferment glucose.
6. The rickettsias and chlamydias are both obligate, intracellular parasites. a. true; b. false.
7. The endospore-forming bacteria are characterized by the formation of reticulate bodies during a stage in their sporulation cycle. a. true; b. false.
8. Certain species of *Staphylococcus* and *Streptococcus* catalase negative and are pyogenic. a. true; b. false.

9. Acid-fast bacteria, which include the genus *Mycobacterium:*
   a. are gram-positive cocci;
   b. include the agents of leprosy and tuberculosis;
   c. are involved in sexually transmitted diseases;
   d. ferment glucose to form lactic acid;
   e. none of the above.
10. The streptomycetes are industrially important bacteria because many of them produce antibiotics. a. true; b. false.

SUPPLEMENTAL READINGS

Atkinson, W. H. and Winkler, H. H. 1985. Transport of ATP by *Rickettsia prowazekii. Journal of Bacteriology* 161(1):32–38.

Daniels, L., Belay, N., Rajagopal, B.S. and Weimer, P.J. 1987. Bacterial methanogenesis and growth from CO_2 with elemental iron as the sole source of electrons. *Science* 237 (4814): 509–511.

Hackstadt, T., Todd, W., and Caldwell, H. 1985. Disulfide-mediated interactions of the chlamydial major outer membrane protein: Role in the differentiation of chlamydiae? *Journal of Bacteriology* 161(1):25–31.

Krieg, N. R., chief ed. and Holt, J. G., ed. 1984. *Bergey's manual of systematic bacteriology.* Vol. 1. 9th ed. Baltimore: Williams & Wilkins.

Krogman, D. W. 1981. Cyanobacteria (blue-green algae)—Their evolution and relation to other photosynthetic organisms. *BioScience* 31:121–124.

Lenette, E. H., Balows, A., Hausler, W. J., and Truant, J. P. eds. 1980. *Manual of clinical microbiology.* Washington, DC: American Society for Microbiology.

Schachter, J. and Caldwell, H. D. 1980. Chlamydiae. *Annual Review of Microbiology* 34:285–311.

Woese, C. R. 1981. The archaebacteria. *Scientific American* 244:98–126.

Woese, C. R. 1987. Bacterial evolution. *Microbiological Reviews* 51(2):221–271.

CHAPTER 9

THE FUNGI

CHAPTER PREVIEW TURKEY-X DISEASE

In 1960 more than 100,000 turkeys died in England due to a previously unknown disease. The disease struck suddenly and without warning. It was not contagious and was not caused by any known microorganism. Since the cause was unknown, the disease was dubbed the "Turkey-X Disease."

Poultry feed was suspected of being the cause of the disease. Studies revealed that the peanut meal in the feed was contaminated with a toxic substance. The toxin could be extracted with methanol, and when injected into ducklings it mimicked the signs and symptoms of Turkey-X Disease.

Eventually, it was discovered that the contaminant in the feed was a fungus called *Aspergillus flavus*. This fungus, when grown under the appropriate cultural conditions, secretes a toxin called **aflatoxin** (**A***spergillus* **fla***vus* **toxin**) that causes liver damage in poultry and liver cancer in laboratory animals.

Aflatoxin has now been shown to be harmful to humans also. For this reason, certain foods (e. g., peanut butter) are tested routinely for the presence of this substance and very strict limits on concentrations of aflatoxin in foods are being enforced.

The reports of Turkey-X Disease heralded the development of modern **mycotoxicology** (the study of fungal toxins and their effects). It is now known that many fungi produce toxins and that these toxins cause a variety of human and animal diseases.

Toxin-producing fungi have a significant impact on humans by causing disease and killing livestock. Fungi also participate in many natural processes that are essential to the survival of many organisms. This chapter introduces the characteristics of the major groups of fungi, their biology, and their impact on humans.

CHAPTER OUTLINE

THE FUNGI ARE A GROUP OF HETEROTROPHIC eukaryotes that are widely distributed in nature. They are considered to be **saprophytes** because they obtain their nutrients from the decomposition of dead organic matter. Although early investigators thought that fungi were related to plants, the fungi are quite different from the green plants and their cells lack chlorophyll. Presently, most biologists recognize this difference and have classified the fungi in a kingdom of their own, the kingdom **Fungi**.

BIOLOGY OF FUNGI

Fungi characteristically form filaments called **hyphae** (fig 9.1). A hypha consists of either a single, elongated cell or a chain of cells whose cytoplasms are generally shared. The hyphae lack chlorophyll and absorb nutrients from the environment. Hyphae also play a role in the reproduction of the fungus, because they can give rise to other hyphae or to specialized reproductive structures called **spores** (fig. 9.2). The hyphae have cell walls consisting of **chitin** or **cellulose**, along with other minor polysaccharides. The fungal colony, or **thallus**, consists of a mass of hyphae called **mycelium** (fig. 9.3). The fungal thallus may originate from a single fungal spore or from a piece of hypha. As simple as the above description of a fungus may be, it is not without exceptions. This is because many microorganisms that are considered to be fungi do not form hyphae. For example, the **yeasts** are single-celled fungi that characteristically form colonies resembling those of bacteria and do not produce demonstrable hyphae. Other fungi produce structures during one stage of their life cycle that are definitely funguslike,

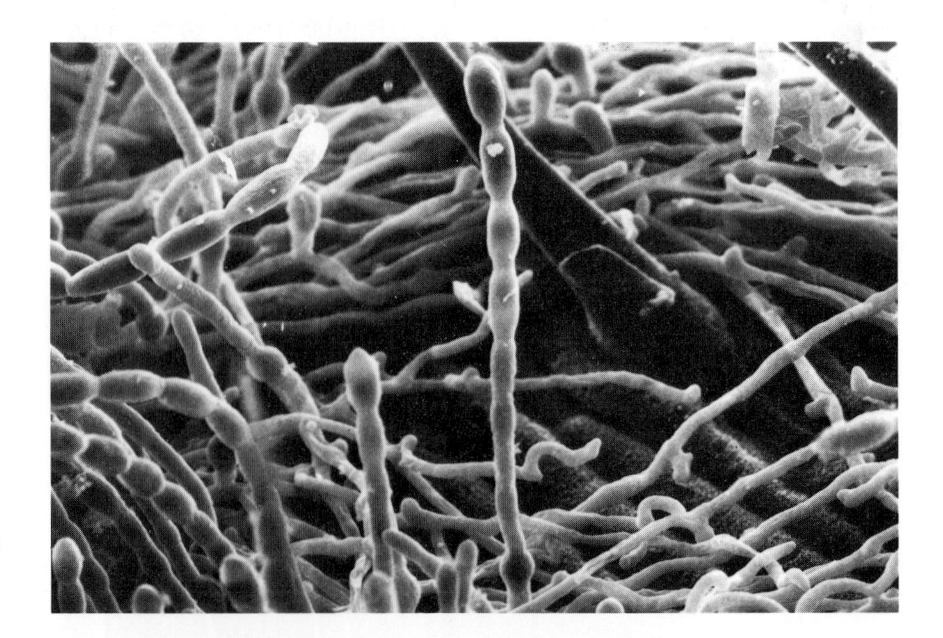

Figure 9.1 Fungal hyphae. Fungal hyphae of *Erysiphe graminis* on a grass leaf as seen with a scanning electron microscope.

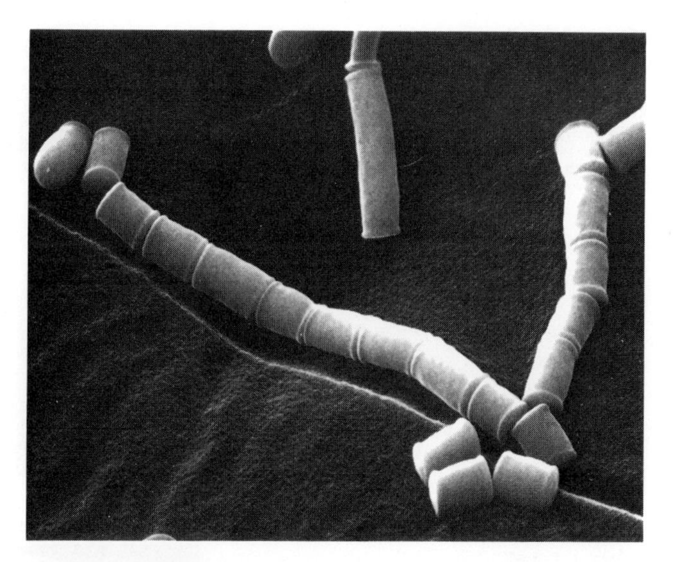

Figure 9.2 **Representative types of fungal spores.** Fungi produce spores of many different shapes. Spores may develop from asexual or sexual processes and are frequently used to identify organisms. Spores efficiently spread fungi throughout the biosphere.

while at another stage they have a protozoalike existence. Organisms that have both fungal and protozoal characteristics are called the **slime molds**.

Materials and nutrients move throughout the hyphae by cytoplasmic streaming. This is possible because hyphae are composed of interconnecting cells. Some hyphae are completely devoid of cross walls and are called **coenocytic hyphae**, while others have distinct cross walls and are called **septate hyphae**. Even septate hyphae can permit the passage of nutrients and cellular structures because the cross walls are perforated at the centers.

The growth of fungi occurs by extension of the hyphal tip (fig. 9.4). As the tips of the hyphae in the thallus grow, they extend outward radially, forming a more or less rounded colony.

The cells that make up the fungal thallus generally possess only one **haploid** nucleus, that is, each nucleus has only one copy of each of its chromosomes. At certain stages of their life cycle, some fungi may

Figure 9.3 Fungal colonies. Colonies of four different fungi as seen growing on agar-solidified media. The colony, also known as a thallus, consists of a mass of fungal hyphae.

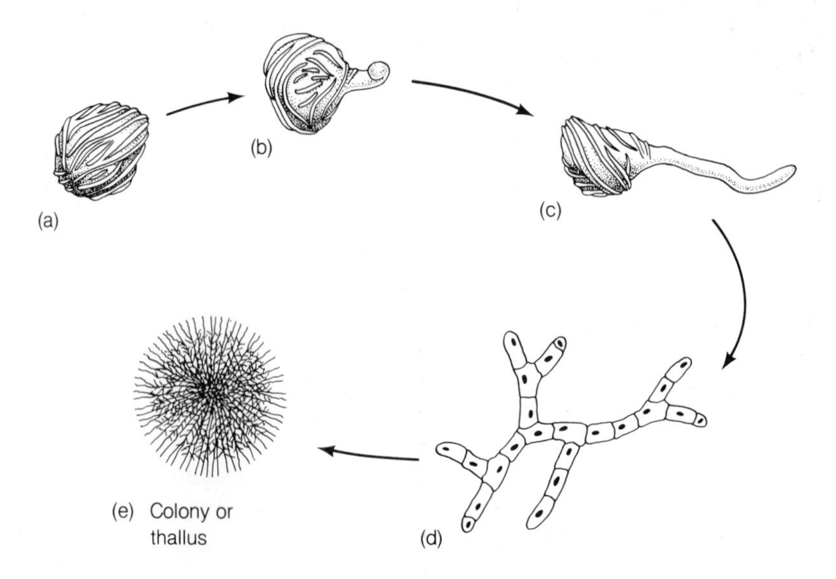

Figure 9.4 Development of a fungal colony. A fungal spore *(a)* germinates producing the beginning of a hypha *(b)*, also known as a germ tube. The hypha elongates *(c)*, branches *(d)*, and eventually forms a mass of hyphae known as mycelium. The mass of mycelium forms the colony or thallus *(e)*. Illustration not drawn to scale. The fungal spore may measure about 5 μm in diameter while the colony or thallus may exceed 5 cm in diameter.

be **diploid** (having a nucleus with two copies of each chromosome). In most fungi, however, diploidy is a brief condition and is usually evident only during certain sexual reproductive stages. The exceptions are the Oomycetes, which are diploid except when forming gametes (sex cells). Some fungi possess more than one haploid nucleus per cell and are said to be **dikaryotic** (di = two, karyon = nucleus) (fig. 9.5).

Fungal Reproduction

The formation of a new thallus (colony) from a preexisting one can take place by either a **sexual** or an **asexual** process. In asexual reproduction, new cells are formed by mitosis and then spread to new sites to form new thalli. This form of reproduction represents the most common form of fungal multiplication during periods of rapid growth and does not require that two compatible fungal elements fuse to form a zygote.

In many fungi, mitotic divisions result in asexual spores (table 9.1) called **conidia**. These structures are often produced in profusion and are carried away by air currents that spread the fungus to other habitats. Many fungi, especially those in the group called the **Fungi Imperfecti (Deuteromycetes)**, produce a large array of different types of conidia, many of which can often be used to differentiate the various species of these fungi.

During sexual reproduction, two compatible nuclei fuse together to form a **diploid zygote**. The zygote subsequently undergoes a meiotic division that results in the production of four or more haploid fungal cells. These cells are generally called **sexual spores**. The sexual spores separate from the fungus and eventually germinate to form new haploid thalli.

Certain fungi can carry out sexual reproduction by fusing two nuclei within the same thallus; that is, they are **self-fertile**. Self-fertile fungi are usually called **homothallic** fungi. Other fungi require a compatible nucleus from another individual in order to undergo sexual reproduction. These fungi are **self-sterile** and are said to be **heterothallic**.

Fungi are unique among microorganisms in that their hyphae can carry two or more genetically different nuclei. These nuclei can arise from the fusion of two genetically different hyphal cells, or from a

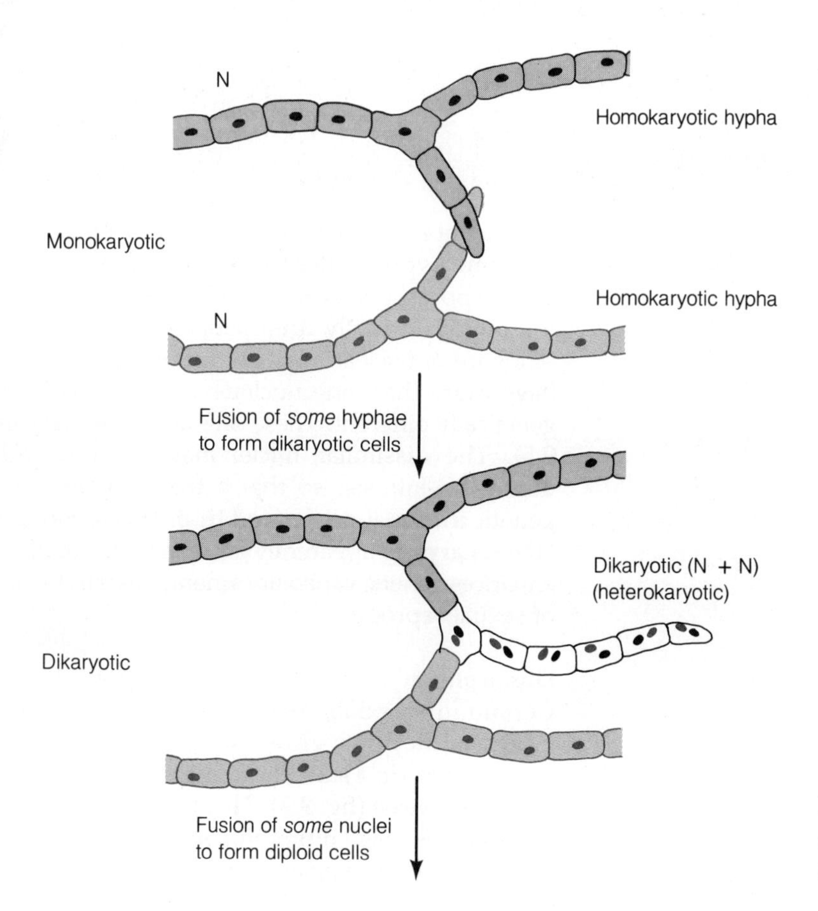

Figure 9.5 The formation of a heterokaryotic hypha from a fusion of two homokaryotic hyphae

TABLE 9.1
VARIOUS TYPES OF FUNGAL SPORES

Fungal group	Distinguishing characteristics	Examples	Asexual spore	Sexual spore
Oomycete	Coenocytic hyphae, diploid nuclei	*Phytophthora*	Flagellated zoospores	Oospores
Zygomycete	Coenocytic hyphae, haploid nuclei	*Rhizopus* *Mucor*	Sporangio-spores	Zygospores
Ascomycete	Septate hyphae, haploid nuclei	*Neurospora* *Eurotium** *Talaromyces** *Arthroderma** *Ajellomyces**	Conidia	Ascospores
Basidiomycete	Septate hyphae, dolipore septum. May form fruiting bodies of various shapes and sizes. Haploid nuclei	*Puccinia* *Amanita* *Boletus* *Filobasidiella**	Conidia Teliospores Pycniospores	Basidiospores
Deuteromycete	Septate hyphae. Haploid nuclei	*Microsporum** *Candida* *Coccidioides* *Sporothrix* *Cladosporium*	Conidia	None

**Eurotium* is the sexual stage of *Aspergillus; Talaromyces* is the perfect stage of *Penicillium; Arthroderma* is the perfect state of *Trichophyton; Ajellomyces* is the perfect state of *Histoplasma; Filobasidiella* is the perfect state of *Cryptococcus;* and *Nanizzia* is the perfect state of *Microsporum.*

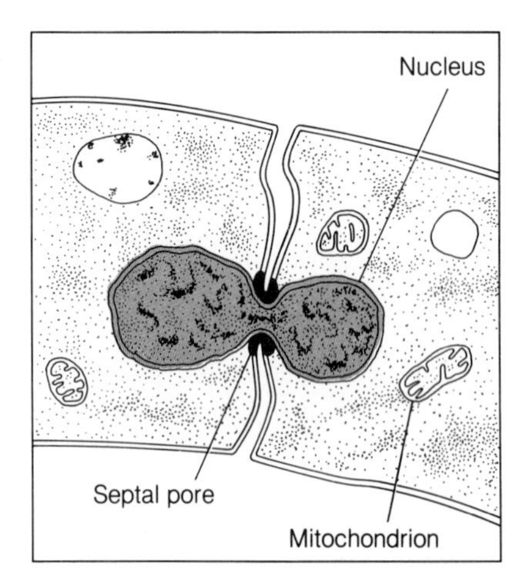

Figure 9.6 Nucleus passing through the septal pore of a hypha of the ascomycete *Neurospora crassa*

mutational event (change in the genetic material) within the same hypha. Since the hyphae of most fungi have septal spores, nuclei apparently can pass from one cell to another virtually unimpeded (fig. 9.6). As a consequence of nuclear migration, a given hyphal cell may have more than one nucleus and the nuclei may be genetically different. These cells are **heterokaryotic** (fig. 9.5). The dissimilar nuclei may multiply independently by mitosis, so that a thallus is created with genetic information derived from two or more nuclei. **Heterokaryosis** apparently is a desirable condition that enhances genetic variability among fungi in the absence of sexual reproduction.

Dimorphism

Certain fungi exhibit two different morphologies. For example, they may grow as filamentous forms in the soil, whereas in a suitable animal host they may reproduce like a yeast (fig. 9.7). This phenomenon of having two forms is called **dimorphism**. Some dimorphic fungi can be made to display their dimorphism simply by

(a)

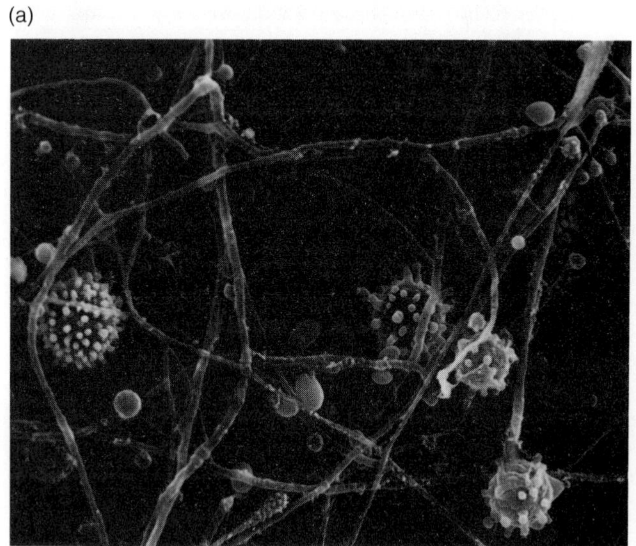

(b)

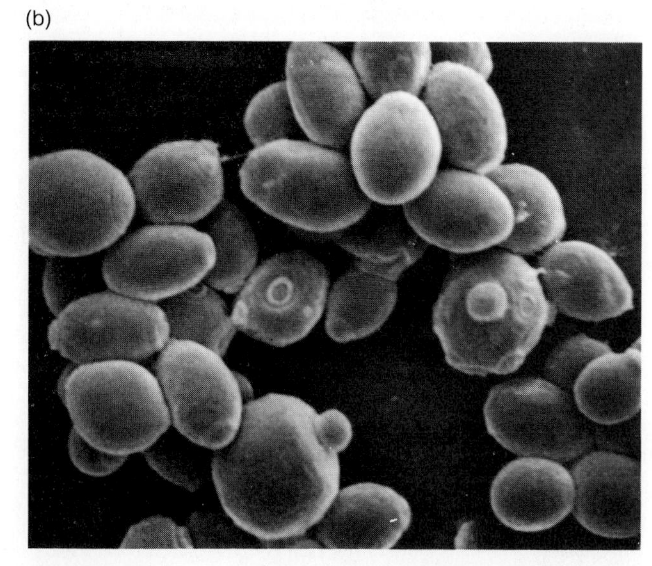

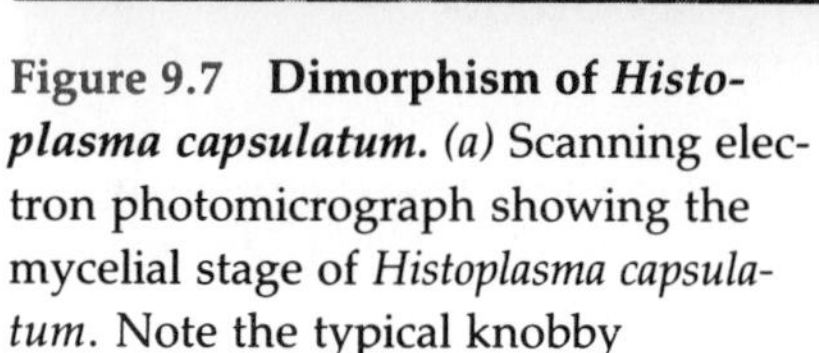

Figure 9.7 Dimorphism of *Histoplasma capsulatum*. *(a)* Scanning electron photomicrograph showing the mycelial stage of *Histoplasma capsulatum*. Note the typical knobby (tuberculate) conidia. They develop when the fungus is cultured at room temperature on suitable culture media *(b)* SEM of the yeast phase of *Histoplasma capsulatum*. This phase develops when the fungus reproduces inside a suitable host or when it is cultured at body temperature on blood-containing agar.

changing the environmental and/or nutritional conditions in which they are grown. For example, the dimorphic fungus *Histoplasma capsulatum*, when grown at room temperature on a culture medium such as Sabouraud Dextrose Agar, grows in a filamentous, **mycelial** form, producing typical large conidia with knobs (tuberculations) on thin hyphae (fig. 9.7a). If however, this fungus is grown on blood agar at 37°C, it will form a single-celled, rounded yeast phase (fig. 9.7b). The same dimorphic change occurs when the mycelial form of the fungus is introduced into a susceptible host, such as a human or a laboratory mouse. This shift is generally called the M to Y shift, because the fungus changes from **m**ycelial form to **y**east form. Another fungus, *Candida albicans*, exists as a yeast at room temperature, but upon inoculation into a susceptible host, or when cultivated under low oxidation-reduction potentials, the fungus forms filaments referred to as **pseudomycelium**. Table 9.2 indicates some important human pathogens that are dimorphic.

Dimorphism is advantageous to fungi because it allows them to reproduce more rapidly in adverse conditions or to fend off immune host responses. Yeast cells generally reproduce faster than filamentous fungi. This rapid growth may also afford the dimorphic fungus a means of obtaining nutrients more rapidly from its environment, thus giving the fungus a competitive edge over other organisms that may be sharing a food resource. The Y to M shift of *C. albicans* allows this fungus to resist host defense mechanisms more effectively. Hence, it is possible that dimorphic changes may also be beneficial to pathogenic fungi because they aid in preventing the host defense mechanisms from destroying the invading fungus.

DISTRIBUTION AND ACTIVITIES OF FUNGI

Fungi have a tremendous impact on our environment. They convert complex organic matter (plant and animal remains) into simple chemical compounds, which

TABLE 9.2
IMPORTANT HUMAN PATHOGENS

Name of fungus	Name of disease	Taxonomic group
Absidia sp.	Zygomycosis	Zygomycetes
**Ajellomyces capsulatus*	Histoplasmosis	Ascomycetes
**Ajellomyces dermatitidis*	Blastomycosis	Ascomycetes
Aspergillus fumigatus	Aspergillosis	Deuteromycetes
**Candida albicans*	Candidiasis	Deuteromycetes
**Coccidioides immitis*	Coccidioidomycosis	Deuteromycetes
Epidermophyton	Ringworms	Deuteromycetes
Filobasidiella neoformans	Cryptococcosis	Basidiomycetes
Fonsecaea	Chromoblastomycosis	Deuteromycetes
Malassezia furfur	Tinea Versicolor	Deuteromycetes†
Microsporum	Ringworms	Deuteromycetes
Mucor sp.	Mucormycosis	Zygomycetes
[1]*Pneumocystis carinii*	Pneumocystis carinii pneumonia	Ascomycetes
Pseudoallescheria	Mycetoma	Ascomycetes
Rhizopus sp.	Zygomycosis	Zygomycetes
**Sporothrix schenckii*	Sporotrichosis	Deuteromycetes
Trichophyton	Ringworms	Deuteromycetest

*Names of organisms preceded by an asterisk indicate that the fungus is dimorphic (see section on dimorphism).
†Some species of *Microsporum* and *Trichophyton* have now been shown to have an Ascomycete state in *Nanizzia* and *Arthroderma*, respectively.
[1]*Pneumocystis carinii* is more closely related to the yeast *Saccharomyces cerevisiae* than to protozoans as shown by rRNA analyses.

can then be used as nutrients by other organisms. Because fungi can act on a large variety of substrates, both living and nonliving, they have been influencing human populations for centuries. Certain fungi are used in making wines, beers, cheeses, breads, and many other foods. Others serve as sources of food (the truffles, morelles, and mushrooms), and still others are the source of life-saving drugs such as antibiotics. The human–fungus association is not always mutually beneficial, however. For example, fungi constitute the single most important cause of disease in plants, resulting in billions of dollars in damage to crops annually throughout the world. Fungi also cause spoilage of foods and destruction of wood and textiles. In addition, certain of these microorganisms can cause diseases in humans and in domestic animals. Fungal diseases of humans result in the loss of millions of dollars due to the lack of productivity. The fungi generally cause disease when they grow within living tissue where they release exoenzymes that destroy tissue, or when they produce toxins that disrupt the proper functioning of living cells.

Mycorrhiza

Mycorrhiza is the fungal growth that appears on the surface and in the cortex of the roots of certain green plants. This association is mutually beneficial and is widespread among terrestrial plants. **Conifers**, such as pine trees and spruces, often form mycorrhizal associations with **basidiomycetes**, such as *Boletus*. Mycorrhizal associations can permit economically important trees to grow on poor soils that normally would not support the tree's growth. The mycorrhizal association is believed to increase the plant's surface area for absorption of nutrients in nutrient-deficient environments and supply the fungus with plant carbohydrates. Many of the mycorrhizal fungi are nutritionally dependent on the plant host (fig. 9.8).

THE MAJOR GROUPS OF FUNGI

Fungi are subdivided into various groups (table 9.3) based upon their structural and reproductive charac-

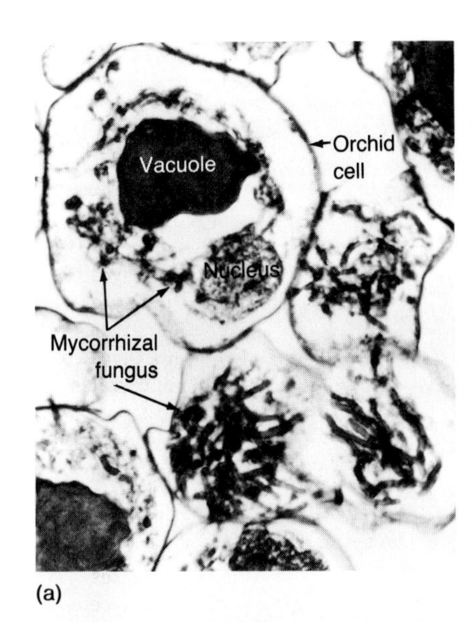

(a)

(b)

(c)

Figure 9.8 Mycorrhizal interactions. *(a)* Mycorrhizal fungus in association with orchid root cells. *(b)* Influence of mycorrhizae on plant growth. The pine tree is growing in a nutrient-poor soil and has not developed a substantial mycorrhizal association. *(c)* The pine tree is growing in the same soil but has a well-developed mycorrhizal association.

teristics. Many of these subdivisions are named after the type of sexual reproductive structure the fungus produces. For example, the basidiomycetes characteristically form **basidia** during sexual reproduction, while the ascomycetes produce **asci** instead. In the following paragraphs some of the major groups of fungi are discussed, with emphasis on their biology and impact on humans.

Zygomycetes

The **zygomycetes** are a group of predominantly terrestrial (land-dwelling) fungi that are widely distributed in nature. The hyphae of these fungi vary in diameter and are characteristically wide, measuring 3–5 μm in diameter. In addition, their hyphae are usually coenocytic (or aseptate). When septa are present, they are often spaced at irregular intervals along the length of the hyphae. Zygomycetes break down organic matter and recycle nutrients. Some species are human pathogens (table 9.2), while others produce industrially useful chemicals. Their life cycle includes both sexual and asexual reproduction. Figure 9.9 illustrates the typical life cycle of the zygomycete *Rhizopus*, a mold that is frequently seen growing on rotting vegetables and starchy food.

The zygomycetes are characterized by the formation of sexual spores called **zygospores**. These spores form after two compatible fungal hyphae fuse: the site where the hyphae fuse becomes enlarged and differentiates into a zygospore. When the zygospore germinates, it gives rise to a **sporangium**, a balloon-shaped structure filled with spores called **sporangiospores** (fig. 9.9). Each sporangiospore can give rise to a new fungal colony. These structures can also develop from hyphae in the absence of sexual reproduction, and constitute a very common means of asexual reproduction in the zygomycetes. A single sporangium can release numerous sporangiospores, which can then be transported for many miles by air and water currents. The sporangiospore represents the major means of dispersal for these fungi. The arrangement of the sporangia and their location along the hyphae sometimes serve to differentiate among the various zygomycetes.

TABLE 9.3
SUMMARY OF THE MAJOR GROUPS OF FUNGI

Classification	Characteristics
Kingdom: Fungi	
Division: Gymnomycota	Naked (wall-less) cells.
Subdivision: Acrasiogymnomycotina Class: Acrasiomycetes (cellular slime molds)	Myxamoeba → pseudoplasmodium (made up of individual amoeba) → sorocarp. Example: *Dictyostelium*
Subdivision: Plasmodiogymnomycotina Class: Protosteliomycetes Class: Myxomycetes	Myxamoeba → plasmodium (a single multinucleated macroscopic cell) → sporangium. Examples: *Physarum, Fuligo*
Division: Mastigomycota	Flagellated cells.
Subdivision: Haploidmastigomyeotina Class: Chytridiomycetes Class: Hyphochytridiomycetes Class: Plasmodiophoromycetes	Haploid hyphae, flagellated zoospores, hyphae coenocytic, many aquatic forms. Examples: *Synchytrium, Allomyces, Rhizidiomyces, Plasmodiaphora.*
Subdivision: Diplomastigomycotina Class: Oomycetes	Diploid hyphae, flagellated zoospores, Hyphae coenocytic. Many aquatic forms. Examples: *Phytophthora* and *Plasmopara*
Division: Amastigomycota	Nonflagellated cells.
Subdivision: Zygomycotina Class: Zygomycetes Class: Trichomycetes	Haploid, coenocytic hyphae. Sexual spore is a zygospore. Asexual spore is the sporangiospore borne inside sporangia. Examples: *Rhizopus* and *Mucor.*
Subdivision: Ascomycotina Class: Ascomycetes	Haploid, septate hyphae with septal pore. Sexual spore is the ascospore borne inside ascus. Asci frequently develop inside ascocarp. Asexual spore is usually the conidium. Examples: *Neurospora, Morchella, Saccharomyces* (yeast), *Ajellomyces,* and *Eurotium.*
Subdivision: Basidiomycotina Class: Basidiomycetes	Hyphae with dolipore septum. Clamp connections seen in certain hyphae. Sexual spores are basidiospores borne on basidia. Examples: *Puccinia* and *Amanita.*
Subdivision: Deuteromycotina Class: Deuteromycetes (Fungi Imperfecti)	Yeastlike or filaments. Hyphae resemble those of the Ascomycetes. Sexual stage unknown. Parasexual cycle in some species. Conidia of various types produced. Examples: *Candida, Trichosporon, Penicillium, Aspergillus, Alternaria,* and *Geotrichum.*

The zygomycetes *Mucor* and *Rhizopus* are capable of very rapid and extensive growth that can cause rapid food spoilage. Fungi in the genera *Rhizopus, Mucor,* and *Absidia* have been implicated in human disease and death. They occasionally invade human tissue in individuals who have been debilitated by diseases such as cancer, immune deficiencies, or diabetes. Their growth and invasion of organs often results in the death of the individual within 2–5 days.

Since the zygomycetes can exhibit a fermentative metabolism, they are used to make fermented food items. Soybean products such as **tempeh** and **sufu** are made using zygomycetes such as *Rhizopus*. Their fermentation of carbohydrates causes the soybean pro-

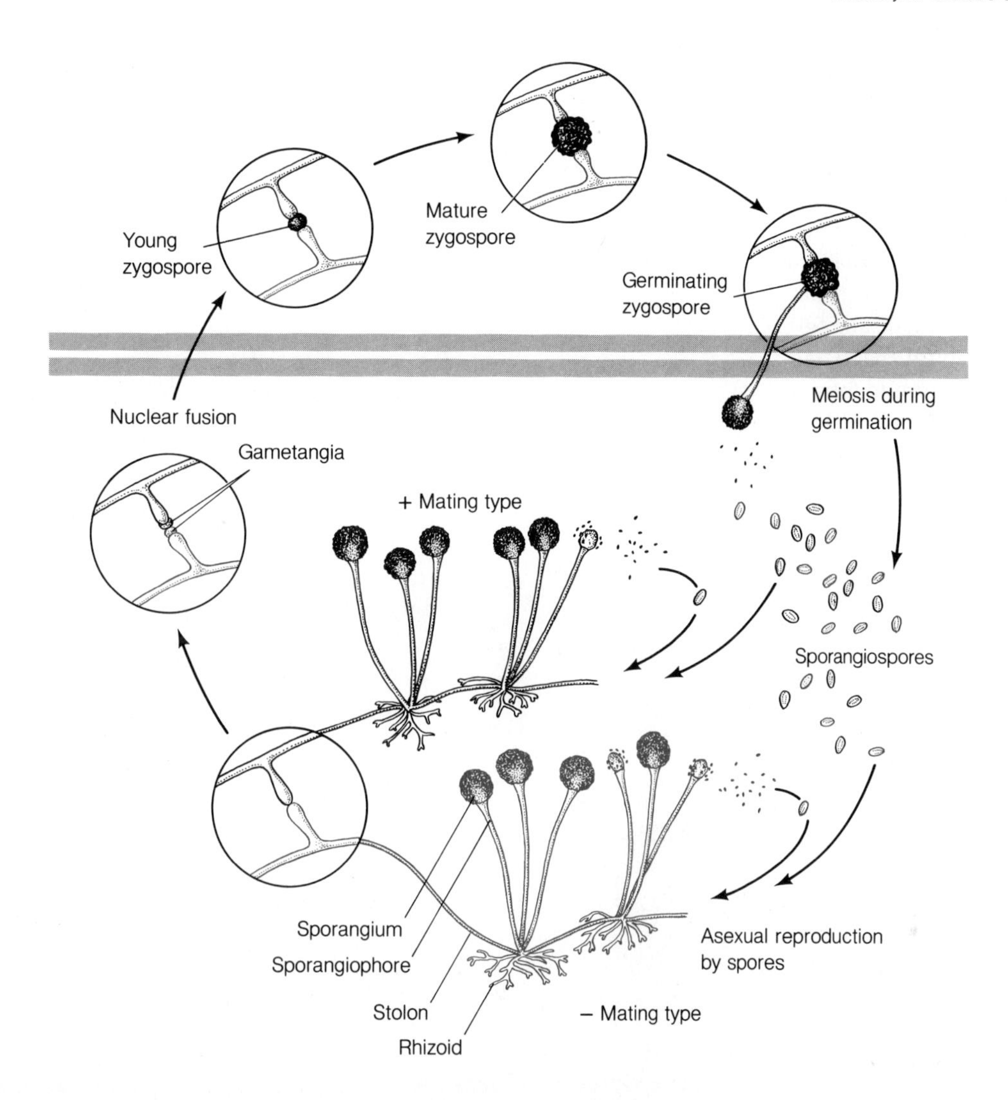

Figure 9.9 Life cycle of *Rhizopus*, a typical zygomycete. Compatible hyphae (+ and −) fuse to form a zygote called the zygospore. The mature zygospore develops a thick coat and can remain dormant for many months. After the dormancy period, the zygospore cracks open, producing a sporangium filled with sporangiospores. The spores can germinate (fig. 9.4), producing new thalli. Vegetative hyphae can also give rise to sporangia by asexual means. When compatible hyphae fuse, the cycle begins again.

tein to curdle and therefore contributes to the formation of the soybean cakes that characterize these food items. Certain zygomycetes are employed in industry for the synthesis of economically important products such as enzymes, organic acids, and alcohols. For example, the enzyme rennin, which is used in the cheese industry to curdle milk as a first step in cheesemaking, is produced by the zygomycete *Mucor pusillus*. Cortisone is another product that is synthesized using zygomycetes, such as *Rhizopus*.

Oomycetes

Until recently, the oomycetes (fig. 9.10) were classified with the zygomycetes in a group called the **phyco-**

Figure 9.10 Life cycle of the oomycete *Saprolegnia*. Asexual reproduction occurs when motile zoospores from sporangia undergo a developmental cycle that leads to the formation of new hyphae. The zoospore may encyst and become dormant for extended periods. Sexual reproduction takes place when male and female structures produce sperm and eggs. These gametes fuse to form a zygote. The zygote eventually develops into oospores. Oospores can then germinate into hyphae and the cycle begins again.

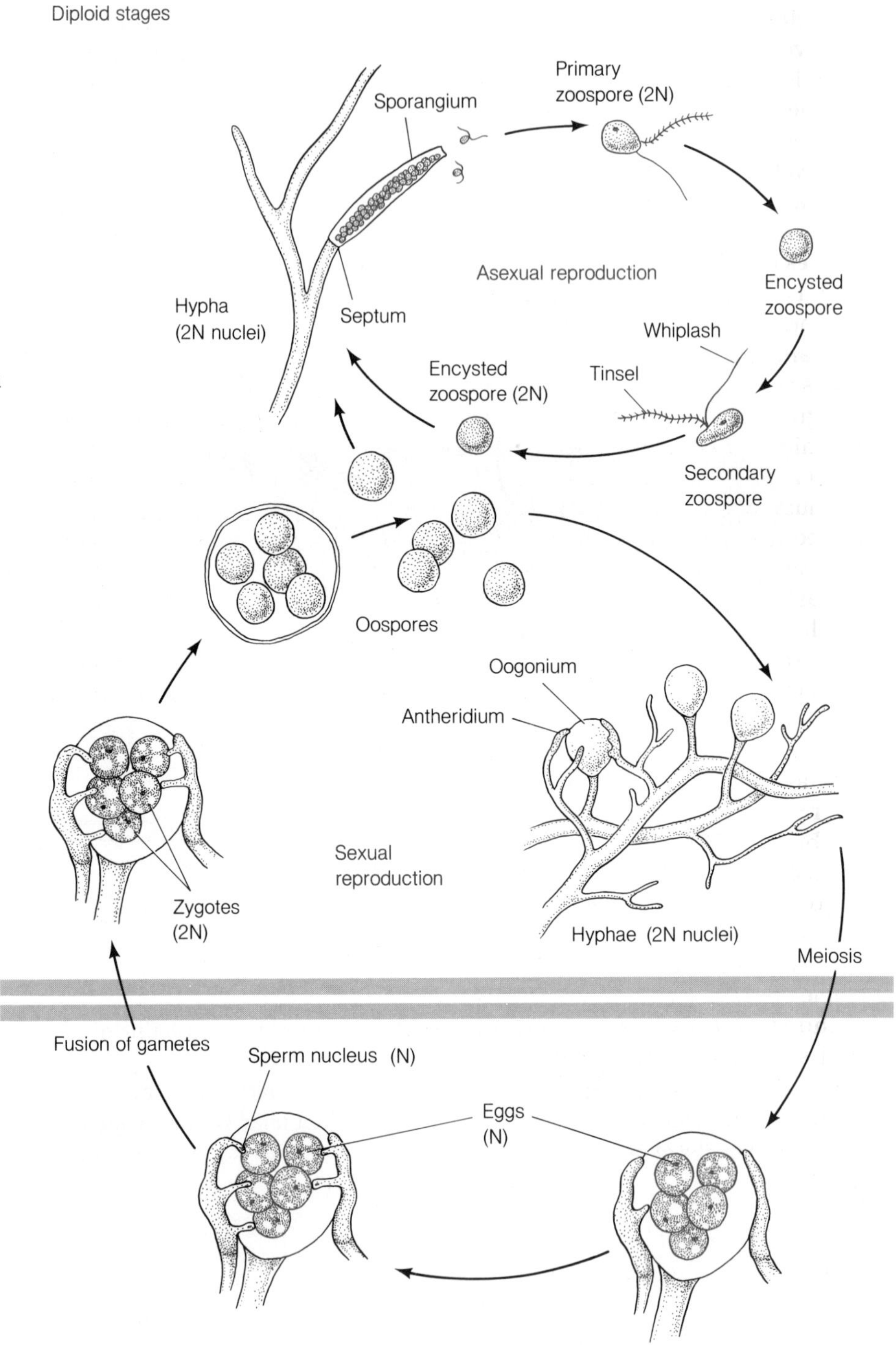

mycetes. Most mycologists believe, however, that the oomycetes are quite different from other fungi and should be classified separately. This argument is based on the facts that the Oomycetes are diploid rather than haploid, that they form motile spores, and that their cell walls have cellulose instead of chitin as the major cell wall polysaccharide. The motile spores develop from mitotic divisions (asexual) and are called **zoospores** (fig. 9.10).

The oomycetes as a group include some very important plant pathogens such as *Phytophthora infestans* and *Plasmopara viticola*. *Phytophthora infestans* causes a disease of potatoes called **late blight**. It was this fungus that destroyed the potato crops in Ireland in the nineteenth century. The destruction of the potato crop led to starvation, disease, and death in Ireland. Because of crop failures, millions of Irish citizens emigrated to various countries, including the United States. *Plasmopara viticola* causes **downy mildew** of grapes, a devastating disease that can result in significant financial losses to viticulturists. Other oomycetes can cause diseases of fish; **ich**, for example, a disease of aquaria fish, can be caused by the oomycete *Saprolegnia*.

Ascomycetes

The ascomycetes produce sexual spores called **ascospores** that are derived from the union of two compatible fungal elements. Ascospores develop inside a sac-like structure called the **ascus**. All the fungi that produce ascospores within asci are called **sac fungi** and are placed in the Class **Ascomycetes**.

The ascomycetes are widely distributed in nature. Some of the ascomycetes parasitize crops, causing tremendous financial losses (billions of dollars annually) and starvation throughout the world. Others cause diseases in humans and other animals such as histoplasmosis and some of the ringworms.

The hyphae of ascomycetes, unlike those of the zygomycetes, are thin and delicate with septations at regular intervals along their length. The hyphal wall may be pigmented, consists primarily of chitin, and generally contains a single nucleus.

The ascomycetes reproduce asexually by producing a variety of spores, predominantly **conidia**. Sexual reproduction in the ascomycetes (fig. 9.11) results in the formation of an **ascus**, which generally occurs within an **ascocarp**. The ascocarp is a complex structure, consisting of hyphae knitted together to form a bed upon which the asci develop (fig. 9.11). The ascomycetes may show male and female differentiation. The male structure can be a conidium, a hyphal fragment, or a nucleus. The female structure is called the **ascogonium** and it has a tubelike structure called the **trichogyne**. When a male structure comes in contact with a trichogyne, it fuses with the trichogyne and a nucleus travels down it to the ascogonium. This initiates the development of specialized hyphae (also known as **ascogenous hyphae**) that eventually develop into asci.

Ascomycetes can cause severe damage to textiles because of the production of enzymes such as **cellulases** and **proteases**, which break down the fibers of cotton, wool, and silk, particularly in humid, warm climates. Some fungi also produce toxic substances that can cause the death of animals and humans. For example, *Claviceps purpurea*, the ascomycete that causes **ergot of rye**, produces toxic alkaloids that cause a disease called **ergotism**. Ergotism can cause the death of anyone who consumes bread made with contaminated rye grain. On the other hand, ergot alkaloids have been used in medicine to stop bleeding, induce smooth muscle contraction, or induce uterine contractions in pregnant women to expedite their labor.

Yeasts

Yeasts are a group of primarily Ascomycete fungi that are typically unicellular and reproduce asexually by budding or by fission (fig. 9.12). In addition to reproducing asexually, many yeasts also exhibit sexual reproduction. The majority of the sexual yeasts are ascomycetes.

Yeasts have a tremendous impact on human activities. For example, yeast infections are commonplace in medical practice. These infections are commonly caused by *Candida albicans*, *Torulopsis glabrata*, and *Cryptococcus neoformans*. Yeast infections are sometimes fatal. Yeasts are also important to humans because of the products they make, such as wine, beer, and bread. The alcohol in wine and the distinct flavors of various wines are due, at least in part, to the action of yeasts on the grape juice. Yeasts are also used as a source of protein for human consumption.

Basidiomycetes

The basidiomycetes constitute a very large group of fungi that are widely distributed in terrestrial habitats. Some of the basidiomycetes have a macroscopic stage

Figure 9.11 Life cycle of a typical ascomycete. Asexual reproduction takes place when conidia, which develop from hyphae, germinate to produce new fungal thalli. During sexual reproduction, two compatible fungal elements come together. The fusion of the fungal elements result in the formation of asci containing four to eight ascospores.

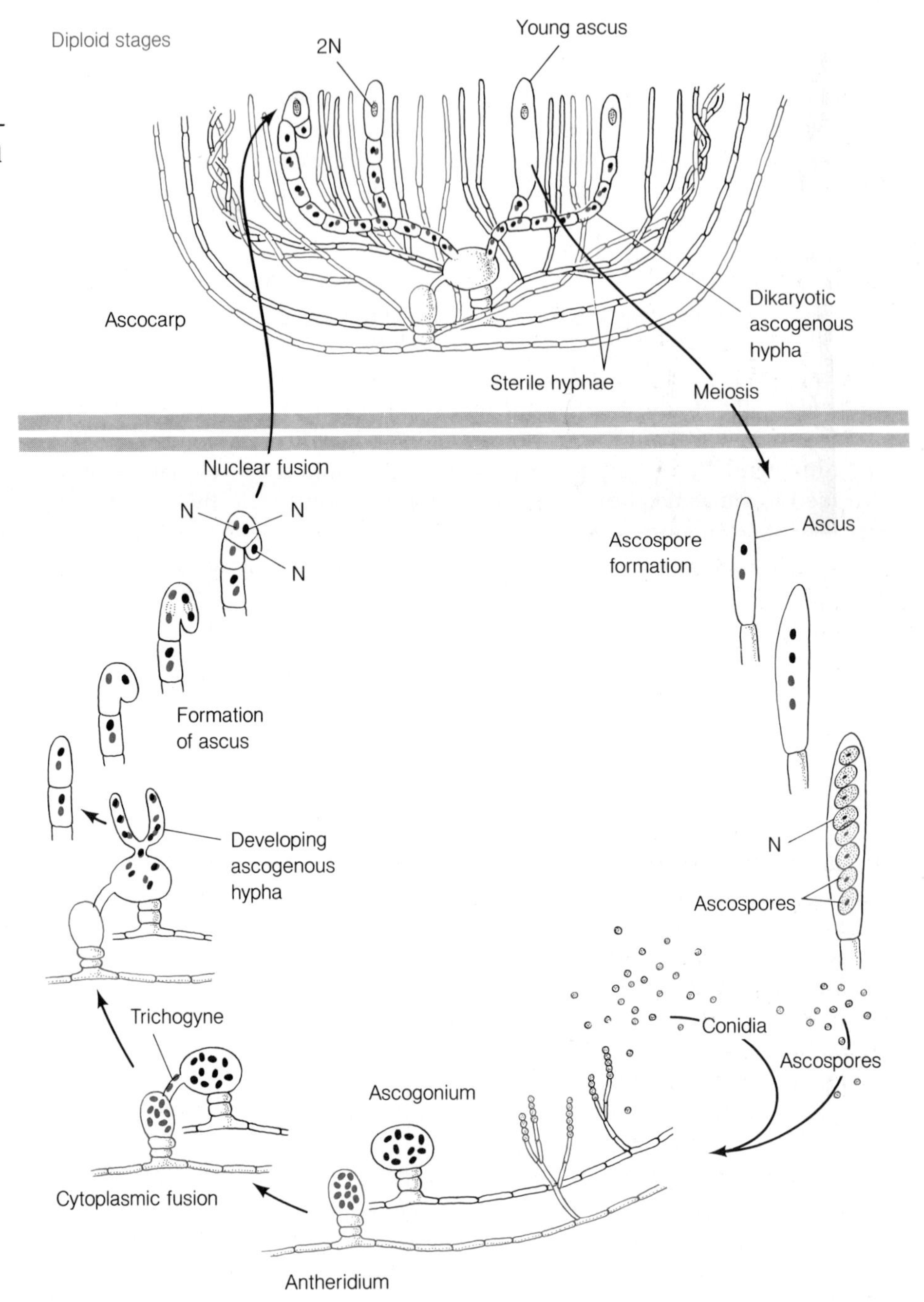

(a)

(b)

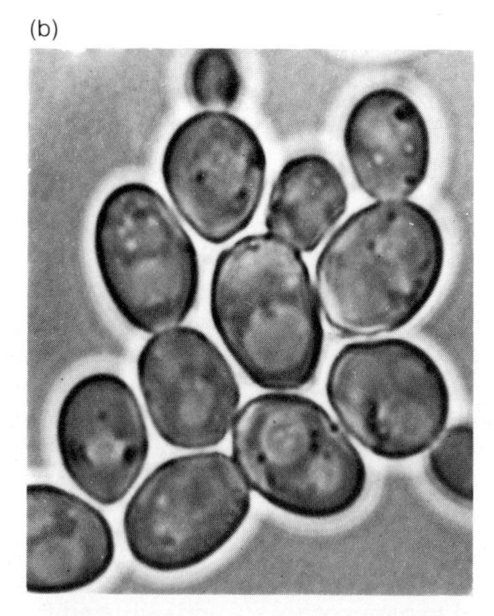

(c)

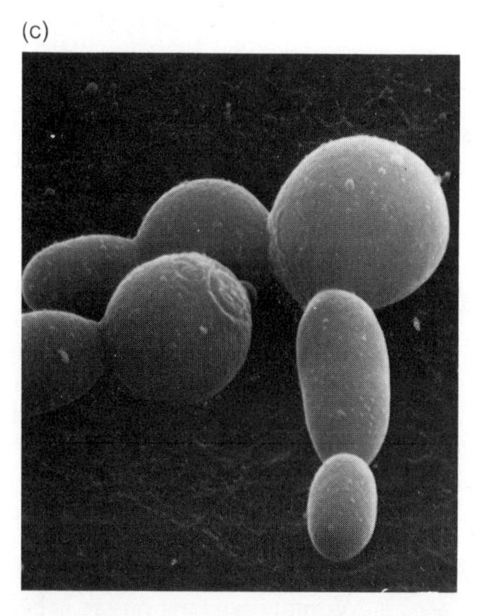

Figure 9.12 Representative yeasts. *(a)* Nomarski interference photomicrograph of the yeast *Schizosaccharomyces* showing vegetative cells and asci containing ascospores. *(b)* Phase contrast photomicrograph of the yeast *Saccharomyces cerevisiae* showing budding yeast cells. *(c)* Scanning electron micrograph of *Candida albicans* showing budding yeast cells.

and are visible with the unaided eye (fig. 9.14). More than 20,000 species of basidiomycetes have been described in the literature. These include **mushrooms, toadstools, rusts, smuts, stink horns, bracket fungi, puffballs, coral fungi**, and **bird's nest fungi**. In this group are some very important plant pathogens, such as *Puccinia graminis*, which causes **wheat rust**, and *Ustilago zeae*, the cause of **corn smut**. Together, these two plant pathogens cause the loss of billions of dollars each year worldwide. Other basidiomycetes are important human pathogens. For example, *Filobasidiella neoformans*, the sexual state of *Cryptococcus neoformans*, infects human nervous tissue, causing a disease called cryptococcosis.

The hyphae of basidiomycetes are tubular structures that consist of typical eukaryotic cells separated by septa. Except for certain rusts and smuts, which have ascomycetelike hyphae, the basidiomycete hypha have a septum called the **dolipore septum** (fig. 9.15).

All basidiomycetes form sexual spores called **basidiospores** (fig. 9.16). The basidiospores develop from supporting structures called **basidia**. The mushroom, or **basidiocarp**, is a fruiting body formed by many basidiomycetes (fig. 9.14). The basidiocarp is a specialized structure that results from the fusion of two compatible hyphae. The hyphae that develop join together to form the basidiocarp. The basidiocarp is composed of three major parts: (a) the **pileus** or cap; (b) the **stipe** or stem; and (c) the **lamellae** or gills. The basidiocarps are found above ground and are generally large enough to be seen easily with the unaided eye. Beneath the ground, the basidiocarp is supported and nourished by vegetative hyphae. On the underside of the cap are the gills, which consist of a thin layer of hyphae that give rise to the basidia and their basidiospores (fig. 9.16). The function of the cap or pileus is to bear the basidia and their basidiospores and to participate in spore dispersal.

When hyphae of two compatible mating strains (for example + and −) fuse, the resulting hypha, containing nuclei from both fungi, is called a **secondary hypha**. Some of the secondary hyphae in the basidiocarp differentiate into basidia (fig. 9.14). The basidia are dikaryotic at first, but as the basidium matures, the nuclei fuse to form diploid nuclei that undergo meiosis almost immediately. The four haploid nuclei that develop from meiosis migrate toward the tip of the basidium and form the basidiospores.

FOCUS WITCHCRAFT OR A CASE OF FOOD POISONING?

The Puritans living in New England during the 1600s and 1700s believed in witches and witchcraft. Almost every year, people were accused of witchcraft, and sometimes the accused witches were tried, found guilty, and hanged. In 1692 in Massachusetts around Salem, approximately 150 persons were accused of witchcraft, over 35 of them were brought to trial, and 12 were hanged. Some spent months in jail, and at least one died there. Most of the witchcraft accusations were made by teenage girls who were afflicted by symptoms that resemble those of persons taking hallucinogenic drugs. About 70% of the accusations were against women, and approximately 60% of those executed were women.

A fungus growing on rye is believed to have been the cause of the numerous strange deaths, bewitched persons, and accusations of witchcraft near Salem in 1692. The years 1690, 1691, and 1692 were cooler than average in eastern New England and the winters were extremely cold. The moist, cool spring along the coast is believed to have favored the growth of the fungus *Claviceps purpurea*, commonly known as **ergot**, on rye and other grasses. One of the fruiting structures of the fungus is called the **sclerotia** (fig. 9.13). When consumed along with the rye it causes food poisoning with symptoms that could be interpreted as bewitchment by people who believe in witchcraft. Ergot sclerotia are a source of lysergic acid diethylamide (LSD), a very potent hallucinogen. The symptoms of this food poisoning, called **convulsive ergotism**, range from mild to severe and can lead to death.

The victims of "bewitchment" in Massachusetts were mainly infants, young children, and teenagers. Seven infants and young children are known to have developed symptoms of convulsive ergotism and died. One boy suffered from a painful urinary difficulty caused by ergotism, and his brother died of a mysterious affliction. It is now known that nursing infants can develop ergotism from their mother's milk. In addition, several cows developed symptoms of food poisoning and died.

The symptoms of "bewitchment" first began in Salem at the end of 1691 and were probably due to contaminated rye harvested in the summer of that year. Records indicate that rye crops in the Salem area frequently remained unthreshed in storage until November and December. This fact may explain why the symptoms did not appear soon after the crops were harvested. Twenty-four of the 30 persons who were "bewitched" in Massachusetts during 1692 had convulsions and feelings of being pinched, pricked, or bitten. Three girls felt that they were being torn to pieces and that their bones were being pulled out of their joints. Symptoms also included temporary blindness, deafness, pain, visions, and flying sensations. A number of girls suffered from hallucinations, fits, and laughing and crying spells.

Since the Salem trials, a number of incidents of convulsive ergotism have occurred in the New England area. Most of the food poisoning epidemics were mild, but a Salem epidemic of convulsive ergotism called "nervous fever" killed more than 30 persons in 1795. Epidemics of ergotism occur throughout the world when cold, damp years weaken rye crops and stimulate the growth of the fungus.

These cases of convulsive ergotism should remind us of the importance of microorganisms that grow in association with animals and plants.

Figure 9.13 **Sclerotia on rye**

Each of these basidiospores is then capable of germinating and giving rise to primary hyphae.

Some of the basidiomycetes are edible. In certain countries, mushroom eating or **mycophagy** is widely practiced. There is always the possibility, however, of ingesting poisonous or hallucinogenic mushrooms instead. This points out the importance of carefully determining the identity of the fungus before eating it, because fungi similar to edible ones can be extremely toxic or even lethal.

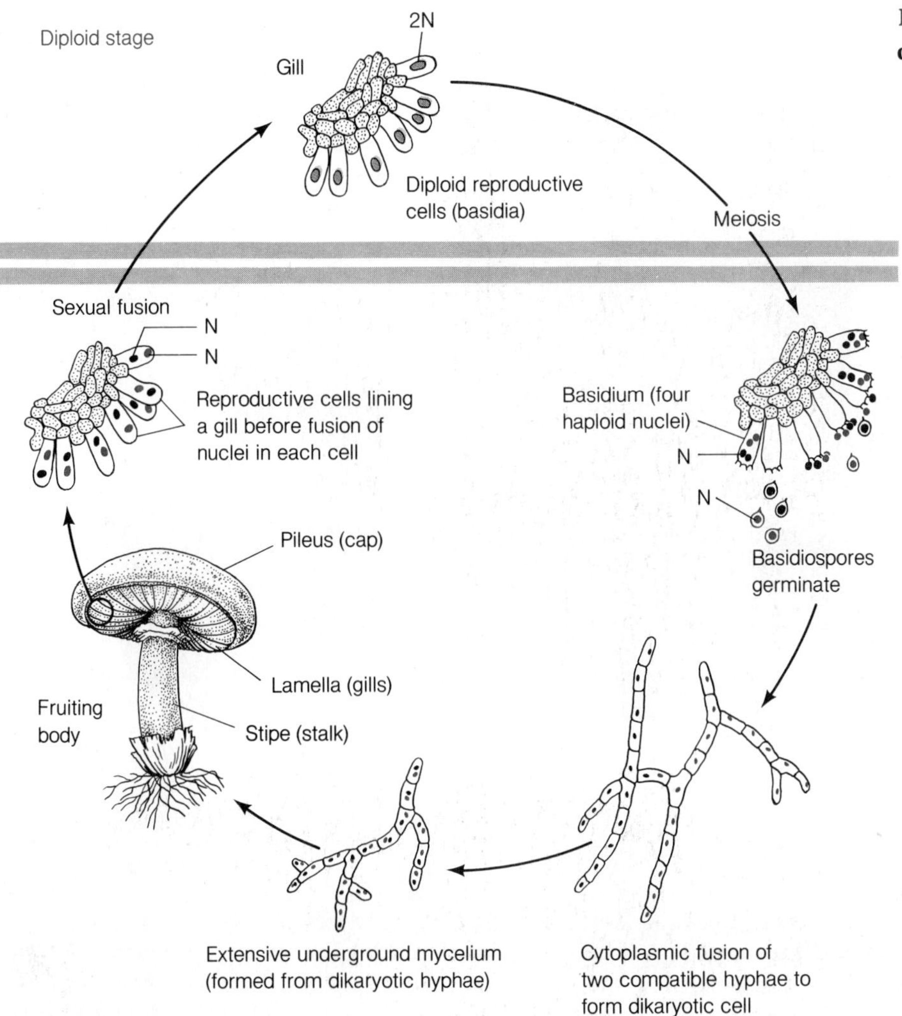

Figure 9.14 Structure and life cycle of a basidiomycete

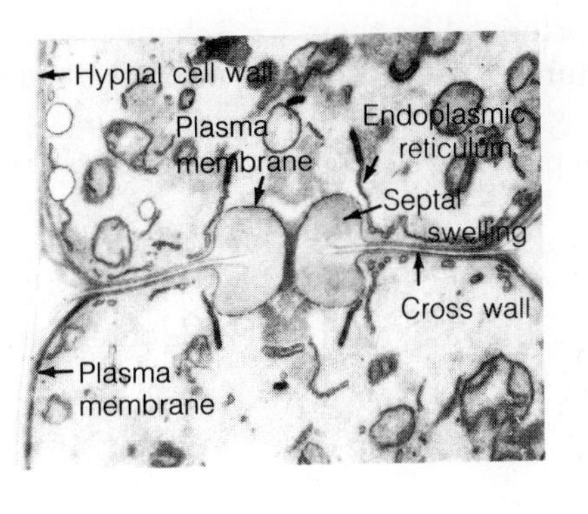

Figure 9.15 The dolipore septum of the secondary hyphae in basidiomycetes. This electron photomicrograph illustrates two hyphal cells of *Rhizoctonia solani* separated by a dolipore septum. The various components of the septum, including the cross walls and septal swellings are indicated.

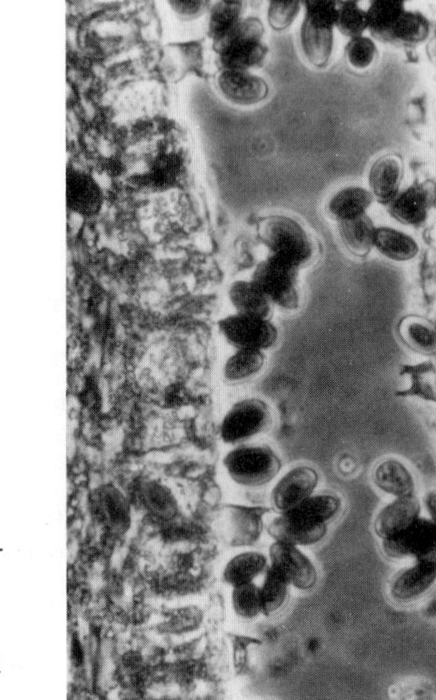

(a)

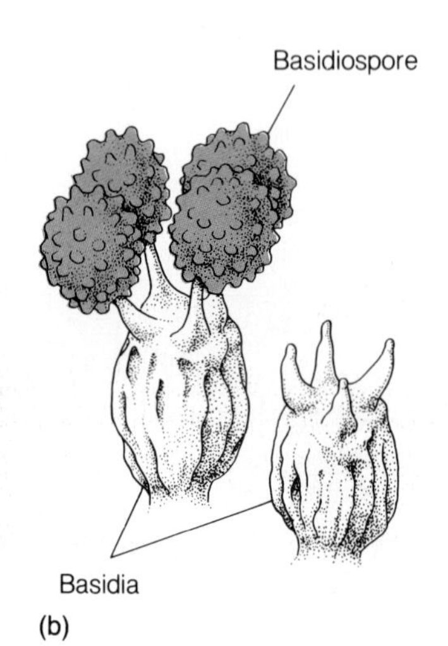

(b)

Figure 9.16 Basidum with basidiospores. (*a*) Stained section of a gill of the basidiomycete *Coprinus*. Numerous, elliptical basidiospores and the club-shaped basidia are seen. (*b*) Interpretive drawing of a scanning electron micrograph of basidia from the gill of a basidiomycete. The basidia are shown with and without attached basidiospores.

The hallucinogenic mushrooms have been used by many ancient cultures. The Maya and Inca civilizations, for example, used hallucinogenic mushrooms in their religious ceremonies. Some of these religious ceremonies involved the consumption of a toxic mushroom called *Amanita muscaria*, which can be hallucinogenic if eaten in minute quantities but lethal if eaten in even slightly greater quantities.

Fungi Imperfecti or Deuteromycetes

The **Fungi Imperfecti**, or **Deuteromycetes**, constitute a very large group of fungi with unknown genetic affinities. As you may have noticed, all major groups of fungi discussed thus far are characterized by their unique sexual reproductive structures. The fungi imperfecti cannot be classified using sexual criteria because they have no known sexual reproductive cycle. Some of the fungi in this group may have the ability to reproduce sexually, but their sexual cycle has not yet been observed; others may have lost this ability altogether. In either case, all fungi lacking a demonstrable sexual cycle are placed in the fungi imperfecti. As **mycologists** (students of fungi) learn more about the fungi and develop new methods for their cultivation and for promoting the sexual reproductive cycles, more of the fungi imperfecti will probably exhibit their sexual cycles. Great strides have already been made in this respect and mycologists have uncovered the sexual cycles of many fungi imperfecti, most of which have turned out to be ascomycetes.

The fungi imperfecti produce a large variety of conidia (from the Greek, meaning "fine dust") that can be used as a means of classifying these organisms (table 9.4). Development of the various types of conidia is controlled by a variety of different genes and is characteristic of species. Mycologists have developed a scheme for classifying and identifying the fungi imperfecti based on how their conidia develop.

Many of the fungi imperfecti are plant and animal pathogens (table 9.2), which can cause disease either by destroying cells during their reproduction or by producing toxins. **Aflatoxin**, which is produced by *Aspergillus flavus*, causes severe liver damage and sometimes death in animals. Other fungi imperfecti are of economic importance because they make useful products. For example, *Pencillium chrysogenum* pro-

TABLE 9.4
CONIDIA PRODUCED BY SOME FUNGI IMPERFECTI

Organism	Illustration of Conidia
Geotrichum candidum *Coccidioides immitis*	
Yeasts *Cladosprorium*	
Alternaria sp. *Dreschlera sp.*	
Sporothrix schenckii	
Scopulariopsis	
Phialophora *Aspergillus* *Penicillium*	

duces the antibiotic penicillin, which is widely used to treat a large variety of infectious diseases caused by bacteria.

Slime Molds

The **slime molds** (table 9.3) are a group of eukaryotic organisms that share characteristics with both the fungi and the protozoa. The slime molds are fungilike during certain stages of their life cycle and protozoalike during others (fig. 9.17).

The slime molds live in cold, moist, shady places, such as under the canopy of forests. They feed upon decaying vegetable matter such as leaves, dead logs, or animal dung. The slime molds are a very important group of decomposers of leaf litter in forests. During the spring months, while it is still cold, one can visit the forest and be awed by the colorful display of fruiting bodies of the organisms on fallen logs and decaying vegetable matter. Some species of slime molds form enormous pseudoplasmodia, measuring several feet in width.

A group of slime molds called the **cellular slime molds**, which belong to the class **Acrasiomycetes**, has been studied extensively because these organisms present a good model for the study of cellular differentiation in eukaryotes. Certain cellular slime molds can be grown easily in the laboratory and therefore serve as a useful organism for research.

When the spores of a cellular slime mold, such as *Dictyostelium discoideum*, are placed in a nutrient medium

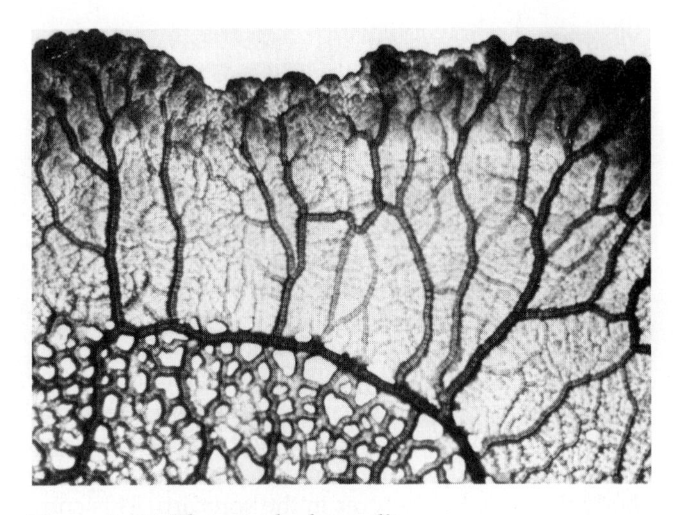

Figure 9.17 The pseudoplasmodium of a slime mold

Figure 9.18 Life cycle of the cellular slime mold *Dictyostelium discoideum*

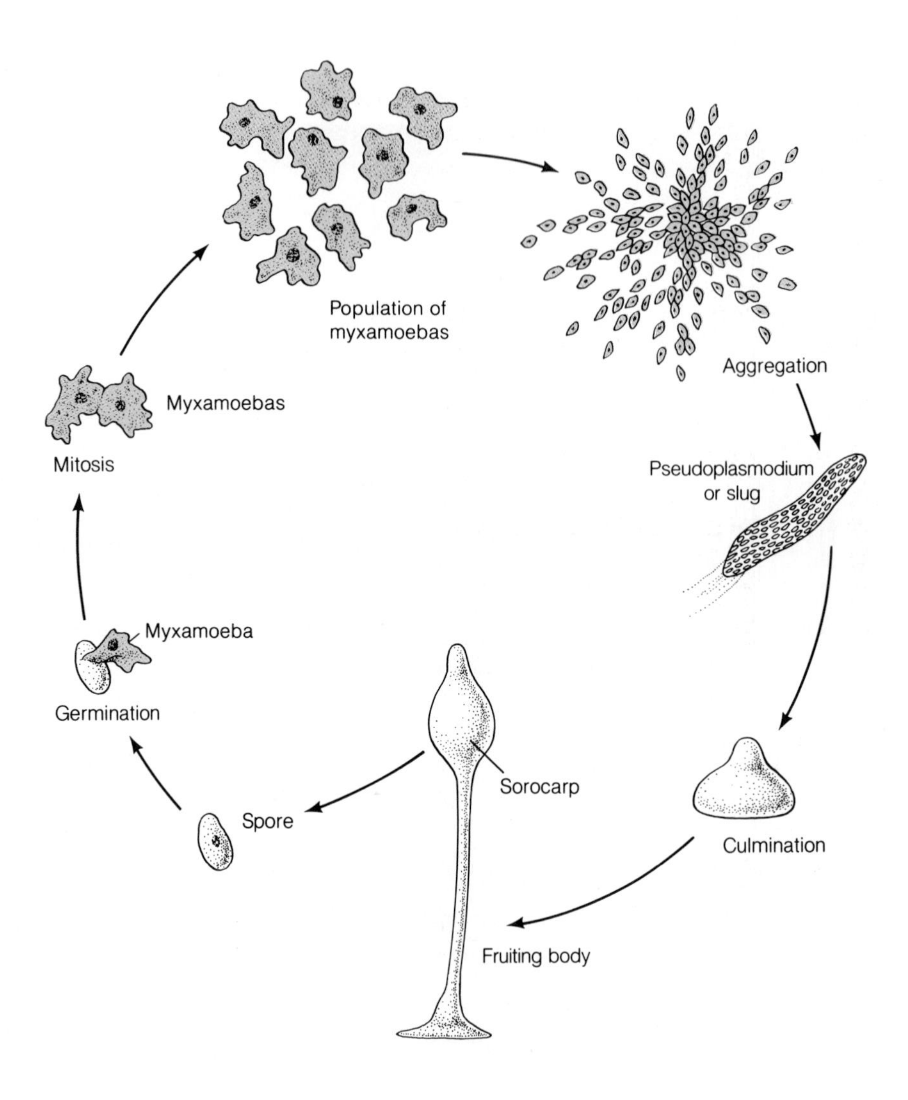

containing *Escherichia coli* cells, the spores germinate and give rise to motile cells known as **myxamoebas** (fig. 9.18). The myxamoebas creep over the surface of the agar and feed upon the bacterial cells. The myxamoebas multiply on the surface of the agar medium by mitosis. When the supply of nutrients is nearly exhausted, the myxamoebas aggregate. They stop feeding and set up **centers of aggregation**, orienting themselves toward these centers. Eventually, they move toward the centers of aggregation, fuse with each other, and develop into what is called a **pseudoplasmodium.**

The pseudoplasmodium develops into a fruiting body that produces spores in the sorocarp. This completes the life cycle of *Dictyostelium*.

SUMMARY

Fungi are heterotrophic, eukaryotic microorganisms that lack chlorophyll.

BIOLOGY OF FUNGI

1. Fungi characteristically form **hyphae** that consist of elongated cells or chains of cells. The hyphae absorb nutrients from the environment.
2. The hyphae of fungi grow by extension of their tips.
3. Yeasts are unicellular fungi and rarely form hyphae.
4. Most fungi can produce sexual and asexual spores. These reproductive structures may be used to differentiate among fungi.

5. Dimorphism is a phenomenon in which certain fungi can have two different morphologies, depending upon the environmental conditions under which they reproduce. Some human pathogens exhibit dimorphism.

DISTRIBUTION AND ACTIVITIES OF FUNGI

1. Fungi play a major ecological role as decomposers. They are largely responsible for the decaying of organic matter and the recycling of nutrients from organic materials in many environments.
2. As decomposers, fungi may cause a variety of diseases in animals (and humans) and in plants.
3. Many metabolic waste products produced by fungi are of industrial importance. For example, some antibiotics (penicillin and cephalosporin), steroids, and ethanol are produced using fungi.

THE MAJOR GROUPS OF FUNGI

1. Fungi are classified on the basis of their reproductive structures. Table 9.3 summarizes the different groups of fungi and their salient characteristics.
2. Fungi reproduce sexually by fusing two compatible haploid cells to form a diploid zygote. The zygote subsequently undergoes **meiosis** to form haploid sexual spores.
3. Fungi reproduce asexually by forming various types of spores by **mitotic** divisions.

STUDY QUESTIONS

1. Describe and diagram a typical fungus.
2. How do yeasts differ from other fungi?
3. Outline the process of hyphal growth. Why are fungal colonies generally circular?
4. Outline the major characteristics of the various groups of fungi.
5. Cite three industrial processes that employ fungi.
6. Describe briefly three fungal activities that are undesirable to humans.
7. What is dimorphism? How can it benefit a fungus? How can microbiologists take advantage of this characteristic to diagnose a fungal disease?
8. What is a conidium? Of what use is it in identifying a fungus? Of what use is it to the fungus? Do sporangiospores serve the same function?
9. Why is sexual reproduction important to the fungi?
10. How do fungi (including yeasts) differ from bacteria?

SELF-TEST

1. The reproductive and nutrient-gathering unit of many fungi consists of tubular, filamentous structures called hyphae. a. true; b. false.
2. The hyphae of ascomycetes and basidiomycetes differ from those of the zygomycetes in that the hyphae of zygomycetes:
   a. are composed of cellulose instead of chitin;
   b. have dolipore septa;
   c. are coenocytic;
   d. form branches;
   e. none of the above.
3. Dimorphism, as illustrated by *Histoplasma*, probably is a survival mechanism of this organism so that it can reproduce faster as a yeast in host tissue and produce large amounts of conidia in its filamentous form in nature. a. true; b. false.
4. *Phytophthora infestans* is an important pathogen in the class Oomycetes. Which of the following diseases does this organism cause?
   a. ergotism;
   b. potato blight;
   c. ringworm of the scalp;
   d. ich;
   e. corn smut.
5. The principal biological function of fungi in nature is the decomposition of organic matter. a. true; b. false.
6. Conidia are fungal structures that arise by asexual processes and represent a form of dispersal of the organism. a. true; b. false.
7. Ascospores and Basidiospores arise after a zygote undergoes the process of mitosis. a. true; b. false.
8. The Deuteromycetes are classified on the basis of:
   a. the type of sexual spores they produce;
   b. the nature of the zygote that develops after gametic fusion;
   c. the type of conidia they produce;
   d. biochemical reactions of the sexual stage;
   e. none of the above.
9. Mushrooms, puff balls, rusts, and smuts are fungi in the class:
   a. Oomycetes;

b. Ascomycetes;
c. Zygomycetes;
d. Deuteromycetes;
e. Basidiomycetes.

10. In general, both the filamentous fungi and the yeasts reproduce by elongation of their cell tips. a. true; b. false.

11. Fungi acquire nutrients via their hyphae by the process of phagocytosis. a. true; b. false.

12. The fungal thallus or colony develops when a spore germinates producing a mass of hyphae that elongate by extension of their tip, forming a more or less round colony. a. true; b. false.

13. A sexual event has occurred in fungi if their haploid spores form after meiosis. a. true; b. false.

14. Which of the following types of spores does not develop by sexual processes?
    a. Ascospores;
    b. Zygospores;
    c. Basidiospores;
    d. Conidia;
    e. Oospores.

15. A fungus is said to be dimorphic when it produces more than one type of spore. a. true; b. false.

SUPPLEMENTAL READINGS

Alexopoulos, C. J. and Mims, C. W. 1975. *Introductory mycology.* 3d ed. New York: Wiley.

American Society for Microbiology, Board of Education and Training. 1978. *Identification of saprophytic fungi commonly encountered in the clinical laboratory.* Washington, DC: American Society for Microbiology.

Christensen, C. M. 1951. *The molds and man: An introduction to the fungi.* Minneapolis: University of Minnesota Press.

Kendrick, B., ed. 1979. *The whole fungus.* Ottawa: National Museums of Canada.

Lennette, E. H., Balows, A., Hausler Jr., W. J., and Shadomy, H. J. 1985. *Manual of clinical microbiology.* 4th ed. Washington, DC: American Society for Microbiology.

McGinnis, M. R., D'Amato, R. F., and Land, G. A. 1982. *Pictorial handbook of medically important fungi and aerobic actinomycetes.* New York: Praeger.

Rippon, J. W. 1982. *Medical mycology, the pathogenic fungi and the pathogenic actinomycetes.* 2d ed. Philadelphia: Saunders.

Rubin, R. H. 1986. Mycotic infections. *Scientific American Medicine* 7–IX: 1–14. New York: Freeman.

Rubin, R. H. 1987. Infection of the immunosuppressed host. *Scientific American Medicine* 7–X: 1–19. New York: Freeman.

Edman, J. C., Kovacs, J. A., Masur, H., Santi, D. V., Elwood, H. J., and Sogin, M. L. 1988. Ribosomal RNA sequences show *Pneumocystis carinii* to be a member of the fungi. *Nature 334:* 519–522.

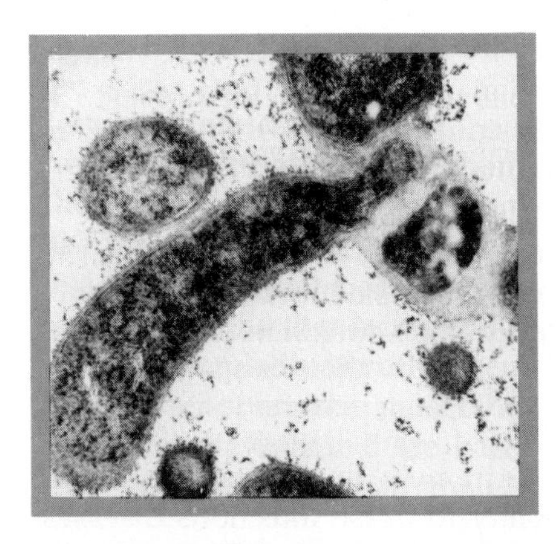

CHAPTER 10

THE EUKARYOTIC PARASITES

CHAPTER PREVIEW *NAEGLERIA FOWLERI:* **AN AGENT OF FATAL MENINGITIS**

Natural bodies of water are inhabited by a myriad of microorganisms, many of which are important to the aquatic community. These microorganisms, which include bacteria, fungi, algae, and protozoa, contribute to the overall "health" of the lake and are prominent members of lakes and streams. Because of them, many lakes and streams are suitable for human recreation (e.g., swimming and fishing) and as reservoirs for drinking water.

Amoebae, single-celled, nonphotosynthetic eukaryotes, abound in these aquatic environments. Like other members of the aquatic community, they contribute positively to the lake and stream ecology by decomposing organic matter and consuming excess microorganisms that may foul the water.

Some of these amoebae, however, may cause severe and often fatal diseases in humans. For example, *Naegleria fowleri*, an amoeba that measures 10 to 20 μm in diameter, reproduces in stagnant waters and can cause severe (and fatal) swelling of the brain and spinal chord (meningoencephalitis).

In 1965, three fatal cases of meningitis were reported in Australia. All three cases were caused by amoebae multiplying in the membranes that envelop the nervous system (meninges). In 1966 and 1968, similar cases were reported in Florida. All the cases had one thing in common: the victims were young people who had been swimming in lakes and ponds about a week before their affliction. By 1973, more than 70 cases of amoebic meningitis caused by *Naegleria fowleri* had been reported from throughout the world.

In 1978 a young English girl was reported dead of amoebic meningitis. The girl was known to frequent the Roman Baths in England. Since the Roman Baths were routinely cleaned and chlorinated, the source of the infection was a puzzle. On close examination, however, the amoebae were found to reproduce in pockets behind the small cracks in the pool's wall, where they were protected from the harmful effects of chlorine.

A large number of cases of amoebic meningitis have been reported from all over the world and in some cases bathing places have been closed or quarantined because they represented public health hazards.

How does *Naegleria* enter the body and infect the central nervous system? Apparently, the amoebae enter through the nose and penetrate by breaching the mucous membranes. From there they enter the brain and multiply at its base. The amoebae may then invade the spinal fluid, from which they can be isolated and cultured. The disease begins with a frontal headache, a fever, and a blocked nose. An altered sense of smell and/or taste, along with other central nervous system disorders, are evident as the disease progresses, until the patient is successfully treated (with a drug called amphotericin B) or dies.

The discovery that free-living amoebae can cause a fatal disease in humans emphasizes the fact that microorganisms influence human lives significantly. Some of their activities are essential to our well-being, while others are detrimental. *Naegleria fowleri* plays both roles. As a free-living organism, this amoeba helps maintain the health of the aquatic environment. As a pathogen, it causes severe disease. In this chapter we discuss the major groups of protozoa and other parasites and emphasize their importance to humans.

CHAPTER OUTLINE

PARASITES ARE ORGANISMS THAT LIVE AT THE expense of their host, usually causing detrimental changes in the host that are sometimes reflected as infectious diseases. There are three prominent groups of parasites: the protozoa, the worms (helminths), and the arthropods (arachnids* and insects). In this chapter we discuss some of the most important members of these groups and describe briefly the diseases they cause or transmit.

INTRODUCTION TO THE PROTOZOA

The protozoa constitute a diverse group of heterotrophic eukaryotic microorganisms widely distributed in nature (table 10.1). They all lack a cell wall and their cytoplasm contains prominent nuclei, mitochondria, Golgi bodies, lysosomes, and vacuoles (fig. 10.1). Protozoa gather nutrients by absorbing them through the plasma membrane or by engulfing particles and microorganisms by phagocytosis.

Reproduction

Protozoa reproduce by asexual and sexual means. Asexual reproduction may consist of binary fission (fig. 10.2) or multiple fission, referred to as **schizogony** (fig. 10.2). Sexual reproduction may involve the transfer of hereditary material between two cells or fusion of two cells or gametes to give rise to diploid zygotes. Both sexual and asexual reproduction may be integrated into a single, complex life cycle.

Importance

The protozoa are extremely important to humans because of their ability to cause infectious diseases. Malaria, African sleeping sickness, and kala-azar are among the many protozoal diseases that afflict millions of people each year. In addition, many parasites, especially in a group called the **coccidians,** are capable of infecting and causing disease in nearly all domestic and wild animals. Protozoa also participate in beneficial associations with humans or other animals. For example, a ciliated protozoan forms a long-lasting association with termites, in which the protozoan digests much of the cellulose consumed by the termites, thus allowing the insect to use wood as a source of nutrients. Ruminants such as cattle and sheep depend on protozoa (among other microorganisms) to digest some of the foods in the rumen. Without these microorganisms, cattle and sheep could not use plant materials efficiently as nutrient sources.

THE VARIOUS GROUPS OF PROTOZOA

There are four major groups of protozoa that are important to humans: the amoebae, the flagellates, the ciliates, and the sporozoans (table 10.2). Their characteristic means of locomotion and morphological characteristics are important criteria used for classification (fig. 10.1).

Amoebae (Subphylum Sarcodina)

In essence, **amoebae** are single-celled microscopic organisms that float or creep in aquatic environments. They resemble eukaryotic animal cells in that they are chemoheterotrophic and lack a cell wall. Amoebae have pseudopodia that serve not only as a means of locomotion but also as a means of gathering food. Nutrients are obtained by phagocytosis (fig. 10.3). During phagocytosis, the pseudopodia extend around food particles and envelop them. The phagocytized food particles

* Arachnids include spiders, ticks, and mites.

TABLE 10.1
CHARACTERISTICS OF THE MAJOR GROUPS OF PROTOZOA

Taxonomic group	Locomotion	Asexual reproduction	Sexual reproduction	Genera of importance
Phylum: Ciliophora Subphylum: Ciliata (ciliates)	Ciliary movement	Transverse fission	Conjugation	*Balantidium* *Paramecium* *Vorticella*
Phylum: Sarcomastigophora Subphylum: Sarcodina (amoebae)	Pseudopodia	Binary fission	Gamete fusion	*Entamoeba* *Amoeba* *Naegleria*
Subphylum: Mastigophora (flagellates)	Flagellar movement	Binary fission	Not seen	*Giardia* *Trypanosoma* *Trichomonas*
Phylum: Apicomplexa Class: Sporozoa (sporozoans)	Usually not motile	Schizogony (multiple fission)	Gamete fusion	*Plasmodium* *Toxoplasma* *Eimeria* *Cryptosporidium* *Pneumocystis**

**Pneumocystis carinii* is more closely related to the yeast *Saccharomyces cerevisiae* than to protozoans such as *Trypanosoma* and *Plasmodium* as shown by rRNA analyses.

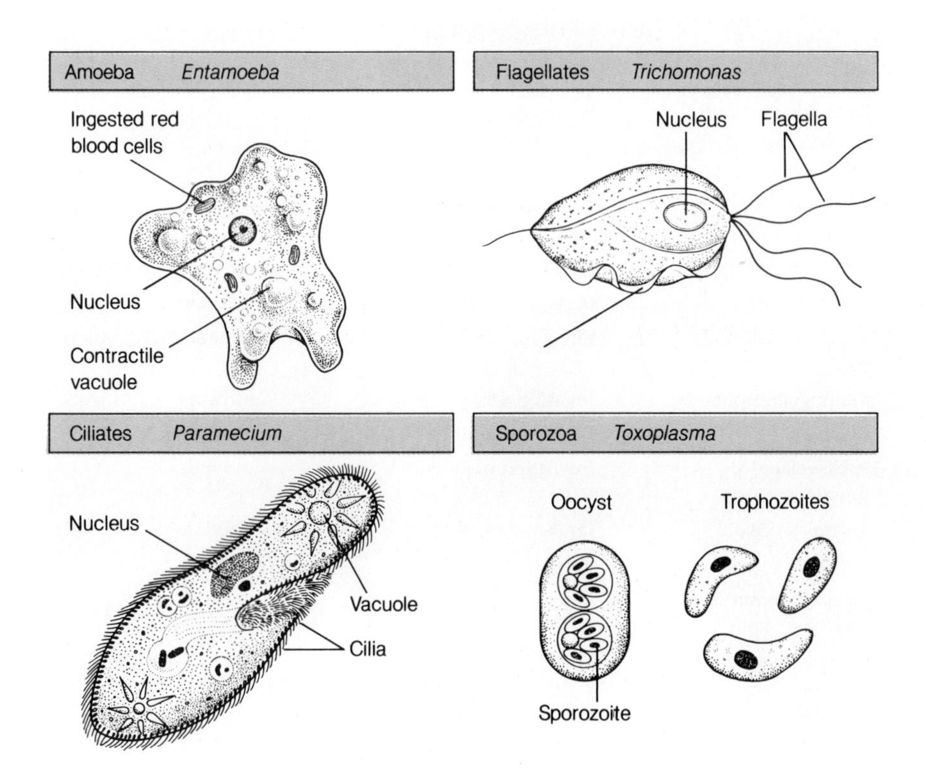

Figure 10.1 Representative protozoa

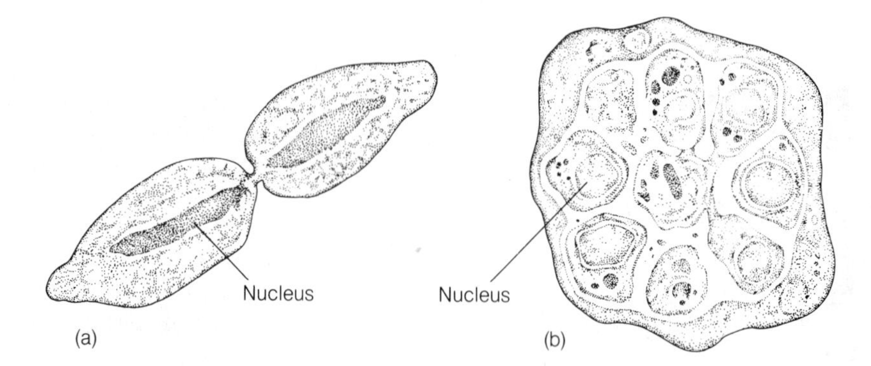

Figure 10.2 Asexual reproduction of protozoa. *(a)* Binary fission by *Paramecium. (b)* Multiple fission by *Plasmodium.*

are surrounded by a membrane within the amoeba, then digested by lysosomal enzymes.

The life cycle of the amoebae is relatively simple. One notable feature is the occurrence of a resting, highly resistant structure called the **cyst** (fig. 10.4). Cysts develop from **trophozoites**, which are the active, feeding, reproducing form of the amoebae. When nutrients are depleted or other adverse changes occur in its environment, the trophozoite differentiates into a cyst. This process is usually called **encystment**. When the cyst is in an environment conducive to growth, it **excysts**, producing a new trophozoite. This life cycle (fig. 10.4) takes place in certain free-living protozoa, as well as in parasitic forms within their hosts.

Amoebae are generally free-living protozoa and

TABLE 10.2
CHARACTERISTICS OF REPRESENTATIVE PARASITIC PROTOZOA AND THE DISEASES THEY CAUSE

Organism	Group*	Disease caused	Mode of transmission	Geography
Babesia sp.	S	Babesiasis (cattle)	Tick bites	Worldwide
Balantidium coli	C	Balantidiasis	Fecal–oral route	Worldwide
Cryptosporidium	S	AIDS-related cryptosporidiosis	Ingestion of cysts	Worldwide
Eimeria tenella	S	Coccidiosis in chickens	Fecal–oral route	Worldwide
Entamoeba histolytica	A	Amoebic dysentery	Fecal–oral route	Worldwide, tropics
Giardia lamblia	F	Gastroenteritis	Fecal–oral route	Worldwide
Leishmania sp.	F	Kala-azar, oriental sore	Bite of sandfly	Tropics and subtropics
Naegleria fowleri	A	Meningoencephalitis	Invasion of nasal mucosa	Worldwide
Plasmodium sp.	S	Malaria	Bite of mosquito	Tropic and subtropics
Pneumocyctis carinii	S	AIDS-related pneumocystosis	Inhalation of cysts	Worldwide
Toxoplasma gondii	S	Toxoplasmosis	Ingestion of cysts	Worldwide
Trichomonas vaginalis	F	Vaginitis	Sexually transmitted	Worldwide
Trypanosoma brucei gambiense	F	Sleeping sickness nagana in cattle	Bite of tsetse fly	Tropical Africa
T. brucei rhodesiense *Trypanosoma cruzi*	F	Chagas' disease	Bite of reduviid bug	Tropical S. America

*A, amoeba; C, ciliate; F, flagellate; S, sporozoan.

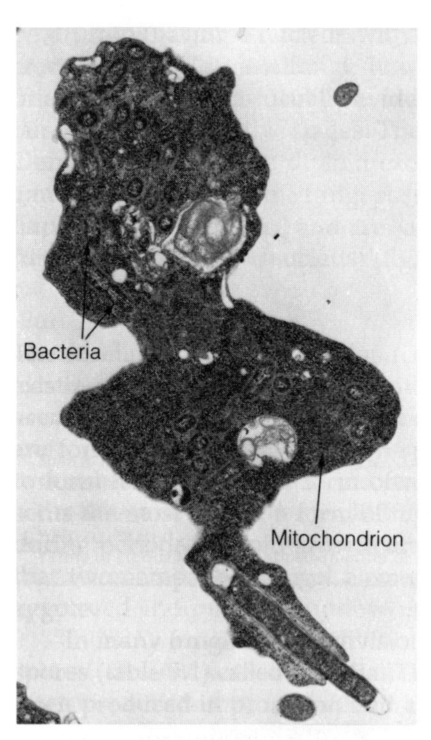

Figure 10.3 Phagocytosis carried out by an amoeba. Transmission electron micrograph of *Vahlkampfa*, (a small soil amoeba) engulfing a bacterium. Notice the large number of bacterial cells (arrows) in phagosomes.

many of them (e.g., radiolarians and foraminiferans) inhabit marine environments. The amoebae participate in the recycling of nutrients by digesting organic matter and by serving as a source of food for other organisms.

Some species of amoebae are capable of colonizing humans and causing disease. For example, *Entamoeba histolytica* is an amoeba that can be ingested along with contaminated food or beverages and can cause a disease called **amoebic dysentery**. This disease is characterized by diarrhea, painful intestinal contractions, and blood and mucus in the stool (feces). Occasionally, this organism can invade other organs such as the liver and cause pus-filled lesions called **abscesses** (fig. 10.4). When this invasion occurs, the outcome of the infection can be very grave. *Entamoeba gingivalis* can be found colonizing the human oral cavity, and *Entamoeba coli, E. histolytica, Endolimax nana,* and *Iodamoeba butschlii* can sometimes be found reproducing in the intestines. Some free-living amoebae, such as *Naegleria fowleri*, may cause serious disease of the central nervous system in humans.

Flagellates (Subphylum Mastigophora)

Flagellates (table 10.2) are single-celled protozoa that move by means of flagella. They are chemoheterotrophic microorganisms that store starch as a reserve material and reproduce by means of longitudinal binary fission (fig. 10.5).

Flagellates are found in aquatic and terrestrial environments as free-living organisms or forming various types of symbiotic (living together) associations with other organisms. Flagellates such as *Giardia lamblia, Trichomonas vaginalis, Leishmania donovani, Trypanosoma cruzi,* and *T. gambiense* can cause severe diseases in humans.

Giardia lamblia (fig. 10.5c) is a common human parasite that inhabits the small intestine and may cause severe infections. Sometimes, *Giardia* infections result in gastrointestinal disorders such as diarrhea, cramps, bleeding, and anorexia. *Giardia lamblia* exists in two forms: the trophozoite stage and the cyst stage. The pear-shaped trophozoite exhibits bilateral symmetry and has a pair of nuclei and two sucking discs that serve to attach the parasite to the intestinal wall. The nuclei, when stained, resemble pairs of eyes, giving the impression that this tiny microorganism is looking back at you through the microscope. The cyst stage is oval with four prominent nuclei. This structure is quite resistant to adverse environmental conditions, and it represents the infectious form of this parasite. The cysts of *Giardia* represent a public health hazard when present in drinking water supplies. They are not killed by the chlorine used to decontaminate water, and can be infectious if ingested along with the water.

Trichomonas vaginalis (fig. 10.1) is another important human parasite. This organism is transmitted by sexual intercourse. In heavily infected females, *T. vaginalis* causes an inflammation of the vagina that is characterized by a profuse, malodorous discharge. The organism differs from *G. lamblia* in that it lacks bilateral symmetry. *Trichomonas* is characterized by a tuft of anterior flagella and undulating membrane associated with a posterior flagellum (fig. 10.1b).

The **trypanosomes** are very important human pathogens. This family includes a rather diverse group

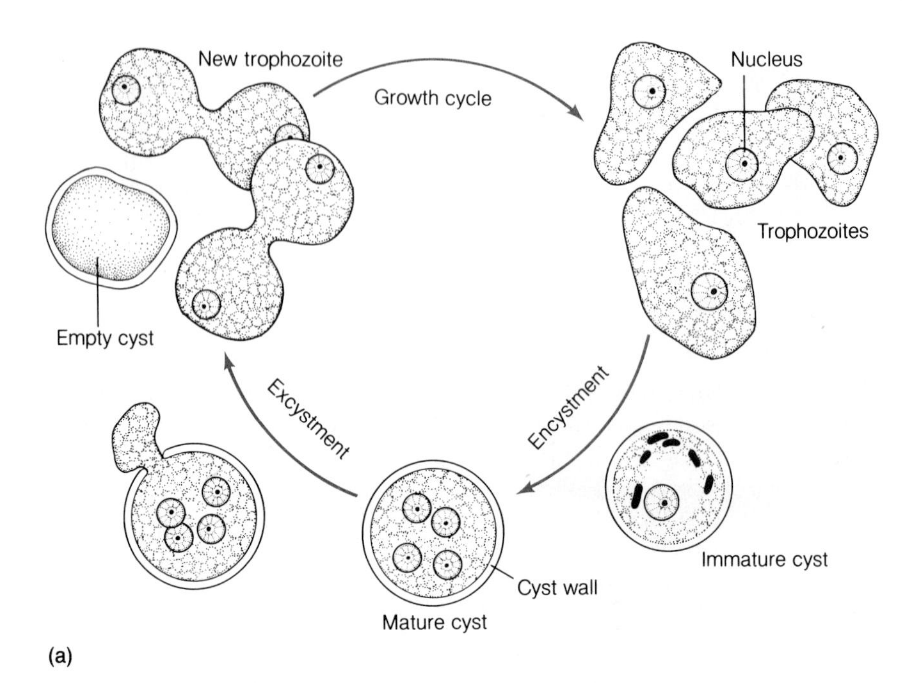

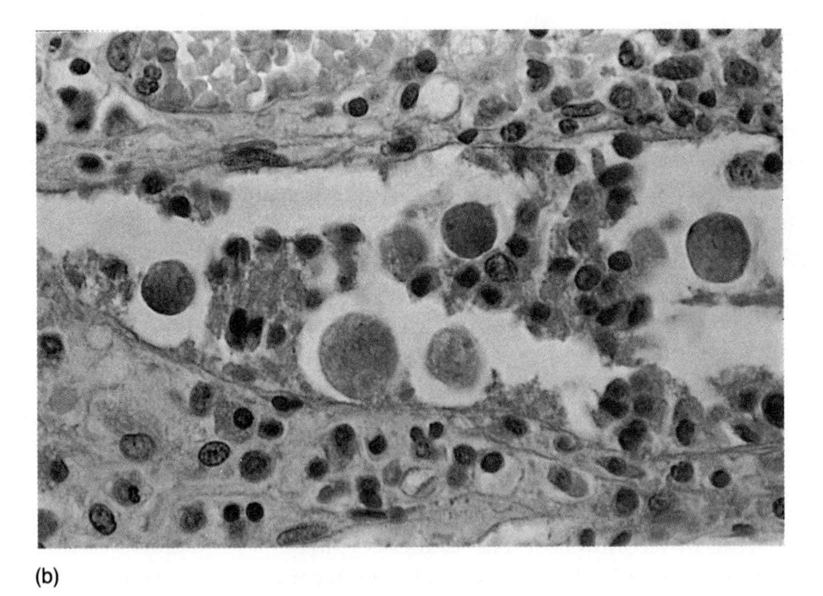

Figure 10.4 Life cycle of *Entamoeba histolytica*. *(a) Entamoeba histolytica* exists in nature predominantly in the cyst form. When humans or other animals consume food contaminated with cysts, they become infected. The cyst germinates (excysts) in the intestines and yields a trophozoite. The trophozoites reproduce by binary fission. Later during infection, the trophozoites may encyst and some of the cysts are passed in the feces. The cysts in the feces may then contaminate foods, which are then infective to other humans or animals. *(b)* Histological section stained with hematoxylin and eosin showing *Entamoeba histolytica* in an eroded region (center) of the intestinal epithelium.

of flagellates that can exist in the latex of plants, the gut of insects, and the blood of most vertebrates. These organisms cause very serious diseases that afflict millions of humans throughout the world, including **Chagas' disease** and **African sleeping sickness**.

Leishmania donovani (fig. 10.5b) causes the often fatal disease **kala-azar**. Inside the human, it exists in a stage called the **amastigote**, oval cells that lack a visible flagellum. The amastigotes of *L. donovani* exist primarily inside cells of the **reticulo-endothelial** system (macrophages). The vector for this protozoan is the sandfly *Phlebotomos*. When it bites an infected host, the sandfly ingests some of the amastigotes in the blood. In the sandfly's midgut, the amastigotes develop into **promastigotes**, elongated forms of the parasite with a short flagellum. The promastigotes divide in the midgut of the fly and migrate to its proboscis (mouth parts). When the fly bites an uninfected individual, it

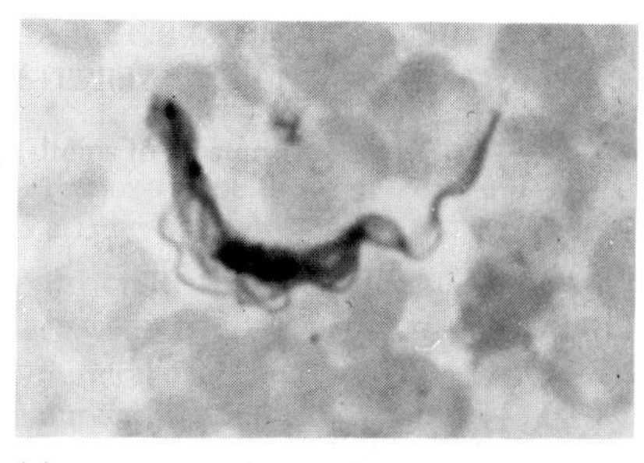

(a)

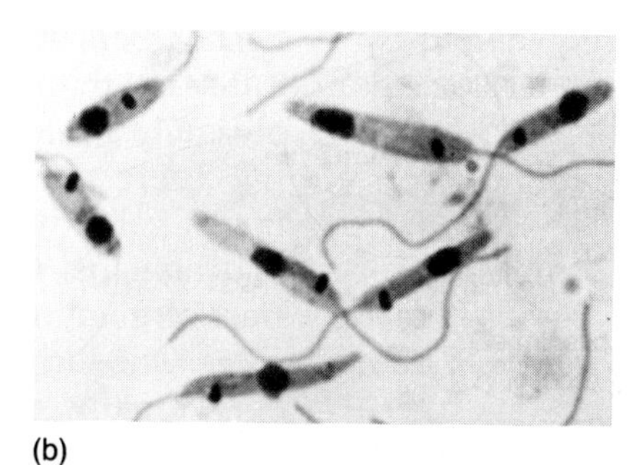

(b)

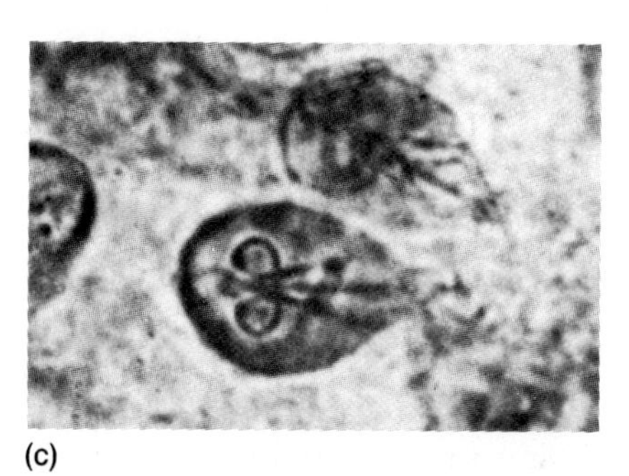

(c)

Figure 10.5 Representative flagellates. *(a) Trypanosoma.* This flagellate causes serious diseases in humans, including African sleeping sickness and Chagas' disease. *(b) Leishmania.* This organism is the cause of kala-azar, a severe infectious disease of the internal organs. *(c) Giardia lamblia.* This is a common cause of traveler's diarrhea.

introduces into the new host numerous promastigotes, which can then initiate an infection in the host. The promastigotes are ingested by phagocytes, within which they again differentiate into amastigotes.

Trypanosoma (fig. 10.5a) causes Chagas' disease and African sleeping sickness (table 10.2). These are primarily parasites that invade the blood, although sometimes they can be found multiplying in brain or heart tissue. The agent of African sleeping sickness, *Trypanosoma brucei*, is transmitted to humans by the bite of the tsetse fly (*Glossina*). The agent of Chagas' disease, *T. cruzi*, is transmitted to humans by reduviid bugs (kissing bugs).

Ciliates (Subphylum Ciliophora)

Ciliated protozoa are predominantly single-celled organisms characterized by the presence of cilia (fig. 10.6). The cilia of these microorganisms not only function in locomotion but also participate in the feeding process. Beating cilia create currents that draw food particles into the oral opening (mouth) of the protozoa.

Some species of ciliates, particularly those that inhabit the intestines of mammals (e.g., *Balantidium coli*), have ciliated trophozoites and nonciliated cysts. Many species of ciliates are free-living organisms that act as decomposers of organic matter. In sewage treatment plants, ciliates help purify water by breaking down the organic matter. Some ciliates, like *B. coli*, are capable of causing human disease. The ciliates are seldom parasitic, however, although they are often living in close association with other organisms. They are

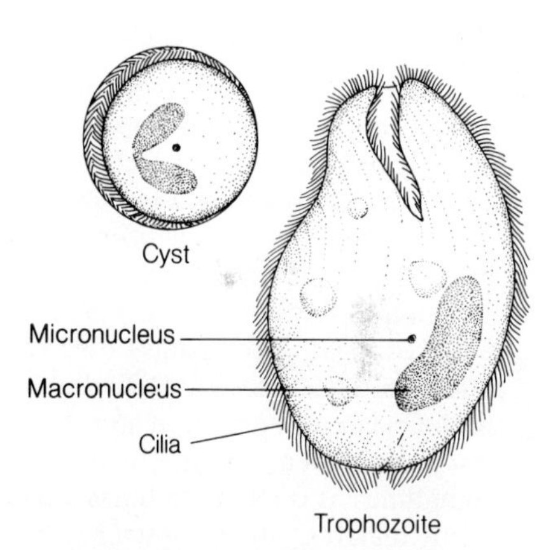

Figure 10.6 *Balantidium coli.* This organism infects humans when they ingest foods or beverages containing cysts. In the intestines, the parasite excysts, releasing a trophozoite. The trophozoite multiplies by transverse fission. Some of the trophozoites may encyst and be passed out in the feces. Encysted organisms may remain dormant for many days until they are consumed by another individual.

found in the rumen (a stomach) of cattle, where they help digest food, and in the gut of termites, where they help in the digestion of cellulose.

Sporozoa (Phylum Apicomplexa)

Sporozoa are protozoa characterized by the production of spores at some stage in their life cycle. Without exception, the sporozoa are parasites of animals. Their life cycles are often quite complex and involve alternating sexual and asexual reproductions in different hosts. Usually, the **final host** (also called the definitive host) harbors the sexual forms while the **intermediate host** harbors the asexual forms. In the case of malaria, a disease that afflicts humans, the final host is the mosquito while the intermediate host is the human.

There are several important species of sporozoa. Some, collectively called **coccidians**, affect livestock and humans. Coccidians reproduce in the digestive tract of many animals, causing severe diarrheal diseases. One such disease is coccidiosis, which can spread very rapidly among a population of animals. For instance, *Eimeria tenella* can destroy an entire flock of chickens in a few days. These sporozoans have a tremendous biotic potential: *one* infectious coccidian ulti-

FOCUS *CRYPTOSPORIDIUM* IS COMMONLY ASSOCIATED WITH AIDS PATIENTS

Cryptosporidium is an infrequent human pathogen related to the coccidians (fig. 10.7). Until recently, *Cryptosporidium* was known to cause disease in animals (other than humans) but the Centers for Disease Control in Atlanta, GA, have been reporting cases of human cryptosporidiosis associated with AIDS (Acquired Immunodeficiency Syndrome) patients. Some of these patients have died as a result of the infection. Patients with AIDS that have been infected with *Cryptosporidium* have a profuse diarrhea that may last more than a month. The diarrhea is very frequent, sometimes as often as 25 times a day. As a consequence of the diarrhea, the patients may become dehydrated from the water loss (1–15 L per day). In patients with normal immune systems, the infection is frequently asymptomatic and much less severe, resulting in abdominal cramps of 2–5 days duration.

Studies conducted by scientists have revealed that cryptosporidiosis can occur in humans but only those who have deficient immune systems (such as AIDS patients) suffer from the severe form of the disease.

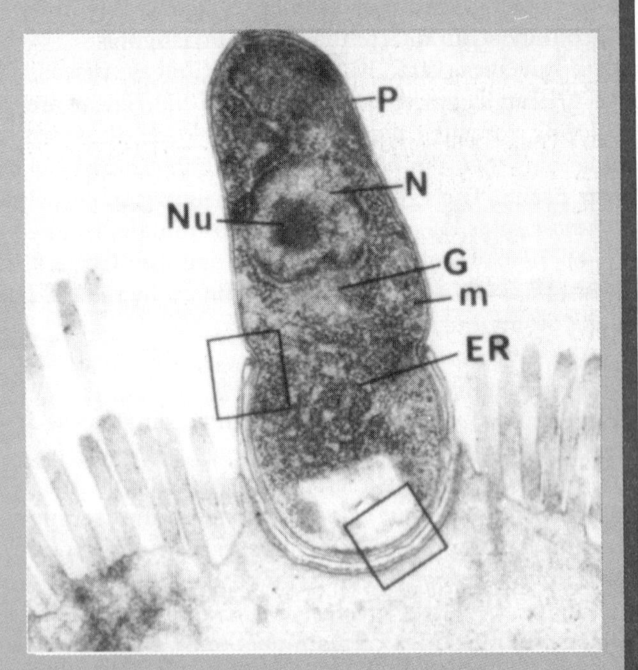

Figure 10.7 ***Cryptosporidium wrairi* penetrating the mucosal cell of a guinea pig.** (From Vetterling, J.M., et al. Reprinted with permission of The Society of Protozoologists, 1971, Vol. 18, p. 255.)

mately results in the formation of over one million infectious cells.

Toxoplasma gondii is another sporozoan (a coccidian) that infects humans. The final host for *T. gondii* is the cat and other felines. In the cat, male and female *T. gondii* gametes fuse to form a zygote that develops into an **oocyst**. The mature oocyst, containing **sporocysts** with **sporozoites**, is shed in the feces. Mature oocysts can last for more than a year in soils, sandboxes, and cat litterboxes. The most common route of infection for humans is by ingestion of the oocyst. Houseflies or cockroaches can serve as a vehicle of transmission for the disease agent. Inside the host, the oocyst releases the sporozoites, which then migrate to the intestinal epithelial cells and infect them. There, the sporozoites transform and reproduce within the infected cells (fig. 10.8). These cells derived from schizogony are called **merozoites** (or trophozoites). The merozoites can invade other intestinal epithelial cells and divide, thereby initiating the cycle once more.

The merozoites can migrate to the brain and heart and develop into cysts in these organs. Merozoites of *T. gondii* can be transferred across the placenta to the fetus, so this organism is of great public health importance. Infected fetuses can be born with congenital malformation as a result of toxoplasmal infection.

To diagnose and treat diseases caused by sporozoa, it is necessary to become familiar with several characteristic stages in their life cycle, which may include many different hosts.

The sporozoan *Plasmodium* (fig. 10.9) causes malaria in humans. The sporozoite, the infective stage for humans, develops from the oocyst in the stomach epithelium of the mosquito. The oocyst forms from the fusion of a male gamete and a female gamete. The sporozoites liberated from the oocyst in the mosquito stomach migrate to the salivary glands and are then injected into the susceptible human host by the bite of the mosquito. The sporozoites leave the blood and immediately enter the liver. There the parasite undergoes asexual reproduction, producing numerous merozoites. Eventually (7–10 days postinfection) the merozoites leave the liver and enter the bloodstream, infecting the red blood cells. In the red blood cells, the malarial parasites continue to reproduce rapidly. When the merozoites complete each of their reproductive cycles in the blood, they burst the red blood cells, releasing toxic substances into the circulation that cause chills and fever, the characteristic symptoms of malaria. The infecting merozoites may also

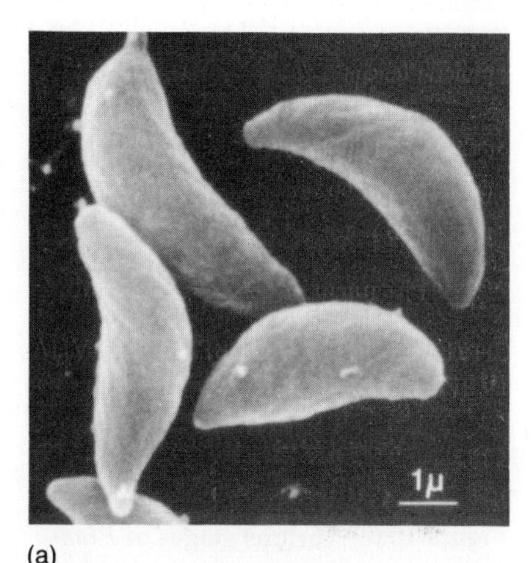

(a)

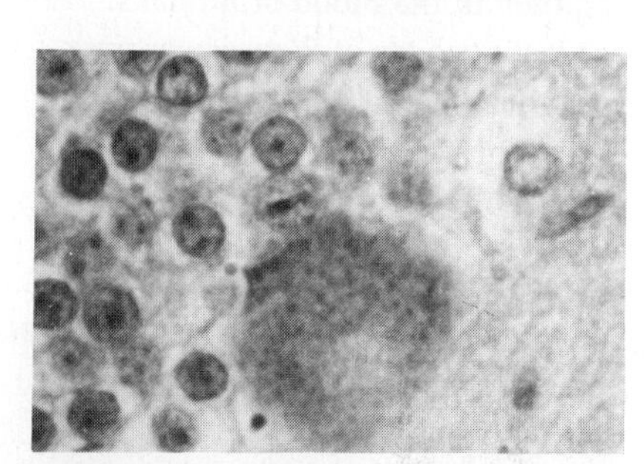

(b)

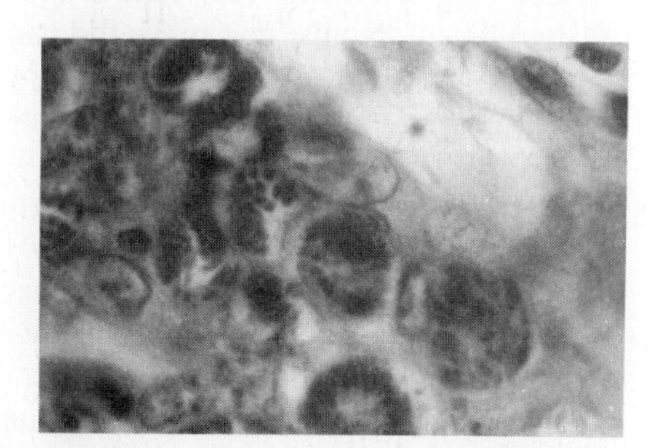

(c)

Figure 10.8 ***Toxoplasma gondii.*** *Toxoplasma gondii* invades many different types of host cells and forms cysts. The cysts represent the site where the parasite is undergoing reproduction. If the infection is very extensive, disease may result. *(a)* SEM of *Toxoplasma gondii*. *(b)* Cyst of *T. gondii* in muscle tissue. *(c)* Cysts of *T. gondii* in intestinal tissue.

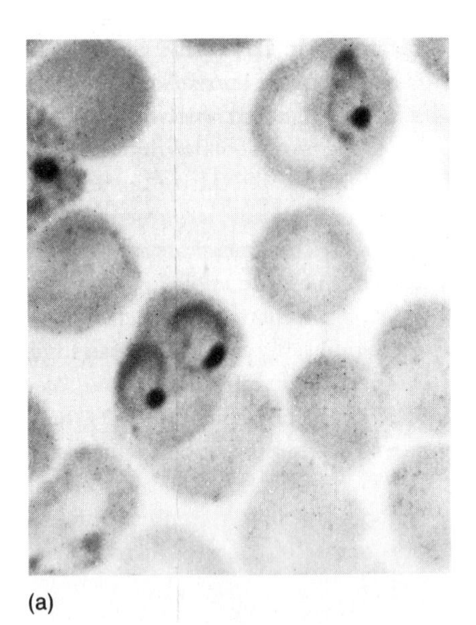

(a)

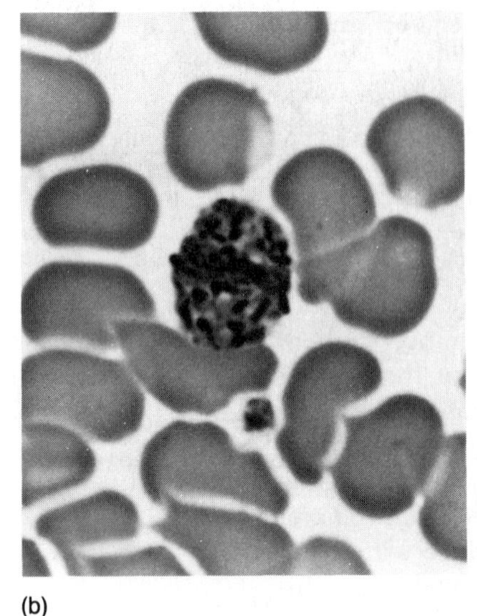

(b)

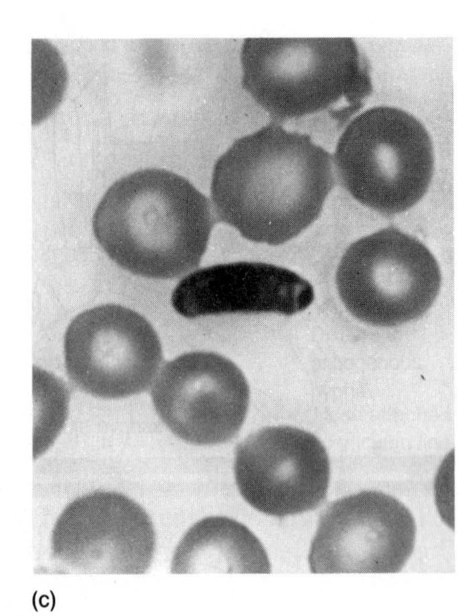

(c)

Figure 10.9 **Various stages of *Plasmodium* in the blood of humans.** *(a)* Trophozoites of *Plasmodium vivax*. *(b)* Schizont of *Plasmodium vivax*. *(c)* Gametocyte of *Plasmodium falciparum*.

FOCUS DINOFLAGELLATES CAN CAUSE SERIOUS FOOD POISONING

Dinoflagellates (fig. 10.10) are single-celled algae with two lateral flagella that make the organisms whirl. While most algae obtain nutrients by absorbing dissolved materials, some dinoflagellates can ingest food particles.

Dinoflagellates are very important sources of food for a variety of aquatic organisms. They have achieved notoriety because some produce toxins that are lethal to fish. During periods of extensive algal growth, fish often die when they feed upon the toxigenic dinoflagellates. In addition, dinoflagellates such as *Gonyaulax excavata* or *G. catanella* are eaten by shellfish. The shellfish are not affected by the dinoflagellate, even though it produces toxins inside them. The most potent of these is the **saxitoxin** produced by *G. catanella*. Saxitoxin affects the central nervous system of many animals and can be lethal to humans. When extensive dinoflagellate growth occurs (a condition also called **red tides**), public health agencies often quarantine the shellfish to prevent their consumption by human populations. The disease caused by this toxin is called **paralytic shellfish poisoning** and is contracted by ingestion of shellfish that have been feeding upon these dinoflagellates.

Figure 10.10 **The dinoflagellate *Ceratium***

develop into gametocytes, which are then released from the red blood cells and may be picked up by another mosquito upon biting an infected host. In the mosquito, the gametocytes differentiate into gametes that mate and give rise to oocysts.

Protists constitute an integral part of our environment. Many of the functions they perform are essential for the survival of organisms in the biosphere. Certain species, however, are capable of producing metabolic products or colonizing human and animal organs, causing disease, death, and/or financial loss. Regardless of the detrimental influences that certain protists have on human populations, these organisms represent a very important part of our environment and must be studied to understand the effects that our actions may have on other organisms sharing this planet.

THE MULTICELLULAR PARASITES

Multicellular parasites are multitissued and multiorganed (metazoan) animals that can live at the expense of other animals or plants (their hosts). The multicellular parasites include animals such as tapeworms, flukes, and roundworms. In addition, a group of arthropods (joint-appendaged invertebrate animals, such as spiders and insects) are considered to be parasites because some can colonize body surfaces and cause much discomfort and disease. Although many of the multicellular parasites (like insects) are not microorganisms, they are frequently studied by microbiologists because many act as carriers of microorganisms that cause human diseases. This section discusses some of the most common multicellular parasites that attack humans, and some of the arthropods

FOCUS RIVER BLINDNESS, A CRUEL PARASITIC DISEASE

Parasitic diseases ravage more than 2 billion people throughout the world. Most of these people reside in Third World countries, although more than 60 million Americans are infected with some kind of worm. It is remarkable, however, that in spite of the magnitude of worm-caused diseases in the world, less than 4% of the funds for research is earmarked for the study and control of these parasitic diseases. Nevertheless, vigorous public health programs are being instituted to eradicate parasitic diseases. For example, the World Health Organization (WHO) has spent more than $125 million trying to control **river blindness** alone.

River blindness, more properly called **onchocerciasis** (ahn-ko-sir-KAI-ah-sis), is a disease caused by the roundworm *Onchocerca volvulus*. This worm infects more than 2 million people in the equatorial regions of Central and South America and Africa, with a potential for infecting an additional 10 million people. The disease ravages entire villages, causing blindness in more than 30% of the inhabitants.

Onchocerca volvulus is transmitted by the bite of bloodsucking female blackflies. The blackfly, upon biting an infected individual, sucks up numerous larval (juvenile worms) into its gut. There the juvenile worms develop into infectious forms, which can then infect another human when the blackfly bites again. The larvae travel throughout the host and eventually develop into adults, which subsequently mate and produce millions of offspring called **microfilariae**. The microfilariae are tiny worms that migrate throughout the body, notably to the skin and eyes. The numerous microfilariae penetrating the retina cause scarring of the tissue and subsequent blindness.

To control river blindness, it is necessary that biologists know intimately the life cycle of the causative organism. For example, it is theoretically possible to control river blindness by controlling the population of blackflies. In practice, such control has proven more difficult than predicted. It is also essential that the anatomy and physiology of the causative organism be known in order to develop effective drugs to cure and/or prevent the disease.

that can transmit infectious agents among human populations or that can themselves cause annoying infestations of the skin and hair.

THE FLATWORMS (PHYLUM PLATYHELMINTHES: CLASSES TREMATODA AND CESTOIDEA)

Flatworms are multicellular, flattened animals in the phylum Platyhelminthes (fig. 10.11). They have primitive digestive systems and generally both sexes are contained in the same worm. One notable exception is the **schistosomes**, which have separate sexes. Many flatworms have very complex life cycles involving more than one host. These parasites are widely distributed in nature and infect millions of humans annually (table 10.3). There are four classes of flatworms, two of which include numerous important human pathogens. These are the **trematodes,** or flukes, and the **cestodes**, or tapeworms.

Trematodes (Flukes)

Trematodes are parasitic worms that can live either inside their host or attached to external surfaces of the host. Generally, the life cycles of these parasites are very complex and may involve two or more different hosts. Most trematodes are parasites of fish, although a few (table 10.3) are important human pathogens. The adult flatworms have powerful suckers with which they adhere to host surfaces and feed upon them (fig. 10.11). Some of the immature stages of trematode parasites have penetrating organs that can pierce human skin and invade the body.

Perhaps the best known of all the flukes are the schistosomes (fig. 10.12). Three species of these organisms, *Schistosoma japonicum*, *S. mansoni*, and *S. haematobium*, cause **schistosomiasis** in humans. This disease is prevalent in more than 70 countries throughout the world and affects more than 200 million people. In countries where schistosomiasis is prevalent, such as Africa and Asia, the disease not only causes much human misery but also inflicts severe financial losses due to lack of productivity as a result of illness and disability. Infections by schistosomes usually result in enlarged and inflamed livers, spleens, kidneys and bladders, depending upon the species of schistosome infecting the host.

The life cycle of all schistosomes (fig. 10.12) is essentially the same, regardless of the species, and

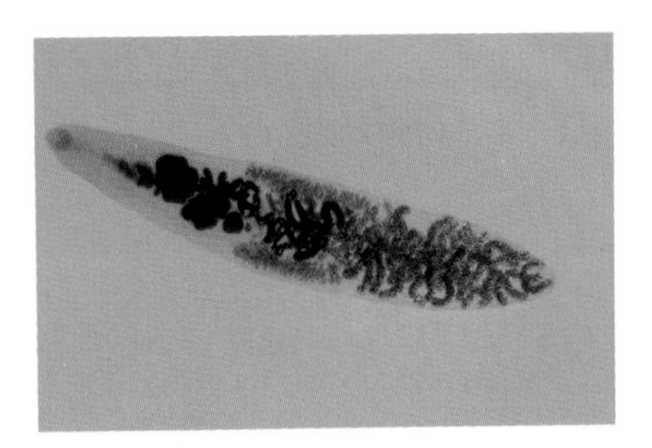

Figure 10.11 **A typical human-parasitic fluke.** Flukes are animals with well-developed sexual reproductive systems. Both male and female organs are present in each animal. This fluke, *Dicrocoelium dendriticum*, is a typical fluke. It infects livestock and humans when they ingest meat containing larval forms of the parasite.

TABLE 10.3
REPRESENTATIVE PARASITIC DISEASES

Disease	Parasite	People affected (in millions)
	Trematodes:	320+
Fasciolopsiasis	*Fasciolopsis* sp.	11
Opistorchiasis	*Opistorchis sinensis*	20
Paragonomiasis	*Paragonimus westermanni*	4
Schistosomiasis	*Schistosoma* spp.	250
	Cestodes:	100+
Tapeworm disease	*Taenia, Hymenolepis, Diphyllobothrium*	69
	Nematodes:	1800+
Ascariasis	*Ascaris lumbricoides*	650
Dracunculiasis	*Dracunculus medinensis*	1
Filariasis	*Loa loa, Wuchereria*	280
Hookworm disease	*Necator, Ancylostoma*	460
Onchocerciasis	*Onchocerca volvulus*	2
Strongyloidasis	*Strogyloides stercoralis*	36
Whipworm disease	*Trichuris trichura*	360

involves a freshwater snail as an intermediate host. The human-infectious stage, called the **cercaria**, develops from a **sporocyst** and matures inside the snail. The cercaria is motile and has a forked tail. After emerging from the snail, the cercaria moves until it either finds a susceptible human or dies. The most common sites of human infection are the feet and legs, because these extremities are usually exposed when the host wades in shallow, cercariae-infested waters. The cercaria penetrates the skin and invades the bloodstream, where it develops into an adult worm. The developing worms travel throughout the body, and upon reaching maturity they mate in the small blood vessels that feed the intestines where the female worm deposits the fertilized eggs. Many eggs erode the blood vessels and work their way into the intestinal lumen or the urinary bladder, where they are shed into the environment in the feces or urine. It is the erosion of host tissue caused by the eggs that is responsible for the signs and symptoms of the disease. If the eggs reach the water, they hatch, each releasing a **miracidium** that swims actively until it finds and infects a suitable snail host. Within the snail, the miracidium develops into a **sporocyst**. The sporocycst, in turn, eventually gives rise to several cercariae and the infectious cycle begins once again.

The schistosomes are not the only important human parasites. Many other flukes, including *Paragonimus westermani* (fig. 10.13) and *Chlonorchis sinensis* afflict more than 100 million people throughout the world. Some of these parasites and the diseases they cause are listed in table 10.3.

Cestodes (Tapeworms)

The cestodes or tapeworms have a segmented body and a head with a holdfast called the **scolex** (fig. 10.14). Each of the body segments, called **proglottids**, has a complete male and female reproductive system. Some of the tapeworms that infect humans may have more than 3000 proglottids attached to the scolex and measure several meters in length. The scolex, which is usually attached to the host, aids in the absorption of nutrients for the rest of the worm. The proglottids, when gravid (after mating), are nothing more than bags filled with fertilized eggs. The eggs and/or the proglottids are periodically shed from the host in the feces. The eggs can then be ingested by susceptible hosts and hence initiate a new infectious cycle.

Tapeworms often affect the health of infected humans. In some countries, it is not uncommon for skinny individuals who consume large quantities of food to be infected with tapeworms. In the early 1900s, some enterprising individuals in the United States sold tapeworm eggs as "weight reduction pills" (would this have worked?). In spite of common belief, many of the tapeworms that infect humans can exist as parasites for extended periods of time without eliciting signs or symptoms of disease.

Three tapeworms that commonly infect humans are the fish tapeworm (*Diphyllobothrium latum*), the pork tapeworm (*Taenia solium*), and the beef tapeworm (*T. saginata*). Each of these parasites can develop long-lasting parasitic associations with humans without causing notable disease.

One of the most common tapeworm parasites of humans is *T. saginata*. This parasite averages 8 m in length (at least one has been reported to measure 25 m). The stage of the parasite that is infective for humans is the **cysticercus** (fig. 10.15). This is an embryonic stage of the parasite that develops in beef tissues and consists of a fluid-filled sac containing an invaginated (inside-out) scolex. Humans become infected when they consume uncooked, "measly" (infected) beef containing cysticerci. Once in the intestine, the scolex emerges from the cysticercus and attaches itself to intestinal epithelium. There it begins to feed upon its host and to develop proglottids. As each proglottid reaches maturity, eggs in the proglottid are fertilized. The eggs are shed in the feces, and if ingested by a suitable intermediate host (cattle or buffalo), the eggs hatch, releasing embryos that migrate throughout the animal and develop into cysticerci. The cycle begins again when a human consumes the measly meat.

The signs and symptoms of tapeworm disease, which are often totally absent, are related to the presence of the worms in the intestine. Intestinal disorders and weight loss are the most common manifestations of tapeworm disease. Occasionally, some of the eggs may be regurgitated by the patient and then swallowed. These eggs may develop into cysticerci, which will affect the cerebrum, cerebellum, meninges, skeletal muscle, or heart, causing a disease called **cysticercosis**. This is a relatively rare occurrence with the beef tapeworm, but occurs more often with *Taenia solium* (the pork tapeworm).

Diseases caused by tapeworms (table 10.3) afflict more than 60 million people throughout the world and are caused primarily by *Taenia saginata*, *Taenia solium*, *Diphyllobothrium latum*, *Hymenolepis nana*, and *H. dimunuta* (table 10.3). The single most important preventive measure for tapeworm disease is to cook all meats well before eating. This precaution will reduce the incidence of tapeworm disease by almost 99%.

Figure 10.12 Life cycle of *Schistosoma mansoni*. *(a)* The adult schistosomes mate and deposit eggs in the capillaries of the intestinal walls. The eggs erode their way into the lumen (cavity) of the intestines and are shed into the environment in the feces. If the feces contaminate freshwater environments, the egg hatches to produce a miracidium, which can then infect a snail host. The miracidium continues to develop within the snail and eventually cercariae emerge from the snail host. The cercariae swim actively until they encounter a suitable human host. The cercariae penetrate the skin of individuals wading in the water and enter the circulation system. The cercaria develop into schistosomula, which eventually mature to adulthood within the human host. Upon reaching the capillaries of the intestines, the adults mate and deposit eggs. *(b)* Ova of *S. mansonii* stained with iodine. *(c)* Adult male and female worms of *S. mansonii* mating in mesenteric capillaries. The term "schistosome," meaning "split body," comes from the observation that the male has folds in his body within which the female places itself during mating. The folds hold the female in place against the flow of blood through the capillaries.

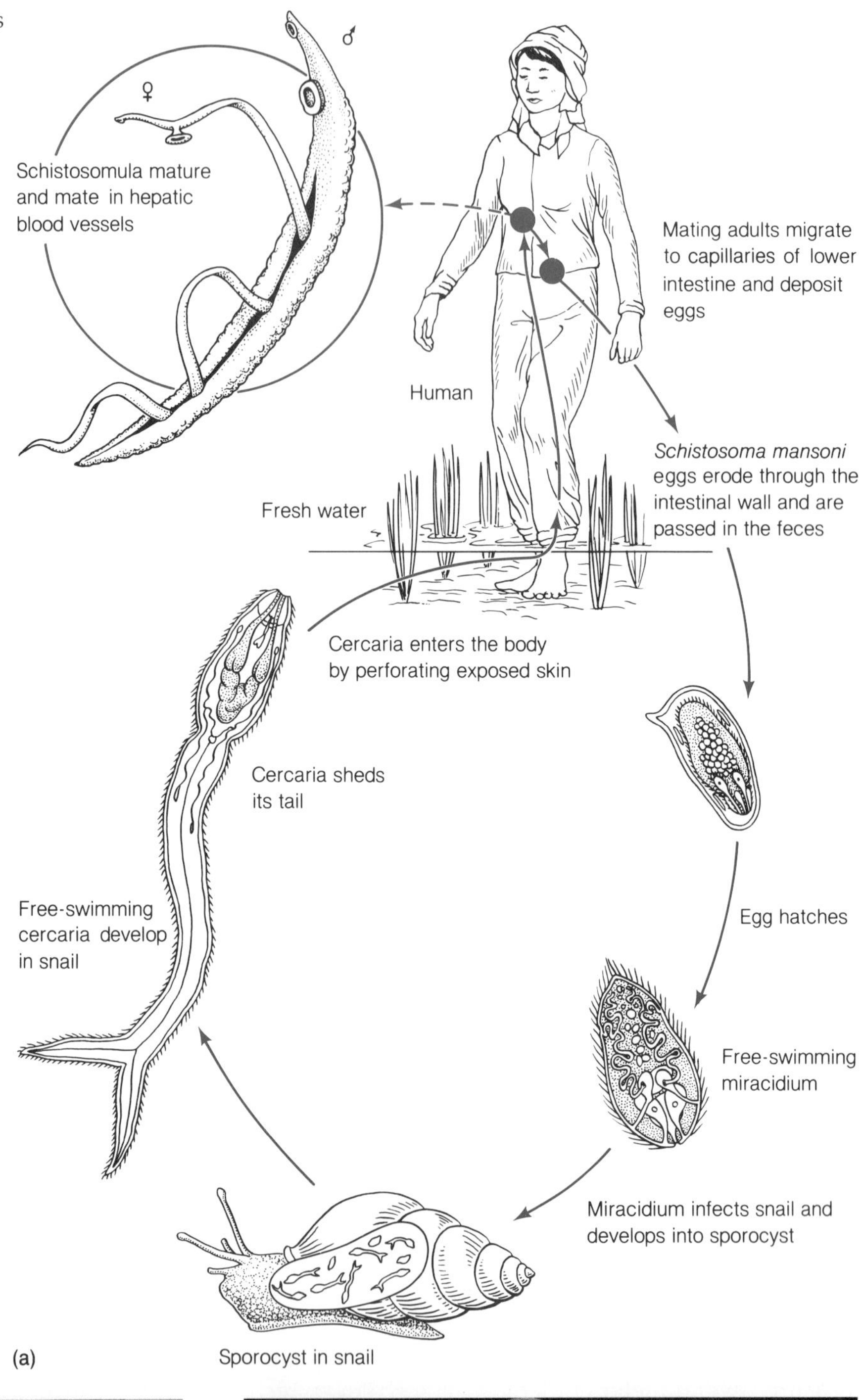

(a)

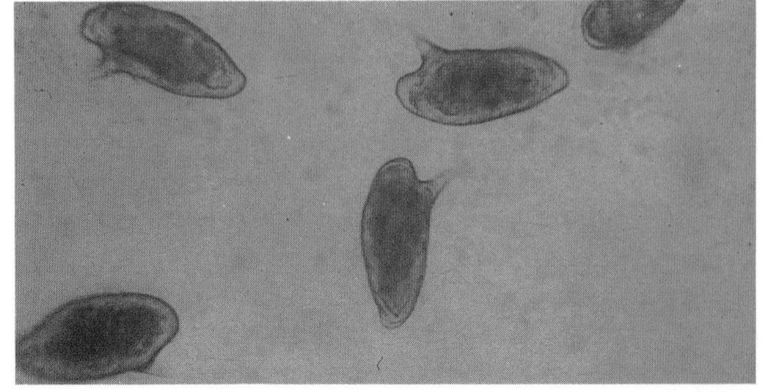

(b)

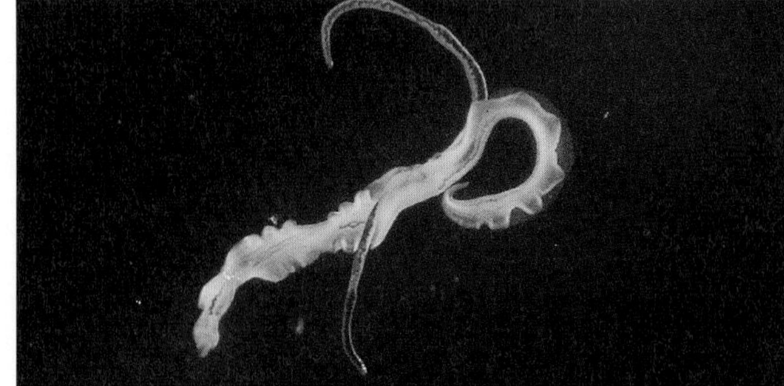

(c)

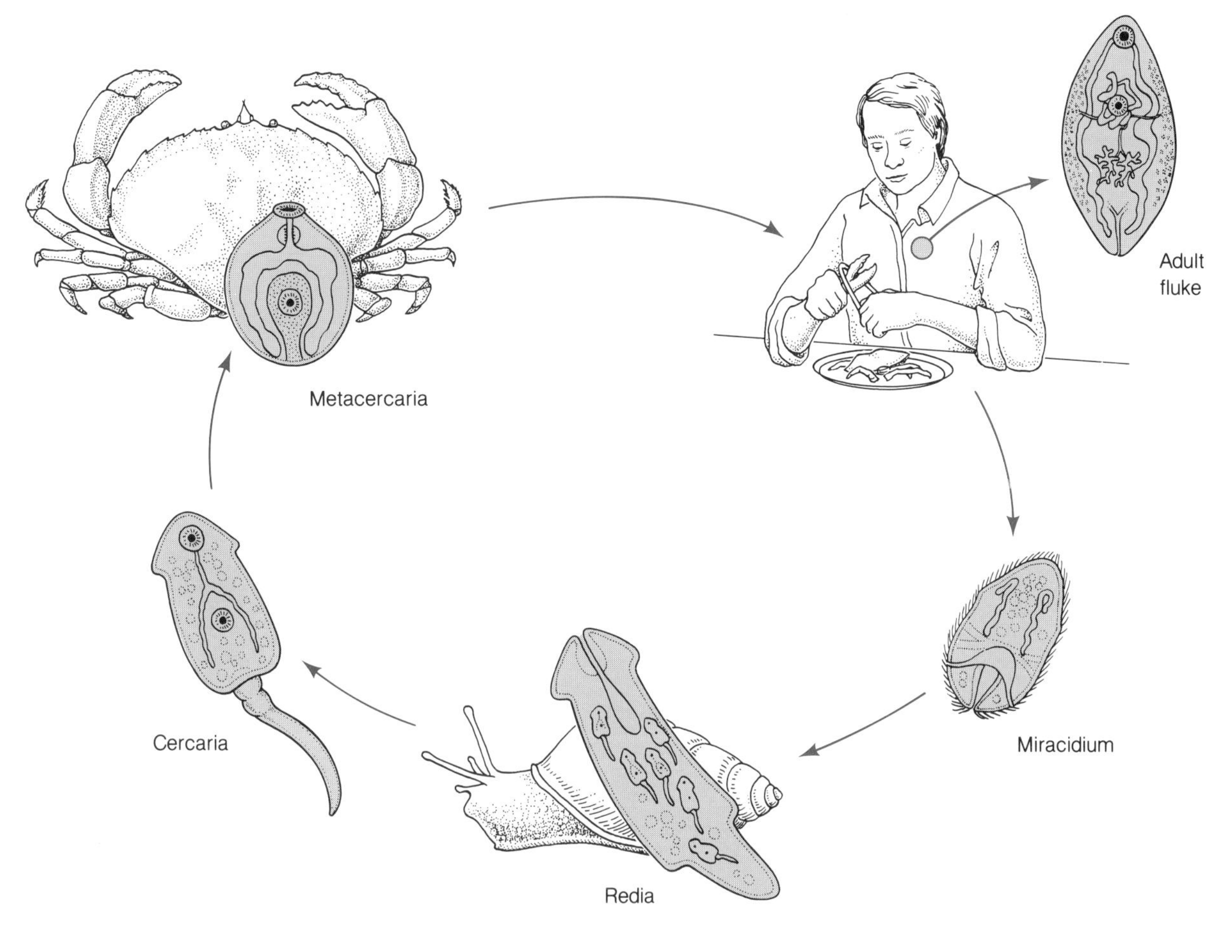

Figure 10.13 Life cycle of the lung fluke *Paragonimus westermani*. Humans become infected with *Paragonimus westermani* when they consume uncooked crab containing encysted metacercaria. The metacercaria make their way to the lungs from the intestines and there they develop into adults. The adults deposit eggs in the lungs and the eggs are shed into freshwater habitats in sputum or feces (after the eggs are swallowed). In freshwater environments the eggs hatch into miracidia, which then infect snails. In the snail, the miracidia develop into cercariae, which then infect freshwater crabs. In the crab meat, the cercariae develop into metacercariae and encyst. The life cycle is completed when humans consume metacercaria-infected crabs.

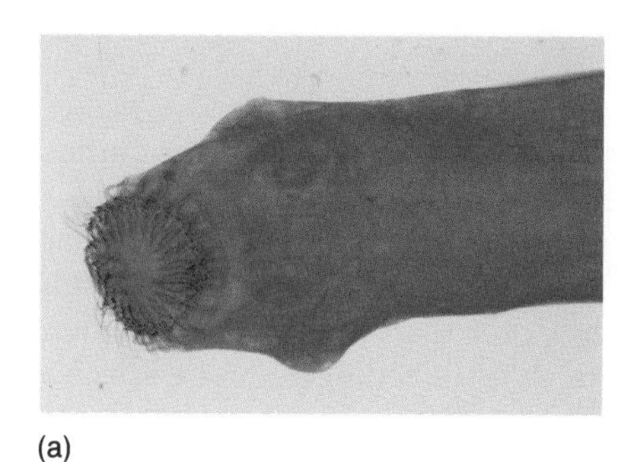
(a)

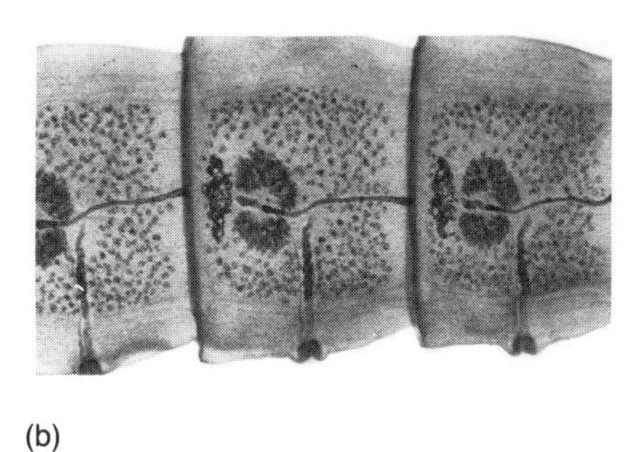
(b)

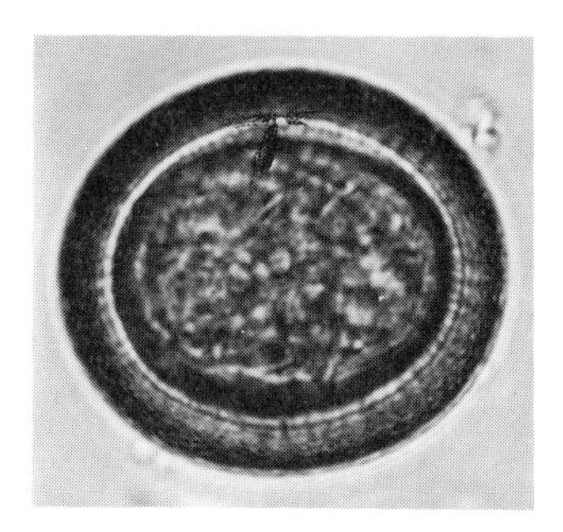
(c)

Figure 10.14 Salient anatomical features of tapeworms. *(a)* The scolex of the beef tapeworm *Taenia saginata*. *(b)* Mature proglottids. *(c)* Egg.

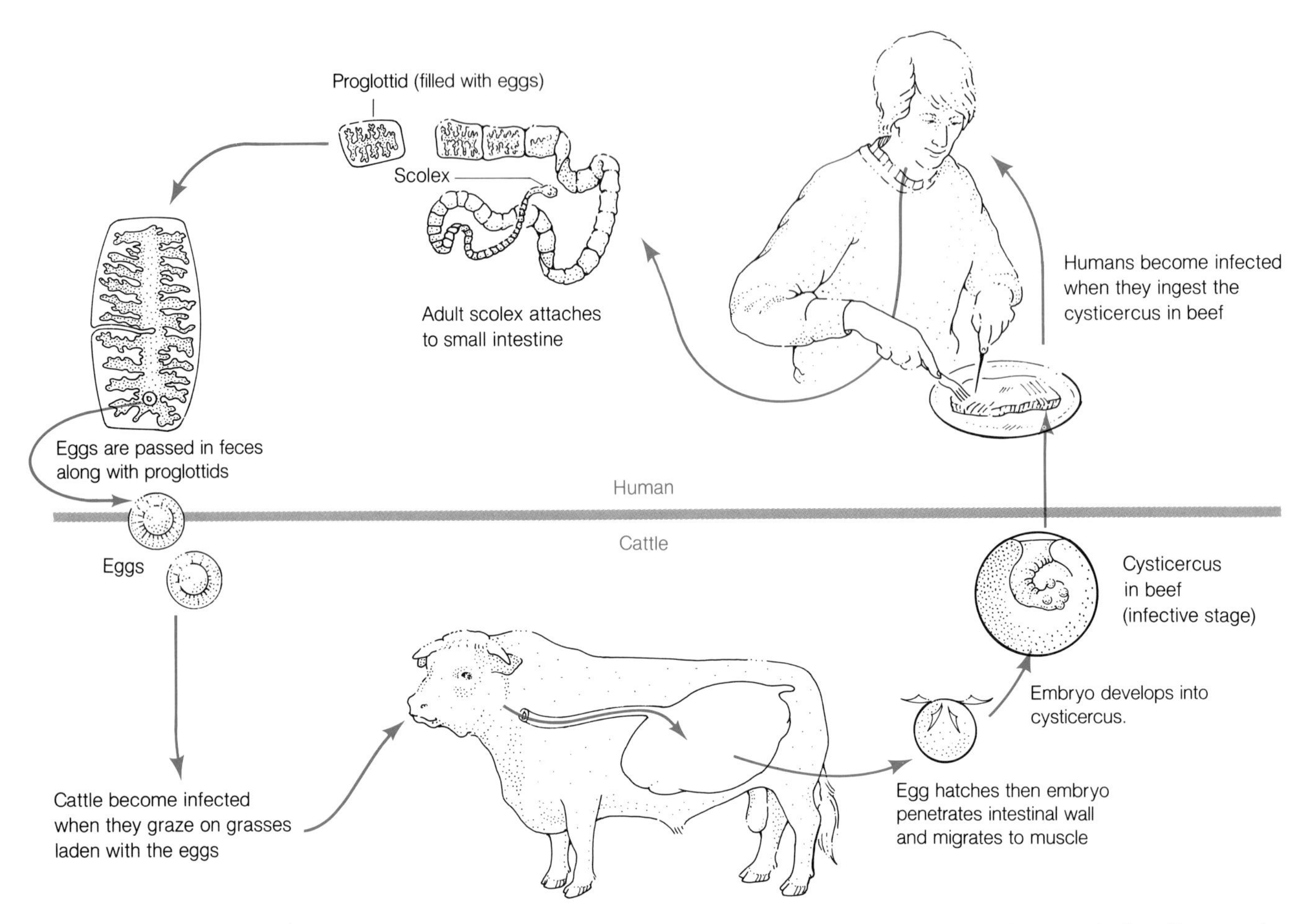

Figure 10.15 Life cycle of *Taenia saginata*. Humans become infected with the beef tapeworm *Taenia saginata* by ingesting rare beef contaminated with tapeworm embryos called cysticerci. The cysticercus yields a scolex, which attaches to the intestines and begins to develop proglottids. Eggs develop in the proglottids and are shed in the feces. The eggs may be ingested by grazing cattle and hatch into oncospheres. The oncospheres penetrate the intestinal wall, enter the blood circulation and migrate to muscles, where they develop into cristicerci. The cycle is completed when measly (contaminated) beef is ingested by humans.

THE ROUNDWORMS (PHYLUM ASCHELMINTHES: CLASS NEMATODA)

Roundworms (nematodes) constitute a very large—possibly the largest—group of animals in the biosphere. Some are found living in water and soil, while many others are parasites of plants or animals. These organisms cause extensive damage to crop plants, inflicting billions of dollars in financial losses to farmers. Similarly, infectious diseases caused by roundworms aflict more than a billion people throughout the world.

Roundworms are generally small (usually smaller than flatworms), round, slender worms with bodies that taper at both ends (fig. 10.16). The mouth is near the anterior portion of the worm and the anus near the posterior. The nematodes have separate sexes, with the female generally larger than the male, and when they mate they produce fertilized eggs. Some species of nematodes, however, like *Trichinella spiralis*, give birth to live young, because the eggs are kept in the uterus of the female until they hatch.

Many nematodes are important human parasites (table 10.3) that cause a variety of diseases, with symp-

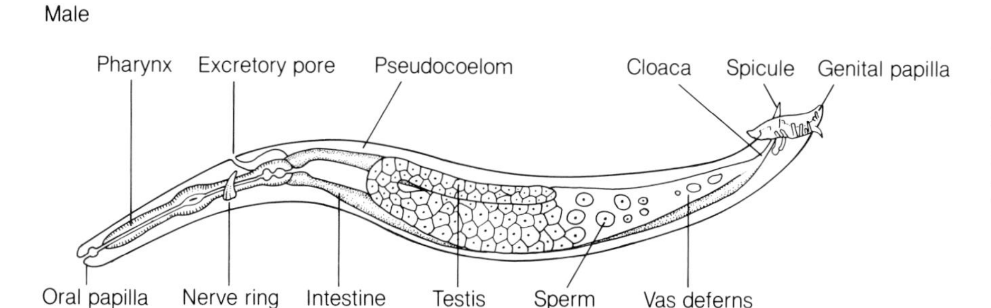

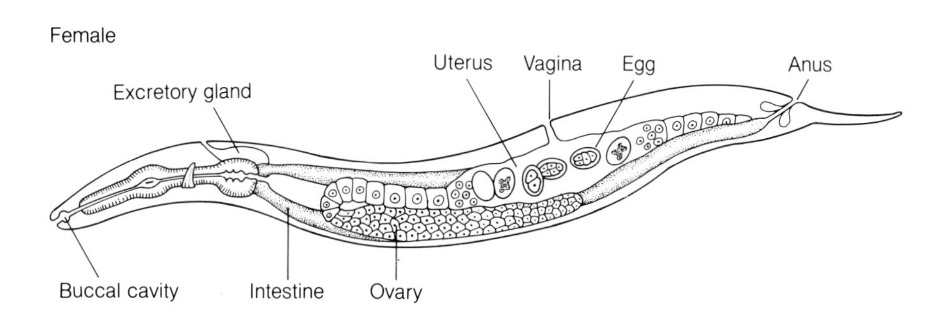

Figure 10.16 Salient anatomical features of roundworms. Internal anatomy of the nematodes. Note that the sexes are separate in these organisms and that they have well-developed reproductive and digestive systems.

toms ranging in severity from an itch to brain damage and death. The life cycles of nematodes are generally simpler than those of cestodes or trematodes, although they do have one or two different larval stages. The simplest life cycle is that of the **human whipworm** *Trichuris trichura* (fig. 10.17). *Trichuris* has a slender anterior end and a broader posterior (fig. 10.17) and measures 30 to 50 mm in length. The males are slightly smaller than the females. Fertilized eggs represent the human-infectious stage of the parasite and are found in moist, shady soils. When these eggs are swallowed, they hatch, releasing larvae that make their way into the intestine, attach to the intestinal wall, and develop into a more mature worm. After a period of development, the whipworms make their way into the lower part of the large intestine and complete their maturation, becoming adult worms. They mate in the large intestine and release as many as 7000 eggs each day. The eggs are shed in the feces and complete their development into infectious larvae in the soil. Heavy infections may lead to intestinal bleeding and anemia, because the adult worms burrow into the intestinal mucosa, where they feed on blood. Bacteria may also infect the site, causing further damage. Inflammation of the colon (colitis) and inflammation of the rectum (proctitis) are common manifestations of heavy parasite burdens. Other symptoms of whipworm disease include insomnia, vomiting, rash, constipation, loss of appetite, and diarrhea.

The **hookworms** *Ancylostoma duodenale* and *Necator americanus* (fig. 10.18) are very important human pathogens. The World Health Organization (WHO) estimates that more than 450 million people are infected with these parasites. It is thought that approximately 2 million people in the United States carry hookworms in their bowels. Because the hookworms attach themselves to the intestinal lining with very sharp "teeth," they lacerate intestinal tissue and cause bleeding. Throughout the course of a hookworm infection, a patient may lose 150–200 ml of blood due to intestinal bleeding. Anemia, abdominal pain, and loss of appetite are common manifestations of hookworm disease. Occasionally, the patient may experience a desire to eat soil (geophagy). In heavy infestations,

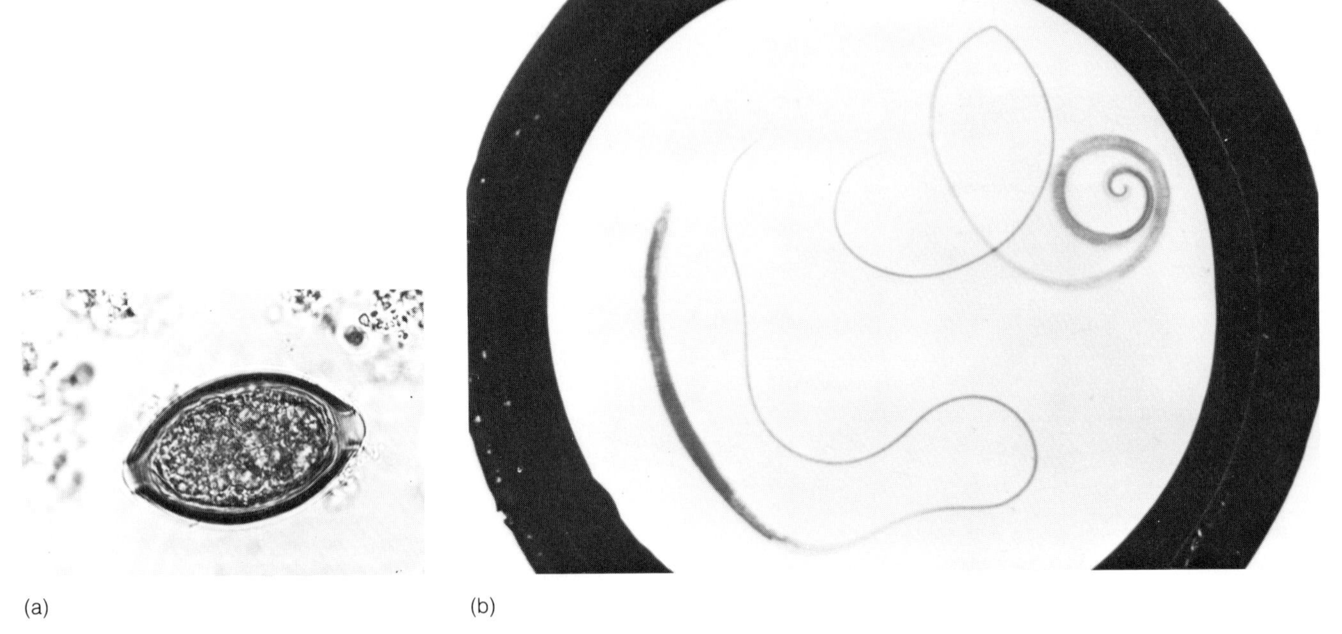

(a) (b)

Figure 10.17 ***Trichuris trichura.*** *(a)* When infective eggs are ingested, they hatch into larvae in the intestines; the larvae penetrate and develop in the mucosa. Adults in the cecum mucosa mate and produce eggs. The eggs are passed in the feces and mature into the infective stage. The life cycle is completed when infective eggs are ingested. *(b) Trichuris trichura* is called the whipworm because it resembles a whip with a broad posterior and a thin anterior.

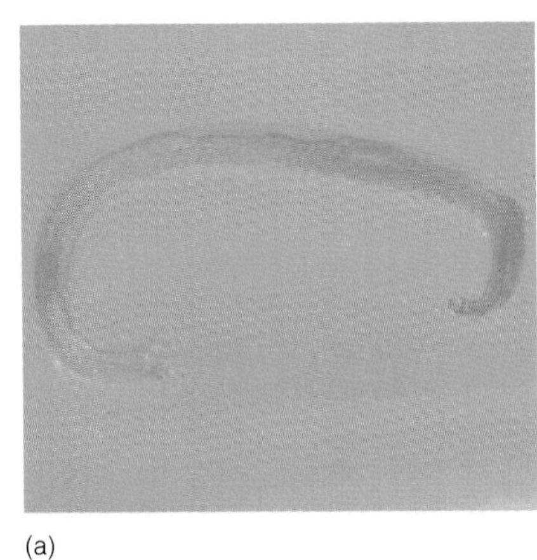

(a)

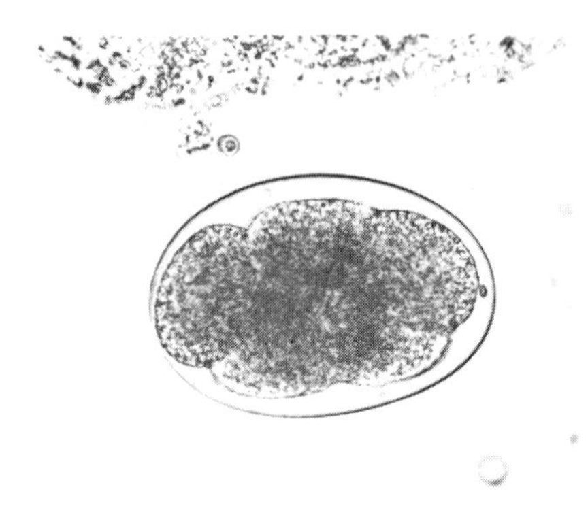

(b)

Figure 10.18 **Hookworms.** *(a)* Photomicrograph of a hookworm. The anterior portion of the worm is used to attach to the intestines. *(b)* Eggs are passed in the feces and develop into larvae. The infective larvae penetrate the skin of the susceptible hosts and migrate to the intestines where they attach, develop into adults, and mate.

the patient may exhibit hair loss, mental dullness, and protein deficiency; in extreme cases, death may result. The life cycle of this parasite involves larval forms that penetrate the skin and travel throughout the body before the adult worms colonize the bowels.

Trichinella spiralis (fig. 10.19) is a very small nematode that causes one of the most important and widespread nematode diseases in the world. The disease is known as **trichinosis**, and it is acquired when a human ingests uncooked meat (primarily pork) laden with encysted larvae (fig. 10.19). The encysted larvae develop into adult worms that live in the host's intestines. There, the adults mate. The male worms die shortly after copulation, but the females live and give birth to live offspring. The young worms penetrate host tissue and enter the venous blood, where they may migrate anywhere in the body. Striated muscle is primarily invaded, although larvae have been found in the stomach, brain, liver, and lungs. In the muscle, the larvae become encysted (fig. 10.19). The migration and encystment of the larvae give rise to the signs and symptoms of trichinosis, which vary because the larvae can invade any organ. Consequently, *Trichinella spiralis* is the "great imitator" because the disease produces signs and symptoms that are very similar to those produced by other organisms.

The **pinworm** *Enterobius vermicularis* (fig. 10.20) is a small nematode that infects more than 300 million people, predominantly in temperate zones. It is estimated that 40 to 50 million humans in the United States, mainly children, are infected. The infection is acquired by ingesting eggs, and the usual route of infection is the fecal–oral route. The ingested eggs hatch in the intestines, where the resulting adults mate. The female carries the eggs to the anus and deposits them there, usually at night, causing the itching and restlessness that are the most prominent manifestations of the disease.

There are many other nematodes that cause human diseases, many of which are deforming or fatal. Some of these organisms and the diseases they cause are listed in table 10.3.

THE ARTHROPODS (PHYLUM ARTHROPODA: CLASSES INSECTA AND ARACHNIDA)

Arthropods constitute a very large group of invertebrate animals that have jointed appendages and segmented bodies. Most arthropods have life cycles that include several immature or embryonic stages.

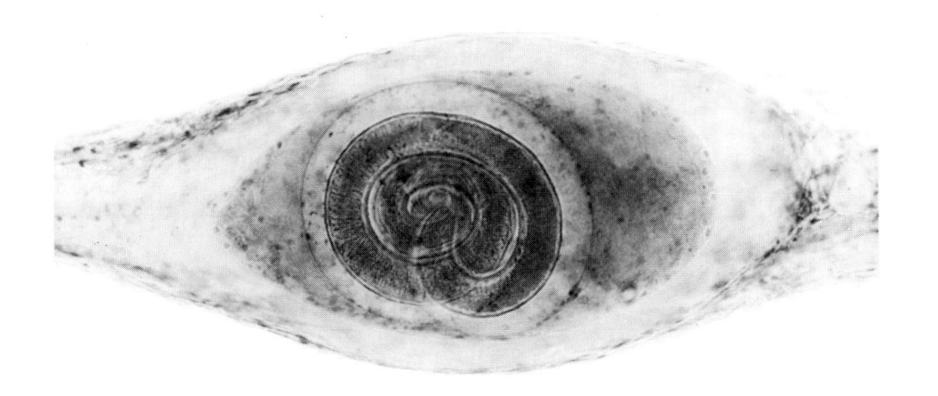

Figure 10.19 **Encysted larva of *Trichinella spiralis*.** Humans become infected when they ingest uncooked pork meat laden with encysted larvae of *Trichinella spiralis*. The larvae excyst and develop into adults in the intestines. Adults mate in the intestines and produce larvae. The larvae penetrate the intestinal tissues and migrate in the blood to skeletal muscle. There they penetrate the muscle fibers and become encysted.

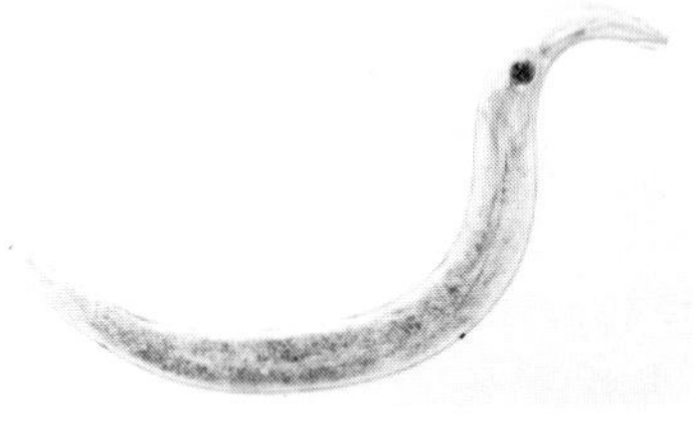

(a) (b)

Figure 10.20 The pinworm, *Enterobius vermicularis*. The pinworm *Enterobius vermicularis (a)* mates in the lumen of the cecum. The gravid females migrate to the anus to deposit the eggs *(b)*. The eggs contaminate clothing or bedding or are picked up on the fingers. The eggs are ingested and develop into larvae in the intestines. The larvae migrate to the cecum and mature into adults.

Arthropods are very important economically because they destroy billions of dollars worth of crops each year and can serve as vehicles for the transmission of infectious diseases. Many species of arthropods, primarily the **insects** (mosquitoes, flies, fleas, and lice) and **arachnids** (a group that includes the spiders, ticks and mites), can cause human and animal disease by parasitizing the skin and other body surfaces. Most arthropods of public health importance participate in the transmission of disease.

FOCUS THE SERPENT ON A STAFF

The serpent on a staff was said to be carried by Aesculapius, the Roman god of medicine. This symbol was adopted by the American Medical Association (AMA) and the Army Medical Corps (fig. 10.21). What was the origin of such an emblem? Was it really a snake on a staff, or was it *Dracunculus medinensis* on a stick?

The female *Dracunculus medinensis* is a rather large, slender roundworm that migrates through human tissue to deposit its eggs near the skin. Sometimes, the skin ulcerates and the worm emerges from the ruptured skin and dangles from the sore. Modern medicine treats these "guinea worms" with drugs or by surgically removing the worm, but many individuals inhabiting Third World countries treat the disease differently: they simply wrap the worm around a stick (it could be a matchstick or something much fancier than that) and wrap the worm a few turns every day until it comes free of the tissue. Undoubtedly, this technique has been passed from generation to generation since ancient times. Woodcuts dating back to the 1600s illustrate the technique for removing guinea worms by winding the worm on a stick. The disease is prevalent in desert areas of the Middle East and Africa.

Whether or not *Dracunculus medinensis* is the "snake" wrapped around the staff in the AMA's emblem will probably never be known for certain, but it is certainly a plausible explanation.

Figure 10.21 Emblem of Army Medical Corps

Parasitic Arthropods

A few arthropods are parasitic on humans (table 10.4). Among these are the itch mite, *Sarcoptes scabiei;* the head or human body louse, *Pediculus humanus;* and the crab louse, *Phthirus pubis* (fig. 10.22). The itch mite is an ectoparasite (parasitizes outer body surfaces) of humans and domestic animals. The adult mites penetrate the host's skin and burrow channels or trenches in the skin (fig. 10.23). In their burrows, the female itch mites deposit eggs that hatch within 5 days. The larvae dig new burrows as they mature into adult mites in 3 to 5 days. The mite infestation causes severe itching, which in turn results in excessive scratching and sores that eventually become covered with scabs. Sometimes the trenches become infected with bacteria that further aggravate the condition. The disease in humans is called **scabies**, but in domestic animals it is called **mange**.

The head or body louse (fig. 10.22) is common in populations where personal hygiene is poor. Some races of lice invade primarily the head, while other races infest the human body. The head louse deposits its eggs (nits) on the hair shaft. The eggs hatch into nymphs that resemble the adult louse. The infestation can be transmitted from person to person by direct contact, in contaminated clothing, or on personal articles. The body louse colonizes the body instead of the hair. People suspected of being infested with lice are diagnosed by making a careful examination of the head, body, and clothing for the presence of nits, nymphs, or adults. The prompt detection of *Pediculus humanus* is important because this ectoparasite spreads rapidly among humans and is capable of transmitting various infectious diseases (table 10.5).

The crab louse, *Phthirus pubis* (fig. 10.22) colonizes primarily the genital areas. The female louse deposits her eggs on pubic hairs. The louse may cause severe itching, and some sensitive hosts develop a rash. The crab louse can also transmit diseases such as trench fever, epidemic typhus, and relapsing fever.

TABLE 10.4
REPRESENTATIVE DISEASES CAUSED BY PARASITIC ARTHROPODS

Disease	Organism
Chigoe flea lesions	*Tunga penetrans* (flea)
Crabs	*Phthirus pubis* (crab louse)
Myiasis	*Dematobia hominis, Callitroga; Sarcophaga, Phaenicia, Phormia* (all flies)
Pediculosis	*Pediculus humanus* (body louse)
Scabies	*Sarcoptes scabiei* (itch mite)

Arthropods as Agents of Disease Transmission

Arthropods, particularly insects and arachnids, are often agents of disease transmission. Some act as **vectors** (table 10.5), that is, animals that harbor a parasitic microorganism and transmit it from one individual to another. For example, the *Anopheles* mosquito (fig. 10.24) is a vector for the transmission of malaria. A portion of the life cycle of the malarial parasites must be completed within the mosquito, which then transmits the parasite by biting a suitable host. Notice that in malaria, as in all vector-borne diseases, the parasite has an intimate (and often necessary) association with the vector. Ticks are also important vectors (table 10.5). For example, the tick *Dermacentor andersonii*

(a)

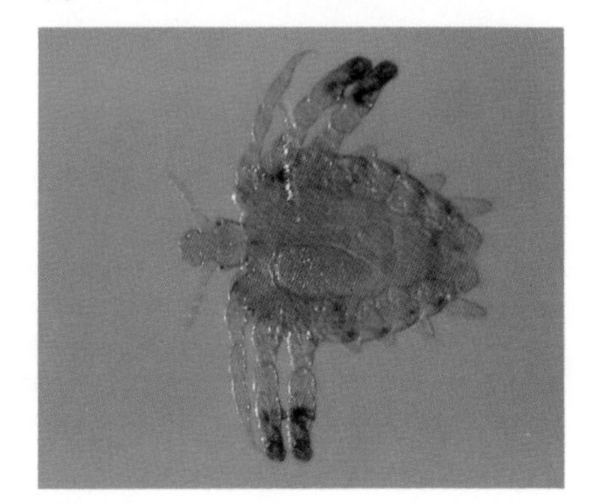

(b)

Figure 10.22 **Representative parasitic arthropods.** *(a) Pediculus humanus* (head or body louse). *(b) Phthirus pubis* (crab louse).

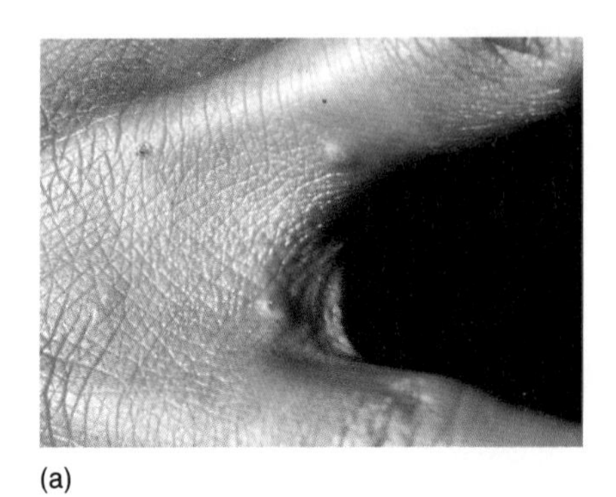

(a)

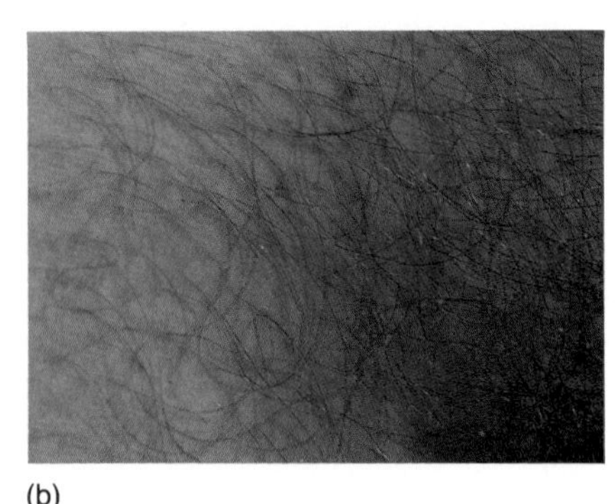

(b)

Figure 10.23 Some human diseases caused by arthropods. *(a)* A case of scabies on the hand caused by *Sarcoptes scabiei*. Courtesy of Professor E. J. Perea. *(b)* Infection of the male pubis (crabs) by *Phthirus pubis*. Courtesy of Professor F. Camacho, Department of Dermatology and Venereology, University Hospital, Seville, Spain.

(fig. 10.24a) is the principal vector for the often fatal disease Rocky Mountain spotted fever.

Some arthropods may act as mechanical means of transmission. For example, a domestic fly may land in a patch of soil contaminated with cat feces containing oocysts of *Toxoplasma gondii*. The fly may pick up some of the oocysts on its feet and legs and transfer them to foods. A human may then consume the food and become infected. The fly did not act as a true vector because the event was an accident. Table 10.5 lists some of the arthropod vectors and the diseases they transmit.

TABLE 10.5
REPRESENTATIVE INFECTIOUS DISEASES THAT ARE TRANSMITTED BY ARTHROPODS

Infectious disease	Geographical distribution	Arthropod vector	Group
African sleeping sickness (nagana in cattle)	Africa	*Glossina* sp. (tsetse)	Flies
Chagas' disease	N. and S. America	*Triatoma, Panstrongylus*	Bugs
Dengue fever	Tropics and subtropics	*Aedes* sp.	Mosquitoes
Diphyllobothriasis	Finland, United States, Russia	*Cyclops* sp.	Copepod
Dracunculosis	Africa, India, Middle East	*Cyclops* sp.	Copepod
Epidemic typhus	Worldwide	*Pediculus humanus*	Lice
Equine encephalitis	N. and S. America	*Culex*	Mosquitoes
Filariasis (bancroftian)	Africa, Asia, Australia, S. Pacific	*Culex, Anopheles, Aedes*	Mosquitoes
Hymenolepiasis	Worldwide	Flour beetles	Beetles
Leishmaniasis	Asia, E. Africa, Mediterranean, C. and S. America	*Phlebotomus* sp.	Sandfly
Malaria	Worldwide	*Anopheles* sp.	Mosquitoes
Murine typhus	Tropics and subtropics	*Xenopsylla cheopis*	Fleas
Onchocerciasis	Africa, Central and S. America	*Simulium* sp.	Flies
Paragonomiasis	Africa, Asia, Phillipines, S. America	Crabs	Crustacean
Plague	Worldwide	*Xenopsylla cheopis*	Fleas
Relapsing fever	Worldwide	*Ornithodorus*	Ticks
Rocky Mountain spotted fever	N. and S. America	*Dermacentor, Ixodes*	Ticks
Scrub typhus	Far East, Phillipines, S. Pacific	*Trombicula* sp.	Mites
Trench fever	Europe	*Pediculus humanus*	Lice
Tularemia	Europe, Japan, N. America	*Dermacentor* sp.	Ticks
Yellow fever	Africa, C. and S. America	*Aedes aegypti*	Mosquitoes

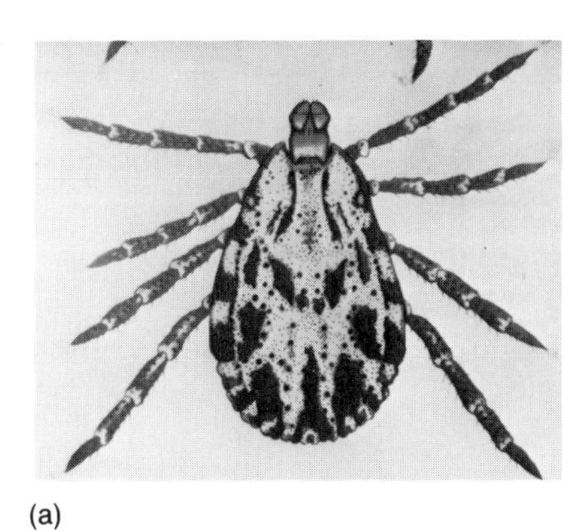

(a)

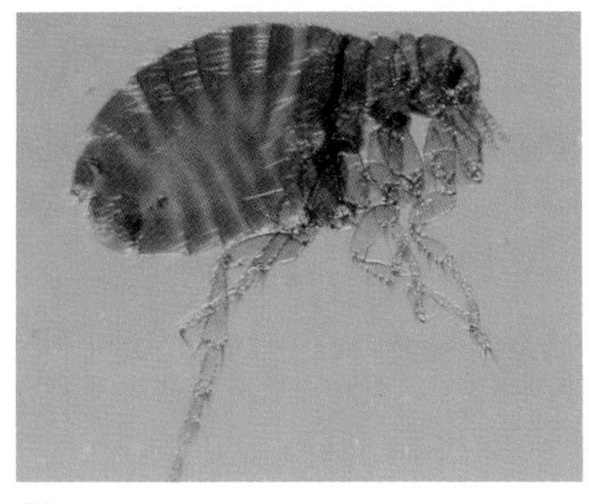

(b)

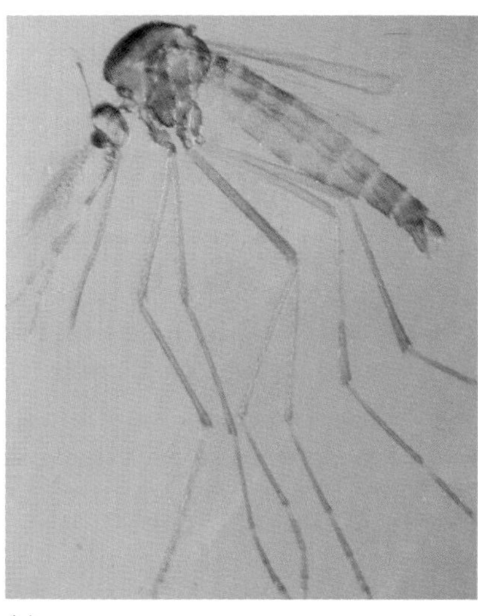

(c)

Figure 10.24 Representative vectors of human disease. *(a) Dermacentor* (tick). *(b) Xenopsylla cheopis* (flea). *(c) Culex* (mosquito).

SUMMARY

Parasites are organisms that live at the expense of their host and can cause serious diseases.

INTRODUCTION TO THE PROTOZOA

1. Protozoa are important organisms because they are responsible for a variety of serious human and animal diseases, many of which affect millions of people throughout the world.
2. The protozoa are a group of single-celled, chemoheterotrophic eukaryotes that are widely distributed in nature. They generally contain nuclei, mitochondria, endoplasmic reticula, and Golgi bodies.
3. In contrast to the algae, which are mostly free-living forms, the protozoa include many important human and animal parasites.

THE VARIOUS GROUPS OF PROTOZOA

1. The protozoa are classified into four major groups, based on their mode of locomotion, mode of reproduction, and chemical and morphological characteristics. These are the amoebae, the flagellates, the ciliates, and the sporozoans.
2. Many protozoa have very complicated life cycles that may involve more than one host.
3. The primary importance of protozoa to humans is their ability to cause disease. Many of the diseases, such as malaria and giardiasis, are very common infections. In addition, diseases caused by coccidians can have a significant impact on domestic and wild animal populations.

THE MULTICELLULAR PARASITES

1. Multicellular parasites, such as trematodes (flatworms such as flukes), cestodes (flatworms such as tapeworms), and nematodes (roundworms), are metazoan animals with the ability to live at the expense of their host.
2. There are four major groups of multicellular parasites that are important to humans: the flukes, the tapeworms, the roundworms, and the arthropods.

THE FLATWORMS (PHYLUM PLATYHELMINTHES: CLASSES TREMATODA AND CESTOIDEA)

1. Flatworms (Platyhelminthes) are parasitic worms that are flattened dorsoventrally.
2. There are two large groups of flatworms: the flukes and the tapeworms.
3. Flukes (trematodes) are flatworms lacking a segmented body and with one or more suckers, which they use to adhere to hosts' surfaces. The flukes cause human diseases such as schistosomiasis, paragonomiasis, and opistorchiasis, which cause millions of deaths each year.

4. Tapeworms (cestodes) are segmented flatworms with a scolex (head) and one or more segments called proglottids. Tapeworm disease in humans may cause severe illnesses and sometimes death.

THE ROUNDWORMS (PHYLUM ASCHELMINTHES: CLASS NEMATODA)

1. Roundworms are slender, cylindrical worms with tapering bodies. Free-living and parasitic roundworms are widely distributed in nature.
2. These organisms are economically important because they cause extensive damage to plants and kill millions of domestic animals and humans each year.

THE ARTHROPODS (PHYLUM ARTHROPODA: CLASSES INSECTA AND ARACHNIDA)

1. Arthropods are invertebrate animals with jointed appendages and segmented bodies.
2. These animals are widely distributed in nature and can cause severe damage to crop plants.
3. Some arthropods, like the itch mite, the head louse, and the crab louse, can cause annoying infestations in animals.
4. Many arthropods, such as mosquitoes, lice, mites, fleas, and ticks, transfer diseases from one person to another. These arthropods are also referred to as vectors.

STUDY QUESTIONS

1. Discuss briefly three different ways in which protozoa affect humans.
2. Outline the life cycle of:
   a. an amoeba;
   b. a trypanosome;
   c. a malaria parasite.
3. Are protozoans important to farmers and ranchers? Explain.
4. Construct a table of 10 important human protozoal pathogens, indicating what disease they cause and how the disease is acquired.
5. Construct a table outlining the differences and similarities among flukes, tapeworms, and roundworms.
6. Using the life cycle of the schistosome illustrated in figure 10.12, discuss two ways in which the spread of schistosomes may be controlled.
7. Using figure 10.15 as an aid, discuss two ways of controlling the spread of tapeworm disease.
8. Using figures 10.22 and 10.24 as aids, construct a table outlining the similarities and differences among mosquitoes, fleas, ticks, and lice.
9. Define the meaning of the term "vector" and give five examples.
10. Explain how you might control the spread of pinworms in human populations.

SELF-TEST

1. Amoebae, like all other protozoal groups obtain their nutrients by the process of phagocytosis. a. true; b. false.
2. The trophozoites of most protozoal parasites are:
   a. a form of sexual reproduction;
   b. involved in the formation of gametes;
   c. involved in the asexual proliferation of the parasite in host tissue;
   d. the resting stage of the parasite;
   e. none of the above.
3. *Giardia lamblia* is a flagellated protozoan that forms a resting stage called the cyst. a. true; b. false.
4. The flagellate *Trypanosoma cruzi* causes the following disease:
   a. Chagas' disease;
   b. sleeping sickness;
   c. traveler's diarrhea;
   d. dysentery;
   e. none of the above.
5. The parasite that causes malaria and the one that causes kala-azar both have an insect host in which the sexual reproductive cycle takes place. a. true; b. false.
6. Which of the following statements *does not* apply to flatworms (Platyhelminthes)?
   a. some species have segmented bodies;
   b. most species have separate sexes;
   c. eggs are produced;
   d. most species have both sexes in the same worm;
   e. most species are flattened dorsoventrally.
7. Which of the following stages in the life cycle of the trematodes is infective for humans?
   a. eggs;
   b. miracidium;
   c. redia;
   d. cercaria;
   e. none of the above.
8. Tapeworms attach to host tissue and obtain their nutrients through the scolex. a. true; b. false.

9. Hookworms cause diseases that often are accompanied by intestinal bleeding because they attach to the intestinal epithelium with powerful hooks that cut deeply into the intestinal wall. a. true; b. false.

10. A sexually transmitted disease that leads to intense itching in the pubic area is caused by:
    a. *Sarcoptes*;
    b. *Pediculus*;
    c. *Phthirus*;
    d. *Aedes*;
    e. *Dermacentor*.

11. The protozoa are divided into groups based on their mode of locomotion, how they reproduce, and features of their life cycle. a. true; b. false.

12. The amoebae are differentiated from other protozoa because they divide by a process called multiple fission. a. true; b. false.

13. Flagellates like *Giardia lamblia* and *Trypanosoma* are transmitted to humans by the bites of insects. a. true; b. false.

14. The final host of sporozoan parasites is always a human or another mammal. a. true; b. false.

15. The parasitic nematodes include the tapeworms and flukes. a. true; b. false.

16. Which of the following groups of animal parasites have segments called proglottids?
    a. Nematodes;
    b. Trematodes;
    c. Cestodes;
    d. Copepods;
    e. Amphipods.

17. This group of animal parasites are flatworms that have separate sexes and the males have a groove that runs lengthwise all through the parasite's body.
    a. Nematodes;
    b. Schistosomes;
    c. Taeniforms;
    d. Fasciolas;
    e. Liver flukes.

18. The flukes are animal parasites that have rounded bodies and separate sexes. a. true; b. false.

19. Vectors are arachnids or insects that are vehicles of disease transmission without themselves being part of the parasite's life cycle. a. true; b. false.

20. Which of the following parasites is responsible for causing scabies in humans?
    a. the itch mite (*Sarcoptes*);
    b. the crab louse (*Phthirus pubis*);
    c. the head louse (*Pediculus*);
    d. *Callitroga* flies;
    e. *Tunga* fleas

21. Mosquitos in the genus *Aedes* can transmit a variety of diseases including Yellow fever and dengue fever.
    a. true; b. false?

SUPPLEMENTAL READINGS

Askew, R. R. 1971. *Parasitic insects*. New York: Elsevier.

Farmer, J. N. 1980. *The protozoa: Introduction to protozoology*. St. Louis: Mosby.

Kolata, G. 1985. Avoiding the schistosome's tricks. *Science*. 227:285–287.

Lennette, E. H., Balows, A., Hausler, W. J., and Shadomy, H. J. 1985. *Manual of clinical microbiology*. 4th ed. Washington, DC: American Society for Microbiology.

Markell, E. K. and Vogue, M. 1981. *Medical parasitology*. 5th ed. Philadelphia: Saunders.

Noble, E. R. and Noble G. A. 1982. *Parasitology: The biology of animal parasites*. 5th ed. Philadelphia: Lea & Febiger.

Richey, H. K., Fenske, N. A., and Cohen, L. E. 1986. Scabies: Diagnosis and management. *Hospital Practice 21(2)*: 124A–124Z.

Rothschild, M. and Clay, T. 1952. *Fleas, flukes and cuckoos*. London: Collins.

Schmidt, G. D. and Roberts, L. S. 1986. *Foundations of parasitology*. 3rd ed. St. Louis: Mosby.

Snow, K. R. 1975. *Insects and disease*. New York: Halsted.

Weller, P. F. 1987. Helminthic infections. *Scientific American Medicine Section* 7–XXXV: 1–17. New York: Freeman.

Weller, P. F. 1985. Protozoan infections. *Scientific American Medicine* 7–XXXIV: 1–18. New York: Freeman.

Whittaker, R. H. and Margulis, L. 1978. Protist classification and the kingdom of organisms. *BioSystems* 10:3–18.

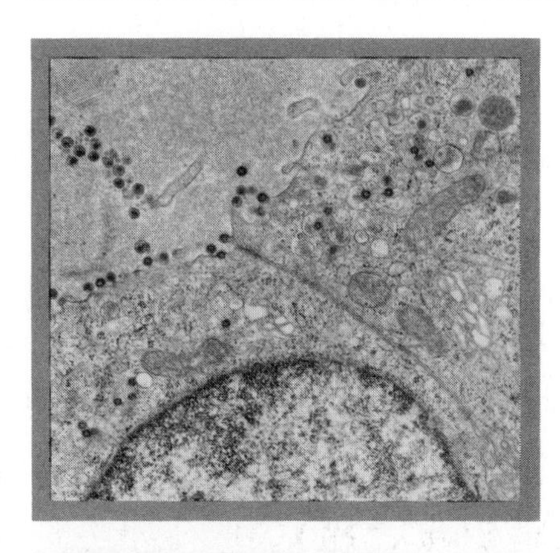

CHAPTER 11

NONCELLULAR INFECTIOUS AGENTS: VIRUSES, VIROIDS, AND PRIONS

CHAPTER PREVIEW A GENETICALLY ENGINEERED VIRUS MAKES A GOOD VACCINE

The vaccinia virus, used to eliminate smallpox from the world, may be used to protect humans from the ravages of a number of other diseases.

Scientists have inserted the genes from a number of pathogenic viruses and from a protozoan into different vaccinia viruses so that the foreign genes are expressed when vaccinia viruses invade host cells. Infections by the genetically engineered, vaccinia viruses have resulted in the following proteins being produced: hepatitis B virus surface antigen (HBsAg), herpes simplex virus glycoprotein D (HSVgD), influenza A virus hemagglutinin (InfHA), vesicular stomatitis virus G protein, rabies virus glycoprotein, and *Plasmodium knowlesi* sporozoite antigen.

Vaccination of separate groups of laboratory animals with one of the different types of recombinant vaccinia viruses resulted in antibodies that eliminated the infectivity of one of the following: hepatitis B, herpes simplex, influenza A, rabies, and vesicular stomatitis viruses. In addition, cytotoxic T-lymphocytes developed against some of the viruses. Most important of all, however, the laboratory animals were protected from these viruses.

The success of individual vaccines has stimulated the development of a recombinant vaccinia virus that carries more than one expressed foreign gene. A recombinant vaccinia virus carrying genes from hepatitis B, herpes simplex, and influenza A viruses has been constructed and tested. The recombinant vaccinia resulted in the production of antibodies against all three virus proteins in one group of animals. The levels of antibodies remained high for over one year in the animals tested.

Because of vaccinia's large genome and large number of nonessential genes, scientists estimate that they should be able to introduce genes from more than 25 organisms. Such genetically engineered vaccinia viruses would be very useful in those situations where multiple forms of a particular pathogen must be attacked to provide immunity. It should become possible to make a single vaccine that would protect against most diseases that afflict humans and animals.

CHAPTER OUTLINE

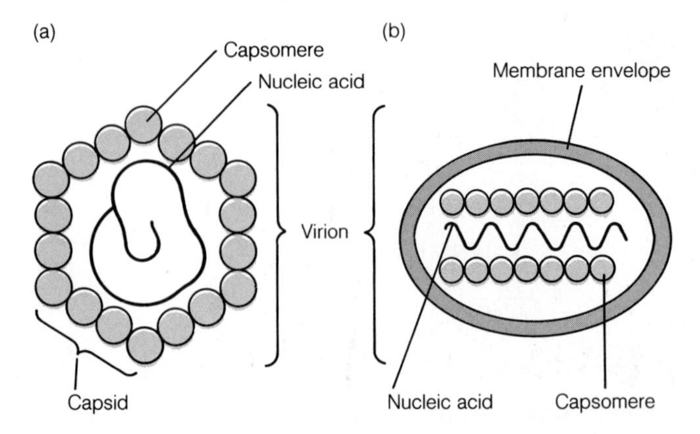

Figure 11.1 Generalized structure of viruses. *(a)* Naked virus, for example, the polio virus. *(b)* Enveloped virus, for example, the AIDS virus (illustrated) or the influenza virus.

BECAUSE OF THEIR SMALL SIZE, NONCELLULAR infectious agents cannot be seen with the light microscope, and they pass through filters that trap bacteria. Viruses were first distinguished and related to each other because of their characteristic of being **filterable infectious** agents. With the development of the electron microscope in the late 1930s, it was possible to see that viruses were very different from bacteria and other cellular organisms. The viruses came in assorted sizes and shapes, but each type maintained its major characteristics as it reproduced. For example, each type of virus has a typical shape, structure, and nucleic acid. Thus, viruses were true breeding infectious agents.

Viruses differ from bacteria and other cellular organisms in a number of ways. Viruses consist of a single type of nucleic acid (DNA or RNA) within a protein coat called a **capsid** (fig. 11.1). The nucleic acid and the capsid are referred to as the **nucleocapsid**. The nucleic acids contain the information for making the proteins found in the viral coat as well as many of the enzymes required to invade the host and to replicate the virus's nucleic acid. Some viruses are covered by a membrane, often called the **envelope**. The term **virion** is used to describe the complete viral particle, including the nucleocapsid and any envelope it may have. None of the viruses is able to replicate in the absence of host cells. Thus, all the viruses are obligate parasites of cellular organisms.

Many viruses produce gross morphological changes in the cells they infect because they transform or destroy them when they multiply. These changes are sometimes referred to as **cytopathic effects** (CPE). For example, when a confluent layer of bacteria is infected by bacterial viruses called **bacteriophages**, morphological changes can be seen (fig. 11.2a). The confluent layer of bacteria, known as a **bacterial lawn**, makes the solidified nutrient plate cloudy. When a bacteriophage reproduces in a bacterium, it destroys

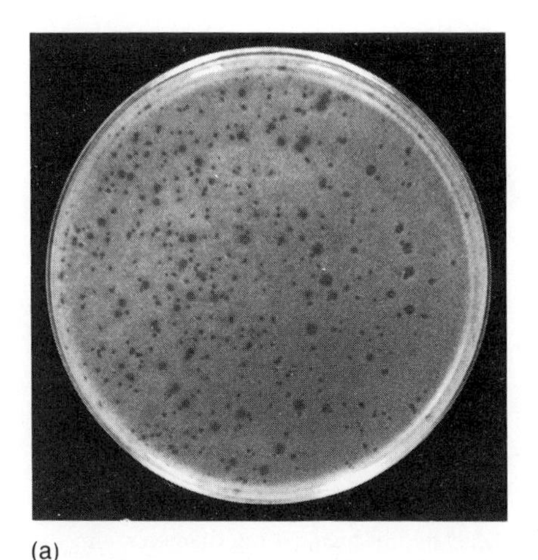
(a)

(b)

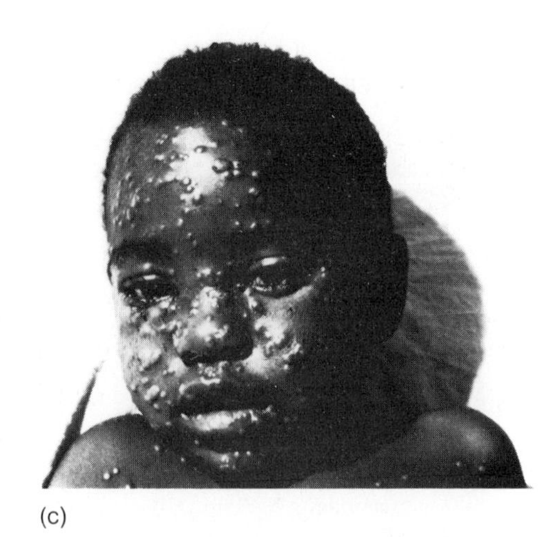
(c)

Figure 11.2 Viral lesions. *(a)* A lawn of bacteria contains clear areas that represent bacteriophage plaques or virus colonies that have developed from a single bacteriophage. *(b)* A leaf with light-colored zones (arrows) that represent tobacco mosaic virus lesions. The leaf can be thought of as a lawn of plant cells. *(c)* A person exhibiting numerous smallpox lesions caused by the smallpox virus.

the cell and the viral progeny spread to adjacent bacteria. Within 10–20 hours, a large area of the bacterial lawn around the original infection becomes lysed, producing a clear area in the lawn. Actually, each clear area represents a virus "colony" that has arisen from a single virus. These virus colonies are known as **plaques**.

Figure 11.2b shows a leaf (which can be thought of as a lawn of plant cells) that has been invaded by **tobacco mosaic virus**. The destruction of large areas of tissue by the virus results in visible lesions. Each lesion probably originated from a single virus.

Figure 11.2c illustrates a human infected by smallpox virus. The smallpox virus is proliferating in the skin cells and causing extensive destruction. The lesions represent virus colonies, and each lesion may have arisen from a single virus. In general, viral infections are detected when their reproduction causes noticable changes in a host's cells.

CLASSIFICATION OF VIRUSES

Bacteriophages

Bacteriophage, or "**phage**" for short, is the term used for the viruses that infect bacteria. The bacteriophages occur in an assortment of shapes and sizes, and they may contain double-stranded DNA (ds-DNA), single-stranded DNA (ss-DNA), double-stranded RNA (ds-RNA), or single-stranded RNA (ss-RNA) (table 11.1). Bacteriophage are generally classified on the basis of their nucleic acid and their structure, which may be **complex**, **polyhedral** (**icosahedral** or 20 sided), **helical** (**cylindrical**) or **enveloped**.

The ds-DNA bacteriophages are very complicated viruses. Their genomes code for 50–200 proteins depending upon the phage. On the other hand, the ss-DNA and ss-RNA bacteriophages are very simple and code for only 3–5 proteins. The ds-RNA bacteriophage $\phi 6$ is one of the largest RNA viruses and codes for aproximately 20 proteins.

Plant Viruses

In contrast to the bacteriophages, almost all the plant viruses are ss-RNA viruses (table 11.2). The protein coat or capsid of plant viruses is generally in the shape of an icosahedron or a cylinder. The plant viruses are generally transmitted from plant to plant by insects, nematodes (roundworms), fungi, or contaminated machinery. Some viruses are also transmitted from parent to offspring in pollen or in seeds.

Many of the plant viruses are unusual in that their genetic information is segmented and packaged into separate capsids. For example, some plant viruses may

TABLE 11.1
CHARACTERISTICS OF COMMON BACTERIOPHAGES

Representative viruses	Shape (morphology)	Type of nucleic acid	Characteristics
T2, T4, T6	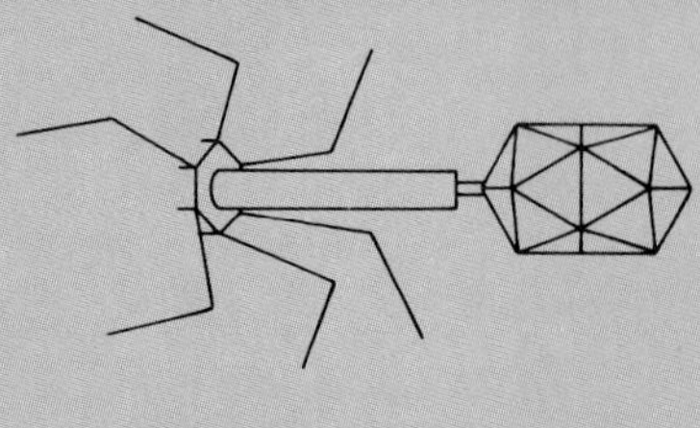	DNA	Prolate icosahedral head 80 × 110 nm, complex tail 110 nm long. Adsorb to wall. Infect *E. coli*
Lambda, Chi	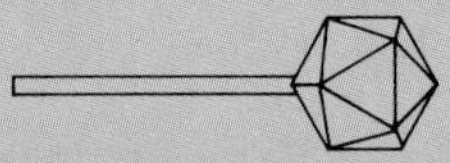	DNA	Regular icosahedral head 54 nm diameter, simple tail 140 nm long. λ adsorbs to wall, X adsorbs to flagella. Infect *E. coli*
T3, T7 P22	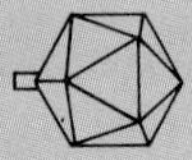	DNA	Regular icosahedral head 60 nm diameter, short tail 20 nm long. Adsorb to wall. P22 infects *Salmonella*, T3, T7 infect *E. coli*
ϕX174		DNA	Icosahedral capsid 30 nm diameter. No tail. Infects *E. coli*
Fd, M13	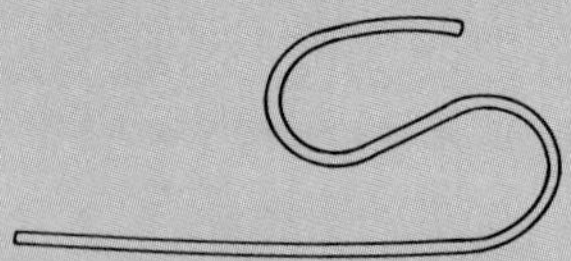	DNA	Filamentous (helical?) capsid 8 × 800 nm. Adsorb to end of pili. Host cell not lysed. Infect *E. coli*
L51		DNA	Helical capsid? enclosed in a membrane. Bullet shaped 14 × 80 nm. Infects *Mycoplasma*
ϕ6		RNA	Icosahedral capsid 70 nm diameter enclosed in a membrane. Infects *Mycoplasma*
R17, MS2, f2 Qβ,		RNA	Icosahedral capsid 25 nm diameter. No tail. Adsorb along the length of pili Infect *E. coli*

package four different genomes into separate capsids, and all four are required for virus multiplication. The genomes of the RNA plant viruses are about the same size as those found in the RNA bacteriophages. The RNA genomes in plants code for 3–15 proteins.

Animal Viruses

Almost all of the animal viruses are either icosahedral or enveloped viruses and have ss-RNA or ds-DNA genomes. Some, however, have ds-RNA or ss-DNA genomes (table 11.3). The DNA animal viruses fall

TABLE 11.2
CHARACTERISTICS OF PLANT VIRUSES

Classification	Shape (morphology)	Major host plant	Mode of transmission
RNA VIRUSES			
Naked, isometric or icosahedral, single-stranded			
Brome mosaic		Broad bean Cowpea	Beetle
Cucumber mosaic		Cucumber	Aphid
Tobacco streak		Tobacco	Seed and pollen
Enveloped, bullet-shaped, single-stranged			
Rhabdovirus		Lettuce	Aphid and leafhopper
Naked, isometric or icosahedral, double-stranded			
Reovirus		Clover Rice	Leafhoppers and Planthoppers
Naked, cylindrical single-stranded RNA genomes			
Tobacco mosaic		Tobacco	Mechanical sap inoculation
Potato X		Potato	Aphid
Beet yellows		Beet	Aphid
DNA VIRUSES			
Naked, icosahedral Double-stranded			
Cauliflower mosaic		Cauliflower	Aphid
Naked, fused capsid, single-stranded			
(Maize streak)		Corn	Whitefly and leafhopper

into two size ranges: those that code for 5–10 proteins and those that code for 30–300 proteins. One of the pox viruses has the largest genome and is believed to code for more than 300 proteins. This, however, is exceptional.

The large percentage of animal viruses that are enveloped (approximately 50%) may reflect the fact that their hosts do not have cell walls and entrance to the host's cytoplasm may be expedited by having an envelope that can fuse with the plasma membrane or

TABLE 11.3
CHARACTERISTICS OF ANIMAL VIRUSES

Classification	Shape (morphology)	Type of nucleic acid	Representative diseases
RNA VIRUSES			
Naked, icosahedral capsid, single-stranded RNA genome			
Picornaviruses		+ ss-RNA	Poliomyelitis, hepatitis A, foot and mouth disease
Naked, icosahedrad capsid, double-stranded, RNA genomes			
Reoviruses		ds-RNA	Respiratory diseases, polyhedrosis disease of insects
Enveloped, icosahedral capsid, single-stranded, RNA genome			
Togaviruses		+ ss-RNA	Encephalitis, yellow fever, rubella (German measles)
Enveloped, cylindrical capsid, single-stranded, RNA genomes			
Orthomyxoviruses	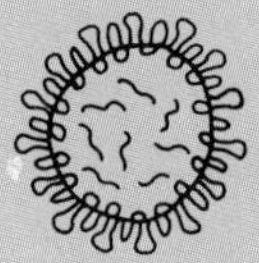	− ss-RNA	Influenza
Bunyaviruses	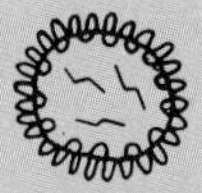	− ss-RNA	Encephalitis
Coronaviruses		+ ss-RNA	Common cold
Arenaviruses		− ss-RNA	Hemorrhagic fevers, lassa fever, lymphocyte choriomeningitis
Paramyxoviruses	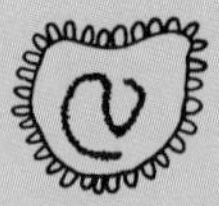	− ss-RNA	Mumps, measles (rubeola), newcastle disease of chickens, distemper in dogs
Retroviruses		+ ss-RNA	Cancers, hepatitis non A non B, AIDS

TABLE 11.3
(CONTINUED)

Classification	Shape (morphology)	Type of nucleic acid	Representative diseases
Enveloped, bullet-shaped, cylindrical capsid, single-stranded RNA genome Rhabdovirus		− ss-RNA	Rabies
DNA VIRUSES			
Naked, icosahedral capsid			
Parvovirus		ss-DNA	Diseases in rodents
Naked, icosahedral capsid			
Papovaviruses		ds-DNA	Warts and tumors
Adenoviruses		ds-DNA	Cancers
Naked, icosahedral double capsid	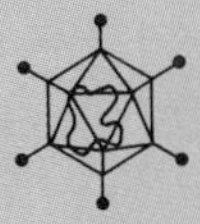		
Hepadnaviruses		ds-DNA	Hepatitis B
Enveloped, icosahedral capsid			
Herpes viruses		ds-DNA	Chicken pox, genital herpes, shingles, mononucleosis, cancer, fever blisters
Iridoviruses		ds-DNA	African swine fever
Complex coats, complex cylindrical capsid			
Pox viruses	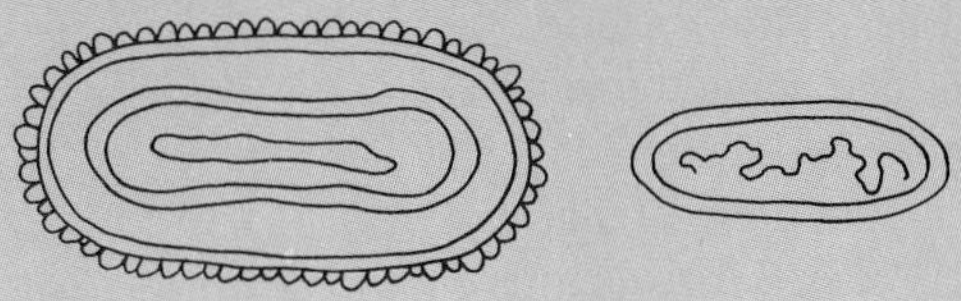	ds-DNA	Smallpox, cowpox

(a)

(b)

(c)

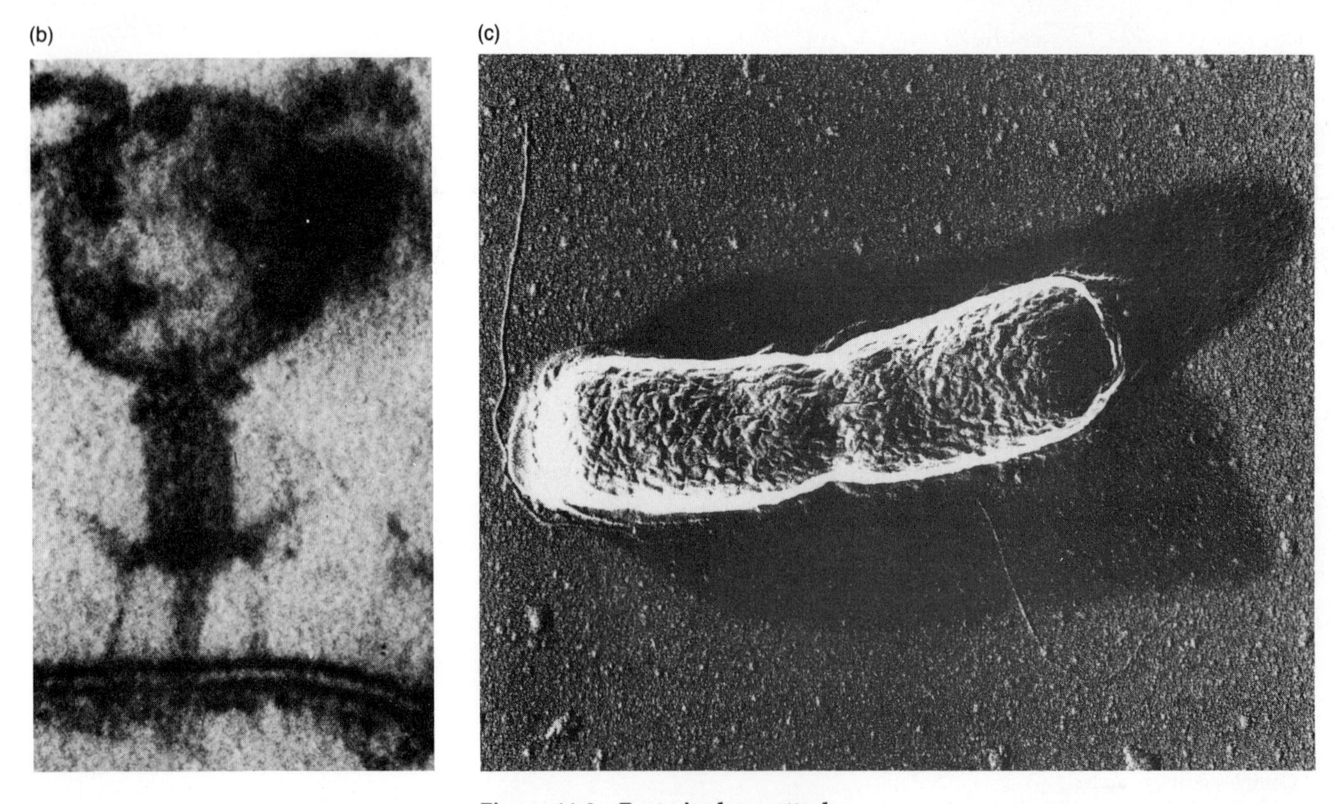

Figure 11.3 Bacteriophage attachment. *(a)* The icosahedral bacteriophage MS2 attached along the length of F pili. *(b)* The complex bacteriophage T4 attached to the bacterial cell wall. *(c)* An M13 bacterial virus infecting an *E. coli* bacterial cell.

a pinocytic vesicular membrane. Alternatively, enveloped animal viruses may reflect the need of a foreign particle to evade the host's immune system. At least two enveloped animals viruses (arenaviruses and bunyaviruses) appear to lack a capsid around their RNAs. Nevertheless, their RNAs are associated with protein. The RNA–protein complex is referred to as the **ribonucleoprotein** core.

REPLICATION OF BACTERIOPHAGES

Viruses lack the necessary enzymes and structures to carry on the metabolism required to reproduce themselves. Reproduction can occur only when viruses are able to get their genetic information into an appropriate host cell and subvert the cell's metabolism to their needs.

Nutrients within the host cell, and frequently host enzymes, are used to synthesize viral enzymes, nucleic acids, and proteins. Once viral structural proteins are synthesized, they aggregate into viral parts. Then viral nucleic acids and viral parts self-assemble into complete viruses.

The life cycle of a virus involves a sequence of steps, usually aimed at penetrating the host cell and reproducing within it. The sequence of steps in an infection by a well-studied bacteriophage will be used as an example of how a virus reproduces.

Attachment

The bacteriophages begin multiplication by attaching to a specific host cell (fig. 11.3). The tailed bacteriophages, such as **T4** and **lambda (λ)**, attach to and puncture the outer envelope of the host's cell wall, while **Chi** attaches along the length of the flagella. The icosahedral bacteriophage **MS2** attaches along the length of F pili, while the cylindrical bacteriophage **M13** binds to the ends of F pili. The enveloped bacteriophages that attack the wall-less mycoplasmas attach to the host cell's plasma membrane.

Genome Penetration

After the virus attaches to its host, it introduces its genetic material into the cell. Bacteriophage genomes enter the bacterial hosts in various ways. **T4** perforates the bacterial cell wall and then inserts the core of its tail into the host cell (fig. 11.4). The DNA genome is then injected into the cell. Chi may inject its hereditary material into the hollow flagella, which must somehow migrate from the flagella to the cytoplasm. MS2 and M13 fuse their capsids with bacterial sex pili. How the virus genomes reach the cytoplasm is unknown; perhaps the virus genomes enter the cytoplasm when the pili are retracted by the cell. The enveloped bacteriophage may enter the wall-less mycoplasmas by pinocytosis or by fusing their envelopes with the host's plasma membrane.

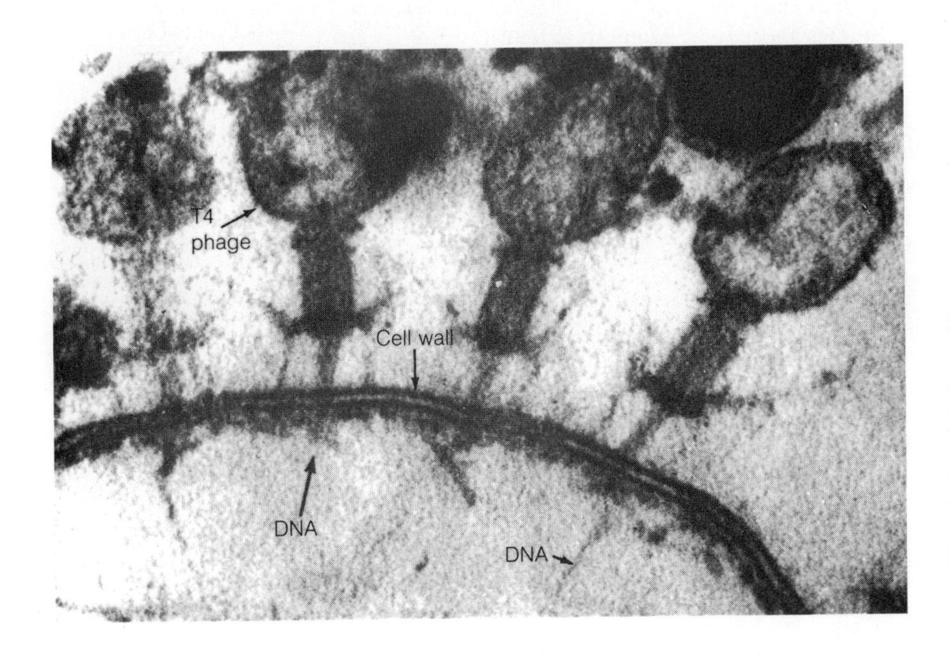

Figure 11.4 Bacteriophage infection. A number of T4 are attached to the cell wall of a bacterium. When the phage sheath contracts, the core is forced through the wall and plasma membrane, and the bacteriophage DNA passes from the phage head into the cytoplasm. Strands of phage DNA can be visualized entering the cell's cytoplasm (arrows).

RNA Synthesis, Protein Synthesis, and Genome Replication

Once the bacteriophage genome enters the cytoplasm, it directs the synthesis of viral mRNA and proteins (fig. 11.5). The viral proteins catalyze the replication of the virus genome and function as structural components of the virus particle.

Assembly

As the virus proteins are synthesized, they self-assemble into viral components such as the head, tail, and tail fibers (fig. 11.5). The assembly of many viral components is catalyzed by viral proteins.

Release

In some infections, viral proteins cause the lysis of the cell so that the assembled particles accumulating in the cytoplasm escape. In a few cases (Fd, Ml3, Fl), the viruses appear to be released from the cell without lysing it. When bacteriophages repeatedly cause the death and lysis of host cells, the infectious cycle is known as the **lytic cycle** (fig. 11.5).

Lysogeny

Some viruses, such as the DNA bacteriophage λ, are able to delay or avoid virus assembly and release by integrating a DNA genome into the host's genome (fig. 11.6). Generally, almost all transcription and translation of viral genes is inhibited when the viral genome integrates into the cell's genetic information. The integrated bacteriophage DNA is called a **prophage**. Since the prophage is replicated along with the cell's genome, each daughter cell carries a prophage. A bacterium that carries a prophage is called a **lysogen**, and this unusual infection is called **lysogeny**.

FOCUS THE DISCOVERY OF VIRUSES

During the "golden age of microbiology" (1875–1915), numerous scientists discovered that some plant and animal diseases were caused by a group of organisms that were much smaller than bacteria and that could not be grown on artificial media. Martinus Beijerinck in the late 1800s hypothesized that plant and animal diseases caused by agents that passed through filters that retained bacteria were not bacteria but something quite different. Eventually these **filterable agents** of disease became known as **viruses**. Because of his studies and insights on these infectious agents, Beijerinck is considered the "father" of virology.

It was not until 1915, however, that virus infections of bacteria were recognized. Frederic Twort observed that bacterial colonies were destroyed by a filterable agent that he hypothesized was either a bacterial virus or a toxin. In 1917, Felix d'Herelle studied the destruction of bacteria by a filterable agent. The virus produced clear holes called **plaques** in bacterial lawns and cleared turbid broth cultures within a few hours, killing most of the bacteria. He reasoned that the infectious agent was reproducing at the expense of the bacteria and consequently causing their destruction. D'Herelle called these bacterial viruses **bacteriophages** because they appeared to consume or eat the bacteria.

In 1935 Wendell Stanley, using techniques for purifying and crystallizing proteins, crystallized tobacco mosaic virus (TMV) and concluded that it was mostly protein. Near the end of 1939, Stanley and Frederic Bawden independently isolated a nucleic acid from the crystallized TMV. This nucleic acid was later shown to be necessary for virus infectivity. Stanley received the 1946 Nobel Prize in chemistry for his crystallization of TMV. Tobacco mosaic virus was the first virus to be observed in the electron microscope in 1939. TMV turned out to be a long, thin, tubelike structure, very much smaller than the smallest bacteria and very different in structure.

In 1939, Emory Ellis and Max Delbruck showed that T4 bacteriophage multiplied within bacteria and that progeny bacteriophages were released when the bacteria lysed. Thomas Anderson and Salvador Luria obtained electron micrographs of T4 bacteriophage in 1942 that clearly indicated it had a "head" and a tubular "tail." In 1952, Martha Chase and Alfred Hershey performed what is now known as the Hershey-Chase experiment, which demonstrated that T4 DNA entered *E. coli* cells but little or no phage protein penetrated. Their experiment showed that the viral genetic information was nucleic acid rather than protein.

Max Delbruck, Salvador Luria, and Alfred Hershey shared the 1969 Nobel Prize for Physiology or Medicine because of their contributions to the understanding of bacteriophage structure, function, and genetics.

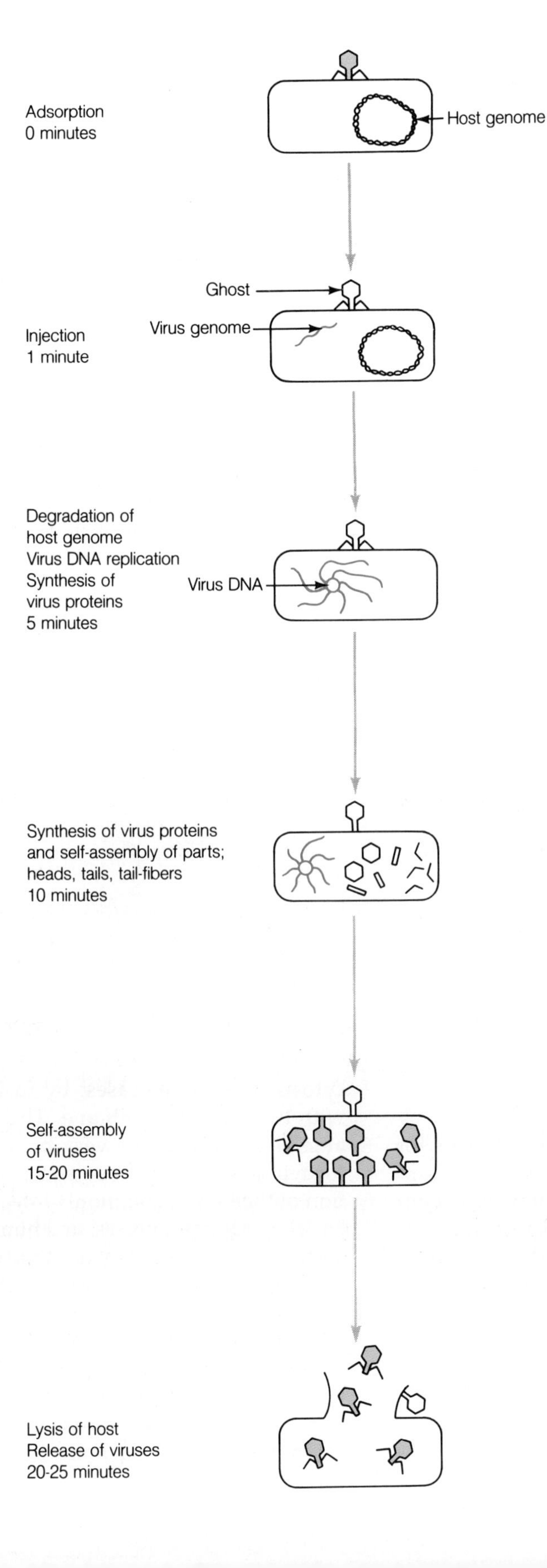

Figure 11.5 **Lytic cycle of T4 bacteriophage.** The cycle begins with the adsorption of the virus to a host cell. Next, the virus genetic information is injected into the host's cytoplasm. Early T4 mRNA and protein synthesis begins approximately 2–3 minutes after infection, while host DNA begins to be degraded approximately 5 minutes after infection. Viral components begin to self-assemble about 10 minutes after infection, and viral components assemble themselves into viruses 15–20 minutes after infection. The viruses escape from their host cells approximately 20–25 minutes after infection by breaking open the cells.

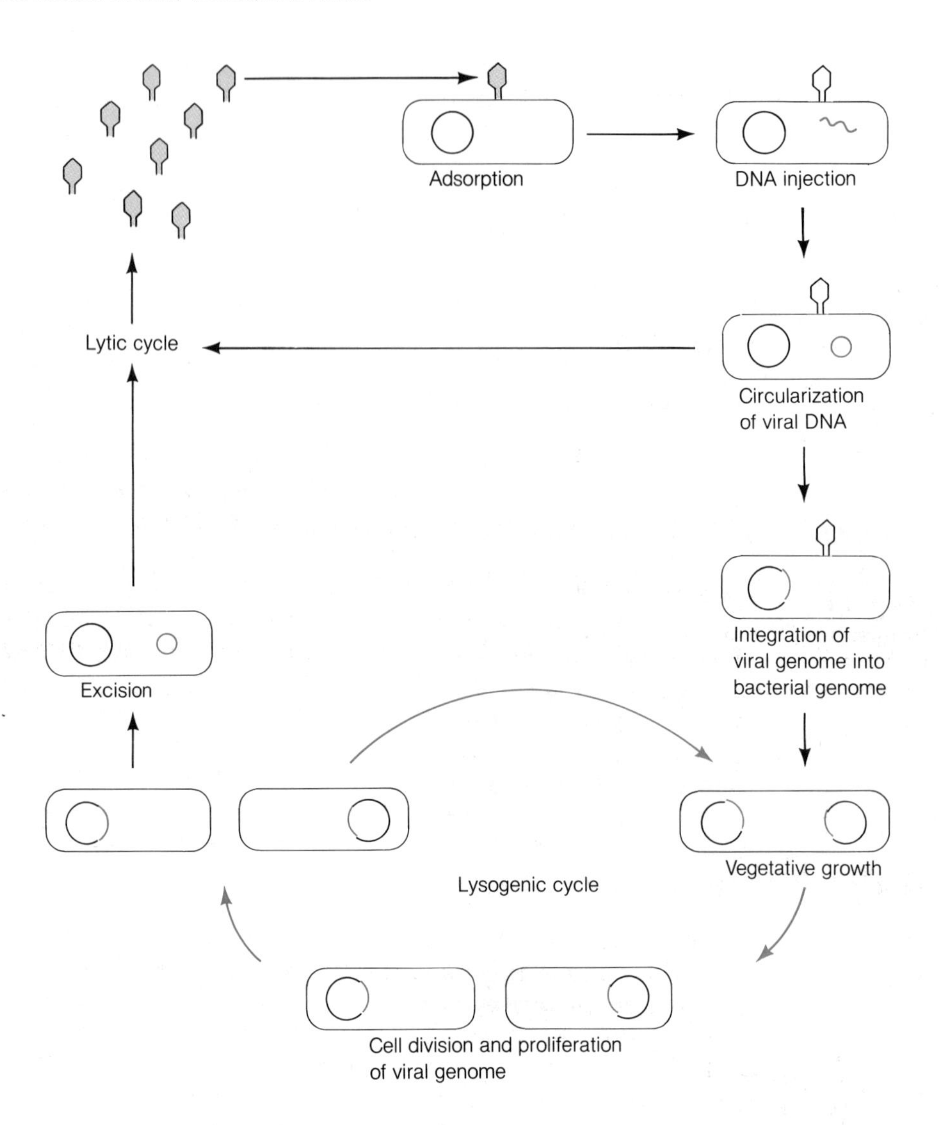

Figure 11.6 **Lysogenic cycle of the lambda bacteriophage.** The lambda bacteriophage frequently goes through a lysogenic cycle rather than a lytic cycle. After adsorption, injection, and genome circularization, the virus genome recombines with the host's genome and becomes part of it. A bacterium that contains an integrated virus genome is called a lysogen. The virus genome is replicated as the lysogen proliferates, so each daughter cell contains a virus genome (prophage). The replication of the viral genome as the lysogen proliferates represents the lysogenic cycle. Generally, when the lysogen's DNA is damaged, the lambda prophage excises itself from the lysogen's DNA and enters a lytic cycle.

REPLICATION OF ANIMAL VIRUSES

Animal viruses rely on a number of mechanisms for penetrating cells, replicating themselves, and escaping from their hosts. For ease of discussion, virus multiplication can be divided into a number of steps: (a) adsorption to cytoplasmic membranes; (b) penetration into the cytoplasm of the cell; (c) uncoating; (d) RNA synthesis, protein synthesis, and genome replication; (e) assembly; and (f) release.

Adsorption to the Plasma Membrane

Animal viruses adsorb specifically to the membranes of host cells and gain entrance either through pinocytosis or, in some cases, by fusing their envelopes with the plasma membrane. The host range of some animal viruses can be very broad. For example, the togaviruses, which cause encephalitis (an inflammation of the brain), commonly infect domestic and wild birds, mosquitoes, horses, and humans. The host range is largely determined by the ability of a virus to attach to the plasma membrane of cells.

Penetration into the Cell's Cytoplasm

In most cases, the entire animal virus (genome, capsid, and envelope) enters the host. The virus is brought into the cell by **pinocytosis.** In most bacteriophages, by contrast, only the genome enter the cell. In the case of some enveloped viruses, only the genome and capsid enter the cell because the envelope fuses with the plasma membrane.

Uncoating

Virus uncoating is believed to take place within pinocytotic vesicles or within the cytoplasm after the virus enter the cell. Envelopes are removed when they fuse with pinocytotic vesicles. In the case of viruses lacking a membrane, the capsid depolymerizes and the nucleic acid is released into the vesicle. The factors stimulating the depolymerization of the capsid are not known.

RNA Synthesis, Protein Synthesis, and Genome Replication

There is a great deal of variety in the way that animal viruses are expressed and replicated. Part of the reason is that virus genomes can be ss-RNA, ds-RNA, ss-DNA, or ds-DNA. In addition, the ss-RNA genomes may be either sense strands (+RNA) that can function as mRNA or nonsense strands (−RNA) that are unable to code for proteins.

Assembly and Release

In most cases, the packaging of animal virus genomes begins when the nucleic acid associates with capsid proteins or with an incomplete capsid.

Many of the animal viruses (herpes, pox, orthomyxovirus, paramyxovirus, rhabodovirus, togavirus, hepatitis B, retrovirus) become enveloped either as they develop within the cell or when they escape from the cell. The herpes virus nucleocapsid becomes enveloped as it passes through the nuclear membrane into the cytoplasm, while the poxvirus obtains its membranes from the cytoplasm of the host cell. Most of the other enveloped viruses obtain their membrane as they pass through the cytoplasmic membrane. Generally, the nonenveloped animal viruses escape from the host cell when it lyses, while the enveloped viruses escape from the host cell when they pinch off from the membrane. Enveloped viruses cause the lysis of the host cell if they disturb the cell's metabolism sufficiently.

REPRESENTATIVE ANIMAL VIRUSES

There is no simple way of discussing a "typical" virus, since each virus has its peculiarities and affects its host in different ways. Consequently, a few noteworthy viruses will be discussed in order to understand how diverse viruses are.

Herpes Viruses

The herpes viruses are double-stranded DNA (ds-DNA) viruses with an icosahedral capsid and a membrane envelope. They are 125–200 nm in diameter and replicate in the nuclei of animal cells (fig. 11.7). The herpes viruses include **herpes simplex** I and II, **herpes zoster** (varicella-zoster), the **cytomegalovirus**, and the **Epstein-Barr virus**. The virus genome exists as a linear ds-DNA molecule and codes for more than 150 proteins. These viruses have a wide host range including mice, guinea pigs, hamsters, rabbits, chickens, monkeys, humans, and many types of cultured animal cells.

The herpes viruses are widely distributed in nature. They cause human diseases that are of public health importance, such as cold sores, genital herpes, chicken pox, shingles, and mononucleosis. One of these diseases, genital herpes, is a sexually transmitted disease of epidemic proportions in the United States. Cytomegalovirus is 125 nm in diameter (slightly smaller than herpes simplex) and replicates in the salivary glands, brain, kidney, liver, and lungs. It is responsible for causing a large percentage of the mental retardation in newborn children each year. The Epstein-Barr virus is about 125 nm in diameter and is responsible for infectious mononucleosis, a disease that generally causes extreme exhaustion and swollen tonsils. The virus replicates in lymphocytes and lymph tissue and can be isolated from pharyngeal secretions of patients with infectious mononucleosis.

Infections caused by herpes viruses usually begin when the virion (nucleocapsid plus envelope) attaches to the cell's plasma membrane through proteins found in the viral envelope (fig. 11.7). The virion enters the cell's cytoplasm by pinocytosis and then moves toward the nucleus within the pinocytotic vesicle. Upon reaching the nucleus, the pinocytotic vesicle fuses with the outer nuclear membrane. The viral envelope then fuses with the inner nuclear membrane (fig. 11.7), and the nucleocapsid is released into the nucleus. At this point, it is believed that the genetic information (core) is uncoated and released (fig. 11.7). DNA replication begins about 3 hours after infection and is completed 12 hours later. Virus particles are evident approximately 18 hours after infection and are released from the cell shortly thereafter (fig. 11.7).

An important characteristic of most herpes viruses is that they can remain latent within cells for many years and reoccur periodically. A good example of this is herpes zoster, which is responsible for both chickenpox (varicella) and shingles. Chickenpox generally subsides with a few days, as a result of the host immune response, and the person never has it again. In some cases, however, the virus that causes chickenpox remains latent for years in neurons. From time to time this latent virus (provirus) is activated to produce virus particles. These viruses travel from the infected nerves to the skin, where they initiate a lytic infection called shingles. The development and/or movement of the virus in the nerves causes severe pain in the infected area. The vesicular eruptions and pain caused by herpes zoster may last as long as a month.

The herpes genome resembles two transposons in series. Because of its structure, it has the potential to integrate into host chromosomes. It is able to make copies of itself that integrate at a number of sites in the host's genome. The ability of the herpes viruses to integrate into host chromosomes and to excise (make copies of) themselves allows the herpes viruses to remain latent for long periods of time, occasionally giving rise to lytic infections. The similarity between transposons (which cause mutations and chromosome breaks) and the herpes genome may shed light on how herpes viruses can transform cells and cause some cancers.

Orthomyxoviruses (Influenza Viruses)

The orthomyxoviruses include all of the influenza viruses that infect humans and animals. These viruses are important because of the widespread illness and death they cause each year in the United States. The orthomyxoviruses are single-stranded segmented RNA viruses that consist of eight nucleocapsids within a membrane or envelope (fig. 11.8). At times the virus may have a filamentous structure.

The eight ss-RNAs code for the 10 virus proteins. The **neuraminidase spikes** on the surface of the virus allow it to cut through glycoproteins in respiratory secretions and bind to host cells rather than to the proteins in the mucus. The **hemagglutinin spikes** on the surface of the virus bind it to specific glycoproteins on the host's plasma membrane. After the virus attaches

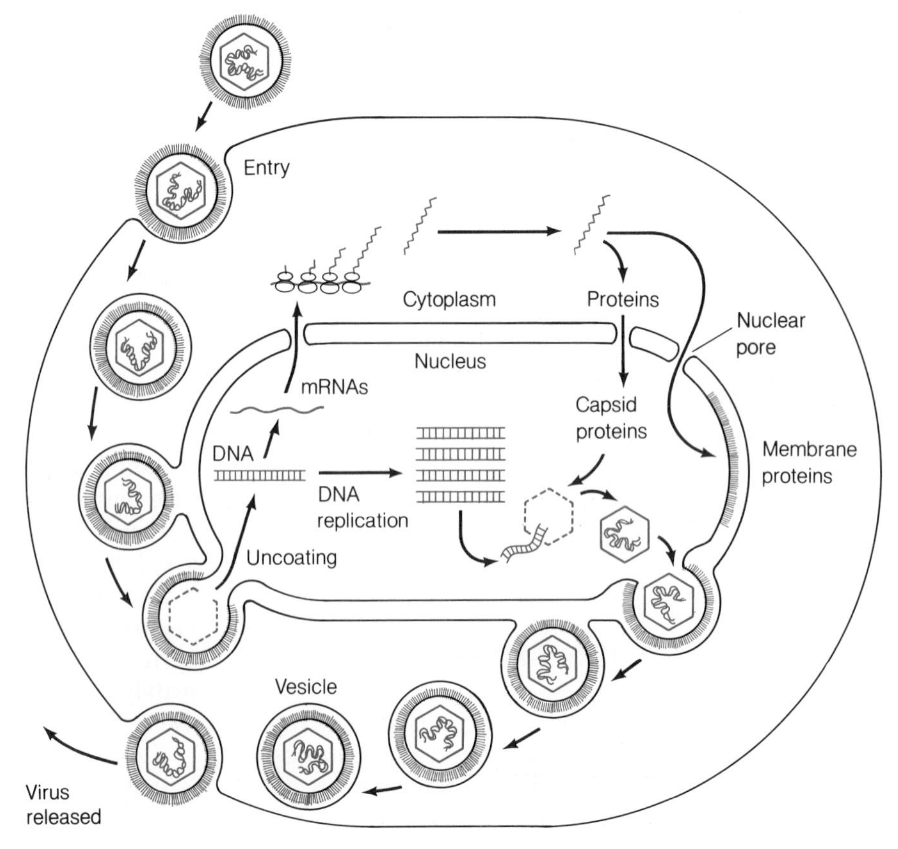

Figure 11.7 Herpes infection. Herpes viruses are enveloped "ds-DNA" animal viruses that enter cells by pinocytosis. After the virus DNA is uncoated in the cytoplasm, it enters the nucleus. DNA replication and transcription occur in the nucleus, but mRNA translation takes place in the cytoplasm. Virus proteins synthesized in the cytoplasm enter the nucleus, where self-assembly of the virus capsid occurs. The virus obtains its envelope when the capsid associates with the inner nuclear membrane. When the enveloped viruses pass through the outer nuclear membrane, they end up in vesicles. The virus escapes from the cell when the vesicles fuse with the plasma membrane.

to the host cell, the host plasma membrane invaginates and the virus enters a pinocytotic vesicle. The virus envelope then fuses with the vesicle membrane and the nucleocapsids are released into the cytoplasm (fig. 11.8). The RNA from the nucleocapsids enters the nucleus, where complementary copies are used as templates to make copies of the virus genome. A virus RNA-dependent RNA polymerase (RNA transcriptase) and a virus replicase are required for multiplication. The RNA transcriptase synthesizes +RNA (mRNA) from the complementary −RNA genomes (fig. 11.8).

Figure 11.8 indicates how the nucleocapsids leave the nucleus and associate with the mass of virus proteins on the plasma membrane. The plasma membrane then buds out, taking with it the eight nucleocapsids. The virus envelope consists of host membrane that has been altered by virus-coded proteins.

Studies of the virus proteins have indicated that there are three distinct types of influenza viruses: influenzas A, B, and C. The hemagglutinin (H) and neuraminidase (N) spikes in influenza A vary slightly through time. This minor variation in virus proteins is called **antigenic drift**, and is due to spontaneous mutations in the RNAs coding for these proteins. Major changes called **antigenic shifts** in the hemagglutinin and neuraminidase occur infrequently and apparently depend upon the recombination of genomes from different virus strains found in widely separated populations or in different hosts. The appearance of mutant viruses in a host population that is partially or totally unfamiliar with the new surface proteins has been responsible for the influenza epidemics that occur every 20 years or so.

Figure 11.9 indicates the major changes that have occurred in influenza A surface antigens through the years. It is believed that the 1889–1890 epidemic that hit Asia, Europe, and America and caused the death of more than 20 million persons worldwide was due to an uncommon virus with a hemagglutinin called H2 and a neuraminidase called N2. In 1900, a less severe epidemic occurred due to an H3N2 influenza virus. This virus was responsible for most of the influ-

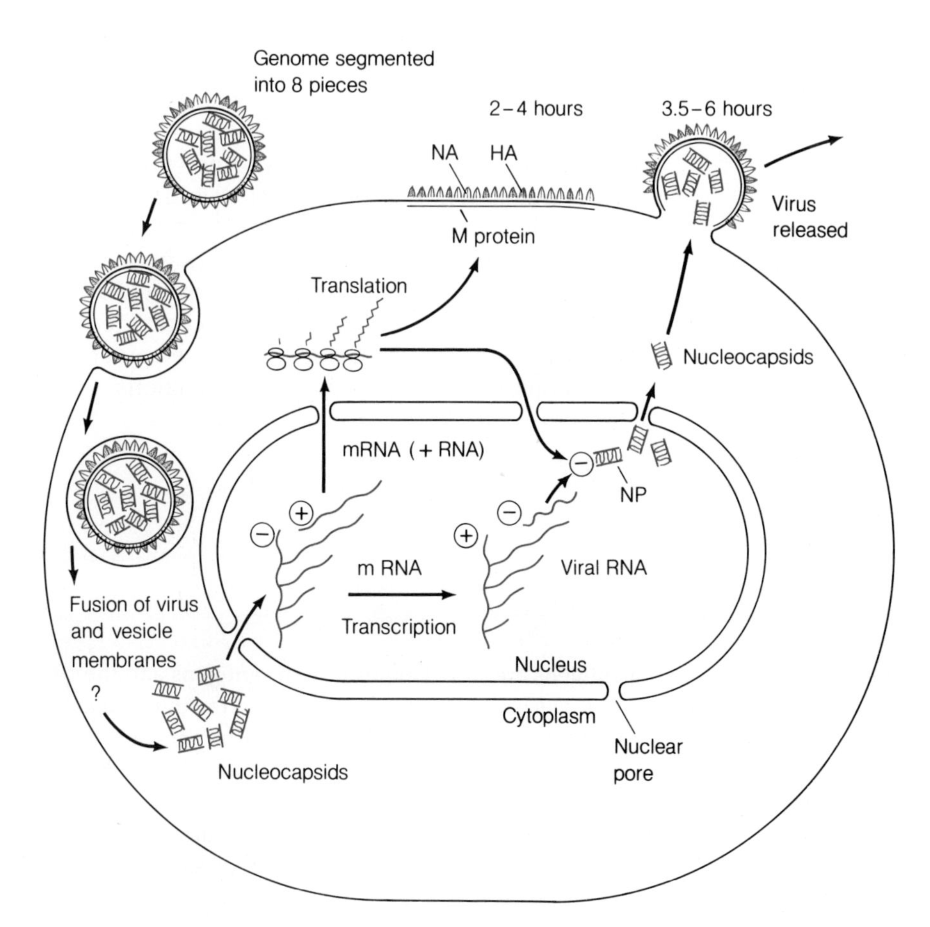

Figure 11.8 Orthomyxovirus infection. Orthomyxoviruses (influenza viruses) are enveloped −RNA animal viruses that enter the cell by pinocytosis. The virus genome is segmented into eight pieces of RNA. After the virus RNAs are uncoated in the cytoplasm, they enter the nucleus. Transcription and RNA replication occur in the nucleus, but mRNA translation takes place in the cytoplasm. A cylindrical capsid forms around each of the eight pieces of the genome. The pieces associate with the cytoplasmic membrane, which undergoes exocytosis. M protein is the matrix protein, NA represents the neuraminidase spike, and HA indicates the hemaglutinin spike.

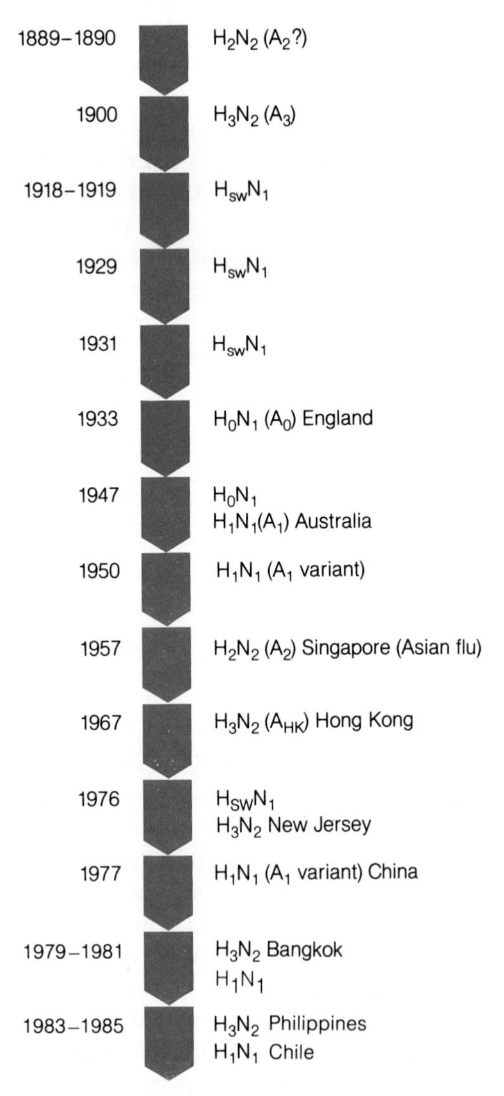

Figure 11.9 Major changes in surface antigens of orthomyxoviruses. In the influenza A virus, hemaglutinin spikes (H) since the 1890s have undergone five major changes (H2, H3, Hsw, H0, H1), while neuraminidase spikes (N) have undergone only two major changes (N2 and N1). The major changes in surface antigens are responsible for the epidemics caused by influenza A virus approximately every 10 to 20 years. When an animal or human population is not immune to altered virus antigens, the virus is able to infect most of the population. Epidemics depend upon major changes in virus antigens and the lack of immunity to the mutated antigens. Influeza B virus generally does not cause major epidemics, because its surface antigens do not undergo major changes. The surface antigens of influenza B change slowly.

enza until 1918, when another epidemic occurred due to a virus that was HswNl (antigenically similar to swine flu). This virus was in turn responsible for most of the influenza until 1933, when a new type known as H0Nl became predominant.

The influenza A viruses that have appeared in the last 100 years have five significantly different hemagglutinins (H0, H1, H2, H3, and Hsw) and two significantly different neuraminidases (N1 and N2). Viruses with different combinations of these surface antigens are believed to reside in animal and human populations by causing mild disease in immunized individuals and virulent infections in susceptible individuals. Previously uninfected individuals may have little or no immunity to infectious agents that are foreign to them, and so the symptoms caused by the influenza virus are severe. Similarly, sick or old individuals frequently have a poorly functioning immune system and consequently are not protected against an influenza virus. The infection in sick or old individuals may be severe enough to cause their death. When the antibodies to surface antigens decrease in a population, the virus with these surface antigens is able to infect, multiply, and spread from person to person. An epidemic occurs when large numbers of individuals are no longer immune to a particular virus.

Influenza B viruses also undergo changes in their envelope proteins but these changes are not as extreme or as frequent as for influenza A. Since influenza B surface antigens show a constant, slow antigenic drift, they do not cause major epidemics.

Morphologically, influenza C differs from A and B by showing a hexagonal surface pattern and by having a large core. Since influenza C surface antigens do not change significantly, it does not cause epidemics. It is responsible for mild respiratory infections.

TABLE 11.4
SELECTED CANCER-CAUSING VIRUSES (ONCOVIRUSES)

Virus group	Oncovirus	Malignant tumors
Adenoviruses (DNA)	Adenovirus IV (12,18,31)	Hamster, rat, and mouse carcinomas
Herpesviruses (DNA)	Herpes simplex virus II (HSV)	Human carcinoma of the cervix
	Epstein-Barr virus (EBV)	Human B-cell leukemia (Burkitt lymphoma)
		Human nasopharyngeal carcinoma
	Herpes Saimiri virus (HSaV)	Monkey T-cell leukemia and reticular cell sarcomas
	Marek disease virus (MDV)	Chicken T-cell luekemia
Papovaviruses (DNA)	Polyoma virus	Mouse leukemia
	Simian virus 40 (SV-40)	Monkey and hamster sarcomas
	Papiloma virus	Rabbit papilloma of the skin
Retroviruses (RNA)	Murine mammary tumor virus (MMTV) Group B	Mouse adenocarcinoma of the mammary
	Rous sarcoma virus (RSV) Group C	Chicken sarcoma
	Murine leukemia virus (MLV) Group C	Mouse leukemia
	Human T-cell lymphotrophic viruses I, II (HTLV-I, HTLV-II) Group C	Human T-cell leukemia
	Feline T-cell leukemia virus (FTLV) Group C	Cat T-cell leukemia

Oncogenic Viruses

Oncogenic viruses cause cancers and tumors in animals (table 11.4). A tumor is a large mass of cells that reproduce abnormally. If the tumor is encapsulated by a layer of cells or by a basement membrane, it is said to be benign. If, however, cells from the tumor spread throughout the body, the tumor is malignant. Cancer is a synonym for a malignant tumor.

DNA viruses A number of DNA viruses can cause cancers. For example, there are several herpes viruses that cause carcinomas and sarcomas in animals. **Carcinomas** are derived from the innermost (endoderm) and outermost (ectoderm) embryonic layers of cells, while **sarcomas** develop from the middle (mesoderm) layer of embryonic cells (table 11.4). The papovaviruses are responsible for sarcomas and carcinomas in animals and in humans, while the adenoviruses and poxviruses generally cause cancers only in animals. Human T-cell lymphotrophic viruses I and II (HTLV-I and HTLV-II) are responsible for two types of leukemia in humans. Retroviruses are the only RNA viruses that are able to cause cancers and tumors in animals and humans.

Retroviruses The retroviruses are RNA viruses that are single-stranded. The genome is unusual in that it is a dimer of two identical RNAs. The RNA is long enough to code for 10 average-size proteins. It is not known exactly how the ss-RNA is coiled within the virus particle, but it is found in association with a **nucleoprotein** and a **reverse transcriptase** (RNA-dependent DNA polymerase). The genome is coated by a capsid protein that is reported to form an icosahedron in some viruses but a cylinder in others. The nucleocapsid (RNA, RNA-associated proteins, and capsid) is covered by an outer protein coat, which in turn is surrounded by an envelope derived from the host cell (fig. 11.10).

The retroviruses adsorb to specific host cell receptors by their glycoprotein knobs. The host cell membrane and the virus envelope subsequently fuse. Once the nucleocapsid enters the cell, the RNA genome is uncoated and the reverse transcriptase begins to polymerize a complementary DNA. Figure 11.10a outlines the reproductive cycle of a retrovirus.

After the double-stranded DNA genome is synthesized, it circularizes and moves into the nucleus. Twenty-four hours after infection, some of the DNA molecules may have integrated into host genomes, becoming **proviruses**. Messenger RNA and progeny RNA are made by transcription of the intergrated proviruses.

Retroviruses with a transforming ability can induce tumors and cancers. The genes that promote transformation and tumors are designated *onc* for **oncogenic genes**. There are at least six classes of oncogenes. These oncogenes usually code for mutated proteins such as protein kinases, hormones, hormone receptors, GTP-binding proteins, and DNA-binding proteins. Each of these proteins is important in the normal functioning of the cell. The following example illustrates the affect an oncogene can have on a cell. An altered **protein kinase** has been identified in mouse sarcoma and in avian sarcoma viruses, which transformed chicken and hamster cells. It is thought that the phosphorylation of Na^+/K^+ – ATPase pumps by the protein kinase disrupts the ionic balance and wastes ATP. The waste of ATP is thought to result in a more active glycolysis. In addition, phosphorylation of membrane proteins that affect cytoskeleton attachment to the plasma membrane can eliminate **contact inhibition** and alter the cell cycle. Contact inhibition is the inhibition of cell division and movement that occurs when normal animal cells come in contact with each other. Cells that do not show contact inhibition can spread through an animal and may represent precancerous or cancerous cells.

Cells transformed with oncogenes often show one or more of the following characteristics: (1) loss of contact inhibition; (2) random orientation in culture; (3) change in chromosome number; (4) infinite number of multiplications; (5) capacity to produce cancer in animals; (6) shorter generation time.

Normal cells are capable of a limited number of generations and are contact-inhibited. Some transformed cells called **cell lines** are able to proliferate indefinitely, but are not cancerous because their multiplication ceases when they come into contact with other cells. Such cells are said to be contact-inhibited. On the other hand, a transformed cell line that reproduces indefinitely and is not contact-inhibited is usually a **cancerous cell**. When the tumor or cancer cells spread through an animal's body, the cancer is said to have **metastasized**. Sarcomas usually spread via the blood system, while carcinomas generally spread through the lymphatic system. A metastasized cancer usually kills the animal.

GROWING VIRUSES IN THE LABORATORY

To study viruses, large quantities of the infectious particles are necessary. This generally means that they must be cultivated in the laboratory. With high concentrations of viruses, it is possible to make detailed studies of their chemistry and biology. The animal and plant viruses, for the most part, are the most difficult organism to cultivate and purify because they require a live host for propagation and often only a few viruses are formed. Of all the viruses, the bacteriophages are the easiest to propagate and purify because they multiply in bacteria that are generally easy to grow and because large numbers of viral progeny are formed.

Bacteriophages

A phage such as P1 can be grown on lawns of *E. coli* proliferating in a layer of soft agar (fig. 11.2). Phage P1 goes through a number of lytic cycles, destroying most of the lawn in about 10 hours. The soft agar containing the phage and surviving bacteria is scraped into a centrifuge tube and shaken with a few drops of chloroform to kill the bacteria. Then the tube is centrifuged to sediment the bacterial debris. The phage remain in the supernatant fluid. The phage concentration, or **titer**, can be determined by mixing dilutions of the phage sample with fresh *E. coli* and soft agar and plating the mixture onto nutrient agar plates. The plaques (virus colonies) that appear in the bacterial lawn 6 to 8 hours later are counted. The number of phages in the undiluted lysate is determined by

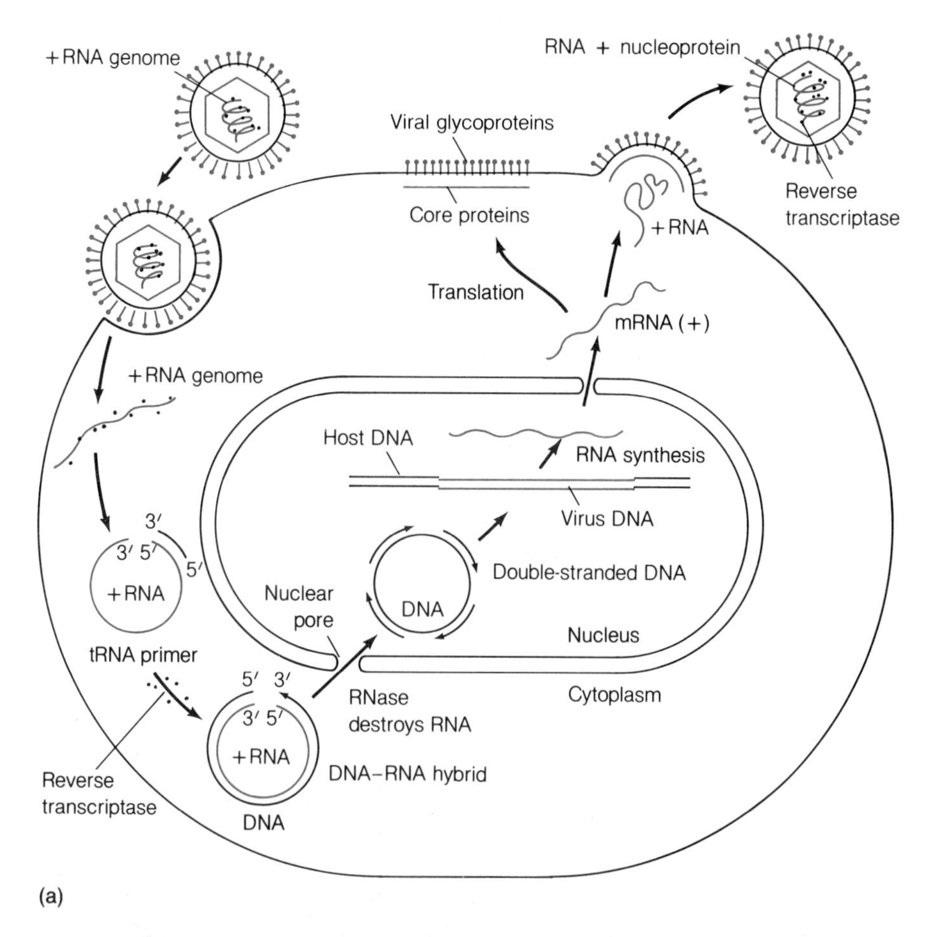

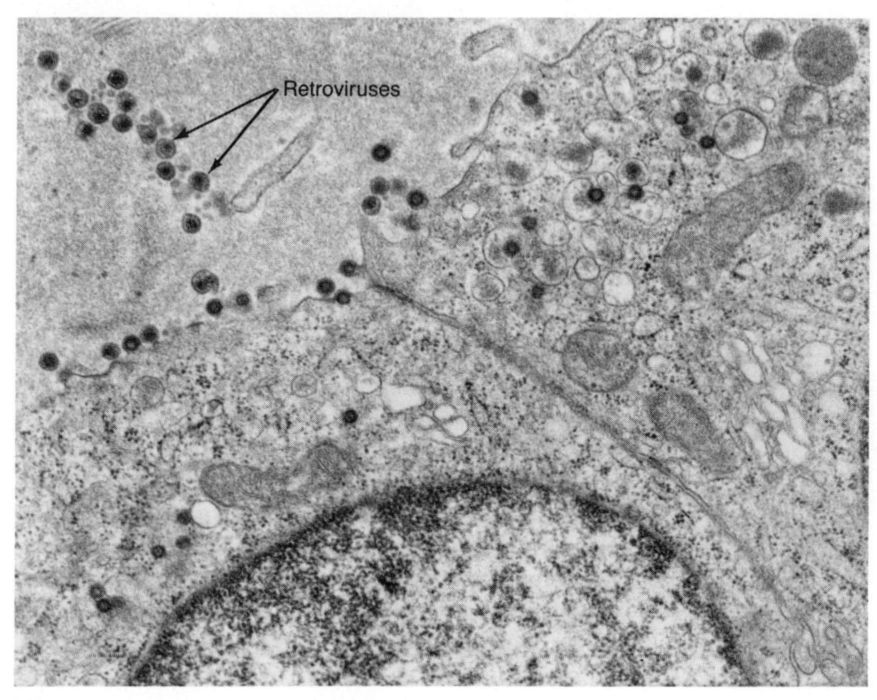

Figure 11.10 Retrovirus infection. *(a)* Retroviruses are enveloped RNA viruses that infect animals. The retroviruses are of interest because they frequently make a DNA copy of their RNA genome and integrate into host chromosomes. In addition, a number of retroviruses carry genes that can transform their host cells into cancerous cells. Retroviruses enter host cells by pinocytosis. The envelope is removed when it fuses with the pinocytotic vescle. After the RNA genome is uncoated, a virus reverse transcriptase makes a DNA copy of the virus RNA. If a lytic cycle occurs, virus capsid protein and RNA associate with the plasma membrane. The virus forms as the plasma membrane undergoes exocytosis. *(b)* C-type retroviruses, exiting the host's cell.

multiplying the plaque number by the inverse of the dilution. The phage titers generally range from 10^8 to 10^{11} phage/ml of phage solution. The phage solution can be stored for many months in the refrigerator without a significant loss of infectivity.

Plant Viruses

Plant viruses are generally grown on living host plants rather than in cell culture. For example, tobacco mosaic virus (TMV) is grown to high titers by rubbing a solution of these viruses into the surface of tobacco leaves attached to healthy plants (fig. 11.2). The rubbing breaks some of the cell walls and allows the viruses to infect. Within two weeks, the plant leaves develop visible lesions that resemble plaques. The TMV can be harvested by grinding the infected leaves with a small amount of water. The grinding releases the TMV into the water.

Animal Viruses

The isolation and identification of viruses in clinical specimens is often accomplished by growing the viruses in an appropriate host. In addition, the production of large titers of animal viruses is of value in making virus vaccines. Animal viruses are generally studied and manipulated by growing them in animal hosts, in eggs, or in cell (tissue) culture.

Bird embryos (fertilized eggs) have been very useful for cultivating a number of animal viruses (fig. 11.11). Five- to ten-day-old duck or chicken embryos developing within their eggs are generally used for growing viruses. Viruses are introduced with a hypodermic needle into the allantoic cavity, chorioallantoic membrane, yolk sac, amniotic cavity, or embryo. The egg is incubated for as long as it takes the viruses to

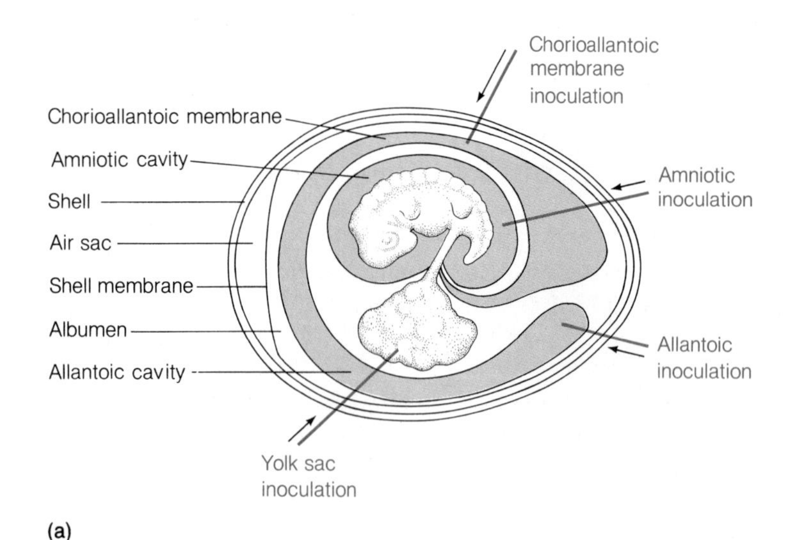

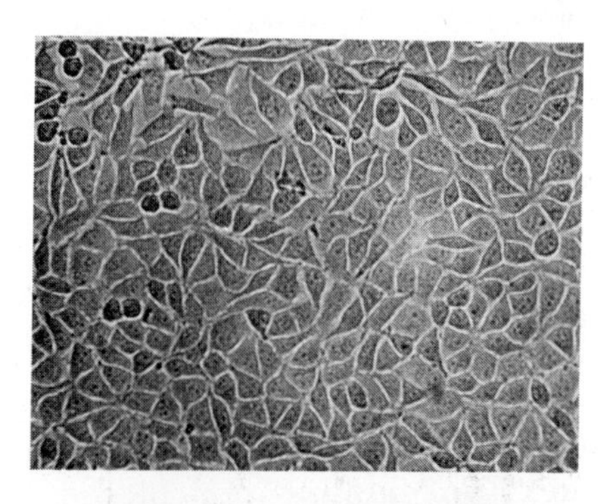

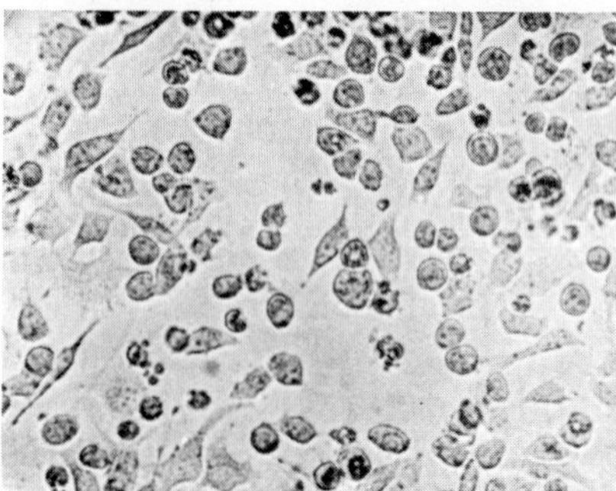

Figure 11.11 **Growing animal viruses** *(a)* Some animal viruses taken from infected tissue are grown on various tissues in chicken eggs. *(b)* Most animal viruses are now proliferated on cells growing in cultures. Mouse L-cells are shown growing in a monolayer. The cells are flattened against the wall of the growth chamber and have a spindle shape. 24 hours after the mouse L-cells have been infected with vesicular stomatitis virus, most of the cells have been transformed. The transformed cells are no longer attached to the wall of the growth chamber and they have become spherical.

reproduce. Inoculation of bird embryos has been used to prepare vaccines against the viruses that cause rabies, poliomyelitis, influenza, smallpox, and yellow fever. Presently, rabies and polio viruses for vaccines are grown in human cell cultures rather than in eggs.

Cell cultures are being used with increasing frequency to grow animal viruses (fig. 11.11). Cell cultures are often started from fresh animal tissue that is separated into individual cells by mechanical and enzymatic treatments. The cells are placed into the bottom of petri dishes, covered with an appropriate growth medium, and cultured in a high CO_2 environment. The cells attach themselves to the plastic petri dishes and grow into a monolayer. Viruses are added to the multiplying animal cells. The destruction or transformation of the cells is referred to as the cytopathic effect (CPE). Not all animal viruses demonstrate cytopathic changes. For example, influenza viruses produce little or no visible change in the cultured cells. Influenza viruses can be detected, however, because red blood cells added to infected cells bind to the surface of the cells.

Unfortunately, cell cultures started from fresh animal tissue do not last very long because the cells multiply only a few times before they die. This means that fresh tissue must be prepared frequently to grow viruses. The longest lasting **primary cell cultures** are derived from human embryo tissue, in which the cells are able to multiply between 50 and 100 times. Occasionally, primary cell cultures become transformed and are able to multiply indefinitely. Transformed cell cultures that do not have a limited number of generations are called **continuous cell lines**. A number of cell lines that are used to grow viruses have been derived from cancerous tissue. The most notable cell line derived from a human tumor is the **HeLa** cell line.

VIROIDS AND PRIONS

Viroids

Viroids are pieces of RNA found in the nucleus of infected plant cells which are able to replicate and spread from one cell to another within the host plant. These pieces of RNA are capable of extensive folding (secondary structure), and consequently form a structure with multiple hairpin loops. The viroids exist as naked ribonucleic acids (fig. 11.12) and are not enclosed in a protein coat. For example, **potato spindle tuber viroids** are single-stranded circular RNAs, 359 bases long, with extensive secondary structure. In electron micrographs, they have the appearance of rods approximately 50 nm long. Viroids differ in the nucleotide sequences of their RNAs. For example, potato spindle tuber viroid and **citrus exocortis viroid** have very different nucleotide sequences, while various strains of potato spindle tuber viroid differ at only a few sites. Analyses of the viroid nucleotide sequences indicate that the nucleotides do not code for proteins. Consequently, it is believed that the viroid RNA does not code for proteins that catalyze its replication. Instead, it is thought that the host provides the necessary RNA polymerase. So far, viroids have not been discovered in animals, the lower eukaryotes, or bacteria, but there is no reason to believe that they will not be found in all these organisms.

Viroids have attracted attention because they cause a number of plant diseases and because they are one of the smallest and simplest biological entities known to proliferate within cells and to spread from cell to cell. Viroids are responsible for a number of plant diseases (table 11.5). The most notable of the diseases are **cadang-cadang** of coconuts, which has killed 12 million coconut trees in the Phillipines, and **potato spindle tuber disease**, which destroys more than $3.5 million worth of potatoes in the United States each year. In the 1950s, a viroid nearly wiped out the chrysanthemum industry in the United States and another disrupted efforts to graft orange and lemon trees to a virus-resistant root stock.

Prions

At one time, scientists believed that "slow viruses" or possibly "DNA viroids" caused several slowly developing brain diseases, such as kuru, Creutzfeldt-Jakob disease, and Gerstmann-Straussler syndrome in humans; scrapie in sheep and goats; chronic wasting disease in mule deer and elk; and transmissible encephalopathy in mink. The slow development of the nervous system diseases suggested to most scientists that slow viruses were involved. On the other hand, the belief in a "DNA viroid" was based on early experiments that indicated the scrapie infectious agent was DNA.

Research from a number of laboratories now indicates that the infectious agent that causes scrapie and the other diseases is a protein called prion (fig. 11.13). Prion protein (PrP) is found as filaments in the brain cells of diseased animals. Researchers are now working on the problems of how PrP reproduces and causes disease. The answer as to how PrP proliferates is slowly

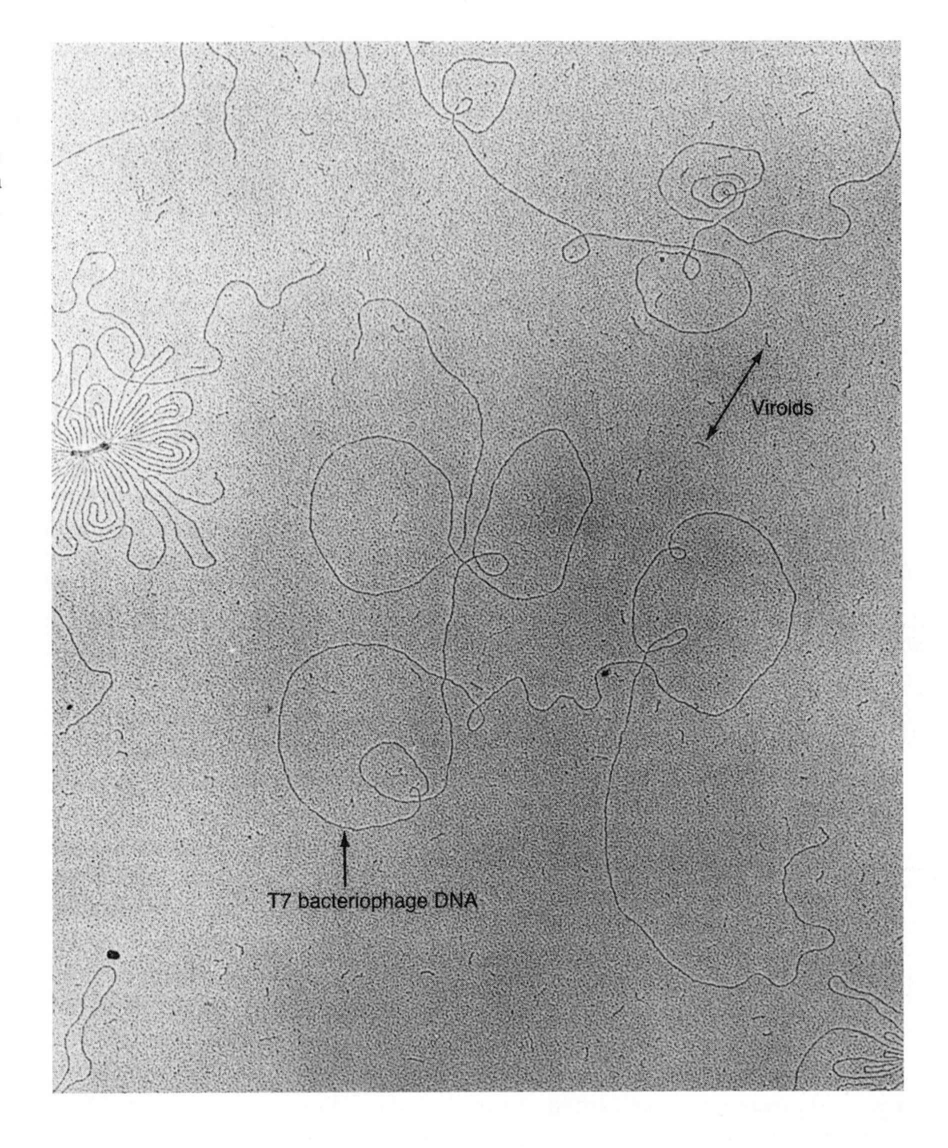

Figure 11.12 Viroid. The potato spindle tuber viroid (arrows) consists of 359 ribonucleotides. The viroid does not appear to code for a protein, since there are no start or termination codons in the viroid RNA. T7 bacteriophage DNA is included in the TEM to show relative sizes.

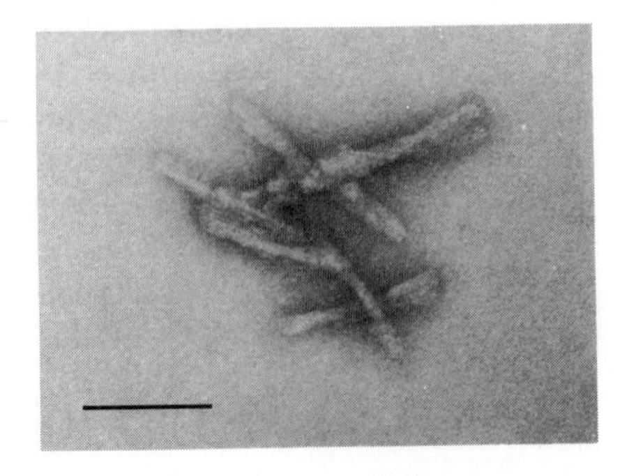

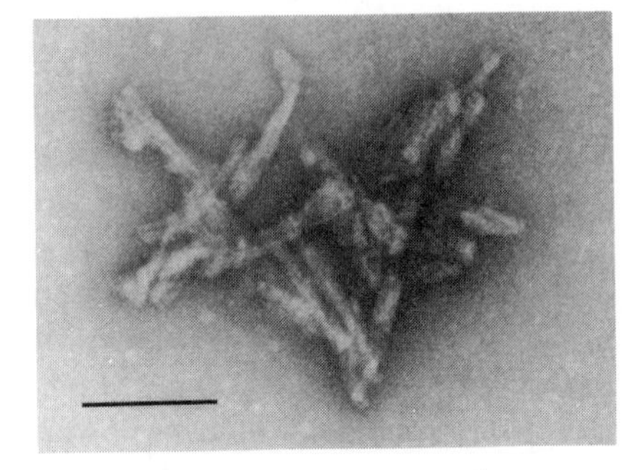

Figure 11.13 Prion. Transmission electron photomicrographs of extensively purified fractions of prions. Bars are 100 nm. Prions are negatively stained with uranyl formate.

TABLE 11.5
THE VIROIDS

Disease	Viroid
Potato spindle tuber disease	Potato spindle tuber viroid (PSTV)
Tomato bunchy top disease	Tomato bunchy top viroid (TBYV)
Citrus exocortis disease	Citrus exocortis viroid (CEV)
Chrysanthemum stunt disease	Chrysanthemum stunt viroid (CSV)
Chrysanthemum chlorotic mottle disease	Chrysanthemum chlorotic mottle viroid (ChCMV)
Coconut cadang-cadang disease	Coconut cadang-cadang viroid (CCCV)
Cucumber pale fruit disease	Cucumber pale fruit viroid (CPFV)
Hop stunt disease	Hop stunt viroid (HSV)

emerging. It has been found that a protein closely related to PrP is coded for by a gene found in the brain cells of normal and diseased animals. The prion-related gene is expressed in healthy animals as indicated by the discovery of mRNA for a prion-related protein. The prion-related protein, if produced in normal tissue, does not form filaments or cause disease.

Presently, it is unclear how the prion acts as an infectious agent. Some scientists have speculated that PrP comes from prion-related protein that may be modified by PrP. This means that PrP must be present initially to cause the formation of more PrP. Unlike prion-related protein, PrP is resistant to proteinase K and appears to be much more stable. The stability of PrP may lead to its accumulation in brain cells, the formation of aggregates and filaments, and the disruption of infected cells.

SUMMARY

CLASSIFICATION OF VIRUSES

1. Bacteriophages are classified on the basis of their structure or shape and type of nucleic acid. Most bateriophages have ds-DNA, ss-DNA, or ss-RNA. Almost all bacteriophages can be characterized as complex, icosahedral, or cylindrical. A few bacteriophages are enveloped (MVL2, ϕ6). In general, the complex viruses have large ds-DNA genomes that code for 50–200 proteins, depending upon the phage. The isosahedral and cylindrical phages are very simple and usually have small ss-RNA, ss-DNA, or ds-RNA genomes that code for 3–5 proteins.
2. Plant viruses are generally classified on the basis of their structure and nucleic acids. Most plant viruses contain ss-RNA. Most are icosahedral and cylindrical.
3. Animal viruses are generally classified on the basis of their structure and nucleic acids. Most viruses contain ss-RNA or ds-DNA and are icosahedral. Many of the animal viruses are enveloped.

REPLICATION OF BACTERIOPHAGES

1. The phage genetic information is known as the genome; the protein coat that directly covers the genome is referred to as the capsid. The individual proteins that make up the capsid are called capsomeres.
2. The infection of a bacterium by a bacteriophage is divided into steps: viral attachment (adsorption); viral genome penetration; viral RNA synthesis, protein synthesis, and genome replication; viral assembly; and viral release.
3. T4 is a complex ds-DNA bacteriophage that undergoes a lytic cycle of infection when it invades *E. coli*.
4. Lambda is a complex ds-DNA bacteriophage capable of both a lytic infection and a lysogenic infection.

REPLICATION OF ANIMAL VIRUSES

1. The replication of animal viruses can be divided into a number of steps: (a) adsorption to the host cell's plasma membrane; (b) penetration of the virus into the cell's cytoplasm; (c) uncoating of the viral genome; (d) viral RNA synthesis, protein synthesis, and genome replication; (e) viral assembly and release.
2. The host range of an animal virus is determined by its ability to attach to specific macromolecules in the plasma membranes of the different host cells.

3. In contrast to bacteriophages, most of the protein associated with animal viruses enters the host cell.
4. The viral capsid depolymerizes within host vesicles.
5. Because of the different types of nucleic acids (+RNA, −RNA, ds-DNA, etc.) associated with animal viruses, the ways in which RNA synthesis and genome replication occur is varied.
6. The assembly of naked viruses and enveloped viruses differs significantly. In the case of most naked viruses, capsid proteins associate with viral nucleic acids in the cytoplasm. In the case of many enveloped viruses, the nucleic acid associates with viral proteins in the host's plasma membrane and the virus forms as the nucleic acid-protein-membrane complex buds from the plasma membrane. An example of this type of virus is the retrovirus responsible for AIDS. In some enveloped viruses, however, the nucleic acid first associates with capsid proteins and then the nucleocapsid associates with a membrane, either endoplasmic reticulum or plasma membrane. An example of this type of virus is the herpes virus.

REPRESENTATIVE ANIMAL VIRUSES

1. Herpes viruses include those that are responsible for cold sores, chicken pox, shingles, and mononucleosis. These viruses are enveloped ds-DNA viruses.
2. Orthomyxoviruses are responsible for influenza. These viruses are enveloped ss-RNA viruses with a −RNA genome. This means that mRNA must be synthesized before viral protein can be produced. When orthomyxoviruses undergo major changes in their two major surface proteins (hemaglutinin and neuraminidase), they often cause epidemics. When minor changes occur in one of the major surface proteins, the viruses are said to undergo antigenic drift. Antigenic drift generally does not result in viruses that are different enough to cause epidemics.
3. Oncogenic viruses are those that are able to cause cancer. There are examples of both DNA and RNA viruses that cause cancer in animals and humans. Human T-cell lymphotrophic viruses-I and II are known to cause different types of leukemia in humans. There is substantial evidence that *Herpes simplex-II* is responsible for human carcinoma of the cervix.

GROWING VIRUSES IN THE LABORATORY

1. Bacteriophages are generally cultured on a lawn of bacteria growing in soft agar. The bacteriophages can be isolated by centrifuging the lysed bacteria and soft agar. The bacteriophages are found in the supernatant fluid.
2. Plant viruses are usually cultured on plants that support the growth of the viruses. The viruses are isolated by grinding up the affected plant tissue and centrifuging the plant debris. The plant viruses are found in the supernatant fluid.
3. Animal viruses may be grown in experimental animals, egg embryos, and cultured cells.

VIROIDS AND PRIONS

1. Viroids are pieces of RNA that cause a number of plant diseases. These pieces of RNA are able to replicate and spread from one cell to another within the host plant. In addition, they can spread from plant to plant through insect vectors or contaminated equipment.
2. Viroids are approximately 360 nucleotides long (about one-third the size of an average gene) and their nucleotide sequence does not appear to code for a protein, since there are no start and stop codons in phase with each other.
3. Viroids do not develop capsids, nor do they become enveloped. Thus, viroids appear to be fundamentally different from viruses.
4. Prions are a new type of infectious agent that is extremely resistant to physical and chemical treatments that inactivate bacteria, viruses, and viroids. Prions are small hydrophobic particles that appear to consist only of protein.

STUDY QUESTIONS

1. What are viruses and how are they classified?
2. What kinds of nucleic acids do bacterial, plant, and animal viruses have?
3. How do viruses differ from cellular organisms?
4. Discuss the lytic cycle of T4.
5. Discuss the lytic cycle and the lysogenic cycle of lambda.
6. What is the lambda genome called when it integrates into the host's genome?

7. What is the host cell called when it contains an integrated viral genome?
8. Compare and contrast a bacteriophage infection with an animal virus infection.
9. Describe a herpes virus infection.
10. Describe an orthomyxovirus infection.
11. Explain what an oncovirus is.
12. Compare and contrast the growth of bacteriophages, plant viruses, and animal viruses in the laboratory.
13. Compare and contrast a viroid with a virus.
14. What is a prion?

SELF-TEST

1. Pathogens that have multiple forms and consequently cannot usually be controlled by a single vaccine may be controlled by:
   a. a genetically engineered vaccinia virus;
   b. a vaccine made of bacteriophage;
   c. antibiotics produced in a genetically engineered vaccinia virus;
   d. a vaccine made from a mixture of different viruses;
   e. none of the above.
2. The individual proteins that make up a viral protein coat are called:
   a. spikes;
   b. envelope proteins;
   c. capsomeres;
   d. nucleocapsids;
   e. virions.
3. A well-isolated plaque on a bacterial lawn is generally assumed to have developed from a single viral particle. a. true; b. false.
4. A plaque may be considered a viral colony. a. true; b. false.
5. If there are 25 well-isolated viral lesions in an animal cell culture of a billion cells, how many viruses are there in the cell culture?
   a. 25;
   b. a billion;
   c. It depends upon how many viruses are released from each cell that was lysed in each of the viral lesions; thus, no definite answer can be given.
6. The smallest RNA viruses generally have how many genes?
   a. 1;
   b. 3;
   c. 20;
   d. 50;
   e. 100.
7. The pox viruses are the largest DNA viruses known. Approximately how many genes do they have?
   a. 40;
   b. 400;
   c. 4000;
   d. 40,000;
   e. 400,000.

For items 8–10, match the description with the infectious agent:
   a. prion;
   b. viroid;
   c. naked virus;
   d. enveloped virus;
   e. unable to match.

8. An infectious agent that is pure RNA.
9. An infectious agent that consists of a nucleic acid, a capsid, and a membrane.
10. An infectious agent that is pure protein.
11. The only part of bacteriophage T4 to enter a host cell is its nucleic acid. a. true; b. false.
12. The only part of the herpes virus to enter a host cell is its nucleic acid. a. true; b. false.
13. What are prophages or proviruses?
   a. infectious agents that are simpler than viruses such as viroids;
   b. the empty capsids on the surface of a host cell;
   c. cells that are infected by viruses;
   d. viral genomes that are integrated into the host's genome;
   e. viral genomes that function like independent plasmids.
14. What is a typical length of time for the lytic cycle of bacteriophage T4?
   a. 3 minutes;
   b. 30 minutes;
   c. 3 hours;
   d. 30 hours;
   e. 1 day.
15. If a single T4 bacteriophage infects a cell, what is a typical number of phage particles produced at the end of one lytic cycle?
   a. 2;
   b. 4;
   c. 100;
   d. a million;
   e. more than a billion.
16. If a single virus genome lysogenizes a bacterium how many lysogenic bacteria will there be after the bacte-

rium undergoes five generations?

a. 1;
b. 5;
c. 32;
d. 500;
e. more than a billion.

SUPPLEMENTAL READINGS

Diener, T. O. 1982. Viroids: Minimal biological systems. *BioScience* 32(1):38–44.

Gallo, R. C. 1987. The AIDS virus. *Scientific American 256(1)*:46–56.

Hendrix, R. W., Roberts, J. W., Stahl, F. W., and Weisberg, R. A. 1983. *Lambda II*. New York: Cold Spring Harbor.

Karpas, A. 1982. Viruses and leukemia. *American Scientist* 70:277–28.

Mathews, C. K., Kotler, E. H., Mosig, G., and Beget, P. B. 1983. *Bacteriophage T4*. Washington, DC: American Society for Microbiology.

Merz, P., Sommerville, R., Wisniewski, H., Manuelidis, L., and Manuelidis, E. 1983. Scrapie-associated fibrils in Creutzfeldt-Jakob disease. *Nature* 306:473–476.

Muesing, M. A., Smith, D. H., Cabradilla, C. D., Benton, C. V., Lasky, L. A., and Capon, D. J. 1985. Nucleic acid structure and expression of the human AIDS/Lymphoadenopathy retrovirus. *Nature* 313:450–457.

Prusiner, S. 1984. Prions. *Scientific American* 251:50–59.

Stuart-Harris 1981. The epidemiology and prevention of influenza. *American Scientists* 69:166–172.

Tiollais, P., Charnay, P., and Vyas, G., 1981. Biology of hepatitis B virus. *Science* 213:406–411.

Varmus, H. 1987. Reverse transcription. *Scientific American* 257(3):56–64.

PART 3

MEDICAL MICROBIOLOGY

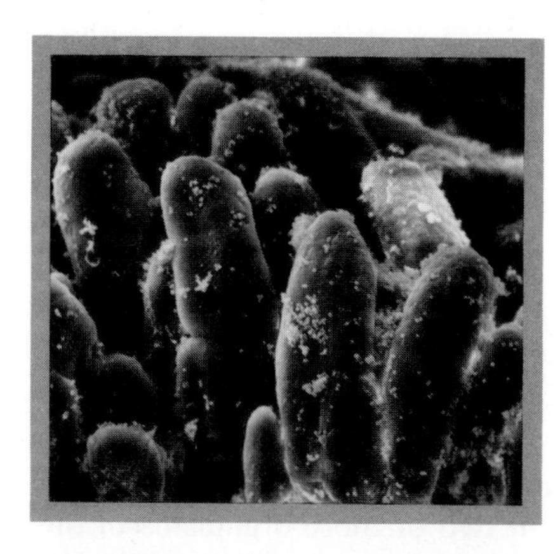

CHAPTER 12

DETERMINANTS OF HEALTH AND DISEASE

CHAPTER PREVIEW LEPROSY OR HANSEN'S DISEASE

Hansen's disease is an infectious disease that disfigures and cripples many millions of people throughout the world. In ancient times, lepers were banished from their villages and towns and forced to roam the countryside begging until they became crippled or died of other diseases. Leprosy evoked such terror in people that many lepers were forced to wear bells so that everyone would know that a leper was near. Even in this century, before the development of antibiotics and drugs for treating leprosy, the disease was so feared that those unfortunate enough to be afflicted by it were ostracized from society and placed in "leper colonies" or locked away for life in institutions or on remote islands. As recently as the 1930s in the United States, leprosy patients at the U.S. Public Health Service hospital in Carville, Louisiana, were restricted to the institution and not allowed to marry or even vote. In many parts of the world, a person with leprosy is still excluded from village life and rejected by family members because of fear of the disease.

There are between 11 and 15 million lepers in the world, about 4000 of whom are in the United States. Although 4000 is a rather small fraction of the leper population, it represents a 500% increase in the incidence of leprosy in the United States since 1960.

Leprosy is caused by the bacterium *Mycobacterium leprae,* a close relative of the organism that causes tuberculosis. Leprosy is contracted by intimate contact with infected individuals or their possessions. Sometimes it may take 15–20 years for the disease to manifest itself.

Leprosy can occur as a mild disease called **tuberculoid** leprosy, characterized by a few inconspicuous lesions, or as a severe disease, called **lepromatous** leprosy, which is manifested by the appearance of numerous skin lesions on the ears, nose, and fingers. The nerves may also be affected resulting in feeling of numbness around the lesions. This often leads to accidental loss of digits, skin, etc. As the lepromatous leprosy develops, nerves are destroyed and the individual becomes crippled. One notable difference between the two forms of leprosy is that lesions of tuberculoid leprosy contain very few bacteria, while those of lepromatous leprosy contain many.

Scientists have discovered that a human host who reacts promptly and efficiently with an immune response will develop the less severe form of the disease. Those who have a sluggish or slow immune response develop the more severe form of the disease. The intensity of the immune response is believed to depend upon factors such as the genetic makeup and overall health of the host, as well as the number of infecting microorganisms and their ability to invade host cells.

Much is yet to be learned about leprosy. We presently know that the severity of the disease and the ultimate recovery of the patient depend upon the invasive properties of the parasite and the defense mechanisms that can be mustered to combat the invading microbe. This chapter illustrates the dynamic interplay between the disease-causing properties of microorganisms and host defense mechanisms.

CHAPTER OUTLINE

WITHIN HOURS AFTER BIRTH, WE ARE COLOnized by a variety of microorganisms from the environment. These microbes colonize all of our body surfaces and cavities, where they live, metabolize, and reproduce. Some populations of colonizing microorganisms become an integral part of our bodies and remain there for our entire lives. These microorganisms and their host become attuned to each other and live together in harmony. Besides these permanent residents, we are colonized by transient microorganisms. Some are harmless, while others have the potential to cause disease. These microorganisms may come from various sources, such as the air, water, food, and soil. Whether or not colonization by these microorganisms results in an infectious disease is a consequence of the genetic ability of the colonizer to cause detrimental changes in the host, the genetic characteristics of the host, and the host's defense mechanisms.

The realm of host–parasite interactions is extremely complicated and involves many factors. It is the sum total of these interactions that determines the outcome of microbial colonization. This chapter discusses some of the ways in which hosts and microorganisms interact in health and in disease.

SOME DEFINITIONS

Before we start discussing host–parasite interactions, it is necessary that we become familiar with a few terms that will serve as a foundation for the following discussion.

Symbiotic Relations

Many organisms have evolved so that they spend their lives together. These close associations are called **symbiotic relations**. The term **symbiosis** means "living together" and describes any interaction, more or less permanent, between two or more organisms of different species. For convenience purposes, biologists have subdivided symbiotic associations into three different types, based upon the impact that one of the members, or **symbiont**, has on the other symbiont.

Symbiotic associations that are mutually advantageous are called **mutualistic** associations. The associations that exist between many microorganisms and their human hosts are often mutualistic. The microorganisms are nourished by the host's secretions and surplus food, while the microoorganisms protect the host from invading microbes by competing for space and food.

Commensalistic associations are symbioses in which one symbiont derives a benefit from the association while the other is unaffected. The word commensalism is derived from a Latin word meaning "eating at the same table." Many microorganisms that are transient colonizers are called **commensals** because they feed upon their temporary host but generally do not cause any harm. When the host becomes debilitated by an illness or the resident flora has been killed after prolonged antibiotic use, some commensals can cause disease and are referred to as **opportunists**.

Parasitic associations are symbiotic relations in which one of the symbionts, the **parasite**, lives at the expense of the other symbiont, the **host**. The parasite is usually smaller than its host. The term parasite can be modified to define the location of the parasitic association. **Ectoparasites** parasitize the host's external surfaces. For example, the ectoparasitic mite *Sarcoptes* parasitizes the human skin, causing irritation and itching. The disease known as scabies is caused by this mite. **Endoparasites** parasitize internal body areas, such as organs and tissues. **Obligate parasites** are organisms that *must* lead a parasitic existence, while **facultative parasites** may obtain their nutrients either from a parasitic association or as a free-living form.

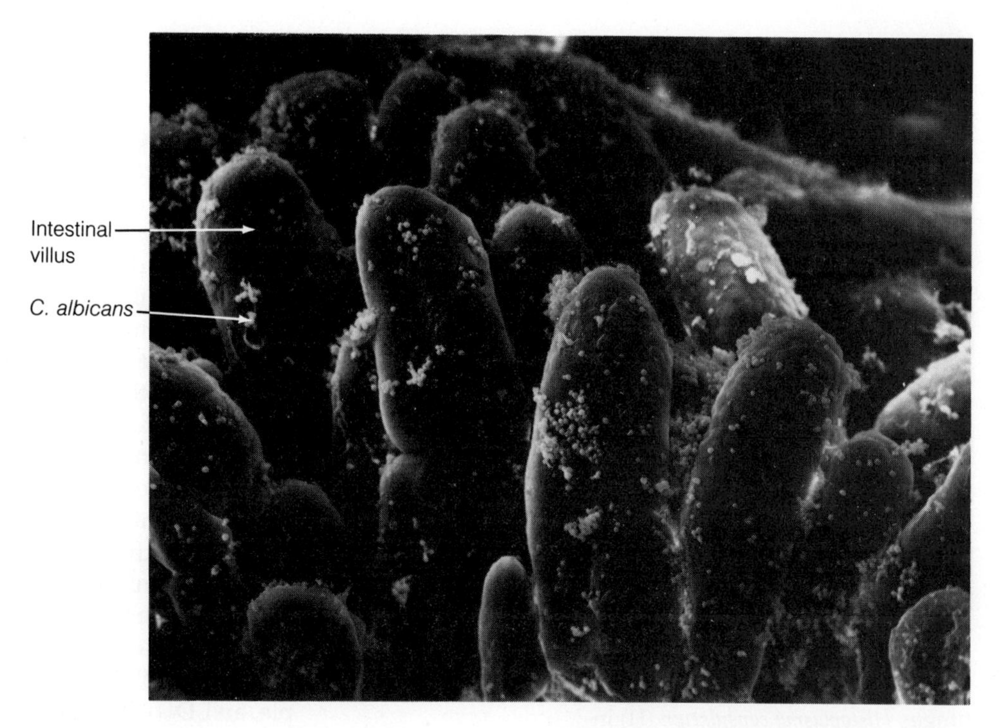

Figure 12.1 ***Candida albicans*** **colonizing intestinal villi.** Experimental infection of a mouse, in which *Candida albicans* (arrow) is reproducing on the surface of intestinal villi, as seen with a scanning electron microscope.

Some microorganisms are well suited for growth on the surface of tissues (fig. 12.1). For example, *Shigella dysenteriae* adheres to the intestinal epithelium and multiplies by using host materials as nutrients. As bacteria grow, they spread to adjacent sites. Some microorganisms invade cells and multiply inside them (fig. 12.2). These **intracellular parasites** can exhibit remarkable adaptations to life inside cells. Extracellular parasites generally cause acute diseases of short duration, such as sore throats, wound infections, and pneumonia, because they can be attacked and eliminated successfully by the host's immune system. Intracellular parasites, on the other hand, are protected from the host's defenses. Often, the intracellular parasite and the host achieve a balance that results in a chronic or long-lasting infection. Certain groups of intracellular parasites have evolved nutrient-gathering mechanisms that make them totally dependent upon their host for survival. These parasites are called **obligate intracellular parasites**. The rickettsiae, the chlamydias, the viruses, and some of the protozoa are examples of obligate intracellular parasites.

The symbiotic association between a parasite and its host may not always result in the ultimate destruction of the host. In fact, a "good parasite" does not kill its host, because in the process it would destroy itself and future generations that might spread to other hosts. In general, parasites tend to evolve mechanisms that ensure their survival within the host. Such mechanisms may involve reduction of irritation so that the host reacts less violently in the presence of the parasite. For the parasite to be able to obtain sufficient nutrients for growth and multiplication and yet not kill its host, both host and parasite must evolve together toward a state of mutual tolerance. Some of these parasitic conditions eventually become mutualistic ones.

Disease versus Infection

Disease is usually defined as a harmful alteration of the host's tissues or metabolic activities. Such an alteration is not always the result of an infection, for there are many noninfectious diseases, such as strokes and

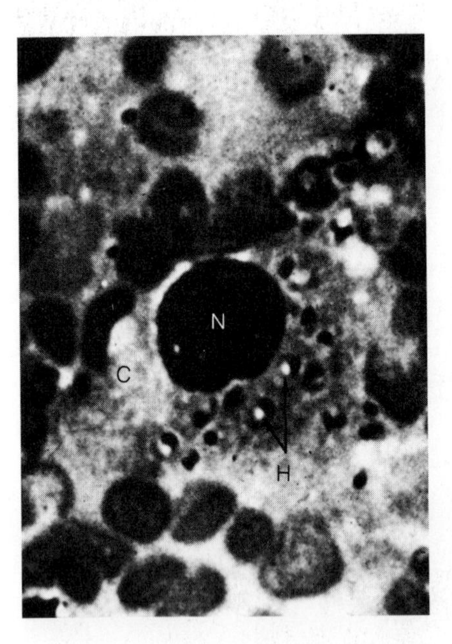

Figure 12.2 ***Histoplasma capsulatum*, an intracellular parasite.** The fungus *Histoplasma capsulatum* (H) invades macrophages and reproduces within them. The photograph illustrates a macrophage infected with reproducing yeast cells of this fungus. Macrophage nucleus (N) and cytoplasm (C) are indicated.

heart attacks. **Infection** is a term that denotes the establishment and proliferation of a microorganism within a host. This multiplication constitutes an infection; through common usage, however, this term is generally restricted to invasion of body organs and tissues by pathogenic (disease-causing) microorganisms. **Disease** is a deleterious alteration of the interaction between the individual and its environment that affects the well being of the host and results in signs and symptoms. The terms **infection** and **disease** should not be used interchangeably. If these terms were synonymous, we would constantly be ill, because we are frequently infected.

Pathogenicity versus Virulence

Pathogenic microorganisms are those microorganisms that have the genetic capacity to cause disease; that is, they produce metabolic products or cause tissue changes that are harmful to the host.

The fact that pathogenic microorganisms such as *Mycobacterium tuberculosis* or *Streptococcus pneumoniae* infect a host does not necessarily mean that the host will become ill. Not all infections by pathogens ultimately result in disease. Even when disease does occur, it may not be of the same severity in all individuals affected. For this reason, the term **virulence** describes the degree of pathogenicity or the **pathogenic potential** of a given microorganism. **Avirulent** strains of *S. pneumoniae* do not cause disease in a susceptible host, while **virulent** strains do. Nearly all pathogenic microorganisms exhibit wide variability in their virulence.

To aid in measuring the degree of virulence of microorganisms, the expressions $\mathbf{LD_{50}}$ and $\mathbf{ID_{50}}$ are used. LD_{50} denotes the *lethal dose* of microorganisms which will kill 50% of the tested population. For example, an LD_{50} of 150 for a bacterium such as *Streptococcus pneumoniae* might mean that a dose of 150 microorganisms per mouse will kill 50% of the mice in a population. A strain of *S. pneumoniae* with an LD_{50} of 25 is much more virulent than another strain with an LD_{50} of 1500. Similarly, an LD_{100} is the dose of microorganisms that will kill the entire population. Figure 12.3 illustrates how the LD_{50} of a hypothetical microorganism and its host is calculated. Along the same lines, the ID_{50} indicates the dose of a given microorganism which will *infect* 50% of a population. These values are extremely important in assessing the disease-causing potentials of pathogens and the impact that they may have on the well-being of a susceptible animal population.

THE INFECTIOUS PROCESS

For microorganisms to initiate an infection in a host, they must first reach those host tissues where microbial growth is favored. Many microorganisms find favorable environments in the respiratory tract and frequently penetrate the host from these tissues. Other microorganisms enter the host through the gastrointestinal and urogenital tracts, and still others penetrate through the skin.

Once a pathogen reaches a suitable site for growth inside a host, it may then become established and begin to reproduce. Many invasions, however, do not

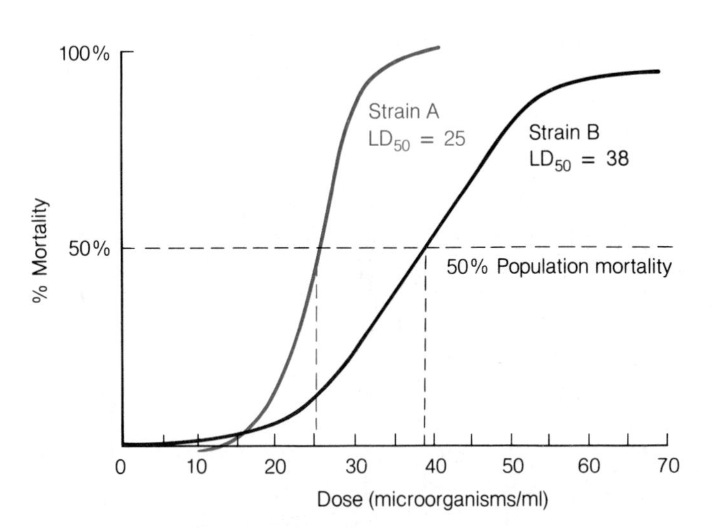

Figure 12.3 Determination of the LD_{50} of a pathogen. The LD_{50} of a population is determined by injecting various doses of the pathogen into members of the population. Deaths are recorded and a graph is constructed. The graph represents the susceptibility of a population to two different strains of the same species of pathogen: strain A is more virulent than strain B. The dose (microorganisms/ml) of pathogen A that killed 50% of the population is 25, while that of pathogen B is 38. Hence the LD_{50} of strain A is 25 and that of strain B is 38.

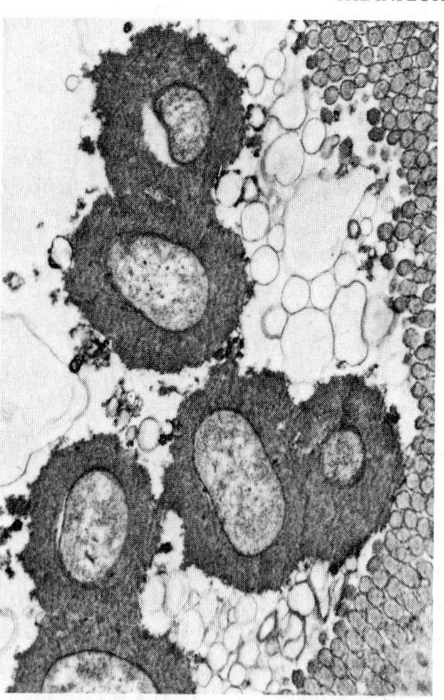

Figure 12.4 Microcolony of *Escherichia coli* enclosed in a glycocalyx. Transmission electron micrograph showing *E. coli* with a thick glycocalyx attached to the microvilli of a calf's ileum. The glycocalyx serves to attach the bacteria to solid surfaces and to shield them from host defense mechanisms.

result in infections. Consider, for example, that every time you inhale a breath of air or eat a morsel of food, hundreds of thousands of microorganisms find their way into your lungs or intestines. But this is as far as many of the invasions go because host defenses prevent the pathogens from becoming established. Similarly, many microorganisms find their way into wounds, but the probability of their initiating a serious infection generally is slight.

For a microorganism to cause an infection, it must have access to nutrients. Microorganisms may obtain nutrients by attaching to tissues. Researchers have established that the attachment often is tissue-specific and is mediated by surface characteristics of the microorganism. One well-studied example is the specific adherence of *Neisseria gonorrhoeae* to the genital tract epithelium. The adherence is mediated by pili, so pilated strains of *N. gonorrhoeae* can adhere to the epithelium and hence initiate an infection.

Adherence may also be due to a **glycocalyx** produced by many bacteria, fungi, and other microorganisms. The glycocalyx is a fibrous matrix of polysaccharide that is found on the surface of many cells. Recent investigations have revealed that, in nature, bacteria grow in glycocalyx-enclosed microcolonies (fig. 12.4). Apparently this covering affords the bacteria a means of binding and channeling nutrients into their cells, as well as a means of protection from phagocytosis by predatory organisms and from infectious virus particles. Certain eukaryotic parasites, such as hookworms, tapeworms, and flukes (fig. 12.5), have evolved structures for attaching to the host. Once a parasite has attached to a suitable nutritive surface and become established there, it can begin to multiply.

PATHOGENIC PROPERTIES OF MICROORGANISMS

Infectious diseases are those that result from a parasite's activities in a host and the host's response to the invading parasite. A microorganism alone, no matter how virulent, cannot cause disease in the absence of a suitable host. In this section we examine the factors that determine the disease-causing potentials of parasites.

Diseases caused by microorganisms can be classified into two distinct categories, based on the level of participation of the pathogen itself in the disease process. These categories are **intoxications** and **infections.** Intoxications are diseases that result from the entrance of a toxin into the body which causes disease in the absence of the toxin-producing microbe.

Infectious diseases result from the physico-chemical alterations of a host as a result of the parasite's growth and extension to adjacent tissues, and often includes the formation and release of toxins as well.

Toxigenicity

Many microorganisms, during their growth and reproduction, release an array of metabolic products. Others, as part of their anatomical makeup, have chemical components that are toxic. Some of these products can be extremely harmful to humans and can cause severe damage in minute quantities (table 12.1). Such noxious substances are called **toxins.** Some toxins interfere with specific physiologic processes, such as protein synthesis or nerve impulse conduction, while others interfere with body defenses.

Microbial toxins are classified into two categories: **exotoxins** (table 12.2) and **endotoxins.** In general, exotoxins are heat-labile (heat-sensitive) proteins produced by a specific microorganism and are not structural components of the cell. Exotoxins have specific and unique modes of action that differ from one species to another. For example, the disease com-

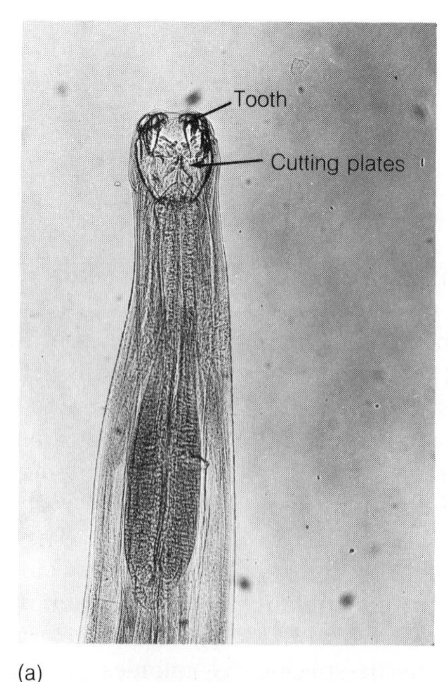

Figure 12.5 Attachment organs of some human-parasitic worms. *(a)* Photomicrograph of the head of a hookworm. The mouth has some prominent teeth that are used to attach to the intestinal epithelium, where the worms feed and reproduce. *(b)* Photomicrograph of a tapeworm. The scolex (head portion) has four suckers, which are used to adhere to the intestinal epithelium.

TABLE 12.1
INFECTIOUS DISEASES CAUSED BY BACTERIA OR FUNGI

Disease	Type* of disease	Microorganism responsible for disease	Factor(s) involved in disease	Mode of action
Anthrax	I/T	*Bacillus anthracis*	Invasive properties, toxin	Affects CNS
Botulism	T	*Clostridium botulinum*	Neurotoxins	Flaccid paralysis
Candidiasis	I	*Candida albicans*	Neuraminidase?, lipase?	
Cholera	T	*Vibrio cholerae*	Choleragen (exotoxin)	Alters intestinal permeability
Coccidioidomycosis	I	*Coccidioides immitis*	Invasive properties	
Diphtheria	T	*Corynebacterium diphtheriae*	Diphtheria toxin	Halts protein synthesis
Dysentery	I	*Shigella* sp.	Invasive properties	
Food poisoning	T	*Staphylococcus aureus*	Enterotoxins	Diarrhea/vomiting
Gas gangrene	I	*Clostridium perfringens*	Various enzymes	
Histoplasmosis	I	*Histoplasma capsulatum*	Invasive properties	
Pharyngotonsillitis	I	*Streptococcus pyogenes*	Various enzymes	
Plague	I/T	*Yersinia pestis*	Invasive enzymes plus plague toxin	Cytotoxic
Pyoderma	I	*Staphylococcus aureus*	Coagulase, other enzymes?	
Salmonellosis	I	*Salmonella* sp.	Invasive properties	
Scarlet fever	I	*Streptococcus pyogenes*	Erythrogenic toxin	Cytotoxic
Tetanus	T	*Clostridium tetani*	Neurotoxin	Spastic paralysis
Toxic shock syndrome	T	*Staphylococcus aureus*	Pyrogenic (fever-causing) exotoxin	Cytotoxic/shock
Tuberculosis	I	*Mycobacterium tuberculosis*	Invasive properties	
Typhoid	I	*Salmonella typhi*	Invasive properties	
Whooping cough	T/I	*Bordetella pertussis*	Various toxins	Cytotoxic

*I, disease caused primarily by invasion of tissue by multiplying bacteria; T, disease caused by ingestion or production of toxins at site of infection.

TABLE 12.2
CHARACTERISTICS OF EXOTOXINS AND ENDOTOXINS

Characteristics	Exotoxin	Endotoxin
Source of toxin	Synthesized by gram-positive and gram-negative as a result of metabolic activity	Gram-negative cells walls
Chemical composition	Proteinaceous materials	Lipopolysaccharides
Resistance to heat (100°C)	Heat sensitive	Heat stable
Toxicity	Extremely toxic in minute doses	Toxic at high doses only
Mode of action	Unique for each toxin	Generally the same regardless of source
Example of action	Spastic paralysis, flaccid paralysis, cytotoxicity, alter protein synthesis, plasma membrane permeability	Fever, shock, and weakness
Immunogenicity (induce immune response)	Very immunogenic	Weakly immunogeneic
Conversion to toxoid (detoxified with formalin)	Readily converted	Not converted

monly known as **botulism** is an intoxication caused by the ingestion of a toxin present in contaminated food. This toxin, called **botulin** or **botulinal toxin**, can cause the disease in the absence of *Clostridium botulinum*. Botulinal toxin typically causes flaccid paralysis because it inhibits the release of **acetylcholine** at the synaptic junction of motor nerve fibers, thus preventing the transmission of a nerve impulse to muscle cells. By contrast, the **diphtheria toxin** inhibits protein synthesis and therefore causes cell death. Another disease, **staphylococcal food poisoning**, results from the ingestions of **enterotoxin** produced by *Staphylococcus aureus*. Tetanus, cholera, and diphtheria result from the production of toxins by the microorganisms involved.

Endotoxins are toxic components of the cell wall of gram-negative bacteria (fig. 12.6). These toxins are not proteins, but **lipopolysaccharides**. All gram-negative bacteria possess endotoxins, which, unlike exotoxins, have similar modes of action (table 12.2): It is the lipid portion, **lipid A** (fig. 12.6), of the lipopolysaccharide which is responsible for the toxicity of the endotoxins. Because we do not normally develop an immunity against lipids, we never acquire immunity to the endotoxin of gram-negative bacteria. The characteristic signs and symptoms that are caused by

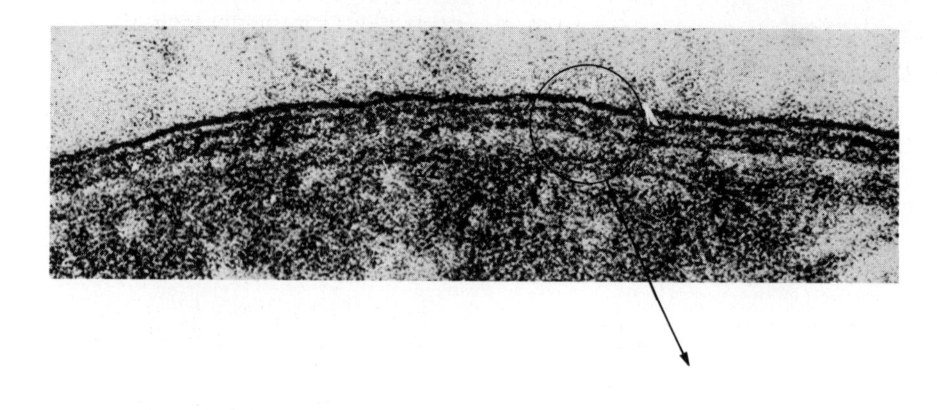

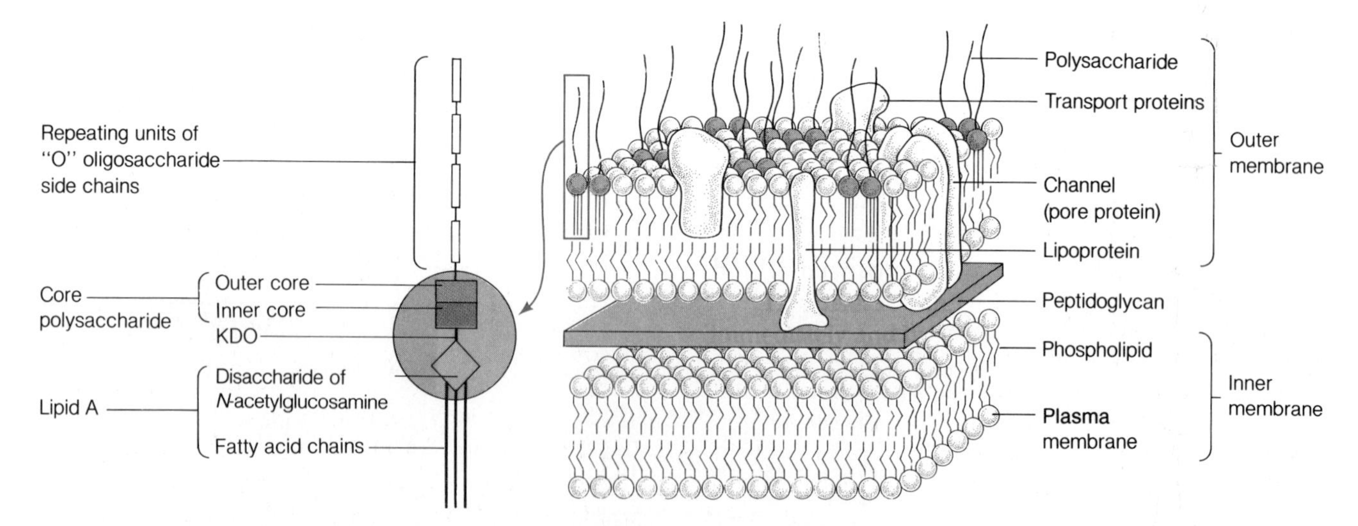

Figure 12.6 The structure and location of endotoxin. Endotoxin is part of the outer membrane of the gram-negative bacterial cell wall. The endotoxin is a lipopolysaccharide that consists of lipid A, a core polysaccharide, and "O"-specific polysaccharide side chains. The polysaccharide core is attached to lipid A by ketodeoxyoctonate (KDO).

endotoxin include fever, shock (1–2 hours after introduction of endotoxin), severe diarrhea, and altered immunity states.

Infectivity

Certain pathogens cause disease as a result of their growth and multiplication within the host. As these microorganisms grow, they produce enzymes and metabolic products that act on host tissues. For example, *Clostridium perfringens,* a causative agent of gas gangrene, causes extensive tissue damage because it produces many exoenzymes when it reproduces within the host. One such enzyme, **collagenase**, breaks down the collagen that provides structural integrity to muscle tissue. The resulting tissue destruction aids the pathogen in its invasion of adjacent muscle tissue. **Hyaluronidase** is produced by *C. perfringens* and by a variety of other microorganisms, including streptococci and staphylococci. This enzyme has been called a **spreading factor** because it breaks down hyaluronic acid of connective tissue, which helps to hold cells together. As we can see from table 12.1, many microorganisms produce enzymes that can play a role in infection. These enzymes act on various tissues and components of the body and therefore can cause the signs and symptoms characteristic of various diseases.

Sometimes a pathogen not only invades adjacent tissues but also erodes its way into blood or lymphatic vessels. Once in circulation, the pathogen can have access to other sites in the host. This grave condition is called **septicemia.** While in the blood, the parasite may release endotoxins, exotoxins, or other metabolic products that cause severe diseases. For example, the parasite that causes **malaria** invades and reproduces inside red blood cells. At the end of its reproductive cycle, it disrupts the red blood cells, releasing its progeny as well as metabolic waste products and cellular debris. This release of toxic wastes results in the characteristic signs and symptoms of malaria, which include chills, fever, and malaise. Of course, some microorganisms owe their virulence to their ability to invade and destroy tissue and to release toxins. Examples include the agents of anthrax and plague.

HOST DEFENSE MECHANISMS

When invaded by a pathogenic microorganism, the host is not passive; rather it confronts the pathogen with a variety of barriers aimed at aborting the infectious process. Vertebrate hosts have evolved remarkably effective mechanisms to deal with invading microbes. Some mechanisms are nonspecific and are effective against a wide variety of potential pathogens, while others are quite specific (table 12.3). An infectious agent is confronted with a well-coordinated barrage of body defenses that will abort the vast majority of infections before they can result in any harm to the host. On the other hand, the host responses can be intense enough to contribute to the signs and symptoms of the disease. In the ensuing discussions we examine various host responses and the impact they have on the overall well-being of the host.

Nonspecific Physical and Mechanical Barriers

Skin and Mucous Membranes The unbroken skin very effectively protects animals against invading microbes, since most microorganisms cannot penetrate this barrier. A few microorganisms, such as the blood flukes and hookworms can penetrate the intact skin by secreting hydrolytic enzymes. For the most part, however, microorganisms penetrate through breaks in the skin, such as cuts, puncture wounds, or insect bites.

The external nares are coated with hairs that can trap some of the microbes in the incoming air before they can enter the host further. The lining of the respiratory tract is coated with mucus, which traps most microorganisms that enter the trachea and bronchi. The ciliary action of the respiratory epithelium brings the microbe-laden mucus into the mouth, where it is swallowed. Many invading microorganisms are removed from the respiratory passages in this manner.

Flushing Mechanisms In addition to physical barriers, the vertebrate host has other mechanisms antagonistic to microbial invasion. Flushing mechanisms, such as coughing, sneezing, ciliary action, urination, peristalsis, lacrimation, and salivation, play central roles in ridding the various organs of infectious microorganisms. Impaired function of any of these mechanisms generally leads to infection. For example, individuals with a partially blocked urethra are more likely to suffer from urinary tract infection than individuals who are capable of normal urination.

Coughing and sneezing are reflexes that serve to clear the respiratory tract of foreign particles. These mechanisms cause the forceful exit of air from the lungs through the mouth and nose, thus serving to maintain the respiratory tract relatively free of infectious agents and irritating particles.

TABLE 12.3
SUMMARY OF DEFENSE MECHANISMS

Defense mechanism	Function
I. Nonspecific Defense Mechanisms	
A. Physical Barriers	
1. Flushing mechanisms	
a. Coughing and sneezing	Expels invaders from respiratory tract
b. Urination	Flushes the urinary tract
c. Lacrimation	Washes out invaders from eyes
d. Ciliary action	Moves microbes out of body
e. Peristalsis	Flushes microbes from intestines
2. Skin and mucous membranes	Prevent entry of pathogens into body
B. Chemical Barriers	
1. pH of body fluids	Inhibits growth of many pathogens
2. Lysozyme	Breaks down cell walls of bacteria
3. Complement	Cell lysis and enhances phagocytosis
4. Interferon	Inhibits viral multiplication
C. Biological Barriers	
1. Natural resistance	Not affected by certain infectious agents
2. Normal flora	Competes with and antagonizes invaders
3. Inflammation	Localizes pathogens and repairs tissue damage
4. Phagocytosis	Engulfs and destroys invaders
II. Specific Defense Mechanisms	
A. Humoral Immunity	Antibodies bind to foreign particles, neutralize toxins and viruses. Also enhance phagocytosis.
B. Cell-Mediated Immunity	Cells of the immune response attack invading microorganisms and cancer cells. Produce cytotoxins and recruit cells to augment the specific immune response.

Periodic intestinal **peristaltic** contractions flush potentially pathogenic microorganisms into the colon and out of the body in the feces.

Urination is another flushing mechanism that plays an important role in the overall health of the vertebrate host. This process keeps the urethra relatively free of microorganisms. As the urine flows through the urethra from the bladder, it washes out the urethral lumen and walls and carries with it many of the microorganisms that have entered the urethra. Similarly, tears and saliva serve to flush the eyes and the mouth, respectively, of potential pathogens.

Nonspecific Chemical Barriers

Vertebrate hosts also have a chemical arsenal with which to combat infectious agents. Body cavities often are coated with acidic secretions that are germicidal or germistatic. For example, the stomach has gastric juices, primarily HCl, with a pH of less than 2. Most microorganisms are killed at this pH level and thus have very little chance of surviving in the stomach. Millions of ingested salmonellae are required to infect a normal human host because most of the bacteria are killed in their passage through the stomach. If the patient is hypochlorohydric (little HCl is produced by the stomach), however, far fewer salmonellae are required to initiate an intestinal infection.

The skin generally is free of harmful microbes, not only because of the dry conditions that exist there but also because of the secretion of fatty acids by sebaceous glands. Some fatty acids, in particular **oleic acid**, are bactericidal for some potentially pathogenic transients.

Saliva and tears function not only as a means of

flushing out invading microbes but also as germicidal agents. Both lacrimal and salivary secretions contain **lysozyme**, an enzyme that weakens or disrupts bacterial cell walls.

Tissue extracts and body fluids possess many antimicrobial agents. Polypeptides with high quantities of lysine or arginine are bactericidal (at least *in vitro*) against the anthrax bacillus, staphylococci, streptococci, and the tubercle bacillus. Blood serum contains **β-lysins**, cationic proteins that are active against a variety of bacterial cells. The β-lysins apparently act by disrupting the plasma membrane. It is believed that the combined action of proteins such as lysozyme and the β-lysins play an important role in the host's defense against microorganisms.

Complement **Complement** is the name given to a group of proteins found in serum. These proteins interact with one another to produce active enzymes and proteins that have several biological functions. Among the most important functions complement proteins carry out are (a) lysis of microorganisms, (b) neutralization of viruses, (c) enhancement of phagocytosis, (d) anaphylatoxin activity, (e) recruitment of phagocytes, and (f) damage to plasma membranes. Complement is generally considered to be part of the immune response and is discussed in greater depth in Chapter 13.

Interferon Isaacs and Lindenmann in the late 1950s described a substance called **interferon** that had antiviral properties. Their search for interferon was initiated by the observation that once an animal was infected with a virus, it was more resistant to an unrelated viral infection than an animal not infected by a virus. In other words, one viral infection "interfered" with another. It is now known that interferon is really a group of **glycoproteins** of related molecular structure with antiviral properties. These molecules are species-specific since human interferons are only effective in humans.

Uninfected animals generally do not contain detectable levels of interferon. Upon infection by a virus, however, the host begins to produce interferon, which "interferes" with viral replication. Because of the action of interferon, many severe viral infections are avoided.

Interferon produced in virus-infected cells initiates the production of molecules in neighboring, uninfected cells, which inhibit transcription or translation of viral mRNA (fig. 12.7).

Nonspecific Biological Barriers

Natural Resistance Every infectious microorganism has a **host range**; that is, it can infect only certain hosts. The host range is a function of both the parasite's characteristics and the host's traits. For example, the first step in the infection of a cell by a virus is its attachment, which is possible only if the host cells have surface structures that can serve as attachment sites for the virus. A simple example is a bacteriophage that attaches to bacterial pili: species of bacteria that lack pili would be **naturally resistant** to this virus. Also, if a given human cell lacks receptor sites that can be recognized by a virus, the cell is naturally resistant to the virus. This fact may explain why humans are resistant to diseases such as distemper and mousepox. It is likely that human cells lack surface receptors for the pathogens that cause these diseases.

Even if a given parasite can attach to a host cell, the attachment may not culminate in an infection. The extracellular and intracellular conditions must be conducive to the parasite's multiplication, or the infection will be aborted. Any number of physiologic or anatomic characteristics of a particular host may make it naturally resistant to a given infection. An example is body temperature. Humans and other mammals are quite susceptible to a disease called anthrax, caused by the invasion of host tissue by *Bacillus anthracis*. Fowl (e.g., chickens), however, are naturally resistant to the disease. Human body temperature is around 36.7°C, while chicken body temperature is closer to 40°C. At 40°C, the bacterium cannot carry out essential metabolic activities and is therefore unable to initiate an infection. At the lower temperature, the disease occurs as a result of the bacterial multiplication. If the chicken's body temperature is lowered to that of humans, it too will succumb to anthrax.

The Normal Flora Immediately after birth, newborns are colonized by many different bacteria, viruses, fungi, and algae. Some of these microorganisms remain associated with the newborn host for only a short time, while others remain throughout the host's lifetime. The microorganisms that colonize body surfaces and cavities are called the **normal flora**.

Each body surface is colonized by its own unique normal flora (table 12.4). For example, the skin and hair have about 10^4 to 10^5 bacteria/cm^2. The most common colonizers of the skin are staphylococci, diphtheroids, and yeasts. The large intestine is normally colonized by enterics, yeasts, and anaerobic bacteria.

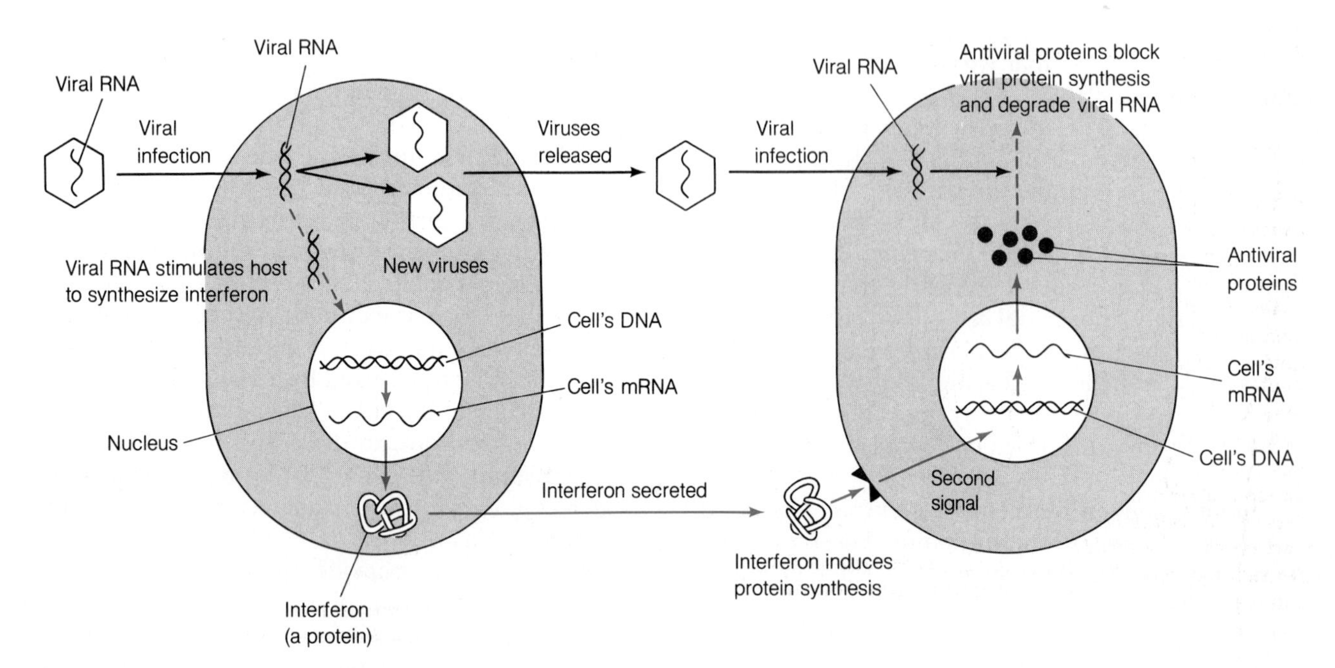

Figure 12.7 Mode of action of interferon. Interferon is a protein synthesized by virus-infected cells which ultimately leads to the inhibition of viral replication. When certain RNA viruses infect cells, they induce the infected cells to synthesize interferon. This protein is secreted by the cell and stimulates other cells to produce antiviral proteins.

The normal flora affords its host a certain amount of protection against invading pathogens. For example, *E. coli, Fusobacterium,* and *Bacteroides,* prominent inhabitants of the human large intestine, make it difficult for enteric pathogens such as *Salmonella* and *Shigella* to colonize the human intestine and cause disease. Evidence that the normal flora protects against invasion by other microorganisms comes from studies using experimental animals. **Gnotobiotic** (of known microbial flora) animals that lack a microbial flora in their intestines are easily infected with *Salmonella* or *Shigella.* These infections usually result in the death of the animal. If these gnotobiotic animals are fed a meal containing *E. coli,* however (which subsequently colonizes the intestines), and are then infected with the enteric pathogens, they are much more resistant to infection.

There are additional lines of evidence supporting the idea that the normal flora protects its host partly by competing with invaders for space and nutrients. Studies have shown that patients treated with antibiotics such as streptomycin become more susceptible to certain intestinal pathogens, apparently because of the destruction of the normal intestinal flora by the antibiotic. Similarly, vaginal infections by certain strains of *Neisseria gonorrhoeae* are inhibited by the presence of *Candida albicans,* which is a common commensal of the human vagina. It is very difficult to ascertain the type of relationship between a host and its normal flora, but some of the colonizers benefit their host, while others may cause disease.

The protective action of normal colonizers such as *E. coli* has been explained in several ways. It is thought that the normal flora consumes all the available nutrients, thus preventing the invaders from obtaining nutrients. It is also suggested that the normal flora inhibits infection by preventing the invaders from attaching to host surfaces. If a given surface area is heavily colonized by the normal flora, the invaders may be unable to bind to specific receptors and may therefore be flushed from the body surface. The pro-

TABLE 12.4
REPRESENTATIVE ORGANISMS FOUND AS RESIDENT FLORA OF HUMANS

Organism	Group	Anatomical site colonized
Actinomyces sp.	Bacterium	Mouth, oral cavity
Bacteroides fragilis	Bacterium	Large intestine
Bacteroides sp.	Bacterium	Mouth and tooth surfaces
Candida albicans	Yeast	Oral cavity, small intestine
Clostridium sp.	Bacterium	Vagina, cervix
Diphtheroids	Bacterium	Skin, oropharynx, nasopharynx, intestines, vagina, cervix
Entamoeba coli	Protozoan	Large intestine
Entamoeba gingivalis	Protozoan	Oral cavity
Enteric bacteria	Bacterium	Vagina
Enterobacter sp.	Bacterium	Large intestine
Enterococci	Bacterium	Small intestine
Escherichia coli	Bacterium	Large intestine
Fusobacteria	Bacterium	Oral cavity large intestine
Group D streptococci	Bacterium	Vagina
Haemophilus influenzae	Bacterium	Oral cavity, oropharynx
Lactobacilli	Bacterium	Small intestine, vagina
Neisseria meningitidis	Bacterium	Oropharynx
Peptostreptococci	Bacterium	Saliva and teeth, large intestine
Pityrosporum	Yeast	Skin
Propionibacterium acnes	Bacterium	Skin
Proteus sp.	Bacterium	Large intestine
Staphylococcus aureus	Bacterium	Skin, nasopharynx
Staphylococcus epidermidis	Bacterium	Skin, nose, mouth, vagina
Streptococcus sp.	Bacterium	Mouth, oropharynx
Treponema sp.	Bacterium	Oral cavity

duction of inhibitory metabolites by the normal flora has also been cited as playing a role in protecting the host.

Competition for available nutrients is a powerful mechanism whereby the normal flora inhibits the growth of pathogens and opportunists. The yeast *Candida albicans,* for example, is normally found in small numbers in the human vagina because of competition with the normal flora for available nutrients. If a patient is subjected to prolonged use of broad spectrum antibiotics, however, very often vaginitis (inflammation of the vaginal lining) caused by *C. albicans* occurs. The antibiotic treatment kills the bacterial vaginal flora and thus eliminates the competition, so *C. albicans* proliferates rapidly, causing vaginitis. Patients with vaginal yeast infections are sometimes treated with douches containing fermented milk or sweet acidophilus milk, which have high concentrations of lactic acid bacteria (e.g., *Lactobacillus acidophilus*) that can recolonize the vagina and reestablish a competition for nutrients. Very often, this treatment is all that is required for the successful control of yeast infections.

Nonspecific Defense Mechanisms Associated with Blood and Lymph

Blood and Its Components The circulatory system serves as a pathway in the delivery of nutrients, oxygen, messenger molecules (hormones), immunity molecules (antibodies and complement), and cells (B and T lymphocytes) from one part of the body to another. It also contains numerous molecules that afford the host an additional dimension in infection defenses. Blood consists of cells suspended in a liquid called **plasma** (table 12.5). If a volume of blood is drawn from a patient, placed in a test tube not containing an anticoagulent, and allowed to stand at room temperature for a few minutes, a **clot** begins to form. The clot eventually will develop into a red mass in the bottom of the tube. The straw-colored liquid above the clot, called **serum**, contains many different proteins such as glob-

TABLE 12.5
BLOOD CELLS, THEIR FUNCTION AND DISTRIBUTION

Type of cell	Morphology	Function of cell
I. ERYTHROCYTES (red blood cells)		Carry oxygen to tissues and exchange CO_2, comprise 35–45 % of blood volume
II. LEUKOCYTES (white blood cells)		
A. Granulocytes	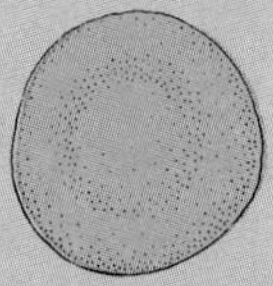	Cells with granules in cytoplasm and lobed nucleus
1. Neutrophils	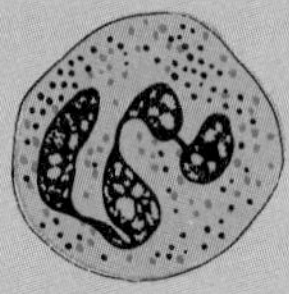	Phagocytosis and digestion of particles and microorganisms, 60% of all white cells
2. Basophils	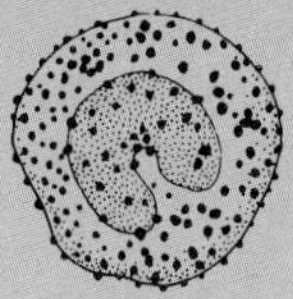	Mediate inflammation, release histamine, 1% of all white cells
3. Eosinophils	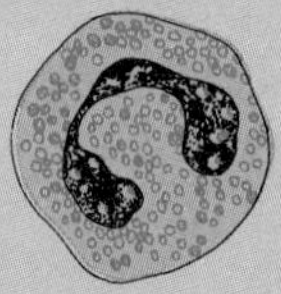	Immunity to some parasites, modulate inflammation, 3% of all white cells
B. Lymphocytes	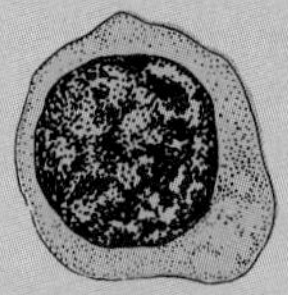	Participate in immune reactions, 30% of all white cells
C. Monocytes	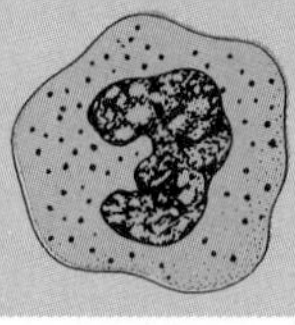	Phagocytosis and digestion of particles, 5% of all white cells

Figure 12.8 **Electron photomicrographs of cells involved in host defense mechanisms.** *(a)* Macrophage. *(b)* Lymphocyte. *(c)* Granulocyte (eosinophil).

ulins, complement, albumens, and interferons. The clot consists of blood cells knitted together by fibers of **fibrin**.

The cells that make up blood are red blood cells (**erythrocytes**), white blood cells (**leukocytes**), and platelets (**thrombocytes**). Each of these **formed elements** has a specific function and is central to the overall health of the individual.

Erythrocytes The function of the erythrocytes is to carry oxygen from the lungs to the rest of the body. Human erythrocytes measure about 7.5 μm in diameter and are biconcave in shape. They develop in the bone marrow and lose their nuclei when they mature. Erythrocytes constitute the largest portion of formed elements in the blood, making up nearly 45% of the blood volume. When the concentration of erythrocytes falls much below this figure, a condition known as **anemia** develops. Associated with the plasma membrane of erythrocytes is **hemoglobin,** a molecule that binds oxygen in the lungs and releases it to cells in other parts of the body. The life span of an erythrocyte is about 120 days. New erythrocytes are constantly being made in the bone marrow to replace the dead erythrocytes.

Leukocytes The blood also contains white cells called leukocytes (fig. 12.8). These cells are nucleated and participate actively in protecting the host against infection. If a blood smear stained using Wright's procedure is examined with a bright field microscope, various types of white blood cells can be seen. Those called **granulocytes** constitute 60–65% of the leukocytes in the blood and contain prominent granules in their cytoplasm. They have a multilobed nucleus, so they are sometimes called **polymorphonuclear** leukocytes or **PMN** leukocytes, although this term is usually reserved to describe only neutrophils.

Granulocytes can be differentiated from one another by their staining properties. Some have granules that react with the acidic dye **eosin** in Wright's stain and stain red, so they are called **eosinophils.** Other granulocytes have granules that take up the basic dye **hematoxylin** and stain blue; these are called **basophils.** There are also granulocytes whose granules stain with neutral dyes as well as with hemotoxylin and eosin. These granulocytes, which stain both red and blue, are called **neutrophils**. Granulocytes are very important in protecting the host against infection. They participate in **inflammation** and in **phagocytosis**, and their numbers increase or decrease in response to various types of infections.

The nongranulated leukocytes constitute the remaining 35–40% of the leukocytes in the blood and include **monocytes** and **lymphocytes**. Monocytes are large leukocytes, 12–15 μm in diameter, with a centrally located, kidney-shaped nucleus. These cells also participate in phagocytosis. The lymphocytes are rounded cells, 10–15 μm in diameter, with a very large ovoid nucleus that takes up most of the cytoplasm. As we shall see in Chapter 13, the lymphocytes play a vital role in specific immunity to infectious agents.

Differential blood counts, which determine the proportions of the various leukocytes in the blood,

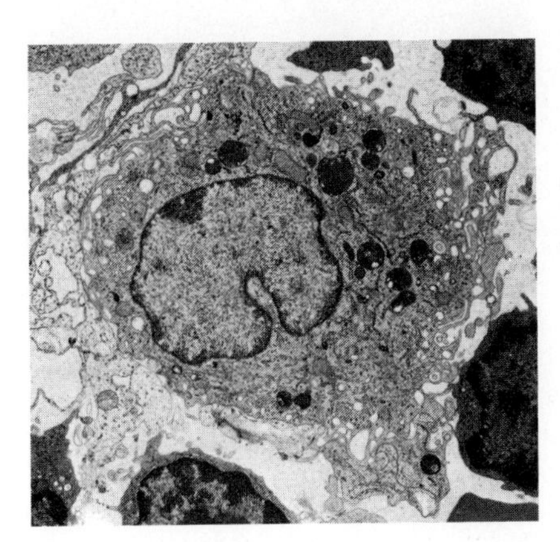

(a)

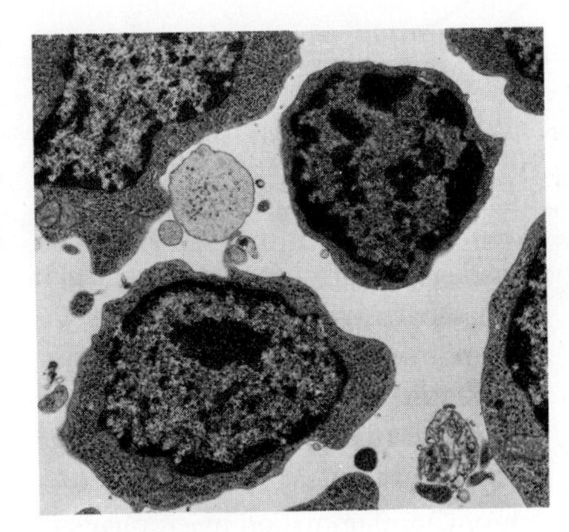

(b)

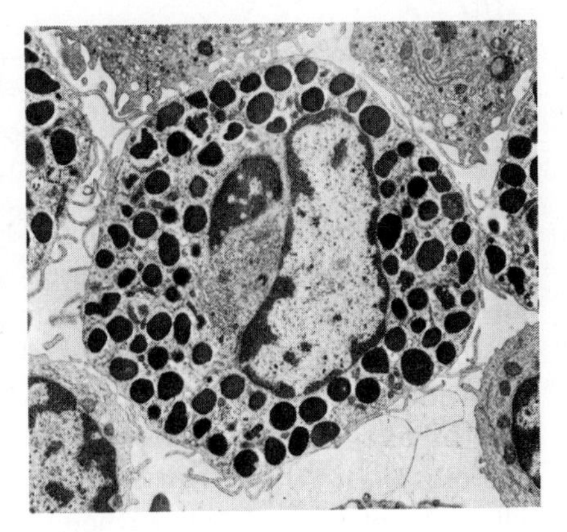

(c)

are important because they serve as an indicator of the state of health of an individual. For example, some diseases characteristically alter the white blood cell count by increasing or decreasing the total number of leukocytes (leukocytosis or leukopenia, respectively). In addition, certain diseases cause an increase in some types of leukocytes but not in others. This information may eventually be of use in determining the cause, treatment, and prognosis of diseases.

The Lymphatic System **Lymph** is a pale fluid with a composition resembling that of blood plasma. This fluid bathes tissues of the body and circulates via the lymphatic vessels. The lymph contains lymphocytes and a type of phagocytic cell called the **macrophage** (fig. 12.8), but no erythrocytes or PMN leukocytes. There is a connection between the circulatory and lymphatic systems via the **thoracic duct**. The lymphatic system consists of **lymphatic vessels**, **lymph nodes**, **lymph**, and **lymphocytes**. Located throughout the lymphatic system are lymph nodes, organs containing numerous lymphocytes and macrophages (table 12.5). Most of the specific immune mechanisms are initiated in these tissues.

Inflammation **Inflammation** is a very effective host defense mechanism that develops in response to tissue injury (fig. 12.9) and may last for as long as the infectious agent is present. The injury may result from mechanical damage such as a puncture wound or scratch, a chemical substance, sunburn, or an infectious agent. The destruction of cells by any of these mechanisms initiates a coordinated sequence of events aimed at the repair of the injury and the removal of the injurious agent. Although this is primarily a protective mechanism, the degree of inflammation can be so extensive that it may be responsible for most of the disease. For example, fever, headaches, and other aches

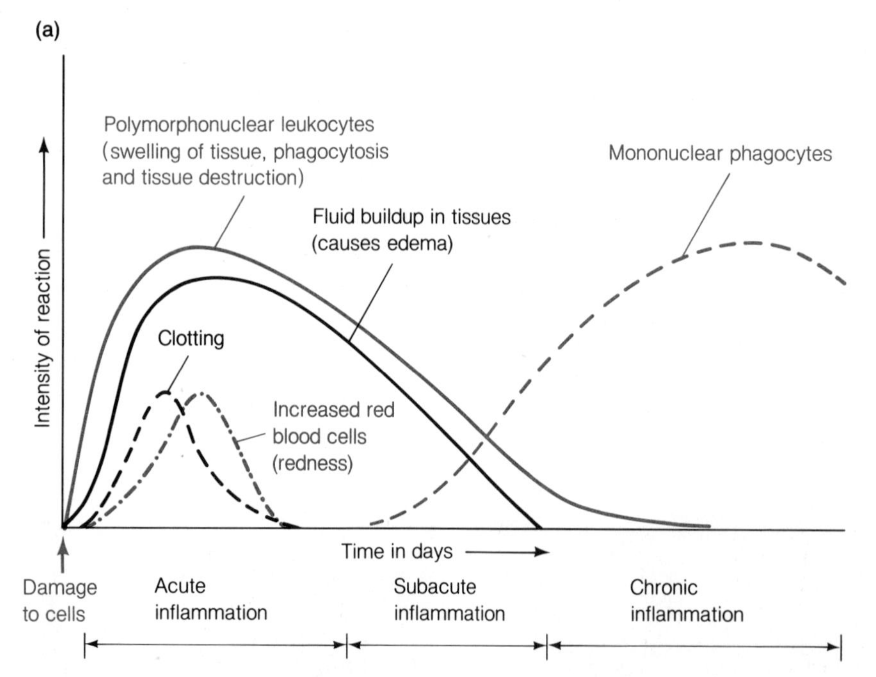

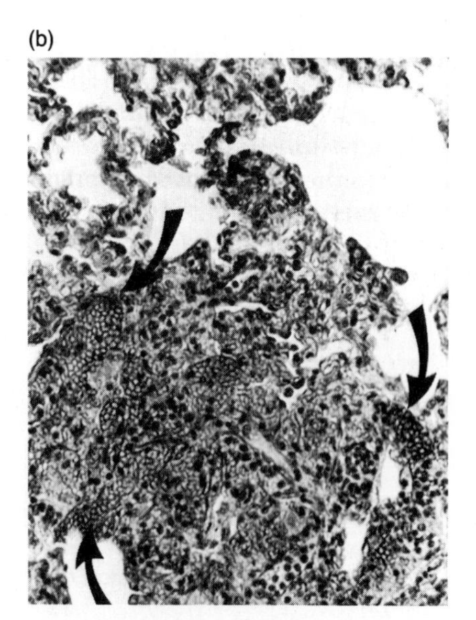

Figure 12.9 Kinetics of inflammation. *(a)* Inflammation is characterized by redness, swelling, and tenderness of the affected area, often accompanied by an increase in body temperature. These signs reflect the cellular events that take place when tissue is damaged. The redness is due to an accumulation of red blood cells in the area. The swelling is due to fluid (plasma) and leukocyte (PMN) accumulation. During acute inflammation, the predominant cell is the PMN, while during long-lasting (chronic) inflammation macrophages predominate. The pain is due partly to the repair of damaged tissue by leukocytes during the healing process. *(b)* Photomicrograph illustrating acute inflammation of lung. Arrows indicate engorged capillaries with many red blood cells.

and pains characteristic of many diseases are directly or indirectly due to inflammation.

Regardless of the cause of inflammation, the sequence of events is essentially the same. Immediately upon damage to tissue, the destroyed cells release cellular materials that initiate the inflammatory response. Initially there is a marked dilatation (widening) of blood vessels near the damaged site which results in an increased blood flow to the area. This event is paralleled by an increase in vascular permeability, which results in the leakage of plasma and a few blood cells to the damaged area. The increased amount of fluid in the damaged area causes a swelling called **edema**. Following the leakage of fluid, leukocytes pass through the blood vessels to the damaged site (fig. 12.9). Initially, the leukocytes are predominantly granulocytes. The neutrophils act primarily as phagocytes, engulfing and digesting microorganisms and damaged host tissue. During this process, the basophils release granules containing **histamine** and **serotonin**. This process, sometimes called **degranulation**, serves to enhance the degree of inflammation. The eosinophils, on the other hand, are thought to play a modulator role in the inflammatory process by releasing antihistaminic substances that counteract, at least in part, the effect of histamine at the site of inflammation.

The presence of endotoxin or other microbial products, or substances produced by the PMN leukocytes themselves, induces fever. These substances, called **pyrogens**, can stimulate the central nervous system (CNS) to elevate the body temperature.

Inflammation is characterized by redness, swelling, heat, and pain. The redness (**rubor**) is due to the increased blood flow to the area of injury. The swelling (**tumor**) is the combined effect of increased extravascular fluid and PMN infiltration in the damaged area. The heat (**calor**) is due to the increased blood flow to

FOCUS METCHNIKOFF ON PHAGOCYTOSIS

During the golden age of microbiology, the concept that infectious diseases were caused by specific microorganisms was established. Scientists throughout the world searched for, and often found, the microbes that were responsible for many of the maladies that afflicted the human race. No sooner was the connection made between infectious diseases and microorganisms than scientists began asking how the survivors of a disease fight off microbes.

One theory, the **phagocytic theory**, was advanced by Elie Metchnikoff in 1884. The Russian-born Metchnikoff theorized that when harmful intruders invade animals, certain host cells move toward the invaders and engulf them. According to Metchnikoff, these cells, called **phagocytes**, are constantly on the lookout for intruders and represent the main line of defense against invading microbes.

Metchnikoff's phagocytic theory grew from his observation that motile cells of starfish larvae move toward and accumulate around rose thorns, which he had used to impale the larvae. The development of Metchnikoff's theory was based on numerous observations in various areas of biology and occupied 25 years of his life.

The most convincing evidence for the phagocytic theory came from studies of a disease that afflicts the water flea *Daphnia*. The disease Metchnikoff studied is caused by a fungus he called *Monospora bicuspidata*. Metchnikoff showed that spores of *Monospora* were quickly surrounded, and often killed, by the water flea's white blood cells (phagocytes). Metchnikoff's study revealed that in some cases, when phagocytes did not respond to the invaders, the fungi would proliferate and kill *Daphnia*. These studies emphasized the importance of phagocytes in protecting the host against disease.

Metchnikoff's studies on phagocytosis created a foundation for the science of *cellular immunology* (immunity conferred by cells). Cellular immunology represents only one part of the complex immune system that protects animals from invading microorganisms.

the area and to the action of pyrogens, while pain (**dolor**) results from the local tissue destruction and irritation of sensory nerve receptors.

The first cells to arrive at the site of inflammation are usually the PMN leukocytes. This kind of inflammation, which occurs within 15 minutes or so after the damage, is called **acute inflammation** (fig. 12.9). As the inflammation progresses as a result of the presence of infectious microorganisms, the PMN leukocytes are gradually replaced by monocytes and lymphocytes, resulting in **chronic inflammation** (fig. 12.9). This type of inflammation is of longer duration and is accompanied by the development of specific immune responses.

Phagocytosis **Phagocytosis** is a very important host defense mechanism that is carried out by phagocytes such as polymorphonuclear leukocytes, monocytes, and macrophages. It rids the body of harmful microorganisms and dead or dying cells (fig. 12.10). Particulate matter, whether dead or alive, that enters host tissues, is taken up by cells (phagocytized) and then destroyed intracellularly.

Some substances enhance the phagocytic ability of PMN leukocytes and macrophages by allowing them to bind more firmly to the particle. These substances (e.g., antibodies and complement) are called **opsonins**. Opsonization is an important antimicrobial mechanism of defense.

Phagocytosis is a continuous process that culminates in the ingestion of a particle and its subsequent intracellular digestion. For convenience, the process has been subdivided into five stages: chemotaxis, attachment (adherence), ingestion, killing, and digestion.

Chemotaxis, the first step in phagocytosis, involves the migration of the phagocyte toward its prey. A random collision between the phagocyte and the prey could occur, but usually the phagocytes are attracted to the prey by chemotactic substances, which may be microbial or host-derived products. As an example, consider the dimorphic fungus *Coccidioides immitis*, which causes a pulmonary disease called valley fever. Inside a host, the fungus develops **endosporulating spherules.** The mature spherules are not particularly chemotactic, but when they rupture they release endospores which, along with other fungus-derived substances inside the spherule, attract PMN leukocytes which then engulf the endospores. Damaged host cells also release chemotactic substances. The influx of PMN leukocytes from the peripheral circulation into the site of injury during inflammation is due to the chemotactic substances released by damaged cells.

Attachment is the necessary next step in phagocytosis: there must be an intimate contact between the phagocyte and the microorganism (or particle) for phagocytosis to take place. In fact, microorganisms that inhibit the attachment of the phagocytes to their cell surfaces generally are resistant to phagocytosis. For example, encapsulated strains of *Streptococcus pneumoniae* are not readily ingested by phagocytes because of their capsules, which do not permit the attachment of the phagocyte to the bacterial surface. *Streptococcus pyogenes* **M protein** inhibits the attachment of phagocytes to their surfaces. To improve the attachment of phagocytes to microbial surfaces, the host produces **opsonins**: proteins, either antibodies or components of complement, that promote the attachment of the phagocyte to its prey.

Ingestion (or engulfment) begins with the extension of pseudopodia by the phagocyte. The pseudopodia surround the prey and totally engulf it. The prey and the membrane that surrounds it are called the **phagosome** (fig. 12.10).

Killing constitutes the most important step in phagocytosis. The neutrophil produces a large number of substances to kill microorganisms. These substances are found in two different types of granules: **azurophilic** granules and **specific** granules. The azurophilic granules are **lysosomes**, which contain acid hydrolases, lysozyme, cationic proteins, collagenase, and myeloperoxidase (MPO). The specific granules have a neutral pH and contain lysozyme, lactoferrin, alkaline phosphatase, collagenase, and vitamin B_{12} binding protein. When the components of these two types of granules are introduced into the phagosome, they are very effective microbiocidal agents.

Phagolysosome Formation When the lysosomes fuse with the phagosomal membrane, the resulting structure is called a **phagolysosome** (fig. 12.11). The lysosomal products act on the ingested microbe or particle. During phagocytosis by PMNs, O_2 uptake increases, a process known as the **respiratory burst**. An enzyme called **NADPH oxidase** reduces the O_2 taken up during the respiratory burst to produce hydrogen peroxide. Other microbiocidal processes produce superoxide ions and hydroxyl radicals that kill the ingested microorganisms.

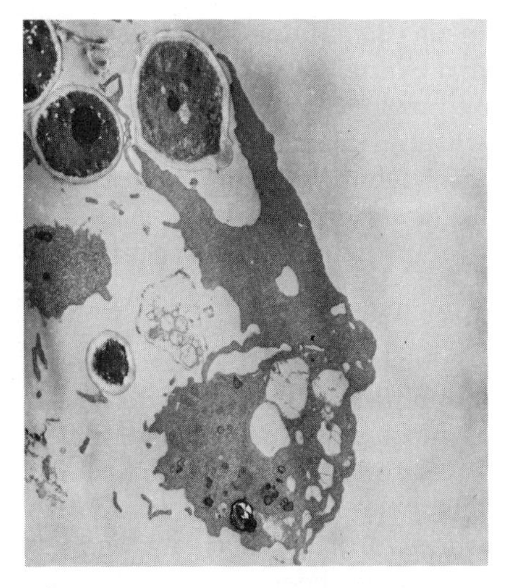
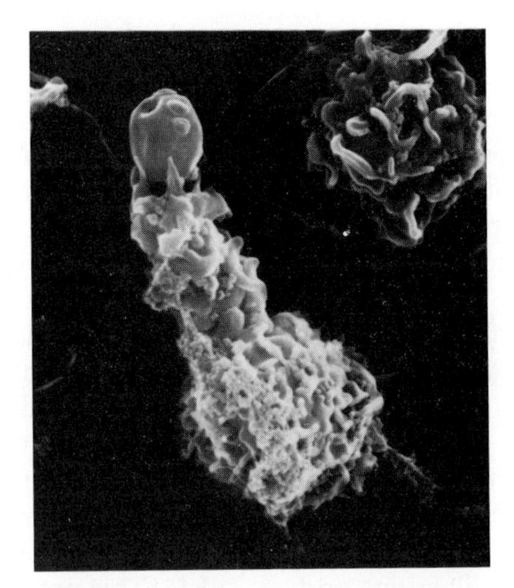

(a)

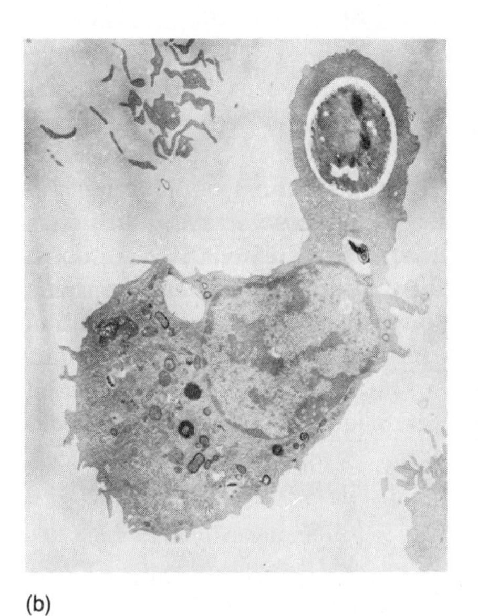
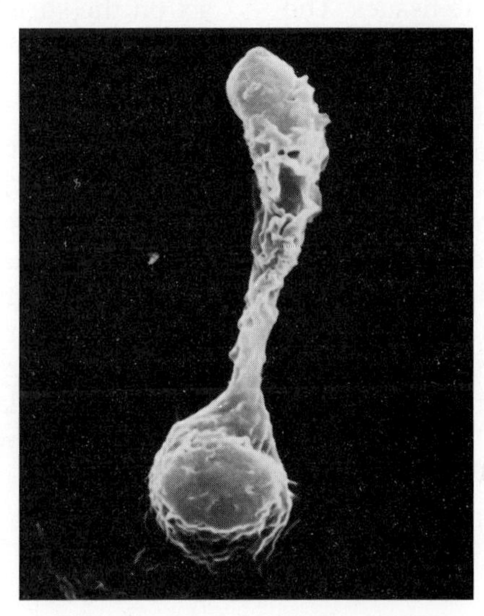

(b)

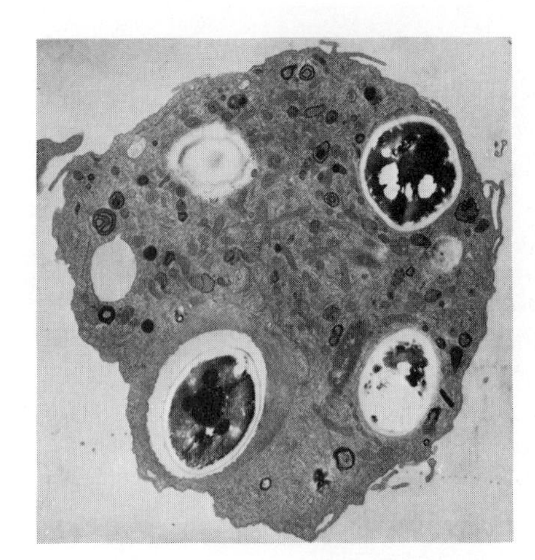

(c)

Figure 12.10 **The process of phagocytosis.** *(a)* Transmission and scanning electron micrographs of alveolar (lung) macrophages attaching to a yeast cell by extended pseudopods. *(b)* Transmission and scanning electron micrographs of alveolar (lung) macrophages of various stages of phagocytosis of the yeast cells. Note the lack of subcellular organelles in the pseudopod and that the pseudopod is being withdrawn into the body of the macrophage. *(c)* Transmission electron micrograph of alveolar (lung) macrophage containing several yeast cells. The yeasts have now been completely drawn into the cell and are digested within phagosomes. Note that the yeasts inside the phagosomes are in various stages of digestion.

The enzyme **myeloperoxidase,** in conjunction with hydrogen peroxide (H_2O_2) and halide ions (e.g., I^- or ClO^-), inactivates enzymes and modifies membrane lipids killing microbes. Enzymes such as **lysozyme** destroy bacterial walls. The protein **lactoferrin** impairs microbial growth within the phagolysosome by sequestering iron, which is needed for microbial growth.

The **digestion** of killed cells within the phagolysosome is carried out by hydrolases, such as nucleases, lipases, elastases, collagenases, and proteases found in lysosomes.

The macrophage is the primary phagocytic cell of the **reticuloendothelial system (RES)**. The RES comprises a group of cells and a network of loose connective tissue (reticulum) that serves to filter out and destroy foreign particles and damaged host material that may be present in the blood or body tissues. The macrophages are either fixed to organ tissues or traveling to sites of infection through the lymphatics and blood. The fixed macrophages are found in organs such as the spleen, liver, tonsils, lymph nodes, bone marrow, and brain. The wandering macrophages are seen throughout the lymphatics, blood, peritoneal cavity, and lungs as they travel to sites of infection. The macrophages have special names, depending upon the site in which they are found. For example, liver macrophages are called **Kupffer cells**; lung macrophages are called **dust cells**; and brain macrophages are also called **microglial cells.**

The mechanism of phagocytosis by the macrophage is similar to that by the PMN. Certain microbes are known to inhibit macrophage degranulation and therefore can live within these phagocytic cells. The **microbiocidal** action of the macrophage is enhanced significantly in the **activated macrophage**. An interaction between the macrophage and specific host defense mechanisms is needed to activate the macrophage and make it an effective microbiocidal agent.

Many parasitic infections are characterized by a condition called **eosinophilia**, an increase in the number of eosinophils in the blood. It has been suggested that the eosinophils may play a role in extracellular digestion. Eosinophils may adhere to the surface of a parasite (such as a roundworm or a fluke) and degranulate. This process, which has been called **exocytosis,** involves the release of lysosomal granules so that they act on the surface layers of the parasite and damage or kill the worm.

Specific Immunity

In addition to the nonspecific host defense mechanisms previously described, the vertebrate host has the ability to attack specifically the invading microbes. This type of immunity, called specific immunity, is mediated by serum proteins called **immunoglobulins** (antibodies), lymphocytes, and plasma cells. The macrophage also plays an important role in this process. Specific immunity mediated by immunoglobulins is

TABLE 12.6
SOME STRATEGIES PARASITES USE TO EVADE HOST RESPONSES

Evasive strategy	Function	Example(s)
Encystation	Forms a resistant outer shell to avoid host attack	*Giardia lamblia* *Entamoeba histolytica*
Multiple forms/stages	Presents the host with a wide array of antigens to which it has to react; allows pathogen to get a foothold on host tissue	Malarial parasites (*Plasmodium*), *Trypanosoma* *Leishmania* sp. dimorphic fungi
Antiphagocytic substances	Inhibits phagocytosis	*Staphylococcus aureus* *Histoplasma capsulatum*
Intracellular parasitism	Protects from host defenses	*Mycobacterium tuberculosis* *Rickettsia rickettsii*
Leukocydin production	Kills phagocytes	*Staphylococcus aureus*
Inhibition of immune response	Prevents development of specific immunity	Malarial parasites (*Plasmodium*)
Changes in antigenic composition	Prevents host from developing an effective immune response	*Trypanosoma brucei gambiense* surface proteins, *Neisseria gonorrhoeae* pili

CLINICAL PERSPECTIVE PROTOZOAN IN THE LIVER?

A 26-year-old sailor on a cruise ship that visited Mexico experienced a short episode of diarrhea followed by fever, chills, and right upper quadrant pain. He was hospitalized when the ship put into Miami, Florida, with a temperature of 102°F, epigastric (adjacent to the stomach) pain, and an epigastric mass that was tender. His eosinophil count was elevated 8%. Stool studies for ova and parasites were positive for *Entamoeba histolytica* cysts but not trophozoites (motile forms). A radionuclide scan of the liver and spleen showed a mass in the left lobe of the liver.

The doctor's diagnosis was amoebic liver abscess. The patient was treated with metronidazole (750 mg 3×/day). Nevertheless, a spiking fever continued and the liver mass did not respond after a week of treatment. The patient should have responded and become afebrile (fever free) in about 48 hours. Further consideration of the increased eosinophil count and the abdominal mass with a rim calcification suggested that the problem might be a cyst caused by a tapeworm. This hypothesis was consistent with the patient's history of having grown up in Greece on a farm where there were dogs and sheep that could serve as the primary and intermediate hosts, respectively, for the tapeworm *Echinococcus*. The cysts can remain asymptomatic for decades and then expand at a rate of 1 cm in diameter per year producing a mass, abdominal pain, and jaundice if the bile duct is obstructed. Cyst expansion may be triggered by a secondary infection. Depending upon the species of *Echinococcus* the disease may be due to a single expanding cyst or numerous budding expanding cysts.

The adult *Echinococcus* tapeworm lives in the small intestine of canines (dogs, wolves, and coyotes), the primary hosts. The eggs are released in the stool and ingested by intermediate hosts (moose, reindeer, sheep, and humans). The embryos penetrate the intestinal wall and enter the portal circulation that carries them to the liver. Infrequently, the embryos will develop into cysts in the lungs, brain, kidneys, and other tissues. In the liver, the larvae develop into cysts that contain the scoleces (heads), brood capsules, and daughter cysts. When intermediate hosts are eaten by canines, the scoleces attach to the canines' intestines and grow into adult worms. Infection of humans occasionally occurs during childhood from infected dogs in sheep-raising communities.

The best cure for an echinococcal cyst is surgery. However, care must be taken to kill the scoleces and the daughter cysts within the cysts with hypertonic saline so that they cannot spread if the cyst ruptures during surgery. Freed scoleces and daughter cysts can form new cysts in the body. In those cases where surgery is not possible, mebendazole (40 mg/kg/day) treatment for 3 months or more is somewhat useful against *E. granulosus* and *E. multilocularis*.

Upon surgery, the left lobe of the patient's liver was seen to have a multilocular lesion. To prevent the seeding of new echinococcal cysts upon resection (removal), the dressings around the cyst were soaked in hypertonic saline. Subsequently, the cavities of the cyst were aspirated and injected with hypertonic saline. After the cyst was resected, it was found to be a multilocular lesion.

Happily, the patient recovered completely and required no futher treatment.

For further details on this problem patient consult the following reference. Schiff, E. R. 1987. A Greek sailor with a hepatic mass. *Hospital Practice* 22(1A):16–22.

called **humoral immunity**, while that mediated by cells is called **cell-mediated immunity**. These types of immunity are the subject matter of Chapter 13.

EVASIVE STRATEGIES OF PARASITES

The development of a parasitic existence requires that the parasite adapt to the host's environment. Not only must the parasite adhere to and penetrate the host, but it must also establish a long-standing relationship with the host. This task is not particularly easy, in view of the vertebrate host's powerful arsenal of protective responses. A parasite must either evolve mechanisms to evade the host response or perish. Table 12.6 summarizes some of the evasive strategies used by parasites. The ensuing discussion attempts to describe some, but not all, of the strategies that successful parasites may use to evade host defenses.

Encystation or the formation of an impervious structure around the cell, serves as a mechanism to avoid host antagonistic responses. For example, the cysts of certain parasitic protozoa, such as *Entamoeba histolytica* and *Giardia lamblia,* have a tough outer layer that makes them resistant to gastric juices and other host-protective responses. Thus, encystment allows them to pass unaffected into the intestines, where they can excyst (come out of their cyst) and initiate an infection.

The **development of a complex life cycle** can afford the infectious agent several avenues for escaping the host response. For example, specific immunity directed toward one stage of the parasite's life cycle will likely not be effective against another stage. This is the case with many protozoans, such as the malaria parasite, and parasitic worms, such as the flukes and roundworms.

Antiphagocytic defenses are also evident among parasites. The polysaccharide capsule of the pneumococcus, the A protein of staphylococci, and the M protein of streptococci are three examples of antiphagocytic strategies. Certain intracellular parasites are readily phagocytosed, but they prevent the formation of phagolysosomes and therefore can live and multiply within the phagosome. In this way, the intracellular parasite can live within host cells and be protected from other host antagonistic actions.

Many parasites secrete substances that can **inhibit or depress the immune response**. The malaria parasite apparently secretes substances that depress the immune response, particularly the immune response mediated by antibodies. This may be an adaptation to its mode of reproduction: since its life cycle involves the infection of blood cells, the invading parasite must make its way into the circulation and be exposed to antimalarial immunoglobulins. By reducing the amount of immunoglobulins produced by an infected host, the parasite increases its chance of reaching and infecting an erythrocyte before being destroyed.

An outstanding example of evasive strategy, in which there are **periodic changes in antigenic composition**, can be illustrated with the flagellated protozoan *Trypanosoma brucei gambiense,* the organism that causes African sleeping sickness. One characteristic feature of this disease is the periodic appearance of the parasite in the blood and an abnormally elevated serum immunoglobulin concentration (IgM). The appearance of the parasite in the blood is paralleled by fever, and its disappearance relieves the fever. Apparently when the parasite appears in the blood it induces a specific immune response, which then drives the parasite out of the blood and into the lymph nodes. There, the parasite changes the composition of its surface layers and reappears in the blood. The immunity against the "previous" parasite is not effective. The parasite then induces an additional humoral immune response. Thus, the appearance and disappearance of the parasite in the blood is accompanied by a change in the chemical composition of its surface layers. In this way, the parasite can withstand long periods of parasitic existence within a host.

SUMMARY

This chapter summarizes the interactions between hosts and parasites and the consequences of such interactions.

SOME DEFINITIONS

1. Symbiosis, or "living together," signifies close association between two or more organisms whose biological activities are closely linked.
2. Mutualistic relations are generally permanent symbiotic associations in which both symbionts benefit from the association.
3. Commensalistic associations are symbioses in which one of the symbionts benefits while the other is unaffected.
4. Parasitic associations are symbioses in which one of the symbionts benefits at the expense of the other.
5. Infections result from the establishment and proliferation of a parasite in or on host tissues.
6. Infectious diseases result when an infecting parasite causes detrimental changes in the host.
7. Pathogenicity is the ability of microorganisms to cause disease. This ability is due, for the most part, to inheritable traits of the microorganism.
8. Virulence is a measure of a microorganism's degree of pathogenicity. Virulent strains cause disease, while avirulent strains do not. The virulence of a strain is usually measured as the LD_{50} or the ID_{50}.

THE INFECTIOUS PROCESS

1. For a microorganism to infect a host, it must enter the host, attach itself to host surfaces, and begin to multiply. Only microorganisms that can successfully carry out these steps are capable of causing an infection.

2. Entrance into a host can be accomplished through a break in the skin or through the mucous membranes of the respiratory, urogenital, or digestive system.
3. Attachment by pathogens to host surfaces can be mediated by various mechanisms, such as recognition of receptor sites on host membranes, pili, capsules, or glycocalyx.
4. Multiplication of the pathogen can take place either inside or outside host cells, depending upon the infecting microorganism.

PATHOGENIC PROPERTIES OF MICROORGANISMS

1. Microorganisms can cause disease either by invading host tissues and causing detrimental changes or by releasing toxic substances.
2. Intoxications are disease resulting from the action of a toxin produced by a microorganism. These toxins could be proteins with specific modes of action, which are called exotoxins, or be lipopolysaccharides, which are called endotoxins.
3. Infectious diseases are those that result from invasion of host tissue by a parasite. Exoenzymes such as collagenases, hyaluronidases, or proteases can be involved in the invasive process.

HOST DEFENSE MECHANISMS

1. The host has numerous mechanisms to prevent infectious diseases. These mechanisms can be specific for the invading microbe or nonspecific and effective against a variety of microorganisms. Defense mechanisms may involve physical barriers, chemical barriers, or biological barriers.
2. Physical barriers include the skin and mucous membranes, and flushing mechanisms such as coughing, sneezing, urinating, and tearing.
3. Chemical barriers are nonspecific defense mechanisms involving chemicals that are active against a variety of microorganisms. These include interferon, complement, and β-lysins.
4. Biological barriers may be nonspecific or specific. The processes of inflammation and phagocytosis are nonspecific defense mechanisms aimed at the removal and destruction of noxious microorganisms.
5. The process of phagocytosis involves the migration of phagocytes to the site where the microorganisms are causing the damage; the engulfment of the parasite by the phagocyte; and the subsequent destruction of the invader within the phagocyte.
6. The normal flora serves a useful function, in that it protects its host from invading pathogens and provides the host with some necessary nutrients such as vitamins and amino acids.
7. Specific immunity can be mediated by antibodies or by certain lymphocytes and the activated macrophage.

EVASIVE STRATEGIES OF PARASITES

Parasites evolve in conjunction with their hosts and develop mechanisms to avoid host defenses. These evasive mechanisms include (a) forming resistant structures, such as cysts; (b) developing multiple life stages; (c) evolving antiphagocytic defenses; (d) attenuating immune responses; and (e) changing antigenic composition.

STUDY QUESTIONS

1. Define symbiosis. Describe three different types of symbiotic associations and give an example of each.
2. Differentiate between:
   a. infection and disease;
   b. pathogenicity and virulence;
   c. LD_{50} and ID_{50};
   d. exotoxin and endotoxin;
3. Explain how a microorganism can cause disease by invading host tissue.
4. Explain how a microorganism can cause disease by producing toxins.
5. Explain how a microorganism can cause disease by eliciting an immune response.
6. Define what is meant by "normal flora" and how it protects the host from infectious diseases.
7. What is inflammation? What role does it play in health? In disease?
8. Describe five different strategies that parasites may employ to evade host defense mechanisms.
9. Describe the process of phagocytosis.
10. What are opportunistic pathogens?

SELF-TEST

1. The term parasite indicates that in a symbiotic relationship both members of the symbiosis profit from the association. a. true; b. false.
2. The degree of disease-causing capacity of a microorganism is known as its:
   a. pathogenicity;
   b. virulence;
   c. commensalism;
   d. biotic potential;
   e. mutualism.
3. The necessary first step before a pathogen in the respiratory passages can initiate a respiratory tract infection is to attach to target cells. a. true; b. false.
4. Which of the following is *not* a nonspecific defense mechanism against infection?
   a. sneezing and urination;
   b. production of interferon;
   c. production of antibodies;
   d. phagocytosis;
   e. inflammation.
5. Exotoxins differ from endotoxins in that exotoxins are proteinaceous substances that have unique detrimental effects on host cell metabolism. a. true; b. false.
6. Diseases that are consequences of exotoxins produced by microorganisms not necessarily present in the affected tissue are called invasive diseases. a. true; b. false.
7. Which of the following nonspecific defense mechanisms act by inhibiting the attachment and proliferation of bacteria in host tissue?
   a. inflammation;
   b. urination;
   c. phagocytosis;
   d. production of interferon;
   e. fixation of complement.
8. Which of the following nonspecific defense mechanisms act by inhibiting viral and foreign DNA replication?
   a. inflammation;
   b. urination;
   c. phagocytosis;
   d. production of interferon;
   e. fixation of complement.
9. The principal function of phagocytosis is to destroy microorganisms before they have an opportunity to cause an inflammatory reaction. a. true; b. false.
10. Parasites evolve mechanisms through which they become more virulent and therefore can destroy additional host tissue, which they can use as sources of nutrients. a. true; b. false.
11. The first line of defense against disease-causing microorganisms is usually the production of specific immune substances. a. true; b. false.
12. Flushing mechanisms are specific defense mechanisms that rid the respiratory and digestive passages of infectious diseases. a. true; b. false.
13. Interferon is an antiviral substance that inhibits viral reproduction in virus-infected cells. a. true; b. false.
14. The normal flora of the body are mutuals that prevent the colonization of normal tissue by potential disease-causing organisms. a. true; b. false.
15. The following are steps in phagocytosis *except:*
    a. chemotaxis;
    b. intracellular multiplication;
    c. ingestion;
    d. digestion;
    e. phagolysosome formation.
16. The inflammatory response is important for the removal of harmful agents and the repair of damaged tissue. a. true; b. false.
17. Symbiotic relations can be of three different types depending upon the nature of the interaction between the two symbionts. a. true; b. false.
18. In symbiotic relations that are considered parasitic, the parasite normally ends up in killing its host. a. true; b. false.
19. Infections of humans by microorganisms usually result in disease. a. true; b. false.
20. Virulence is a measure of the disease-causing capacity of a microorganism and can be quantified using values such as LD_{50} and ID_{50}. a. true; b. false.

SUPPLEMENTAL READINGS

Burnet, M. and White, D. B. 1972. *Natural history of infectious diseases.* 4th ed. London: Cambridge University Press.

Boxer, G. J., Curnutte, J. T., and Boxer, L. A. 1985. Polymorphonuclear leukocyte function. *Hospital Practice* 20(3): 69–90.

Dautry-Varsat, A. and Lodish, H. 1984. How receptors bring proteins and particles into cells. *Scientific American* 250(5): 52–58.

Donelson, J. E. and Turner, M. J. 1985. How the trypanosome changes its coat. *Scientific American* 252(2):42–52.

Hood, L. E., Weissman, I. L., and Wood, W. B. 1986. *Immunology.* 2nd ed. Menlo Park, CA: Benjamin/Cummings.

Horwitz, M. A. 1982. Phagocytosis of microorganisms. *Review of Infectious Diseases* 4:104–108.

Iglewski, W. J. 1984. Critical roles for mono (ADP-ribosyl) transferases in cellular regulation. *ASM News* 50(5):195–198.

Lewin, R. 1984. New regulatory mechanism of parasitism. *Science* 226(4673):427.

Mechnikov, I. 1908. *The prolongation of life*. Vol. 5. *Lactic acid as inhibitor of intestinal putrefaction*. New York: Putnam.

Middlebrook, J. L. and Dorland, R. B. 1984. Bacterial toxins: Cellular mechanisms of action. *Microbiological Reviews* 48(3): 199–221.

Ranney, D. F. 1985. Targeted modulation of acute inflammation. *Science* 227:182–184.

Root, R. K. and Cohen, M. S. 1981. The microbiocidal mechanism of human neutrophils and eosinophils. *Review of Infectious Diseases* 3:565–568.

Smith, H., Skehel, J. J., and Turner, M. J. 1980. *The molecular basis of microbial pathogenicity.* Deerfield Beach, FL: Verlag Chemie.

Taylor, P. W. 1983. Bacteriocidal and bacteriolytic activity against gram-negative bacteria. *Microbiological Reviews* 47:46–65.

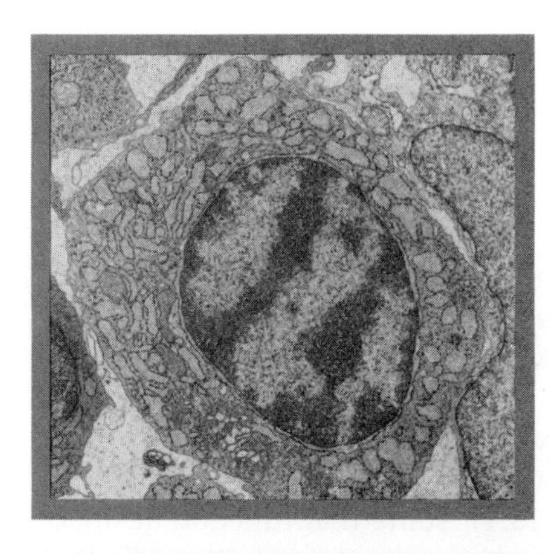

CHAPTER 13

THE IMMUNE RESPONSE

CHAPTER PREVIEW IMMUNOLOGY, A BENEVOLENT SCIENCE

The science of immunology (the study of immune systems) developed from the experiments conducted by scientists such as Edward Jenner. Jenner's experiments in 1798 demonstrated that vaccination of individuals with the benign cowpox virus could protect vaccinated individuals from smallpox, a disease that scarred and killed many humans until quite recently. Approximately 80 years later, Pasteur showed that the bacterium that causes anthrax (a disease that killed many farm animals as well as humans) could be tamed (attenuated) and used to induce protective immunity to the disease-causing bacterium in vaccinated animals. After this discovery, Pasteur developed vaccines against swine erysipelas, chicken cholera, and rabies. Largely through the technique of vaccination we have been able to eliminate the risk of acquiring smallpox and poliomyelitis and significantly reduce the risk of death due to diseases such as influenza, diphtheria, whooping cough, and tetanus.

Immunology has grown considerably in importance since its origin in the eighteenth century. Discoveries regarding the initiation and expression of the immune response are of fundamental importance in biology because they have led to the improvement of human health and the treatment of infectious diseases. The development of immunological tests has been extremely important in medicine because these tests allow physicians to diagnose infectious diseases before debilitating or fatal illnesses have resulted. Immunological tests are easy to perform and can give results that are fast and accurate. Through the use of such tests, the time it takes to diagnose a disease has been reduced from days to a few hours. In many cases immunological methods have helped physicians diagnose diseases that could not have been diagnosed using other methods. The rapid diagnoses of diseases through the use of immunological methods have saved countless lives.

The treatment of cancer is also possible through the use of immunology. Scientists have been experimenting with anticancer drugs that are attached to antibodies. The antibodies will seek the cancerous cells and deliver the anticancer drugs to the target cells without exposing normal cells to the toxic effects of the anticancer drugs.

It is as true now as it was in the nineteenth century: immunology is a pertinent science that has a direct impact on human welfare and health. In this chapter we discuss the basic concepts of immunology and how these concepts can be put to practical use in the detection and prevention of disease.

CHAPTER OUTLINE

IMMUNOLOGY IS THE SCIENCE THAT STUDIES the immune system. The immune system consists of a number of different types of cells and their products, whose primary function is to protect animals from invading microorganisms, their toxic products, and cancerous cells. An intact and functioning immune system is essential for the health of an animal and even short lapses in the proper functioning of the immune system could lead to serious disease or even death.

The purpose of this chapter is to acquaint you with the various cells of the immune system and the substances they produce to block the multiplication and spread of microorganisms and cancerous cells in the body.

TYPES OF IMMUNITY

The term immunity refers to the resistance that an individual has against one or more infectious agents, their products, or cancerous cells. Immunity occurs when antimicrobial substances are produced by the body to combat a pathogen (disease-causing organism), or when various types of immune cells attack and inactivate infectious agents. Immunity to certain diseases may also occur as the result of certain genetic traits.

The immunity of an organism to disease results from the coordinated action of a number of protective mechanisms (fig. 13.1), some that function against a wide variety of infectious agents and others that are specific for one type of infecting agent. The specific type of immunity, also known as acquired immunity, is expressed in two distinct, but interrelated ways: humoral immunity and cell-mediated immunity.

Humoral immunity is effected by a group of complex proteins found in serum and certain mucous membranes called **antibodies**. Antibodies are formed in response to the presence in the body of foreign substances or particles (which may be toxins, microorganisms, or cancerous cells). Those substances or particles that can induce the formation of antibodies are known as **antigens**. We describe antigens and antibodies in greater detail later in this chapter.

Cell-mediated immunity is effected by a variety of immune cells whose function is to recognize and inactivate infectious agents and cells.

Immunity is often referred to as either **native** or **acquired immunity**. Native immunity, the resistance that an animal has at birth, is not dependent upon antibodies or immune cells. Native immunity is often due to the chemical makeup or physiological properties of an animal that make it impossible for a microorganism to invade and reproduce in that animal. For example, humans are not subject to infections by bacteriophages or the tobacco mosaic virus because the plasma membranes of human cells lack the necessary receptor sites to which these viruses attach. Even if these viruses could enter human cells, it is unlikely that they could reproduce within them because human enzymes for nucleic acid replication, transcription, and translation do not function with these viral nucleic acids. Another example of native immunity is the natural resistance of many birds against infections by the anthrax bacillus. The bacterium that causes anthrax (*Bacillus anthracis*) cannot reproduce at the body temperatures of birds and hence cannot infect them. Even within a species, some individuals are immune to a

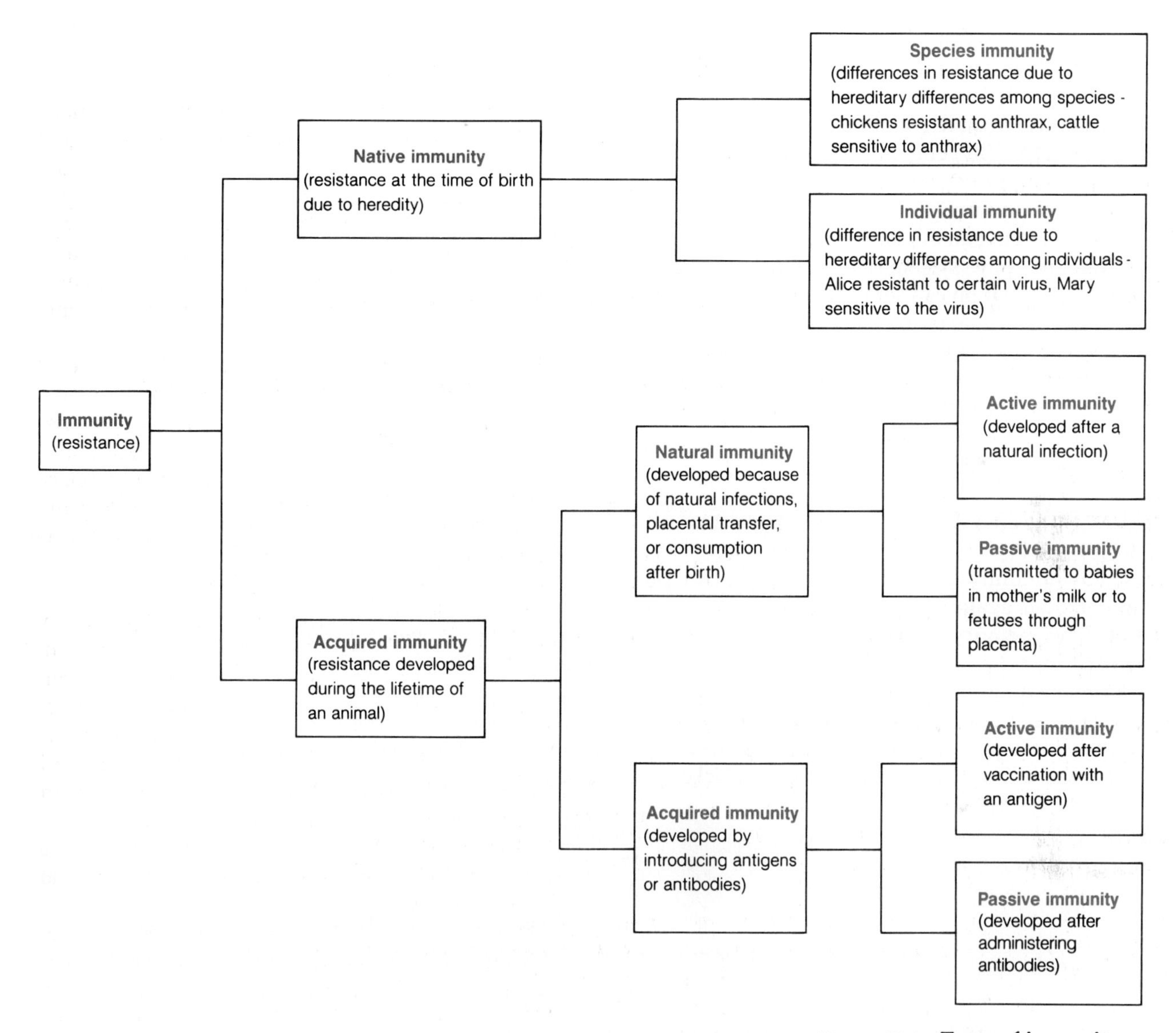

Figure 13.1 **Types of immunity**

disease while others are not, because of slight differences in their physiologies. Factors that affect the physiology of an animal include its sex, age, nutrition, genetic background, and overall health.

Acquired immunity is the specific immunity that develops after the host obtains antibodies or immune cells against specific antigens. Acquired immunity may develop after contact with antigens that induce an immune response or through the transfer of antibodies or cells that developed in other hosts.

Active immunity is the type of acquired immunity that develops in a host after coming in contact with infectious agents or after vaccination. Active immunity is of long duration and sometimes hosts can be immune to certain infectious agents for as long as they live.

Passive immunity results after a host receives antibodies or immune cells from another individual. This type of immunity is of short duration. Passive immunity can develop temporarily after the passage of antibodies from a mother to her fetus or from her milk to the baby. Babies are frequently resistant to many infectious diseases because mothers transfer some antibodies and other antimicrobial substances to their offspring in the milk. Passive immunity can also develop after the injection of antibodies that developed in other animals. Some infectious diseases such as botulism and tetanus, which result from the detrimental action of potent exotoxins on host cells, are frequently treated by administering antibodies against the toxins (hyperimmune serum) to the patient and thus giving protection until the host can develop its own active immunity.

Persons lacking an immune system at birth or with a damaged immune system due to exposure to large quantities of radiation can sometimes be given a functional immune system by an injection of bone marrow. The bone marrow (which must come from close relatives of the recipient) contains immune cells and can therefore help the body regain the capacity to develop its own immunity to antigens.

CHARACTERISTICS OF ACQUIRED IMMUNITY

Acquired immunity is extremely important in protecting animals against infectious diseases. It is characterized by memory, specificity, and recognition of "non-self."

Memory allows the immune system to respond swiftly and vigorously to an infectious agent that has previously been encountered. For example, the organisms that cause childhood diseases such as chickenpox and mumps seldom attack an individual more than once. An individual develops an effective, long-lasting immune response when first infected, and this first response is largely responsible for the clinical cure of the disease. When infected with the same agent at a later date, the body "remembers" the infectious agent, because it developed long-lived "memory" cells that stimulate an immediate, powerful secondary immune response (fig. 13.2). This secondary immune response, sometimes called the **anamnestic immune response**, is characterized by a more rapid and intense response than the primary response.

The anamnestic response can be studied by following the production of antibodies when an animal is injected with an antigen. When an antigen such as the tetanus toxoid (tetanus vaccine) is injected into an animal, there is a lapse of several days before any antibodies against the toxoid can be detected in the blood's serum. The concentration of antibodies in the serum gradually increases, peaks out, and then begins to decline (fig. 13.2). After 3 to 4 weeks, the level of antibodies is very low. If, however, the same animal is injected again 5 to 10 weeks after the first injection (a "booster" shot), the concentration of antibodies in the serum will increase rapidly within 2 or 3 days and will reach a much higher level than after the first injection (fig. 13.2).

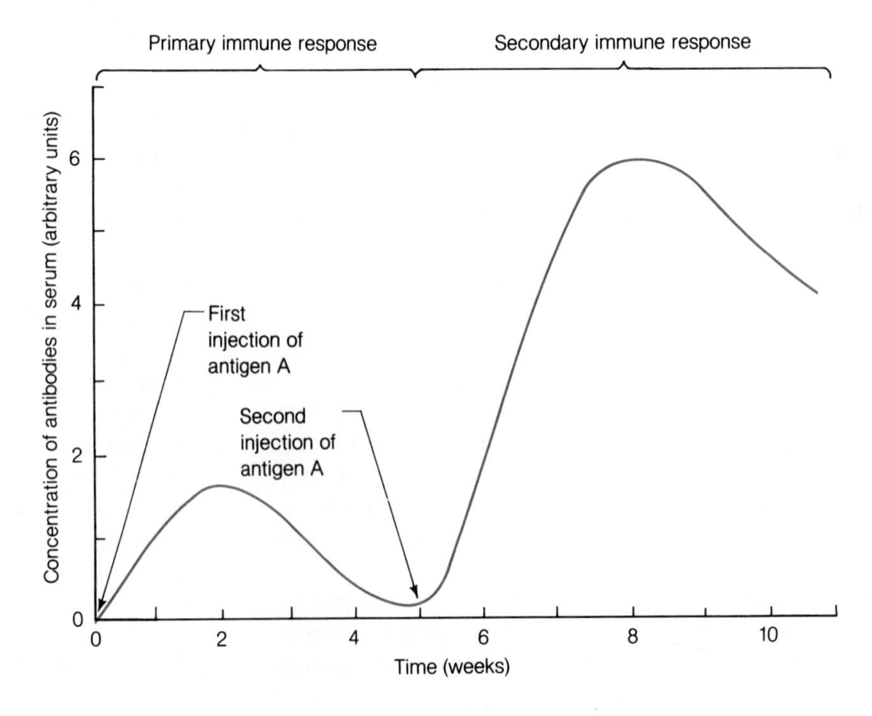

Figure 13.2 **Kinetics of the immune response.** Upon first encounter with a given antigen, the immune response is not detected immediately. The immune response develops over a period of 1 to 3 weeks. As time goes on, the level of immunity gradually declines. When the immune system encounters the same antigen for the second time, it reacts quickly and vigorously. More antibodies are produced and for a much longer period of time than after the first encounter with an antigen. The second encounter with antigen is referred to as the secondary or anamnestic response.

The dynamics of antibody production help us understand why the primary invasion of the human body by a pathogen, such as chickenpox or mumps viruses, can result in disease, while the secondary invasion of these pathogens generally has little effect on a host. Pathogens that reproduce rapidly in a non-immune host may kill the host within a week after infection. The reason is that the immune system in these individuals does not have a chance to develop since antibody levels generally do not reach effective serum concentrations for at least 2 weeks. When a previously sensitized host is invaded for the second time by the same pathogen, however, the antibodies in the blood reach effective levels within a short period of time (less than a week), aborting the infection. This is an important reason why people are resistant to those diseases against which they have been vaccinated.

Another characteristic of the acquired immune response is its **specificity**. When an individual is vaccinated against a disease such as mumps, the immune response that develops is specific against the mumps virus antigens. This is because the memory cells that arise after the first bout of mumps respond only to the mumps virus. If an individual who had mumps at an earlier date becomes infected by another virus, such as the rubella virus (which causes German measles), the immunity against mumps will not be of use against rubella and the individual may contract the disease. Any subsequent encounter with the rubella virus, however, results in the development of a rapid, specific, and protective immune response.

The normal immune response **distinguishes between self and non-self**. Since the immune response can be induced by a variety of large molecules, it is

TABLE 13.1
CELLS OF THE IMMUNE SYSTEM AND THEIR FUNCTIONS

Cell type	Protein produced	Function of protein
Neutrophils	Lysozymes Interferon	Digest cells and degrade macromolecules Inhibits virus replication
Monocytes	Lysozymes	Digest cells and degrade macromolecules
Macrophages	Lysozymes Complement components (C4, C2) Genetically related macrophage factor (GRF) Lymphocyte activation factor (LAF) (Interleukin-1)	Digest cells and degrade macromolecules Create holes in plasma membranes (stimulates release of histamine) Stimulates T-lymphocytes (T_H) to secrete immune response proteins Stimulates T-lymphocytes (T_H) to multiply
Plasma cells (B-lymphocytes)	Immunoglobulins (antibodies): IgM, IgD, IgG, IgA, IgE	Bind to and agglutinate foreign cells and debris, neutralize toxins and viruses, enhance phagocytosis
T-lymphocytes	Interferon Chemotactic factors Migration inhibitory factors Macrophage activation factors	Inhibits virus replication Attract macrophages and other cells Inhibit departure of macrophages Stimulate macrophages to fuse lysosomes with food vacuoles
T_C and T_K	Killing factor	Kills foreign cells, cancerous cells, and infected cells
T_H	Helper factor Interleukin-2 Interferon	Stimulates development of lymphocytes, initiates immune response Stimulates T-cell growth and division Inhibits virus replication
T_S	Suppressor factor	Inhibits development of lymphocytes and immune response. Inhibits immune response against self-antigens.
Fibroblasts	Interferon	Inhibits virus replication
Epithelial gastrointestinal mucosa	Complement components	Lyse plasma membranes, kill microorganisms
Liver cells	Complement components	Lyse plasma membranes, kill microorganisms

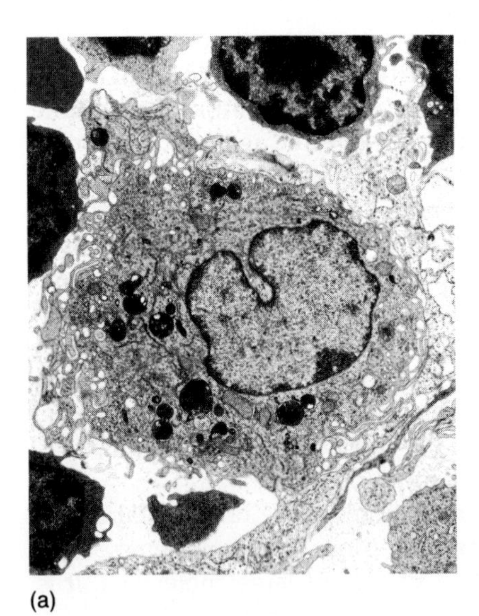
(a)

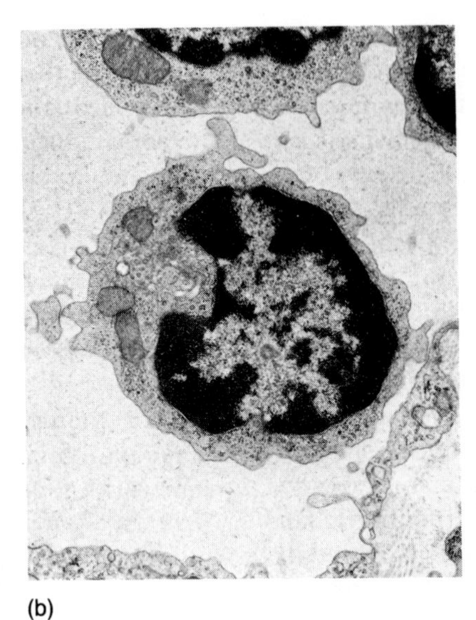
(b)

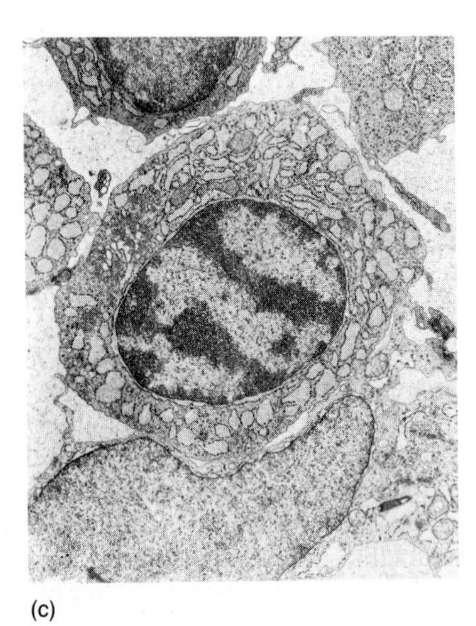
(c)

Figure 13.3 Cells of the immune response. Cells involved in the immune response arise from multipotent stem cells found in the bone marrow. The stem cells mature in various organs of the body and become effectors of special branches of the immune system. Thymus-processed lymphocytes (T-lymphocytes) are effectors of cell-mediated immunity; bone marrow- and gut-processed lymphocytes (B-lymphocytes) are involved in antibody-mediated immunity; plasma cells are involved in antibody production; and macrophages carry out phagocytosis. *(a)* Macrophage. *(b)* Lymphocyte. *(c)* Plasma cell.

theoretically possible that the immune system could attack the host and cause severe damage throughout the body. For this reason, the immune system becomes nonresponsive to self antigens. It is believed that this nonresponsiveness develops during fetal life. Sometimes, however, the mechanisms that make the immune system unresponsive to self do not function properly and the individual develops an immune response to some part of the body. This response results in what is known as an **autoimmune disease**. Some diseases, such as systemic lupus erythematosus and rheumatic fever result when the body's defenses are directed against self antigens. The next chapter is concerned with some of the most common autoimmune disorders that afflict humans.

CELLS OF THE IMMUNE SYSTEM

Several types of specialized cells are responsible for an animal's immunity (table 13.1). These cells recognize the presence of antigens, interact with other cells to initiate an immune response, produce substances that interact specifically with the antigen, or attack the antigen directly. The most important cells of the immune system are lymphocytes and macrophages, although certain types of granulocytes also play a role in immunity (fig. 13.3). The cells of the immune system begin to develop before the animal is born (fig. 13.4). In the embryo, cells associated with the yolk sac

Figure 13.4 Development of blood cells. Humans have red and white blood cells. The red cells, or erythrocytes, carry oxygen to cells. White blood cells include granulocytes, lymphocytes, and monocytes. There are three types of granulocytes, known as neutrophils, basophils, and eosinophils. The two types of lymphocytes are B- and T-lymphocytes. ▶

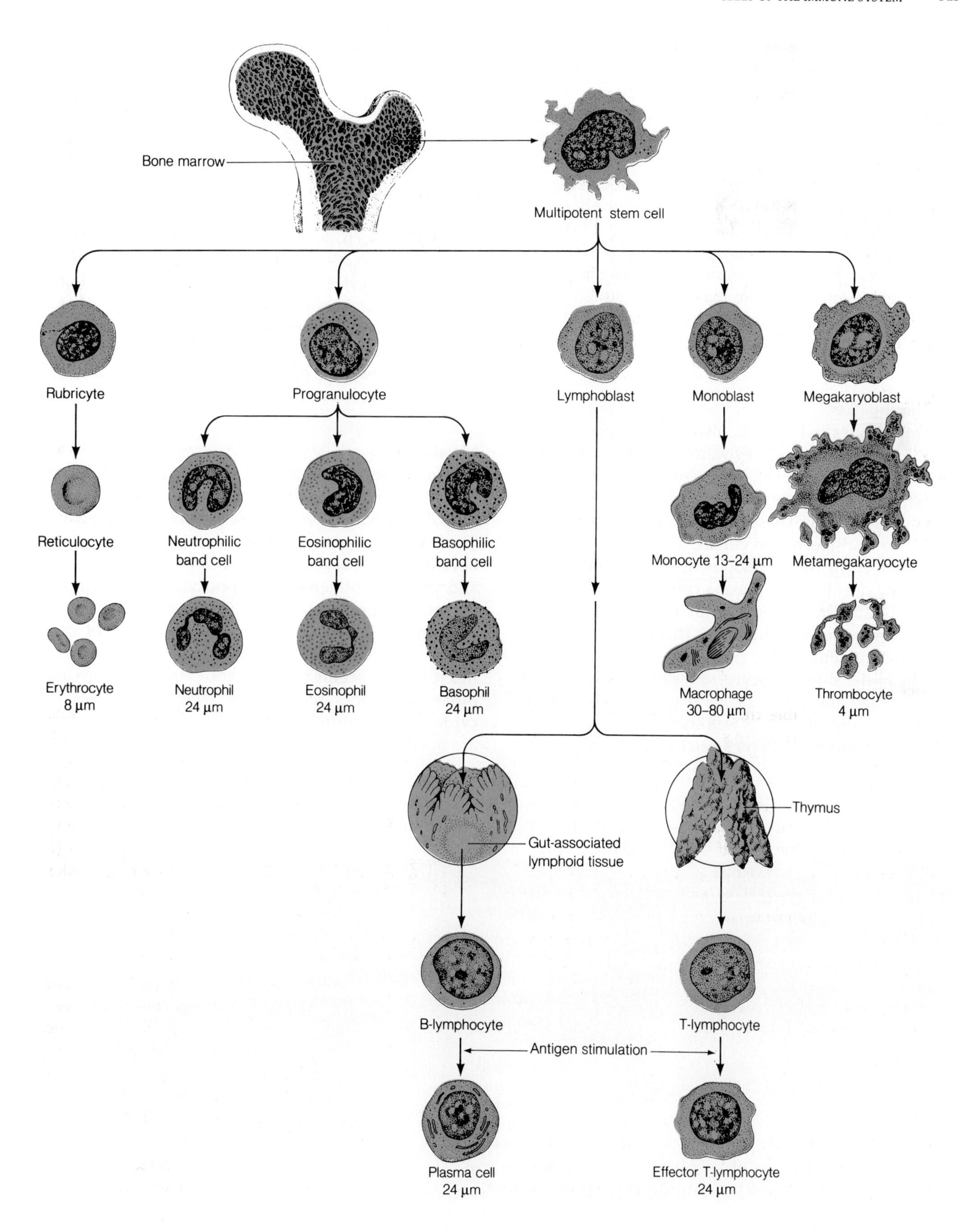
Bone marrow
Multipotent stem cell
Rubricyte
Progranulocyte
Lymphoblast
Monoblast
Megakaryoblast
Reticulocyte
Neutrophilic band cell
Eosinophilic band cell
Basophilic band cell
Monocyte 13–24 μm
Metamegakaryocyte
Erythrocyte 8 μm
Neutrophil 24 μm
Eosinophil 24 μm
Basophil 24 μm
Macrophage 30–80 μm
Thrombocyte 4 μm
Gut-associated lymphoid tissue
Thymus
B-lymphocyte
T-lymphocyte
Antigen stimulation
Plasma cell 24 μm
Effector T-lymphocyte 24 μm

and blood islets in the circulatory system migrate to the fetal liver, where they develop into **stem cells.** Stem cells, in turn, migrate to other fetal tissues such as the thymus, spleen, gut, lymph nodes, and bone marrow, where they differentiate into the cells of the immune system (fig. 13.4). For example, a population of lymphocytes (fig. 13.3) called T-lymphocytes completes its maturation in the thymus. B-lymphocytes, granulocytes, monocytes, and macrophages develop primarily in the bone marrow, but are also found in the spleen and gut. In mammals, various stem cells that migrate into the bone marrow are stored there and give rise to fresh stem cells, which can repopulate the spleen, gut, and lymph nodes during the animal's life.

B-Lymphocytes

B-lymphocytes (table 13.2), also called bursa-derived lymphocytes (because they were first discovered in a gut-associated organ called the **bursa of Fabricius** in birds), make up 10–20% of the lymphocytes found in the human blood. These cells are also found in the bone marrow, spleen, and gut (fig. 13.5). B-lymphocytes are involved in the development of humoral immunity; they develop into **plasma cells** (fig. 13.3c) when stimulated by the presence of an antigen.The plasma cells, in turn, secrete antibodies that react specifically with the antigen.

In birds, B-lymphocytes develop in a lymphoid organ called the bursa of Fabricius. The lymphoid organs of mammals are not well characterized, although the bone marrow, the **Peyer's patches** in the intestines, and the **mesenteric** (intestinal) lymph nodes are the lymphoid organs thought to be the sites of B-lymphocyte maturation.

T-Lymphocytes

The thymus-derived lymphocytes, called T-lymphocytes, make up approximately 80% of the lymphocytes in the human blood. They mature in the thymus before the fetus is born, where they acquire the characteristics they will express during the life of the individual.

T-lymphocytes perform a variety of functions (table 13.2) involved in the protection of the host against disease. Some of these functions include the induction of the immune response, the amplification of immunity, the modulation of the immune response, and the inactivation of harmful cells or microorganisms.

Macrophages

Macrophages (fig. 13.3a) are scavenger cells found in the blood, lymph, and various organs of the body. The **reticuloendothelial system** consists of wandering and fixed macrophages. Macrophages are important in the development of specific immunity because they process and present antigens to lymphocytes, an important first step in the development of immunity to antigens. Macrophages are also important effectors of cell-mediated immunity. The engulfing of antigens by macrophages seems to be enhanced by antibodies and other substances released by lymphocytes. **Monocytes** are immature macrophages. Blood monocytes are not able to destroy infectious agents.

TABLE 13.2
CHARACTERISTICS OF T- AND B-LYMPHOCYTES

Characteristic	T-lymphocyte	B-lymphocyte
Organ where they mature	Thymus	Bone marrow
Function	Cell-mediated immunity Antigen recognition Modulate immune response	Humoral immunity
Proportion of lymphocytes	70–80%	20%
Membrane antigens (Markers)	Theta antigen Ly antigen	Fc receptor
Differentiate into:	Helper cells Memory cells Suppressor cells Cytotoxic cells Killer cells	Plasma cells Memory cells
Products	Lymphokines Cytotoxin	Immunoglobulins

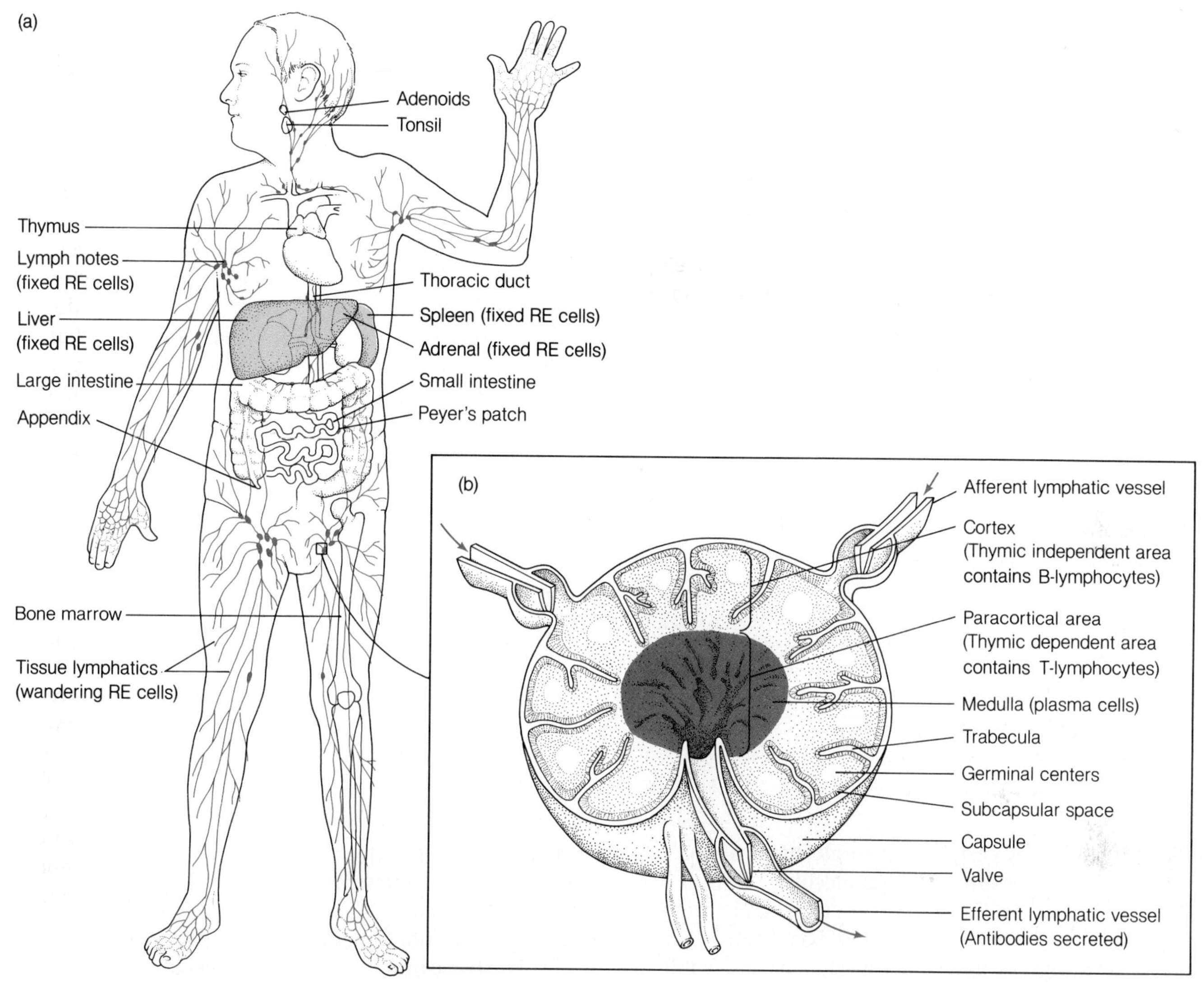

Figure 13.5 **The lymphatic and reticuloendothelial systems.** *(a)* The lymphatic system is a network of vessels, lymph nodes (see illustration b), and organs that function in the initiation and maintenance of the immune response. The **reticuloendothelial system** consists of the fixed and wandering phagocytic cells distributed throughout the body. They monitor and rid the body of infectious agents and cancerous cells. *(b)* Structure of the lymph node. The germinal centers of the lymph node contain primarily B-lymphocytes, while the thymic-dependent areas are composed of T-lymphocytes. The very center of the lymph node, the medulla, contains primarily antibody-forming cells or plasma cells.

ANTIGENS AND ANTIBODIES

What Are Antigens?

Any molecule that induces a specific immune response is called an **antigen**. Antigens include microbial components such as toxins, cell walls, flagella, and capsules, as well as microorganisms, viral particles, and cancerous cells. Antigens are generally found to be macromolecules such as polypeptides, polysaccharides, lipopolysaccharides, and glycoproteins. The chemical group in an antigen that specifically stimulates the immune response is called the **antigenic determinant** or **epitope** (fig. 13.6). The antigenic

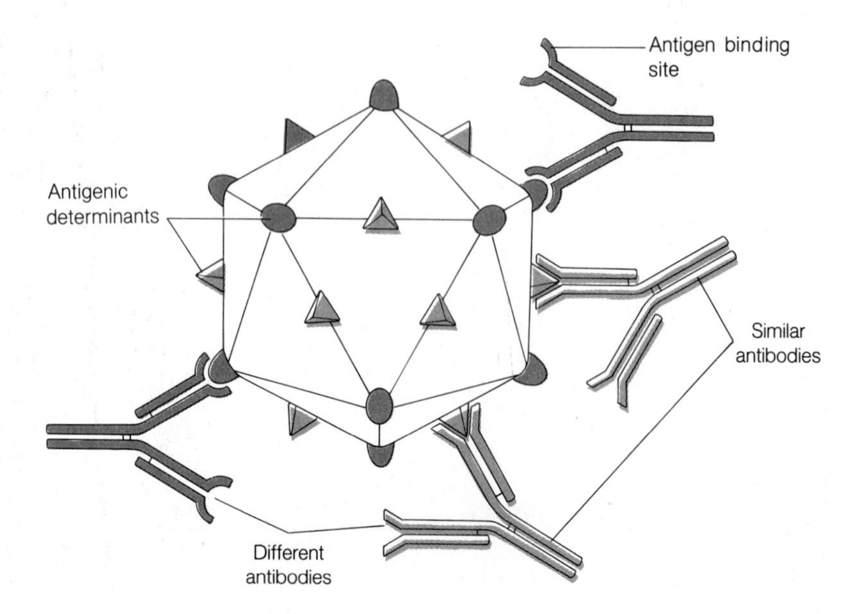

Figure 13.6 Antigens and antigenic determinants. The antigen represented by a virus particle has two different types of antigenic determinants, each recognized by antibodies of different specificities. The specificity of each antibody molecule is represented by the shape of the antigen-binding site of the molecule, which has a shape complementary to that of the antigenic determinants (pyramids and half spheres). When an antigen, such as a virus, is injected into an animal, it induces the animal to produce various types of antibodies against the antigen. Each type of antibody will react with a different antigenic determinant.

determinant may consist of a single sugar or a group of amino acids or other small molecules. Thus, a bacterial cell or a flagellum may contain several antigenic determinants. The term **hapten** has been used to describe antigenic determinants that are incapable of stimulating an immune response by themselves. Almost any molecule that is not antigenic by itself can be made antigenic by covalently joining it to macromolecules such as polypeptides or polysaccharides, which are referred to as **carrier molecules**. Since many of the subunits in a polypeptide or polysaccharide can stimulate the development of specific antibodies, an antigen generally stimulates the production of a family of antibodies with different specificities that bind different antigenic determinants (fig. 13.6).

Sometimes two unrelated antigens have antigenic determinants that are so similar that they bind to the same antibody molecule. Antibodies that bind to more than one antigen are said to be **cross-reactive**. Cross-reacting antibodies are sometimes a problem when microbiologists use antigen-antibody reactions to diagnose infectious diseases. It is always possible that an observed antigen-antibody reaction was due to cross-reactivity, and this observation may lead to a misdiagnosis.

What Are Antibodies?

Antibodies or **immunoglobulins** are a family of complex glycoproteins found primarily in serum and mucous secretions that react with antigens.

The basic immunoglobulin molecule is composed of four polypeptides joined together by disulfide bonds. The four polypeptides consist of two identical heavy (H) chains, containing approximately 440 amino acids each, and two identical light (L) chains with approximately 220 amino acids each. Each heavy chain is covalently bonded to a light chain so that each immunoglobulin molecule is made up of two identical halves.

Both the H and L chains are composed of two regions: the variable (V) region and the constant (C) region. The constant regions of the heavy and light chains contain approximately 320 and 105 amino acids, respectively, and do not vary much among various immunoglobulin molecules. The variable regions of both the H and L chains show considerable variation in their amino acid sequence and are responsible for the specificity of each type of antibody.

The immunoglobulins perform a number of functions. The arms, called **Fab** regions, bind to antigens,

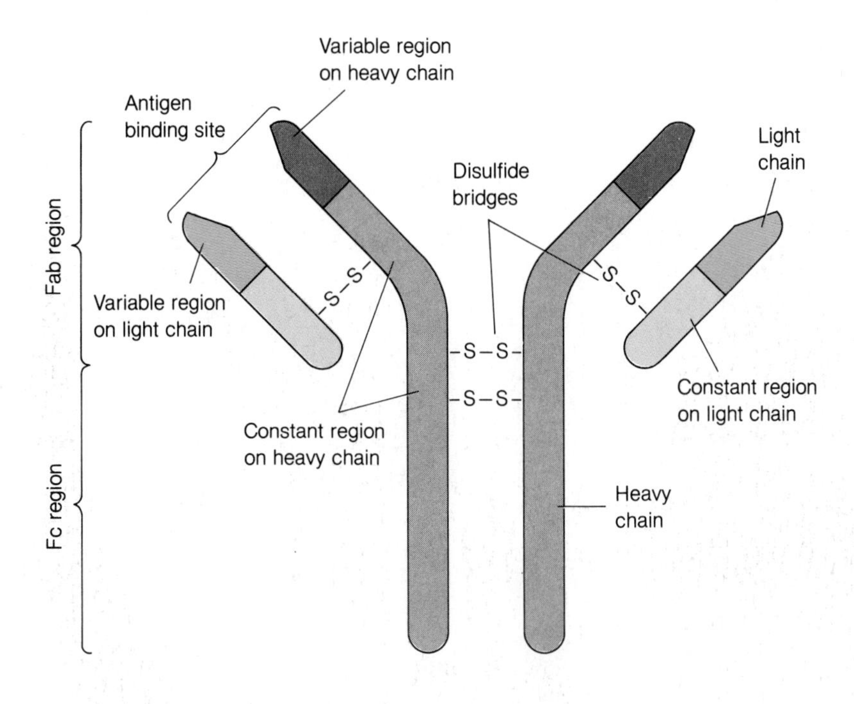

Figure 13.7 **Basic structure of immunoglobulins.** Immunoglobulins are proteins composed of four polypeptides joined by disulfide bonds (–S–S–). Two of the polypeptides are large molecules, referred to as heavy chains: the other two polypeptides are smaller and are called light chains. Each heavy chain is bonded to a light chain. The two halves are joined to form a bivalent antibody molecule. The anterior portion of each light and heavy chain has a variable region. The variable regions of the heavy and light chains form the antigen binding site (Fab region), which confers specificity on the molecule. The three-dimensional shape of the Fab region is complementary to that of the antigen with which it reacts. The Fc region is made up of the constant regions of both heavy chains and may bind to cell surfaces and activate certain components of complement.

while the stem, known as the **Fc** region (fig. 13.7), may bind to various types of cells or be involved in activating complement. The binding between immunoglobulins and antigens is due to hydrogen bonding, electrostatic bonding, and Van der Waals attraction. The basic immunoglobulin molecule is said to be **bivalent** because its two Fab regions can attach to two antigen molecules.

The specificity with which antibodies bind to antigens is related to the three-dimensional structure of both the antigen and the antibody. The three-dimensional structure of the Fab region is complementary to the shape of the antigenic determinant (fig. 13.6). Any antigenic determinant with a shape that fits within the Fab region of an antibody will react with the antibody. Cross-reacting antibodies have binding sites that are able to fit a number of antigenic determinants.

Immunoglobulin Variability

An animal is able to produce millions of different immunoglobulins that can recognize and bind specifically to millions of antigens. It is believed that millions of different lymphocytes are constantly developing in animals, and that each produces one of the millions of possible immunoglobulins. If an antigen is present that can bind to some of the immunoglobulins on the surface of B-lymphocytes, the B-lymphocytes are stimulated to proliferate and differentiate (fig. 13.8). The B-lymphocytes become plasma cells, which secrete antibodies that bind to the antigen that stimulated the B-lymphocyte to become a plasma cell.

Clonal Selection Theory

How does the antigen "know" to stimulate the appropriate lymphocyte? This question has intrigued scientists since the nineteenth century. It is currently thought that the genetic information necessary to synthesize an antibody is expressed as antibody molecules on the plasma membranes of lymphoid cells. The antigen stimulates the multiplication of each type of lymphoid cell when it combines specifically with the antibody molecules present on the surface of the lymphoid cells. Thus the antigen-stimulated cell becomes an immunologically committed cell (fig. 13.8). This theory, called the **clonal selection theory**, was advanced originally by Sir Macfarlane Burnett in the

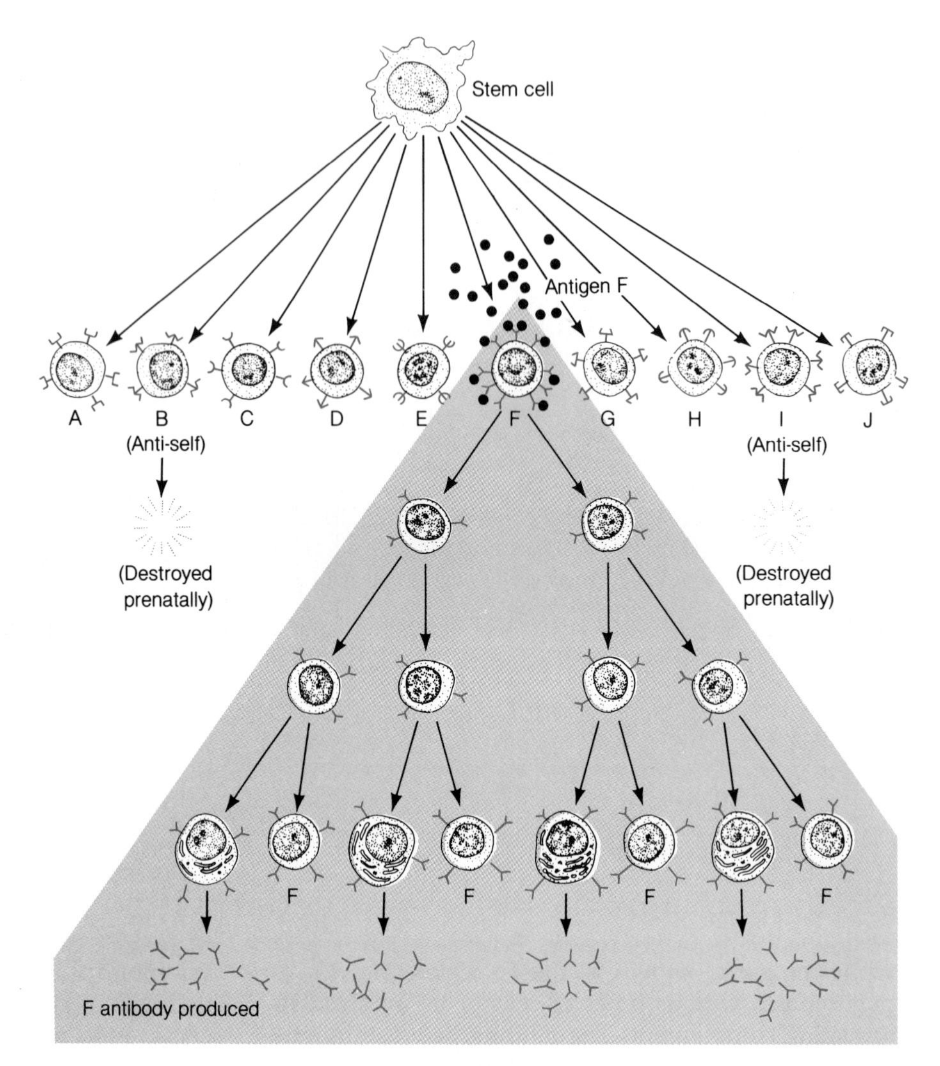

Figure 13.8 The clonal selection hypothesis of antibody production. The clonal selection hypothesis states that the body possesses numerous small clones of immunocompetent lymphocytes, and each clone is able to recognize a different antigenic determinant. This specificity of recognition is expressed on the plasma membrane in the form of antibody receptors. When antigen F enters the body, it binds to clone F and induces the clone to proliferate. This response leads to the differentiation of clone F into additional memory cells and plasma cells that secret antibodies of the same specificity. During prenatal life, certain clones arise that recognize self antigens. These clones are eliminated before birth.

early 1960s and has withstood considerable scientific scrutiny throughout the years.

The clonal selection theory suggests that each lymphocyte has the genetic information to synthesize a particular antibody molecule, and that these molecules can be found "displayed" on the surface (plasma membrane) of each immunocompetent lymphocyte. Each lymphocyte has a different **antibody receptor** on its surface, and when an antigen is introduced into the body, it will bind only to those receptors that have a complementary shape. The reaction between the antigen and the surface receptor stimulates lymphocytes to develop and multiply into a **clone** of cells synthesizing the same type of antibody. Not all the cells in the clone will become antibody-forming cells; some will become memory cells. On first exposure to an antigen, few of the cells will have receptors complementary to the antigen, so there will be a long time lapse between stimulation and the presence of detectable levels of antibodies. On second exposure to the antigen (anamnestic response), however, a clone of memory cells exists that can immediately multiply, giving rise to numerous antibody-forming cells as well as more memory cells. For this reason, the secondary immune response appears much faster and is stronger than the primary immune response.

The clonal selection theory also explains how the immune response differentiates self from nonself. Apparently, during fetal development, lymphocytes carrying surface receptors that recognize self antigens are destroyed. Thus, at birth humans normally do not have lymphocytes with surface receptors that can recognize self antigens. If new lymphocytes develop that attack self antigens, these, too, are eliminated or inhibited by T_S-lymphocytes. As we age there is an ever increasing amount of autoantibody suggesting that this system "makes mistakes" as we grow older.

Classes of Immunoglobulins

There are five classes of immunoglobulins (table 13.3), each of which is constructed from very similar proteins. The various types of immunoglobulins are classified according to the amino acid sequences in the constant region of the heavy chain. These immunoglobulins, which include IgG, IgM, IgA, IgE, and IgD, differ not only in their heavy chains but also in their biological and chemical properties (table 13.3). For example, serum IgM is a pentameric (five basic antibody molecules joined) molecule that appears first during the primary immune response. This immunoglobulin is eventually replaced by the monomeric immunoglobulin IgG. IgA molecules are most common in mucous secretions, where they interact with foreign antigens associated with the mucous membranes of the nasal passages, intestines, urethra, and vagina. The physical and biological properties of the various classes of immunoglobulins are summarized in table 13.3.

Monoclonal Antibodies

Cesar Milstein and Georges Koehler in 1975 developed a technique to produce "immortal" clones of antibody-producing cells that secrete antibodies of a single specificity. For their work, Milstein and Koehler received the Nobel Prize in 1984, along with Niels K. Jerne. This procedure consists of fusing antibody-forming cells with a tumorous culture of B-lymphocytes (myeloma). The resulting cell hybrid is called a **hybridoma.** All antibodies secreted by the clone are specific for a single antigenic determinant. By contrast, live animals produce a spectrum of different antibodies against a variety of antigenic determinants, and as many as 90% of these antibodies do not specifically bind to a particular antigen.

Monoclonal antibodies have been used extensively in research and in diagnostic microbiology. It is now possible to develop monoclonal antibodies to very specific microbial antigens. Monoclonal antibodies are used frequently to diagnose diseases by immunological methods and to detect antigens in tissues and other materials. Experimental drugs consisting of monoclonal antibodies and anti-cancer agents are being studied.

IMMUNE RESPONSE TO ANTIGENS

When an infectious agent or a foreign antigen enters the body, it sets in motion a series of events aimed at ridding the body of the invaders. These responses, called specific immune responses, can be classified as **humoral** or **cell-mediated** immune responses. The humoral immune response involves the synthesis and release of immunoglobulins into body fluids, such as the blood or the mucous secretions. The antibodies act by combining with microbial toxins or viral particles and neutralizing them, or by binding to microbial cells and promoting their phagocytosis. The process whereby antibody molecules promote or enhance phagocytosis is known as **opsonization**.

The level of antibodies (often referred to as the **titer**) produced during an immune response is dependent upon the concentration of antigen. Some materials, when introduced into a host along with an antigen, are able to stimulate the immune response to a greater degree than if the antigen were introduced by itself. Those compounds that are able to enhance the immune response to an antigen are called **adjuvants**.

The cell-mediated immune response is effected by "sensitized" lymphocytes and by macrophages (fig. 13.9). The lymphocytes mediate cellular immunity by releasing a variety of substances that are microbiocidal (or cytocidal), or that stimulate other cells to become microbiocidal. For example, when a macrophage becomes activated by **lymphokines**, polypeptides released by T-lymphocytes (table 13.4), it becomes very aggressive and effective in killing microbial pathogens. The activated macrophage is an important effector of cell-mediated immunity, playing a central role in protecting mammals against intracellular parasites, such as the tubercle bacillus and viruses, by engulfing and subsequently killing them. Other cells that take part in cell-mediated immunity are various types of T-lymphocytes (see table 13.2). The T-lymphocytes act by destroying cancerous cells, virus-infected cells, or infectious microorganisms as a result of toxic substances that they produce.

Humoral and cell-mediated immune mechanisms work in unison to rid the body of harmful microorganisms and cells. If either of these two branches of the specific immune system is faulty, serious disease usually results.

Induction of the Immune Response

When an antigen enters the body, it is promptly attacked by macrophages and partly digested. A portion of the

TABLE 13.3
CHARACTERISTICS OF THE FIVE MAJOR CLASSES OF HUMAN IMMUNOGLOBULINS

Class	Diagram	Sedimentation constant	Relative abundance in serum (%)	Subclasses	Chain subunits			Arrangement of chains	Main characteristics
					Light	Heavy	Others		
IgG		7S	80	IgG_1 IgG_2 IgG_3 IgG_4	Kappa or lambda	Gamma		Monomer	Present in blood and lymph. Half-life is 23 days. Active against microbes and toxins.
IgM		19S	6	IgM	Kappa or lambda	Mu	J	Pentamer	Present in blood and lymph. Half-life is 5 days. Appears early in infection. Effective against microorganisms.
IgA		7S 11S	13	IgA_1 IgA_2	Kappa or lambda	Alpha	J	Monomer Dimer	Present in secretions and lymph. Half-life is 6 days. Protects body's external surfaces. Found in tears, saliva, vagina mucous membranes.
IgE		8S	0.002	IgE	Kappa or lambda	Epsilon		Monomer	Plays role in allergic reactions. Half-life is 3 days.
IgD		7S	1	IgD	Kappa or lambda	Delta		Monomer	Present on surface of lymphocytes. Half-life is 3 days.

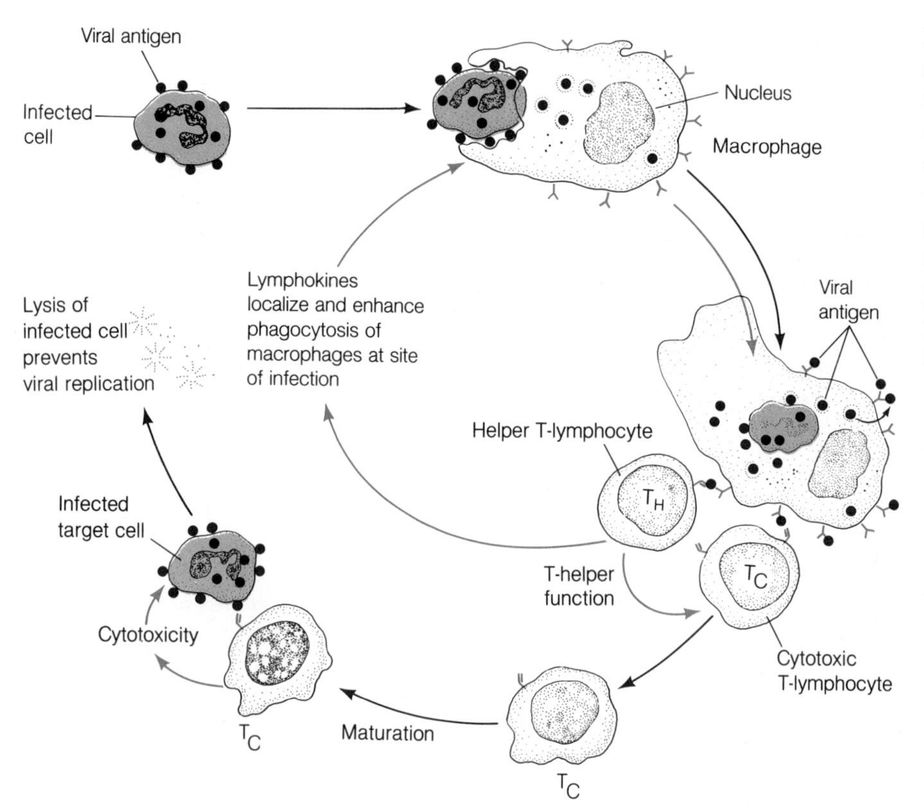

Figure 13.9 Cell-mediated immune mechanisms. T-Lymphocytes and macrophages are the effector cells of cell-mediated immunity. T-Lymphocytes may develop into cytotoxic cells, which can then destroy virus-infected cells or tumor cells or may produce substances called lymphokines. Lymphokines have various functions (see table 13.4) aimed at mounting a strong immune response.

TABLE 13.4
PROPERTIES OF VARIOUS LYMPHOKINES PRODUCED BY T-CELLS

Lymphokine	Cell affected	Mode of action
Macrophage activation factor	Macrophage	Enhances killing by macrophages
Chemotactic factor	Macrophage	Attracts macrophages to site of infection
Migration inhibition factor	Macrophages	Inhibits migration of macrophages from site of infection
Mitogenic factor	Lymphocyte	Promotes cell division
Chemotactic factor	Lymphocyte	Attracts lymphocytes to antigen
Leukocyte chemotactic factors	Eosinophils	Attracts eosinophils
	Basophils	Attracts basophils
	Neutrophils	Attracts neutrophils
Lymphotoxins	Various cells	Kills cells
Colony stimulating factor	Various cells	Promotes mitosis
Interferon	Various cells	Inhibits viral replication

FOCUS IS THERE A LINK BETWEEN THE BRAIN AND THE IMMUNE SYSTEM?

You have no doubt heard of "miraculous" cures of patients with terminal diseases. These cures can baffle physicians and kindle speculation of faith healing or of "willing" the disease away. Apparently, the cure of diseases using willpower to combat them has recruited a faithful following and the concept of positive thinking has emerged. The use of positive thinking as a medical treatment is not looked upon with favor by the American Medical Association, largely because of the unorthodox methods that are practiced and because the many physical and chemical links between the brain and the immune system have only recently begun to be elucidated.

A new group of medical scientists called psychoneuroimmunologists (PNI) claim that positive thinking and the maintenance of strong positive emotions can be used in the treatment of diseases. They assert that the brain regulates the immune defenses by sending signals to cells of the immune system, which are then stimulated to multiply and kill microorganisms. The brain–immune system connection, they contend, is bidirectional in that the immune system acts like a sixth sense in "informing" the brain of infections. It is claimed that the changes in body temperature, heart rate, and sleep in persons afflicted by infectious diseases are a direct result of the stimulation of the brain centers by the immune system.

How do the brain and the immune system communicate with each other? PNI scientists say that the communication is carried out in three different ways: by hormones, by a direct connection between nerve fibers and lymphocytes, and by neurotransmitters called **neuropeptides**. Evidence of a direct connection between the brain and the immune system is seen in the fact that many nerve fibers pass through areas where numerous lymphocytes are found. These fibers, they suggest, are linked directly with the lymphocytes so that there is physical as well as chemical contact between them.

The most intriguing means of communication between the brain and the immune system involves neurotransmitters called neuropeptides. It has been shown that neuropeptides control the emotions of the individual and PNI scientists suggest that they also induce immunological changes in the cells of the immune system. Hence, one's emotions and attitudes could favorably or adversely affect the outcome of infections or cancers by stimulating the secretion (or undersecretion) of certain neuropeptides.

The claims of the PNI scientists are being scrutinized very carefully and the scientific and medical communities are far from convinced that positive thinking and emotions really work. However, PNI scientists have at least gone to the laboratory to test their ideas, instead of simply talking about them.

antigen is incorporated into the plasma membrane of the macrophage so that it can be recognized by other cells of the immune system. These macrophages with the incorporated antigen interact with helper T-lymphocytes (table 13.2) which are then stimulated to react (fig. 13.10). The helper T-lymphocytes induce B-lymphocytes to develop into antibody-forming cells (plasma cells) or into memory cells. The B-lymphocytes must also interact with the antigen to be stimulated to become plasma cells. A plausible mechanism of the induction of the immune response is outlined in figure 13.10.

COMPLEMENT

Complement is a term given to a group of serum proteins and their cleavage products that are involved in disrupting plasma membranes. In addition, some complement components stimulate the release of histamine from leukocytes and so are indirectly involved in the inflammatory process. Other complement proteins stimulate chemotaxis of neutrophils, monocytes, and eosinophils, as well as enhance phagocytosis (opsonization) by phagocytes and platelet-stimulated neutralization of viruses (fig. 13.11). Complement

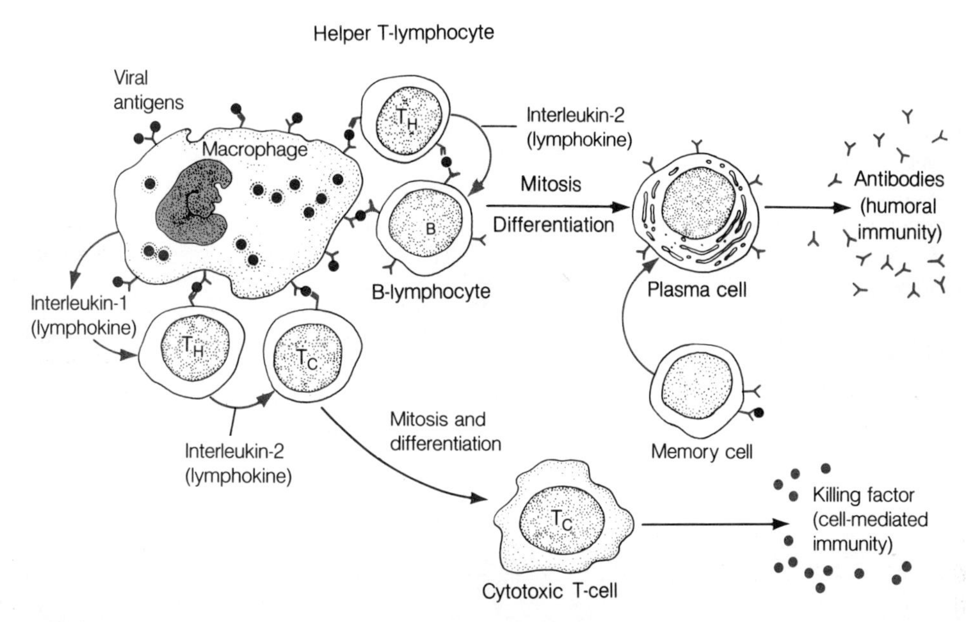

Figure 13.10 The induction of the immune response. The immune response is induced by the cooperative interaction among B-lymphocytes, helper T-lymphocytes, and macrophages. The macrophages process antigen and stimulate helper T-lymphocytes. These, in turn, interact with B-lymphocytes. B-lymphocytes differentiate into plasma cells, which secrete antibodies. Soluble factors (proteins? hormones?) are also involved in the induction process. Helper T-lymphocytes may also interact with other T-lymphocytes to induce cell-mediated immunity.

components are thought to be produced by various cells, including those that make up the epithelium of the gastrointestinal mucosa.

The Classical Pathway of Complement Activation

Complement proteins in the blood and lymph are not active until they interact with antigen-bound immunoglobulins (immune complexes). Not all immunoglobulins can **fix** (initiate the complement reaction) complement; only IgM and IgG can do so. A single IgM molecule attached to an antigen can activate complement, but it takes two adjacent molecules of IgG to activate complement (fig. 13.12) by the classical pathway. The complement cascade begins when C1 components of complement bind to the Fc region of the antigen-bound IgG or IgM immunoglobulins (fig. 13.12). The activated C1 complex, in turn, activates C4 and C2. Two products of this activation, C2a and C4b, form an enzyme that binds to membranes and activates the C3 component of complement. Activated C3a component is an extremely potent peptide that causes the release of chemicals from certain leukocytes, leading to inflammation, pain, and the contraction of smooth muscle. C3a also induces chemotaxis in a variety of phagocytes. Another C3 product, C3b, associates with the membrane-bound complement components, C4b and C2a, and promotes the activation of other components that attach to the membrane of target cells and disrupts them. The complex of complement components disrupts the cell by creating pores in the plasma membrane. This alteration leads to cell death, because the internal physiology of the cell is irreversibly altered.

The Properdin Pathway

The **properdin pathway** of complement activation (fig. 13.13), also called the alternative pathway of complement activation, is independent of immunoglobulins.

Figure 13.11 Functions of complement. Complement has various protective functions in host responses to infection. (*a*) Its main function is to destroy foreign cells by damaging their plasma membranes. (*b*) Some components of complement can also enhance binding of phagocytes to target cells (opsonization) and hence improve the efficiency of phagocytosis. (*c*) Complement can also stimulate basophils and mast cells to release histamine (inflammatory response) and (*d*) attract various cells of the immune system (chemotaxis).

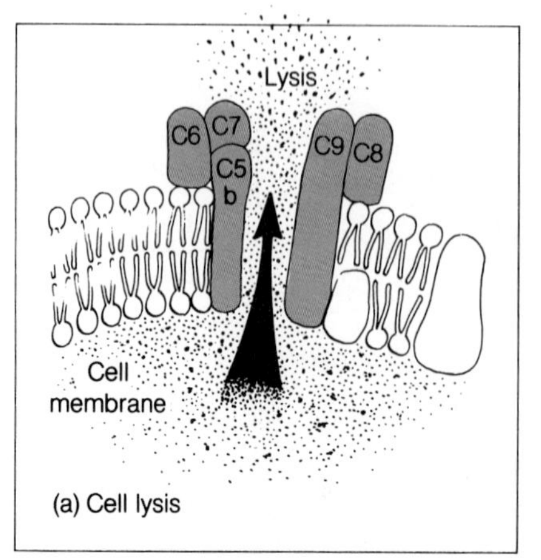

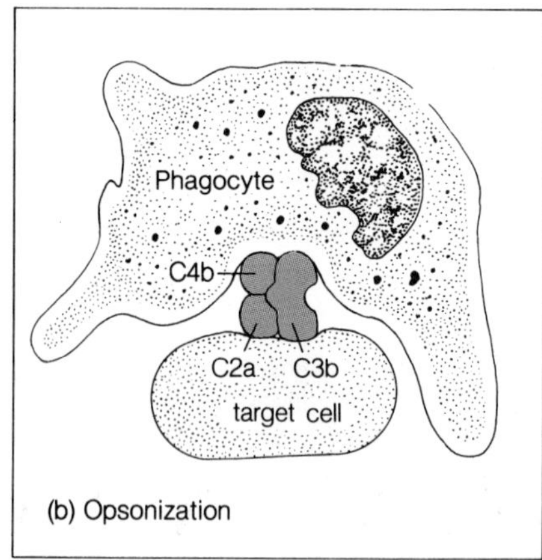

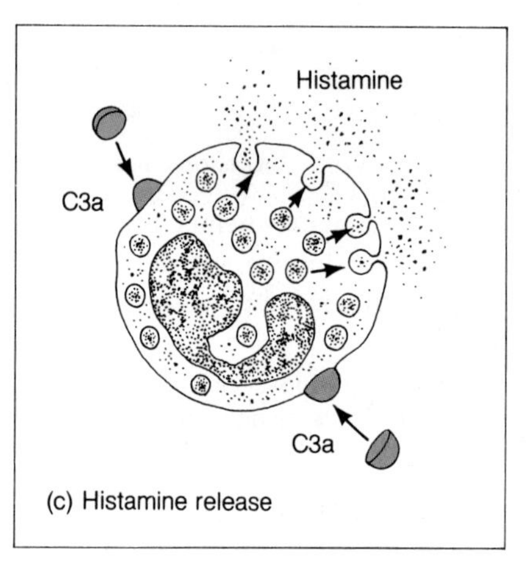

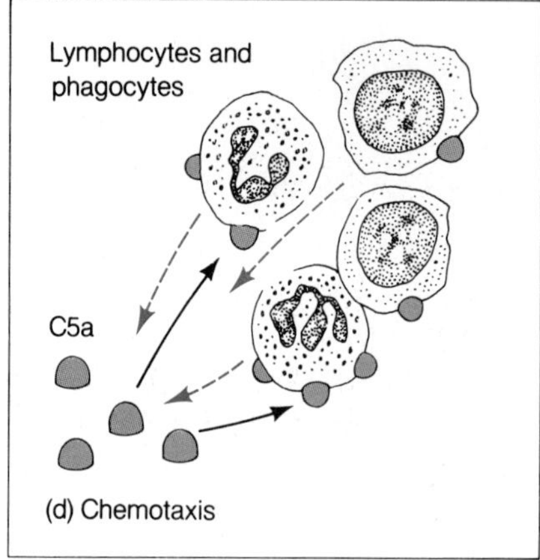

Insoluble polysaccharides and other molecules from bacteria and yeast can activate C3 in the absence of C1, C2, and C4. This antibody-independent activation is an important initial defense against various types of infection.

REACTIONS BETWEEN ANTIGENS AND ANTIBODIES

The immune response is specific for the stimulating antigen. Hence, if we become infected with a particular microbe, we probably will develop an immune response against the microbe. By detecting the presence of immunity to infectious agents using antigen–antibody reactions it is possible to diagnose infectious diseases. The field of immunology that deals with antigen–antibody reactions is called **serology.** **Serological tests** are widely used in the diagnosis of infectious diseases because they are rapid and, if properly controlled, yield reliable results.

Figure 13.13 The properdin pathway of complement activation. *(a)* The alternative, or properdin, pathway is triggered when complement component C3 is stimulated to split when it binds to certain polysaccharides on cell surfaces. C3b attaches to the membrane. *(b, c)* Component B then attaches to C3b and is activated by a serine esterase, D. *(d)* Bb catalyzes the formulation of more C3b, which then attaches to the membrane. As more B attaches to the C3b, a chain reaction occurs. This results in an amplification of the complement reaction. The C3bBb complex is stabilized by properdin (P). The enzyme called convertase Bb also activates C5. This leads to the complement cascade illustrated in figure 13.12 (steps d through f). ▶

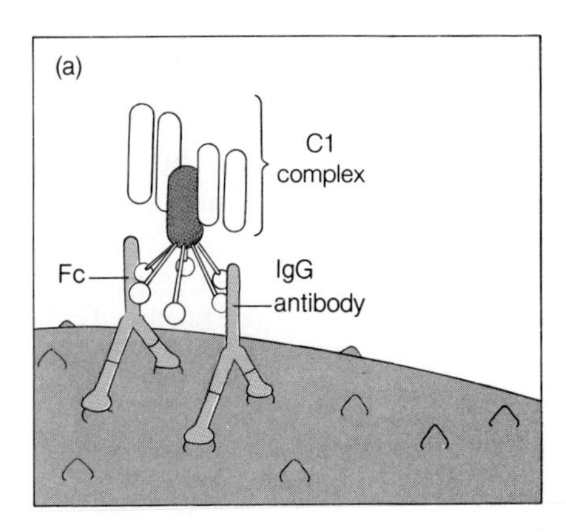

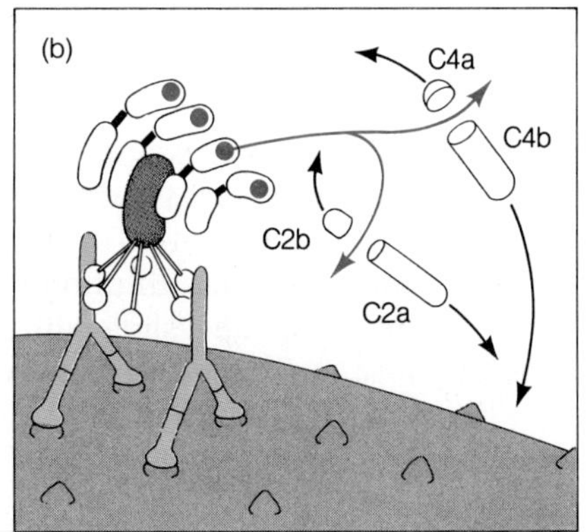

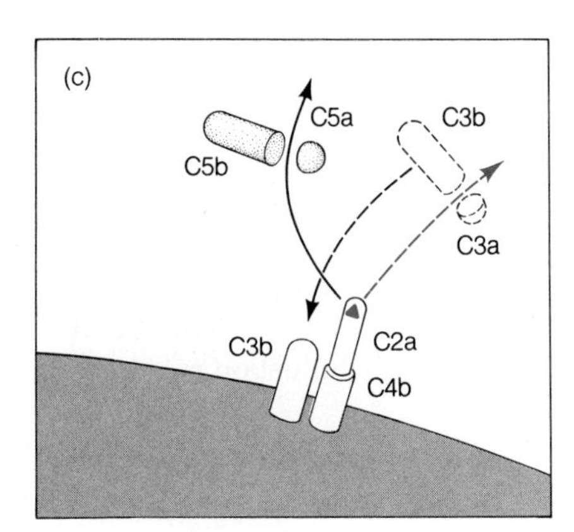

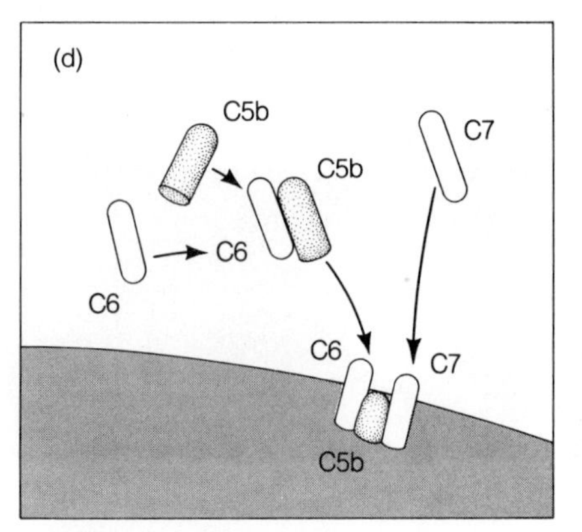

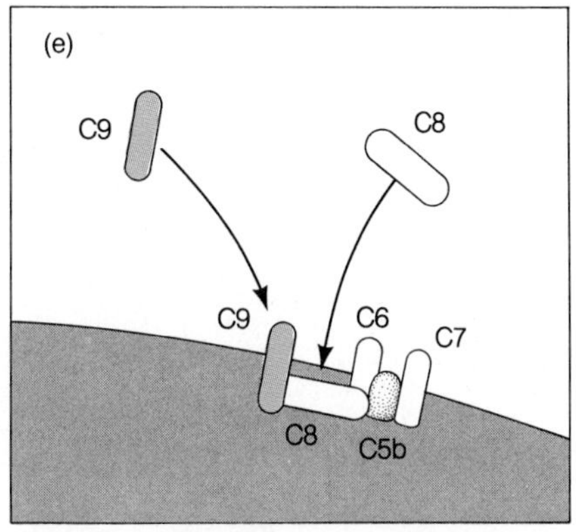

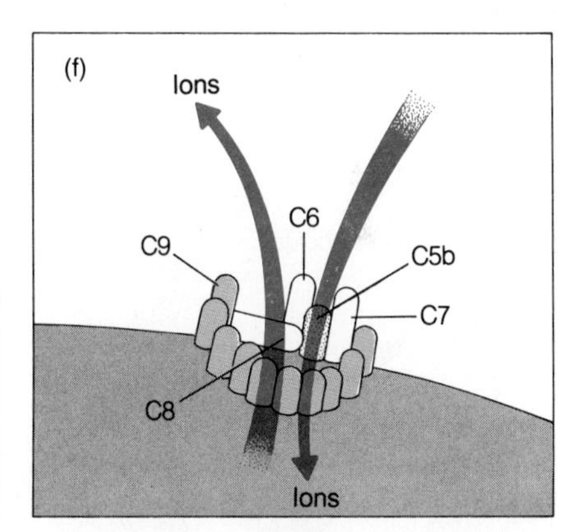

Figure 13.12 **The complement cascade.** (*a*) At the start of the classical complement activation pathway, antibodies recognize and bind to the antigen (e.g., foreign cells or bacteria). The complement component C1, binds to two adjacent IgG antibody molecules. Binding by C1 subunit to the antibody alters its configuration, which results in conformational changes in C1, exposing proteolytic sites. (*b*) The C1 now cleaves two other complement components, C4 and C2, allowing the C4b fragment to attach to the cell membrane. C2a then binds to C4b, producing $\overline{\text{C4b2a}}$ (C3 convertase), which then cleaves C3. C3b can attach to the cell membrane. (*c*) When C3b attaches itself close to the C3 convertase, it alters the specificity of the enzyme site on the C2a, which can now split C5, and is thus called a C5 convertase. (*d*, *e*) C5b binds to C6 and C7 and attaches to the cell surface. C8 and C9 now attach to the $\overline{\text{C5b67}}$. The insertion of this late complex damages the cell membrane and permits a rapid flux of ions that ultimately leads to cell lysis.

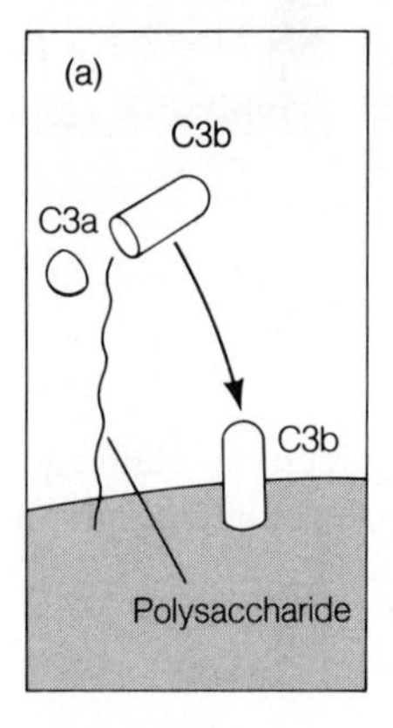

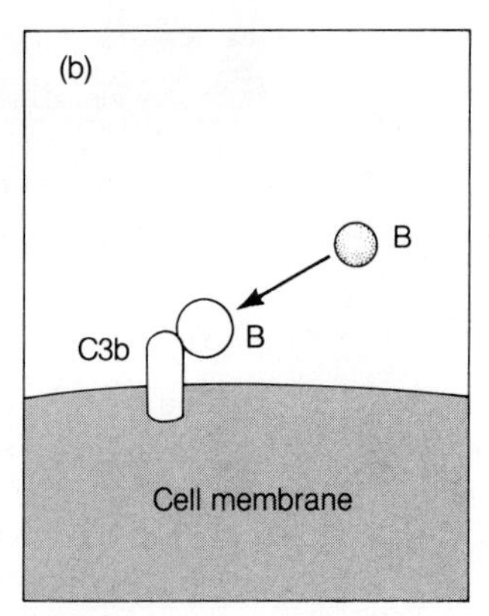

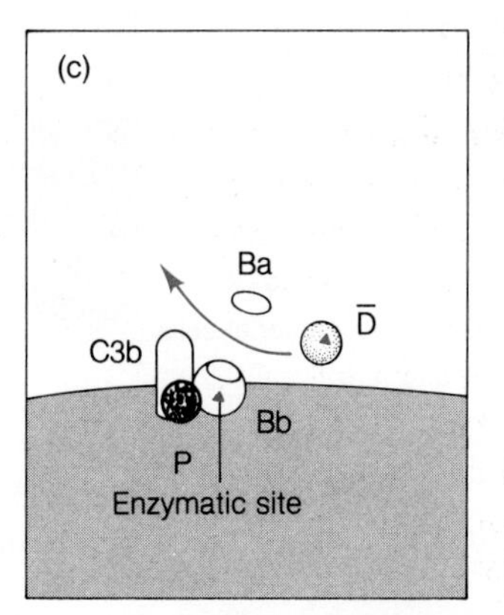

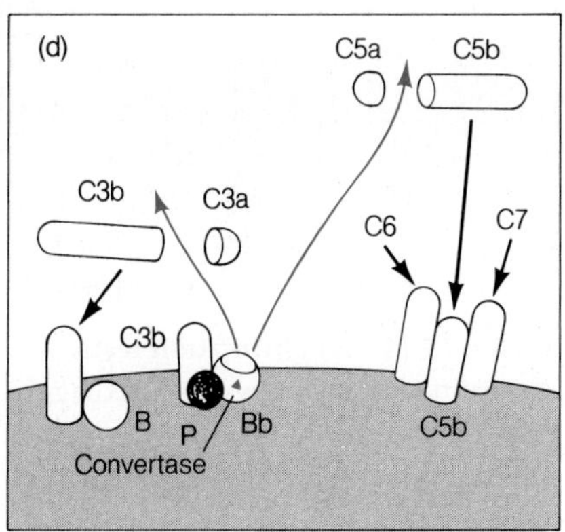

Agglutination Reactions

Agglutination reactions generally involve whole cell antigens (particulate), such as bacterial cells or red blood cells. These antigens contain many different antigenic determinants on their surfaces. Specific antibodies bind selectively to the various antigenic determinants so that cells are joined together by antibodies forming a lattice or network. Some of the most important agglutination reactions are discussed below.

Bacterial Agglutination Bacterial agglutination tests, sometimes called **Widal** tests, are used routinely in many laboratories for diagnosing infectious diseases and identifying bacterial isolates. These procedures include tube and slide agglutination tests (fig. 13.14). To carry out such a test, the patient's serum (or a dilution of the serum) is mixed in a test tube or on a slide with a standard suspension of bacterial cells (the antigen). The bacteria–antibody mixture is allowed

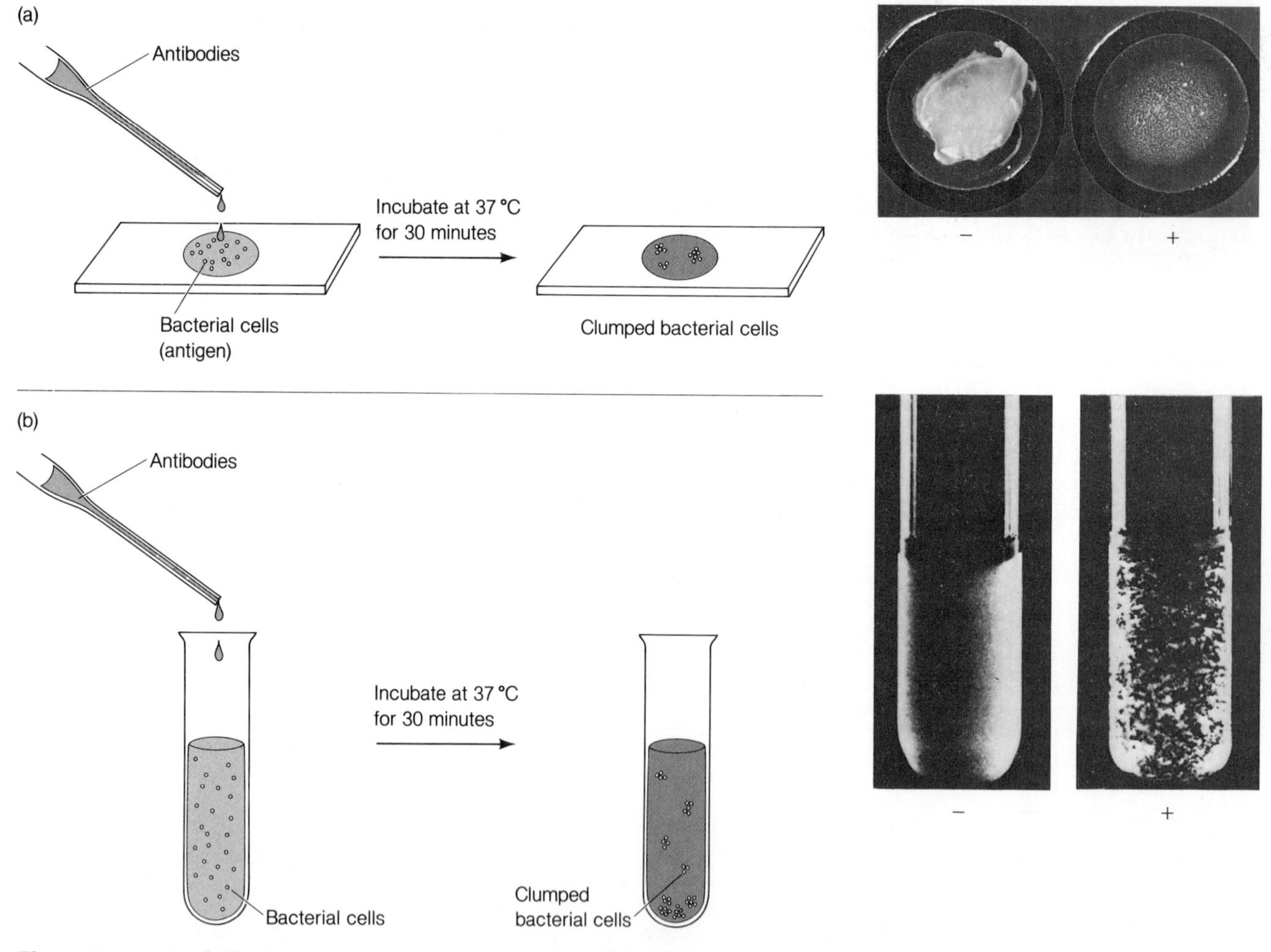

Figure 13.14 Agglutination tests. Agglutination tests are used to diagnose certain infectious diseases or to identify unknown microorganisms. A drop of antigen (bacterial cells) is mixed with a standard volume of antiserum and mixed well. Positive results are detected by evidence of clumping of the antigen. *(a)* Slide agglutination test. *(b)* Tube agglutination test.

to react at a constant temperature for a specified amount of time and is then read for agglutination. Test bacteria that clump (agglutinate) indicate that the patient's serum contains antibodies against the bacteria and hence, has been previously exposed to the agents.

Hemagglutination Reactions Hemagglutinations are agglutination reactions using red blood cells (fig. 13.15). Red blood cells are used because various antigens can be easily attached to their surface, the tests are easy to read, and the results obtained are very reliable. Many infectious diseases are diagnosed using red blood cells with antigens from the infectious agent attached to their surface.

Blood typing is a hemagglutination test that is carried out like a slide agglutination test. This procedure, used to type blood for transfusions, is based on the fact that red blood cells have one or two antigens, called A and B antigens, on their plasma membrane. These antigens are glycolipids (molecules composed of a polysaccharide portion and a lipid portion) that can induce antibody production and can react with these antibodies. The blood-typing procedure consists of mixing the patient's blood with antisera (serum containing antibodies) against the A and B antigens. After a brief period of incubation, the drops are examined for evidence of clumping. The clumping indicates that the antiserum has agglutinated the red blood cell. The pattern of clumping determines the blood type of the individual. Figure 13.15 illustrates the agglutination patterns of the four ABO blood groups.

Indirect hemagglutination techniques are widely used in clinical microbiology to diagnose viral and other infectious diseases. The procedure is based on the fact that many antigens, such as proteins and polysaccharides, can be adsorbed (attached) to red blood cells (RBCs). Thus, the RBCs serve as an easily observed indicator system for serological reactions. When RBCs suspended in a fluid such as physiological saline are left undisturbed, they settle to the bottom of the test tube, forming a distinct red button (fig. 13.16). If, however, these RBCs are agglutinated by a serological reaction between antibodies and the antigens adsorbed to the surface of the RBCs, the button does not form. Instead, a diffuse layer of RBCs is formed in the bottom of the tube (fig. 13.16). Agglutination of antigen-coated red blood cells by the patient's serum normally indicate that antibodies against the antigen are present in the blood.

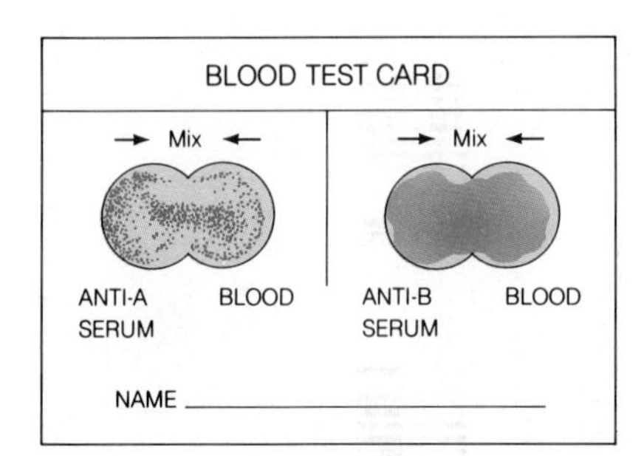

Group A

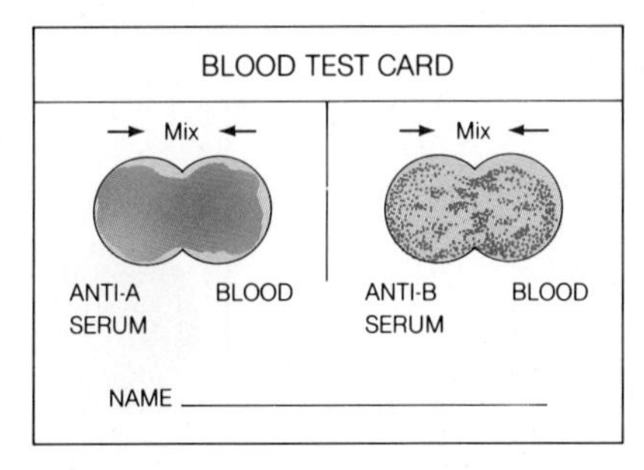

Group B

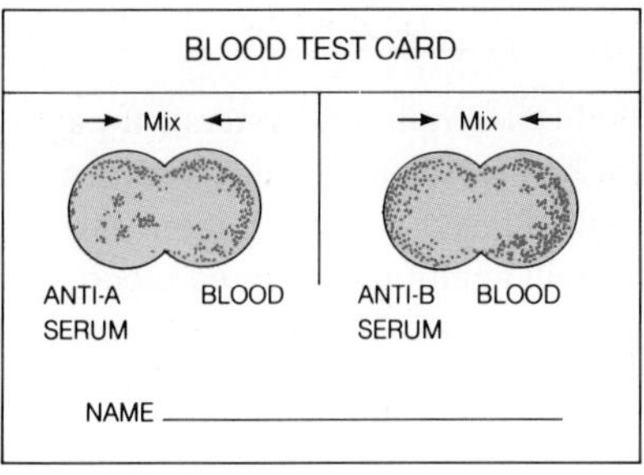

Group AB

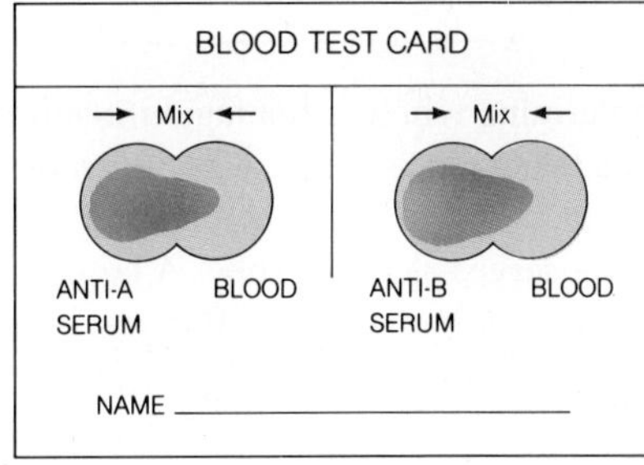

Group O

Figure 13.15 Agglutination patterns of human ABO blood groups. Blood typing is carried out by mixing known antisera against A and B antigens that are present on the surface of red blood cells. Blood that is agglutinated with anti-A antiserum only is A type blood. Blood agglutinated by anti-B antiserum only is type B blood. Blood agglutinated by both anti-A and anti-B antiserum is blood type AB. Blood that is not agglutinated by either antiserum is type O blood.

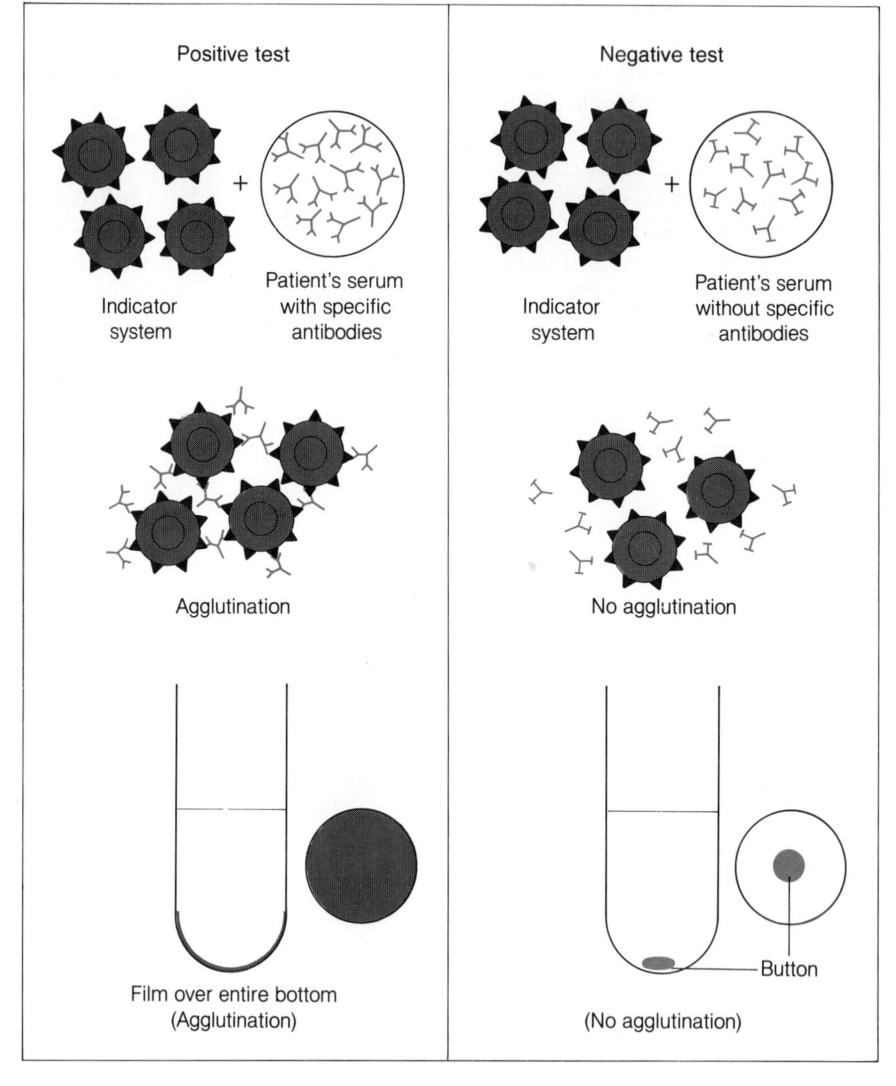

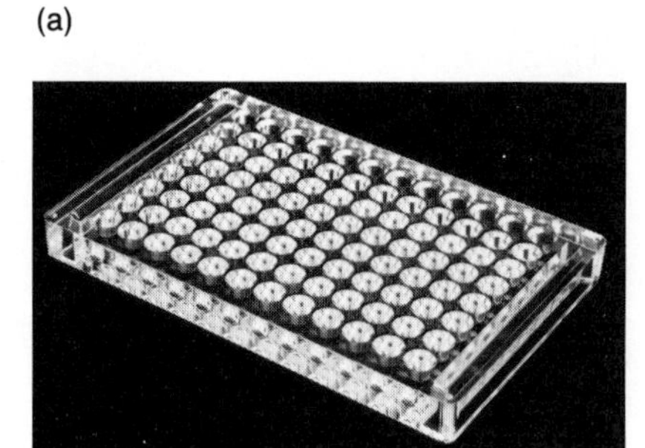

Figure 13.16 Hemagglutination test. When red blood cells are allowed to settle in a tube, they form a small button at the bottom of the tube. When blood cells are aggluntinated by antibodies (or viruses), the clumps settle and form a thin layer all over the bottom of the tube. By viewing the bottom of the tube, one can easily determine whether hemagglutination has taken place or not. *(a)* A microtiter plate. Each well in the microtiter plate serves as a tiny tube. A large number of hemagglutination tests can be carried out and examined quickly. *(b)* Diagrammatic representation of a passive hemagglutination test. Antigen is adsorbed to the surface of red blood cells (indicator system). If patient has antibodies against antigen A (positive test), then the patient's serum will agglutinate the red blood cells: otherwise (negative test) a button will develop as an indication of no hemagglutination.

Immunofluorescence

Immunofluorescence techniques are serological methods for detecting antigens or antibodies in serum or clinical materials. These techniques make use of special dyes, called **fluorochromes**, that emit visible light when stimulated with ultraviolet light. Two of the most popular fluorochromes are fluorescein isothiocyanate (FITC) and rhodamine isothiocyanate (RITC). These dyes are normally attached to antibodies, which can then be made to react with the specimen. The results are detected by observing the specimen with a microscope equipped with a fluorescent light source. If the specimen has antigens that are bound by the fluorochrome-labeled antibody, it will be lit up when examined with the fluorescence microscope. Figure 13.17 illustrates two methods of detecting antigens using immunofluorescence.

Precipitin Tests

Precipitin tests involve antigen–antibody reactions in which the antigens are soluble substances (fig. 13.18). When a volume of serum known to contain antibodies against a soluble antigen is mixed with an aqueous solution of the antigen, these two components will react with each other in a fashion similar to that of agglutination. Eventually, after a period of minutes to hours, the antigen and the antibodies will form a cross-linked matrix that will precipitate out of solution. For

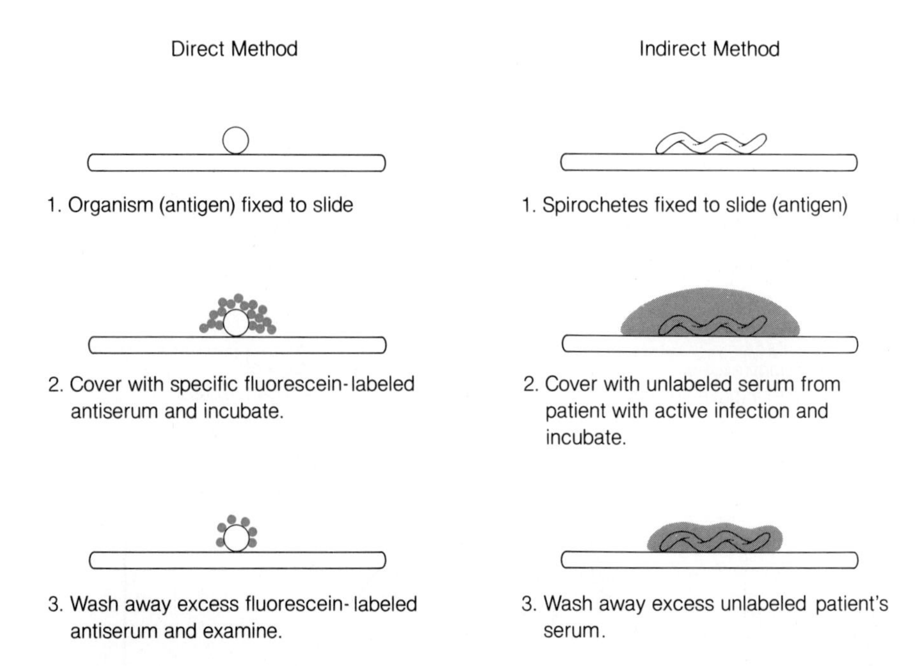

Figure 13.17 **Fluorescent antibody test.** The fluorescent antibody test is performed by adding antibodies labeled with a fluorescent dye to a specimen. The antibodies bind to the antigen, which can then be viewed with a fluorescence microscope. Positive results will be detected because the specimens will fluoresce. *(a)* Direct fluorescent antibody test. *(b)* Indirect fluorescent antibody test.

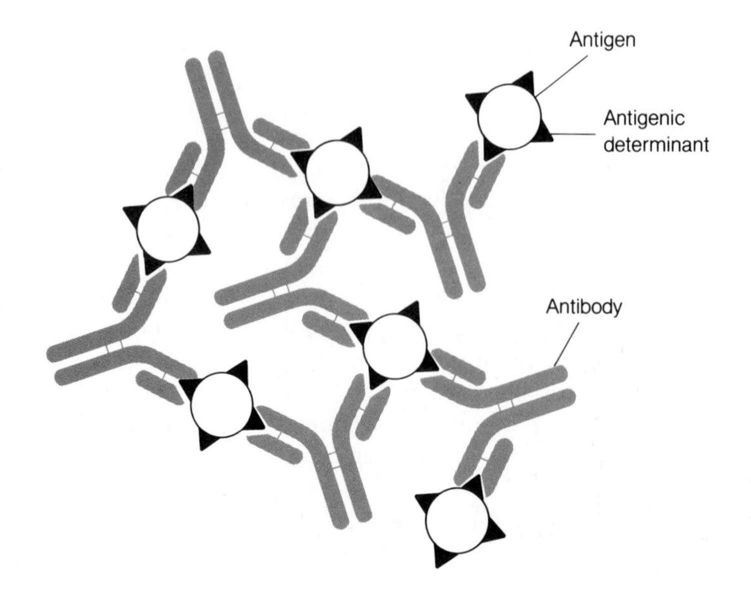

Figure 13.18 Cross-linking in antigen-antibody reactions. Antibodies are polyvalent molecules (i.e., they can bind to two or more antigen molecules). Antigen–antibody reactions can lead to the formation of lattice networks created by antibodies binding to antigens. The clumps can be so large that they settle out. Inside the host, these clumps can be phagocytosed by PMNs or macrophages.

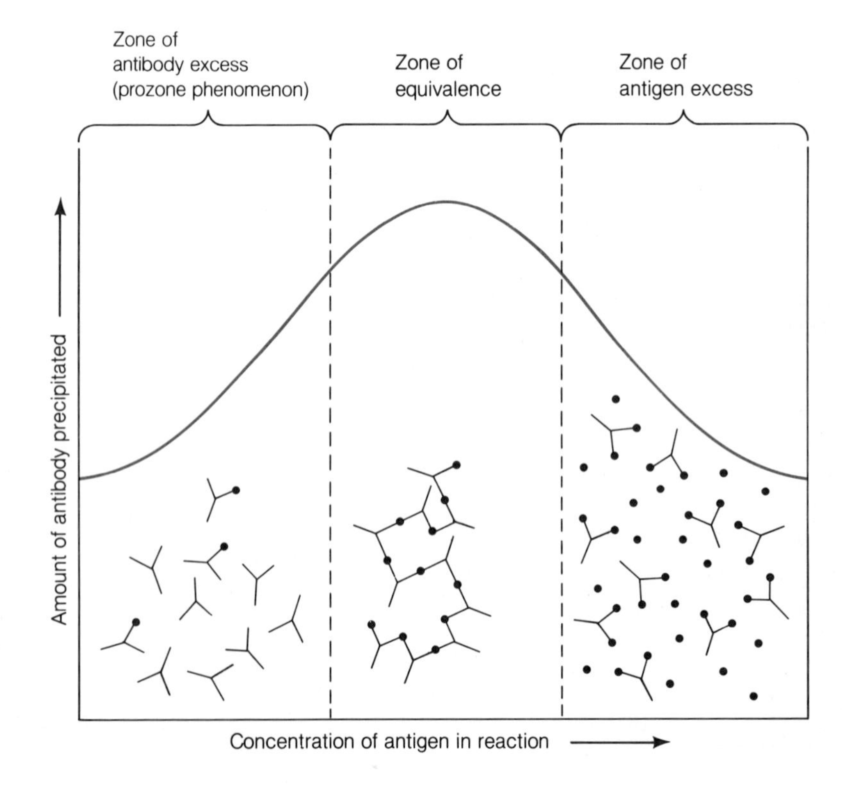

Figure 13.19 The classical precipitin reaction. The degree of cross-linking in antigen–antibody reactions is determined largely by the relative concentrations of antigen and antibody present. When increasing concentrations of antigen are reacted with a standard concentration of antibody and the amount of precipitated antigen–antibody complexes measured, the results resemble those illustrated. At low antigen concentration, there is an excess of antibody, and very little cross-linking occurs. The precipitate (representing antigen–antibody complexes) is small and the supernatant fluid (containing unbound reagents) has a large concentration of antibody. At optimal concentrations of antigen, extensive cross-linking occurs and forms a large precipitate. Little or no free antigen or antibody will be left in the supernatant fluid. At high concentrations of antigen, again little cross-linking takes place.

example, if a volume of tetanus toxin is mixed with a suitable amount of antitoxin in a capillary tube, a line of precipitation will form where the antigen and the antibodies combine in optimal proportions. This procedure can also be carried out in small test tubes with similar results; in this case, it is called the tube precipitin test (fig. 13.19).

Gel Precipitation Techniques All serological tests involving precipitin reactions should be closely monitored to insure that the antigen and the antibody are present in optimal proportions. Several dilutions of the reagents must be made so that optimal proportions are achieved. For this reason, various modifications of the precipitin test using gels have been developed. The purpose of the gel is to provide a medium for diffusion of antigen and antibody, thereby achieving optimal proportions in a single tube and obviating the need for multiple dilutions.

The **Ouchterlony immunodiffusion** (ID) test is the most widely used of the agar gel techniques. In a common procedure, a volume of liquefied agarose (a purified agar) is poured into a 60-mm petri dish and allowed to gel on a cool, level surface. Once the agarose has hardened, a pattern of wells (fig. 13.20) is cut into it. The antigen is placed in one well (usually the center well), and the antisera are placed in peripheral wells. Both antigen and antibody diffuse radially, forming a concentration gradient from a high concentration in the well to a low concentration at the edge of the diffusion pattern. The antigen and antibody form a precipitate where they meet in optimal proportions (fig. 13.20).

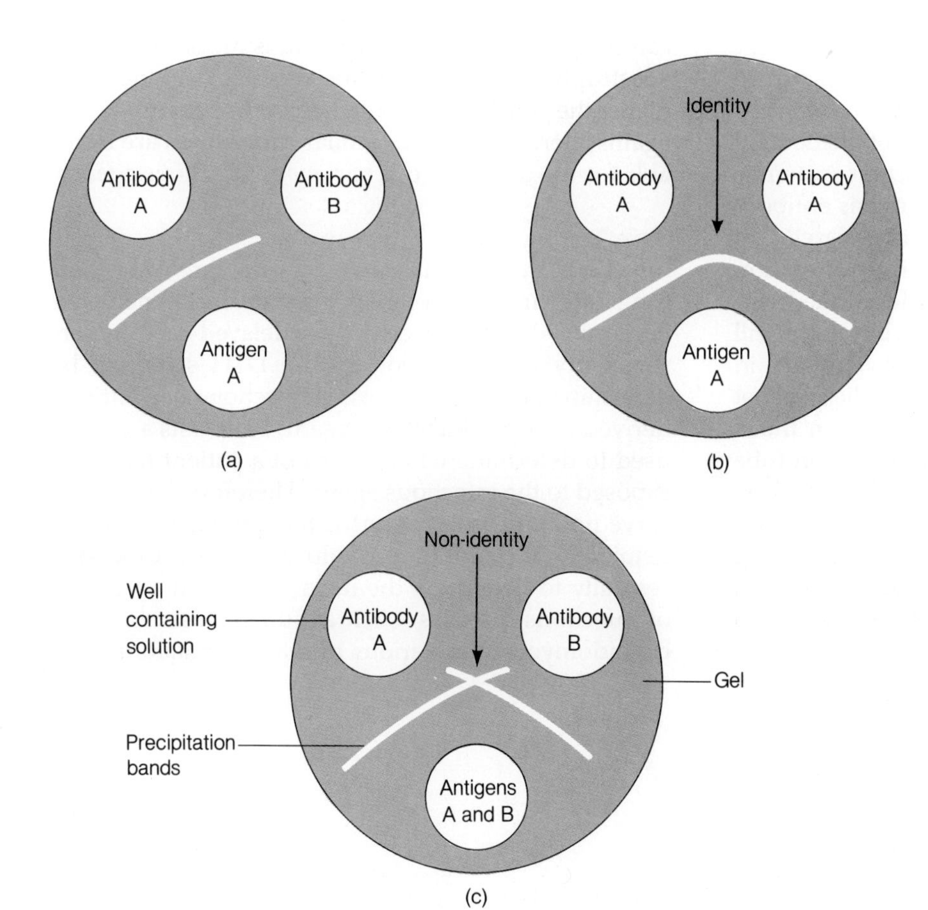

Figure 13.20 The immunodiffusion test. Antigen is placed in a well in an agar plate and the antiserum in another. During an incubation period, both antigen and antibody diffuse. Where the zones of diffusion meet in optimal proportions, they will form a band. Each antigen reacts with its corresponding antibody and forms a line of precipitation.

The ID technique can separate various antigens based on differences in diffusion rates exhibited by various components of the antigen preparation. In cases where there is more than one antigen in the well, a different line of precipitation forms for each antigen (fig. 13.20). Immunodiffusion techniques are widely used as screening tests for a variety of infectious diseases. They have the advantage of being relatively easy to set up, and the results can be obtained within 24 hours.

Complement Fixation Test

The **complement** fixation test (fig. 13.21) detects the presence of specific antibodies in the patient's serum by observing whether or not red blood cells are lysed (disrupted) by complement. The complement fixation test is carried out by using an **indicator** system and a **test** system. The indicator system consists of sheep red blood cells and antibodies (called **hemolysins**) against these cells. The indicator system provides a visible means of detecting the fixation of complement. The test system consists of the patient's serum, a known soluble antigen, and a known amount of complement (obtained from guinea pig serum) so that all of it is used up when an antigen-antibody reaction occurs.

In a complement fixation test, the patient's serum is made to react with the known antigen in the presence of complement. After a period of incubation (usually 37°C for 30 minutes), the indicator system is added (fig. 13.21). If the patient's serum has antibodies against the antigen, the immune complexes that form will fix all of the complement in the reaction tube. When the indicator system is added, the hemolysin will react with the erythrocytes, but there will be no complement left to fix. As a consequence, in cases where the patient's serum has antibodies against the test antigen, the erythrocytes will remain intact and form a button at the bottom of the reaction tube (fig. 13.21). If the patient's serum lacks antibodies, however, no immune complexes will form. Thus, when the indicator system is added, the hemolysin binding to the erythrocytes will fix complement and the red cells will be lysed, turning the medium red due to the hemoglobin released from the blood cells.

Enzyme Immunoassays

Enzyme immunoassays (EIA), also called **enzyme-linked immunosorbent assays** or ELISA tests, are widely used in the laboratory for the diagnosis of infectious diseases. The advantage of this type of procedure is that it is well adapted to automation. Most EIA procedures currently on the market are designed to detect the presence of specific antibodies against disease agents such as herpes virus, rubella virus, cytomegalovirus, and protozoa, as well as a variety of bacterial and fungal pathogens.

Let us examine an ELISA procedure that is commonly employed for the detection of herpes simplex antibodies (fig. 13.22). All the necessary reagents can be purchased in a kit, which also provides microtiter plates with the viruses adsorbed to the bottom of each well. Upon addition of the patient's serum, any antibodies against the virus which are present will attach to the viruses adsorbed to the wells. The unbound antibody is subsequently rinsed off and a measured amount of anti-human immunoglobulin, conjugated with an enzyme such as alkaline phosphatase, is added to the wells. During this period, the enzyme-conjugated anti-immunoglobulin binds to the patient's antibodies, which are attached to the virus (fig. 13.22). After an incubation period, the contents of the cuvette are rinsed several times and the substrate for the enzyme, in this case *p*-nitrophenyl phosphate (PNPP), is added to the wells. The enzymatic breakdown of the PNPP is allowed to take place for a specified amount of time, after which the amount of substrate degraded can be seen as a color change and is measured with a spectrophotometer. The more substrate degraded, the higher the antibody concentration in the patient's serum against herpes simplex. Similar procedures are available for the detection of a variety of infectious agents.

Skin Testing

Infections by many bacteria, viruses, protozoa, and fungi often induce a long-lasting state of cell-mediated immunity that is reflected by a delayed-type hypersensitivity (DTH) response. The DTH state can be determined by the intradermal injection of an antigen derived from the infectious agent. Skin tests are often used to determine whether or not a patient has been exposed to the infectious agent. Therefore, these tests serve not only as a diagnostic tool but also as an epidemiological tool. The procedure has been used successfully to determine the incidence and distribution of infectious diseases, such as tuberculosis and coccidioidomycosis, in various human populations.

Figure 13.21 **The complement fixation test** ▶

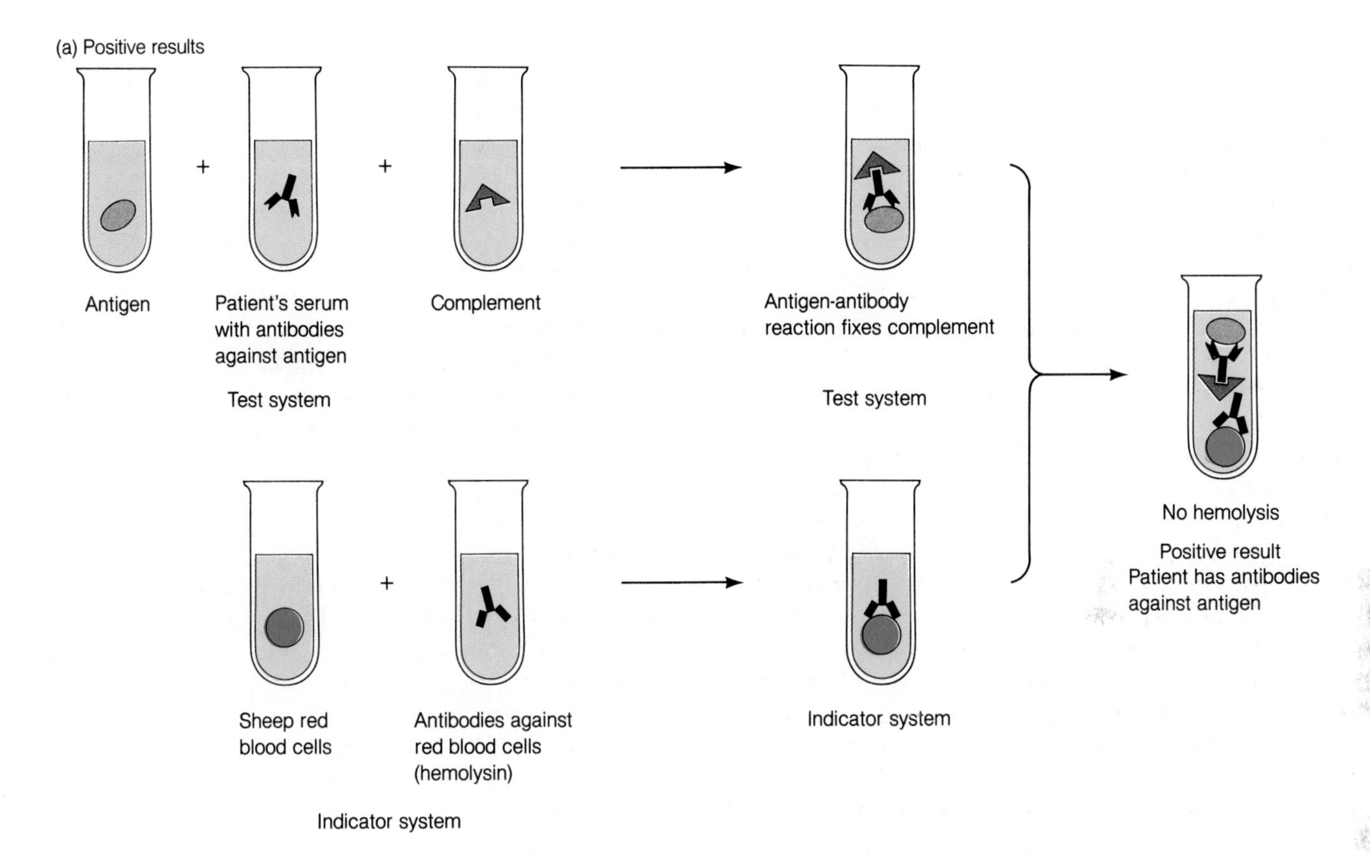
(a) Positive results
Antigen
Patient's serum with antibodies against antigen
Complement
Test system
Antigen-antibody reaction fixes complement
Test system
Sheep red blood cells
Antibodies against red blood cells (hemolysin)
Indicator system
Indicator system
No hemolysis
Positive result
Patient has antibodies against antigen

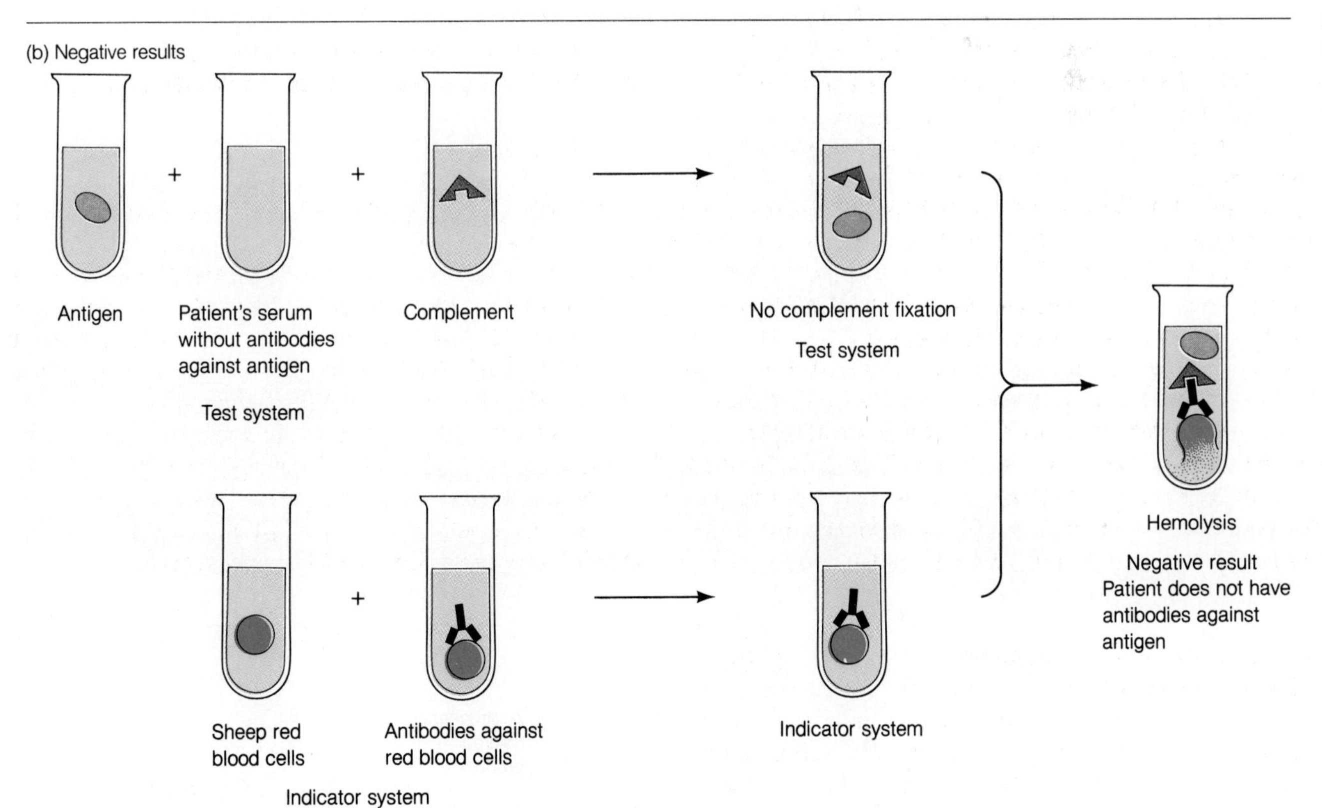
(b) Negative results
Antigen
Patient's serum without antibodies against antigen
Complement
Test system
No complement fixation
Test system
Sheep red blood cells
Antibodies against red blood cells
Indicator system
Indicator system
Hemolysis
Negative result
Patient does not have antibodies against antigen

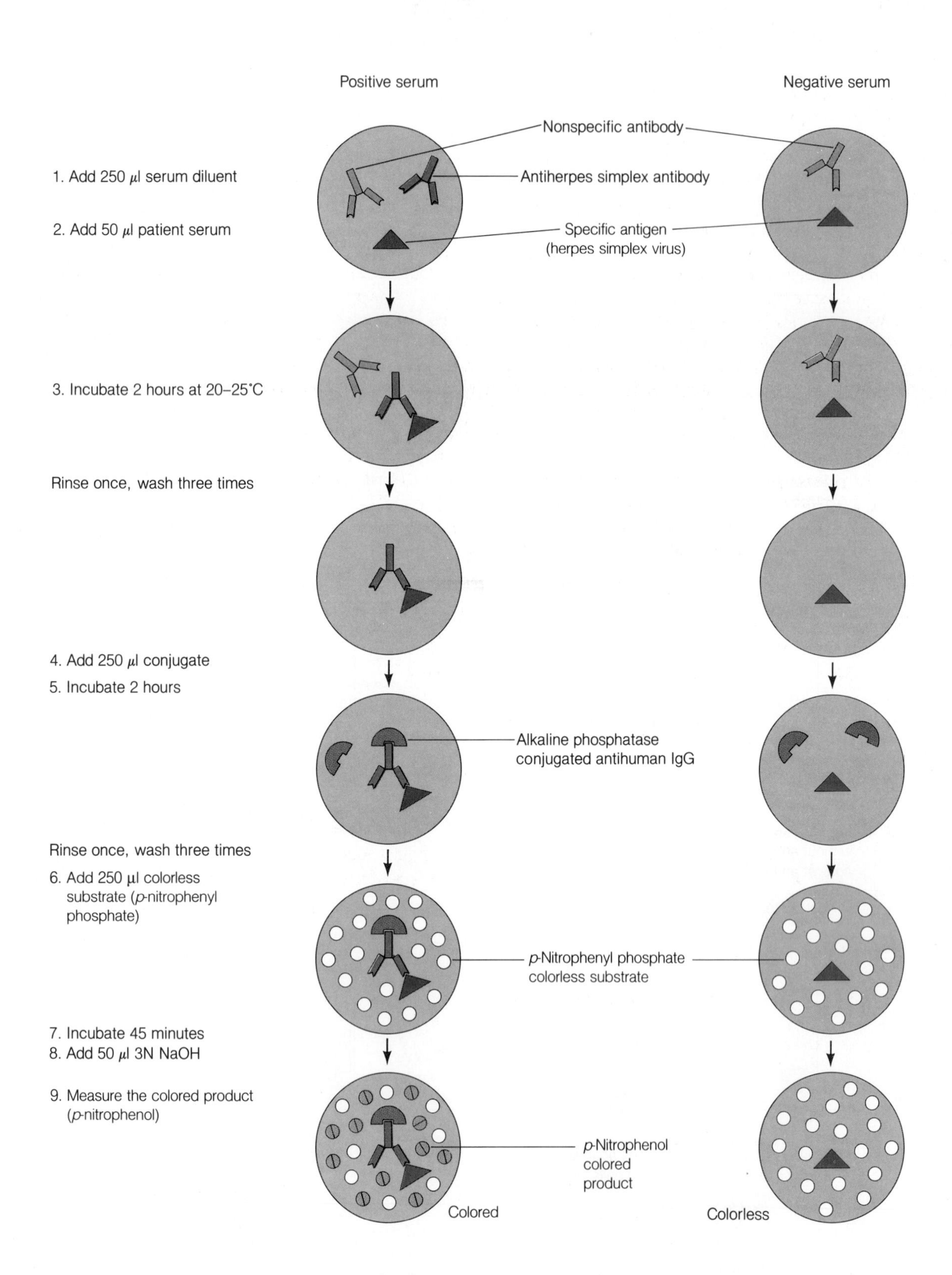
Positive serum
Negative serum
Nonspecific antibody
1. Add 250 μl serum diluent
Antiherpes simplex antibody
2. Add 50 μl patient serum
Specific antigen
(herpes simplex virus)
3. Incubate 2 hours at 20–25°C
Rinse once, wash three times
4. Add 250 μl conjugate
5. Incubate 2 hours
Alkaline phosphatase
conjugated antihuman IgG
Rinse once, wash three times
6. Add 250 μl colorless
substrate (p-nitrophenyl
phosphate)
p-Nitrophenyl phosphate
colorless substrate
7. Incubate 45 minutes
8. Add 50 μl 3N NaOH
9. Measure the colored product
(p-nitrophenol)
p-Nitrophenol
colored
product
Colored
Colorless

CLINICAL PERSPECTIVE CONTROLLING PAIN BY CONTROLLING THE IMMUNE RESPONSE

A 32-year-old male Philippine emigrant was hospitalized with lower right quadrant pain and multiple erythematous (reddish) nodules on the face, abdomen, and legs. He had a temperature of 101°F and a white blood cell count of 17,000/ml (normal 7500). For 2 years he had suffered from painful, erythematous nodules of the skin that frequently developed necrotic (containing dead cells) centers yet subsided spontaneously over a period of a week. He had used corticosteroids occasionally to speed recovery.

The lower right quadrant pain indicated appendicitis and an inflamed and necrotic appendix was removed. At the same time the skin nodules were biopsied. Histological study of the skin specimens indicated that the man was also suffering from lepromatous leprosy and erythema nodosum leprosum (an inflammation of blood vessels with subcutaneous skin nodules).

Soon after the man's operation, he developed shaking chills and a temperature of more than 104°F that subsequently oscillated between 100° and 103°F. Skin lesions, some of which developed necrotic centers, increased. In addition, the patient's testes became swollen and painful (orchitis), and he suffered from extreme nerve pain (neuritis). To control the temperature, skin nodules, swelling, and pain, thalidomide and prednisone treatment was initiated. Within 48 hours, the patient's temperature had dropped to normal, his neuritis was resolved, and skin nodules had begun to regress.

Thalidomide (the same chemical that caused birth defects) is very effective in eliminating the skin nodules that are due to a hypersensitive (allergic) reaction of the blood vessels and skin against *Mycobacterium* antigens. The mechanism of action of thalidomide is unknown. Prednisone, a corticosteroid, alleviates inflammation as well as pain and was used to speed up the resolution of the skin nodules and neuritis. It was discontinued nine days after the patient entered the hospital when dapsone and rifampin treatment was started to clear up the leprosy. Thalidomide therapy was continued with the antileprosy drugs.

For further information on this case see the following reference: Bullock, W. E. 1986. *Hospital Practice* 21(3):102E–102X.

◀ Figure 13.22 **Enzyme-linked immunosorbent assay (ELISA).** The wells in the microtiter plate are coated with the antigen (herpes simplex viruses). The patient's serum is dispensed into the well and allowed to react with the antigen. After a period of reaction, the wells are washed, and any unbound antibody is removed from the well. At this point, the reagent, which consists of anti-human immunoglobulin antibodies conjugated with alkaline phosphatase (the enzyme), is added. The reagent reacts with the bound antibody and excess reagent is washed away by rinsing. Then the substrate for the enzyme is added (*p*-nitrophenyl phosphate) and the results are assessed by looking for a colored product. If the patient has antibodies against herpes simplex virus, the well will appear colored. If the patient's serum lacks those antibodies there will be no color reaction.

SUMMARY

The immune response protects animals from invading microorganisms and cancerous outgrowths.

TYPES OF IMMUNITY

1. Humoral immunity is due to circulating antibodies, while cell-mediated immunity is due in part to lymphocytes and macrophages.
2. Native immunity is the resistance that an organism has at birth because of its distinct physiology, chemistry, or structure.
3. Acquired immunity is that obtained after exposure to foreign substances.

CHARACTERISTICS OF ACQUIRED IMMUNITY

1. Acquired immunity is characterized by memory, specificity, and recognition of self.
2. Memory allows the immune system to respond swiftly and vigorously to an infectious agent that has been previously encountered.

3. The anamnestic response, which involves the rapid and intense response to antigens that have previously been encountered, is based on the ability of the immune system to "remember" the antigen and the presence of memory cells that can respond rapidly.
4. Repeated vaccinations with the same antigen increase the anamnestic response.
5. The specificity of the immune response allows it to differentiate among antigens.
6. The immune response is capable of differentiating between self and foreign antigens, so that the immune response does not cause damage to the host.

CELLS OF THE IMMUNE SYSTEM

1. In mammals, the B-lymphocytes, granulocytes, macrophages, and monocytes develop primarily in the bone marrow, while the T-lymphocytes develop in the thymus.
2. When stimulated by an antigen, specific B-lymphocytes differentiate into plasma cells (antibody-secreting).
3. Specific T_H, T_S, and T_C multiply in response to lymphokines and antigens. T_C releases a killing factor, T_H participates in helping the development and maintenance of immunity, and T_S supresses immune responses.
4. The neutrophils and macrophages are the major phagocytic cells in the immune system.

ANTIGENS AND ANTIBODIES

1. An antigen is any molecule that can stimulate an immune response. Antigens generally contain numerous antigenic determinants or chemical groups that are recognized by different antibodies.
2. A hapten is a chemical group that is not immunogenic in itself, but when attached to a large molecule, called a carrier, can induce the production of specific antibodies.
3. Antibodies, also called immunoglobulins, are glycoproteins found in blood serum which are produced by cells in the immune system in response to the presence of antigen.
4. Antibodies are bivalent molecules consisting of two light chains and two heavy chains. The arms (Fab regions) of the antibody molecule bind to antigen, while the tails (Fc region) bind to some cells of the immune system and can also fix complement.
5. There are five classes of immunoglobulins: IgG, IgM, IgD, IgA, and IgE.
6. Immunologic tolerance is due to the inability of T-lymphocytes and B-lymphocytes to respond to antigen.

IMMUNE RESPONSE TO ANTIGENS

1. Antibodies protect by combining with antigen and inactivating it or by enhancing its phagocytosis and destruction.
2. Adjuvants are chemicals that enhance the immune response to antigen.
3. Cell-mediated immunity is effected by various subpopulations of lymphocytes and macrophages.
4. The immune response is induced by the interaction of antigen with macrophages, B-lymphocytes, and T-lymphocytes. Antigen stimulates immune cells, and their proliferation and differentiation into effector cells.
5. The clonal selection theory explains how antigens select and stimulate the proper immunocompetent cells to become immune committed.

COMPLEMENT

1. Complement is the name for a number of serum proteins involved in immune responses.
2. Complement proteins frequently are activated when they associate with antibodies attached to antigens.
3. The complement cascade ultimately results in the disruption of plasma membranes of target cells. Certain components of complement, such as C3a and C5a, have other biological properties. Some function as chemotactic factors, allergic factors, and opsonins.

REACTIONS BETWEEN ANTIGENS AND ANTIBODIES

1. Antigens react with antibodies to form lattices of cross-linked molecules, thus causing the precipitation of these molecules. These reactions, called serological reactions, can be used as a means of diagnosing infectious disease.
2. In the classical precipitin reaction, soluble anti-

gens react with antibodies to form a visible precipitate when they combine in optimal proportions.

3. Serological methods for diagnosing infectious diseases are desirable when the suspected agent is difficult or expensive to grow in the laboratory, or when the results are desired quickly.
4. Serological tests commonly used in the laboratory are based on the fact that an individual infected by a microorganism produces antibodies against the infectious agent, and the antibodies can be detected.
5. Serological methods routinely used in the laboratory include agglutination, precipitation, complement fixation immunofluorescence, and skin tests.

STUDY QUESTIONS

1. Comment on the following statements:
   a. An antibody molecule can bind only one type of antigen.
   b. A given antigen may bind to various antibodies.
   c. Antibody production is based on antigen-antibody reactions.
   d. Natural immunity is based on genetic traits.
2. Define the following terms:
   a. hapten;
   b. carrier;
   c. antigen;
   d. antibody;
   e. immunogenic determinant.
3. Describe in detail the structure of an:
   a. antigen;
   b. lgG molecule;
   c. IgM molecule;
   d. IgA molecule.
4. How does the clonal selection theory account for the ability of an animal to respond immunologically to thousands of different antigens?
5. Describe the three characteristics of the immune response.
6. Describe the various types of cells involved in immunity.
7. Describe the process of induction of antibody formation by plasma cells.
8. What would be the consequence of defects in:
   a. B-lymphocyte populations;
   b. T-lymphocyte populations;
   c. macrophage populations.
9. What would happen if the body produced too many T_S-lymphocytes in response to a pathogenic microorganism?
10. What are vaccines and how do they work?
11. Describe the kinetics of antibody formation in response to a first and second exposure to the same antigen.
12. What is the basis of antigen-antibody reactions?

SELF-TEST

1. The type of immunity that develops in an individual after the administration of hyperimmune serum is known as:
   a. innate immunity;
   b. individual immunity;
   c. natural immunity;
   d. active immunity;
   e. passive immunity.
2. Antigens that have similar, although not identical three-dimensional shapes are said to be cross-reactive if they can be bound by the same type of antibody. a. true; b. false.
3. A molecule that by itself cannot induce an immune response but can do so when linked to a large molecule is called:
   a. an antigen;
   b. a carrier;
   c. a hapten;
   d. an antigenic determinant;
   e. an immunoglobulin.
4. The IgM molecule is found in serum in the pentameric form and is the first antibody that appears after a first encounter with antigen. a. true; b. false.
5. Which of the following is a function of macrophages?
   a. production of antibodies;
   b. development into cytotoxic cells;
   c. phagocytization and processing of antigens;
   d. production of interferon;
   e. release of antigens.
6. The main function of supressor T-cells is to inhibit the immune response against self antigens and thus prevent autoimmune diseases. a. true; b. false.
7. Antibodies can agglutinate bacterial antigens because they are bivalent and can bind to two different antigen molecules at once. a. true; b. false.
8. Acute glomerulonephritis and rheumatic fever are diseases that result from the destruction of host tissue by the immune system of the host. These are examples of:
   a. active immunity;
   b. passive immunity;

c. natural immunity;
d. autoimmunity;
e. none of the above.

9. The indirect immunofluorescence test uses labeled antibodies against the specific antigen. a. true; b. false.

10. If horse antibodies are injected into rabbits the rabbits will produce antibodies because the horse antibodies act as antigens in the rabbit. a. true; b. false.

11. The type of immunity that develops after an individual is infected with chickenpox virus is:
a. passive immunity;
b. innate immunity;
c. active immunity;
d. individual immunity;
e. none of the above.

12. The type of immunity that develops after an individual with diphtheria receives horse hyperimmune serum is:
a. passive immunity;
b. innate immunity;
c. active immunity;
d. individual immunity;
e. none of the above.

13. The success of vaccination is based on the fact that a strong acquired immunity develops after a single exposure to antigen. a. true; b. false.

14. The anamnestic response is possible because of the formation of "memory" cells after a first encounter with antigen. a. true; b. false.

15. Because of the specificity of the immune system, a person vaccinated against the rubella virus is not protected against any other virus. a. true; b. false.

16. Autoimmune diseases result when the immune system reacts with foreign substances in the patient's body. a. true; b. false.

17. All of the following are cells involved in the development of a specific immune response *except:*
a. macrophages;
b. T-lymphocyte;
c. B-lymphocyte;
d. plasma cells;
e. neutrophils.

18. Which of the following cells produce and secrete antibodies?
a. macrophages;
b. T-lymphocyte;
c. B-lymphocyte;
d. plasma cells;
e. neutrophils.

19. This cell functions in specific immunity as a helper cell:
a. macrophages;
b. T-lymphocyte;
c. B-lymphocyte;
d. plasma cells;
e. neutrophils.

20. The macrophage and the T-lymphocyte are two important cells in the induction of the specific immune response to microorganisms. a. true; b. false.

21. After the stimulation of B-lymphocytes by antigen, they divide and become plasma cells. a. true; b. false.

22. Cell-mediated immunity is effected by the production of lymphokines and cytotoxic lymphocytes. a. true; b. false.

23. Molecules that induce the development of specific immunity are called immunoglobulins. a. true; b. false.

24. A given antigen may possess two or more different antigenic determinants. a. true; b. false.

25. Antibodies are lipopolysaccharides that are composed of two light chains and two heavy chains. a. true; b. false.

26. The antibody molecule is responsible for binding to antigens in the:
a. Fc region;
b. Fab region;
c. S–S bonds;
d. constant regions of the light chain;
e. constant regions of the heavy chain.

27. According to the clonal selection theory, all the different immune cells develop and are found in small clones after the birth of the fetus. a. true; b. false.

28. The principle upon which serological reactions are based is that antibodies bind specifically to the antigen that induced its formation. a. true; b. false.

SUPPLEMENTAL READINGS

Ada, G. L. and Nossal, G. 1987. The clonal-selection theory. *Scientific American* 257(2):62–69.

Centers for Disease Control, 1985. Provisional public health service inter-agency recommendations for screening donated blood and plasma for antibody to the virus causing acquired immunodeficiency syndrome. *Morbidity and Mortality Weekly Report* 34(1):5–7.

Delmee, M., Homel, M., and Wauters, G. 1985. Serogrouping of *Clostridium* difficile strains by slide agglutination. *Journal of Clinical Microbiology* 21(3):323–327.

Dinome, M. A. and Young, J. D. 1987. How lymphocytes kill tumor and other cellular targets. *Hospital Practice* 22(5A):59–66.

Golub, E. S. 1982. *The cellular basis of the immune response.* Sunderland, MA: Sinauer Assoc.

Grimont, P. A. D., Grimont, F., Displaces, N., and Tchen, P. 1985. DNA probe specific for *Legionella pneumophila*. *Journal of Clinical Microbiology* 21(3):431–437.

Hood, L. E., Weissman, I. L., and Wood, W. B. 1986. *Immunology*. 2d ed. Menlo Park, CA: Benjamin/Cummings.

Jaret, P. and Nilsson, L. 1986. Our immune system: the wars within. *National Geographic* 169(6): 702–735.

Kapp, J. A., Pierce, C. W., and Sorensen, C. M. 1984. Antigen-specific suppressor T-cell factors. *Hospital Practice* 19(8):85–99.

Lennette, E. H., Balows, A., Hausler, W. J. Jr., and Shadomy, H. T., eds. 1985. *Manual of clinical microbiology*. 4th ed. Washington, DC: American Society for Microbiology.

Manser, T., Huang, S. Y., and Gefter, M. L. 1984. Influence of clonal selection on the expression of immunoglobulin variable region genes. *Science* 266(4680):1283–1288.

Merigan, T. C. 1985. Viral vaccines, immunoglobulins and antiviral drugs. *Scientific American Medicine* Section 7–XXXIII. New York: Freeman

Pál, T., Pácsa, A. S., Emödy, L., Vörös, S., and Selley, E. 1985. Modified enzyme-linked immunosorbent assay for detecting enteroinvasive *Escherichia coli* and virulent *Shigella* strains. *Journal of Clinical Microbiology* 21(3):415–418.

Sabin, A. B. 1981. Evaluation of some currently available and prospective vaccines. *Journal of the American Medical Association* 246:236–241.

Unanue, E. R. and Allen, P. M. 1987. The immunoregulatory role of the macrophage. *Hospital Practice* 22(4):87–104.

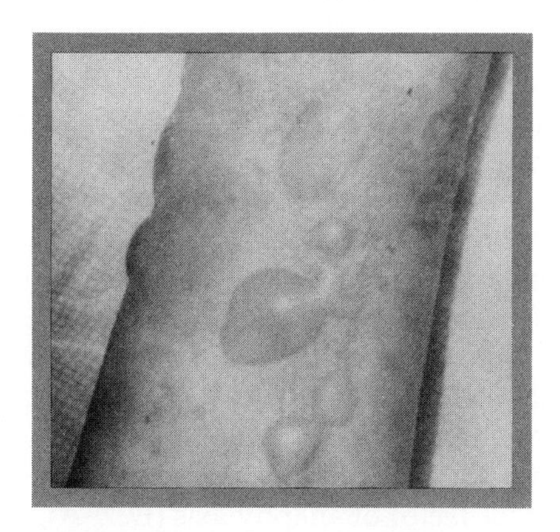

CHAPTER 14

DISORDERS ASSOCIATED WITH THE IMMUNE RESPONSE

CHAPTER PREVIEW THE BOY IN THE PLASTIC BUBBLE

In 1971 a boy named David was born with **combined immune deficiency**, a hereditary condition in which T-lymphocytes and B-lymphocytes are missing. Without these cells, humans develop lethal viral, bacterial, and fungal infections. To save David from certain death, he was placed in a germ-free environment. Until his death in 1984, he was the longest-living survivor of combined immune deficiency.

David lived his entire life (except for the last two weeks) in a plastic bubble of one sort or another, in order to exclude microorganisms (fig. 14.1). His bubble consisted of three large chambers that occupied most of his family's living and dining rooms. Everything that entered the bubble had to be sterilized. Air was forced into his bubble through high-efficiency filters that trapped viruses, bacteria, and fungi. David's food was heat-sterilized and stored in jars, while his paper, books, toys, and other heat-sensitive items were sprayed with a bactericide. Materials were passed to David through a double air lock. Inside the chamber, he prepared his own food. David's garbage, dirty dishes, and soiled clothing were slipped out the air locks.

During his life in the bubble, David never touched another human or an animal. He could touch people and animals indirectly using long rubber gloves that extended outside his bubble. David's schooling was via a closed-circuit TV and telephone hookup with a local school. Teachers visited him occasionally for special tutoring, and classmates joined David at his home for class get-togethers. David was an intelligent child, in the upper 20% of all students tested. When not studying or playing he spent his time watching TV and talking to his friends on the telephone. In spite of his genetic disorder and the fact that he could never leave his bubble, David was a very healthy, intelligent, well-adjusted child. David never had an infectious disease during the first twelve years of his life.

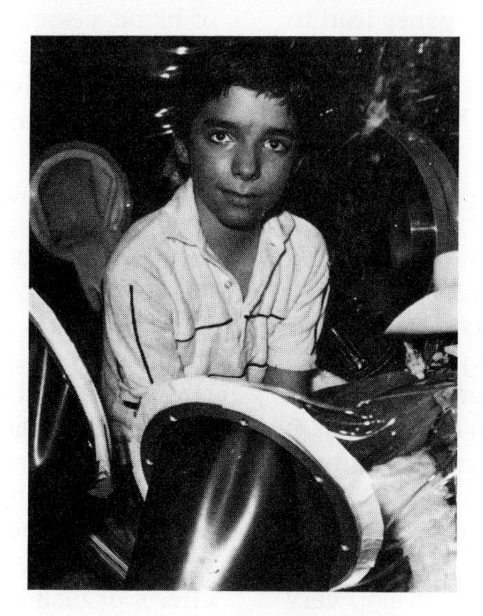

Figure 14.1 **David in his plastic bubble**

In 1983, David underwent a bone marrow transplant to give him an immune system. It was hoped that this operation would allow him to live a normal life in the outside world. Bone marrow cells from his sister were used. Unfortunately, some of the transplanted cells were infected with Epstein-Barr virus (a herpes virus) that caused the B-lymphocytes to become cancerous. David developed B-cell leukemia approximately three months after the transplant. David's first illness was characterized by vomiting and diarrhea. He also developed a stomach ulcer and internal bleeding. In order to be treated, he was removed from his protective bubble and placed in a "sterile" hospital room. He was treated with the drug acyclovir to counteract the Epstein-Barr virus. David died of heart failure two weeks after he emerged from his germ-free plastic bubble.

David's rare condition was due to a defect in his immune system. A great many people suffer because their immune systems do not work as they should. Because of the importance of some disorders associated with the immune system, this chapter is devoted to discussing the relationship between disease and the immune system.

CHAPTER OUTLINE

IN MOST SITUATIONS THE IMMUNE SYSTEM efficiently protects animals and humans from invading microorganisms and toxic substances. Sometimes, however, an immune system does not do its job in a way that is beneficial and can be responsible for troublesome allergies, chronic degenerative diseases, and even an attack upon the very individual that it is supposed to protect. Any of these conditions, in its most severe form, can result in the death of the individual. When the immune system reacts strongly or inappropriately to an antigen and pathological changes result, the individual is said to be **hypersensitive** or **allergic**. The most common of the allergic states are summarized in table 14.1. The purpose of this chapter is to explain how microorganisms and noninfectious agents can lead to an immune response that is detrimental to the host. In addition, there is a discussion on how defective genes lead to defective immune responses and to disease.

IgE-MEDIATED (IMMEDIATE-TYPE) HYPERSENSITIVITY

Some antigens, when reintroduced into a sensitized person, result in **anaphylaxis** (Latin, *ana* = against + *phylaxis* = protection), a situation that often harms rather than protects. **Localized anaphylaxis** is exemplified by such conditions as hay fever, asthma, and hives. **Systemic** (throughout the body) **anaphylaxis** is characterized by dilation of blood vessels, swelling of tissue, contraction of smooth muscle, and pain. These changes in the body often lead to shock, asphyxia, and (if not reversed in time) death. The IgE-mediated **atopic** (foreign or out of place) allergies develop very rapidly and occur after frequent exposures to certain antigens.

The antigens that stimulate the production of IgE antibodies are responsible for immediate-type (IgE-mediated) hypersensitivity (fig. 14.2), so called because the effects of the allergy occur shortly after exposure to an antigen. During immediate-type hypersensitivity, the IgE molecules bind to **mast cells** (located in the respiratory tract lymph, and lining the blood vessels and capillaries) and to **basophils** (found in the blood) through their Fc regions. When an antigen **(allergen)** enters a hypersensitive individual, it binds to specific IgE on mast cells and basophils. The antigen-antibody complexes stimulate these cells to release a number of chemicals, such as histamine, serotonin, eosinophil chemotactic factor of anaphylaxis (ECF-A), heparin, platelet activating factors (PAF), slow-reacting substance of anaphylaxis (SRS-A, also called **leucotriene**), and serotonin (table 14.2). These chemicals are responsible for most of the symptoms of an immediate hypersensitive reaction. Immediate hypersensitive reactions that are not excessive are generally beneficial because they promote the movement of antibodies and white blood cells from the blood into damaged tissue and also stimulate blood clotting. In addition, immediate hypersensitive reactions may have benefits such as ridding an infected gut of worms.

Histamine induces smooth muscle contraction, the release of mucus, and the dilation of blood vessels (fig. 14.2). Excessive smooth muscle contraction in the trachea and bronchi closes these air passages and causes **asphyxia** (lack of air intake) and death if not reversed. Severe uterine contraction during pregnancy can lead to abortion of the fetus. Histamine-induced dilation of blood vessels causes redness of the affected area and allows the escape of fluid from the blood into the tissues. Extensive loss of fluid into the tissues causes swelling and can lead to shock, which occurs when the volume of blood flowing through the body is decreased to the extent that tissues are not properly oxygenated or cleared of CO_2. Under these conditions, organs begin to fail and death may rapidly follow.

The effects of excessive histamine release can be partially reversed by **antihistamines**, such as diphenhydramine and chlorpheniramine maleate. Antihistamines are believed to compete with histamines by inhibiting histamine binding to tissue. Antihistamines help alleviate the symptoms of hay fever, since this allergy is IgE-mediated. **Epinephrine** (adrenaline) is a hormone that is also frequently used to reverse the effects of histamine. Epinephrine stimulates smooth muscle dilation, and this effect leads to the relaxing of the trachea and bronchi and consequently prevents

TABLE 14.1
MAJOR TYPES OF HYPERSENSITIVITY REACTIONS

Antibody-mediated (immediate-type)*			Cell-mediated (delayed-type)**	
I Anaphylactic-type reactions	II Immune-complex reactions	III Cytotoxic reactions	IV Allergy of infection	V Transplant rejection
IgE binds to mast cells or circulating basophils. Antigen combines with bound antibody to trigger the release of various vaso-active amines, such as histamine, serotonin, and slow reacting substance of anaphylaxis.	Antibody combines with a large dose of antigen to produce antigen-antibody complexes that trigger the release of histamine and other mediators. Blood vessels are damaged, resulting in inflammation and necrosis of tissue.	Antigen elicits the formation of antibody, which then combines with target cells. The attached antibody causes the cells to be destroyed by various means (phagocytosis, complement-mediated tissue lysis, and destruction by lymphoid cells).	Antigens of the infectious agent sensitize the infected individual, causing a proliferation of lymphoid cells. Sensitized lymphoid cells react to the presence of the antigens by releasing lymphokines that mediate the hypersensitivity reaction. Humoral antibody not involved.	Tissue from a donor is grafted onto a recipient. A population of sensitized lymphoid cells develops and is instrumental in the destruction of the graft by mechanisms similar to those involved in type IV allergies. Antibody-mediated hypersensitivity can also be involved in graft rejection.
Examples: Food allergies and anaphylactic shock due to bee stings or penicillin injections.	Examples: Serum sickness, rheumatoid arthritis, systemic lupus erythematosus.	Examples: Drug reactions, agranulocytosis, hemolytic anemia, thrombocytopenic purpura, transfusion reactions, and erythroblastosis fetalis.	Examples: Tuberculin reaction and similar manifestations.	Example: Graft rejection.

*Time of onset of reaction occurs within 5–12 hours after contact with antigen.
**Time of onset of reaction is delayed 24–48 hours after contact with antigen.

asphyxiation. The action of epinephrine is very rapid, so individuals undergoing anaphylactic shock can be saved by prompt administration of this hormone.

Serotonin (5-hydroxytryptamine) functions both as a neurotransmitter in the brain and as a chemoeffector of smooth muscle. In some situations, however, serotonin has been found to be 10,000 times more effective than histamine in causing smooth muscle contraction. In addition to causing asphyxiation, serotonin increases the respiratory rate, decreases central nervous system activity, produces pain, and stimulates histamine release. A large amount of serotonin is found in blood platelets and is released when platelets are aggregated and lysed by platelet activating factors (PAFs).

Bradykinin, a small peptide released in minute amounts by mast cells and basophils, also causes smooth muscle to contract. Bradykinin enhances capillary permeability and leukocyte migration outside the blood vessels. In addition, bradykinin may stimulate peripheral nerves, since it induces pain.

Slow-reacting substance of anaphylaxis (SRS-A) causes capillary dilation and smooth muscle contraction, and consequently causes many of the same symptoms as histamine, bradykinin, and serotonin. This mediator of anaphylaxis is also known as a **leucotriene**.

Platelet activating factors (PAFs) cause the aggregation and lysis of platelets, and these effects, in turn, aid clot formation and the release of serotonin.

Eosinophil chemotactic factors of anaphylaxis (ECF-A) draw eosinophils to the inflamed site and are thought to promote some complement fixation reactions. Eosinophils are believed to release antihistaminic substances and help reverse some of the deleterious effects of anaphylaxis (fig. 14.2).

Heparin blocks the formation of thrombin, which converts fibrinogen into fibrin in the blood. Thus heparin decreases blood coagulation.

As you can see, histamine alone does not cause all of the annoying signs and symptoms of hay fever. This is why antihistamines, which do not inhibit the

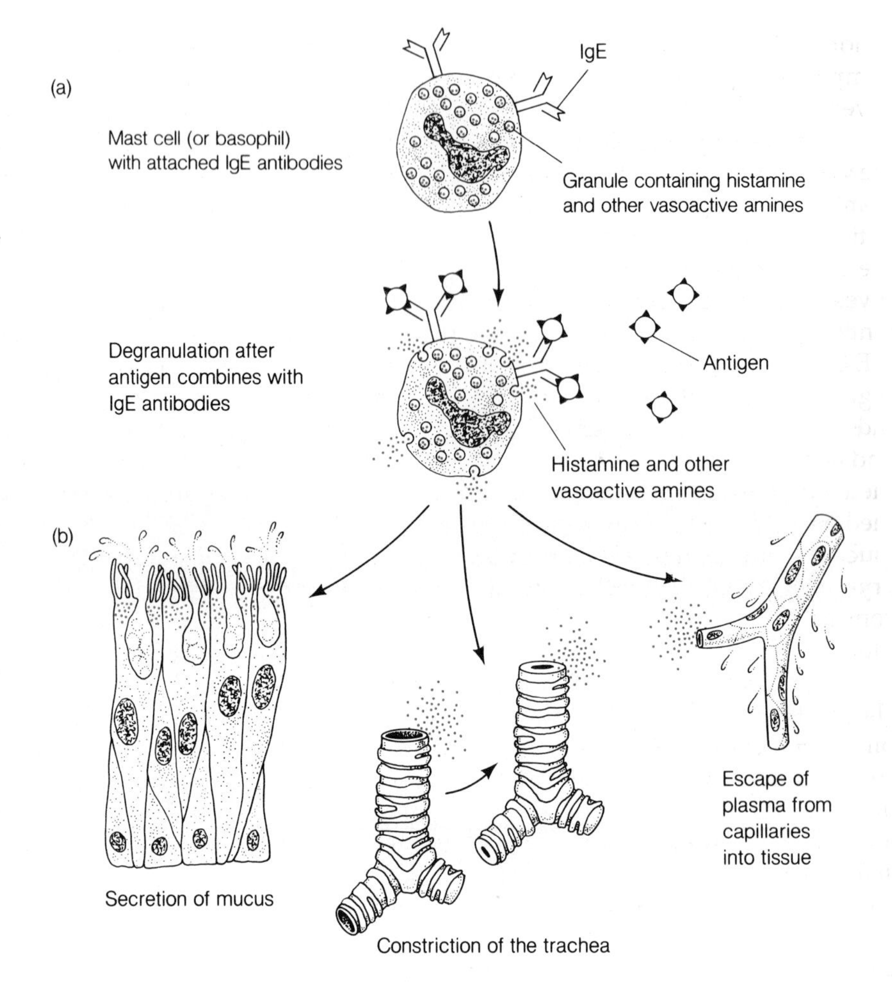

Figure 14.2 Anaphylactic reaction. *(a)* IgE are produced in response to certain antigens. These antibodies attach to mast cells in the lymph and basophils in the blood by their stems (Fc portion). When antigen binds to the arms of the cell-bound antibodies, the mast cells and basophils are stimulated to release histamine and other effector molecules (see table 14.2). *(b)* The effector molecules cause the release of plasma from the capillaries, the release of mucus from cells, and the contraction of smooth muscle in the trachea and intestines.

TABLE 14.2
MEDIATORS OF ANAPHYLAXIS

Mediator	Comment and function
Histamine	Found in mast cells and basophils. Increases blood capillary permeability and smooth muscle contraction of bronchial tubes and attracts eosinophils. Increases gastric secretions of hydrochloric acid.
Eosinophil chemotactic factor of anaphylaxis (ECF-A)	Found in mast cells and basophils. Attracts eosinophils and enhances complement reactions.
Heparin	Found in mast cells and basophils. Inhibits coagulation.
Platelet-activating factors (PAFs)	Found in mast cells and basophils. Aggregate and cause the lysis of platelets, which release serotonin.
Slow-reacting substance of anaphylaxis (SRS-A)	Formed and secreted by mast cells and basophils after they are activated. Increases blood capillary permeability and smooth muscle contraction of bronchial tubes.
Serotonin	Found in blood platelets. Increases blood capillary permeability.

action of bradykinin, serotonin, and SRS-A, do not completely alleviate hay fever, but simply reduce its severity.

Localized IgE-mediated hypersensitive reactions are very common in humans. Allergies to fungal spores, plant antigens (many of which are pollens), animal antigens (dander), insect venoms, and sometimes foods are responsible for hay fever, asthma, eczema, and hives. Figure 14.3 illustrates how an insect's venom (antigen) can stimulate the production of IgE, and how IgE bind to mast cells and basophils through their Fc regions. When antigen is reintroduced in a sensitized individual, it binds to the Fab regions of the antibody and induces the release of histamine and other cell mediators. An atopic allergy results when the cell mediators bind to goblet cells, causing the release of mucus; to capillaries, causing **edema** (swelling) and **erythema** (redness); and to smooth muscle, causing contraction of the trachea, bronchi, intestines, and uterus, as well as pain.

Hay fever (allergic rhinitis) is an IgE-mediated allergic reaction that generally affects the upper respiratory tract and includes such symptoms as nasal congestion, draining sinuses, puffiness and swelling in the nasal passages and around the eyes, itching, and sneezing. Antihistamines alleviate most hay fever symptoms, apparently because histamine is the major mediator of these allergies.

Asthma is an allergic reaction that results in the contraction of the trachea and bronchi. The inability to fill the lungs, accompanied by wheezing and the lack of response to antihistamines, is a sure sign of this type of allergic reaction. Antihistamines are not effective in reversing the smooth muscle contraction because its main mediators are leucotrienes (SRS-A) and serotonin. This condition can be reversed by epinephrine and aminophylline.

Hives (urticaria) is an IgE-mediated allergic reaction that results in wheals on the skin (fig. 14.3). **Wheals** are swollen white to pink blotches that appear on the skin sometimes within minutes after contact with the allergen. Allergens (antigens that induce allergies) that do not touch the skin, such as those in foods, can also cause hives. Food allergies can cause diarrhea, vomiting, and severe pain in the intestines instead of hives. Eczema is a very severe IgE-mediated allergic reaction of the skin and is characterized by inflammation, itching, and the formation of scaly skin.

Cutaneous anaphylaxis is an IgE-mediated allergic reaction of the skin to an intradermal injection of an allergen (an antigen that induces allergies). Within a few minutes, the allergen causes the formation of a flat, pale wheal surrounded by an inflamed red area. This type of hypersensitivity is sometimes called a **wheal and flare** reaction. **Generalized anaphylaxis** is an IgE-mediated allergic reaction throughout the body caused by an injected allergen in a sensitized individual. The allergen causes constriction of the respiratory passages, which leads to suffocation and loss of fluid from the blood. Either one of these conditions can cause death within a few minutes.

Generalized anaphylaxis is sometimes observed in people and animals stung by bees or injected with penicillin. Penicillin by itself is incapable of inducing the production of specific IgE; it is thought that penicillin functions as a hapten and becomes antigenic when it combines chemically with host protein, as occurs in approximately 4% of penicillin injections. Antibodies can then be made against the "penicillin" antigen. As mentioned earlier, the production of antibodies against penicillin requires approximately two to three weeks. Thus, no adverse reaction occurs unless penicillin is still being injected, but the person becomes sensitized to penicillin.

If penicillin is injected into a sensitized person, it binds to the Fab portions of the IgE attached to mast cells and basophils. This antibody–hapten complex induces the release of the cell mediators (histamine, etc.) throughout the body. If high concentrations of the cell mediators are released, extensive dilation of blood vessels and excessive loss of fluid from the blood causes shock. In addition, there is a complete constriction of the trachea and bronchi, which results in suffocation. These reactions occur so rapidly that death can result within 10 to 15 minutes.

Generalized anaphylaxis can be treated with an injection of epinephrine. Animals or humans who demonstrate mild allergic reactions to penicillin must not be given this antibiotic, because there is a good chance that a severe reaction will occur. We shall see later that penicillin can also cause a cell-mediated (delayed-type) hypersensitivity. Even skin contact with penicillin can result in lesions of the skin and mucous membranes.

People who are allergic to common plant, animal, and fungal antigens can sometimes be **desensitized** to the antigens by receiving injections of small amounts of antigen over a long period of time so that high titers of IgG antibodies against them are produced. These IgG are called **blocking antibodies** because they bind to the antigens so that they are unable to attach to IgE. Some scientists claim, however, that desensitization

Figure 14.3 Bee sting and the allergic reaction. IgE, produced in response to bee venom, attach to mast cells and basophils. If the bee venom enters a sensitized person, it binds to the IgE cell-bound antibodies and stimulates the release of histamine and other effector molecules. I ocal reactions (wheals) and/or systemic reactions (anaphylactic shock) may occur.

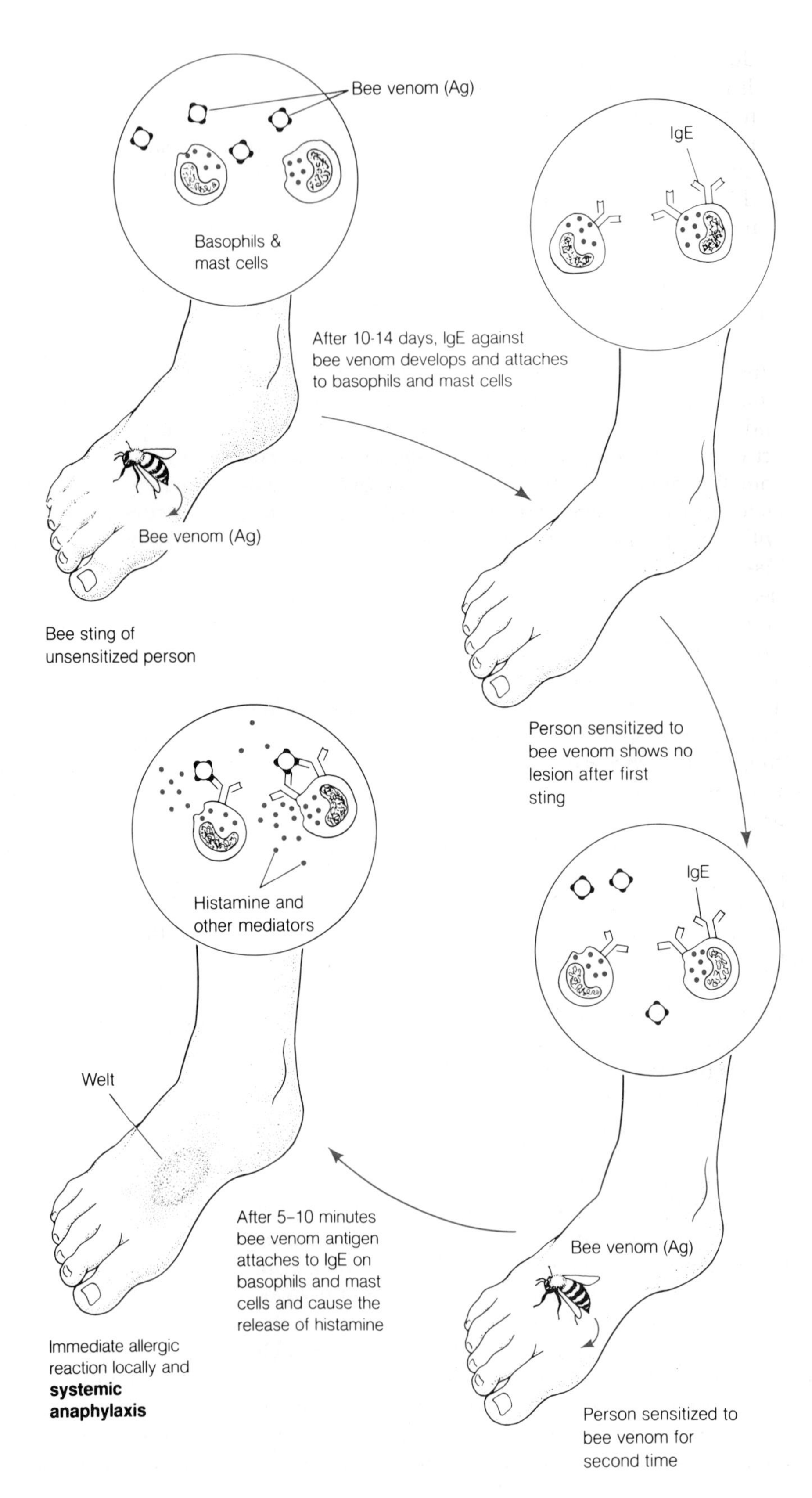

is due to a preferential stimulation of suppressor T-cells rather than blocking antibody. The allergen (antigen or hapten) is identified by injecting it intradermally (under the skin) into the sensitized person or animal. The development of a wheal and flare reaction within a few minutes indicates an allergy to the antigen or hapten.

IgG-COMPLEMENT-MEDIATED HYPERSENSITIVITY

When antigens are introduced repeatedly into a sensitized animal, they react with IgG and IgM antibodies and complement is activated. Frequent complement activation causes severe tissue damage. This situation sometimes occurs in the treatment of a disease such as tetanus by passive immunization against the toxin with immunoglobulins made in domestic animals. Tissue damage in this case is known as **serum sickness**. Serum sickness is characterized by fever, rash, and swelling of the lymph nodes, ankles, and face. The immune reaction, a complement-mediated hypersensitivity, can cause serious internal problems, such as kidney dysfunction, degeneration of vascular and cardiac tissue, and lesions in the joints (arthritis). The complement fixed affects the body in a number of ways. C3a and C5a, called **anaphylatoxins**, cause the release of cell mediators such as histamine from mast cells and basophils. These mediators cause the fever, rash, and swelling characteristic of serum sickness. The lysis of host cells by complement also contributes to the symptoms. The antibody–antigen complexes that are fixing complement in blood vessels and heart, in the glomeruli of the kidney, and in the joints can lead to vascular disease, glomerulonephritis, and arthritis, respectively.

A subcutaneous injection of foreign antigens into a sensitized animal can result in what is called the **Arthus reaction**: antibody-antigen complexes can also occur in any tissue where the antibody-antigen-complement complexes accumulate. Dilation of capillaries results from the anaphylatoxin activity of C3a and C5a and results in tissue swelling. C5a also contributes to Arthus reactions by attracting PMN leukocytes (see fig. 13.11). IgG-complement-mediated immune reactions play an important role in such diseases as **rheumatic fever, erythoblastosis fetalis**, some **drug-induced diseases**, and some **autoimmune diseases**.

Rheumatic fever is caused by the immune system's reaction to a chronic throat infection by *Streptococcus pyogenes*. The bacterial antigens sensitize the infected person. IgG and IgM form antigen-antibody-complement complexes with the bacterial antigens that are always present in a chronic infection. When these complexes localize in the heart, they cause **endocarditis**; in the kidney glomeruli, they cause **glomerulonephritis**; and in the joints they cause **arthritis**. Some IgG and IgM made against M-protein from the bacterial cell wall cross-react with host antigens found in the heart. In this case, immunoglobulins bind to heart cells and activate complement, which results in the lysing of heart cells. When an immune system attacks its owner, we have what is known as an **autoimmune reaction** (see Autoimmune Diseases).

CELL-MEDIATED (DELAYED-TYPE) HYPERSENSITIVITY

Viruses, bacteria, fungi, and noninfectious agents that continually stimulate the immune system induce T-lymphocytes and macrophages to activity.This activity occasionally leads to **delayed-type hypersensitivity**, so named because it appears 24 to 48 hours after the host encounters the antigen. It is characterized by **erythema** (redness), due to blood flow to the damaged area, and **induration** (hardness), due to the accumulation of macrophages and lymphocytes in the affected area. The activated T-lymphocytes and macrophages, in their attempt to eliminate the foreign antigens, can cause severe tissue destruction.

Much of the discomfort and damage associated with infections by organisms such as *Schistosoma japonicum* (schistosomiasis), *Mycobacterium tuberculosis* (tuberculosis), *M. leprae* (leprosy), *Treponema pallidum* (syphilis), and *Histoplasma capsulatum* (histoplasmosis) are due to long-term immune reactions against these organisms by macrophages and cytotoxic T-cells. For example, the organism that causes syphilis is an intracellular bacterium that usually causes little damage to the cells it infects and is not readily destroyed by the host's immune system, because it generally resides inside cells. The constant release of *Treponema* antigens from the infected tissue sensitizes the animal, however, and leukocytes, macrophages, and T-cells are drawn to the site of the infection.

Macrophages and neutrophils actively phagocytize and destroy the free *Treponema* at the site of infection. T-cells also release **cytotoxic factors** that kill the *Treponema* and nearby host cells. If the infection is a chronic one, there is continual destruction of tissue. The **chancres** (lesions of the primary stage), **rashes**

FOCUS AN ALLERGIC REACTION TO A VACCINE

In the United States, most babies and children under the age of two receive a series of three to four inoculations of the diphtheria–pertussis–tetanus (DPT) vaccine to protect them from these diseases. Before entering school, most children are required to have a recent booster shot. Thus, most people in the United States receive four or five inoculations of the DPT vaccine early in life. Without inoculations against diphtheria, whooping cough (pertussis), and tetanus, epidemics of these diseases would occur frequently.

There is a problem, however, with the pertussis component of the DPT vaccine. Studies on the effects of DPT indicate that some children vaccinated with DPT suffer severe reactions including crying, fever, and drowsiness. These reactions occur 3–4 hours after injection and last for approximately 12 hours. More severe allergic reactions also occur: approximately 1 in 700 babies and children who receive DPT injections suffers convulsions and goes into shock. Brain and motor nerve damage that leads to severe retardation and paralysis occurs at a frequency of 1 in 50,000 babies and children vaccinated. Over the years, the pertussis component in the DPT vaccine has produced many thousands of mentally retarded and crippled children and adults in the United States because of allergic reactions.

Presently, there are approximately 2000 cases of pertussis and an average of six deaths from pertussis each year in the United States. If the pertussis toxoid were left out of the DPT vaccine, the nation would risk epidemics of whooping cough in which millions would become ill and thousands would die each year. In 1934, before the DPT vaccine was used in the United States there were more than 7500 deaths. In England, were the pertussis vaccination is no longer given to all babies and young children, whooping cough has recently become epidemic: in one year there were more than 100,000 cases and about 30 deaths.

The dangers of a whooping cough epidemic in the United States far outweigh the dangers of allergic reactions. This knowledge is little consolation, however, to those parents whose children become mentally retarded and crippled by the vaccine. Stimulated by public outcry and numerous lawsuits against drug firms that prepare the DPT vaccine, Congress is attempting to make the Public Health Agencies of the U.S. Government and the American Medical Association warn doctors and parents of the risks of the vaccine, so that babies who have allergic reactions to the first vaccination can be excused from further vaccinations.

(characteristic of the secondary stage), and **gummas** (lesions of the tertiary stage) of syphilis result partly from tissue destruction by T-cells and macrophages at sites of infection. Since primary chancres appear 3 to 5 weeks after infection, they may be due not only to extensive *Treponema* replication but also to cell-mediated hypersensitivity. Arthus-type (IgG-complement-mediated) allergic reactions and cell-mediated allergic reactions are responsible for the extensive rash and ulcers that are part of secondary syphilis and generally begin a month or so after the initial infection. IgG-complement-mediated allergic reactions and cell-mediated allergic reactions are responsible for the symptoms of tertiary syphilis, which can include brain, circulatory system, kidney and liver damage, as well as arthritis and bone deformation. Death from a chronic syphilis infection in some cases does not occur for 10–20 years after the initial infection.

Chronic infections of the brain by viruses, such as those that cause chickenpox and measles, are believed to cause one type of **multiple sclerosis**. These viruses are thought to damage the **glial cells** that form the myelin sheaths around nerve cells, as well as the capillaries that bring blood to the brain. Myelin protein released from the damaged cells passes from the brain into the blood because of a breakdown in the blood–brain barrier. Consequently, myelin protein which normally does not enter the circulatory system, is recognized as a foreign antigen and stimulates the immune system against the brain myelin. Macrophages (fig. 14.4) and T-lymphocytes infiltrate the brain and further destroy the myelin sheaths around the nerve cells. Because nerve cells in the brain do not function normally without a myelin sheath surrounding them, a person suffering from this induced autoimmune disease often loses his or her sight and

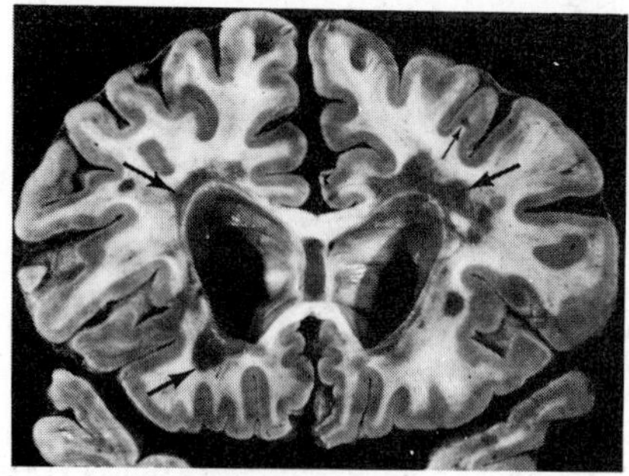
(a)

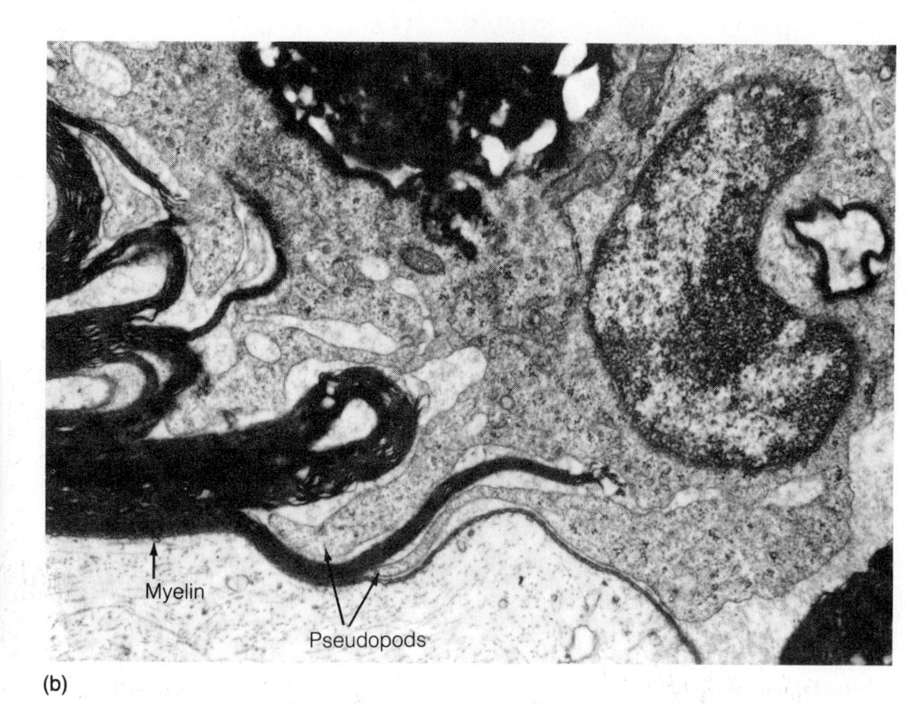

(b)

Figure 14.4 Multiple sclerosis and cell-mediated immune reaction. *(a)* The brain from a man who died of multiple sclerosis shows numerous dark areas where the nerves have lost their myelin sheaths. Arrows indicate location of lesions caused by the autoimmune response. *(b)* A macrophage is destroying the myelin sheath in the nervous system of a rabbit that was sensitized to myelin. Macrophage pseudopods are separating the myelin from an axon.

muscular coordination. A person suffering from multiple sclerosis eventually becomes unable to speak and wastes away because his or her muscles are no longer stimulated. Because the host's own immune system attacks self antigens, this type of multiple sclerosis is considered an autoimmune disease (see Autoimmune Diseases).

Allergic contact dermatitis is a cell-mediated allergic reaction that occurs when certain chemicals come into contact with the skin. For example, penicillin and catechols from poison ivy or poison oak leaves, rubbed on the skin of a sensitized individual, can cause severe blistering of the skin (figs. 14.5 and 14.6). Destruction of the epidermis by T-lymphocytes and macrophages is accompanied by severe itching. The small molecules that cause allergic contact dermatitis react with skin proteins and consequently become antigens that trigger host immune reactions. The first time an individual comes into contact with the allergen, no dermatitis results, but memory T-lymphocytes can quickly respond to the chemical if exposure is repeated.

AUTOIMMUNE DISEASES

An autoimmune disease occurs when the immune system becomes sensitized to self antigens, or when it attacks new antigens on cells and then proceeds to destroy extensive amounts of tissue.

One type of arthritis, **rheumatoid arthritis**, is caused by IgM (rheumatoid factor) that bind to IgG associated with the **synovial membrane** in the joint (table 14.3). Complement fixation leads to inflammation, stiffness, and deformity of the joints. It is hypothesized that IgM, possibly made against some foreign antigen, cross-reacts with human IgG.

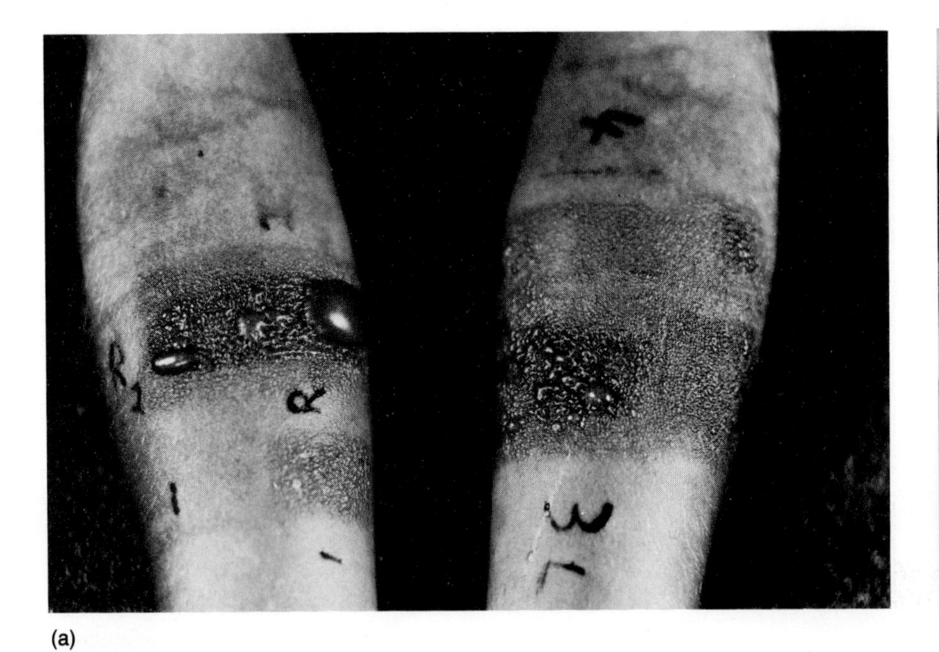

(a)

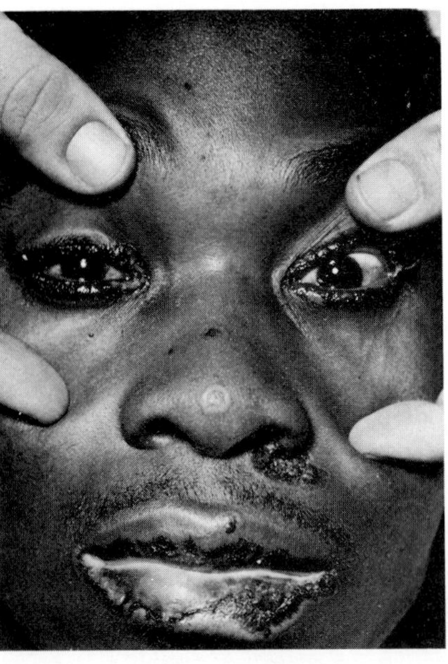

(b)

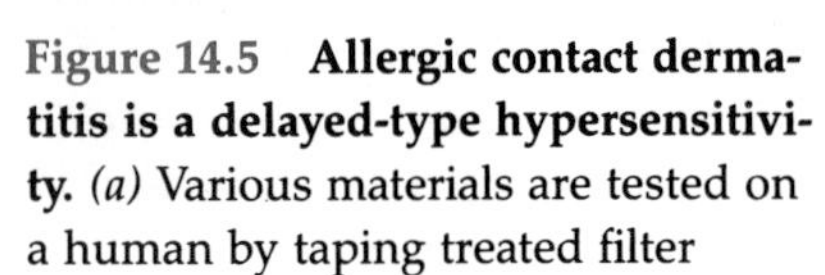

Figure 14.5 **Allergic contact dermatitis is a delayed-type hypersensitivity.** *(a)* Various materials are tested on a human by taping treated filter paper patches to the skin. If a person is sensitized to a material, blisters appear within a day or so after contact. *(b)* This man demonstrates allergic contact dermatitis to penicillin. He has developed blisters on his eyelids, nasal mucosa, and lips.

TABLE 14.3
AUTOIMMUNE DISEASES

Disease	Part of immune system involved	Tissue affected (specific autoantigen)
Encephalomyelitis & Choriomeningitis	T-cells and macrophages	Brain cells
Diabetes mellitus (juvenile)	T-cells	Pancreas
Multiple sclerosis	Macrophages and T-cells	Brain cells (myelin protein)
Myasthenia gravis	Antibodies	Nerve and muscle (receptor for acetylcholine)
Goodpasture glomerulonephritis	Antibodies and complement	Kidney
Hashimoto's thyroiditis	T-cells and antibodies	Thyroid gland
Hemolytic anemia	Antibodies and complement	Red blood cells
Thrombocytopenic purpura	Antibodies	Blood platelets
Pernicious anemia	Antibodies and complement	Intestines (receptor for B_{12} absorption)
Addison's disease	T-cells	Adrenal gland
Sympathetic ophthalmia	Antibodies	Eye
Ulcerative colitis	T-cells	Intestines
Rheumatic fever	Antibodies	Systemic; in particular heart, kidneys, joints
Rheumatoid arthritis	Antibodies	Systemic; in particular joints (mutant IgG associated with synovial membranes)
Sytemic lupus erythematosus	Antibodies	Systemic (DNA from many tissues)
Guillain-Barré	T-cells	Myelin sheaths of peripheral nerves

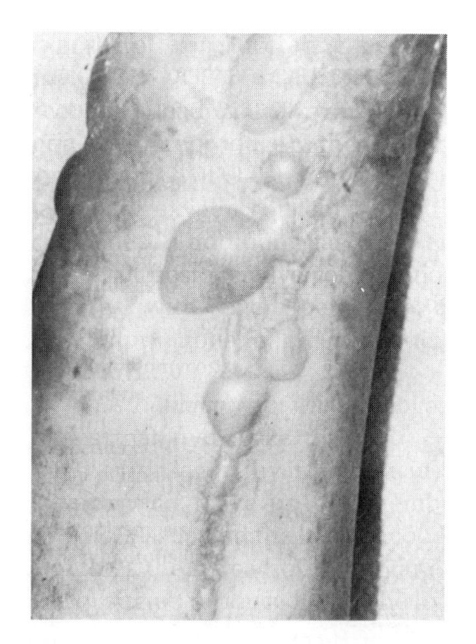

(a)

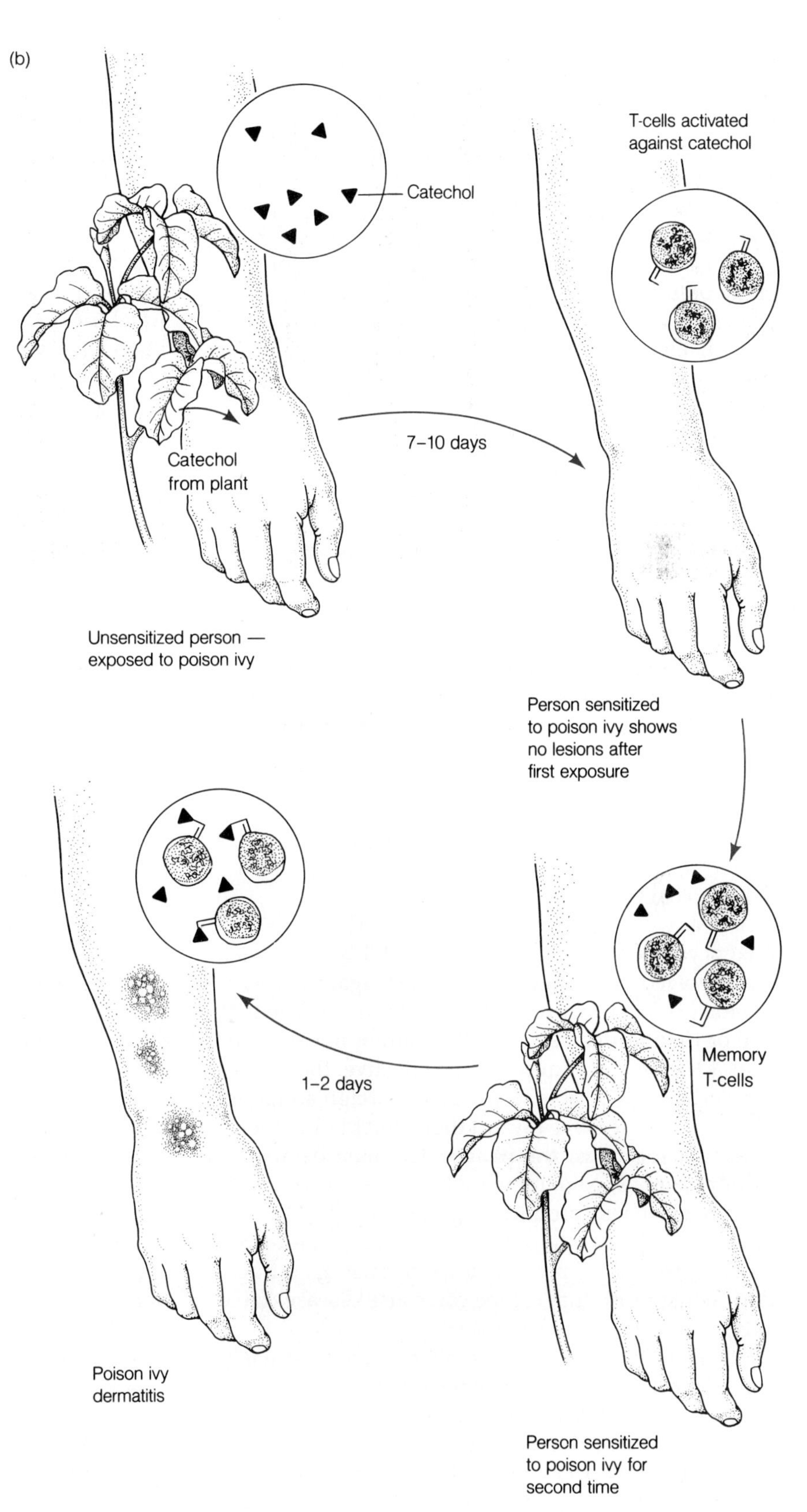

Figure 14.6 Allergy to poison ivy is a delayed-type hypersensitivity. *(a)* Typical lesion of poison ivy dermatitis. *(b)* Exposure to catechol molecules in poison ivy results in the formation of a catechol–skin protein complex that becomes immunogenic. The initial contact results in the development, in approximately one to two weeks, of T-lymphocytes that are sensitized to the catechol molecules, but no visible lesions develop. Upon subsequent contacts with the catechols, sensitized T-lymphocytes will respond immunologically to the antigen and initiate a delayed-type hypersensitivity reaction that leads to a dermatitis at the site of contact with the catechols.

Many chemicals and drugs can react with the surfaces of cells and hence create new antigens that stimulate an immune response against the cells. This is what happens in **thrombocytopenic purpura** (table 14.3).

A number of enveloped viruses, such as those that cause influenza, measles, mumps, and chickenpox, insert their antigens into the plasma membranes of the cells they infect. This activity results in an attack by macrophages and T-lymphocytes on virus-infected cells. If the virus remains inside the host cells, the immune system destroys the virus-containing cells in an attempt to clear the infection. This causes further damage to the host. It is believed that **juvenile diabetes mellitus** is caused by a virus infection of the cells found in the pancreatic islet of Langerhans. The virus is thought to modify these cells so that they are destroyed by macrophages and T-lymphocytes. Islet of Langerhans cells produce the essential hormone **insulin**, which controls the level of glucose in the blood. Without appropriate amounts of the hormone, a person slowly starves and various tissues degenerate. Some of the diseases that result from immune responses directed against foreign and self antigens are listed in Table 14.3.

IMMUNODEFICIENCIES

Phagocytic cells, B-lymphocytes, and T-lymphocytes are involved in protecting the host against microorganisms that can cause harmful infections. The first line of cellular defense against invading microorganisms is the phagocytic cells. To be effective, the phagocytic cells must be able to recognize foreign antigens. Then they must move toward them and bind to them. After phagocytosis, the phagocytes must be able to fuse **lysosomes**, containing hydrolytic enzymes, with the phagosomes; produce toxic O_2^- and H_2O_2; and ultimately kill the pathogen. Those organisms destroyed by this process are prevented from causing an infection, while those that escape can cause disease (table 14.4).

A fatal childhood disease called **Chediak-Higashi syndrome** is due to abnormal phagocytes whose lysosomes fuse slowly with the phagosomes. Because bacteria are not effectively destroyed, they grow inside the phagocytes and spread throughout the body. Another fatal childhood disease, called **chronic granulomatous disease**, results from the inability of lysosomes to produce H_2O_2 and their inability to iodinate microorganisms. Phagocytized bacteria grow inside the phagocytes and stimulate an inflammatory reaction. The phagocytes and surrounding fibroblasts (cells found in most tissues) develop into a chronic inflammation known as a **granuloma**.

Some people lack T-lymphocytes, or their T-lymphocytes are ineffective. Consequently, these people are especially susceptible to viral and intracellular bacterial infections. For example, a vaccination with live attenuated viruses, such as those commonly used in the Sabin polio vaccine, can be lethal to persons who lack functional T-lymphocytes. Some of these defects are inherited, while others are not.

Persons lacking B-lymphocytes or having nonfunctional B-cells are especially sensitive to bacterial infections of the skin and the respiratory tract. In the case of **acquired agammaglobulinemia**, the individual has normal levels of B-cells, but the B-cells are unable to develop into antibody-releasing plasma cells because of an excess of suppressor T-lymphocytes, which suppress B-cell development.

AIDS

Acquired immune deficiency syndrome (AIDS) is caused by a retrovirus called HIV (human immunodeficiency virus) that attacks and destroys helper T-lymphocytes and macrophages, and therefore hampers the host's ability to combat disease. Because of the virus-induced destruction of the immune system, cancers and infectious agents ravage the body and eventually cause death (table 14.5). Thus, AIDS can be considered a group of diseases that are associated with a defective immune system. People are diagnosed as having AIDS if they have a suppressed immune system and also fall into any of the following categories: (a) suffering from characteristic opportunistic infections (such as *Pneumocystis carinii* pneumonia); (b) afflicted by characteristic cancers (such as Kaposi's sarcoma fig. 14.7); or (c) having antibodies against the AIDS virus and one or more severe chronic infections (e.g., hepatitis, amebiasis, fungal infections).

In the first 9 years after AIDS appeared in the United States (1978–1986), there were more than 30,000 cases with a mortality of over 50%. During this time, less than 1% of the persons with AIDS survived more than 5 years after they were diagnosed.

Another group of individuals of public health importance are those with **AIDS-related complex (ARC).** This group is made up of people with anti-

TABLE 14.4
IMMUNODEFICIENCY DISEASES

Disease	Defect	Consequence
Selective dysgammaglobulinemia	Deficiencies of IgM, IgG, or IgA.	IgA deficiency results in gastrointestinal problems such as diarrhea and poor absorption of fats and fat-soluble vitamins. IgM and IgG deficiencies lead to bacterial infections.
Acquired agammaglobulinemia	Suppressor T-cells inhibit development of B-lymphocytes into plasma cells.	Lack of plasma cells results in serious bacterial infections of the skin and lungs.
Congenital immune deficiency syndrome (CIDS)	Most B-cells, plasma cells, and T-cells are missing (inherited autosomal mutation).	Lack of plasma cells and T-cells results in virtually no protection against infection.
Acquired immune deficiency syndrome (AIDS)	Helper T-cells are depressed or missing because they are destroyed by HIV.	Lack of helper T-cells leads to multiple opportunistic infections that are eventually fatal.
DiGeorge syndrome	Thymus does not develop correctly. Consequently, T-cells are absent (not inherited).	T-cell deficiency leads to fatal viral and intracellular bacterial infections.
Hodgkin's disease	Cancer of lymph nodes leads to serum factors that block T-cell receptors (acquired).	T-cell deficiency results in serious viral and intracellular bacterial infections (herpes, tuberculosis).
Bruton's disease	Total lack of B-cells (inherited, X-linked).	B-cell deficiency leads to serious bacterial infection of the skin and respiratory tract.
Cushing's disease	Excess secretion of cortisone and cortisol leads to lysis of most T- and B-cells and decreased levels of blood monocytes (developed).	Extreme susceptibility to viral and intracellular bacterial infections.
Hereditary angioneurotic edema	Deficiency of C1 esterase inhibitor (inherited).	Unchecked complement activation leads to acute inflammation and possible death by suffocation.
Complement deficiencies	Deficiency in C3 or C5 (inherited).	Susceptibility to bacterial infections.
Complement deficiencies	Deficiency in C1r, C1s, C2, or C4 (inherited).	Hypersensitivity.
Chediak-Higashi syndrome	Defective phagosome formation (inherited).	Fatal bacterial infections.
Chronic granulomatous disease	Defective lysosomes. They lack hydrogen peroxide and iodination enzymes (inherited).	Fatal bacterial infections and inflammation.

bodies against HIV, who have persistent, generalized, swollen lymph nodes (lymphoadenopathy), and who are sick but do not yet suffer from a characteristic opportunistic infection, cancer, or severe chronic diseases. At the end of the first 9 years of the AIDS outbreak in the United States, there were more than 200,000 individuals suffering from ARC and over a million more that had been exposed to the virus. The data gathered up to 1987 indicate that most of those with ARC will develop AIDS and about 20% of those exposed will develop AIDS within 5 years. This means that sometime in 1991, 400,000 persons will have acquired AIDS (fig. 14.8). Almost all of the million people with HIV will progress to ARC or AIDS within the next 10 years.

During the first 9 years of the AIDS epidemic, approximately 2000 children were born to mothers infected by HIV. Between 50 and 75% of these children were born with congenital ARC. More than 350 soon developed AIDS. Congenital ARC is indicated by (a) small heads with boxlike foreheads, (b) flattened nose bridges, and (c) wide-set eyes with bluish whites.

TABLE 14.5
DISEASES ASSOCIATED WITH HIV INFECTIONS

Characteristic opportunistic infections and cancers	Severe chronic infections or diseases	Prodromal AIDS (AIDS related complex)
Kaposi's sarcoma of the skin (4.3–36%)*	Gonorrhea	**ADULTS**
Pneumocystis carinii pneumoniae (62.4%)	Syphilis	Tuberculosis
Mycobacterium avium-intracellulare and *M. tuberculosis* pulmonary (4–65%)‡ disseminated infections (50%)	Hepatitis B	Lymphopenia
CNS Toxoplamosis (2.9–27%)†	Amebiasis	Impaired T-cell function
Salmonellosis (45%)	Giardiasis	Persistent generalized lymphadenopathy
Cytomegalovirus pneumonia (6%) disseminated infections (4.9–25%)	Generalized herpes simplex (3.6%)	**NEWBORNS**
Herpes encephalitis (2%)	Oral candidiasis (thrush)	Small heads with box-like foreheads
Aseptic meningitis (5.5%)	Full-body tinea corporis (ringworm)	Flattened nose bridge
Subacute encephalitis (11%)	Granuloma inguinale	Wide-set eyes with bluish whites
Cryptococcal meningitis (5.1%), pneumonia (1%)	Full-body granulomatous dermatitis	
CNS lymphoma (2.7%)	Epstein-Barr induced oral hairy leukoplakia	
Pulmonary legionellosis (1%)	B-cell lymphoma	
Esophageal, bronchial, or pulmonary candidiasis (10.8%)	Oral papillomas	
Aspergillosis	Ulcerative gingivitis	
Nocardiosis		
Cryptosporidiosis (3.3%)		
Generalized herpes simplex (3.6%)		
Prolonged varicella zoster (chickenpox)		

*Kaposi's sarcoma (KS) occurs in 36% of homosexual men with AIDS but only in 4.3% of IV drug abusers with AIDS. The average is 24.4%.
†Central nervous system toxoplasmosis occurs in about 27% of Haitians but in less than 2.9% of other groups.
‡*Mycobacterium tuberculosis* and *Mycobacterium avium-intracellulare* pulmonary infections occur in 65% and disseminated infections occur in 40% of Haitians with AIDS. Tuberculosis occurs in about 4% of the homosexual men with AIDS.

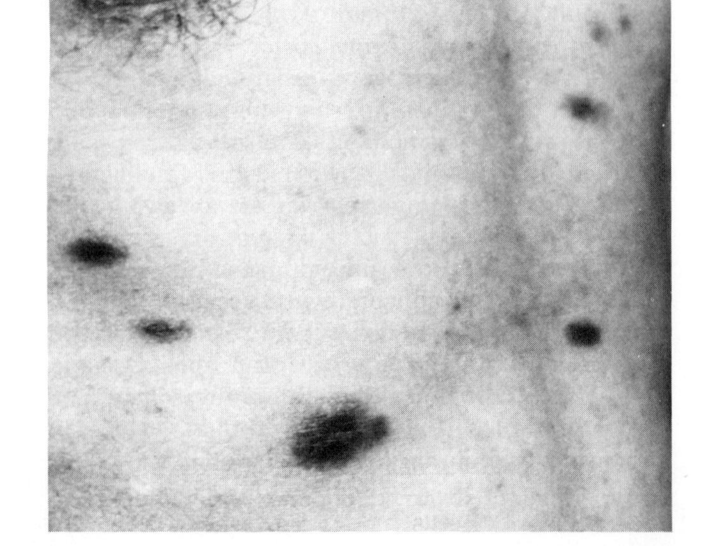

Figure 14.7 **Patient with Kaposi's sarcoma.** This AIDS patient is afflicted with Kaposi's sarcoma. Courtesy of Professor F. Camacho, Department of Dermatology and Venereology, University Hospital, Seville, Spain.

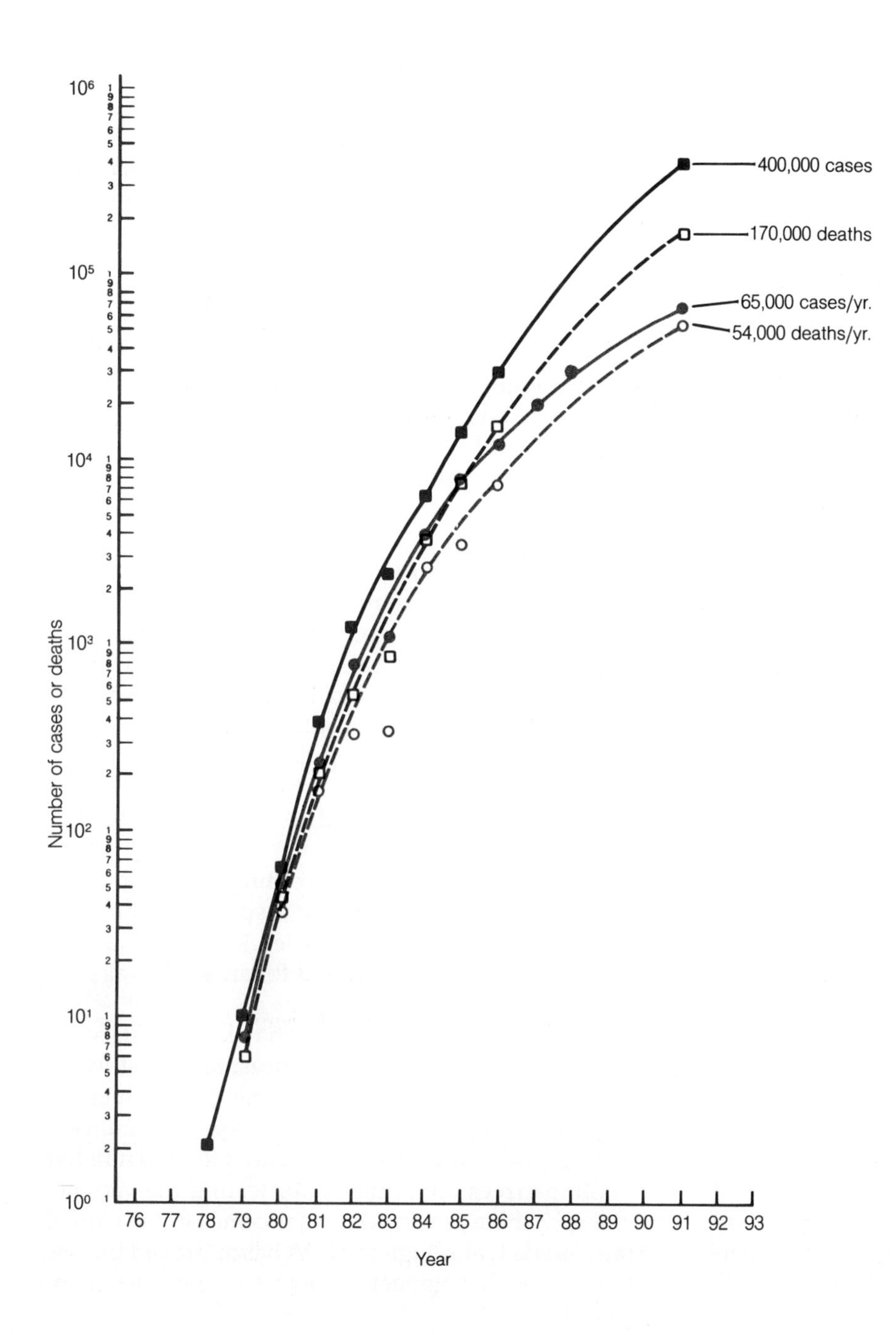

Figure 14.8 AIDS morbidity and mortality. In 1978, only 2 cases of AIDS were reported in the United States. Recent epidemiological studies indicate that by the end of 1991, a total of 400,000 cases and 170,000 deaths will have been reported in the 14 years of the epidemic. At the end of 1991 the rate of new cases is expected to be 65,000 cases/year and the rate of new deaths 54,000 deaths/year Total cases, ■; total deaths, □; cases/year, ●; deaths/year, ○.

Cause The suppression of the immune system that leads to AIDS and ARC is due to the destruction of immune cells by HIV. HIV causes much of its damage to the immune system when it infects helper T-lymphocytes and rapidly reproduces within these cells. This causes the cells to fuse with each other, to become inactive as helper cells, and to die. Without helper T-lymphocytes, the immune response to infectious agents cannot be initiated.

The AIDS virus can also invade the macrophages of the reticuloendothelial system, which also are essential in the expression of immunity to infectious diseases. Macrophages in the brain that are infected cause deterioration of the central nervous system such

TABLE 14.6
AIDS HIGH RISK GROUPS

Group	United States (%)
Homosexual and bisexual men	72.5
Heterosexual IV drug abusers	17.0
Heterosexuals from Haiti or Africa	2.0
Heterosexuals having sexual intercourse with persons with AIDS	2.0
Transfusion recipients	1.5
Hemophilia patients	1.0
Children born to mothers with AIDS	1.0
Undetermined because of incomplete records or do not fall into those groups listed above.	3.0

Source: Norman, C. 1986. Sex and Needles, Not Insects and Pigs, Spread AIDS in Florida Town. *Science* 234: 415–417.

as memory loss, mental retardation, paralysis, and death. In many of those infected with HIV, mental symptoms appear before any of the opportunistic infections or cancers. Memory loss and strong emotions are the first symptoms.

Transmission AIDS is, for the most part, a sexually transmitted disease. In the United States, HIV is generally transmitted from one individual to another by vaginal or anal sex (77%), by sharing needles used to inject intravenous drugs (17%), and by the transfusion of blood or injection of blood products (2%). The risk of acquiring AIDS from casual contact with saliva, tears, urine, and fecal matter from patients with AIDS is too small to be measured. Similarly, the risk of contracting AIDS from biting insects is unmeasurable. However, AIDS has been acquired by at least three people who had contact with HIV-contaminated blood. Extensive contact with contaminated blood represents a small risk to medical and laboratory personnel. Most of the sexually transmitted cases of AIDS have occurred in bisexual and homosexual men. However, because of the large number of intravenous drug users and bisexual men that engage in sexual activities with heterosexuals, the disease is spreading to the general (heterosexual) population (Table 14.6).

Prevention A study of over 500 dentists and dental hygienists in San Francisco and New York, two cities that have very high rates of AIDS, showed that none had antibodies against HIV in 1986. In addition, none of the many thousands of people that cared for AIDS patients and came in contact with their saliva, tears, urine, and fecal matter developed antibodies against HIV. A conservative analysis of this and other epidemiological data indicates that casual contact cannot be responsible for more than 1 in 10,000 cases (<0.01%). Thus, after 9 years of the AIDS epidemic, less than 3 out of 30,000 cases of AIDS might have been acquired through casual contact. Consequently, isolation or quarantine of the more than 1,500,000 people who have been exposed to HIV is not a practical means of prevention. Further infections could be reduced significantly by abstaining from anal, oral, or vaginal intercourse with individuals in the high risk group (see table 14.6) and by using condoms when engaging in sexual intercourse.

For persons in high risk groups abstinence is not a realistic expectation. Consequently, other behavioral changes should be made. This must include the use of condoms during intercourse.

Intravenous drug abusers and their sexual contacts, for the most part, are irresponsible and will not alter their behavior without intervention. They are, and will be responsible for continually spreading AIDS throughout the general population. Legislation and practical measures to reduce the number of intravenous drug users is essential to prevent the further spread of HIV.

Prostitutes are and will continue to be another important reservoir of HIV that is spreading throughout the general population. They should be closely monitored and regularly tested for antibodies against HIV.

Ultimately, vaccination of high-risk individuals will be the most effective preventive measure against AIDS. One of the problems in the vaccine development is that only chimpanzees, not other apes or animals, develope the human AIDS, and only a limited number of chimpanzees are available for testing the vaccine. Presently, there are no licenced vaccines in the United States market, although the FDA has approved the use of a genetically engineered vaccine in controlled clinical trials.

Treatment Presently, a handful of drugs have been discovered that alleviate the signs and symptoms of AIDS. AZT (3′-azido-3′-deoxythymidine) is one of the most promising drug available. In patients tested, AZT inhibits the replication of HIV throughout the body, including the brain. The results are an improvement in immune function, weight gain, an increase in helper T-lymphocytes, a reversal of mental retardation, a regression of cancers, a decrease in the rate of opportunistic infections, and a decrease in the death rate. Figure 14.9 demonstrates the decrease in AIDS-associated events in patients treated with AZT.

AZT blocks viral reverse transcriptase as well as cellular DNA polymerases, and thus inhibits the viral replication in infected cells. The most common side effect seen in AZT users is the development of anemia because the drug also affects the synthesis and development of red blood cells.

Another drug that shows promise is AL721. It apparently strips cholesterol from the outer membrane of the AIDS virus so that it is unable to attach to host cells and multiply in them. If proven effective, this drug may be more desirable than AZT because it does not cause serious side effects.

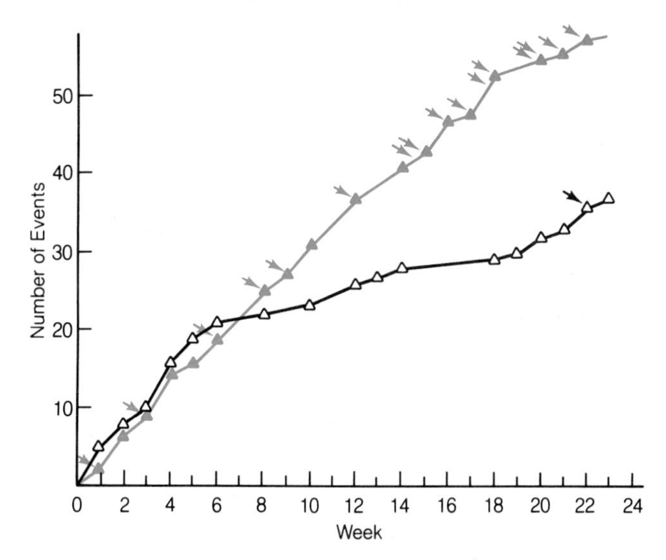

Figure 14.9 **AZT-treated ARC and AIDS patients.** Forty-five percent of the patients had ARC, while the remainder had AIDS. Five of the ARC patients and 11 of the AIDS patients who were in the placebo group died. 3′-Azido-3′-deoxythymidine (AZT) was given to 145 patients with ARC and AIDS, while a placebo was given to 137 patients in a double blind experiment. During 23 weeks (almost 6 months) of treatment, the number of events (opportunistic infections, cancers, and deaths) in the AZT group increased slowly in contrast to the control group. Most significant was the difference in the number of deaths during the 23 weeks of treatment, one compared to 16. ▲, Placebo; △, AZT; ↓ deaths.

TISSUE TRANSPLANTATION AND BLOOD TRANSFUSIONS

Often it is desirable to transfer or transplant tissue or organs from one site to another or from one individual to another. For example, treatment of severe burns often requires that skin be transplanted from an intact region of the body to the burn site so that the wound can heal more quickly. When tissue is transplanted (**grafted**) from one site to another in the same individual, the transplant is called an **autograft**. To save an individual from death in the case of kidney disease, a kidney from another person is transplanted into the diseased individual. When tissues are transplanted from one individual to another, the tissue is known as an **allograft**. If the transplant is between identical twins or highly inbred animals, the transplant is called an **isograft**. The tissue transplanted between different species, such as between a cat and a dog, is known as a **xenograft**. It is found that autografts and isografts of tissues such as skin generally are successful, because the immune system does not recognize the tissue antigens as foreign. On the other hand, allografts and xenografts generally are not successful because of cell-mediated immunity. T-lymphocytes, which recognize foreign antigens, are responsible for the rejection of grafts. The mechanism of allograft rejection also involves humoral antibodies and complement.

BLOOD TYPE INCOMPATABILITIES

Erythroblastosis fetalis is a disease of the fetus and newborn, caused by IgG from the mother attacking the fetus's red blood cells. During development the fetus suffers from anemia, and at birth it suffers from both anemia and jaundice. Often a woman who has

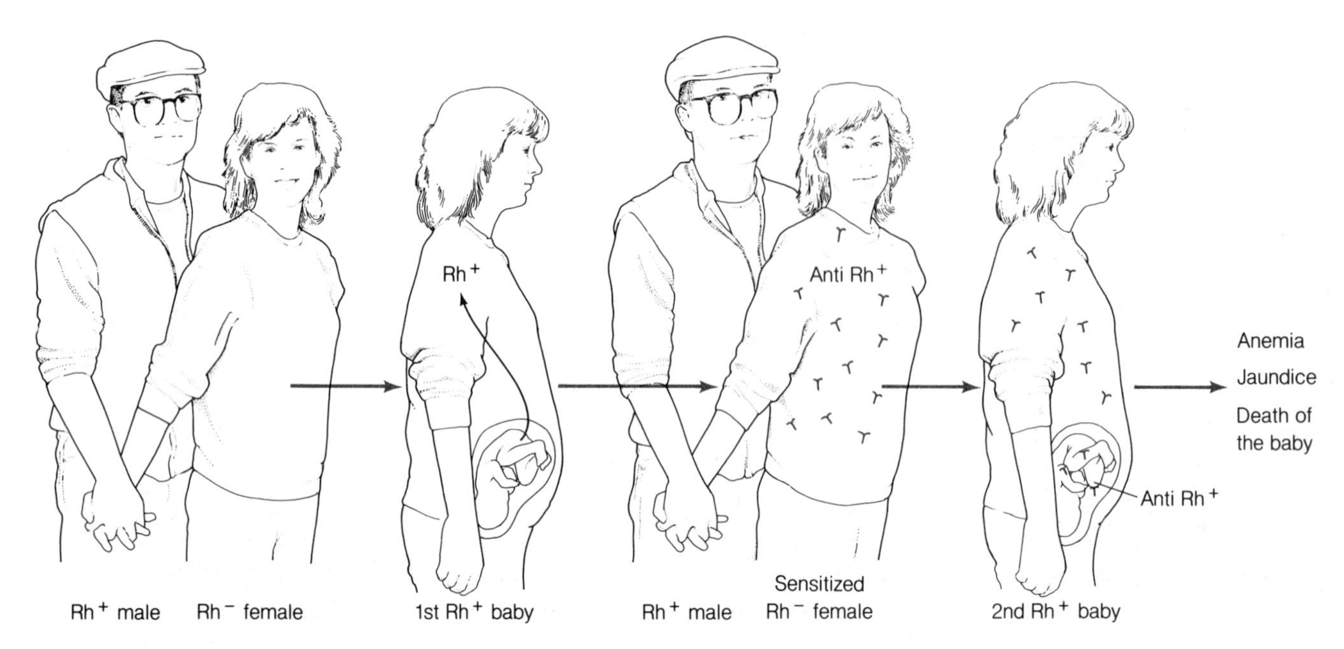

Figure 14.10 The development of blood incompatibility. The child of an Rh^+ male and an Rh^- female may be Rh^+. When the Rh^+ baby is born, some of its blood enters the mother and sensitizes her against the Rh^+ blood. Nothing happens to the first Rh^+ baby, but subsequent Rh^+ babies are in danger of being attacked by antibodies from the sensitized Rh^- mother. The mother's antibodies against Rh^+ blood pass through the placenta, enter the baby's blood system, and attack the baby's blood cells. This attack results in anemia, jaundice, and often death of the baby. Those babies that survive the attack on their blood have the disease known as erythroblastosis fetalis.

Rh^- type blood becomes sensitized to Rh^+ type blood after the birth of her first Rh^+ child. At birth, some of the newborn child's Rh^+ red blood cells enter the mother and sensitize her to these foreign antigens. Subsequently, IgG that passes from her blood system through the placenta into the fetal blood system will attack the blood of any fetus that has Rh^+ blood (fig. 14.10). The IgG binds to the fetal red blood cells, complement is fixed, and the cells are lysed. The fetus responds by releasing immature red blood cells, called **erythroblasts**, which do not efficiently remove CO_2 or transport O_2. Thus, the fetus suffers from anemia and hypoxia. Most of the byproducts of red blood cell destruction are removed by the mother's circulatory system, so jaundice is mild in the fetus. At birth, however, the mother's circulatory system no longer removes red blood cell debris and the cholesterol from the red blood cells. The cholesterol is turned into bile, which accumulates in the baby's blood, liver, skin, and urine. The bile turns the baby yellow.

Since jaundice develops rapidly after birth and is due to the destruction of the baby's RBCs by the mother's circulating antibodies, the condition is also known as **hemolytic disease of the newborn**. If the anemia and jaundice are not alleviated, they result in the death of the infant. These conditions can be reversed by the complete removal of the baby's blood and its replacement with type O blood. This eliminates most of the mother's circulating antibodies and prevents further hemolysis of the baby's erythroblasts and red blood cells. Slowly, the baby's RBCs replace the transfused type O red blood cells. Type O blood is used so that the baby does not attack its new blood with antibodies against A and B antigens.

Erythroblastosis fetalis can be prevented by passive immunization of the Rh^- mother with antibodies against Rh^+ blood at the time she gives birth to her first Rh^+ baby. Any Rh^+ baby cells that might enter the mother's circulatory system at birth are rapidly destroyed by the injected immunoglobulins and so do not stimulate the mother's immune system. Thus, the next child is protected, because passive immunization blocks the development of the mother's natural immunity against the baby's foreign antigens.

CLINICAL PERSPECTIVE ANAPHYLAXIS TO A BEE STING

The patient is a 47-year-old female who had a bee sting in July 1975 and a yellow jacket sting in August. The latter sting was followed by generalized erythema (redness) and what she described as a semiconscious state. The patient was immediately transported to an emergency room, where she was treated for shock.

The patient had positive skin tests to bee and yellow jacket venoms and was placed on immunotherapy. The immunotherapy consisted of injections of extracts of ground, whole bodies of stinging insects. This treatment continued from late 1975 until 1978. Serum analyses also revealed that the patient had IgE antibodies against honeybee, yellow jacket, and hornet venoms in her serum.

When the patient moved from her home in Texas to California, she was reevaluated. Upon skin testing with a variety of venoms, she was found to react strongly to yellow jacket, hornet, and wasp venoms. By 1979, the therapy for people with potentially life-threatening allergies to insect stings had changed and consisted of injections with extracts of sting insect venoms. This type of treatment is called **venom immunotherapy** (VIT). The patient was placed in a VIT program that included those venoms to which she reacted strongly.

Shortly after this, an assay (a method) became available for measuring not only venom-specific IgE antibodies, but also venom-specific IgG **blocking** antibodies. The production of blocking antibodies is thought to be an important reason immunotherapy increases tolerance to the venoms. While on monthly (maintenance) injections the patient had the first of what became yearly measurements of venom-specific IgG and IgE (allergic) in her serum to determine the efficacy of the immunotherapy (which should result in an increase in venom-specific blocking IgG antibodies and a decrease in IgE). The tests showed that her serum had low levels of IgE antibodies to hornet and wasp venoms and adequate levels to yellow jacket venom. The patient's dose of VIT was increased by about 30% and was found to increase significantly the levels of IgG antibodies to all three insect venoms.

In 1983, the patient's venom-specific IgE antibodies dropped to insignificant levels while her IgG levels persisted. She was, therefore taken off VIT after a total of 5 years in the program.

Recent studies of patients whose VIT has been stopped after five years of treatment and whose venom-specific IgE has diminished to insignificant levels, show a good record of continued protection against potentially life-threatening reactions to insect stings. The patient was, nevertheless cautioned to continue to carry her insect-sting emergency kit (containing antihistamine tablets and a syringe of epinephrine for self-administration) to further insure her safety.

Clinical history courtesy of Robert J. Holzhauer, M.D., Diplomate, American Board of Allergy and Immunology.

SUMMARY

1. The immune system can at times function improperly or overreact and cause serious diseases.
2. An individual is said to be hypersensitive or allergic to an antigen when pathological changes occur.
3. A sensitized individual is one that has a memory of an antigen and is able to mount a rapid immune response to the antigen.

IgE-MEDIATED (IMMEDIATE-TYPE) HYPERSENSITIVITY

1. In a sensitized individual, antigens binding to IgE attached to mast cell and basophils cause these cells to release cell mediators, such as histamine.
2. Histamine causes the contraction of smooth muscle in the trachea and bronchi, dilation of blood vessels (and consequently the outpouring of plasma from capillaries), and the release of mucus from goblet cells.
3. Excessive contraction of the bronchi and trachea and excessive loss of plasma from the blood lead to asphyxiation and shock, a condition known as anaphylaxis.
4. Anaphylactic shock can occur within 5 to 10 minutes after an antigen enters a sensitized animal, so this type of allergy is known as an immediate-type (IgE mediated) hypersensitivity.
5. Antihistamines are drugs that can inhibit histamine activity; epinephrine is a hormone that

reverses some of the symptoms histamine produces.

6. Since numerous cell mediators, in addition to histamine, are released during an allergic reaction, antihistamines are unable to alleviate all the symptoms of an allergy.
7. Hay fever, asthma, eczema, and hives are all immediate-type (IgE-mediated) hypersensitivities. These conditions are due to the different types of cell mediators that predominate.
8. People who are allergic to common antigens can sometimes be desensitized by injecting antigen over a long period of time. This treatment may result in the production of blocking antibodies (IgG or IgA), which inactivate antigen before it can bind to IgE.

IgG-COMPLEMENT-MEDIATED HYPERSENSITIVITY

1. Antigens injected into a sensitized animal generally bind to IgG and IgM, and this leads to complement activation.
2. Repeated injection of antigen can cause serum sickness, a condition due to activated complement components that cause anaphylaxis and lyse host cells.
3. Subcutaneous injection of antigen can cause an Arthus reaction, a condition due to activated complement component, which causes a localized allergic reaction and lysis of host cells.
4. Some forms of endocarditis (heart damage), nephritis (kidney damage), and arthritis (joint damage) are due to IgG-complement-mediated hypersensitivities. It is thought that antibody-antigen-complement complexes develop in the heart, kidney, and joints, and so lead to immune damage.
5. IgG-complement-mediated hypersensitivities play an important part in rheumatic fever. Rheumatic fever results from a chronic infection by *Streptococcus pyogenes* which continually stimulates the formation of antigen-antibody-complement complexes.
6. Some IgG made against *Streptococcus* M-protein are known to cross-react with proteins in the heart. Thus, rheumatic fever is also partly an autoimmune disease.

CELL-MEDIATED (DELAYED-TYPE) HYPERSENSITIVITY

1. Antigens not only stimulate the production of antibodies but also activate T-lymphocytes and macrophages.
2. Activated macrophages bind, phagocytize, and consume antigens that they recognize as foreign. Activated T-lymphocytes bind and attack antigens they recognize as foreign.
3. The destruction of nerve cells in the brain (one type of multiple sclerosis) is due to an attack by macrophages and T-lymphocytes on glial cells (myelin sheaths), which cover nerve cells.
4. The destruction of tissue cells that occurs in tuberculosis or leprosy is largely due to an attack by macrophages and T-lymphocytes on infected host cells. Arthus reactions may also play a role in tissue destruction.
5. Some of the destruction of tissue cells that occurs in syphilis is due to macrophages and T-lymphocytes attacking infected host cells. Arthus reactions may be important in tissue destruction.
6. Allergic contact dermatitis is a cell-mediated reaction to some chemicals that contact the skin.
7. The hapten penicillin is a very reactive molecule that often reacts with animal proteins and becomes an antigen. In a sensitized animal, simple contact with penicillin can result within one or two days in blistering of the skin and tissue destruction by macrophages and T-lymphocytes.
8. Catechols are very reactive haptens that frequently react with human proteins and become antigens. In a sensitized person, simple contact with catechols results within one or two days in blistering of the skin and tissue destruction by macrophages and T-lymphocytes.

AUTOIMMUNE DISEASES

1. An autoimmune disease occurs when the immune system attacks self antigens.
2. Many autoimmune diseases are initiated by an infection, by chemicals that modify antigen on host cells, or by viruses that introduce or modify antigens on host cells.
3. Rheumatic fever is an example of an autoimmune

disease caused by antibodies made against *Streptococcus* M-protein, which cross-reacts with heart tissue.

4. Juvenile diabetes mellitus is an autoimmune disease caused by macrophages and T-lymphocytes attacking pancreatic cells that have been altered by a virus.
5. One type of multiple sclerosis is an autoimmune disease caused by macrophages and T-lymphocytes attacking glial cells in the brain. Chickenpox and measles viruses have been implicated in the sclerosis of the brain.

IMMUNODEFICIENCIES

1. Immunodeficiences are due to one or more components of the immune system not functioning correctly. An immunodeficiency can be genetic, or it can be the result of an infection.
2. Persons missing or having defective cytotoxic T-lymphocytes (or killer T-cells) are particularly sensitive to viruses, intracellular bacterial infections, fungal infections, protozoal infections, and cancer.
3. Persons missing or having defective B-lymphocytes or helper T-lymphocytes are especially sensitive to bacterial infections.
4. Those missing or having defective suppressor T-lymphocytes may develop an autoimmune disease.
5. Combined immune deficiency is a hereditary condition in which T-lymphocytes and B-lymphocytes (plasma cells) are missing.
6. A depressed immune system allows virus infections that can lead directly to cancer.
7. A depressed immune system makes virus infections, bacterial infections, fungal infections, and protozoan infections more likely.
8. AIDS (acquired immune deficiency syndrome) is caused by the retrovirus HIV (human immunodeficiency virus).
9. HIV infects T-lymphocytes and macrophages. The destruction of these cells blocks the development of immune responses to a variety of microorganisms and cancers.
10. Diseases that frequently develop in HIV infected persons are cancers (Kaposi's sarcoma), pneumonia *(Pneumocystis carinii)*, and tuberculosis *(Mycobacterium tuberculosis* and *M. avium-intracellulare)*.
11. The virus that causes AIDS is transmitted from an infected person to another by sexual intercourse, and by the transfer of contaminated blood or blood products that occurs with IV drug abuse and transfusions.
12. AIDS may take longer than 5 years to develop after infection. Nearly all people who develop AIDS die from the consequences of the disease.
13. ARC (AIDS related cases) are individuals who have antibodies against HIV and persistent swollen lymph nodes, but do not suffer from a characteristic infection or cancer. Most ARC progress to AIDS within 5 years.
14. The spread of AIDS in a population can be slowed by sexual abstinence, monogamous relationships, and the use of condoms.
15. In the United States, the high risk groups for AIDS are homosexual and bisexual men, IV drug abusers, people from East and Central Africa, people from Haiti, and people receiving blood transfusions or blood products.
16. AZT (azidodeoxythymidine) is a base analogue that alleviates many of the AIDS symptoms and prolongs life. It does not cure the disease.

TISSUE TRANSPLANTATION AND BLOOD TRANSFUSIONS

1. Tissues can be transplanted from one site to another in the same individual or between individuals.
2. The transplantation of tissues from one site to another in the same individual is called an autograft.
3. The transplantation of tissues from one individual to another is called an allograft.
4. Allograft rejections are usually due to cell-mediated immunity.

BLOOD TYPE INCOMPATIBILITIES

Erythroblastosis fetalis is a disease of the fetus and newborn caused by the mother's IgG attaching to the fetus's RBCs.

STUDY QUESTIONS

1. Define the term allergy.
2. What are the salient characteristics of:
   a. anaphylactic reactions;
   b. immune complex reactions;
   c. cytotoxic reactions;
   d. delayed reactions.
3. What types of conditions can result from:
   a. an immediate-type hypersensitivity?
   b. a complement-mediated hypersensitivity?
   c. a delayed-type hypersensitivity?
4. What is an autoimmune disease?
5. How might an autoimmune disease develop?
6. What would be a consequence of an immunodeficiency in:
   a. phagocytic cells?
   b. B-lymphocytes?
   c. T-lymphocytes?
7. Define the following:
   a. autograft;
   b. isograft;
   c. allograft;
   d. anaphylaxis;
   e. serum sickness.
8. Explain how the HIV causes AIDS.

SELF-TEST

1. The immune response to an infectious agent can sometimes cause allergies or serious diseases in the host it is supposed to protect. a. true; b. false.
2. Which of the following classes of immunoglobulins is involved in anaphylactic reactions?
   a. IgA;
   b. IgD;
   c. IgE;
   d. IgG;
   e. IgM.
3. Which of the following classes of immunoglobulins is involved in Arthus-type reactions?
   a. IgA;
   b. IgD;
   c. IgE;
   d. IgG;
   e. none of the above.
4. Immediate-type hypersensitivities are easily treated with antihistamines because this drug counteracts the action of all the mediators such as histamine, serotonin, and SRS-A. a. true; b. false.
5. Epinephrine is very effective in the treatment of serum sickness because this condition results from the action of vasoactive amines released by mast cells. a. true; b. false.
6. Vasoactive amines are released by certain cells after they have been stimulated by an antigen-antibody reaction on their plasma membrane. Choose, from the following list, the cell type most commonly associated with the release of these substances:
   a. neutrophils;
   b. basophils;
   c. eosinophils;
   d. macrophages;
   e. lymphocytes.
7. An antigenic substance that elicits a hypersensitivity reaction in a host is called an allergen. a. true; b. false.
8. Select from the following list of allergic conditions the one that is caused by a delayed-type hypersensitivity:
   a. hay fever;
   b. serum sickness;
   c. hives;
   d. rheumatic fever;
   e. allergic contact dermatitis.
9. Select from the following list of allergic conditions the one that is caused by a complement-mediated hypersensitivity:
   a. hay fever;
   b. gummas;
   c. hives;
   d. rheumatic fever;
   e. allergic contact dermatitis.
10. The process of desensitization to an allergen involves frequent injections of small quantities of the allergen to promote the production of IgG antibodies instead of IgE. a. true; b. false.
11. The cells that are involved in delayed-type hypersensitivities are macrophages and lymphocytes instead of mast cells or neutrophils. a. true; b. false.
12. While the typical lesions associated with immediate-type hypersensitivities exhibit erythema (redness) and edema (swelling due to fluid), those of delayed-type hypersensitivities exhibit erythema and induration (hardness due to cell accumulation). a. true; b. false.
13. Allergic contact dermatitis is an allergic reaction that results when certain chemicals come in contact with the skin. The chemicals become allergenic only when they bind to skin proteins. Therefore, these chemicals are:
    a. carriers;
    b. haptens;
    c. antibodies;
    d. vasoactive amines;
    e. none of the above.

14. The following represents a list of autoimmune diseases that commonly afflict humans. Of the five diseases listed, which one is caused by antibodies directed against the patient's nucleus (and DNA)?
    a. Guillain-Barré syndrome;
    b. myasthenia gravis;
    c. hemolytic anemia;
    d. systemic lupus erythematosus;
    e. multiple sclerosis.
15. Some viral infections result in autoimmune disease because the host develops an immune response against virus-infected cells. a. true; b. false.
16. The following represents a list of immunodeficiences that can afflict humans. Of the five diseases listed, which one is caused by a deficient phagocytosis system?
    a. acquired agammaglobulinemia;
    b. Chediak-Higashi syndrome;
    c. AIDS;
    d. Hodgkin's disease;
    e. Cushing's disease.
17. AIDS is primarily transmitted in salivary secretions and by casual contact with infected individuals: a. true; b. false.
18. The following group of individuals comprise five AIDS risk groups. Which one of them has the highest incidence of disease?
    a. persons from Haiti;
    b. hemophiliacs;
    c. newborn babies;
    d. prostitutes;
    e. homosexual and bisexual men.
19. AIDS differs from ARC in that AIDS patients are infected by the HIV while ARC patients are not. a. true; b. false.
20. Tissue transplantations from one identical twin to another are called:
    a. homografts;
    b. autografts;
    c. allografts;
    d. isografts;
    e. xenografts.
21. Erythoblastosis fetalis results when an Rh^+ mother produces antibodies against the Rh^- baby. a. true; b. false.

SUPPLEMENTAL READINGS

Arnason, B. G. W. 1982. Multiple sclerosis: Current concepts and management. *Hospital Practice* 17:81–89.

Austen, K. F. 1984. The heterogeneity of mast cell populations and products. *Hospital Practice* 19:135–146.

Barnes, D. M. 1986. Strategies for an AIDS vaccine. *Science 233:* 1149–1153.

Berkman, S. A. 1984. The spectrum of transfusion reactions. *Hospital Practice* 19:205–219.

Centers for Disease Control. 1985. Acquired immuno-deficiency syndrome—Europe. *Morbidity and Mortality Weekly Report* 34(11):147–156.

Dixon, F. J. and Fisher, D. W. eds. 1983. *The Biology of Immunologic Disease.* Sunderland, MA: Sinauer Associates, Inc.

Fayer, R. and Ungar, B. L. P. 1986. *Cryptosporidium* spp. and Cryptosporidiosis. *Microbiological Reviews* 50(4):458–483.

Gallo, R. C. 1987. The AIDS virus. *Scientific American* 256(1):46–56.

Herberman, R. B. 1982. Natural killer cells. *Hospital Practice* 17(4):93–103.

Hirsch, M. S. and Kaplan, J. C. 1987. Treatment of human immunodeficiency virus infections. *Antimicrobial agents and Chemotherapy* 31(6):839–843.

Hood, L. Weissman, I. W. Wood. 1978. *Immunology.* Menlo Park, CA: Benjamin/Cummings.

Koffler, D. 1980. Systemic lupus erythematosus. *Scientific American* 243(1):52–61.

Leder, P. 1982. 1979. The genetics of antibody diversity. *Scientific American* 246(5):102–115.

Medley, G. F., Anderson, R. M., Cox, D. R., and Billard, L. 1987. Incubation period of AIDS in patients infected via blood transfusions. *Nature* 328:719–721.

Muller-Eberhard, H. J. 1977. Chemistry and function of the complement system. *Hospital Practice* 12:33–43.

Muller-Eberhard, H. J. 1978. Complement abnormalities in human disease. *Hospital Practice* 13:65–76.

Norman, C. 1986. Sex and Needles, Not Insects and Pigs, Spread AIDS in Florida Town. *Science* 234: 415–417.

Rose, N. R. 1981. Autoimmune diseases. *Scientific American* 244(2):80–103.

Selwyn, P. A. 1986. AIDS: What is now known, II. Epidemiology. *Hospital Practice* 21(6):127–164.

Zinkernagel, R. M. 1978. Major transplanion antigens in host responses to infection. *Hospital Practice* 13:83–92.

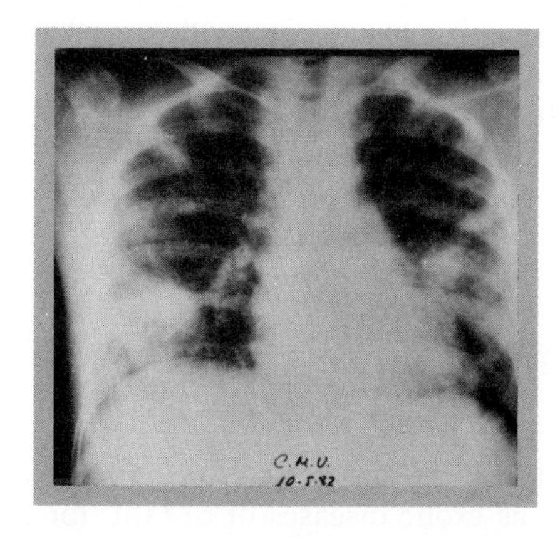

CHAPTER 15

EPIDEMIOLOGY OF INFECTIOUS DISEASES

CHAPTER PREVIEW PERSPECTIVE: JOHN SNOW ON CHOLERA

Cholera is an infectious disease that has ravaged populations for many centuries. The cause of the disease, *Vibrio cholerae*, is spread in drinking water. The disease is characterized by a profuse diarrhea that has the consistency of rice water (rice water stool). The water loss due to diarrhea is so great that patients can lose 10–15% of their body weight in less than a day. Death due to cholera usually results from kidney failure or low blood volume, both consequences of fluid loss.

We now have a clear understanding of the cause of cholera, the disease process, and how it spreads in human populations. This was not the case as recently as 150 years ago. In fact, cholera was thought to be a curse of the gods as punishment for our transgressions. Much of what we know about the spread of cholera is based on studies carried out by the British physician and anesthesiologist John Snow. His studies clearly showed that cholera spreads in human populations through drinking water polluted with sewage. This research was done even before the discovery of *Vibrio cholerae* in 1883 by Robert Koch.

Snow's knowledge of cholera was acquired during a series of outbreaks in London in 1853 and 1854. At that time, the drinking water was supplied to Londoners by two different purveyors: the Southwark & Vauxhall Company (S & V) and the Lambeth Company.

Snow's studies led him to suspect that contaminated water might be the source of cholera. He conducted a survey of those houses in which cholera patients lived and found that the S & V company supplied the drinking water to most of those homes. He also discovered that the S & V company obtained its water from the Thames River, right in the heart of London—the very same place where Londoners discharged their sewage. In contrast, the Lambeth company obtained its water from the Thames River before it reached the city. Thus, Snow concluded that cholera was spread by drinking water contaminated with raw sewage (which contained the cholera agent). He also pointed out the need for water purification to alleviate this malady.

The studies conducted by Snow in the 1850s are considered to be classic studies in epidemiology. He employed methods that included both interviews of patients and experimentation. The results he obtained allowed him to decipher the epidemiology of cholera. In this chapter we discuss some of the characteristics of epidemics, how epidemics are detected, and how infectious diseases are spread among humans.

CHAPTER OUTLINE

EPIDEMIOLOGY IS THE SCIENCE THAT STUDIES the prevalence and distribution of disease in a population. This information can enable physicians to make prompt and accurate diagnoses. For example, certain infectious diseases, such as influenza, elicit signs and symptoms such as headache, malaise, fever, and achy joints, which are characteristic of a variety of diseases. Under ordinary conditions, a physician may need additional information to make a diagnosis of influenza based on these signs and symptoms. If the same physician is aware that an epidemic of influenza is taking place, however, the diagnosis is facilitated, because the signs and symptoms exhibited by the patient are consistent with those of influenza. Also, by studying groups of patients, enough information can be generated to answer basic questions such as how to prevent a given disease. For instance, epidemiological studies of patients with myocardial infarctions (heart attacks) and of normal patients have helped physicians to determine that factors such as personality, social pressures, smoking, obesity, and lack of exercise contribute to heart attacks. Therefore, heart attacks can be prevented by reducing these risk factors.

Increasing population densities and rapid worldwide travel must be considered in epidemiological studies. No longer is it possible to think of infectious diseases such as leprosy, malaria, schistosomiasis, and hemorrhagic fever as exotic diseases of distant, foreign lands. A patient may contract the disease 10,000 miles from home and not show its signs and symptoms for a number of weeks after returning. For example, the number of cases of Hansen's disease (leprosy) in the United States has increased consistently for the last two decades (fig. 15.1). The increase can be attributed to foreign travel as well as to increased immigration of infected individuals. Furthermore, with increasing population densities, the risk of outbreaks of infectious diseases also increases.

EPIDEMICS AND PANDEMICS

An **epidemic** or outbreak is usually a short-term increase in the occurrence of a disease in a particular population. Three cases of diphtheria in a week in a small town of 1000 inhabitants, which normally has one such case every 5 to 10 years, is definitely an epidemic. Three cases of influenza in a day in a large metropolis of 1 million inhabitants, however, is certainly not an epidemic. Figure 15.2 illustrates a typical epidemic graph. Notice that this epidemic represents a significant increase in the number of cases above the number expected.

An **endemic disease** is one that occurs in a population or geographic area on a constant basis. One speaks of venereal diseases as being endemic in the United States, or histoplasmosis as being endemic in the Mississippi River Valley. On the other hand, **sporadic** cases occur in a population without any type of periodicity. They appear in and disappear from a population at unpredictable rates.

Sometimes the term **pandemic** is used to describe a long-term increase in the incidence of a disease in a very large population. The duration is usually years, and the spread of the disease goes beyond international boundaries. For example, the most recent pan-

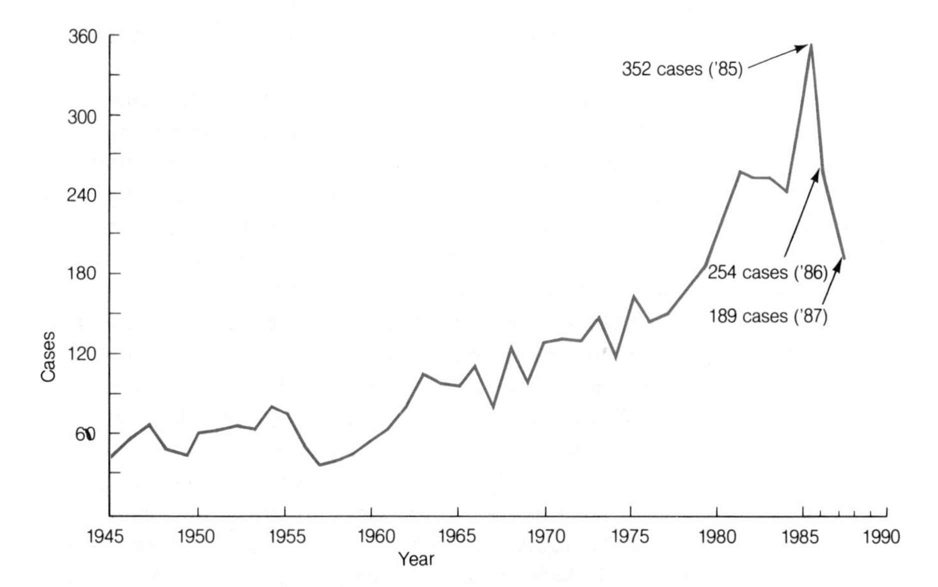

Figure 15.1 Yearly incidence of Hansen's disease in the United States. The incidence of Hansen's disease (leprosy) in the United States has been on the rise since 1957. This increase is due primarily to the increased immigration of individuals from areas where leprosy is prevalent.

demic of cholera, originated in Indonesia in the early 1960s and spread to parts of Europe, Africa, and the Middle East within 10 years.

PARAMETERS USED BY EPIDEMIOLOGISTS TO MEASURE EPIDEMICS

To make sense of data obtained from studies of infectious diseases, epidemiologists determine the number of cases in a particular group. Simple counts are often not very informative because they do not reveal the extent of an outbreak. In our examples of diphtheria and influenza, the counts were the same—3 cases. But 3 cases/1000 (300/100,000) is much greater in significance than 3 cases/1,000,000 (0.3/100,000). Therefore, rates or proportions can provide valuable information, while simple counts generally do not.

Rates

Prevalence rate, case rate, or **cumulative incidence** quantify the total number of events in a population at a specified time. The population may be a school, a hospital, a city, a state, or a country. Rates for a disease are calculated by dividing the number of individuals in the population suffering from the disease by the total number of individuals in the population. Consider, for example, the following statement: "On August 8 there were 150 cases of influenza in a city of 13,000 inhabitants." The rate of influenza in the population is 150/13,000. Usually, the rate is converted to the

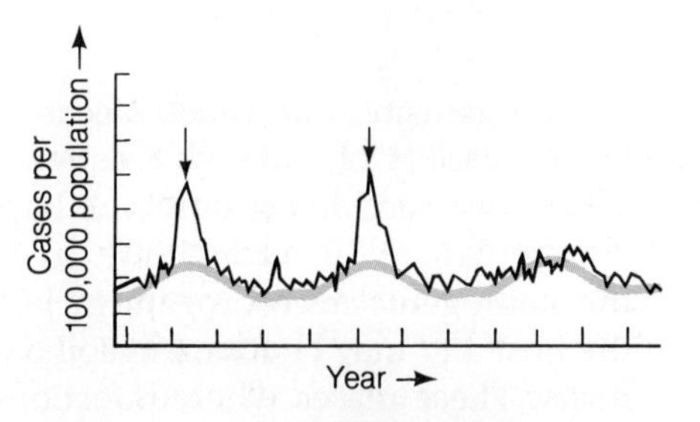

Figure 15.2 Graph illustrating an epidemic. The color lines indicate the expected number of cases; the black lines indicate the actual number of cases. Epidemics (arrows) are sharp increases in the number of cases of a disease above that expected.

number per 100,000 so that rates can be compared:

$$150/13{,}000 \rightarrow 1154/100{,}000.$$

Sometimes, the rate is presented as a percentage:

$$(150/13000) \times 100 \rightarrow 1.15\%.$$

Prevalence rate can be measured for any time interval, such as a day, a month, or a year.

Mortality, Morbidity, and Case Fatality Rates

The **mortality rate** measures the number of individuals dying in a population, either from a specific cause or from all causes:

$$[\text{number dead/total population}] \times 100 = \text{death rate}.$$

Morbidity rates measure the number of individuals who become ill as a result of a particular infection:

$$[\text{number ill/total population}] \times 100 = \text{illness rate}.$$

The **case fatality rate** measures the number of deaths among individuals suffering from a particular disease:

$$[\text{number dead/number ill}] \times 100 = \text{case fatality rate}.$$

Together, the morbidity and case fatality rates give epidemiologists an indication of the intensity of exposure and the virulence of the strain causing a particular disease.

SOURCES OF INFECTIOUS DISEASES

Microorganisms that cause disease can enter a host from a variety of sources. Frequently, a human host is merely an accidental host who happens to be in the wrong place at the wrong time. In cases such as this, the pathogenic microoroganism leads a life outside the host and may be found in soil or water or in other hosts. These places where infectious agents are perpetuated in nature are called **reservoirs**. For example, the fungal disease called valley fever (coccidioidomycosis), which afflicts thousands of individuals living in endemic areas, has *soil* as its reservoir. The fungus reproduces in the soil, and when it becomes airborne in wind currents it can infect susceptible individuals. On the other hand, the natural reservoir for rabies is small wild mammals, primarily skunks. Rabies is transmitted from skunk to skunk in the wild, but occasionally an infected skunk finds its way to the suburban environment and quarrels with a domestic animal, which can then become infected. These animals, as well as skunks, can transmit the disease to humans. Such diseases, which are endemic in animal populations, are called **zoonoses**. Typhoid, by contrast, which is caused by *Salmonella typhi,* is transmitted from person to person by the fecal-oral route, by ingestion of foods or water contaminated with human fecal matter from individuals carrying the typhoid bacterium. Humans represent the only known reservoir for this disease. The dangerous aspect of typhoid resides in the fact that certain infected individuals harbor the infectious agent without themselves becoming ill. These asymptomatic individuals, called **carriers**, serve to perpetuate the disease in the human population.

MODES OF TRANSMISSION

Infectious diseases result when an individual comes in contact with an infectious agent. This contact may come about by one or more of the following routes: (a) person to person contact; (b) contact with infected animals; (c) contact with vectors; (d) contact with inanimate objects (fomites); (e) consumption of contaminated food or water; (f) inhalation of airborne particles.

Person to Person Contact

Many infectious diseases are transmitted from one individual to another by direct contact. Venereal diseases, for example, are contracted by sexual intercourse or other sexually related activities. Touching infected lesions, such as boils and ulcers may transfer the infectious agent to a susceptible host. Other diseases can be transmitted by direct contact, such as touching hands or kissing. Infectious mononucleosis ("mono") can be transmitted by kissing, but it can also be transmitted by indirect means. The common cold can be transmitted from person to person by contact with hands contaminated with nasal secretions of infected individuals.

Contact with Infected Animals

Diseases such as rabies, tularemia, and ringworm can be acquired by contact with infected animals. For example, ringworm can be contracted from cats, dogs,

FOCUS CONTROL OF MARY MALLON: A CARRIER OF THE TYPHOID FEVER BACILLUS

Presently, 400–500 cases of typhoid fever and 5–25 deaths from this disease are reported each year in the United States. In the early 1900s, however, there were many thousands of cases and 11,000–25,000 deaths each year. Most of these cases of typhoid fever were due to drinking water contaminated with sewage and eating foods handled or prepared by persons shedding the typhoid fever bacterium. One of the most famous carriers of the typhoid bacillus was Mary Mallon.

Mary Mallon's story began in 1906, when the New York City Department of Health hired a sanitary engineer named George Soper to investigate an outbreak of typhoid fever in a family in Oyster Bay, Long Island. Since neither the water drunk by the family nor the food in the home was contaminated, Soper suspected that the family cook, who had quit her job when the outbreaks of typhoid fever occurred, might have had the disease and spread it to the family. The family cook was Mary Mallon.

In his attempt to find Mary Mallon, Soper discovered that she had worked in seven homes from 1896 to 1906. In these homes, there had been 28 cases of typhoid fever during the time Mary had worked as a cook. In every case, Mary had left the job when persons in the homes came down with typhoid fever. In 1907, Soper caught up with Mary Mallon in Manhattan, where she was working for a family. He confronted her with his suspicions that she was a typhoid carrier and that she was responsible for the outbreaks of typhoid fever in the homes where she had worked. Soper asked for blood, urine, and fecal specimens to determine conclusively whether or not she was shedding the bacteria and promised her free medical treatment. Instead of accepting Soper's offer, Mary became belligerent and threatened Soper with a carving fork. Mary did not believe in bacteria nor in the possibility that she was the cause of the typhoid outbreaks where she had worked.

The New York City Health Department soon ordered Mary arrested. She was brought fighting and screaming to a city hospital for patients with infectious diseases. Examination of Mary's stool samples demonstrated that she was shedding large numbers of the typhoid bacilli, but Mary showed no external symptoms of typhoid fever. An article in the *Journal of the American Medical Association* of 1908 referred to her as Typhoid Mary, an epithet by which she is known today. Doctors believed that the bacteria were growing in her gall bladder, and recommended that it be removed in order to eliminate her carrier condition. Since there was disagreement as to exactly where the bacteria were growing in her body, she refused the operation. Mary realized that the operation might kill her because bacteria would spread into her abdominal cavity when the infected tissue was removed. Operations of this type were very dangerous because there were no antibiotics then.

In 1910, Mary was released because of public pressure on the Health Department. The public felt it was not right to imprison persons who had not committed crimes but were simply carriers of disease-causing organisms. Mary remained at large until 1915, when the Public Health Department connected a 1914 outbreak of typhoid fever at a New Jersey sanatorium and a 1915 outbreak at a New York hospital to a cook matching Mary's description. In 1915 she was returned to her island hospital prison in the middle of the New York River. She spent the next 23 years at the hospital, where she died in 1938 at the age of 70. During her life in prison, she worked in the hospital laboratory but was forced to eat alone. Mary was imprisoned for 26 years of her life because she was a danger to the community and was responsible for a number of deaths: at least 53 cases of typhoid fever and three deaths were positively connected with her.

Control of carriers is a serious moral and public health issue even today. For example, there are approximately 800,000 carriers in the United States—and more than 200 million carriers throughout the world—of the virus that causes hepatitis B. This virus is generally transmitted by sexual contact, by blood transfusions, and through needle pricks among drug addicts. What is the responsibility of Public Health Departments in controlling carriers who are prostitutes or sexually promiscuous, or who want to sell their blood to blood banks? Should these carriers be held accountable if illness and death result from their behavior? This is an ethical problem for which there may be no simple solution.

and horses. Tularemia, a zoonosis that afflicts chiefly rabbits, is not uncommon among rabbit hunters in endemic areas. Rats, mice, and other rodents are reservoirs for a variety of infectious diseases, many of which can be transmitted to humans. Bats can transmit diseases such as rabies and histoplasmosis. It was once thought that only vampire bats attacked animals for their blood, but it is now known that other bats, if rabid, may attack and bite animals or humans and thereby transmit the rabies virus.

Vectors

Vectors are a special group of animals that can transmit infectious agents to other animals and to humans. Vectors are defined in various ways by different authors. In the strict sense, however, a vector is an arthropod that has a long-lasting association with a disease-causing microorganism and can transmit the pathogen to susceptible individuals. For example, the mosquito *Anopheles* is a vector for malaria; that is, it can transmit the malaria parasite by biting a susceptible host while feeding. The sexual stage of the life cycle of the malarial parasite takes place in the mosquito, so the mosquito is a necessary host for the completion of the life cycle of the parasite.

A smiliar situation exists with the vector that transmits Rocky Mountain spotted fever. The tick *Dermacentor andersonii* can harbor *Rickettsia rickettsii*, a bacterium that multiplies within the tick and can be transmitted from generation to generation of ticks in the eggs. The tick, upon biting a suitable host, transmits the infectious agent.

Arboviruses are a group of medically and economically important viruses that are transmitted by vectors. These viruses cause a variety of diseases, including encephalitis (inflammation of the brain), that afflict humans and livestock. The viruses are transmitted by the bite of arthropods, such as ticks and mosquitoes.

Now let us examine another situation in which an arthropod is not strictly a vector, but does transmit an infectious organism. Toxoplasmosis is an infectious disease caused by *Toxoplasma gondii*, a protozoan that commonly infects cats and occasionally is transmitted to humans. The parasite is excreted in the cat's feces into soil, sandboxes, or "kitty litter" boxes. Flies and roaches, for example, can land on contaminated soil or sand and pick up the parasite with their feet and legs. The insect may then land on food and deposit the parasite, and a human ingesting the contaminated food may become infected. The insect, although it served as a **vehicle of transmission**, is not considered a vector, because the association between the parasite and the insect was accidental and transient.

Fomites

Fomites are inanimate objects such as drinking cups, toothbrushes, towels, combs, toilet seats, and soap that serve to transmit a disease agent but do not support its multiplication. Ringworm of the scalp in preadolescents is most often caused by a fungus called *Microsporum audouinii*. This fungus can be transmitted among children by direct contact with one another, but it can also be transmitted through fomites, such as combs or backrests of theater seats. Backrests, especially those upholstered with piled fabrics such as corduroy or velveteen, serve as thousands of tiny "inoculating needles" that can pick up spores from the heads of infected individuals and "inoculate" others. Infectious mononucleosis can be transmitted among individuals by using contaminated glasses, cups, or eating utensils.

Food and Water as Vehicles of Infection

Food and water can serve to transmit a variety of infectious diseases. Poliomyelitis, hepatitis, cholera, typhoid, and various forms of food poisoning are transmitted by contaminated food or water. Improperly cooked foods can transmit disease agents with which they are contaminated. For example, trichinosis and salmonellosis are frequently transmitted by improperly cooked foods.

The **fecal-oral** route is a common mode of transmission for a variety of infectious agents. Such diseases are transmitted when feces of an infected individual enter a potable (drinkable) water supply

Figure 15.3 **Possible sources of food contamination.** *(a)* Sewage contaminates fish, shellfish, and crustaceans. *(b)* Carriers and microorganisms in the environment contaminate prepared foods. *(c)* Undercooked or raw eggs contaminate prepared foods such as custard, eggnog and mayonnaise. *(d)* Improperly pasteurized milk from diseased animals contaminates cheeses and milk products. *(e)* Undercooked fowl or meat contains pathogens. ▶

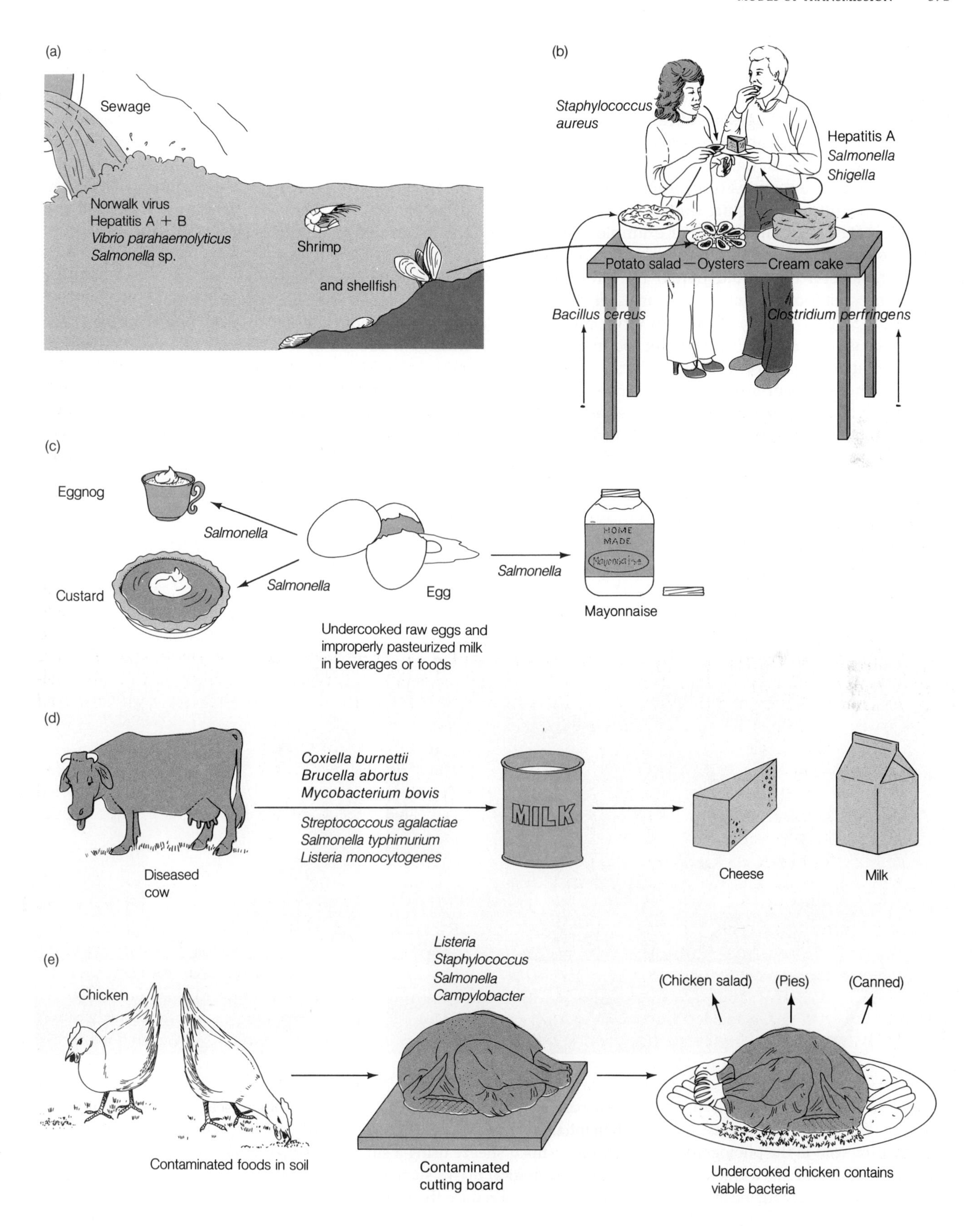
(a)
Sewage
Norwalk virus
Hepatitis A + B
Vibrio parahaemolyticus
Salmonella sp.
Shrimp
and shellfish
(b)
Staphylococcus aureus
Hepatitis A
Salmonella
Shigella
Potato salad—Oysters—Cream cake
Bacillus cereus
Clostridium perfringens
(c)
Eggnog
Salmonella
Custard
Salmonella
Egg
Salmonella
HOME MADE
Mayonnaise
Undercooked raw eggs and improperly pasteurized milk in beverages or foods
(d)
Coxiella burnettii
Brucella abortus
Mycobacterium bovis
Streptococcous agalactiae
Salmonella typhimurium
Listeria monocytogenes
MILK
Diseased cow
Cheese
Milk
(e)
Chicken
Listeria
Staphylococcus
Salmonella
Campylobacter
(Chicken salad)
(Pies)
(Canned)
Contaminated foods in soil
Contaminated cutting board
Undercooked chicken contains viable bacteria

that is subsequently consumed by other individuals. Diseases such as typhoid, cholera, poliomyelitis, dysentery, and various worm infections are transmitted by the fecal-oral route. Ingestion of contaminated foods, water, and beverages can be the cause of a large number of human diseases (fig. 15.3). These diseases may arise from either infections or intoxications.

Health authorities require that food be monitored for the presence of pathogens. For example, shellfish such as oysters and mussels are closely monitored for indicator bacteria that could herald the presence of pathogenic microorganisms, such as *Campylobacter, Salmonella, Shigella, Vibrio,* hepatitis A virus, and polio virus. Because of this risk, shellfish are routinely tested for the presence of coliforms as indicators of fecal pollution of the shellfish beds. The procedure used to test for the microbiological quality of shellfish is similar to that used to test water, but also includes a standard plate count of the shellfish meat to determine the total number of microorganisms present.

Foods implicated in outbreaks of food poisoning are tested for the presence of specific pathogens. For example, if a particular food is suspected of being the source of staphylococcal enterotoxin during an outbreak of staphylococcal food poisoning, it is highly desirable that live *Staphylococcus aureus* be isolated from the food in large numbers to support the suspicion.

Airborne Particles

The air is laden with numerous microorganisms, many of which are harmless; however, the flushing mechanisms of coughing and sneezing can serve to introduce infectious agents into the air (fig. 15.4). An unstifled sneeze can discharge into the atmosphere numerous drops of mucus and saliva, and if the individual is currently infected, numerous infectious agents

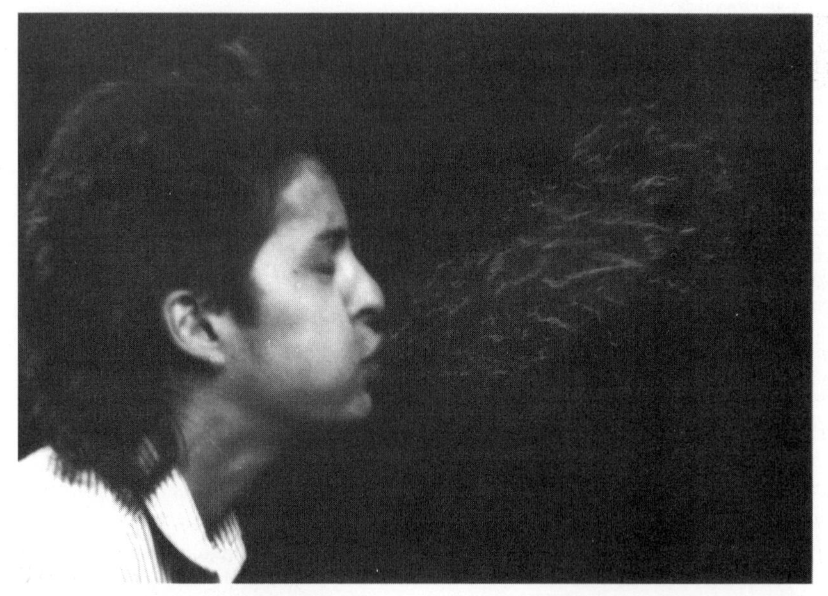
(a)

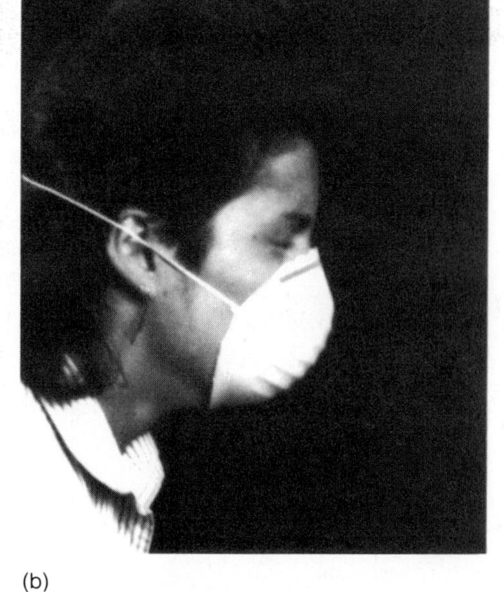
(b)

Figure 15.4 Aerosols created by sneezing. *(a)* An unstifled sneeze. The particles seen in the photograph are bits of saliva and mucus secretions laden with microorganisms. These particles may be infectious when inhaled by susceptible individuals. *(b)* A stifled sneeze using a surgical mask. Surgical masks such as the one illustrated reduce the formation of aerosols during sneezing to negligible levels, thus reducing considerably the risks of infection in hospitals and surgical suites.

are also released. Many of these infectious agents, trapped in large drops, fall to the ground and dry out and may subsequently be redistributed in the dust. The dust particles can then become airborne and infect other individuals. Tiny droplets, containing one or two infectious agents, dry up before they reach the ground. These droplets are surrounded by a dehydrated film of oral discharge and are called **droplet nuclei**. They can be airborne for long periods of time and yet retain their infectivity.

The danger of airborne infections is maximized in hospital environments and in large buildings where individuals are concentrated and droplet nuclei have a good chance of being inhaled (fig. 15.5). Air conditioning systems are potential hazards for transmission of airborne infectious agents. For example, the organism that causes Legionnaires' disease (legionellosis), primarily *Legionella pneumophila* grows well in aquatic environments, including the water of air-cooling towers that are part of air conditioning systems. The bacteria become aerosolized into the air ducts and are distributed throughout the building, resulting in a potential source of an epidemic.

FACTORS INFLUENCING EPIDEMICS

Infectious diseases result from complex interactions among the host, the parasite, and the environment (fig. 15.6). The virulence of the pathogen, the susceptibility of the host, and environmental factors make each epidemic unique. Epidemics differ in character even when they are caused by the same agent, as can be ascertained from various parameters such as the incidence rate and the case fatality rate.

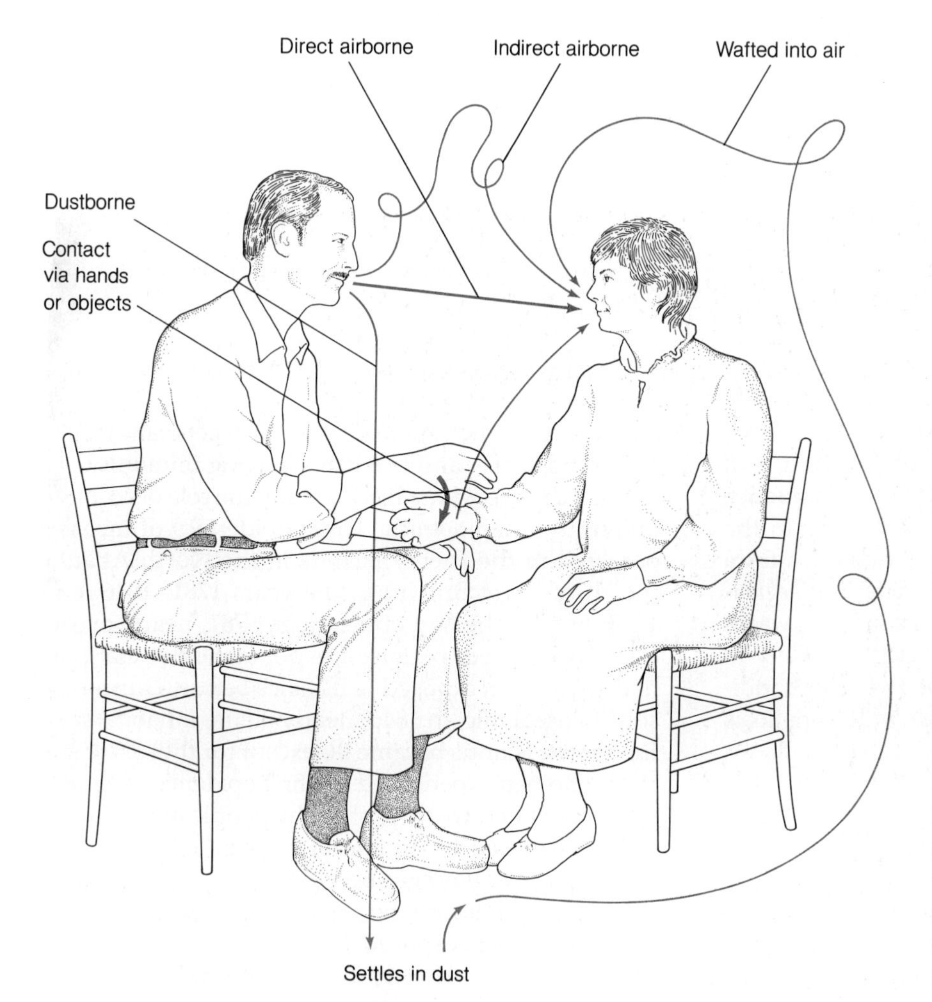

Figure 15.5 The dispersal of airborne infectious agents. Airborne microorganisms can be transmitted directly by coughing and sneezing. Indirect transmission can occur by inhaling contaminated dust or by touching contaminated objects. Insects may also transmit infectious agents when they carry microorganisms on their appendages.

Figure 15.6 Interactions that promote an infectious disease. Infectious diseases result from complex interactions among the host, the pathogen, and the environment.

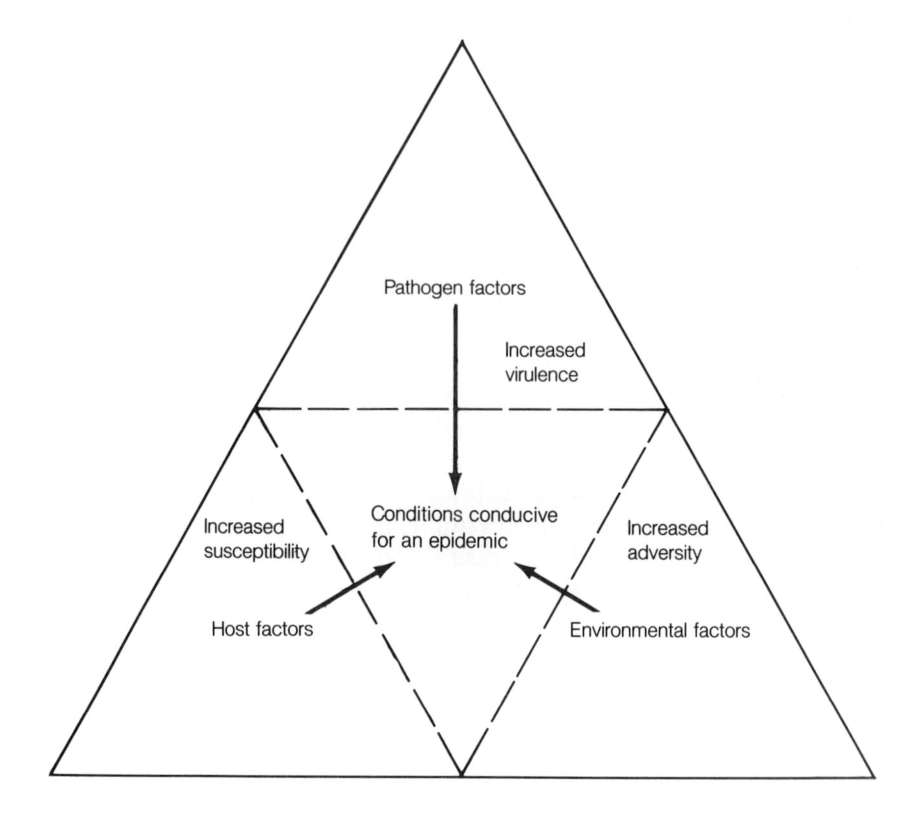

Environmental Factors

The physical environment plays a role in determining the size and severity of an epidemic. Environmental conditions such as crowding, temperature, wind, and relative humidity can drastically alter the incidence and case fatality rates of an infectious disease in the population. For example, Legionnaires' disease, a form of severe pneumonia caused by the bacterium *Legionella pneumophila,* is most common during the warm months, when air conditioning systems are likely to be used extensively (fig. 15.7). Thus, weather conditions contribute to the high incidence of legionellosis in a population.

Host Factors

Acquired Immunity Whether or not an infectious agent causes a few cases of a disease or an epidemic often depends upon the immunological state of the population. A population that has been subjected to many infections by a particular pathogen or that has been vaccinated against the pathogen will have a low morbidity rate.

A disease such as measles—which generally results in overt disease, and produces lifelong immunity—is an ideal disease with which to study the role of acquired immunity in epidemics. Three epidemics of measles occurred in the Faeroe Islands in the North Atlantic (north of Great Britain) in the years 1781, 1846, and 1875. During the first epidemic in 1781, nearly everyone was afflicted, because the population at that time had never been exposed to measles before. During the second epidemic, in 1846, nearly all the inhabitants of the Faeroe Islands became ill, except for those few still alive who had experienced the first epidemic. The 1875 epidemic is noteworthy: of all the people in the Faeroe Islands exposed to the measles virus, only those 29 years of age or younger became ill. These individuals were not alive during the 1846 epidemic and therefore had never been exposed to measles. The rest were already immune. Hence, the immune state of the pop-

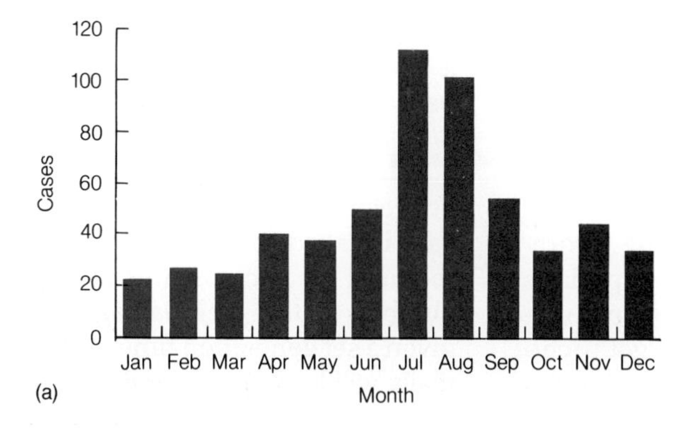

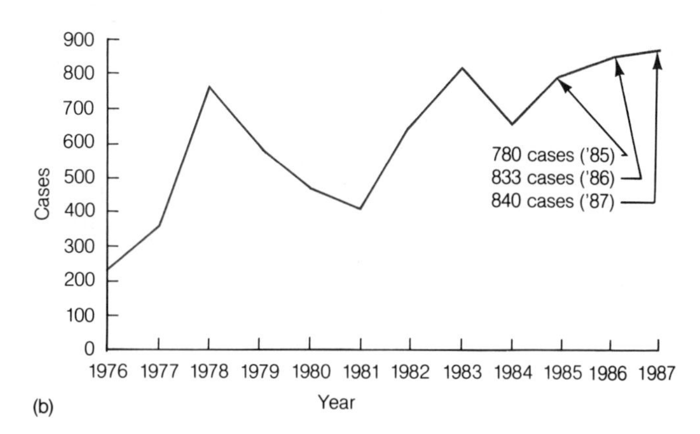

Figure 15.7 Incidence of legionellosis in the United States. *(a)* Peak incidence of legionellosis in the United States occurs during the hot months, when air conditioning systems are more likely to be used. The plot is for the year 1979. *(b)* Yearly incidence of legionellosis in the United States.

ulation at risk (**herd immunity**) largely influences the number of individuals that will become ill during an outbreak of a disease.

Genetic Background If we analyze the distribution of genetic traits (different gene frequencies) in a population, it becomes apparent that certain traits are common in one subpopulation but not in others. Some of these characteristics may render a group less susceptible to a particular disease.

Malaria occurs more often in Caucasians than in persons of Black African ancestry. About 70% of this group have mutated red blood cells that inhibit the replication of *Plasmodium* (malaria) and are thus genetically immune to it. Malaria cannot take place in the absence of erythrocyte infection by the merozoite.

It is known that individuals who are heterozygous for the allele that causes sickle-cell anemia *(s)* are more resistant to malaria than those individuals who carry only the normal allele *(S)*. Individuals with **sickle-cell anemia** *(ss)* have defective erythrocytes, and their oxygen transport is greatly diminished. Individuals with **sickle-cell trait** *(Ss)*, however, have a functional but fragile red blood cell. When these erythrocytes are infected with merozoites, they "sickle up," causing the death of the parasite within the erythrocyte. The sickled erythrocyte eventually is ingested by cells of the reticuloendothelial system. As a consequence, individuals with sickle-cell trait generally are more resistant to malaria than normal *(SS)* individuals. Sickle-cell trait is much more common in some Black populations than in Caucasian populations. Therefore, the incidence and case fatality rates in a malaria outbreak are much lower in the Black population than in the Caucasian.

Health of Host Emotional stress and malnutrition are detrimental to an animal's health because the immune system does not function efficiently when emotional stress and malnutrition are problems. Individuals and populations are more susceptible to pathogens under these types of stress.

Cultural Factors Religious or ethnic practices may affect the transmission of disease agents. For example, *Trichinella spiralis* is a roundworm that causes a disease called **trichinosis**. It is contracted by eating poorly cooked meats of several domestic animals, especially pork. Individuals who, for religious or personal reasons, choose not to eat pork are unlikely to get trichinosis. In Jewish populations the incidence of trichinosis is very low because they usually do not eat pork.

In New Guinea, members of one tribe formerly handled and ate the brains of their dead ancestors. Because of this practice, a high proportion of the population suffered from **kuru**, a disease of the central nervous system, which causes a general debilitation that often leads to death. Apparently, the infectious agent responsible for this disease was infecting the brains of many people in the tribe and was transmitted when the brains were handled or consumed. High incidence rates of kuru will obviously be greater in populations that consume and handle human brains than in populations that do not follow these cannibalistic practices.

Pathogen Factors

It has been found that epidemics caused by the same disease agent usually differ in morbidity and case fatality rates. Some epidemics have case fatality rates as high as 70%, while others have much lower rates, even in the same population. In many cases, it appears that the virulence of the infectious agent varies; i.e., it is possible that the LD_{50} of one strain was much lower than that of the other and that this difference is reflected in the higher case fatality rate.

Characteristics That Make a Good Pathogen The characteristics that make a good pathogen were discussed in Chapter 14. These factors may include the pathogen's (1) ability to attach to the host, (2) invasiveness, (3) antigenic variation so as to escape the host's defenses, (4) ability to exist as an intracellular parasite, and (5) production of toxins.

Dose of Infectious Agent The number of individuals that become ill in a given outbreak may depend on the number of infective doses present. For example, *Coccidioides immitis* normally has a prevalence rate of about 10 cases of valley fever per year in San Luis Obispo County, California; but a dust storm may cause 15–20 cases in a single month. The sudden increase is a direct result of the increased number of infectious particles present in the air during the dust storm.

Incubation Period The period of time that elapses between infection and clinical manifestation of a disease is called the **incubation period**. This is the time required for microorganisms to multiply to sufficient numbers so that their damage is noticeable. Obviously, given a set generation time, large infectious doses normally will lead to shorter incubation periods than small infectious doses. Infectious diseases have characteristic incubation periods, however, that reflect the growth habits of the pathogen. Hence, epidemiologists can obtain clues as to the identity of a pathogen by determining the incubation period of the disease. Knowledge of incubation periods for various pathogens can also help the epidemiologist to establish approximately when the patient came in contact with the infectious agent and to trace the source of an epidemic.

DETERMINATION OF THE INFECTIOUS AGENT

To control an epidemic, its exact cause must be known. It is sometimes necessary to differentiate between the endemic cases and those that are part of an epidemic, particularly when the disease outbreak is caused by a pathogen that does not cause characteristic signs and symptoms.

Let us look at a hypothetical outbreak of diarrhea (gastroenteritis) to see the importance of determining the exact cause of an epidemic. Suppose that during a period of two weeks, 50 cases of gastroenteritis are reported to the local health agency by a hospital infection control officer. Upon examination of the stool specimens, 41 yield *Salmonella*, 1 yields *Shigella sonnei*, and 8 yield *Campylobacter jejuni*. These data indicate that *Salmonella* might be the cause of this epidemic. Further investigation of all the people (41) who yielded positive stool cultures for *Salmonella* indicated that 35 of them were at a catered affair while the other 6 were not. This information indicates that the 35 individuals at the catered affair are probably part of an outbreak, while the other 6 were infected from endemic sources. This hypothesis can be supported further if we find that the same species of *Salmonella* is responsible for the 35 cases of gastroenteritis. Any unrelated cases may likely be caused by a different species. Current methodology for the identification of *Salmonella* is based on biochemical and serological procedures (e.g., agglutination). Let us assume that these procedures were carried out, and that 34 isolates were identified as *S. montevideo* and 1 as *S. heidelberg*.

For all practical purposes, the study should concentrate on *S. montevideo* as the common cause of the outbreak. The importance of this identification lies in the fact that now we can trace both the immediate source of the infection and possibly the carrier. In such an outbreak, all people who attended the catered affair should be questioned as to whether they were ill or not, which food(s) they ate, and which food(s) they prepared. It is likely that all the ill patients ate the same food and that their stools would yield *S. montevideo*. Once the vehicle of infection is known (in this case a food item), the epidemiologist can begin to trace the source of the infectious agent and ultimately eliminate it. This hypothetical study shows that the precise identification of a causative agent allows the determination of the size of the epidemic, the source of infection, and any previously unrecognized carriers.

THE CONTROL OF EPIDEMICS

The successful control of epidemics requires an understanding of the factors that lead to them. Control measures vary depending upon the disease, but generally fall into several related categories. Some are short-term measures designed to control the current outbreak, while others are long-term measures designed to prevent future outbreaks.

Reduction of the Source of Infection

Prompt implementation of measures that will reduce the source of infection in an ongoing epidemic is essential to its control. Thus, it is essential that cases and carriers be recognized and isolated from the susceptible population. Most hospitals have an isolation ward that serves to house individuals with **contagious** diseases. Strict rules are followed in these wards to prevent the escape of infectious agents. Another method of separating infectious agents from the susceptible population is by **quarantine**. This is a procedure similar to isolation, except that it generally involves individuals, animals, or materials that are suspected of being contaminated with an infectious agent, but that do not show signs or symptoms of the disease.

Reporting outbreaks promptly is important so that control measures can be implemented as soon as possible. In the case of cruise vessels that stop at many foreign ports, information on outbreaks can prevent the spread of disease from port to port. The master of the vessel is required to report the number of cases of diarrhea to the quarantine station portside within 24 hours of arrival so that health authorities can control the introduction and spread of diseases such as cholera, plague, and dysentery.

International travel poses a significant obstacle to the successful control of infectious diseases. As previously mentioned, an individual might travel by jet to a distant land where an infectious disease (e.g., yellow fever) is prevalent, become infected, and introduce a "new" disease upon returning home. In cooperation with the **World Health Organization** (WHO), health authorities throughout the world enforce quarantine measures to control diseases such as plague, cholera, and yellow fever. Along with the isolation and quarantine measures, suspected patients are treated with suitable drugs to reduce their infectivity.

In outbreaks in which the source of infectious agent is contaminated soil, soil sterilants or disinfectants are used to kill the pathogen. In agricultural areas in which valley fever (coccidioidomycosis) is endemic, farmers sometimes spray the soil with fungicides before plowing to reduce the number of infectious agents that may become airborne.

In the case of animal vectors or reservoirs, a campaign may be initiated to destroy the animals. In a recent **epizootic** (an outbreak of an infectious disease in animal populations) of rabies in California, control measures included trapping and destroying skunks to reduce the size of the reservoir host and therefore reduce the chances of human infection.

Disruption of the Chain of Transmission

Standards for water, milk, and food quality are established and enforced by local, state, and federal health agencies to prevent epidemics. Drinking water supplies, milk, and dairy products are closely monitored to evaluate their suitability for human consumption. Chlorination of water supplies, pasteurization of milk and dairy products, and inspection of restaurants are very important epidemic control measures designed to break the chain of transmission of infectious diseases.

Many diseases such as malaria, Rocky Mountain spotted fever, typhus, western equine encephalitis, and yellow fever are transmitted by vectors. Interruption of the transmission cycle is achieved by pest control. The use of insecticides is one common way of controlling vector populations in endemic areas. Schistosomiasis (which affects more than 1 million individuals annually) can be controlled by snail eradication programs, since the parasite that causes this disease requires a snail as an intermediate host to complete its life cycle.

Immunization

The immune state of the population, also known as herd immunity, plays a central role in determining the extent and course of an epidemic. Transmission of infectious agents occurs only among susceptible individuals; thus, in a population with low herd immunity, an infectious agent can spread with ease, because most individuals are able to "catch" the disease. In a population with high herd immunity, spread of the disease is slow. A well-implemented vaccination program can reduce the incidence of disease (fig. 15.8). Table 15.1 summarizes some of the vaccinations that are carried out routinely in the United States.

Travel to countries that are endemic areas for diseases such as cholera, yellow fever, and typhus could represent a health hazard. For this reason, vaccinations are recommended. Table 15.2 outlines some of the vaccinations recommended for travel to various countries. It is also common practice to vaccinate a population at risk during a severe epidemic to break the chain of transmission. The impact of vaccination on a disease can be clearly seen in figure 15.8, which represents the number of reported cases of poliomyelitis in England and Wales for the years 1945–1987. The vaccination program was initiated in 1957; after that date, the annual number of polio cases dropped from about 7000 in 1946 to fewer than 10 in 1980. No such decline has occurred, however, in Central and South America where an extensive vaccination program has not been carried out.

Vaccination

It has been known (or at least suspected) for the past 2000 years that persons who have recovered from an illness such as smallpox, plague, or typhus are very resistant or completely immune to a subsequent infection by the same agent. In the late 1700s Edward Jenner carried out an experiment showing that inoculation of a person with the cowpox "germ", which was innocuous for humans, would protect that person from smallpox. Because of Jenner's work, the inoculation

TABLE 15.1
IMMUNIZATION USED IN THE UNITED STATES

Vaccine	Contents of vaccine	Who should receive	Frequency*
Diphtheria–pertussis–tetanus (DPT)	Adsorbed diphtheria and tetanus toxoids, killed *Bordetella pertussis* cells	Infants: unimmunized children < 7 yrs. old	IM 2, 4, and 6 mos. Boost at 1½ and 4–6 yrs.
Tetanus–diphtheria (Td)	Adsorbed tetanus and diphtheria toxoids	Unimmunized children 7 yrs. and older	IM 2 mos. apart, then 1 dose 6–12 mos. later. Booster every 10 yrs.
Mumps	Attenuated live virus	Children 12 mos. and older	1 dose SC
Measles	Attenuated live virus	Children 15 mos. and older	1 dose SC
Rubella	Attenuated live virus, grown in human diploid cells	Children 12 mos. and older	1 dose SC
Polio (OPV-1, 2, 3)	Trivalent oral, attenuated live virus	Children 2 mos. to 18 yrs.	1 dose O at 2 and 4 mos. Booster at 1½ and 4–6 yrs.
Hemophilus b (HbPV)	Polysaccharide	Children 18–24 mos.	1 dose SC
Rabies	Inactivated virus	High risk professions	As needed
Influenza A	Attenuated live virus	High risk individuals	As needed
Hepatitis B	Surface antigen	High risk individuals	As needed

*IM, intramuscular; SC, subcutaneous; O, oral.

TABLE 15.2
VACCINATIONS OR DRUG TREATMENTS RECOMMENDED FOR FOREIGN TRAVEL

Disease	Countries	Disease prevention
Cholera	Mozambique, Albania, Angola, Cape Verde, Dominican Republic, Egypt, India, Korea, Pakistan	Vaccination, every 6–8 mos.
Yellow fever	Benin, Cameroon, Congo, Mali, Nigeria, Mauritania, Ivory Coast, Uganda, Senegal, South America	Vaccination, every 10 years
Malaria	Africa, Southeast Asia, Central and South America, India, Mexico	Chemoprophylaxis (Chloroquine or pyrimethamine + sulfadoxine)
Plague	Southeast Asia, Africa, South America	Vaccination, every 6 mos.
Typhoid	Tropical countries	Vaccination
Hepatitis A	Developing countries, Southeast Asia, Africa	Injection of immunoglobulins
Hepatitis B	Developing countries	Vaccination
Traveler's diarrhea *(Escherichia, Shigella, Salmonella)*	Developing countries	Large doses of bismuth subsalicylate (e.g., Pepto-Bismol®)
Measles	Worldwide	Vaccination
Tuberculosis	Developing countries	BCG or INH prophylaxis
Typhus	Ethiopia, Mexico, Ecuador, Bolivia, Peru, Asia, Burundi	Vaccination, every 6 mos.

of individuals with cowpox pus to protect against smallpox became known as **vaccination** (from the Latin *Vacca*, for cow). We now use the term vaccination to describe an inoculation of any antigen (or microorganism) into an animal to induce a protective immune response. The material injected is often referred to as a **vaccine**.

Vaccines may consist of living or dead microorganisms or extracts of these microorganisms. Some vaccines consist of purified materials, such as protein toxins or polysaccharides. During the preparation of these vaccines, organisms generally are **attenuated** or **inactivated**. Organisms are inactivated by heating or by treating them with chemicals such as Merthiolate, acetone, or formalin. Attenuated microorganisms are live organisms that are unable to cause a serious infec-

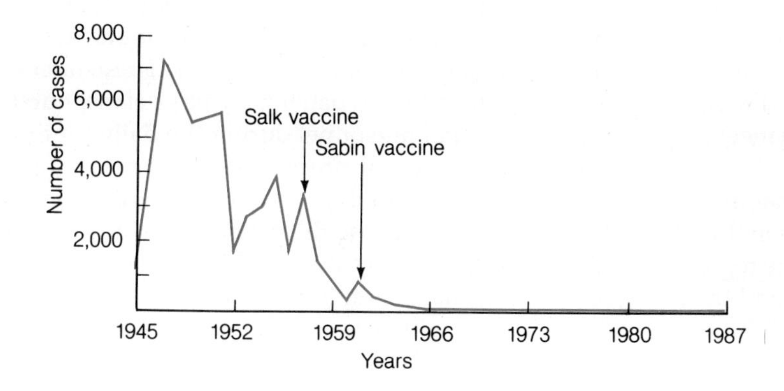

Figure 15.8 **Effect of vaccination on the incidence of poliomyelitis.** The introduction of the Salk vaccine (inactivated poliovirus) in the late 1950s and the Sabin vaccine (attenuated poliovirus) in the early 1960s reduced the number of cases in England and Wales from 4000–7000 cases per year to fewer than 10 cases annually.

tion, but can still induce an immune reaction. For example, the Sabin polio (oral) vaccine consists of an attenuated poliomyelitis virus. This virus does not cause serious disease in humans, although it can induce a lifelong immunity in the human host. The living organisms are attenuated by various procedures, such as repeated cultivation in laboratory media until the vaccine is no longer able to cause serious disease, or by selection of avirulent mutants.

The exotoxins from several different bacteria, including those produced by *Clostridium tetani* and *Corynebacterium diphtheriae,* are used to make important vaccines. The exotoxins are denatured or inactivated by treatment with formalin, and in this form they are known as **toxoids**. Toxoids do not cause detrimental changes in the host, although they can initiate an immune response that will neutralize the active toxin. To make some toxoids more immunogenic, they are sometimes absorbed to adjuvants (substances that enhance the immune response), like aluminum hydroxide or aluminum phosphate. Such a complex, when injected into an animal, releases the toxoid slowly so that the immune system becomes stimulated over a period of a few days, resulting in an enhanced immune reaction.

Viral vaccines may be made by treating the virus with formalin or with ultraviolet light, or by attenuating the virus using special culture techniques. Formalin-killed viruses are used in the Salk polio vaccine, developed by Jonas Salk and his coworkers during the early 1950s for use against the polio viruses. The vaccine consists of the three known types of polio viruses that have been inactivated by treatment with formalin.

The Sabin vaccine, developed by Albert Sabin in the late 1950s and early 1960s, consists of attenuated strains of the three major types of polio viruses. These viruses apparently grow to a limited extent in the epithelium of the throat and intestines, eliciting a strong and long-lasting immune response. Large numbers of memory cells can rapidly yield high levels of IgA in the respiratory tract, throat, and intestines, as well as high levels of IgG and IgM, when the vaccinated host encounters the virulent form of the poliovirus. The cell-mediated immune system is also stimulated by the live vaccine and is better prepared to meet a challenge from virulent poliomyelitis viruses.

Tables 15.1 and 15.2 list bacterial and viral diseases that can be averted by vaccination. In the United States, only a few of these vaccines are regularly used. These include the oral polio vaccine and the DPT vaccine, which includes the toxoids of the diphtheria, tetanus, and whooping cough organisms. Booster vaccinations are required with the DPT vaccination to build up the antibody titers against the toxins. Booster shots capitalize on the anamnestic response to increase the level and duration of antibodies in the serum.

Vaccination is one of the most effective yet simple means of protecting populations against deadly or debilitating diseases. It represents the results of the efforts of many scientists working toward unveiling the basic mechanisms of the immune response and the application of these principles.

Recombinant DNA technology (genetic engineering—see Chapter 24) has recently led to the development of a vaccine against the hepatitis-B virus, which causes a serious liver infection affecting 200,000 to 300,000 people in the United States each year. The hepatitis B vaccine consists of a purified viral envelope protein that must be injected in three doses over a 6-month period. This vaccination appears to induce an immune response that lasts up to 5 years.

A vaccine against an agent of pneumonia, *Streptococcus pneumoniae,* has also been developed and is used primarily in older persons or those with a compromised immune system. This vaccine consists of a mixture of the capsular materials from the most common strains that infect humans. The vaccine was released in 1977 and has played a central role in reducing the number of deaths due to pneumonia in older individuals. This disease ranks among the 10 leading causes of death due to infection in the United States.

CONTROL OF HOSPITAL (NOSOCOMIAL) INFECTIONS

Hospitals are ideal environments for the spread of infectious diseases (fig. 15.9). Many of the pathogens are opportunists and flourish in hospital environments. In addition, the high population density brings individuals into close proximity, thus favoring the transmission of infectious agents. Infections acquired in hospitals and similar institutions are called **nosocomial infections** (table 15.3). These infections can be transmitted to patients by physicians, nurses, or other hospital personnel during the daily routine of patient care, surgical procedures, or therapy (fig. 15.10). Illnesses caused by nosocomial infections tend to be severe, because the patient is already compromised or debilitated by the condition that required his or her hospitalization in the first place.

Eight percent of all patients entering general hos-

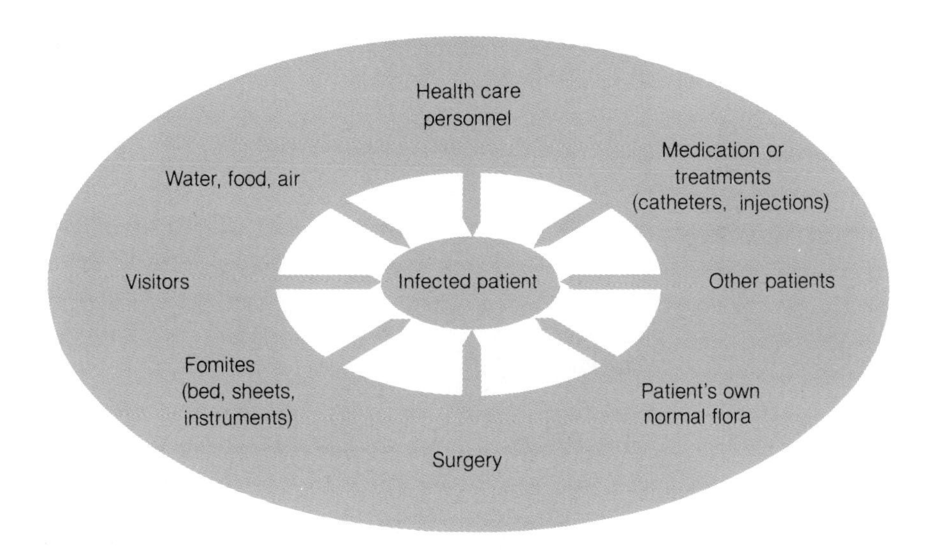

Figure 15.9 Spread of nosocomial infections. Institutionalized individuals, such as hospital patients, may become infected from health care personnel, other patients, or visitors. Occasionally, the patient's own resident flora may cause the nosocomial infection.

pitals in the United States contract infections while at the hospital. Since there are approximately 2.5 million hospital patients each year, this amounts to 200 thousand patients that acquire nosocomial infections. A conservative estimate is that 10% of these people die (20,000) as a direct result of the nosocomial infection and that of an additional 25% that die (50,000), the nosocomial infection contributed to their death but was not the sole cause. Patients with terminal illnesses are at a high risk of acquiring nosocomial infection. About 25% of terminal cases acquire nosocomial infections in comparison with 2.5% of patients hospitalized for nonfatal illnesses.

Nosocomial infections can be caused by a variety of microorganisms and affect various tissues and organs (fig. 15.11). Table 15.3 summarizes some common nosocomial infections, as well as some of the most common pathogens involved.

Every hospital accredited by the Joint Commission for the Accreditation of Hospitals must have an infection control committee (ICC). The ICC usually includes representatives from the various departments within the hospital, such as nurses, physicians, housekeepers, clinical laboratory technologists, engineers, and administrators. These individuals evaluate laboratory reports and the patients' charts to determine any increases in a particular disease or isolation of microorganisms that seems to be out of the ordinary.

Control Measures

Control measures vary from one hospital to another, but certain factors must be considered. For example, all personnel involved with the care of patients must be familiar with basic infection control measures. These should include (a) the isolation policies of the hospital; (b) handwashing, aseptic techniques, and the proper way of handling medical equipment and supplies; (c) the use of closed-system urinary catheters and intravenous injections; (d) surgical wound care; and (e) monitoring antimicrobial usage policies. An important aspect of infection control is the constant sur-

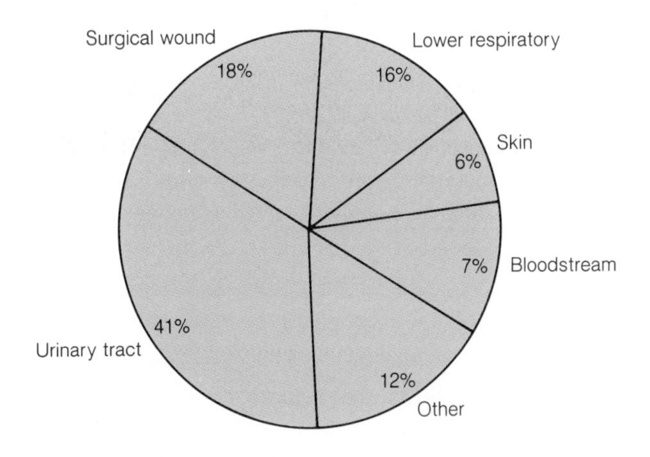

Figure 15.10 Incidence of hospital-acquired infections, according to site. (Courtesy of the Centers for Disease Control)

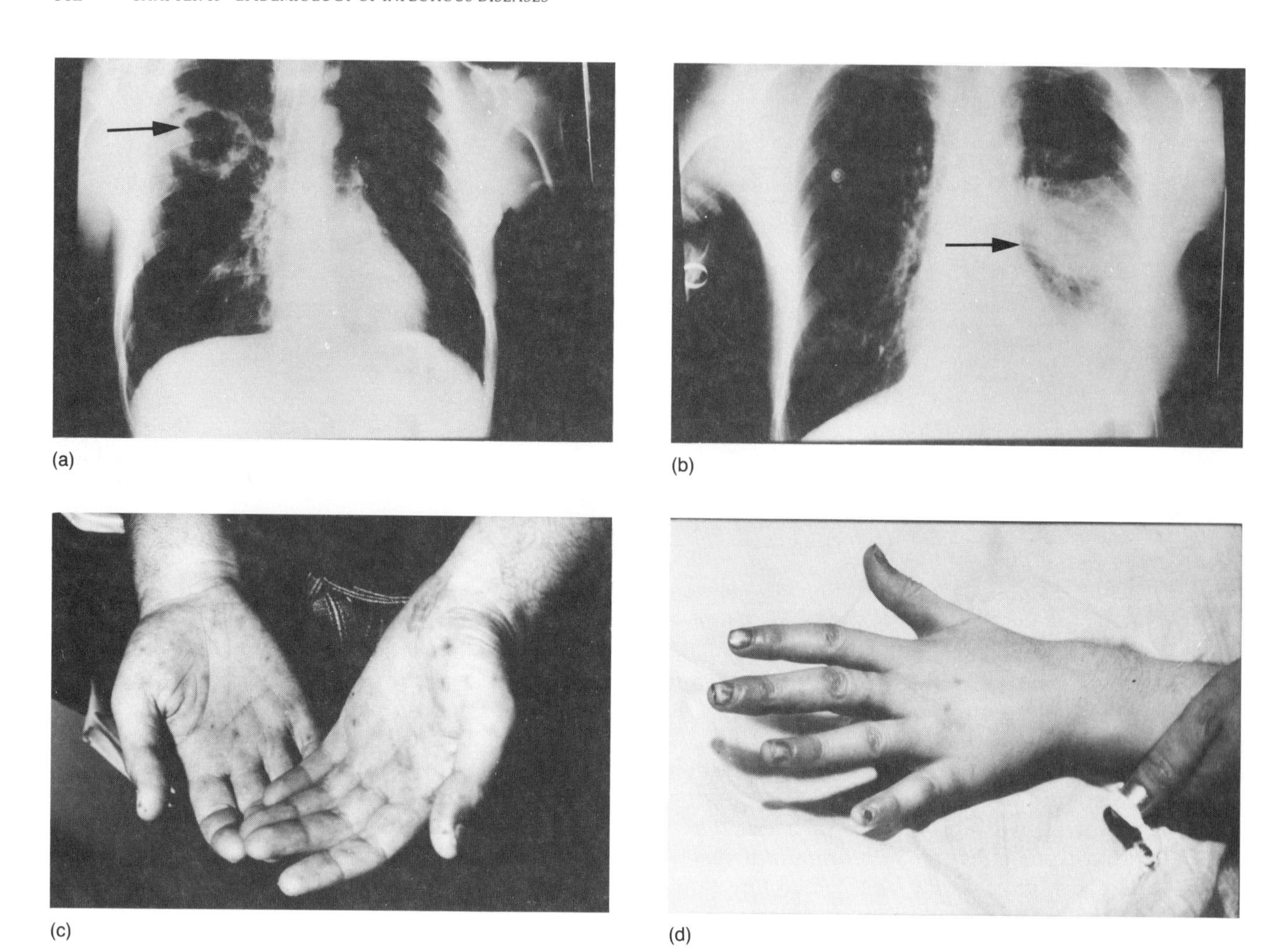

Figure 15.11 Clinical manifestations of some nosocomial infections. *(a)* Lung X-ray of a patient with an abscess caused by *Staphylococcus aureus.* Courtesy of Pediatrics Department, Division of Infectious Diseases, University Hospital, Seville, Spain. *(b)* Lung abscess that developed after surgery in a child caused by *Klebsiella.* Courtesy of Pediatrics Department, Division of Infectious Diseases, University Hospital, Seville, Spain. *(c)* Hand lesions reflecting a hospital-acquired endocarditis. Courtesy of Dr. J. L. Corral Arias. *(d)* Peeling of fingertips of child reflecting a sepsis caused by *Staphylococcus aureus.* Courtesy of Pediatrics Department, Division of Infectious Diseases, University Hospital, Seville, Spain.

veillance of patients. Surveillance regarding the frequency, distribution, symptomatology, and other characteristics of common nosocomial infections, as well as their etiology, can be an invaluable aid in controlling these infections. Through control measures it is possible to reduce the number of nosocomial infections and therefore patient morbidity, mortality, and expense.

The Centers for Disease Control (CDC)

Many of the epidemiological studies conducted in the U.S. are monitored or directed by personnel affiliated with or employed by the National Centers for Disease Control (CDC). CDC, a branch of the United States Public Health Service (USPHS) located in Atlanta, Georgia, provides laboratory support to state and regional health

FOCUS NOSOCOMIAL INFECTIONS AND LEGIONNAIRES' DISEASE

Infections acquired in hospitals not only debilitate and kill many thousands of people each year in the United States, they also add about four days to a typical hospital stay at a cost of over $1.5 billion.

The most common cause of nosocomial infections is the use of catheters in the urinary tract and blood system. When catheters are pushed into the urethra or into blood vessels, they often force microorganisms on the skin into the wounds produced or provide a pathway for organisms that come from hospital personnel that change the catheters. Surgical wounds that become infected by organisms at the site of surgery, or by organisms on personnel that check on the surgery, as well as respiratory infections due to contaminated breathing equipment or personnel also contribute significantly to nosocomial infections.

One organism that causes a significant number of nosocomial pneumonias in patients and hospital personnel is *Legionella pneumophila*, the cause of Legionnaires' disease (fig 15.12). A study of one hospital where Legionnaires' disease was endemic showed that 22% of all pneumonias were due to *L. pneumophila* and 69% were acquired in the hospital. In the study of another hospital where the disease had not been reported, 14% of the nosocomial pneumonias were due to *L. pneumophila* and 64% of the hospital's water distribution sites were contaminated with the bacterium. The bacteria are usually acquired from aerosols created by contaminated air conditioning units and rapidly moving water in spas, basins, and showers.

The case fatality rate from *L. pneumophila* pneumonias generally ranges from 5 to 15%. The seriousness of the disease in hospitals is indicated by the 1985 outbreak of *Legionella* in the Stafford General Hospital in England. In one month, 37 people died and 6 nurses contracted pneumonia. The bacterium was found in water droplets associated with the air conditioning system.

Nosocomial infections by *Legionella* could be reduced if hospitals would periodically heat their water supplies to 77°C (171°F). Diagnosis is achieved by the indirect fluorescent antibody technique that discovers antibodies against the pathogen. Infections by *L. pneumophila* can be treated with erythromycin, tetracycline, and rifampin.

Other species of *Legionella*, such as *L. micdadei*, which cause acute suppurative pneumonia (Pittsburgh pneumonia), *L. feelei*, which causes a nonpneumonic febrile disease (Pontiac fever), and *L. bozemanii*, which causes a pneumonia, are responsible for a small number of additional nosocomial infections.

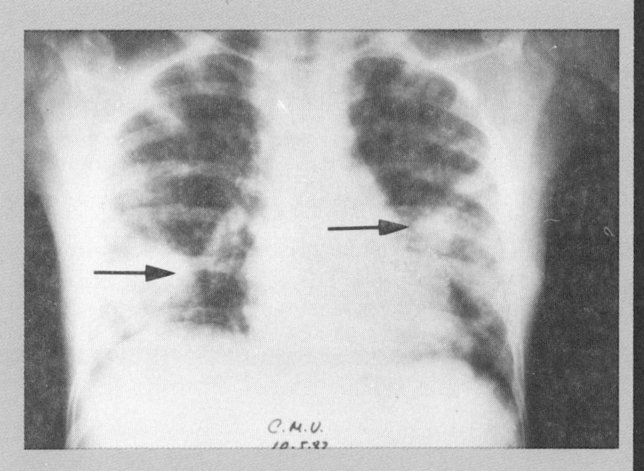

Figure 15.12 Pneumonia caused by *Legionella pneumophila*. Courtesy of Dr. J. L. Corral Arias.

agencies in the area of disease control and prevention. CDC also sets criteria for the safe and accurate performance of microbiological procedures, as well as training personnel and conducting updating seminars.

The National Centers for Disease Control also gathers and collates epidemiological data. Significant disease occurrences and national disease trends are published weekly in a publication entitled *Morbidity and Mortality Weekly Report (MMWR)*. The extensive laboratory facilities of CDC and its highly trained personnel makes CDC one of the premier laboratories in the world.

CLINICAL PERSPECTIVE A GENETIC PREDISPOSITION TO A DISEASE

In 1962, a United States warship of the Mediterranean fleet visited a port where shigellosis was endemic. On the last day in port, food served at a picnic for the crew was contaminated by two cooks who had contracted a mild dysentery. Less than 18 hours after the picnic over 500 cases of bacillary dysentery affected the crew sending 50% to sick bay. By the time the epidemic was over 602 crew members had suffered with dysentery. An equally large number of the crew escaped the dysentery epidemic. About 10 days after recovering from the bacillary dysentery, nine crew members developed inflamed eyes, a mild urethritis (inflammation of the urethra), and arthritis (inflammation of the joints) in one or more joints of the lower extremities. This unusual malady, known as Reiter's syndrome, did not go away in most of the men.

A followup study 13 years later of five members of the crew affected by Reiter's syndrome showed that four had persistent aggressive disease which included in the most extreme case crippling arthritis and blindness. The most severe case had developed in the oldest crew member who was 36 years old at the time of the followup study. He was blind in the left eye because of a recurrent inflammation of the pigmentary layer of the eye (**uveitis**) and subretinal hemorrhages. The right eye was deteriorating more slowly and was blurring his remaining vision. He had widespread arthritis with limited movement in the left wrist and right elbow. Inflammation of the urethra and of the glans penis (**balanitis**) had reoccurred for 13 years. His clinical state had rendered him unemployable.

The four sailors in their early thirties with Reiter's syndrome were shown to have a specific genetic marker (HLA-B27) on their cells. The sailor who had recovered from Reiter's syndrome and had been symptom-free for two years lacked HLA-B27. This genetic marker predisposes a person to Reiter's syndrome after certain microbial infections. A study of Finnish victims of a *Shigella* epidemic demonstrated that more than 80% with Reiter's syndrome carried the genetic marker HLA-B27. Because the HLA-B27 marker is less prevalent in non-Caucasians, they are affected at a much lower rate. Women are rarely affected, indicating that the disease is sex-influenced.

Reiter's syndrome is diagnosed by the combined signs and symptoms of arthritis, uveitis, and urethritis. It is said to be the most common form of arthritis in military patients under 40. It was named after Professor Hans Reiter, who described a Prussian lieutenant suffering from the disease in 1916. Reiter incorrectly guessed at the bacte-

TABLE 15.3
SELECTED NOSOCOMIAL INFECTIONS AND THEIR CAUSE

Nosocomial infection	Common pathogens	Nosocomial infection	Common pathogens
Gastroenteritis	*Escherichia coli, Salmonella, Shigella, Campylobacter jejuni,* viruses, *Cryptosporidium*	Burns	*P. aeruginosa, E. coli, S. aureus*
Upper respiratory infections	*Haemophilus influenzae, Streptococcus pyogenes, S. pneumoniae,* viruses	Wounds	*E. coli, P. aeruginosa, S. aureus, Streptococcus* Group A, enterococcus, *Klebsiella, Bacteroides*
Lower respiratory infections	*S. pneumoniae, Pseudomonas aeruginosa, Klebsiella pneumoniae, Legionella pneumophila,* influenza viruses, *Pneumocystis carinii*	Urinary tract infections	*E. coli, P. aeruginosa, Proteus, Enterobacter aerogenes,* enterococcus
		Hepatitis	Hepatitis B, hepatitis A, hepatitis non A non B, and delta viruses
Septicemias	*E. coli, P. aeruginosa, Staphylococcus aureus, Candida albicans,* viruses	Acquired immune deficiency syndrome (AIDS)	Human immunodeficiency virus (HIV)

rial cause. A French journal of 1916 reported that Reiter's syndrome (inflammation of the joints, eyes, and urinary tract) followed an outbreak of dysentery among French soldiers. This hinted at the initiator of the disease. Today, it is believed that *Shigella flexneri* is the major initiator of Reiter's syndrome in persons with the appropriate genetic background. Other bacteria in the genera *Shigella, Salmonella, Yersinia,* and *Mycoplasma* as well as myxoviruses are also implicated as initiators in the disease.

The onset of Reiter's syndrome is marked by diarrhea and fever. Two to four weeks after recovery from the diarrhea, and fever, arthritis, uveitis, and urethritis develop. Superficial ulcers may form on the palms and soles. Treatment generally involves the use of tetracycline to treat any crytic infection and phenylbutazone to relieve pain and inflammation in the joints. Recovery from the infection is possible but immunological disturbances may occur for years to come.

For an interesting discussion of Reiter's syndrome in military personnel and in Christopher Columbus the following reference is suggested: Weissmann, G. 1986. They all laughed at Christopher Columbus. *Hospital Practice* 21(1): 29–41.

SUMMARY

Epidemiology is the science that studies the incidence and distribution of disease in a population as well as the factors that influence them.

EPIDEMICS AND PANDEMICS

1. An epidemic is a short-term increase in the number of cases of a disease in a population.
2. The endemic rate of a disease represents the "normal" number of cases in a given population.
3. A pandemic is a very large epidemic of long duration, which may include populations in several different countries.

PARAMETERS USED BY EPIDEMIOLOGISTS TO MEASURE EPIDEMICS

1. Prevalence rates measure the usual number of cases in a population.
2. Incidence rates measure the number of *new* cases of a disease in a population.
3. Case fatality rates measure the number of fatalities among those who become ill with a particular disease.
4. Mortality rates measure the number of individuals dying in a population.

SOURCES OF INFECTIOUS DISEASES

1. Infectious diseases can be acquired from a variety of sources: (a) humans; (b) animals; (c) water; (d) food; and (e) air.
2. Reservoirs are places, materials, or animals where infectious agents are perpetuated in nature.
3. Carriers are asymptomatic individuals who are capable of transmitting infectious diseases to other individuals.

MODES OF TRANSMISSION

1. Person to person transmission occurs by direct contact between two or more individuals. Sexually transmitted diseases are examples of person to person transmission.
2. Contact with infected animals may result in an infectious disease. Those diseases that are perpetuated in nature in animal reservoirs are called zoonoses. Rabies is an example.
3. Fomites (inanimate objects) serve as passive vehicles for the transmission of infectious agents. Many diseases are transmissible through fomites.
4. Food and water serve as vehicles for the transmission of infectious diseases such as salmonellosis, typhoid fever, dysentery, cholera, and poliomyelitis. This route of transmission is sometimes called the fecal-oral route.
5. The inhalation of airborne particles containing infectious agents is a very common way of becoming infected. Many microorganisms are transmitted in this way. Many infectious diseases, particularly those that affect the respiratory system, are acquired by inhalation of airborne infectious agents.

FACTORS INFLUENCING EPIDEMICS

1. Infectious diseases result from a complex interaction among the host, the parasite, and the environment.
2. Acquired immunity determines which members of a population will be able to "catch" a disease.
3. Genetic background sometimes determines the susceptibility of an individual (or a population) to a particular infectious disease.
4. Cultural factors, including religious, ethnic, and behavioral practices, may predispose some populations to contracting an infectious disease. Certain behaviors may protect other populations from diseases.
5. Virulence may determine how devastating an epidemic is. Particularly virulent strains of microorganisms may cause many deaths in a population. The virulence of a strain may be reflected in the incidence or case fatality rates in the afflicted population.
6. The number of pathogens that infect an individual can also determine, at least in part, how severe an epidemic is.

DETERMINATION OF THE INFECTIOUS AGENT

1. To control an epidemic, its cause and the source of infection must be known.
2. Identification of a pathogen can be accomplished by cultural, serological, and biochemical methods.

THE CONTROL OF EPIDEMICS

1. To control epidemics the factors that contribute to the web of causation must be reduced or eliminated.
2. Reduction of the source of infection by isolation, quarantine, prophylactic chemotherapy, and reservoir eradication programs are important measures in the control of epidemics.
3. Disruption of the chain of transmission can be done by vector and intermediate host eradication programs, quarantine measures, and reservoir eradication.
4. Immunization is an effective measure in controlling epidemics.

CONTROL OF HOSPITAL (NOSOCOMIAL) INFECTIONS

1. Hospitals are ideal environments for the spread of infectious diseases.
2. Diseases that are acquired in hospitals are called nosocomial.
3. The major cause of nosocomial infections is contaminated catheters that initiate urinary tract and bloodstream infections.
4. Control of nosocomial infections involves knowledge of isolation policies in the hospital, the proper use of equipment, practicing aseptic techniques, and continuing education programs.

STUDY QUESTIONS

1. Define the following terms:
   a. epidemic;
   b. endemic;
   c. pandemic;
   d. morbidity;
   e. mortality.
2. Why is the science of epidemiology of central importance in public health?
3. Describe five different sources of infectious agents and give an example of each.
4. How can infectious diseases be transmitted? Are animals important in the transmission of disease? Explain.
5. Define the following terms:
   a. vector;
   b. fomite;
   c. reservoir;
   d. carrier.
6. Briefly discuss the relationship between the host, the parasite, and the environment in epidemics.
7. What factors influence the case fatality and morbidity rates of an epidemic? Explain.
8. Why is it important to identify precisely the pathogen responsible for an epidemic?
9. Cite five methods of infraspecies classification.
10. What are nosocomial infections? Of what significance are they?

SELF-TEST

1. A disease that appears and disappears in an unpredictable manner is best described as:
   a. pandemic;

b. epidemic;
c. endemic;
d. sporadic;
e. nosocomial.

2. A disease that is always in a population is best described as:
a. pandemic;
b. epidemic;
c. endemic;
d. sporadic;
e. nosocomial.

3. If there are 24 cases of a disease in a population of 10,000, what is the rate per 100,000?
a. 12;
b. 24;
c. 48;
d. 240;
e. none of the above.

4. If city A has 500 cases in a population of 1 million while city B has 30 cases in a population of 10,000, which city is having the most severe epidemic? a. A; b. B.

5. If there are 80 deaths for every 100 cases of a disease, what can be said about the disease?
a. the mortality rate for the population is high;
b. the morbidity rate for the population is high;
c. the case fatality rate is high.

6. Diseases that are endemic in animal populations and that can be transferred to humans are known as:
a. nosocomial;
b. pandemic;
c. zoonoses;
d. carried;
e. none of the above.

7. Symptomatic or asymptomatic animals or humans that harbor a pathogen would be best described by the term:
a. vector;
b. reservoir;
c. carrier;
d. fomite;
e. phoront.

8. Which is *not* a fomite?
a. mosquito;
b. tick;
c. flea;
d. fly;
e. all of the above.

9. Which is *not* a vector?
a. cups;
b. eating utensils;
c. towels;
d. handkerchiefs;
e. all of the above.

10. Epidemics of Legionnaires' disease are generally due to bacteria that become aerosolized in air conditioning systems that use tanks of water for cooling the air.
a. true; b. false.

For items 11–18, match the disease with the manner in which it is generally transmitted:
a. vectors;
b. fomites;
c. infected animals;
d. infected people;
e. food and water.

11. Cholera

12. Hepatitis A

13. Common cold

14. Mononucleosis

15. Herpes

16. Salmonellosis

17. Rocky Mountain spotted fever

18. Ringworm

19. What factors affect the incidence rate of a disease?
a. environmental characteristics;
b. host characteristics;
c. pathogen characteristics;
d. all of the above;
e. none of the above.

20. Which host characteristic is important in determining whether or not a population will be affected by a few cases or by an epidemic?
a. acquired herd immunity;
b. genetic background;
c. host health;
d. behavior;
e. all of the above.

21. Which pathogen characteristic is important in determining whether or not a host population will be affected by a few cases or by an epidemic?
a. infectious dose;
b. genetic background;
c. water content;
d. temperature;
e. a and b.

22. Specifically, which genetically determined characteristics of a pathogen make for a virulent organism?
a. ability to attach to the host;
b. invasiveness and ability to reproduce within cells;
c. antigenic variation;
d. production of toxins;
e. all of the above.

23. It is important to identify the organism responsible for

an epidemic in order to control the outbreak because knowing the identity of the pathogen allows public health workers to:
a. determine the size of the epidemic;
b. find carriers;
c. trace the immediate source of the pathogen;
d. effect changes that eliminate the pathogen, reservoir, vector, or transmission vehicle;
e. all of the above.

24. Which is *not* a method used to control *large* epidemics?
a. immunization;
b. quarantine;
c. eliminating the chain of transmission;
d. eliminating the vector, transmission vehicle, or reservoir;
e. using drugs to eliminate the pathogen.

25. When is it recommended that the diphtheria–pertussis–tetanus immunization first be given?
a. 2, 4, and 6 months after birth;
b. 5 years;
c. 10 years;
d. only if there is an epidemic;
e. only to the elderly.

26. When is it recommended that the mumps, measles, and rubella vaccines first be given?
a. 2, 4, and 6 months after birth;
b. 1 year and older;
c. 5 years;
d. 10 years;
e. only if there is an epidemic.

27. When is it recommended that the polio vaccine first be given?
a. 2 and 4 months and 5 years after birth;
b. anytime;
c. 5 years;
d. 10 years;
e. only if there is an epidemic.

28. What is *not* a major cause of nosocomial infections?
a. arthropod vectors;
b. catheters and breathing equipment;
c. visitors, patients, nurses, and doctors;
d. fomites;
e. surgery.

29. Approximately 20,000 people die as a direct cause of nosocomial infections, and nosocomial infections contribute to the death of 50,000 per year in U.S. hospitals.
a. true; b. false.

30. The site of most nosocomial infections is:
a. lower respiratory tract;
b. bloodstream;
c. surgical wounds;
d. urinary tract;
e. skin.

SUPPLEMENTAL READINGS

Bopp, C. A., Birkness, K. A., Wachsmuth, I. K., and Barrett, T. J. 1985. In vitro antimicrobial susceptibility, plasmid analysis, and serotyping of epidemic-associated *Campylobacter jejuni. Journal of Clinical Microbiology* 21:4–7.

Brunton, J., Clare, D., and Meier, M. A. 1986. Molecular epidemiology of antibiotic resistance plasmids of *Haemophilus* species and *Neisseria gonorrhoeae. Reviews of Infectious Diseases* 8(5):713–724.

Cliff, A. and Haggett, P. 1984. Island epidemics. *Scientific American* 250(5):138–147.

Cohen, M., Tauxe, R. 1986. Drug-resistant *Salmonella* in the United States: an epidemiological perspective. *Science* 234: 964–969.

Friedman, G. D. 1974. *Primer of epidemiology.* New York: McGraw-Hill.

Harris, A. A., Levin, S., and Trenholme, G. 1984. Selected aspects of nosocomial infections in the 1980s. *The American Journal of Medicine* 77(1B):3–11.

Liñares, J., Sitges-Serra, A., Garau, J., Pérez, J. L., and Martin, R. 1985. Pathogenesis of catheter sepsis: A prospective study with quantitative and semiquantitative cultures of catheter hub and segments. *Journal of Clinical Microbiology.* 21(3):357–360.

May, C. W. ed. 1980. *Microbial diseases.* Los Altos, CA: William Kaufmann.

Roueche, B. 1967. *Annals of epidemiology.* Boston: Little, Brown.

Saxinger, W. C., Levine, P. H., Dean, A. G., Thé, G., Lange-Wantzin, G., Moghissi, J., Mei Hoh, F. L., Sarngadharan, M. G., and Gallo, R. C. 1985. Evidence for exposure of HTLV-III in Uganda before 1973. *Science* 227:1038–1040.

Simon, H. B. and Swartz, M. N. 1987. Legionnaire's disease. *Scientific American Medicine* 7(xx):9–12.

Stamm, W. E. 1981. Nosocomial infections, etiologic changes, therapeutic challenges. *Hospital Practice* 16(8):75–88.

Weissman, G. 1986. They all laughed at Christopher Columbus. *Hospital Practice* 21(1):29–41.

Zinsser, H. 1935. *Rats, lice and history.* Boston: Bantam.

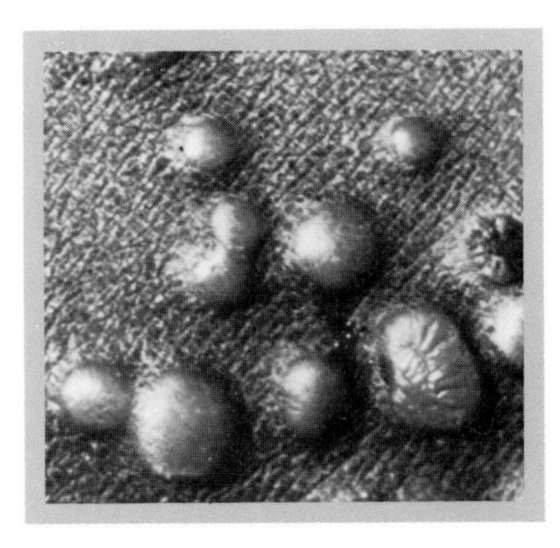

CHAPTER 16

DISEASES OF THE SKIN

CHAPTER PREVIEW THE CONQUEST OF SMALLPOX

Smallpox, a frequently fatal disease, is characterized by the development of skin eruptions that eventually form a crust. When the crust falls off, it leaves scars called "pockmarks." In the severe form of the disease, 15–45% of those infected die. The disease, caused by a virus, spreads from person to person in aerosols, making it a particularly contagious disease. This characteristic, together with the high degree of infectivity of the virus, makes smallpox a fearsome disease.

Smallpox has ravaged human populations since antiquity. Pharaoh Ramses V is believed to have died of smallpox in 1137 B.C. During one epidemic in the early 1520s, more than 3 million Aztecs died, and it is believed that this outbreak contributed significantly to the conquest of the Aztecs by Cortez.

The United States was not spared from smallpox. For example, during the 18th century in Boston, almost 60% of a population of about 1000 became ill with smallpox within 10 months of the introduction of the disease. Of those who became ill, more than 50% died of the disease.

Because this disease had such an impact on the welfare of human populations, it prompted a great deal of interest that ultimately resulted in its eradication. The first significant advance in smallpox control occurred when a **vaccination** was discovered. Later, with the intervention of the World Health Organization, the disease was eradicated. The last naturally occurring case of smallpox occurred in 1975, and the disease was declared "eradicated" on October 26, 1979.

Brilliant work by scientists throughout the world, capitalizing on knowledge gathered during centuries of suffering, ultimately resulted in the eradication of smallpox from the biosphere. This success was achieved because humans were the only natural reservoir of the virus, and scientists were knowledgeable about the biology of the virus, the disease process, its epidemiology, and means of prevention and treatment. In this chapter we discuss some of the most common human diseases of the skin, their epidemiology, pathogenesis (disease process), and treatment.

CHAPTER OUTLINE

MANY INFECTIOUS AGENTS CAUSE DISEASES that are manifested on the skin. Many of these microorganisms cause disease that affect the skin directly, while others cause disease elsewhere in the body and then spread to the skin. For example, coccidioidomycosis and syphilis sometimes cause skin diseases as a result of invasion of skin tissue or allergic reactions. Diseases such as these are discussed in the chapters dealing with the organs in which the primary or most commonly recognized form of the disease occurs. This chapter discusses some of the most important infectious diseases that affect the skin, and those with signs and symptoms that are primarily cutaneous (of the skin).

STRUCTURE AND FUNCTION OF THE SKIN

The skin is an organ that covers all of the body surfaces exposed to the external environment. The human skin (fig. 16.1) consists of an outer layer, or epidermis,

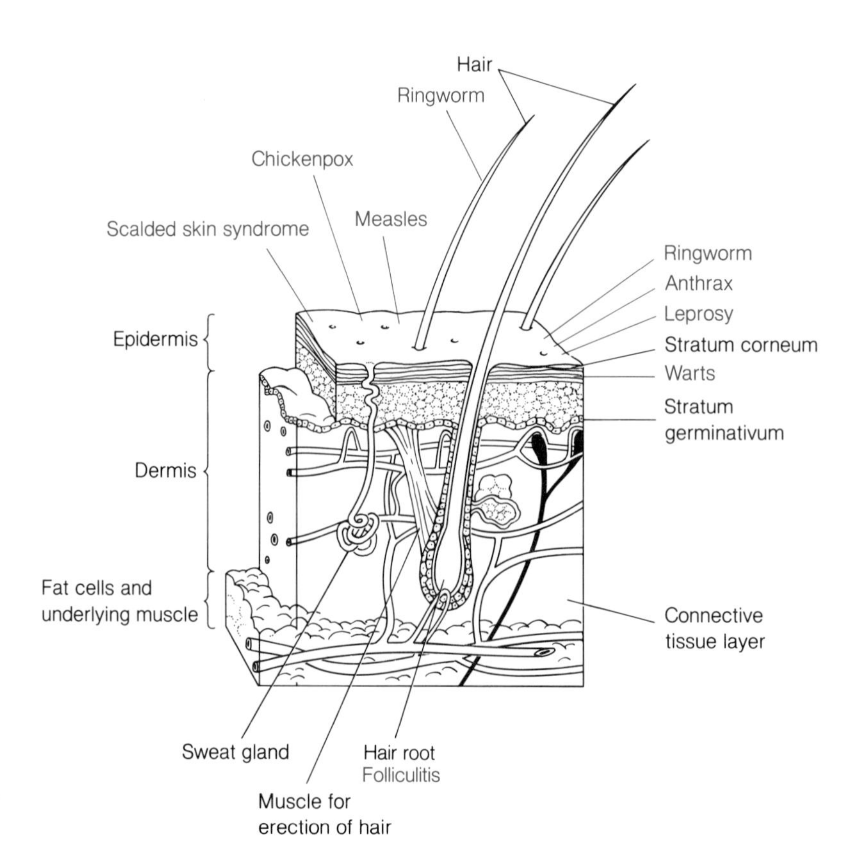

Figure 16.1 Diagram of the various components of the human skin. The diagram indicates the various components of the skin and some of the diseases that affect the skin (in color).

containing keratin, and an inner layer called the dermis. In addition, the skin has a variety of integumentary glands (sweat and sebaceous glands), blood vessels, nerves, and muscle cells and other connective tissue. In addition, hair and nails are also considered to be part of the skin.

The skin performs many important functions. It protects the individual from harsh environmental influences and integumentary pigments such as melanin protect the vertebrate animal from damage due to solar radiation. Skin also has numerous nerve endings that are sensitive to touch, pressure, heat, cold, and chemical conditions. These nerve endings collectively permit the organism to monitor the environment constantly and react accordingly.

The intact skin is a formidable barrier against infection. When its physiology or integrity is altered, however, infectious diseases can occur. The breach of normal skin by abrasions, punctures, or cuts allows microorganisms to enter the subcutaneous tissues and initiate an infection.

Conditions that alter the proper functioning of the skin also predispose the host to disease. For example, the **stratum corneum** (fig. 16.1) is continuously being shed and replaced. In this way, microorganisms present on the skin are removed along with the old stratum corneum. The rate of exfoliation (shedding of old stratum corneum) is important in the overall health of the skin. Conditions that retard the shedding of old stratum corneum can also lead to extensive colonization by microorganisms. Excessive moisture or steroid treatment, for example, can slow down the rate of exfoliation and can predispose individuals to certain diseases, such as candidiasis (yeast) and tinea versicolor (mold). Thus, the proper functioning of the skin is essential to the health of the individual.

THE NORMAL FLORA OF THE SKIN

The healthy skin is a rather harsh environment for many microorganisms because of its lack of moisture and its acidity. Although the skin is constantly exposed to a myriad of microorganisms, only a few of these are able to establish permanent residence (table 16.1). These microorganisms colonize primarily the stratum corneum and the hair follicles. The resident flora often protect the skin against infection by pathogenic microorganisms, by competing for nutrients and space and by releasing toxic substances that are inhibitory to many microorganisms. Bacteria that occupy hair follicles constitute a reservoir of microorganisms that reseed the skin surfaces as they are exfoliated or washed away.

The predominant flora of the skin consist of bacteria and fungi. The distribution and types of microorganisms on the skin vary according to the anatomical site. Skin on the forearm may harbor *Staphylococcus epidermidis*, diphteroids, and some yeasts, while the skin of the perineum (area between the anus and the genitals) may harbor these organisms in addition to enterics and coliforms. The number of microorganisms per cm^2 of skin may also vary from about 1000 in dry skin areas to about 10 million in moist areas.

Some members of the normal flora may become pathogenic if the opportunity arises. For example, the fungus *Malasezzia furfur*, under conditions of excessive moisture or slow skin exfoliation, may cause a "cosmetic" disease called tinea versicolor. Acne, a com-

TABLE 16.1
REPRESENTATIVE MICROORGANISMS COMMONLY ISOLATED FROM THE HUMAN SKIN

Microorganism	Microbial group	Average frequency of isolation (%)*
Candida albicans	Yeast	<10
Candida parapsilosis	Yeast	8
Diphtheroids (aerobic)	Gram + rods	60
Enteric bacteria	Gram − rods	< 5
Pityrosporum spp.	Yeast	35
Propionibacterium acnes	Gram + rods	72
Staphylococcus aureus	Gram + coccus	20
Staphylococcus epidermidis	Gram + coccus	92

*Numbers indicate the proportion of individuals in a population carrying the organisms in their skin.

mon condition in teenagers, can be caused by some bacterial members of the normal flora.

BACTERIAL DISEASES OF THE SKIN

Anthrax

Anthrax, a disease caused by *Bacillus anthracis*, is characterized by flu-like symptoms and the formation of a **malignant pustule** (fig. 16.2) at the site of infection. The lesion consists of a central black eschar (scab) and a conspicuous ring of edema (swelling due to fluid accumulation) and erythema (redness due to increased blood flow to the area). A pulmonary disease called **woolsorters' disease** is acquired by the inhalation of *B. anthracis* endospores, which germinate in the lungs and cause a severe, often fatal pulmonary infection.

Bacillus anthracis is an aerobic, gram-positive, endospore-forming rod. It grows well on a variety of laboratory culture media. On blood agar containing 5% sheep blood, it causes complete hemolysis (lysis of red blood cells) and forms large, gray colonies of rough texture. Examination of the colony with a dissecting scope shows that the edges of the colonies resemble "medusa heads" (fig. 16.2). Virulent strains of *B. anthracis* synthesize a polyglutamic acid (polypeptide) capsule and produce at least 3 exotoxins. The capsule protects the dividing bacterium against phagocytosis. One of the exotoxins released by the virulent bacilli affects the central nervous system, causing respiratory failure and sometimes death. The other exotoxins promote the growth and penetration of the bacteria in the host.

Endospores commonly enter the skin through cuts or scratches and then germinate in the wound. Three to five days after infection, a small papule (raised lesion) develops, which later becomes filled with a dark, gelatinous fluid and is surrounded by a ring of inflammation. At this stage, the lesion is called a **malignant pustule**. The bacteria multiply locally at the site of entry, producing the exotoxin, and subsequently spread via the bloodstream or lymphatics. The signs and symptoms of anthrax are similar to those of an acute respiratory infection, with cough, malaise, fever, and achy muscles. These symptoms are probably attributable to the exotoxin.

Livestock grazing on *B. anthracis*-contaminated pastures are usually infected through abrasions in the oral mucosa and mouth acquired while grazing. Infected animals generally die soon after the signs and symptoms of anthrax appear. Humans usually contract the disease from infected animals or their prod-

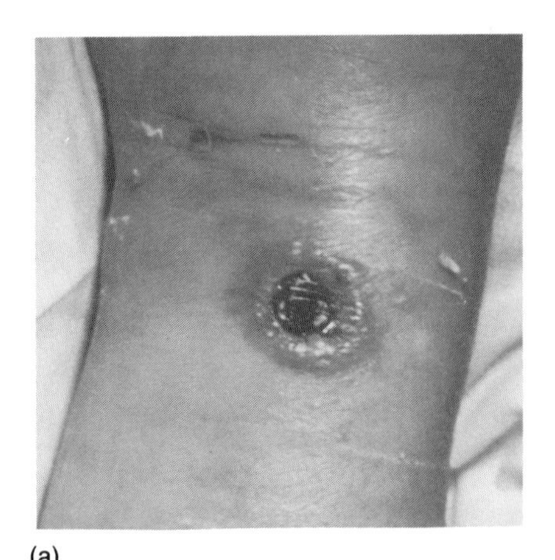

(a)

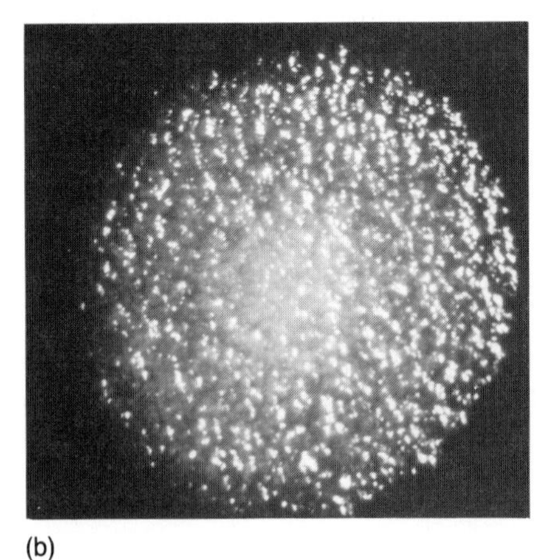

(b)

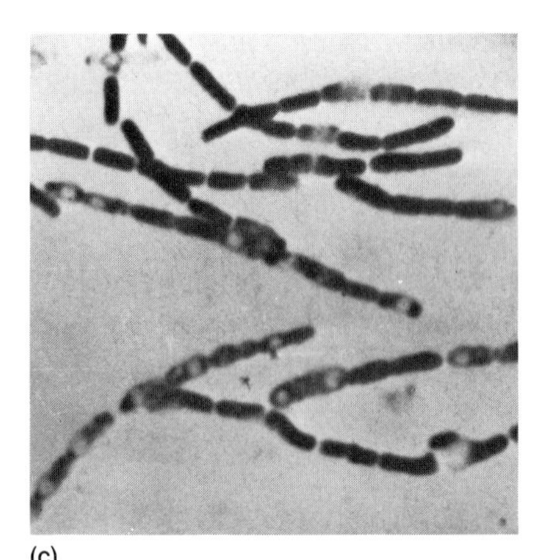

(c)

Figure 16.2 Clinical manifestations and cultural characteristics of *Bacillus anthracis*. *(a)* The characteristic lesion caused by anthrax at the site of infection consists of a central region of dead tissue (necrosis) and a surrounding ring of swelling and redness. Anthrax bacilli can be isolated from the lesion. *(b)* Colony of *Bacillus anthracis* on blood agar exhibiting the characteristic "Medusa Head" appearance. Courtesy of Professor E. J. Perea. *(c)* Gram stain of *Bacillus anthracis*. Notice the chain of bacterial cells containing endospores (clear circular areas inside cells).

ucts. Most reported cases of anthrax in humans are acquired by individuals intimately associated with domestic animals, such as veterinarians, farmers, hunters, and slaughterhouse workers. Another group of individuals sometimes affected includes textile and felt mill workers, who get the disease by handling contaminated wool, goat's hair, or other animal-derived materials.

Vaccination with endospores from a nonencapsulated strain of *B. anthracis* prevents the disease in animals and in humans. The treatment of choice for anthrax is penicillin, which is particularly effective when given early in the course of the disease. The effectiveness of penicillin in controlling the infection diminishes as the disease progresses; this fact emphasizes the need for prompt and accurate diagnosis of anthrax. In cases in which the patient is allergic to penicillin, tetracycline can be used successfully.

Hansen's Disease (Leprosy)

Hansen's disease, or **leprosy**, is characterized by the development of multiple lesions of the skin, often accompanied by loss of sensory perception of the affected areas. The disease dates back to antiquity. Ancient Hindu and Hebrew writings make reference to it. Accounts of the skin lesions and the social attitude toward this disease are commented upon in the Old Testament. Nowadays, the disease is better understood, and although the stigma of old remains, social acceptance has improved. In fact, individuals suffering from leprosy have formed a society and publish a journal called "The Star," in which they report cases and advances in the treatment and biology of the disease. The causative agent of Hansen's disease, *Mycobacterium leprae,* was first seen by Armauer Hansen in 1878. In his report he made reference to the abundance of rod-shaped bacteria in the specimens. To date, there has been no successful cultivation of this bacterium in laboratory media.

Leprosy is a highly variable disease that ranges from benign **tuberculoid** leprosy to a severe, extensive form of the disease called **lepromatous** leprosy (fig. 16.3). In tuberculoid leprosy, a single **macule** (discolored lesion) develops, measuring about 5–20 cm in diameter. The color of the skin on the lesion is generally lighter than the uninfected skin. The skin lesions that develop in tuberculoid leprosy are the result of cell-mediated immune reactions that cause tissue damage. Many patients with tuberculoid leprosy undergo spontaneous cures. Lepromatous leprosy is a severe form of the disease which includes multiple lesions on the skin. The lesions result from the dep-

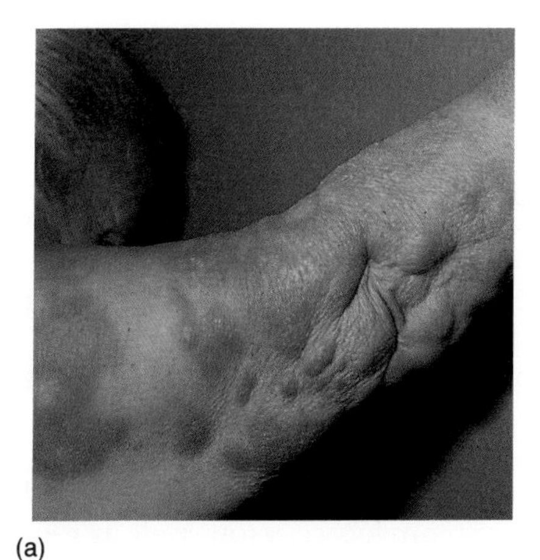

(a)

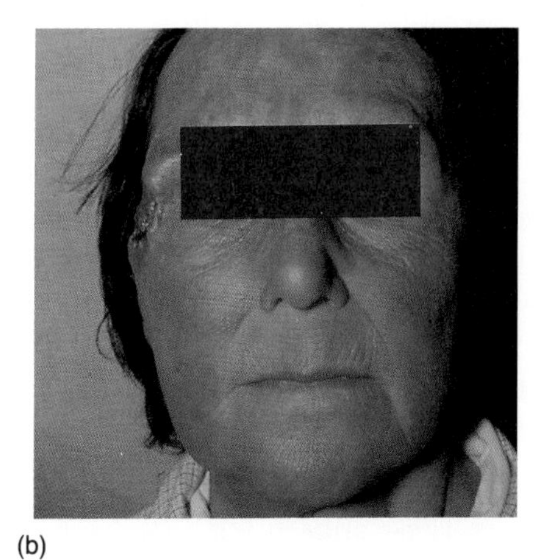

(b)

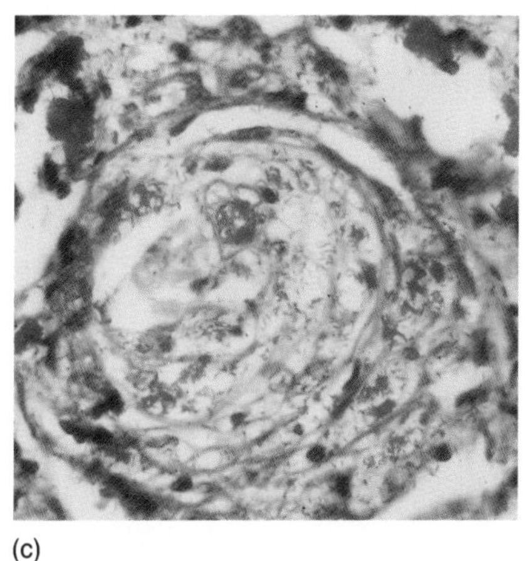

(c)

Figure 16.3 Characteristics of Hansen's disease (leprosy). *(a)* Patient with tuberculoid leprosy of the arm. Courtesy of Professor F. Camacho, Department of Dermatology and Venereology, University Hospital, Seville, Spain. *(b)* Patient with lepromatous leprosy of the face. Courtesy of Professor F. Camacho, Department of Dermatology and Venereology, University Hospital, Seville, Spain. *(c)* Acid-fast stain of a biopsy sample from a patient with lepromatous leprosy. The magenta-red rods are the acid-fast bacilli of *Mycobacterium leprae.*

osition of immune complexes at the site where damage occurs. In lepromatous leprosy, nerve involvement occurs, often with concomitant loss of sensory perception in the affected area. Intermediate forms of leprosy also occur and are referred to as **borderline** cases.

Mycobacterium leprae is an acid-fast bacillus that can be seen inside infected epithelial cells called **lepra cells**. Little is known about the physiology of this organism, because it has not been grown on artificial media in the laboratory. Studies in experimental animals, principally mice, armadillos, and monkeys, suggest that the bacteria grow very slowly, with a generation time of about 12–14 days. The bacillus grows best on areas that are usually below body temperature (~37°C). For this reason, hands, ears, nose, and face, areas that are not usually clothed, and are therefore cooler, are the sites most often affected by lesions of Hansen's disease. The incubation period for this disease averages more than 2 years, with a range of 12 weeks to 40 years.

The bacteria generally penetrate the skin through small cuts and abrasions, then divide slowly inside human skin and nerve cells. Humans are very resistant to infection by *M. leprae*; experimental infections using human volunteers are seldom successful. This resistance may be due to the fact that normal individuals develop prompt and strong cell-mediated immune (CMI) responses that inhibit the growth of the pathogen. Tuberculoid, borderline, and lepromatous forms of the disease apparently are related to the time of onset of the cell-mediated immune response following infection. The later the CMI response develops, the longer the parasite can reproduce inside host cells and the more severe the infection is. In tuberculoid leprosy, evidence of a strong CMI response can be seen early in the infection as an extensive infiltration of lymphocytes and giant cells (large, multinucleated cells derived from macrophages) at the site. As a consequence, few acid-fast bacilli can be demonstrated (fewer than 100,000/cm^2). In lepromatous leprosy, little cell-mediated immunity is evident and numerous acid-fast bacilli (more than 100 million/cm^2) can be found inside macrophages. Persons with lepromatous leprosy have a high titer of antibodies against *M. leprae*. These antibodies are not protective. In fact, antigen-antibody complexes (immune complexes) fixing complement may be responsible for much of the tissue damage in lepromatous leprosy.

Leprosy is prevalent in tropical areas of Africa, Southeast Asia, and South America. The prevalence rate for leprosy in these areas is about 2000–4000 per 100,000 population. Most cases of leprosy in the United States are traceable to immigrants.

Dapsone (a sulfone), which is a sulfa drug, has been effective in treating leprosy, but lepromatous leprosy requires prolonged treatment, sometimes for the rest of the patient's life. In cases in which sulfones cannot be used because of resistance to the drug, rifampin and clofazimine are moderately useful. A combined regiment of rifampin, clofazimine, and dapsone has been successful in treating patients in endemic areas. A vaccine currently used in Europe against *M. tuberculosis*, consisting of live *M. bovis* and known as BCG (Bacillus Calmette-Guerin), may have some prophylactic value against *M. leprae* but has not yet been fully evaluated. India has carried out a large-scale test of an anti-leprosy vaccine that uses gamma-ray-inactivated *M. avium*. This bacterium stimulates a strong immune response against various species of *Mycobacterium*. The large-scale test will determine which vaccine (BCG, BCG and *M. leprae*, or *M. avium*) is most effective.

Pyoderma

Pyoderma is a diagnostic term used to define any acute inflammatory infection of the skin with **purulent** (pus containing) exudates (fig. 16.4). Pyoderma can occur as a result of primary invasion of the skin by pathogenic bacteria, or as a secondary invasion by opportunistic bacteria following diseases such as scabies or after cuts and scratches.

Staphylococcal Skin Infections The majority of the suppurative (festering, pus-discharging) skin diseases encountered in clinical practice are caused by *Staphylococcus aureus* (fig. 16.5). Staphylococcal skin lesions or **abscesses** are acute inflammatory lesions due to host cell destruction caused by the invading bacteria. *Staphylococcus aureus* is a gram-positive coccus measuring approximately 1 μm in diameter. When this bacterium divides, it forms characteristic grape-like clusters that can be seen with the aid of a microscope in smears of both clinical specimens and laboratory cultures. Staphylococci grow well in routine laboratory media (e.g., nutrient agar or blood agar) forming small, opaque, white to yellow colonies.

Different strains of *Staphylococcus aureus* produce various enzymes and toxins that have harmful effects on the human host. Some of these enzymes aid the bacterium in spreading to adjacent tissues by degrad-

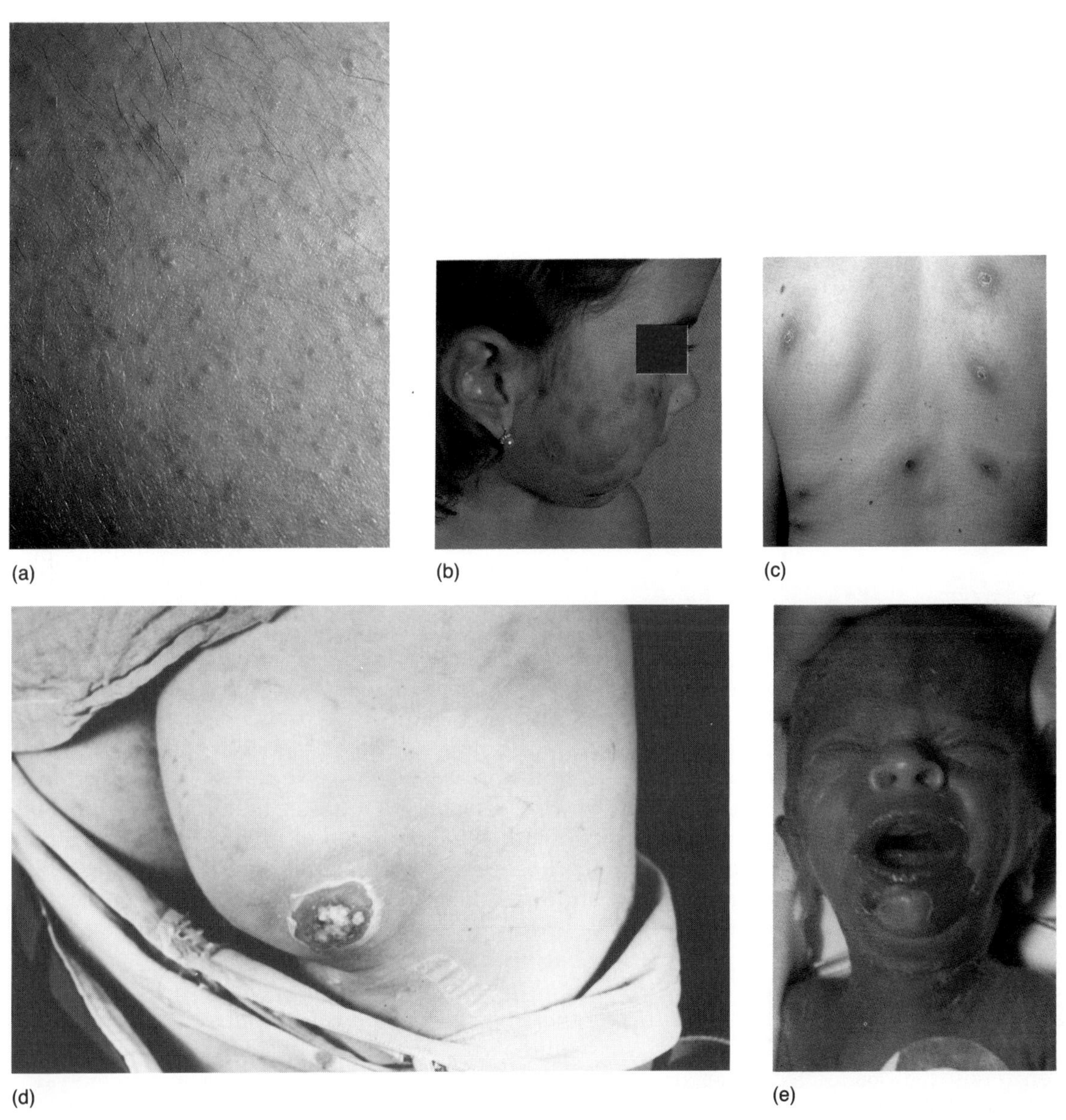

Figure 16.4 Diseases caused by *Staphylococcus aureus*. *(a)* Folliculitis (pimples). *(b)* Child with impetigo on the face. Courtesy of Professor F. Camacho, Department of Dermatology and Venereology, University Hospital, Seville, Spain. *(c)* Patient with numerous furuncules on the back and shoulder region. Courtesy of Professor F. Camacho, Department of Dermatology and Venereology, University Hospital, Seville, Spain. *(d)* Carbuncle on buttocks. *(e)* Child with scalded skin syndrome (note peeling of skin around mouth, neck, and forehead).

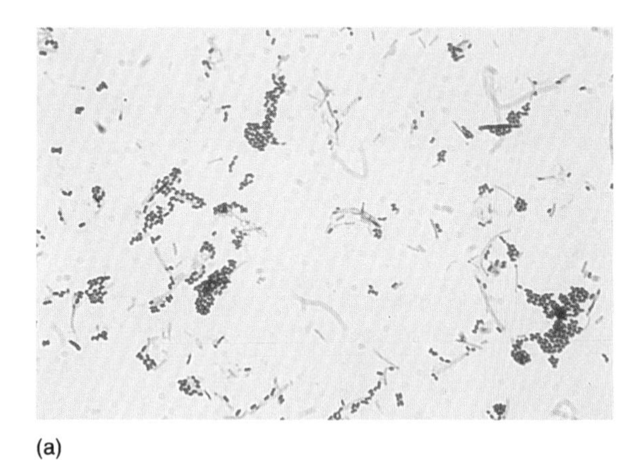

(a)

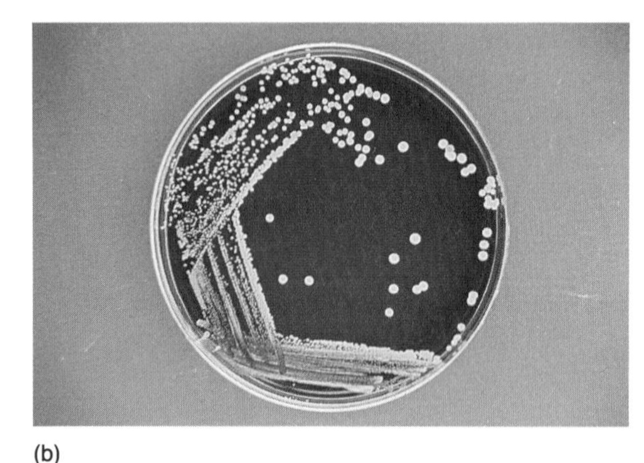

(b)

Figure 16.5 Cultural and staining characteristics of *Staphylococcus.* *(a)* Gram stain of *Staphylococcus epidermidis* exhibiting gram-positive cocci arranged in grape-like clusters. Courtesy of the Biological Sciences Department, California Polytechnic State University, San Luis Obispo. *(b) Staphylococcus epidermidis* growing on blood agar. Characteristic whitish, round, convex colonies appear on the agar surface.

ing host cells and by interfering with host defense mechanisms.

Coagulases are enzymes that clot plasma by a mechanism resembling that of normal clotting. Coagulase activity is used by microbiologists to distinguish between pathogenic and nonpathogenic staphylococci. Although coagulase production is not required for pathogenicity, this enzyme is generally viewed as a good indicator of the pathogenic potential of *Staphylococcus aureus.* It is postulated that coagulase-producing staphylococci produce this enzyme so as to form a fibrin lattice around themselves to avoid the host's attack. It is also suggested that coagulase activity is at least partly responsible for the hard periphery seen in most staphylococcal lesions.

β-lactamases are enzymes that break down penicillins and cephalosporins, thus rendering these antibiotics useless in fighting off infections. The genes coding for these enzymes, may be located either on plasmids or on the bacterial chromosome. Most strains of *Staphylococcus aureus* isolated from hospital environments and some strains isolated from the general population carry β-lactamase genes in their genomes. Although β-lactamases do not participate directly in the disease process, they are important because they protect the bacteria from antibiotics used in treating the infection.

In addition to these enzymes, *Staphylococcus aureus* may produce an array of other enzymes that take an active part in the disease process. For example, **hydrolases** are enzymes secreted by staphylococci that break down various types of cellular material, which helps the pathogens invade host tissue (fig. 16.6). **Staphylokinase** is a protein-degrading enzyme (protease) that breaks down fibrin. It is thought that this enzyme dissolves clots and so permits the spread of *Staphylococcus aureus* to adjacent tissues. **Hyaluronidases** break down the hyaluronic acid that holds cells together and consequently facilitates the spread of the pathogen in the host tissue. **Lipases** hydrolyze membrane lipids and fats, causing the lysis of host cells. The production of lipases by *Staphylococcus aureus* correlates well with its ability to cause pimples and boils.

Staphylococcus aureus also produces a variety of exotoxins that may be responsible for the signs and symptoms of some staphylococcal skin diseases. Cytolytic toxins, such as **leukocydin** and **hemolysin**, disrupt the plasma membrane of host cells and play important roles in the disease process. Leukocydin, a toxin produced by most pathogenic staphylococci, destroys PMN leukocytes (e.g., neutrophils) and macrophages by attacking the phospholipids in their plasma membrane. This toxin is important in the disease process because it protects the invading staphylococci from cellular defense mechanisms. Hemolysins are

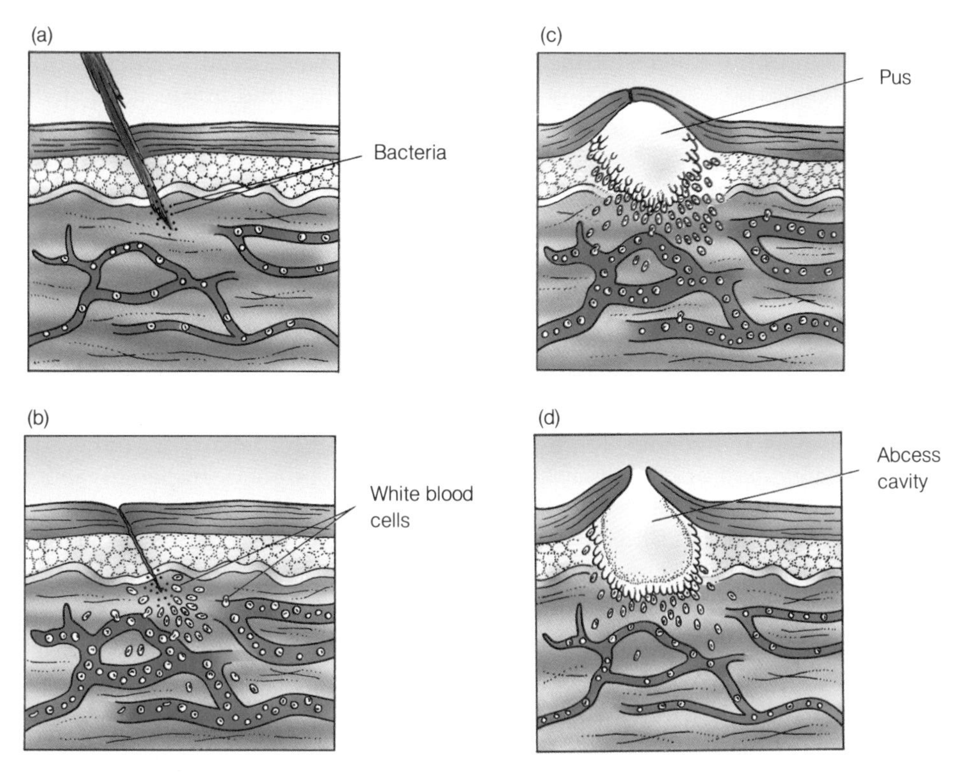

Figure 16.6 The formation of an abscess. *(a)* When a pyogenic (pus-forming) microorganism such as *Staphylococcus aureus* is introduced into the skin, it begins to multiply at the site of infection. *(b)* The developing colony produces a variety of proteins (aggressins) that cause tissue destruction and allow the colony to spread. *(c)* The damage to tissue caused by the multiplying bacteria induces an inflammation, so that the area becomes swollen, reddened, and painful to the touch. *(d)* Eventually, the developing abscess perforates through the skin, draining pus (dead cells, bacteria, and body fluids) to the outside. The erosion of tissue could continue if the infection is allowed to fester without treatment or removal.

toxins that destroy red blood cells. One such toxin is the alpha toxin, which destroys not only red blood cells, but also skin, muscle, and kidney cells.

Enterotoxins are important exotoxins that cause various signs and symptoms in the human host. The most common signs and symptoms due to enterotoxins are nausea, vomiting, and diarrhea, which become apparent about 4 hours after ingestion of foods contaminated with the toxin. One type of enterotoxin, Enterotoxin-F, apparently is responsible for the signs and symptoms associated with toxic shock syndrome (see Chapter 19).

A cell wall component that may also play a role in the pathogenesis of staphylococcal infections is **protein A**. It is believed that this protein is a major factor contributing to the virulence of *Staphylococcus aureus* because it allows the staphylococci to adhere to host cells and to evade the immune responses of the host. Protein A binds to the Fc regions of antibodies (IgG), inactivating them. It also inhibits phagocytosis and has been found to cause allergic reactions in mammals.

Staphylococcal infections generally involve localized invasion of the skin and soft tissue. Clinical entities such as folliculitis (inflammation of the hair follicles) and impetigo fall in this category. These infections are rarely hazardous, although at times they cause a great deal of discomfort. Localized infections can become disseminated, however, and cause endocarditis, osteomyelitis, bacteremia, arthritis, pneumonia, and meningitis. When infections are caused by **toxigenic** (toxin-producing) strains, diseases such as toxic shock syndrome or scalded skin syndrome can result (see fig. 16.4).

How the host protects itself from staphylococcal infections is poorly understood. Apparently, both cell-mediated and antibody-mediated (humoral) immunity are important. Cell-mediated immunity is believed to be carried out by macrophages that actively phagocytize the pathogens and by cytotoxic T-lymphocytes that inactivate the cells. Antibodies enhance the phagocytosis by PMN leukocytes and neutralize the toxins produced by the pathogen.

Infections by *Staphylococcus aureus* are difficult to control because the organism is indigenous to the human skin and external nares, and is carried by a large proportion of healthy individuals. Staphylococci present a particularly difficult problem in hospitals, where they cause a large number of **nosocomial** (hospital acquired) infections. They can be brought into hospitals by patients and health care personnel who carry *Staphylococcus aureus* in their nasopharynx or have small, draining lesions (such as pimples) and pass them on to patients. Most of the hospital-acquired strains of *Staphylococcus aureus* are resistant to penicillin, a factor that makes the treatment of nosocomial infections even more difficult.

Penicillin or its derivatives are the drug of choice for the treatment of staphylococcal infections. This is particularly true for community-acquired infections since many strains of the pathogen are still sensitive to penicillin. In cases where β-lactamase producing strains are causing the disease, lincomycin, erythromycin, and novomycin have been shown to be effective.

Impetigo is a form of pyoderma, characterized by the development of pustules that become encrusted and rupture (fig. 16.4). The external nares, face, and extremities are the most common areas involved. This condition is most prevalent among children, possibly because of the high propensity they have for scratches and scrapes. In addition, person to person contact is more frequent among school-age children than among adults, so children are more likely to acquire and transmit virulent staphylococci or streptococci, either of which can cause impetigo. When toxigenic strains of *S. aureus* that produce the toxin **exfoliatin** are the cause of impetigo, the lesions are blisterlike and are known as **bullous** impetigo. The blisters eventually rupture and develop a crust.

Scalded skin syndrome (SSS) is the name given to a condition that results in necrosis of the epidermis with little or no dermal involvement (fig. 16.4). The name "scalded skin" is given because the afflicted skin appears to have been scalded with boiling water. Although this condition can result from drug reactions and other causes, SSS is caused by exfoliatin-producing strains of *S. aureus*. Scalded skin syndrome can be considered a generalized impetigo. It is most often seen in the newborn human child and may involve up to 80% of the body. Constitutional symptoms such as fever, general malaise, and prostration can accompany scalded skin syndrome.

Streptococcal Skin Infections Streptococcal infections are quite common in humans. The commonest are upper respiratory tract infections such as **strep throat**, but **erysipelas**, **impetigo**, and **scarlet fever** are also common. Like staphylococcal infections, streptococcal infections cause pyogenic (pus-forming) lesions. Generally speaking, streptococcal skin infections result from invasion of skin lesions such as scratches, insect bites, cuts, or sores. Occasionally, superficial streptococcal infections result in nonsuppurative (no pus produced) complications, such as **rheumatic fever** or **acute glomerulonephritis**.

The genus *Streptococcus* (table 16.2) contains several species that are pathogenic for humans. All the bacteria in this genus are gram-positive, fermentative cocci, typically arranged in chains and in pairs. In this genus, *S. pyogenes, S. agalactiae, S. faecalis,* and *S. pneumoniae* are the most important pathogens (table 16.2).

Streptococci are classified into groups based on their surface antigens. Rebecca Lancefield in 1933 classified the streptococci on the basis of serological reactions involving surface polysaccharides and placed them into groups (A–H, J, and K). Of these, Group A and B streptococci contain the most important human pathogens. Their pathogenicity resides in virulence factors such as the **M-protein, hemolysins**, and **erythrogenic toxin.**

The **M-protein** of Group A streptococci is a structural component of the cell wall. It helps *Streptococcus pyogenes* bind to respiratory epithelial cells and inhibit phagocytosis and therefore protects the bacteria from ingestion by neutrophils.

Streptococci also produce hemolysins and are classified into three groups based on their hemolysin activity (fig. 16.7). **Alpha hemolytic streptococci** growing on blood agar break down erythrocytes and produce a zone of green discoloration around the colonies. **Beta hemolytic streptococci** produce a large, clear zone of hemolysis surrounding the colony; **gamma hemolytic streptococci** do not hemolyse blood. Two

TABLE 16.2
SALIENT CHARACTERISTICS OF STREPTOCOCCI THAT COMMONLY AFFLICT HUMANS

Species	Group*	Hemolysis	Isolated from	Diseases caused	Salient features
Aerobic					
S. pyogenes	A	Beta	Nasopharynx, skin	Strep throat, others	Major pathogen, toxigenic
S. agalactiae	B	Beta	Nose, genital tract	Endocarditis, sepsis	Also causes mastitis
S. equi	C	Beta	GU and GI tract, skin	Wound infections	Toxigenic
S. faecalis	D	Gamma (beta)	GI and GU tracts	Peritonitis, UTI	6.5% NaCl resistant
S. salivarius	K	Gamma	Oropharynx	Sinusitis, abscesses	Grow at 45°C
S. sanguis	H	Alpha	GI tract, oropharynx	Abscesses, meningitis	Resistant to bile
S. mutans	NG	Alpha	Oropharynx	Caries, endocarditis	Grows in dental plaques
S. pneumoniae	NG	Gamma (alpha)	Oropharynx	Pneumonia, meningitis	Optochin sensitive
Anaerobic					
Peptostreptococcus	NG	Gamma	Digestive system	Empyema, sinusitis	Anaerobe, causes abscesses

*Streptococci are grouped based on serological reactions involving cell surface polysaccharides. NG, not groupable; UTI, urinary tract infection; GU, genitourinary; GI, gastrointestinal.

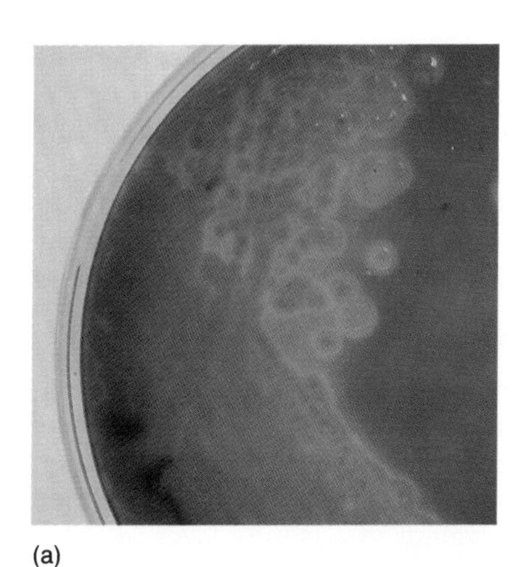
(a)

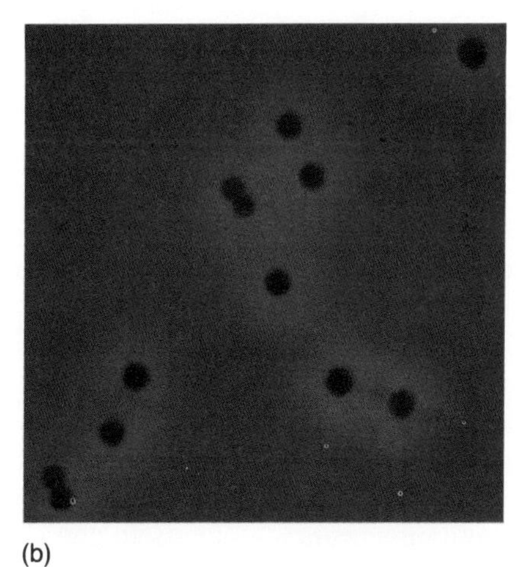
(b)

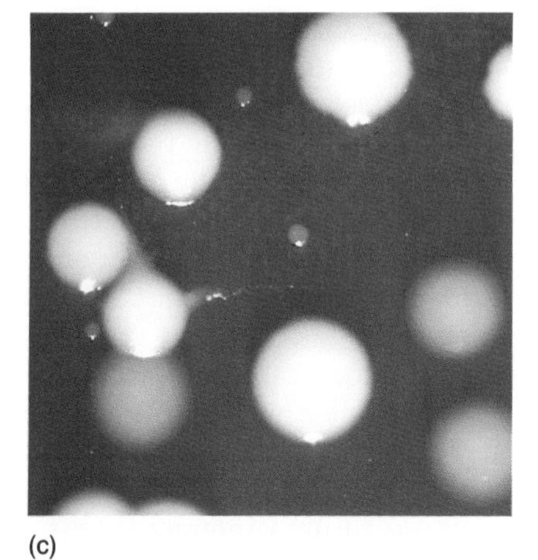
(c)

Figure 16.7 Hemolytic patterns of various streptococci grown on blood agar. *(a) Streptococcus pneumoniae* exhibiting its typical alpha hemolysis, which is characterized by a greening discoloration of the medium surrounding the colony. *(b) Colonies of Streptococcus pyogenes* exhibiting beta hemolysis. This type of hemolysis is characterized by the complete clearing of the medium around the hemolytic colony. Courtesy of Microbiology Department, University Hospital, Seville, Spain. *(c)* Gamma hemolysis of nonhemolytic *Streptococcus salivarius.* Courtesy of Microbiology Department, University Hospital, Seville, Spain.

hemolysins, **streptolysin O** (SLO) and **streptolysin S** (SLS), are important. Streptolysin O is an oxygen-sensitive enzyme produced by most Group A streptococci. This enzyme is active not only against RBCs but also against leukocytes and myocardial cells. In contrast, streptolysin S is oxygen-tolerant and contributes to the beta hemolysis of Group A streptococci.

Erythrogenic toxins (A, B, and C) are toxins that cause damage to the blood vessels underlying the skin. These toxins are produced by strains of group A, C, and G streptococci that carry a provirus with the genetic material coding for erythrogenic toxins. Other enzymes, such as DNases, ATPases, neuraminidases, hyaluronidases, and streptokinases, are produced by various streptococci. **Streptokinase** breaks down fibrin and other proteins. Together, these cell products bring about the pyogenic infections by streptococci.

In general, streptococci can cause disease in three different ways: (a) direct invasions; (b) intoxications; and (c) destructive immune responses. Direct invasions usually involve attachment of the streptococci to epithelial cells, with the subsequent production of toxic substances. Streptolysin O, SLS, and various exoenzymes may account for the invasive properties. Since both SLO and SLS have antileukocytic activities, pyogenic lesions may be due to the action of these enzymes on migrating leukocytes. In addition, the M-protein plays a major role in the pathogenesis of the disease, by helping the cells bind to the host and by inhibiting phagocytosis and hence augmenting the chances of survival for the streptococci. Skin diseases such as impetigo and erysipelas are consequences of direct invasion by streptococci. Intoxications due to erythrogenic toxins are responsible for the signs and symptoms characteristic of scarlet fever. Postinfection sequelae, such as rheumatic fever or acute glomerulonephritis, are most commonly complications of untreated upper respiratory tract streptococcal infections.

Streptococcal infections are commonly transmitted by droplet nuclei. They originate from asymptomatic carriers or from individuals with active streptococcal infections. While "strep throat" is most common during winter months, pyodermal infections are very common in temperate climates during the warm months. Pyodermal diseases caused by Group A streptococci are usually transmitted by direct contact, and the bacteria colonize the breached skin. Transmission of streptococci can also involve fomites or insects. Nosocomial infections caused by streptococci are most commonly surgical wound infections; asymptomatic hospital personnel are most frequently implicated as the source of infection.

Unlike staphylococci, which have developed resistance to penicillin, streptococci are still quite sensitive to this drug. Thus, penicillin remains the drug of choice. Patients allergic to penicillin can be treated successfully with lincomycin, erythromycin, or clindamycin. Third-generation cephalosporins are also effective against streptococci.

Impetigo, erysipelas, and scarlet fever are common streptococcal diseases of the skin.

Streptococcal impetigo is similar to staphylococcal impetigo, except that the former disease is seldom of the bullous type; rather, a thick crust develops around the lesion. Beta hemolytic Group A streptococci are the most common cause.

Erysipelas is an acute febrile disease resulting form a deep streptococcal infection, with localized inflammation and erythematous (reddish) skin. Face and head lesions are common and are often accompanied by fever, headache, nausea, and vomiting. This disease must be treated promptly, for untreated infections may result in septicemia (dissemination to the blood), multiple abscesses, nephritis, or rheumatic fever. It responds well to penicillin or erythromycin.

Scarlet fever was once a common disease resulting from streptococcal infections, but since the advent of penicillin and erythromycin its incidence has declined significantly in the United States. The typical fine, red "sandpaper" rash associated with the disease results from the action of **erythrogenic toxin** on the blood vessels of the skin. This is a common consequence of streptococcal tonsillitis, but it also occurs after skin infections. The initial rash may occur on the chest, but it spreads quickly to other parts of the body. Fever, vomiting, and prostration can also accompany the rash. During the healing process, desquamation of the skin and "strawberry tongue" are characteristic.

Trachoma **Trachoma** is a highly contagious conjunctivitis (an inflammation of the membrane that covers the eye, which is called the conjunctiva) caused by *Chlamydia trachomatis* (fig. 16.8). The infection often leads to the development of a cloudy layer over the eye and, in severe cases, to scar tissue formation in the infected area of the conjunctiva. The disease is uncommon in the United States (fewer than 300 cases reported annually), but it affects more than 400 million people worldwide, primarily in Africa and Asia. The infection may lead to the formation of corneal ulcers and to blindness. It has been estimated that

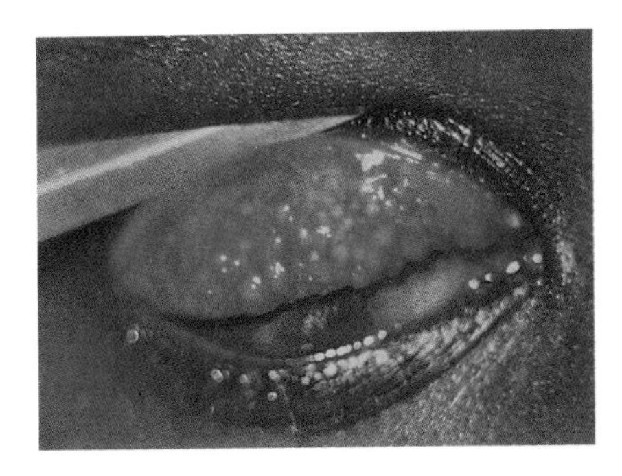

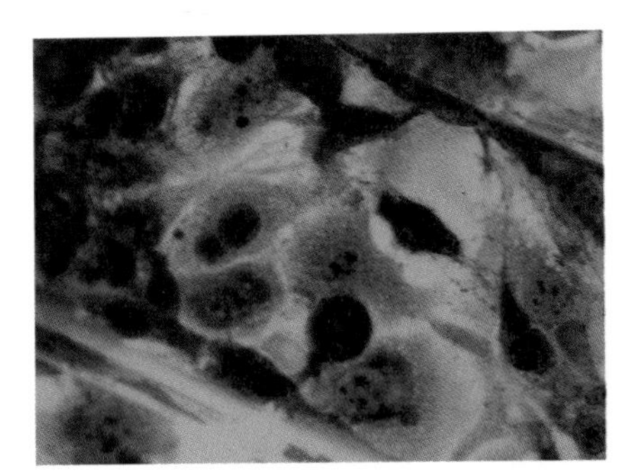

Figure 16.8 **Trachoma.** *(a)* Inflammatory nodules of the eye characteristic of trachoma. Courtesy of Professor E. J. Perea. *(b)* HeLa cells infected with *Chlamydia trachomatis,* the cause of trachoma. The inclusion bodies of *C. trachomatis* are the dark staining structures within the cells. Courtesy of Dr. J. Aznar.

more than 20 million people have been blinded by this disease. It is easily transmitted from person to person by direct contact with eye secretions or fomites contaminated with the bacteria.

Chlamydia trachomatis is an obligate intracellular parasitic gram-negative bacterium. The infectious particle, also referred to as the elementary body, infects host cells and differentiates into a reticulate body. After several rounds of binary fission, many elementary bodies arise. The elementary bodies are released following disruption of the host cell and can then infect other cells. Extensive cell disruption can occur from the infection, leading to a destructive inflammation of the affected tissue.

The etiology of the disease is usually determined by demonstrating the pathogen in inflammatory exudates, using staining techniques of fluorescent antibodies. Eye ointments containing tetracycline or erythromycin are effective in controlling the infection. Avoiding contact with infected persons or contaminated materials to prevent self-inoculation, and sanitation of contaminated materials prevents infection. Prompt treatment with antibiotics is required to stop the spread of the disease in human populations.

FOCUS LYME DISEASE

Lyme disease, first noted in 1975 in Connecticut, consists of an infectious arthritis that is characterized by a large, ringlike lesion on the skin. The lesions measure between 5 and 14 cm in diameter. Fever, headache, and muscle aches may also accompany the infection. Investigators have found that a spirochete is responsible for Lyme disease and have named it *Borrelia burgdorferi.* The disease is effectively treated with penicillin, erythromycin, or tetracycline, especially when the medications are administered early in the course of infection.

The disease occurs primarily in the coastal regions of the northeastern United States (Connecticut, Rhode Island, New York, Maryland, Massachusetts), the western states (California, Utah, Nevada, Oregon), and the Northern Midwest (Minnesota and Wisconsin). Lyme disease is transmitted by the bite of **ixodid ticks,** and the endemic area corresponds to those areas where the ticks proliferate. Although these regions are the hotbeds for Lyme disease, it can be found in any region where ixodid ticks occur.

The incidence of Lyme disease has been increasing considerably since 1975. In 1980, the Centers for Disease Control in Atlanta, Georgia, reported 226 cases. In 1983 there were more than 500 cases while in 1985 there were more than 1000 cases reported. The reason for the increase in Lyme disease is most likely due to the increase in hiking into endemic areas. In addition, an increased awareness of the disease by physicians has contributed greatly to the rise in reported cases.

Many dogs and horses in endemic areas are developing debilitating joint problems. Thus, Lyme disease is also of concern in veterinary medicine.

FUNGAL DISEASES OF THE SKIN

Fungi cause many diseases that involve the skin (fig. 16.9). Some fungi, such as *Malasezzia furfur* and *Exophiala werneckii,* cause superficial skin diseases that involve only the outermost layers of the stratum corneum (fig. 16.1). Others, such as *Sporothrix schenckii, Phialophora verrucosa,* and *Pseudoallescheria boydii,* are capable of invading deep tissues of the skin and penetrating to the bone. In addition, many pulmonary infections caused by fungi (such as cryptococcosis), upon dissemination from the lung, can cause skin lesions as well. The most common fungal infections, however, are the **ringworms** and infections caused by *Candida albicans.*

Ringworm

Ringworm, tinea, or dermatophytosis are names given to infections of the skin and scalp caused by a group of fungi called the **dermatophytes**. The name "ringworm" was coined because these fungi characteristically form concentric lesions on the hairless skin. The rings represent the spread of the agent outward from the site of infection, as well as an inflammatory response. Ringworm can occur in almost any part of the body. The site of infection determines the name of the disease. For example athlete's foot is called **tinea pedis** while ringworm of the scalp is called **tinea capitis**.

The disease process in all the ringworms is generally the same, and different symptoms arise as a result of the virulence of the strain, the susceptibility of the host, and the site of the infection. The infection begins on the stratum corneum and spreads outward, giving rise to the typical ringworm lesion. As a consequence of this growth, itching, patchy scaling, inflammation, or eczema develop from toxic wastes or tissue damage produced by the fungi. The lesions are restricted to the keratinized layers of the epidermis. The inflammatory and allergic responses are generally due to waste products released by the growing fungus. Rarely does a dermatophyte invade the dermis; when this does occur, it is most likely *Trichophyton rubrum* in an immune-compromised individual.

When *Microsporum audouinii* infects the human scalp, there is a period of incubation during which the fungus grows on the keratinized layers of the scalp. When the hyphae reach hair follicles, they grow into them on the hair shafts, where arthroconidia develop (fig. 16.10). After a period of spread, there is a refractory period, characterized by profuse arthrospore formation and little or no follicle invasion. After this period, infected hairs are sloughed off and other hairs return to normal.

The dermatophytes cause a variety of clinical manifestations, depending upon the species and the anatomical site of infection. Let us look briefly at tinea capitis, an infection of the scalp caused by a variety

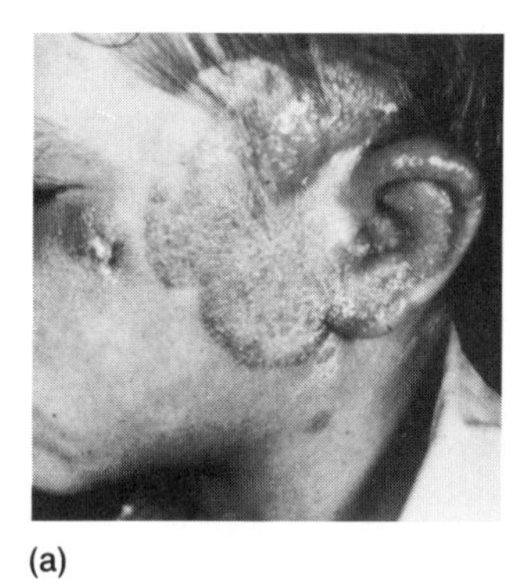

(a)

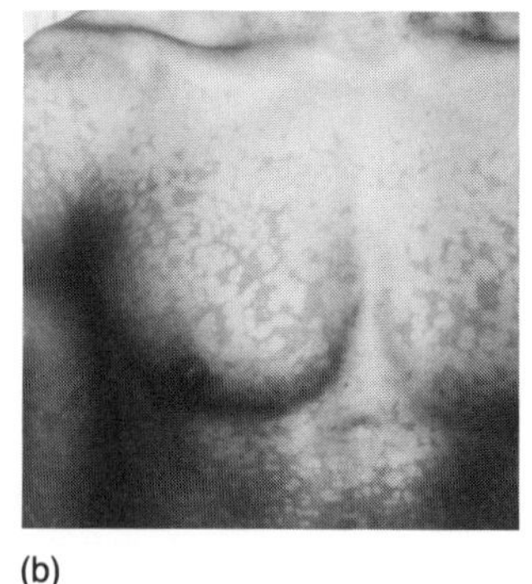

(b)

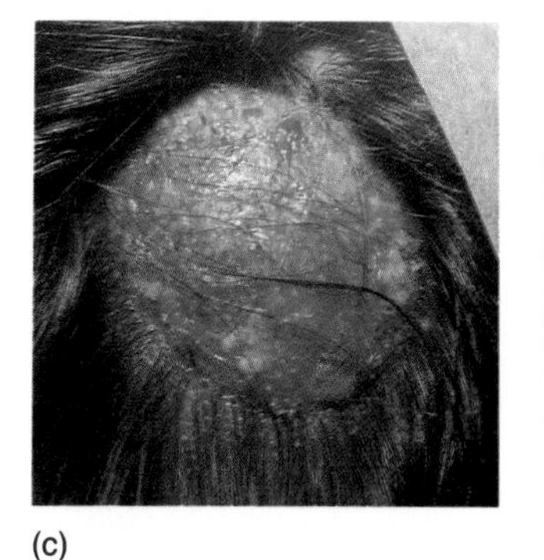

(c)

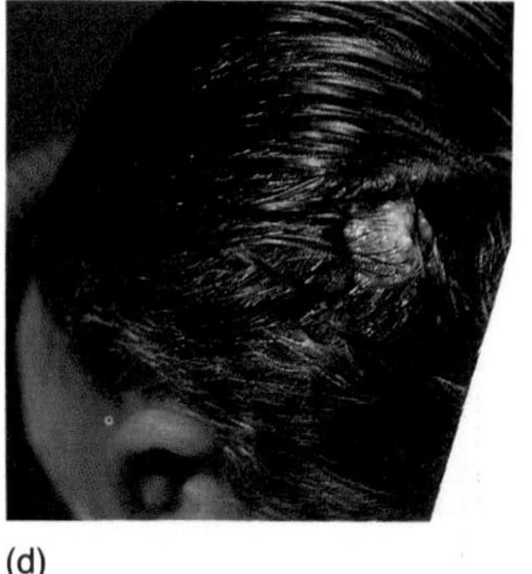

(d)

Figure 16.9 Diseases caused by fungi. *(a)* Child with tinea capitis (ringworm of the scalp) caused by *Microsporum audouinii. (b)* Child with tinea versicolor. The disease is caused by the yeast *Malasezzia furfur. (c)* Kerion (inflammatory lesion of the scalp) resulting from ringworm of the scalp (tinea capitis) caused by *Microsporum canis.* Courtesy of Professor F. Camacho, Department of Dermatology and Venereology, University Hospital, Seville, Spain. *(d)* Tinea capitis caused by *Trichophyton mentagrophytes.* Courtesy of Professor F. Camacho, Department of Dermatology and Venereology, University Hospital, Seville, Spain.

of dermatophytes including *Microsporum audouinii*, *Trichophyton tonsurans*, and *T. schoenleinii*. Each of these dermatophytes causes a different variety of tinea capitis. *Microsporum audouinii*, which generally infects children, causes a ringworm of the scalp called "gray patch." This ringworm is due to a noninflammatory disease in which the affected hairs become lusterless and break off a few millimeters above the scalp, giving the appearance of a gray patch of hair. The black dot variety of tinea capitis is usually caused by *T. tonsurans*, a fungus that invades the inside of the hair shaft. The hair breaks at the follicular region, leaving a stub of hair "loaded" with spores. In dark-haired individuals, the scalp appears laden with black dots. *Trichophyton schoenleinii* causes a ringworm of the scalp called **favus** or **tinea favosa**, a chronic, noninflammatory infection. The affected portions of the scalp have **scutules**, which are flakes consisting of hyphae and scalp material knitted together.

Treatment of ringworm is generally accomplished by using ointments containing miconazole or other imidazoles. Unresponsive cases can be treated by administering the oral antibiotic griseofulvin for extended periods of time. This antibiotic accumulates in the skin and inhibits fungal growth.

Figure 16.10 **Infectious spore of a dermatophyte in human hair**

Candidiasis

Candidiasis is a name given to diseases caused by yeasts in the genus *Candida*. The clinical manifestations of candidiasis vary from benign integumentary (skin and nail) diseases to severe renal (kidney) and myocardial infections. The skin diseases usually involve skin eruptions, which are papular or pustular inflammatory lesions with redness, swelling, and scaling of the skin. *Candida* infections may also involve the mucous membranes of the mouth and vagina (fig. 16.11). These forms of candidiasis are also called **thrush**. Individuals with defective immune systems may acquire a disease called chronic mucocutaneous candidiasis (fig. 16.11). The infection is recurrent and may involve both the integument and the mucous membranes.

Two species of *Candida* are most commonly associated with human diseases: *C. albicans and C. tropicalis; C. albicans* is far more common. Both are dimorphic, forming yeast cells on routine culture media (e.g., Sabouraud glucose agar) and pseudohyphae in host tissue or under reduced oxygen tensions. Both yeasts (like all other yeasts) are differentiated on the basis of their carbohydrate assimilation and fermentation patterns.

Candida albicans generally is unable to penetrate the intact skin, and breaking or softening of the skin is a prerequisite for infection. Virulence of *C. albicans* is believed to be related to its ability to form pseudohyphae and to produce the enzyme phospholipase. Pseudohyphae are more resistant than the yeast cell to phagocytosis by the PMN leukocytes and can penetrate host cells, possibly because of the phospholipase activity concentrated at the tips of the pseudohyphae. Phospholipase apparently breaks down plasma membrane phospholipids. The candidal infection causes abscesses that contain remnants of skin cells, leukocytes, and fungal elements.

Yeasts in the genus *Candida* are common inhabitants of the human body and are found in the skin, oropharynx, vagina, and digestive tract. Infections occur when the host's physiology allows the extensive growth of the yeast, generally as a consequence of predisposing factors such as diabetic acidosis, leukemias, excessive moisture, and obesity. An infection of the oral or vaginal mucosa often occurs after prolonged use of antibacterial antibiotics that cause a reduction of the normal flora in these areas.

The most effective treatment for candidiasis is correction of the predisposing factors. For example, prolonged broad-spectrum antibiotic treatment depletes the normal bacterial flora and allows *Candida albicans* to proliferate, since it is not susceptible to these antibacterial antibiotics. Stopping the treatment or recolonizing the infected area with normal flora can control

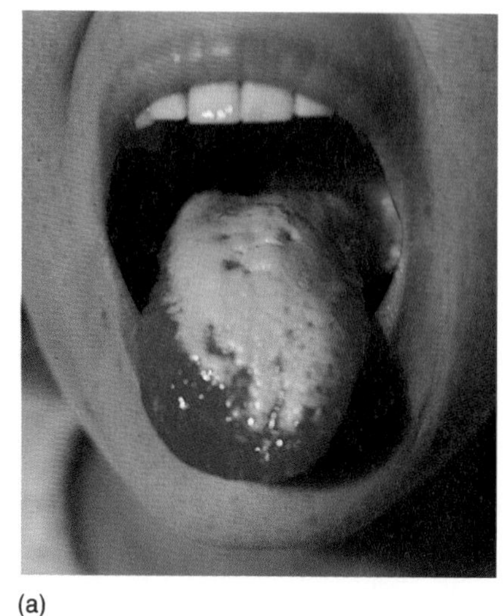

(a)

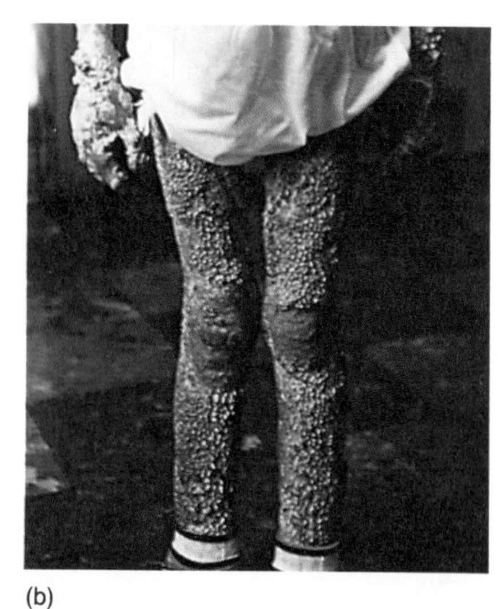

(b)

Figure 16.11 Candidiasis of skin and mucous membranes. *(a)* Patient with oral candidiasis. The disease, also known as "thrush," is characterized by the development of a cream-colored pseudomembrane over the afflicted area. Courtesy of Professor F. Camacho, Department of Dermatology and Venereology, University Hospital, Seville, Spain. *(b)* Chronic granulomatous candidiasis. The disease often occurs in individuals with various types of leukocyte dysfunction.

TABLE 16.3
SUMMARY OF INFECTIOUS DISEASES THAT AFFECT THE HUMAN SKIN AND EYE

Infectious disease	Causative agent	Microbial group
Anthrax	*Bacillus anthracis*	Gram + spore forming rod
Chickenpox (varicella)	Varicella-zoster virus	Herpesvirus
Conjunctivitis	*Neisseria gonorrhoeae*	Gram − coccus
	Newcastle disease virus	
	Streptococcus spp.	Gram + coccus
Erysipelas	*Streptococcus pyogenes*	Gram + coccus
Folliculitis	*Staphylococcus aureus*	Gram + coccus
Furunculosis	*Staphylococcus aureus*	Gram + coccus
Gas gangrene	*Clostridium perfringens*	Gram + sporogenous rod
Impetigo	*Streptococcus pyogenes*	Gram + coccus
Impetigo	*Staphylococcus aureus*	Gram + coccus
Inclusion conjunctivitis	*Chlamydia trachomatis*	Chlamydia
Leprosy	*Mycobacterium leprae*	Acid-fast rod
Lyme disease	*Borrelia burgdorferi*	Spirochete
Measles (Rubeola)	Measles virus	Paramyxovirus
Molluscum contagiosum	Togavirus	Togaviruses
Piedra, Black	*Piedraia hortae*	Fungus
Piedra, white	*Trichosporon* sp.	Yeastlike fungus
Pink eye	*Haemophilus aegypticus*	Gram − rod
Pyoderma	*Pseudomonas aeruginosa*	Gram − rod
	Streptococcus pyogenes	Gram + coccus
	Staphylococcus aureus	Gram + coccus
Ringworm (Tineas)	*Microsporum* spp.	Dermatophytic fungi
	Trichophyton spp.	Dermatophytic fungi
	Epidermophyton floccosum	Dermatophytic fungi
Rubella (German measles)	Rubella virus	Togaviruses
Scarlet fever	*Streptococcus pyogenes*	Gram + coccus
Smallpox (variola)	Smallpox virus	Poxviruses
Syphilis	*Treponema pallidum*	Spirochete
Tinea nigra	*Exophiala werneckii*	Fungus
Tinea versicolor (pityriasis versicolor)	*Malasezzia furfur*	Yeast
Trachoma	*Chlamydia trachomatis*	Chlamydia

the infection. Ointments containing the antifungal agent Nystatin may be helpful on skin lesions. In severe infections, the use of amphotericin B, or other antifungal drugs such as miconazole and ketoconazole, is necessary. Chronic mucocutaneous candidiasis presents a special problem because it results from an immunodeficiency. Some patients can be treated by administering transfer factor, a substance released by sensitized T-lymphocytes which recruits other immune cells to the site of infection. Apparently, the deficiency of these patients lies in the inability of their sensitized lymphocytes to muster a strong enough cell-mediated immune response. Transfer factor obtained from normal individuals helps in amplifying the specific anticandidal response and attaining at least a temporary cure.

VIRAL DISEASES OF THE SKIN

Viral infections involving the skin are common (see table 16.3). Childhood diseases such as chickenpox (varicella) and German measles (rubella) have cutaneous manifestations. Some of the viral infections (e.g., warts) involve only the skin, while others, such as herpes and smallpox (variola) viruses, also involve blood vessels and nervous tissue (fig. 16.12).

Dermotropic viruses, or viruses that have clinical manifestations affecting the skin, include pox, measles, varicella-zoster, herpes simplex, rubella, human wart (papilloma), some coxsackie viruses, and some echoviruses. Although these viruses can enter the host through the skin, some can also enter via the respiratory tract and others through the alimentary tract.

The life cycles of many viruses involve the disruption of the host cell as a result of viral multiplication. The newly released virus particles can, in turn, infect other cells and thereby initiate another round of infection. Eventually, the cellular destruction is so extensive that it is noticeable. The first few waves of infection cause enough cellular destruction to initiate an inflammatory response, which, in turn, contributes to tissue destruction and the subsequent formation of a viral lesion.

Specific antiviral mechanisms may also be involved in the formation of a lesion. When cells are infected by a virus, the individual develops antibody and cell-mediated immunity against it. Much of this immunity is effective in aborting viral infections, but sometimes it is responsible for some of the signs and symptoms of a disease. For example, some viruses alter the host cell surface by causing display of antigens; thus the host's immune system, in an attempt to abort the viral infection, recognizes viral antigens on the cell surface and destroys the cell. This response might appear suicidal, but it is a protective mechanism: the host destroys an infected cell before the virus has had a chance to multiply. This process, together with the inflammatory response, is partly responsible for the sores caused by some viruses.

Herpes Simplex Infections

Infections by the herpes simplex virus may involve the skin, eye, mucous membranes, and nervous system. These infections can be quite severe, even to the point of causing death. The infectious agent is usually acquired by direct contact through the broken skin or the mucous membranes. The infections eventually develop into lesions as a result of host- and viral-mediated tissue destruction.

One important characteristic of herpes simplex is its tendency to remain quiescent in the ganglia of sensory neurons that innervate the infected area. Because these foci of infection persist in the host, recrudescences (reappearances) of the disease often occur at the same site. A tingling or burning sensation precedes the eruption, and many patients "feel" a cold sore coming on. Recurrent herpetic lesions often develop after a febrile disease, UV irradiation (sunbathing), or stressful situations. The latent virus, when reactivated in ganglionic cells, moves along the axon

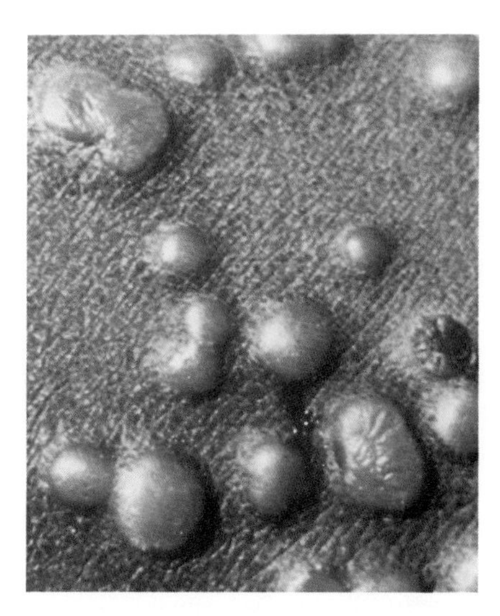

Figure 16.12 **Characteristic rash of smallpox**

FOCUS DO TOADS GIVE YOU WARTS?

In folklore, warts, or **papillomas**, have been attributed to various causes, such as touching a toad.

In reality, warts are contracted by direct contact with individuals or objects carrying a type of virus called the **papillomavirus**. These are small, double-stranded DNA viruses that are highly specific for their host. The human papillomavirus, for instance, infects only humans.

In essence, human warts are benign tumors caused by the virus reproducing in host skin (fig. 16.13). Viral reproduction causes a localized, excessive proliferation of skin and an increased deposit of keratin. Human warts generally do not become malignant (cancerous) and frequently go away on their own.

There are many "recipes" for curing warts, none of which work at all. In fact, simply leaving the warts alone may eventually be the best treatment of all. The more tenacious warts can be removed surgically, by freezing, or by applying ointments that dissolve keratin. One fact about warts is certain: toads won't give you warts, unless they are carrying the human papilloma virus.

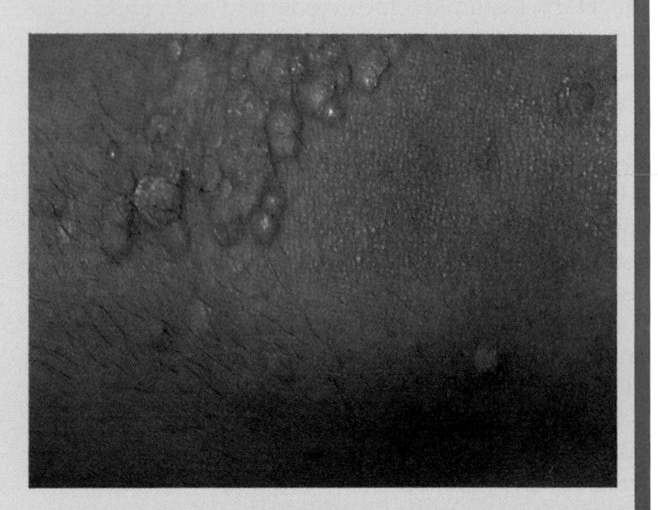

Figure 16.13 **Knee with warts.** Courtesy of Professor F. Camacho, Department of Dermatology and Venereology, University Hospital, Seville, Spain.

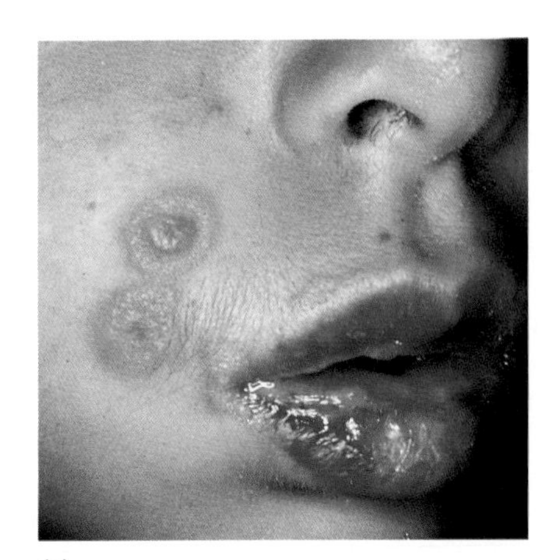

(a)

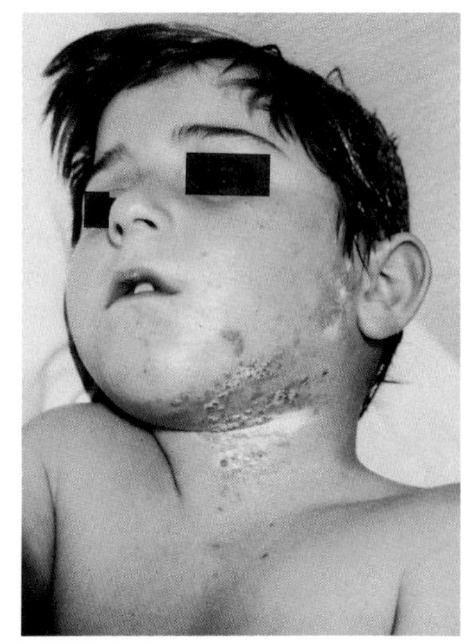

(b)

Figure 16.14 **Characteristic lesions of Herpes Simplex I infections.** *(a)* Herpetic lesions of the face and lips in a young female patient. Courtesy of Pediatrics Department, Division of Infectious Diseases, University Hospital, Seville, Spain. *(b)* Herpes simplex I infection of the chin and neck in a young boy. Courtesy of Pediatrics Department, Division of Infectious Diseases, University Hospital, Seville, Spain.

to the skin and initiates another infection. The painful lesions (fig. 16.14) initially are blisterlike, but after they rupture, they develop into a scab.

Herpes simplex virus is worldwide in distribution, and nearly everyone eventually becomes infected. There are two types of herpes simplex virus (I and II) that generally cause different diseases. Herpes simplex type I is most commonly associated with disease of the mouth, respiratory tract, skin, and central nervous system, while herpes simplex type II is an important cause of genital infection, although infection of the genital tract caused by herpes simplex type I has been described.

Few drugs are currently available for the control of herpetic infections. Topical 5-iodo-2′-deoxyuridine (IDU) is partially effective against eye infections but not against superficial skin infections. Intravenous, intramuscular, and oral acyclovir (sold under the trade name Zovirax) reduce the severity and duration of skin and systemic herpetic infections. Zovirax in ointment form is ineffective against genital or oral herpes.

Varicella-Zoster Virus Infections

Upon first exposure to a varicella-zoster virus, the patient develops the typical rash characteristic of **chickenpox** (fig. 16.15a). This infection is generally mild, lasting 5–10 days in children. Approximately 14 days after infection, a papular rash (raised spots) appears on the skin and mucous membranes. The papules eventually become vesicular and filled with a watery fluid. Fever, headache, and malaise may be evident during the **prodromal period** (initial stage of a disease). Primary varicella-zoster infections in adults generally are more severe than in children, often resulting in pneumonia. Mortality is high (15%) in adults. Chickenpox seems to increase in incidence (fig. 16.16) between the months of March and May.

Adults who have had chickenpox may experience a reappearance of varicella-zoster called **shingles** (fig. 16.15b). The recrudescence involves a return of the dormant virus to the innervated skin along sensory neurons. The reactivation of the latent virus and its movement in the axons cause severe pain in the infected area. Vesicular lesions resembling those of chickenpox also occur; however, these lesions are distributed only in the skin innervated by the infected neurons. Unlike chickenpox, the vesicular lesions of shingles may last up to a month, and the pain can persist even longer.

Measles (Rubeola)

Measles (also called **rubeola**) is a highly infectious disease caused by a pseudomyxovirus. The virus infects and multiplies in respiratory tract epithelium and in draining lymph nodes. The prodromal period has signs

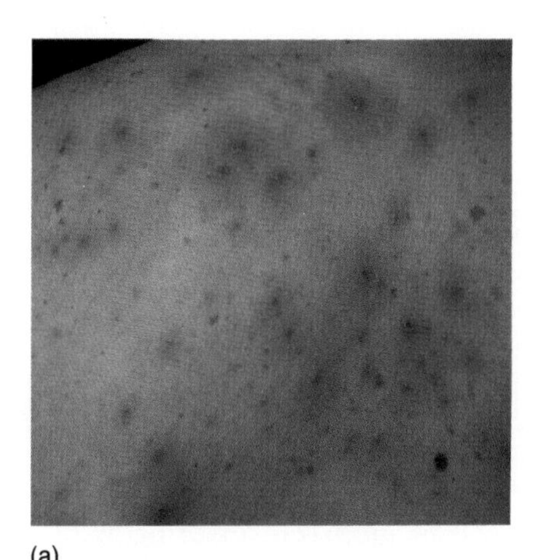

(a)

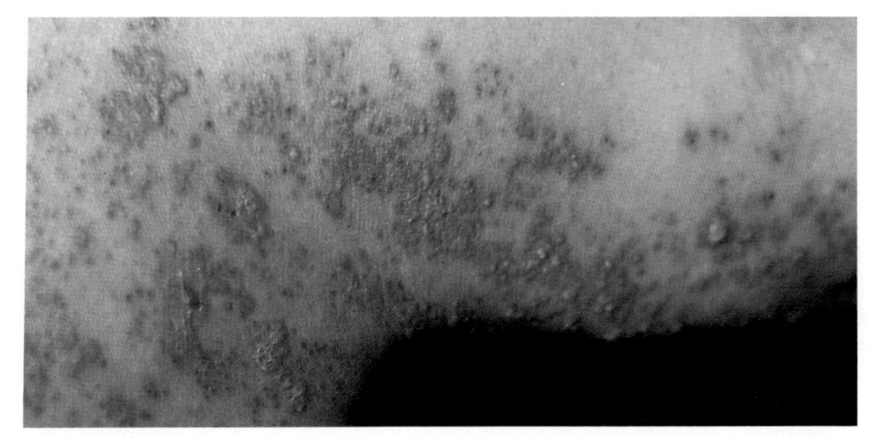

(b)

Figure 16.15 Diseases caused by Varicella-Zoster virus. *(a)* Typical case of chickenpox in young boy. Courtesy of Professor F. Camacho, Department of Dermatology and Venereology, University Hospital, Seville, Spain. *(b)* Characteristic lesions of shingles in the arm and shoulder region. Courtesy of Professor F. Camacho, Department of Dermatology and Venereology, University Hospital, Seville, Spain.

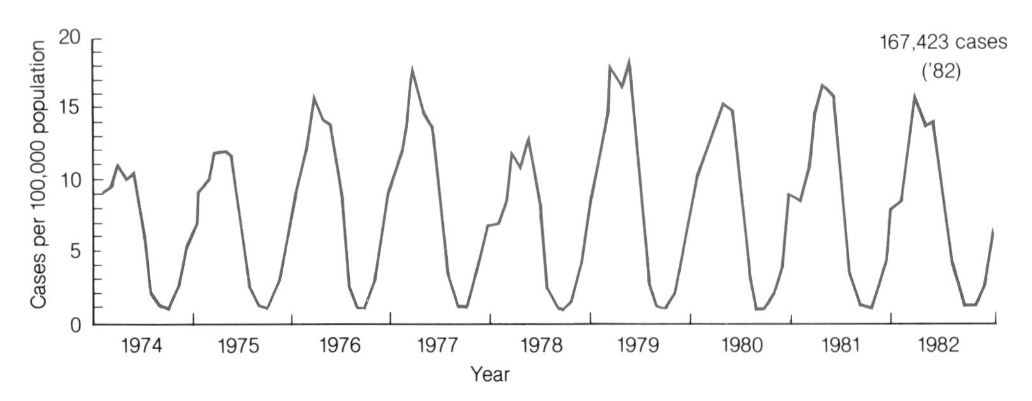

Figure 16.16 Incidence of chickenpox in the United States. Chickenpox has appeared with predictable frequency for the past 10 years. The incidence rate is 10–20 cases/100,000 inhabitants. The majority of cases occur in children between the months of March and May.

and symptoms resembling those of the common cold, with sore throat, low-grade fever, and cough. Allergiclike symptoms (e.g., nasal congestion) are sometimes present, as is conjunctivitis. With the spread of the virus in the blood, a typical rash appears (fig. 16.17), usually heralded by the development of small red dots with blue-white centers (**Koplik dots**) located on the lateral oral mucosa. This sign is considered to be diagnostic for measles. The maculopapular rash appears first on the face and neck and later spreads throughout the body. The incubation period for the disease is 10–21 days, and the acute stage of the disease with the rash may last 4–5 days (fig. 16.18).

Measles is transmitted from person to person in respiratory secretions and aerosols. Because humans are the only reservoir and, once infected, are immune for long periods of time, the disease tends to disappear from small communities until it is reintroduced by travellers. It reoccurs in epidemic proportions every 5 to 7 years, or as soon as a large number of nonimmune children have been born and the immunity has declined in a sufficient number of adults (fig. 16.19). Serious consequences of the disease are encephalomyelitis and subacute sclerosing panencephalitis. The disease is common in children, especially during December and May. Immunity to measles may be transferred passively from mother to offspring, which explains why children under the ages of 5 or 6 months seldom are afflicted with measles. In the early 1980s, there were an average of 4000 cases per year in the United States (fig. 16.19).

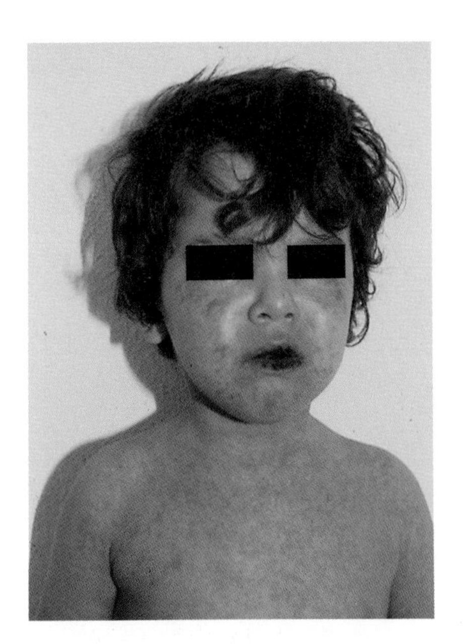

Figure 16.17 Measles. Young boy exhibiting the characteristic rash of measles. Courtesy of Pediatrics Department, Division of Infectious Diseases, University Hospital, Seville, Spain.

Rubella (German measles) is caused by a togavirus and is characterized by a macular rash accompanied by a low-grade fever. It spreads from person to person by direct contact, through nasal secretions, and by aerosols. The virus initially infects the mucous membranes of the upper respiratory tract and then spreads to other organs of the body. Fever, nasal discharge, and lymph node enlargement are early signs and symptoms of rubella. The characteristic rash is apparently due to a hypersensitivity (allergic response) to circulating viral antigens.

Rubella is an important disease (fig. 16.20) because during the first trimester of pregnancy the virus can infect the fetus and cause major destruction of fetal tissue. Such infection can result in abortion or in developmental abnormalities, including brain damage and mental retardation, deafness, congenital cataracts and blindness, heart defects, bone lesions, enlarged spleen (splenomegaly), enlarged liver (hepatomegaly), and low birth weight.

Rubella is diagnosed almost exclusively on a serologic basis, because this method is rapid and results can be attained the same day. Rubella testing has been implemented in many states in family planning and maternity clinics to advise potential mothers on the risk of rubella during pregnancy. Prospective mothers who have never been exposed to rubella should be vaccinated to avoid the possibility of becoming infected with the rubella virus during pregnancy. In 1984, a total of 745 cases were reported in the United States.

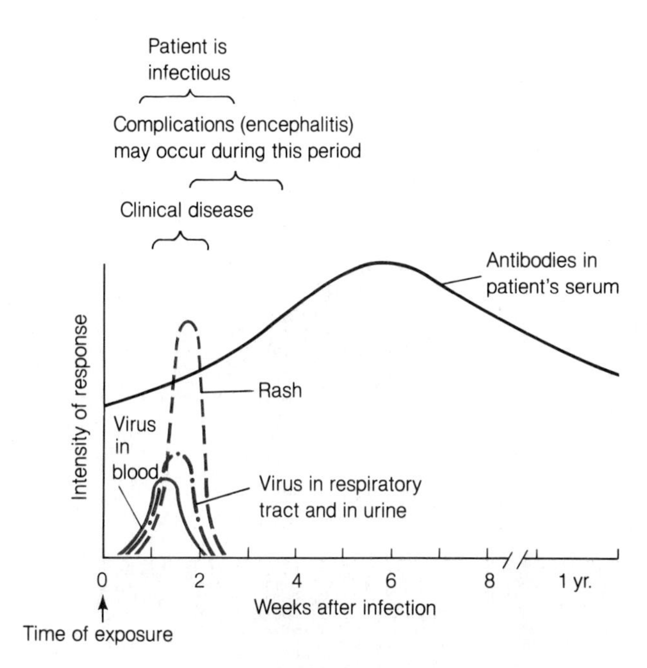

Figure 16.18 Course of infection by the measles (rubeola) virus. The characteristic rash of the disease appears 10–14 days after infection and is accompanied by a rise in serum antibodies. During the period of clinical illness, patients are contagious because they are shedding viruses in saliva and urine.

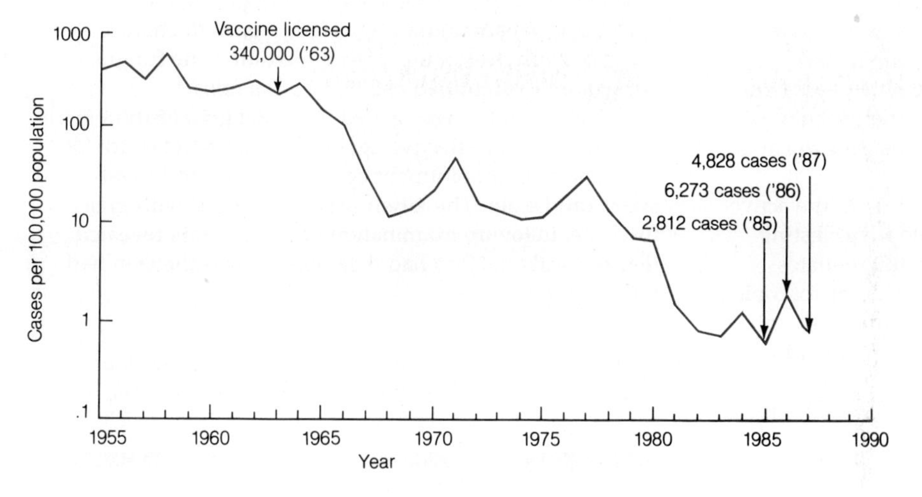

Figure 16.19 Incidence of measles in the United States. Before the introduction of the vaccine in 1963, there were about 250 cases/100,000 individuals. The present number of annual cases ranges between 1 and 5/100,000.

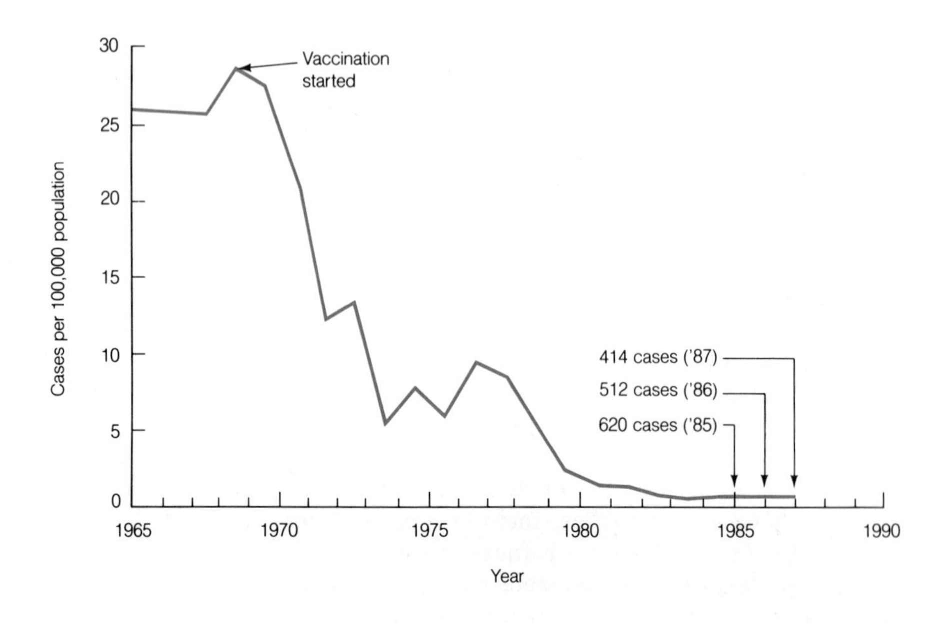

Figure 16.20 Incidence of German measles in the United States

CLINICAL PERSPECTIVE A CASE OF RINGWORM FROM A KITTEN

A 32-year-old female and her 5-year-old daughter visited the doctor's office with a two-week history of multiple, round, red, scaly patches, 1–2 cm in size, on the arms and chest. The little girl also had two small patches on her right cheek. Two weeks prior to the appearance of the scaly patches, the family had received a new kitten from a friend. The kitten had some scaling, balding spots on its body. When the patients were seen, a tentative diagnosis of **tinea corporis** (ringworm of the body) was made.

Skin scrapings from the edge of the lesion (the active site) were taken and cultured on Dermatophyte Testing Medium (a type of culture medium that differentiates between ringworm fungi and other fungi). A microscopic examination of part of the sample was also made by mixing the sample in a drop of 10% potassium hydroxide (KOH). This substance dissolves much of the tissue and allows the visualization of fungal material. The KOH wet mount revealed the presence of branching hyphae and spores characteristic of ringworm-causing fungi (dermatophytes), verifying the diagnosis of ringworm. The cultures of skin scrapings exhibited several fungal colonies that had turned the medium bright red, a growth characteristic of dermatophytes on the testing medium. The fungus was subsequently identified as *Microsporum canis*.

The patients were treated with oral **griseofulvin** (a drug that is very effective against dermatophytes) for 28 days and with an ointment containing the antifungal agent **imidazole**. The kitten was also treated with griseofulvin. A followup examination of the patients revealed that the scaly patches had disappeared and the skin had returned to normal.

Courtesy of Dr. Charles B. Fishman, M.D., Diplomate of the American Boards of Dermatology and Dermatopathology.

SUMMARY

Many infectious agents are capable of causing diseases of the skin. The most frequently-encountered skin pathogens are bacteria, viruses, and fungi.

STRUCTURE AND FUNCTION OF THE SKIN

1. The skin covers the body and provides an effective barrier against infections by microorganisms.
2. Conditions that alter the proper structure of the skin predispose the host to infectious diseases.

THE NORMAL FLORA OF THE SKIN

1. The normal flora of the skin colonize the stratum corneum and hair follicles.
2. The normal flora protect the host against pathogenic bacteria and fungi by competing for space and nutrients and by changing the skin environment so that it is hostile to invading pathogens.

BACTERIAL DISEASES OF THE SKIN

1. Many bacteria may invade the human skin and cause various diseases.
2. Anthrax is an often fatal disease characterized by the formation of a malignant pustule at the site of infection. The pathogen spreads from that site and produces an exotoxin. The disease is the result of the combined action of the spreading bacterium and the toxin.
3. Hansen's disease is a disease of the reticuloendothelial system. *Mycobacterium leprae* invades host cells, where it multiplies. The severity of the disease is related to the time of onset and strength of cell-mediated immunity. In the absence of cell-mediated immunity, the more severe form of the disease develops, lepromatous leprosy.
4. Pyoderma includes any acute inflammatory infection of the skin that results in the formation of pus. It is most commonly caused by *Staphylococcus* and *Streptococcus*.
5. Pyoderma includes diseases such as impetigo, folliculitis, carbuncles, and furuncles.
6. The pathogenic properties of *Staphylococcus aureus* reside in the ability of this bacterium to produce an array of exoenzymes as well as exotoxins.
7. Group A and B streptococci are the most important human pathogens. These microorganisms produce many exoenzymes and exotoxins that are responsible for the disease process.
8. Trachoma is a serious eye infection caused by *Chlamydia trachomatis*. The infection causes an extensive destructive inflammation that often leads to permanent blindness.

FUNGAL DISEASES OF THE SKIN

1. Fungi cause a variety of diseases that involve the skin. The most common of these are the ringworms and candidiasis.
2. Ringworms, such as athlete's foot and tinea capitis, are caused by a group of fungi called dermatophytes that grow on keratinized tissues. These are superficial infections that may elicit severe inflammatory responses, leading to permanent hair loss and the formation of scar tissue.
3. *Candida albicans* causes infections of the skin and mucous membranes. These infections cause primarily inflammatory reactions that vary in severity from mild to fatal.
4. Candidal infections often result when the host's physiology is altered or when the normal bacterial flora, which normally keep the yeasts in check, become depleted.

VIRAL DISEASES OF THE SKIN

1. Many viruses cause skin infections.
2. Dermotropic viruses have clinical manifestations that affect the skin and include smallpox, varicella-zoster, herpes simplex, rubella, and human wart viruses.
3. Most of the viral diseases result from host cell disruption by the reproducing virus, or from protective host responses.
4. Herpes simplex infections are characterized by the formation of a sore at the site of entry. Repeated episodes of "cold sores" result because the virus may remain quiescent in nerve ganglia that innervate the infected area.
5. Some herpetic infections are very serious and may involve the eye and central nervous system.
6. Chickenpox and shingles are caused by the varicella-zoster virus. Chickenpox is a benign skin infection that results in the formation of vesicular

lesions. The virus may remain quiescent for many years, after which it may cause the disease called shingles.

7. Measles, or rubeola, is caused by a pseudomyxovirus. The typical disease is characterized by a rash and by the formation of Koplik dots on the lateral oral mucosa. This virus can cause very serious diseases, such as encephalomyelitis and subacute sclerosing panencephalitis.
8. Rubella, or German measles, is caused by a togavirus. The disease is characterized by a rash and low-grade fever. Rubella in pregnant women, especially in the first trimester, can lead to severe fetal deformations or fetal death.

STUDY QUESTIONS

1. Describe the salient features of the skin and the function(s) it performs. Indicate which areas of the skin are colonized by microorganisms.
2. Discuss how the normal skin flora protect the host.
3. Discuss the various mechanisms employed by microorganisms to enter the human skin and cause disease.
4. Construct a table of bacterial skin diseases, including their etiology, characteristic signs and symptoms, and mode of transmission.
5. Construct a table of fungal diseases of the skin, including their etiology, characteristic signs and symptoms, and mode of transmission.
6. Construct a table of viral diseases of the skin, including their etiology, characteristic signs and symptoms, and mode of transmission.
7. Discuss various ways of preventing skin infections.
8. Discuss the pathogenesis (disease process) of:
   a. leprosy;
   b. impetigo;
   c. scarlet fever;
   d. anthrax;
   e. shingles.
9. Why are staphylococci and streptococci considered to be pyogenic bacteria? How are exoenzymes and toxins produced by these bacteria related to their ability to cause pyogenic lesions?

SELF-TEST

1. Which disease has the World Health Organization eliminated from human populations?
   a. measles;
   b. chickenpox;
   c. German measles;
   d. smallpox;
   e. no disease has ever been eradicated from human populations.
2. Why has it been possible to control smallpox and not other diseases?
   a. humans are the only natural reservoir for the pathogen;
   b. the pathogen did not demonstrate significant genetic drift so one vaccine was effective;
   c. a very effective vaccine was developed that conferred long-lasting immunity;
   d. the disease was so feared that all nations actively contributed to the effort to immunize all persons;
   e. all of the above.
3. Which organism is isolated from peoples' skin more than 90% of the time?
   a. *Candida albicans;*
   b. *Escherichia coli;*
   c. *Streptococcus pyogenes;*
   d. *Staphylococcus epidermidis;*
   e. *Staphylococcus aureus.*
4. Anthrax can be acquired when endospores:
   a. get into wounds or cuts;
   b. get into the lungs;
   c. are consumed in undercooked contaminated meat;
   d. all of the above;
   e. none of the above.
5. The vaccine for anthrax is made from:
   a. a crude cell extract of *Bacillus anthracis;*
   b. purified *B. anthracis* capsules;
   c. endospores from a nonencapsulated strain of *B. anthracis;*
   d. formalin-treated *B. anthracis;*
   e. none of the above.
6. Individuals who do not develop an effective immune response to *Mycobacterium leprae* most likely have what type of leprosy? a. tuberculoid; b. lepromatous.
7. What is the immune response found in persons with tuberculoid leprosy?
   a. strong and rapid response;
   b. thymus-derived lymphocytes are very active;
   c. macrophages are very active;
   d. all of the above;
   e. there is a poor response.
8. What is dapsone?
   a. an antibiotic;
   b. a type of penicillin;
   c. a type of sulfa drug;
   d. similar to cephalosporin;
   e. none of the above.

9. What helps *Staphylococcus aureus* to be a successful pathogen?
   a. staphylokinase;
   b. β-lactamase;
   c. hydrolases;
   d. protein A;
   e. all of the above.

10. The importance of discovering that *Staphylococcus aureus* has a coagulase is that it generally indicates that the organism is:
   a. a pathogen;
   b. resistant to penicillin;
   c. not a pathogen;
   d. able to break out of clots;
   e. resistant to sulfa drugs.

For items 11–14, match the hydrolases to their functions:
   a. breaks down fibrin protein;
   b. breaks down nucleic acids;
   c. breaks down polysaccharides;
   d. breaks down phospholipids;
   e. converts blood proteins to fibrin.

11. Hyaluronidase
12. Lipase
13. Staphylokinase
14. Coagulase

15. The bacterium that causes scalded skin syndrome is unusual because it produces:
   a. coagulase;
   b. exfoliatin;
   c. hyaluronidase;
   d. collagenase;
   e. DNase.

16. When *Staphylococcus aureus* causes impetigo with blisterlike lesions, what toxin is it producing?
   a. coagulase;
   b. exfoliatin;
   c. hylauronidase;
   d. collagenase;
   e. DNase.

17. Which organism, other than *S. aureus*, is often the cause of impetigo?
   a. *Candida albicans;*
   b. *Escherichia coli;*
   c. *Streptococcus pyogenes;*
   d. *Staphylococcus epidermidis;*
   e. *Bacillus anthracis.*

18. Which disease is *not* caused by a *Streptococcus?*
   a. rheumatic fever;
   b. scarlet fever;
   c. erysipelas;
   d. impetigo;
   e. scalded skin syndrome.

For items 19–20: streptococci produce hemolysins that allow them to be divided into three groups based upon their activity on blood agar plates. Match the activity to the classification:
   a. alpha hemolysis;
   b. beta hemolysis;
   c. gamma reaction.

19. These organisms cause large zones of clearing in the blood agar around colonies.

20. These organisms turn the blood around colonies a greenish brown. There is usually some lysis also.

For items 21–24, match the classification to the organism:
   a. group A;
   b. group B;
   c. group D;
   d. pneumococci;
   e. viridans.

21. *Streptococcus faecalis.*
22. *Streptococcus salivarius* and *S. mutans.*
23. *Streptococcus pyogenes.*
24. *Streptococcus agalactiae.*

25. Which is primarily responsible for the signs and symptoms of scarlet fever?
   a. erthrogenic toxin;
   b. DNase;
   c. M-protein;
   d. streptokinase;
   e. exfoliatin.

26. The Lancefield groups are determined by the type of:
   a. hemolysin;
   b. erythrogenic toxin;
   c. activity on blood agar;
   d. serological reaction;
   e. all of the above.

27. What characteristic of dermatophytic fungi enables them to penetrate and grow in hair and skin?
   a. production of keratinase;
   b. production of collagenase;
   c. production of lipase;
   d. all of the above;
   e. none of the above.

For items 28–30, match the pathogen with the disease:
   a. pox virus,
   b. rubella,
   c. varicella-zoster;
   d. rubeola;
   e. papilloma or polyoma.

28. German measles.
29. Chickenpox.
30. Measles.

SUPPLEMENTAL READINGS

Habicht, G. S., Beck, G., and Benach, J. L. 1987. Lyme disease. *Scientific American* 257(1):78–83.

Henderson, D. A. 1976. The eradication of smallpox. *Scientific American* 235:25–33.

Hirsch, M. S. 1984. *Cutaneous viral diseases. Scientific American Medicine* Section 7-XXX New York: Freeman.

Joklik, W. K., Willett, H. P., and Amos, D. B. eds. 1984. *Zinsser microbiology.* 18th ed. New York: Appleton-Century-Crofts.

Marples, M. J. 1969. Life on the human skin. *Scientific American* 220:108–115.

Rosebury, T. 1969. *Life on man.* New York: Berkeley.

Simon, H. B. 1984. *Gram-positive cocci. Scientific American Medicine* Section 7-I New York: Freeman.

Wolf, R. H., Gormus, B. J., Martin, L. N., Baskin, G. B., Walsh, G. P., Meyers, W. M., and Binford, C. H. 1985. Experimental leprosy in three species of monkeys. *Science* 227:529–530.

CHAPTER 17

DISEASES OF THE RESPIRATORY TRACT

CHAPTER PREVIEW **AIRBORNE INFECTIONS IN DAY-CARE CENTERS**

Increases in the cost of housing and living have forced both parents in many households to seek employment in order to make ends meet. As a consequence, more and more preschool children are going to day-care centers while their parents are at work. This rapid increase in the number of children attending child-care centers has created a new public health hazard: overcrowding and less than optimum hygienic facilities. This makes a larger number of children vulnerable to a vast array of infectious diseases. Of importance are diseases that are spread in airborne particles or by direct contact. This hazard is compounded by the fact that preschool children have no concept of how diseases spread or how to prevent them. Hence, unstifled sneezes, coughs, and nasal secretions, as well as the close contact that exists among the children, create an environment in which children are continually exposed to respiratory disease agents.

Transmission of common colds, influenza, cytomegaloviruses, *Cryptosporidium* and meningitis increases dramatically under these conditions. Infected children can then pass on the infectious agents to their families and neighborhood playmates.

Public health officials need to know the biology and mode of transmission of infectious agents of the respiratory tract in order to control respiratory diseases in institutions such as day-care centers. This chapter discusses some of the most common diseases of the human respiratory tract, their cause, treatment, and control.

CHAPTER OUTLINE

An average individual, when breathing quietly, inhales about 7500 ml of air every minute along with numerous air contaminants, many of which are potentially pathogenic microorganisms. In light of this fact, it is not surprising that many infectious diseases are acquired and transmitted through the respiratory route. Since there is such a potential for lung infection, the body has evolved an array of specific and nonspecific mechanisms for protecting the lungs. Coughing, sneezing, mucus secretions, and ciliary action expel inhaled particles and microorganisms from the respiratory tract. In addition, macrophages and immunoglobulins (IgA) help destroy microorganisms that remain in the lungs.

The respiratory tract is divided into two regions: the **upper respiratory tract** (URT) and the **lower respiratory tract** (LRT). URT infections are generally benign and can be treated effectively, although some URT infections, such as diphtheria and acute epiglottitis, can be life-threatening. Conversely, LRT infections often are serious, and must be treated promptly and effectively or serious complications or death will result. In this chapter we examine some of the most common respiratory tract infections, including their clinical and biological nature as well as their treatment.

STRUCTURE AND FUNCTION OF THE RESPIRATORY TRACT

The primary function of the respiratory tract is to bring air into contact with a respiratory surface that allows oxygen to diffuse into the blood and carbon dioxide to escape. The respiratory surface is highly vascular, not only because gases must be exchanged efficiently during breathing, but also because cells of the immune system must be delivered promptly. The respiratory system includes a system of air-transporting tubes that bring the air to the respiratory surface (fig. 17.1). The tubes in the upper respiratory tract consist of the **nasal cavity, eustachian tubes**, and throat, while those in the lower respiratory tract include the pharynx, larynx, trachea, and bronchi. These structures are much too thick to permit the free exchange of air with the blood, and they function solely as air passages to the respiratory **bronchioles, alveolar ducts**, and **alveolar sacs** (fig. 17.1).

The alveoli or air sacs are the site of exchange of gases between the blood and the environment. These thin-walled sacs, made up of a single layer of flat epithelial cells (fig. 17.1), have numerous capillaries. The alveolar sacs and capillaries exchange gases with each other, and the capillaries may also deliver cells of the immune system to the air sacs. The alveolar surface area that functions in gas exchange in adult humans is about 50–70 m^2 (1 $m^2 \approx$ 1 square yard) and consists of more than 300 million alveoli. Any condition that interferes with the proper exchange of gases is harmful to the individual. In pneumonia, for example, the damaged alveoli fill with fluids from alveolar capillaries. The fluid in the alveoli reduces the efficiency of gas diffusion and blood aeration.

THE NORMAL FLORA OF THE UPPER RESPIRATORY TRACT

The microbial flora of the upper respiratory tract (table 17.1) consists mainly of bacteria and fungi. Staphylococci, streptococci, diphtheroids, actinomycetes, gram-negative cocci, coccobacilli, aspergilli, yeasts, and zygomycetes may all be present in the upper respiratory tract.

The indigenous flora of the upper respiratory tract may contain commensals as well as potential pathogens. In addition, the normal flora of the upper res-

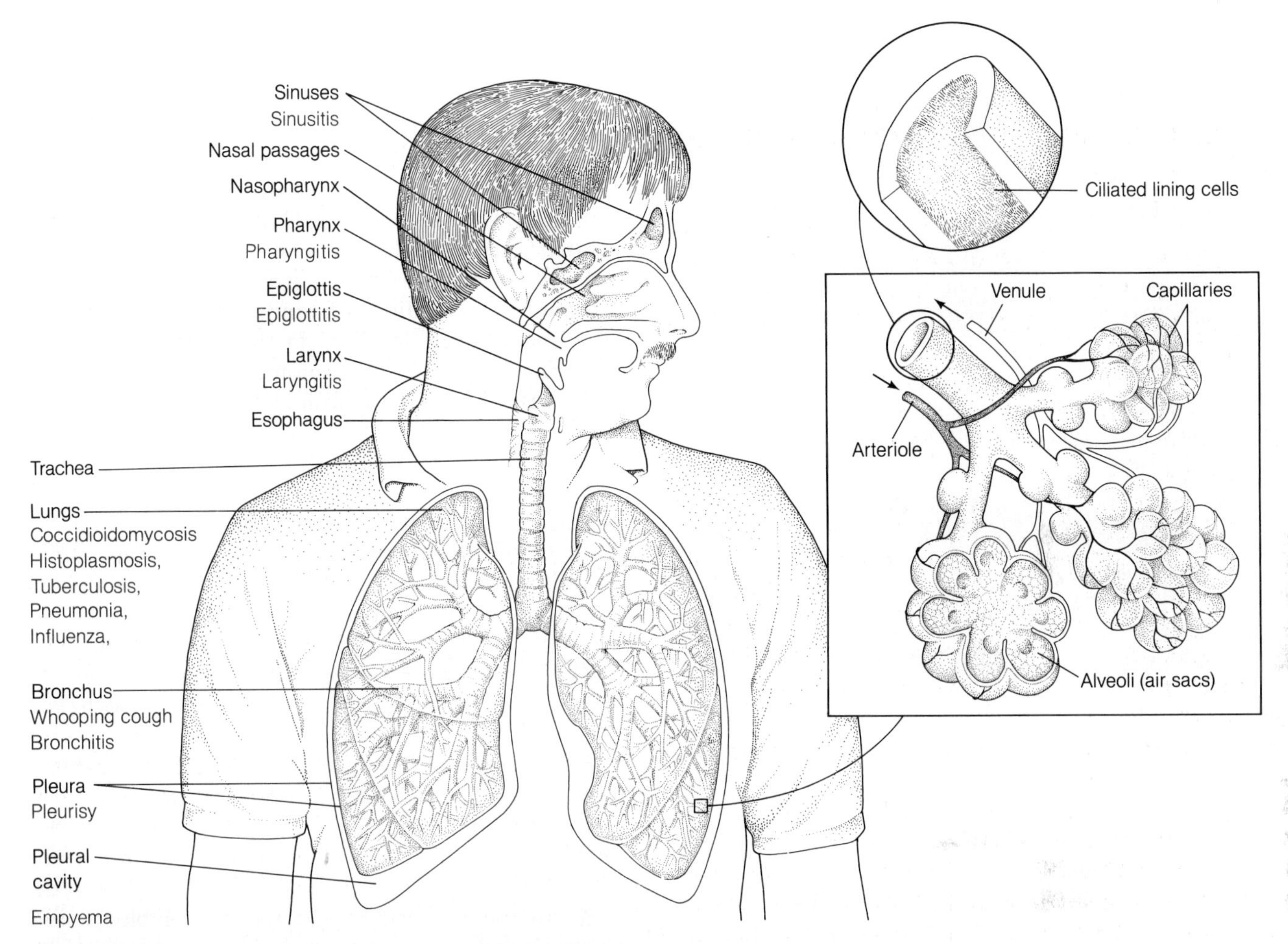

Figure 17.1 Anatomy of the human respiratory system. *(a)* Upper respiratory system. The colored print indicates infectious diseases that affect the anatomical site. *(b)* Lower respiratory tract. Insert illustrates the structure of alveoli. Notice that respiratory tubes are lined by ciliated cells, which serve to flush out infectious agents and dust particles before they reach the alveoli. The colored print indicates infectious diseases that affect the anatomical site.

piratory tract contains microorganisms that protect the host from colonizing pathogenic bacteria. For example, certain indigenous streptococci of the throat, especially those in the **viridans group**, inhibit the growth of Group A streptococci and therefore protect the host from a possible attack of "strep throat" (**pharyngotonsillitis**). Antibiotics or other inhibitory substances may be produced by the viridans streptococci, or they may deplete the available nutrients.

BACTERIAL DISEASES OF THE UPPER RESPIRATORY TRACT

Throat Infections

Pharyngotonsillitis (table 17.2) is an inflammation of the pharynx, usually accompanied by throat pain, malaise, fever, and postnasal secretions. The throat is scarlet red, with exudative pyogenic (pus-containing) material.

TABLE 17.1
REPRESENTATIVE MICROORGANISMS FOUND IN THE UPPER RESPIRATORY TRACT

Microorganism	Oral	Nasopharynx	Sinuses
BACTERIA			
Bacteroides sp.	+	−	−
Borrelia sp.	+	+	−
Branhamella sp.	+	+	−
Diphtheroids	+	+	−
Fusobacterium sp.	+	−	−
Haemophilus sp.	+	+	−
Mycoplasma sp.	−	−	+
Neisseria sp.	+	+	−
Peptostreptococcus sp.	+	+	−
Staphylococcus aureus	+	+	−
S. epidermidis	+	+	−
Streptococcus sp.	+	+	+
Veillonella sp.	+	−	−
FUNGI			
Aspergillus sp.	+/−	+/−	−
Candida albicans	+	+	+/−
VIRUSES			
Adenoviruses	−	+	+
Coxsackieviruses	+	+	+
Epstein-Barr virus	−	+	−
Herpes simplex	+	−	−
Influenza viruses	−	+	+
Rhinoviruses	+	+	+

+, Organism commonly isolated from site.
−, Organism infrequently or never isolated from site.

Group A streptococci, a group of β-hemolytic streptococci such as *Streptococcus pyogenes* (see table 16.2), are the most common cause of bacterial pharyngotonsillitis. Occasionally, Groups C (S. *equi*) and G (S. *anginosus*) streptococci can also cause pharyngotonsillitis. Streptococcal pharyngotonsillitis (or "strep throat") is most common in school-age children (5–15 years) and occurs most often during the winter months. The streptococci are transmitted in aerosols and occasionally in foods. Other bacteria that have been cited as causes of pharyngotonsillitis, albeit infrequently, are *Staphylococcus aureus, Haemophilus influenzae,* and *Corynebacterium diphtheriae.* In spite of a common belief that most pharyngotonsillitis is due to streptococci, most cases (70%) are in fact caused by adeno-, coxsackie-, rhino-, Epstein-Barr, influenza, parainfluenza, and respiratory syncytial viruses.

Sequelae to Pharyngotonsillitis

A small but significant percentage of individuals suffering from streptococcal pharyngotonsillitis, if they receive no treatment, may end up with nonsuppurative (no pus produced) sequelae. The most common of these are rheumatic fever and acute glomerulonephritis.

Rheumatic fever is a serious sequela that occurs in about 3% of human streptococcal infections. It is responsible for 10,000–20,000 deaths each year in the United States. The disease, also called rheumatic heart disease, is characterized by arthritis (inflammation of the joints) and carditis (inflammation of heart muscle). Rheumatic fever develops in three stages: (a) acute nasopharyngitis; (b) asymptomatic carriage of Group A streptococci in the throat; and (c) electrocardiographic changes, accompanied by fever and inflammation of the joints and heart muscle.

Rheumatic fever patients characteristically have antibodies against hyaluronic acid, streptolysin O, DNase, and M-protein. It is thought that the cardiac lesions and the joint inflammation may be due to antistreptococcal antibodies that cross-react with the heart tissue. The fact that streptococcal M-protein, when injected repeatedly into laboratory animals, is able to cause rheumatic fever-like diseases, supports this idea.

Acute glomerulonephritis is another complication of streptococcal infections that occurs in an average of 7% of individuals suffering from a streptococcal infection. The disease is an acute inflammation of the glomeruli in the kidney, characterized by blood in the urine (hematuria) and hypertension (high blood pressure). Delirium and coma may accompany severe cases.

The pathogenesis of the disease appears to be related to the formation of circulating immune complexes (antigens bound by antibody). These complexes are deposited in the glomeruli, fixing complement. The complement-mediated kidney tissue destruction gives rise to the signs and symptoms of the disease. Epidemics of this disease have been documented among school-age children. The disease does not tend to recur in afflicted individuals, and deaths are rare.

Acute epiglottitis is a fulminant (rapid in onset and very severe), grave inflammation of the epiglottis (fig. 17.1). In children, the disease progresses quickly, causing fever, a painful sore throat, extreme difficulty in swallowing, and a progressive enlargement of the epiglottis. Difficulty in breathing becomes increasingly pronounced as the disease advances. If the airway is not maintained patent (open), apnea (lack of breathing) and death soon follow. The most common cause of epiglottitis is *H. influenzae*, although *S. pneumoniae*, streptococci, staphylococci, parainfluenza virus,

TABLE 17.2
COMMON INFECTIOUS DISEASES OF THE UPPER RESPIRATORY TRACT

Name of disease	Causative agent	Group of organisms
Common cold	Rhinoviruses	Virus
	Coronaviruses	Virus
Croup	Respiratory syncytial virus	Virus
	Myxoviruses	Virus
	Adenoviruses	Virus
Diphtheria	*Corynebacterium diphtheriae*	Bacteria (gram + rods)
Epiglottitis	*Haemophilus influenzae*	
	Cytomegaloviruses	Virus
	Branhamella catarrhalis	
Otitis Media	*Haemophilus influenzae*	Bacteria (gram − rods)
	Staphylococcus aureus	Bacteria (gram + coccus)
	Streptococcus pneumoniae	Bacteria (gram + coccus)
Pharyngotonsillitis (strep throat)	*Streptococcus pyogenes*	Bacteria (gram + coccus)
	Staphylococcus aureus	
Sinusitis	*Haemophilus influenzae*	
	Bacteroides sp.	Bacteria (gram − rod)
	Streptococcus pneumoniae	
Zygomycosis	*Rhizopus* sp.	Fungi (zygomycetes)

and respiratory syncytial virus have also been isolated from patients with the disease. The treatment involves prompt reestablishment of the airway and massive intravenous antibiotic treatment with chloramphenicol and penicillin. Moistened, cool air also helps to alleviate the inflammation.

Diphtheria is an infectious disease characterized by fever, headache, malaise, sore throat, and eventually the formation of a pseudomembrane on mucous membranes of the throat and nasal passages. In pharyngeal diphtheria, the airway is often obstructed due to bacterial growth and the pseudomembrane. Death is due to the action of the exotoxin (diphtherotioxin) on the heart and peripheral nerves. **Myocarditis** (inflammation of the heart muscle) manifests itself late in the course of the disease and represents a grave turn of events.

The cause of diphtheria is *Corynebacterium diphtheriae*, a gram-positive, nonmotile, non-spore-forming pleomorphic rod. The individual cells are generally club-shaped with a tendency to form V's, M's, L's, the so-called Chinese lettering, and palisades. This organism produces a toxin, called **diphtherotoxin**, an exotoxin that inhibits protein synthesis and is responsible for the signs and symptoms of the fatal disease. The production of toxin by *C. diphtheriae* is due to a **lysogenic conversion** of the bacterium by a bacteriophage. The lysogenic bacteriophage β apparently contains as part of its genome the gene coding for diphtherotoxin. In a prophage state (viral DNA incorporated in the bacterial chromosome) within the host, the viral phenotype is expressed. Only lysogenized *C. diphtheriae* are capable of producing the toxin.

The signs and symptoms of diphtheria are due to the bacteria growing in the throat where they release the diphtherotoxin. The exotoxin is absorbed through the mucous membranes and then distributed to organs such as the heart by the circulatory system. Myocardial lesions characteristic of the severe disease result from the destruction of the heart muscle cells by the exotoxin. If the degree of myocardium destruction is sufficiently extensive, the disease is fatal. Nervous system involvement results from the destruction of neurons (nerve cells) by the toxin. The severity of the disease is directly proportional to the amount of cell-bound diphtherotoxin in the host.

Corynebacterium diphtheriae is transmitted by direct contact with asymptomatic carriers, by inhalation of aerosols, or via fomites. Epidemics of diphtheria have been reported from economically deprived urban areas of various countries. Individuals in these areas had not been immunized—a fact that demonstrates the importance of immunization for preventing potentially devastating epidemics. The highest incidence of diphtheria occurs during the winter months in patients 15 years of age or older. This is a departure from past observations, in which individuals between the ages of 6 months and 10 years showed the highest incidence. Protection against this disease is by vaccination (DPT shots), which is compulsory in the United States. The vaccine preparation or **toxoid** consists of modified, nontoxic, but antigenic toxin (fig 17.2).

The most effective treatment consists of the prompt administration of antitoxin and treatment with erythromycin. Passive immunization with equine (horse) antidiphtherotoxin antibodies is the only specific treatment for the severe form of the disease. In addition, treatment with erythromycin is very effective in eliminating the carrier state.

Ear Infections

Otitis is an inflammation of the ear that may or may not be caused by an infectious agent. **Otitis externa** is a benign, inflammatory condition of the external auditory canal. The condition, very common in human populations, can be initiated by excessive moisture in the outer ear or by damage (e.g., from cotton swabs or hairpins) to the auditory canal. The most common complaints are pain and itching. Crusting and purulent discharges are also common manifestations. The etiology (cause) of otitis externa is varied and includes gram-positive cocci, gram-negative bacilli, and fungi. Infections of the external ear by *Pseudomonas aeruginosa* are characterized by the formation of a green pus that is discharged from the inflamed area. Central nervous system complications have been reported as a consequence of perforation of the ear by *P. aeruginosa* and require prompt treatment with carbenicillin and tetracycline.

Otitis media is a common childhood disease, but it is relatively rare in adults. This condition usually results from the invasion of the middle ear by bacteria from a pharyngeal infection. The signs and symptoms of this disease are pain, fever, and temporary hearing loss due to the acute inflammation and the resulting edema. A characteristic sign is a convex tympanic membrane, bulging outward because of the fluid buildup and pressure in the middle ear. A myriad of bacteria can cause the disease, however, pyogenic cocci, gram-negative bacteria, *Haemophilus influenzae*, and *Branhamella catarrhalis* are often implicated.

Sinus Infections

Sinusitis is an inflammation of the sinuses, especially the paranasal sinuses (fig. 17.1). Any condition that impedes the proper drainage of sinuses can lead to this disease. Nasal septum deviation, polyps, or rhinitis are common predisposing factors. The signs and

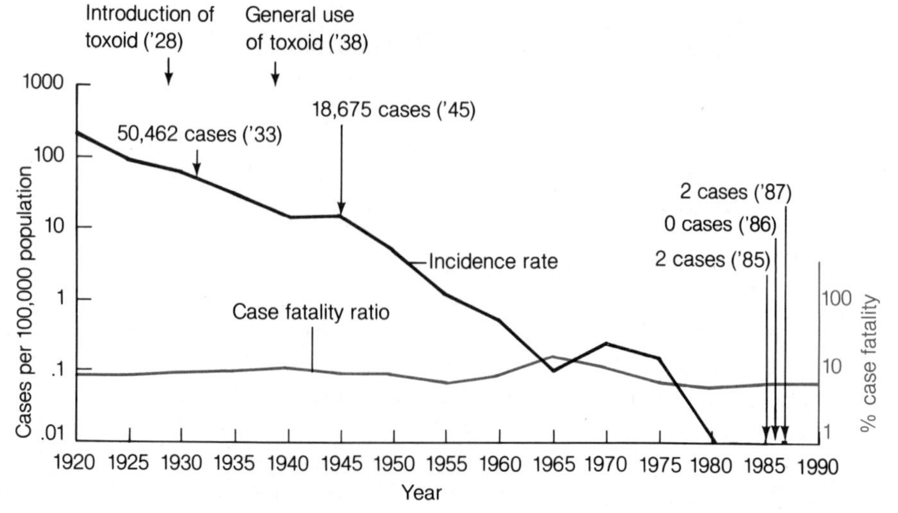

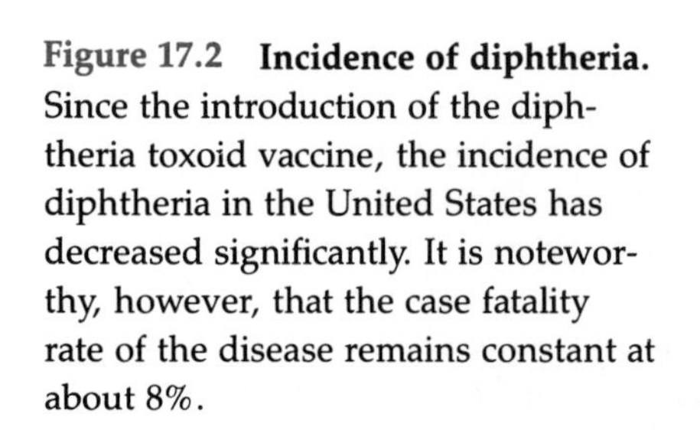
Figure 17.2 Incidence of diphtheria. Since the introduction of the diphtheria toxoid vaccine, the incidence of diphtheria in the United States has decreased significantly. It is noteworthy, however, that the case fatality rate of the disease remains constant at about 8%.

symptoms of sinusitis depend upon which sinuses are involved. Paranasal sinusitis, the most common form of sinusitis, results in pain and tenderness in the forehead, with purulent (pus-containing) drainage. It is usually caused by *Haemophilus influenzae*, *Streptococcus pneumoniae*, and Group A streptococci. Treatment usually entails draining the sinuses and administering antibiotics.

Although not strictly a sinusitis, a fungal disease called **zygomycosis** sometimes affects the sinuses. The disease is infrequent but often fatal. It is caused by opportunistic zygomycetes such as *Rhizopus* and *Mucor.* These opportunistic pathogens cause a rapidly developing disease in individuals with diabetic acidosis or leukemias. These fungi grow well in arterial blood and invade the vessels, causing thrombi (blood clots). The lesions initiate in the paranasal sinuses, creating a black, mucoid exudate (fig. 17.3) consisting of necrotic tissue, fungal elements, and clotted blood. The fungi spread rapidly into the brain, causing severe central nervous system disorders or death. The treatment of choice is correction of the predisposing factor(s) and prompt administration of antifungal antibiotics such as amphotericin B or imidazoles.

VIRAL DISEASES OF THE UPPER RESPIRATORY TRACT

The Common Cold

The **common cold** is a mild illness of the upper respiratory tract characterized by a "stuffy" nose, "scratchy" throat, general malaise, headache, sore throat, sneezing, and watery discharge from the nose. The nasal discharge often thickens and becomes yellowish. This is usually an afebrile (fever-free) disease with an incubation period of 1–4 days. The illness runs its course within 4–7 days. Sinusitis and otitis media are common complications, since drainage of the sinuses and middle ear is blocked.

The common cold results from a number of viral infections of the upper respiratory tract. Various types of viruses, including rhino-, corona-, myxo-, coxsackie-respiratory syncytial, adeno-, and enteroviruses have been isolated from nasal secretions of patients suffering from the common cold (table 17.3).

Common colds occurring in the fall and spring are usually caused by the rhinoviruses, while those in the winter months are often caused by coronaviruses. The rhinoviruses cause approximately 30–35% of all colds.

Since the signs and symptoms of the common cold resemble those of allergic rhinitis, it is possible that at least some of the symptoms of the common cold are due to allergic reactions to viral antigens. The virus is known to adhere to and multiply in the mucous membrane of the nose and induce the formation of antibodies. In tissue culture, it also causes cytopathic effects, thus illustrating the cell-destruction potential of the virus.

Humans are the only natural hosts for the rhinoviruses and coronaviruses. Although chimpanzees can be infected with the virus and develop specific antibodies, the infection in these animals does not usually culminate in disease.

At one time it was thought that coughs and sneezes

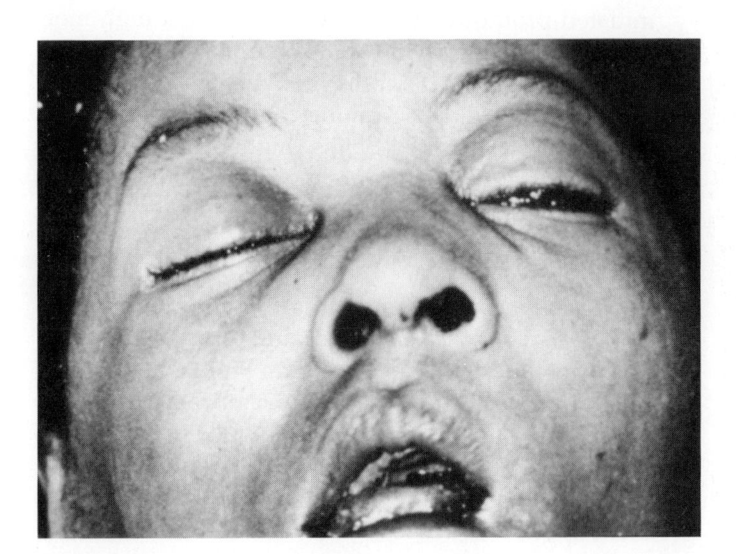

Figure 17.3 **Rhinocerebral zygomycosis.** The dark discharge evident in the nostrils, eyes, and mouth of the patient is characteristic of the disease and consists of fungal elements, dead tissue, and coagulated blood.

TABLE 17.3
RESPIRATORY DISEASES CAUSED BY VIRUSES

	Virus importance						
	Influenza A	Rhino-virus	Corona-virus	Adeno-virus	Respiratory syncytial virus	Coxsackie virus	Echo virus
Adult pneumonia	+++	+	−	−	+	+++	++
Childhood pneumonia	++	++	−	+	++++++	+++	+++
Adult colds	++	+++++	+++	++	+	+	+
Childhood colds	++	+++	++	+++	++++	++	++
Adult pharyngitis	+++	+++	−	++	++	++	++
Childhood pharyngitis	+++	+++	−	+++	+++	++	++
Croup	+++	++	−	+	+++	++	+

created aerosols that other individuals might inhale and thereby become infected. Recent evidence suggests, however, that person-to-person—or person-to-fomite-to-person—contacts are the primary modes of transmission. For example, viruses in nasal secretions contaminating the infected person's hands can spread by hand contact to other individuals, who then inoculate themselves.

Prevention of the common cold by vaccination is unlikely in the near future, because there are so many antigenically different viruses (more than 113 different serotypes) that can cause the common cold. Thus, it would be extremely difficult to develop a polyvalent vaccine that would contain all the common antigenic types. The problem is further compounded by the fact that common cold viruses induce specific immunity of short duration; therefore, repeated, frequent vaccinations would be needed. Behavioral modification, such as frequent hand-washing, would reduce the chances of becoming infected. There is no useful treatment at the present time, nor is there a preventive measure. The old-fashioned remedy of drinking lots of fluid (including chicken soup) and bed rest appears to be the most comforting of available treatments.

Acute Laryngotracheobronchitis (Croup)

Croup is an acute infectious disease of children (mostly under the age of 3 years) characterized by coughing, hoarseness, high-pitch sounds, and fever. Severe symptoms may lead to cyanosis (due to apnea), convulsions, and sometimes death. The disease is caused by a variety of viruses, most commonly parainfluenza viruses, respiratory syncytial viruses, and some adenoviruses.

BACTERIAL DISEASES OF THE LOWER RESPIRATORY TRACT

Lung Infections

Pneumonia (table 17.4) is an inflammation of the lungs, with accompanying fluid buildup in the alveolar sacs that can result from infectious or noninfectious processes. Bacterial pneumonia caused by pneumococci, pyogenic cocci, and bacilli appears suddenly and is characterized by high fever, chest pain, chills, and a purulent (pus-containing) cough. If therapy is not initiated promptly, these infections have a high mortality rate. During 1980–1985 the mortality rate stabilized at about 23 cases per 100,000 (fig. 17.4); pneumonias still rank among the ten leading causes of death in the United States, higher than any other infectious disease. There are more than 1 million cases of pneumonia each year in the United States, and the case fatality rate is about 5–15%.

Pneumococcal Pneumonia Pneumonias caused by *Streptococcus pneumoniae* account for more than 50% of all pneumonias and more than 90% of all bacterial pneumonias. The disease is characterized by an acute onset of fever, chills, dyspnea (difficulty in breathing), pleurisy (inflammation of the pleura), and productive cough. The sputum has a purulent discharge and is often spotted with blood.

TABLE 17.4
COMMON DISEASES OF THE LOWER RESPIRATORY TRACT

Name of disease	Causative agent(s)	Group of organisms
Aspergillosis	*Aspergillus* sp.	Fungi
Blastomycosis	*Blastomyces dermatitidis*	Fungi
Coccidioidomycosis	*Coccidioides immitis*	Fungi
Cryptococcosis	*Cryptococcus neoformans*	Fungi
Histoplasmosis	*Histoplasma capsulatum*	Fungi
Influenza	Influenza viruses	Orthomyxoviruses
Hydatid cyst disease	*Echinococcus granulosus*	Cestodes
Paragonomiasis	*Paragonimus westermanni*	Trematodes
Pneumonia	*Streptococcus pneumoniae*	Bacteria (gram + cocci)
(Legionellosis)	*Legionella pneumophila*	Bacteria (gram − rods)
	Klebsiella pneumoniae	Bacteria (gram − rods)
	Streptococcus pyogenes	Bacteria (gram + cocci)
	Haemophilus influenzae	Bacteria (gram − rods)
(Atypical primary)	*Mycoplasma pneumoniae*	Bacteria (wall-less)
	Influenza viruses	Orthomyxoviruses
	Respiratory syncytial	Paramyxoviruses
	Adenovirus	Adenoviruses
	Coccidioides immitis	Fungi
	Histoplasma capsulatum	Fungi
	Blastomyces dermatitidis	Fungi
	Pneumocystis carinii	Protozoa
Pontiac fever	*Legionella* sp.	Bacteria (gram − rods)
Psittacosis	*Chlamydia psittaci*	Chlamydia (gram − rods)
Q fever	*Coxiella burnetti*	Rickettsia (gram − rod)
Tuberculosis	*Mycobacterium tuberculosis*	Bacteria (acid fast rod)
	Mycobacterium sp.	Bacteria (acid fast rod)
Whooping cough	*Bordetella pertussis*	Bacteria (gram − rods)

FOCUS NOSE DROPS TO CURE THE COMMON COLD?

The common cold affects nearly everyone. The multiplication of rhinoviruses (a common cause of colds) on the nasal mucosa causes the characteristic signs and symptoms of the disease. Nasal secretions are an important characteristic of the disease, because they are laden with infectious viruses and serve as the main vehicle of transmission to other individuals.

The traditional treatment of common colds, by drinking plenty of fluids and getting sufficient rest, does little to reduce the nasal secretions and hence the transmissibility of the viruses. Scientists at the University of Virginia and the University of Rochester medical centers apparently have developed a mode of treatment that arrests viral shedding (in the nasal secretions) and viral multiplication. This treatment is effective in reducing the severity of the disease and at the same time eliminating its contagiousness.

The treatment involves the administration of human interferon directly into the nasal cavity. In these studies, the interferon was administered either as a nasal spray or as drops. Either way, the human interferon either prevented the disease altogether or reduced significantly its signs and symptoms. Also, the shedding of infectious viruses in nasal secretions was significantly reduced.

This study represents another example of how physicians and microbiologists can treat an infectious disease more effectively once they know the mode of action of antimicrobials and the way in which microorganisms cause disease.

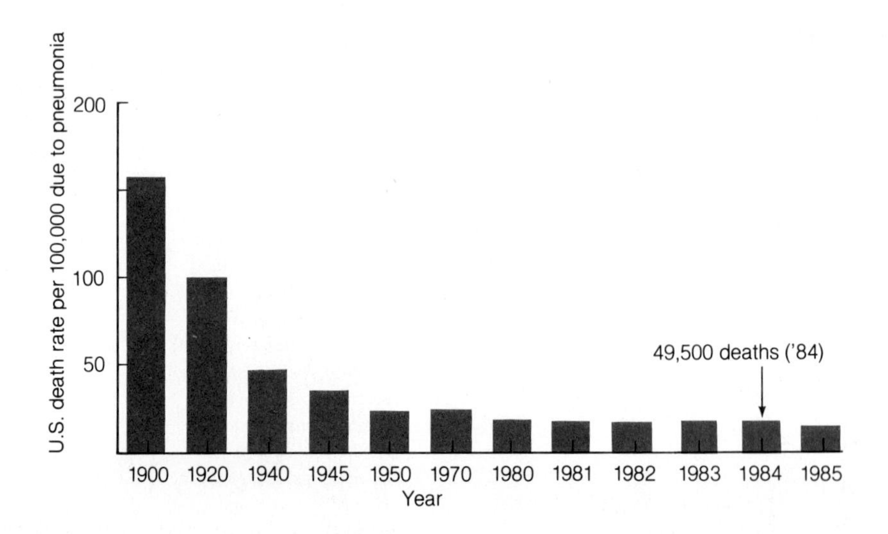

Figure 17.4 Deaths due to pneumonia in the United States. Although the number of deaths due to pneumonia has decreased considerably in the United States since the 1920s, the disease still ranks among the 10 most common causes of death in the United States. The decrease in the number of deaths is due primarily to improved patient care and the use of antibiotics. There are about 1,000,000 cases of pneumonia each year in the United States.

Streptococcus pneumoniae is a gram-positive coccus, characteristically arranged in lancet-shaped diplococci. Not infrequently, it forms chains. Virulent strains of pneumococci are encapsulated. On blood agar, the pneumococcus forms small, mucoid colonies with a central depression and a characteristic alpha hemolysis (fig. 17.5). Laboratory identification of *S. pneumoniae* is based on colony morphology, hemolysis pattern, sensitivity to surfactants, and microscopic and staining characteristics.

The normal human host is generally resistant to pneumococcal infections. Studies show that up to 40% of normal adults carry various strains of pneumococci in their throats. Even when these bacteria gain access to the alveoli, they rarely cause disease in the normal human host, but alcoholics, patients under the influence of anesthesia, and debilitated individuals are highly susceptible. Pneumonia accompanying influenza is sometimes encountered.

Once an infection is initiated in the alveoli, the disease develops along a fairly predictable path (fig. 17.5a). The pneumococci multiply extracellularly within the alveoli. Phagocytized pneumococci that are nonencapsulated are readily destroyed by polymorphonuclear (PMN) leukocytes, but encapsulated organisms are not. The capsule of *S. pneumoniae* inhibits phagocytosis and therefore is a virulence factor.

The reproducing pneumococci produce a variety of enzymes, including a hemolysin called **pneumolysin**, which may also play a role in the pathogenesis of pneumonia. An inflammatory reaction ensues following pneumococcal multiplication and the release of toxic substances into the alveoli. The edema fluid within the alveoli is rich in nutrients and serves to support further growth of the bacterium. Following the edema period, the area becomes infiltrated with PMN leukocytes. Eventually, the alveoli become crowded with bacteria and PMN leukocytes attempting to destroy the bacteria. Some of the bacteria, however, may escape phagocytosis and pass through the thin wall of the alveoli to adjacent areas. The **pleura** (membrane that lines the lungs) may be invaded, causing empyema (pus in the lungs) and pleurisy. In severe cases, the pneumococci may penetrate the walls of capillaries surrounding the alveoli, thus invading the bloodstream. From the blood, the pneumococci may invade other organs of the body, causing meningitis, encephalitis, or myocarditis. Recovery from pneumonia is associated with the development of anticapsular antibodies. These antibodies act as opsonins, enhancing phagocytosis of these bacteria by PMN leukocytes and macrophages.

Most cases of pneumococcal pneumonia are due to endogenous bacteria originating in the upper respiratory tract that spread from the URT into the lungs. The remaining cases of pneumococcal pneumonia are due to intimate contact with infected persons. Since *S. pneumoniae* readily dies when it dries out, aerosols and fomites are unimportant in the spread of this bacterium in human populations. A vaccine consisting of capsular material from the 14 most common serotypes (>75% of all the cases) has been developed and is successful in reducing the incidence rate of pneumonia, especially among high-risk individuals.

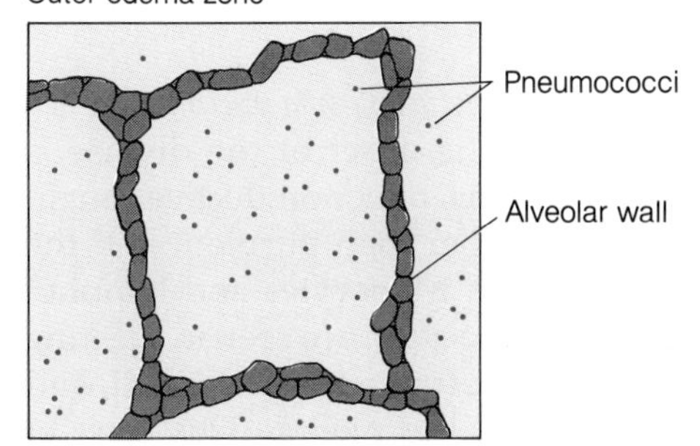

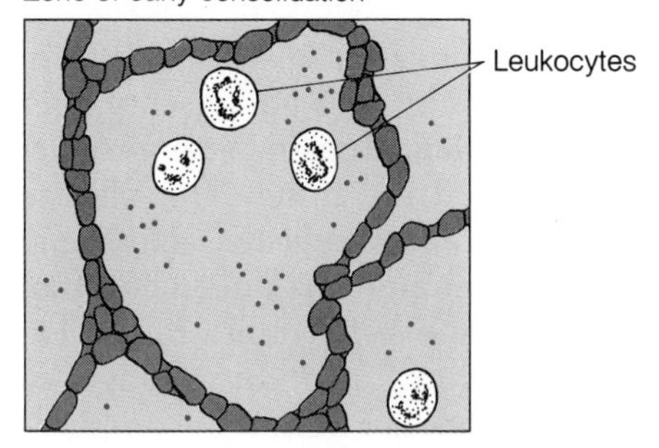

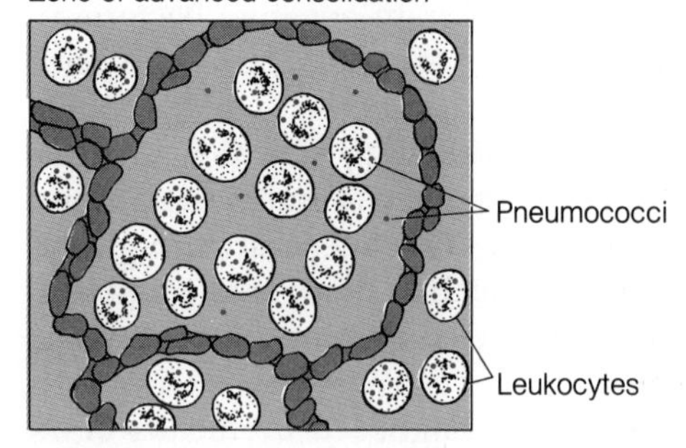

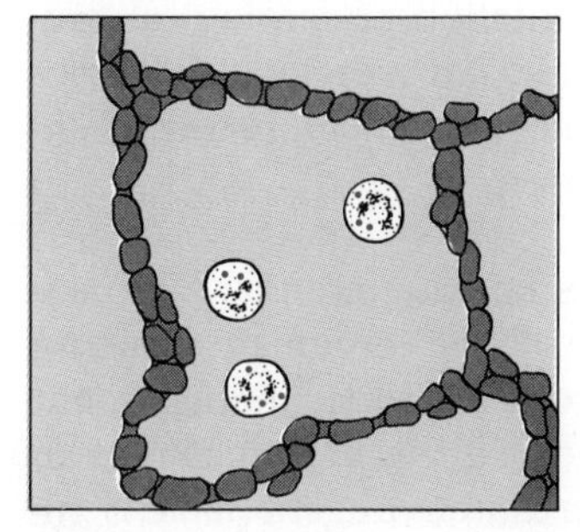

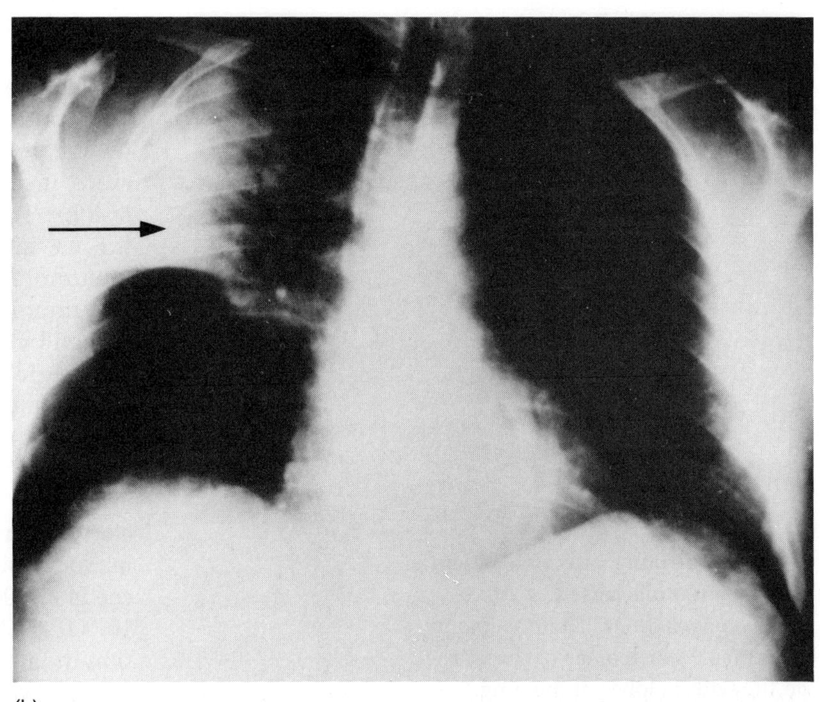

(c)

Figure 17.5 Pneumococcal pneumonia. *(a)* The pneumonia spreads from the site of infection outward until it involves the entire alveolus and adjacent alveoli. At the peak of infection, the alveoli are filled with leukocytes and fluids due to the inflammatory response that results from the damage caused by the invading pneumococci. *(b)* Lung x ray of a patient exhibiting pneumonia. The pneumonia is seen as a whitish region on the right side of the lung. Courtesy of Dr. J. L. Corral Arias. *(c)* Gram stain of *Streptococcus pneumoniae.* Notice the chains of gram-positive cocci that are formed by the pneumococcus. Courtesy of Microbiology Department, University Hospital, Seville, Spain.

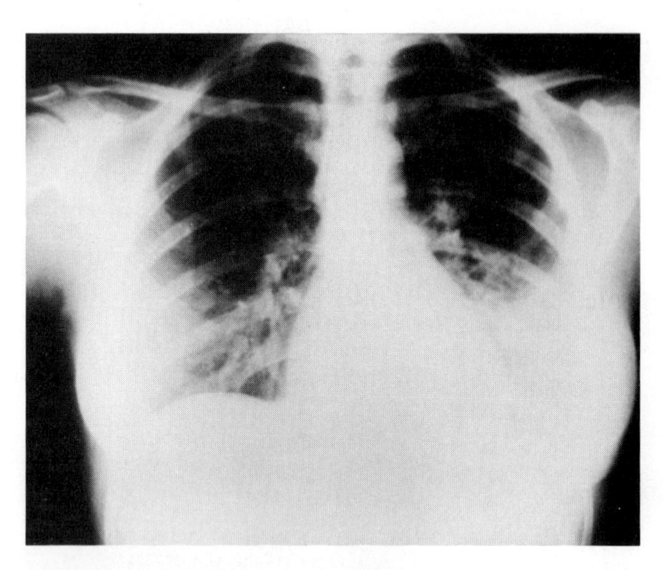

Figure 17.6 Lung x ray of a patient with pneumonia caused by *Mycoplasma pneumoniae*. The pneumonia is seen as whitish areas of the x ray on the bottom lobes of the lung. Courtesy of Dr. J. L. Corral Arias.

Pneumonias Due to *Haemophilus, Klebsiella, Mycoplasma,* and *Legionella* In addition to the pneumococcus, gram-negative bacteria, such as *Haemophilus influenzae* and *Klebsiella pneumoniae*, are capable of causing pneumonia (table 17.4). The pneumonias caused by these bacteria are similar to that caused by the pneumococcus.

Haemophilus influenzae pneumonia is rare in adults and most commonly develops in children after a bout of pharyngitis or a viral infection (e.g., influenza). In 1918, during the great influenza epidemic, *H. influenzae* was isolated from many people with pneumonia. At that time, the viral etiology of influenza was not yet known and so it was thought that this bacterium was the cause of the disease (hence, the name *H. influenzae*).

Klebsiella pneumoniae is a gram-negative bacterium that can cause primary pneumonia in humans. The pneumonia caused by this opportunistic bacterium is associated with chronic alcoholics. The course of the pneumonia caused by *K. pneumoniae* is very similar to that caused by *S. pneumoniae*, although the disease generally is more destructive. For this reason, the mortality rate for untreated cases can exceed 50%.

Pneumonia caused by *Mycoplasma pneumoniae* is a mild disease (fig. 17.6). The onset of the disease is gradual, with fits (paroxysms) of nonproductive cough. The sputum is mucoid rather than purulent and the fever rarely exceeds 102° F. Monocytes and lymphocytes, instead of PMN leukocytes, are seen in the lung lesions. These lesions are restricted to the alveoli, and the bacteria rarely invade the pleura. This type of pneumonia is sometimes called primary atypical pneumonia, and in many ways it resembles those caused by the viruses.

Legionellosis, or Legionnaires' disease, is a form of pneumonia characterized by an acute onset of high fever, bed-shaking chills, pleuritic pain, a nonproductive cough, and accelerated breathing. The disease is usually heralded by a mild headache, achy muscles, and malaise. This pneumonia is unusual in that leukocytosis (elevated white cell count) rarely exceeds 16,000 leukocytes/mm^3, and an average of 10,000–12,000 is seen. Shock is a common complication in fatal cases.

The cause of legionellosis is a group of fastidious (nutritionally demanding) gram-negative bacteria in the family Legionellaceae, of which *Legionella pneumophila* is the best known member.

Epidemiological studies suggest that *L. pneumophila* is an aquatic microorganism found in natural and artificial bodies of water. The agent appears to be transmitted by aerosols, such as those created by ventilation systems that are cooled by contaminated waters in cooling towers.

Legionellosis can be diagnosed by means of an indirect fluorescent antibody test, which detects the presence of anti-legionellae antibodies in the patient's serum. Since the implementation of serological procedures to detect legionellosis, an average of 600 cases per year in the United States have been recorded. In 1984, 651 cases were reported.

Management of Bacterial Pneumonia The isolation and identification of the organism causing pneumonia are essential for the proper management of the disease. This is because the proper choice of antibiotic depends upon the causative agent (table 17.5). Pneumococcal pneumonia can be treated successfully with penicillin, but pneumonias caused by gram-negative bacteria cannot. Pneumonias caused by *H. influenzae* respond well to treatment with chloramphenicol or ampicillin, while those caused by *K. pneumoniae* are best treated with gentamycin and tetracycline. Mycoplasmal pneumonias cannot be treated with penicillin, since they lack a cell wall, and are

TABLE 17.5
ANTIBIOTIC TREATMENT OF PNEUMONIAS

Organism	Antibiotic
S. pneumoniae	Penicillin
K. pneumoniae	Gentamycin Tetracycline
M. pneumoniae	Erythromycin
H. influenzae	Chloramphenicol Ampicillin
L. pneumophila	Erythromycin Tetracycline

best treated with erythromycin. Erythromycin and tetracycline are the drugs of choice for the treatment of legionellosis.

Pertussis A highly infectious disease, characterized by fits of coughing with a "whooping" sound and an incubation period of about 10 days (5–21 days) is known as **whooping cough** or **pertussis**. During the prodromal period (before the characteristic signs and symptoms appear), which lasts one to two weeks, the patient has coldlike symptoms, after which the characteristic cough appears. Each fit of coughing may last 10–20 seconds. The shortness of breath due to prolonged coughing forces the patient to inhale air forcibly, making the whooping noise as the air rushes in through the epiglottis.

The cause of whooping cough is *Bordetella pertussis*, a small, gram-negative coccobacillus. This bacterium produces a capsule that is partly responsible for its virulence. *Bordetella pertussis* also produces a toxin known as **tracheal cytotoxin**. This toxin is a peptide fragment from the *B. pertussis* cell wall. A number of bacterial enzymes cut the peptide from the cell wall during log phase growth. When exposed to the peptide toxin, ciliated cells lining the trachea cease to beat rhythmically, DNA synthesis is inhibited, cell metabolism is extensively modified, and the ciliated cells become detached from the epithelium. The inhibition and loss of ciliated cells causes the deep continuous cough characteristic of pertussis and blocks the movement of mucus. The mucus and inflammation block the airways and promote secondary pulmonary infections. Because the ciliated cells of the trachea are inhibited and the epithelium is disrupted for many days after exposure to the peptide toxin, coughing persists for up to a week after elimination of the pathogen.

The infection begins when inhaled *B. pertussis* attaches to the ciliated epithelium of the bronchi (fig. 17.7). This attachment may be mediated by pili. At the site of attachment, the bacteria multiply, causing the prodromal signs and symptoms. As the bacteria multiply locally, they elaborate their toxin(s). During the period of paroxysmal (fits of) coughing, there is a marked lymphocytosis (increase in the number of lymphocytes in the blood), subepithelial necrosis, and inflammation of the bronchi.

Humans, the only reservoir of *B. pertussis*, transmit the disease by aerosols created during coughing. The disease is worldwide in distribution, with no apparent seasonal variation. Since the implementation of vaccination (DPT shots), the incidence of this disease in the United States has dropped to an average of fewer than 4000 cases/year and a mortality rate of $<1\%$ (<30 deaths/year). The vaccine usually consists of a crude extract of *Bordetella*.

Bordetella pertussis is sensitive to erythromycin, the drug of choice. An understanding of the disease process is essential for its proper management, since the antibiotics kill the bacteria but do not inactivate the toxins already present. Therefore, prompt therapy reduces the severity of the disease but does not eliminate the signs and symptoms entirely, since any exotoxin present, no matter how little, has some effect on the host's tissue. Administration of **human pertussis immune serum globulin** can offset the effects of the toxins.

Tuberculosis **Tuberculosis** is a chronic infectious disease of the lower respiratory tract. In its most common form, the disease is characterized by a chronic cough, low-grade fever, and malaise. The long-lasting nature of this syndrome can lead to prostration. Patients with tuberculosis are constantly shedding infectious agents in aerosols.

Tuberculosis caused by *Mycobacterium tuberculosis*, remains the number 1 killer disease in the world. There are an estimated 1.5 billion infected individuals throughout the world; of those, 20 million are shedding the bacteria, causing 3–5 million new cases annually. The mortality world-wide from tuberculosis is about 600,000 deaths/year; however, countries with a high standard of living have a relatively low incidence of the disease. The morbidity in the United States is about 30,000 cases per year, with approximately 10% mortality (fig. 17.8).

(a)

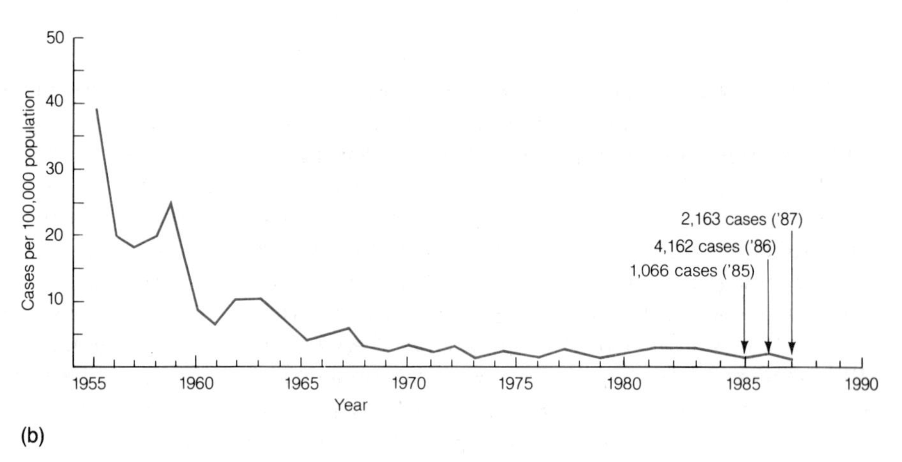

(b)

Figure 17.7 Characteristics of whooping cough. *(a) Bordetella pertussis* reproducing amidst ciliated epithelium. The bacteria multiply locally, releasing toxic substances that cause the characteristic disease. *(b)* Cases of pertussis in the United States. Since the 1970s the rate has been between 1–3 cases/100,000 population (2000–6000 cases/year in the United States). The sharp reduction in cases since the early 1950s is due largely to the compulsory use of the DPT vaccine.

Humans are capable of mounting a significant and effective immune response against the highly infectious **tubercle bacillus.** Unfortunately, however, the host's immune response is responsible for much of the tissue destruction associated with some forms of tuberculosis.

Mycobacterium tuberculosis is a slender, acid-fast rod containing a high concentration of lipids in its wall. The bacterium is an obligate aerobe, and in host tissues it multiplies inside cells (intracellular parasite). Some strains have a tendency to stick together and form colonies, showing a **serpentine growth** on artificial media. The tubercle bacillus does not produce

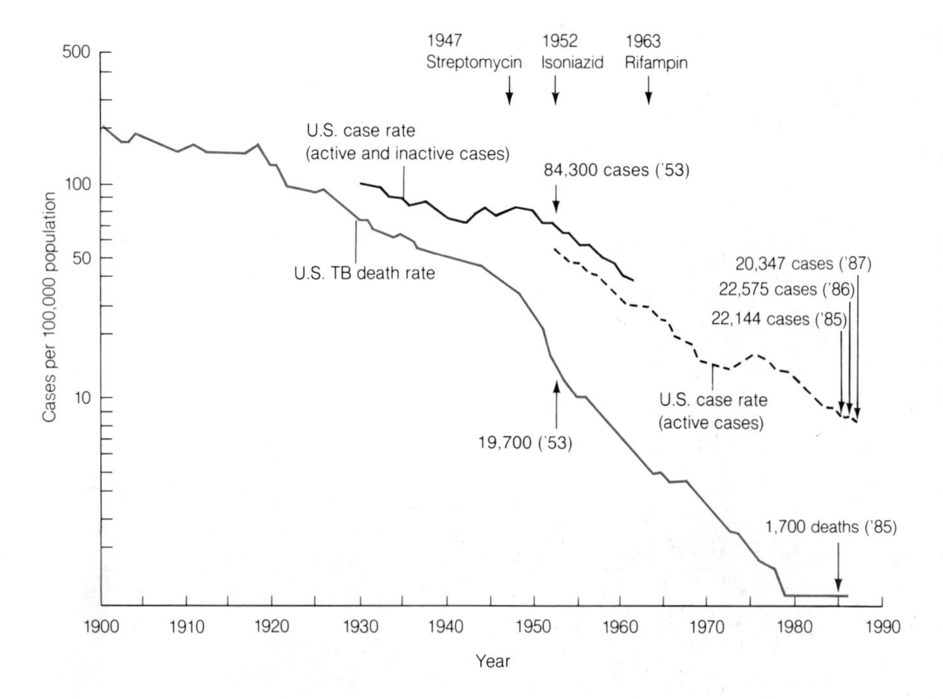

Figure 17.8 Tuberculosis in the United States. The graph illustrates that tuberculosis is still prevalent in the United States, although the number of cases/year is declining. The salient feature of the graph is that the death rate due to tuberculosis is declining at an encouraging rate. The rapid decline in the death rate began in the early 1950s when combined drug therapy with streptomycin and isoniazid was initiated.

invasive enzymes or toxins, and its pathogenicity resides in its ability to induce a strong immune response and to avert these responses by growing inside mononuclear phagocytes.

The transmission of tuberculosis is primarily through aerosols created by people with the active disease. During the initial stages of tuberculosis, the tubercle bacillus multiplies inside alveolar macrophages in the lungs. After this initial period of multiplication, the microorganisms enter the lymphatics and are transported to the regional lymph nodes, where they continue to multiply intracellularly. After a period of multiplication, the bacteria exit the lymph nodes and enter the circulation through the thoracic duct. From there, they spread to other organs, including the lungs. At this stage, the infected host may not show any signs or symptoms. If symptoms are evident, they are due to inflammatory responses to tissue damage during bacterial multiplication. If the lung is the site of the primary infection, there may be a mild, transient pneumonia. Up to this point, the organism has been virtually unchallenged by the host. Now, the host develops a cell-mediated immunity that is tuberculocidal, and the bacilli are attacked by T-lymphocytes and activated macrophages. The cell-mediated immunity in a healthy individual can efficiently control the spreading infection.

Soon after cell-mediated immunity is detected, the host responds to the infection by walling off the pathogen from the rest of the body within multinucleated giant cells surrounded by lymphocytes and fibroblasts. These structures, which can be seen on x-ray examinations, are called **tubercles** (fig. 17.9). The tubercle bacilli become localized in foci, primarily in the lungs and in other organs where the oxygen concentration is high (spleen and liver). Dormant bacteria are able to remain viable in the tubercles for many years. The tubercles may eventually undergo **caseation necrosis**, acquiring a cheesy consistency, and may ultimately become scarred or calcified. These structures are called **Ghon complexes** (fig. 17.9).

Reactivation tuberculosis is the clinical manifestation that most people associate with tuberculosis. It occurs in individuals who have cell-mediated immunity to *M. tuberculosis* antigens and who suffer a temporary lapse in their immunity or are reinfected. In either case, the bacteria in the tubercles begin to multiply, stimulating the anti–*M. tuberculosis* immune response in the host. This response causes caseation necrosis of the tubercles. This partial destruction of

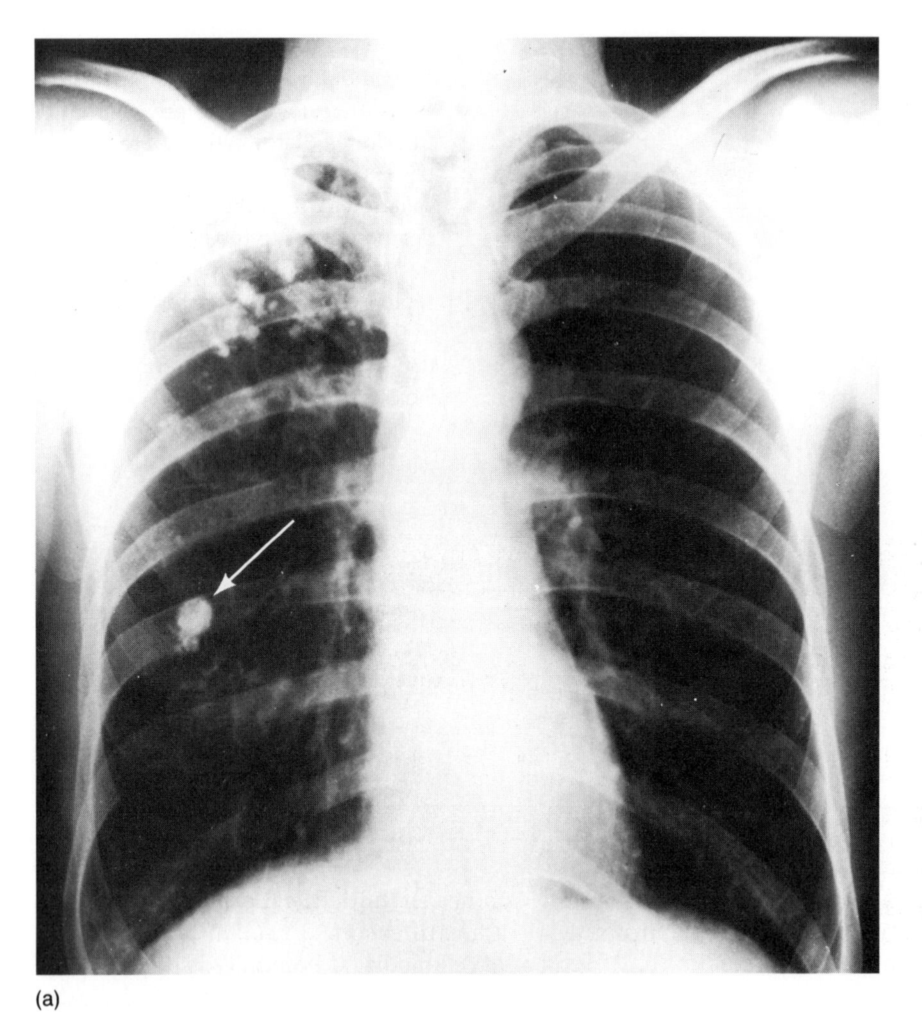

(a)

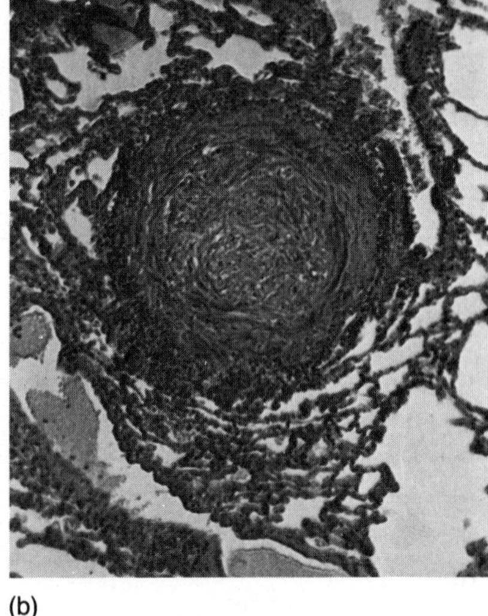

(b)

Figure 17.9 Clinical manifestations of tuberculosis. *(a)* X ray showing a large tubercle (arrow) and several smaller ones in the right lung. The left lung appears normal. (b) Histological section of a tubercle stained hematoxylin and eosin. The center of the tubercle contains *Mycobacterium tuberculosis* and epithelioid cells.

the tubercles provides a suitable environment for the bacteria to multiply, and this multiplication further stimulates the immune response. As the reactivation disease progresses, the lesion(s) enlarge and eventually erode through the bronchi, which then fill with the liquid caseum. The fluid in the bronchi, laden with infectious mycobacteria, induces the characteristic cough. The ruptured tubercles provide a cavity in which the bacilli continue to multiply. Tubercles that develop in the spleen, liver, or other organs can also undergo similar caseation necrosis, thus damaging the organ.

Vaccination of human volunteers with **BCG** (Bacillus Calmette-Guerin) has reduced the incidence of tuberculosis by as much as 80% in some developing countries. Presumably, BCG induces a strong cell-mediated immune response and promotes the activation of macrophages, which play a central role in the control of the disease. It has been suggested that BCG vaccination be initiated in endemic areas to reduce the number of cases of tuberculosis.

Patients with positive skin tests generally are treated with isoniazid (INH) daily for up to one year to prevent reactivation disease. Active infections are treated with a combination of agents such as INH, ethambutol, rifampin, streptomycin, and para-amino salicylic acid (PAS), because of the possibility that the tuberculosis has been caused by drug-resistant bacteria. In active cases, prolonged bed rest and wholesome nutrition help expedite recovery.

VIRAL DISEASES OF THE LOWER RESPIRATORY TRACT

Lung Diseases Due to Viruses

Viral pneumonia (table 17.3) is an acute systemic disease that can be caused by a variety of viruses, including influenza, adeno-, and respiratory syncytial viruses. The disease occurs most often in adults with chronic lung and heart diseases, or as a complication of other infections such as influenza or chickenpox. After influenza-like symptoms, the patient develops a fever (about 102°F), experiences difficulty in breathing, and has a cough. The sputum is usually frothy and tinged with blood. The severity of the disease varies, depending upon the individual infected and the virulence of the virus. A diagnosis of viral pneumonia is usually made after attempts to isolate and identify bacterial and fungal pathogens have failed. It has been estimated, using the above criteria, that 15–30% of all pneumonias are of viral origin.

Influenza (grippe), or "flu," is an acute infection of the lower respiratory tract characterized by a sudden onset of fever, chills, myalgia (muscle ache), and headache. Coldlike symptoms, such as coryza (nasal inflammation with profuse discharge), sore throat, and cough, are also common. Generally speaking, the disease is benign, self-limiting, and of short duration. On occasion, "stomach flu" symptoms may appear: loss of appetite, diarrhea, and vomiting. **Reye's syndrome**, a degenerative disease of the central nervous system, has been implicated with certain strains of influenza and parainfluenza viruses. This syndrome may be initiated by ingestion of aspirin.

Influenza is caused by a group of RNA viruses called the orthomyxoviruses. The surface of the virion contains two major components: **hemagglutinin** and **neuraminidase receptors**. These components play very important roles in the pathogenensis and epidemiology of the disease.

The virus attaches to epithelial cells of the respiratory tree via the hemagglutinin receptors on its surface. Local multiplication of the virus results in the destruction of the ciliated epithelium. The virus spreads to adjacent sites and eventually causes considerable denudation (destruction of ciliated epithelium) of the trachea and bronchi; irritation of the airways probably induces the cough. After a period of local multiplication, the virus may spread hematogenously (via the blood). The viremia (virus in the blood) may be responsible for the fever, myalgia, and malaise associated with the illness. Viruses are shed in the cough during the early stages of the clinical disease and disappear shortly after the signs and symptoms disappear. The recovery of the patient is due to a rise in humoral immunity, cell mediated immunity, and interferon activity.

Influenza is transmitted from person to person by virus-laden aerosols created by coughing and sneezing. High incidence rates occur in institutions that harbor individuals in close proximity. Such outbreaks occur in army barracks, school classrooms, and hospitals. High attack rates are more common among children than adults, perhaps because children are immunologically naive with respect to the virus. Influenza epidemics are common during the winter months, possibly because susceptible individuals spend more time indoors and in close proximity with infectious individuals.

The periodicity of influenza epidemics is believed to be due to major antigenic shifts in the hemagglutinin (H) and neuraminidase (N) receptors of the viral coat. When a new HN type of influenza virus appears, the majority of the population is susceptible to the virus. After infection, immunity is developed against the new virus antigens, and eventually the population becomes resistant to that virus. When another HN type appears, much of the population has no antibodies against it and therefore is again susceptible to influenza. Treatment of influenza is generally supportive to alleviate the signs and symptoms. In cases in which secondary bacterial infection is suspected or expected, antibiotics are administered.

FUNGAL DISEASES OF THE LOWER RESPIRATORY TRACT

San Joaquin Valley Fever (Coccidioidomycosis)

Coccidioidomycosis is a fungal infection of the lungs. Many of the infections (about 60%) are asymptomatic (no signs or symptoms) and are detected by indirect means such as routine x rays, skin testing, or serology. Approximately 35% of the people who become infected have symptoms that are quite similar to (and often mistaken for) influenza. Sometimes, during the later stages of the disease, rashes or rheumatism-like symptoms appear. In some cases, full recovery may take many months. The remaining 5% or so who become infected develop chronic pulmonary disease, which may disseminate to the joints, bones, central nervous system, and skin. Chronic disease is characterized by a low-grade fever, anorexia (weight loss), weakness, a productive cough, and occasionally hemoptysis (blood in the sputum). Since coccidioidomycosis is not a reportable disease in all states, it is difficult to estimate how many cases there are per year, but it is estimated that between 500 and 1000 new cases occur per year in endemic areas with a mortality rate of 1–5%.

The cause of San Joaquin valley fever is *Coccidioides immitis*, a fungus with a complex life cycle that is commonly found in the soil in a filamentous form

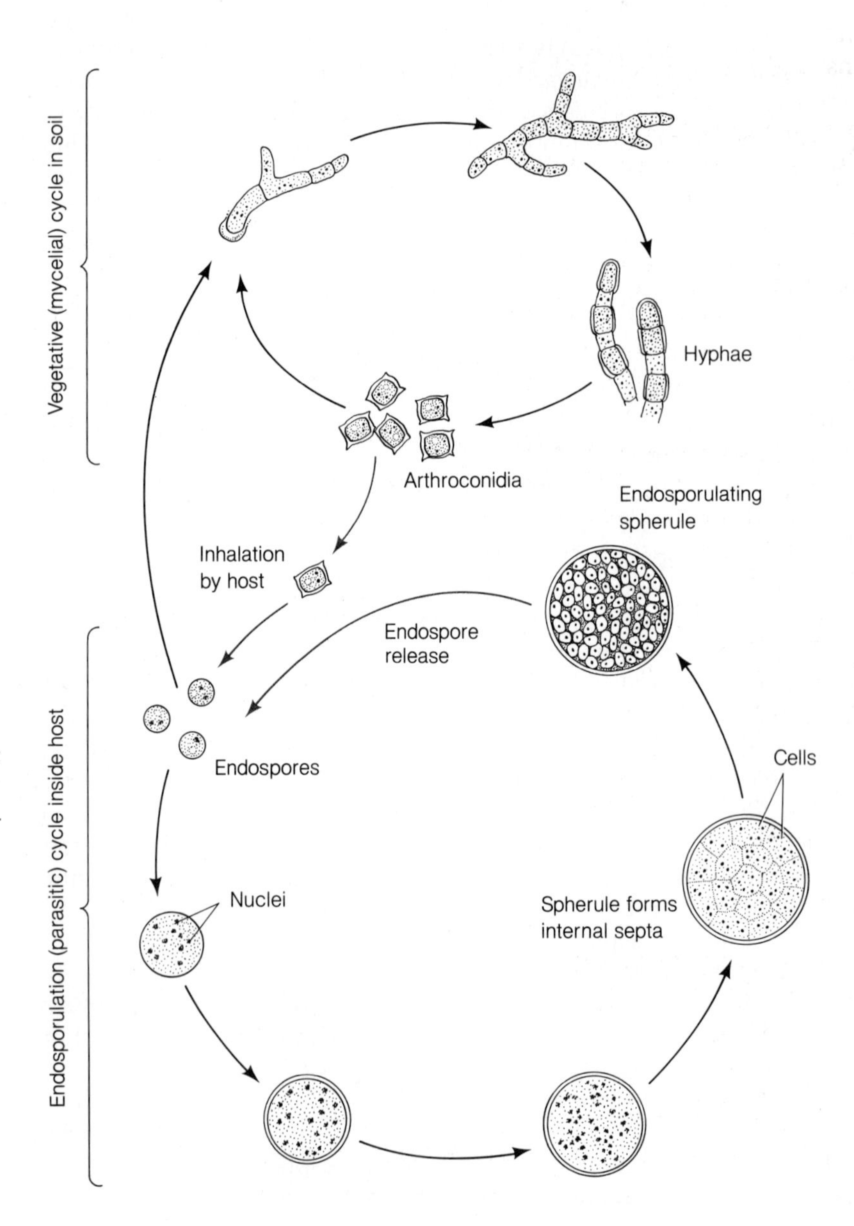

Figure 17.10 Life cycle of *Coccidioides immitis*. *Coccidioides immitis* has two life cycles: a vegetative cycle in soils (mycelial form) and a parasitic cycle in susceptible mammalian hosts (endosporulating spherule form). Human infection occurs when arthroconidia of the fungus are inhaled. Inside the host's lung, the arthroconidia develop into endospores and eventually into endosporulating spherules. The endosporulating cycle is perpetuated inside the host. When endospores are released into the external environment, they develop into vegetative hyphae and the vegetative cycle takes place.

(fig. 17.10). The complex life cycle of this fungus may be of significance in the pathogenesis of the disease. The constantly changing surface components may serve as an evasive mechanism to avert the host immune responses. In addition, the various stages of the life cycle of *C. immitis* produce soluble components that may participate in the disease process as well.

Nearly everyone who comes in contact with *Coccidioides immitis* becomes infected, yet 60% of those who become infected show no symptoms and 35% have flulike symptoms and nothing more. These facts suggest that humans are highly resistant to serious infections by this fungus. Studies have revealed that activated macrophages play a central role in protecting against dissemination. Most of the patients who exhibit the severe form of the disease have either a temporary or a long-lasting depression of their cell-mediated immunity (CMI).

Arthroconidia (a type of spore) adhering to lung tissue initiates an inflammatory response that may be responsible for the flulike symptoms of the early stages of the disease. Although polymorphonuclear leukocytes migrate rapidly to the site of infection, they are unable to destroy all the arthroconidia. The surviving arthroconidia rapidly differentiate into **endosporulating spherules** (fig. 17.10), which eventually rupture and release numerous **endospores**. This event, in turn, induces the migration of yet more PMN leukocytes to the foci of infection. Some of the endospores are killed by the PMN leukocytes. Eventually, the PMN leukocytes are replaced by monocytes and macrophages. Endospores are transported to the regional lymph nodes, where they stimulate the development of cell-mediated immunity. This development leads to a chronic inflammation in the affected portions of the lung, characterized by the abundance of lymphocytes and macrophages.

Once cell-mediated immunity has developed, the infection is checked and recovery usually follows. The lesions eventually heal and are reabsorbed by the body. The formation of cavities with caseation necrosis, similar to those seen in tuberculosis, are sometimes observed in valley fever. "Desert bumps," rashes, and rheumatism that sometimes accompany valley fever are believed to be due to allergic reactions to circulating fungal antigens.

Coccidioides immitis (fig. 17.11) is endemic to the southwestern United States and arid regions of Mexico and California. These areas are characterized by arid or semiarid conditions, hot summers, and mild winters, with rainfall between 10 and 20 inches annually. The fungus is transmitted by air currents carrying arthroconidia. On windy days, large numbers of arthroconidia become airborne and can lead to serious consequences for animals and humans. In December 1977, a severe dust storm blowing from the southeast carried arthroconidia to northern and central coastal California. An increase of coccidioidomycosis was noted in areas not usually colonized by the fungus. Coccidioidomycosis in humans and animals (including a sea otter) was reported in California coastal communities. Presumably, burrowing mammals such as gophers, ground squirrels, and kit foxes may also serve as reservoirs for the infectious agent in nature.

Most cases of coccidioidomycosis recover without antifungal treatment; however, 60–70% of all cases of untreated, disseminated coccidioidomycosis are fatal. Hence, treatment with amphotericin B is required to reduce the mortality rate. The imidazole antifungal agents, such as miconazole and ketoconazole, have been used in clinical trials and appear to be as effective as amphotericin B but less toxic.

Pneumonia Due to *Histoplasma capsulatum*

Histoplasmosis is a systemic disease characterized by infection of cells of the reticuloendothelial system by the dimorphic fungus *Histoplasma capsulatum*. The clinical outcome of the host-parasite relationship is quite varied. More than 75% of those who become infected show either no symptoms or those of influenza. The other 25% show various degrees of severity in their illness. The most common syndrome is a pneumonia of short duration. In these patients, a nonproductive cough, pleurisy, dyspnea (difficulty in breathing), myalgia (muscle ache), arthralgia (joint pain), fever, and night sweats are common symptoms.

The mycelial form of *H. capsulatum* grows in soils, especially those enriched with bird droppings or bat guano. On laboratory culture media such as Sabouraud glucose agar at room temperature, *H. capsulatum* forms characteristic macro- and microconidia. When the microconidia or other fungal elements are inhaled into the lungs, they develop into yeast cells, facultative intracellular parasites that multiply inside macrophages (fig. 17.12). The fungi are transported in the macrophages to the lymph nodes, where they multiply further. Many escape the lymph nodes, however, and enter the circulatory system, disseminating

Figure 17.11 Characteristics of coccidioidomycosis (valley fever). *(a)* Cutaneous manifestations of disseminated coccidioidomycosis. *(b)* Vegetative phase of *Coccidioides immitis.* Notice the characteristic barrel-shaped spores (called arthroconidia) that alternate with an empty cell in the hyphae. *(c)* Stained spherule (tissue phase) of *Coccidioides immitis* as normally seen in the lungs of infected individuals.

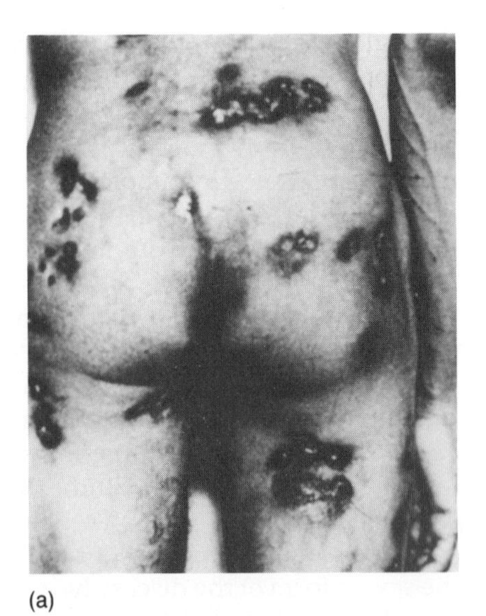

(a)

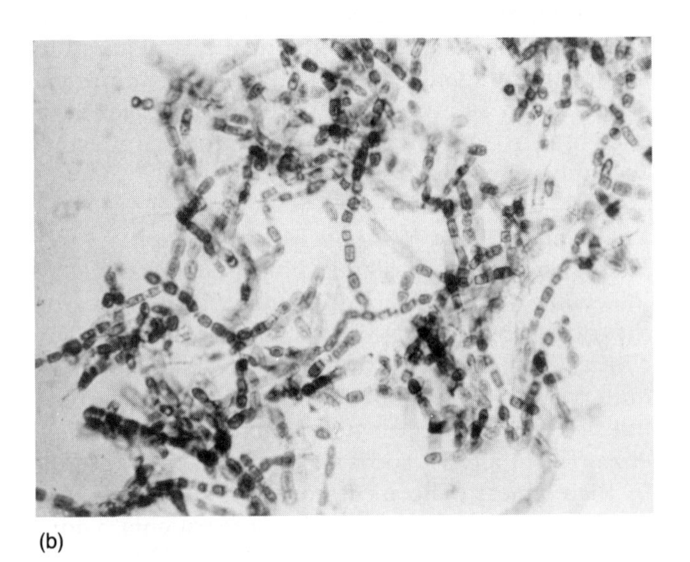

(b)

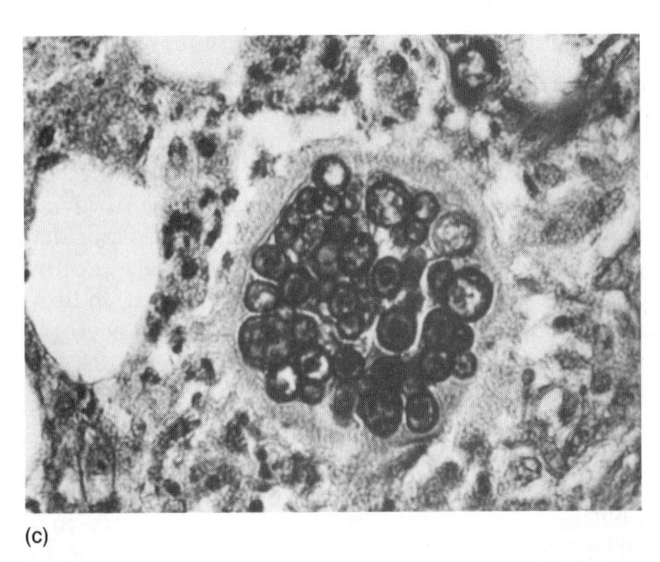

(c)

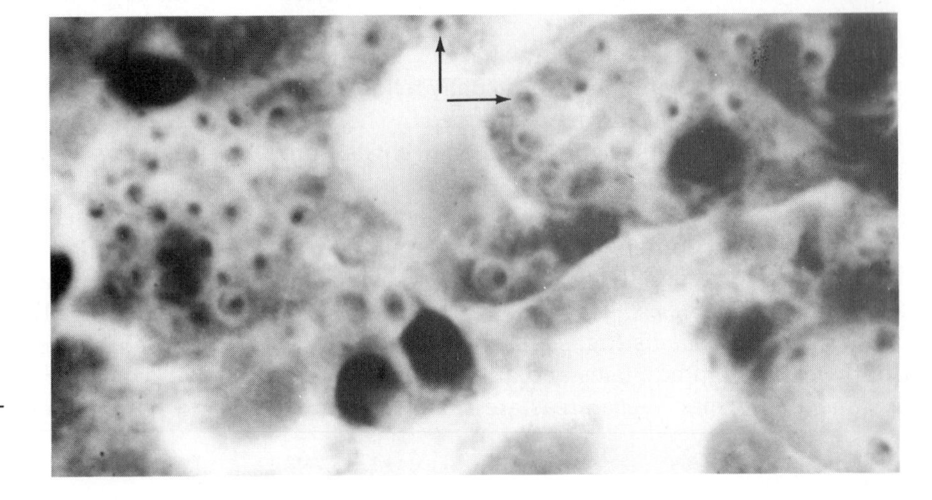

Figure 17.12 *Histoplasma capsulatum* in histiocyte. Hematoxylin and eosin stain of lesion, showing histiocyte filled with *Histoplasma capsulatum.* Courtesy of the Biological Sciences Department, California Polytechnic State University, San Luis Obispo.

throughout the body. Within a week or so, the host develops a cell-mediated immunity that may involve fungicidal T-lymphocytes and activated macrophages. The development of cell-mediated immunity is paralleled by a chronic inflammation around foci of infection. The fungi are quickly destroyed, and the lesions eventually disappear or become calcified. On occasion, cavities develop, containing dormant fungi that may become reactivated at a later date. Reactivation histoplasmosis resembles tuberculosis.

More than 500,000 people in the United States become infected every year with *Histoplasma capsulatum*, most of them in endemic areas: moist, humid regions such as the Mississippi and Missouri river valleys. The infectious stage is found in soils that have been enriched with chicken, bird, or bat feces. In urban environments, where there are many pigeons, *Histoplasma capsulatum* can be found in bird droppings. Thus, persons feeding pigeons or disturbing their roosts are in danger of contracting the disease. Construction and demolition workers who disturb contaminated soil, creating dust clouds, are also at risk. Airborne fungi can enter buildings through windows or air conditioning systems and infect people inside. Severe cases of histoplasmosis are treated with amphotericin B. Miconazole and ketoconazole are being used experimentally to treat the severe infections, with encouraging results.

CLINICAL PERSPECTIVE RESPIRATORY DISTRESS AND MORE

A 33-year-old woman came to the emergency room of a hospital because of shaking chills, fever, a nonproductive cough, and chest pain. Upon examination, the woman was found to have a fever of 103°F, a pulse of 135 (normal 72), blood pressure of 110/60 (normal 120/80), and to be in severe respiratory distress with 40 respirations per minute (normal 15). An electrocardiogram indicated a rapid heart beat (tachycardia) without any wave changes. A Gram stain of her sputum showed a large number of polymorphonuclear cells and gram-positive diplococci.

Intravenous penicillin was administered after sputum and blood cultures were obtained. Nevertheless, the patient did not respond to antibiotic treatment after 36 hours. The patient remained feverish and had difficulty breathing. A chest puncture gave a purulent fluid with gram-negative rods. The preliminary sputum culture and the blood cultures were positive for *Haemophilus influenzae* type B. These observations suggested that a bacterial pneumonia had become a septicemia (blood infection) and a purulent pericarditis (inflammation of the membrane surrounding the heart). Upon determination of the pathogen, penicillin treatment was replaced by gentamicin, and cefotaxime therapy.

On the third day of treatment, the patient showed signs of a worsening condition. Aspiration of 240 ml of a purulent yellow fluid from the chest changed her pulse from an irregular 140 to a regular 95 (normal 72). A pericardial drainage tube was inserted to draw off accumulating fluid around the heart. With the discovery that the *Haemophilus* was sensitive to gentamicin, cefotaxime, and ampicillin, treatment was changed to ampicillin alone.

After four days (hospital day seven) of ampicillin treatment, improvement was seen. Chest tubes were removed after 14 days in the hospital. However, on hospital day 18, a diffuse rash developed on the patient. To eliminate this, the ampicillin was replaced by cefotaxime. After 28 days of therapy, the patient was discharged in good condition and remained asymptomatic for more than six months.

This patient illustrates that purulent pericarditis can develop from pneumonia. In this case *Haemophilus influenzae* within an infected lung may have spread to the adjacent pericardium. Control of the disease requires immediate antimicrobial treatment with a broad-spectrum antibiotic until the culture sensitivities can be determined. The chest should be drained in cases of purulent pericarditis.

For further information on pneumonia and purulent pericarditis consider the following article. Danowitz, J.S. and Reza, R. (1986). A healthy young woman in respiratory distress. *Hospital Practice* 21(6): 200–202.

SUMMARY

STRUCTURE AND FUNCTION OF THE RESPIRATORY TRACT

1. The respiratory tract, consisting of the upper and lower tracts, brings air into contact with blood vessels, where an exchange of oxygen and carbon dioxide takes place.
2. Because the potential for infection via the respiratory tract is great, mammals have evolved an array of host defense mechanisms designed to abort infections.

THE NORMAL FLORA OF THE UPPER RESPIRATORY TRACT

1. The upper respiratory tract is richly colonized with microorganisms such as bacteria and fungi. Certain of these colonizers are potential pathogens that can initiate an infection if conditions are conducive to their extensive growth.
2. The normal flora protects the host from infections by inhibiting potential pathogens.
3. The lower respiratory tract is generally free of microorganisms, and microorganisms entering the lungs have a great potential for causing disease.

BACTERIAL DISEASES OF THE UPPER RESPIRATORY TRACT

1. Pharyngotonsillitis or "strep throat" is an infectious disease usually caused by Group A streptococci. The disease is characterized by a sore throat, malaise, and fever, which result from an acute inflammation due to the destruction caused by the reproducing bacterium.
2. Untreated streptococcal infections may give rise to autoimmune diseases, such as rheumatic fever and acute glomerulonephritis.
3. Diphtheria is a serious infectious disease characterized by fever, headache, malaise, sore throat, and the development of a pseudomembrane in the throat. It is caused by *Corynebacterium diphtheriae*, which produces a toxin called diphtherotoxin. The toxin, as well as the damage caused by the reproducing bacterium, cause the characteristic signs and symptoms of diphtheria.
4. Sinusitis is an inflammation of the sinuses as a result of hampered drainage. It is generally caused by *Haemophilus influenzae, Streptococcus pneumoniae*, and Group A streptococci.
5. Acute epiglottitis is a serious inflammation of the epiglottis as a consequence of microbial growth. The onset of the disease is rapid, and it can lead to death due to asphyxiation. It is most commonly caused by *Haemophilus influenzae.*

VIRAL DISEASES OF THE UPPER RESPIRATORY TRACT

1. The common cold is a mild illness characterized by headache, malaise, stuffy nose, sore throat, and nasal discharge.
2. The common cold is caused by a variety of different viruses, including rhino-, corona-, myxo-, coxsackie-, respiratory syncytial, and adenoviruses.
3. The signs and symptoms of the common cold result from both allergic reactions to viral antigens and tissue destruction caused by reproducing viruses.
4. Transmission of the common cold is generally via person to person contact or contact with objects contaminated with nasal secretions of infected individuals.
5. Viruses such as the adeno- and respiratory syncytial viruses cause a variety of other diseases of the upper respiratory tract. One such disease is croup, which leads to difficulty in breathing and can have serious complications.

BACTERIAL DISEASES OF THE LOWER RESPIRATORY TRACT

1. Pneumonia is an inflammation of the lungs, accompanied by a buildup of fluids, and ranks in the top ten of the killer diseases in the United States.
2. *Streptococcus pneumoniae* is the most common cause of bacterial pneumonia. It causes a fulminant pneumonia characterized by a rapid onset of high fever, chills, dyspnea, and productive cough.
3. Most cases of pneumococcal pneumonia are caused by endogenous pneumococci that spread from the upper respiratory tract.
4. Other bacteria such as *Haemophilus influenzae, Klebsiella pneumoniae*, and *Mycoplasma pneumoniae* cause pneumonias as well.
5. Legionellosis is a form of pneumonia characterized by a rapid onset of fever, bed-shaking chills, pleuritic pain, nonproductive cough, and accel-

erated breathing. It is caused by several species in the genus *Legionella*. The bacteria enter the lower respiratory tract via aerosols.

6. Whooping cough is a highly infectious disease characterized by fits of coughing. It is caused by *Bordetella pertussis*. The disease is caused by exotoxins produced by the bacteria while attached to the ciliated epithelium of the respiratory tract.
7. Tuberculosis is a chronic infectious disease caused by *Mycobacterium tuberculosis*. It is characterized by a chronic cough, low-grade fever, and malaise. The infectious bacteria are constantly being shed in the cough of patients.
8. The tubercle bacillus multiplies inside cells of the reticuloendothelial system and induces a strong immune response.
9. The immune response toward the tubercle bacillus is responsible for the control of the infection. Upon reinfection, however, it is the immune response that causes the characteristic signs and symptoms of the disease.

VIRAL DISEASES OF THE LOWER RESPIRATORY TRACT

1. Viral pneumonia is an acute infection of the lungs caused by a variety of viruses, including influenza, adeno-, and respiratory syncytial viruses.
2. Influenza (grippe) is an acute infection of the lower respiratory tract, resulting in a rapid onset of fever, chill, achy muscles, and coldlike symptoms.
3. Influenza is caused by orthomyxoviruses, primarily Influenza A and B. The influenza A virus has hemagglutinin (H) and neuraminidase (N) receptors that promote its pathogenesis.
4. Influenza is transmitted from person to person via droplet nuclei.
5. The periodicity of influenza epidemics is believed to be due to major antigenic shifts in the H and N receptors of the viral membrane envelope.

FUNGAL DISEASES OF THE LOWER RESPIRATORY TRACT

1. Coccidioidomycosis is a fungal infection of the lungs caused by *Coccidioides immitis*. The disease is generally mild with flulike symptoms, although at times the fungus may spread from the lungs, causing serious and sometime fatal disease.
2. *Coccidioides immitis* is a dimorphic fungus that enters the lungs as arthrospores via inhalation of dust particles. The arthrospores develop into endosporulating spherules.
3. Histoplasmosis is a fungal infection caused by *Histoplasma capsulatum*, a dimorphic fungus that invades cells of the reticuloendothelial system and multiplies intracellularly as a yeast.
4. *Histoplasma capsulatum* enters the lungs via aerosols and invades phagocytic cells. It spreads throughout the body inside these cells and causes a disease similar to tuberculosis.

STUDY QUESTIONS

1. Construct a table containing the causative agent, the symptoms, and the mode of transmission of:
   a. five diseases of the upper respiratory tract;
   b. five diseases of the lower respiratory tract.
2. Most infectious diseases of the respiratory tract are characterized by a cough. Of what importance is this clinical manifestation to the host? to the pathogen?
3. Given the following signs or symptoms of respiratory tract disease, explain the host-parasite interaction that may have taken place and resulted in the clinical manifestation.
   a. cough;
   b. malaise;
   c. fever;
   d. pneumonia;
   e. scarlet-red throat;
   f. chills.
4. Briefly describe several methods that may be useful in reducing the incidence of:
   a. influenza;
   b. pneumococcal pneumonia;
   c. diphtheria;
   d. tuberculosis;
   e. common colds.
5. Why are the common cold and influenza more common during the winter months?
6. Briefly discuss the cause of rheumatic fever.
7. Discuss the value of vaccination against the common cold and influenza. Is it feasible? Why?
8. Zygomycosis occurs primarily in individuals who are somehow compromised by an underlying condition. What does that fact indicate about the pathogenic potential of these organisms? Would you predict that zygomycosis is very common in human populations?

SELF-TEST

1. Which is most likely to be responsible for pharyngotonsillitis? Streptococci of:
   a. viridans group;
   b. pneumoccocal group;
   c. group A;
   d. group B;
   e. group C.
2. The major cause of the common cold in adults is:
   a. fungi;
   b. protozoans;
   c. bacteria;
   d. viruses;
   e. allergins such as pollen.
3. What may result from a *chronic* pharyngotonsillitis caused by *Streptococcus pyogenes*?
   a. acute epiglottitis;
   b. diphtheria;
   c. rheumatic fever;
   d. acute glomerulonephritis;
   e. c and d.
4. The diphtheria toxoid vaccine since the late 1930s has steadily reduced the case fatality ratio (the number of deaths/number of cases). a. true; b. false.
5. Middle ear infections usually arise from microorganisms that work their way up the eustachian tube from nasopharyngeal infections. a. true; b. false.
6. Which organism is frequently associated with middle ear infections?
   a. *Streptococcus pyogenes;*
   b. *Haemophilus influenzae;*
   c. *Branhamella catarrhalis;*
   d. all of the above;
   e. none of the above.
7. Organisms commonly associated with sinusitis are:
   a. *Streptococcus pyogenes;*
   b. *Haemophilus influenzae;*
   c. *Streptococcus pneumonia;*
   d. all of the above;
   e. none of the above.
8. Which bacterium accounts for 90% of all bacterial pneumonias?
   a. *Klebsiella pneumoniae;*
   b. *Legionella pneumophila;*
   c. *Haemophilus influenzae;*
   d. *Mycoplasma pneumoniae;*
   e. *Streptococcus pneumoniae.*
9. In epidemics of legionellosis, what is the usual contributing factor that causes the epidemic?
   a. human carriers;
   b. animal carriers;
   c. wind storms that pick up the bacteria from the soil;
   d. a depressed immune system;
   e. air conditioning systems where the air is cooled by contaminated waters in cooling towers.
10. The identification of the agent responsible for bacterial pneumonias is not essential because they are all sensitive to the same antibiotics. a. true; b. false.

For items 11–15, match the agent that causes pneumonia with the antibiotic of choice:

   a. penicillin;
   b. chloramphenicol;
   c. erythromycin;
   d. sulfanilamide;
   e. gentamycin.

11. *Mycobacterium pneumoniae.*
12. *Streptococcus pneumoniae.*
13. *Klebsiella pneumoniae.*
14. *Legionella pneumophila.*
15. *Haemophilus influenzae.*
16. The effect of pertussis toxin can be offset by the use of:
   a. the antibiotic erthromycin;
   b. the whooping cough vaccine;
   c. human pertussis immune serum globulin;
   d. all of the above;
   e. none of the above.
17. The toxin produced by *Bordetella pertussis* causes the death of epithelial cells in the bronchi where the bacterium grows. a. true; b. false.
18. Which is the antibiotic of choice used to treat *Bordetella pertussis* infections?
   a. penicillin;
   b. chloramphenicol;
   c. erythromycin;
   d. sulfanilamide;
   e. gentamycin.
19. What are characteristics of *Mycobacterium tuberculosis*?
   a. intracellular parasite of macrophages;
   b. acid-fast positive;
   c. transmitted primarily by aerosols;
   d. infections treated with streptomycin, isoniazid, and rifampin;
   e. all of the above.
20. What are tubercles?
   a. colonies of *M. tuberculosis* growing on the lungs;
   b. masses of tissue that have walled up *M. tuberculosis* in the lungs;
   c. any type of necrosis in lung tissue;
   d. stained bacteria in the lungs;
   e. all of the above.

21. There is a vaccine called Bacillus Calmette-Guerine (BCG) that protects persons from tuberculosis. a. true; b. false.

22. The damage to lung tissue caused by *M. tuberculosis* is due to the:
    a. exotoxin it produces;
    b. lipopolysaccharide it releases from its wall;
    c. waxes it releases from its wall;
    d. chronic allergic reaction that the bacterium stimulates;
    e. formation of tubercles.

23. A person may have tubercles but not be infective. a. true; b. false.

24. Which virus is responsible for the majority of childhood viral pneumonias?
    a. influenza A virus;
    b. coxsackie virus;
    c. respiratory syncytial virus;
    d. echovirus;
    e. adenovirus.

25. Which viruses are responsible for the majority of adult viral pneumonias?
    a. influenza A and coxsackie viruses;
    b. rhino- and respiratory syncytial viruses;
    c. echo- and coronaviruses;
    d. rhino- and adenoviruses.

26. Rhinoviruses and coronaviruses are the major cause of the common cold in adults. a. true; b. false.

For items 27–28: this fungus is found in arid and semiarid regions of California, Arizona, New Mexico, and Mexico. In a fungal infection of the lungs, 60% of those infected have very mild or no symptoms, 35% of those infected have symptoms that resemble influenza, while the remaining 5% develop disseminated chronic infections that severely cripple and frequently lead to death.

27. What is the name of the disease?
    a. aspergillosis;
    b. San Joaquin valley fever;
    c. candidiasis;
    d. histoplasmosis;
    e. none of the above.

28. What is the name of the fungus?
    a. *Aspergillis;*
    b. *Candida;*
    c. *Coccidioides;*
    d. *Histoplasma;*
    e. none of the above.

For items 29–30: this fungus grows in soils that are enriched by chicken, bird, or bat feces. The disease it causes is prevalent in moist, humid regions such as the Mississippi and Missouri river valleys. The fungus infects macrophages in the lungs and is carried to the lymph nodes. In about 25% of the infections, the fungus spreads to other tissues.

29. What is the name of the disease?
    a. aspergillosis;
    b. San Joaquin valley fever;
    c. candidiasis;
    d. histoplasmosis;
    e. none of the above.

30. What is the name of the fungus?
    a. *Aspergillis;*
    b. *Candida;*
    c. *Coccidioides;*
    d. *Histoplasma;*
    e. none of the above.

SUPPLEMENTAL READINGS

Fraser, D. W. and McDade, J. E. 1979. Legionellosis. *Scientific American* 241:82–99.

Gwaltney, J. M., Jr., Moskalski, P. B., and Hendley, J. O. 1978. Hand-to-hand transmission of rhinovirus colds. *Annals of Internal Medicine* 88:463–466.

Kaplan, M. M. and Webster, R. G. 1977. The epidemiology of influenza. *Scientific American* 237:88–106.

Merigan, T. C. 1984. Respiratory viral infections of adults. *Scientific American Medicine* Section 7-XXV. New York: Freeman.

Middlebrook, J. L. and Dorland, R. B. 1984. Bacterial toxins: Cellular mechanisms of action. *Microbiological Reviews* 48(3):199–221.

Rytel, M. W. 1987. Influenza and its complications: recognition and prevention. *Hospital Practice* 22(1):102A–102V.

Simon, H. B. 1984. Mycobacteria. *Scientific American Medicine* Section 7-VIII New York: Freeman.

Stuart-Harris, C. 1981. The epidemiology and prevention of influenza. *American Scientist* 69:166–172.

Turck, M. 1985. An AIDS patient who died too soon. *Hospital Practice* 20(1):77–80.

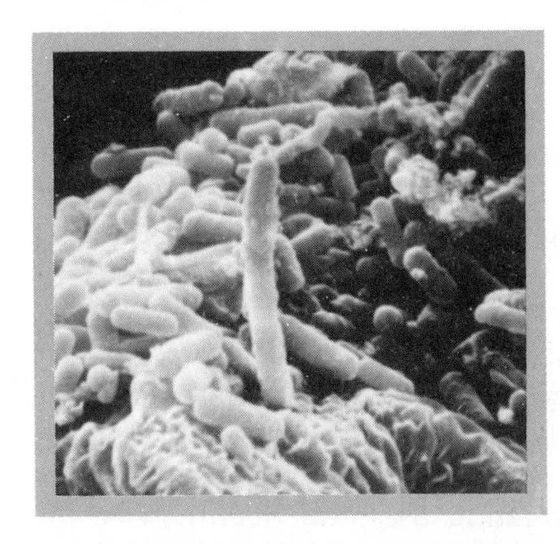

CHAPTER 18

DISEASES OF THE DIGESTIVE SYSTEM

CHAPTER PREVIEW KIYOSHI SHIGA AND THE AGENT OF DYSENTERY

The discovery by Koch that microorganisms were responsible for infectious diseases sparked an intensive search for the agents of dreaded diseases such as diphtheria, cholera, plague, typhoid, and dysentery. The span of time during which this search took place is referred to as the "golden age of microbiology," and it represents a bright period in the history of microbiology. During the golden age of microbiology, many scientists working in the laboratories of Pasteur and Koch achieved worldwide recognition for their contributions to the subjugation of infectious diseases. One such scientist was Kiyoshi Shiga.

Shiga became interested in dysentery—a disease characterized by bloody diarrhea laden with mucus, intestinal pain, and painful intestinal contractions—because the disease was very common in his native land, Japan. In one of his many publications on dysentery, entitled "The agent of dysentery in Japan" (title is a translation from the original German publication), he indicated that between June and December 1897 there were 89,400 cases of dysentery in Japan, of which approximately 21,500 were fatal. His studies of the disease led him to believe that dysentery was a distinct clinical entity that was not caused by any known agent (e.g., typhoid or cholera bacteria). Based on this assumption, he studied specimens from 36 dysentery patients, using observational and cultural techniques made popular by Koch in his studies of anthrax and tuberculosis. Shiga's efforts culminated in the isolation of a bacterium that he named *Bacillus dysenteriae.* This name was later changed to *Shigella dysenteriae* in his honor.

Shiga's discovery was a significant contribution to the ever-increasing mass of knowledge about infectious diseases. His contributions did not end there; once he knew what caused dysentery, he directed his efforts toward the prevention and treatment of this disease. Again he was successful: he developed a vaccine that was useful in preventing the disease. He did this while working with Kitasato at Koch's laboratory in Germany.

Shiga's many successes with dysentery gained him prestige and earned him valuable posts in his native country, including a directorship at the Institute of Infectious Diseases in Tokyo, a post in which he served for 16 years, and a deanship of the Medical Faculty of Keijo Imperial University.

The successes of Kiyoshi Shiga came only after he spent countless hours studying the various aspects of dysentery to gain a thorough understanding of its clinical and epidemiological aspects. In this chapter we discuss some of the important infectious diseases that afflict human digestive systems, the disease processes, their causes, and their control.

CHAPTER OUTLINE

THE FOOD AND WATER WE CONSUME ARE often laden with microorganisms; many of these are harmless, but a few are capable of causing disease. This chapter discusses where these pathogens come from, how they cause disease, their impact on human populations, and their control and treatment.

STRUCTURE AND FUNCTION OF THE DIGESTIVE SYSTEM

The digestive system consists of the **alimentary** or **gastrointestinal tract** (GI) and the accessory organs. The GI tract is composed of the tubular organs of the system, extending from the mouth to the anus, and includes the mouth, pharynx, esophagus, stomach, small intestine, large intestine, rectum, and anal canal. The anatomical relationships among these organs can be seen in figure 18.1.

Food is ground down by chewing and is partially digested by enzymes in the mouth. Acid and acid-tolerant enzymes continue to digest food in the stomach. In the intestines, the food is further digested enzymatically, and water and small molecules are absorbed into the blood and the lymph. Materials not digested or assimilated are released from the body through the anus.

Accessory organs involved in digestion include the teeth, tongue, salivary glands, liver, gall bladder, and pancreas (fig. 18.1). These organs participate in digestion by macerating the food (teeth); releasing hydrolytic enzymes (salivary glands and pancreas); producing or storing chemicals that aid in absorption of nutrients (gall bladder, pancreas, and liver); and functioning as nutrient storage areas (liver).

The GI tract is composed of four distinct layers or tunics: the mucosa, the submucosa, the muscularis, and the serosa. The **mucosa**, or mucous membrane, is the innermost lining of the tract and is composed of two layers; an epithelial lining and an underlying layer of connective tissue called the **lamina propria** (fig. 18.2). The lamina propria is rich in glands that have digestive functions. The major functions of the GI mucosa are to absorb nutrients and to secrete various substances needed for digestion. The epithelium consists of a single layer of cells, some of which absorb nutrients while others secrete substances such as mucus. The lamina propria is made of loose connective tissue and is rich in blood and lymph vessels, which absorb nutrients that have come through the epithelium. The blood and lymphatics of the mucosa also contain immunologically competent cells to fight off infectious agents that may have penetrated the epithelium.

The **submucosa** is made of loose connective tissue and is highly vascularized to provide nourishment to adjacent tissues of the GI tract. This tissue is also innervated and contains an autonomic nerve network.

The **muscularis** layer consists of two sheets of smooth muscle running perpendicular to each other. The smooth muscle is responsible for periodic contractions known as peristaltic movements.

The **serosa** is a layer of connective tissue and epithelial tissue that covers the outer layers of most organs of the digestive system. The **peritoneum** is an important serosa layer, because sometimes it becomes infected and allows the spread of microorganisms into the peritoneal cavity and subsequently to internal organs.

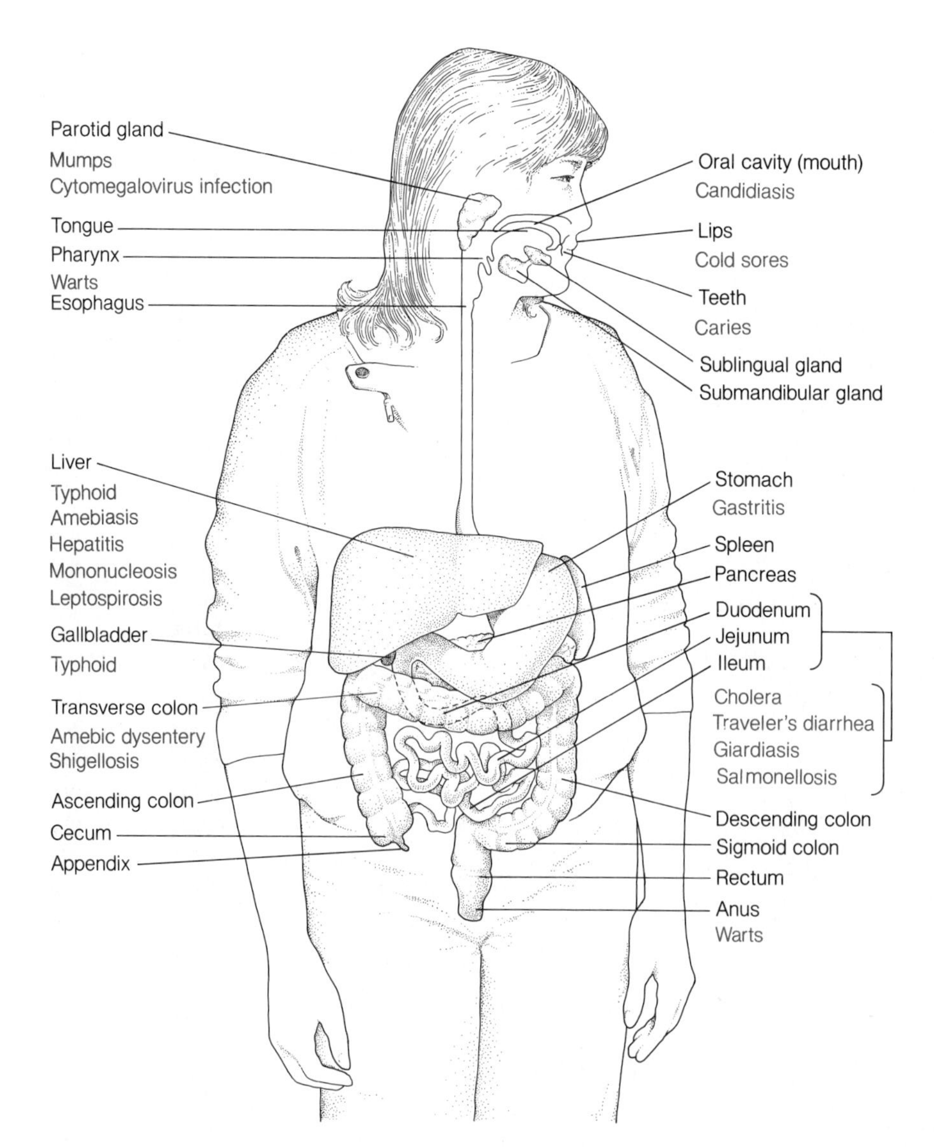

Figure 18.1 Anatomy of the digestive system. Colored print indicates the diseases that afflict the anatomical site.

Infection and inflammation of the peritoneum is called **peritonitis**.

NORMAL FLORA OF THE DIGESTIVE SYSTEM

The normal flora of the oral cavity is established shortly after birth and consists of streptococci, veillonellae, actinomycetes, spirochetes, and many other bacteria (table 18.1). Fungi and protozoa may also colonize the oral cavity. Microorganisms in the oral cavity derive their nutrients from food particles adhering to the dentition and the gums, soluble material in the saliva, and metabolic products from other colonizing microorganisms.

The upper GI tract (esophagus and stomach) contains few bacteria, mostly transients because foods and liquids are constantly washing the esophagus and the acid in the stomach does not allow the survival of bacteria there.

The duodenal area of the small intestine contains bacteria, mostly lactobacilli and enterococci. This flora gradually changes in the jejunum, where an increased number of coliforms and anaerobic bacteria are found. The ileal portion of the small intestine contains enteric gram-negative bacteria, as well as yeasts and anaero-

TABLE 18.1
MICROORGANISMS COMMONLY ISOLATED FROM THE HUMAN DIGESTIVE SYSTEM

Genus of microorganism	Group of microorganisms	Oral cavity	Small intestine	Large intestine	Feces (%)
Actinomyces	Gram + rod	+	−	+/−	
Bacteroides	Gram − rod	+	+	+	25–30
Borrelia	Spirochete	+	−	−	
Branhamella	Gram − coccus	+	−	−	
Candida	Yeast	+	+	+	
Clostridium	Gram + sporogenous rod	−	+	+	
Corynebacterium	Gram + rod	+	−	−	
Diphtheroids	Gram + rods	+	+	−	
Entamoeba	amoebic protozoa	+	−	−	
Escherichia	Gram − rod (coliforms)	−	+	+	0.05–0.1
Eubacterium	Gram − rod	−	+/−	+	15–18
Fusobacterium	Gram − rod	+	−	+	5–8
Giardia	Flagellated protozoa	−	+/−	−	
Lactobacillus	Gram + rod	+	+	+	
Neisseria	Gram − rod	+	−	−	
Peptococcus	Gram + cocci (anaerobic)	+/−	−	+	8–10
Proteus	Gram − rod (enterics)	−	+	+	
Salmonella	Gram − rod (enterics)	−	−	+	
Shigella	Gram − rod (enterics)	−	−	+	
Staphylococcus	Gram + coccus	+	−	+	
Streptococcus	Gram + coccus	+	+	+	0.1–0.2
Treponema	Spirochete	+	−	−	
Veillonella	Gram − coccus (anaerobic)	+	−	−	

*Percent of microbial flora that constitute the feces.
+/−, infrequently present.
Data from Moore & Holdeman, *Applied Microbiology* 27:916, 1974.

bic bacteria. These include coliforms, *Bacteroides*, *Eubacterium*, and bifidobacteria.

The large intestine contains the vast majority of microorganisms in the GI tract (fig. 18.2). These bacteria, which make up 20–60% of the mass of feces, are primarily *Bacteroides* and bifidobacteria. Other microorganisms in the large intestine include lactobacilli, coliforms, clostridia, streptococci, and yeasts. Every gram of feces contains approximately 100 billion microorganisms.

The normal flora protects the host against certain pathogens by competing for space and nutrients. The normal flora may also prevent certain infections by releasing inhibitory substances, such as bacteriocins. The contribution of the normal flora to its host may extend beyond protection; studies involving **gnotobiotic** animals (animals with known microbial flora) indicate that the normal flora is essential to the good health of the host because it stimulates the host's immune system and participates in the nutrition of the host. There is some evidence that much of the vitamin K required by animals is synthesized primarily by bacteria in the gut.

DISEASES OF THE ORAL CAVITY

The oral cavity is the site of entry for large numbers of microorganisms in food and water. Since there is an ample supply of nutrients from food, many microorganisms live attached to the teeth, gums, cheeks, and tongue. As a consequence, these tissues are sometimes subject to infection and disease.

Tooth Decay

Dental caries, or cavities, result from the gradual decay and disintegration of teeth (fig. 18.3). Dental caries are not limited to the nonliving portions of the tooth,

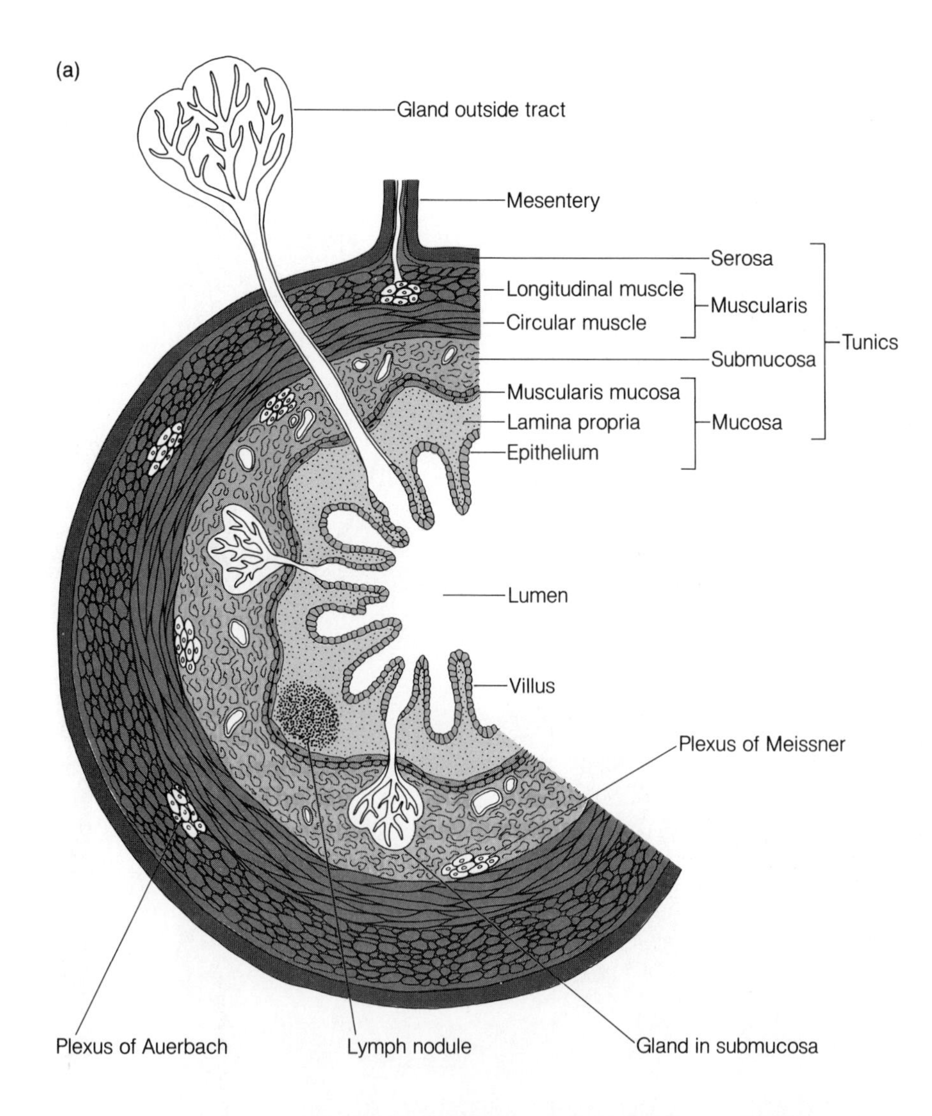

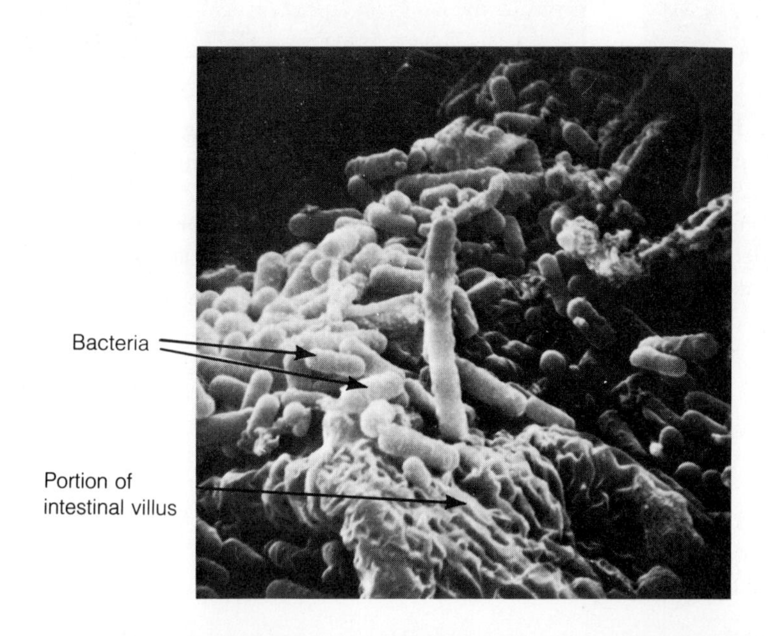

Figure 18.2 **The gastrointestinal tract.** *(a)* The gastrointestinal tract consists of a series of tubes that conduct food and its byproducts. The tubes consist of a number of layers called tunics, each with a function (see text for details). The mucosa is the site of many infectious diseases, although other tunics may also become involved. *(b)* Scanning electron photomicrograph of bacteria adhering to intestinal villi.

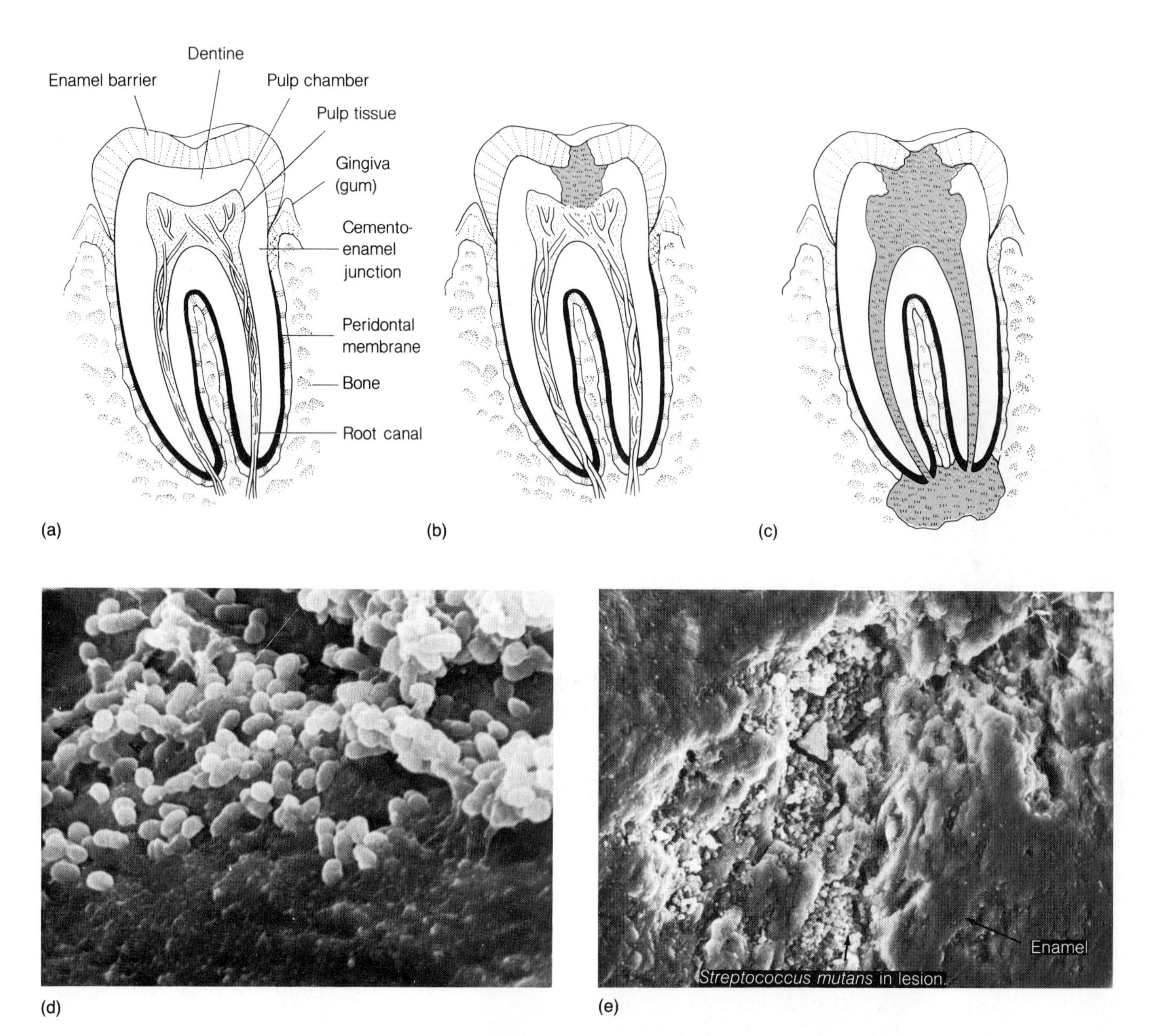

Figure 18.3 Caries. *(a)* Normal tooth. *(b)* The development of tooth decay begins when acidogenic, plaque-forming bacteria colonize a tooth and weaken the enamel. Illustration shows decay involving the enamel, dentine and pulp. Notice that the tooth is inflamed, as evidenced by the dilated vessels of the pulp. *(c)* Complete involvement of pulp. The decay is spreading into the bone, and causing degeneration of bone tissue. *(d)* Scanning electron micrograph of *Streptococcus mutans* attached to dental enamel. *(e)* Scanning electron micrograph of carious lesions showing *Streptococcus mutans*.

but may involve the pulp and the nerves within the pulp. The gradual decay and disintegration of teeth is caused by acids produced by some of the bacteria attached to the teeth. Apparently, acids produced when sugars are fermented by the tooth flora demineralize the hydroxyapatite that makes up the enamel (hard portion of the tooth). The bacterium that is largely responsible for enamel decay is *Streptococcus mutans,* which is able to attach in large numbers to the surfaces of teeth because of the dextrans (polymers of glucose) it synthesizes from sucrose. The attached bacteria form **dental plaques** that can cover large areas of a tooth and continuously bathe these regions in acid.

It is unlikely that *S. mutans* is the only microorganism involved in tooth decay. Dental caries is a progressive disease involving the destruction of the enamel, the dentine, and the pulp. The initial demineralization of the enamel is effected by acidogenic bacteria, but as soon as the dentine is reached, proteolytic bacteria (many of them anaerobes) begin to multiply. These anaerobic bacteria may be responsible for some of the damage seen in tooth decay.

Microorganisms growing on the teeth are the cause of dental caries, but whether or not they will cause disease in a particular instance depends on a number of factors such as dental hygiene, host immune defenses, tooth anatomy, arrangement of the dentition, and diet. Diet drastically affects the growth of cariogenic (caries-causing) bacteria on the teeth. Dental caries are most common in children and adolescents, who consume large amounts of candy and sweets that contain sucrose. Adults, in general, have fewer caries than children because of pH changes of oral secretions that occur during adulthood and reduced consumption of sucrose-containing sweets.

The trauma of tooth decay and repair can be significantly reduced by a diet low in sucrose and by proper dental hygiene. Daily brushing and flossing of the teeth results in fewer cariogenic plaques. Fluoridated drinking water and toothpaste have been reported to reduce cavities by as much as 50%. Cavities are treated by removing the decayed portion of the tooth and replacing it with an inert filling material.

Gum Disease

Periodontal disease is a chronic inflammation of the anchoring and supporting tissue of the teeth (fig. 18.4). **Gingivitis** (inflammation of the gums) is a periodontal disease that develops slowly as a consequence of dental plaque formation near the gums. An inflammatory response occurs in the gums that may be due to the host's immune response to growing microorganisms or to their products (e.g., lipopolysaccharide) on the dental plaque. The inflammation causes swelling of the tissue and interdentitional crevices. Bacteria later colonize the inflamed tissue and cause more damage. The initial host response to the accumulating dental plaque is a PMN leukocyte infiltration. This acute inflammatory reaction is replaced by a chronic inflammation, containing lymphocytes and monocytes.

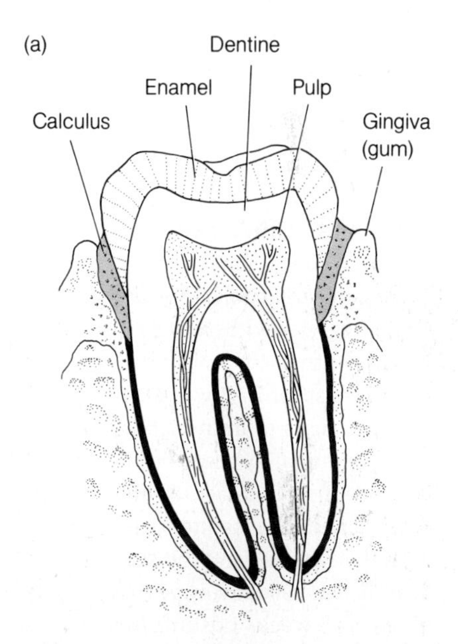

(b)

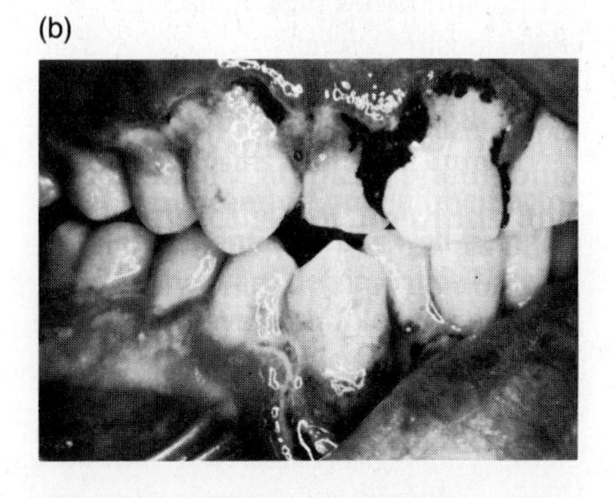

Figure 18.4 Periodontal disease. *(a)* Diagram illustrating gingivitis. Notice the inflamed gums and heavy deposits of plaque and calculus between tooth and gum. *(b)* Photograph showing a severe case of gingivitis.

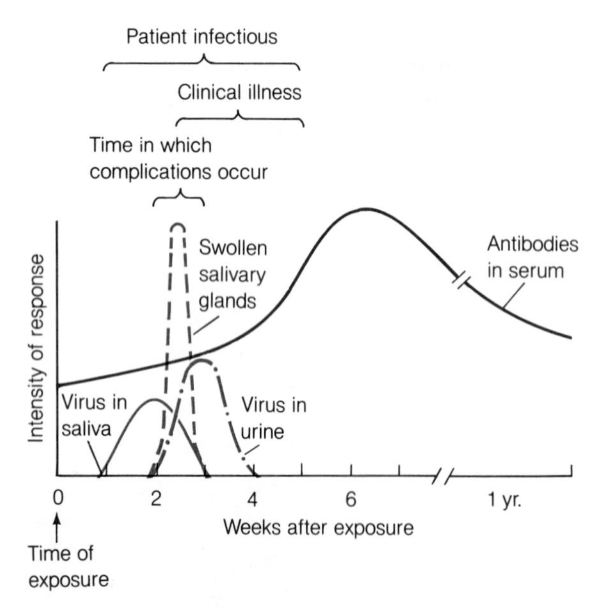

Figure 18.5 Clinical course of mumps. Clinical illness occurs 1–3 weeks after infection. This is the most contagious period because the virus is present in the saliva. Clinical symptoms persist for 2–3 weeks. The characteristic signs of mumps, inflamed parotid glands, appear about 2 weeks after infection and last for about a week. Serious consequences, such as inflamed testicles and meninges or encephalitis, usually appear (albeit rarely) 3–5 weeks postinfection. Serum antibodies peak at about 5–6 weeks postinfection.

In gingivitis, a general increase in the number of microorganisms is seen, predominantly actinomycetes. In periodontitis, a destructive gingivitis, the microbial flora is complex and includes *Bacteroides*, *Capnocytophaga* (*Bacteroides ochraceus*), *Eikenella*, and *Vibrio*. *Actinomyces*, *Actinobacillus*, and *Bacteroides* in high numbers are an indication of active periodontal disease. Because of their importance, gene probes have been developed for their identification and to detect their presence in plaques.

Infection of the Salivary Glands

Mumps is a disease characterized by fever and swelling of the salivary (parotid) glands. It is caused by a paramyxovirus known as the mumps virus. The disease occurs primarily in childhood, although some adults contract it. Complications such as orchitis (inflammation of the testes), meningitis (inflammations of the membranes lining the brain), encephalitis (inflammation of the brain), or pancreatitis (inflammation of the pancreas) occur mainly in adults. Orchitis, the commonest complication occurs in 25–30% of all adult male patients. Transmission of the mumps virus is from person to person via salivary or respiratory secretions.

The infection begins when a virus attaches to upper respiratory tract epithelium. There is a period of local multiplication in the epithelium and cervical lymph nodes after which the virus spreads into the bloodstream (viremia). Infection of the parotid glands results from this viremia. In the parotid glands, the virus multiplies, causing damage to the cells and an inflammatory reaction. Neutralizing antibodies and cell-mediated immunity play a role in blocking viral multiplication. Figure 18.5 illustrates the pathogenesis of mumps.

The highest incidence of mumps is found in individuals between 5 and 10 years old. This group of children is the most important source of infectious viruses. Adults usually get the disease from infected children. Consequently, vaccination of children eliminates (or reduces) the source of infectious virus and reduces the number of susceptible individuals in a population (fig. 18.6). Adults may be treated with hyperimmune serum containing IgG. This treatment is directed toward the prevention of orchitis (inflammation of the testes) and possible ensuing sterility in adult males.

INFECTIOUS DIARRHEAS CAUSED BY BACTERIA

Gastritis and **enteritis** are inflammations of the stomach and small intestines, respectively, while **gastroenteritis** is an inflammation of both the stomach and the small intestines. The inflammations may be due to diet, toxins, infections, drugs, or psychological factors. An inflammation of just the large intestine is defined as **colitis**. The small and large intestines are the most common sites of intestinal disease. **Diarrhea**, a symptom of many gastrointestinal disorders, is characterized by a frequent passage of watery stool. Diarrhea, which is caused by infectious agents, such as viruses, bacteria, and protozoa, still accounts for a

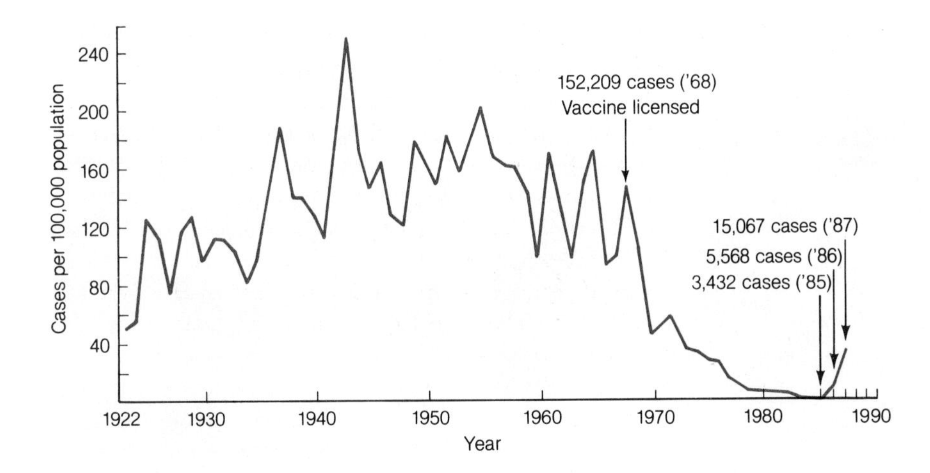

Figure 18.6 Reported cases of mumps in the United States. The number of cases of mumps in the United States waxed and waned until 1967, when the vaccine was approved. Before 1967, there was an average of about 160 cases/100,000. With the introduction of the vaccine, the number of cases dropped sharply. Presently <2 cases/100,000 are reported. About 70% of all cases are reported in children between the ages of 5 and 14. Adults (20+) are least often afflicted.

significant proportion of morbidity and mortality in developing countries.

Enteritis

Infections of the small intestine generally result in a profuse, watery discharge, resulting from an alteration of the ionic balance in the intestinal epithelium. In some infections, the mucosa and submucosa tunics are not altered pathologically and no inflammatory response is noted. Generally, the invading microorganisms produce toxins that induce the profuse secretion of fluids from the epithelium. In other infections, the mucosa and submucosa are damaged directly by the invading microorganisms, causing an inflammation of the small bowel which may also cause diarrhea.

Colitis

Infections of the large intestine often cause mucosal erosion accompanied by abdominal cramps, painful spasmodic contraction of the bowel (tenesmus), and inflammatory exudate (pus, PMN leukocytes, and blood) in the stool. These signs and symptoms characterize a condition known as **dysentery**.

Cholera

An acute infection of the human small intestine caused by the bacterium *Vibrio cholerae* and known as **cholera** is characterized by a profuse, watery discharge. The diarrhea is brownish at first, but soon turns a pale whitish color. This is the characteristic "rice water" stool of cholera. Vomiting may also accompany the diarrhea.

Vibrio cholerae is a gram-negative motile rod exhibiting a characteristic curved or comma shape although the curved appearance of the rods may disappear in laboratory cultures. The cholera vibrios generally enter the small intestine in contaminated water and food. They adhere to the microvilli of epithelial cells (fig. 18.7), where they multiply and release a powerful enterotoxin called **choleragen**. This toxin causes the efflux of interstitial and cellular water and electrolytes (chloride and calcium ions) into the lumen of the small intestine (fig. 18.8). The amount of water lost is so great (4–5 gallons/day) that it cannot be reabsorbed by the large intestine. This tremendous loss of fluid causes dehydration, leading to diminished blood volume. The diminished blood volume leads in turn to **hypovolemic shock**, since organs and tissues are not oxygenated or nourished properly. Death soon follows if this condition is not reversed.

Humans are the most important reservoir for *V. cholerae*. Infected individuals shed the vibrios into sewage, which can contaminate water supplies. The spread of the disease is by the fecal-oral route via contaminated water. Carriers harbor the bacteria in their gall bladders. A vaccine has been developed that induces the development of an immunity that acts directly on the vibrio, inhibiting its attachment to intestinal epithelium rather than neutralizing the toxin. Some countries still require travellers to be immunized with the cholera vaccine, although the efficacy of the vaccine is under debate.

Vibrio cholerae is found in nearly all large bays and rivers associated with densely populated areas. In September 1978, for example, the Public Health Ser-

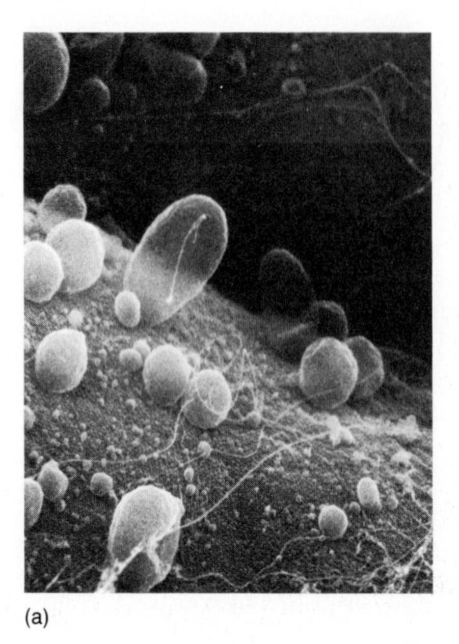

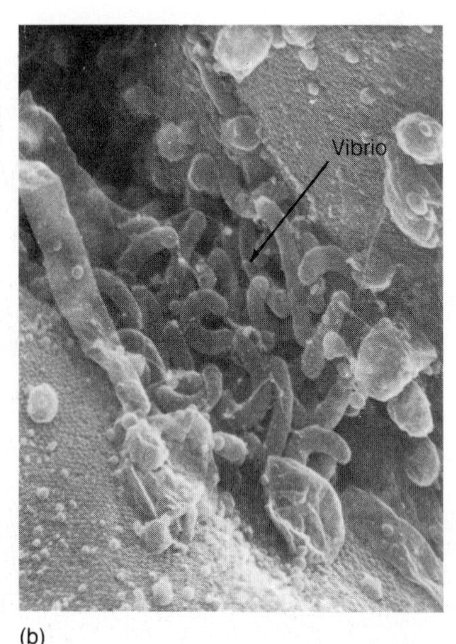

(a) (b)

Figure 18.7 Vibrio cholerae in intestines. *(a)* Scanning electron micrograph of mouse ileum showing normal flora on microvilli. *(b)* Intestine extensively colonized by *Vibrio cholerae.* Cholera is caused by the effect of an exotoxin on intestinal epithelium.

vice reported the isolation of *V. cholerae* in 11 persons from Louisiana. Improperly cooked crab was the source of the bacterium. Since then, however, the organism has been isolated from local shrimp as well as crab.

The most effective treatment for cholera is the prompt replacement of lost fluids and electrolytes. This treatment alone reduces the mortality of cholera to less than 1%. Intravenous solutions containing Na^+, K^+, Cl^-, and HCO_3^-, and oral administration of 5% glucose in water constitute a common method of treatment. Tetracycline reduces the number of vibrios present in the small intestine and helps in eliminating the carrier state.

Typhoid Fever

Typhoid fever, an acute infectious disease of the GI tract caused by *Salmonella typhi*, is characterized by a high fever (104°F), headache, diarrhea, and rose spots and tenderness in the abdomen. The incubation period for typhoid fever is 1–3 weeks (average, 2 weeks). The early symptoms include nosebleeds, general weakness, mild headache, and malaise. After these general symptoms, the characteristic tenderness and rose spots on the abdomen appear, along with a high fever, splenomegaly (an enlargement of the spleen), and diarrhea. Thin bowel movements generally occur 3–7 times per day. Shock caused by endotoxin may also occur if large numbers of *S. typhi* are in the blood. Mortality of untreated cases may be as high as 40%, but generally is between 15 and 25%.

Salmonella typhi, a facultative intracellular parasite of macrophages, is a gram-negative rod in the family Enterobacteriaeceae. The microorganism enters the human small intestine in contaminated food and water. During the first 5–7 days of infection, *S. typhi* attaches to the surface of the intestinal lining, where it multiplies. When it penetrates the lamina propria and submucosa, some of the invading bacteria enter the blood and lymph, causing a bacteremia. The early symptoms (fever, malaise, and lethargy) are caused by the penetration into the intestinal wall and the bacteremia. The symptoms may be aggravated by the presence of endotoxin in the blood.

When *S. typhi* enters the circulation, it is distributed throughout the body. The bacteria are phagocytized, but not killed, by macrophages in the lymph nodes and other organs of the reticuloendothelial system. They multiply within the macrophages and escape into the lymph and blood in very large numbers. The high fever of typhoid is correlated with this bacteremia. Endotoxin and the host's fever-producing substances (**endogenous pyrogens**) released by the phagocytic cells also contribute to the symptoms. At this stage, the bacteria enter the gall bladder and then reinfect the intestines. Diarrhea and abdominal tenderness and rose spots parallel the reinfection of the

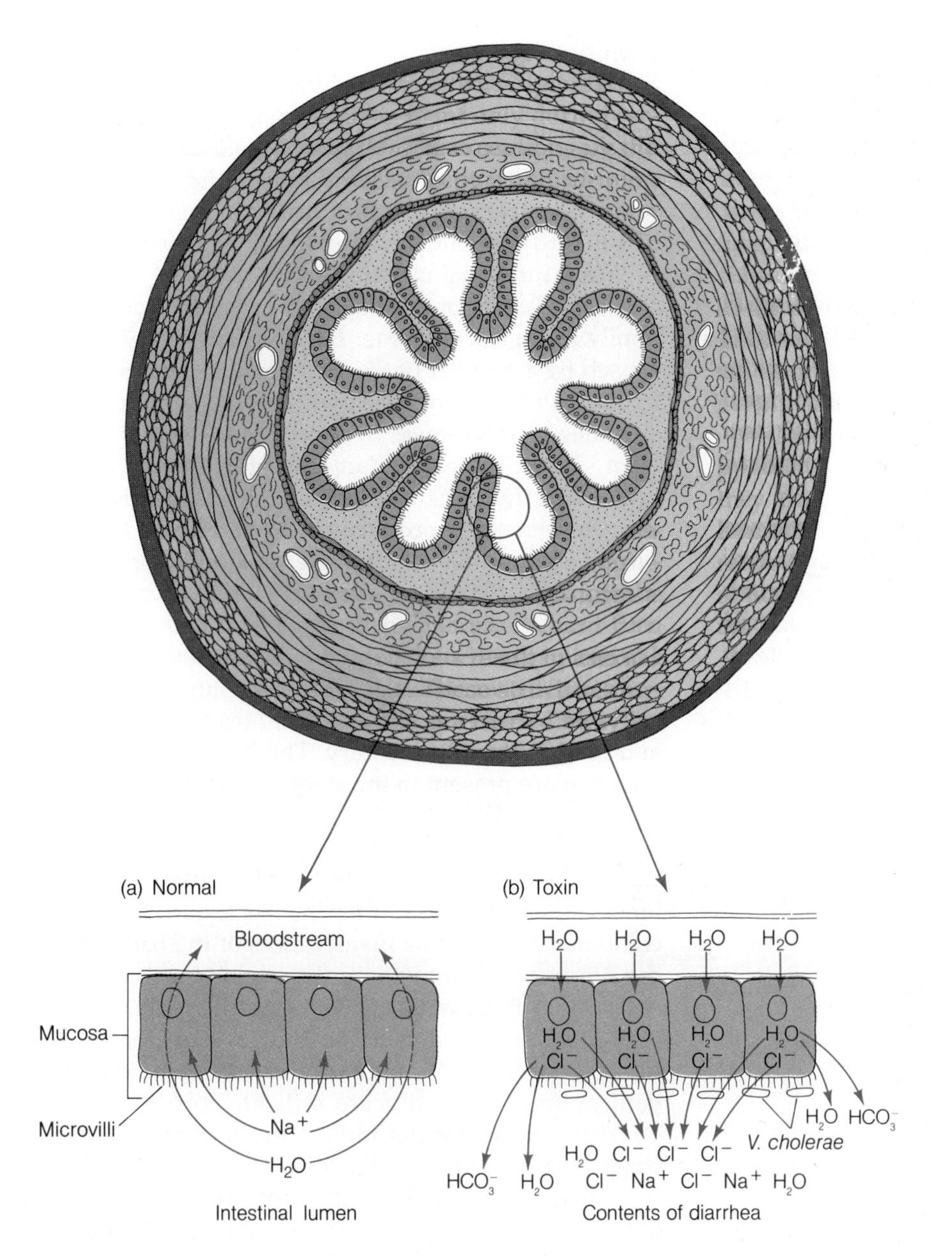

Figure 18.8 Mode of action of the cholera toxin. Cholera toxin causes the excessive flow of cellular fluids and electrolytes into the lumen of the intestines. *(a)* Normal movement of fluids and sodium in the intestines. Water flows from the lumen of the intestines into the bloodstream, along with sodium ions. *(b) Vibrio cholerae* colonization of intestines and production of exotoxin. The toxin acts by stimulating adenyl cyclase to convert ATP into cAMP. This nucleotide blocks the flow of sodium ions into the bloodstream, which causes chloride and bicarbonate ions to be excreted from the blood. As a consequence, large volumes of water exit the blood and tissues into the lumen and are excreted, along with the electrolytes, as diarrhea.

intestines by *S. typhi*. With the development of cell-mediated immunity, activated macrophages destroy the invading bacteria. This event generally leads to the recovery of the patient; however, 1% of these individuals die and 10–15% become chronic carriers, harboring the bacteria in the gall bladder.

In the last few years, there have been fewer than 500 reported cases of typhoid fever per year in the United States and about 5 deaths annually. Approximately 45–75 new carriers enter the population each year (fig. 18.9).

Typhoid fever is transmitted by the fecal-oral route. Chronic carriers are the most common source of infectious agents, since their excrements contaminate food and water with virulent bacteria. The Public Health Service (PHS) maintains a very close surveillance of these carriers, because it would be dangerous for them to handle foods in such places as food processing plants, restaurants, and bakeries. Most cases of typhoid fever in the United States are traced to carriers who contaminate food or water. In most cases of typhoid fever in which water is the vehicle, the source is traced to faulty sewage systems.

Prevention of typhoid involves the surveillance and control of carriers. In addition, water supplies should be properly chlorinated and reservoirs and water pipes maintained. In areas where typhoid fever is prevalent, all foods should be steaming hot when eaten, and only beverages from unopened bottles should be consumed. Foods should be cooked and/or refrigerated to prevent growth of salmonellae.

To date, no vaccine has been developed that gives full protection against typhoid fever, although some vaccines provide partial protection. Vaccines containing the capsular antigen have given encouraging results in clinical trials. Ampicillin and chloramphenicol can be used to treat typhoid fever, but ampicillin-resistant trains of *S. typhi* have evolved. Thus, sensitivity tests should be conducted on isolates so that the best antibiotic can be used to treat the infection.

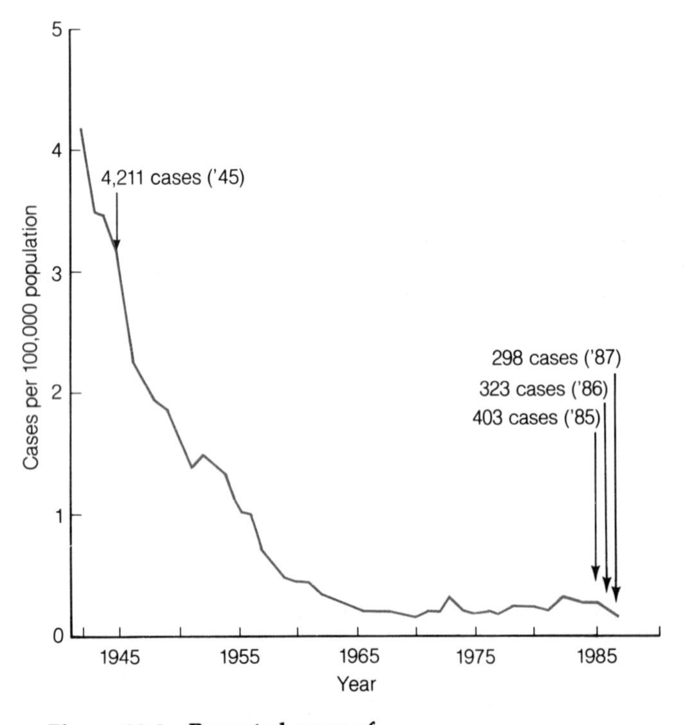

Figure 18.9 Reported cases of typhoid in the United States

Bacterial Dysentery

Shigella is responsible for **bacillary dysentery**, an acute infectious disease. It is characterized by diarrhea, tenesmus (distressing but ineffectual urge to defecate), and cramps, along with mucus and blood in the feces. The disease involves an inflammation of the ileum or the colon which causes sloughing of mucosal cells and intestinal ulceration.

Shigella is a gram-negative, nonmotile rod in the family Enterobacteriaceae. Bacillary dysentery can be caused by a number of *Shigella*. Small numbers of this bacterium are able to cause dysentery in humans. Three shigellae, *S. dysenteriae, S. sonnei*, and *S. flexneri*, have been shown to produce an **enterotoxin** that appears to work in the same way as choleragen. It is believed that these bacteria produce their enterotoxin early in the infection and that the enterotoxin is responsible for the early diarrhea. The characteristic common to all shigellae, however, is their ability to adhere to and invade the intestinal epithelium of the ileum and colon. The invasion of the intestinal epithelium by *Shigella* causes an acute inflammation that leads to the signs and symptoms of dysentery. The blood mucus, and cells that are present in the stool are a result of inflammation caused by *Shigella*.

The characteristic lesion of the intestinal mucosa is an ulcer covered by a layer of PMN leukocytes, bacteria, cellular debris, and fibrin. The lesions arise as a combined effect of the invasiveness of the bacterium, the release of endotoxin, and an acute inflammatory reaction on the part of the host. As the bacteria penetrate the intestinal mucosa, they release metabolic wastes, enterotoxin, and endotoxin, which cause cell damage and elicit an acute inflammatory response. Together, these factors induce the formation of the characteristic ulcer and diarrhea.

Dysentery caused by *S. dysenteriae* type I, also called the Shiga bacillus, produces an exotoxin called **neurotoxin**. The Shiga exotoxin causes bleeding and paralysis apparently by inhibiting protein synthesis. The role of the exotoxin in the pathogenesis of dysentery has not been clearly defined, but it appears that the neurotoxin is responsible for the severity of the dysentery caused by the Shiga bacillus as compared to the other species. Circulating antibodies appear during the infection, but they are relatively ineffective in stopping the invasion. Cell-mediated immunity is thought to be responsible for protective immunity and recovery of the patient.

Dysentery is transmitted by the fecal-oral route, and the most common source of infection is other humans. In countries with high standards of hygiene, where human wastes are properly discarded, the disease is infrequent. The majority of the cases are laboratory-associated infections and outbreaks in institutions. In institutions where coprophagy (consumption of feces) sometimes occurs, bacillary dysentery and hepatitis are common. In developing countries, where sanitary conditions and public health measures frequently are marginal, bacillary dysentery is endemic. The disease is most often transmitted in foods contaminated by infected individuals. Although the carrier stage is rare, inapparent infections do occur, and the infected person can serve as a temporary source of contamination.

Preventive measures for dysentery include the chlorination of water supplies and the maintenance of reservoirs and water pipes, as well as the treatment of raw sewage (table 18.2). Dairy foods have been reported as a source of infection, and therefore it is essential that milk be properly pasteurized before drinking and that the consumption of raw milk be avoided. There are no vaccines available with which to immunize the public. In the last few years, between 15,000 and 20,000 cases of dysentery have been reported each year in the United States. The number of deaths among these cases has been reported to be between 20 and 30.

Severe infections generally are treated with ampicillin. Chloramphenicol and trimethoprim sulfamethoxazole are used for those infections caused by ampicillin-resistant strains. The majority of cases are not treated at all, however, since the infection is self-limiting and of short duration. Supportive therapy is recommended to alleviate the discomfort and to replace the fluids and electrolytes lost in the diarrhea.

Bacterial Gastroenteritis

Gastroenteritis is an inflammation of the epithelial linings of the stomach and intestines. The inflammation can be manifested in a variety of signs and symptoms, although abdominal pain, nausea, vomiting, and diarrhea are most frequently seen.

***Salmonella* gastroenteritis** is a disease characterized by a sudden onset of diarrhea, headache, abdom-

TABLE 18.2
REPORTED CASES OF DISEASE AND DEATHS ASSOCIATED WITH PUBLIC DRINKING WATER

Bacteria	Average per year outbreaks	Average per year cases	Average per year deaths
Shigella spp.	2.3	324	0.3
Salmonella spp.	1.6	838	0.1
Campylobacter spp.	0.2	207	0
Toxigenic *E. coli*	0.2	52	0.2
Vibrio spp.	<0.1	0.7	0
Yersinia spp.	<0.1	0.7	0
Viruses			
Hepatitis A. virus	2.2	70	<0.1
Norwalk virus	0.7	172	0
Rotavirus	<0.1	76	0
Protozoans			
Giardia spp.	3.7	995	0
Entamoeba spp.	0.1	2	<0.1
Chemicals			
Inorganic (metals, nitrate)	1.3	39	0
Organic (pesticides, herbicides)	1.0	118	0.3
Unidentified agents	11.6	3771	0

inal pain, vomiting, and fever. The disease is commonly called *Salmonella* food poisoning, but it is not really a "poisoning," because large numbers of bacteria, not exotoxins, are responsible for the signs and symptoms of the disease. *Salmonella enteritidis* and *Salmonella cholerasuis* are the species of *Salmonella* most frequently isolated from patients with gastroenteritis. Although the diseases caused by these two organisms can be very similar, *Salmonella cholerasuis* normally causes a septicemia (invasion of the blood) while *Salmonella enteritidis* does not.

The salmonellae almost invariably enter the GI tract within contaminated foods. Unlike shigellae, of which fewer than 200 can initiate an infection, more than 100,000 salmonellae are needed for a successful infection. The salmonellae that survive the harsh stomach environment and reach the intestines adhere to the intestinal epithelium and initiate foci of infection in both the small and large intestines. In contrast to enteric fevers (e.g., typhoid), the salmonellae causing gastroenteritis rarely invade the blood, and their growth is limited to the GI tract. When the infecting bacteria penetrate the epithelium and enter the lamina propria, they are ingested by phagocytes. Some of the bacteria are transported to the regional lymph nodes, where they stimulate specific immunity that eventually eliminates the infectious agent from the intestines.

The growth of salmonellae causes pathological changes that are thought to account for most of the symptoms of the disease. An acute inflammation is believed to be responsible for focal ulceration and microabscesses seen in biopsy materials from infected persons. Gastroenteritis caused by salmonellae is generally not treated with antibiotics, since it is a self-limiting disease.

The infection is generally acquired by ingesting food contaminated with salmonellae. Leaving contaminated food in a warm place for longer than 4 hours may allow the multiplication of salmonellae to infectious levels in the food. In contrast to *S. typhi*, which generally infects only humans, *S. enteritidis* and *S. cholerasuis* also infect animals; consequently, animals can serve as a source of infection. Outbreaks have been traced to chickens, pet turtles, pigs, and many other domestic animals. More than 1400 serotypes of salmonellae have been associated with gastroenteritis. In the last few years, the number of cases of *Salmonella* gastroenteritis reported in the United States has increased drastically (fig. 18.10). The morbidity is over 30,000 cases and the mortality is between 60 and 80 cases each year. Since it is a rather mild disease, however, many of these cases are not reported, and it is estimated that fewer than 10% of all cases of *Salmonella* gastroenteritis are reported to health agencies.

Diarrheas Caused by *Escherichia coli*

Escherichia coli is a gram-negative, glucose-fermenting rod that has long been known to colonize the human intestine. This organism is part of the normal flora in a mutualistic relationship, but it can also become a pathogen. For example, approximately 90% of urinary tract infections are caused by *E. coli*. Three distinct types of *E. coli* have been described that cause diarrheal diseases in humans. These three types of pathogenic *E. coli* are not normally part of the human intestinal flora.

Strains of *E. coli* known as **enteropathogenic** cause a diarrheal disease that resembles *Salmonella* gastroenteritis. This group of *E. coli* serotypes is most frequently responsible for outbreaks of diarrhea in institutions such as nurseries and children's hospitals. They do not produce any kind of enterotoxin, nor are they able to invade the lamina propria of the intestines. How they cause disease is not known.

Invasive strains of *E. coli* cause a disease that resembles bacillary dysentery. These *E. coli* are capable of invading epithelial cells of the intestines. Adults and infants are equally susceptible.

Toxin-producing (enterotoxigenic) strains of *E. coli* cause choleralike diseases in humans. These strains have been implicated as the cause of "travelers' diarrhea." Enterotoxigenic *E. coli* strains also cause diarrhea in piglets and calves. The disease can be very serious in children because it appears suddenly, causing a severe dehydration. These *E. coli* cause the disease by elaborating a powerful **heat-labile** (LT) **enterotoxin** that acts in a manner similar to the cholera toxin. Pili are important in the pathogenesis of toxin-producing *E. coli*, since they attach the bacteria to the intestinal epithelium. This results in the multiplication of the bacteria and the subsequent production of enterotoxin.

Pathogenic strains of *E. coli* are sensitive to a variety of drugs, including sulfonamides, tetracyclines, and ampicillin. The treatment is directed toward eliminating the bacteria and, in cases of severe diarrhea, replacing electrolytes and fluids, as for cholera patients.

The best way to control these diseases is by avoiding contaminated drinking water by drinking bottled beverages or water. If contaminated H_2O must be drunk, it should be boiled or treated in such a way as

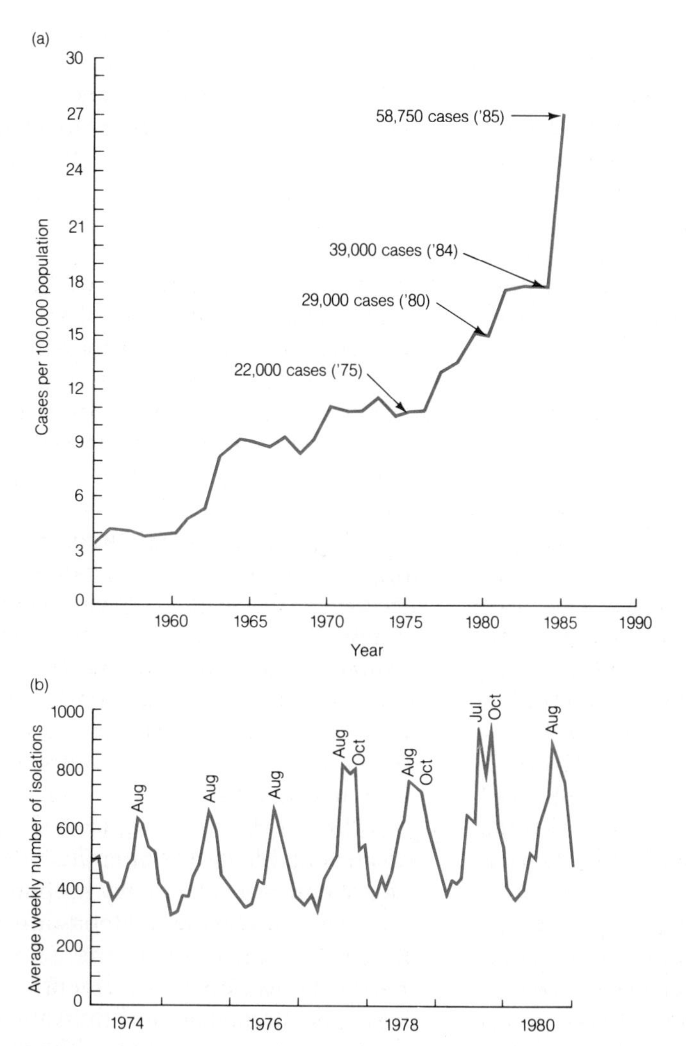

Figure 18.10 Characteristics of salmonellosis. *(a)* Incidence of *Salmonella* gastroenteritis in the United States. *(b)* Reported isolation of *Salmonella*. It is noteworthy that most isolates occur during the summer months, when group picnics and outings are most frequent.

to kill the pathogens. Raw vegetables in salads or "munchies" have also been implicated in diarrheal diseases caused by *E. coli*. Oral sulfonamides have been recommended for travellers as a prophylactic measure for travellers' diarrhea. Bismuth subsalicylate (Pepto-Bismol) has been shown to be effective in reducing the effect of toxin-producing *E. coli* strains.

Other Bacterial Diarrheas

Campylobacter jejuni has recently been discovered to be a common cause of gastroenteritis. It is a gram-negative, flagellated, curved rod that is isolated from approximately 9% of all diarrheas and from 3% of formed stools. It requires a microaerophilic environment and grows best at 40–42°C. The disease is characterized by fever and malaise, followed by diarrhea and severe cramping that may mimic acute appendicitis or ulcerative colitis. The bacteria appear to penetrate the mucosal epithelium, since inflammatory cells and blood are often present in the stool. *Campylobacter jejuni* causes hemorrhagic necrotic foci in the jejunum and ileum. The disease usually lasts less than a week and is self-limiting; however, fatal cases have been reported.

Domestic animals such as cows, swine, sheep, dogs, and fowl serve as reservoirs for *C. jejuni*. The infectious agent is generally acquired by ingesting contaminated water or raw milk, or by contact with infected ducks and geese or humans. All ages are susceptible to the infection, which has a seasonal peak during the summer. The number of annual cases of *C. jejuni*-caused diarrheas in the United States is estimated to be at least 100,000. This bacterium is now considered to be one of the most common causes of diarrhea in the United States, as well as in developing countries throughout the world.

Yersinia enterocolitica is a gram-negative bacillus that causes a severe **enterocolitis**. It is not a frequent cause of infectious diarrhea in the United States and is more prevalent in western Europe and Scandinavian countries. The symptoms include fever, diarrhea, and abdominal pain. Animals are believed to serve as reservoirs for *Y. enterocolitica*. The bacterium has been isolated from water, food, and unpasteurized dairy products. This pathogen invades the wall of the small intestine and may spread to the regional lymph nodes. *Yersinia's* behavior may be responsible for the abdominal pains and appendicitislike symptoms of the enterocolitis it produces. The enterotoxin produced by this bacterium is believed to be responsible for the diarrhea.

Clostridium difficile has recently been implicated in **antibiotic-associated pseudomembranous enterocolitis** (AAPE). The disease is characterized by diarrhea and ulcerative lesions similar to those caused by *Shigella*. The pathogenesis of the microorganism is related to the production of a cytolytic enterotoxin that is believed to cause the ulcerative colitis and the diarrhea. The symptoms are often associated with the administration of antibiotics; clindamycin and some other antibiotics at subinhibitory concentrations cause the production of enterotoxin by *C. difficile*. It is believed that the antibiotic suppresses the normal flora and hence allows the extensive multiplication of *C. difficile*. Clindamycin has also been shown to act as an inducer of the enterotoxin.

FOOD INTOXICATIONS

So far we have been discussing diseases that are caused by infectious agents multiplying in the gastrointestinal tract. The diseases they cause are attributed to their growth, metabolism, and spread. The term "food poisoning" is commonly used to describe diseases caused by ingesting infectious agents or foods contaminated with toxic substances. In this section we discuss diseases of the GI tract that are caused by the ingestion of toxins produced by microorganisms multiplying in foods (table 18.3). Botulism, a very serious form of food poisoning, is not discussed in this chapter because it involves the nervous system rather than the GI tract.

Staphylococcal Gastroenteritis

Staphylococcal gastroenteritis is thought to be the most common cause of food poisoning in the United States. The disease is caused by ingesting foods containing *S. aureus* enterotoxins and is characterized by a rapid onset of nausea, vomiting, and diarrhea within an average of 4 hours after consuming the contaminated food. The severity of the intoxication varies depending on the amount and type of enterotoxin ingested. Foods such as custards, hams, hollandaise sauce, and creamy fruit salads, tuna, chicken, and macaroni salads contaminated by food handlers frequently become sources of *S. aureus* enterotoxin when the foods are not kept refrigerated. Large amounts of enterotoxin can be produced in a few hours when the food is at room temperature.

The symptoms of staphylococcal gastroenteritis are attributable to the enterotoxin(s). Since the toxins are not denatured readily by heating, disease can result even when contaminated foods are cooked. The mode of action of the enterotoxins is unknown; however, they are thought to act on emetic (vomiting) receptors in the intestine, to impede fluid absorption by the gut, and to promote the efflux of fluids into the lumen of the intestines. The toxins are also believed to be pyrogenic (fever causing), a characteristic that may account for the fever observed in some cases of food poisoning.

The source of contamination of foods is usually chronic carriers who come into contact with the food. Approximately 10% of the United States population are carriers for enterotoxigenic staphylococci. Refrigerating foods until they are served and promoting safe food handling procedures insure that staphylococci do not grow and produce the enterotoxin in the food. Cases of staphylococcal food poisoning generally are not treated, because the disease is rarely fatal and the symptoms are of short duration. In severe cases, replacement of fluids and electrolytes is recommended.

Clostridial Gastroenteritis

Clostridium perfringens causes a mild form of food poisoning characterized by abdominal pain, cramps, and

TABLE 18.3
FOOD POISONING ASSOCIATED WITH THE PRODUCTION OF TOXINS

Disease	Causative agent	Symptoms or disease	Foods implicated	Time of onset (Hours)
Botulism	*Clostridium botulinum* (botulinal exotoxins A-E)	Flaccid paralysis, cardiac paralysis	Meat, fish, low-acid foods, canned vegetables	24–48
B. cereus food poisoning	*Bacillus cereus* (toxins)	Nausea, vomiting, colic pain, diarrhea	Unrefrigerated starchy foods	8–12
Enterococcus food poisoning	*Streptococcus faecalis* (toxins)	Nausea, vomiting, diarrhea	Unrefrigerated foods	8–12
Mycetismus	*Amanita* spp. (mushroom) (Amanitin)	Nervous disorders, death	Poisonous mushrooms	24–36
Mycotoxicosis	*Aspergillus flavus* (Aflatoxin)	Liver disorders, death	Improperly stored grains and damaged grains in the field	12–64
"Perfringens" food poisoning	*Clostridium perfringens* (exotoxins)	Abdominal pain, diarrhea	Unrefrigerated cooked meats	10–20
"Staph" food poisoning	*Staphylococcus aureus* (enterotoxins A-F)	Sudden nausea, vomiting, diarrhea	Potato salad, cream-filled pastries, dry skim milk, ham	4–12

diarrhea. It differs from staphylococcal gastroenteritis in that nausea and vomiting are uncommon symptoms. The onset of the disease occurs 10–20 hours after ingestion of food contaminated with *C. perfringens* enterotoxins. The symptoms may last 10–20 hours. This type of food poisoning ranks second only to staphylococcal food poisoning in incidence in the United States. The cramps and diarrhea are caused by the enterotoxins produced.

The disease is most often caused by eating foods that have been cooked in bulk and then stored at temperatures that allow the bacteria to multiply. Pinto beans and other legumes are excellent substrates for the growth of these bacteria. When the foods are served without reheating to above 165°C, the toxin is not inactivated. For this reason, most outbreaks of *C. perfringens* food poisoning occur at large gatherings and in institutions, where bulk preparation of food is a necessity.

Clostridium perfringens is a gram-positive, anaerobic, endospore-forming rod. It is widely distributed in soils, sewage, water, the intestinal tract of mammals, raw meats, poultry, and fish. The organism elaborates a wide variety of enzymes, many of which cause extensive tissue damage. In addition to food poisoning, *C. perfringens* is one of several agents involved in a mixed infection of connective tissue called **gas gangrene.**

PROTOZOAL INFECTIONS OF THE DIGESTIVE SYSTEM

Protozoal diseases, such as those caused by *Entamoeba histolytica* and *Giardia intestinalis*, are quite common, especially in developing countries and in areas where poor sanitation is sometimes practiced. These organisms cause diarrheal diseases resembling those caused by bacteria. The symptoms range from mild to severe, and the infections may include accessory organs such as the liver and gall bladder.

Amoebic Dysentery

Amebiasis is a disease caused by *Entamoeba histolytica*. The classical clinical presentation of this disease is a dysentery, also called **amoebic dysentery**. At the onset, frequent bowel movements and abdominal pain are experienced. Typically the stools are watery or loose and accompanied by blood and mucus. Rectal pain and tenesmus are also common manifestations of amebiasis. This symptom reflects an acute inflammatory response due to the invasion of the colonic

Figure 18.11 Amoebic ulcers. *(a)* Liver section showing amoebic abscess. Arrows indicate abscess area of liver with many trophozoites of *Entamoeba histolytica*. *(b)* Diagram illustrating the formation of an intestinal abscess. The invasive trophozoites of *Entamoeba histolytica* erode the mucosa and submucosa.

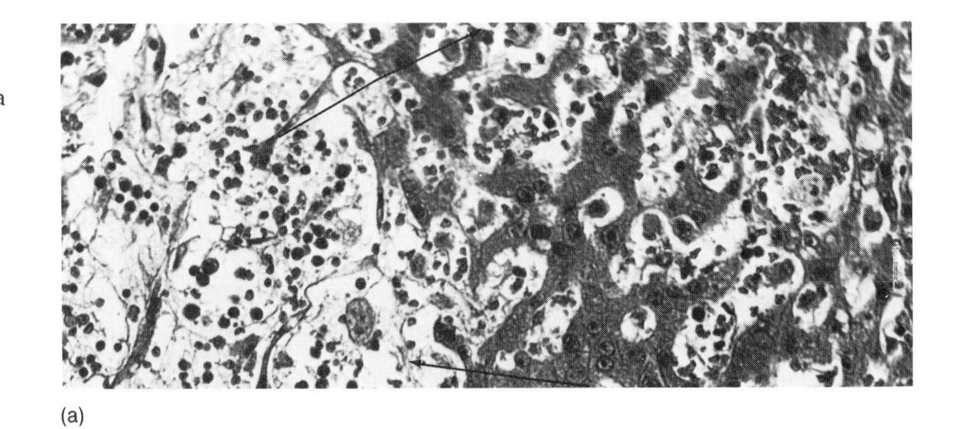

(a)

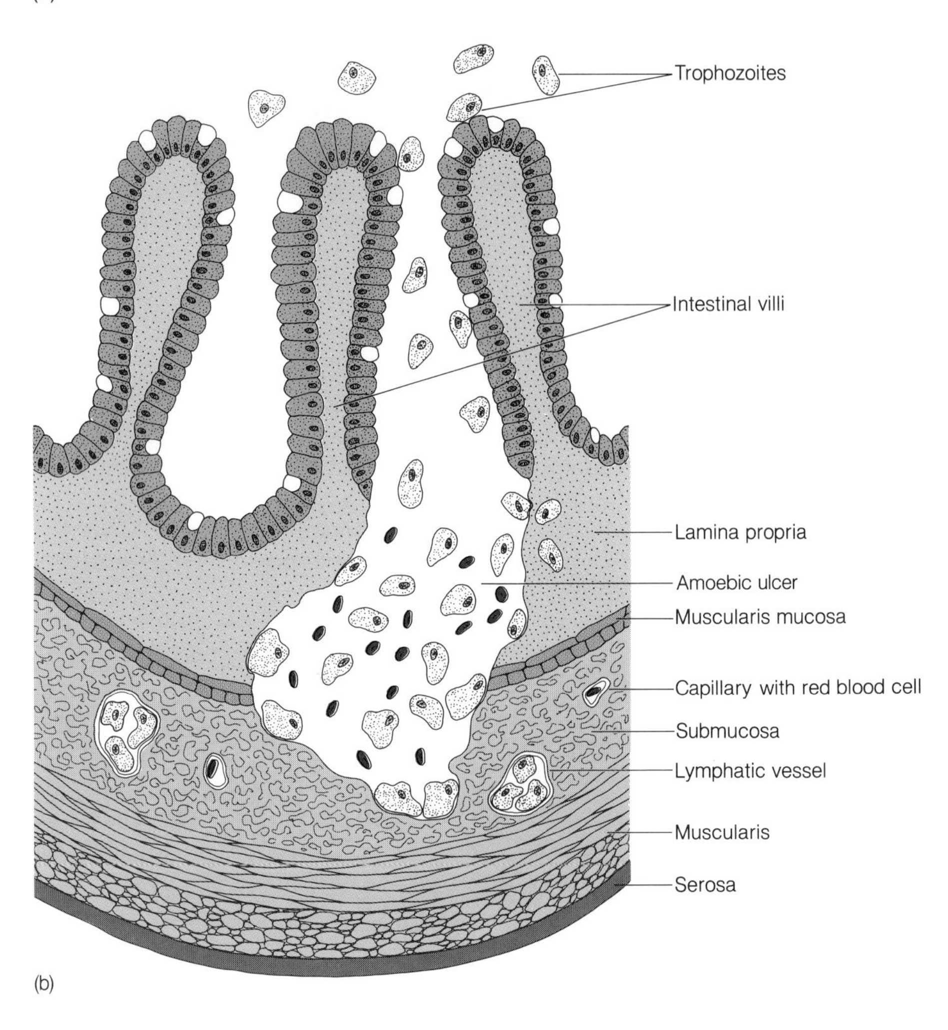

(b)

epithelium by the amoebae. The large bowel may develop an ulcer (fig. 18.11). Amebiasis may also be seen as a mild diarrhea, with no ulceration of the colon and little or no abdominal pain. The acute diarrhea is characterized by the presence of **trophozoites** while in chronic infections the **cyst** is most common. Liver abscesses occur when the amoebae spread to the liver; pain in the right upper quadrant is the most common complaint with this condition, and hepatomegaly (an enlargement of the liver) is often felt on palpation. Peritonitis can develop from either the rupture of the abscess or the perforation of an intestinal ulcer.

An infection is initiated by ingestion of cysts, either in food or in water. The amoebae excyst in the large intestine and then adhere to the mucosal epithelium, where they multiply and derive nutrients from cellular secretions and extracellular materials. They may remain in this state for a prolonged period of time without causing disease. It is not known why certain amoebae become invasive, but it is at this stage that the dysentery syndrome appears.

Prevention is achieved by avoiding untreated water or foods contaminated with human feces. Amoebicides such as metronidazole (Flagyl), diloxadine furoate (Furamide), or diiodohydroxyquin are used successfully to control amebiasis. In the last few years, the number of reported cases of amebiasis in the United States has slowly increased to over 5000 per year. The number of fatalities has decreased to around 20 per year, however, probably due to the availability of better amoebicidal drugs.

Figure 18.12 ***Giardia*** **in mouse intestines.** Scanning electron photomicrograph of *Giardia muris* on mouse jejunal mucosa. The parasite adheres to the mucosa, using its sucking disk. Lesions caused by the sucking disks are indicated. This adherence may be responsible for some of the signs and symptoms of giardiasis.

Giardia Enterocolitis

Giardiasis is an infectious disease caused by the flagellated protozoan *Giardia lamblia* (fig. 18.12). This flagellate infects the small intestine and causes diarrhea, abdominal cramps, nausea, vomiting, flatulence, and greasy stools. The infection apparently inhibits absorption of fats and certain vitamins by the human host.

The parasite enters the host in the form of a cyst in food or water. In the duodenum, it excysts and develops into a trophozoite. The trophozoites attach to the crypts (folds of intestinal mucosa) of the duodenum and jejunum with their ventral sucking disks (fig. 18.12) and multiply there. The ability of the trophozoites to adhere to the intestinal epithelium may account for the fact that the trophozoites are rarely found in stools and can be seen only in the severest of diarrheas. The disease might be confused with amebiasis, but is characterized by the foul-smelling, greasy stool and flatulence (intestinal gas).

The symptoms of the disease may be explained only in part by tissue destruction caused by the sucking disks. In fact, examination of biopsy materials reveals little, if any, invasion of the mucosa. On closer examination with an electron microscope, however, one can see that the organism may damage microvilli. This damage may result in some abnormality in the small intestine so that it does not absorb nutrients and electrolytes efficiently. The electrolyte imbalance may be the cause of the diarrhea.

Giardia lamblia is a cosmopolitan microorganism and is a common cause of travellers' diarrhea. Most of the infections are acquired by drinking contaminated water (table 18.2). Like those of *E. histolytica*, the cysts of *G. lamblia* are resistant to chlorine, so chlorinated water may still contain infectious organisms.

FOCUS INFANTS MAY BE PROTECTED FROM *GIARDIA* BY MOTHER'S MILK

Human milk contains an array of antimicrobial agents that protect the infant from infectious diseases. These antimicrobial agents include IgA and macrophages. In addition, there are many still-undefined antimicrobial agents in milk. One of them protects infants from parasitic protozoal diseases, such as **amebiasis** and **giardiasis.** Both of these diseases are common waterborne diseases in the United States and other parts of the world.

Giardiasis is particularly common in the United States, and infected children tend not to thrive. Some infants, however, recover rapidly and without treatment. Some investigators propose that because these infants are breast-fed that the anti-*Giardia* substances in the milk protect them from infection.

To test their hypothesis, investigators incubated *Giardia intestinalis* with normal human milk. In the presence of even low concentrations of human milk, the parasite was rapidly killed. The antimicrobial substance, believed to be a **bile salt-stimulated lipase,** is very effective in killing *Giardia.*

Although this phenomenon has been demonstrated only in culture systems, it is believed that it may also work *in vivo* and that it may be responsible for the protection of breast-fed infants against giardiasis.

Outbreaks in the United States are usually associated with the consumption of raw surface water or potable (drinking) water supplies contaminated with sewage. Although this protozoan is resistant to chlorination, proper sedimentation, flocculation, and filtration of potable water effectively remove the cysts. Giardiasis is now in epidemic proportions in some counties of California and it presents a hazard for backpackers and other wilderness campers unless they filter their drinking water.

The drug of choice for the treatment of giardiasis is quinacrine HCl (Atabrine). Because numerous side effects have been documented with this drug, however, metronidazole (Flagyl) has been used as an effective alternative. This observation is noteworthy because Flagyl is also used to control infections caused by anaerobic bacteria. Thus, the disease known as giardiasis may be a reflection of a secondary bacterial infection, as well as the primary infection by *Giardia lamblia*. In the last few years, about 11,000 cases of giardiasis have been reported annually to health authorities in the United States, with deaths averaging fewer than 1 per year.

VIRAL DISEASES OF THE DIGESTIVE SYSTEM

Viral Hepatitis

Hepatitis is an inflammation of the liver, manifested by anorexia (weight loss), jaundice, hepatomegaly (enlargement of liver), fever, and other constitutional disorders (fig. 18.13). Hepatitis may be caused by at least three different viruses: hepatitis A virus (HAV), hepatitis B virus (HBV) and non-A and non-B hepatitis viruses (NABV).

The hepatitis A virus traditionally was considered to be the agent of infectious hepatitis, acquired by the fecal-oral route, while the hepatitis B virus and the non-A, non-B hepatitis viruses were considered to be the agents of serum hepatitis, acquired by parenteral injections or by puncture wounds with contaminated utensils (table 18.4). Blood transfusions using blood containing HBV and NABV and the use of contaminated syringes for injecting drugs have been cited as common ways of acquiring serum hepatitis. Recent studies have revealed, however, that HBV and NABV can be acquired by the fecal–oral route as well. HBV disease has a much longer incubation period (50–160 days) than that caused by HAV (15–50 days).

The host-parasite interaction in hepatitis acquired by the fecal-oral route, regardless of the agent, is essentially the same. Upon ingestion, the hepatitis virus enters the GI tract and multiplies there, probably in the cells of the mucosal epithelium. During this time, the virus may be shed in the feces. Eventually, the viruses enter the circulatory system, causing a viremia, and some of the viruses lodge in the liver parenchymal cells, where they multiply further. The virus continues to be shed in the feces. Liver damage may result from the multiplication of the pathogen or from

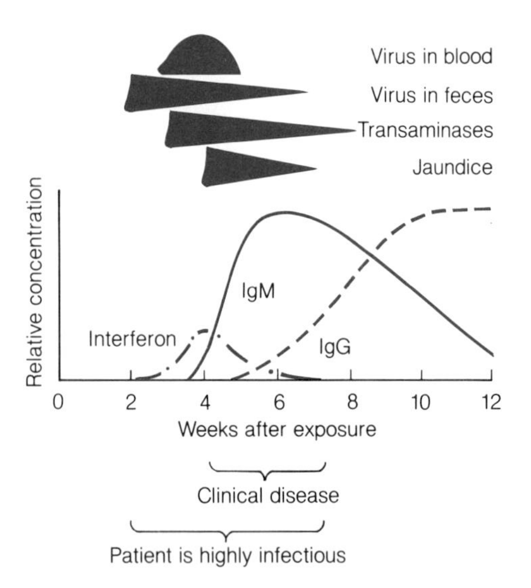

Figure 18.13 Course of hepatitis A virus infection in humans. Clinical disease caused by hepatitis A virus, which includes jaundice and liver dysfunction, is evident 3–4 weeks after infection. From week 2 to week 6, the patient is infectious and is shedding the virus in the feces. The peak of viremia (virus in the blood) is paralleled by an increase in interferon production. Immunoglobulins are not detectable until after 4 weeks post-infection.

immune reactions to viral antigens. Whether the damage results from humoral or cell-mediated immunity is still uncertain. Recovery, however, appears to coincide with the development of immunity and the inhibition of viral multiplication.

In epidemics, the source of the infection can almost always be traced to either water or food. Direct spread may also occur. Epidemics of hepatitis have been documented in which mentally handicapped children have acquired the infection by coprophagy. Shellfish and crustaceans living in contaminated water may serve as a source of infectious agent. Since there is no specific treatment available for hepatitis, the best course is prevention. No vaccines are available against HAV or NABV. However, there is a vaccine against HBV that is used for high-risk groups such as homosexuals and biomedical researchers. Each year in the United States there are 55,000–60,000 reported cases of hepatitis and 500–600 deaths (fig. 18.14). The actual number of hepatitis cases is estimated to be closer to 500,000/year and the number of deaths more than 5000/year (table 18.4).

Gastroenteritis of Viral Origin

Scientists estimate that fewer than 30% of the gastroenterites are of bacterial origin. What about the other 70%? Some of them are caused by protozoans, but the vast majority are caused by viruses. It is generally agreed that benign, self-limiting gastroenteritis with no definite bacterial etiology is of viral origin. The diseases are characterized by nausea, vomiting, diarrhea, fever, cramps, headache, and prostration.

In the average clinical laboratory, there are no facilities for isolating these agents from the specimens: therefore, diagnosis is made on epidemiological and clinical grounds and on the absence of bacterial or protozoal etiology. Viruses suspected of causing gastroenteritis in humans include picorna-, echo-, parvo-, and rotaviruses. Some of these viruses can cause waterborne and foodborne outbreaks. Outbreaks have also been traced to contaminated serving utensils.

A group of double-stranded RNA viruses, called the **rotaviruses**, are becoming increasingly important as agents of viral diarrheas. These viruses are the most common cause of a widespread disease called **infantile enteritis**. The rotaviruses cause considerable vomiting and diarrhea, which lead to dehydration. The

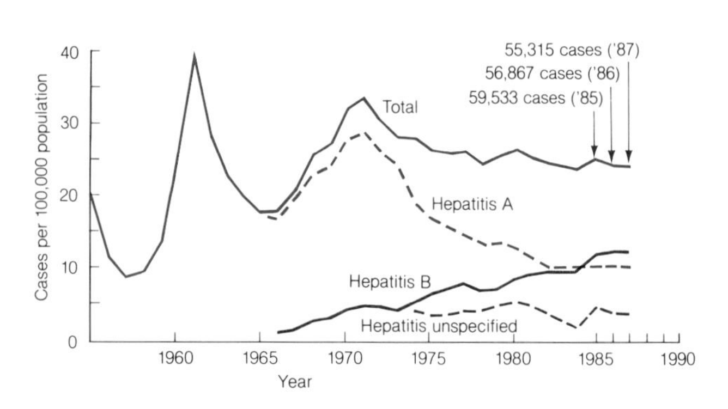

Figure 18.14 Reported cases of hepatitis in the United States

TABLE 18.4
COMPARISON OF HEPATITIS VIRUSES

Characteristic	Hepatitis A virus	Hepatitis B virus	Hepatitis Non-A Non-B
Incubation period	Short (15–50 days)	Long (50–160 days)	Short–long (15–140 days)
Most common route of infection	Fecal-oral	Injection, fecal-oral	Injection, fecal-oral
Nosocomial infection	Rare	Frequent	Frequent
Onset	Sudden	Insidious (gradual)	Insidious (gradual)
Serum transaminase (SGOT)	Temporary elevation	Prolonged elevation	Fluctuations
Humans most susceptible	Children	Adults	Adults
Time of year	Fall and winter months	Year-round	Year-round
High fever	Common	Rare	Rare
Serum immunoglobulin (IgM)	Elevated	Normal	?
Prevention using gamma-globulin	Successful	Doubtful	?
Viral antigens	Fecal antigen Liver antigen	Surface antigen (HB_b-Ag) Core antigen (HB_c-Ag)	?
Serum sickness	Rare (15%)	Rare (15%)	Very rare
Cases/year in U.S.	>100,000	>200,000	>200,000
Carriers/year in U.S.	Very rare	10,000	25,000
Deaths/year in U.S.	100	1000	4000

TABLE 18.5
SUMMARY OF INFECTIOUS DISEASES OF DIGESTIVE SYSTEM

Infectious diseases	Cause of disease	Microbial group
Amebiasis	*Entamoeba histolytica*	Amoebic protozoa
Ascariasis	*Ascaris lumbricoides*	Nematode
Bacillary dysentery	*Shigella* spp.	Gram − rods
Balantidiasis	*Balantidium coli*	Ciliated protozoa
Brucellosis	*Brucella melitensis*	Gram − rod
Cholera	*Vibrio cholerae*	Gram − curved rod
Diphyllobothriasis	*Diphyllobothrium latus*	Tapeworm (cestode)
Enterocolitis	*Clostridium difficile*	Gram + anaerobic sporeformer
Fascioliasis	*Fasciola hepatica*	Liver fluke (trematode)
Gastroenteritis	*Campylobacter fetus*	Gram − curved rod
Gastroenteritis	*Cryptosporidium*	Protozoa
Gastroenteritis	echoviruses	Picornaviruses
Gastroenteritis	parvoviruses	Parvoviruses
Gastroenteritis	rotaviruses	Reovirus
Giardiasis	*Giardia lamblia*	Flagellated protozoa
Hepatitis A	Hepatitis A virus	Picornavirus
Hepatitis B	Hepatitis B virus	Complex virus?
Hepatitis non-A non-B	Hepatitis non-A non-B virus	Retroviruses
Hookworm disease	*Ancylostoma* or *Necator*	Nematodes
Opistorchiasis	*Opistorchis sinensis*	Liver fluke (trematode)
Pinworms	*Enterobius vermicularis*	Nematode
Poliomyelitis	Poliovirus	Picornavirus
Salmonellosis	*Salmonella enteritidis*	Gram − rods
Schistosomiasis	*Schistosoma* spp.	Blood flukes (trematodes)
Taeniasis (tapeworm)	*Taenia solium* or *T. saginata*	Tapeworms (cestodes)
Traveler's diarrhea	*Escherichia coli*	Gram − rod
Traveler's diarrhea	*Giardia lamblia*	Flagellated protozoa
Trichinosis	*Trichinella spiralis*	Nematode
Typhoid fever	*Salmonella typhi*	Gram − rod
Weil's disease	*Leptospira* sp.	Spirochete
Whipworm disease	*Trichuris trichura*	Nematode
Yersiniosis	*Yersinia enterocolitica*	Gram − rod

infected children, most commonly 6–25 months of age, may also exhibit fever and upper respiratory tract involvement. Infected adults generally are asymptomatic. Most cases of rotavirus diarrhea occur in the winter and last 2 weeks. Apparently, the rotaviruses are widely distributed in human populations, and are usually transmitted by the fecal-oral route. The treatment of rotavirus-caused diarrheas is administration of fluids and electrolytes to replace those lost during vomiting and diarrhea.

Nematodes, tapeworms, flukes, bacteria, protozoans, and viruses are responsible for a large variety of diseases of the digestive system. Some of these are summarized in table 18.5.

CLINICAL PERSPECTIVE AN ALCOHOLIC WITH PERSISTENT DIARRHEA AND VOMITING

A 56-year-old alcoholic man was admitted to the hospital with persistent diarrhea and vomiting. His abdomen was enlarged and he was suffering from anorexia. The diarrhea, which consisted of several loose, yellow stools each day, and his abdominal swelling began approximately two weeks before he entered the hospital. The vomiting began two days before he sought medical attention.

The man's medical history indicated that he had been treated surgically two years earlier for lung cancer (squamous cell carcinoma) and was currently a heavy smoker. Six months earlier he was found to have fatty metamorphosis and fibrosis of the liver due to his chronic alcoholism.

Upon examination, the distended abdomen was found to be soft and tender. A distended abdomen is usually the result of an inhibition of peristalsis in the digestive tract and the accumulation of materials. This paralytic obstruction is known as adynamic ileus. Fluid and electrolytes are secreted into the intestinal lumen, but are not moved along by peristalsis. Consequently, the intestines become distended with accumulated materials. In addition, there was a large accumulation of fluid (ascites) into the body cavity. Significantly, chemical tests indicated minute amounts of blood (occult blood) in the stool. X-rays supported the idea of a mechanical bowel obstruction or adynamic ileus.

A diagnosis of spontaneous bacterial peritonitis and alcoholic cirrhosis could explain the major signs: adynamic ileus and ascites, and possibly the diarrhea. This initial diagnosis was weakened when the ascites and blood cultures were found to be free of bacteria and fungi. A study of the man's stool for parasites turned up a large number of filariform larvae of the nematode *Strongyloides stercoralis*. An explanation for the adynamic ileus, the abdominal swelling, diarrhea, and vomiting was edema of the intestines due to chronic nematode penetration of the intestinal mucosa.

The discovery of nematode larvae prompted the doctors to treat the patient with thiabendazole (25 mg/kg twice daily). Over a period of five days the patient's signs and symptoms gradually improved. A week after his discharge from the hospital, stool samples were negative for ova and parasites.

Strongyloides stercoralis is an intestinal nematode endemic in the southeastern United States where this man lived the first 26 years of his life. This nematode is able to reproduce both in the human host and in soil. It is hypothesized that a filariform larva in the soil entered the patient when he was a young boy or man through a lesion in the skin. It entered the blood system and reached one of the lungs which it penetrated. From the lung it migrated to the glottis where it was swallowed. In the intestines it invaded the mucosa of the duodenum or jejunum where the larva developed into a hermaphroditic female that laid eggs. These eggs develop into larvae that can be excreted in the stool or to filariform larvae that can penetrate the intestinal mucosa and the circulatory system. The cycle of autoinfection is believed to have gone on for as long as 30 years until the flare-up that brought the patient to the hospital. The flare-up was probably brought on by alcoholism, which compromised the patient's immune system. Flare-ups by the nematode are very serious because they can lead to hyperinfection syndrome that results from bacteria sepsis of the damaged intestines and the spread of the bacteria into the blood.

Strongyloidiasis should be suspected in any patient with persistent, unexplained enteric bacteremia and in chronically debilitated or immunocompromised patients who develop nausea, vomiting, diarrhea, abdominal pain, or adynamic ileus. As an indication of how important this pathogen is in some parts of the country, it has been observed that *Strongyloides stercoralis* is the most common intestinal parasite of patients in one university hospital in Kentucky.

For more details on this clinical perspective see the following article. Nahman, N. S. and Salt, W. B. 1983. *Hospital Practice* 18(9): 249–254.

SUMMARY

The digestive tract serves as a common route of human infection by microorganisms.

STRUCTURE AND FUNCTION OF THE DIGESTIVE SYSTEM

1. The digestive system is made up of the gastrointestinal tract and accessory organs. The GI tract is a system of tubes through which food passes and nutrients are absorbed into the blood and lymph. The accessory organs aid in the digestion and storage of nutrients.
2. The GI tract is made up of four tunics: the mucosa, the submucosa, the muscularis, and the serosa.

NORMAL FLORA OF THE DIGESTIVE SYSTEM

1. The digestive system is colonized by a variety of microorganisms. The distribution of these microorganisms varies, depending upon the anatomical site.
2. The normal flora of the digestive system is established soon after birth. It lives on excess host nutrients or mucous exudates.
3. The normal flora protects the host from invading pathogens, and some of these organisms supply the host with essential chemicals.

DISEASES OF THE ORAL CAVITY

1. The most important diseases of the oral cavity are dental caries, periodontal disease, and mumps.
2. Dental caries result after acidogenic microorganisms in dental plaque produce acid metabolites that break down the enamel and dentine. Various microorganisms invade the pulp of the tooth. Preventing the formation of dental plaques by brushing and flossing is a successful means of avoiding caries.
3. Periodontal disease results from an inflammatory reaction caused by microorganisms colonizing dental plaques near the gums.
4. Mumps is a childhood disease characterized by an inflammation of the salivary glands and is caused by the mumps virus. The disease, which can have serious consequences in adults, is prevented by vaccination.

INFECTIOUS DIARRHEAS CAUSED BY BACTERIA

1. Diarrhea, a symptom of many infections of the GI tract, is characterized by frequent, watery stools.
2. Infections of the small intestine generally result in profuse, watery discharge, while infections of the large intestine are accompanied by mucosal erosion and inflammatory exudate.
3. Cholera is caused by *Vibrio cholerae* and is characterized by profuse, whitish, watery diarrhea that can lead to dehydration, shock, and death. The disease is caused by the toxin choleragen, which is produced by the bacteria reproducing in the small intestine. The transmission of the disease is by the fecal-oral route. The disease is treated by replacement of body fluids and electrolytes. Tetracycline helps in alleviating the disease because it reduces the number of vibrios in the intestines.
4. Typhoid fever is caused by *Salmonella typhi*, and is characterized by high fever, headache, diarrhea, and rose spots. The disease results when the bacteria invade the blood and spread to the liver, multiplying inside macrophages and causing inflammatory responses. The disease is transmitted by the fecal-oral route and is treated with ampicillin or chloramphenicol.
5. Bacillary dysentery is caused by several species of the genus *Shigella*. It is an inflammation of the large bowel, resulting in tenesmus, diarrhea cramps, and inflammatory exudate. The disease is transmitted by the fecal-oral route. Severe infections are treated with ampicillin, chloramphenicol, or trimethoprim sulfamethoxazole.
6. *Salmonella* gastroenteritis is a diarrheal disease accompanied by headaches, abdominal pain, vomiting, and fever. The disease results from the multiplication of the bacteria in the small intestine and is acquired by the fecal-oral route. Treatment of the disease generally is not necessary.
7. Certain strains (or serogroups) of *Escherichia coli* are responsible for diarrheal diseases that resemble those caused by *Vibrio*, *Salmonella*, and *Shigella*. These diseases have epidemiologies similar to those of other diarrheal diseases.
8. *Campylobacter jejuni*, *Yersinia enterocolitica*, and *Clostridium difficile* are human pathogens that also cause diarrheal diseases.

FOOD INTOXICATIONS

1. Food intoxications are diseases that result from the ingestion of foods containing toxins produced by microorganisms.
2. Staphylococcal gastroenteritis is a very common

form of food poisoning, characterized by a rapid onset of vomiting and diarrhea. It is caused by the action of *Staphylococcus aureus* enterotoxins on the epithelium of the digestive tract.

3. Toxins produced by *Clostridium perfringens* cause a form of gastroenteritis consisting of abdominal pain, cramps, and diarrhea. The disease is acquired by ingestion of foods containing clostridial enterotoxins.

PROTOZOAL INFECTIONS OF THE DIGESTIVE SYSTEM

1. Amebiasis, a disease caused by *Entamoeba histolytica*, is characterized by dysentery. The disease results from the multiplication of the pathogen in the large bowel causing an acute inflammation. The cysts of *E. histolytica* enter the GI tract in contaminated water or foods. Treatment is by administration of Flagyl or Furamide.
2. Giardiasis is caused by *Giardia lamblia*. It infects the small intestine, causing diarrhea, nausea, vomiting, flatulence, and greasy stools. The characteristic disease, which results from the multiplication of the flagellate on intestinal epithelium, results in malabsorption of lipids and certain vitamins. Treatment of giardiasis is achieved by administration of Atabrine or Flagyl.

VIRAL DISEASES OF THE DIGESTIVE SYSTEM

1. Hepatitis is an inflammation of the liver which is caused by hepatitis viruses. It is characterized by jaundice, hepatomegaly, and fever. The viruses, HAV, HBV, and non-A non-B virus, enter the host either by the fecal-oral route or parenterally (e.g., by injections or blood transfusions). The disease is caused by the virus replicating in GI epithelium and liver parenchyma. A delayed-type hypersensitivity, which results from prolonged viral infection, may be responsible for some of the signs and symptoms of the disease.
2. Gastroenterites caused by viruses are quite common in human populations. They are generally benign and self-limiting, with nausea, vomiting, and diarrhea. Rotaviruses and parvoviruses are very common causes of viral diarrheas.

STUDY QUESTIONS

1. What is food poisoning? How does it differ from a food intoxication?
2. Describe the epidemiology of:
   a. cholera;
   b. typhoid;
   c. infantile diarrhea;
   d. hepatitis;
   e. giardiasis;
   f. travelers' diarrhea;
   g. bacillary dysentery.
3. For each of the above-mentioned diseases, describe the causative agent and discuss the way in which it causes the disease.
4. How do infections of the small intestine differ from those of the large intestine? Why?
5. Compare and contrast the diseases caused by *Shigella*, *Vibrio cholerae*, and *Salmonella typhi*.
6. How could infectious diseases of the GI tract be diagnosed? How about food intoxications?
7. Define the following:
   a. diarrhea;
   b. tenesmus;
   c. staphylococcal food poisoning;
   d. viral gastroenteritis;
   e. enterocolitis.
8. Explain the role that the following play in the transmission of enteric pathogens:
   a. food;
   b. water;
   c. aerosols;
   d. fomites;
   e. domestic animals.
9. Describe the role(s) of the normal flora of the GI tract.

SELF-TEST

1. Where in the digestive tract is the greatest mass of bacteria found?
   a. mouth;
   b. esophagus;
   c. stomach;
   d. small intestine;
   e. large intestine.
2. *Escherichia coli* is the predominant bacterium in the feces. a. true; b. false.
3. *Bacteroides*, *Bifidobacterium*, *Fusobacterium*, and *Eubacterium* represent more than 50% of the bacteria in the feces. a. true; b. false.
4. Of what benefit are the bacteria in the intestines?
   a. they produce some of the vitamin K necessary for blood clotting;
   b. they inhibit yeast growth in the intestines;

c. they stimulate the host's immune system;
d. all of the above;
e. there is no evidence that bacteria in the intestines are of benefit to the host.

5. The bacterium that is believed to be responsible for most enamel decay in children and adults is:
a. *Bacteroides* spp.;
b. *Streptococcus mutans;*
c. *Staphylococcus aureus;*
d. *Lactobacillus lactis;*
e. *Actinobacillus* and *Actinomyces.*

6. Periodontal disease is due to bacteria that cause pockets to form in the gums around the teeth. The pockets fill with bacteria and further deterioration of the ligaments and bone occurs. This process eventually results in tooth loss. a. true; b. false.

7. Which bacteria are believed to be responsible for periodontal disease?
a. *Escherichia coli;*
b. *Streptococcus mutans;*
c. *Staphylococcus aureus;*
d. *Lactobacillus lactis;*
e. *Actinomyces, Bacteroides, Actinobacillus.*

8. The paramyxovirus that causes mumps may infect and cause signs and symptoms in the:
a. testes;
b. meninges;
c. cerebrum;
d. salivary glands;
e. all of the above.

For items 9–12, match the definitions below with the following:
a. colitis;
b. enteritis;
c. gastritis;
d. gastroenteritis;
e. enterocolitis.

9. An inflammation of *only* the small intestine.

10. An inflammation of *only* the large intestine.

11. An inflammation of *only* the stomach.

12. An inflammation of *both* the small and large intestine.

For items 13–14, match the definitions below with the following:
a. diarrhea;
b. gastritis;
c. dysentery.

13. A frequent passage of watery stool.

14. A frequent bloody stool accompanied by abdominal cramps and painful spasmodic contraction of the bowel.

15. The cholera toxin affects the small intestine by blocking the uptake of sodium ions, which in turn results in the movement of water, chloride, and bicarbonate ions from the blood into the intestines. a. true; b. false.

16. Cholera is considered a dysentery rather then a diarrhea. a. true; b. false.

17. The organism that causes typhoid fever attaches to the small intestine and becomes an intracellular parasite of macrophages. a. true; b. false.

18. Typhoid fever is considered a dysentery rather than a diarrhea. a. true; b. false.

19. The inflammation and disease caused by *Shigella* is considered a dysentery rather than a diarrhea because this pathogen invades the large intestine and causes abdominal cramps, painful spasmodic contractions of the bowel, and a frequent bloody stool. a. true; b. false.

20. *Shigella* produces an enterotoxin that works in much the same way as the cholera enterotoxin. a. true; b. false.

21. *Shigella* produces a neurotoxin that causes bleeding and paralysis apparently by blocking protein synthesis. a. true; b. false.

22. *Shigella* releases an endotoxin that causes an allergic reaction. This contributes to the acute inflammatory reaction in the large intestine. a. true; b. false.

23. Which bacterium is required in the hundreds of thousands to cause disease? a. *Shigella;* b. *Salmonella.*

24. There are strains of *Escherichia coli* that are invasive and cause a dysentery similar to that caused by *Shigella dysenteriae.* a. true; b. false.

25. There are strains of *Escherichia coli* that produce an enterotoxin, which cause a disease known as "travelers' diarrhea." a. true; b. false.

For items 26–27, match the food intoxication to the organism that causes it:
a. *Clostridium botulinum;*
b. *Staphylococcus aureas;*
c. *Clostridium perfringens;*
d. *Bacillus cereus;*
e. *Aspergillus flavus.*

26. Rapid onset of nausea, vomiting, and diarrhea develop about 4 hours after consuming food (frequently potato salad, cream-filled pastries, and other prepared foods that are not refrigerated) contaminated by the enterotoxin produced by this bacterium. Duration depends upon the amount of toxin consumed.

27. Cramps and diarrhea begin more than 10 hours after consuming food (frequently unrefrigerated cooked meats) contaminated by the enterotoxin produced by

this bacterium. Symptoms last for 10–20 hours. Nausea and vomiting are rare.

28. Which hepatitis virus is generally transmitted through the fecal–oral route, generally does not result in a chronic infection, and causes about 100 deaths/year in the United States?
    a. hepatitis A;
    b. hepatitis B;
    c. hepatitis non A non B;
    d. hepatitis delta;
    e. does not fit the characteristic of any known hepatitis virus.
29. Which hepatitis virus is presently a problem in blood transfusions, is a retrovirus, flares up many years after first contracted, and is responsible for as many as 4000 deaths/year in the United States?
30. The majority of gastroenterites are caused by viruses such as the rotaviruses. a. true; b. false.

SUPPLEMENTAL READINGS

Binder, H. J. 1984. The pathophysiology of diarrhea. *Hospital Practice* 19:107–129.

Brandt, L. J. 1987. Colitis in the elderly. *Hospital Practice* 22(6): 165–188.

Cukor, G. and Blacklow, N. R. 1984. Human viral gastroenteritis. *Microbiological Reviews* 48(2):157–179.

Elwell, L. P. and Shipley, P. L. 1980. Plasmid-mediated factors associated with virulence of bacteria to animals. *Annual Review of Microbiology* 34:465–496.

Hollinger, F. B. 1987. Serologic evaluation of viral hepatitis. *Hospital Practice* 22(2):101–114.

Loesche, W. J. 1986. Role of *Streptococcus mutans* in human dental decay. *Microbiological Reviews* 50(4):353–380.

Mandel, I. D. 1979. Dental caries. *American Scientist* 67:680–687.

Marshall, B. J. 1987. Peptic ulcer: an infectious disease. *Hospital Practice* 22(8):87–96.

Middlebrook, J. L. and Dorland, R. B. 1984. Bacterial toxins: Cellular mechanisms of action. *Microbiological Reviews* 48(3):199–221.

Mirelman, D. 1987. Ameba-bacterium relationship in amebiasis. *Microbiological Reviews* 51(2):272–284.

Murphy, A. M., Borden, E.C., and Crewe, E. B. 1977. Rotavirus infections of neonates. *Lancet* 2:1149–1152.

O'Brien, A. D. and Holmes, R. K. 1987. Shiga and Shiga-like toxins. *Microbiological Reviews* 51(2):206–220.

O'Brien, A. D., Newland, J., Hiller, S., Holmes, R., Smith, H., and Formal, S. 1984. Shiga-like toxin-converting phages from *Escherichia coli* strains that cause hemorrhagic colitis or infantile diarrhea. *Science* 226:694–696.

Rubin, R. H., Hopkins, C. C., and Swartz, M. N. 1985. Infections due to gram negative bacilli. *Scientific American Medicine* Section 7-II. New York: Freeman.

Savage, D. C. 1977. Microbial ecology of the gastrointestinal tract. *Annual Review of Microbiology* 31:107–133.

Ward Moser, P. 1987. It must have been something you ate. *Discover* 8(2):94–100.

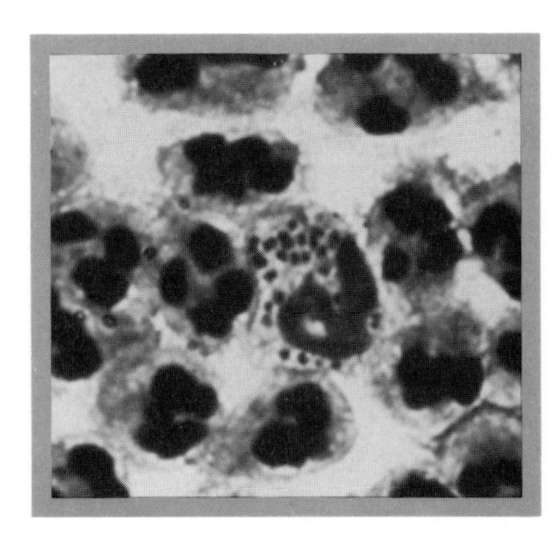

CHAPTER 19

DISEASES OF THE UROGENITAL TRACT

CHAPTER PREVIEW TOXIC SHOCK SYNDROME AND THE USE OF TAMPONS AND CONTRACEPTIVES

During 1980 and 1981 there were numerous outbreaks of toxic shock syndrome (TSS). Most of these cases were shown to be associated with the use of superabsorbent tampons during menstruation and the practice of not changing the tampons frequently. Because many women were uncertain about the safety of tampons, they looked for other ways of absorbing their menstruation. Some women turned to sea sponges, menstrual cups, and contraceptive sponges in the hope that these would reduce the risk of toxic shock syndrome. Unfortunately, there have been cases of toxic shock syndrome associated with the use of some of these materials.

In 1983, vaginal sponges made of polyurethane and impregnated with the spermicide nonoxynol-9 were released as contraceptives. The manufacturer's label recommended that the sponges not be left in place for more than 30 hours. Late in 1983, four cases of toxic shock syndrome were associated with the use of vaginal contraceptive sponges. One of the patients had been unsuccessful in removing the sponge because it fragmented. Another patient had not removed her sponge for 32 hours, and a third patient had the sponge in place for about 5 days. By keeping sponges in place for longer than about 8 hours, women increase the risk of *Staphylococcus aureus* growth and toxic shock syndrome.

A study has shown that women who wear birth control diaphragms for 24 hours, the maximum recommended by manufacturers, promote the growth of *Staphylococcus aureus* on the cervix and in the vagina. Normally, bacteria and their toxins are shed from the body by menstruation and vaginal secretions. The use of contraceptive diaphragms and sponges inhibits the normal shedding process and creates a favorable growth environment for the bacteria. Approximately 10 women per year risk acquiring toxic shock syndrome because of their use of contraceptive diaphragms.

A minimum estimate of the danger of contracting toxic shock syndrome from contraceptive sponges is 10 cases/100,000. Since approximately 250,000 women currently use the sponges each year, this amounts to 25 cases. In comparison, 2 cases/100,000 women who use tampons are expected each year. Because there are more than 25 million women in the United States who use tampons, however, most of the cases of toxic shock syndrome, about 500 per year, are associated with tampon use. The death rate due to toxic shock syndrome is about 3%.

Toxic shock syndrome is a much better understood disease now than it was in the early 1980s, largely because of the interest that it generated due to its severity. In this chapter we discuss some of the most common diseases of the urogenital tract, their causes, epidemiology, prevention, and treatment.

CHAPTER OUTLINE

DISEASES OF THE UROGENITAL TRACT MAY BE so mild as to be asymptomatic, or so severe as to cause long-lasting pain, tissue degeneration, or death. Some of these diseases, particularly sexually transmitted diseases, still carry a social stigma that makes patients reluctant to seek medical help or to reveal the identity of their contacts. This stigma makes it extremely difficult to evaluate the extent of the disease in a population or to control it. Some sexually transmitted diseases have reached pandemic proportions in human populations, and urinary tract infections (UTI) afflict nearly every human being at some time in his or her life.

Pathogens such as *Chlamydia trachomatis* and herpes simplex are now known to be the cause of some very common sexually transmitted diseases. Because of the increase in sexual freedom in the past 20 years, classical sexually transmitted disease agents have increased in prevalence, and microorganisms not usually transmitted sexually, have also emerged as important pathogens. Cytomegalovirus, hepatitis B, Group B streptococci, *Giardia lamblia*, and *Entamoeba histolytica* have now been associated with sexual transmission. Human immunodeficiency virus (HIV) is a newly evolved pathogen that causes acquired immune deficiency syndrome (AIDS). Because of its effect on the immune system, this sexually transmitted disease is discussed in Chapter 14.

STRUCTURE AND FUNCTION OF THE UROGENITAL TRACT

The Urinary System

The principal role of the urinary system is to eliminate wastes from the blood and to maintain a steady-state equilibrium between the fluid and the solutes of the blood, by voiding (urinating) wastes and excess fluids. In addition, the urinary system controls the volume of the blood by the amount of water it allows to be passed in the urine.

The human urinary system consists of two **kidneys**, two **ureters**, a **urinary bladder**, and the **urethra** (fig. 19.1). Each kidney empties fluid that has been removed from the blood into its ureter, which in turn empties into the urinary bladder. The urethra, which originates in the bladder, serves as a conduit for transporting the urine out of the body. This flow of urine dislodges many of the microorganisms in the urethra.

On gross examination, the kidney consists of a thin surface layer, the capsule, a cortex, and a medulla. The cortex contains primary **glomeruli** (fig. 19.1), in which diffusion of fluids and solutes from the blood into the kidney tubules takes place. The electrolytes and fluids that pass across the glomeruli enter the convoluted tubules (fig. 19.1) and then drain into the papillary ducts, which merge into the calyx of the medulla. The glomerulus, convoluted tubules, and papillary duct make up the **nephron**, the functional unit of the kidney.

The Reproductive System

The reproductive system generates gametes, provides a means for fertilizing eggs, and nurtures the resulting embryo. In humans, as in other mammals, male and female sexes are well differentiated. Figure 19.2 illustrates the male and female reproductive systems.

The male reproductive system includes the **testes**, which produce the male gametes, or **sperm**, and a number of ducts that serve to store the sperm and transport them to the outside. In addition, there are several accessory glands that secrete substances to protect the sperm and form the semen.

The female reproductive system includes the **ovaries**, which produce the female gametes or **ova**; the **uterus**; the **fallopian tubes**, which serve as conduits for the ova from the ovaries to the **uterus**; and the **vagina**. The uterus is a muscular organ that serves as a receptacle for the fertilized ovum or zygote. Fetal development takes place in the uterus. The vagina serves as a receptacle for the penis during coitus or

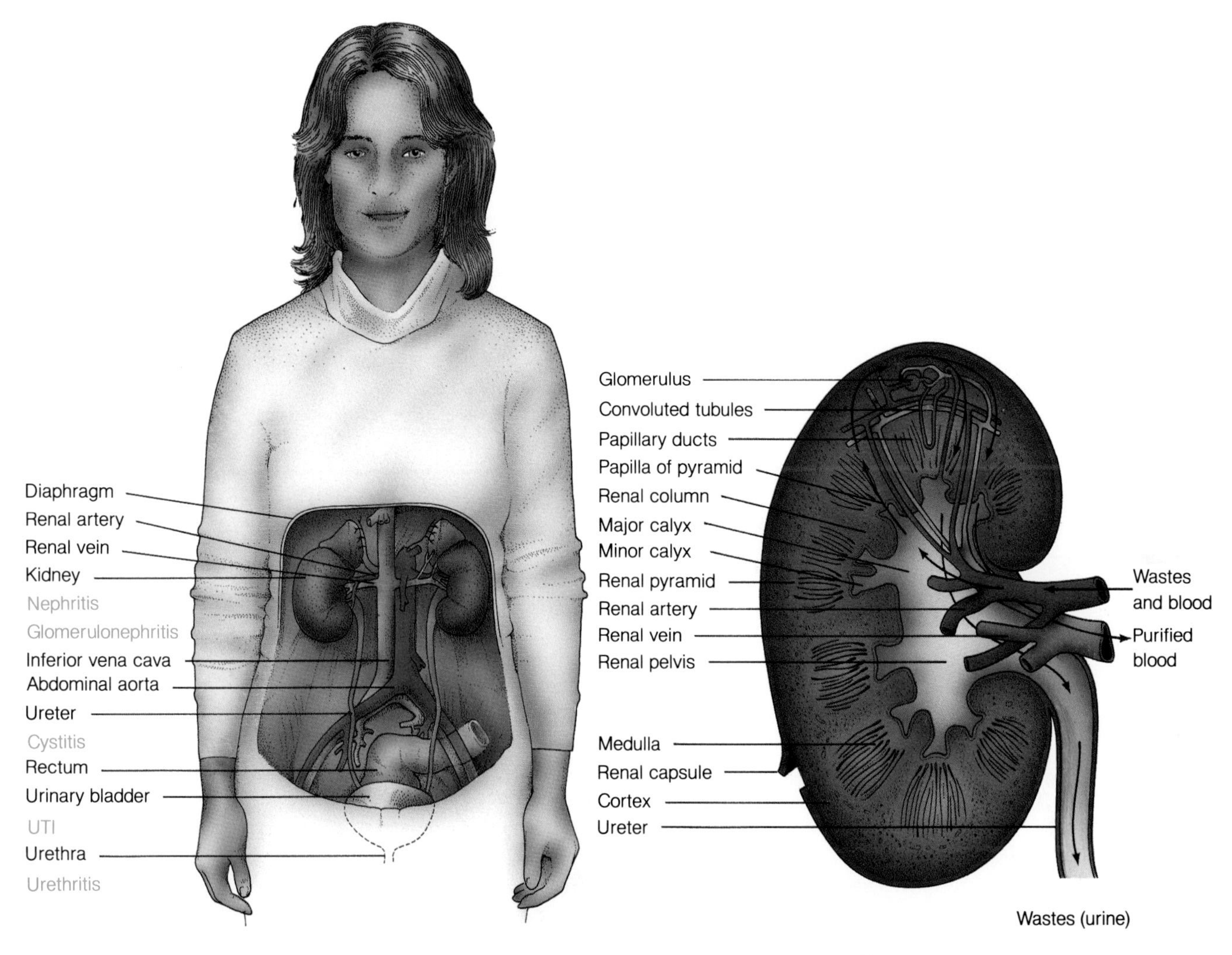

Figure 19.1 **Anatomy of the urinary system.** Colored print indicates diseases that afflict the anatomical site.

sexual intercourse and also as an exit for the developed fetus.

NORMAL FLORA OF THE UROGENITAL SYSTEM

Most of the urinary tract, except for the anterior urethra near the external orifice, normally is free of microorganisms because of urine flow. The anterior urethra, however, contains an array of bacteria, including staphylococci, diphtheroids, and gamma hemolytic streptococci (table 19.1). The female urethral opening may also contain lactobacilli.

The vagina and cervix have a very complex and dynamic flora (table 19.1). Shortly after birth, the female baby's vagina is colonized predominantly by lactobacilli, because progesterone from the mother promotes the multiplication of the resident bacteria in the baby's vagina. After the maternal hormonal influences wane in the baby, the lactobacilli decline in numbers and the vagina becomes alkaline. During the time the vagina is alkaline, the normal flora consists of diphtheroids,

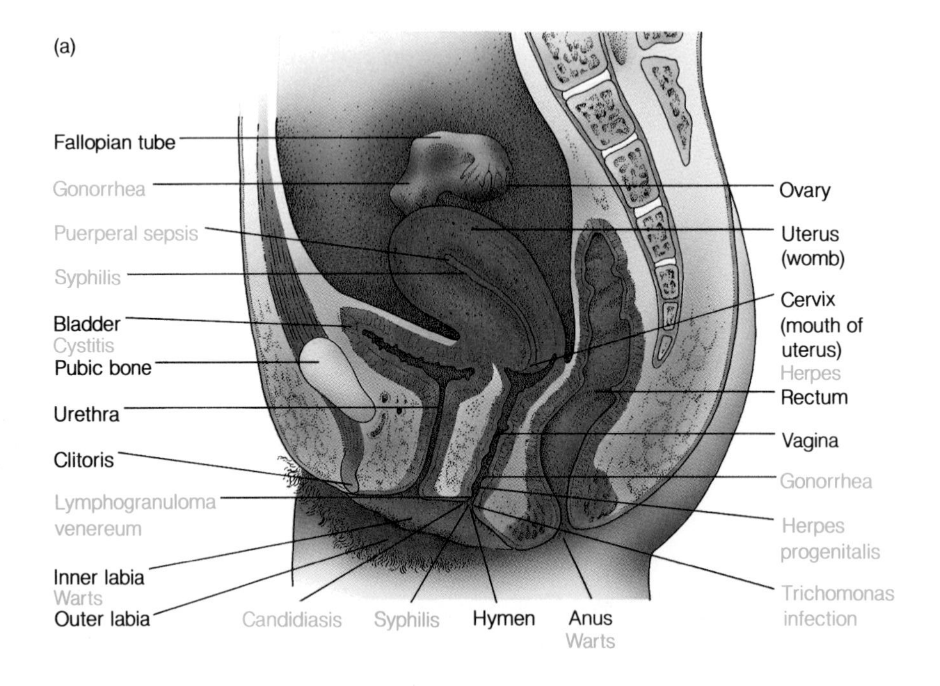

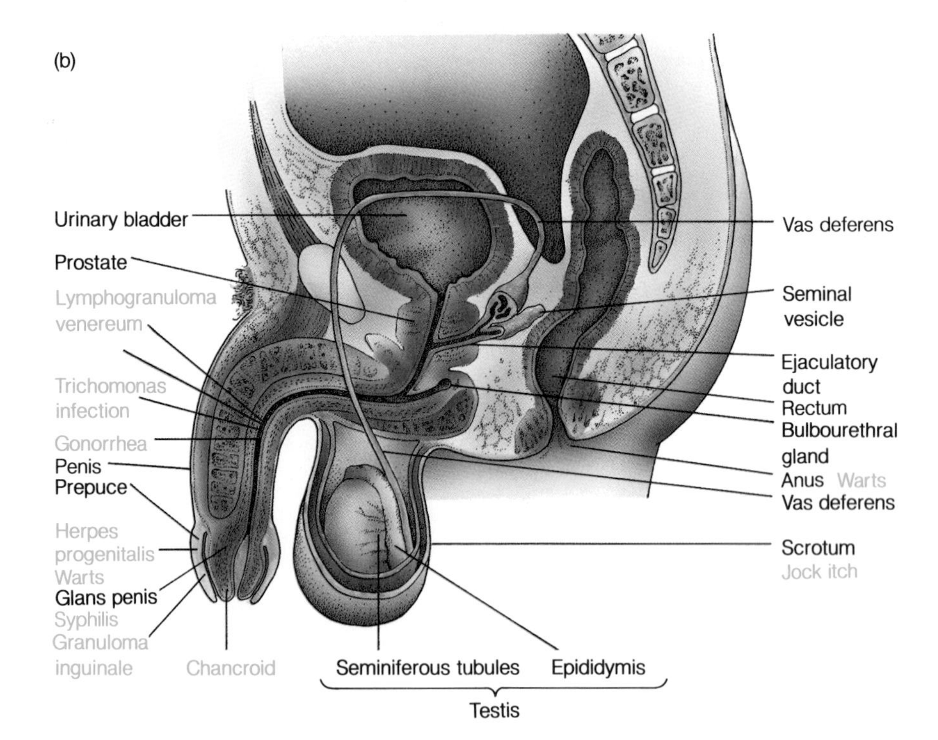

Figure 19.2 The human reproductive system. Colored print indicates infectious diseases that afflict the anatomical site. *(a)* Female reproductive system. *(b)* Male reproductive system.

FOCUS *UREAPLASMA* **LINKED TO HUMAN INFERTILITY**

It is estimated that 8 million or 15% of all couples in the United States are infertile, and that half of them are infertile because of infections. The mycoplasma *Ureaplasma urealyticum* is responsible for approximately 2.4 million or 60% of the infections that make couples infertile. Although *Ureaplasma* infects the genitals in both men and women, it is not clear how it causes infertility.

When both partners are treated with appropriate antibiotics, the bacteria are eliminated in over 80% of the couples. Sixty percent of the treated couples who become free of *Ureaplasma* are then able to have children, compared to only 5% of the untreated, *Ureaplasma* infected couples.

Chlamydia trachomatis (the agent of nonspecific urethritis and lymphogranuloma venereum) and *Neisseria gonorrhoeae* (the agent of gonorrhea) are believed to account for much of the remaining infertility (40%) due to infectious agents. Long-term *Chlamydia* and *Neisseria* infections destroy the fallopian tubes and consequently make women infertile.

Yearly, approximately 1–2.5 million people in the United States suffer from *Ureaplasma* infections, 1 million from *Neisseria* infections, and 3 million from *Chlamydia* infections.

TABLE 19.1
MICROORGANISMS COMMONLY ISOLATED FROM THE UROGENITAL SYSTEM

Microorganism	Microbial group	Average frequency of isolation
Actinomyces sp.	Actinomycetes	25%
Bacteroides sp.	Gram − rods	70%
Bifidobacterium sp.	Gram − rod (anaerobe)	<10%
Candida albicans	Yeast	40%
Clostridium sp.	Gram + spore formers	25%
Diphtheroids	Gram + rods (aerobe)	60%
Enteric bacteria	Gram − rods	25%
Enterococci	Gram + cocci	70%
Eubacterium sp.	Gram − rod (anaerobe)	<10%
Gardnerella vaginalis	Gram − coccobacilli	30%
Lactobacillus sp.	Gram + rods	75%*
Mobiluncus sp.	Gram − curved rod	25%
Peptostreptococcus sp.	Gram + cocci (anaerobe)	35%
Staphylococcus sp.	Gram + cocci	75%
Trichomonas vaginalis	Flagellated protozoa	17%

*Most adults carry *Lactobacillus*, but prepubertal children do not often carry these organisms.

enterococci, and anaerobes, such as *Bacteroides* and *Peptostreptococcus*. Hormonal influences at the onset of menstruation (10–15 years) cause the lactobacilli to reappear and become predominant because of the new supply of glycogen. They consume most of the nutrients in the vagina and release acids, thus maintaining a low pH in the vagina which inhibits many microorganisms from proliferating and causing disease.

An important function of the indigenous flora is to keep potential pathogens in check. For example, competition for nutrients, combined with the low pH resulting from the fermentation of carbohydrates in the adult vagina by the lactobacilli, restricts the population size of the yeast *Candida albicans*. If for any reason the lactobacilli are reduced in numbers, the increased nutrients available stimulate the *C. albicans* population to increase in size and cause vaginitis. Interestingly enough, *C. albicans* is capable of inhibiting the proliferation of *Neisseria gonorrhoeae*, the cause of gonorrhea, in the normal human vagina.

DISEASES OF THE URINARY TRACT

Urinary tract infections are very common infections in humans. They range from asymptomatic infections to full-blown kidney disease that can result in shock and death. Symptoms vary, depending upon the agent causing the infection and the anatomical site that is invaded. Symptoms frequently include **cystitis** (inflammation of the urinary bladder) with **dysuria** (difficulty in urination), urinary frequency, **hematuria** (blood in the urine), low back pain, and fever.

The diagnosis of a urinary tract infection is generally based on the isolation of a large number of bacteria from the urine. An infection is clearly indicated when 100,000 or more bacteria are found per milliliter of clean-catch voided urine. Mixed infections have been reported, but more than 90% of all urinary tract infections are caused by a single species (table 19.2).

Gram-negative enterics from stool, primarily *Escherichia coli* and *Proteus mirabilis,* account for most of the cases of urinary tract infections. These bacteria colonize the anterior urethra and from there may reach the bladder. The kidneys may also be colonized by bacteria that infect the bladder, often causing a disease called **pyelonephritis**. The bacteria reach the kidneys most commonly via the ureters. It has been estimated that more than 95% of urinary tract infections result from an ascending infection (i.e., urethra → bladder → ureters → kidney). Another route, albeit considerably less common, is through the bloodstream. These infections result from invasion of the urinary tract from other sites via the blood. These infections are caused most often by salmonellae and staphylococci.

Urinary tract infections often start because of toilet habits, incontinence, or sexual habits. Since *E. coli, P. mirabilis, Klebsiella pneumoniae,* and *Enterobacter aerogenes* represent a minority of the total bacteria in the feces, it is noteworthy that they are responsible for 95% of all infections of the urinary tract. The pathogenicity of these organisms is believed to be due to their ability to attach to urinary tract epithelium. This ability permits the bacteria to multiply within the urinary tract without being subject to the flushing action of urine. In *E. coli,* the pili and the capsule both appear to influence the ability of the bacteria to cause disease. The pili are mainly responsible for attaching the bacteria to the epithelium. Although the capsule may also have adhesive properties, its primary role is thought to be that of protecting the bacteria from host defenses such as immunoglobulins or phagocytes. Similar mechanisms probably account for the virulence of *P. mirabilis*.

TABLE 19.2
SIX MOST COMMON AGENTS OF URINARY TRACT INFECTIONS IN HUMANS

Causative agent	Frequency of isolation from patient's urine
Escherichia coli	90%
Proteus mirabilis	3%
Klebsiella pneumoniae	2.5%
Streptococcus faecalis	2%
Enterobacter aerogenes	1%
Pseudomonas aeruginosa	0.5%

About 5–10% of pregnant women are subject to urinary tract infections with **bacteruria** (bacteria in the urine). These cases may be due to the fact that, during pregnancy, the fetus sometimes partly blocks the urethra and reduces the flow of urine. Thus, bacteria in the urethra are not flushed out efficiently and the bladder retains residual amounts of urine, which serves as a suitable culture medium for the growth of the bacteria.

Host defenses against urinary tract infections are both local and systemic. The urinary mucosa, predominantly that of the bladder, has many antibacterial substances that inhibit microbial growth. Associated with the urinary tract epithelium is secretory IgA that may bind to pili and inhibit the attachment of potential pathogens. Anticapsular antibodies may also be present. Cell-mediated immunity is also evident in the urinary tract.

The prevalence of urinary tract infection in human populations varies depending on the age and sex of the individuals. During infancy, males are almost twice as susceptible as females to urinary tract infection. This situation quickly changes, however, and the susceptibility of females to urinary tract infection increases with age. For example, preschool girls are 10 times as susceptible as boys of the same age, but adult females are almost 50 times as susceptible as adult males. The high incidence of urinary tract infection in females is partly due to the much shorter urethra and its proximity to the anus. In old age, males and females are equally susceptible, possibly due to a decreasing immune response or hormonal changes in both males and females.

Control of urinary tract infection involves toilet practices and sexual habits that prevent contamination of the urethral opening with feces. Management of urinary tract infections involves treatment with antibiotics. Amoxicillin, tetracycline, cephalosporin, or trimethoprim sulfamethoxathole generally are used to treat urinary tract infections.

DISEASES OF THE GENITAL TRACT

Many microorganisms are capable of colonizing the human genital tract, and some of these can cause diseases that are transmitted sexually or by intimate person to person contact. Not all infectious diseases of the genital tract are transmitted sexually, however; some are caused by commensals that multiply rapidly under certain conditions, causing acute inflammations. This section discusses both sexually transmitted and opportunistic pathogens (table 19.3).

Candida albicans Vaginitis

Vaginitis caused by *Candida albicans*, and less fre-

TABLE 19.3
SEXUALLY TRANSMITTED DISEASES OF THE UROGENITAL TRACT

Infectious disease	Causative agent	Group	Comments
Actinomycosis	*Actinomyces israelii*	Actinomycete	Associated with use of IUDs*
AIDS	HIV	Retrovirus	Common in homosexual men and intravenous drug abusers
Amebiasis	*Entamoeba histolytica*	Amebic protozoa	Occurs in homosexual men
Chancroid	*Haemophilus ducreyi*	Gram − rods	Develop soft chancre
Condyloma acuminatum	Papilloma virus	Papilloma viruses	Warts of external genitalia
Crabs (pediculosis)	*Phthirus pubis*	Louse	Causes severe itching
Cytomegalovirus infection	Cytomegalovirus	Herpesvirus	Fetal infection during pregnancy
Cystitis	*Escherichia coli*	Gram − rods	Commonest cause of UTI**
Genital herpes	Herpes simplex virus	Herpesviruses (HSV-2)	>1,000,000 cases/yr
Giardiasis	*Giardia lamblia*	Flagellated protozoa	Occurs in homosexual men
Glomerulonephritis	*Streptococcus pyogenes*	Gram + cocci	Nonpyogenic sequelae
Gonorrhea	*Neisseria gonorrhoeae*	Gram + cocci	More than 1 million cases yearly
Granuloma inguinale	*Calymmatobacterium* sp.	Gram − rods	Also known as Donovan bodies
Hepatitis A	Hepatitis A virus	Picornavirus	Common in homosexual men
Hepatitis B	Hepatitis B virus	Complex virus	Common in homosexual men
Hepatitis non-A non-B	Hepatitis non-A non-B virus	Retrovirus	Common in homosexual men
Hepatitis D	Hepatitis delta virus	Virus	Defective virus that requires hepatitis B virus to reproduce
Lymphogranuloma venereum	*Chlamydia trachomatis*	Chlamydia	Also causes trachoma
Nonspecific urethritis	*Pseudomonas*, coliforms	Gram − rods	Caused by various bacteria
Nonspecific vaginosis	*Gardnerella vaginalis*	Gram − coccobacilli	May be due to mixed flora infection, including *Mobiluncus*
Pinta	*Treponema carateum*	Spirochetes	Prevalent in the tropics
Proctitis	*Campylobacter jejuni*	Gram − curved rods	Common in homosexual men
Puerperal sepsis	*Peptostreptococcus* sp.	Gram + cocci (anaerobe)	Also caused by other bacteria
Pyelonephritis	*Bacteroides*, enterics	Gram − rods	Inflamed kidney & pelvis
Reiter's syndrome	*Chlamydia* sp.	Chlamydias	Conjunctivitis + arthritis + urethritis
Shigellosis	*Shigella* sp.	Gram − rods (enterics)	Common in homosexual men
Syphilis	*Treponema pallidum*	Spirochete	35,000 cases annually
Thrush	*Candida albicans*	Yeasts	Also causes balanitis***
Trichomoniasis	*Trichomonas vaginalis*	Flagellated protozoa	A form of vaginitis
Urethritis, nongonococcal	*Ureaplasma urealyticum* *Chlamydia trachomatis*	Mycoplasmas Chlamydia	>1,000,000/yr afflicted 30–50% of NGUs

*IUD, Intrauterine device.
**UTI, Urinary tract infection.
***Lesions of the penis.

quently by *C. tropicalis* and *Torulopsis glabrata,* is characterized by a thick, cream-colored vaginal discharge. The vaginal mucosa may be spotted with patchy lesions, which are gray-white pseudomembranes consisting of host tissue and fungal material. In more serious cases, the lesions may also involve the perineum and inguinal area (fig. 19.3). Acute inflammatory reactions resulting from the infection lead to swollen, erythematous (reddened) lesions that may itch severely.

Diabetes, prolonged antibiotic treatment, and pregnancy often promote the growth of the yeast and therefore predispose to vaginitis. Conditions that promote an increase in the progesterone level, such as the use of some oral contraceptives, also promote the growth of *C. albicans,* possibly because the increase in glycogen deposition in the vagina provides additional nutrients for the yeasts. A condition similar to vaginitis in males is **balanitis**, a disease characterized by inflammatory lesions of the penis. This is an uncommon condition and is usually acquired through sexual activities with an infected individual.

The primary host response to *C. albicans* infections is an acute inflammation. The PMN leukocytes effectively control most infections by this yeast; however, the lesions develop as a result not only of the infection but also of the inflammation. Secondary colonization of the lesions by bacteria is common, and they can be seen in abundance in stained clinical specimens. These bacteria may be transient or permanent residents of the vagina.

Since *C. albicans* can be considered to be an opportunist, control of the infection usually involves measures to eliminate the predisposing factor. When diabetes is the predisposing factor, control of the condition with insulin will generally help resolve the infection. If the cause of vaginitis is a depleted *Lactobacillus* flora, replacement of the normal flora can often solve the problem. Douches containing buttermilk or acidophilus milk can be administered to reintroduce the lactic acid bacteria into the vagina. Ointments containing nystatin (mycostatin) may also be applied to the lesions to control the infection.

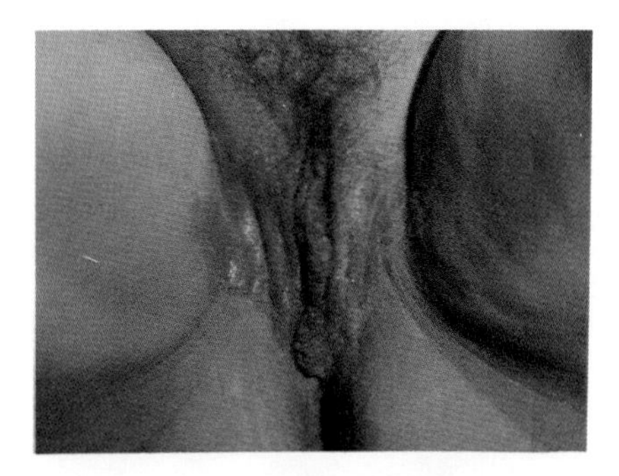

Figure 19.3 ***Candida* vaginitis.** Infection of the vagina and inguinal area by *Candida albicans.* The reddening of the inguinal area is a result of an inflammatory response and reflects an active infection by the yeast. Courtesy of Professor F. Camacho, Department of Dermatology and Venereology, University Hospital, Seville, Spain.

Trichomonas vaginalis Vaginitis

Vaginitis caused by *Trichomonas vaginalis* is a common condition in females and consists of burning and itching of the vulvar area, accompanied by a profuse, whitish exudate (pH 5–6). *Trichomonas vaginalis,* which may be found in profusion in the exudates, is a flagellated protozoan that colonizes the mucosal epithelium of male and female genitalia (fig. 19.4). This organism is widely distributed, and it is estimated that 20–40% of females may be infected worldwide.

In males, many infections are asymptomatic and the incidence of the clinical disease is 5–10% of those infected. *Trichomonas vaginalis* in males is found in the ureters, bladder, prostate gland, epididymis, and urethra. It is thought that males serve as a source of infection for females. The infection is transmitted primarily by sexual intercourse, but infections can also be acquired by contact with fomites such as contaminated toilet articles and clothing.

The growth of the parasite and the production of toxic metabolic products on the epithelium may cause an inflammation that results in the symptoms of the disease. Diagnosis of trichomoniasis is usually made by noticing the characteristic discharge and by demonstrating the parasite in wet mounts of the discharge. Oral metronidazole (Flagyl) is very effective in controlling the infection. Best results are obtained when both sexual partners are treated simultaneously.

Bacterial Vaginosis

Bacterial vaginosis is an infection of the human vagina characterized by a homogeneous, malodorous (often "fishy") vaginal discharge with a pH of 4.5–5.0. Epithelial cells (called **clue cells**) with adhering coccobacilli, presumably the etiologic agent, are frequently seen in wet mounts of the discharge.

The cause of nonspecific vaginosis is still in question, although a gram-negative coccobacillus called *Gardnerella vaginalis* is found in large numbers in the exudates of about 50% of patients with the disease. Approximately 20% of asymptomatic women also carry this bacterium in low numbers. Anaerobic bacteria such as *Bacteroides*, *Peptostreptococcus*, and *Mobiluncus* are also found in the human vagina and have been implicated in the disease process, along with *G. vaginalis*. At present, few laboratories routinely culture for *G. vaginalis* in clinical specimens. On rich culture media with 5% human blood at an elevated (5%) carbon dioxide atmosphere, *G. vaginalis* forms small, raised colonies with a distinct zone of diffuse beta hemolysis which is characteristic. Metronidazole (Flagyl) or broad-spectrum antibiotics are used to treat the infection successfully.

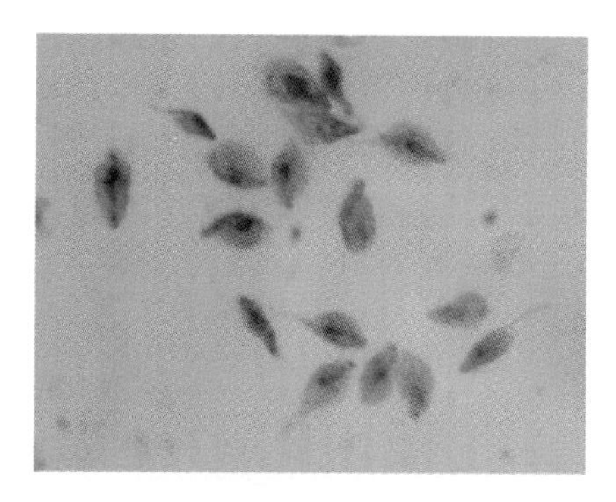

Figure 19.4 ***Trichomonas vaginalis.*** Wet mount of a vaginal exudate from a patient with *Trichomonas vaginalis*. Notice the flagellated cells with a pointed end (axostyle) that is characteristic of *Trichomonas*. Courtesy of Dr. R. Cano.

Bacterial Toxins Released into the Vagina Cause Disease

Toxic shock syndrome is characterized by hypotension (low blood pressure), shock, fever, rash, sore throat, conjunctival infection, muscle aches, and a variety of other symptoms, such as diarrhea and vomiting. The rash resembles that seen in scarlet fever, but is usually followed by peeling (desquamation) of the hands and soles. The disease gained notoriety in 1980, when a number of cases were associated with the use of superabsorbent tampons (fig. 19.5). Even though this condition has been found primarily in menstruating individuals, it has also occurred in persons with furuncles, carbuncles, impetigo, and other forms of pyoderma. The disease has been studied extensively since 1980 and the evidence points to toxigenic strains of *S. aureus* as the cause of toxic shock syndrome. Two different toxins, **pyrogenic exotoxin C** and **enterotoxin F**, have been implicated in the disease process.

Toxigenic strains of *S. aureus* colonize tampons and damaged tissue and release the toxin(s). The toxins cause the characteristic signs and symptoms of toxic shock syndrome. Before the introduction of the superabsorbent tampon in about 1977, only 2 to 5 cases of toxic shock syndrome were reported in the United States each year. The number of cases jumped dra-

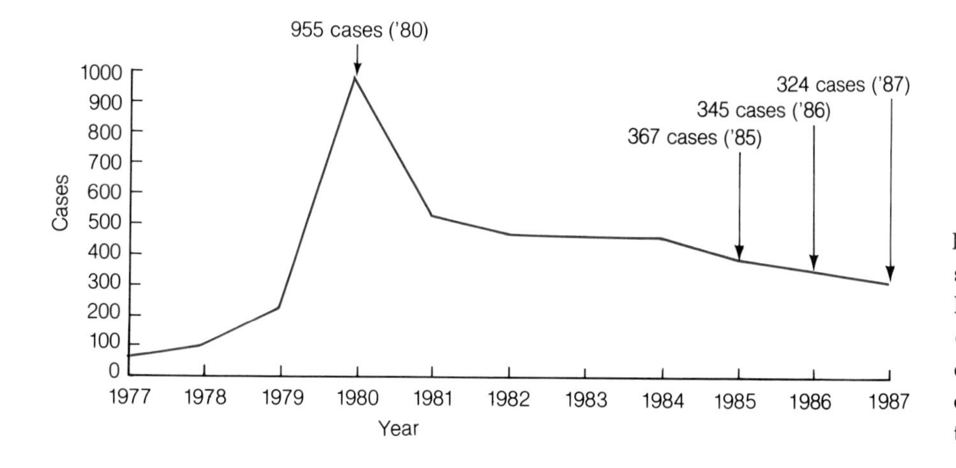

Figure 19.5 Reported cases of toxic shock syndrome in the United States. Most cases of toxic shock syndrome occur in menstruating females. These data led investigators to deduce that certain tampons were associated with the disease.

matically each year from 1978 to 1980, when over 900 cases were reported with a mortality of about 3%. In 1984, 461 cases were reported to CDC.

Because some of the superabsorbent tampons are kept in place for extended periods of time, they concentrate nutrients and provide a rich, moist environment for the growth of *S. aureus*. The chances of contracting the disease can be reduced by not using superabsorbent tampons, by changing tampons three times a day, and by not using them at night.

Treatment of the disease requires the immediate replacement of fluids and electrolytes to reverse hypotension and shock. Abscesses, if present, must be drained, and antibiotics such as cephalosporin and β-lactamase-resistant penicillins must be started as soon as possible.

SEXUALLY TRANSMITTED DISEASES (STD)

The incidence of **sexually transmitted diseases** (table 19.3) has increased significantly in the last 20 years. The increase has included not only the classical venereal diseases, such as syphilis and gonorrhea, but also some newly recognized diseases such as nongonococcal urethritis, genital herpes, and nonspecific vaginitis. In addition, there has been an alarming increase in the number of cases of diseases that normally affect the gastrointestinal tract. For example, amebiasis, giardiasis, and hepatitis can sometimes be acquired by the sexual route. These and the more conventional diseases transmitted by sexual activities are summarized in table 19.3.

Gonorrhea

Gonorrhea, commonly known as clap, the drip, the dose, and the strain, is a sexually transmitted disease caused by *Neisseria gonorrhoeae* and characterized by an acute inflammation of the genital mucosal epithelium (both male and female), and a purulent discharge. The bacterium that causes the disease is a gram-negative diplococcus that is normally seen in the purulent discharge inside polymorphonuclear leukocytes, where it sometimes multiplies.

When introduced into a suitable host, the gonococcus attaches to the mucosal epithelium, presumably through its pili. Only pilated strains of *N. gonorrhoeae* are thought to cause gonorrhea, at least in natural infections. However, pili alone do not make the gonococcus virulent. A cell wall component called IgA protease functions in the pathogenesis of gonorrhea by cleaving antigonococcal IgA antibodies. As we know, IgA is associated with mucosal secretions and provides local immunity. Once *N. gonorrhoeae* is attached to the epithelium, it begins to multiply, causing localized tissue damage and eliciting an inflammation.

Gonorrhea is present in epidemic proportions in our society (fig. 19.6). Among those diseases that physicians are required to report, gonorrhea is the most

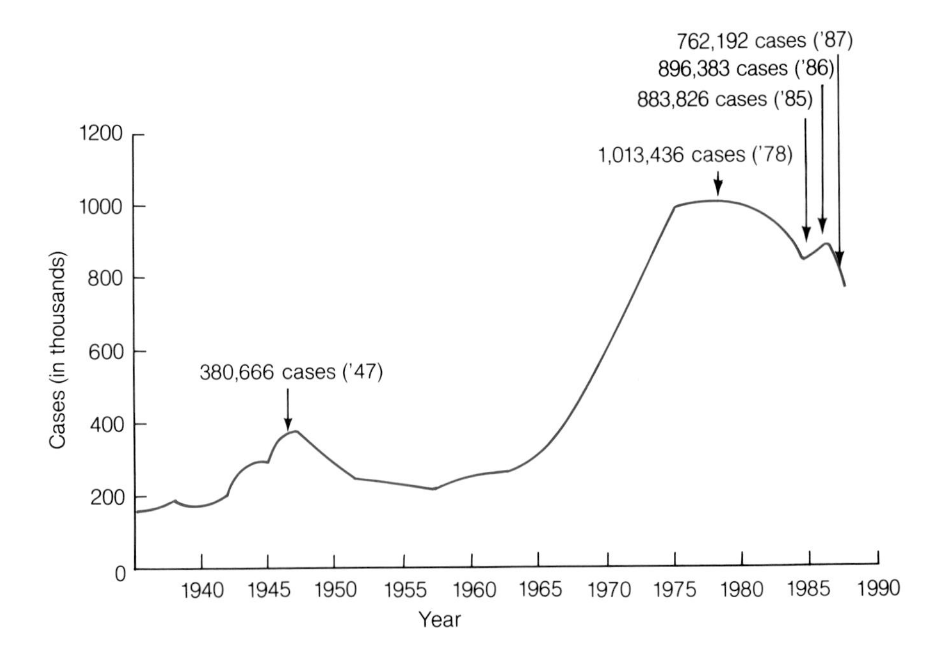

Figure 19.6 Civilian incidence of gonorrhea in the United States. In the past five years, the military incidence of gonorrhea has averaged about 23,000 cases per year.

common. In 1978 and 1979, more than 1 million new cases per year of gonorrhea were reported in the United States. Of these, young adults (20–24 years old) and teenagers (14–19 years old) were responsible for 65–70% of all cases, about 750,000 cases. The number of deaths due to gonorrhea has been 1–10 per year for the last few years. Although gonorrhea is a reportable disease, it has been estimated that no more than 50% (some estimate 25%) of all actual cases are reported to health authorities. If this is an accurate estimation, the true number of cases in the United States during the last few years has ranged between 2 and 4 million per year. The cost of screening for gonorrhea exceeds $1 billion per year.

Gonorrhea normally is acquired through sexual intercourse. During coitus with an infected female, the **gonococcus** is frequently transmitted to the male's urethra. The vast majority of the infected males (90%) have the typical signs and symptoms of gonorrhea, which include a purulent urethral discharge accompanied by painful urination (fig. 19.7). These individuals play a relatively minor role in the transmission of gonorrhea because the discomfort of gonoccocal urethritis prevents sexual activity. Also, these males seek medical help to alleviate the symptoms. The remaining males who become infected (10%), however, show no evident signs or symptoms of gonorrhea. They play a very important role in the transmission of the disease, because they continue to have sexual intercourse and consequently infect their sexual partners. Assuming that 2 million cases per year is a realistic estimate of the actual number of new cases of gonorrhea in the United States, and that these include equal numbers of males and females, there will be approximately 100,000 new male carriers annually who can spread the disease.

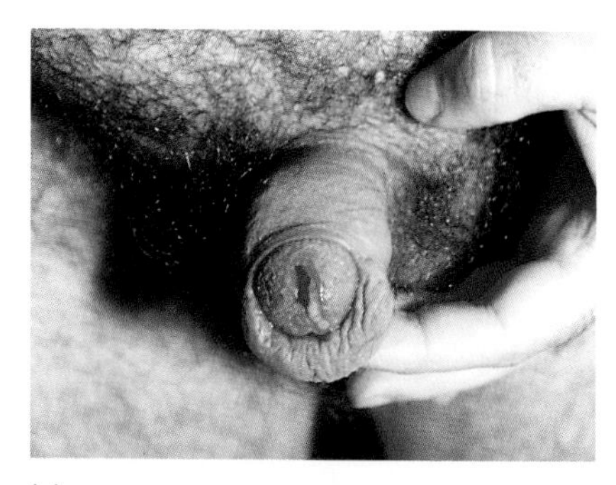

(a)

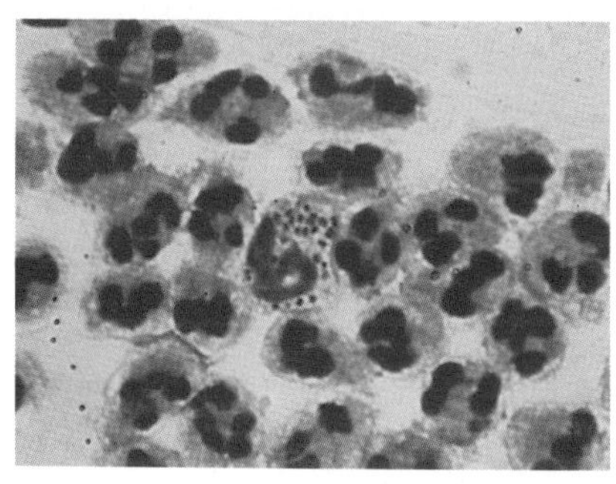

(b)

Figure 19.7 Characteristics of gonorrhea. *(a)* Thick, purulent urethral exudate, along with painful urination are characteristic signs and symptoms of gonorrhea in males. Blood may also be evident in some exudates. Courtesy of Professor F. Camacho, Department of Dermatology and Venereology, University Hospital, Seville, Spain. *(b)* Gram stain of exudate, revealing gram-negative diplococci inside leukocytes.

Among females, the epidemiology of gonorrhea is quite different. More than 50% (perhaps closer to 75%) of the women infected are asymptomatic or have only mild, transient symptoms. These individuals rarely seek medical help on their own, so they constitute a vast reservoir for the infectious agent. The remaining 25–50% seek medical help, because they have a recognizable disease. Thus, there are probably over 500,000 new female carriers per year in the United States. Another picture of transmission is emerging among homosexuals: the incidence of anorectal and pharyngeal gonorrhea is increasing every year.

In heterosexual men, gonorrhea is characterized by an inflammation of the urethra, called **anterior urethritis**. The incubation period is generally 2–4 days, after which the characteristic purulent exudate and the painful urination become obvious. The exudate contains leukocytes, cellular debris, and gonococci (fig. 19.7). Even though anterior urethritis is the most common form of the disease in heterosexual men, about 2–5% of these individuals have **gonococcal pharyngitis**. Since the advent of antibiotics, formerly common complications such as urethral stricture, **epididymitis** (inflammation of epididymis), and **prostatitis** (inflammation of the prostate glands) have all but disappeared.

In homosexual men with gonococcal urethritis, involvement of the anal canal is also common. Studies have revealed that 50% of homosexual males who have gonorrhea have **anorectal gonorrhea**. Pharyngitis is also common. Unfortunately, about 20% of individuals carrying *N. gonorrhoeae* in the pharynx are asymptomatic and serve as reservoirs and sources of infection.

Gonorrhea in females most often involves the cervix and results in a purulent exudate similar to that seen in males. Urethritis, anal canal involvement, and pharyngitis are also common.

Infants and children may suffer from gonorrhea as well. In infants, the disease is acquired from the infected mother as the newborn passes through the birth canal. The bacteria most often infect the eyes, causing a disease sometimes called gonococcal **ophthalmia neonatorum**. For this reason, all newborn babies are treated with eye drops containing silver nitrate, tetracycline, or erythromycin, to kill any pathogen that may be present in the eyes. Other areas, such as the pharynx, the umbilical cord, or the anus, may also be infected. Gonococcal infections of prepubertal children resemble those of adults; sexual molestation or precocious sexual activity serve to transmit the disease in this age group.

Penicillin remains the drug of choice for the treatment of gonorrhea, but certain strains of penicillinase-producing *Neisseria gonorrhoeae* are found to be resistant to penicillins. These strains are being isolated from patients at a rate that is alarmingly high and increasing (fig. 19.8). In these instances, tetracycline, spectinomycin, and the quinolones have been found useful. Sexually transmitted disease, family practice, premarital, and prematernity clinics are of value because they screen the population to identify carriers and to treat those infected.

Syphilis

Syphilis is a sexually transmitted disease caused by the spirochete *Treponema pallidum* (fig. 19.9). It is characterized by the formation of a firm, ulcerative lesion called a **chancre** at the site of infection, by repeated rashes and lesions all over the body, and finally, by the development of ulcers known as **gummata**.

Treponema pallidum is a spirochete that measures 5–20 μm in length and about 0.2 μm in diameter (fig. 19.9). Because of its small diameter, this microorganism cannot be resolved with the bright field microscope, and special techniques such as phase contrast or dark field microscopy must be used to see it in clinical specimens. At present, human pathogenic treponemes cannot be cultured in laboratory media so little is known about their virulence factors.

Treponema pallidum enters the host through the mucous membranes or abraded skin. The most common sites of infection include the genitals, lips, anus, breasts, tonsils, and fingers. Once inside the host, the

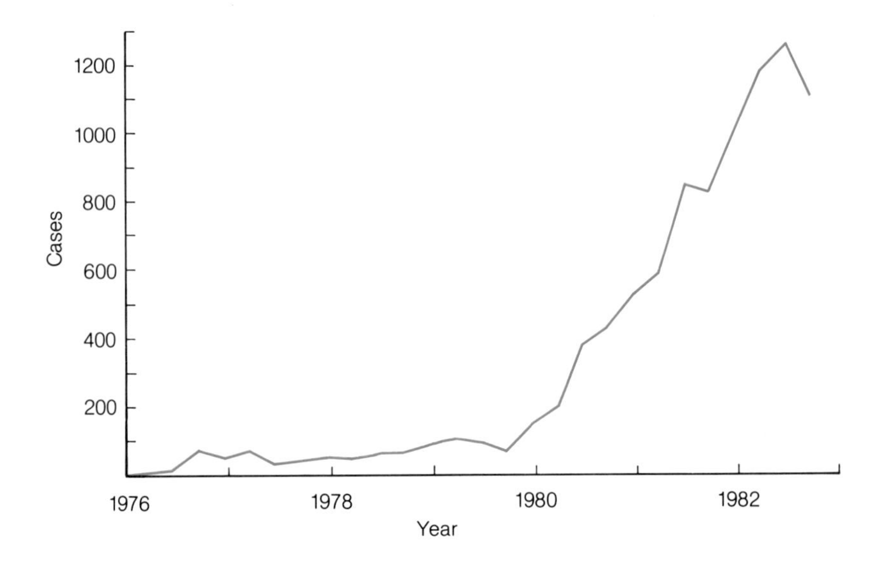

Figure 19.8 **Incidence of gonorrhea caused by penicillin-resistant *Neisseria gonorrhoeae*.** Note the sharp increase in penicillin-resistant *N. gonorrhoeae* isolated since 1979.

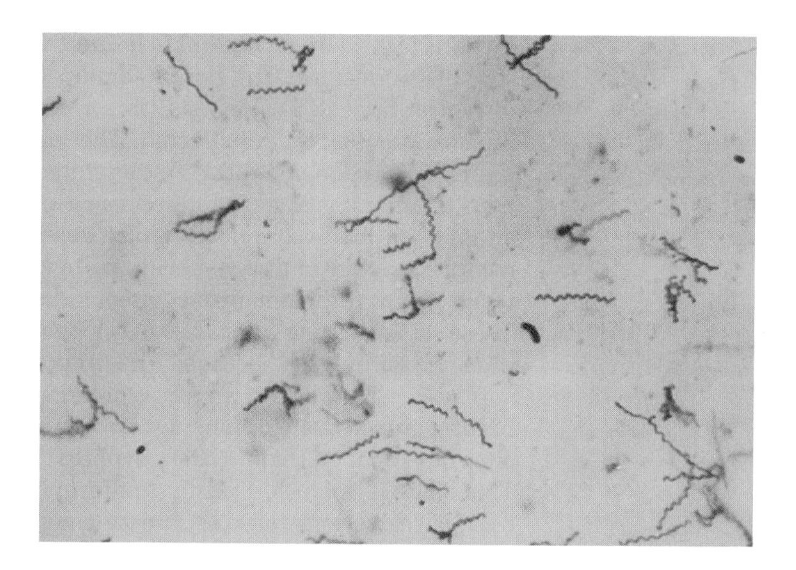

Figure 19.9 ***Treponema pallidum.*** Silver stain of *Treponema pallidum* obtained from a primary chancre of a syphilitic patient. Notice the characteristic spiral shape of the treponemes.

spirochetes multiply locally (extracellularly and intracellularly) and then spread to the regional lymph nodes.

Syphilis has traditionally been subdivided into three stages, based upon differences in signs and symptoms: primary, secondary, and tertiary. **Primary syphilis** is characterized by the appearance of a chancre at the initial site of infection (figs. 19.10 and 19.11). The chancre, an ulcer with a firm border, appears about three weeks after the infection and is laden with infectious spirochetes. The chancre is accompanied by an enlargement of the regional lymph nodes (lymphadenopathy), reflecting the spread of the parasite and its multiplication in the lymph nodes. After 5–10 days, the chancre heals spontaneously. If the infection is not treated during the chancre stage, **secondary syphilis** follows after a period of weeks or months. This stage

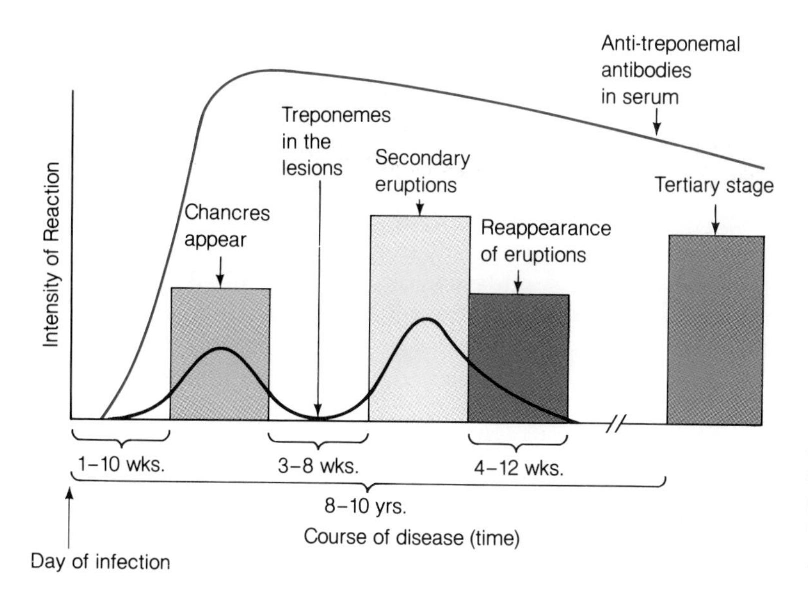

Figure 19.10 Clinical course of the disease in syphilis. In primary syphilis a chancre develops at the site of infection. The chancre has numerous treponemes and can infect sexual partners. The chancre heals after 2–3 weeks. 3–8 weeks later, a rash may appear, heralding the onset of secondary syphilis. These lesions may also contain treponemes and hence are infectious. Tertiary syphilis may appear 8–10 years after infection. Severe damage to organs may occur. The lesions are due to allergic responses and are generally free of treponemes. Antitreponemal antibodies appear early in the disease and remain high for long periods of time. These antibodies are of diagnostic value.

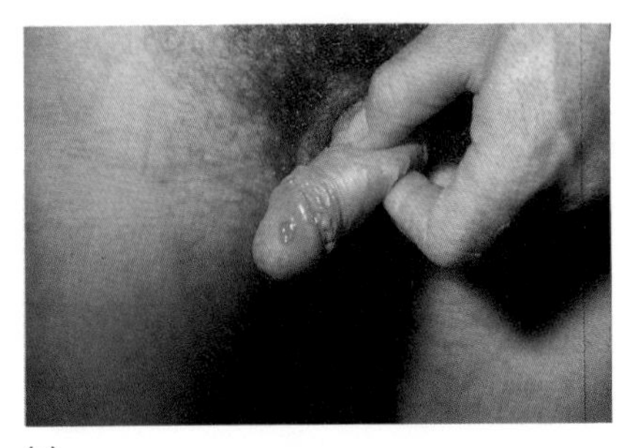

(a)

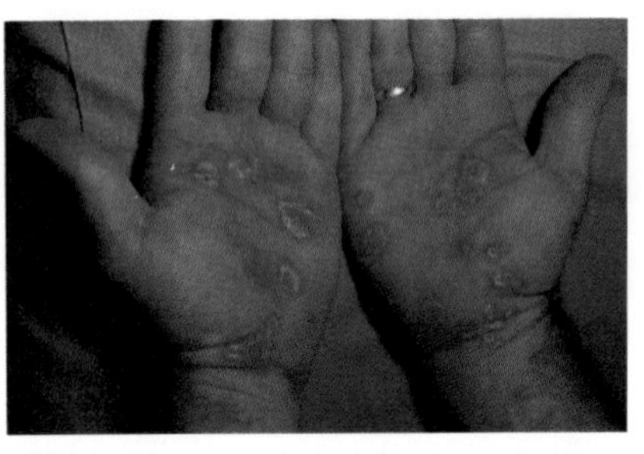

(b)

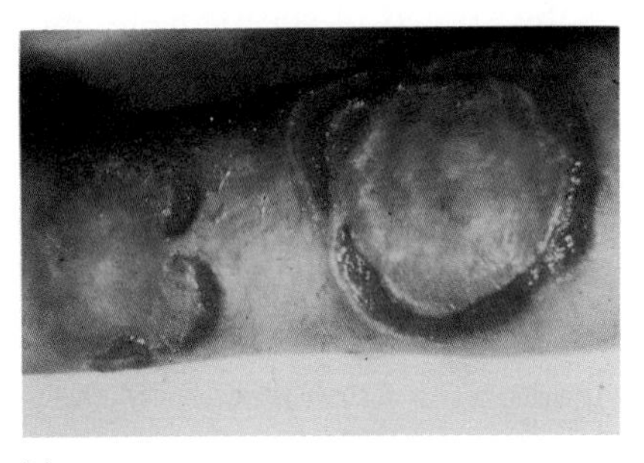

(c)

Figure 19.11 Clinical manifestations of syphilis. *(a)* Primary chancre of the penis in a syphilitic patient. The primary lesion is usually a single open sore with redness and swelling in its periphery. Courtesy of Professor F. Camacho, Department of Dermatology and Venereology, University Hospital, Seville, Spain. *(b)* Secondary syphilitic rash on the hands. Courtesy of Professor F. Camacho, Department of Dermatology and Venereology, University Hospital, Seville, Spain. *(c)* Gummata sometimes seen in patients with tertiary syphilis.

is characterized by multiple skin lesions, rashes, fever, sore throat, malaise, and **lymphadenopathy** (swelling of lymph nodes). Lesions also occur at this stage in internal organs that are affected. The lesions contain numerous treponemes and are therefore highly infectious. During this stage, antitreponemal antibodies and antigen–antibody complexes can be detected. If these complexes lodge in the glomeruli, kidney disease may result from complement-mediated kidney damage. These same immune complexes may lodge in the joints, causing arthralgias. Because the treponemes have spread to other parts of the body, organ systems often become infected. If syphilis remains untreated, years after the initial infection **tertiary syphilis** may develop. This stage is characterized by the formation of localized dermal lesions called gummata (singular, gumma) (fig. 19.11). These lesions contain few, if any, treponemes and arise as a result of cell-mediated immunity to circulating *T. pallidum* antigens. Tertiary syphilis may also be reflected in cardiovascular diseases that are often fatal.

In pregnant syphilitic patients, *T. pallidum* may cross the placenta and infect the fetus. The treponemes enter the unborn child via the fetal circulation and multiply in fetal tissue. **Congenital syphilis** can cause severe damage in the fetus and lead to congenital malformation (fig. 19.12). Although routine screening and treatment of pregnant, syphilitic individuals is widely practiced in the United States, the incidence of congenital syphilis is about 0.4%.

Figure 19.13 illustrates the incidence rate of primary and secondary syphilis from 1920 to 1988. As you can see, syphilis reached a peak in 1943, when over 400 cases per 100,000 population were reported. At this time, antibiotics such as penicillin were introduced, and the incidence of the disease declined rapidly to 150 cases per 100,000 in 1950. Since the 1950s, the disease has continued to decline; however, the reported cases of syphilis represent only the tip of the iceberg, since it is estimated that only 25% of cases are reported. If this estimate is correct, then there are presently over 100,000 new cases of primary and secondary syphilis in the United States each year.

The control of syphilis requires that carriers and new cases be discovered and treated with penicillin. This procedure reduces the number of infected individuals in the population. Other control measures are similar to those employed in the control of gonorrhea. The treatment of choice is penicillin for active cases of syphilis since penicillin resistance has not developed in *T. pallidum*.

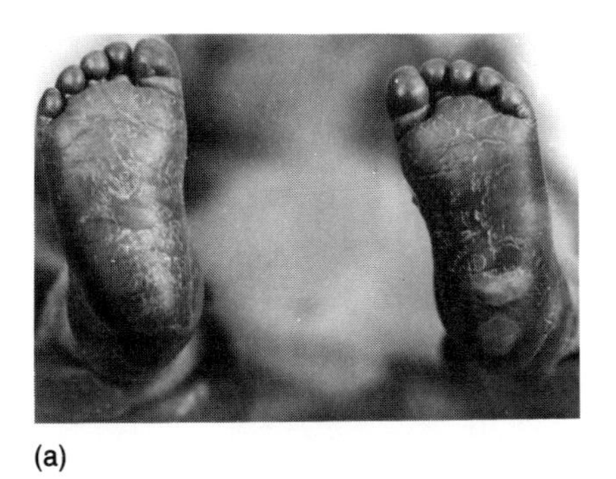
(a)

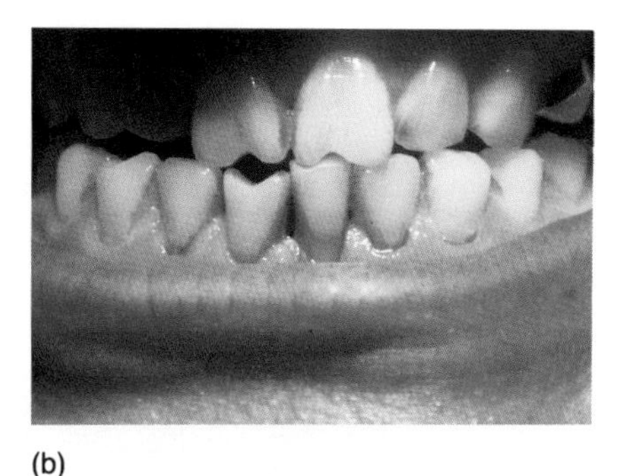
(b)

Figure 19.12 Congenital syphilis. *(a)* Superficial skin erosion in child with congenital syphilis. *(b)* Notched incisors characteristic of congenital syphilis.

Herpes Virus Infections

Genital herpes is a sexually transmitted disease characterized by the development of small blisters on the skin and mucous membranes of the genitourinary tract, which ulcerate after 4–5 days. The outbreak of blisters may be accompanied by fever and exhaustion. These lesions recur at frequent intervals and cause considerable discomfort and distress. The disease is caused by the herpes simplex virus, which can become latent by infecting nerve cells. There are two types of herpes simplex (I and II) that can cause genital herpes, but type II is far more common. Infected individuals are infected for life. They appear to transmit the virus to sexual partners only when the blisters and ulcers are present (fig. 19.14).

The fetus in pregnant individuals can be infected by the circulating viruses during a genital herpes episode, thus resulting in severe damage to the fetus or in stillbirth. In addition, the virus can be transmitted to the newborn as the child passes through the birth canal if the mother is having a herpes attack. This infection can result in very severe and often fatal disease. Approximately 200 newborn babies die each year in the United States because of systemic herpes infections and more than 200 suffer physical or mental impairment. As if these complications were not enough, females with genital herpes have an increased incidence of cervical cancer.

Genital herpes is an infectious disease of considerable public health significance because of the large number of people infected for life and capable of transmitting the infection. Some studies have revealed that genital herpes is more prevalent in human populations than gonorrhea. It reached epidemic proportions during the 1970s, and continues to be the leading cause of morbidity among sexually transmitted diseases. It is estimated that there are over 1 million new cases of genital herpes every year in the United States.

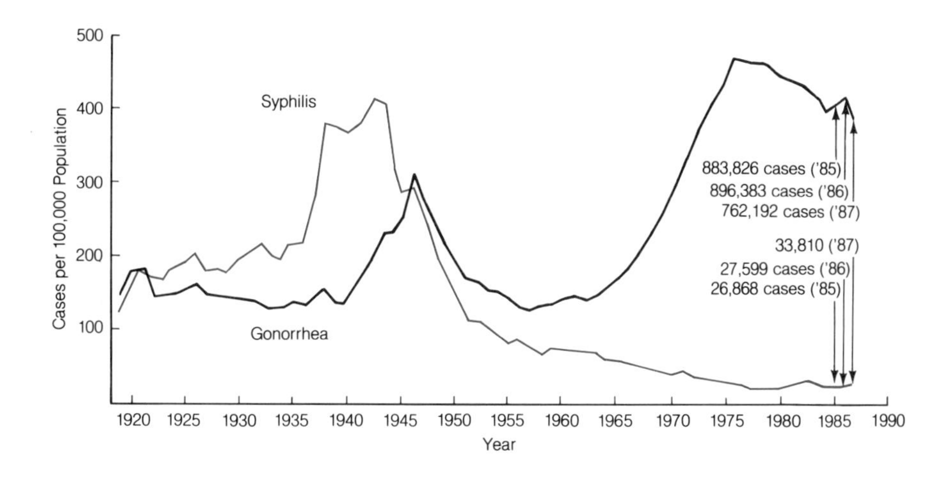

Figure 19.13 Comparison of civilian syphilis and gonorrhea in the United States. Graph illustrating the incidence of primary and secondary syphilis in the United States. Of the more than 27,000 cases of syphilis in the United States in 1984, fewer than 100 were congenital syphilis. The incidence of military syphilis during the last five years has averaged at about 370 cases per year.

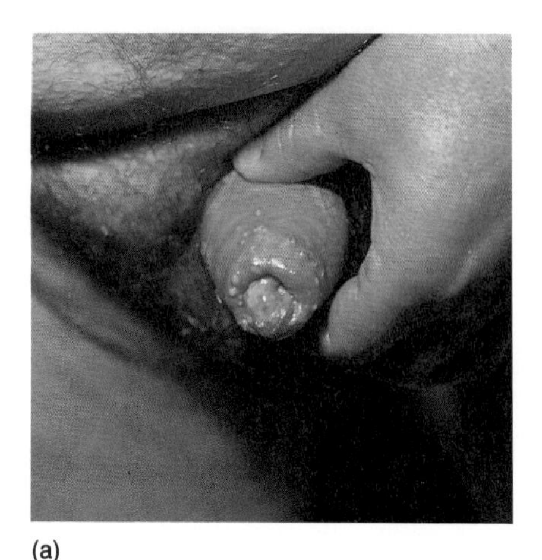

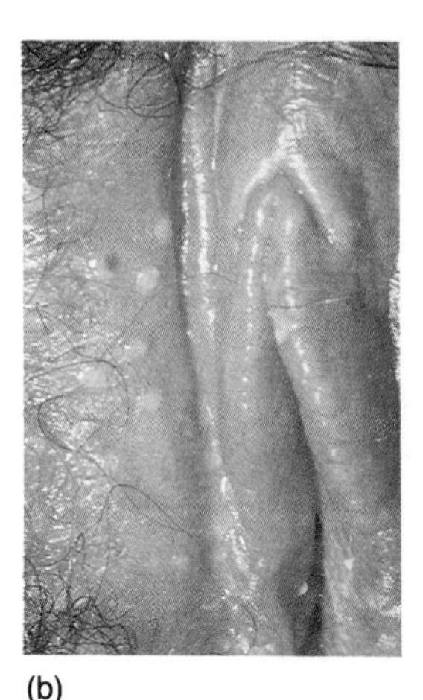

(a) (b)

Figure 19.14 Genital herpes. *(a)* Herpetic lesion of the penis. Courtesy of Professor F. Camacho, Department of Dermatology and Venereology, University Hospital, Seville, Spain. *(b)* Herpes simplex II lesion of the vaginal area. Notice the moist, blister-like lesions that are characteristic of the disease. Courtesy of Professor E. J. Perea.

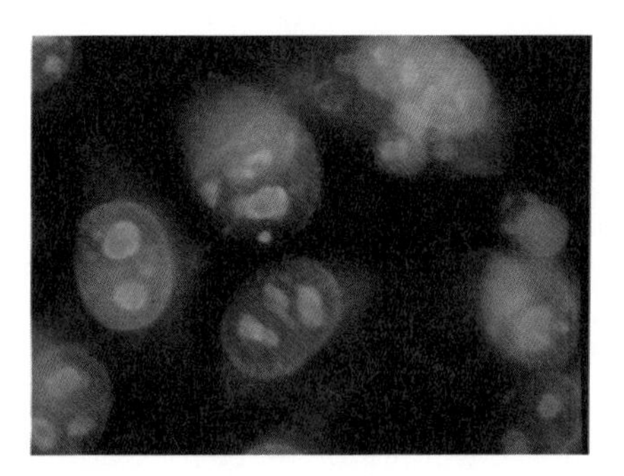

Figure 19.15 *Chlamydia trachomatis.* Immunofluorescence stain of *Chlamydia trachomatis*-infected HeLa cells. This is a common method of diagnosing chlamydial infections. Clinical samples are inoculated into tissue cultures and incubated. After the incubation period, the infected cells are stained with fluorescent antibodies and examined using a fluorescence microscope. *C. trachomatis*-infected cells are characterized by orange-fluorescing inclusion bodies. Courtesy of Dr. J. Aznar.

The ability of the virus to become latent, and the lack of drugs that can destroy the latent virus, make it almost impossible to eradicate this disease from human populations. The disease can be prevented, however, by avoiding sexual contact during the time when the blisters are developing and ulcers are present and by the use of condoms during sexual intercourse.

There is no known cure for genital herpes. In 1982, however, the FDA permitted the release of a topical drug called **acyclovir** (Zovirax) that inhibited the replication of the virus. However, the ointment was useless since it did not eliminate the virus or reduce the severity of genital herpes attacks. The oral form of acyclovir does help to reduce the incidence of recurrences as well as the severity and duration of the episodes. Intravenous acyclovir is even more effective than oral acyclovir in reducing the incidence, severity, and duration of outbreaks.

Urethritis

Nongonococcal urethritis (NGU) is a sexually transmitted disease characterized by an inflammation of the urethra accompanied by a discharge of pus. It is estimated that in the United States 30–50% of the 2 million annual cases of acute urethritis (other than gonococcal) are caused by *Chlamydia trachomatis* (fig. 19.15). This organism is also responsible for the more serious pelvic inflammatory diseases. Other microorganisms, such as the mycoplasma *Ureaplasma urealyticum,* can also cause NGU. The diagnosis of NGU is usually made when it is discovered that the etiology is not *N. gonorrhoeae.*

FOCUS GENITAL WARTS

One venereal disease that appears to have increased in frequency in the last few years is genital and anal warts, which are benign tumors of the skin. Genital and anal warts are caused by a number of viruses, such as the papilloma and polyoma viruses. Additionally, it is suspected that the human papilloma virus may be involved in the increasing number of vaginal and cervical cancers being diagnosed. The viruses are spread by sexual contact when infected tissue breaks down and sheds the virus.

Warts develop from infected epithelium that grows into slender cones or fingerlike projections. The warts may develop singly or in dense groups (fig. 19.16).

Individual warts on the penis or vulva or around the anus are often removed by electrodesiccation. The epidermis is subsequently treated with various chemicals that denature the viruses, so that viruses escaping from the diseased tissue do not initiate new infections. When the warts have developed in thick groups, chemicals such as podophylen in alcohol are applied, causing an often painful inflammatory reaction that sloughs off the diseased epidermis. Vaginal warts frequently are treated with triple sulfonamide urea cream.

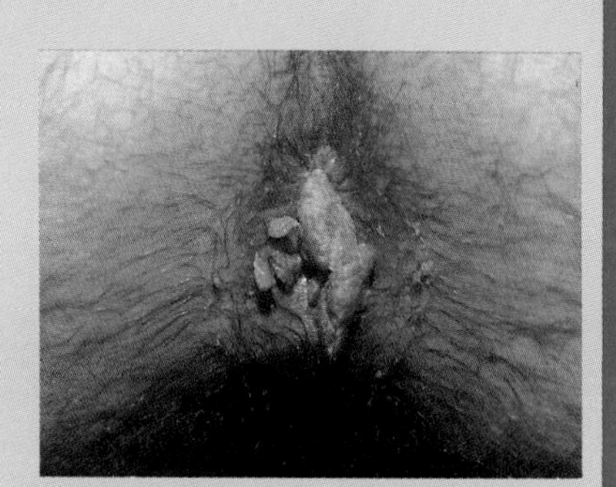

Figure 19.16 **Perianal warts.** Courtesy of Professor F. Camacho, Department of Dermatology and Venereology, University Hospital, Seville, Spain.

Chlamydia trachomatis (fig. 19.17) is an obligate intracellular parasite that causes, in addition to NGU, a variety of other diseases including lymphogranuloma venereum, salpingitis (an inflammation of the fallopian tubes), **Reiter's syndrome** (urethritis, arthritis, and conjunctivitis), **cervicitis** (an inflammation of the cervix), and trachoma.

Chlamydia trachomatis is transmitted by sexual activity or during childbirth. Elementary bodies of *C. trachomatis* are taken up by susceptible host cells in the urethra (or other organs) by a process resembling phagocytosis. The elementary bodies then differentiate into reticulate bodies, which, in turn, undergo a series of binary fissions inside the urethral cells. About 36–72 hours later, this process gives rise to elementary bodies. One reticulate body can give rise to as many as 1000 elementary bodies. The parasitized host cell bursts, releasing the elementary bodies. Several such cycles can cause considerable damage to urethral epithelium and therefore incite an inflammatory reaction. The chlamydiae are thought to produce a toxin that contributes to the cell damage and subsequent inflammation.

Chlamydial infections are not routinely diagnosed, although more and more clinical laboratories are beginning to check for *C. trachomatis* in clinical specimens. Diagnostic procedures usually involve the growth of the suspected agent in tissue culture. The infected tissue culture cells are then viewed using special staining techniques or fluorescent antibodies. Routine screening for this pathogen is desirable in cases of urethritis, so that the appropriate treatment can be carried out and sexual contacts can be informed of their possible infection and subsequently treated. Active cases are treated effectively with tetracyclines or with erythromycin.

A sexually transmitted disease known as **granuloma inguinale** is distinguished by skin ulcers and the

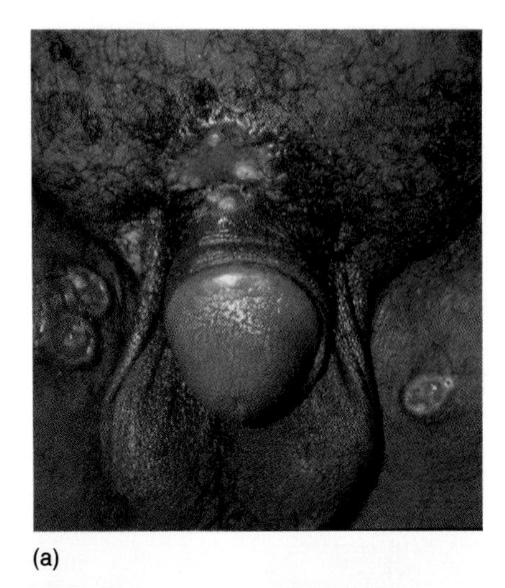

(a)

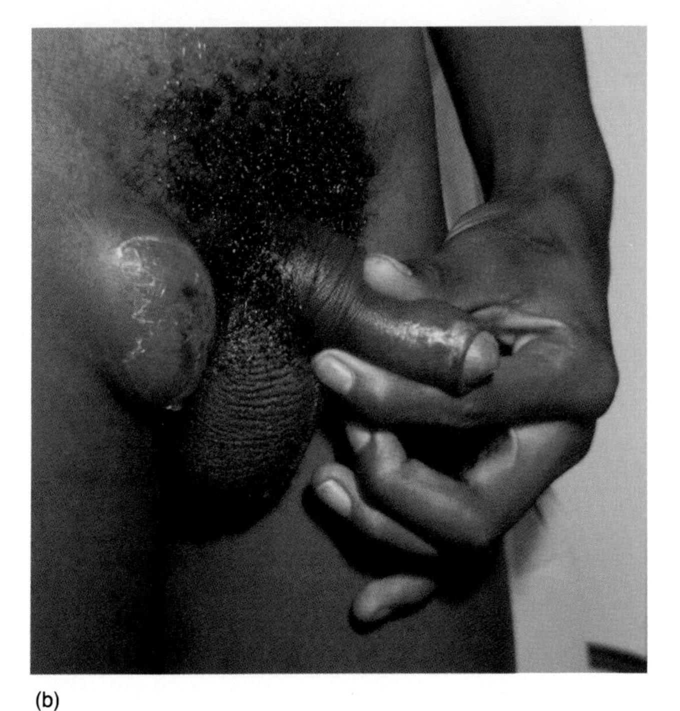

(b)

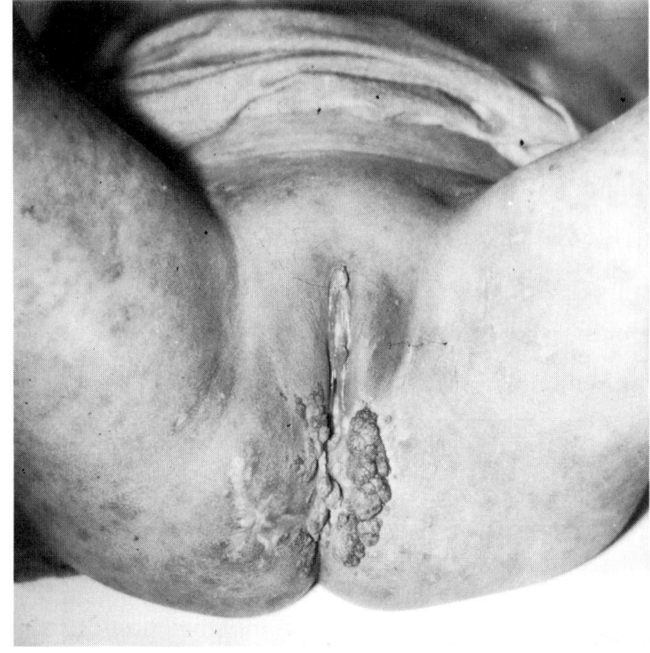

(c)

Figure 19.17 **Clinical manifestations of selected sexually transmitted diseases.** *(a)* Chancroid with ulcers in 39-year-old male. This disease is caused by *Haemophilus ducreyii*. *(b)* Lymphogranuloma venereum caused by *Chlamydia trachomatis* in a male and *(c)* in a female. Notice that swollen lymph nodes are characteristic of the disease in males while distortion of tissue and rectal narrowing are characteristic in females.

development of subcutaneous tissue in the groin and genitals. The cause of the disease is a small gram-negative bacillus called *Calymmatobacterium granulomatis*. This disease is being found in some persons with acquired immunodeficiency syndrome (AIDS). If untreated, the lesions spread and become deeper and infected with other organisms. Streptomycin is used to treat the disease.

CLINICAL PERSPECTIVE A CASE OF SYPHILIS

A 30-year-old white male visited the County Health Department Sexually Transmitted Disease Clinic complaining of an open sore on the penis (a **chancre**), which he had had for about four days. A sample was taken from the chancre and immediately examined with a dark field microscope. The microscopic examination revealed spirochetes with the characteristic motility of *Treponema pollidum*.

A diagnosis of **primary syphilis** was made and the patient was treated with 2.4 million units (m.u.) of the antibiotic **bicillin** by intramuscular (IM) injection. Blood samples were also drawn from the patient to measure the levels of antibodies against syphilis antigens in the blood. Both the Rapid Plasma Reagin (RPR) test (a screening test for syphilis) and the VDRL (Veneral Disease Research Laboratory) test (a test used to verify the results of the RPR test), were negative. Another test for syphilis, the MHATP, was inconclusive. These results are to be expected because the blood serum does not contain antitreponemal antibodies until about five to seven days after infection.

The patient was interviewed by the Communicable Disease Investigator to determine the number of sexual contacts he had during the last 90 days. For a primary case of syphilis, contacts made during the last 90 days plus the duration of the symptoms (since the incubation period for syphilis is 10–90 days) need to be investigated. Three contacts were named.

One of the contacts was a female with whom the patient had had sexual relations several times over a period of six months. She was located and examined. She gave no history of clinical syphilis and had no signs or symptoms upon examination. Her blood tests (RPR and VDRL) were positive (said to be reactive) and she was diagnosed as an **early latent syphilis** case (less than one year of duration). The female contact was treated with 2.4 m.u. bicillin IM.

The other two females named in the interview, both of whom had had sexual relations with the patient, were located, examined, and their blood drawn for the RPR and VDRL tests. Both these females were found to be nonreactive and treated preventatively with 2.4 m.u. bicillin IM.

A cluster interview (interview of the sexual contacts the females and their other sexual partners had had) was done. A man was named as a cluster because he had had sex with one of the same women. He was located and examined, and his blood drawn. He was preventatively treated; his blood was negative for syphilis.

Case analysis indicates that the woman who was diagnosed as an early latent case was the source of infection for the original patient with the chancre. Her interview revealed further contacts, all of which were located, interviewed, and treated.

Preventive measures of future infection were explained to the patient, of which he chose the use of condoms as the preferred method.

Responding to symptoms in the future was also discussed. He was admonished to get an examination as soon as possible and not have sexual relations if he had a sign or symptom or noticed a sign of a sexually transmitted disease on a partner. Followup blood tests were recommended at one month, three months, six months and then at one-year intervals.

Courtesy of the San Luis Obispo County Health Department.

SUMMARY

1. Diseases of the urogenital tract are common in human populations.
2. Urinary tract infections caused by *Escherichia coli* are the most common infections in humans.
3. Sexually transmitted diseases, in general, are currently on the rise.
4. Pathogens such as *Entamoeba histolytica* and *Giardia lamblia* not normally considered to be sexually transmitted diseases, are now being encountered as the cause of some venereal diseases.

STRUCTURE AND FUNCTION OF THE UROGENITAL TRACT

1. The principal roles of the urinary tract are to maintain homeostasis in the blood and to void body wastes.
2. The flow of urine from the bladder through the urethra helps cleanse the urethra of adhering microorganisms.

3. The reproductive system serves to generate gametes and to nurture the embryo.
4. Secretions of the reproductive system, particularly that of the female, help maintain a population of microorganisms that may protect the host from infections.

NORMAL FLORA OF THE UROGENITAL SYSTEM

1. The urinary tract, except for the anterior urethra, normally is maintained free of microorganisms by body defenses. The anterior urethra contains an array of bacteria, which include gram-positive cocci and diphtheroids.
2. The vagina and cervix have a complex flora composed primarily of lactobacilli.
3. The indigenous flora helps keep potential pathogens from causing disease.

DISEASES OF THE URINARY TRACT

1. Urinary tract infections (UTI) are very common in human populations. They are caused primarily by *Escherichia coli* and *Proteus mirabilis*.
2. UTIs are characterized by cystitis, urinary frequency, hematuria, low back pain, and fever.
3. Virulence factors, such as pili and capsular antigens, play a role in pathogenesis in that they help the pathogen adhere to epithelial surfaces or to be protected from host defense mechanisms.
4. UTIs can be controlled by preventing urethral contamination with feces. Large doses of amoxicillin, kanamycin, or trimethoprim sulfamethoxathole are used to treat UTIs.

DISEASES OF THE GENITAL TRACT

1. *Candida albicans* is an endogenous pathogen that causes a form of vaginitis sometimes called thrush. The disease is characterized by a thick, cream-colored vaginal discharge. Predisposing factors such as diabetes, hormonal imbalance, or depletion of the normal flora usually account for the abnormal proliferation of the opportunistic yeast.
2. *Trichomonas vaginalis*, a flagellate, causes a vaginitis that consists of a profuse, whitish exudate accompanied by burning and itching. The disease is transmitted sexually and is treated with Flagyl.
3. Nonspecific vaginitis is an inflammation of the vagina that generally is not caused by the usual pathogens. The exudate is homogeneous and malodorous and usually contains "clue cells." It is thought that the disease is caused by *Gardnerella vaginalis*, and it can be treated with broad-spectrum antibiotics or with Flagyl.
4. Toxic shock syndrome (TSS) is an intoxication caused by certain strains of *Staphylococcus aureus*. The disease usually ensues in menstruating individuals who use tampons and keep them in place for long periods of time. The bacteria multiply at the expense of nutrients that have been absorbed by the tampon and produce a toxin. The intoxication is characterized by hypotension, shock, fever, a rash, and several other constitutional symptoms. Treatment is by removing the tampon and immediately replacing electrolytes to reverse shock and hypotension.

SEXUALLY TRANSMITTED DISEASES (STD)

1. Gonorrhea, caused by *Neisseria gonorrhoeae*, is characterized by an acute inflammation of the genital mucosa with purulent discharge.
2. Virulence factors such as pili and IgA proteases are important in the pathogenicity of the bacterium.
3. The disease occurs in epidemic proportions in human populations, with more than 1 million cases reported annually in the United States alone.
4. Gonorrhea is treated successfully with penicillin, although some penicillinase-producing strains have been isolated. These are treated with tetracycline or spectinomycin.
5. Primary syphilis, caused by *Treponema pallidum*, is characterized by the formation of a firm, ulcerative lesion called a chancre.
6. The pathogenesis of the disease involves both the invasive properties of the pathogen and its ability to elicit allergic responses in the host.
7. Primary and secondary syphilis are due primarily to the invasion of host tissues by the parasite, while tertiary syphilis is due to delayed hypersensitivity responses to treponemal antigens.
8. There are an average of 35,000 new cases of syphilis reported in the United States each year. This figure, however, may represent only about 25% of the actual number of cases.
9. The treatment of choice for active cases of syphilis is penicillin.

10. Genital herpes is caused by herpes simplex type I or II, although the latter is much more frequent. The disease is characterized by a first episode of painful, prolonged ulceration of the genital area, with fever and a burning sensation with frequent recurrences of the lesions.
11. Genital herpes has reached epidemic proportions in the United States and has influenced sexual practices in this country.
12. If the treatment for genital herpes is symptomatic, Zovirax has shown promising results when administered orally or intravenously.
13. Nongonococcal urethritis afflicts more than 1 million persons in the United States each year. *Chlamydia trachomatis* and *Ureaplasma urealyticum* are the most common causative agents. *Chlamydia* is also responsible for lymphogranuloma venereum, while *Ureaplasma* is implicated in many cases of infertility.
14. Genital warts are caused by a number of viruses, in particular papilloma and polyoma viruses.

STUDY QUESTIONS

1. Explain the pathogenesis of urinary tract infections caused by *Escherichia coli*.
2. Describe the epidemiology of:
   a. UTI;
   b. candidiasis;
   c. TSS;
   d. genital herpes;
   e. syphilis.
3. Describe the role(s) of the host's immune response in the following disease processes:
   a. syphilis;
   b. gonorrhea;
   c. genital herpes;
   d. UTI;
   e. candidiasis;
   f. TSS.
4. Outline a program of prevention for the following diseases:
   a. TSS;
   b. candidiasis;
   c. syphilis;
   d. gonorrhea;
   e. genital herpes;
   f. STD.
5. Describe the role (if any) of the carrier population in the following diseases:
   a. UTI;
   b. gonorrhea (both males and females);
   c. syphilis (both males and females);
   d. candidiasis;
   e. genital herpes (both males and females);
   f. trichomoniasis.
6. Describe the typical signs and symptoms of the following diseases, and explain the mechanisms by which they occur:
   a. candidiasis;
   b. gonorrhea;
   c. tertiary syphilis;
   d. TSS;
   e. genital herpes.
7. Why is genital herpes of such public health importance?
8. Construct a table of diseases of the urogenital tract containing the causative agent, epidemiology, and treatment of choice.

SELF-TEST

1. What are the predominant bacteria in the vagina of the prepubescent child?
   a. enterococci, diphtheroids, *Peptostreptococcus*, and *Bacteroides;*
   b. lactobacilli;
   c. *Neisseria;*
   d. enterics;
   e. *Bifidobacterium*.
2. Cystitis is an inflammation of the:
   a. kidneys;
   b. ureters;
   c. bladder;
   d. urethra;
   e. vagina.
3. Ninety percent of all urinary tract infections are due to:
   a. *Pseudomonas;*
   b. *Proteus;*
   c. *Klebsiella;*
   d. *Eschericha;*
   e. *Enterobacter.*
4. What is the path that *Escherichia* usually takes from the anus to the kidneys?
   a. anus → urethra → bladder → ureters → convoluted tubules → glomerulus;
   b. rectum → intestine → bladder → kidneys;
   c. rectum → intestines → renal veins → kidneys;
   d. anus → rectum → bladder → kidneys;
   e. anus → capillaries → renal vein → glomerulus.
5. What is believed to be the reason *Escherichia* is able to

efficiently reproduce in the urinary tract and not be flushed out by the urine?
a. high concentration of nutrients in the urine;
b. pili that attach the bacterium to the epithelium;
c. capsule that protects the bacterium from the immune system;
d. glycocalycx that attaches the bacterium to the epithelium;
e. production of a growth factor.

6. Adult females are more than 50 times more susceptible to urinary tract infections than adult males because females:
a. urinate less than males;
b. have bacteria in their vagina;
c. have shorter ureters;
d. have shorter urethras that are near the anus;
e. have babies that block the ureters.

7. The pathogen that causes *Candida* vaginitis is a:
a. virus;
b. bacterium;
c. fungus;
d. protozoan;
e. nematode.

8. The pathogen that causes *Trichomonas* vaginitis is a:
a. virus;
b. bacterium;
c. fungus;
d. protozoan;
e. nematode.

9. *Trichomonas vagininalis* is usually found in a female's:
a. anus;
b. urethra;
c. ureters;
d. bladder;
e. vagina.

10. *Trichomonas vagininalis* is usually found in a male's:
a. anus;
b. urethra;
c. ureters;
d. bladder;
e. vagina.

11. What is the major cause of nonspecific vaginitis?
a. *Chlamydia;*
b. *Bacteroides;*
c. *Gardnerella;*
d. *Staphylococcus;*
e. *Ureaplasma.*

12. Toxic shock syndrome is caused by which organism?
a. *Chlamydia;*
b. *Bacteroides;*
c. *Gardnerella;*
d. *Staphylococcus;*
e. *Ureaplasma.*

13. What are contributing factors that lead to toxic shock syndrome?
a. using superabsorbent tampons;
b. leaving tampons in the vagina for a day or more;
c. carrying certain strains of *Staphylococcus aureus* that are lysogenic for a toxin producing prophage;
d. undergoing menstruation;
e. all of the above.

14. If the mortality of toxic shock syndrome is about 3% of the morbidity, how many women die each year in the United States from TSS?
a. 1;
b. 10;
c. 100;
d. 1000;
e. 10,000.

For items 15–20, match the pathogen with the disease:
a. *Neisseria;*
b. *Treponema;*
c. *Chlamydia;*
d. *Ureaplasma;*
e. *Haemophilus.*

15. Nongonococcal urethritis.

16. Gonorrhea.

17. Lymphogranuloma venereum.

18. Syphilis.

19. Chancroid.

20. Infertility due to bacteria.

21. Which organisms are frequently responsible for genital warts?
a. polyma and papilloma viruses;
b. *Treponema;*
c. *Chlamydia;*
d. *Ureaplasma;*
e. *Haemophilus.*

SUPPLEMENTAL READINGS

Arego, D. E. and Koch, S. J. 1986. Bacteruria in patients with spinal chord injury. *Hospital Practice* 21(3A):87–114.

Fayer, R. and Ungar, B.L.P. 1986. *Cryptosporidium* spp. and cryptosporidiosis. *Microbiological Reviews* 50(4):458–483.

Gallo, R. C. 1987. The AIDS virus. *Scientific American* 256(1):46–56.

Karchmer, A. W. 1982. Sexually transmitted diseases. *Scientific American Medicine* Section 7-XXII.

Karchmer, A. W. 1983. Infections due to *Neisseria*. *Scientific American Medicine* Section 7-III.

Persson, E. and Holmberg, K. 1984. Clinical evaluation of preciptin tests for genital actinomycosis. *Journal of Clinical Microbiology* 20(5):917–922.

Rubin, R. H. 1984. Infections of the urinary tract. *Scientific American Medicine* Section 7-XXIII.

Selwyn, P. A. 1986. AIDS: What is now known, II. Epidemiology. *Hospital Practice* 21(6):127–164.

Selwyn, P. A. 1986. AIDS: What is now known, IV. Psychosocial aspects, treatment prospects. *Hospital Practice* 21(10):125–164.

Sewell, D. L. and Horn, S. A. 1985. Evaluation of a commercial enzyme-linked immunosorbent assay for the detection of Herpes Simplex virus. *Journal of Clinical Microbiology* 21(3):457–458.

Turck, M. 1980. Urinary tract infections. *Hospital Practice* 15:49–58.

Wroblewsky, S. S. 1981. Toxic shock syndrome. *American Journal of Nursing* 81:82–85.

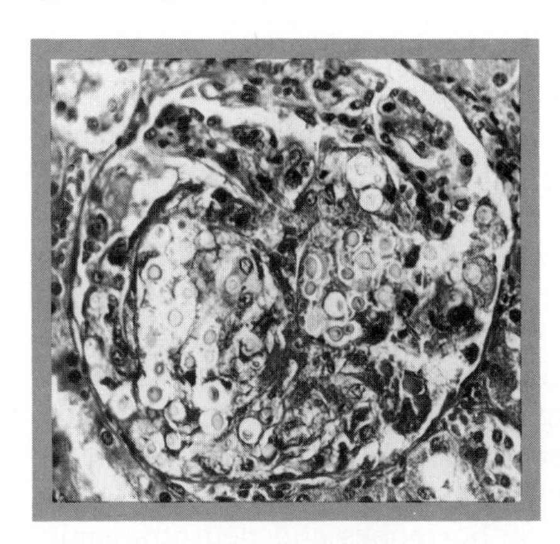

CHAPTER 20

DISEASES OF THE NERVOUS SYSTEM

CHAPTER PREVIEW **WHERE DO THE SLOW VIRUSES HIDE?**

Certain viruses infect animals and humans for many years before the damage they cause becomes apparent. Since the incubation periods are sometimes many decades, these viruses are known as **slow viruses.** Scientists are interested in knowing how these viruses escape the host's immune system and where they reside during their long incubation periods.

A study of the slow virus that causes **visna** in sheep and goats has indicated that it hides in macrophages (cells that usually destroy invading microorganisms), where it replicates and accumulates in cytoplasmic vacuoles. Surprisingly, no cytopathic effects are visible in the macrophages.

Damage in the host occurs only after the infected macrophages fuse with cells in the brain, lungs, or joints. The visna viruses begin to replicate in the fusion cells, lysing them. The virus replication and cell destruction draw cells of the immune system to the affected tissue. The attack on the visna virus by the cells of the immune system eventually causes most of the damage to the animal. As the disease develops, the animal becomes paralyzed and emaciated and dies.

Subacute sclerosing panencephalitis is a slowly progressing, fatal human disease of the central nervous system caused by the long-term presence of measles virus antigens. Virus parts are seen in the nucleus and cytoplasm of infected brain cells. The infectious measles virus is not present because one of the proteins required for its maturation is not synthesized, but all the other virus proteins are synthesized. The measles antigens associated with brain cells stimulate the immune system to attack the brain tissue.

The visna virus and the measles virus illustrate two ways in which viruses reside in host cells and cause slowly progressing diseases of the central nervous system. The visna virus accumulates quietly in macrophages until the macrophages fuse with other cells, while the measles viruses fail to mature in brain cells. Since virus multiplication is delayed, the destruction of nervous tissue progresses slowly under the onslaught of the immune system.

In this chapter we are introduced to other pathogens that cause neurological disorders. We discover that microorganisms are responsible for a number of diseases of the nervous system that account for many thousands of deaths each year in the United States.

CHAPTER OUTLINE

DISEASES OF THE NERVOUS SYSTEM ARE caused by a variety of bacteria, viruses, protozoa, and fungi. These diseases often result when the infectious agent or some toxin spreads from the initial site of infection to nerve cells. Since infections and diseases of the nervous system are often fatal, they require prompt treatment. In this chapter we discuss diseases in which the most obvious symptoms result from infections or disturbances of the nervous system.

ORGANIZATION OF THE NERVOUS SYSTEM

The nervous system is generally divided into two parts: the **central nervous system** and the **peripheral nervous system**. The central nervous system includes all of the nerve cells, supporting cells, cerebrospinal fluid, and membranes that make up and are associated with the brain and spinal cord (fig. 20.1). The peripheral nervous system consists of all the nerve cells and supporting cells that innervate the body and that connect to the spinal cord or directly to the brain. The **sympathetic** and **parasympathetic nerves** are those in the autonomic nervous system that innervate smooth muscles, heart muscle, and glands.

The brain and spinal cord are protected by three membranes called **meninges** (singular, meninx) and by cerebrospinal fluid. The meninx that covers the surface of the brain and spinal cord is called the **pia mater**. This meninx is, in turn, covered by the **arachnoid**. Cerebrospinal fluid produced by brain cells flows between the pia mater and the arachnoid in what is called the **subarachnoid space**. The third meninx, called the **dura mater**, encloses another layer of cerebrospinal fluid around the brain, called the **superior sagittal sinus**.

The brain and spinal cord are protected by the skull bones and the vertebrae, which make up the backbone. There are four interconnecting cavities within the brain called **ventricles**, which are filled with cerebrospinal fluid. Specialized cells and neurons lining the ventricles release hormones and neurotransmitters into the fluid. It is thought that the brain, spinal cord, cerebrospinal fluid, and blood are integrated into one neuroendocrine unit by hormones and neurotransmitters.

Even though small molecules such as hormones and neurotransmitters can pass back and forth between the blood and the central nervous system, large molecules (such as antibodies) and microorganisms (such as viruses and bacteria) are unable to penetrate the barriers between the brain capillaries and the blood system. Consequently, unlike most of the body, the central nervous system is immunologically isolated and there is a "blood-brain barrier" that inhibits the movement of some drugs and antibiotics into the brain. Thus, infections of the brain may be very severe because components of the immune system and many drugs and antibiotics are unable to fight infections in this organ.

BACTERIAL DISEASES OF THE NERVOUS SYSTEM

Infections of the Meninges

Acute pyogenic meningitis is a severe inflammation of the meninges accompanied by the formation of pus, which can be caused by a number of bacteria (table 20.1). The bacteria causing meningitis in babies one month old or less are *Escherichia coli* (40%) and Group B *Streptococcus* (30%). In children 15 years old or younger, the major causes of meningitis are *Haemophilus influenzae* (50%), *Neisseria meningitidis* (30%), and *Streptococcus pneumoniae* (15%). In adults, meningitis is due mainly to *Streptococcus pneumoniae* (40%), *Neisseria meningitidis* (25%), and *Staphylococcus aureus* (10%).

Meningococcal meningitis (caused by *Neisseria meningitidis*) is the only kind of bacterial meningitis that causes epidemics and small outbreaks (fig. 20.2). The outbreaks are common in crowded military camps

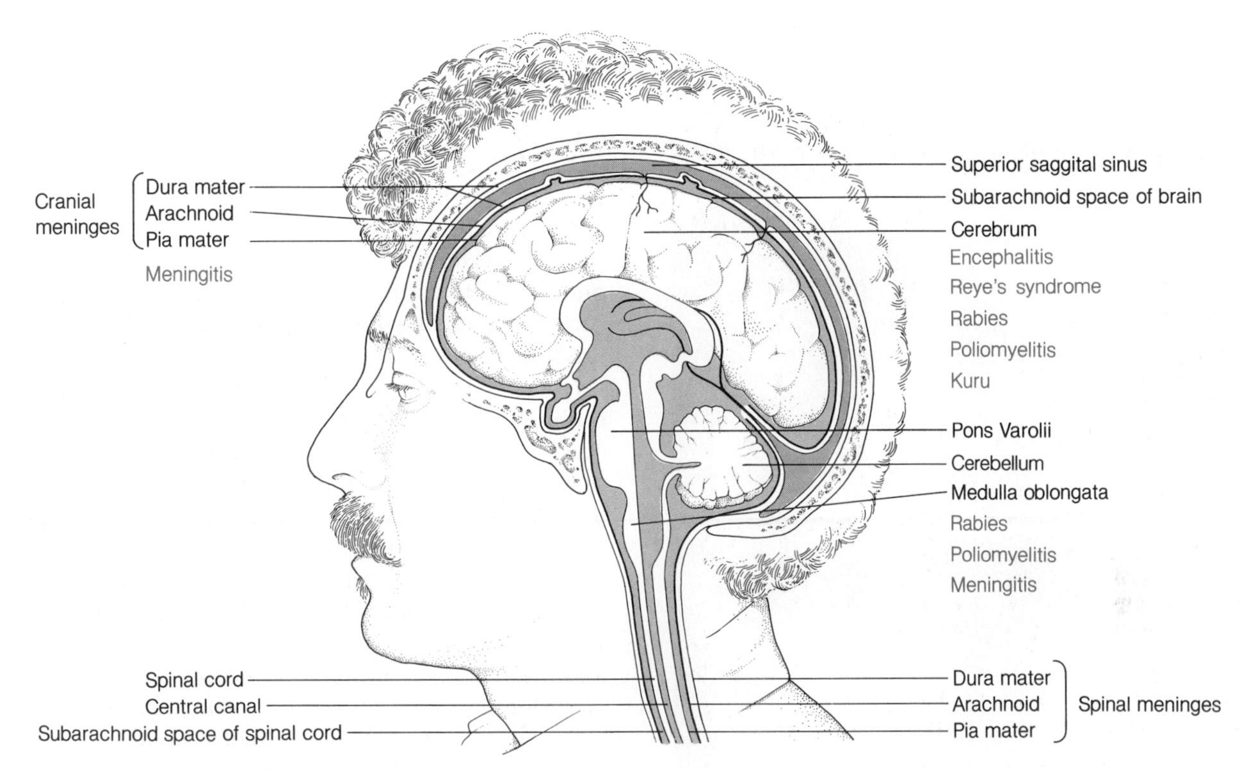

Figure 20.1 The central nervous system. The central nervous system consists of the brain, spinal cord, and meninges. Color print indicates the diseases affecting the anatomical site.

TABLE 20.1
FREQUENCY OF BACTERIAL CAUSES OF MENINGITIS BY AGE GROUP

	Neonates (one month of age or less)	Children (one month to 15 years of age)	Adults (older than 15 years of age)
Streptococcus pneumoniae	0–5%	10–20%	30–50%
Neisseria meningitidis	0–1%	25–40%	10–35%
Haemophilus influenzae	0–3%	40–60%	1–3%
*Streptococcus agalactiae**	20–40%	2–4%	1–5%
Staphylococci	1–5%	1–2%	5–15%
Listeria	2–10%	1–2%	5%
Gram-negative rods**	50–60%	1–2%	1–10%

*Almost all streptococci isolated in neonatal meningitis are Group B.
***Escherichia coli* accounts for approximately 40% of all bacterial neonatal meningitis. *Enterobacter, Klebsiella, Proteus, Pseudomonas,* and *Serratia* are responsible for the remaining cases caused by gram-negative bacteria.

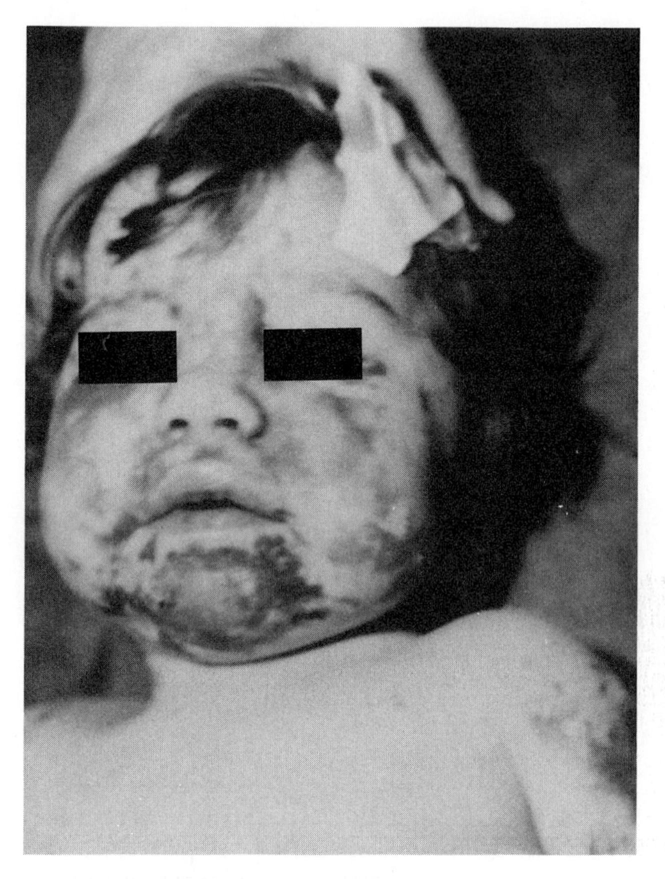

Figure 20.2 **Meningococcal septicemia.** Meningococcal septicemia is characterized by petechial (spots) hemorrhages throughout the body. The photo shows a child exhibiting petechial hemorrhages on the face and trunk due to *Neisseria meningitidis* septicemia. Courtesy of Pediatrics Department, Division of Infectious Diseases, University Hospital, Seville, Spain.

and in crowded slum areas. In the United States the number of reported cases per year averages about 2500 (1.5/100,000) (fig. 20.3). In 1943–1944 there was an epidemic of meningococcal meningitis in the United States related to the large number of recruits gathered together during the war. At least 35,000 persons suffered from meningococcal meningitis during those two years.

Pneumococcal meningitis (caused by *Streptococcus pneumoniae*) occurs most frequently in persons with ear infections (acute otitis media) and mastoid infections (mastoiditis) (30%); in persons with pneumonia (30%); in persons suffering from alcoholism and cirrhosis of the liver (20%); and in persons with severe nonpenetrating head injuries (10%).

The bacteria that cause meningitis are spread from one person to another in aerosols and on fomites. The bacteria generally enter the nasopharyngeal region and the respiratory tract, where they establish infections. Infections of the sinuses (sinusitis), mastoids (mastoiditis), and middle ear (otitis) can provide a direct access to the meninges. Infections of the lungs (pneumonia), bronchi (bronchitis), and pleura (pleuritis) can lead to bacteremia, which provides another route to the meninges. Frequently, the bacteria migrate from nasopharyngeal venules (blood vessels) to the meninges. The bacteria infecting the meninges rapidly spread throughout the subarachnoid space (fig. 20.1) and infect the ventricles (ventriculitis).

Meningitis caused by *N. meningitidis*, *H. influenzae*, and *S. pneumoniae* is usually preceded by an upper respiratory tract infection, an ear infection, or a lung infection. Headache, stiff neck, muscle pain (myalgia), back pain, and vomiting are common early symptoms before meningitis occurs. Meningitis begins when the bacteria spread to the meninges. Symptoms then include muscle ache, backache, stiff neck, fever, drowsiness, confusion, and unconsciousness. The pus and blood clots that accumulate in the brain block the flow of cerebrospinal fluid out of the brain, causing the subarachnoid space to swell. The increased pressure on the brain can cause seizures, paralysis, and coma.

Meningococcal (*Neisseria meningitidis*) infections are generally accompanied by a purpuric (purplish) rash all over the body (fig. 20.2). The purple color is due to hemorrhaging capillaries just under the skin. This rash may be caused by a **bacterial protease** that is known to hydrolyze IgA, or it may be due to damage to blood vessels caused by endotoxin released by the gram-negative *Neisseria meningitidis* (Shwartzmann phenomenon). A purpuric rash rarely occurs during a *Streptococcus pneumoniae* bacteremia and meningitis; however, the purpuric rash of echovirus **aseptic meningitis** (an infection of the meninges that does not involve bacteria) looks very much like the meningococcal meningitis rash.

The risk of meningococcal meningitis can be reduced by avoiding crowded quarters with poor air

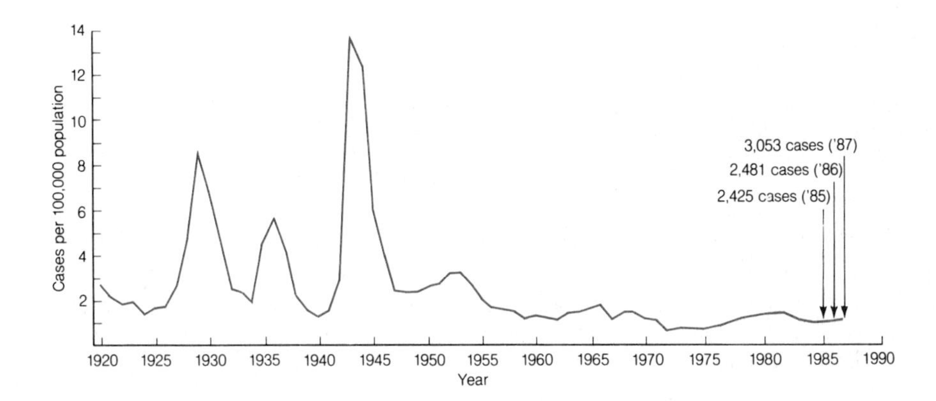

Figure 20.3 Incidence of meningococcal infections. During the last few years (1985–1987), approximately 2500 cases of meningococcal cases have been reported per year. The last major epidemic of meningococcal disease in the United States occurred during the Second World War, when large numbers of young soldiers were brought together in army camps. Today most cases occur in infants less than one year old.

circulation in which large numbers of persons gather, such as crowded infant day-care centers or military barracks. Meningitis caused by *S. pneumoniae*, *N. meningitidis*, and *H. influenzae* can be controlled by early diagnosis and treatment of ear infections, mastoid infections, and pneumonia. In the last few years, the incidence of meningococcal infections in the United States has been about 2500 cases/year, while the number of deaths has ranged between 300 and 700 per year (fig. 20.3).

Treatment of bacterial meningitis involves the use of antibiotics, but shock can be brought on by antibiotic treatment during acute bacteremia (blood infection). The shock is due to an allergic reaction against the lipopolysaccharide (endotoxin) released by the lysing gram-negative bacteria (*Neisseria*, *Hemophilus*). Lipopolysaccharide from the membranous envelope that surrounds gram-negative bacteria reacts with IgE, sometimes resulting in a severe anaphylactic reaction and shock known as the Shwartzmann phenomenon. The shock often is severe enough to kill the individual being treated.

Acute cerebral swelling generally is treated by intravenous infusion of mannitol and the use of corticosteroids (such as dexamethasone sodium phosphate). Cerebellar pressure is manifested by enlargement of the pupils and by rigid extension of the feet (bilateral extensor plantar responses). Seizures, which may cause fluids to go down the trachea and choke a patient, can be controlled by anticonvulsants.

Tetanus

Tetanus, commonly known as lockjaw, is a disease characterized by painful cramps, convulsions, labored breathing, and spastic paralysis, caused by a neurotoxin produced by the gram-positive bacillus *Clostridium tetani*. The neurotoxin binds to central nervous system nerve cells and inhibits nerve signals. *Clostridium tetani*, an anaerobic endospore former, is ubiquitous and lives in most soils in the warmer parts of the world.

When the endospores of *C. tetani* enter into a host tissue as a result of a deep puncture wound, they are able to germinate into vegetative cells because the damaged tissue provides an anaerobic environment. As the vegetative cells multiply, there is local tissue destruction and necrosis (tissue death) due to the exoenzymes released. The bacteria remain within the necrotic tissue, because they are unable to reproduce in normal tissue due to the high O_2 concentration. The bacteria multiply slowly for 7–14 days before any symptoms appear. As some of the bacteria die, they autolyse and release one of the most potent neurotoxins known to humans: less than 2.5×10^{-9} g/kg body weight of this neurotoxin is required to kill a human. Thus, a 150-pound (70-kg) person can be killed by only 1.75×10^{-7} g (0.175 μg) of the neurotoxin after it spreads throughout the body. It may also travel inside peripheral nerves to the central nervous system; the neurotoxin has an affinity for the axons of neurons, to which it binds (fig. 20.4).

Nerve cells transmit signals by releasing the neu-

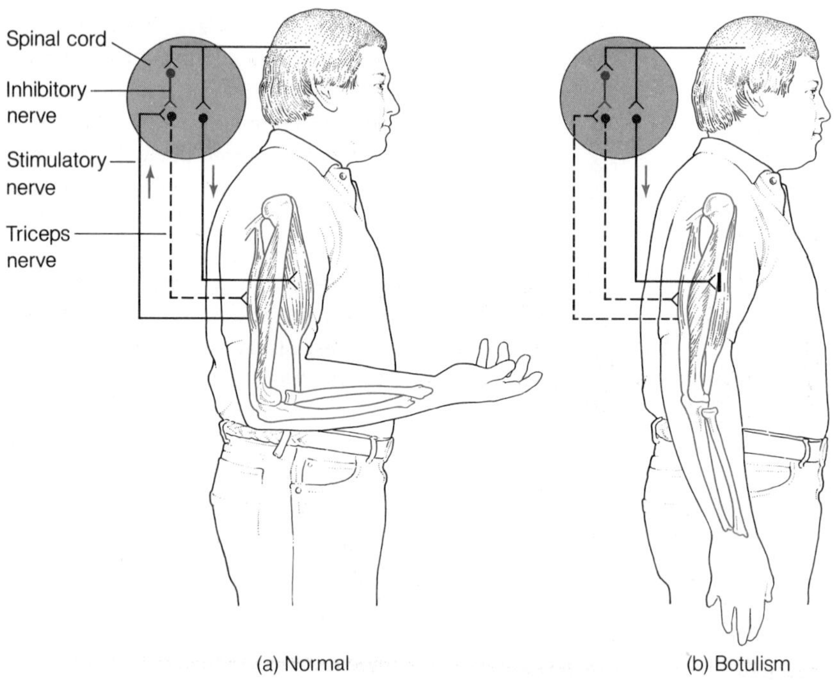

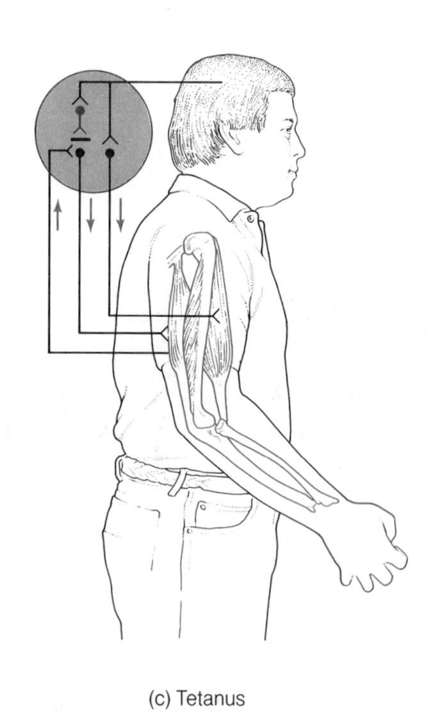

Figure 20.4 **How botulism and tetanus toxins work.** *(a)* Nervous control of the biceps and triceps are illustrated. Nerve impulses from the brain result in the contraction of the biceps. The stretching of the triceps sends a stimulatory signal to the triceps to prevent the stretching. This signal is inhibited by the inhibitory nerve, which prevents contraction of the triceps. If nerve impulses to the triceps were not inhibited, the triceps and biceps would work against each other. *(b)* The botulism toxins bind to nerves at the nerve-muscle junctions and block the release of acetylcholine. The biceps (and triceps) cannot be stimulated to contract, resulting in flaccid paralysis. *(c)* The tetanus toxins bind to the inhibitory nerve and block the release of inhibitory neurotransmitters. The triceps, which are normally inhibited, are now able to contract against the biceps, resulting in cramps known as tetanus.

rotransmitter acetylcholine from the ends of their axons. The neurotransmitter diffuses across the synaptic junction to the dendrite of another nerve cell, where it causes a transient collapse of the membrane potential. This effect initiates a nerve signal, which rapidly moves the length of the nerve. Then the neurotransmitter is rapidly destroyed by an enzyme called acetylcholinesterase, so that the nerve can reestablish its membrane potential and be ready to receive new signals.

It has been found that the tetanus neurotoxin binds mainly to the axons of inhibitory nerve cells and blocks the release of acetylcholine (fig. 20.4). Consequently, the nerves that go directly to the muscles constantly send signals to opposing muscles. **Tetanus** occurs when opposing muscles are constantly stimulated to contract. Muscles pulling against one another result in painful cramps all over the body (fig. 20.5). Overactivity of the sympathetic nerves produces drastic changes in blood pressure and heart rate which may result in death. The tremendous use of energy and the severe pain lead to exhaustion, shock, and death.

Tetanus is prevented by vaccination with **tetanus toxoid**. Tetanus toxoid is a denatured form of the tetanus neurotoxin which produces no disease but is able to stimulate the immune response. Tetanus toxoid generally is given at the same time that the vaccines for diphtheria and whooping cough (pertussis) are administered. The combined vaccine is known as DPT (diphtheria-pertussis-tetanus vaccine). Immunization of children at 2, 4, 6, and 18 months of age is recommended. Booster shots at 5 years, and every 10 years

Figure 20.5 Tetanus victim. A soldier wounded at the Battle of Corunna in 1809 is a victim of the effects of the tetanus toxin. The drawing is by a Scottish surgeon and anatomist, Sir Charles Bell, and was published in 1832. The soldier is in a state of spastic paralysis or tetanus. His muscles are working against one another, locking his jaw, bending his back, and curling his toes. The painful and exhausting cramps eventually lead to shock and death.

thereafter, supply ample protection against tetanus. Persons who acquire deep puncture wounds, however, or who have serious compound fractures that break through the skin, should receive a booster shot of tetanus toxoid.

Persons who have never been vaccinated for tetanus, or who have not had a booster shot for more than 10 years, should be given **antiserum** as well as the **tetanus toxoid** if they get a deep puncture wound. Antiserum contains antibodies against the tetanus neurotoxin and is obtained from humans or animals (generally horses, sheep, or rabbits) that have been vaccinated with tetanus toxoid. The antiserum provides passive immunity against the tetanus neurotoxin until sufficient antibody titer can develop after vaccination.

The number of reported cases of tetanus per year in the United States has declined steadily from approximately 450 in 1955 to 60 in 1986. The decline in tetanus cases is due to the increased incidence of vaccination since the 1950s. The lack of adequate vaccination is blamed for the current incidence of tetanus. In the last few years, the mortality in the United States due to tetanus has ranged between 20 and 50 per year.

Once an individual shows signs of tetanus (spontaneous contractions of the muscles, convulsions, cramps, and labored breathing), there is little that can be done to alleviate the symptoms. Antiserum is not very effective, because it cannot remove neurotoxin already bound to the nerve cells. It can, however, inactivate newly synthesized neurotoxin. Cleaning of the wound and penicillin treatment can also block the production of new neurotoxin. Limiting tetanus neurotoxin production may prevent death in the less severe cases.

Botulism

Botulism is a foodborne disease that involves the progressive loss of muscle activity all over the body due to a number of neurotoxins produced by the anaerobic gram-positive bacillus *Clostridium botulinum*. The bacterium and its endospores can be found in most soils, on plants, on animals, and in water. Consequently, these bacteria commonly contaminate foods. Under anaerobic conditions, the vegetative cells may produce a neurotoxin. If ingested, the neurotoxin causes disease. The neurotoxins bind to nerve cells, thus blocking the signals between nerve cells and between nerve and muscle cells (fig. 20.4).

Botulism in animals and humans is caused by the consumption of neurotoxins in food in which *Clostridium botulinum* was able to grow. The botulinal toxins may be produced in canned and bottled foods that have been improperly sterilized. The anaerobic environment and nutrients in canned and bottled foods promote the germination of *Clostridium* endospores that have escaped sterilization. Growth of *Clostridium* can also occur in sausages, fish, and hams that do not contain bacterial inhibitors.

Vegetative cells produce and release the botulism toxin into the food. When the food is consumed, the toxin passes from the intestines into the circulatory system and spreads throughout the body. The botulism toxin has strong affinity for nerve cells; it binds to the axons and prevents the release of acetylcholine, thus blocking nerve impulses to the muscle cells so that they are not stimulated to contract (fig. 20.4). Thus, botulism is a type of food poisoning that results in progressive flaccid paralysis. Twelve to 48 hours after ingestion of the botulism toxin, the first signs of the disease appear: blurred vision, progressive weakness throughout the body, and difficulty in chewing and swallowing. Death, generally due to respiratory or cardiac paralysis, occurs in 25–50% of those who show signs and symptoms of the disease. In the United States there is an average of 16 cases per year in humans, but the range is 5–80 cases annually (fig. 20.6). The number of deaths each year ranges between 1 and 5.

Many animals also suffer from botulism. For example, in parts of South Africa, where the soils are deficient in phosphorus, cattle frequently chew on animal bones and associated flesh to make up for their phosphorus deficiencies. Often the cattle die from consuming botulism toxin that has been produced in the carcasses. Any animal eating remains of this type ingests spores that germinate in its intestines after death. The vegetative cells then invade the rest of the carcass. Whole populations of wild waterfowl, mostly ducks, frequently are killed by botulism. Often, as many as 10,000–50,000 birds die per year. "Western duck sickness" occurs in the Western United States and Canada and is responsible, in some years, for the deaths of millions of wild waterfowl. It generally occurs when lakes and marshes are filled with decaying vegetation. Anaerobic conditions prevail at the bottom of these bodies of water, and this situation promotes the growth of *C. botulinum*. The waterfowl feeding in these environments pick up the bacteria and the toxin.

Botulism can be prevented by destroying spoiled canned foods, hams, and sausages. Spoiled food should not be given to animals, since they can also be affected by the botulism toxins. To prevent bovine botulism, carcasses should be disposed of immediately by burning or burial. To prevent fowl botulism, wild birds should be herded away from contaminated lakes and ponds. The contaminated areas should be drained, and feed should be distributed to keep wild birds from returning to these areas. Immunization of cattle, mink, and pheasants with toxoids is of some use in controlling outbreaks in wild animal populations. Humans are not vaccinated against the botulinal toxins, but persons who have eaten tainted foods can be treated with a botulism antiserum. The antiserum contains antibodies that inactivate the circulating botulism toxin,

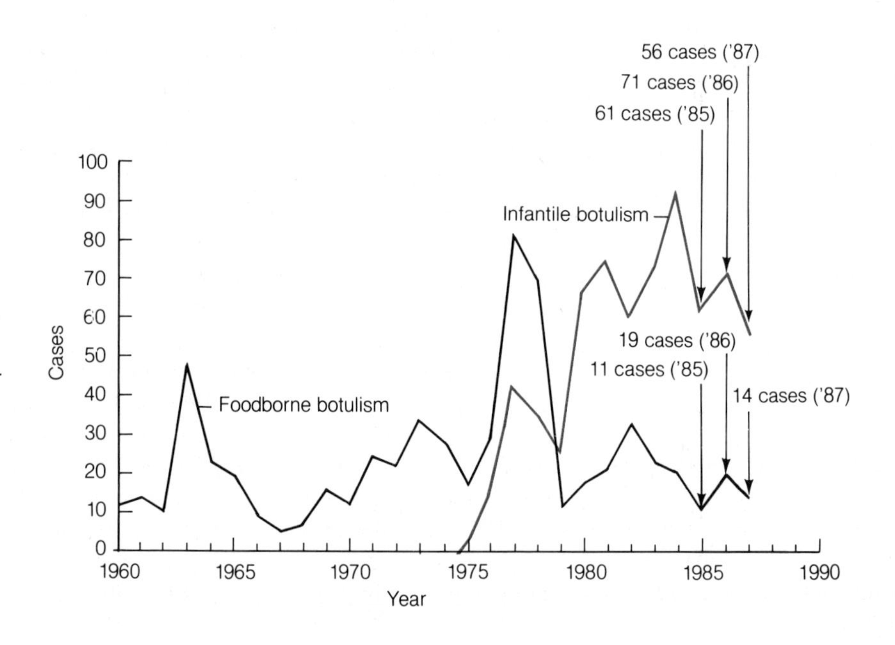

Figure 20.6 Morbidity due to botulism. The number of cases of foodborne botulism reported annually in the United States averages 16 per year. Every few years there is a drastic increase: 1951 (33), 1963 (45), 1973 (33), 1977 (80), and 1978 (65). During the last few years, the number of reported deaths has averaged about 5 per year. The number of reported cases of infant botulism in the last few years (1980–1985) in the United States has averaged about 70 per year. Infant botulism generally occurs in 30- to 40-week-old infants of both sexes. There is an average of one death per year, but some scientists believe that there are actually many more cases and deaths due to infant botulism than those reported. The actual number per year may be more than 400.

but has little effect on botulism toxin attached to nerve cells. Consequently, once the disease symptoms appear, botulism antiserum is only partially effective.

Infant Botulism

Infants from birth to about 12 months of age are particularly susceptible to **infant botulism**. Apparently, *C. botulinum* endospores are able to germinate in babies' intestines, and the vegetative cells are able to produce neurotoxins there. The first nerves inhibited by the toxin are those that control the muscles in the neck and head. The infant suckles poorly, has difficulty in swallowing, and is unable to hold up its head. Thus, this syndrome is known as the "floppy baby syndrome." As the disease progresses, the child's arms and legs become paralyzed, and respiratory paralysis may lead to death. The number of reported cases each year in the United States (most are in California) is about 70 (fig. 20.6). Reported deaths in the United States due to infant botulism have been 1–2 per year; however, some studies have indicated that 5% of the infants reported to have died of respiratory failure, and classified as **sudden infant deaths**, were infected with *Clostridium* or contained botulism toxin. Since there are approximately 7000 infants per year in the United States who die of sudden infant death, the actual number of infant botulism deaths may be more than 350 per year.

Infant botulism can be prevented by avoiding foods that contain large numbers of spores (such as honey) and those that promote germination of spores (such as spinach). Treatment of infant botulism includes administration of the appropriate antiserum and respiratory support units to help the infant continue breathing. *Clostridium botulinum* vegetative cells and toxins are found in the feces of infants who have recovered from the disease. Eventually, the clostridia and toxins disappear as a result of the competing microorganism in the feces and the children recover fully.

VIRAL DISEASES OF THE NERVOUS SYSTEM

Viral diseases of the nervous system are often due to viruses that initiated their attack elsewhere in the body. The initial infection generally occurs at the site of an animal bite or insect bite, or in the upper respiratory tract, which then spreads to the nervous system.

Hydrophobia (Rabies)

Rabies is a disease of the central nervous system that results in the deterioration of the brain and eventually death. It is caused by the growth of a bullet-shaped **rhabdovirus** in the spinal cord and brain and is characterized by paralysis, acute renal failure, coma, and death. **Rabies viruses** are found all over the world, infecting wild animals such as skunks, raccoons, foxes, coyotes, bobcats, and some bats. Because the rabies viruses grow exceedingly well in the salivary glands, they are found in large numbers in the saliva of an infected animal. Rabies generally is spread from one animal to another, or from animals to humans, through bites that inject the virus into the tissue. Consequently, many domestic animals become infected when they are bitten by a rabid animal. Cattle, sheep, and oxen are frequently infected when they are bitten by a rabid bat, skunk, fox, coyote, or wolf. Dogs and cats become infected when they attack a rabid animal or disturb its carcass. Rabid domestic animals, in turn, can transmit the disease to humans through bites. It is believed that the rabies virus can also be transmitted by virus aerosols; in this case, the virus begins its infection in the epithelium lining the nasal passages and the mouth.

Once the rabies virus enters the body, it begins to replicate in host cells at the site of infection. It appears to proliferate most efficiently in peripheral nerve cells. Interferon or other components of the immune system generally do not inhibit the rabies virus infection, so the virus spreads slowly from the site of infection along nerve cells, eventually reaching the spinal cord or brain. If the bite is on the hand, it might take the virus 6–12 weeks to reach the brain (fig. 20.7). If the infection occurs on the head, however, it might take the virus only 2–4 weeks to reach the brain. In humans, death occurs about 2–3 weeks after symptoms appear. No symptoms are observed until the virus begins to grow in the spinal cord and brain. The clinical course of rabies in humans involves malaise, fever, weakness, paralysis, acute renal failure, coma, and death (fig. 20.7).

The danger of rabies to humans and pets can be reduced by vaccination of domestic animals and of humans with high-risk occupations (skunk trappers, lab workers, veterinarians, dogcatchers) with a rabies virus vaccine. Vaccination of dogs and cats in rural areas is extremely important. The incidence of rabies can also be reduced by restricting dogs so that they do not run wild and attack rabid wild animals (fig. 20.8).

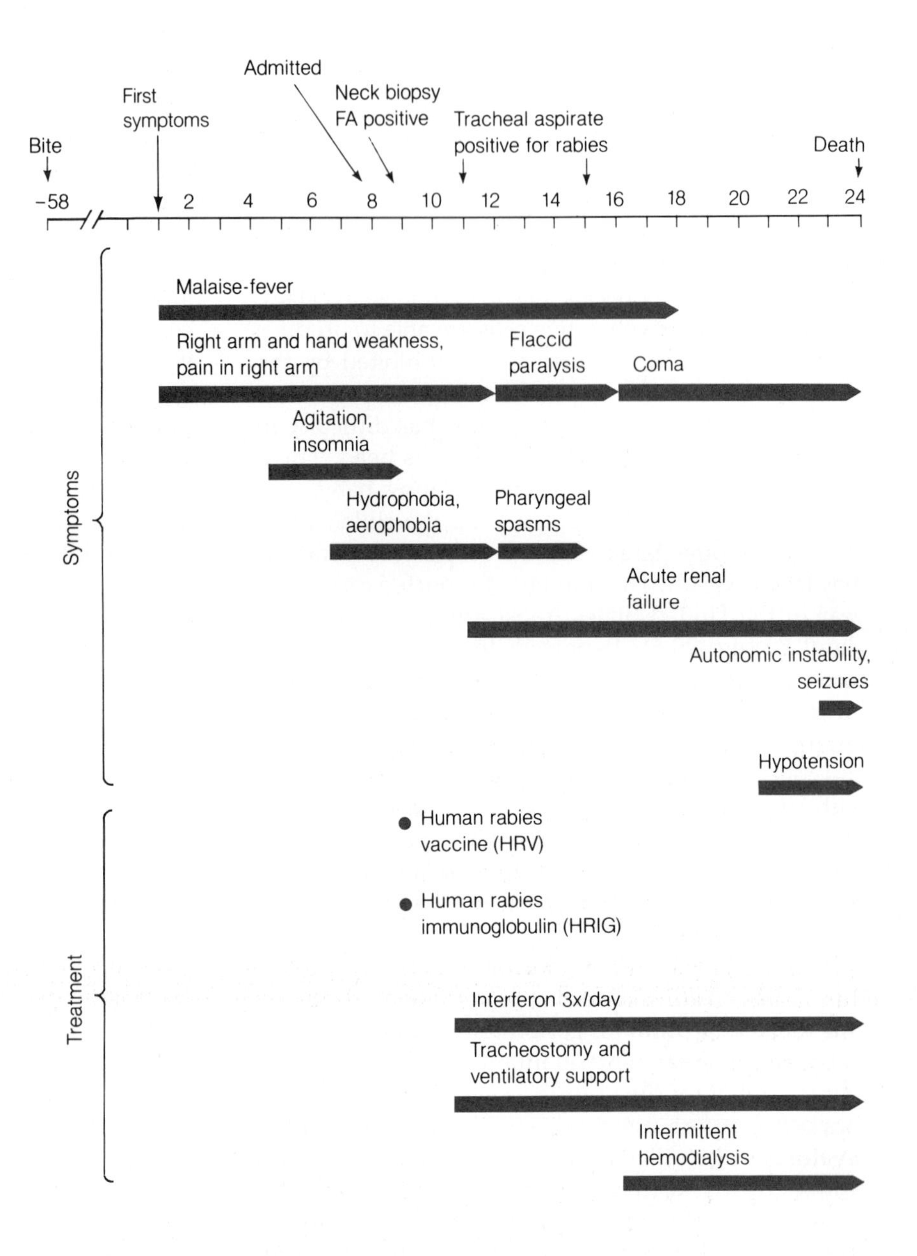

Figure 20.7 Clinical course of a human rabies case. This illustration schematizes the clinical course and management of a well-studied case of rabies. Bilateral bronchopneumonia, acute renal failure, hypotension, autonomic instability, and seizures caused sudden death due to respiratory and vascular collapse. Postmortem examination indicated fluorescent antibody positive rabies antigen in kidney parenchyma, bladder nerve tissue, skin of neck, skin from bite site, occipital nerve, pharynx, choroid plexus, hippocampus, pons, cervical spinal nerve, and dorsal root ganglion.

There are a number of rabies vaccines for animals that require only one vaccination per year. The viruses may be grown in tissue culture, extracted, and heat- and phenol-inactivated. Live virus vaccines are prepared in chicken or duck eggs. The growth of the viruses in chicken embryos leads to their attenuation. Caution must be used with live vaccines, however, because they can revert to virulent strains. Some are used only in dogs, because they become pathogenic in other animals.

Active immunization against rabies in humans requires six inoculations of chemically inactivated viruses known as the **human diploid cell** vaccine. Five intramuscular injections are given over a 30-day period, and one injection is given 2 months later. High-risk individuals, such as veterinarians and persons working with the virus, are immunized prophylactically. Persons in the general population who are exposed to the virus are passively immunized with **human rabies immune globulin** or with **horse antirabies serum**, and

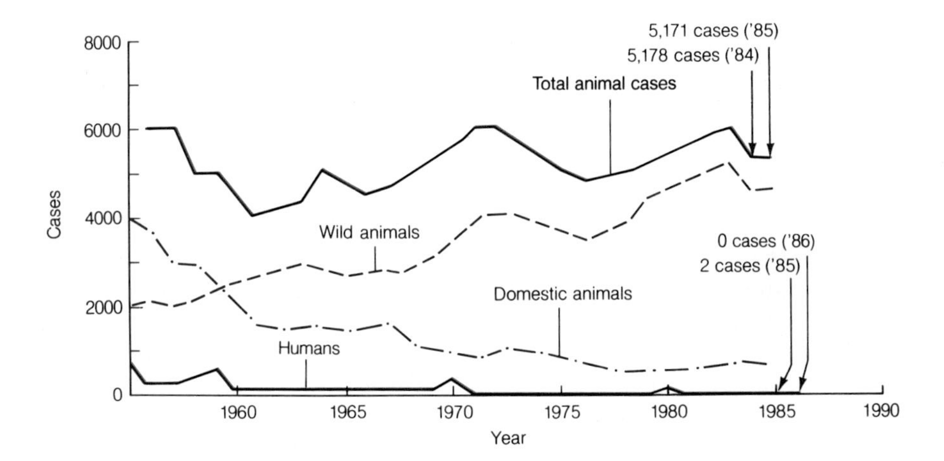

Figure 20.8 Incidence of rabies. In the last few years the number of rabies cases in wild and domestic animals has averaged about 5500 per year and the number of rabies cases in humans has averaged about 1 per year.

are then actively immunized with **human diploid cell vaccine** or with the older, infrequently used **duck embryo vaccine**. Human diploid cell vaccine is the vaccine of choice, since it requires only six intramuscular injections and it does not produce severe allergic reactions. Rabies is the only disease in which immunization started after exposure to the virus can prevent the disease. Protection against the rabies virus can develop before the disease symptoms appear because it takes such a long time for the viruses to reach the central nervous system. Once symptoms of the disease appear, however, rabies is nearly 100% fatal, so there is no treatment.

Poliomyelitis (Polio)

Poliomyelitis, in its most severe form, is a disease of the central nervous system characterized by destruction of nerve tissue and muscle paralysis. It is caused by the growth of a small, nonenveloped, icosahedral virus (**picornavirus**) in the brain and spinal cord. The destruction caused by the virus in the central nervous system can lead to paralysis of various parts of the body and to death. In those who survive the disease, some part of the body is often permanently paralyzed. The paralyzed part often wastes away because the muscles are not being stimulated, and this in turn leads to deformity (fig. 20.9).

Humans appear to be the only host species for the polio virus. Because the virus often multiplies in the throat and intestines and causes mild symptoms such as sore throat, headache, fever, and nausea, the virus is readily spread from one person to another. Antibodies against the polio virus are found in

Figure 20.9 Ancient Egyptian with polio. This Egyptian hieroglyph illustrates a man with an atrophied leg, which closely resembles the results of a paralytic poliomyelitis infection.

approximately 99% of the human population, indicating that most people have come in contact with the virus at some time (either naturally or by vaccination). It is transmitted from one person to another in water or food contaminated with saliva or feces. Public swimming pools are a good reservoir of polio viruses, which are not usually inactivated by the chlorine concentrations used to control other microorganisms.

The virus initially infects the throat, tonsils, intestines, and lymph nodes of the ileum (opening into the large intestine). From these tissues it spreads into the blood, causing what is called **viremia**. Usually the infection does not proceed beyond viremia and the patient recovers fully. Occasionally, however, the viruses in the blood manage to infect nerve cells of the spinal cord or brain. The replication of the viruses in nerve cells causes their death, which in turn leads to paralysis. The virus does not infect or destroy peripheral nerves. Destruction of large numbers of nerve cells in the brain results in death.

The severe symptoms of "polio" are more prevalent in developed countries, where the level of sanitation is high, than in underdeveloped countries. One reason is that, in countries with poor sanitation, polio viruses generally infect infants soon after they are born when they are still protected by their mother's circulating antibodies. These infections stimulate the infant to develop an immune response against the polioviruses. It also appears that infants less than a year old are not very good hosts for the virus; thus, infections of infants lead to the mild form of the disease.

In countries where hygiene is highly developed, immunization against polio is imperative. In the United States, vaccination against polio was initiated on a massive scale in 1955 (fig. 20.10). The incidence of paralytic polio in the United States in the 1970s and 1980s ranged between 5 and 10 cases per year, and the number of deaths averaged less than 1 per year. Most cases of paralytic poliomyelitis in the United States during that period were due to the oral polio vaccine reverting to virulent or were introduced by immigrants. The average death rate for those who contract the severe form of polio is 10%, but it may be as high as 50% in some epidemics. Ten to 50% of those who survive have residual paralysis.

In 1955, formalin-inactivated polio viruses were

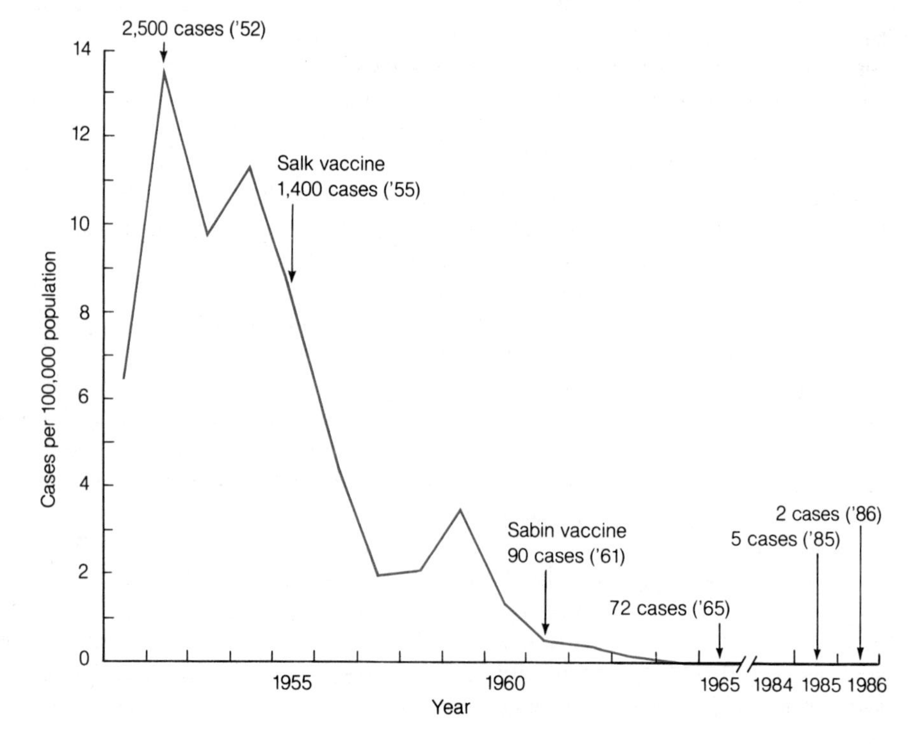

Figure 20.10 **Incidence of polio.** The incidence of polio reached a peak in the early 1950s. The Salk inactivated virus vaccine was introduced in 1955. Within 5 years, it was responsible for decreasing the number of cases more than twentyfold. In 1961 the Sabin oral vaccine using attenuated polioviruses was introduced. The Sabin vaccine was responsible for reducing the number of cases in the United States to an average of 5 cases per year.

used in a polio vaccine known as the **Salk vaccine**. Jonas Salk is given credit for the development of this vaccine, which is injected into the body. The dead viruses stimulate an immune reaction; however, a number of injections are required over a few months to develop a strong immune response to prevent viral infection, and booster shots are required every 5 years to maintain a high level of immunity. On the other hand, there have been no reported cases of polio due to this vaccine and it is the vaccine used in most countries outside of North America.

In 1961, attenuated live polio viruses were used in a polio vaccine called the **Sabin oral polio vaccine**. Albert Sabin is given credit for the development of this vaccine, which is simply swallowed. The attenuated polio viruses invade the intestinal epithelium and multiply, but because of their attenuated form, they are unable to spread efficiently to other parts of the body. The growth of the viruses in the intestines is believed to stimulate the production of IgA in the intestines as well as a generalized immune response. IgA in the intestines inhibits the growth of the wild-type polio viruses, thus reducing their prevalence in the immunized population.

The Sabin oral polio vaccine is very simple to administer, because it requires only the swallowing of a few drops of virus-laden syrup. Two separate inoculations are required over a period of a month to insure a strong immune response. The two disadvantages of the Sabin vaccine are its possible reversion to a virulent form and its transmissibility to others, which could lead to an outbreak of polio. Both of these problems account for about 50% of the cases of paralytic poliomyelitis in the United States.* Polio is contracted from vaccines at a rate of about one in 5 million immunizations (0.2×10^6). In the 1980s approximately 5–10 people contracted paralytic poliomyelitis in the United States because of the Sabin oral polio vaccine.

Once polio is contracted, treatment with vaccine may help to boost the immune system; however, there is no effective treatment for polio.

Complications Following Viral Infections

Reye's syndrome is a complication that can follow influenza A, influenza B, and varicella (chickenpox) infections and is characterized by a swelling of the brain and a fatty degeneration of the liver. Approximately 95% of the cases occur in children 14 years old or younger. Reported cases in the United States range from 200 to 400 cases per year, or 1 to 2 cases per 100,000 children per year.

In a typical case, a child with influenza or chickenpox (varicella) who is beginning to recover suddenly begins to vomit repeatedly. Within a day, the child becomes lethargic, disoriented, and irritable. Soon afterward, the child may become unconscious, and death follows in about 25% of the cases (50 to 100 deaths/year in United States). Apparently, the influenza or varicella viruses infect the liver and brain. Growth of the viruses in the brain appears to cause an acute noninflammatory encephalopathy. The brain swelling is due to the accumulation of cerebrospinal fluid around the brain. As the volume of cerebrospinal fluid in the brain increases, there is an increase in pressure on the brain, which damages it. Some children who survive the disease have permanent brain damage. Death from Reye's syndrome can sometimes be avoided by controlling the swelling of the brain with drugs.

Guillain-Barré syndrome is a complication that can follow some virus infections. A reversible paralysis is associated with the Guillian-Barré syndrome, and death may result if the lung diaphragm is paralyzed. Approximately 500 cases are reported each year in the United States. Because Guillain-Barré syndrome occurred in some people after they received the swine flu vaccine in 1978, the vaccination program against swine flu was halted. A study of the situation eventually indicated, however, that Guillain-Barré syndrome was not associated with influenza vaccinations.

It is believed, however, that influenza viruses infect and kill Schwann cells, which myelinate peripheral nerves, thus causing a demyelination of these nerves. This effect results in the paralysis of some muscles, since many peripheral nerves do not transmit nerve impulses without a myelin sheath. Death can be prevented by artificially stimulating breathing. Since Schwann cells can proliferate and wrap around demyelinated nerves, nerves can be made to function once more, and recovery from the Guillain-Barré syndrome is generally complete.

Slow Viruses and Prion Diseases

There are several diseases of the brain and spinal cord that often take many years to develop after the infectious agent invades the nervous system. Some of the infectious agents that cause these "slow virus diseases" are conventional viruses, while others are non-

Source: Postgraduate Medicine 75(1):171–175 (1984).

conventional pathogens that appear to be fundamentally different from either viruses or cellular microorganisms (table 20.2). There is now evidence, albeit controversial, to suggest that some of the nonconventional pathogens are **prions**, proteins that stimulate their own synthesis. Allergic reactions and inflammation stimulated by the conventional viruses are responsible for some viral symptoms. The nonconventional pathogens, by contrast, do not seem to stimulate an immunologic response or inflammation.

Kuru is a progressive degenerative disease of humans that is due to a nonconventional pathogen believed to be a prion. Kuru is distinguished by erratic jerking, tremors, loss of muscular control, inability to walk or stand, tissue wasting, and death 9–12 months after the onset of the signs and symptoms.

Kuru is found infrequently today, but it did infect a large number of individuals in a primitive New Guinea tribe who practiced headhunting and cannibalism of brains. Kuru attacked primarily young boys and girls and adult women, but almost never adult men. The reason was that the women prepared infected brain tissue from persons who had died of kuru, and the children and women ate this tissue. Tribal laws prohibited the men from touching or consuming human tissues, so they infrequently developed the disease.

Creutzfeldt-Jakob disease in humans is due to a nonconventional pathogen thought to be a prion. The disease usually affects people between 50 and 60 years of age and is distinguished by a progressive mental deterioration, jerking, muscle tremors, and eventual death. The nonconventional pathogen causes lysis of nerve cells and the proliferation of astrocytes, so that a spongiform encephalitis develops.

A high incidence of this disease is found in Libyan Jews, who frequently consume ovine eyes and brains. Creutzfeldt-Jakob disease can infect recipients of corneal transplants when the donor is infected. Neurosurgeons can become infected while working on diseased patients. There is a high incidence of Creutzfeldt-Jakob disease in people who have had neurosurgery or eye surgery, indicating improperly sterilized equipment (brain electrodes) or surgical tools.

A reservoir for the nonconventional pathogen that causes Creutzfeldt-Jakob disease may be prion-infected sheep that have **scrapie**. Human infections could come from eating scrapie-infected sheep tissue. The scrapie agent is extremely resistant to inactivation by heat or chemicals, so it could be transmitted during preparation of infected meat or in undercooked lamb.

Alzheimer's disease is distinguished by a progressive mental deterioration of persons in their 50s, 60s, and 70s characterized by loss of memory and apathy. It is an important disease because it affects over 3 million people in the United States and cases are occurring at a rate of approximately 100,000 per year. At this time it is not clear whether Alzheimer's disease is hereditary or due to a nonconventional agent. A number of researchers believe that it is caused by a prion. However, proteins isolated from the brain lesions

TABLE 20.2
DISEASES DUE TO SLOW VIRUSES AND PRIONS

Disease	Host	Cause	Characteristic	Inflammation
Scrapie	Sheep	Prion	Lysis of nerve cells	0
Mink encephalitis	Mink	Prion?		0
Visna	Sheep	Virus	Demyelination	+
Aleutian disease of mink	Mink	Virus	Demyelination	+
Kuru	Human	Prion?	Lysis of nerve cells	0
Creutzfeldt-Jakob disease	Human	Prion?	Lysis of nerve cells	0
Subacute sclerosing panencephalitis	Human	Rubeola virus	Demyelination	+
Progressive rubella panencephalitis	Human	Rubella virus	Demyelination	+
Multiple sclerosis	Human	Rubeola virus	Demyelination	+
Amyotrophic lateral sclerosis	Human	Retrovirus	Lysis of motor neurons	+
Progressive multifocal leukoencephalopathy	Human	Picornavirus	Demyelination	+/0
Alzheimer's disease	Human	Defective gene	Lysis of nerve cells	0

are not infectious. Also a gene has been detected in chromosome 21 that might be responsible for the disease.

Alzheimer's disease destroys neurons and causes atrophy of the cerebral cortex and neurofibrillar tangles. A typical case of Alzheimer's disease might be diagnosed at 60, progress for 10 to 15 years, and result in a demented, mute, bedridden individual.

Subacute sclerosing panencephalitis is a neurologic disease of young children and adolescents, rarely affecting people older than 14. It is initially characterized by mental deterioration, jerking, and the inability to control muscles. As the disease progresses, it leads to dementia, stupor, rigidity, and death. Serum and cerebrospinal fluid have high titers of antibodies against the measles virus. The virus causes the destruction of the myelin sheath around brain nerve cells, and it stimulates the proliferation of astrocytes and the invasion of the brain by mononuclear cells.

Subacute sclerosing panencephalitis is due to a mutant form of the rubeola (red measles) virus, while **progressive rubella panencephalitis** is due to a persistent infection by the rubella virus. For the viruses to cause these diseases, apparently they must infect the host early in life: before the age of two, in the case of the rubeola virus, and before birth, in the case of the rubella virus. High titers of antibodies that diminish host cell-mediated immune responses appear to select for mutant forms of the rubeola and rubella viruses which are able to persist in the brain.

Patients with congenital rubella infections may undergo progressive mental deterioration, muscle spasms, lack of muscle control, and seizures. The damage to the brain resembles that caused by subacute sclerosing panencephalitis.

Multiple sclerosis is a general term describing a type of degeneration of the brain or spinal cord which involves the destruction of myelin sheaths around nerve cells. This results in muscular weakness, trembling, slurred speech, diminished vision, and paralysis. Some types of multiple sclerosis are initiated by virus infections; the measles virus is a favorite candidate.

The normal brain or spinal cord appears white when sectioned, because of the large amount of myelin surrounding the nerve cells. If the **oligodendroglial cells** that form the myelin sheaths are destroyed, however, an abnormal growth of astrocytes surrounds the nerve cells and produces a discoloration and hardening (sclerosis) of the tissue.

It is hypothesized that a virus (measles, influenza, etc.) infects the brain during an illness. The virus infection is thought to damage the oligodendroglial cells and the "blood-brain barrier," so that myelin proteins, which normally are found only in the brain, pass into the body and initiate an autoimmune reaction. Apparently, stimulated macrophages and lymphocytes can penetrate the "blood-brain barrier" and attack the myelin sheath that covers the nerve cells in the brain and spinal cord. An experimentally induced multiple sclerosis has been produced in guinea pigs simply by injecting a single dose of white matter from other guinea pigs. Consequently, it is believed that multiple sclerosis is due to an autoimmune reaction that is initiated by a virus infection of the brain or spinal cord. This fatal disease in humans is often of very long duration, taking 15–20 years to kill the individual.

Rapidly Developing Infections of the Central Nervous System

Viral encephalitis (swelling of the brain) is caused mainly by **enteroviruses** (mumps, polio-, coxsackie-, and echoviruses) and by **arboviruses** (eastern equine encephalitis [EEE], western equine encephalitis [WEE], Venezuelan equine encephalitis [VEE], dengue fever, Japanese B encephalitis, Saint Louis encephalitis, California encephalitis, and yellow fever viruses). These viruses infect the brain, meninges, and spinal cord. This leads to fever, headache, chills, vomiting, encephalitis, and sometimes death. All of these viruses not only infect humans but have a number of animal hosts as well. The arboviruses (*ar*thropod *bo*rne *viruses*) occasionally infect horses and cause widespread illness and death among herds. The Saint Louis encephalitis, California encephalitis, and Japanese B encephalitis viruses, by contrast, usually infect humans. In the United States, there are about 1200 reported cases of human encephalitis per year (fig. 20.11), with a 10% mortality. In 1975 there was an epidemic of Saint Louis encephalitis in the United States which struck over 4000 individuals and killed 387 of them.

The viral reservoirs are believed to be mainly wild and domestic fowl, rodents, horses, and the mosquito vector. The bite of the infected mosquito introduces the viruses into the blood, where they can spread to all parts of the body and multiply in susceptible cells. The systemic replication and spread of the virus causes fever, headache, chills, and vomiting. If the virus spreads into the brain, it causes sufficient damage so that the brain swells. This encephalitis can result in loss of consciousness and in death.

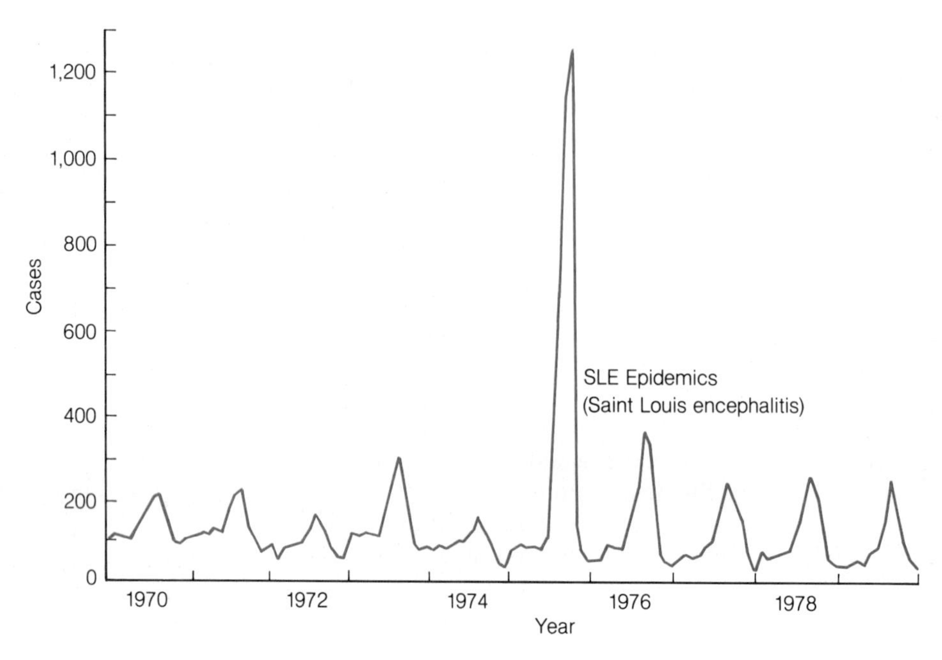

Figure 20.11 Incidence of viral encephalitis (arbovirus and enteroviruses). Viral encephalitis reaches a peak each year in the late summer (July–Sept.), when people are outdoors and likely to be infected by mosquito bites and in contact with each other.

Encephalitis caused by arboviruses can be controlled by eliminating the mosquito's breeding grounds and by using insecticides where breeding sites cannot be destroyed. There are vaccines that can be used to protect humans against the EEE, WEE, and VEE viruses. In addition, there are vaccines for horses against these viruses, and these are recommended during outbreaks of the diseases; mortality in horses is about 20–50% for WEE, as high as 75% for VEE, and over 90% for EEE. Two intradermal injections of vaccine 10 days apart give immunity that lasts approximately a year.

FOCUS THE INFLUENZA VIRUS MAY BE RESPONSIBLE FOR MAJOR NEUROLOGICAL DISEASES

Scientists at the Centers for Disease Control (CDC) in Atlanta, Georgia, have proposed that the influenza virus causes not only pneumonia but also encephalitis and chronic heart diseases that may affect people many decades after they contracted influenza.

The more than 20 million people who died worldwide in the 1918 influenza epidemic were only the first victims of the virus. During the years 1919–1928, approximately 500,000 people died worldwide of **encephalitis lethargica**, a severe neurological disease caused by influenza A virus. In addition, the CDC researchers propose that much of the heart disease in elderly people could be due to the 1918 influenza virus having damaged the nerves that regulate the heart. In the 1970s, there was a decline in the number of cases of chronic mental disorders and heart disease. The CDC researchers postulate that this was because many of the people alive during the 1918 influenza epidemic had died because of old age and/or disease.

Aseptic meningitis is an inflammation of the meninges that may be caused by the mumps, polio-, coxsackie-, or echoviruses. The term "aseptic" refers to the fact that bacteria cannot be cultured from the diseased meninges or spinal fluid. The echoviruses are responsible for the vast majority of aseptic meningitis cases. These viruses apparently spread from the throat and intestines into the blood, producing a viremia. From the blood, they penetrate to the meninges and infect these membranes. Outbreaks of aseptic meningitis occur in July, August, and September each year, the months during which large numbers of people are most likely to use pools, lakes, and streams that are contaminated by fecal matter.

FUNGAL AND PROTOZOAL DISEASES OF THE CENTRAL NERVOUS SYSTEM

Fungal Infections of the Meninges

Cryptococcosis is a disease caused by the fungus *Cryptococcus neoformans* that is usually accompanied by fever, cough, and pleural pain. In some cases the disease can become systemic and involve the brain (fig. 20.12).

Cryptococcus neoformans is a yeastlike fungus often found growing in soils and in pigeon droppings. The yeast has a thick polysaccharide capsule that is clearly visible in a capsule stain and aids in the yeast's identification. The yeast can be blown into the air on windy days and thus transmitted to humans. The inhaled *Cryptococcus* often is able to grow in the lungs; when it does so, it leads to a fever, a severe cough, and pleural pain. In relatively healthy humans, cryptococcosis is self-limiting. In older persons and immunosuppressed individuals, however, the yeast may manage to spread from the lungs into the blood and reach various organs. If the meninges are infected by the yeast, a chronic meningitis results that is fatal if untreated. Cryptococcosis can be treated effectively with amphotericin B.

Protozoal Infections of the Central Nervous System

African sleeping sickness, or **trypanosomiasis,** is a disease that is characterized by attacks of fever over a period of several months and then a progressive disorientation, slurred speech, and difficulty in walking. The last stages of the disease include convulsions, paralysis, mental deterioration, extended periods of sleep, coma, and then death. These last stages of the disease generally occur over a number of months. The disease is caused by the protozoa *Trypanosoma brucei gambiense* and *T. brucei rhodesiense*.

Trypanosoma brucei gambiense reproduces in humans and in the tsetse fly, while *T. brucei rhodesiense* additionally reproduces in wild and domestic animals. Trypanosomes in the salivary gland of the tsetse fly are injected into animals by the bite of the fly. The

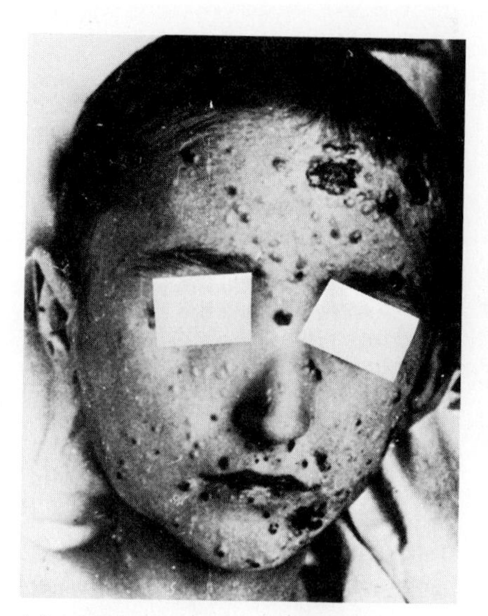
(a)

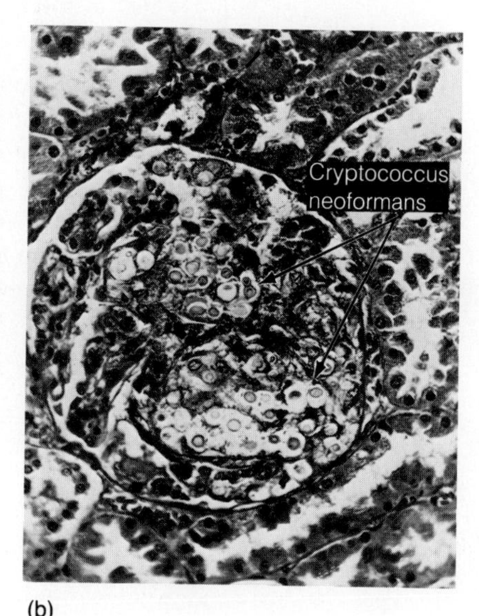

(b)

Figure 20.12 Disseminated cryptococcosis. The fungus *Cryptococcus neoformans* invades the lungs, and in severe cases disseminates to the central nervous system and other organs, including the skin. *(a)* Child with disseminated cryptococcosis. Fungi have spread to the brain and skin from the lungs. *(b)* Large number of *Cryptococcus neoformans* cells in kidney glomerulus.

Figure 20.13 ***Trypanosoma brucei gambiense*** **in a blood smear**

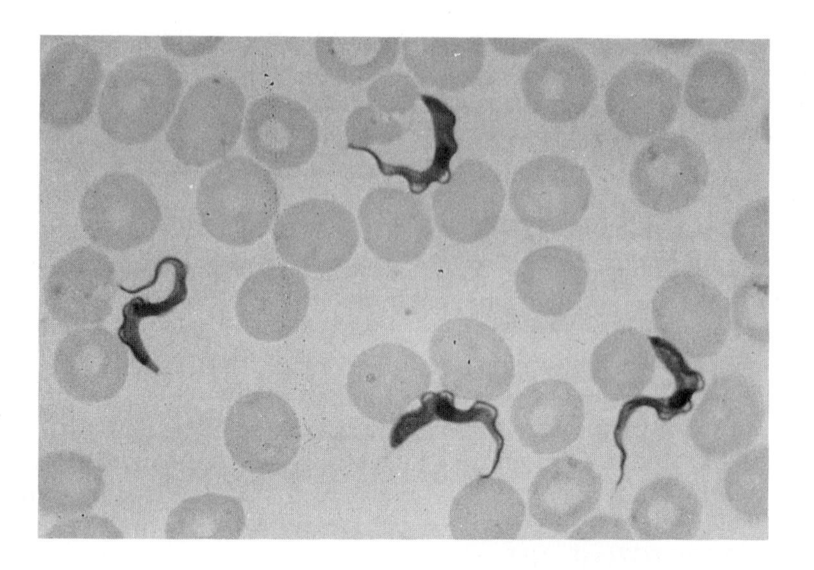

trypanosomes develop in the blood and lymph and may become polymorphic (fig. 20.13). Multiplication of the organisms in the lymph cause periodic fevers 2–3 weeks after infection. Eventually, the trypanosomes invade the central nervous system, producing a mild meningitis and encephalitis that produces headaches, general weakness, loss of appetite, nausea, slurred speech, and mental deterioration. The infection may last for as long as a year before the individual becomes comatose and dies.

The trypanosomes are generally numerous enough in infected humans and animals that a biting tsetse fly will become infected. The trypanosomes reproduce in the fly's midgut and then migrate to its salivary glands. Tsetse flies with trypanosomes in their salivary glands act as the vectors for the disease.

TABLE 20.3
DISEASES OF THE NERVOUS SYSTEM

Disease	Causative agent	Microbial group
Meningitis	*Neisseria meningitidis* *Streptococcus pneumoniae* *Haemophilus influenzae* *Escherichia coli*	Gram − cocci Gram + cocci Gram − rods Gram − rods
Aseptic meningitis	Viruses	Various groups
Tetanus	*Clostridium tetani*	Sporogenous rod
Botulism	*Clostridium botulinum*	Sporogenous rod
Rabies	Rabies virus	Rhabdoviruses
Poliomyelitis	Polioviruses	Picornaviruses
Reye's syndrome	Influenza viruses Varicella-zoster virus	Orthomyxoviruses Herpesvirus
Encephalitis	Arboviruses	Arboviruses
Cryptococcosis	*Cryptococcus neoformans*	Yeast
Sleeping sickness	*Trypanosoma gambiense* *Trypanosoma rhodesiense*	Protozoa Protozoa

The best ways to prevent the disease in Africa are to eradicate the tsetse fly's breeding sites wherever possible, use insect repellant, wear clothing that protects against fly bites, and take prophylactic drugs such as chloroquine during times of high risk. At this time, there are no effective vaccines against the trypanosome because the protozoan changes its antigenic coat during the course of infection. However, this is a project under study by the World Health Organization, and a vaccine may be developed in the near future.

Trypanosomiasis can be diagnosed by observing the parasites in the blood, in the lymph nodes, or in the spinal fluid. Once African sleeping sickness is contracted, it can be treated with the drugs **pentamidine isothionate** and **melarsoprol**. Treatment of *T. rhodesiense* infections also relies on the drug **suramin**.

Microorganisms cause many diseases that involve the human nervous system. The main characteristics of some of these diseases are summarized in table 20.3.

CLINICAL PERSPECTIVE HUMAN RABIES CASE DUE TO AN INADEQUATE TEST

A 40-year-old man suspected of having rabies was admitted to the hospital eight days after first developing numbness of his right hand and then suffering from pain in his right arm, fever, and malaise (fig. 20.7). Two days before being admitted the patient had experienced throat spasms when drinking water. Hydrophobia became so severe because of the resulting painful throat spasms that the patient had avoided taking a shower. The patient was highly agitated and was suffering from insomnia.

The patient's history indicated that he was probably exposed to the rabies virus when he was bitten on his right hand by his dog nearly two months before he first began to experience the rabies symptoms. After the bite, the dog had been killed and the head shipped to a health laboratory to check for rabies. The dog's brain showed no Negri bodies with Sellers' stain indicating the dog was free of rabies. In addition, the dog had had a recent rabies vaccination. Because of the negative test for rabies and the dog's vaccination history, antirabies vaccine was not given to the man.

A neck biopsy was taken from the patient and sent to the Centers for Disease Control (CDC). Direct fluorescent antibody testing of the tissue confirmed the presence of rabies viruses.

Treatment was initiated with a single dose each of human diploid cell rabies vaccine and human rabies immune globulin. On the second day after being admitted, human leukocyte interferon was given twice a day intramuscularly and once a day intraventricularly (into the brain ventricles).

The patient did not respond to treatment. He developed bilateral bronchopneumonia, acute renal failure, autonomic instability, and flaccid paralysis. He went into a coma and a few days later died suddenly of respiratory and vascular collapse.

Serum and cerobrospinal fluid that were collected daily for the 16 days of hospitalization before the patient died were negative for antibodies against rabies. A tracheal aspirate used to inoculate a mouse demonstrated the presence of rabies virus.

The Sellers' stain used to identify Negri bodies that are diagnostic for rabies is known to give false-negative results in 10% of specimens from dogs that are clinically ill with rabies. Because of this large error, direct fluorescent antibody staining of brain material is the recommended procedure for identifying rabies viruses in animal tissue.

The single dose of human rabies immunoglobulin and the interferon treatments were not successful. In animal studies, protection against the rabies occurs only if the interferon is administered before or shortly after virus infection. In addition, the human rabies vaccine did not stimulate the patient to develop antibodies against the rabies virus in the 16 days before he died.

For more information on the case discussed, see the following reference: Sibley, W.A., et. al. 1981. Human rabies acquired outside the United States from a dog bite. *Morbidity and Mortality Weekly Report* 30(43): 537–540.

SUMMARY

ORGANIZATION OF THE NERVOUS SYSTEM

1. The central nervous system includes the brain, spinal cord, cerebrospinal fluid, and meninges (the membranes that cover the brain and spinal cord).
2. The peripheral nervous system consists of all the nerve cells from the body that connect to the spinal cord or brain.
3. The "blood-brain barrier" inhibits the movement of antibodies and some drugs from the blood into the spinal cord and brain. Consequently, the brain is not directly protected by the immune system, and some drugs and antibiotics are useless in treating spinal cord or brain infections.

BACTERIAL DISEASES OF THE NERVOUS SYSTEM

1. Acute pyogenic meningitis is caused by *Escherichia coli* and Group B streptococci in babies younger than 1 month; by *Haemophilus influenzae, Neisseria meningitidis,* and *Streptococcus pneumoniae* in children less than 15 years old; and by *Streptococcus pneumoniae, Neisseria meningitidis,* and *Staphylococcus aureus* in adults. There are 2500 cases/year and 300–700 deaths/year in the United States due to meningococcal meningitis.
2. *Neisseria meningitidis* is the only bacterium that causes meningitis epidemics.
3. Meningitis is preceded by upper respiratory tract infections, ear infections, or lung infections.
4. Meningococcal infections, in contrast to pneumococcal infections, are generally accompanied by a rash due to hemorrhaging capillaries.
5. Tetanus or lockjaw is due to a neurotoxin produced by the anaerobic bacterium *Clostridium tetani.*
6. The tetanus neurotoxin binds primarily to inhibitory nerve cells and blocks the release of acetylcholine. Thus, nerves send signals to opposing muscles and painful cramps develop all over the body.
7. The tetanus neurotoxin also blocks neuromuscular transmissions, so that muscles become paralyzed.
8. Tetanus is prevented by vaccination with tetanus toxoid, a denatured form of the neurotoxin.
9. There are 50–100 human cases/year and 20–50 deaths/year in the United States due to tetanus.
10. Botulism is due to a neurotoxin produced by the anaerobic bacterium *Clostridium botulinum.*
11. The botulism neurotoxin blocks neuromuscular tranmissions so that muscles become paralyzed.
12. Deaths due to botulism are generally due to respiratory or cardiac paralysis. There are 5–80 adult cases annually and 1–5 deaths each year.
13. Flocks of wild waterfowl, mostly ducks, are often wiped out by botulism.
14. Infant botulism may kill 70–350 infants/year in the United States.

VIRAL DISEASES OF THE NERVOUS SYSTEM

1. Rabies is caused by a rhabdovirus that infects the central nervous system.
2. The rabies virus causes death because of its growth in the brain.
3. The morbidity and mortality of rabies in the United States is approximately five each year.
4. Rabies in wild animals has been increasing for a number of years, and has reached over 5000/year in the United States.
5. Poliomyelitis is caused by three strains of a picornavirus that initially infect the throat and intestines and subsequently infect the central nervous system.
6. The Salk vaccine is made from formalin-inactivated polioviruses and was first used in 1955. The Sabin oral polio vaccine is made from attenuated live polioviruses and was introduced in 1961.
7. The polio vaccines have dramatically reduced the number of polio cases from approximately 2000 cases/year in the early 1950s to fewer than 10 cases/year in the 1980s.
8. Reversion of one of the polioviruses to a virulent form in the Sabin oral polio vaccine is responsible for 5 to 10 cases of polio each year.
9. Reye's syndrome is characterized by a swelling of the brain and fatty degeneration of the tissue and is due to a complication following infections by influenza A, influenza B, and varicella (chickenpox). There are between 200 and 400 cases of Reye's syndrome/year in the United States. Death occurs in 25% of the cases.
10. Guillain-Barré syndrome is characterized by a reversible paralysis that occurs after some virus

infections. There are approximately 500 cases/year in the United States.

11. Slow virus diseases include Aleutian disease of mink, subacute sclerosing panencephalitis, progressive rubella panencephalitis, and multiple sclerosis.
12. Sheep scrapie, mink encephalitis, Kuru, and Creutzfeldt-Jakob disease may be caused by prions. Alzheimers disease may be due to a faulty gene.
13. Viral encephalitis is caused by enteroviruses and arboviruses that infect the meninges, spinal cord, and brain. The infection causes the brain to swell. Death occurs in 10% of cases.
14. There are approximately 1200 cases of human encephalitis and 150 deaths/year in the United States.
15. Arbovirus reservoirs are found mainly in wild and domestic fowl, rodents, horses, and the mosquito vectors.
16. Aseptic meningitis is an inflammation of the meninges caused by the enteroviruses. There are approximately 1200 cases/year in the United States. The echoviruses are responsible for most of the cases of aseptic meningitis.

FUNGAL AND PROTOZOAL DISEASES OF THE CENTRAL NERVOUS SYSTEM

1. African sleeping sickness is caused by the protozoans *Trypanosoma gambiense* and *T. rhodesiense.*
2. The trypanosomes are spread by the bite of the tsetse fly, which functions both as a host and as a vector.
3. Cryptococcosis is due to the yeastlike fungus *Cryptococcus neoformans.* This fungus sometimes achieves a systemic infection that leads to meningitis.

STUDY QUESTIONS

1. Explain what the central nervous system, peripheral nervous system, and autonomic nervous system are.
2. How does the "blood-brain barrier" protect the spinal cord and brain? What is a disadvantage of this barrier?
3. Which bacteria are responsible for acute pyogenic meningitis in babies less than a month old; in children less than 15 years old; and in adults?
4. Which bacterium, of the ones that cause meningitis, usually causes a rash?
5. Give the genus and species of the bacteria that cause tetanus and botulism.
6. How does the tetanus neurotoxin affect a human, and what is the cause of death?
7. How does a person acquire tetanus?
8. How does the botulism neurotoxin affect a human, and what is the cause of death?
9. How does a person acquire botulism?
10. What type of organism causes rabies, and how is the disease generally contracted?
11. Who developed the first rabies vaccine, and how was it prepared?
12. What is the difference between the Pasteur treatment and the modern treatment for rabies?
13. What type of organisms cause poliomyelitis, and how is the disease generally contracted?
14. What is the difference between the Salk and the Sabin polio vaccines?
15. How effective have the polio vaccines been in reducing the incidence of the disease?
16. What potential problem is there with the Sabin vaccine?
17. What kind of organisms cause Reye's syndrome, and how are they contracted?
18. What kind of organisms cause Guillain-Barré syndrome, and how are they contracted?
19. Scrapies, kuru, and Creutzfeldt-Jakob disease are due to what kind of infectious agent?
20. Compare and contrast subacute sclerosing panencephalitis, progressive rubella panencephalitis, multiple sclerosis, and viral encephalitis.
21. What kind of organisms are believed to be responsible for multiple sclerosis? How do they cause the disease?
22. What kind of organisms lead to viral encephalitis? How do they cause the disease?
23. Aseptic meningitis is caused by what type of viruses? How many cases are there each year in the United States?
24. What kind of an organism causes African sleeping sickness? How does the organism cause the disease?

SELF-TEST

1. The inner two meninges enclose what important tissue?
   a. blood;

b. lymph;
c. cerebrospinal fluid;
d. peripheral nervous system;
e. none of the above.

2. The brain and spinal chord are considered the:
a. sympathetic nervous system;
b. parasympathetic nervous system;
c. autonomic nervous system;
d. central nervous system;
e. peripheral nervous system.

3. Normally, antibodies and cells of the immune system are unable to enter the brain and spinal chord to protect them from microorganisms that sometimes infect. a. true; b. false.

4. What important tissue is enclosed by the dura mater?
a. blood;
b. lymph;
c. cerebrospinal fluid;
d. peripheral nervous system;
e. none of the above.

5. The majority of cases of bacterial meningitis in newborn babies are caused by *Neisseria meningitidis*. a. true; b. false.

6. The majority of cases of meningitis in newborn babies and adults are caused by *Haemophils influenzae*. a. true; b. false

7. The purplish rash that accompanies meningococcal infections is due to:
a. a pigment released by the brain;
b. a pigment released into the cerebrospinal fluid;
c. ruptured capillaries and coagulating blood;
d. antigen-antibody-complement complexes;
e. lgE-mediated allergic reactions.

8. Approximately, how many deaths occur each year in the United States due to meningococcal infections?
a. 5;
b. 50;
c. 500;
d. 5000;
e. 50,000.

9. Any bacterium on rusty nails can cause lockjaw (tetanus). a. true; b. false.

10. The neurotoxin that causes lockjaw (tetanus) causes paralysis because it:
a. enters muscle cells and blocks contraction;
b. enters muscle cells and stimulates contraction;
c. blocks inhibitory nerves so that complimentary muscles are stimulated;
d. stimulates nerves so that complimentary muscles are inhibited;
e. blocks the neuromuscular junction so that no muscles are stimulated.

11. The neurotoxin that causes botulism causes flaccid paralysis because it:
a. enters muscle cells and blocks contraction;
b. enters muscle cells and stimulates contraction;
c. blocks inhibitory nerves so that complimentary muscles are stimulated;
d. stimulates nerves so that complimentary muscles are inhibited;
e. blocks the neuromuscular junction so that no muscles are stimulated.

12. When is the vaccine to protect against tetanus generally first given in the United States?
a. 2–18 months old;
b. 5–10 years old;
c. 10–20 years old;
d. only when needed;
e. none of the above.

13. How many *reported* cases of infant botulism are there per year in the United States?
a. 7;
b. 70;
c. 700;
d. 7000;
e. 70,000.

14. How can infant botulism be prevented?
a. infant vaccination with toxoid;
b. breast-feed infants;
c. avoid foods with large numbers of endospores and those that promote the germination of endospores;
d. use only cooked foods;
e. change diapers frequently.

15. There are more cases of hydrophobia (rabies) in domestic and wild animals than in humans. a. true; b. false.

16. Who developed the first successful vaccine for rabies?
a. Erlich;
b. Koch;
c. Pasteur;
d. Sabin;
e. Salk.

17. Who developed the first successful vaccine for polio?
a. Erlich;
b. Koch;
c. Pasteur;
d. Sabin;
e. Salk.

18. What is the difference between the injected and oral polio vaccine?
a. one virus/variety of viruses;
b. inactivated nonproductive viruses/attenuated reproducing viruses;
c. variety of viruses/one virus;
d. developed in brain stems/developed in human cell cultures;

e. only one injection required/many separate doses required.

19. Viral encephalitis is caused almost exclusively by:
    a. retroviruses;
    b. herpes viruses;
    c. pox viruses;
    d. enteroviruses and arboviruses;
    e. none of the above.

20. What are examples of enteroviruses?
    a. mumps virus;
    b. polio viruses;
    c. echoviruses;
    d. coxsackieviruses;
    e. all of the above.

For items 21–25, match the disease with its cause:
    a. viroid;
    b. bacterium;
    c. virus;
    d. prion;
    e. fungus.

21. Multiple sclerosis.

22. Kuru.

23. Progressive rubella panencephalitis.

24. Creutzfeldt-Jakob disease.

25. Subacute sclerosing panencephalitis.

26. The vector associated with African sleeping sickness is:
    a. *Trypanosoma gambiense;*
    b. *Trypanosoma rhodesiense;*
    c. mosquito *Anopheles;*
    d. Tsetse fly;
    e. mosquito *Aedes.*

For items 27–30, match the pathogen with the disease.
    a. *Haemophilus influenzae* and *Neisseria meningitidis;*
    b. influenza viruses and varicella-zoster virus;
    c. *Escherichia coli* and *Streptococcus agalactiae;*
    d. *Streptococcus pneumoniae* and *Neisseria meningitidis;*
    e. enteroviruses and arboviruses.

27. The major cause of meningitis in the newborn.

28. The major cause of meningitis in children under 15.

28. The major cause of meningitis in adults.

30. The cause of Reye's syndrome.

SUPPLEMENTAL READINGS

Bennett, D. D. 1985. Like sheep virus, AIDS virus infects brain. *Science News* 127:22.

Hirsch, M. S. 1984. Acute viral central nervous system diseases. *Scientific American Medicine* Section 7-XXVII. New York: Freeman.

Holland, J. S. 1974. Slow, inapparent, and recurrent viruses. *Scientific American* 230:32–40.

Joklik, W. K. and Willett, H. P. (eds.). 1984. *Zinsser Microbiology.* New York: Appleton Century Crofts.

McKinstry, D. W. 1983. The slow visna virus: where it hides. *Research Resources Reporter* 7:1–3.

Merigan, T. C. 1984. Slow virus diseases. *Scientific American Medicine* Section 7-XXXII. New York: Freeman.

Shaw, G. M., *et al.* 1985. HTLV-III infection in brains of children and adults with AIDS encephalopathy. *Science* 227:177–182.

U.S. Department of Health and Human Services. 1980–1987. *Morbidity and Mortality Weekly Reports.* Washington, D.C.: U.S. Department of Health and Human Services.

Van Heyningen, W. E. 1968. Tetanus. *Scientific American* 218:69–77.

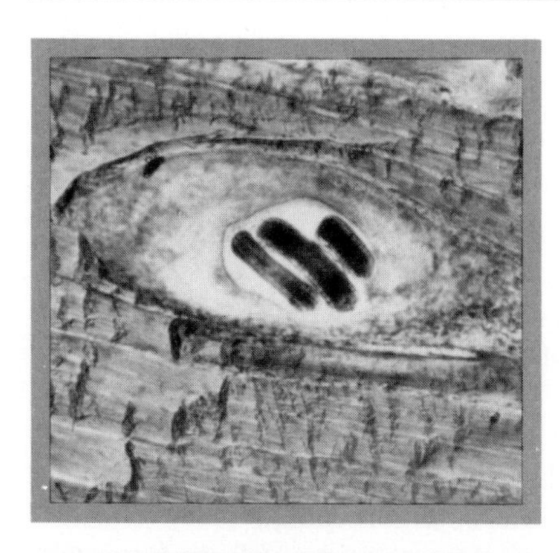

CHAPTER 21

DISEASES OF THE BLOOD, LYMPH, MUSCLE, AND INTERNAL ORGANS

CHAPTER PREVIEW THE BLACK DEATH

The most severe plague epidemic, known as the Black Death, took place in Europe during the years 1347–1350. Millions of people died from a blood infection after suffering from a type of pneumonia (pneumonic plague) and/or swollen lymph nodes (bubonic plague). Eighty percent of those who came down with the plague died in great pain within two or three days. Studies indicate that the plague was introduced from Asia by rats or diseased sailors at one or more ports in southern France and Italy in 1347 and transmitted to humans by the bite of the rat flea. The plague spread rapidly throughout Europe.

Records indicate that the population of Europe in 1347 was about 84 million, but by the end of 1350 only 64 million people remained. Europe lost one-fourth of its population, while England is estimated to have lost one-half of its population. After the European pandemic of 1347–1350, there were recurrent outbreaks all over Europe every 10 years or so which reduced the population even further. After 1450, the epidemics killed fewer people and the population increased rapidly. Between 1347 and 1600, 25 epidemics of plague occurred in Venice and 20 in London. Between 1630 and 1670 serious epidemics in Europe broke out again and are estimated to have killed more than 10 million persons.

In the late 1700s, the recurrent epidemics of plague in Europe came to an end, possibly due to the selection of more resistant hosts and vectors and/or to the selection of less virulent bacterial strains.

Plague is an example of a disease that involves the blood, lymph, and lungs. The bacterium that causes the disease is found in many wild rodents and is spread by fleas among animals and to humans. In this chapter we learn more about plague and other diseases that affect the blood, lymph, muscle, and internal organs.

CHAPTER OUTLINE

SUPERFICIAL INFECTIONS OF THE BODY, SUCH as those that sometimes accompany cuts, insect bites, scratches, and tooth decay, can result in life-threatening diseases if microorganisms spread to internal organs such as the heart, brain, lungs, liver, spleen, kidneys, or bone. The purpose of this chapter is to discuss some of the mechanisms by which microorganisms infect blood, lymph, muscle, and internal organs and cause disease. As we will see, many of these infections are initiated by vectors that introduce the microorganisms into the blood and lymph. Most of the diseases occur because toxins and microorganisms are disseminated throughout the body by the lymphatic and circulatory systems.

STRUCTURE AND FUNCTION OF THE CIRCULATORY AND LYMPHATIC SYSTEMS

The circulatory system is a network of vessels that supplies all cells of the body with nutrients and oxygen and removes wastes and toxic substances (fig. 21.1). Part of the blood plasma passes through blood capillaries and bathes the tissue cells in what is called **interstitial fluid** (fluid between cells).

The **lymphatic system** consists of **lymph vessels** that drain the tissues supplied by blood capillaries. The interstitial fluid, also called lymph, is picked up by the lymph vessels and carried to the **lymph nodes** (fig. 21.1), where the immune system reacts with toxins and microorganisms that may have been drawn into the lymph capillaries. After passing through the lymph nodes, the lymph flows into the blood.

During infections, most of the toxins and microorganisms that escape from the initial site of infection enter the lymphatic system, where they are attacked by the immune system. Toxins and microorganisms that escape from the lymph nodes enter the venous blood and are pumped directly to the heart. From there, they are pumped to the lungs and then to other organs and tissues.

The immune system attacks toxins and microorganisms in the blood; for example, leukocytes and monocytes destroy microorganisms, while circulating antibodies and complement inactivate toxins and also kill microorganisms. Microorganisms and toxins that escape destruction can initiate infections or cause damage in those tissues they reach through the blood system. Thus, some diseases of the heart (endocarditis), lungs (pleurisy), liver (hepatitis), kidney (glomerulonephritis), brain (meningitis, encephalitis), and bone (osteomyelitis) arise from superficial infections.

BACTERIAL DISEASES CAUSED BY THE INVASION OF THE LYMPH AND BLOOD

Bacteremia is the situation in which bacteria are found in the blood. The terms bacteremia, **septicemia**, and **bacterial sepsis** are often used interchangeably, but septicemia and bacterial sepsis are most commonly used when the infection of the blood produces signs

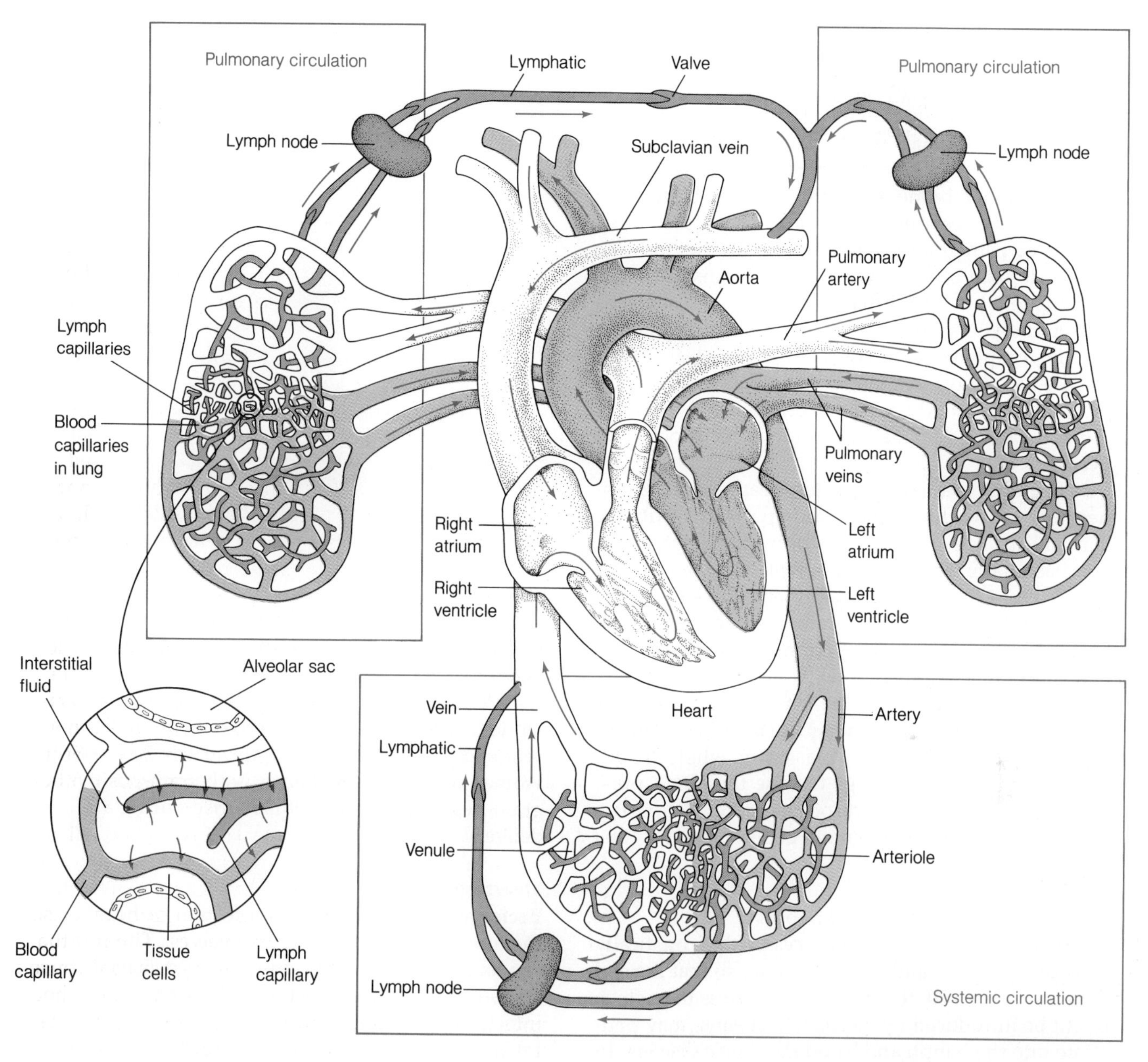

Figure 21.1 Relationship between the blood and lymphatic circulation

and symptoms. Signs and symptoms of septicemia include fever, chills, low blood pressure (hypotension), and shock.

Normal flora from some part of the body entering the lymph and blood is responsible for the bacteremia. It often penetrates the normal barriers in people suffering from leukemia, solid tumors, and multiple myelomas. However, urinary catheterizations, intravenous cannulas, surgical procedures, and burns are usually responsible for the entrance of bacteria into the lymph and blood. There are also some bacterial pathogens that cause septicemia after they invade the blood by eroding from the tissue they are infecting into the circulatory system.

Since the introduction of antibiotics in the late 1940s and early 1950s, gram-negative bacteria have become responsible for about 80% of the estimated 300,000 septicemias/year in the United States. Of these, approximately 100,000 result in death (table 21.1).

The lipopolysaccaride or endotoxin associated with the outer membrane of gram-negative bacteria is thought to be responsible for much of the pathology associated with gram-negative sepsis. Studies indicate that endotoxin causes the release of kinins, activates complement by the alternative pathway, and activates the fibrinolytic system and the coagulation systems. Kinins, complement, and histamines are released during the blood invasion, and intravascular coagulation throughout the body results in chills, fever, blockage in microvascular circulation, anaphylactic shock, and death.

The antibiotic of choice for treatment of gram-negative septicemia is the broad-spectrum antibiotic gentamicin. It is ineffective, however, against anaerobic bacteria such as *Bacteroides, Peptostreptococcus,* and *Peptococcus.* Clindamycin and carbenicillin are used to treat these anaerobes. The gram-positive bacteria are best treated with a cephalosporin or a semisynthetic penicillin, in combination with an aminoglycoside (streptomycin) to inhibit the enterococci. Shock is treated by maintaining an adequate blood volume, by replacing lost plasma with plasma or whole blood, and by administering drugs to raise the blood pressure without producing vasoconstriction. Isoproterenol and dopamine are now being used for this purpose.

TABLE 21.1
MOST COMMON CAUSES OF BACTERIAL SEPSIS

Organisms	Relative frequency (%)
Gram-negative bacteria	81%
Escherichia coli	35
Klebsiella pneumonia and *Enterobacter aerogenes*	16
Pseudomonas aeruginosa	9
Bacteroides sp.	8
Proteus sp.	4
Serratia marcescens	4
Haemophilus sp.	2
Gardnerella vaginalis	2
Providencia and *Citrobacter*	1
Gram-positive bacteria	19%
Staphylococcus aureus	8
Streptococcus pneumoniae	5
Group D streptococci	3
Viridans group of streptococci (other than group D)	1
Group A streptococci	1

Childbed Fever or Puerperal Sepsis

Childbirth and abortion are processes that drastically disrupt the uterus and the birth canal, so that normal flora associated with these tissues, as well as that which might be introduced by mechanical means, may penetrate into the lymph and blood and cause disease. In the early 1800s, the incidence of **puerperal sepsis** (childbed fever) among new mothers often reached 50% and the death rate ranged between 5% and 15% (fig. 21.2). The morbidity (incidence) and mortality (death) due to puerperal sepsis in the late 1800s were reduced to less than 15% and 0.5%, respectively, when the doctors and midwives began to wash their hands and sterilize surgical equipment used on women giving birth. Nevertheless, a 0.5% death rate amounts to 500 deaths/100,000 women giving birth. This death rate is 25–50 times higher than that of today. Before the use of antibiotics, puerperal sepsis was due mainly to the gram-positive bacteria *Streptococcus* and *Staphylococcus.*

Presently, the incidence of puerperal sepsis in the United States is estimated to be about 30,000 cases/year and the death rate is approximately 300–600/year. Women die from infections by both gram-positive and gram-negative bacteria, which arise during and after delivery.

Heart Infections

Bacterial endocarditis is an infection of heart endothelium, usually in one of the valves. The infection results in the constant shedding of organisms that may spread to other parts of the body via the blood and infect internal organs such as the kidneys, spleen, and brain. Clinical symptoms include severe weakness and a persistent fever that lasts for months. The patient may have chills and sweats, difficulty in breathing, lesions on the lower legs, an enlarged spleen and heart, and a blowing sound associated with the aortic heart valve. If treatment is not initiated or is not effective, aortic insufficiency increases because of damage to the aortic valves, and the heart becomes unable to pump blood efficiently through the body. Consequently, the lymph does not pick up interstitial fluids and return them to the heart, so fluids accumulate in the tissue, particularly in the lungs. This situation is known as **congestive heart failure.**

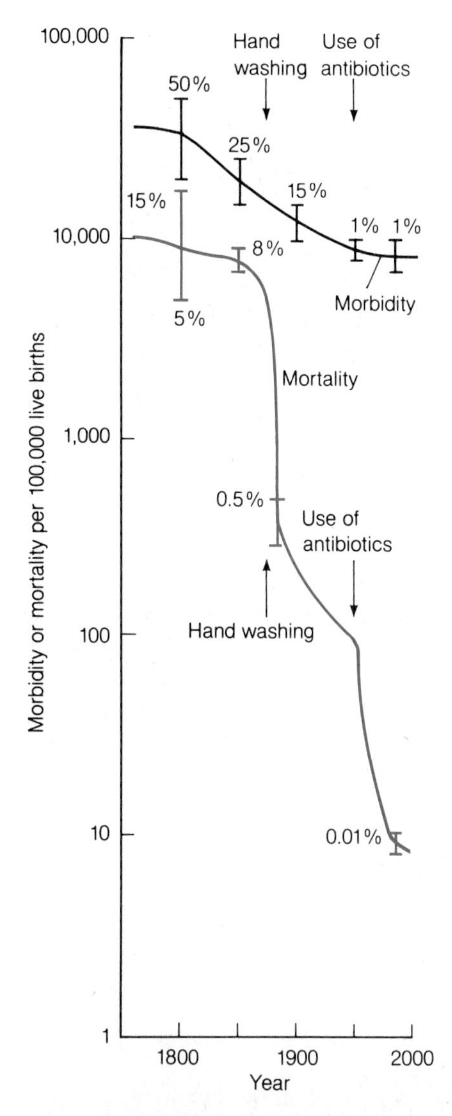

Figure 21.2 **Morbidity and mortality of puerperal sepsis in the United States and Europe**

FOCUS IGNATZ SEMMELWEISS (1816–1865) AND CHILDBED FEVER

Before Pasteur's and Koch's research showing that microorganisms were responsible for decomposition and disease, much of the medical profession was unaware of the importance of microorganisms in causing infection. Doctors frequently attended diseased patients and performed autopsies and then assisted in births, without washing their hands or changing their clothes. Because of ignorance about infection, many new mothers became infected following childbirth and a high percentage died of septicemia.

In the late 1840s, Ignatz Semmelweiss, working in the Vienna General Hospital, noticed that childbed fever was much more common in wards where physicians attended new mothers than in wards where midwives aided in births. Semmelweiss hypothesized that the high incidence of puerperal sepsis and death might be due to the doctors transferring decaying material from corpses to expectant mothers. The midwives did not perform autopsies, and they washed their hands occasionally during their shifts. After Semmelweiss insisted that midwives and other personnel in his obstetric ward wash their hands in chlorinated lime before they attended new mothers, the rate of puerperal sepsis and death in his ward dropped significantly.

Even after Semmelweiss's demonstration that hand washing could reduce the mortality and morbidity of puerperal sepsis, doctors at the Vienna General Hospital refused to institute aseptic procedures such as hand washing, and they continued to ignore the possibility that they were responsible for most of the cases of puerperal sepsis in their hospital. Semmelweiss's hypothesis and accusation stirred such hostility against him by doctors in the hospital that he was discharged. Nevertheless, within 25 years most doctors came to accept the validity of his observation. Ironically, Semmelweiss died of a streptococcal infection resulting from an accidental nick during an autopsy.

Subacute endocarditis generally progresses for 2 to 3 months before it flares up (table 21.2). Eighty percent of cases of subacute endocarditis are due to α-hemolytic streptococci (viridans streptococci), such as *Streptococcus mutans.* Infection of the heart tissue usually occurs in abnormal or damaged epithelium. The remainder of cases of subacute endocarditis appear to be caused by group D enterococci, such as *Streptococcus faecalis;* and anaerobic bacteria, such as *Propionibacterium acnes, Bacteroides, Fusobacterium,* and *Clostridium.*

Acute endocarditis usually reaches a fulminant

TABLE 21.2
CHARACTERISTICS OF BACTERIAL ENDOCARDITIS

	Acute endocarditis	Subacute endocarditis
Length of disease	<6 weeks	6 weeks or longer
Predisposing factors	Normal or prosthetic valve	Rheumatic or congenital heart disease; prosthetic valve
Microorganisms responsible	50% Viridans streptococci 35–40% *Streptococcus pyogenes* *Neisseria gonorrhoeae* *Pseudomonas aeruginosa* 10–15% *Staphylococcus aureus* *S. epidermidis*	80% Viridans streptococci 20% *Streptococcus faecalis* Anaerobic streptococci and rods
Treatment	Penicillin and streptomycin	Penicillin and streptomycin

stage within a month after the heart is infected (table 12.2). Fifty percent of cases are attributed to α-hemolytic steptococci, and 10–15% are due to *Staphylococcus aureus* and *Staphylococcus epidermidis.* The Group A, such as *Streptococcus pyogenes,* and the gram-negative bacteria *Neisseria gonorrhoeae* and *Pseudomonas aeruginosa* account for the remaining cases.

Dental surgery and tooth decay can introduce into the bloodstream α-hemolytic streptococci, β-hemolytic streptococci, and some of the anaerobic species mentioned above. Inflammation, ulcers, or cancers of the intestinal tract generally introduce *S. faecalis* and anaerobic bacteria, while boils and wounds usually introduce *S. aureus.*

Gangrene

Gas gangrene is a disease that results from a mixed infection involving *Clostridium perfringens* in conjunction with other clostridia (e.g., *C. septicum* and *C. novii*) as well as streptococci and peptococci. The disease is characterized by an invasion of the lymph and blood, rotting away (necrosis) of tissue, the presence of ulcerating lesions, and gas bubbles often caught under the skin or within the muscle tissue (fig. 21.3).

Clostridium perfringens, the organism present in almost all cases of gas gangrene, can be isolated from the skin of 45% of noninfected individuals and from the gastrointestinal tract of 35% of normal persons. Invasion of the bloodstream by *C. perfringens* can result from serious wounds that are contaminated by soil

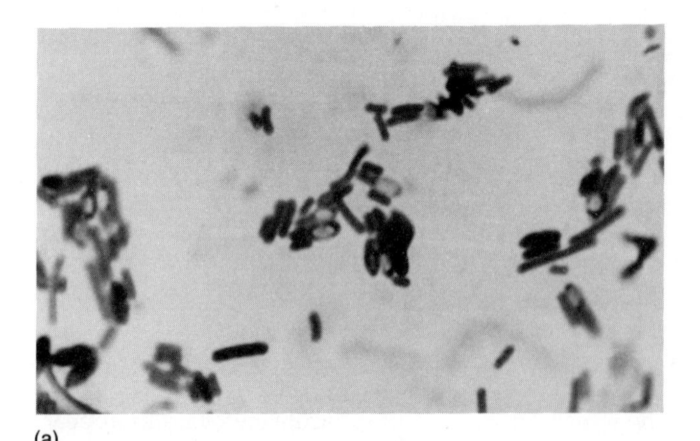
(a)

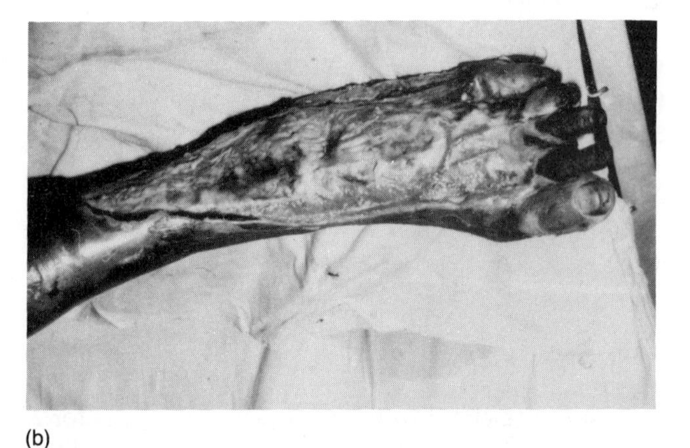
(b)

Figure 21.3 **Clinical characteristics of gas gangrene.** *(a)* Gram stained smear of a biopsy material from a patient with gas gangrene. The endospore forming, gram-positive rods are those of *Clostridium perfringens,* a common cause of gas gangrene. Courtesy of Microbiology Department, University Hospital, Seville, Spain. *(b)* Patient with gas gangrene of the foot.

bacteria, or from ulcerating lesions of the gastrointestinal tract. Malignant tumors of the gastrointestinal tract often lead to *C. perfringens* bacteremia. The bacteria do not reproduce in the blood because of the high concentration of O_2. If, however, they enter tissue that has a low oxygen concentration, they can cause gas gangrene. In addition, gas gangrene often occurs as a result of postoperative complications in patients with diabetes, with severe peripheral arteriosclerosis, or with amputations that result in decreased blood flow to the area.

Clostridium perfringens produces numerous exoenzymes that hydrolyze the matrix that holds tissues together (collagen, hyaluronic acid) and the lipids and proteins that make up cell membranes. The breakdown of tissue and the lack of blood flow into the area promotes the spread of the organism. The fermentation of sugars produced from the hydrolysis of cellular polysaccharide is the source of the gas caught in the decaying tissues.

Gas gangrene can be treated effectively with antibiotics (such as penicillins and cephalosporins) and hyperbaric (high oxygen tension) therapy. Hyperbaric therapy increases oxygen pressure around infected tissues so as to increase the oxygen concentration in the damaged tissues. This extra oxygen concentration inhibits the growth and exotoxin release by the clostridia.

The Plague (The Black Death)

The plague, a disease caused by the small gram-negative bacterium *Yersinia pestis*, can involve the lymphatic system, the circulatory system, and the lower respiratory tract. When the infection develops in the lymph, the disease is known as **bubonic plague**. The lymph nodes in the groin and in the neck swell to about the size of golf balls; these swollen lymph nodes are known as **buboes**, hence the name bubonic plague. During the early stages of bubonic plague there is fever, delirium, and swelling of the lymph nodes. Eventually, a septicemia develops and hemorrhagic, blackened lesions appear. The hemorrhaging is the reason this disease is often called the "black death."

Yersinia pestis, sometimes spread from the blood into the lungs and initiate a **pneumonic plague**, which is generally fatal. In the early stages of pneumonic plague there is a high fever, a heavy cough that disperses the bacteria into the air, and thick mucus, often with blood. Bacteria in the air can be inhaled by others, so pneumonic plague is highly contagious. As the disease progresses, a septicemia or blood infection develops and hemorrhagic lesions may be visible in the mouth and on the lips.

Yersinia pestis is ubiquitous in rodent populations in the temperate parts of the world. From time to time, when rodent populations are stressed by malnutrition, overcrowding or disease, *Yersinia pestis* causes the plague in animals. The bacteria are spread from diseased and dead animals to other animals by fleas, principally *Xenopsylla cheopis*, which feed on the blood of diseased animals and become infected with the bacteria. *Yersinia pestis* is able to multiply in the flea, so if the flea bites another animal before it dies of the plague, it inoculates the animal with the plague bacterium. Diseased fleas can rapidly spread the plague through a rodent population and to other animals and humans. **Epizootic plague** is indicated when numerous rodents such as ground squirrels or rats are found dead due to *Y. pestis* infections.

When infected fleas bite animals or humans, the bacteria are injected subcutaneously. Granulocytes, monocytes, and macrophages attack the bacteria, but some escape destruction by the phagocytes and enter the blood and lymph. Because the blood and lymph are filtered by lymph nodes, *Y. pestis* ends up in the nodes, where it is attacked by macrophages. Macrophages often do not destroy the bacteria, however, so they are able to multiply in the nodes. *Yersinia pestis* also produces a powerful **necrotizing exotoxin** that kills many of the phagocytes. The toxic materials released during the multiplication of *Y. pestis* in the lymph nodes cause them to swell and also elicit a fever. If the immune system does not control the infection, *Yersinia* spreads from the swollen lymph nodes into the lymph and then into the blood, causing a bacteremia, which causes fever, delirium, shock, and eventually death. The necrotizing exotoxin may cause hemorrhaging in various parts of the body.

The plague can be prevented by eliminating slums and garbage heaps that support large populations of rats and by limiting the size of ground squirrel populations that come into contact with humans. Vaccination of persons at high risk, such as veterinarians and personnel monitoring ground squirrel and rodent populations, with a heat- or chemically-inactivated strain of *Y. pestis* can prevent the spread of plague among humans. Treatment of plague includes the rapid diagnosis of the disease and the use of streptomycin or tetracycline.

Currently, there is an average of 15 reported cases of human plague in the United States each year, with about 1 death per year (fig. 21.4).

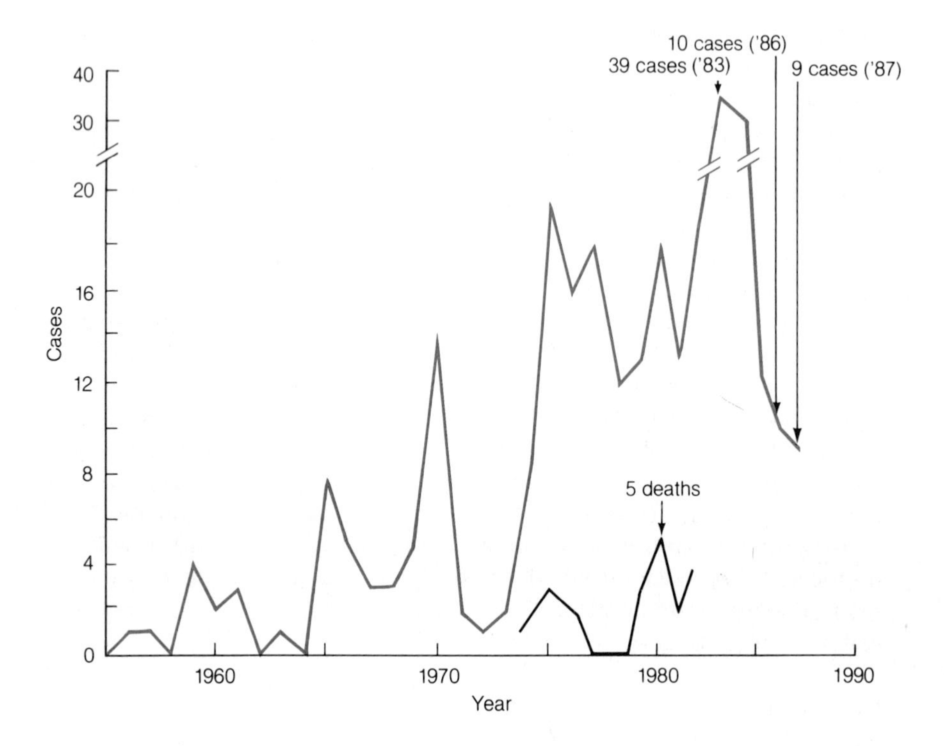

Figure 21.4 **Occurrence of plague in the United States.** The incidence of plague (color) exhibited a sudden increase during the 1970s. But now it is in the decline again. Most of the cases are due to contact with wild animals. The number of deaths (black) fluctuates between 2 and 5 each year.

Relapsing Fever (Tick Fever)

Relapsing fever is a disease caused by the spirochetes *Borrelia hermsii* and *B. recurrentis* (fig. 21.5). The disease is characterized by recurrent periods of high fever that alternate with periods of normal body temperature. The infection causes headache, muscular pains, chills, nausea, and septicemia, and there may be a rash of rose-colored spots. The symptoms subside after a few days, but a day or so later they return in a milder form. Recoveries and relapses may occur three or four times, but eventually the symptoms disappear completely. Diagnosis of the disease is made by observing *Borrelia* in the blood.

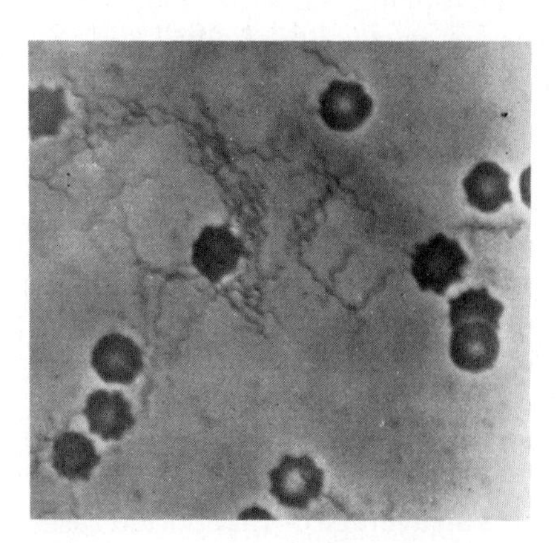

Figure 21.5 ***Borrelia recurrentis*, the cause of relapsing fever**

The disease is found primarily in the tropics. Reservoirs of *B. recurrentis* include ticks, rodents, and humans; ticks and lice function as vectors. In ticks, **transovarial** passages of *Borrelia* from the infected female to her eggs allows the bacteria to multiply wherever the tick is found. Thus, ticks are an important reservoir. Most infections in the United States are caused by tick bites, but lice are responsible for spreading *Borrelia* from person to person during epidemics.

The disease can be controlled by removing ticks immediately and by maintaining a high level of hygiene to control lice. Once the disease is contracted, it can be treated with chloramphenicol or tetracycline.

Brucellosis (Undulant or Malta Fever)

Brucellosis is a disease often found in cattle, goats, swine, and sheep. The disease is caused in the United States primarily by *Brucella abortus* and *B. suis;* both are gram-negative coccobacilli. In animals, brucellosis is characterized by abortion in the female, infertility in both sexes, a bacteremia (blood infection), and an undulating fever that drops during the day. In humans, the disease is characterized by recurring bouts of fever followed by periods of normal body temperature, enlarged lymph nodes, enlarged spleen, and muscle aches. The disease is often severe in humans and may linger for a year without antimicrobial treatment.

Many herds have carrier animals infected by *Brucella*. In bulls with infected seminal vesicles, testicles, and epididymides, semen is contaminated with *Brucella*. If contaminated semen is deposited in the uterus of a cow, she can become infected. Thus, it is important that semen used in artificial insemination come from disease-free bulls. In the cow, *Brucella* infects the uterus and developing fetus, causing abortion. The rate of abortion in an infected herd often reaches 50%. *Brucella* also infects the udders and is shed in the milk, sometimes for the life of the animal. The disease is frequently spread by ingestion of organisms that are present in milk, contaminated feed, and contaminated water. Feed and water are contaminated by uterine discharges, milk release, and feces. The licking of aborted fetuses is another important way that animals become infected.

It is believed that the bacteria multiply in the mucous membranes of the gastrointestinal tract (or the uterus) and then spread into the lymphatic system and blood, from which they infect many tissues and organs. Long-term infections are caused by the bacteria that establish a symbiotic relationship inside cells of the reticuloendothelial system. Carriers in a population generally produce 20% less milk than normal animals.

Humans can also become infected by *Brucella:* approximately 200 cases of brucellosis in humans (Malta fever) are reported each year in the United States, and 60% of these cases occur in butchers and workers from meat processing plants. About 30% occur in ranchers and hunters, while fewer than 10% occur in persons who consume unpasteurized dairy products. Mortality in humans is low, with an average of 1 death/year in the United States.

In herds of cattle, the disease can be eliminated by vaccination of calves, annual booster injections of adults, and slaughtering and disposal of infected animals. The vaccines use killed or attenuated organisms. The most effective measures for controlling brucellosis in humans are inspection of herds, elimination of diseased animals, and pasteurization of dairy products. Chlortetracycline and streptomycin are effective for curing valuable animals, but antibiotic treatment is expensive in large animals and is rarely used to treat a herd. Tetracycline and ampicillin are used to treat human cases of brucellosis.

Tularemia (Rabbit Fever)

Tularemia, a disease caused by *Franciscella tularensis*, is characterized by fever, fatigue, coughing, diarrhea, and enlarged lymph nodes, spleen, and liver. Most often, however, an infection due to *F. tularensis* produces no clinical symptoms. The disease is endemic in the United States in wild animals such as rabbits, squirrels, deer, and birds. The gram-negative coccobacilli are often transmitted to domestic animals, such as cattle, sheep, goats, hogs, horses, dogs, and cats, by ticks that feed on infected wild animals. Tularemia is transmitted to humans by bites of infected ticks and deerflies. Hunters who skin infected rabbits and deer can acquire the organism if it gets into cuts or onto mucous membranes of the mouth or nose.

The pathogens usually multiply at the site of entry and produce an ulcer (fig. 21.6). The bacteria are engulfed by phagocytes, but are not effectively eliminated; thus, *F. tularensis* often spreads to the lymph and lymph nodes in dead and dying phagocytic cells. From there the bacteria spread into the blood, causing a septicemia, and into various tissues and organs, such as the liver and spleen. About 200 human cases of tularemia are reported each year in the United States.

Prevention of tularemia is through the control of ticks infesting herds or flocks. Domestic animals (cattle, sheep, goats, hogs, horses, dogs, and cats) can be dip-washed or sprayed with chemicals such as arsenic, rotenone, toxaphene, or malathion. The disease can be treated effectively with the antibiotic oxytetracycline.

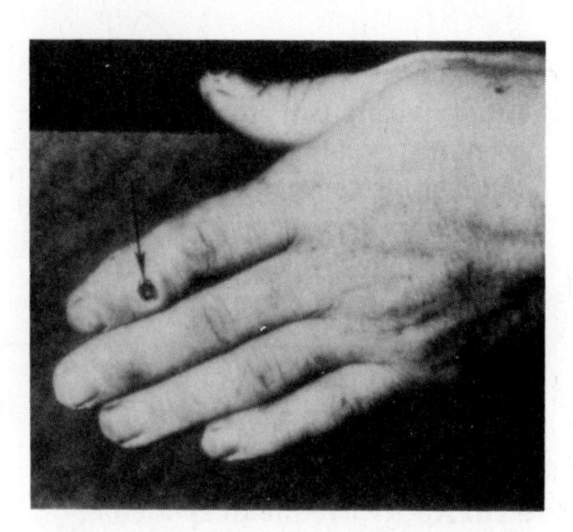

Figure 21.6 **Characteristic lesions of tularemia caused by *Francisella tularensis***

Rocky Mountain Spotted Fever

Rocky Mountain spotted fever is characterized in humans by a fever and by a rash that starts at the extremities and then develops on the rest of the body. The rash is caused by damage to peripheral blood vessels, which leads to inflammation, hemorrhaging, and escape of blood into surrounding tissues. The disease is caused by the bacterium *Rickettsia rickettsii,* an obligate intracellular coccobacillus with a gram-negative type of cell wall.

Rocky Mountain spotted fever is prevalent in the South Atlantic states (60% of reported cases), as well as in the Rocky Mountain states (fig. 21.7). There are approximately 1150 cases of Rocky Mountain spotted fever per year in the United States with a 5% fatality rate (about 55 deaths/year).

The reservoirs for *Rickettsia rickettsii* are rodents, rabbits, and ticks. Infected ticks transmit the organism through the egg to offspring and also function as the vector. The bacteria are transmitted to animals and to humans almost exclusively by tick bites: When the ticks feed, the bacteria enter the lymph and blood and then reproduce in neutrophils and in the cells lining the peripheral blood vessels, causing **thrombosis** (blood clots) and **extravasation** (loss of blood into the tissues). The rash is due to the destruction of the endothelial cells that line the superficial blood vessels.

A vaccine consisting of formalin-killed bacteria has been developed for humans; it diminishes the severity of the disease but does not stop it. There are no vaccines for domestic animals. Control is best achieved by prompt removal of ticks. Generally, ticks must feed for several hours before they transmit virulent rickettsia, and so their removal twice a day can prevent the disease. The tick bite should be treated with tincture of iodine to prevent secondary bacterial infection. The use of tick repellents and tight, overlapping clothing can prevent tick bites. Treatment of the disease includes bed rest and the administration of antibiotics such as tetracycline.

Typhus Fevers

Typhus fevers are infectious diseases caused by various species of *Rickettsia.* These diseases spread by the bites of arthropods such as lice, fleas, ticks, and mites. Human typhus is characterized by septicemia, high fever, headache, haziness, confusion, and a rash, often associated with hemorrhaging beneath the skin.

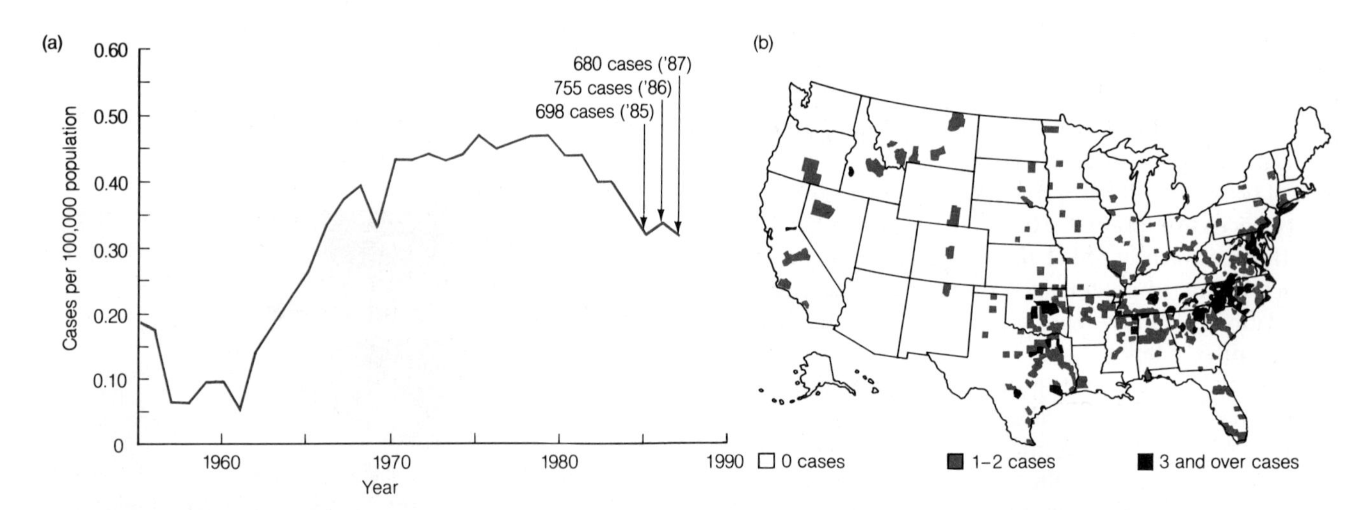

Figure 21.7 Incidence and distribution of Rocky Mountain spotted fever. *(a)* Reported cases of Rocky Mountain spotted fever caused by *Rickettsia rickettsii. (b)* Distribution of cases of Rocky Mountain spotted fever in the United States. It is noteworthy that the vast majority of the cases occur in the eastern United States, not in the Rocky Mountains in the west. The name of the disease was coined because the spotted fever was first known to occur in the Rocky Mountains. The incidence of the disease parallels well the distribution of the tick vectors.

Endemic typhus, also known as **murine typhus**, caused by *R. typhi*, is widely distributed in warm climates and is transmitted to humans in the feces of infected fleas. The bacteria enter the body through a flea bite or other break in the skin. They invade neutrophils and endothelial cells of capillaries, where they reproduce, causing the symptoms and signs of typhus. There are approximately 60 cases of endemic typhus every year in the United States and a 5% mortality rate (fig. 21.8). The reservoir for murine typhus is rodents, such as rats and squirrels.

Epidemic typhus, caused by *R. prowazekii*, occurs in cool climates throughout the world and is transmitted to humans in the feces of infected body lice. The bacteria may enter through bites or other breaks in the skin. Epidemics of this disease generally occur in times of war and famine, when overcrowding and unsanitary conditions prevail. The mortality rate can reach 40%. Very few cases of epidemic typhus occur in the United States.

Humans are an important reservoir for epidemic typhus, because the organisms can exist in a latent form for many years. In some instances, infected persons who are clinically cured may exhibit a relapse of the disease many years later. This recurrent form of typhus is called **Brill-Zinsser** disease.

Typhus fevers can be controlled by spraying with pesticides and by practicing personal hygiene and good sanitation. Pesticides and cleanliness eliminate body lice, mites, and fleas, the vectors for *Rickettsia*, while sanitation (garbage removal and deposition) can eliminate some of the reservoir animals, such as mice and rats. A vaccine of chemically treated microorganisms is effective against both epidemic typhus and murine typhus. Treatment of the disease with tetracycline or chloramphenicol effectively destroys the organisms and generally eliminates the carrier state in epidemic typhus.

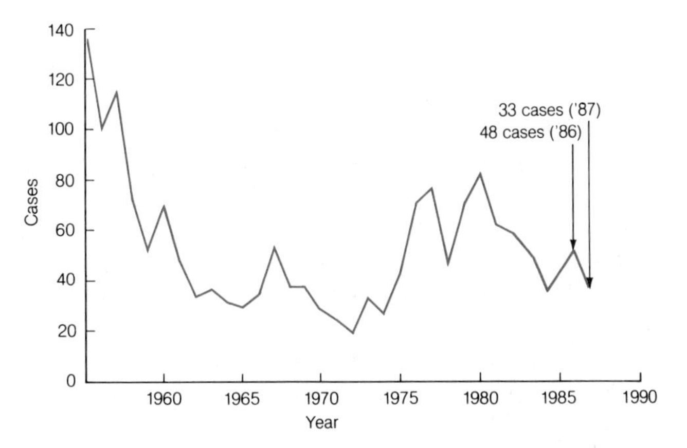

Figure 21.8 **Cases of endemic typhus in the United States.** Murine rodents are the reservoirs for the agent of murine typhus, *Rickettsia typhi*. A total of 58 cases were reported in 1982, of which 41 occurred in Texas and 9 in Hawaii.

VIRAL DISEASES CAUSED BY INVASION OF THE LYMPH AND BLOOD

Colorado Tick Fever

Colorado tick fever in humans is an illness characterized by fever, chills, headache, and muscle aches, and it may be accompanied by a rash. The disease is caused by an arbovirus and is transmitted to humans from infected wild rodents by tick vectors (fig. 21.9). The disease occurs in western North America, and can be diagnosed by detecting antibodies against the virus or by isolating the virus in clinical specimens (blood).

Control of the disease is through the use of pesticides and tick repellents to eliminate the vector population. There is no treatment for this infection, which generally subsides after running its course. Since this is a non-notifiable disease in the United States, it is difficult to determine how many cases there are; in 1980, however, there were approximately 200 reported cases in the United States, none of which was fatal. A vaccine is prepared by growing the virus in chick embryos so that they are attenuated. Infection with either the virulent strain or the vaccine strain produces a long-lasting immunity.

Yellow Fever

Yellow fever is caused by an arbovirus and is spread among humans, or from monkeys to humans, by the mosquito *Aedes aegypti* (fig. 21.9). The disease is found in the tropical regions of South and Central America, India, Southeast Asia, and West, Central, and East Africa. It is characterized by fever, internal hemorrhaging, jaundice (due to extensive liver destruction), and vomiting. When mosquitoes bite, they inject the

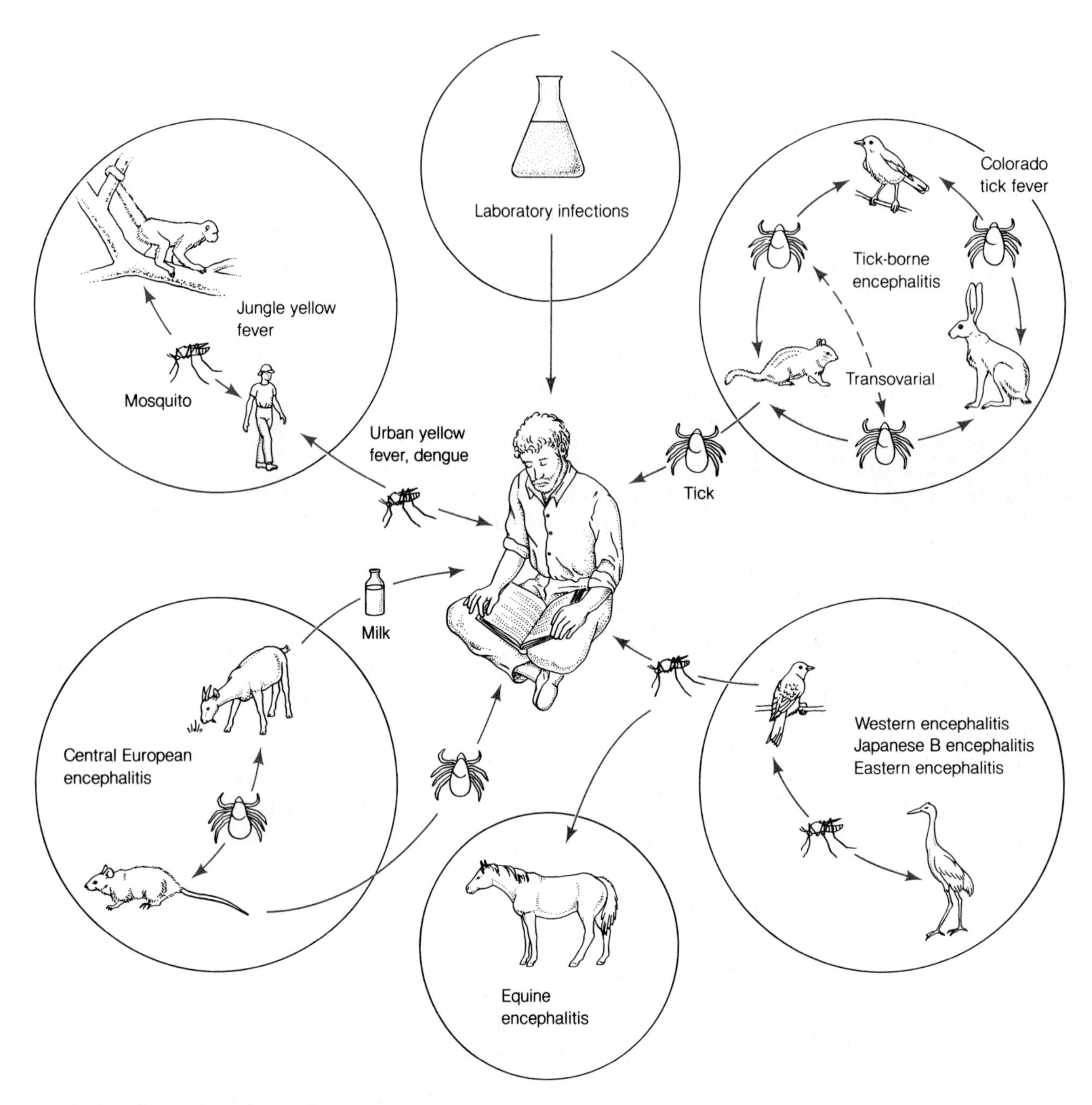

Figure 21.9 The epidemiology of arbovirus-caused encephalitis

virus into the blood and lymph of the host. The viruses infect and multiply within macrophages, monocytes, and reticulocytes in lymph nodes and in the spleen, and then spread into the blood, where they invade the endothelial lining of the blood vessels and cause hemorrhaging. They also spread to the brain, causing an encephalitis. Infections of the internal organs cause hepatitis, nephritis, endocarditis, and osteomyelitis. Proliferation of the virus causes fever, chills, rash, headache, and muscle pains.

In the Mississippi Valley in 1878, an epidemic of yellow fever killed approximately 13,000 persons. Although there has not been a reported case of yellow fever in the United States since 1924, suitable mosquito vectors can be found in the United States, and therefore the possibility always remains that an infected immigrant will initiate an outbreak such as that of 1878. The yellow fever virus is of importance because it kills many people each year throughout the world. Those persons traveling to South America and Africa, where the disease is endemic, should be vaccinated against the virus.

Dengue Fever

Dengue (hemorrhagic) fever is a disease defined by fever, joint pain, skin rash, and mental depression. These signs and symptoms may be accompanied by severe hemorrhaging. It is caused by an arbovirus similar to the one that causes yellow fever. The disease is found in tropical and subtropical areas in Southeast Asia, the Caribbean, Central and South America, and the Philippines. The dengue fever virus is transmitted by females of the mosquito *Aedes aegypti,* which also act as a reservoir for the virus (fig. 21.9). In Malaysia, monkeys are an important reservoir.

Occasionally, when a new serotype of the dengue virus infects an individual sensitized to another serotype, antibodies cross-react with the new virus but do not inactivate it. In fact, the antibodies appear to aid the adsorption of the virus to blood leukocytes and, consequently, the multiplication of the virus. Virus-antibody complexes sometimes activate 80% of the serum C3 and C5, causing intravascular coagulation and inflammatory reactions throughout the body. The disease caused by the immune complex is known as **dengue hemorrhagic shock syndrome**.

Dengue is not common in the continental United States, but in Puerto Rico there has been an average of 11,000 cases and over 100 deaths reported/year in the last few years.

Mononucleosis

Infectious mononucleosis is caused by the Epstein-Barr virus (a herpes virus). The prodromal (early, nonspecific) symptoms include exhaustion and a general poor feeling. The disease itself is characterized by extreme exhaustion, fever, sore throat, pharyngitis, tonsillitis, and swollen lymph nodes and spleen. The presence of more than 10% atypical T-lymphocytes (called **Downey cells**) in the blood, and a total WBC count of about 15,000–20,000 cells/mm^3 (5000 is normal), are further signs of the disease. Recovery generally takes 4–8 weeks, during which extreme exhaustion after only a few hours of activity is one of the symptoms of the disease.

The human population serves as a reservoir for the Epstein-Barr virus (EB-virus). In areas where sanitation is poor, nearly all people in a population have been infected by the age of five without recognizable clinical disease. Eighty to 95% of Africans, for example, have antibodies against EB-virus by the age of two, but infectious mononucleosis is extremely rare. In contrast, in industrialized areas where sanitation practices are better, antibodies against EB-virus are found in only 25% of those under two and in less than 50% of those under 20 years of age. Thus, infection by the EB-virus is delayed until adolescence and early adulthood in high socioeconomic populations, and 50% of those infected develop the signs and symptoms of mononucleosis.

It is believed that infected adolescents in Western countries spread the EB-virus within their age group when they begin to share drinks and cigarettes and when kissing first becomes common. EB-virus may be spread from infant to infant in saliva-contaminated breasts or communal food in those populations where individuals are infected by the age of two.

The virus initially infects the throat and reproduces in epithelial cells of the throat and pharynx, then it spreads to the lymph and blood. Viruses are shed into the throat and pharynx beginning approximately two weeks after exposure and for up to two years. Excretion of the EB-virus can occur in the absence of symptoms of mononucleosis.

After infecting the oral pharynx, the EB-virus invades B-lymphocytes and circulates in a noninfectious form in these cells. In this form, the virus DNA is found within the B-lymphocytes, but progeny viruses are not produced. Cell-mediated immunity is carried out by Downey cells, which increase in the blood a week after infection. Downey cells account for 10–80% of the total white blood cells, and their numbers remain at a high level for 4–5 weeks. Humoral immunity against the viral capsid is mediated by IgM, which is found in 90% of adults at the beginning of an infection and lasts for a month or so.

The disease may be very difficult to prevent in industrialized countries, since prevention would require adolescents and young adults to avoid kissing. A vaccine has not been developed so far, and there are no drugs that can be used as prophylactics or for treatment. Mononucleosis is generally self-limiting; complications such as ruptured spleen, Guillain-Barré syndrome, and hemolytic anemia occur infrequently. Even though infectious mononucleosis is a nonreportable disease, approximately 20,000 cases of mononucleosis are reported per year in the United States, along with an average of 20 deaths per year.

It is believed that the EB-virus may also cause two human cancers: **African Burkitt's lymphoma** and **nasopharyngeal carcinoma**. Burkitt's lymphoma is a malignant lymphoma of children which causes jaw tumors in more than 50% of cases. Its incidence is highest in Africa and in New Guinea, where 95% of the children have antibody titers against the EB-virus. Nasopharyngeal carcinoma occurs with an incidence of 24/100,000 per year in China, as compared with 1/100,000 per year in the United States. Again, the incidence of nasopharyngeal carcinoma may be highest in persons subjected to an EB-virus in infancy. Thus, vaccines containing viral DNA cannot be used until it has been learned if there is any danger from infant vaccination.

PROTOZOAL DISEASES CAUSED BY INVASION OF THE LYMPH AND BLOOD

Malaria

Malaria is due to an infection of the red blood cells (RBCs) by the protozoans *Plasmodium falciparum* (50%), *P. vivax* (43%), *P. malariae* (6%), or *P. ovale* (1%) (fig. 21.10.) One stage of the parasite infects RBCs, causing their lysis. This process leads to periodic chills and fever that leave the affected person drenched in sweat and exhausted. Hemolysis results in anemia and jaundice (yellowish pigmentation of the skin due to excessive bile). Infected RBCs, because they are no longer pliable, stick in small capillaries and cause blood clots. Anoxia (lack of oxygen) and electrolyte imbalance result and can lead to death. The liver and spleen enlarge to handle the debris from the RBCs; the spleen can become so large and fibrous that sometimes it ruptures spontaneously. In cases of acute hemolysis, large amounts of hemoglobin are released into the blood and plug

FOCUS THE DANGERS OF TRANSFUSION

Many accident victims, persons undergoing major surgeries, and hemophiliacs require blood transfusions. Unfortunately, each year hundreds of persons in the United States acquire serious diseases because of blood donated by individuals carrying infectious organisms.

One of the most serious diseases transmitted to patients during transfusion is hepatitis caused by a non-A non-B virus. Post-transfusion non-A and non-B hepatitis occurs in 40–50% of patients who receive more than six units of paid-donor blood, and in 6–10% of patients receiving volunteer blood. The problem with non-A non-B hepatitis is that 40–50% of the cases become chronic, and about half of these develop cirrhosis of the liver 5–10 years later. Death results from liver hemorrhaging, infection, and liver failure. It is estimated that there are over 100,000 cases of non-A non-B hepatitis per year and more than 4000 annual deaths.

At least two distinct viral agents may be responsible for non-A non-B hepatitis. One type of non-A non-B hepatitis has an incubation period of 15–45 days, produces a mild disease, and can be transmitted by the fecal-oral route as well as by transfusions and injections. Ten percent of the cases become chronic. The second type has an incubation period of 90–180 days and is transmitted only by transfusions and injections. A majority of the cases become chronic.

At present, there is no way of screening blood donors to see if they are carriers for non-A non-B. In addition, there is no serological test that can identify non-A non-B hepatitis agents in blood. The only precaution used to avoid non-A non-B is to exclude donors with serological evidence of hepatitis A or B.

Another disease transmitted through blood transfusions is AIDS (acquired immune deficiency syndrome). The agent that causes this disease is the human immunodeficiency virus (HIV). In the last few years, a number of persons who have had transfusions have acquired the disease. Most of the victims have been hemophiliacs and young children. In 1985 the ELISA screening test was licensed to detect the presence of antibodies against HIV. This test detects 92–98% of the blood samples with anti-HIV antibodies. Because this test does not detect all possible blood donors that might have AIDS and because some of the positive results may be false positives, further tests (western-blot) are necessary to ascertain whether or not the donor indeed has AIDS.

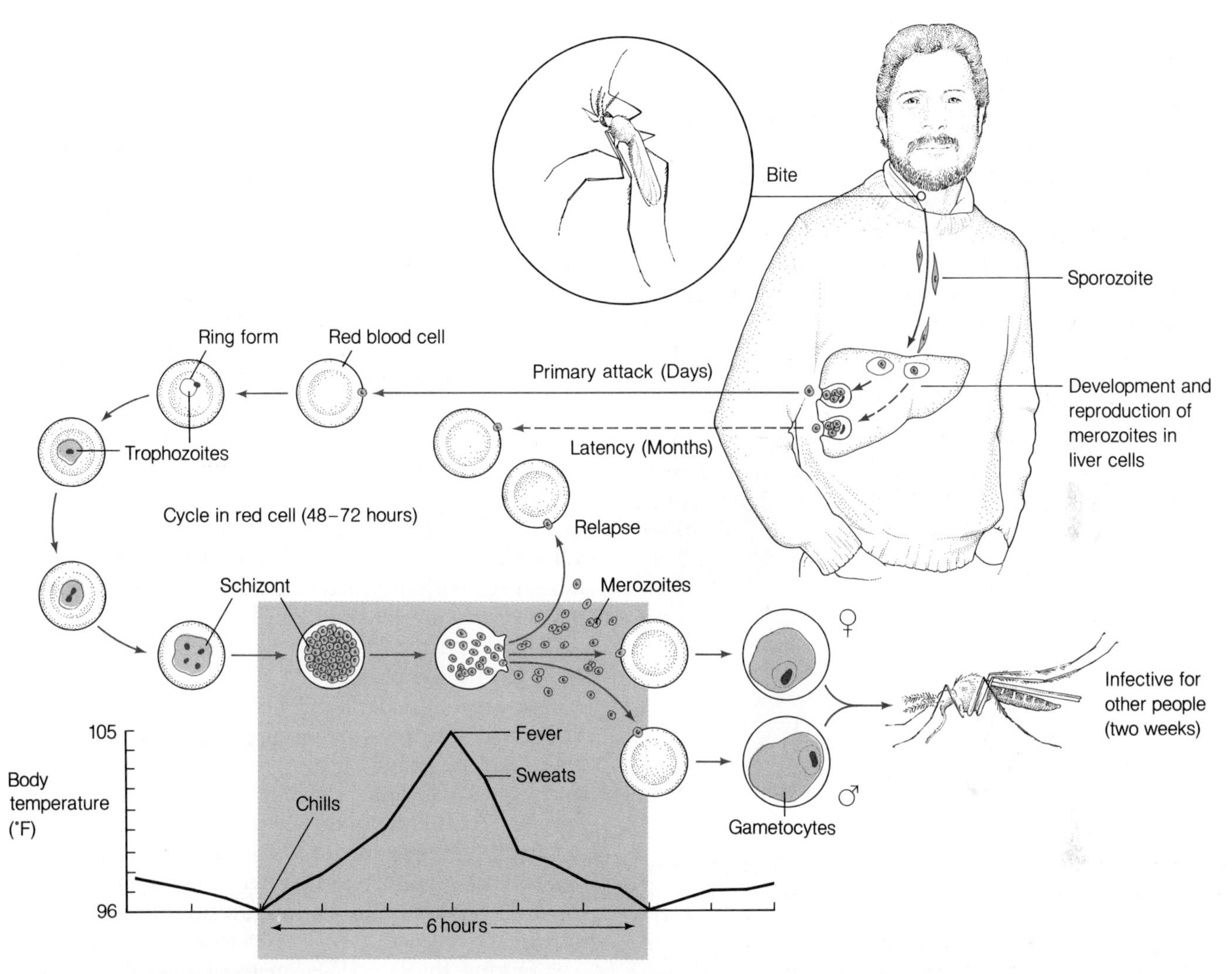

Figure 21.10 Life cycle of *Plasmodium vivax* and its relationship to human malaria. Malaria is characterized by recurring bouts of fever and chills (paroxysms). The paroxysms begin with severe chills, followed by progressive increase in body temperature (104–105°F). The fever is followed by sweats, with a concurrent decline in body temperature. The paroxysms last about 6 hours (vivax malaria). The frequency of the paroxysms is related to the life cycle of the malaria parasite in the blood: the paroxysms coincide with the time during which the parasite is in schizogony.

up the kidneys, leading to renal failure. It is estimated that there are 100 million new cases of malaria/year and about 1 million deaths/year worldwide.

The major reservoir of the malarial parasites is humans, although some monkeys can also carry the protozoan. The female *Anopheles* mosquito is the vector for the disease, spreading it from human to human. When the female *Anopheles* bites, it releases crescent-shaped **sporozoites** (15 μm long) into the lymph and blood (fig. 21.10). The sporozoites migrate directly to the liver, where they invade hepatic cells. Within the liver cells, the sporozoites multiply and differentiate into oval **merozoites**, which are released about 10 days after infection into blood capillaries. The merozoites spread throughout the blood and invade RBCs and develop into ring-shaped **trophozoites**. The trophozoites differentiate into **schinzonts** when they undergo multiple fission inside the red blood cells to form more merozoites. The RBCs lyse at this point, and the merozoites are released into the blood; they may invade other RBCs or return to the liver.

The debris from the hemolysis of RBCs, and from the merozoites that are attacked and destroyed by white blood cells (WBCs), leads to the fever and chills of malaria. The reproduction of *Plasmodium vivax* in the blood takes approximately 48 hours and therefore causes spells of chills and fever every third day (tertian malaria), while *P. malariae* requires 72 hours to complete its cycle (quartan malaria).

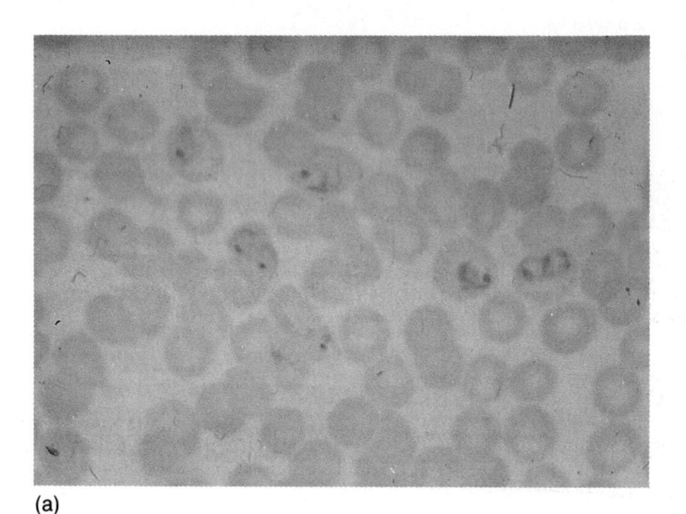

(a)

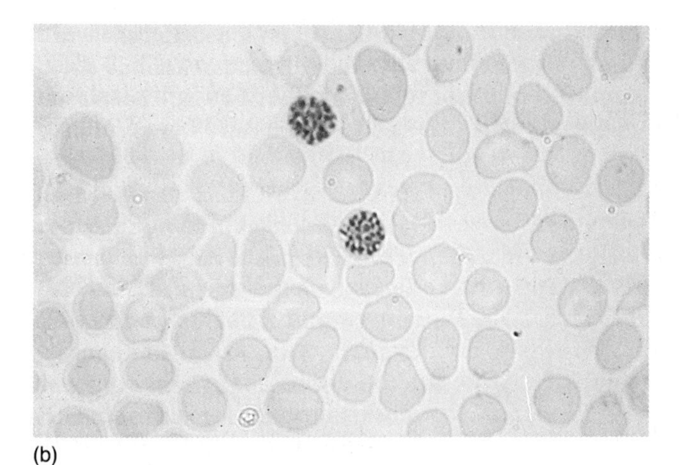

(b)

Figure 21.11 Blood stages of malarial parasites. *(a) Plasmodium vivax* trophozoites (ring forms) in red blood cells. *(b) Plasmodium vivax* schizonts (parasite undergoing multiple fission) in blood smear.

The most severe malaria (pernicious) is caused by *P. falciparum*. In this type of malaria, the patient's fever does not subside completely between cycles of multiplication. Anoxia and kidney failure due to the breakdown of red blood cells frequently lead to shock and death. After a series of attacks, an individual sometimes gets well, apparently because the immune system has eliminated all the infected RBCs and merozoites. It is believed, however, that sporozoites lie dormant within hepatic cells. Thus, relapses can occur when dormant sporozoites become active and form merozoites. Ultimately, when all sporozoites introduced by the original mosquito bite have differentiated and all merozoites are destroyed, relapses no longer occur. Relapses lose their strength over a number of years.

Diagnosis of malaria may be confirmed by observing malarial parasites in Wright- or Giemsa-stained RBCs (fig. 21.11). The parasites show up as blue rings with attached red chromatin dots, or they may completely fill the RBC. Fluorescent antibodies that bind specifically to the malaria parasite are also used to find the protozoa.

Control of malaria can be achieved by screening doors and windows, by using mosquito netting over beds, and by using clothing that protects from mosquito bites. Malaria is also prevented by draining, spraying, or oiling ponds where mosquitoes breed, or by planting fish in the ponds to eat the mosquito larvae. It is also important to detect and cure carriers of malaria, especially if there are mosquitoes that could spread the disease.

During the late 1970s in the United States, the number of cases of malaria increased from about 375 in 1975 to approximately 2000 in 1980 (fig. 21.12). About 80–85% of these cases were found in immigrants from Southeast Asia. Deaths in the United States average 5 per year. Malaria is treated with chemicals such as quinine and chloraquine, a derivative of quinine. Prophylactic treatment with such drugs is recommended for persons visiting endemic areas since at present there is no vaccine commercially available.

Toxoplasmosis

Toxoplasmosis is an infection by the protozoan *Toxoplasma gondii* characterized by mild fever, headache, muscle aches, and swollen lymph nodes and spleen. There may also be lesions of the gastrointestinal tract, myocarditis, necrosis of the liver, central nervous system disorders, or abortion. Cats are important reser-

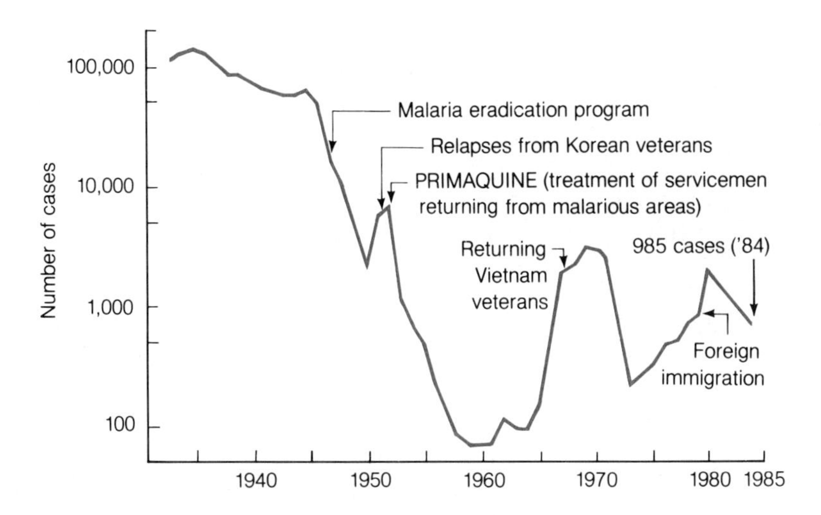

Figure 21.12 Incidence of malaria in the United States

voirs for *T. gondii,* although many wild and domestic animals and birds often become infected. For example, latent toxoplasmosis may occur in up to 50% of pigs, 50% of sheep, and 10% of cattle. This disease is an important cause of abortions and stillbirths in sheep.

Humans can become infected by accidentally ingesting dried particles of cat feces contaminated with oocysts (fig. 21.13). Subclinical infections in pregnant women can often result in the infection of the fetus, since the parasites can cross the placenta. Severe brain damage, blindness, and death of the fetus may result. Using serological procedures, it has been estimated that about 30% of the population has been infected with *T. gondii,* but most of these infections are asymptomatic.

Toxoplasmosis in humans can be prevented by avoiding contact with cat feces, by covering children's sandboxes, and by carefully disposing of cat litter boxes. Thoroughly cooking lamb, pork, and beef (hamburger meat) also reduces the likelihood of picking up these parasites. Cats can be kept relatively free of the protozoan by feeding them only cooked meats and not allowing them to eat mice and birds that they might catch. No vaccine is available against this organism. The drugs **sulfadiazine** and **pyrimethamine** (folinic acid antagonist), which affect only the trophozoites, are used to treat toxoplasmosis.

Trypanosomiasis

African sleeping sickness is a trypanosomiasis transmitted to humans by the tsetse fly. The organisms causing the disease are *Trypanosoma brucei gambiense* and *T. brucei rhodesiense.* The symptoms, epidemiology, prevention, and treatment are described in Chapter 20.

Chagas' disease is found throughout Central and South America and is caused by *Trypanosoma cruzi.* The protozoan is spread among humans by insects commonly called "kissing bugs" (Reduviidae). The disease is characterized by fever, exhaustion, lack of appetite, muscle aches, a puffy face, heart irregularities, and swelling of the lymph nodes, liver, and spleen. The trypanosome often causes an infection of the eyes in children, resulting in swelling of the eyelids (Romaña's sign). Organisms in the circulatory system infect the spleen, lymph nodes, liver, heart, and digestive tract. In chronic cases, cysts develop within heart tissue, leading to necrosis of the tissue, heart failure, and death. Infections of the esophagus interfere with swallowing, while those in the intestines can result in severe constipation. Esophagus and colon disease may require surgery to correct the damage. Nifurtimox is used to treat infections that have not progressed too far. Suramin is the drug of choice in the treatment of early trypanosomiasis.

There are a great number of reservoir hosts for *Trypanosoma cruzi,* including opossums, armadillos, rodents, racoons, monkeys, deer, cats, and dogs. Important vectors of *Trypanosoma cruzi* are hemipterans and other blood-sucking insects (bedbugs, kissing bugs, lice, etc.). These insects generally ingest the trypanosomes during a blood meal on an animal. The trypanosome multiplies in the insect's gut, attaches to the rectal region, and flows out in the feces. Since the

Figure 21.13 Life cycle of *Toxoplasma gondii*. *(a) Toxoplasma gondii* infects cats that consume animals infected with cysts of *Toxoplasma*. The parasite undergoes sexual reproduction in the cat's intestine, and the oocysts are passed in the feces. Children playing in sandboxes may become infected with the oocysts. Adults become infected when they consume the oocysts on their hands or the cysts in foods. *Toxoplasma* infections are dangerous when they occur in pregnant females because the parasite can invade the fetus through the placenta and cause severe damage. It has also been noted that 52% of those AIDS patients that have serious neurological damage have *Toxoplasma* cysts in the brain. *(b)* Scanning electron micrograph of *T. gondii* invading a mammalian cell.

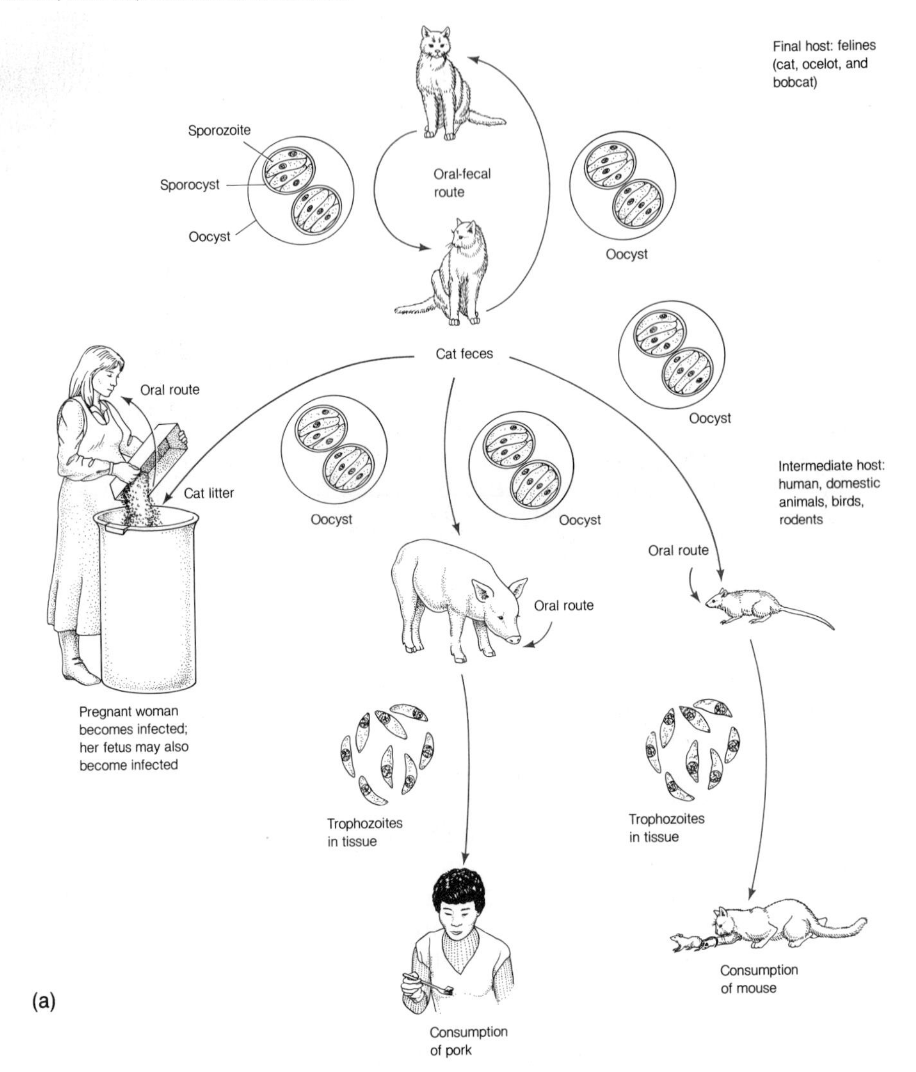

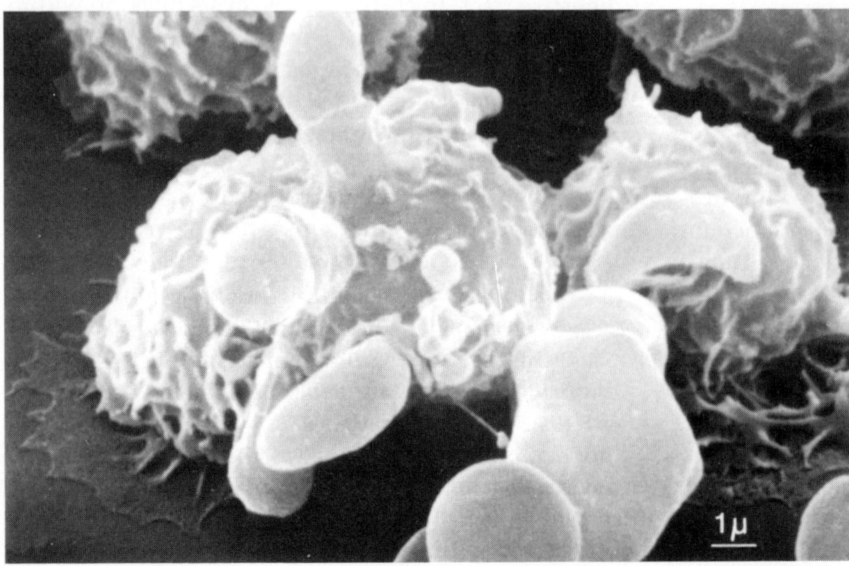

(b)

insect vectors generally defecate while they are biting and feeding, the bite often becomes infected. The insects sometimes bite children on the eyelids and consequently cause an infection of the eyelids.

DISEASES CAUSED BY MULTICELLULAR PARASITES

River Blindness and Loaisis

Onchocerciasis, or river blindness, is caused by the parasitic nematode *Onchocerca volvulus* and is spread by blackflies. The disease afflicts more than 150 million people worldwide. Most of the cases are in West Africa, but infections are also reported in east Africa, Yemen, Mexico, Central America, and South America.

An African disease called **loaisis** is caused by a nematode similar to *Onchocerca,* called *Loa loa,* and is spread by the bite of a horsefly or mango fly. The infection is similar to river blindness. The *Loa loa* microfilariae are easily seen crawling over the eye just under the corneal conjuctiva (fig. 21.14).

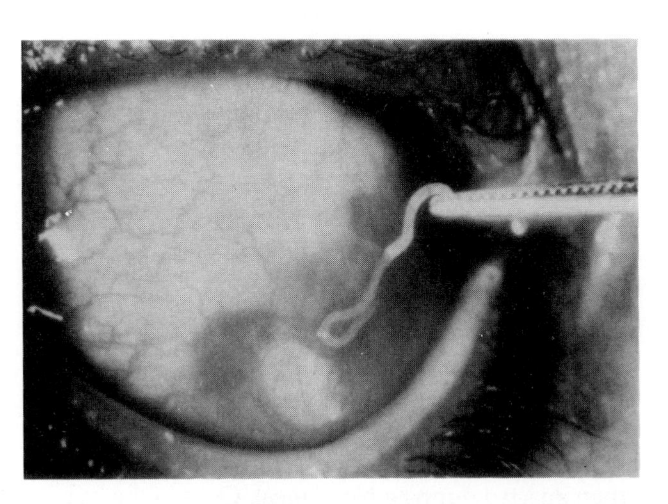

Figure 21.14 ***Loa loa* in the eye.** Surgical removal of *loa loa* worm.

Filariasis (Elephantiasis)

Filariasis is a disease caused by the nematode *Wuchereria bancrofti* (fig. 21.15). It is characterized by **microfilariae** infecting the lymph vessels and lymph nodes and obstructing the flow of lymph. The microfilariae also cause an inflammatory reaction in the infected tissue. Connective tissue is stimulated to multiply, so that the skin of the lower extremities and external genitalia becomes enlarged, coarse, and thickened (fig. 21.15). Legs often become so swollen and altered that they resemble the legs of elephants, hence the name **elephantiasis.** Filariasis is very prevalent in tropical

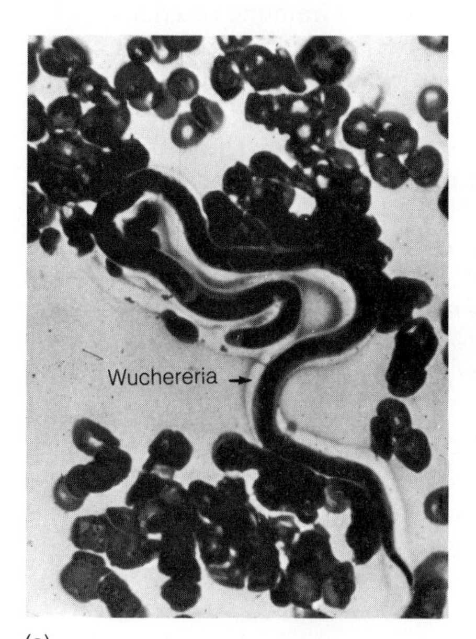

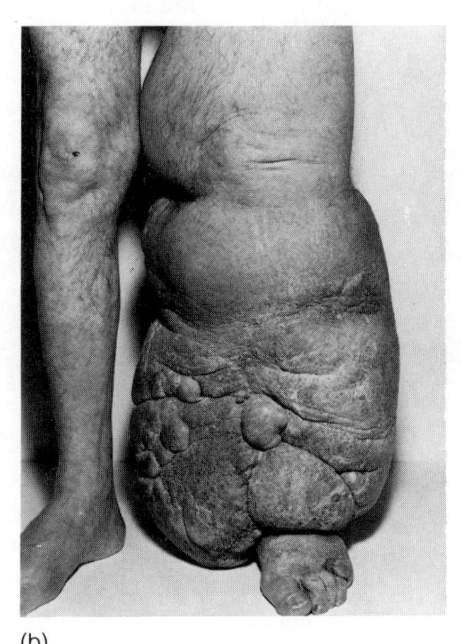

Figure 21.15 **Elephantiasis.** *(a)* Microfilaria of *Wuchereria bancrofti* in blood smear. *(b)* Patient with extensive elephantiasis of the leg.

regions all over the world and extends into some subtropical areas. Humans are the only known reservoir for this nematode, and the mosquito *Culex* is the vector for the filarial parasite.

Female (90 × 0.25 mm) and male (40 × 0.10 mm) *Wuchereria* reside in the lymph nodes. The female gives birth to threadlike embryos that develop into microfilariae (250 × 10 μm) in the lymph and blood. The microfilariae lodge in capillary walls and in subcutaneous tissue, but migrate into the peripheral blood at night, the time that the mosquito vector feeds. Thus, the mosquito sucks in microfilariae along with blood from an infected person. The larvae develop in the mosquito into a form that successfully infects humans when injected during a blood meal. Once in the human host, the larvae develop into adults that reside in lymph node sinuses, where they mate. The allergic reactions to the microfilariae and larvae growing in the subcutaneous tissue, lymph, and blood are responsible for many of the signs and symptoms of the disease.

Control of the disease can be achieved by spraying or draining pools of water where mosquitoes breed. Alternatively, the pools can be stocked with fish that eat the mosquito larvae. If mosquito vectors do not take in the microfilariae, the larvae die in the human host. Apparently, the development of the larvae in the mosquito is an essential part of the developmental process of these parasites.

Figure 21.16 ***Trichinella spiralis* in muscle tissue.** Encysted larva of the nematode *Trichinella spiralis* as seen in infected human muscle tissue.

Trichinosis

Trichinosis is a disease of muscular tissue caused by the nematode (roundworm) *Trichinella spiralis*. About 24 hours after consuming *Trichinella*-infected meat, the person develops a nonspecific gastroenteritis that may include abdominal pain, nausea, vomiting, diarrhea, and fever. Approximately a week after infection, and for several weeks thereafter, the infected person suffers from muscle pains, low-grade fever, dyspnea (painful or difficult breathing), edema (swelling of tissues), skin eruptions, anorexia (loss of appetite), and emaciation. After the disease becomes chronic, muscular pains persist for months. Trichinosis is found worldwide in animals and in humans who consume raw or undercooked meat, especially pork and bear.

Infection generally occurs by consuming meat with *Trichinella* cysts embedded between muscle fibers (fig. 21.16). Stomach acids and intestinal enzymes digest the cyst wall and free the nematode larva, which takes up residence between the cells of the duodenal and jejunal mucosa. In two days or so the larvae develop into sexually mature adults, which then mate. The females (3–4 mm long) burrow further into the intestinal lining, where they give birth about a week later to 500–1000 small larvae (0.1 mm long) over a 2- to 6-week period. The adults die and are digested, but the larvae migrate to all parts of the body through the connective tissue and through the lymphatic and circulatory systems. In most of the organs of the body, the larvae are destroyed by immune reactions; in the muscles, however, the larvae burrow between muscle cells, grow to 1 mm in length, coil up, and become encysted. **Myositis** (inflammation of the muscle), rigidity, and loss of strength occurs. The nematode can remain viable within the cyst for a number of years.

Control can be achieved by cooking garbage and meat fed to pigs, cats, and dogs. Humans can avoid the disease by thoroughly cooking meats. Once the disease is contracted, treatment consists of alleviating the pain and discomfort with adrenocorticotrophic hormone (ACTH) and treating the infection with thiabendazole.

In the United States, for the past 10 years, an average of 150 cases of trichinosis have been reported annually with a very low mortality. An average of less than 1 death/year has been reported in the last 10 years in the United States.

TABLE 21.3
DISEASES OF THE BLOOD, LYMPH, MUSCLE, AND INTERNAL ORGANS

Disease	Name of pathogen	Reservoir	Vector	How acquired
Acquired immune deficiency syndrome (AIDS)	Human immunodeficiency virus (HIV)	Humans	—	Sexual intercourse, inoculation or injection of virus-laden body fluids/blood products
African sleeping sickness	*Trypanosoma brucei gambiense* *T. brucei rhodesiense*	Humans	Tsetse fly	Fly bite
Brill-Zinsser disease	*Rickettsia prowazekii*	Humans	—	Latent typhus
Brucellosis	*Brucella abortus*	Cattle, goats, sheep	—	Ingestion, tissue rupture
Cat scratch fever	*Chlamydia*	Cats	—	Cat scratch
Chagas' disease	*Trypanosoma cruzi*	Rodents, cats, raccoons, deer, dogs	Bedbugs, lice	Bite
Colorado tick fever	Arbovirus	Rodents	Tick	Tick bite
Dengue fever	Arbovirus	Mosquito	Mosquito	Mosquito bite
Elephantiasis (filariasis)	*Wuchereia bancrofti*	Humans	Mosquito	Bite
Endemic (murine) typhus	*Rickettsia mooseri*	Rodents	Flea	Flea bite
Endocarditis	*Streptococcus* *Staphylococcus*	Skin	—	Tissue rupture, infection
Epidemic typhus	*Rickettsia prowazekii*	Humans	Lice	Louse bite
Gas gangrene	*Clostridium perfringens*	Soil	—	Tissue rupture, infection
Loiasis	*Loa loa*	Humans	Horsefly	Fly bite
Malaria	*Plasmodium falciparum* *P. vivax* *P. malariae* *P. ovale*	Humans, monkeys	Mosquito	Mosquito bite
Mononucleosis	Herpes virus (Epstein-Barr)	Humans	—	Kissing, ingestion
Plague	*Yersinia pestis*	Rodents	Fleas	Flea bite
Puerperal sepsis	*Escherichia coli*	Intestines	—	Tissue rupture
Q fever	*Coxiella burnetii*	Cattle	—	Ingestion
Rat bite fever	*Spirillum minor* *Streptobacillus moniliformis*	Rats, turkeys, weasels	—	Rat bite
Relapsing fever	*Borrelia recurrentis* *B. hermsii*	Rodents, ticks, lice, opossums	Tick, louse	Tick or louse bite
Rickettsialpox	*Rickettsia akari*	Rodents	Mites	Mite bite
River blindness	*Onchocerca volvulus*	Humans	Blackfly	Fly bite
Rocky Mountain spotted fever	*Rickettsia rickettsii*	Rodents, ticks	Tick	Tick bite
Schistosomiasis	*Schistosoma mansoni (japonicum) (haematobium)*	Human, snails	—	Burrow through skin
Scrub typhus	*Rickettsia tsutsugamushi*	Rodents	Mites	Mite bite
Toxoplasmosis	*Toxoplasma gondii*	Cats, rodents, pigs, sheep	—	Ingestion, inhalation
Trench fever	*Rochalimaea quintana*	Humans	Lice	Louse bite
Trichinosis	*Trichinella spiralis*	Pigs, many carnivores	—	Ingestion
Tularemia	*Francisella tularensis*	Rabbits, deer	Tick, deerfly	Ingestion, bites, & tissue rupture
Weil's disease (leptospirosis)	*Leptospira interrogans*	Rodents, skunks, dogs, racoons, cattle, swine	—	Ingestion, through skin
Yellow fever	Arbovirus	Humans, monkeys	Mosquito	Mosquito bite

CONCLUDING REMARKS

The great majority of infections that affect blood, lymph, muscle, and internal organs are initiated by insect vectors that inject the microorganisms into the lymph and blood (table 21.3). Some internal infections are due to microorganisms that penetrate intestinal or respiratory mucous membranes after being consumed or inhaled. A few organisms penetrate via animal bites (rabies) or scratches (cat scratch fever), open cuts (gangrene and sometimes tularemia), or by direct penetration (schistosomiasis).

Table 21.3 summarizes a number of characteristics of bacterial, viral, protozoal, and helminthic diseases that affect the lymph, blood, muscle, and internal organs. Notice that many of the bacteria are transmitted by fleas (plague), lice (typhus), and ticks (Rocky Mountain spotted fever). Viruses are transmitted by mosquitoes (yellow fever) and ticks (Colorado tick fever); protozoans are spread by flies (African sleeping sickness), bugs (Chagas' disease), and mosquitoes (malaria); and animals such as nematodes are disseminated by flies (river blindness) and mosquitoes (filariasis).

Fig. 21.17 illustrates the seriousness in the United States of selected diseases of the muscles, lymph, blood, and internal organs, by showing the yearly morbidity and mortality rates of these diseases.

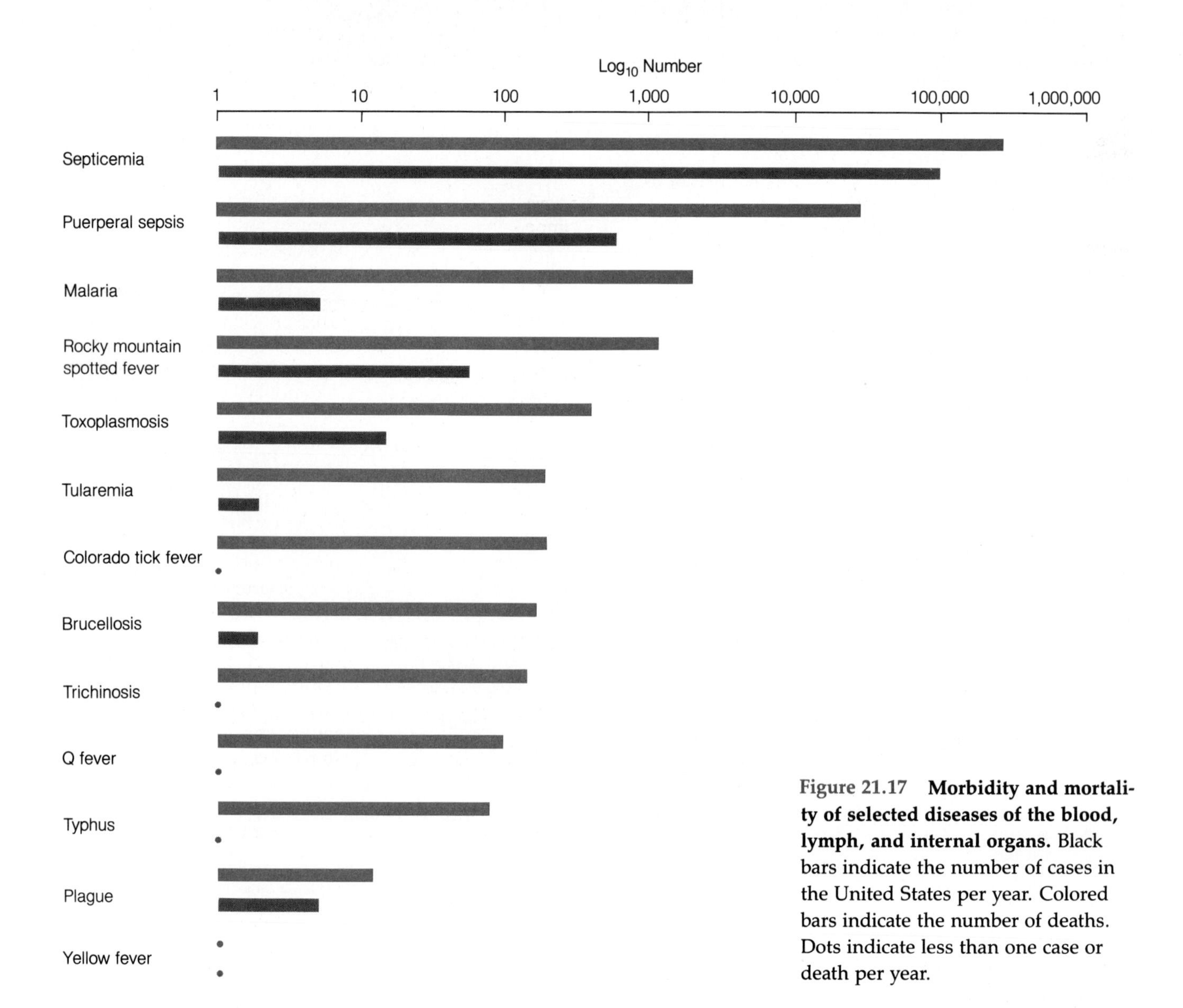

Figure 21.17 Morbidity and mortality of selected diseases of the blood, lymph, and internal organs. Black bars indicate the number of cases in the United States per year. Colored bars indicate the number of deaths. Dots indicate less than one case or death per year.

CLINICAL PERSPECTIVE STAPHYLOCOCCAL ENDOCARDITIS AND BACTEREMIA

A 35-year-old woman entered the hospital because of a cough that yielded approximately 0.5 ml of bright red blood a number of times each day. She complained of moderate bilateral chest pain and shortness of breath. Her feet and hands were swollen and she had a fever with chills. The patient's history indicated that she had been a heavy user of intravenous drugs for more than 15 years and for the past year had been injecting cocaine exclusively. There were multiple hyperpigmented track marks on both forearms and a hemorrhage on the right little finger. She did not smoke and she did not suffer from hepatitis. There was no indication of any recent surgery.

A chest x-ray revealed patchy infiltrates in both lungs and a slight enlargement of the cardiac silhouette. X-rays over a number of days showed an increase in patchy bilateral infiltrates and enlargement of the cardiac silhouette. The patient had the following signs: temperature 100°F (normal 98.6), pulse 100 (normal 72), respiration 24 (normal 15) and shallow, and blood pressure 100/70 (normal 120/80). Most significant was evidence of right ventricular failure: neck vein distention, tender hepatomegaly, and marked swelling (edema) of the legs. An electrocardiogram showed an increase in heart rate and nonspecific wave changes. The evidence indicated that the patient was experiencing right heart failure.

The white blood cell count was 17,300/ml (normal 7500). The nonpurulent sputum contained no acid-fast bacilli and Gram stains showed no predominant organism. A test for rheumatoid factor, lgM directed against lgG, was positive at a serum dilution of 1:320 and indicated that high levels of immune complexes were circulating in the bloodstream. Blood in the urine (hematuria) and urea as well as other nitrogenous compounds in the blood (azotemia) reflected the fact that immune complexes were being deposited in the kidneys and causing glomerulonephritis. Kidney disease is a common development in patients with endocarditis.

The chills, fever, muscle pain, joint pain, and increased white blood cells indicated an acute infection. The distribution of lung lesions seen in x-rays and the absence of purulent sputum suggested that the patient was suffering from a hematogenous septic embolization. *Streptococcus pneumoniae* (rusty sputum) and *Klebsiella pneumoniae* (currant-jelly sputum) are unlikely causes because they generally cause lobal pneumonia and grossly purulent sputum. Pulmonary tuberculosis and histoplasmosis were not completely ruled out until the sputum sample showed no acid fast bacteria or fungi and blood cultures were found to contain a penicillin-resistant methicillin-, oxacillin-, and nafcillin-sensitive *Staphylococcus aureus*.

Hematogenous septic embolization frequently occurs because of seeding of the lungs from a right-sided (tricuspid valve) endocarditis. The diagnosis of endocarditis would also explain the patient's right heart failure, renal dysfunction, and systemic symptomatology (swelling). The discovery of *S. aureus* in blood cultures rules out other common causes of endocarditis: viridans streptococci, enterococci, coagulase-negative stahylococci, *Pseudomonas aeruginosa*, and *Candida*. The diagnosis is of clinical importance since patients require more prolonged antibiotic therapy when they have endocarditis than when they have "simple" bacteriemia.

Staphylococcus aureus resistant to the semisynthetic penicillins (methicillin, nafcillin, oxacillin), as well as to cephalosporins may be treated with vancomycin. Since the patient was infected with a penicillin-resistant methicillin-sensitive *S. aureus* and had no history of allergic reactions to penicillin, she was started on gentamicin and nafcillin intravenously. The gentamicin was given for the first two weeks while nafcillin was given for eight weeks. Gentamicin was used for two weeks rather than one because soon after treatment began the patient's temperature rose to 102°F and the lesions in her right lower lobe progressively consolidated, forming cavities. Nevertheless, the patient made a gradual recovery after the problem developments and was discharged after eight weeks. Her heart was no longer failing, but there was a persistent systolic murmur. The patient was told of the danger of further intravenous drug abuse and any future surgical procedures that might introduce microorganisms that could reinfect the heart. The prognosis for the patient was considered poor because she was expected to reinfect herself.

A more detailed discussion of this patient and bacterial endocarditis can be found in the reference Bisno, A.L. (1986). Staphylococcal endocarditis and bacteremia. *Hospital Practice* 21(4): 139–158.

SUMMARY

STRUCTURE AND FUNCTION OF THE CIRCULATORY AND LYMPHATIC SYSTEMS

1. A number of serious diseases can occur when bacteria penetrate the skin or mucous membranes and invade the lymph and blood. Access to the lymph and blood generally occurs because of wounds, microorganism-induced ulcers, or animal or insect bites.
2. The blood system, at times, provides a pathway for infecting microorganisms to spread throughout the body.
3. Infecting microorganisms that are not destroyed in the lymph nodes may escape from the lymph nodes and flow into the blood and then into the heart.
4. Infecting microorganisms that are not destroyed in the blood by the immune systems can spread to the internal organs and cause disease.

BACTERIAL DISEASES CAUSED BY INVASION OF THE LYMPH AND BLOOD

1. Septicemia, or bacterial sepsis, is a blood infection characterized by fever, chills, decreased blood pressure, shock, and death in 30–50% of cases. As many as 100,000 persons die each year in the United States from septicemia. Eighty percent of the infections are due to gram-negative bacteria, such as *Escherichia*, *Klebsiella*, and *Enterobacter.*
2. Puerperal sepsis, or childbed fever, is a septicemia that results when a woman who has recently given birth becomes infected. Today, as many as 600 women die each year in the United States from puerperal sepsis. Before the extensive use of antibiotics in the late 1940s and early 1950s, puerperal sepsis was due mainly to the gram-positive bacteria *Streptococcus pyogenes*, *Staphylococcus aureus*, and *Staphylococcus epidermidis*. Presently, most of the cases of puerperal sepsis are due to gram-negative bacteria such as *Escherichia*, *Klebsiella* and *Enterobacter.*
3. Endocarditis is an infection of heart endothelium, generally in one of the valves that is damaged or abnormal. This infection leads to a septicemia and heart failure. Subacute endocarditis progresses for 2 to 3 months before it fulminates (produces severe symptoms). The gram-positive α-hemolytic streptococci, such as *Streptococcus mutans*, are responsible for 80% of the cases of subacute endorcarditis. Acute endocarditis generally flares up within a month after the infection begins. The gram-positive α-hemolytic streptococci account for 50% of the cases.
4. Gas gangrene results from an infection of the muscles or internal organs by any of a number of clostridia. The disease is characterized by the production of gas and necrotic tissue. The infection generally occurs in tissues that have poor blood circulation. Serious wounds, operations, arteriosclerosis, and diabetes predispose toward gas gangrene. *Clostridium perfringens* produces a number of exoenzymes (exotoxins) that are responsible for the degradation of the matrix and tissue cells. The bacterium ferments the hydrolyzed polysaccharide and produces gas, which often gets caught in the necrotic muscle and under the skin.
5. The plague, or black death, is caused by the bacterium *Yersina pestis* and is spread from animal to animal and from animal to human by fleas. When the bacteria multiply in the lymph nodes and cause them to swell, the plague is known as bubonic plague. When the bacteria multiply in the lungs, the disease is called pneumonic plague. Pneumonic plague is extremely contagious because it can spread from person to person in respiratory secretions and aerosols.
6. Relapsing fever is caused by the spirochete *Borrelia recurrentis*. The bacterium is usually spread by ticks, which are also the reservoir for this pathogen.
7. Brucellosis, or undulant fever, generally occurs in cattle, goats, swine, and sheep and is due to various species of the bacterium *Brucella*. In dairy herds, the disease frequently results in abortion. The bacterium can be spread to humans in unpasteurized milk from diseased animals, but butchers, workers in meat processing plants, ranchers, and hunters account for most of the brucellosis in humans (90%).
8. Tularemia, or rabbit fever, is caused by the bacterium *Francisella tularensis*. Tularemia is transmitted to humans when infected ticks and deerflies bite. The bacteria sometimes enter small cuts or penetrate the mucous membranes when infected animals are skinned by hunters.

9. Rocky Mountain spotted fever is due to the obligate intracellular bacterium *Rickettsia rickettsii*. The reservoirs for this bacterium are rodents, rabbits, and ticks. The ticks also act as vectors.

10. Typhus fevers are caused by various *Rickettsia*. Epidemic typhus is caused by *R. prowazekii* and is transmitted to humans in the feces of infected body lice that bite. Endemic typhus is caused by *R. mooseri* in the feces of infected fleas when they bite. *Rickettsia* are obligate intracellular parasites.

VIRAL DISEASES CAUSED BY INVASION OF THE LYMPH AND BLOOD

1. Colorado tick fever is due to an arbovirus and is transmitted to humans from infected wild rodents by tick vectors.
2. Yellow fever is caused by an arbovirus and is spread among humans, or from monkeys to humans, by mosquito vectors. There have been no cases of yellow fever in the United States since 1924. The disease is endemic in Africa and the Far East.
3. Dengue fever is due to an arbovirus that is transmitted by a specific mosquito, which acts as a reservoir. There are few cases of dengue in the United States, but recently in Puerto Rico an average of 11,000 cases has been reported/year.
4. Infectious mononucleosis is caused by a herpes virus called the Epstein-Barr virus. About 20,000 cases of mononucleosis, and 20 deaths, are reported per year in the United States.

PROTOZOAL DISEASES CAUSED BY INVASION OF THE LYMPH AND BLOOD

1. Malaria is caused by the protozoans *Plasmodium falciparum* (50%), *P. vivax* (43%), *P. malariae* (6%), and *P. ovale* (1%). The major reservoir of *Plasmodium* is humans, although some monkeys can carry the protozoan. A specific mosquito is responsible for spreading the organisms from human to human when it bites.
2. Toxoplasmosis is due to the protozoan *Taxoplasma gondii*. Cats are an important reservoir for this protozoan, although wild and domestic animals (pigs, sheep, and cattle) and birds often become infected. Humans become infected when they ingest dried particles of cat feces contaminated with the *Toxoplasma* oocysts.
3. African sleeping sickness is caused by *Trypanosoma brucei gambiense* and *T. brucei rhodesiense*, which are spread from human to human by the bite of the tsetse fly. Humans, animals, and the tsetse fly act as reservoirs for the protozoan.
4. Chagas' disease is caused by *Trypanosoma cruzi*, which is spread among humans by biting insects such as kissing bugs, bedbugs, and lice.

DISEASES CAUSED BY MULTICELLULAR PARASITES

1. River blindness, or onchocerciasis, is caused by the parasitic nematode *Onchocerca volvulus*, which is spread by blackflies. The disease afflicts more than 150 million people worldwide, mostly in West Africa. Humans appear to be the nematode's only reservoir.
2. Loiasis, a disease similar to onchocerciasis, is caused by nematode *Loa loa* that is spread by the bite of a horsefly or mango fly.
3. Filariasis, or elephantiasis, is caused by the nematode *Wuchereria bancrofti*. Humans are the only known reservoir for the nematode, which is spread from human to human by a specific mosquito.

STUDY QUESTIONS

1. For each of the following diseases, list (a) the name of the pathogen, (b) reservoirs, (c) vectors, and (d) how acquired.
   a. Puerperal sepsis
   b. Endocarditis
   c. Gas gangrene
   d. Plague
   e. Relapsing fever
   f. Brucellosis
   g. Tularemia
   h. Epidemic typhus
   i. Endemic typhus
   j. Tick fever
   k. Yellow fever
   l. Dengue fever
   m. Mononucleosis
   n. Malaria
   o. Toxoplasmosis
   p. Chagas' disease
   q. Sleeping sickness
   r. River blindness
   s. Loiasis
   t. Filariasis
2. From which activity could a person become infected by the pathogen listed?
   a. eating undercooked meat

b. injecting drugs intravenously
c. changing cat litter
d. drinking unpasteurized milk
e. butchering rabbits or deer
f. sleeping without mosquito netting
g. wading or swimming in waters that contain flukes and their snail hosts
h. collecting dead rats
i. walking in heavy brush with bare legs and bare upper body
j. infrequently washing clothes and body
k. kissing or using contaminated cups or utensils
l. using bird or cat feces as fertilizer

Francisella tularensis
Brucella melitensis
Yersinia pestis
Clostridium perfringens
Rickettsia rickettsi
Rickettsia prowazekii
Yellow fever virus
Dengue fever virus
Epstein-Barr virus
Plasmodium falciparum
Toxoplasma gondii
Trypanosoma gambiense
Trypanosoma cruzi
Onchocerca volvulus
Wuchereria bancrofti
Trichinella spiralis
Schistosoma mansoni
Histoplasma capsulatum

3. Most septicemias are due to which group of bacteria? How many people in the United States die from septicemias each year?
4. Puerperal sepsis is caused mostly by which bacteria? How many women die each year in the United States because of puerperal sepsis?
5. Which group of bacteria is responsible for most cases of subacute endocarditis?
6. Which bacteria are responsible for most cases of acute endocarditis?
7. What is the difference between subacute and acute endocarditis?
8. Lung infections would be expected to occur most readily from endocarditis in which part of the heart?
9. Which of the following diseases do *not* occur naturally in the United States? Give at least one reason for this.
   a. malaria;
   b. plague;
   c. yellow fever;
   d. puerperal sepsis;
   e. relapsing fever;
   f. brucellosis;
   g. Q-fever;
   h. tularemia;
   i. Rocky Mountain spotted fever;
   j. epidemic typhus;
   k. endemic typhus;
   l. Colorado tick fever;
   m. dengue fever;
   n. mononucleosis;
   o. toxoplasmosis;
   p. onchocerciasis;
   q. elephantiasis (filariasis);
   r. trichinosis;
   s. schistosomiasis;
   t. histoplasmosis;
   u. blastomycosis.

SELF-TEST

1. Toxins and microorganisms that escape the lymph nodes enter the venous blood. The first major organ that they pass through is the:
   a. liver;
   b. kidneys;
   c. heart;
   d. lungs;
   e. brain.
2. Approximately how many cases of septicemia are there per year in the United States?
   a. 300;
   b. 3000;
   c. 30,000;
   d. 300,000;
   e. 3 million.
3. Indicate the approximate number of deaths per year due to septicemia:
   a. 100;
   b. 1000;
   c. 10,000;
   d. 100,000;
   e. 1 million.
4. Which of the following bacteria accounts for the fewest cases of septicemia?
   a. *Escherichia;*
   b. *Staphylococcus;*
   c. *Klebsiella;*
   d. *Enterobacter;*
   e. *Proteus.*
5. In the 1880s, what drastically reduced the number of deaths due to childbed fever in European and American hospitals?

a. antibiotics;
b. better surgeons;
c. the development of disinfectants;
d. the use of surgical masks;
e. hand washing.

6. Which bacteria are most likely to be the cause of endocarditis?

a. viridans streptococci;
b. *Streptococcus pyogenes;*
c. *Streptococcus faecalis;*
d. *Staphylococcus aureus* and *S. epidermidis;*
e. *Pseudomonas aeruginosa.*

7. Gas gangrene is due mostly to the bacterium *Pseudomonas aerugenosa.* a. true; b. false.

8. *Yersinia pestis* is the cause of:

a. gas gangrene;
b. puerperal sepsis;
c. endocarditis;
d. plague;
e. tick fever.

9. What do endemic typhus, epidemic typhus, scrub typhus, Rocky Mountain spotted fever, Brill-Zinsser disease, and a type of pox have in common?

a. the vector is a mosquito;
b. the reservoir is rodents;
c. the cause is *Rickettsia* spp.;
d. the vector is a flea;
e. the pathogen is a virus.

10. Trench fever is spread by a louse and caused by the bacterium *Rochalimaea quintana.* a. true; b. false.

For items 11–20, match the major vector associated with the disease:

a. flea;
b. Tsetse fly;
c. tick;
d. mosquito;
e. louse or mite.

11. Relapsing fever
12. Malaria
13. Plague
14. Epidemic typhus
15. Rabbit fever (tularemia)
16. Rocky Mountain spotted fever
17. Elephantiasis (filariasis)
18. Yellow fever
19. Endemic typhus
20. Scrub typhus

21. What do African Burkitt's lymphoma, nasopharyngeal carcinoma, and mononucleosis have in common?

a. nothing;
b. the vector is a mosquito;
c. the reservoir is farm animals;
d. they are caused by Epstein-Barr virus;
e. the vector is a bacterium.

22. Post-transfusion non-A non-B hepatitis occurs in what percentage of patients receiving six or more units of paid-donor blood?

a. 1%;
b. 5%;
c. 15%;
d. 25%
e. 45%.

23. What is responsible for the chills, fever, and sweats that occur during the recurring bouts of malaria?

a. allergic reaction to mosquito saliva;
b. the presence of sporozoites in the blood;
c. the lysis of red blood cells and release of merozoites into the blood;
d. the presence of trophozoites in red blood cells;
e. all of the above.

24. Where do the zygotes of *Plasmodium* develop into sporozoites?

a. mosquito;
b. liver cells;
c. red blood cells;
d. white blood cells;
e. any of the above.

25. Which is not a common reservoir for *Toxoplasma gondii*?

a. mice;
b. rats;
c. cats;
d. birds;
e. humans.

26. What form of *Toxoplasma* is infective when undercooked pork is involved?

a. merozoite;
b. trophozoite;
c. oocyst with sporozoites;
d. all of the above;
e. none of the above.

27. What type of organism is responsible for river blindness, loaisis, trichinosis, and filariasis?

a. fungus;
b. nematode;
c. virus;
d. bacterium;
e. mnone of the above.

28. Filariasis is transmitted by eating contaminated undercooked meat. a. true; b. false.

29. In the United States which disease is responsible for the most deaths?
    a. malaria;
    b. Rocky Mountain spotted fever;
    c. typhus;
    d. plague;
    e. septicemia.

30. Arboviruses, which cause encephalitis, are spread to humans by:
    a. ticks;
    b. mosquitos;
    c. contaminated milk;
    d. laboratory accidents;
    e. all of the above.

SUPPLEMENTAL READINGS

Goldstein, E. J. C. and Richwald, G. A. 1987. Human and animal bite wounds. *American Family Physician* 36(1):101–111.

Kolata, G. 1984. The search for a malaria vaccine. *Science* 226:679–682.

Komaroff, A. L. 1987. The chronic mononucleosis syndromes. *Hospital Practice*. 22(5A):71–75.

Lange, W. R. 1987. Viral hepatitis and international travel. *American Family Physician* 36(1):179–189.

Langer, W. L. 1964. The black death. *Scientific American* 210:114–121.

Merigan, T. C. and Rubenstein, E. 1985. Viral zoonoses. *Scientific American Medicine* 7(XXXI):1–10.

Paterson, P. Y. 1982. Bacterial endocarditis. Chap. 50 in *Microbiology: Basic science and medical applications* (A. Braude, C. Davis, and J. Fierer, Eds.). Philadelphia: Saunders.

Rabinowitz, S. G. 1982. Bacterial sepsis and endotoxic shock. Chap. 36 in *Microbiology: Basic science and medical applications* (A. Braude, C. Davis, and J. Fierer, Eds.). Philadelphia: Saunders.

Rubin, R. H. 1985. Infection in the immunosuppressed host. *Scientific American Medicine* Section 7-X. New York: Freeman.

Sommers, H. M. 1982. Diseases due to anaerobic bacteria. Chap. 48 in *Microbiology: Basic science and medical applications* (A. Braude, C. Davis, and J. Fierer, Eds.). Philadelphia: Saunders.

Van Heynigen, W. E. 1968. Tetanus. *Scientific American* 218:69–77.

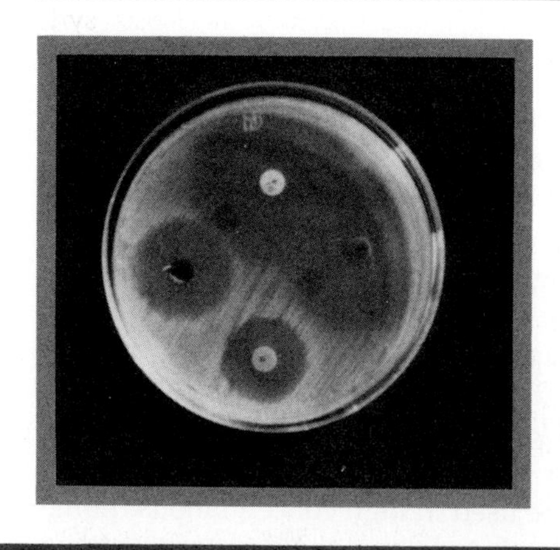

CHAPTER 22

CHEMOTHERAPY OF INFECTIOUS DISEASES

CHAPTER PREVIEW CHEMOTHERAPEUTIC DRUGS FROM THE SEA

Humans have experimented with plants for thousands of years to find drugs that might alleviate or cure their ills. The most notable discoveries have been the ipecacuanha root from Brazil, which yields **emetine**, a chemical that kills *Entamoeba histolytica,* the cause of amebic dysentery, and the cinchona bark from Peru, which yields **quinine**, an alkaloid that kills *Plasmodium*, the protozoa that causes malaria. Fungi and bacteria also produce chemicals that help alleviate or cure ills. For example, the fungus *Penicillium* produces the antibiotic **penicillin** and bacteria in the genus *Streptomyces* produce an array of antibiotics that includes **streptomycin** and **tetracycline**.

Now scientists are finding useful chemicals and drugs derived from sea life. Marine biologists have discovered a number of drugs that may be effective against various cancers. One species of animal that forms mats on rocks and on the hulls of ships, known as "false coral," produces a compound called **bryostatin-1** that doubles the lifespan of mice with leukemia. As little as 0.1 μg is effective in prolonging the life of these animals. Marine biologists have also discovered that soft corals and mollusks that lack an external shell produce toxic chemicals to avoid being eaten by predators. More than 30 chemicals that inhibit cell division have been isolated from soft sea corals and mollusks. Some of these chemicals may be useful in treating cancerous growths and parasitic infections. Soft corals also produce peptide hormones called **pseudopterosins**, which inhibit the release of prostaglandins. The hormones reduce pain, inflammation, and fever. The pseudopterosins, unlike other analgesic compounds such as morphine, are not addictive or toxic at the doses that are effective in humans.

A compound called **stypoldone,** which repels fish, has been extracted from a Caribbean seaweed. Because stypoldione also inhibits cell division, it is being investigated to see if it might be useful in treating some types of cancer or parasitic infections.

For more than 40 years, scientists have isolated many hundreds of drugs from bacteria and fungi. In fact, most useful drugs are produced by bacteria and fungi that live in the soil. The discovery of drugs produced by sea creatures will undoubtedly stimulate scientists to begin an extensive search for new chemicals. The main task ahead is to characterize these chemicals and to determine how they might be useful to humans in preventing and treating disease.

In this chapter we will be introduced to some chemicals that are used as chemotherapeutic agents. We will also learn what factors are important in a good drug and how various drugs function.

CHAPTER OUTLINE

THE PROPER MANAGEMENT OF INFECTIOUS diseases requires an understanding of how chemotherapeutic agents work and of their limitations. With this information,physicians are able to choose the correct drug and dosage to treat a disease optimally and reduce the risk of side effects. It is also important to know how effective levels of antimicrobials are reached in the body. In addition, the doctor must know whether or not the pathogen is resistant to the chemotherapeutic agent. In this chapter you will study some of the most commonly-used modern antimicrobials, their mode of action, and how their effectiveness against microorganisms are measured in the laboratory. You will also study some of the mechanisms whereby microorganisms develop resistance to antimicrobials and thus render them ineffective.

PERSPECTIVE

Since ancient times, humans have used medicines and potions discovered by chance to treat their diseases. Scientific chemotherapy began in 1910 after a methodical search for an antimicrobial, when Paul Ehrlich announced his discovery of an arsenic compound he called **salvarsan** (fig. 22.1). This compound, also known as Ehrlich's "magic bullet," was useful in treating yaws, syphilis, and relapsing fever. It was Ehrlich's belief that chemicals could be synthesized that would specifically bind and kill microorganisms. Unfortunately, salvarsan was not specific, and it was extremely toxic because it had to be used in high concentrations. Long-term treatment at reduced dosages led to serious brain and nerve damage. At recommended doses, the patients became nauseated and weak and often developed unusual allergies. salvarsan and neo-salvarsan (which was less toxic) were nevertheless much better than the folk cures for syphilis being used at the time, and Ehrlich's arsenic compounds cured many people of this disease.

The Discovery of Sulfanilamide

In 1935, Gerhard Domagk reported that **prontosil rubrum**, a red azo dye combined with a sulfonamide group (fig. 22.2), was effective against bacterial infections in experimental animals and in humans. He discovered its usefulness in humans when he used the chemical on his daughter, who was dying of septicemia: the drug effected a rapid and miraculous recovery. Although prontosil had no effect on bacteria growing in test tubes, it did have an effect *in vivo*, where it was cleaved into a sulfa drug called **sulfanilamide** (fig. 22.2).

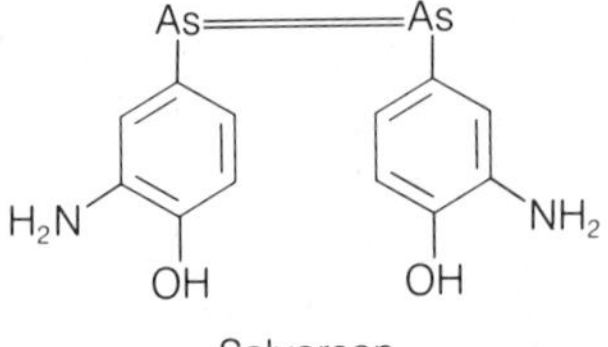

Figure 22.1 Salvarsan. Compound 606, salvarsan, was Paul Ehrlich's "magic bullet." This arsenic-containing compound was effective in combating syphilis in experimental animals.

Prontosil

Sulfanilimide

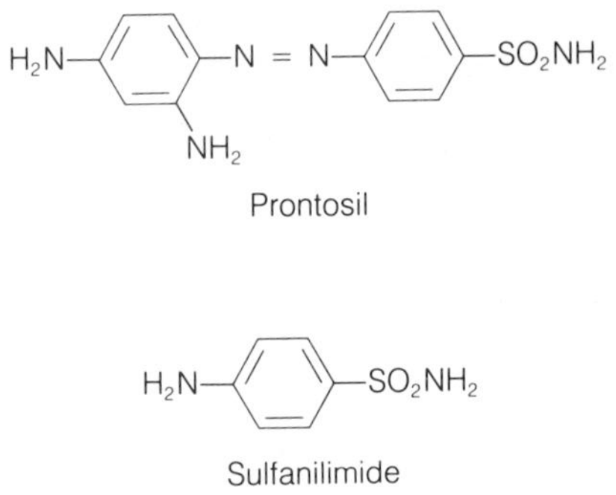

Figure 22.2 Prontosil and sulfanilamide sulfa drugs

Sulfa drugs are effective only against organisms that synthesize the coenzyme folic acid from para-aminobenzoic acid (PABA). Sulfanilamide competes with PABA for the enzyme that joins PABA to another compound in the synthesis of folic acid. Because the sulfa drug is unable to take the place of PABA, the formation of the coenzyme is blocked (fig. 22.3). Humans and animals do not synthesize their own folic acid but instead require it preformed, so they are not disturbed by the sulfa drug.

Sulfonamides such as sulfanilamide, sulfapyridine, and sulfathiazole were found to be effective against a variety of bacterial infections. Sulfa drugs are effective against *Streptococcus pyogenes* (skin and throat infections), *Neisseria meningitidis* (meningitis), *Escherichia coli* (urinary tract infections), *Chlamydia trachomatis* (trachoma), *Haemophilus ducreyi* (chancroid), *Nocardia asteroides* (nocardiosis), and *Actinomyces israelii* (actinomycosis). Domagk was awarded the 1939 Nobel Prize for medicine and physiology, but for political reasons he declined the honor and money.

The Discovery of the First Antibiotics

In 1939, Rene Dubos discovered that the soil bacterium *Bacillus brevis* was producing a chemical he called **tyrothricin**, which killed *Streptococcus* in a liquid solution. Tyrothricin and another antibiotic, **gramicidin**, produced by another species of *Bacillus*, proved to be too toxic to be taken internally by humans and so they were used only for curing skin infections.

Selman Waksman, stimulated by Rene Dubos's successes, began a search for actinomycetes that might produce useful antimicrobial agents. Waksman coined the term **antibiotic** to describe the natural antimicrobial agents produced by microorganisms. Today, antimicrobial agents produced by microorganisms are called antibiotics to distinguish them from manmade chemicals, such as salvarsan and sulfanilamide. In 1940, Waksman and coworkers discovered an actinomycete that produced **actinomycin**. Although the antibiotic proved very useful in research, it was too toxic to be used to treat animal infections. By 1943, another actinomycete called *Streptomyces griseus* was

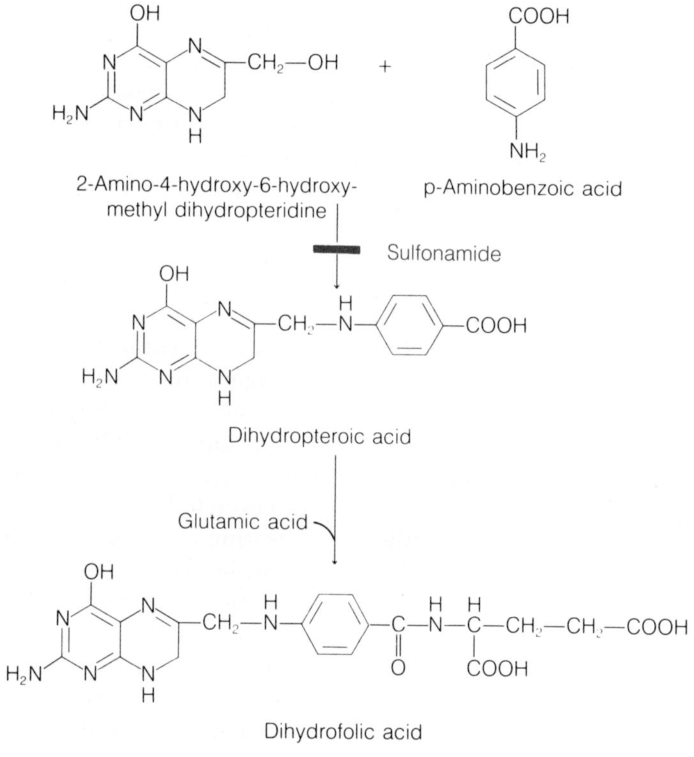

2-Amino-4-hydroxy-6-hydroxymethyl dihydropteridine + p-Aminobenzoic acid

Sulfonamide

Dihydropteroic acid

Glutamic acid

Dihydrofolic acid

Figure 22.3 Mode of action of sulfa drugs. Sulfa drugs inhibit the synthesis of folic acid, a coenzyme needed for the synthesis of nucleic acids. The synthesis of folic acid requires that para-aminobenzoic acid (PABA) be present. Sulfa drugs like sulfonamide compete with PABA in the synthesis of folic acid. When sulfonamide is used instead of PABA, the resulting product is inactive, and hence the synthesis of nucleic acids is inhibited.

discovered by Albert Schatz, one of Waksman's assistants. It produced the antibiotic called **streptomycin** (fig. 22.4). Streptomycin turned out to be effective against *Mycobacterium tuberculosis* (tuberculosis), *Francisella tularensis* (tularemia), *Neisseria meningitidis*, *Streptococcus pneumoniae* (pneumonia), and *Escherichia coli*. Waksman and his colleagues are also credited with the discovery of **erythromycin**, **neomycin**, and **candicidin**. For his discovery, isolation, and testing of streptomycin, Waksman received the Nobel Prize for physiology and medicine in 1952.

One of the most significant developments in chemotherapy was the discovery of the antibiotic **penicillin G** (fig. 22.5), produced by the fungus *Penicillium notatum*. The discovery of penicillin occurred in 1928, but its purification and use did not take place until the early 1940s. Sir Alexander Fleming is credited with the discovery of penicillin; Howard Florey and Ernst Chain are responsible for its isolation and purification, and they won the Nobel Prize in 1945.

The discovery and successful use of penicillin during the last part of World War II was important, because it stimulated the search for other natural substances produced by fungi and soil bacteria which could inhibit infectious microorganisms. The search for new antibiotics was also spurred by the discovery that penicillin was not a panacea: the fungi, protozoans, and some bacteria are not affected by this antibiotic. Until the early 1940s, plant substances (quinine), poisonous arsenic compounds (salvarsan), and sulfonamides (sulfanilamide) were the only substances that had been discovered after years of searching for antimicrobial agents. Suddenly, after the isolation of penicillin, researchers were checking all types of soil microorganisms to find other antibiotics that would work where penicillin did not.

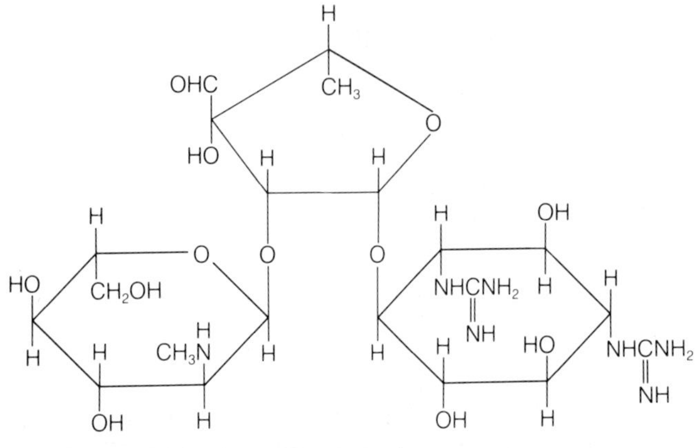

Streptomycin

Figure 22.4 Streptomycin

FUNDAMENTALS OF CHEMOTHERAPY

An antimicrobial drug is useful in the treatment of infectious diseases only if it inhibits or kills the infectious agent without harming the host. This desirable property of antimicrobial agents is called **selective toxicity**; that is, it is toxic to the pathogen but not to the host. For example, the antibiotic penicillin has selective toxicity. It inhibits the synthesis of new cell wall murein by bacteria. Without murein, growing bacterial cells would burst. Since human cells do not have cell walls, much less murein-containing cell walls, they will not be affected adversely by penicillin. On the other hand, the arsenic compound salvarsan, prepared by Paul Ehrlich to combat syphilis, had very little selective toxicity: it was extremely toxic and made the patients severely ill.

Even antimicrobial agents that show a selective toxicity can be toxic if taken in large quantities or for extended periods of time. For this reason, all antimicrobials are defined by their **therapeutic ratio**. This ratio is the highest amount of the antimicrobial that can be tolerated by the patient divided by the minimum amount of antimicrobial that will kill or inhibit the infectious agent. Antimicrobials such as ampicillin, cephalosporins, and quinolones, which can be taken in large quantities and are microbicidal at very low concentrations, have high therapeutic ratios. By contrast, Ehrlich's arsenic compounds had very low therapeutic ratios.

A number of factors besides selective toxicity affect the therapeutic ratio of antimicrobials. For example, an antimicrobial agent that is very reactive (e.g., penicillin), and that sometimes stimulates a life-threatening allergic reaction, has no value on sensitized individuals. Similarly, a drug that is insoluble in water or that is rapidly excreted from the body, is generally of little use in treating internal infections.

The nature of the pathogen must also be considered when selecting an antimicrobial for treatment. Antimicrobial agents used against bacteria that are resistant to the agent will not clear up the infection, even if very high doses are used. For this reason, the sensitivity of a pathogen isolated from a patient is nor-

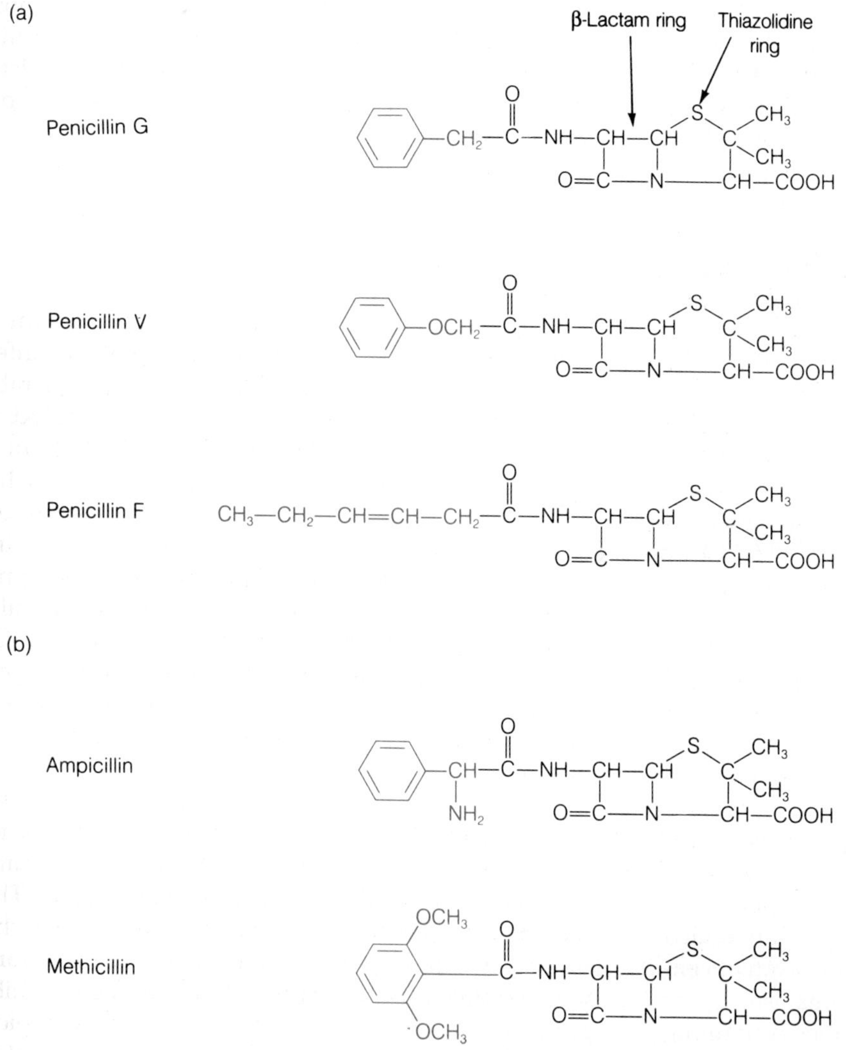

(a)

β-Lactam ring Thiazolidine ring

Penicillin G

Penicillin V

Penicillin F

(b)

Ampicillin

Methicillin

Figure 22.5 Chemical structure of penicillins. All penicillins, both naturally occurring *(a)* and synthetic *(b)*, include a β-lactam ring. The side chains (in color print), which make each penicillin type distinct, influence the activity and range of activity of the molecule.

mally tested against a variety of antimicrobials to determine the best course of treatment for the infection.

Antimicrobials are also defined based on the spectrum of microorganisms that they inhibit or kill. Those antimicrobial agents that are effective against a small number of microorganisms are called **narrow-spectrum** antimicrobials, while those drugs that are effective against a large number of very different organisms are called **broad-spectrum** antimicrobials (table 22.1). For example, the antibiotic penicillin is most effective against gram-positive organisms and a few special gram-negative bacteria, so it is classified as a narrow-spectrum antibiotic. On the other hand, tetracycline is active against most gram-negative and gram-positive organisms and so is classified as a broad-spectrum antibiotic. If the presence of an infectious agent has not been determined before chemotherapy is initiated, it is recommended that a broad-spectrum antimicrobial be used until the diagnosis is made and a more appropriate drug can be prescribed (if necessary).

DETERMINATION OF DRUG AND ANTIBIOTIC SENSITIVITIES

It is essential in treating an infection to determine what antimicrobial agent is most effective. The sensitivity of an infectious agent can be determined in a number of ways. One method for determining sensitivities to

TABLE 22.1
BREADTH OF ACTIVITY OF CERTAIN ANTIBIOTICS*

Antibiotic	Broad	Narrow	Organisms affected
Tetracycline	X		Most bacteria
Chloramphenicol (Chloromycetin)	X		Most bacteria
Democlocycline (Declomycin)	X		Most bacteria
Oxytetracycline (Terramycin)	X		Most bacteria
Kanamycin (Kantrex)	X		Most bacteria
Ampicillin	X		Most bacteria
Cephalothin (Keflin)	X		Most bacteria
Gentamycin	X		Most bacteria
Penicillin		X	Gram + bacteria, *Neisseria*
Streptomycin		X	*Streptococcus*, some gram − bacteria
Erythromycin (Ilotycin)		X	Gram + bacteria
Polymyxin B		X	Gram − bacteria
Nystatin (Mycostatin)		X	Yeasts
Griseofulvin (Grisactin)		X	Dermatophyte fungi
Rifampin		X	*Mycobacterium*
Amphotericin B (Fungizone)		X	Systemic fungi

*Names in parentheses are trade names of antibiotics.

drugs is the **tube dilution test** (fig. 22.6). In the tube dilution test, a number of tubes are inoculated with a dilute suspension of microorganisms and then varying concentrations of the antimicrobial agent are added to the tubes. In the tubes containing very low concentrations of the antimicrobial agent, the organisms grow and turn the medium turbid. In the tubes containing intermediate to high concentrations, however, no growth is visible. The **minimum inhibitory concentration (MIC)** of an antimicrobial agent can be determined from the tube with the lowest concentration that shows no turbidity (no growth). Similar procedures with other agents allow a comparison between antimicrobials. The antimicrobial agent with the lowest MIC is generally the most effective.

The **minimum killing concentration** is established by transferring the organisms in the clear tubes into fresh, antimicrobial-free media. The tube with the lowest concentration of antimicrobial agent that shows no growth upon transfer to fresh media indicates the minimum killing concentration (fig. 22.7). Since these measurements are generally done on bacteria, the minimum killing concentration is often called the **minimum bactericidal concentration.**

By definition, if the concentration of an antimicrobial agent needed to kill a population of organisms is 10 × higher than the amount necessary to inhibit the organisms, the agent is said to be **biostatic** (bacteriostatic, fungistatic, etc.). On the other hand, if the concentration needed to kill a population of organisms is less than 5 × the amount required to inhibit the organism, the agent is said to be **biocidal** (bactericidal, fungicidal, etc.).

To determine the dosage and how often the antimicrobial agent should be given, it is necessary to know not only the minimum killing concentration but also the concentration of the agent in the blood and tissues and how it changes with time. One procedure for determining *in vivo* concentrations of drugs involves sampling the blood several times before and after administration of the drug. The blood samples are placed in wells punched into an agar plate seeded with sensitive microorganisms (fig. 22.8). Various known concentrations of the antibiotic are placed in other wells, so that a standard curve can be developed (fig. 22.8), using the diameter of inhibition around the holes as a measure of the concentration of the antibiotic in the blood. A graph relating the concentration of antibiotic in the blood to time can be constructed (fig. 22.8). This type of curve indicates what the dose should be and how often it should be given to maintain a minimum killing concentration.

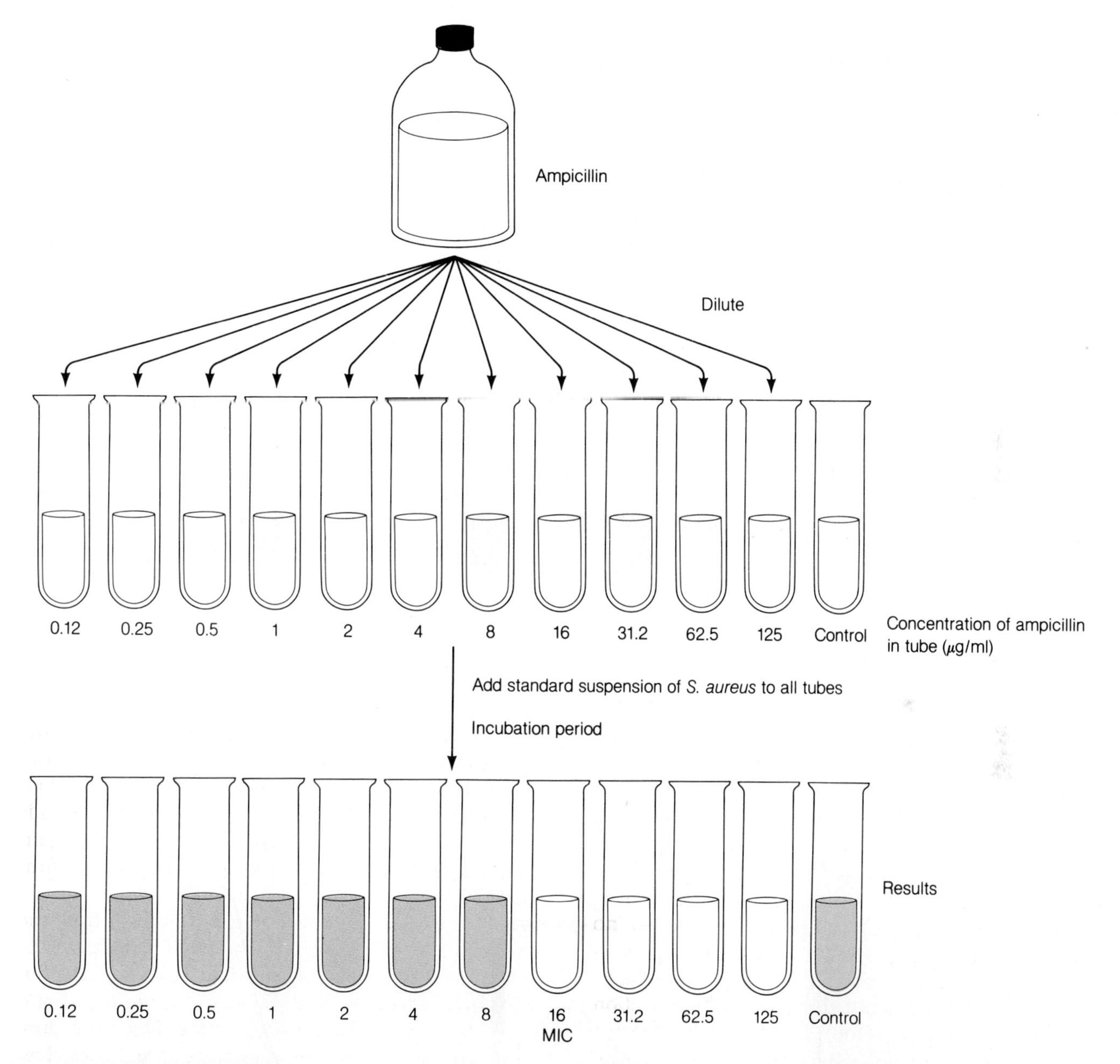

Figure 22.6 Determination of the minimum inhibitory concentration of an antimicrobial agent. The minimum inhibitory concentration (MIC) of an antimicrobial consists of the lowest concentration that will inhibit the growth of a test organism. The diagram illustrates a common procedure for determining the MIC of ampicillin for *Staphylococcus aureus*. Various dilutions of ampicillin (ranging from 125 μg/ml to 0.12 μg/ml) in a liquid broth are inoculated with a standard suspension of *S. aureus* and incubated at 35°C. After the incubation period, the tubes are examined for growth. The highest dilution of ampicillin that inhibits bacterial growth (clear tube) is the MIC. In this example, the MIC is 16 μg/ml.

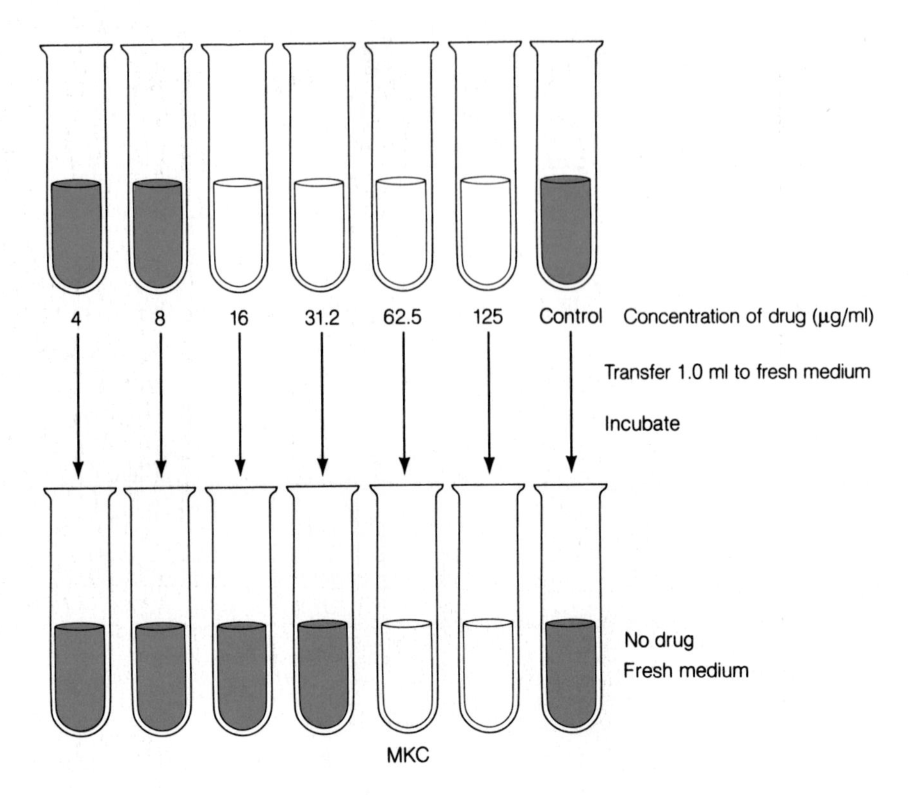

Figure 22.7 **Determination of the minimum killing concentration.** The minimum killing concentration (MKC) is the lowest concentration that will kill a standard population of bacteria. In the example given, samples from tubes of the MIC test (fig. 22.6) are inoculated into antibiotic-free broths and incubated at 35°C. After an incubation period, the tubes are examined for growth. The MKC is the lowest concentration of antibiotic that will kill the bacterial population. In this example, the MKC is 62.5 μg/ml. Notice that, although the MIC is 16 μg/ml, the MKC is 62.5 μg/ml, indicating that at concentrations of 16 and 31.2 μg/ml the bacteria are merely inhibited from multiplying, but not killed.

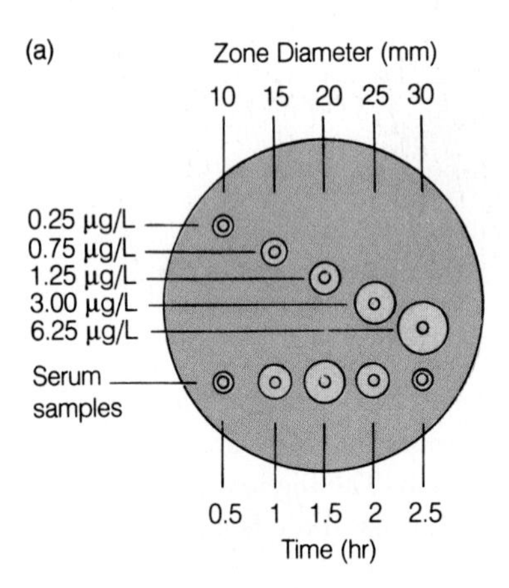

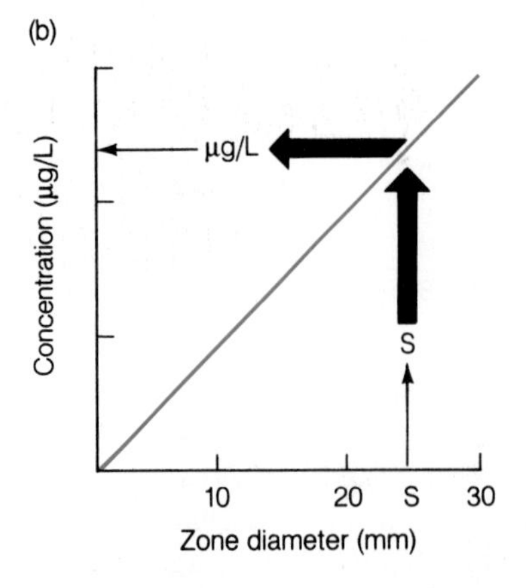

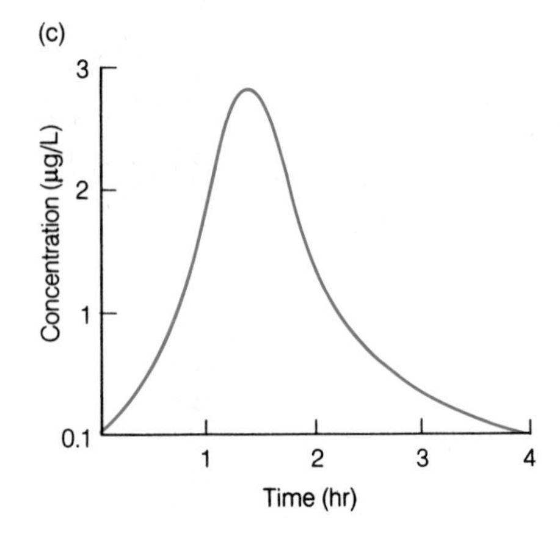

Figure 22.8 **Determination of *in vivo* concentrations of antimicrobial agents.** *(a)* Agar plates are seeded with test bacteria and wells are cut into the agar. The wells are filled with various concentrations of an antimicrobial agent or with serum samples extracted at various times. The plates are incubated overnight at 35°C. *(b)* The diameters of the zones of inhibition are plotted against the concentration of antimicrobials to obtain a standard curve. The antimicrobial concentration in serum can be determined by using the standard curve. *(c)* The concentration of an antimicrobial agent in serum varies with time. For example, suppose the zone diameter for the serum is S. Then the μg serum concentration is μg/L. The serum concentration μg/L and the time the serum was taken is used to plot one point in curve c.

A rapid procedure for determining the degree of sensitivity or resistance to a number of antibiotics is the **disk diffusion** test, also known as the **Kirby-Bauer** antibiotic susceptibility test. This test was introduced in 1966 by Fred Bauer, William Kirby, J. C. Sherris and M. Turck. In the Kirby-Bauer test, a lawn of the organism in question is seeded onto an agar petri dish (usually Müller-Hinton agar). Antibiotic-impregnated filter disks are then placed onto the developing lawn and incubated for 16 hours (fig. 22.9). During the incubation period, each of the antibiotics in the disks will diffuse in the agar, creating a concentration gradient, from high to low, as it diffuses radially from the disk. The bacteria on the plate will multiply throughout, except in those areas where the antibiotic's concentration is high enough to inhibit microbial growth and produce a zone of no growth, called the **zone of inhibition**. The diameter of the zone of inhibition around the antibiotic disks is related to minimum inhibitory concentrations and minimum killing concentrations. Table 22.2 lists a number of antimicrobial agents and the data that indicate whether or not the agent would be effective against the organism tested. (The values in the table are for a specific strain of *Staphylococcus aureus* and give no information as to how another organism might respond.)

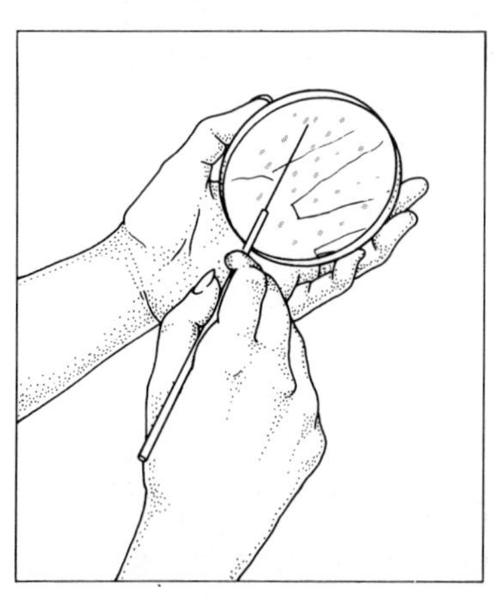
(a) Pick colony

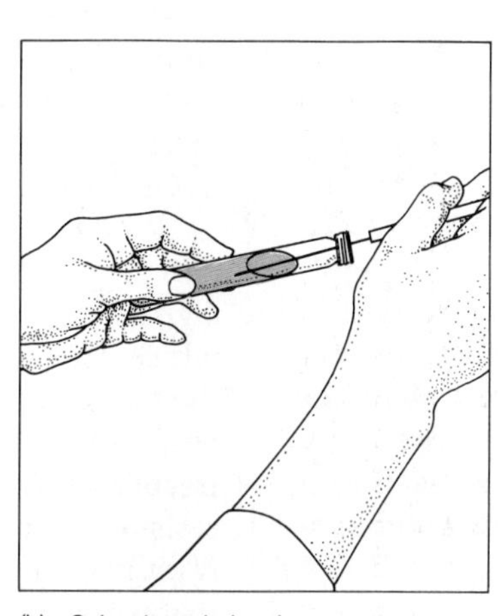
(b) Subculture in broth

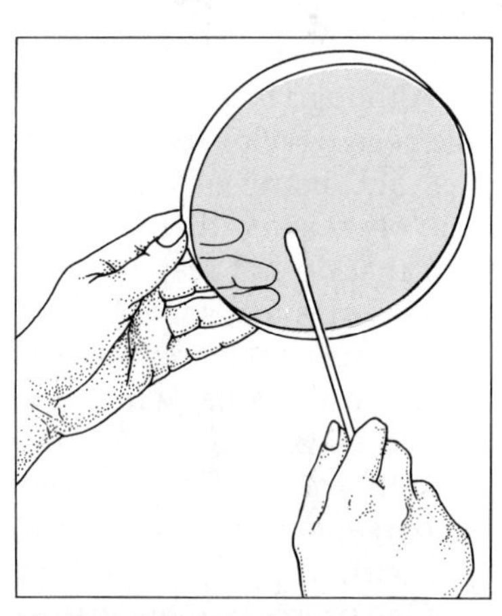
(c) Spread diluted culture over agar plate; allow to dry for a few minutes

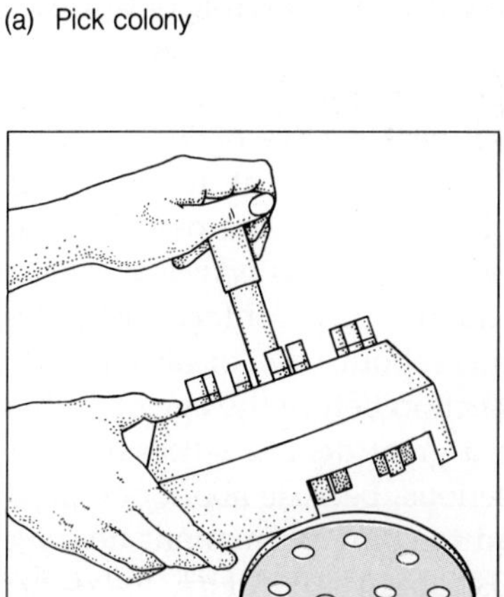
(d) Apply antibiotic impregnated disks to plate

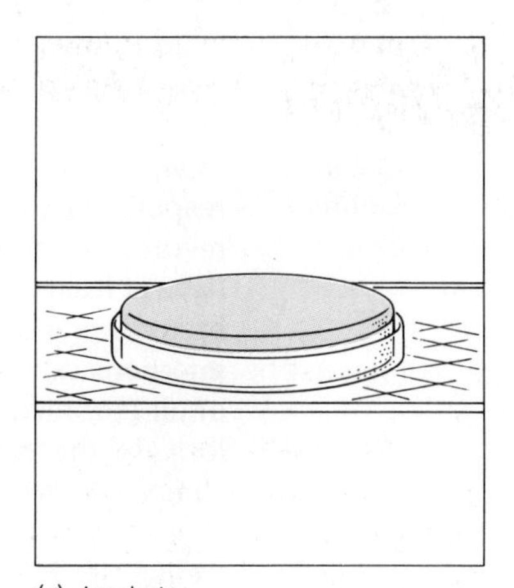
(e) Incubate

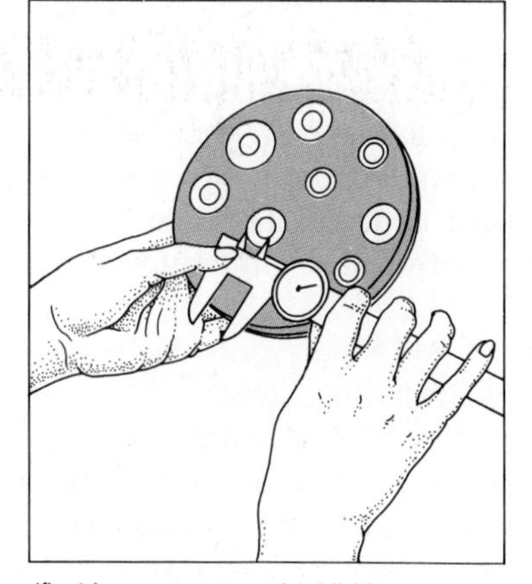
(f) Measure zones of inhibition

Figure 22.9 The Kirby-Bauer method

TABLE 22.2
ZONES OF INHIBITION FOR CERTAIN ANTIMICROBIALS USING *STAPHYLOCOCCUS AUREUS* AS THE TEST ORGANISM

Antimicrobial drug*	Zone of inhibition (mm) Sensitive	Resistant
Ampicillin (10-μg disk)	>28	<21
Chloramphenicol (30-μg disk)	>17	<13
Colistin (10-μg disk)	>10	< 9
Kanamycin (30-μg disk)	>17	<14
Penicillin (10 unit disk)	>28	<21
Sulfadiazine (300-μg disk)	>16	<13
Tetracycline (30-μg disk)	>18	<15

*Values in parentheses indicate the concentration of the antibiotic in the paper disk.

Although many hundreds of antimicrobial agents have been developed and extracted from microorganisms, and many of these are very effective against some organisms, there are still many infectious diseases that do not respond to treatment. One important reason is that some individuals have a depressed immune response and are therefore unable to eliminate the organisms, which may only be inhibited or reduced in number by the treatment. Another very common reason for persistent infections is that a microorganism has become resistant to the antimicrobial agent.

One problem in the use of antimicrobial agents is that toxins may be released by the infecting agent when it is killed. In the case of septicemia by a gram-negative bacterium such as *E. coli*, an antibiotic that kills all the organisms can lead to death because of an allergic reaction precipitated by the release of large quantities of endotoxin. In some cases, the antimicrobial agent is unable to get into infected tissues at sufficiently high concentrations to kill the organisms. Consequently, these antibiotics are not very useful.

DRUG RESISTANCE

Because of the extensive use of many antibiotics all over the world, numerous bacteria are becoming resistant to one or more antibiotics. In the 1940s, when penicillin was first introduced to treat gonorrhea, a single dose of 200,000 units of penicillin was sufficient to clear up all *Neisseria gonorrhoeae* infections. In the mid 1950s, partially resistant strains began to appear that required higher doses of penicillin. By 1976, 4.8 million units of penicillin were required to eliminate most *N. gonorrhoeae*. A few outbreaks also occurred in which the bacteria were completely resistant to penicillin. The source of these penicillin-resistant strains is found in those countries where penicillin has been used as a prophylactic and as a treatment for many different infections. Penicillinase-producing *N. gonorrhoeae* (PPNG) is endemic in many countries that have large camps of foreign military or businessmen and the associated prostitution. Penicillinase-producing *N. gonorrhoeae* have been isolated in Korea (U.S. military), the Philippines (U.S. military), Vietnam (U.S. military), Japan (U.S. military), Hong Kong (British military and international businessmen), Singapore (British military and international businessmen), Africa Australia, Great Britain, Norway, Canada, and the United States. Spectinomycin is often used in penicillin-resistant cases, but unfortunately some spectinomycin-resistant *N. gonorrhoeae* have also been reported in Great Britain and in the United States.

Similarly, many drug-resistant *Neisseria meningitidis* strains are appearing. Over 70% of the *N. meningitidis* now isolated are resistant to the sulfa drugs, which are the antimicrobial agents of choice because they work better than penicillin against meningitis. Since the 1960s, more and more cases have had to be treated with penicillin. It is feared that penicillin drug resistance will be transferred from *N. gonorrhoeae* to *N. meningitidis*. Since penicillin is one of the few drugs that works on *N. meningitidis*, an epidemic of untreatable meningitis is expected.

Since 1972, the bacterium *Haemophilus influenzae*, which causes infant meningitis at a rate of 40,000 cases/year, has been found to be resistant to penicillin (ampicillin and methicillin) because of a penicillinase (β-lactamase) it produces. *Haemophilus influenzae* is also responsible for 30% of all middle ear infections (otitis media) in children. Since about 95% of all children have at least one ear infection before the age of 5, this bacterium accounts for a great deal of suffering. Ear infections can be very serious, because about one-third of all chronic cases lead to some permanent hearing loss. In the early 1980s, about 40% of *H. influenzae* strains checked turned out to be resistant to ampicillin.

Multiple drug resistance is being observed more frequently in bacteria, especially among the gram-negative bacteria such as *Escherichia* and *Pseudomonas*. Multiple drug resistance is due to genes that are carried on plasmids, often referred to as **R plasmids** because they confer drug resistance. The genes that make an organism resistant to a drug or antibiotic generally code for an enzyme that modifies the antimicrobial agent. Mutant **permeases** and **drug-resistant enzymes** are common strategies for creating a resistant state. Many soil microorganisms normally carry drug-resistant genes to protect themselves from antibiotics in the soil and also from antibiotics they may produce. For example, *Bacillus circulans* produces the aminoglycoside antibiotic **butirosin**, as well as an enzyme that inactivates butirosin and several other aminoglycosides such as neomycin. Plasmids specifying β-lactamase have been found in *Staphylococcus* strains isolated before the use of penicillin. Similarly, plasmids conferring resistance to tetracycline are found in several species of soil bacteria. When plasmids were first discovered, they specified resistance to one or two antibiotics. Now, however, many confer resistance to as many as 10 different antibiotics, drugs, heavy metals, and serum components.

MECHANISMS OF DRUG ACTION

Chemicals and antibiotics are characterized as to whether they affect viruses, bacteria, fungi, protozoans, or metazoans. In recent years, the modes of action of many chemicals and antibiotics have been worked out. Table 22.3 lists some of the ways in which chemicals and antibiotics affect microorganisms. As we can see, bacteria can be inhibited or killed by blocking DNA replication (naladixic acid, quinolones, novobiocin), RNA synthesis (rifampin, actinomycin D), or protein synthesis (streptomycin, chloramphenicol, tetracycline); by disrupting membranes (valinomycin, gramicidin); by poisoning electron transport chains (rotenone, antimycin A); by poisoning ATPases (rutamycin); by blocking cell wall synthesis (cycloserine, bacitracin, vancomycin, penicillin); or by interfering with a specific step in metabolism (sulfanilamide). Although many of the drugs and antibiotics used to treat infections have been claimed to be "wonder drugs," there is potential danger in their improper use. Many of the "wonder drugs" can produce very serious side effects. For example, some enter mitochondria and inhibit the bacterial-like metabolism in these organelles.

Most viral infections have been difficult to treat with drugs because most of the steps in a virus infection involve normal cellular components. Attempts to inhibit viral infections also disrupt the cell. The drugs that have been useful in treating some viral infections are analogs of nucleotides or chemicals that block DNA and RNA synthesis. These drugs inhibit viral DNA and RNA synthesis differentially because viruses replicate much more rapidly than eukaryotic cells. Drugs that inhibit virus uncoating and that block virus protein synthesis have also had limited usefulness.

Many fungal infections are particularly difficult to treat because of the toxicity of effective chemicals. Most chemicals that inhibit the fungi also inhibit the host's cells, since both are eukaryotic cells; however, a number of drugs have been found that inhibit and kill fungi differentially.

Antibacterial Drugs

Drugs That Inhibit Bacterial Cell Wall Synthesis The pencillins kill growing bacteria by binding to and inhibiting enzymes that join the peptidoglycan polysaccharides into a rigid layer of the cell wall (fig. 22.10). Penicillin binds to approximately six different proteins in the bacterial membranes, including a transpeptidase which is involved in cross-linking peptidoglycan polysaccharides. If the peptidoglycan polysaccharides are not cross-linked in growing bacteria, autolysins (self-digesting enzymes) are activated, the plasma membrane balloons out through the weakened wall, and the membrane eventually gives way. Penicillin has little effect on bacteria that are not dividing since it does not cause the breakdown of the bacterial cell wall but only interferes with its synthesis. Penicillin also has no effect on plant or animal cells, because plant cells have very different enzymes operating in the synthesis of their cellulose walls, and animal cells lack cell walls.

One type of penicillin resistance occurs when the transpeptidase connecting peptidoglycan polysaccharides is replaced by a mutated enzyme that does not bind penicillin. The mutated transpeptidase continues to join the peptidoglycan polysaccharides together because it is not efficiently inhibited by penicillin. Mutated transpeptidases are believed to be responsible for most of the low-level resistance to penicillin.

Resistance to high levels of penicillin is due to the production of enzymes that modify or break down penicillin so that it is no longer capable of inhibiting transpeptidases. Enzymes that modify penicillin are called **penicillinases**. The most frequently encountered penicillinase is known as a **β-lactamase** because it cleaves the lactam ring of the penicillin molecule. The β-lactamases are also active against cephalosporins because they posses a β-lactam ring. The hydrol-

TABLE 22.3
MODE OF ACTION OF SELECTED ANTIMICROBIAL DRUGS

Antimicrobial	Mode of action: Inhibits cell wall synthesis	Inhibits protein synthesis	Inhibits nucleic acid synthesis	Inhibits enzyme activity	Damages membrane	Poisons electron transport system
Penicillin	X					
Cephalosporin	X					
Cycloserine	X					
Ristocetin	X					
Bacitracin	X					
Vancomycin	X					
Chloramphenicol		X				
Lincomycin		X				
Erythromycin		X				
Tetracycline		X				
Streptomycin		X				
Azidodeoxythymidine (AZT)			X			
Rifampin			X			
Rifamycin B			X			
Nalidixic acid			X			
Quinolones (Ofloxacin)			X			
Trimethoprim			X			
Mitomycin			X			
Fluorocytocine			X			
Novobiocin			X			
Actinomycin D			X			
Sulfanilamide				X		
Isoniazid				X		
Nitrofuran				X		
Para-aminosalicylic acid				X		
Rutamycin				X		
Polymyxin B					X	
Valinomycin					X	
Nystatin					X	
Amphotericin B					X	
Tryocidine					X	
Gramicidin					X	
Rotenone						X
Antimycin A						X

ysis of penicillin G yields penicilloic acid, which has no antibacterial activity. These enzymes are coded for by genes (*bla*) found in plasmids.

There are a number of antimicrobial agents that block cell wall synthesis in bacteria, in addition to penicillin (fig. 22.11 and table 22.3).

Cephalosporins (fig. 22.11) are antibiotics produced by the fungus *Cephalosporium acremonium.* The cephalosporins block the final step of peptidoglycan synthesis in growing bacteria, but turn out to be bacteriostatic. These antibiotics are broad spectrum, affecting many gram-positive and gram-negative bacteria. Some of the derivatives can be taken orally, but others must be injected because they are inactivated by stomach acid as is penicillin G. The cephalosporins do not cause allergic reactions in persons sensitized to penicillin, and consequently are useful for treating patients who are hypersensitive to penicillin or who

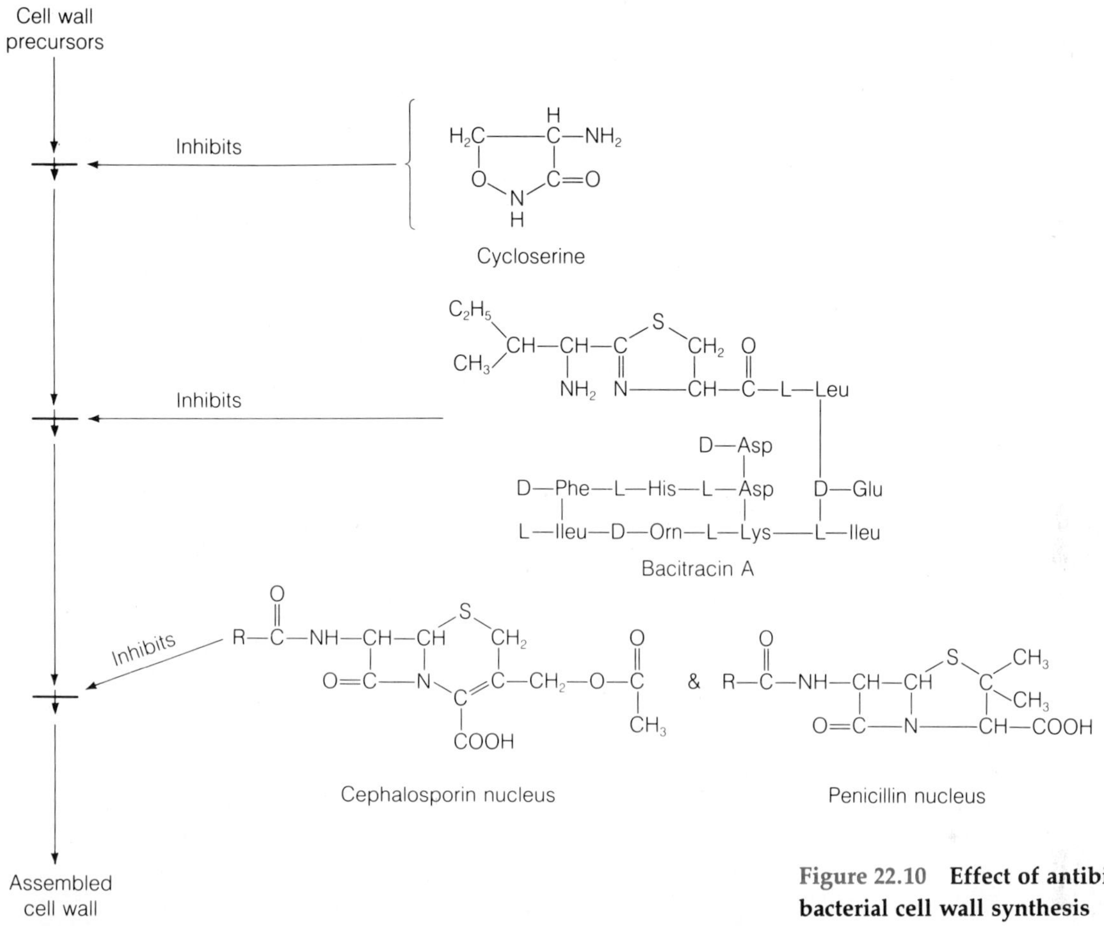

Cell wall precursors

Inhibits

Cycloserine

Inhibits

Bacitracin A

Inhibits

Cephalosporin nucleus

Penicillin nucleus

Assembled cell wall

Figure 22.10 Effect of antibiotics on bacterial cell wall synthesis

are infected by a penicillin-resistant bacterium. The cephalosporins can cause thrombophlebitis, a serum sickness, and occasional gastrointestinal disturbances.

The antibiotic **cycloserine** (oxamycin) is bactericidal, like penicillin. Cycloserine blocks the synthesis of the pentapeptide necessary for peptidoglycan synthesis. Side effects of cycloserine include confusion, coma, occasionally liver damage, folate deficiency, peripheral nervous system disorders, and malabsorption syndrome.

Bacitracin is a polypeptide antibiotic that blocks the attachment of peptidoglycan polysaccharide to the membrane. It does this by inhibiting the dephosphorylation of a membrane phospholipid to which the peptidoglycan polysaccharide must attach (fig. 22.10). Bacitracin is most effective against gram-positive bacteria, such as *Streptococcus* and *Staphylococcus,* but because it is so toxic to the kidneys it is used only topically, in creams and ointments to treat skin infections. Used internally, bacitracin may cause local pain at the injection site, gastrointestinal disturbances, and renal damage.

Antibiotics That Block Bacterial Protein Synthesis One of the best-studied groups of antibiotics includes those that block protein synthesis in bacteria (table 22.3). Most of these antibiotics are synthesized by various species of soil bacteria in the genus *Streptomyces.* Most of these antibiotics bind to some component of the bacterial ribosome and thereby inhibit a specific step in protein synthesis (fig. 22.11). Resistance to the antibiotics can arise in a number of ways: (1) an alteration of the cell's membranes can keep the antibiotic out of the cell; (2) an enzyme can alter the antibiotic so that it is unable to bind to the ribosome; or (3) the antibiotic binding site on the ribosome can

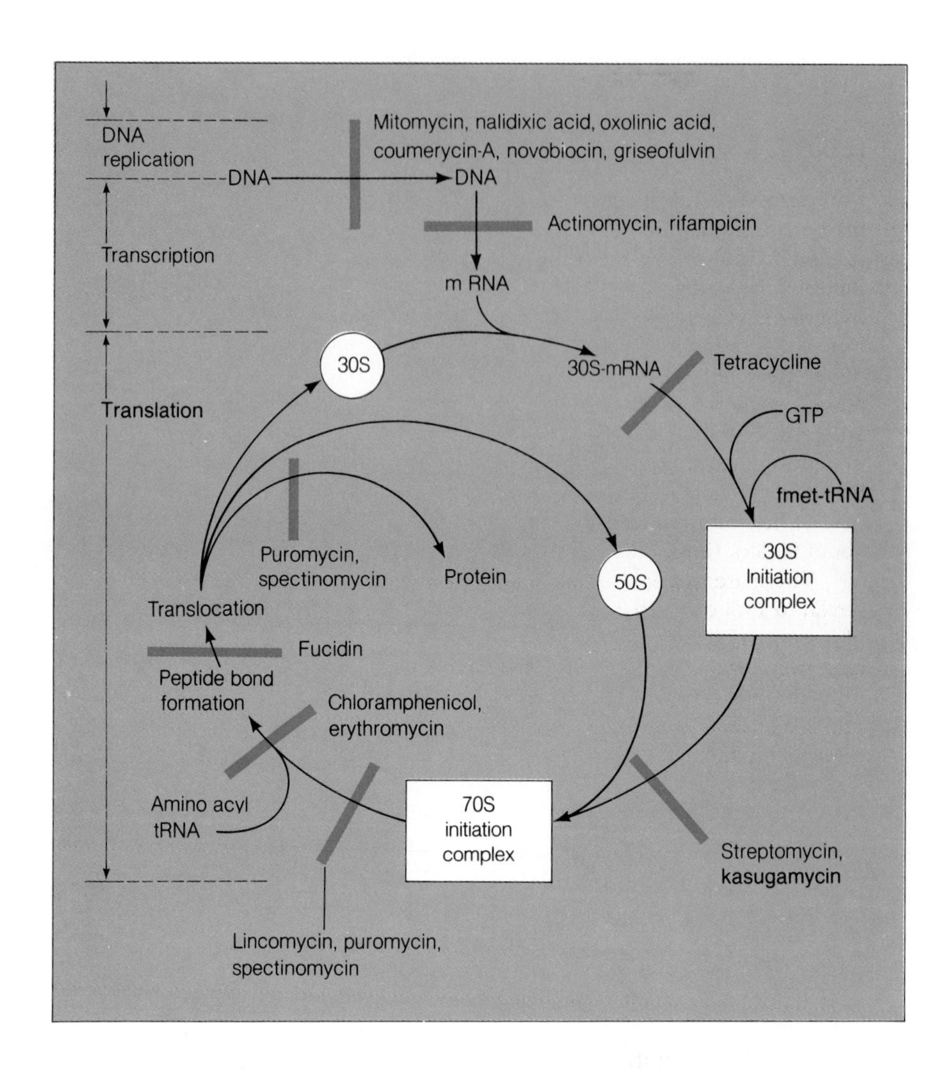

Figure 22.11 Mode of action of drugs that affect bacterial protein synthesis and nucleic acid synthesis

be altered by mutations so that the antibiotic cannot bind.

Tetracycline, chlortetracycline (aureomycin), and **oxytetracycline** (terramycin) are believed to bind to the 30S ribosomal subunit at the initiation of protein synthesis and block the attachment of the first aminoacyl-tRNA (*N*-formylmethionine-tRNA) to the 30S-mRNA complex (fig. 22.11). The tetracyclines are bacteriostatic. Resistance to these antibiotics occurs when membrane permeability is altered or when enzymes inactivate the antibiotics. The tetracyclines may cause gastrointestinal disturbances and bone lesions, as well as deformed and stained teeth in children up to 8 years of age. Newborn infants frequently have bone lesions and stained teeth when the antibiotics are given to the mother after the fourth month of pregnancy.

Aminoglycoside antibiotics contain a number of glycosidic bonds and include **streptomycin**, **neomycin**, **kanamycin**, **paromycin**, and **gentamicin**. These antibiotics bind to a number of different proteins in the 30S ribosomal subunit. For example, streptomycin binds strongly to one part of the 30S subunit, causing extensive misreading by the 70S ribosome. It also causes false initiation of protein synthesis at internal sites in the mRNA. At high concentrations, streptomycin binds to a second site on the 30S subunit and blocks all initiation complexes on the mRNA. Resistance to the antibiotic is due to mutations in one of the 30S proteins and to enzymes that modify the antibiotic so that it is unable to bind to the 30S subunit.

The aminoglycosides are bactericidal because they remain tightly attached to the bacterial ribosome long

FOCUS SYMMETREL FIGHTS INFLUENZA

Influenza, a disease that is responsible for the death of many thousands of elderly persons each year, can be prevented and treated with the synthetic drug Symmetrel (amantadine) (fig. 22.12). Studies indicate that Symmetrel is very effective in preventing influenza A infections. In fact, it appears that this drug may be more effective than the flu vaccines because it affects all influenza A type viruses regardless of their surface antigens. Symmetrel prevents the influenza virus from penetrating cells, and so blocks the spread of the virus. Some researchers claim that vaccines protect only about 50% of those immunized, while Symmetrel protects about 90% of those taking the drug.

There are minor adverse reactions to Symmetrel. Seven to 10% of the persons taking the drug complain of insomnia, jitteriness, slurred speech, difficulty in concentrating, lethargy, and dizziness. These symptoms diminish, however, after the drug is used for a number of days and disappear after the drug is discontinued.

Doctors familiar with Symmetrel recommend that influenza vaccination be coupled with the use of 200 mg/day of the drug at the first signs of the flu, in order to prevent the thousands of deaths each year. Although Symmetrel has been produced since the early 1970s, it has not been very popular with doctors.

Amantadine hydrochloride (Symmetrel)

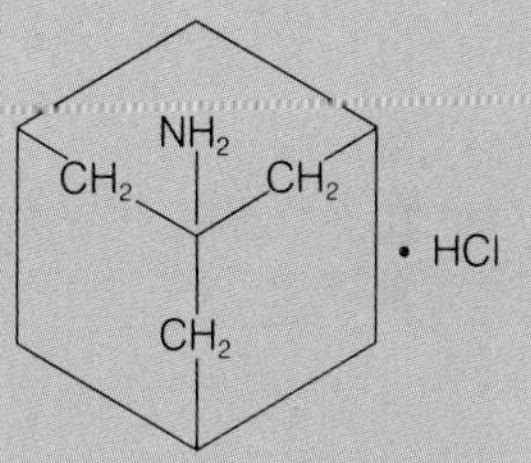

NH_2 CH_2 CH_2 CH_2 • HCl

Figure 22.12 Chemical structure of the antiviral drug symmetrel

after levels of antibiotic drop. Even one ribosome blocking an mRNA can completely block translation. Streptomycin is bactericidal for many gram-negative and gram-positive bacteria; in particular, it kills *Mycobacterium tuberculosis.* The antibiotic is ineffective against anaerobic bacteria, however, and is very poorly absorbed when taken orally and consequently must be injected. One drawback to streptomycin is that it has a cumulative effect when given for long periods of time. It can destroy nerves in the inner ear (eighth cranial nerve), affecting balance and hearing. Occasionally there is renal damage.

Chloramphenicol binds to the 50S subunit of the 70S ribosome and inhibits the enzyme called **peptidyltransferase**, which catalyzes the formation of peptide bonds between amino acids. Resistance is due to a mutated 50S protein that blocks chloramphenicol binding, to enzymes that modify chloramphenicol, or to an altered membrane permeability. A bacterial acetyltransferase that adds acetyl groups to chloramphenicol makes an organism resistant to the antibiotic. Chloramphenicol is a bacteriostatic agent against a broad spectrum of bacteria. It is the antibiotic most frequently used to treat difficult cases of typhoid fever. Chloramphenicol has been reported to cause occasional gastrointestinal disturbances, blood clumping, and **gray syndrome** in infants.*

Erythromycin binds to the 50S ribosomal subunit, apparently to the peptidyltransferase, and blocks the formation of peptide bonds (fig. 22.11). The antibiotic is bacteriostatic and is most active against gram-positive bacteria. It is frequently used as an alternative to penicillin, when penicillin-resistant organisms are involved or when the patient is allergic to penicillin. Erythromycin has been reported to cause mild damage to the liver, gastrointestinal disturbances, and transient hearing loss at high doses or when used for prolonged periods.

* Vomiting, lack of sucking response, rapid and irregular respiration, cyanosis and abdominal distension in newborn infants treated with chloramphenicol. Infants appear ashen-gray in color. Death occurs in 40% of the patients.

Drugs and Antibiotics That Inhibit Bacterial DNA Replication A number of drugs and antibiotics have been discovered that specifically block DNA replication (fig. 22.11, table 22.3).

Four drugs are known to affect bacterial DNA gyrase, the enzyme involved in unwinding DNA. They are **nalidixic acid**, **ofloxacin**, **novobiocin**, and **coumerycin-A**. Nalidixic and ofloxacin are synthetic chemicals that bind to the α-subunit of bacterial DNA gyrase. The binding of these chemicals inhibits the activity of the enzyme, which is required for bacterial DNA synthesis. Resistance to nalidixic acid occurs when the α-subunit undergoes a mutation, and also when the transport system that brings the drugs into the cell is altered. Naladixic acid is most effective against gram-negative enteric bacteria and is used to treat urinary tract infections.

Novobiocin and coumerycin-A bind to the β-subunit of DNA gyrase and inhibit the enzyme's activity. Novobiocin treatment results in side effects that include

TABLE 22.4
ANTIMICROBIALS EFFECTIVE AGAINST EUKARYOTES AND VIRUSES

Antimicrobial drug	Microorganisms affected			
	Viruses	Fungi	Protozoa	Helminths
Amantidine	X			
Azidodeoxythymidine (AZT)	X			
Vidarabine	X			
Acyclovir	X			
Interferon	X			
Idoxuridine	X			
Methisazone	X			
Cytosine arabinoside	X			
2-Deoxy-*d*-glucose	X			
Griseofulvin		X		
Nystatin		X		
Ketoconazole		X		
Myconazole		X		
Natamycin		X		
Clotrimazole		X		
Tolnaftate		X		
Haloprogin		X		
Amphotericin B		X		
Quinine			X	
Chloroquine			X	
Amodiaquine			X	
Dihydroemetine			X	
Diiodohydroxyquin			X	
Melarsoprol			X	
Nifurtimox			X	
Primaquine			X	
Quinacrine			X	
Stibogluconate			X	
Suramin			X	
Emetine			X	
Metronidazole			X	
Niclosamine				X
Pyrivinium				X
Mebendazole				X
Pyrantel pamoate				X
Piperazine citrate				X
Avermectin				X

gastrointestinal problems, jaundice, neonatal hyperbilirubinemia, and allergic reactions. Occasionally, severe blood clumping is induced by novobiocin.

Drugs That Block Bacterial RNA Synthesis Some drugs and antibiotics specifically block RNA synthesis in bacteria (table 22.3). **Rifampicin** (rifampin) is a semisynthetic antibiotic produced from the antibiotic rifamycin B. It binds to the β-subunit of bacterial RNA polymerases and inhibits the initiation of RNA synthesis (fig. 22.12). Resistant bacterial mutants arise when the β-subunit of the RNA polymerase undergoes an appropriate mutation. Rifampin is a broad spectrum antibiotic that is effective against *Mycobacterium* and can be taken by mouth.

Drugs That Block Bacterial Metabolism **Para-aminosalicylic acid** (table 22.3) is a synthetic drug that inhibits the synthesis of folic acid by competing with para-aminobenzoic acid. Para-aminosalicylic acid and the antibiotic streptomycin are frequently used to treat tuberculosis.

Isonicotinic hydrazide (or isoniazid) is a synthetic drug that resembles pyridoxin (vitamin B_6). Because it is an analog of vitamin B_6, isonicotinic hydrazide interferes with some of the reactions that require this vitamin (table 22.3). It is believed that it may also be incorporated into NAD or NADP. Isonicotinic hydrazide is bacteriostatic for *Mycobacterium tuberculosis,* but has little effect on other bacteria. This drug is generally used in conjunction with streptomycin and rifampin to treat tuberculosis in humans.

Drugs That Affect Bacterial Membranes A number of drugs and antibiotics interact with bacterial membranes and thereby kill the bacteria (table 22.3). **Polymyxin** antibiotics, produced by *Bacillus polymyxa,* are used to treat topical infections by gram-negative bacteria, in particular *Pseudomonas.* Polymyxin B destroys bacteria by interacting with the envelope and cytoplasmic membrane, disrupting transport and ionic balance.

Drugs That Are Effective Against Eukaryotes The treatment of infections caused by eukaryotes requires a completely different group of drugs and antibiotics than those used to treat bacterial infections (table 22.4). The most useful antibacterial agents are those that specifically interfere with prokaryotic metabolism and have little effect on eukaryotic organisms, except when used at toxic concentrations. The drugs and antibiotics that are useful against fungi and protozoans are those that affect these organisms more adversely than they do the patient's cells (table 22.5).

Drugs That Are Effective Against Fungi
Nystatin is an antibiotic used to treat nonsystemic fungal infections. This fungicidal agent is a **polyene** antibiotic that interacts with sterols in the cytoplasmic membrane of many fungi, disrupting the ionic balance within the cells. The leakage of Ca^{2+} into the cells and of K^+ and Mg^{2+} out of the cells drastically alters cell physiology so that RNA and protein synthesis is blocked. Some algae, protozoans, and mammalian cells are also sensitive to nystatin; however, there seems to be little toxicity associated with nystatin when it is used to treat intestinal, vaginal, or superficial candidiasis. The discovery of nystatin was reported in the 1950s by Elizabeth Hazen and Rachel Brown. Polyenes have no effect on bacteria, except for some of the mycoplasmas that contain sterols.

Another polyene antifungal agent is **amphotericin B**. The drug is believed to disrupt fungal cells in the same manner as nystatin. It is used to treat superficial and systemic candidiasis, as well as systemic aspergillosis, coccidioidomycosis, histoplasmosis, blastomycosis, and cryptococcosis. Unfortunately, it has a low therapeutic ratio and causes damage to kidney tissue and to blood-forming tissues. Symptoms

TABLE 22.5
ADVERSE REACTIONS TO ANTIMICROBIAL AGENTS

Adverse reaction	Drugs that may be involved
Hypersensitivities (skin eruptions, fever, and anaphylaxis) and immunothrombocytopenias	Penicillins, streptomycin, isoniazid, sulfonamides, stibophen, chloroquine
Kidney dysfunction	Aminoglycosides, polymyxins, amphotericin B
Damage to bone marrow	Chloramphenicol, amphotericin B
Damage to nervous system	Streptomycin, gentamycin, neomycin, kanamycin, amikacin, aminoglycosides
Liver damage	Isoniazid, rifampin, PAS, pyrazinamide
Diarrheas and enterocolitis	Erythromycin, clindamycin, tetracycline, chloramphenicol, neomycin

CLINICAL PERSPECTIVE PENICILLIN USE IN THE FACE OF ALLERGIC REACTIONS

A 77-year-old man was admitted to the hospital because of blood in the urine (hematuria) and acute or chronic inflammation of the prostate gland (prostatism). For a number of years, he had experienced decreasing force in his urinary stream and for six months he suffered intermittent pain with hematuria upon urination. Cysts in the right kidney and multiple bladder calculi were found a month before admission. There was a history of allergic reactions to oral penicillin 25 years previously.

The patient underwent an examination of the bladder and biopsy of tumors (cystolithotomy) as well as an operation to remove part of the prostate (prostatectomy). Seven days after the operation, the patient developed a fever of 40°C (febrile), had a leukocyte count of 18,200/mm^3, and had a urinary tract and blood infection. An enterococcus was isolated from urine and blood cultures. An echocardiogram indicated enlarged ventricles, a small amount of pericardial fluid, and vegetation on the aortic valve. In other words, the patient was suffering from endocarditis, an extremely serious complication of urinary tract and prostate operations.

It was found that the enterococcus was very sensitive *in vitro* to penicillin but less so to gentamicin, vancomyucin, and streptomycin. The drug of choice was penicillin. However, a skin test to determine whether the patient was still allergic to penicillin indicated that he was sensitive to penicilloyl-polylysine, but not to penicillin G. Thus, to avoid an allergic reaction, intravenous vancomycin and gentamicin were used to treat the bacterial endocarditis.

Severe chemical phlebitis and renal deterioration (elevated creatinine) occurred in response to vancomycin and gentamicin treatment. Thus, an alternative treatment involving the use of penicillin was decided upon. However, before penicillin use was undertaken, a desensitization program was perfoımed. The patient received intravenous penicillin every 30 minutes starting with a low dose (0.01 μg/ml) and building up to the full dose 100,000 μg/ml). Penicillin concentrations were maintained at a consistently high serum level to suppress any potential IgE-mediated allergic reaction. During desensitization there was no allergic response and so a penicillin and streptomycin treatment replaced the vancomycin and gentamicin treatment.

Five days after the penicillin and streptomycin treatment was initiated the patient developed malaise, high fever, and a diffuse pruritic (itching) maculopapular (spotted raised solid lesion) rash. His white blood count was 6200/mm^3 and his serum complement was normal. The reaction was diagnosed as serum sickness. Serum sickness is caused by penicillin-IgG or -IgM complexes in the blood that activate serum complement and the production of anaphylatoxins C3a and C5a. These complement proteins cause mast cells or basophils to release mediators that damage cells.

Because of the necessity to treat the endocarditis with an effective antibiotic, penicillin treatment was continued even though allergic reactions were occurring. Prednisone, a corticosteriod was given to the patient to block the development of the allergic response. Corticosteroids are effective in providing relief of serum sickness within 24 to 48 hours. Corticosteroids are not usually given because they inhibit the immune system needed to destroy the infectious agent and clear the body of circulating immune complexes. Drug induced serum sickness generally has a good prognosis and recovery is complete several days to a week after withdrawal of the offending drug.

For a more complete discussion of penicillin desensitization and its use in patients who are allergic to the antibiotic, the following reference is suggested: Rocklin, R. (1986). Penicillin reactions desensitize or discontinue? *Hospital Practice* 21 (1A): 75–89.

due to the treatment include nausea, vomiting, abdominal pain, convulsions, anemia, and cardiac arrest. Because of the toxicity of this drug, it is necessary that the patient be hospitalized for treatment.

Griseofulvin is a fungistatic antibiotic that is generally used to treat fungal skin infections, such as ringworm. It is taken orally, but must be taken for many weeks to eliminate the fungus. Griseofulvin is thought to accumulate within the fungal cell, where it disrupts components of the cytoskeleton, causing aberrant cell growth and mitosis. It is mildly toxic, occasionally causing fatigue, gastrointestinal disturbances (such as nausea, vomiting, and diarrhea), allergies, skin lesions, and photosensitive reactions.

Tolnaftate is a synthetic antifungal agent first introduced in 1965 for the treatment of athlete's foot (tinea pedis), jock itch (tinea cruris), and body ringworm (tinea corporis). Since this chemical is used topically, its toxic effects are very mild.

Drugs That Are Effective Against Protozoans

Bacteria and fungi are not the only living forms that can produce antimicrobials. The tropical plants **ipecacuanha** and **cinchona** produce emetine and quinine, respectively. Since very early times, American Indians treated malaria simply by chewing the bark of the cinchona tree, and Europeans as early as the 1800s extracted quinine from cinchona bark to treat malaria.

Chloroquine, a synthetic derivative of quinine, was produced in 1934 for treatment of malaria. It kills the malaria parasite, apparently by blocking DNA and RNA synthesis and by creating frame shift mutations in the mRNA. Chloroquine is used as a prophylactic drug to avoid malaria when traveling through infested areas. A dose is taken weekly, beginning 2 weeks before departure and continuing for 6 weeks after leaving the infested area. *Plasmodium* residing in the liver must be treated with **primaquine phosphate** (table 22.4). Chloroquine is also effective against *Entamoeba histolytica,* which causes amoebic abscesses and dysentery. Quinine causes headache, blurred vision, ringing in the ears, and blood disorders. Chloroquine causes vomiting, blurred vision, hair loss, and anemia, while primaquine phosphate may cause nausea and blood disorders. **Emetine** and **dihydroemetine** are used to treat *Entamoeba histolytica* infections.

Chemotherapeutic Agents Used to Treat Viral Infections

Most viral infections are difficult to treat with antimicrobial agents because many of the enzymes used by viruses in their replication are host enzymes. Thus, a good viricidal or viristatic agent must inhibit a process or enzyme that is associated only with the virus. So far, most of the antiviral agents that have been developed (table 22.4) block DNA or RNA synthesis differentially because the viruses replicate much more rapidly than the host cells. These drugs are generally very toxic to the patient, however.

Vidarabine is a nucleoside that resembles adenosine but has an attached sugar called arabinose (adenine arabinoside). The antibiotic is isolated from *Streptomyces antibioticus.* Vidarabine binds to the DNA polymerase used by the virus and inhibits its activity. It is most effective against herpes that causes encephalitis. In treating encephalitis, vidarabine is taken intravenously and is able to reduce the incidence of death by about 50%. Vidarabine is also used topically for treating herpes infections of the cornea.

AZT (3′azido-3′-deoxythymidine) is a synthetic nucleoside analog that shows promise in inhibiting the replication of the AIDS virus (HIV). It has significantly reduced the signs and symptoms of ARC and AIDS patients.

Idoxuridine is a synthetic antiviral nucleoside, also called 2′-deoxy-5-iodouridine, which is incorporated into some viral DNAs instead of thymidine. This substitution inhibits the replication of viral DNA and results in noninfective DNA. The idoxuridine is used topically to treat herpes simplex I infections of the eye. It is ineffective on herpes simplex II, which is the cause of most genital infections. Idoxuridine also appears to have little effect on recurrent herpes infections.

Acyclovir (Zovirax) is a synthetic antiviral agent released in 1982 to treat genital herpes, which is estimated to afflict 20 million people in the United States. Acyclovir can also be used to treat herpes infections of the eye. Oral acyclovir and intravenous injection inhibit herpes simplex replication, shorten the healing time, and reduce pain. The drug attacks the virus selectively and so is not toxic to the patient. Unfortunately, the drug does not appear to eliminate the recurrent infections of the virus.

SUMMARY

PERSPECTIVE

1. Paul Ehrlich discovered an arsenic compound called salvarsan that was effective against *Treponema pallidum,* the bacterium that causes syphilis.
2. Gerhard Domagk is credited with the discovery of prontosil, a dye combined with a sulfonamide. In animals, prontosil is cleaved into the dye and sulfanilamide.
3. Sulfonamides block the synthesis of the coenzyme folic acid and consequently inhibit those organisms that synthesize the coenzyme.
4. Sulfonamides are active against a number of bacteria.
5. Selman Waksman discovered the various species of the soil bacterium *Streptomyces* produced anti-

biotics, such as streptomycin, erythromycin, and neomycin. These antibiotics block protein synthesis in bacteria.

6. Alexander Fleming discovered that the fungus *Penicillium* produced an antibiotic he called penicillin. This antibiotic blocks cell wall synthesis in gram-positive bacteria.

FUNDAMENTALS OF CHEMOTHERAPY

1. The best antimicrobial agents have a high selective toxicity (they harm only the parasite) and a high therapeutic ratio (they can be used at high concentrations without harm to the patient, and low concentrations are effective against the pathogen).
2. Narrow-spectrum antimicrobials affect only a small group of organisms, while broad-spectrum antimicrobials affect a large number of different organisms.

DETERMINATION OF DRUG AND ANTIBIOTIC SENSITIVITIES

1. The minimum inhibitory concentration (MIC) of an antimicrobial agent can be determined from a tube dilution test. The MIC is the lowest concentration of an antimicrobial that does not allow growth.
2. The minimum killing concentration (MKC) is the lowest concentration that kills an organism.
3. In the Kirby-Bauer test, paper disks with various concentrations of antimicrobials are placed on a lawn of bacteria to determine the organism's sensitivity.
4. The zone of inhibition around antimicrobial disks is related to the *in vivo* minimum killing concentration.

DRUG RESISTANCE

1. Some strains of *Neisseria gonorrhoeae* are completely resistant to penicillin because they produce an exoenzyme called penicillinase. *Neisseria gonorrhoeae* are also becoming resistant to spectinomycin.
2. *Neisseria meningitidis* strains resistant to sulfa drugs are appearing. Sulfa drugs and penicillin are the most effective drugs against *Neisseria meningitidis*.
3. *Haemophilus influenzae* strains resistant to ampicillin are frequently being isolated.
4. Multiple drug resistance is due to genes carried on drug-resistant plasmids called R plasmids.
5. The extensive use of antibiotics all over the world has apparently selected for organisms that have R plasmids with multiple drug-resistant genes.

MECHANISMS OF DRUG ACTION

1. DNA replication in bacteria is blocked by naladixic acid and novobiocin.
2. RNA replication in bacteria is blocked by rifampin and actinomycin D.
3. Protein synthesis in bacteria is inhibited by streptomycin, chloramphenicol, and tetracycline.
4. Membrane potentials are disrupted by valinomycin, gramicidin, and polymyxins.
5. The electron transport system is blocked by rotenone and antimycin.
6. ATPase activity is inhibited by rutamycin.
7. Cell wall synthesis in bacteria is inhibited by cycloserine, bacitracin, vancomycin, and penicillin.
8. Nystatin and amphotericin B are polyene antibiotics used to kill fungi.
9. Griseofulvin is an antibiotic used extensively to treat fungal skin infections.
10. Tolnaftate is a synthetic antifungal agent used to treat athlete's foot, jock itch, and ringworm.
11. Quinine is one of the oldest antimicrobials used to treat malaria.
12. Chloroquine is a synthetic derivative of quinine that blocks DNA and RNA synthesis in *Plasmodium* (malaria parasite) and in *Entamoeba* (amoebic dysentery parasite).
13. Most drugs used to treat viral infections are analogs of nucleotides and consequently inhibit DNA and RNA replication.
14. Idoxuridine and acyclovir are used to treat herpes simplex I and II infections. Acyclovir appears to be the more effective of the two drugs.

STUDY QUESTIONS

1. What is salvarsan? Which diseases were treated with it? What problems were there with salvarsan? Who is credited with its discovery?
2. How do sulfonamides inhibit bacteria? Sulfonamides have no effect on what type of bacteria?

3. How does streptomycin inhibit bacteria? By what organism is it produced?
4. How does penicillin inhibit bacteria? What organism produces penicillin?
5. Explain the difference between an antibiotic and a man-made antimicrobial.
6. What was one of the first antibiotics used by humans?
7. What contributions did the following people make to the science of chemotherapy? Paul Ehrlich, Gerhard Domagk, Rene Dubos, Selman Waksman, Alexander Fleming.
8. Is a drug that shows selective toxicity a good one or a poor one? Explain.
9. Is a drug that has a high therapeutic ratio a good one or a poor one? Explain.
10. What is the difference between a bacteriostatic and a bactericidal agent?
11. Explain what a broad-spectrum antimicrobial is.
12. Explain how the minimum inhibitory concentration (MIC) of an antimicrobial is determined.
13. Explain how the minimum killing concentration (MKC) is determined.
14. How is it determined whether an antimicrobial is bacteriostatic or bactericidal?
15. Describe a typical Kirby-Bauer test.
16. What is one reason that penicillin-resistant *Neisseria* are increasing?
17. Which antibiotic is used to treat penicillin-resistant *Neisseria?*
18. Explain what multiple drug resistance is and what is causing it in bacteria.
19. How do the penicillins kill bacteria?
20. Describe the two ways that a bacterium can become resistant to penicillin.
21. Compare and contrast the action of cephalosporins and penicillin. Which type of organism produces cephalosporins?

SELF-TEST

1. Sulfa drugs have no effect on bacteria that have a growth factor requirement for folic acid. a. true; b. false.

For items 2–5, match the drug with the person credited with its discovery:
   a. Alexander Fleming;
   b. Gerhard Domagk;
   c. Paul Ehrlich;
   d. Selman Waksman;
   e. Rene Dubos

2. Salvarsan
3. Gramicidin
4. Penicillin
5. Prontosil
6. Streptomycin
7. A drug with a high therapeutic ratio is one that has no effect on the host at a high concentration but inhibits or destroys the infectious agent at a much lower concentration. a. true; b. false.
8. Salvarsan has a high therapeutic ratio. a. true; b. false.

For items 9–13, match the drug with the organism it is used to eliminate:
   a. quinine;
   b. emetine;
   c. azidodeoxythymidine;
   d. acyclovir;
   e. penicillin.

9. Herpes virus
10. AIDS virus
11. gram-positive bacteria
12. *Entamoeba histolytica*
13. *Plasmodium falciparum*
14. Indicate which is a broad-spectrum antibiotic:
    a. penicillin;
    b. streptomycin;
    c. erythromycin;
    d. rifampin;
    e. ampicillin.
15. If it takes 1000 times more of an antimicrobial agent to kill a pathogen than to inhibit the pathogen, the antimicrobial agent is said to be microbicidal. a. true; b. false.

For items 16–17: Eight tubes of broth contain, 0, 1, 5, 10, 20, 40, 80, and 100 μg of an antimicrobial agent. Tubes 0, 1, 5, and 10 show growth. Tubes 0, 1, 5, 10, 20, and 80 show growth when very small amounts of the culture are transferred to broth with no antimicrobial agent: a. 10, b. 20, c. 40, d. 80, e. 100.

16. What is the minimum inhibitory concentration of the antimicrobial?
17. What is the minimum killing concentration of the antimicrobial?
18. What do cycloserine, bacitracin, cephalosporin, and penicillin have in common?

a. similar chemical structures;
b. inactivated by penicillinases;
c. inhibit protein synthesis;
d. inhibit cell wall synthesis;
e. disrupt plasma membrane.

19. Multiple drug resistance genes are transferred together when bacteria conjugate if the genes are located on:
a. the bacterial chromosome;
b. transposons;
c. R plasmids;
d. a, b, and c;
e. none of the above.

20. Long-term use of antibiotics tends to select for drug-resistant microorganisms. a. true; b. false.

SUPPLEMENTAL READINGS

American Medical Association. 1971. *AMA drug evaluation 1971.* 1st ed. Chicago: American Medical Association.

Barbieri, E. J. 1987. Leukotriene antagonists: New antihistaminic drugs. American Family Physician 36(1):204–208.

Bopp, C. A., Birkness, K. A., Wachsmuth, I. K., and Barrett, T. J. 1985. In vitro antimicrobial susceptibility, plasmid analysis, and serotyping of epidemic-associated *Campylobacter jejuni. Journal of Clinical Microbiology* 21:4–7.

Cartwright, F. and Biddiss, M. 1972. *Disease and history.* New York: T. Y. Crowell.

Collier, J. R. and Kaplan, D. A. 1984. Immunotoxins. *Scientific American* 251(1):56–64.

Culliton, B. J. 1976. Penicillin-resistant gonorrhea: New strain spreading worldwide. *Science* 194:1395–1397.

Gold, D. and Corey, L. 1987. Acyclovir prophylaxis for herpes simplex infections. *Antimicrobial Agents and Chemotherapy* 31(3):361–367.

Hammond, S. M., et al. 1987. A new class of synthetic antibacterials acting on lipopolysaccharide biosynthesis. *Nature* 327(6124):730–732.

Klein, J. O. 1975. Shifts in microbial sensitivity: Implications for pediatrics. *Hospital Practice* 10:81–88.

Langone, J. 1985. AIDS: The quest for a cure. *Discover* 6:75–77.

Lyon, B. R. and Skurray, R. 1987. Antimicrobial resistance of *Staphylococcus aureus:* Genetic basis. *Microbiological Reviews* 51(1):88–134.

Maugh, T. H. 1981. A new wave of antibiotics. *Science* 214:1225–1228.

Merigan, T. C. 1985. Viral vaccines, immunoglobulins, and antiviral drugs. *Scientific American Medicine* Section 7-XXXIII.

Neu, H. C. 1984. Changing mechanisms of bacterial resistance. *The American Journal of Medicine* Section 77(1B):11–24.

Sobell, H. M. 1974. How actinomycin binds to DNA. *Scientific American* 231:82–91.

PART 4

ENVIRONMENTAL AND APPLIED MICROBIOLOGY

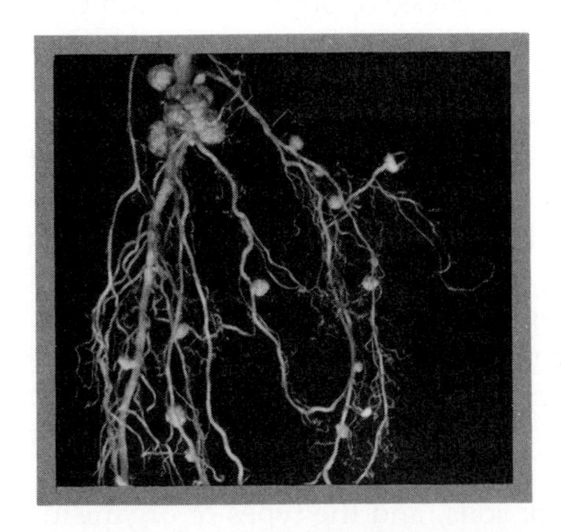

CHAPTER 23

INTRODUCTION TO ENVIRONMENTAL MICROBIOLOGY

CHAPTER PREVIEW CAN TERRESTRIAL LIFE FORMS COLONIZE MARS?

In August 1975, the Viking 1 and 2 spacecrafts began transmitting data about Mars to Earth. Scientists at NASA-Ames Research Center, Moffett Field, California, examined the data to determine whether the Martian environment could support humans and other terrestrial life forms, and to develop a model Martian community that could make Mars habitable by humans.

Scientists found that the present Martian atmosphere has no oxygen and very little water vapor, but some carbon dioxide. The water on the planet is mostly ice in the polar ice caps and in the permafrost (frozen water below the surface of the soil). The average Martian temperature is well below freezing (0°C) and stays that way for much of the Martian year.

It is also known that the Martian surface is heavily bombarded with ultraviolet radiation and whipped by winds exceeding 200 miles per hour. In essence, the present Martian environment is quite hostile to terrestrial life and in some ways resembles the conditions encountered in dry valleys in Antarctica on the earth.

Despite these harsh conditions, it was generally agreed by the researchers at NASA-Ames Research Center that certain terrestrial microorganisms could reproduce in the Martian environment. Furthermore, it was thought that these microorganisms could fill the atmosphere with oxygen, raise the ozone level high enough to filter out much of the ultraviolet irradiation, elevate the planet's temperature, and increase the availability of liquid water. It was their proposal to seed the Martian soil with a **community** of terrestrial microorganisms that could change the Martian environment within 100,000 years so that humans could colonize the planet.

Which microorganisms would be suitable for life on Mars, and where on earth could they be found? Since the Martian environment resembles in some ways the dry valleys on Antarctica, the microbial communities reproducing in these terrestrial habitats could be likely candidates for the colonization of Mars.

Any pioneer Martian community must include photosynthetic organisms, such as the cyanobacteria or the lichens, because these organisms are not entirely dependent on oxygen for survival. They produce oxygen from their photosynthesis which would accumulate in the Martian atmosphere and create an aerobic environment. The accumulation of oxygen in the atmosphere would also result in the formation of an ozone layer that would shield the surface of the planet from irradiation. As the photosynthetic organisms multiplied, the amount of organic matter would accumulate on the planet's soils, providing nutrients for other pioneer populations of microorganisms. Once the organic matter had built up, other pioneer populations could recycle these nutrients and create a steady state equilibrium for biologically important chemical elements such as nitrogen, oxygen, carbon, and sulfur.

The studies conducted by NASA scientists, using computer-simulated models and laboratory experiments, were based on the fact that microorganisms can drastically alter the environment as a result of their metabolism. In this chapter, many of the concepts that served as foundations for this NASA project will be discussed. As you read this chapter, it will become obvious that without microorganisms life on the planet Earth would be impossible.

MICROORGANISMS, AS THEY GROW AND reproduce, obtain nutrients from their environment and release metabolic by-products. In the process, microbial populations change their environment. Some of the by-products of microbial metabolism or the changes microorganisms make in their environment can be used by humans to improve their living conditions. In this chapter we will study some of the ways in which microbial activities recycle nutrients, provide foods, and purify the environment.

MICROORGANISMS AND THE RECYCLING OF NUTRIENTS

Microoranisms have played an important part in shaping terrestrial habitats. On the primitive earth, cyanobacteria, over a period of about 2 billion years (3.5 to 1.5 billion years before the present), fixed CO_2 into organic compounds and released O_2 into the atmosphere. The organic compounds and O_2 stimulated the evolution and growth of aerobic, terrestrial organisms. Part of the organic compounds that microorganisms made was sequestered under sediments and eventually became oil and natural gas. Another portion of the photosynthesized carbon remained tied up in the living mass (**biomass**) of organisms accumulating on the earth.

In the last 500 million years, plants and eukaryotic algae have replaced the prokaryotes as the major producers of organic materials and O_2. Some of the plant material has been sequestered (tied up) under sediments, where part of it has become or will eventually become coal, oil, and natural gas. Although plant material is the major source of organic carbon in soils, soil microorganisms are largely responsible for converting the plant material into CO_2 and simple organic compounds that can be utilized by a variety of heterotrophs.

The fertility of soils, and the types of organisms that are able to colonize a particular habitat, are determined partly by the physical environment of the area and partly by the microbial composition of the soils.

Nutrients in the environment are shared and recycled by microbial populations. The recycling also takes place as a consequence of the metabolism of plants and animals. Most nutrients are recycled by microorganisms and thus made available to all organisms. In the following sections, we will discuss some of the ways in which microorganisms participate in nutrient recycling in nature and the impact that they have on human populations.

Carbon Cycle

Biological molecules are made up mostly of carbon. Consequently, this element is of central importance to all living organisms. Most of the carbon in nature is found as CO_2 and in biological molecules. Organisms that convert carbon from one form to another provide various communities of organisms in an ecosystem with the form of carbon that they need to survive. When CO_2 in the atmosphere is fixed by photosynthetic organisms, O_2 is released into the atmosphere and carbon is incorporated into biological compounds. The organic compounds resulting from the CO_2 fixation are oxidized by chemoheterotrophic organisms to produce CO_2. As can be seen from figure 23.1 carbon is recycled among autotrophic and heterotrophic organisms; the autotrophs convert the gaseous CO_2 into biological molecules, while the heterotrophs return the CO_2 to the atmosphere.

Not all carbon is cycled immediately. Some of it, for example, accumulates as coal, oil, or natural gas. Based on pool sizes and utilization rates, it has been calculated that each molecule of CO_2 in the atmosphere is likely to be fixed by photosynthetic organisms every 300 years.

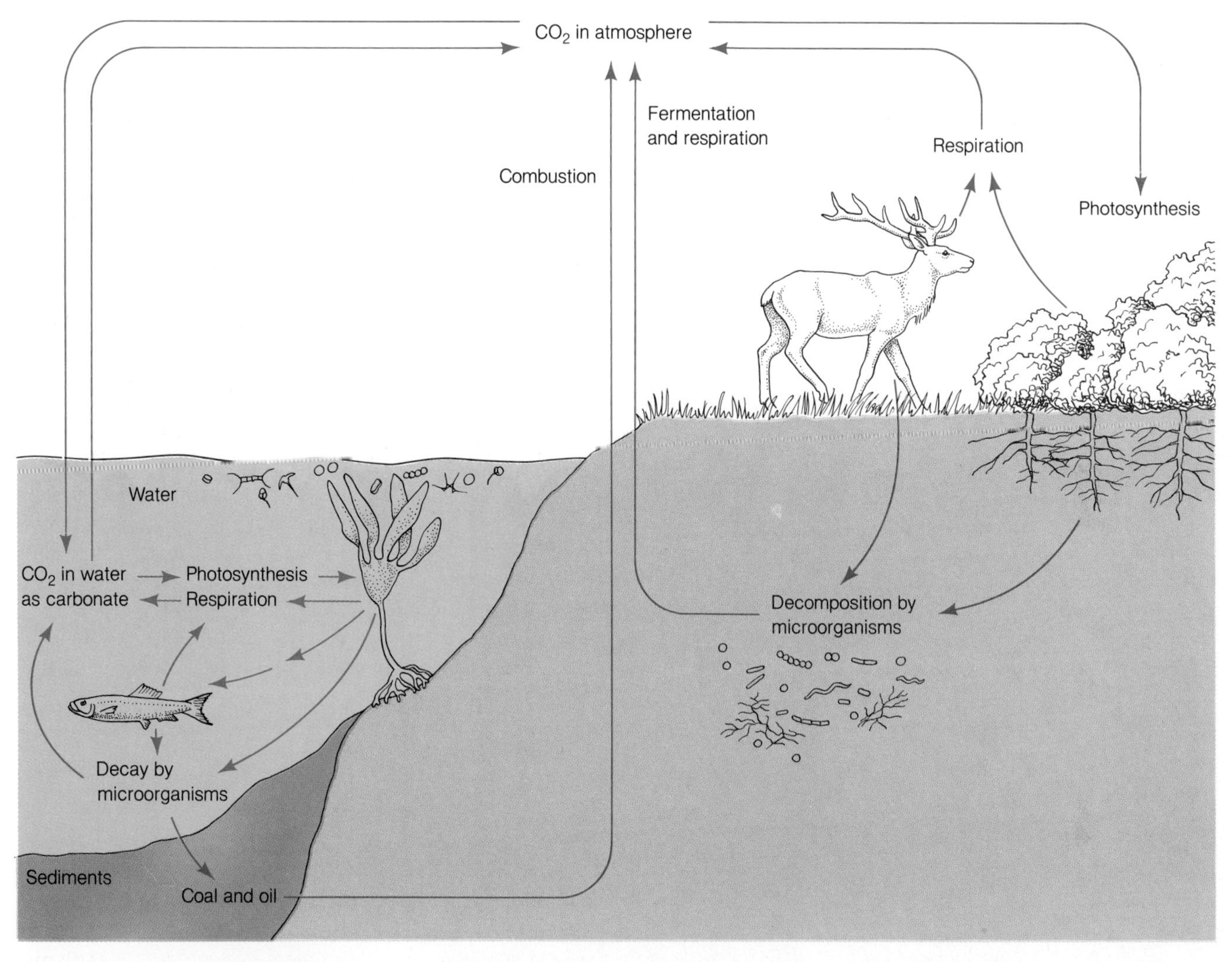

Figure 23.1 The carbon cycle. Much of the carbon available to living organisms is in the form of CO_2. The CO_2 is incorporated into organic matter through photosynthesis. The organic carbon, as part of living organisms, is oxidized to CO_2 during respiration and returned to the atmosphere. Some of the organic carbon may be in the form of fossil fuels, which are oxidized to CO_2 during combustion.

Nitrogen Cycle

Organisms require substantial amounts of nitrogen to synthesize many of their cellular constituents, including proteins and nucleic acids. Nitrogen may be found in soils, as inorganic salts of nitrogen (ammonium or nitrate), or as organic matter. In addition, a major source of nitrogen is the atmosphere, which consists of 78% nitrogen gas (N_2). This latter form of nitrogen is largely inaccessible to organisms, however, except for those that can incorporate this gas into organic molecules. The nitrogen-containing compounds serve not only as a source of nutrients for soil microorganisms but also as energy sources or electron acceptors. The conversion of nitrogenous compounds from one form to another by various communities results in what is called the **nitrogen cycle** (fig. 23.2).

Nitrogen Fixation **Nitrogen fixation** is a process that involves the conversion of atmospheric nitrogen (N_2) into organic nitrogen. This process is carried out by many different types of bacteria. The bacterial genus *Rhizobium*, when it forms a symbiotic association with

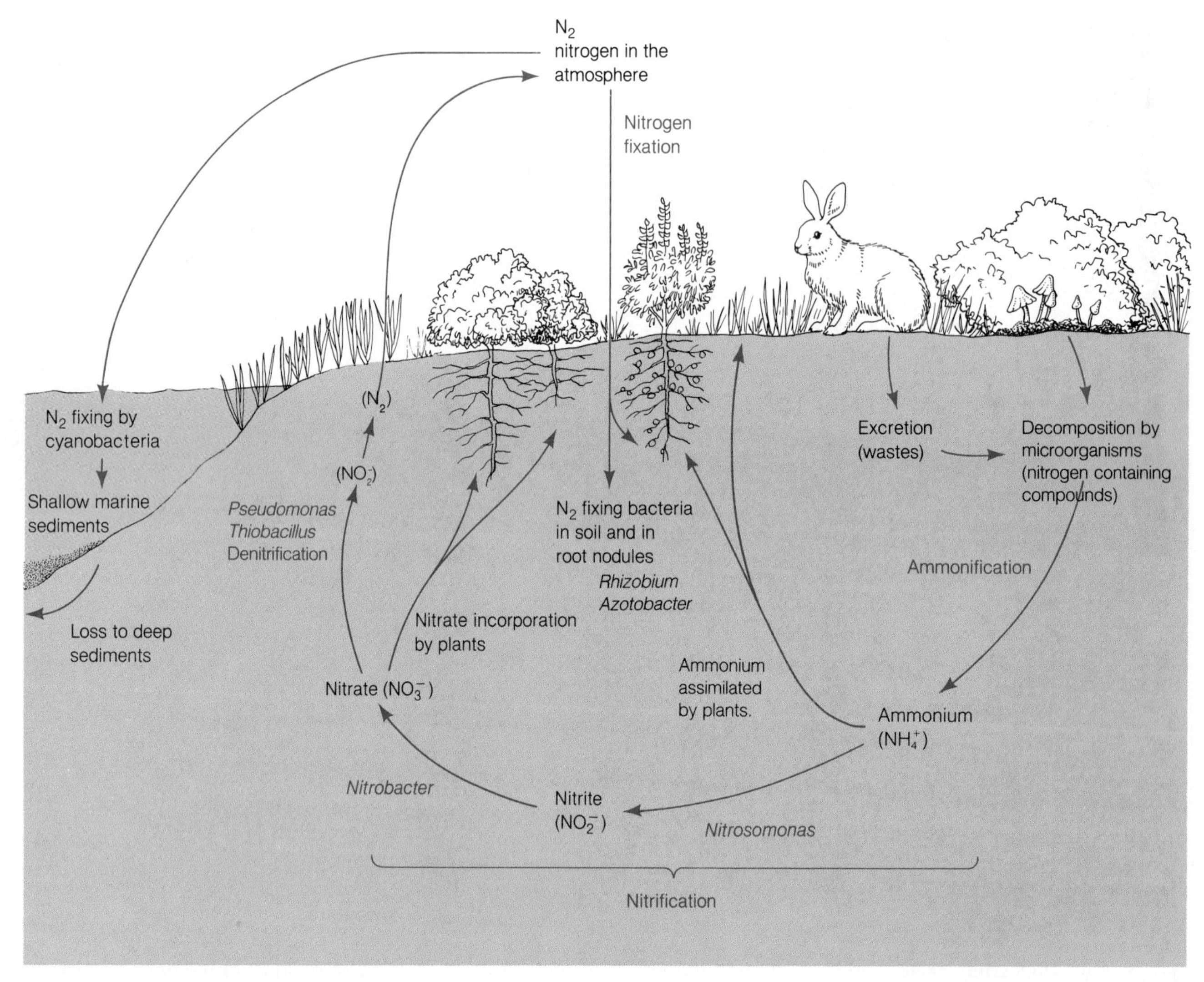

Figure 23.2 The nitrogen cycle. The nitrogen cycle is an important process that takes place in nature and insures an adequate supply of nitrogenous compounds to living organisms. The cycle is a sequence of transformations of various forms of organic and inorganic nitrogen compounds carried out by microorganisms. Nitrification involves the aerobic respiration of reduced nitrogen compounds, while denitrification occurs when nitrogen compounds are used as electron acceptors in anaerobic respirations.

leguminous plants (e.g., peas and clover), is responsible for much of the nitrogen fixed on land. For example, species of *Rhizobium* associated with alfalfa plants may fix more than 100 kg (1 kg = 2.2 pounds) of nitrogen/acre/year. In contrast, *Azotobacter,* a free-living, nitrogen-fixing bacterium, fixes less than 1 kg of nitrogen/acre/year in the same types of soils. Some actinomycetes fix nitrogen in symbiotic relationships with trees such as alder and bayberry. In aquatic environments, cyanobacteria such as *Anabaena*, *Rivularia*, and *Nostoc* fix as much as 10 kg of nitrogen/acre/year.

The soil bacterium *Rhizobium* can be isolated from tumorous growths (fig. 23.3) on the roots of leguminous plants such as alfalfa, clover, peas, lentils, and soybeans. These tumors, called **root nodules**, constitute the site of symbiotic nitrogen fixation. The bacteria induce the formation of root nodules when they invade plant root cells (fig. 23.4).

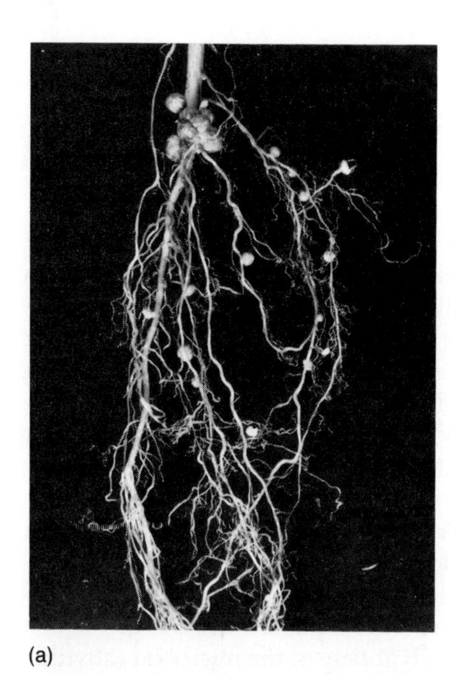

(a)

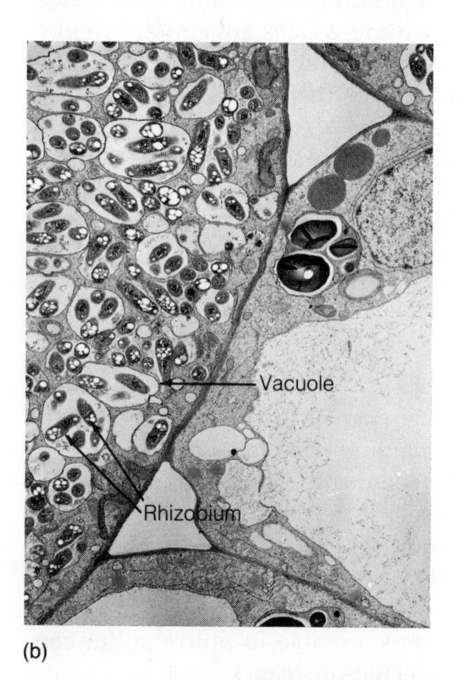

(b)

Figure 23.3 Root nodules. *(a) Rhizobium* root nodules on a leguminous plant. The nodules are the site where nitrogen fixation takes place. *(b)* Transmission electron photomicrograph of root nodule in cross section. Numerous vacuoles filled with *Rhizobium* can be seen in the plant cell on the left.

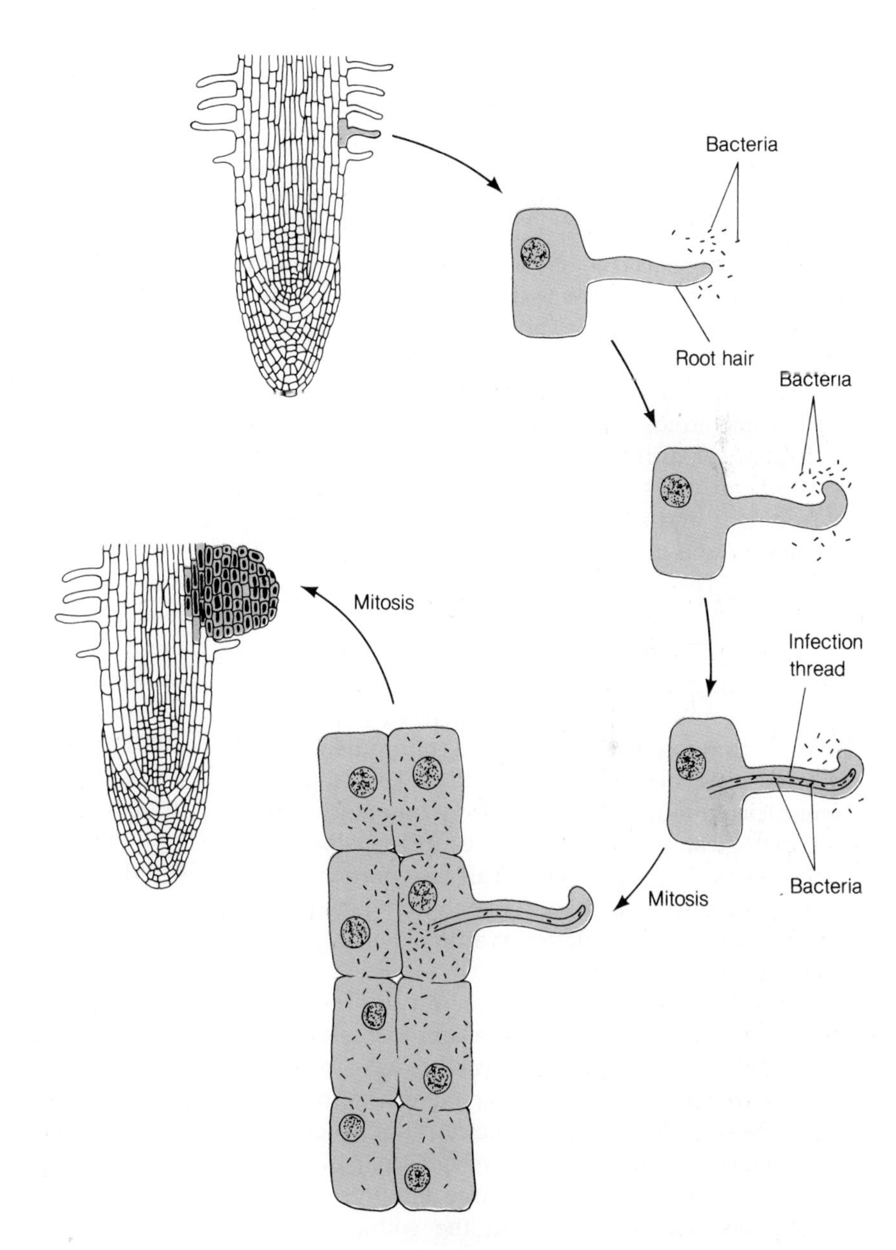

Figure 23.4 Sequence of events leading to the formation of a *Rhizobium*-induced root nodule

The development of the root nodule begins when *Rhizobium* growing around the plant roots penetrates the plant cell wall around the root hairs and resides between the cell wall and plasma membrane. The infection stimulates the root hair to develop a groove called the **infection thread**. The bacteria migrate down the infection thread to the plant cell body, where they invade the cytoplasm. The bacteria multiply in the plant cell and spread as the plant cell divides. The plant cells are stimulated to multiply uncontrollably by the infecting bacteria. It is this uncontrolled growth that is responsible for the formation of the root nodules (fig. 23.4). Within the nodules, many of the bacteria become oddly shaped resembling X's, Y's, and T's. These oddly shaped cells are called **bacteroids**.

The bacteroids convert N_2 to NH_3, a first step in the fixation of nitrogen into organic molecules. The plant cell provides the bacteroids with an anaerobic environment and all the nutrients required for their metabolism and growth.

Ammonification and Nitrification When nitrogen-containing organic molecules (amino acids, nucleotides, and urea) in excretions or cellular material are decomposed (catabolized), ammonium is generally released into the environment. This is called **ammonification** (fig. 23.2). Ammonification is a common biological process that takes place wherever microorganisms metabolize organic nitrogenous compounds. Many of the characteristic odors emanating from decaying organic matter or animal wastes result from ammonification.

A few bacteria also use ammonium as a source of energy for their metabolism. For example, the aerobic bacteria *Nitrosomonas* and *Nitrosococcus* oxidize (remove electrons from) ammonium ions, converting them to nitrite. These bacteria run the electrons obtained from the oxidation of ammonium through their electron transport system, thus making large amounts of ATP. Nitrite does not accumulate in the soil, however, because other aerobic soil bacteria, such as *Nitrobacter* and *Nitrococcus*, convert the nitrite to nitrate by removing more electrons from the nitrogen atom. These bacteria also pass the electrons through an electron transport system and make ATP. The conversion of ammonium into nitrate by microorganisms is called **nitrification**. The process of nitrification is quite common in well-aerated soils and provides a large quantity of nitrate for plant growth.

Denitrification Nitrate ions accumulate in aerobic soils because of nitrification and generally are assimilated by organisms in the terrestrial ecosystem. Bacteria such as *Pseudomonas* reduce nitrate ions to nitrite, nitrous oxide, and eventually molecular nitrogen. In essence, nitrate reduction by anaerobic bacteria depletes the soil of its nitrates. The reduction of environmental nitrate and its subsequent loss to the atmosphere is called **denitrification**.

The nitrogen cycle (fig. 23.2) clearly shows that microorganisms are intimately involved in nitrogen fixation, ammonification, nitrification, and denitrification. These processes, which involve the reuse of a finite resource, demonstrate that the waste products of one population can serve as a nutrient or energy source for another population, which in turn provides nutrients for yet another population. Together, these organisms recycle nitrogen atoms among the various populations of the soil or aquatic community.

Farmers take advantage of the microbial activities that occur in soils. For example, plowing fields before fertilization and planting creates an aerobic environment that promotes the growth of ammonifiers and nitrifiers and discourages the growth of denitrifiers. Plowing helps to achieve a maximum amount of ammonium and nitrate in the soil. Inexpensive ammonium-containing fertilizers such as $(NH_4)_2SO_4$ can be applied to soils because the nitrifying bacteria in the soil will convert the ammonium to nitrate, thus promoting lush plant growth.

Sulfur Cycle

Sulfur is an essential element that is required for the synthesis of certain amino acids (methionine and cysteine), coenzymes (coenzyme A), and sulfate-containing polysaccharides. Sulfur is also used as a source of electrons (energy) and as an electron acceptor by various microorganisms. Sulfate is believed to be the preferred form of sulfur for assimilation, even though many microorganisms are able to utilize sulfur-containing organic molecules instead.

The conversion of sulfur from one form to another by microorganisms is called the **sulfur cycle** (fig. 23.5). The microbial community that participates in the recycling of sulfur is made up of various populations.

Hydrogen sulfide, sodium thiosulfate, and hydrogen gas are used by photosynthetic bacteria and nonphotosynthetic bacteria, such as *Thiothrix* and *Beggiatoa* for converting CO_2 to glucose. Both groups of

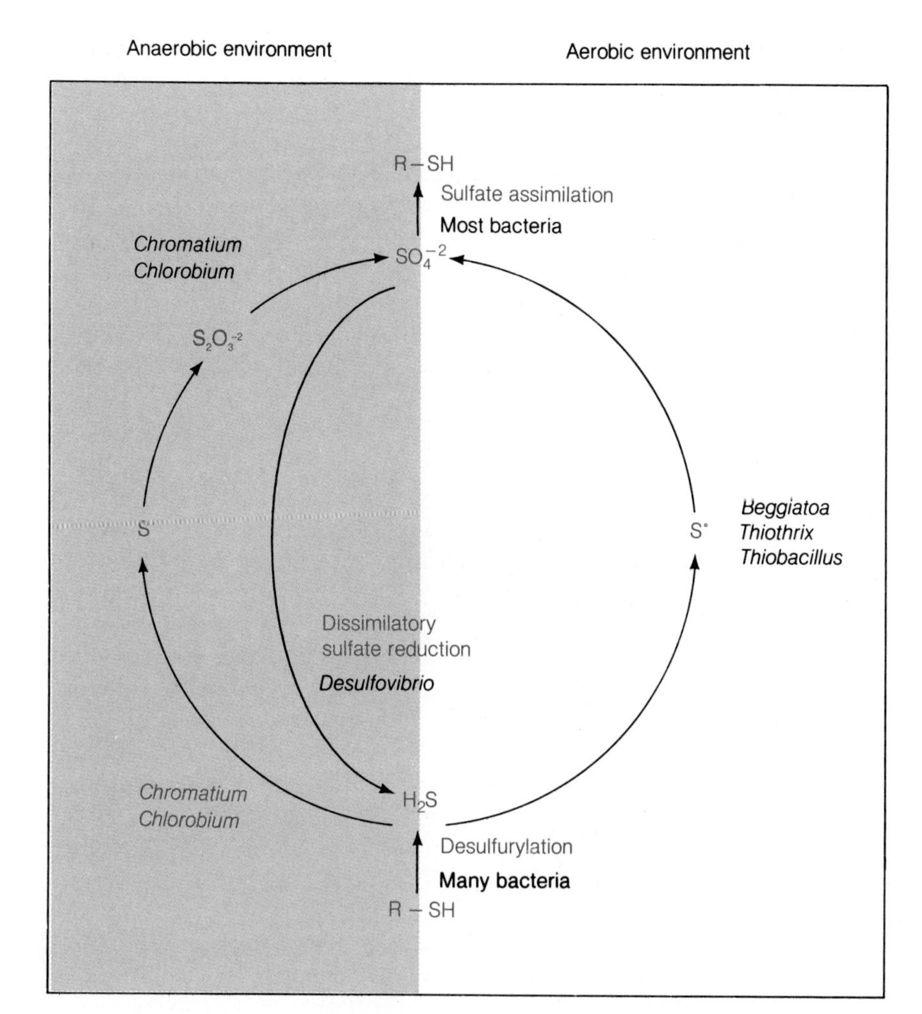

Figure 23.5 The sulfur cycle. The sulfur cycle is an important process that takes place in nature. It consists of a sequence of transformations of sulfur-containing compounds achieved by the metabolic activities of microorganisms.

microorganisms release sulfate as a waste product of their metabolism. The sulfate does not accumulate in the soil, but instead is assimilated by a variety of soil inhabitants.

One of the most interesting soil bacteria, which is active in the sulfur cycle as well as in the nitrogen cycle, is *Thiobacillus*. Under aerobic conditions, this organism can remove electrons from sulfur (S) and use them to make ATP, producing sulfuric acid as a by-product. Under anaerobic conditions, however these bacteria reduce nitrate to nitrite. Some species obtain their electrons from organic compounds, others from sulfur compounds.

Some chemoheterotrophic bacteria, such as *Proteus* and *Salmonella*, remove the sulfide from the amino acids methionine and cysteine and release H_2S. Another bacterium, called *Desulfovibrio*, in the absence of oxygen and in the presence of sulfate ions will carry out anaerobic respiration to produce hydrogen sulfide. These bacteria use sulfate ions as the terminal electron acceptor and consequently reduce the sulfate to hydrogen sulfide. The hydrogen sulfide may then react with iron in the soil to form ferrous sulfide, which makes anaerobic soils black. Since extensive ferrous sulfide forms in marshes and stagnant ponds, the soil in these areas is usually black. *Desulfovibrio* causes problems because its metabolic products can corrode iron pipes and foundations in sulfate-containing soils. Under anaerobic conditions, this bacterium produces H_2S, which then reacts with the $Fe(OH)_3$ layer on iron,

FOCUS UNDERWATER MARINE GARDENS

The bottom of the ocean normally is very cold (<5°C) and dark where no light penetrates and hence no photosynthesis takes place. Most organisms that live at these depths must depend entirely on food from the ocean's surface. Thus, many sea bottom communities, like surface communities, are dependent on photosynthetic organisms as their source of food. In 1977, an oceanographic expedition, headed by Dr. John B. Corliss of Oregon State University, discovered a new type of submarine ecosystem based, not on light, but on chemical reactions carried out by chemolithotrophic bacteria.

The expedition—aboard the submarine *Alvin* from the Woods Hole Oceanographic Institute—discovered this new ecosystem at a place called the Galapagos Rift, located between two adjacent ocean bottom plates (fig. 23.6). At the rift, water seeps into fissures at the ocean's bottom and into the earth's mantle, where it becomes heated to temperatures above 350°C (662°F). The heated water also picks up inorganic materials like sulfur and hydrogen sulfide from the earth's crust. These inorganic materials are used by bacteria such as *Thermodiscus* (S) and *Thiobacillus* (H_2S) as a source of energy. The bacteria serve as the primary producers for the ecosystem. The biomass created by these bacteria is the nutrient source for a complex food chain that includes gigantic tubeworms and many marine animals.

This type of chemosynthetic community was found not to be unique to the Galapagos Rift; apparently it is commonplace in mid-ocean ridges where molten rock from the earth's mantle comes in contact with ocean waters, resulting in underwater oases where the primary producers are bacteria. It is now thought that mid-oceanic rifts not only nourish a chemosynthetic ecosystem, but also may be largely responsible for the chemical composition of seawater.

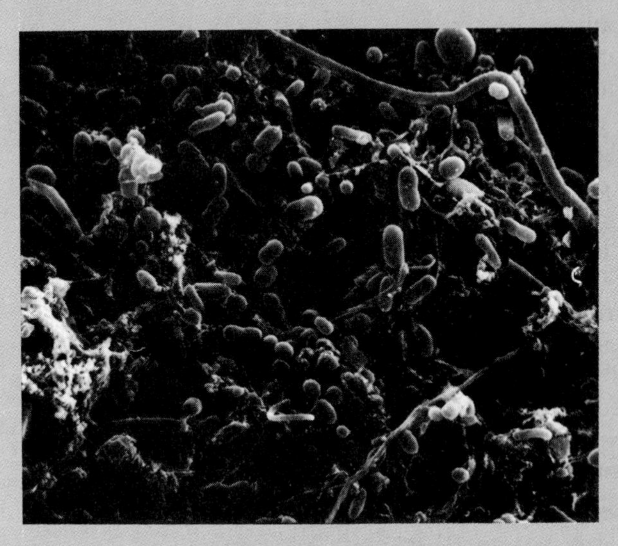

Figure 23.6 **The Galapagos Rift community.** Scanning electron photomicrographs of bacteria that serve as the primary producers for the Galapagos Rift ecosystem.

converting it to FeS. The black FeS flakes off the iron, a process that leads to the disintegration of the pipe or foundation.

MICROORGANISMS AND ANIMAL NUTRITION

Microorganisms have established relationships not only with plants but also with animals. In some cases, both organisms evolved into new life forms that exist in harmony with one another. Examples of animals that harbor essential symbionts are the termites and the ruminants. Termites are unable to exist without their gut protozoans who are, in turn, hosts for bacteria. Similarly, ruminants cannot exist without the microorganisms that are found in their rumen, and many of these microorganisms cannot live outside the rumen.

Intestinal bacteria in animals and humans provide nutrients that promote good health, but both the bacteria and the animals can usually exist without each other. In this section we will consider a few examples of how microorganisms foster good health or induce disease in animals.

Intestinal Bacteria

The microbial flora in the gastrointestinal tract of many animals provides the animal with acids, polysaccharides, proteins, and vitamins that it requires. Such microorganisms also aid in digestion and protect the intestines from many potentially pathogenic microorganisms. Animals raised in the absence of microorganisms develop abnormally. For instance, the reticuloendothelial organs that provide cells for the immune system do not develop, probably because they are not stimulated by the great number of microbial antigens. In gnotobiotic rabbits, mice, and guinea pigs, the cecum is greatly enlarged (fig. 23.7). This enlargement may be a response to the lack of bacterial proliferation and lack of nutrients produced by the bacteria; by increasing the organ's size, the absorption of nutrients may be improved. In addition, the entire intestine has a thin wall and does not respond normally to mechanical stimuli. Vitamin K must be supplied to gnotobiotic animals, because miroorganisms such as *Escherichia coli,* which synthesize much of the vitamin K that animals require, are not present.

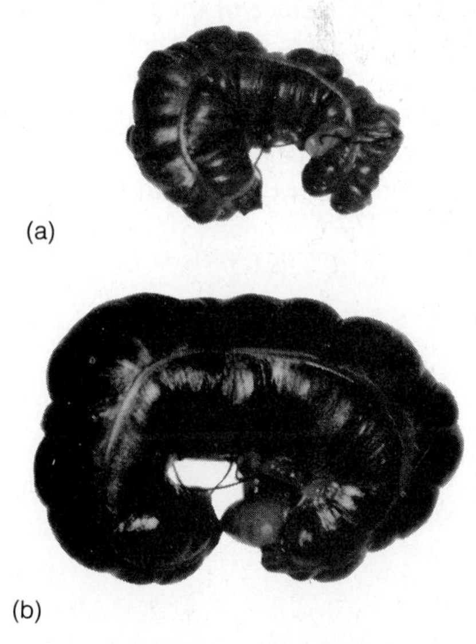

Figure 23.7 Enlarged cecum of germ-free animal. *(a)* Cecum from a normal rodent. *(b)* Cecum from a germ-free rodent.

Ruminants

The ruminant intestine includes a rumen in which large numbers of microorganisms (bacteria, protozoans, and fungi) break down cellulose and hemicellulose. These organisms convert plant material into acids that can be absorbed by the animal:

$$\text{Sugar} \rightarrow \text{acetate} + \text{propionate} + \text{butyrate} + CO_2 + H_2 + CH_4$$

The microorganisms in the rumen also produce vitamins, proteins, and sugars that are utilized by the ruminant when the organisms are digested in the stomach. The partially digested organisms are moved to the small intestine and large intestine, where they are further digested and sugars, amino acids, short peptides, fats, vitamins, and other nutrients are absorbed into the blood.

MICROORGANISMS AND SEWAGE TREATMENT

Sewage consists of human and domestic wastes as well as industrial and agricultural discharges. These liquid wastes, if not treated properly, find their way into natural bodies of water and cause very serious detrimental changes in the ecosystem. Rivers, lakes and artificial reservoirs are irreplaceable, valuable resources that must be saved at all costs and protected from pollution. They provide the water needed for drinking, bathing, laundering, irrigating, and manufacturing.

The amount of organic material in sewage is important because it increases the rate of growth of algae as well as other microorganisms and the death of animals in lakes and estuaries receiving the polluted waters. The degeneration of lakes and estuaries due to increased amounts of organic matter and loss of O_2 is referred to as **eutrophication**.

Primary Treatment

Primary treatment of sewage involves the removal of solid wastes from raw sewage. As raw sewage (influent) enters the sewage treatment plant, it is passed through a series of screens that remove large objects such as marbles, paper, rocks, dentures, and glasses (fig. 23.8). The screened water is then diverted into settling tanks or basins, where the suspended solids are allowed to settle to the bottom. In many cases, skimmer arms cruise the surface of the **settling tanks** to remove any floating objects such as feathers, grease, cellophane, or wood. The settled wastes are collected and digested or composted. Primary treatment removes much of the organic material in the water (30–40%), but a significant amount remains.

Secondary Treatment

In **secondary treatment** of sewage, much of the remaining organic matter in the waste is oxidized by microorganisms before the wastewater is discharged from the plant. **Trickling filter** and **activated sludge** (fig. 23.8) systems are common installations used to expedite the oxidation of organic wastes. In both of these systems, the liquid sewage is well aerated to speed up the oxidation of organic matter by aerobic organisms. During this process, the organic matter is converted to inorganic molecules such as ammonium, nitrates, and sulfates, which can be sold as "organic" fertilizers. If the secondary treatment is the final purification process, the water is drained off, chlorinated, and returned to the environment. After secondary treatment, 85–90% of the organic matter has been removed from the sewage.

Tertiary Treatment

Although effluent waters from secondary treatment have little organic material, they contain minerals that

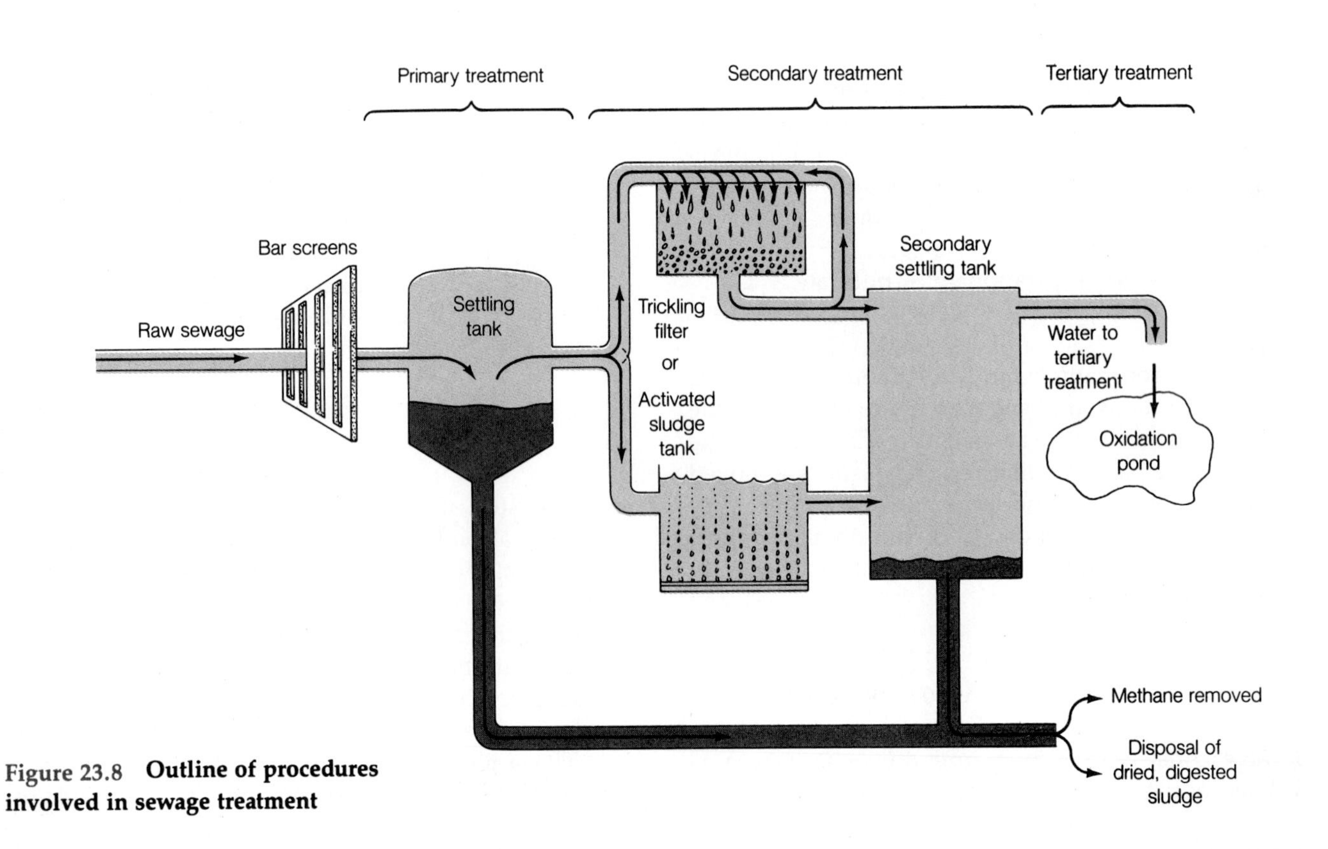

Figure 23.8 Outline of procedures involved in sewage treatment

are serious pollutants. When minerals such as NH_4^+, NO_3^-, and PO_4^{-3} are introduced into receiving waters, they promote the growth of autotrophic microorganisms (e.g., cyanobacteria), leading to eutrophication. To reduce the rate of eutrophication in receiving waters, effluents from secondary treatment are subjected to a third stage of purification. The **tertiary treatment** removes much of the ammonium, nitrate, and phosphate from the water, as well as toxic organic substances, such as pesticides and herbicides, and any nonbiodegradable organic matter.

Many communities process their secondary treatment waste in **oxidation ponds** or shallow lagoons (3–5 feet deep), in which microbial activities remove much of the organic and inorganic materials remaining after secondary treatment.

MICROORGANISMS AND FOOD SPOILAGE

Spoilage is any disagreeable change or departure from the normal state in the food that can be detected with the senses of smell, touch, taste, or vision. The changes that occur depend on the food composition and the microorganisms that are present (table 23.1) and are the result of chemical reactions in the food or metabolic activities of the microorganisms.

Not all foods are equally susceptible to spoilage, and various physical, chemical, and biological factors play a role in spoilage. For example, lipolytic microorganisms, such as pseudomonads or certain fungi, growing in sweet butter (mostly fats) cause a form of spoilage called **rancidity**. This type of spoilage results from the microbial breakdown of butterfats to produce glycerol and fatty acids, which are responsible for the smell and taste of rancid butter. Similarly, proteolytic bacteria growing on hamburger meat (mostly protein) will break down the muscle protein and release products such as putrescine and cadaverine, which give the rotting meat its characteristic odor. This type of spoilage, called **putrescence**, results from the incomplete microbial metabolism of amino acids.

Another form of food spoilage is **sour spoilage**.

TABLE 23.1
MICROORGANISMS INVOLVED IN FOOD SPOILAGE

Foods affected	Type of spoilage	Microorganisms involved	Microbial group
Bacon, sausage	Greening	*Lactobacillus*	Gram + rod
		Leuconostoc	Gram + coccus
Bread	Ropiness (stringy)	*Bacillus subtilis*	Gram + rod
	Moldy	*Aspergillus, Rhizopus*	Fungi
Canned goods	Moldy	*Byssochlamys*	Yeast
Eggs	Rotting, greening	*Pseudomonas, Proteus*	Gram − rods
Fresh fish	Rotting (fishy)	*Pseudomonas, Serratia*	Gram − rods
		Flavobacterium	Gram − rod
		Micrococcus	Gram + coccus
	Discoloration	*Micrococcus*	Gram + coccus
		Pseudomonas, Serratia	Gram − rods
		Yeasts and molds	Fungi
Fresh meats	Slimy	*Flavobacterium, Pseudomonas*	Gram − rods
		Yeasts	Fungi
Milk	Souring	*Lactobacillus*	Gram + rod
		Streptococcus	Gram + coccus
	Ropiness	*Alcaligenes*	Gram − rod
	Gassy	*Bacillus, Clostridium*	Gram + rods
Preserves	Moldy	*Aspergillus*	Fungi
		Penicillium, yeasts	

For example, if milk is allowed to stand undisturbed, either in the refrigerator or outside, eventually it will sour. The sour spoilage results from the growth of bacteria in the milk. These bacteria obtain their energy and carbon by fermenting the milk sugar (lactose) to lactic acid and other acids, which give milk an unpalatable, sour taste.

MICROORGANISMS AND FOOD PRODUCTION

Microorganisms, as they grow and reproduce, utilize nutrients from their environment and release wastes. These metabolic activities in foodstuffs can result in very desirable changes in texture, aroma, and taste. The resulting product is referred to as a **fermented food**. Foods such as cheeses, breads, yogurt, buttermilk, and sauerkraut are made by the intervention of microorganisms such as bacteria and yeasts (table 23.2).

People throughout the ages have learned to capitalize on the fermentation processes of microorganisms to preserve foods. For example, almost all cultures have allowed a desirable fermentation of milk, called **lactic acid fermentation**, to preserve the easily spoiled fresh milk protein as cheeses and fermented milk beverages. Lactic acid released by bacteria from the metabolism of lactose (milk sugar) acidifies food and thus makes it more resistant to microbial spoilage because the acid inhibits the growth of many bacteria. Additionally, some ethnic groups, which have lactase (an enzyme that degrades lactose) deficiencies, have utilized fermentation of milk products to compensate for their deficiency because bacteria that ferment lactose have enzymes necessary to break-down the sugar.

TABLE 23.2
FOODS PREPARED USING MICROORGANISMS

Fermented food	Starting material(s)	Microorganisms involved in fermentation	Microbial group
Pickles	Cucumbers	*Lactobacillus* sp., *Pediococcus* sp.	Gram + rod, coccus
Acidophilus milk	Pasteurized milk	*Lactobacillus acidophilus*	Gram + rod
Bread	Wheat flour, sugars	*Saccharomyces cerevisiae*	Yeast
Bread, sourdough	Wheat flours	*Saccharomyces exiguus* *Lactobacillus sanfrancisco*	Yeast Gram + rod
Buttermilk	Pasteurized milk	*Lactobacillus bulgaricus*	Gram + rod
Koumiss	Mare's milk	*Lactobacillus bulgaricus* *Torula, mycoderma*	Gram + rod Yeasts
Kefir	Fresh whole milk	*Streptococcus* sp., *Lactobacillus* sp., *Candida, Saccharomyces*	Gram + coccus, rod Yeasts
Yogurt	Pasteurized milk	*Lactobacillus bulgaricus,* *Streptococcus thermophilus*	Gram + rod Gram + coccus
Olives	Fresh olives	*Leuconostoc mesenteroides,* *Lactobacillus plantarum*	Gram + coccus Gram + rod
Poi	Taro roots	*Lactobacillus* spp.	Gram + rod
Shoyu (soy sauce)	Rice, soy beans, rice	*Lactobacillus delbrüeckii,* *Aspergillus oryzae,* *Saccharomyces rouxii*	Gram + rod Fungi Yeast
Cheeses	Curdled milk	Various bacteria & fungi	
Beer	Grains	*Saccharomyces carlsbergii*	Yeast
Wine	Grape juice	*Saccharomyces cerevisiae,* *S. champagnii*	Yeast Yeast
Cured hams and sausage			
Sausages	Pork and beef	*Pediococcus cerevisiae*	Gram + rods
Cured hams	Pork	*Aspergillus, Penicillium*	Fungi

The effects of microorganisms on foods are evaluated subjectively by humans. For example, fermentation of cabbage by lactic acid bacteria to produce sauerkraut may be considered a definite improvement of the cabbage by some individuals while others may consider the product to be spoiled cabbage. This illustrates the point that microorganisms do not set out to make a food more palatable or to spoil it. Instead, the changes noted and evaluated by humans are simply the result of the growth and reproduction of microorganisms.

Buttermilk

Buttermilk results from the souring of low-fat milk by lactic acid bacteria. The texture of buttermilk results from a broken curd and the aroma and the flavor are due to **diacetyl, acetaldehyde**, and other metabolic products released by the fermenting bacteria. Modern cultured buttermilk is made commercially utilizing a lactic acid culture consisting of *Streptococcus cremoris, Streptococcus diacetylactis,* and *Leuconostoc cremoris*. The nature of the starter culture varies among different manufacturers, and some use *Lactobacillus bulgaricus* to make Bulgarian buttermilk. The production of acid and the formation of the curd are due to *S. cremoris,* while the aroma and the flavor are due to the metabolism and multiplication of the other two bacteria in the curdled milk.

To make buttermilk, skim or lowfat milk that has been homogenized and pasteurized is inoculated with 1% starter culture and allowed to ferment at 18–22°C for about 14 hours. After the fermentation, the resulting product is shaken vigorously to break up the curd, cooled to 4°C, and packaged in milk containers. The final product is a homogeneous, thickened liquid that is slightly effervescent (due to carbon dioxide production), with an acid flavor and a buttery aroma.

Yogurt

Yogurt is a fermented milk with a pudding-like consistency that results from the action of *Streptococcus thermophilus* and *Lactobacillus bulgaricus*. Traditionally, the milk was heated for several hours to evaporate some of the water and increase the proportion of milk solids to liquid. After the evaporation step, the milk was cooled to about 108°F (40–42°C) and inoculated with a previous batch of yogurt. After an overnight incubation in a warm place, the product was cooled. The bacteria produce diacetyl, acetaldehyde, and a variety of other metabolic products that impart the characteristic flavor and aroma of yogurt. Today, the starting milk is thickened by adding powdered milk to pasteurized milk rather than by evaporating away the liquid. Thickeners such as gelatin or carrageenan may also be added.

Cheese

A simple cheese is made by briefly scalding milk so as to kill most milk spoilage microorganisms and then adding a starter culture to the cooled milk. The starter culture usually consists of lactic acid bacteria such as *Streptococcus lactis, Streptococcus cremoris, Leuconostoc citrovorum,* and *Leuconostoc dextranicum*. The seeded milk is allowed to ferment at 18°C for approximately 24 hours so that a curd is produced. In some cases, the curd is produced by adding proteases such as **rennin**. The liquid, or whey, can be removed by draining the curd in cheesecloth. The curd is then salted to inhibit further microbial growth. *Leuconostoc* releases diacetyl, a compound synthesized from citric acid, which is responsible for the aroma and flavor associated with cheeses. Other bacterial products also contribute to the flavors and aromas of various cheeses. Cheeses may be incubated for long periods to allow the maturation of the curd and the growth of microorganisms to add flavors. Some of the steps involved in the production of cheeses are illustrated in figure 23.9.

Soft cheeses contain 50–80% water. They are consumed soon after they are made. Cottage cheese, an unripened soft cheese, is very moist, not heavily salted, and not very acidic since it is made using rennin. Because of its low acid content, it spoils relatively rapidly even when refrigerated. On the other hand, **Camembert** is a ripened soft cheese that does not spoil rapidly. The fungus *Penicillium camemberti* is grown on the surface of the curd to introduce the characteristic flavor and aroma of this cheese.

Hard cheeses have a water content of less than 40% and are generally ripened with bacteria or fungi. For example, Swiss cheese is made with *Streptococcus lactis, Streptococcus thermophilus, Propionibacterium shermanii, Propionibacterium freudenreich, Lactobacillus helveticus,* and *Lactobacillus bulgaricus*. The propionic acid bacteria give Swiss cheese its characteristic flavor and produce CO_2 pockets (the holes) in the curd.

Bread

Bread is one of the earliest processed foods made by humans. In the British Museum there are samples of bread made by Egyptians before 2000 B.C. Bread is made by mixing flour, water, salt (sometimes sugar),

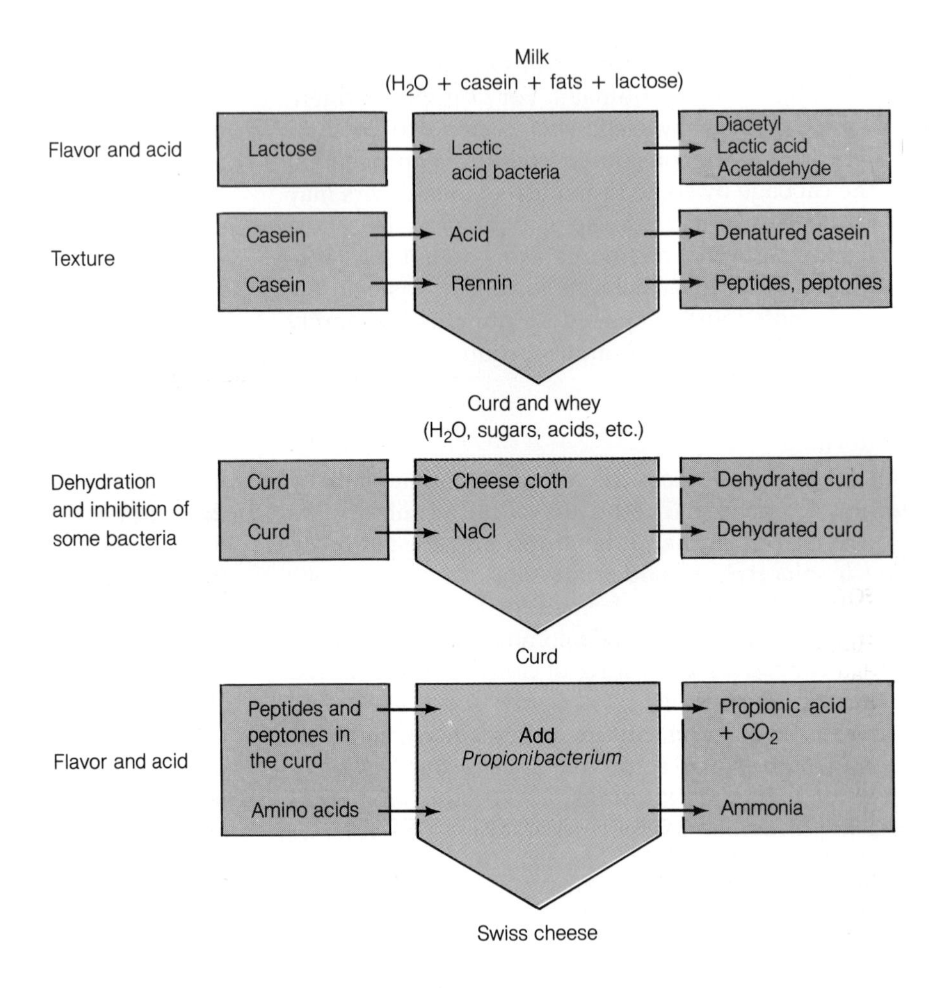

Figure 23.9 Steps involved in the preparation of hard cheese. Outline of the preparation of Swiss cheese.

and yeast to make dough. The yeast most commonly used is *Saccharomyces cerevisiae.* It ferments the carbohydrates (sugars) in the dough and produces carbon dioxide and ethanol. With refined flour, sugar must also be added, because the yeasts are unable to synthesize amylases that break down the grain starch into fermentable carbohydrates. The carbon dioxide causes the bread to rise (leaven), and gives the final product the light, porous texture characteristic of leavened bread. The bread is then baked and the ethanol evaporates. Although the yeasts are killed during the baking, metabolic products remain in the bread and add flavor and vitamins to it.

Sourdough breads are made in much the same way as other leavened breads, except that the microorganisms contributing to the flavor also include lactic acid bacteria (*Lactobacillus sanfrancisco*). These bacteria produce lactic acid, acetic acid, ethanol, and carbon dioxide from their energy-producing chemical activities. The carbon dioxide causes the dough to rise, while the acids (in particular acetic acid) give the sourdough bread its characteristic sour and tart flavor.

Sauerkraut (Acid Cabbage)

The making of sauerkraut in northern Europe dates back to the 1500s. Sauerkraut results from the fermentation of salted cabbage by lactic acid bacteria. The cabbage, which contains about 2% carbohydrate (sugar), is usually shredded and mixed with 2–3% salt. The salted cabbage is then packed firmly into vats to eliminate any air pockets and allowed to ferment for about three weeks. During the fermentation process, at least three successive species of lactic acid bacteria, which are resident flora on the cabbage leaves, participate in the production of sauerkraut. The salt serves as a selective agent and promotes preferentially the growth of these bacteria. The fermentation is carried out at room temperature.

Leuconostoc mesenteroides is the first population of microorganisms to appear in high numbers (fig. 23.10).

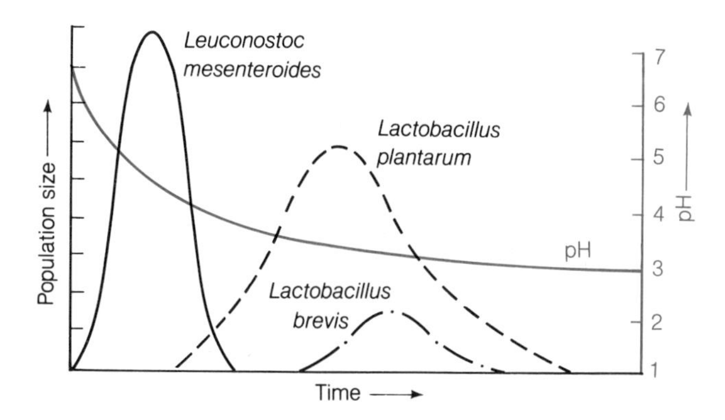

Figure 23.10 Microorganisms involved in the fermentation of cabbage into sauerkraut. *Leuconostoc mesenteroides* initiates the lactic acid fermentation of the cabbage carbohydrates, and the resulting acid lowers the pH of the brine. This change provides a stimulus for *Lactobacillus* to reproduce and proceed with the fermentation of the cabbage to produce sauerkraut.

FOCUS BEER- AND BREAD-MAKING ARE ANCIENT INDUSTRIES

Humans have been using microorganisms since the dawn of history to make useful products. For example, the Sumerians and Babylonians used "natural" yeast to make beer as far back as 6000 B.C., while the Egyptians, as long ago as 4000 B.C., used brewer's yeast to leaven their bread. The Babylonians also knew how to convert the ethanol in their beers into acetic acid (vinegar).

Egyptian tomb reliefs dated at about 2400 B.C. indicate that the preparation of leavened bread and beer was not just a family project, but probably a major industry employing many men and women (fig. 23.11). The top panel of the relief illustrates how the Egyptians manually separated the stalks from the heads of grain and how they pounded the heads of grain to separate the chaff. Flour was produced by grinding the grain between two rocks. An Egyptian depicted in the center relief is believed to be soaking the pounded, winnowed grain in a basket, so that it will sprout (malt) and yield starch-degrading enzymes and subsequently sugar for natural or added yeasts to ferment.

The central relief in the second panel indicates that some of the malted grain was added to flour to produce leavened dough. The malted grain contained sugar that yeast could ferment to ethanol and CO_2. The CO_2 caused the dough to rise. Two Egyptians are shown kneading the dough to stimulate the growth of yeast and the production of CO_2. Above the Egyptians are spherical loaves being allowed to rise before baking. Another individual is shown tending the oven where the bread was baked.

In the bottom panel, two individuals are preparing beer in the leftover malted grain in the basket. The Egyptian may be pouring warm water over the malted grain to wash sugar and natural yeasts through the basket into the clay fermenting vat below. The grain in the basket may have been used to feed cattle. The sugar extract (wort) was converted to ethanol and CO_2 by the yeast in the clay vat over a period of a few days. When the alcoholic beverage was ready, it was poured into numerous deep pottery jars that were sealed with clay.

Many of the steps in beer production today are similar to those used by the Egyptians over 4000 years ago.

Figure 23.11 Egyptian tomb relief

It multiplies well on the salted cabbage, rapidly producing lactic acid and lowering the pH within three days to about 4.5. This acidity discourages the growth of contaminating microorganisms. Acetic acid, mannitol, and ethanol are produced during this stage in addition to lactic acid. The mannitol, which would give a bitter flavor to the sauerkraut, is fermented by other populations of lactic acid bacteria (*Pediococcus cerevisiae, Lactobacillus brevis,* and *Lactobacillus plantarum*). These bacteria continue to produce lactic acid, lowering the pH to about 3.5. After 21 days, the process is completed, possibly because the fermentable carbohydrates are depleted, and the sauerkraut has 2% salt, a pH of 3.0–3.5, and a little bit of ethanol.

SUMMARY

MICROORGANISMS AND THE RECYCLING OF NUTRIENTS

1. Microorganisms participate in nutrient recycling by utilizing available nutrients and releasing metabolic wastes. The wastes can then be used as nutrient sources by other populations in the community.
2. Microorganisms and plants are important in establishing and changing the earth's environment because they are able to promote chemical conversions of large amounts of carbon, oxygen, nitrogen, phosphorus, sulfur, and metals.
3. The earth's organic matter, CO_2, H_2O, and O_2 are closely linked and interconnected by the processes we call photosynthesis and respiration.
4. O_2 has accumulated and CO_2 has decreased to present atmospheric levels because of the sequestering of organic compounds under sediment. If all the organic sediments should suddenly be oxidized, the oxygen would virtually disappear from the atmosphere and the carbon dioxide would increase drastically.
5. All living organisms are dependent upon the organic nitrogen produced by a few groups of microorganisms and their plant partners.
6. Organic nitrogen is created when microorganisms fix atmospheric nitrogen gas into organic compounds. This process is called nitrogen fixation. Organic nitrogen can also be formed when microorganisms and plants assimilate ammonia or nitrates into their organic compounds.
7. Organic nitrogen can be converted to ammonium by numerous organisms, by a process called ammonification.
8. Ammonium ions can be converted to nitrite ions, and these in turn can be converted to nitrate ions, by a small group of bacteria. This process is called nitrification.
9. Nitrogen can temporarily be lost to organisms when it is converted into N_2 gas and escapes to the atmosphere. The process of converting nitrate to N_2 is called denitrification.
10. Sulfur is required to make two amino acids, a coenzyme, and sulfate-containing polysaccharides. Various chemical forms of sulfur are interconvertible by microorganisms. Some organisms use H_2S as an electron source and convert it to S and this in turn to SO_4^{2-}. Other microorganisms use SO_4^{2-} as an electron acceptor to create H_2S.

MICROORGANISMS AND ANIMAL NUTRITION

1. Microorganisms colonize various parts of an animal and may cause changes in the animal's metabolism. Many animals develop long-lasting associations with microorganisms that provide digestive enzymes or nutrients.
2. Termites have a gut flora that helps the termite digest cellulose.
3. Ruminants nurture a complex community of anaerobic microorganisms in their rumen which provides the animal with nutrients from the cellulose they digest.

MICROORGANISMS AND SEWAGE TREATMENT

1. Sewage treatment is aimed at reducing the organic material before the effluent is discharged into receiving waters.
2. Primary treatment of sewage involves the removal of suspended soils and lowers the organic material by 30–40%.
3. Secondary treatment involves the biological oxidation of dissolved organic materials in the wastewater, which reduces this material by about 90%. This treatment can be accomplished by using trickling filters or activated sludge facilities.
4. Tertiary treatment is a process aimed at removing inorganic ions resulting from secondary treat-

ment. It reduces the organic material by more than 98%.

MICROORGANISMS AND FOOD SPOILAGE

1. Food spoilage is considered to be any undesirable change that takes place in the food.
2. Microorganisms multiplying in food can cause various types of changes in the food, thus spoiling it.

MICROORGANISMS AND FOOD PRODUCTION

1. Fermented foods are made when microorganisms multiply in the food, changing the flavor and texture of the original product. The changes are caused by metabolic products released by the microorganisms while fermenting a food substrate.
2. Using milk as a starting material, various dairy products can be made by allowing lactic acid bacteria to ferment the milk sugars and form a curd.
3. The yeast *Saccharomyces*, which produces ethanol and CO_2, is used to make bread rise. The CO_2 inflates the dough.
4. Lactic acid bacteria such as *Lactobacillus sanfrancisco* are used to make sourdough breads. These bacteria produce lactic acid and acetic acid. The acids give the bread its sour taste.
5. Naturally occurring lactic acid bacteria are responsible for the acidification of cabbage and the formation of sauerkraut (acid cabbage). A succession of bacteria are responsible for dropping the pH from about 6.5 to 3 within 2 to 3 weeks.
6. The yeast *Saccharomyces*, which produces ethanol and CO_2, is used to make alcoholic beverages.

STUDY QUESTIONS

1. Compare the microbial processes that take place in the nitrogen cycle with those in the sulfur cycle.
2. What happens to fertile, well-aerated soils when excessive rains make the soils waterlogged?
3. How might soil microorganisms be used in industrial processes?
4. Where in the nitrogen cycle do the following processes take place? Name one organism that carries out this process, and state if the process requires the input of energy or is energy-producing.
   a. anaerobic respiration;
   b. aerobic respiration;
   c. chemolithotrophic metabolism;
   d. nitrate reduction;
   e. N_2 reduction.
5. What useful function(s) do soil microorganisms perform in nature?
6. What characteristics are shared by all the various biogeochemical cycles discussed in this chapter?
7. What is sewage? How is it treated?
8. What is primary sewage treatment? Secondary sewage treatment? Tertiary sewage treatment?
9. How might microorganisms affect foods?
10. Describe three different ways in which microorganisms can spoil foods.
11. Describe how microorganisms participate in the production of the following foods:
    a. cheese;
    b. sauerkraut;
    c. yogurt;
    d. buttermilk;
    e. bread.

SELF-TEST

1. Which of the following are involved in the recycling of nutrients by microorganisms?
   a. aerobic respiration;
   b. anaerobic respiration;
   c. photosynthesis;
   d. fermentation;
   e. all of the above.
2. Both chemoheterotrophic and photoautotrophic organisms participate actively in the carbon cycle. a. true; b. false.
3. Which of the following bacteria participate in the nitrogen cycle by fixing nitrogen?
   a. *Nitrosomonas;*
   b. *Nitrobacter;*
   c. *Proteus;*
   d. *Pseudomonas;*
   e. *Rhizobium.*
4. Microorganisms that participate in the process of nitrification differ from those involved in denitrification in that the nitrifying organisms respire anaerobically while those that denitrify are strict aerobes. a. true; b. false.

5. The aim of sewage treatment is to remove as much of the organic matter from the water as possible so that the process of eutrophication is slowed down. a. true; b. false.

6. Secondary treatment of sewage consists of the:
   a. removal of particulate organic matter in settling tanks;
   b. anaerobic digestion of sludge by microorganisms;
   c. oxidation of dissolved organic matter by microorganisms;
   d. removal of inorganic ions by chemical means;
   e. chlorination of treated water after oxidation.

7. Which of the following microbial activities is most often associated with sour spoilage of foods?
   a. fermentation of lactose;
   b. putrefactive processes;
   c. aerobic respiration;
   d. hydrolysis of fats;
   e. hydrolysis of DNA.

8. Bacteria spoil foods by mechanisms that are very different from those observed in laboratory culture media. a. true; b. false.

9. Which one of the following fermented foods is prepared through the intervention of a fungus?
   a. buttermilk;
   b. yogurt;
   c. leavened bread;
   d. poi;
   e. olives.

10. The sour taste that is characteristic of fermented milks such as yogurt and buttermilk is due to the lactic acid produced during the fermentation of milk sugar by the lactic acid bacteria. a. true; b. false.

11. In an aerobic respiration of glucose, what serves as the final electron acceptor?
    a. nitrate;
    b. sulfate;
    c. pyruvic acid;
    d. ferric iron;
    e. none of the above.

12. In a complete anaerobic respiration of glucose, using SO_4^{-2} as the final electron acceptor, what are the waste products?
    a. H_2SO_4 and lactic acid;
    b. H_2SO_4 and CO_2;
    c. H_2S and CO_2;
    d. CO_2 and H_2O;
    e. none of the above.

13. In a complete anaerobic respiration of glucose, N_2 and CO_2 are the waste products. What is the most likely electron acceptor?
    a. pyruvic acid;
    b. NO_3^-;
    c. H_2O and CO_2;
    d. lactic acid;
    e. CO and NH_4^+.

14. Which organisms fix nitrogen?
    a. some cyanobacteria;
    b. a few fungi;
    c. all symbiotic bacteria;
    d. some protozoans;
    e. many microscopic animals.

15. Which organism fixes nitrogen only when it forms an endosymbiotic relationship with legumes?
    a. *Clostridium;*
    b. *Azotobacter;*
    c. *Anabaena;*
    d. *Klebsiella;*
    e. none of the above.

16. Which process requires aerobic conditions?
    a. decarboxylation;
    b. ammonification;
    c. nitrification;
    d. denitrification;
    e. nitrogen fixation.

17. Which process is promoted by anaerobic conditions?
    a. ammonification;
    b. nitrification;
    c. denitrification;
    d. nitrogen fixation;
    e. all of the above.

18. Which process decreases the fertility of a soil?
    a. ammonification;
    b. nitrification;
    c. denitrification;
    d. nitrogen fixation;
    e. none of the above.

19. In the nitrogen cycle, when organisms convert ammonia to nitrate, they are engaged in what type of metabolism?
    a. assimilatory;
    b. dissimilatory;
    c. catabolic;
    d. anabolic;
    e. all of the above.

20. Which organisms carry out nitrification?
    a. *Pseudomonas* and *Thiobacillus;*
    b. *Anabaena* and *Nostoc;*
    c. *Rhizobium* and *Azotobacter;*
    d. *Nitrosomonas* and *Nitrobacter;*
    e. none of the above.

21. If the denitrifying bacteria were to suddenly all die and all other types of bacteria were to survive, what would happen?

a. all forms of life would increase greatly within a few years;
b. the amount of nitrogen available to make life would decrease to very low levels and all organisms would die within a few years except microorganisms that could assimilate nitrates from dissolving rocks;
c. N_2 and CO_2 and O_2 would increase in the atmosphere;
d. there would be no significant change in the atmosphere;
e. there would be no significant change in the number or types of organisms.

22. In the sulfur cycle, when organisms convert extracellular SO_4^{-2} to H_2S, they are using the sulfur:
a. to make sulfur containing compounds;
b. as a final electron acceptor;
c. as a source of reducing electrons;
d. as a source of electrons for making ATP;
e. none of the above.

23. In the sulfur cycle, when nonphotosynthetic organisms convert extracellular H_2S to S and then to SO_4^{-2}, they may be using the sulfur:
a. to make sulfur-containing compounds;
b. as a final electron acceptor;
c. as a source of reducing electrons;
d. as a source of electrons for making ATP;
e. both c and d.

24. What is not part of the definition of lactic acid bacteria?
a. gram-negative;
b. rod-shaped or spherical;
c. produce lactic acid when fermenting sugars;
d. catalase-negative;
e. all of the above.

For items 25–27, indicate which organisms are usually involved in making the food or beverage indicated.
a. lactic acid bacteria;
b. filamentous fungi;
c. yeasts;
d. acetic acid bacteria;
e. coliforms.

25. Beer, wine, and bread

26. Yogurt and cheese

27. Sauerkraut

28. In yogurt production, what is responsible for the formation of the curd?
a. caseinase;
b. amylase;
c. gelatinase;
d. acetylmethylcarbinol (acetoin);
e. none of the above.

29. What is responsible for the sour taste of yogurt?
a. lactose;
b. lactic acid;
c. acetylmethylcarbinol;
d. ethanol;
e. none of the above.

30. Which amine contributes to spoilage of meats?
a. 2,3 butanediol;
b. indole;
c. hydrogen sulfide;
d. acetate;
e. putrescine.

31. The organisms that initiate the spoilage of meat are:
a. lipolytic and proteolytic bacteria and fungi;
b. coliforms;
c. lactic acid bacteria;
d. fermentative;
e. antibiotic producers.

32. The most likely thing about an organism that initiates the spoilage of meat is that it releases:
a. hydrogen peroxide;
b. indole;
c. antibiotics;
d. exoenzymes;
e. acids.

33. The greening of meat is due to lactic acid bacteria that release hydrogen peroxide which interacts with hemoglobin and possibly cytochromes in the meat. a. true; b. false.

34. *Serratia* gives off a yellow-green pigment and is responsible for the greening of milk. a. true; b. false.

35. Lipolytic fungi and bacteria cause spoilage by breaking down fats to glycerol and fatty acids. The fatty acids are responsible for a type of spoilage called:
a. souring;
b. blackening;
c. greening;
d. rancidity;
e. none of the above.

36. Which type of bacteria are usually responsible for the formation of the curd in sour milk?
a. nitrifying bacteria;
b. denitrifying bacteria;
c. lactic acid bacteria;
d. coliforms;
e. fungi and pseudomonads.

SUPPLEMENTAL READINGS

American Public Health Association. 1976. *Recommended methods for the microbiological examination of foods*, 2nd ed. New York: American Public Health Association.

American Public Health Association. 1985. *Standard methods for the examination of water and wastewater,* 16th ed. Washington, DC: American Public Health Association.

Atlas, R. M. and R. Bartha. 1981. *Microbial ecology. Fundamentals and applications.* Menlo Park, CA: Addison-Wesley.

Ayres, J. C., Mundt, J. O., and Sandine, W. E. 1980. *Microbiology of foods.* San Francisco: W. H. Freeman.

Baross, J. A., Dahm, C. N., Ward, A. K., Lilley, M. D., and Sedell, J. R. 1982. Initial microbiological response to the Mt. St. Helens eruption. *Nature,* 296:49–52.

Block, E. 1985. The chemistry of garlic and onions. *Scientific American* 252(3):114–119.

Brill, W. J. 1977. Biological nitrogen fixation. *Scientific American* 236:68–81.

Childress, J. J., Felbeck, H., and Somero, G. N. 1987. Symbiosis in the deep sea. *Scientific American* 256(5),114–120.

Hardy, R. W. F. and U. D. Havelka. 1975. Nitrogen fixation research: A key to world food. *Science* 188:633–643.

Kolata, G. 1985. Testing for trichinosis. *Science* 227:621–624.

Ourisson, G., Albrecht, P., and Rhomer, M. 1984. The microbial origin of fossil fuels. *Scientific American* 251(2):44–51.

Payne, W. J. 1983. Bacterial denitrification: Asset or defect. *BioScience* 33:319–325.

Rose, A. H. 1981. The microbiological production of food and drink. *Scientific American* 245:140–154.

CHAPTER 24

BIOTECHNOLOGY

CHAPTER PREVIEW GENETIC ENGINEERING IN NATURE

Many isolates of the soil bacterium *Agrobacterium tumefaciens* contain a plasmid known as the Ti-plasmid (tumor-inducing plasmid). When this bacterium infects wounds in plants such as grapes, tobaccos, tomatoes, and begonias, it stimulates a "cancerous growth" or tumor at the site of infection. The plant tumors generally develop on the stems, and the disease is known as crown gall. It becomes noticeable three to four weeks after infection.

Agrobacterium attaches to plant cells within a lesion, and Ti-plasmids from the bacterium gain entrance to the plant cells. Part of the Ti-plasmid, called T-DNA, recombines with the plant's genome and transforms the plant cell. The transforming bacterial DNA is believed to cause an imbalance in the production of plant hormones. This imbalance results in the uncontrolled growth of plant cells and the formation of a crown gall.

The effects of hormones on plant cells in tissue cuture give us some insight as to how hormones act in whole plants. Most normal plant cells will not grow in tissue culture without plant hormones. Crown gall tissue free of *Agrobacterium* can be cultivated on simple media without the addition of hormones, evidence that some of the plant cells from the crown gall have been transformed.

In addition to causing the uncontrolled growth of plant cells, T-DNA directs plant cells to produce unusual nitrogenous compounds called **opines**, which are released into the soil around the crown gall. Agrobacteria in the soil that carry the Ti-plasmid use these opines as a source of energy, nitrogen, and carbon. The opines induce *Agrobacterium* to conjugate and spread the Ti-plasmid throughout the population. It has been reported that *Agrobacterium* conjugates very infrequently in the absence of opines.

The Ti-plasmids from *Agrobacterium* are of interest because they are expressed in plant cells. Genetic engineers have succeeded in splicing various bacterial and plant genes to the controlling sites in the T-DNA, introducing these genes into plant cells, and having them expressed. The genetically modified Ti-plamids may, in the future, allow biologists to introduce a number of genes that would improve the plant. Genetic engineers are looking into the possibility of introducing genes that would allow plants to fix their own nitrogen, increase their nutritive value, and make them more resistant to plant pathogens and dessication.

Genetic engineers have succeeded in introducing a gene that codes for an enterotoxin in tobacco plant cells. This enterotoxin, from the bacterium *Bacillus thuringiensis*, makes the tobacco plants that arise from the treated cells resistant to certain insect larvae: when the larvae eat the tobacco leaves, they become sick and die.

In this chapter we will be introduced to some procedures used by genetic engineers to create recombinant DNAs, to make genes, to clone genes, and to make commercially important products.

BIOTECHNOLOGY IS AN APPLIED AREA OF biology that uses organisms to produce medically or industrially important products such as enzymes, drugs, antibiotics, foods, and beverages. Some of the organisms commonly utilized include bacteria, yeasts, fungi, and algae as well as animal and plant cells.

The first step in biotechnology is the selection of a suitable organism to provide the product. Often, the appropriate source is a naturally occurring microorganism. For example, the biotechnological processes employed in the production of antibiotics use soil bacteria (streptomycetes) or fungi (*Penicillium*) to produce the antibiotics. Sometimes, however, it becomes necessary to genetically manipulate an organism so that it acquires the capacity to produce the desired product. For example, human **interferon**, a substance that is used for the treatment of hairy-cell leukemia and certain viral infections, is produced in strains of *Escherichia coli* that have received a human gene that codes for interferon. This genetic manipulation of organisms so that they acquire novel characteristics is known as **genetic engineering**.

Once the appropriate biological source of the product has been selected, it is necessary to culture the organism in volumes so that large quantities of

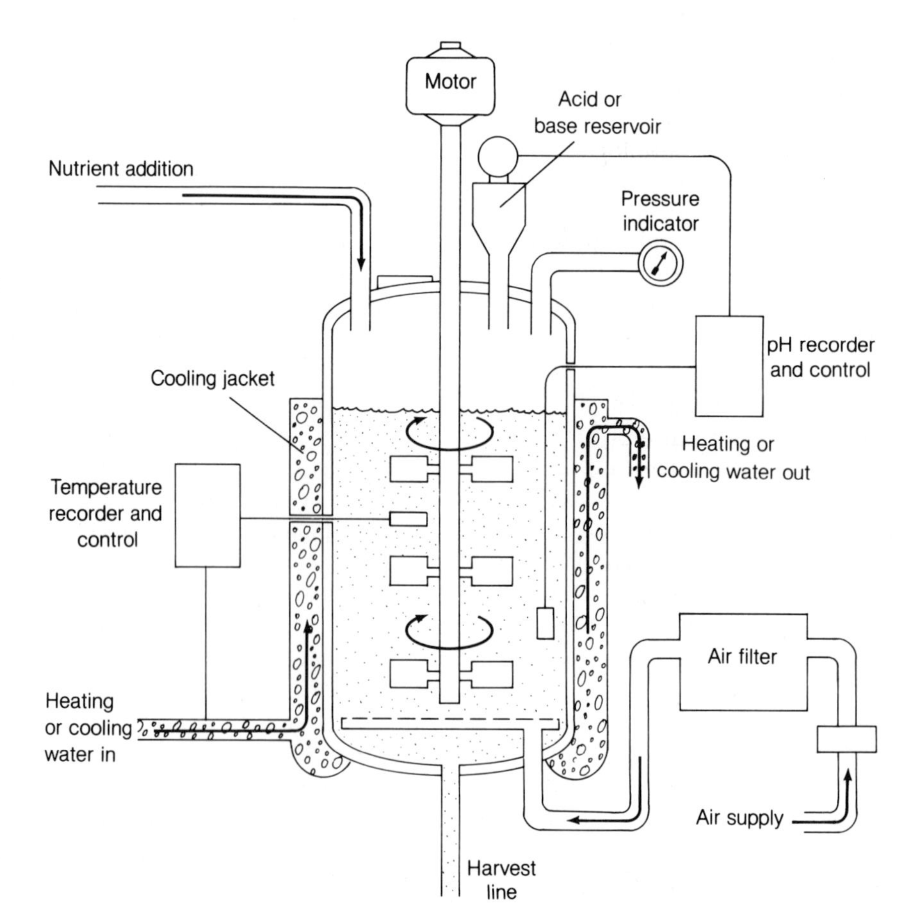

Figure 24.1 Fermenter. A batch fermenter is used to grow many industrially important microorganisms. The chamber of the fermenter and all lines leading into and out of it are sterilized with steam. The sterile medium enters the fermenter through the nutrient addition line. Microorganisms may be added to the nutrient before it enters the chamber, or they may be added through a port. Filtered air, if required, enters through the air supply line. Paddle wheels mix the nutrients and microorganisms to insure rapid growth. The temperature is controlled by the flow of warm or cold water around the fermenter. The pH is controlled by the release of acid or base from a reservoir. The microorganisms and any substances they have released are harvested through the harvest line.

the desired substance are produced. This is usually accomplished in **fermenters** (fig. 24.1) or in fermentation tanks (fig. 24.2).

The last, and very important step in the biotechnological process is the extraction and purification of the desired product so that it can be packaged and distributed commercially. This step involves the cooperation of microbiologists, chemists, and engineers.

This chapter will consider some of the various steps involved in biotechnology, utilizing specific examples to illustrate a concept.

Figure 24.2 Fermentation tanks. These fermentation tanks are used to grow microorganisms that produce antibiotics. The ports and gauges are used to monitor and maintain optimal growth conditions for the production of antibiotics.

GENETIC ENGINEERING

The manipulation of genes and cells to create new organisms or inexpensive biological products is referred to as genetic engineering. Genetic engineering is a relatively new area of microbiology that developed in the 1970s from the many advances in microbial physiology and genetics. A clear understanding of DNA and how it is expressed is the basis of genetic engineering. However, the discovery and purification of various bacterial endonucleases that cleave DNA precisely within certain sequences as well as the use of modified bacterial plasmids and host bacteria were the most important factors in the development of genetic engineering.

Presently, microbiologists are able to insert genes into viruses, bacteria, and eukaryotes. By growing the transformed viruses or cells, billions of copies of the inserted genes can be produced. In addition, appropriate manipulation of the genes results in valuable products being made in large quantities (table 24.1 and table 24.2). For example, human hormones such as insulin and growth hormone are being produced in microorganisms and are being used to treat genetic deficiencies. Certain enzymes such as streptokinase and urokinase, which dissolve blood clots, also are being produced in bacteria. In addition, the enzyme tissue plasminogen activator, which converts plasminogen to plasmin that dissolves blood clots, has been tested in people suffering from heart attacks and has been very successful in preventing severe damage to the heart. Proteins from various viruses (foot and mouth disease, herpes, hepatitis, HIV, etc.) made in genetically engineered cells are being tested to determine whether they can be used as effective vaccines. In addition, the human interferons are being studied to determine if they will be useful in inhibiting some cancers and virus infections.

While the term genetic engineering generally implies that there is manipulation of purified DNA and a change in an organism's genome by inserting or removing DNA, **bioengineering** usually indicates the making of a new organism either through genetic engineering or through other techniques, such as cell fusion, or breeding (table 24.3). For example, in the late 1970s, microbiologists learned how to make **hybridomas**, cells formed from the fusion of antibody-producing cells and cancer cells. Hybridomas have turned out to be little factories that make very useful antibodies (monoclonal antibodies) that can be used for the diagnosis of infectious diseases, as vaccines, to purify proteins, and for tissue typing.

THE BASIS OF GENETIC ENGINEERING

Genetic engineering is dependent upon a number of important discoveries and techniques that were made in the 1960s, 1970s, and 1980s: (a) restriction endonucleases, which cut DNA at specific sites; (b) DNA

TABLE 24.1
PRODUCTION OF USEFUL PROTEINS

Product	Field	Specific use
Human insulin	Medicine	Treatment for diabetes
Human interferons	Medicine	Treatment for virus infections and certain cancers
Human growth hormone	Medicine	Prevention of growth deficiencies
Human thymosin-alpha 1	Medicine	Treatment for certain cancers
Human interleukin-2	Medicine	Stimulates growth of lymphocytes and the immune system
Human interleukin-3	Medicine	Stimulates growth of many cell types in the immune system
Granulocyte colony-stimulating factor	Medicine	Stimulates growth of granulocytes
Macrophage colony-stimulating factor	Medicine	Stimulates growth of macrophages
Granulocyte-macrophage colony-stimulating factor	Medicine	Stimulates growth of granulocytes and macrophages
Human somatostatin	Medicine	Inhibits release of insulin and growth hormone
Tissue plasminogen activator	Medicine	Stimulates destruction of blood clots
Urokinase	Medicine	Treatment for dissolving blood clots
Hoof and mouth virus proteins	Medicine	Vaccine for hoof and mouth disease
Hepatitis B virus proteins	Medicine	Vaccine for hepatitis B
Colibacillus proteins	Medicine	Vaccine for calf and piglet disease
AIDS virus core protein	Medicine	Radioactive core protein that binds to core of AIDS virus and is used to detect the presence of the virus
Monoclonal antibodies	Industry	Purification of proteins
	Medicine	Tissue typing and diagnosis of infectious diseases
Antiendotoxin*	Medicine	Protects against endotoxins released during blood infections
Anti T-3*	Medicine	Destroys T-cells involved in tissue rejection
Anti cancer cells*	Medicine	Destroys specific cancer cells

*Monoclonal antibodies

ligases, which splice DNA fragments; (c) vectors (viruses and plasmids) used to introduce DNA fragments into recipient cells; (d) host selection (cloning); and (e) requirements for gene expression.

Today we know that each species or strain of bacterium has **restriction endonucleases** that bind DNA and cut it specifically at one or more points (fig. 24.3). Restriction endonucleases help a cell maintain its genetic integrity by protecting it from foreign DNA that might be acquired during conjugation, transduction, or transformation. More importantly, restriction endonucleases may protect a cell from invading DNA viruses.

If two pieces of DNA are cut by a restriction

TABLE 24.2
PRODUCTION OF USEFUL NUCLEIC ACIDS

Product	Field	Specific use
DNA probe for retinoblastoma	Medicine	Detects gene that causes retinoblastoma and osteosarcoma
DNA probe for cystic fibrosis	Medicine	Detects gene that causes cystic fibrosis
DNA probe for Huntington's chorea	Medicine	Detects gene that causes Huntington's chorea
DNA probes for specific organisms	Medicine	Detect and identify specific organisms
Hemoglobin genes	Medicine	Replace defective genes in hemophilia and β-thalassemia patients

TABLE 24.3
PRODUCTION OF USEFUL MICROORGANISMS

Product	Field	Specific use
Ice-minus bacteria (*P. Fluorescence, P. syringae*)	Agriculture	Protect commercial crops (potatoes and strawberries) from frost damage
Attenuated pseudorabies virus	Agriculture	Vaccine for swine against pseudorabies virus
Vaccinia virus with herpes virus genes	Medicine	Vaccine for humans against herpes simplex I and II
Vaccinia virus with rabies virus genes	Agriculture	Vaccine for livestock against rabies
Hybridomas	Medicine, Industry	Production of monoclonal antibodies

endonuclease that creates staggered ends, the pieces of DNA will hybridize with each other because of the complementary nucleotide sequences in the staggered ends (fig. 24.4). Thus, hybrid or recombinant DNAs can be created quite easily by mixing the DNA pieces together and repairing the nicks in the DNA. The enzyme that repairs DNA nicks is known as **DNA ligase**.

One gene or one region of DNA is impossible to use and not very valuable in making proteins. Consequently, it is necessary to have a way of making many copies of the gene. One way that a large number of copies can be made is to insert the DNA of interest into a virus or plasmid and subsequently proliferate the infectious agent or the plasmid in a cell (fig. 24.5). The virus or plasmid that carries the spliced DNA is referred to as the **vector**.

After regions of DNA have been spliced into a vector, they must be introduced into an appropriate host cell to reproduce them. To select the cells con-

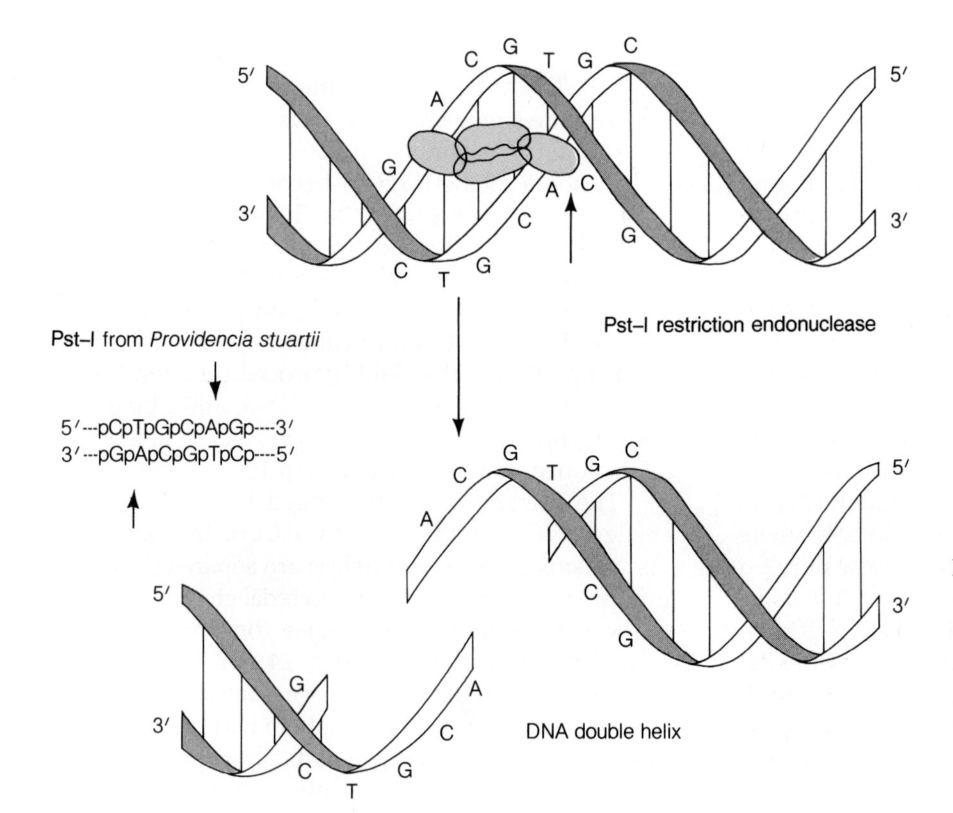

Figure 24.3 **DNA cuts made by restriction nucleases.** Each restriction endonuclease cuts DNA at a precise location. Pst–I cuts the DNA so that it has staggered ends. T, thymine (thymidine); C, cytosine (cytidine); A, adenine (adenosine); G, guanine (guanosine).

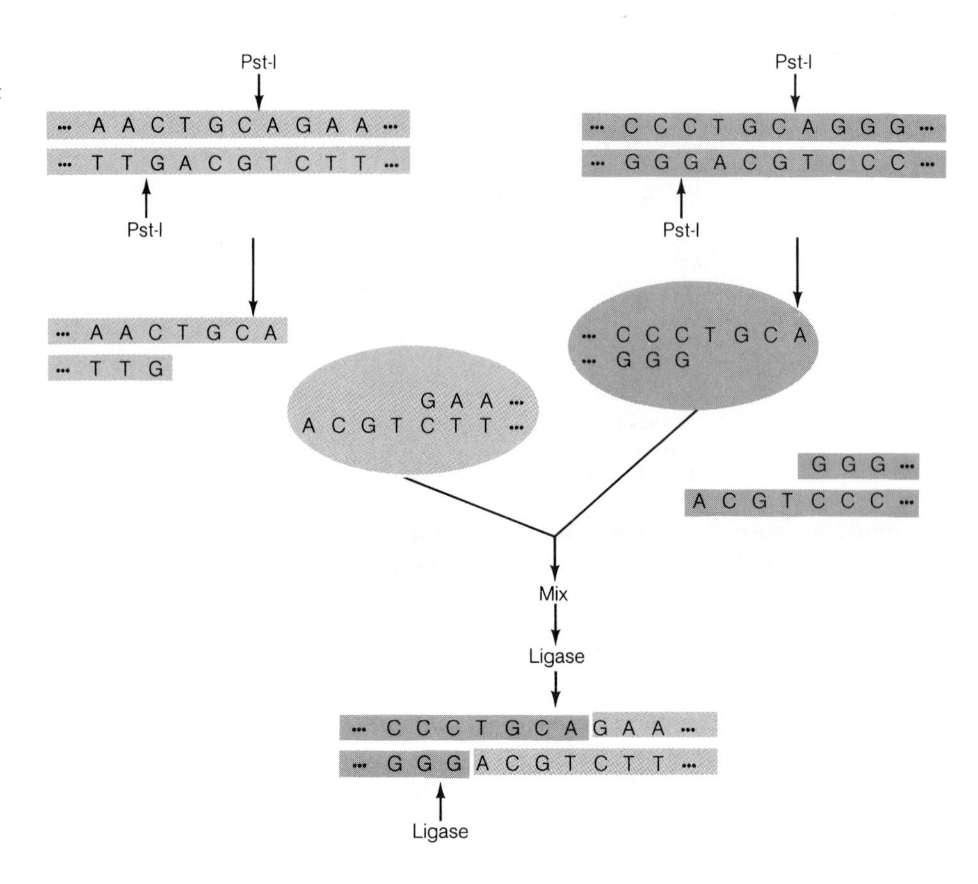

Figure 24.4 Genetic engineering. The same endonuclease is used to cut two different DNA molecules. Since the fragments have compatible overlapping regions they are able to recombine. Ligase repairs the cuts in strands that recombine.

taining the plasmids of interest, plasmids have been engineered with antibiotic-resistance genes. These genes make it possible to detect the presence of the plasmids and select for them. The proliferation and selection of specific regions of DNA in cells is known as **cloning** and the colonies of cells as **clones**.

A METHOD FOR CLONING GENES

One way of cloning a gene is to cut it out of an organism's genome with a restriction endonuclease, introduce it into a vector cut with the same endonuclease, and then insert the recombinant vector into cells where it can proliferate. Figure 24.5a illustrates how purified plasmids carrying genes that confer tetracycline resistance are cut with a restriction endonuclease (Pst-I), and how foreign genes with ends that match the cut plasmid are inserted into the middle of the plasmid. Since Pst-I creates staggered or overlapping ends, foreign DNAs with staggered ends that are complementary to the staggered ends in the plasmid bind to the plasmid DNA because of the hydrogen bonding between the single-stranded ends.

When the recombinant plasmids are mixed with *E. coli,* some of the bacteria become transformed to tetracycline resistance (fig. 24.5b). Many bacteria do not take up the plasmid and consequently remain tetracycline-sensitive. If the bacteria are grown on a medium containing tetracycline, only those with the plasmid having the tetracycline-resistance gene will grow (fig. 24.5c). The cloning procedure allows us to splice a single piece of foreign DNA and obtain billions of copies within a day. It is very easy to grow *E. coli* to concentrations greater than 10^9 bacteria/ml).

The bacteria are concentrated by centrifugation and gently lysed so as not to disrupt the plasmids. The plasmids and *E. coli* debris are separated by centrifugation. The plasmids and bacterial genomes remain in the **supernatant fluid** because they are very light in comparison to the debris (fig. 24.5d). The plasmids and bacterial genomes can be separated by centrifugation in a cesium chloride gradient because they have different densities. The plasmids are concentrated in a band that is easily drawn off with a syringe. The

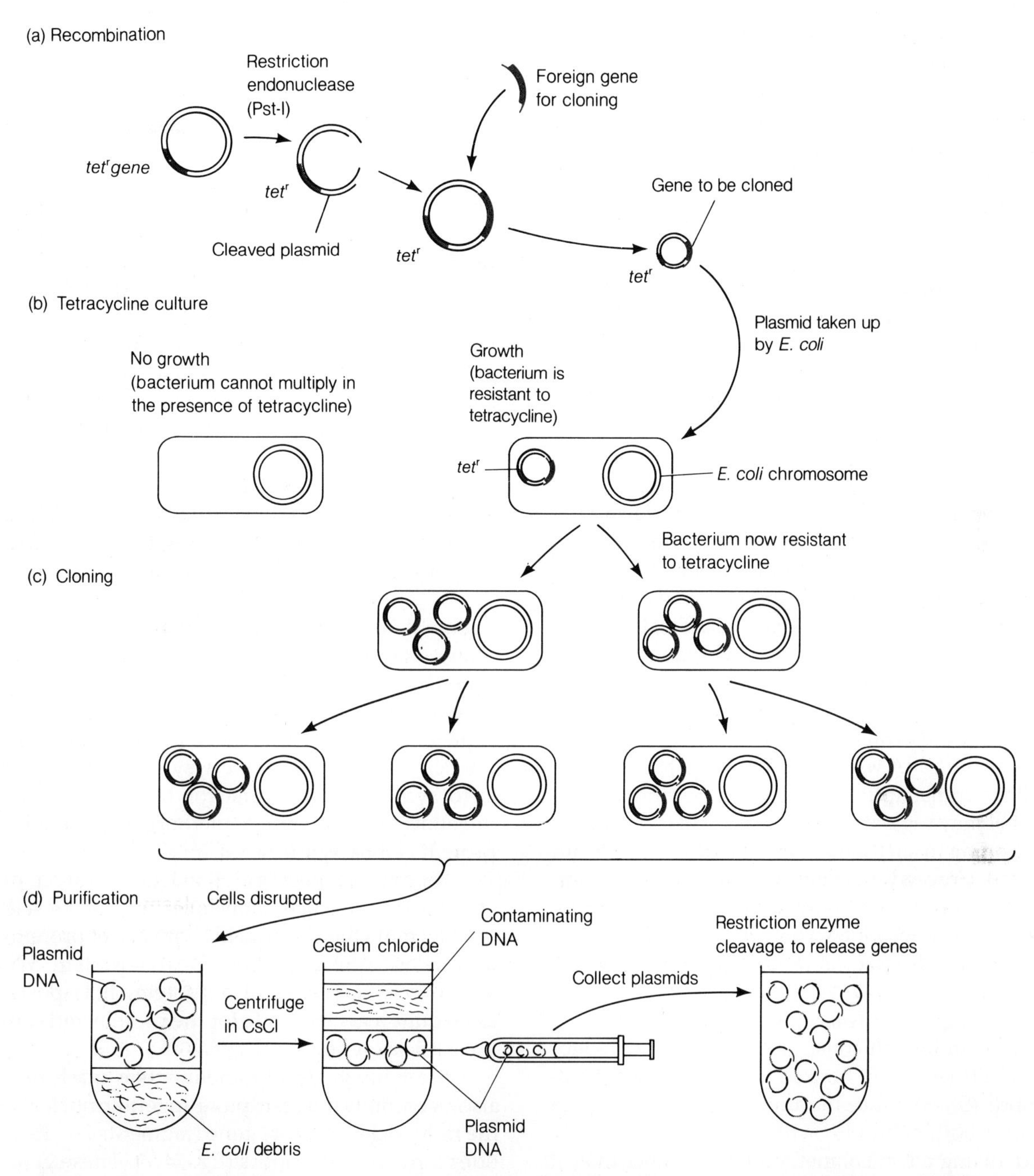

Figure 24.5 A method for cloning DNA. *(a)* Bacterial plasmids that can confer resistance to tetracycline are cut with a restriction endonuclease and foreign genes are spliced into the plasmids. *(b)* The bacteria are plated on a tetracycline-containing medium. Only bacteria that contain plasmids and are resistant to tetracycline multiply and form colonies. *(c)* Consequently, all the bacteria in the final population contain the plasmid with the foreign gene. *(d)* The bacterial plasmids can be isolated from the bacteria and purified. The foreign genes can be extracted from the plasmids by using restriction endonucleases.

plasmids may then be cleaved with a restriction endonuclease so that the foreign DNA is released.

METHODS FOR PRODUCING MEDICALLY IMPORTANT PROTEINS

A number of useful proteins, such as human insulin, human interferon, and human growth hormone, are already being produced in bacteria. **Human insulin** obtained from genetically engineered cells will eventually replace bovine (cow) and porcine (pig) insulin for treating diabetes. Use of human insulin will be advantageous because it does not have the side effects that the bovine and porcine insulins do. **Human growth hormone** for treating children with growth hormone deficiencies, and various **human interferons** for inhibiting some cancers and virus infections, have already been produced in bacteria and tested clinically. No doubt, these and many other human products obtained from genetic engineering will soon be available to the public at a reasonable price (table 24.1). The following section will discuss some of the techniques used to produce valuable proteins and genes in bacteria.

To produce a protein, more than just the gene that codes for the protein is required. First, there must be a site on the DNA where RNA polymerase can bind and initiate transcription of the gene. Second, all elements that may regulate the expression of a gene must be functioning in such a manner that transcription and translation proceed normally. After this is accomplished, it is necessary to have the technology to purify the desired gene product. The production of human insulin by biotechnological processes is an example.

Genetically Engineered Human Insulin

Active human insulin consists of two short polypeptide subunits (A and B) hooked together by two disulfide bonds (S—S). Since the amino acid sequences are known for both subunits, synthetic DNAs can easily be made using chemical methods. Remember that, if we know the amino acid sequence of a polypeptide, it is possible to determine from the genetic code a nucleotide sequence for the RNA and DNA that code for the protein.

The synthetic genes for the insulin polypeptides are then introduced into *E. coli* plasmids (e.g., pBR322) that contain the gene for resistance to the antibiotic ampicillin. The artificial A and B genes are inserted at the end of a gene in separate plasmids and cloned separately to produce stable protein-insulin molecules.

Transcription of the plasmid DNA and translation of the mRNA result in the accumulation of a protein-insulin hybrid, also known as **fusion protein**. This protein is isolated by concentrating the cells and lysing them gently under pressure. The fusion protein is somewhat insoluble and consequently it precipitates after low-speed centrifugation (fig. 24.6). The treatment of the fusion proteins with cyanogen bromide cleaves the A and B subunits away from the other unwanted protein (fig. 24.6e). Since the unwanted protein fragments are much larger than the insulin peptides, the two are easily separated. The sulfhydryl groups (SH) on the A and B peptides are chemically altered so that they will react with each other. When the peptides are mixed they form disulfide bonds with each other and active human insulin results. Active human insulin has been produced in large amounts from the insulin subunits produced in *E. coli*. It is now cheaper to produce insulin using *E. coli* than to extract the hormone from the pancreas of animals.

Automated Genetic Engineering

Many valuable proteins are extremely difficult or expensive to isolate in more than minute quantities. One of the goals of genetic engineering is to produce proteins in large amounts and economically for use in medicine or industry. One way to produce a protein is to synthesize the gene for the protein and then introduce the gene into a microorganism in which the protein can be synthesized.

Recently, a number of machines called **protein sequencers** have been developed that allow scientists to determine the amino acid sequence of proteins with as little as a few μg (10^{-6} grams) of protein. With the amino acid sequence of a protein, it is possible to determine a genetic code for the protein and construct an artificial gene.

After much development, **gene machines** have also been built that can piece together nucleotides in the right order to create gene fragments 20–30 nucleotides long in 10–15 hours (fig. 24.7). These gene fragments can then be spliced together in the right order to create a gene (sometimes 1,000 or more nucleotides long), or they can be radioactively labeled and used as a probe to detect cellular mRNA from which a complementary DNA can be synthesized. Once the gene has been synthesized, it can be spliced into a vector and cloned in a host cell, where a large number of copies can be synthesized.

The protein sequencer and the gene machine should facilitate the production of hundreds of med-

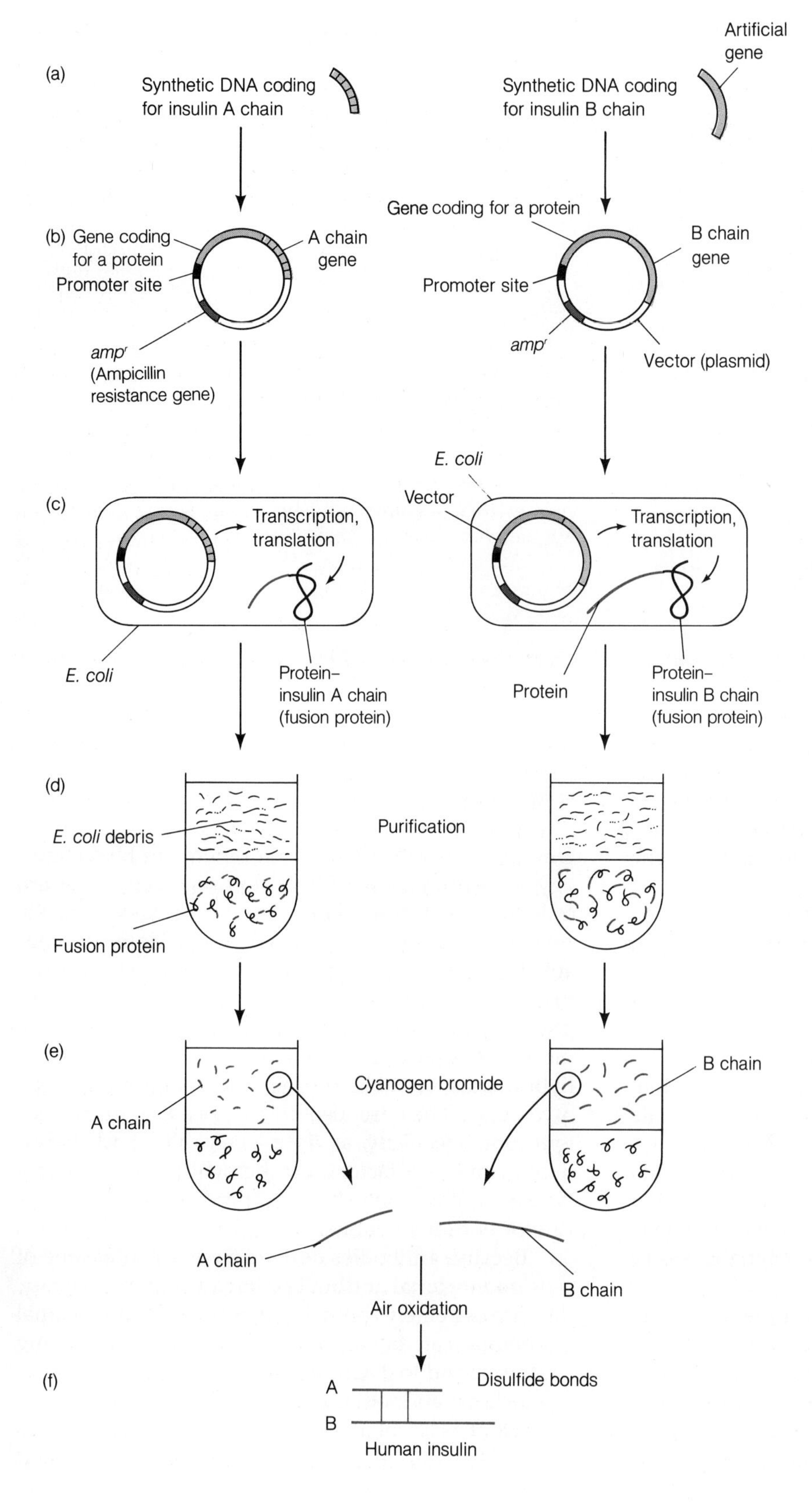

Figure 24.6 A method for cloning insulin genes and producing insulin. *(a)* DNA coding for the insulin chains is synthesized chemically. *(b)* The insulin DNA is spliced into plasmids in such a way that it will be expressed. The insulin DNA is spliced near the end of the gene for a cellular protein. This gene can easily be regulated. If it is expressed, the gene for the insulin subunits will also be expressed. *(c)* Each of the plasmids with the two different insulin genes is mixed with separate populations of bacteria. Some transform the bacteria to ampicillin resistance. The bacteria are grown on an ampicillin-containing medium to select for bacteria that contain the plasmids. After the ampicillin-resistant populations have been selected, the gene for the cellular protein is induced. The resultant products are fusion proteins that consist of the cellular protein and the insulin subunits. *(d)* Next, the fusion proteins are isolated from their respective host bacteria. Then, the fusion proteins are split into cellular protein and the insulin subunits. Subsequently, the insulin subunits are purified. *(f)* Finally, the two different insulin subunits are used to make insulin.

FOCUS THE USE OF CLONING AND GENETIC ENGINEERING TO RECREATE EXTINCT ANIMALS

In parts of northern Europe and Russia, numerous extinct animals have been discovered frozen. The best-known finds are those of complete Siberian mammoths in 1938 and in 1977. In recent years, Russian cytologists and molecular biologists have been attempting to revive some of the frozen cells from mammoth carcasses, so that nuclei from the revived cells might be exchanged with nuclei in fertilized elephant eggs. The altered elephant egg might then be implanted in an elephant, in the hope that it would develop into a baby mammoth. The Russian scientists have so far been unsuccessful in reviving any of the mammoth cells in tissue culture. Most likely, the mammoths were not frozen rapidly enough or maintained at low enough temperatures to avoid cellular damage.

Another approach to recreating a mammoth might be to isolate nuclei from the frozen mammoth cells and introduce them into elephant cells in tissue culture. If this could be achieved, then the next step would be to introduce the nuclei into fertilized elephant eggs and implant these in elephants. If this second procedure fails, genetic engineering might be tried to create a mammoth out of an elephant cell.

Most of the cells from the frozen mammoths probably contain undamaged DNA from which a mammoth DNA library could be created by cloning the DNA in *E. coli.* The mammoth DNA could be compared to elephant DNA and the differences determined. Then elephant cells in tissue culture could be transformed with the mammoth DNA that differed from the elephant DNA, and the transformed elephant nuclei could be transferred to anucleated, fertilized elephant eggs.

Even if it turns out to be impossible to recreate a mammoth by cloning and genetic engineering, these studies should reveal much about the evolutionary relationship between mammoths and modern-day elephants and the relationship between DNA and an animal's final form.

ically and industrially important proteins, such as interferons and vaccines. Already, medically important genes are being synthesized using the gene machine.

The protein sequencer and gene machine will facilitate the mapping of animal and plant genes and the determination of their nucleotide sequence. Because it is impossible (or extremely difficult) to select for mutations in numerous genes, these genes cannot be mapped, nor can their boundaries be determined. By making labeled DNA probes from the proteins that these genes code for, however, it is possible to locate the genes in the chromosomes.

After genes have been located and sequenced, accurate copies can be synthesized to establish a "genetic library" of animal and plant genes. A genetic library has many uses. For example, a genetic library for humans would allow the screening of fetuses for the presence of abnormal genes. A sample of a fetus's genes obtained from fetal cells in the amniotic fluid could be compared with the genetic library of normal DNAs. Prospective parents could then be counseled in cases where it was discovered that there were deviations in the fetus's genes that would produce birth defects.

HYBRIDOMAS

One of the most recent developments in biotechnology is the production of hybridomas. Hybridomas are cells that result from the fusion of normal eukaryotic cells with cancerous eukaryotic cells. The most valuable hybridoma cells produced so far have been from fusions between plasma cells (differentiated B-lymphocytes that produce antibodies) and cancerous plasma cells called **myeloma** cells. The hybrid cells, or hybridomas, are able to proliferate indefinitely, as do myelomas. They are also able to produce antibodies with one specificity, as do plasma cells. Each hybridoma clone is a factory for the synthesis of a single type of antibody, which can be produced in tissue culture or within an animal.

Because antibodies derived from a single clone of cells (**monoclonal antibodies**) have a single specificity, they are extremely useful in microbiology. Monoclonal antibodies can be used to destroy disease-causing organisms and to diagnose the diseases that they cause. Monoclonal antibodies are becoming more and more important as a tool in the identification of blood types (ABO, Rh, MN, etc.), transplantation antigens, and cancer cells. Researchers are presently employing

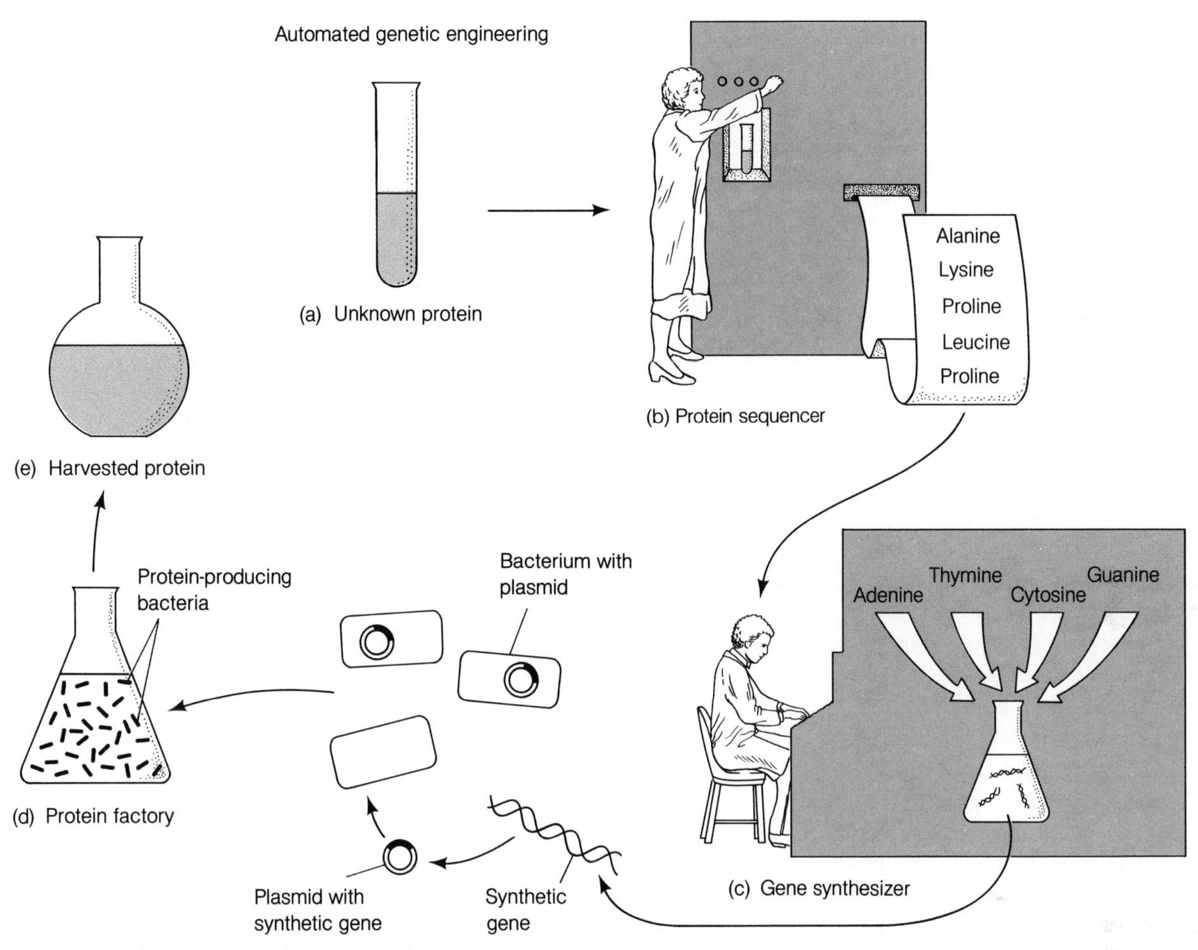

Figure 24.7 Protein sequencer and gene machine. The protein sequencer and gene machine allow scientists to automatically make a gene from minute samples of an unknown protein. This in turn makes it possible to synthesize large amounts of the protein for study. The making of a gene and then protein begins when minute samples of an unknown protein (a) are put into the amino acid sequencer. (b) This machine gives the sequence of amino acids in the protein. (c) To make a gene that will code for the protein, a scientist types a genetic code into the gene synthesizer that codes for the protein's amino acid sequence. The gene machine then chemically hooks together the correct nucleotides to make the gene. (d) By other chemical means the DNA is integrated into a bacterial plasmid which is taken up by a population of bacteria. (e) The bacteria function like a factory and produce the protein.

monoclonal antibodies to identify, quantify, and purify membrane proteins involved in regulating the physiology and development of eukaryotic cells. Valuable proteins such as interferons, neurotransmitters, and hormones, which are produced naturally in minute quantities, can be purified using single-specificity antibodies.

Passive immunization against toxins (e.g., snake venom, bee and wasp toxins, tetanus and botulism toxins) and disease-causing organisms can be greatly improved by using monoclonal antibodies. Monoclonal antibodies made against specific tumor antigens can be used to treat many types of cancers. The antibodies may aid the patient's immune system to destroy the cancerous tissue naturally, or they may be used to direct toxic drugs at cancerous tissue and selectively destroy the abnormality.

Figure 24.8 summarizes the immune response in

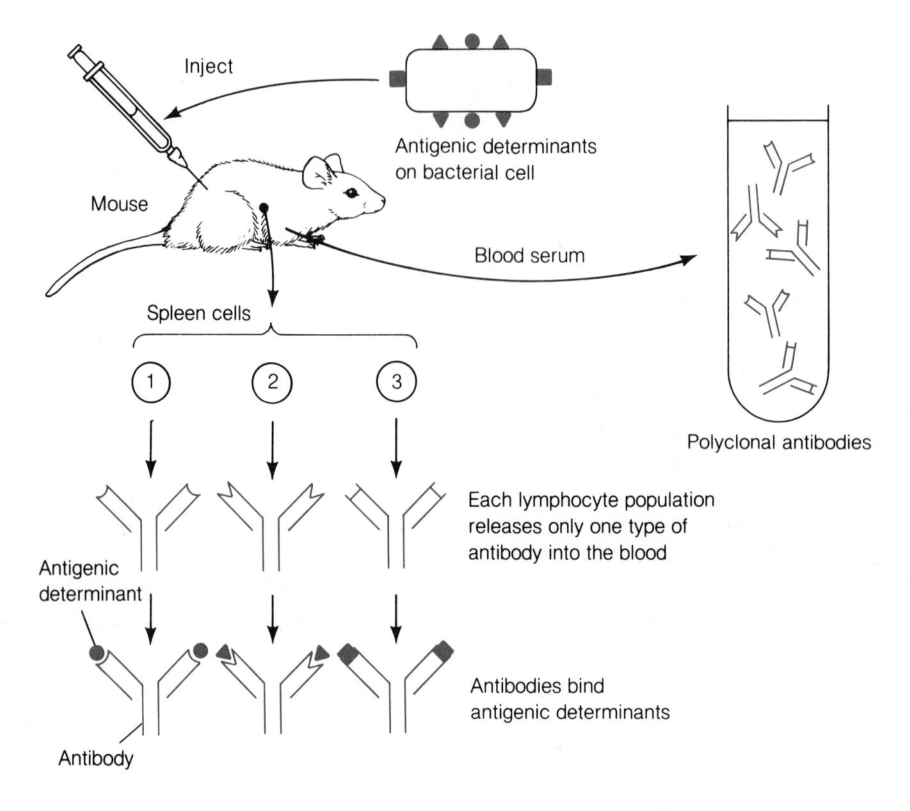

Figure 24.8 Production of polyclonal antibodies. Complex molecules such as polysaccharides and proteins, when injected into animals, are capable of inducing an immune response. These molecules are called antigens. The immune response is usually against a number of sites on the antigen known as antigenic determinants. In this illustration, a bacterium is shown with three different antigenic determinants. In a typical immune response, clones of B-lymphocytes multiply and release antibodies that bind to the antigenic determinants on the antigen that stimulated their proliferation and production of antibodies. Each clone of B-lymphocytes produces only one type of antibody, which generally binds to only one type of antigenic determinant. A mixture of antibodies that bind to an antigen can be isolated from the blood serum of a stimulated animal. The antibodies are known as gamma globulins or immunoglobulins. Their range of diversity is illustrated.

an animal and shows how **polyclonal** (mixed) sera are obtained, while figure 24.9 outlines how hybridomas are made and how monoclonal antibodies are produced. A cell (or a large molecule) with a number of different chemical groups is called an **antigen**, while the different chemical groups are known as **antigenic determinants** or **epitopes** (fig. 24.8). When an antigen is injected into an animal, numerous different B-lymphocytes, which produce immunoglobulin against the antigenic determinants of the antigen, are stimulated to develop into plasma cells. Each different population of plasma cells (which arose from a single B-lymphocyte stimulated by one of the antigenic determinants) produces only one immunoglobulin, and this immunoglobulin binds specifically to the antigenic determinant that induced its production. These lymphocytes are said to be **committed**. The committed populations of lymphocytes concentrate in the animal's spleen and lymph nodes, where they produce large amounts of immunoglobulins. The mixture of immunoglobulins moves from the lymphatic system into the animal's blood, where the immunoglobulins bind to foreign antigenic determinants that induced their production.

Hybridomas that produce only one of the many antibodies made against an antigen can be produced by mixing mouse spleen cells with myeloma cells from a plasma cell tumor. A mixed monolayer of B-lymphocytes and myeloma cells is treated for 1 minute with a solution consisting of polyethylene glycol (PEG) and balanced salt solution (BSS) to stimulate cell fusion (fig. 24.9). After 24 hours, the cells are transferred to a medium that selects for hybridomas. The various hybridoma clones are assayed for immunoglobulin production. Those clones that are of value are recloned and then stored for future projects by freezing in liquid nitrogen. Some of the cells can be grown in mass culture or in mice, and the immunoglogulins produced can be harvested.

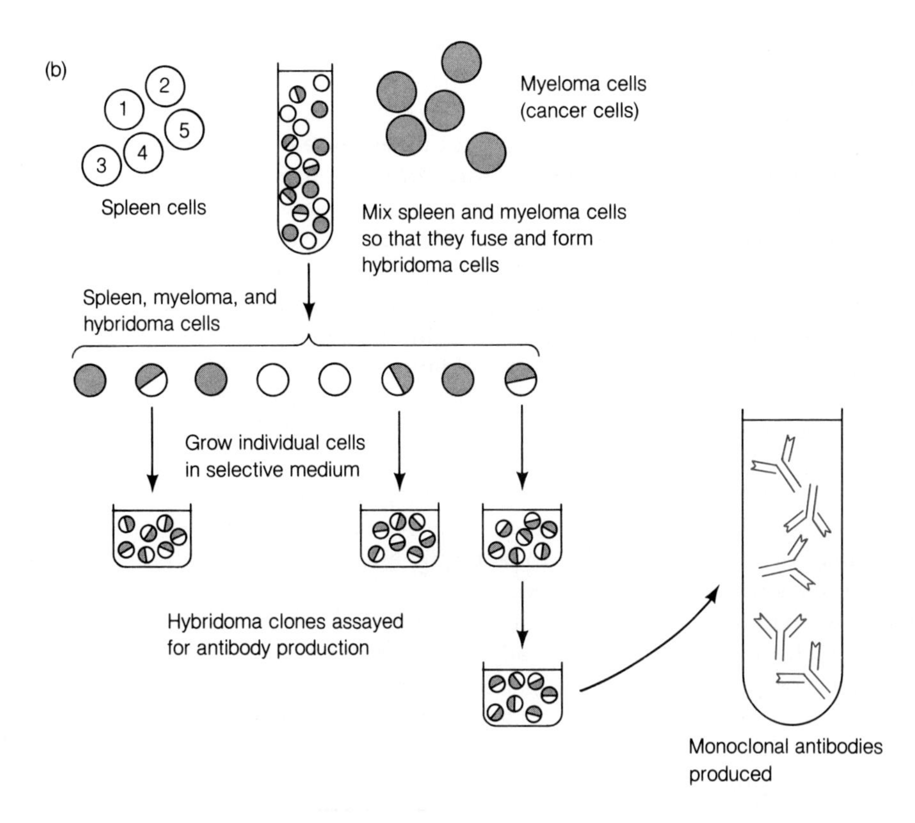

Figure 24.9 Production of monoclonal antibodies. B-lymphocytes isolated from the spleens of antigen stimulated rats and cell cultured cancerous B-lymphocytes (myeloma line) are mixed together in polyethylene glycol to stimulate cell fusion. The fused B-lymphocytes, known as hybridomas, and the parent cells are placed in a special medium (HAT medium) that selects for the hybridomas. B-lymphocytes and myeloma cells die in the HAT medium. Each of the hybridomas (many thousands) is grown in its own well so that a clone develops that produces only one type of antibody. The hybridoma clones are assayed to determine whether they produce specific antibodies against the antigen used to stimulate the rat. The clones may be injected into rats where they induce myelomas that secrete only one type of antibody. Alternatively, the clones may be grown in mass culture where they secrete only one type of antibody.

BIOENGINEERING, SAFETY, AND POLITICS

With the development of genetic engineering, many scientists and laypersons became concerned about the possibility that new infectious agents and microorganisms might be released into the environment and play havoc with plants, animals, and humans.

In 1971, Paul Berg and his colleagues at Stanford University School of Medicine were persuaded not to grow a λ DNA and SV40 DNA hybrid virus in *Escherichia coli* because of the chance that the λ-SV40-infected *E. coli* might infect researchers. SV40 is a dangerous virus because it can transform numerous mammalian cell types into cancerous cells.

To avoid contamination by the organisms used to clone potentially dangerous DNAs and at the same time reduce the chances of dangerous hereditary material propagating uncontrollably, cloning organisms were developed that could not grow in the absence of very specialized laboratory media and conditions. To protect themselves and the public, microbiologists have developed numerous safety rules and various types of containment laboratories to deal with suspected dangers.

In recent years bioengineers and government agencies that are supposed to regulate the release of bioengineered organisms have come under considerable attack because of their attempts to release into the environment new microorganisms. U.S. military research into biological weapons slowed down by an injunction blocking the U.S. Army from building a planned $300 million biological aerosol test facility at Dugway Proving Ground in Utah until an environmental impact statement had been developed. The Foundation on Economic Trends, an organization run by Jeremy Rifkin, which is against biotechnology, is fighting the Army's biological weapons research by asking for environmental impact statements. A $100,000 defense fund for microbiologists who publicize illegal biological weapons research has also been established.

The opposition to the production of new living forms and their release into the environment believes that the organisms cannot be tested adequately to insure that there is no danger to plants, animals, and humans. In support of the idea that a cautious approach should be taken is the observation that some experimental organisms may not be benign.

In 1985, Advanced Genetic Sciences of Oakland, California, which had developed a *Pseudomonas* that reduces frost damage to certain crops, tested the bacterium on trees on their rooftop to see if the genetically engineered *Pseudomonas* caused damage. They tested before they received a permit and most importantly, they did not report to the U.S. Government agency that approves such releases that damage occurred to some of the inoculated trees. When the Environmental Protection Agency found out about the unauthorized experiment and the damage to the trees, they withdrew the permit and fined the company $20,000.

Also in 1985, the Department of Agriculture's Animal and Plant Health Inspection Service, apparently with little consideration of environmental release and with no permit, encouraged a firm in Omaha, Nebraska, to produce and test a genetically engineered vaccine against the herpes-like virus that causes pseudorabies in swine. The disease affects most of the nation's swine herds causing many pig deaths and millions of dollars of loss. Previously produced pseudorabies vaccines were made by growing the virus in chick cell cultures until attenuated mutants were isolated that did not cause disease. The genetically engineered virus is attenuated by removing a region of its genome necessary for replication. However, no experiments were done to determine what the reversion rate in nature might be compared to that in the attenuated viruses obtained from cell cultures.

In 1986, the U.S. Department of Agriculture licensed the Omaha firm to make the pseudorabies vaccine (Omnivac). Omnivac became the first live product of genetic engineering licensed for use in the environment. The few tests performed showed the vaccine to be highly effective and safe to the extent it was tested.

The general public is concerned about the release of genetically engineered organisms into the environment. It is not clear to the public what the danger or risk is to agricultural crops, farm and domestic animals, or humans because testing has been done poorly or not at all. Some microbiologists are aware of the dangers when large amounts of usually nonpathogenic organisms are released into a population. Those engaged in genetic engineering experiments and those businesses and government agencies promoting the

FOCUS *SERRATIA MARCESCENS* **AND GERM WARFARE**

In the late 1940s, the U.S. Army was involved in research to determine the feasibility of using microorganisms in war and the impact these microorganisms might have when used on U.S. cities. In 1950, the U.S. Army released aerosols of *Serratia marcescens* near the Golden Gate Bridge to determine how the bacteria would spread over San Francisco. *Serratia* was used in the experiment because it was considered nonpathogenic and because it produces a bright red pigment when grown at 18°C. The pigment allowed investigators to distinguish easily between *Serratia* and other bacteria in the air that might fall onto solid nutrient media placed around the city.

As it turns out, *Serratia marcescens* is not as innocuous as was once thought. *Serratia* is a common cause of diarrhea, pulmonary infection, and urinary tract infections. Records gathered at the time of the experiment indicate that 11 patients at a San Francisco hospital developed urinary tract infections and one patient developed a pneumonia due to *Serratia* during the five months after the release of the bacteria. Apparently, the bacteria did a good job of contaminating the hospital. How many other infections in San Francisco were caused by the Army experiment will probably remain unknown forever. This experiment points out, however, that uncontrolled release of microorganisms can be very dangerous and cause adverse effects over a long period of time.

use and release of genetically engineered organisms have not been politically active. The public has become suspicious and mistrustful because of the incomplete manner in which the organisms have been tested. No environmental report or adequate public education has been forthcoming.

THE IMPORTANCE OF MICROORGANISMS IN INDUSTRY

Natural products synthesized by microorganisms are becoming increasingly important in world markets. Anticoagulants, antidepressants, vasodilators, herbicides, insecticides, plant hormones, enzymes, and enzyme inhibitors have all been isolated from microorganisms (table 24.4). Microorganisms are being used increasingly in the production of enzymes, such as **amylases** (breakdown starches), used in brewing, baking, and textile production, and **proteases** (breakdown proteins), used for tenderizing meats, preparing leathers, and making detergents and cheeses.

The food, petroleum, cosmetic, and pharmaceutical industries are also using microorganisms to make polysaccharides. For example, the bacterium *Xanthomonas campestris* produces the polysaccharide called **xanthan**, which is used to stabilize and thicken foods, as part of drilling muds, as a base for cosmetics, as a binding agent in many pharmaceuticals, and for textile printing and dyeing. *Leuconostoc mesenteroides*, when grown on sucrose, produces **dextran**, a polysaccharide that is used to extend blood plasma.

The chemical industry produces tons of amino acids, nucleotides, vitamins, and organic acids, which are sold to many different types of laboratories and to health food stores. L-Lysine, which is prescribed by some doctors to treat herpes infections, is synthesized by the bacterium *Corynebacterium glutamicum*. Vitamins such as B_{12} (cyanocobalamin) and B_2 (riboflavin)

TABLE 24.4
COMMERCIALLY IMPORTANT MICROBIAL PRODUCTS

Product	Use	Source
Phialocin	Anticoagulant	*Phialocephala repens*
1,3-Diphenethylurea	Antidepressant	*Streptomyces* sp.
Avermectin	Antihelminthic	*Streptomyces avermitilus*
Vitamin B_{12}	Antipernicious anemia	*Streptomyces griseus*
Naematolin	Coronary vasodialator	*Naematoloma fasciculare*
Monascin	Food pigment	*Monascus* sp.
Herbicidin	Herbicide	*Streptomyces saganonensis*
Fusaric acid	Lowers blood pressure	*Fusarium* sp.
N-Acetylmuramyl tripeptide	Immune enhancer	*Bacillus cereus*
Pericidin	Insecticide	*Streptomyces mobaraensis*
Tetranactin	Miticide	*Streptomyces aureus*
Gibberillic acid	Plant hormone	*Gibberella fujikuroi*
Lactic acid	Chemical reagents	*Lactobacillus delbruekii*
Acetic acid	Vinegar, solvent	*Acetobacter* sp.
Citric acid	Food preservative	*Aspergillus niger*
Glutamic acid	Growth factor, preservative	*Micrococcus glutamicus*
Lysine	Growth factor	*Micrococcus glutamicus*
Valine	Growth factor	*Escherichia coli*
Amylase	Hydrolysis of starch (brewing)	*Aspergillus* sp.
Pectinases	Degrading pectin	*Aspergillus* sp. & *Rhizopus* sp.
Streptokinase	Degradation of blood clots (thrombolytic)	*Streptococcus pyogenes*
Proteases	Proteins to amino acids (tenderizing meats)	*Bacillus subtilis, B. licheniformis*
Invertase	Sucrose degraded to glucose & fructose	*Saccharomyces cerevisiae*
Penicillinase	Degradation of penicillin	*Bacillus subtilis*
Glycerol	Preserving foods	*Saccharomyces cerevisiae*
2,3-Butanediol	Chemical laboratories	*Bacillus polymyxa*
Butanol, acetone	Chemical laboratories, manufacturing	*Clostridium acetobutylicum*
Ethanol	Chemical laboratories	*Saccharomyces* sp.

are synthesized by the bacterium *Pseudomonas denitrificans* and the yeast *Ashbya gossypii*, respectively.

Microorganisms themselves are valuable because of the jobs they can carry out for humans. For example, microorganisms such as the bacterium *Pseudomonas putida* are being used to degrade various components of oil in order to clean up oil spills.

The bacterium *Bacillus thuringiensis* is being used successfully as a pesticide and as a pesticide producer. *Bacillus thuringiensis,* when consumed by many insect larvae, produces an enzyme that degrades the polysaccharide that holds cells together. This enzyme severely damages the digestive tracts of the larvae and kills them while showing no adverse effects on other organisms. In addition, a toxin has been isolated from *Bacillus thuringiensis* that can be sprayed on plants to kill insect larvae in the absence of the bacteria.

Microorganisms are being used to leach low-grade ores so as to extract valuable metals. For example, copper and uranium can be leached from low-grade ores by bacteria such as *Thiobacillus*. First, *Thiobacillus* frees ferrous ions (Fe^{2+}) in FeS ores by oxidizing the ferrous ion to the ferric ion (Fe^{3+}). Then, the ferric ion oxidizes the sulfur in pyrite (FeS_2), the yellow mineral often called fool's gold, so that sulfuric acid is produced. In addition, the ferric ion oxidizes insoluble CuS to the soluble $CuSO_4$, which can easily be collected.

The Production of Pharmaceuticals

Microorganisms are involved in the production of more than $1.5 billion worth of drugs in the United States each year. When you compare this to the $7.5 billion of prescription drugs sold each year in the United States, it is clear that biotechnology is big business. Worldwide sales of just four groups of antibiotics—the tetracyclines, the penicillins, the cephalosporins, and erythromycin—total over $4 billion per year. The industrial production of antibiotics relies upon a small

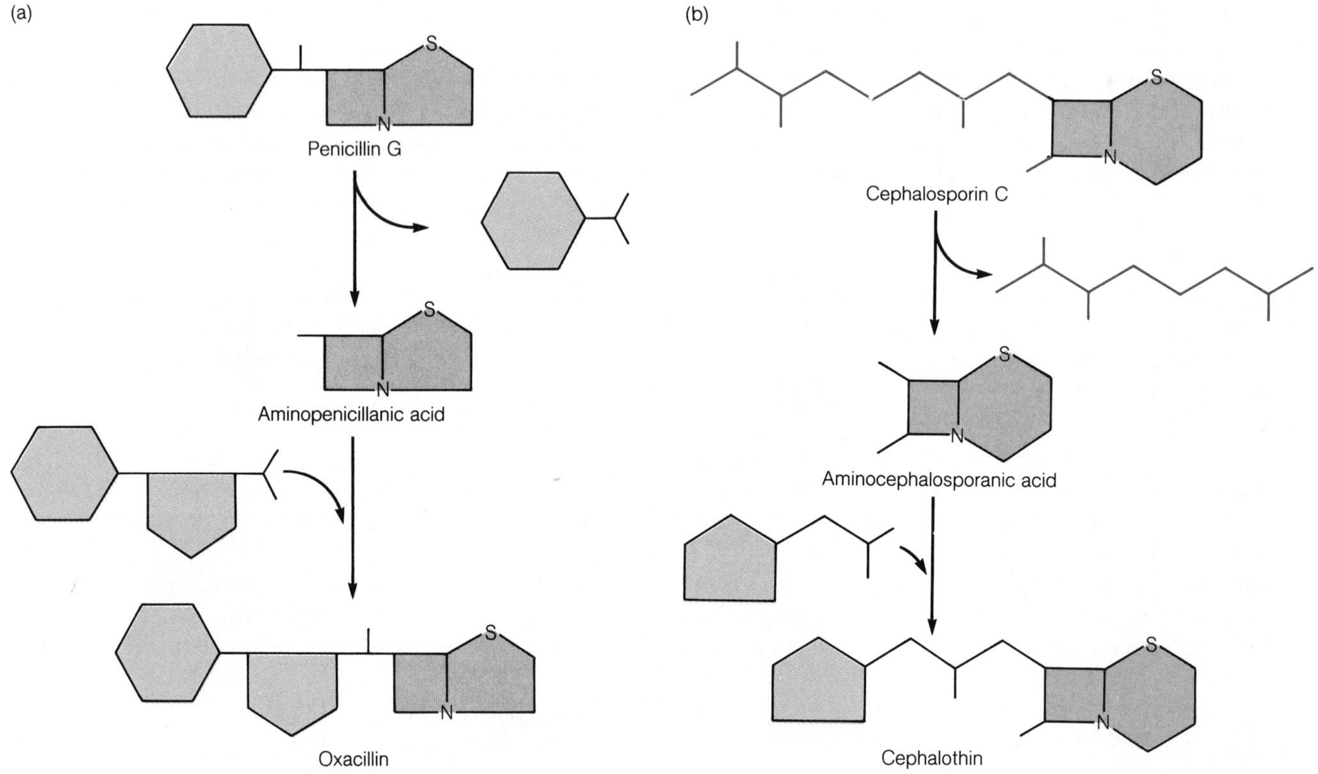

(a)
Penicillin G
Aminopenicillanic acid
Oxacillin

(b)
Cephalosporin C
Aminocephalosporanic acid
Cephalothin

Figure 24.10 The production of synthetic penicillins and cephalosporins

group of spore-forming microorganisms. Various species of the fungus *Cephalosporium* synthesize the cephalosporins; the fungus *Penicillium* produces the natural penicillins; and species of the bacterium *Streptomyces* synthesize numerous antibiotics, such as streptomycin, tetracycline, erythromycin, and chloramphenicol.

In the last few years, the natural penicillins and cephalosoporins have been modified chemically to produce semisynthetic antibiotics that are effective against a broader spectrum of bacteria, and that are not inactivated as easily by bacterial enzymes.

The manufacture of antibiotics takes place in large growth tanks (fig. 24.2). For example, *Penicillium chrysogenum* may be grown in 100,000-liter (26,000-gallon) fermenters for approximately 200 hours (8 days). The removal of the fungus from such large amounts of growth media takes approximately 15 hours. The fungus grows on sugar and phenylacetic acid, which are added continuously. The phenylacetic acid provides the group side of penicillin G (fig. 24.10). Penicillin G is extracted from the filtrate and crystallized. To make the semisynthetic penicillins, penicillin G is mixed with a bacterium that secretes enzymes that remove the side group from penicillin G and so convert it to **aminopenicillanic acid**. Aminopenicillanic acid is the core molecule that is used to make other penicillins. Various chemical groups are added to aminopenicillanic acid (fig. 24.10). Similarly, cephalosporin C has an amide side group removed and then new groups are added to the core (aminocephalosporanic acid) (fig. 24.11).

Wine and Vinegar Production

Winemaking Wine is the aged product of the alcoholic fermentation of grape juice by yeasts. The term "wine" has been applied loosely to any alcoholic beverage that results from the fermentation of fruit juices by yeasts.

Winemaking is an industrial process involving several steps (fig. 24.11). It begins with the crushing and destemming of the grapes. The crushed grapes, called the **must**, are treated with sodium metabisulfite (or another form of sulfur dioxide) to kill wild yeasts, bacteria, and fungi. The must consists of grape juice (90–95%), grape skins, and seeds. It is placed in fermentation tanks, inoculated with wine yeasts (*Saccharomyces cerevisiae* or *Saccharomyces ellipsoideus*), and allowed to ferment under carefully controlled conditions of temperature (27–30°C) and humidity.

During the fermentation step, red, rosé, or white wines are made. The skins and seeds of red grapes contain dyes and chemicals, which, if allowed to stay in the fermentation tanks for 5–10 days, will make the wine red and flavor it. If the grape skins are removed shortly after the fermentation begins (1–2 days), the wine is pink and is called rosé wine. If white grapes are used or the skins from red grapes are excluded from the must altogether, white wines will result.

The fermentation is allowed to take place until the desired alcohol concentration is achieved, usually ranging from about 8–9% in sweet wines to about 12–14% in dry wines. Dry wines are those in which all of the available sugar has been fermented to ethanol and carbon dioxide.

During the fermentation of most wines, the carbon dioxide is allowed to escape; however, sparkling wines are prepared so that they retain some of the carbon dioxide. The carbon dioxide is responsible for the bubbles in sparkling wines, such as champagne, spumante, and cold duck.

After the fermentation has reached the desired point, the fermented grape juice is clarified (fining), filtered, pasteurized, and placed into casks for aging. The fermented grape juice is allowed to age for months to years before the resulting wine is bottled. The aging process is essential to winemaking because it is during this step that the subtle changes or **secondary fermentations** take place, making the fermented grape juice into wine. As a general rule, red wines are aged for much longer periods of time than white wines (two to ten years for red versus one to two years for white) before they are bottled. After aging and bottling, the wines are distributed to the consumer or aged further in the bottle. Sparkling wines, for example, are allowed to ferment further in the bottle (secondary fermentation of sugars) so that a small amount of carbon dioxide is produced while in the bottle, giving the wine its sparkle.

Vinegar Production More than 1.4 million tons of acetic acid, with a market value of $500 million, are synthesized each year in the United States. This acid is used in the manufacture of rubber, plastics, dyes, and insecticides, and as a substrate for the production of amino acids. Presently, very little of the industrial acetic acid is made by microorganisms converting ethanol to acetic acid. Much of the acetic acid used as vinegar, however, is made from ethyl alcohol under aerobic conditions by the bacteria *Acetobacter aceti*, *A. orleanense*, and *A. schutzenbachii*, which are widely distributed in the soil and on plant material.

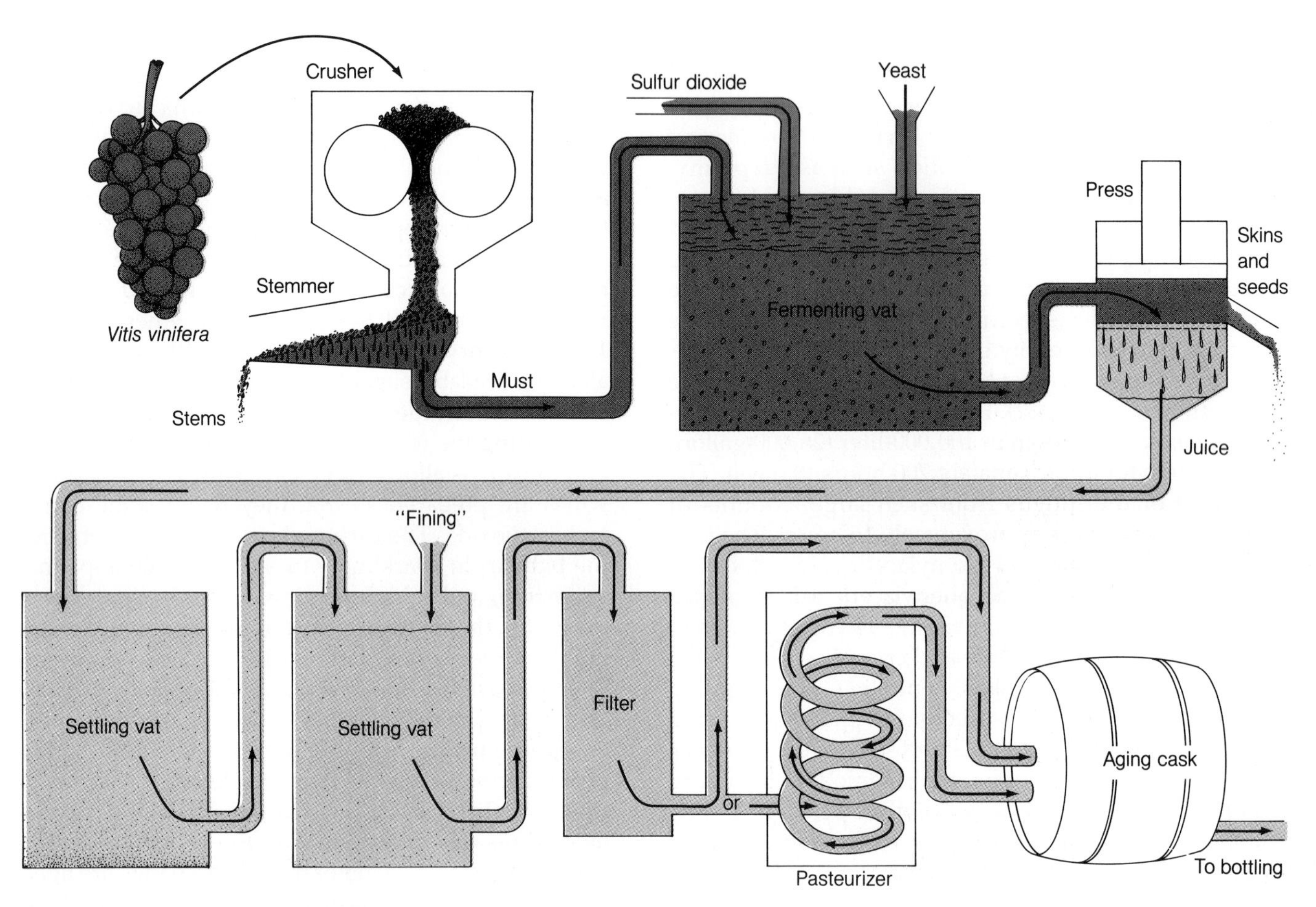

Figure 24.11 Winemaking. White wines are usually made from white grapes, while red wines are made from red grapes. Grapes are crushed to release the juice and the stems are removed. The juice, skins, and seeds (called the must) are treated with sulfur dioxide (SO_2) at about 100 ppm to inhibit undesirable yeasts and filamentous fungi. The must is transferred to a fermenting vat, where yeasts such as *Saccharomyces cerevisiae* are added. The juice is separated from the grape skins and seeds after a few days of fermentation in the press. The time at which skins and seeds are removed determines the color of the wine and the concentration of tannins. The red wines are fermented at about 22°C for 3–5 days with the skins; the light red wines have the skins removed earlier; the white wines are fermented at about 15°C for 7–14 days without the skins. Dry wines contain little or no sugar, while sweet wines contain sugar. Settling vats are used to allow the sedimentation of yeast, proteins, and tannins. The process of adding chemicals to promote settling of particles is called fining. The wine is then filtered and pasteurized. The aging of wine occurs as chemical reactions change some of the alcohol and remaining organic acids.

Commercial vinegar is made in large tanks that contain beechwood shavings (fig. 24.12). Alcohol is trickled from the top of the tank through wood shavings that have *Acetobacter* growing on them. The alcohol that accumulates at the bottom of the tank is recirculated and allowed to trickle through the shavings a number of times. Air enters from the bottom of the tank to provide oxygen for the bacteria. The tanks are kept at a temperature 25–30°C for 8–10 days. Oxidation of the alcohol produces a 4–5% solution of vinegar. The vinegar is often concentrated by distillation.

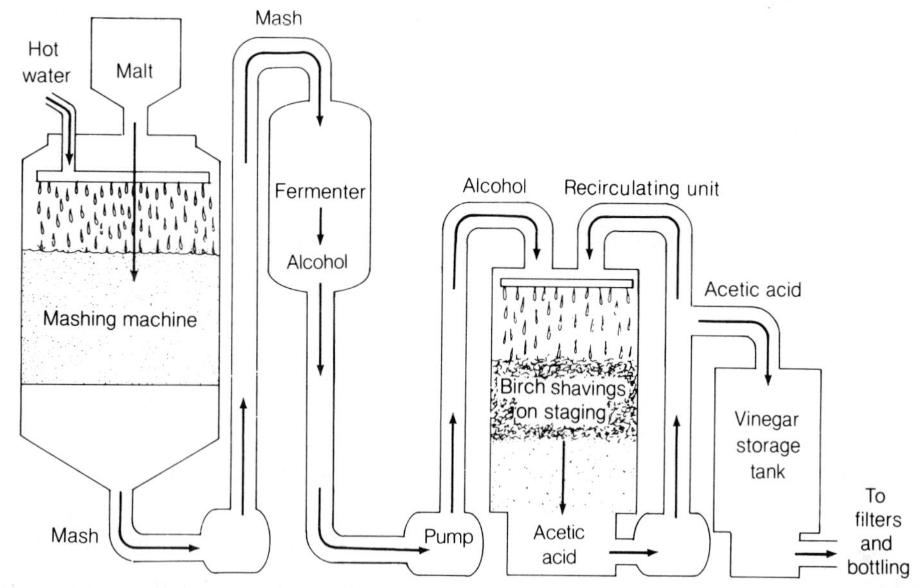

Figure 24.12 Manufacture of vinegar. Barley malt (sprouted barley seeds) can be the starting material used to make acetic acid (vinegar). The sprouted barley seeds produce enzymes that can break down the starches in the seeds to glucose and maltose. The malted barley is crushed in the mashing machine to release the enzymes that break down the starch. After the enzymes have broken down the starch, hot water is added to denature the enzymes and to create a sugar medium that will support the growth of fermenting yeasts. The mash is pumped to a fermenter, where yeasts *(Saccharomyces)* ferment the sugar to ethanol and CO_2. The alcoholic product may be stored for a number of months before being used to make vinegar. The ethanol is pumped to a tower and allowed to trickle down through shelves of beechwood shavings upon which *Acetobacter* is growing. *Acetobacter* oxidizes the alcohol to acetic acid. The ethanol and acetic acid are recycled until there is approximately 5% vinegar. The vinegar is stored in large tanks before it is filtered and bottled.

SUMMARY

GENETIC ENGINEERING

1. When new organisms are made either through breeding experiments or by fusing cells, this is known as bioengineering.
2. Genetic engineering is the alteration of an organism's hereditary information by introducing new pieces of DNA.

THE BASIS OF GENETIC ENGINEERING

1. The discovery, isolation, and characterization of restriction endonucleases that cleave DNA molecules at specific sites has allowed researchers to recombine DNA molecules from different organisms *in vitro*.
2. DNA molecules of interest are generally spliced into vectors, such as bacterial plasmids, bacteriophages, and hybrid eukaryotic and prokaryotic viruses.

A METHOD FOR CLONING GENES

1. Genetically engineered bacterial plasmids (vectors) carrying human, animal, and chemically synthesized genes are taken up by *E. coli*. The bacteria are proliferated in a medium that selects for the vectors. The selection and proliferation of a particular genetically engineered vector in a host cell is known as cloning.
2. Presently, genes for human insulin, interferons, growth hormone, urokinase, and tissue plasminogen activator have been spliced into genetically engineered bacterial plasmids.

METHODS FOR PRODUCING MEDICALLY IMPORTANT PROTEINS

Genes can be expressed in the host cell if they are placed next to a promoter site that the host cell's RNA polymerase can recognize. Genes that are not under the control of a promoter will not be expressed.

HYBRIDOMAS

1. Hybridomas are eukaryotic cells formed from the fusion of a normal plasma cell and a cancerous plasma cell.
2. Hybridomas are of use because they produce only one type of antibody (monoclonal antibody) against a specific antigenic determinant.
3. Monoclonal antibodies can be used (a) as a diagnostic tool; (b) to identify, quantify, and purify valuable proteins; (c) as passive vaccines against toxins and disease-causing organisms; and (d) to treat many types of cancers.
4. In the years ahead, genetic engineering will play an important role in the development of useful genes and proteins.

BIOENGINEERING, SAFETY, AND POLITICS

1. Microorganisms and viruses, when released into the environment in large numbers, may cause disease in plants or animals where no disease existed previously. Consequently, all genetically engineered organisms, viruses, and vaccines should be tested thoroughly under carefully controlled conditions to determine the consequences of their use and release into the environment.
2. In a number of cases governments, scientists, and regulatory agencies have carried out or allowed experiments that were not carefully controlled. These few cases have undermined public confidence in bioengineering and may make it difficult for valuable and safe developments to rapidly reach the public.

THE IMPORTANCE OF MICROORGANISMS IN INDUSTRY

1. Microorganisms are now being used in a number of industries to produce valuable products such as enzymes, polysaccharides, amino acids, hormones, and monoclonal antibodies.
2. Microorganisms are also being used to degrade toxic chemicals, disintegrate oil spills, serve as pesticides, and aid in mining.
3. Enzymes are used to tenderize meat, prepare leathers, produce detergents, and make cheeses.
4. Polysaccharides are used to stabilize and thicken foods, as part of drilling muds, as cosmetic bases, as binding agents in many pharmaceuticals, as extenders of blood plasma, and as molecular sieves.
5. Hormones such as insulin and growth hormone are used to treat people who have genetic defects.
6. The fungi *Penicillium* and *Cephalosporium* produce the β-lactam antibiotics pencillin G and cephalosporin C, respectively. Bacteria in the genus *Streptomyces* also produce antibiotics.
7. Penicillins and cephalosporins inhibit the synthesis and assembly of peptidoglycan cell walls in growing bacteria.
8. Wines are generally made from the juice of grapes.
9. Yeasts of the genus *Saccharomyces* ferment the grape sugar, turning it into ethanol.
10. The skins and seeds from red grapes contain various chemicals and pigments that flavor and color the ferment.

11. If the skins and seeds are totally excluded from the fermentation, white wine results.
12. Alcohols produced by microorganisms are used as solvents, antifreeze, fuel (gasohol), and substrates for vinegar production.
13. *Acetobacter* species are used to make vinegar from ethanol under aerobic conditions.

STUDY QUESTIONS

1. Define and explain the term restriction endonuclease.
2. Show with a drawing why two pieces of DNA that have been cut with Pst-I will join together.
3. Explain what cloning means.
4. What does the term vector mean to the molecular biologist? What does this term mean to the epidemiologist?
5. Why are plasmids, bacterial viruses, and animal viruses used as vectors rather than any piece of DNA?
6. Explain how vectors and the spliced DNA are selected for in bacteria.
7. Outline the procedure used to produce the protein subunits of human insulin in *E. coli*.
8. Define and explain the term, hybridoma.
9. Define a monoclonal antibody.
10. List the uses for monoclonal antibodies.
11. What is the difference between penicillin G and semisynthetic penicillins?
12. How do the penicillins and cephalosporins differ chemically and in the manner in which they affect bacteria?
13. Which human proteins are being produced in genetically engineered bacteria?
14. Which microorganisms are used to make ethanol and vinegar?
15. What materials and microorganisms are generally used to make wines?
16. What procedure accounts for red, rosé, and white wines?

SELF-TEST

1. Which of the following have been produced by genetic engineering?
   a. interleukin-2;
   b. growth hormone;
   c. interferon;
   d. insulin;
   e. all of the above.
2. Hybridomas produce what useful molecule?
   a. interleukin-2;
   b. growth hormone;
   c. interferon;
   d. insulin;
   e. monoclonal antibodies.
3. What are nucleic acid probes?
   a. monoclonal antibodies used to identify pieces of DNA;
   b. plasmids used as vaccines;
   c. pieces of DNA used to discover identical regions in the genomes of organisms;
   d. pieces of DNA used as vaccines;
   e. none of the above.
4. Bioengineering has produced which of the following useful organisms?
   a. vaccinia virus that produces rabies proteins;
   b. vaccinia virus that produces herpes proteins;
   c. genetically modified pseudorabies virus;
   d. *Pseudomonas* that does not catalyze the formation of ice;
   e. all of the above.
5. What cuts DNA so that staggered ends are produced?
   a. ligase;
   b. endonuclease;
   c. DNA polymerase;
   d. methylase;
   e. ATPase.
6. Two pieces of DNA can always recombine with each other if they are cut with different endonucleases. a. true; b. false.
7. Pieces of DNA to be cloned are generally inserted into a:
   a. chromosome;
   b. plasmid;
   c. virus;
   d. all of the above;
   e. b and c only.
8. Most vectors can easily be selected and cloned if they make the host cell resistant to some drug. a. true; b. false.
9. How many "genes" are necessary to code for the proteins in genetically engineered insulin?
   a. 1;
   b. 2;
   c. 3;
   d. 4;
   e. 5.

10. A bacterium has how many antigenic determinants?
    a. 0;
    b. 1;
    c. 2;
    d. many;
    e. it depends upon the size of the culture.

11. When microorganisms are used to make materials such as steroid hormones, antibiotics, amino acids, or ethanol, which term best describes the industries?
    a. bioengineering;
    b. genetic engineering;
    c. biotechnology;
    d. fermentation;
    e. genetic recombination.

12. When new organisms are made either through breeding experiments or by fusing cells, which term best describes the science?
    a. bioengineering;
    b. genetic engineering;
    c. biotechnology;
    d. fermentation;
    e. genetic recombination.

13. When an organism's hereditary information is altered by introducing new pieces of DNA, which term best describes the procedure?
    a. bioengineering;
    b. genetic engineering;
    c. biotechnology;
    d. fermentation;
    e. genetic recombination.

14. Which enzyme is used to cut DNA at specific locations?
    a. DNA methylase;
    b. DNA polymerase;
    c. DNA ligase;
    d. reverse transcriptase;
    e. restriction endonuclease.

15. In genetic engineering, which is frequently used as a vector?
    a. cell clone;
    b. genome;
    c. hybridoma;
    d. plasmid;
    e. mosquito.

16. If two separate DNAs are cut by the same endonuclease, what does this indicate?
    a. each DNA has an identical restriction endonuclease site;
    b. the two DNAs are identical;
    c. recombinant DNAs can be made by mixing and ligating;
    d. all of the above;
    e. a and c only.

17. What makes it possible to clone a particular plasmid in a medium containing an antibiotic?
    a. antibiotic sensitivity gene on the plasmid;
    b. antibiotic resistance gene on the chromosome;
    c. antibiotic sensitivity gene on the chromosome;
    d. antibiotic resistance gene on the plasmid;
    e. any of the above.

18. The various penicillins and cephalosporins are made by using strains of fungi and bacteria that produce the different antibiotics. a. true; b. false.

19. What important function does the yeast *Saccharomyces* play in wine production?
    a. colors the wine;
    b. makes the wine acidic;
    c. produces ethanol;
    d. prevents spoilage;
    e. produces tannins.

20. What material does *Acetobacter* use to make vinegar?
    a. ethanol;
    b. lactic acid;
    c. sucrose;
    d. lactose;
    e. none of the above.

21. A cell formed from the fusion of a plasma cell and a cancerous cell is called a:
    a. myeloma;
    b. hybridoma;
    c. monoclonal granulocyte;
    d. activated K-lymphocyte;
    e. monoclonal antibody.

22. Antibodies formed from a single isolate of a hybridoma are known as:
    a. antigenic determinants;
    b. antiserum;
    c. antigens;
    d. monoclonal antibodies;
    e. none of the above.

SUPPLEMENTAL READINGS

Abelson, J. 1983. Biotechnology: An overview. *Science* 219:611–613.

Abraham, E. P. 1981. The beta-lactam antibiotics. *Scientific American* 244(6):76–86.

Aharonowitz, Y. and Cohn, G. 1981. The microbiological production of pharmaceuticals. *Scientific American* 245(3): 140–152.

Brill, W. J. 1985. Safety concerns and genetic engineering in agriculture. *Science* 227:381–384.

Caskey, T. C. 1987. Disease diagnosis by recombinant DNA methods. *Science* 236(4806):1223–1228.

Chilton, M. 1983. A vector for introducing new genes into plants. *Scientific American* 248:50–59.

Clark, S. C. and Kamen, R. 1987. The human hematopoietic colony-stimulating factors. *Science* 236(4806):1229–1237.

Collier, J. R. and Kaplan, D. A. 1984. Immunotoxins. *Scientific American* 251(1):56–64.

Demain, A. L. and Solomon, N. A. 1981. Industrial microbiology. *Scientific American* 245(3):66–75.

Eveleigh, D. E. 1981. The microbiological production of industrial chemicals. *Scientific American* 245(3):154–178.

Gaden, E. L. 1981. Production methods in industrial microbiology. *Scientific American* 245(3):180–196.

Gilbert, W. and Villa-Komarooff, L. 1980. Useful proteins from recombinant bacteria. *Scientific American* 242:74–94.

Hirsch, A. M., Drake, D., Jacobs, T. W., and Long, S. R. 1985. Nodules are induced on alfalfa roots by *Agrobacterium tumefaciens* and *Rhizobium trifolli* containing small segments of *Rhizobium meliloti* nodulation region. *Journal of Bacteriology* 161(1):223–230.

Hopwood, D. A. 1981. The genetic programming of industrial microorganisms. *Scientific American* 245(3):90–102.

Kolata, G. 1987. Immune boostera. *Discover* 8(9):68–74.

Macario, E. and Macario, A. 1983. Monoclonal antibodies for bacterial identification and taxonomy. *ASM News* 49:1–7.

Milstein, C. 1980. Monoclonal antibodies. *Scientific American* 243:66–74.

Novick, R. 1980. Plasmids. *Scientific American* 243:102–127.

Nowinski, R., Tam, M., Goldstein, L., Stong, L., Kuo, C., Caorey, L., Stamm, W., Handsfield, H., Knapp, J., and Holmes, K. 1983. Monoclonal antibodies for diagnosis of infectious diseases in humans. *Science* 219:637–644.

Old, R. and Primrose, S. 1985. *Principles of gene manipulation*. Berkeley: University of California Press.

Phaff, H. 1981. Industrial microorganisms. *Scientific American* 245(3):76–89.

Rose, M. J. 1981. The microbiological production of food and drink. *Scientific American* 245(3):126–138.

Smith, J. E. 1985. *Biotechnology principles: Aspects of microbiology.* American Society for Microbiology. Washington, D.C.

Vaeck, M., Reynaerts, A., Höfte, H., Jansens, S., De Beuckeleer, M., Dean, C., Zabeau, M., Van Montagu, M., and Leemans, J. 1987. Transgenic plants protected from insect attack. *Nature* 328(6125):33–37.

Webb, A. D. 1984. The science of making wine. *American Scientist* 72:360–367.

APPENDIX A

DISEASE REFERENCE GUIDE

ACQUIRED IMMUNODEFICIENCY SYNDROME

Cause(s) Human Immunodeficiency virus (HIV)
Signs and Symptoms Inability to develop a strong immune response to infectious agents. Susceptibility to many infectious agents.
Mode of Transmission Sexual contact. Blood transfusions. Intravenous drug abuse.
Preventive Measures Avoid intimate contact with infected individuals. Receive blood that has been screened for HIV
Treatment of Choice AZT

AMEBIASIS

Cause(s) *Entamoeba histolytica*
Signs and Symptoms Dysentery-like symptoms. Fever, malaise, diarrhea with blood and mucus, painful intestinal contractions.
Mode of Transmission Fecal-oral route
Preventive Measures Sanitize foods and water before consumption
Treatment of Choice Quinacrine hydrochloride, metrodinazole

BACILLARY DYSENTERY

Cause(s) *Shigella* sp., *Campylobacter jejuni*, *Escherichia coli*
Signs and Symptoms Abdominal pain and cramps, vomiting, diarrhea with blood and mucus, fever, and malaise
Mode of Transmission Fecal-oral route
Preventive Measures Cook foods well; sanitize water
Treatment of Choice Ampicillin, trimethoprim-sulfamethoxazole, gentamycin

BOTULISM

Cause(s) *Clostridium botulinum*
Signs and Symptoms Fatigue, weakness, dizziness, nausea, vomiting, diarrhea, and cardiac and respiratory paralysis
Mode of Transmission Ingestion of foods containing botulinal toxin
Preventive Measures Cook foods well, especially canned goods
Treatment of Choice Administration of polyvalent antiserum

BRUCELLOSIS

Cause(s) *Brucella abortus*, *B. melitensis*, and *B. suis*
Signs and Symptoms Fever, often fluctuating; body aches and anorexia (loss of appetite)
Mode of Transmission Ingestion of contaminated dairy products or other foods; skin abrasions or inhalation
Preventive Measures Sanitize foods
Treatment of Choice Tetracycline and streptomycin administered together

CANDIDIASIS

Cause(s) *Candida albicans*
Signs and Symptoms Development of a pseudomembrane and a yellowish exudate
Mode of Transmission Disease usually endogenous. Some forms (vaginitis and balanitis) are transmitted sexually
Preventive Measures Maintain overall good health
Treatment of Choice Mycostatin, Amphotericin B, miconazole

CHANCROID

Cause(s) *Haemophilus ducreyi*
Signs and Symptoms Pustule or ulcer at site of infection. Scar remains after healing. Progresses rapidly
Mode of Transmission Sexually transmitted
Preventive Measures Avoid sexual contact with infected individuals
Treatment of Choice Sulfonamides, chloroamphenicol, tetracyclines

CHOLERA

Causes(s) *Vibrio cholerae*
Signs and Symptoms Violent vomiting, copious rice water stool (diarrhea). Hypovolemic shock and death in severe cases
Mode of Transmission Fecal-oral route

Preventive Measures Vaccination; sanitize water before drinking
Treatment of Choice Tetracyclines, and replacement of fluids and electrolytes

COCCIDIOIDOMYCOSIS

Cause(s) *Coccidioides immitis*
Signs and Symptoms Chronic cough, fever, chest pains, malaise, weight loss
Mode of Transmission Inhalation of conidia
Preventive Measures Vaccination (experimental); wear face masks during dust storms in endemic areas
Treatment of Choice Amphotericin B, ketoconazole

COMMON COLD

Cause(s) Rhinoviruses, coronaviruses, and adenoviruses
Signs and Symptoms Cough, runny nose, fever, malaise
Mode of Transmission Hand-to-nose, mouth or eye. Some via aerosols
Preventive Measures Wash hands; avoid contact with infected individuals
Treatment of Choice None

CONJUNCTIVITIS

Cause(s) *Haemophilus influenzae, Streptococcus pneumoniae, Moraxella lacunata,* or *Chlamydia trachomatis*
Signs and Symptoms Tearing, itching, and inflammation of the conjunctive. Sometimes blindness
Mode of Transmission Hand-to-eye, aerosols
Preventive Measures Wash hands; avoid contact with infected individuals
Treatment of Choice Chloroamphenicol, penicillin, tetracyclines

CROUP

Cause(s) Respiratory syncytial virus, parainfluenza virus, and Influenza A and B viruses
Signs and Symptoms Laryngitis, loss of voice, deep, noisy cough, difficult breathing, rapid pulse, and fever
Mode of Transmission Aerosols
Preventive Measures Avoid contact with infected individuals
Treatment of Choice Hot compresses on throat, humidify air; establishment of airway if necessary

CYSTITIS

Cause(s) *Escherichia coli, Proteus mirabilis, Enterobacter aerogenes, Klebsiella pneumoniae*
Signs and Symptoms Painful and frequent urination, hematuria
Mode of Transmission Endogenous infection
Preventive Measures Practice good hygiene
Treatment of Choice Gentamycin or tobramycin

DERMATOPHYTOSIS

Cause(s) Fungi in the genera *Microsporum, Trichophyton,* and *Epidermophyton*
Signs and Symptoms Athlete's foot, jock itch, ringlike scalp and skin lesions
Mode of Transmission Contact with humans, animals, or fomites
Preventive Measures Avoid contact with animals or objects carrying infectious agents
Treatment of Choice Oral griseofulvin

DIPHTHERIA

Cause(s) *Corynebacterium diphtheriae*
Signs and Symptoms Fever, headache, malaise. Sore throat, with yellowish membrane in throat and tonsils. Adenitis. Death due to cardiac failure
Mode of Transmission Direct contact with human carrier, or via fomites
Preventive Measures Vaccination
Treatment of Choice Erythromycin and administration of diphtheria antitoxin

EPIGLOTTITIS

Cause(s) *Haemophilus influenzae*
Signs and Symptoms Sore throat, fever, cough, drooling, cyanosis, coma, and (if airway is not reestablished) death
Mode of Transmission Endogenous or by aerosols
Preventive Measures Avoid contact with infected individuals
Treatment of Choice Establish airway and then treat with chloroamphenicol

GASTROENTERITIS

Cause(s) *Salmonella* sp., *Campylobacter jejuni, Shigella* sp., *Escherichia coli*
Signs and Symptoms Diarrhea, vomiting, fever, and malaise
Mode of Transmission Fecal-oral route
Preventive Measures Sanitize or cook foods and beverages before consuming
Treatment of Choice Usually none necessary; ampicillin, chloroamphenicol, gentamycin, or tobramycin for severe cases

GENITAL HERPES

Cause(s) *Herpes simplex virus*
Signs and Symptoms Painful, recurring blisters on genitals
Mode of Transmission Sexually transmitted
Preventive Measures Avoid sexual contact with actively infected individuals
Treatment of Choice Zovirax (topical, oral, or parenteral)

GONORRHEA

Cause(s) *Neisseria gonorrhoeae*
Signs and Symptoms In males: Yellow, mucopurulent discharge from urethra with painful urination. In females: urethral or vaginal discharge; painful, frequent urination
Mode of Transmission Sexually transmitted
Preventive Measures Avoid sexual contact with infected individuals
Treatment of Choice Penicillin or ampicillin

GIARDIASIS

Cause(s) *Giardia lamblia*
Signs and Symptoms Malodorous flatus, greasy stools, diarrhea
Mode of Transmission Fecal-oral route
Preventive Measures Sanitize water
Treatment of Choice Metrodinazole, quinacrine

HEPATITIS, VIRAL

Cause(s) Hepatitis A, B, or non-A non-B viruses
Signs and Symptoms Loss of appetite, jaundice, fever, weakness
Mode of Transmission Fecal-oral route or parenteral
Preventive Measures Sanitize foods or beverages; practice good hygiene
Treatment of Choice Treat symptoms to relieve discomfort

HISTOPLASMOSIS

Cause(s) *Histoplasma capsulatum*
Signs and Symptoms Fever, malaise, cough, chest pain
Mode of Transmission Inhalation of conidia or hyphal fragments
Preventive Measures Wear masks during dust storms and demolition of old buildings
Treatment of Choice Amphotericin B or miconazole

IMPETIGO

Cause(s) *Staphylococcus aureus* or *Streptococcus* group A
Signs and Symptoms Development of pustules on the skin, which become encrusted and rupture
Mode of Transmission Direct contact with nasal secretions of carriers or exudates of infected individuals
Preventive Measures Avoid contact with infected individuals
Treatment of Choice Penicillin or erythromycin

INFECTIOUS MONONUCLEOSIS

Cause(s) Epstein-Barr virus
Signs and Symptoms Increased number of mononuclear white blood cells; weakness, fever, sore throat, and lymphoadenopathy
Mode of Transmission Kissing infected individuals, or contact with contaminated utensils or glassware
Preventive Measures Sanitize glassware, avoid contact with infected individuals
Treatment of Choice Treat symptoms, bed rest

INFLUENZA

Cause(s) Myxoviruses
Signs and Symptoms Fever, chills, headache, cough, runny nose, muscle aches, and sore throat
Mode of Transmission Aerosols
Preventive Measures Vaccination; avoid contact with infected individuals
Treatment of Choice Amantadine, bed rest, and symptomatic treatment

LEGIONELLOSIS

Cause(s) *Legionella pneumophila* and other legionellae
Signs and Symptoms Fever, chills, pneumonia, cough, and chest pain. Onset very rapid
Mode of Transmission Aerosols created by air-conditioning systems
Preventive Measures Sanitize water in air-cooling systems
Treatment of Choice Erythromycin or rifampin

LEPROSY

Cause(s) *Mycobacterium leprae*
Signs and Symptoms Anesthesia of affected parts of the skin, paralysis, ulceration, and deformity of affected areas
Mode of Transmission Direct contact with infected individuals or fomites
Preventive Measures Avoid close personal contact with infected individuals
Treatment of Choice Dapsone

LYME DISEASE

Cause(s) *Borrelia burgdorferi*
Signs and Symptoms Fever, joint pain, and a ringlike rash at site of infection
Mode of Transmission Bites of ticks
Preventive Measures Wear long sleeves and head coverings when venturing into tick-infected areas
Treatment of Choice Tetracyclines

MALARIA

Cause(s) *Plasmodium vivax, P. falciparum, P. malariae, P. ovale*
Signs and Symptoms Bouts of alternating fever and chills every 2–3 days; sweating and prostration
Mode of Transmission Bite of female *Anopheles* mosquito
Preventive Measures Prophylaxis of chloroquine phosphate once a week
Treatment of Choice Chloroquine phosphate and quinine sulfate

MENINGITIS

Cause(s) *Neisseria meningitidis, Haemophilus influenzae, streptococcus pneumoniae.* Viruses, protozoa, and fungi also can cause meningitis
Signs and Symptoms Fever, loss of appetite, hyperesthesia, delirium, convulsions, and coma
Mode of Transmission Direct contact, aerosols
Preventive Measures Prophylaxis with antibiotics; isolating active cases
Treatment of Choice Ampicillin, penicillin, kanamycin, chloroamphenicol, and gentamycin

MUMPS

Cause(s) Mumps virus
Signs and Symptoms Malaise, chills, headache, swelling of parotid glands. Orchitis is a complication
Mode of Transmission Aerosols or contact with secretions contaminated with the mumps virus
Preventive Measures Vaccination
Treatment of Choice Rest and analgesics

NONGONOCOCCAL URETHRITIS

Cause(s) *Chlamydia trachomatis, Ureaplasma urealyticum*
Signs and Symptoms Yellowish urethral discharge, with painful urination
Mode of Transmission Sexually transmitted
Preventive Measures Avoid sexual contact with infected individuals
Treatment of Choice Tetracyclines, erythromycin

NONSPECIFIC VAGINOSIS

Cause(s) *Gardnerella vaginalis* and certain anaerobes
Signs and Symptoms Profuse, malodorous vaginal discharge; clue cells (Epithelical cells with gram-variable rods attached) present.
Mode of Transmission Sexually trasmitted
Preventive Measures Avoid sexual contact with infected individuals
Treatment of Choice Metrodinazole or ampicillin

PHARYNGOTONSILLITIS

Cause(s) *Streptococcus pyogenes*
Signs and Symptoms Sore throat, fever, and swollen lymph nodes of neck region
Mode of Transmission Contact with carriers, or ingestion of contaminated foods
Preventive Measures Avoid contact with infected individuals or eating foods handled by them
Treatment of Choice Penicillin or erythromycin

PLAGUE

Cause(s) *Yersinia pestis*
Signs and Symptoms High fever, restlessness, confusion, prostration, inflammation of lymph nodes (buboes), delirium, shock, and death
Mode of Transmission Bite of infected fleas
Preventive Measures Vaccination; removal of fleas and rodents
Treatment of Choice Streptomycin

PNEUMONIA

Cause(s) *Streptococcus pneumoniae, Klebsiella pneumoniae, Mycoplasma pneumoniae,* Influenza viruses
Signs and Symptoms Chills, high fever of rapid onset, chest pain, purulent exudate from cough, blood in sputum
Mode of Transmission Endogenous of from aerosols
Preventive Measures Vaccination, or avoid contact with infected individuals
Treatment of Choice Penicillin, gentamycin, erythromycin

PSEUDOMEMBRANOUS ENTEROCOLITIS

Cause(s) *Clostridium difficile*
Signs and Symptoms Diarrhea, vomiting, abdominal pain, and plaque-formation in colon mucosa
Mode of Transmission Complication following treatment with certain antibiotics
Preventive Measures None that are effective
Treatment of Choice Penicillin

PUERPERAL SEPSIS

Cause(s) *Streptococcus pyogenes* or *Clostridium* sp.
Signs and Symptoms Fever above 38°C on two consecutive days after parturition or abortion
Mode of Transmission Contamination from gloves or surgical instruments during delivery or abortion
Preventive Measures Practice of aseptic techniques during delivery or abortion
Treatment of Choice Penicillin, erythromycin

Q-FEVER

Cause(s) *Coxiella burnetii*
Signs and Symptoms Headache, fever, myalgia, weight loss, and malaise
Mode of Transmission Drinking contaminated milk; inhalation of cells; handling infected animals
Preventive Measures Pasteurize milk; avoid handling infected animals
Treatment of Choice Tetracyclines

RABIES

Cause(s) Rabies virus (a rhabdovirus)
Signs and Symptoms Malaise, depression, hydrophobia, hyperesthesia, aggressiveness, and death
Mode of Transmission Bite of rabid dog, skunk, bat, or other animals
Preventive Measures Vaccination
Treatment of Choice Vaccination (Pasteur treatment) immediately after bite

RHEUMATIC FEVER

Cause(s) *Streptococcus pyogenes*
Signs and Symptoms Fever, arthralgia, myocarditis, following streptococcal infection
Mode of Transmission Allergic response to previous streptococcal infection
Preventive Measures Prophylaxis with penicillin; treat streptococcal infections promptly
Treatment of Choice Penicillin, with bed rest

ROCKY MOUNTAIN SPOTTED FEVER

Cause(s) *Rickettsia rickettsii*
Signs and Symptoms Rash, fever, headache, myalgia
Mode ot Transmission Bite of infected ticks
Preventive Measures Wear protective clothing when visiting tick-infested areas
Treatment of Choice Tetracyclines

RUBELLA

Cause(s) Rubella virus
Signs and Symptoms Upper respiratory tract disease, accompanied by a rash. Fetal defects or miscarriages occur if mother is infected during first trimester of pregnancy
Mode of Transmission Aerosols
Preventive Measures Vaccination
Treatment of Choice Treat symptoms

RUBEOLA

Cause(s) Measles virus
Signs and Symptoms Upper respiratory tract disease, accompanied by mouth lesions called "Koplik spots"
Mode of Transmission Aerosols and person-to-person contact
Preventive Measures Vaccination
Treatment of Choice Treat symptoms

SCALDED SKIN SYNDROME

Cause(s) *Staphylococcus aureus*
Signs and Symptoms Blistering and shedding of outermost layer of skin
Mode of Transmission Occurs mostly in infants. Contact with persons or objects carrying exfoliatin-producing pathogens
Preventive Measures None that are effective
Treatment of Choice Penicillin

SCARLET FEVER

Cause(s) *Streptococcus pyogenes*
Signs and Symptoms Complication of "strep throat." Development of rash, roughening of the skin, and a "strawberry" tongue. Vomiting may also occur.
Mode of Transmission Direct contact with carriers; ingestion of contaminated foods
Preventive Measures Avoid contact with carriers or foods handled by infected individuals
Treatment of Choice Penicillin or erythromycin

SYPHILIS

Cause(s) *Treponema pallidium*
Signs and Symptoms Development of a chancre (at site of treponemal entrance). Healing of chancre followed by a rash, runny nose, and pronounced tearing.
Mode of Transmission Sexually transmitted
Preventive Measures Avoid sexual contact with infected individuals
Treatment of Choice Penicillin

TAENIASIS

Cause(s) *Taenia saginata* (beef tapeworm), *Taenia solium* (pork tapeworm)
Signs and Symptoms Abdominal pain, weight loss, and weakness. Cysticercosis may develop after *Taenia solium* infections
Mode of Transmission Eating uncooked meats with cysticerci of tapeworms
Preventive Measures Cook meats well
Treatment of Choice Niclosamide, Avermectins

TETANUS

Cause(s) *Clostridium tetani*
Signs and Symptoms Stiffness of jaw and neck muscles. Rigidity of jaw, alteration of voice, tetanic contractions of muscles of back and extremities, arching the body
Mode of Transmission Traumatic implantation of endospores
Preventive Measures Vaccination
Treatment of Choice Booster shot, injection of antitoxin, and treatment with penicillin

TOXIC SHOCK SYNDROME

Cause(s) *Staphylococcus aureus*
Signs and Symptoms Fever, diarrhea, low blood pressure, myalgia, and a rash, which may be followed by peeling of skin; shock
Mode of Transmission Edogenous, or in contaminated tampons
Preventive Measures Change tampons frequently; practice good hygiene
Treatment of Choice Penicillin, and treatment to reverse shock

TOXOPLASMOSIS

Cause(s) *Toxoplasma gondii*
Signs and Symptoms Swellng of lymph nodes, inflammation of the eye, chorioretinitis, and sometimes neurological symptoms
Mode of Transmission Ingestion of oocysts of *Toxoplasma gondii,* in foods or from fingers.
Preventive Measures Wear gloves when changing "kitty litter"; wash hands
Treatment of Choice Pyrimethamine and sulfonamides

TRICHINOSIS

Cause(s) *Trichinella spiralis* (a nematode)
Signs and Symptoms Nausea, vomiting, constipation, or abdominal pain, followed by fever, myalgia, facial edema, and hemorrhaging in eye or tongue
Mode of Transmission Eating poorly cooked pork or bear meat (or other meats) with encysted larvae
Preventive Measures Cook meats well (especially pork and bear)
Treatment of Choice Serious cases are treated with prednisone and thiabendazole

TUBERCULOSIS

Cause(s) *Mycobacterium tuberculosis* or *M. bovis*
Signs and Symptoms Persistent cough, sputum production, weakness, prostration, weight loss. X-rays reveal many lung lesions
Mode of Transmission Inhalation of aerosol particles, or ingestion of contaminated milk
Preventive Measures Avoid contact with infected individuals; pasteurize milk
Treatment of Choice Streptomycin, Para-amino salicylic acid, rifampin

TYPHOID FEVER

Cause(s) *Salmonella typhi*
Signs and Symptoms High fever, abdominal pain, headache. Accompanying lesions on Peyer's patches and spleen

Mode of Transmission Ingestion of contaminated water or foods
Preventive Measures Vaccination; sanitize foods and beverages
Treatment of Choice Chloramphenicol

TYPHUS

Cause(s) *Rickettsia prowazekii*
Signs and Symptoms Severe headache, backache, prostration, high fever, and formation of a rash. Bronchopneumonia is a complication
Mode of Transmission Bite of lice
Preventive Measures Vaccination; avoid contact with louse-infested individuals
Treatment of Choice Tetracyclines

TRICHOMONIASIS

Cause(s) *Trichomonas vaginalis*
Signs and Symptoms Profuse, whitish, frothy vaginal discharge
Mode of Transmission Sexually transmitted
Preventive Measures Avoid sexual contact with infected individuals
Treatment of Choice Metrodinazole

VARICELLA (CHICKENPOX)

Cause(s) Varicella-zoster virus
Signs and Symptoms Fever, headache, malaise, and a vesicular rash that later encrusts
Mode of Transmission Aerosols, or contact with infected individuals
Preventive Measures Avoid contact with infected individuals
Treatment of Choice Bed rest and treatment of symptoms

WHOOPING COUGH

Cause(s) *Bordetella pertussis*
Signs and Symptoms Bouts of violent coughing that last for almost a minute, causing the patient to suck air forcibly producing a "whooping" sound. Vomiting and convulsions
Mode of Transmission Aerosols
Preventive Measures Vaccination
Treatment of Choice Administration of pertussis immune human globulin

APPENDIX B

ANSWERS TO SELF-TESTS

CHAPTER 1

1. a
2. a
3. c
4. d
5. e
6. a
7. c
8. d
9. b
10. d

CHAPTER 2

1. a
2. b
3. a
4. a
5. a
6. d
7. c
8. d
9. d
10. c
11. d
12. e
13. a
14. e
15. e

CHAPTER 3

1. b
2. c
3. a
4. b
5. a
6. e
7. a
8. a
9. d
10. e
11. a
12. b
13. d
14. b
15. a
16. c
17. a
18. a
19. b
20. b

CHAPTER 4

1. a
2. b,d
3. d,e
4. d
5. a
6. a
7. a
8. b
9. b
10. a,c,d
11. a
12. a
13. c
14. b
15. a
16. c
17. e
18. c
19. b
20. a
21. a
22. c
23. c
24. e
25. c
26. b
27. d
28. b
29. a
30. e

CHAPTER 5

1. d
2. c
3. b
4. a
5. e
6. d
7. b
8. d
9. c
10. b
11. a or e
12. a or e
13. c
14. b
15. e
16. e
17. e
18. d
19. c
20. b
21. a
22. b
23. b
24. a

CHAPTER 6

1. c
2. a
3. c
4. b
5. b
6. a
7. a
8. a
9. d
10. a
11. b
12. d
13. a
14. c
15. b
16. c
17. c
18. b
19. b
20. a
21. a
22. a

CHAPTER 7

1. a
2. c
3. b
4. d
5. a
6. e
7. a
8. c
9. a
10. a

CHAPTER 8

1. c
2. b
3. b
4. b
5. d
6. a
7. b
8. b
9. b
10. a

CHAPTER 9

1. a
2. c
3. a
4. b
5. a
6. a
7. b
8. c
9. e
10. a
11. b
12. a
13. a
14. d
15. b

CHAPTER 10

1. a
2. c
3. a
4. a
5. a
6. b
7. d
8. a
9. a
10. c
11. a
12. b
13. b
14. b
15. b
16. c
17. b
18. b
19. b
20. a
21. a

CHAPTER 11

1. a
2. c
3. a
4. a
5. c
6. b
7. b
8. b
9. d
10. a
11. a
12. b
13. d
14. b
15. c
16. c

CHAPTER 12

1. b
2. b
3. a
4. c
5. a
6. b
7. b
8. d
9. b
10. b
11. b
12. b

13. a
14. a
15. b
16. a
17. a
18. b
19. a
20. a

CHAPTER 13

1. e
2. a
3. c
4. a
5. c
6. a
7. a
8. d
9. b
10. a
11. c
12. a
13. a
14. a
15. a
16. b
17. e
18. d
19. b
20. a
21. a
22. a
23. b
24. a
25. b
26. a
27. b
28. a

CHAPTER 14

1. a
2. c
3. d
4. b
5. b
6. b
7. a
8. e
9. d
10. a
11. b
12. a
13. b
14. d
15. a
16. b
17. b
18. e
19. b
20. d
21. b

CHAPTER 15

1. d
2. c
3. d
4. b
5. c
6. c
7. c
8. e
9. e
10. a
11. e
12. e
13. b,d
14. b,d
15. d
16. e
17. a
18. b,c,d
19. d
20. e
21. e
22. e
23. e
24. b
25. a
26. b
27. a
28. a
29. a
30. d

CHAPTER 16

1. d
2. e
3. d
4. d
5. c
6. b
7. d
8. c
9. e
10. a
11. c
12. d
13. a
14. e
15. b
16. b
17. c
18. e
19. a
20. b
21. c
22. e
23. a
24. b
25. a
26. d
27. d
28. b
29. c
30. d

CHAPTER 17

1. c
2. d
3. e
4. b
5. a
6. d
7. d
8. e
9. e
10. a
11. c
12. a
13. e
14. c
15. b
16. c
17. a
18. c
19. e
20. b
21. a
22. d
23. a
24. c
25. a
26. a
27. b
28. c
29. d
30. d

CHAPTER 18

1. e
2. b
3. a
4. d
5. b
6. a
7. e
8. e
9. b
10. a
11. c
12. e
13. a
14. c
15. a
16. b
17. a
18. b
19. a
20. a
21. a
22. a
23. b
24. a
25. a
26. b
27. c
28. a
29. c
30. a

CHAPTER 19

1. a
2. c
3. d
4. a
5. b
6. d
7. c
8. d
9. e
10. b
11. c
12. d
13. e
14. b
15. c,d
16. a
17. c
18. b
19. e
20. a,c,d
21. a

CHAPTER 20

1. c
2. d
3. a
4. a
5. b
6. b
7. c
8. c
9. b
10. c
11. e
12. a
13. b
14. c
15. b
16. c
17. e
18. b
19. d
20. e
21. c
22. d
23. c
24. d
25. c
26. d
27. c
28. a
29. d
30. b

CHAPTER 21

1. a
2. d
3. d
4. e
5. e
6. a
7. b
8. d
9. c
10. a
11. c
12. d
13. a
14. e
15. c
16. c
17. d
18. d
19. a
20. e
21. d
22. e
23. c
24. a
25. e
26. b
27. b
28. b
29. e
30. e

CHAPTER 22

1. a
2. c
3. e
4. a
5. b
6. d
7. a
8. b
9. d
10. c
11. e
12. b
13. a
14. e
15. b
16. b
17. e
18. d
19. c
20. a

CHAPTER 23

1. e
2. a
3. e
4. b
5. a
6. c
7. a
8. b
9. c
10. a
11. e
12. c
13. b
14. a
15. e
16. c
17. c
18. c
19. b
20. d
21. b
22. b
23. d
24. a
25. c
26. a
27. a
28. e
29. b
30. e
31. a
32. d
33. a
34. b
35. d
36. c

CHAPTER 24

1. e
2. e
3. c
4. e

5. b
6. b
7. e
8. a
9. b
10. d
11. c
12. a
13. b
14. e
15. d
16. e
17. d
18. b
19. c
20. a
21. b
22. d

GLOSSARY

Acetobacterium (ah-SEE-toe-back-TER-ee-um) Gram negative, anaerobic, ellipsoid bacterium that may oxidize hydrogen gas and reduce carbon dioxide or ferment carbohydrates to acetic acid (vinegar).

Acetoin (as-SEE-toe-in) The same as acetylmethylcarbinol. An intermediate in the 2, 3-butanediol fermentation which is used to detect the pathway. See Voges-Proskauer test.

Acetylmethylcarbinol The same as acetoin.

Acetone-butanol fermentation A fermentation pathway in which the major waste products are carbon dioxide, acetone, and butanol. Species of *Clostridium* carry out this type of fermentation.

Acid dyes Dyes in which the colored portion is negatively charged. Used to stain positively charged structures within cells or the background.

Acid-fast characteristic The property, shared by most mycobacteria and some nocardiae, of retaining heated carbofuchsin, even after treating with an acidified decolorizing (leaching) agent.

Acidophile (as-SID-doe-file) An organism that grows at (or requires) low pH values.

Acinetobacter calcoaceticus (ah-sin-ET-toe-BAK-ter kal-koh-a-SIT-tee-kus) Gram negative, aerobic, short bacilli in pairs and short chains, which oxidize carbohydrates.

Acrasiomycetes (ah-KRAY-see-oh-MY-seats) A class of fungi known as cellular slime molds. During part of their life cycle, they resemble amebae. Upon nutrient deprivation, the amebae congregate and form a multicellular structure called a pseudoplasmodium. The pseudoplasmodium forms a fruiting body that contains spores. The spores germinate into amebae. The cells may develop cellulose cell walls—unusual for fungi, which generally have chitin cell walls.

Actinomycetes (ACK-tin-no-MY-seats) Gram positive, irregularly-staining, rod-shaped, diphtheroid or branched, nonmotile, aerotolerant or anaerobic bacteria. In the 8th edition of Bergey's there were eight families in the order Actinomycetales: Actinomycetaceae, Mycobacteriaceae, Frankiaceae, Actinoplanaceae, Dermatophilaceae, Nocardiaceae, Streptomycetaceae, and Micromonosporaceae. In the 9th edition the actinomycetes have been broken up and placed in a number of sections (16, 17, 27, 28, 29, and 30).

Adenosine triphosphate ATP.

Adjuvant (ADD-jew-vant) Compound that increases the efficiency of antigens to induce an immune response.

***Aedes* sp.** (EH-des) Mosquito vectors for encephalitis (EEE, WEE, yellow fever).

Aedes aegypti (eh-GIP-tee) Mosquito vector for the yellow fever virus and the Dengue fever virus.

Aerobe (EH-robe) An organism that requires molecular oxygen for life. Examples are *Micrococcus* and *Pseudomonas*.

Aerotolerant anaerobe (eh-row-TOL-er-ant AN-eh-robe) Usually, a fermentative organism that is able to grow in the presence of molecular oxygen. Examples are *Streptococcus* and *Lactobacillus*.

Aeromonas hydrophila (eh-row-MO-nas hi-dro-FILL-la) Gram negative, polarly flagellated, facultatively anaerobic, oxidase-positive, rod-shaped bacterium.

Aeromonas shigelloides (she-gehl-LOYD-dees)

Agglutinins Antibodies that bind cells together, causing them to clump.

Aflatoxins (AFF-la-tox-ins) Toxins harmful to animals and humans, produced by fungi growing on wheat, rye corn, and other grains. Some of the toxins cause cancer in animals and humans.

Agrobacterium tumefaciens (ag-grow-back-TEAR-ee-um too-ma-FAY-see-ens) Gram negative, peritrichously flagellated, aerobic, oxidase positive, plant parasitic, rod-shaped bacterium. Causes a disease of plants called crown gall.

AIDS Acquired immunodeficiency syndrome. A virus (HIV) caused disease which results from the inability of helper T-lymphocytes to initiate an immune response.

Akinetes (A-kin-neats) Thick-walled, elongated spores formed by cyanobacteria and some green algae.

Alcaligenes (al-ka-LI-jen-eez) Gram negative, peritrichously flagellated, aerobic, oxidase positive coccobacilli.

Alga [pl. algae] (AL-gah [pl. should be AL-geh but always pronounced AL-gee]) Includes both microscopic, single-celled, prokaryotic and eukaryotic photosynthetic organisms and macroscopic, multicelled, eukaryotic photosynthetic organisms.

Alkalinophile Also spelled Alkalophile. An organism that grows under basic conditions or at a high pH.

Alcoholic (ethanolic) fermentation A fermentation in which the major waste products are carbon dioxide and alcohol (ethanol).

Allele (ah-LEEL) An alternate form of a gene.

Allergens Antigens that initiate a hypersensitive (allergic) reaction.

Allergy An overactive response by the immune system to an antigen which results in a diseased state (e.g. asthma, rash, welts). An allergy may be mediated by humoral antibodies or T-lymphocytes.

Alpha hemolysis A greening and partial lysis of red blood cells around certain colonies growing on blood agar plates. See Beta hemolysis and Gamma reaction.

Aminoacyl-tRNA (Ah-ME-no-A-seal-tee-R-N-A) An amino acid–tRNA complex.

Aminoacyl-tRNA synthetase (SIN-tha-tayz) An enzyme that hooks together a specific tRNA and a specific amino acid.

Ammonification (ah-MOAN-ni-fi-KAY-shun) The release of ammonium from organic molecules.

Amoeba (ah-MEE-ba) A protozoan that moves by extending pseudopodia and retracting terminal portions of the cell.

Amoeboid motion (ah-MEE-boyd) Movement due to the extension of one edge of a cell and the retraction of another edge.

Anabaena (an-na-BEH-na) A filamentous cyanobacterium.

Anabolism (an-NEH-bow-liz-um) The synthesis of compounds within a cell. See also Catabolism.

Anaerobe (an-EH-robe) An organism that is inhibited or killed by the presence of molecular oxygen and obtains its energy by fermentative processes or anaerobic respiration.

Anaerobic respiration Respiration in which the final electron acceptor is an inorganic molecule (sulfate or nitrate) other than molecular oxygen.

Anamnestic response (ah-nam-NESS-tik) Rapid immune response against a specific antigen by B-cells or T-cells because they have previously been sensitized to the antigen.

Anaphylactic shock (AN-eh-feh-LAK-tik) Severe contraction of smooth muscle and excessive loss of fluid from the blood into the tissues due to an immediate allergic reaction (atopic hypersensitivity) mediated by IgE.

Anaphylatoxin (AF-feh-lah-TOX-in) The C3 and C5 complement proteins that cause basophils and mast cells to release various molecules, such as histamine, which cause inflammation.

Annealing of DNA The joining together, through hydrogen bonding, of complementary strands of DNA.

Anopheles **sp.** (ah-NOF-fell-lees) Mosquito vectors for the malaria parasite and the western equine encephalitis (WEE) virus.

Antibiotic Organic compounds produced by microorganisms which inhibit or kill other organisms.

Antibodies Immunoglobulins. Serum proteins that are synthesized by plasma cells in response to stimulation by antigens. They react specifically with the inducing antigen.

Antigens Compounds that are able to elicit an immune response.

Antimicrobial agent Any physical or chemical agent that inhibits or kills microorganisms.

Antiseptic A chemical agent that inhibits or kills microorganisms and that can be used safely on the surface of an animal.

Antitoxins Antibodies that inactivate toxins.

Arboviruses Viruses carried by arthropod vectors which multiply in the arthropod and in the vertebrate host. The arboviruses cause various types of encephalitis (EEE, VEE, WEE, St. Louis) and yellow fever.

Archaebacteria (are-cheh-back-TEAR-ee-ah) Prokaryotes that lack the usual peptidoglycan (murein) in their cell walls, and that have phospholipids with ether bonds rather than the usual ester bonds. Many of the archaebacteria are chemoautotrophic methanogens, reducing carbon dioxide with electrons from hydrogen. Another large group are the sulfur oxidizing chemoautotrophic extreme thermophiles. The archaebacteria are generally found in extreme environments where high temperatures, high salinity, and high acid content are the rule. Examples of archaebacteria are *Sulfolobus*, *Thermoplasma*, *Halobacterium*, and *Methanobacterium*.

Arthrobacter (are-throw-BACK-ter) Gram positive, aerobic coccobacilli, showing great variation in size; may be flagellated.

Arthropods Invertebrate animals with jointed legs, such as the insects and arachnids.

Ascomycetes A class of fungi that produces sexual spores called ascospores inside a sack known as the ascus. Ascomycetes produce asexual spores called conidia. Examples include the mold *Neurospora* and the yeast *Saccharomyces*.

Asepsis (a-SEP-sis) The absence of contaminating or infecting microorganisms. Asepsis does not imply sterility. See Sepsis.

Aseptic technique Procedures that reduce the risk of contaminating materials or infecting patients.

Aspergillus (as-per-JILL-us) A fungus belonging to the class Deuteromycetes.

Assimilation The uptake of nutrients and their conversion to cellular material.

Atopic hypersensitivity IgE-mediated immediate allergic reaction.

Autoimmunity A situation in which an animal's immune system destroys its own tissues.

Autotroph An organism that uses carbon dioxide as its source of carbon.

Auxotroph A mutant that cannot synthesize simple compounds, such as amino acids, nucleosides, and vitamins. A medium must be supplemented with these growth factors if the organism is to proliferate.

Axenic culture (ax-ZEE-nik) Pure culture of microorganisms.

Azotobacter (ah-zo-toe-BACK-ter) Gram negative, aerobic, oval cocci; may be peritrichously flagellated; carries out nonsymbiotic nitrogen fixation.

B-cells B-lymphocytes that circulate in the lymph and blood and mature in the bone marrow and spleen. B-cells, when stimulated by foreign materials, differentiate into plasma cells and release antibodies into the lymph and blood.

Bacillus (ba-SILL-lus) A rod-shaped bacterium. When capitalized and italicized (or underscored), the term refers to a bacterial genus.

Bacillus (ba-Sill-lus) Gram positive, rod-shaped, flagellated or nonmotile, aerobic or facultatively anaerobic, endospore-forming bacteria.

Bacteriocins (back-TER-ee-oh-sins) Proteins produced by bacteria which kill sensitive members of related bacterial species. A special class of antibiotics.

Bacteriophage A bacterial virus, usually referred to as a phage.

Bacteroids Irregularly shaped bacteria, found within the plant cells that make up root nodules.

Bacteroides (back-ter-OYE-dees) Gram negative, rod-shaped, nonmotile, anaerobic, catalase negative (or weakly catalase positive) bacterium.

Balantidium coli (bal-lan-TID-dee-um CO-lee) Ciliated protozoan that may cause diarrhea and dysentery.

Basic dye A dye in which the colored portion is positively charged. A dye that stains cells.

Basidiomycetes (ba-sid-dee-oh-MY-seats) A class of fungi that produces sexual spores, called basidospores, from a structure called the basidium. The meadow mushroom, *Agaricus compestris*, is an example of a Basidiomycete.

Basophil White blood cell that releases histamines and other compounds that act on smooth muscle and other tissues.

Beta hemolysis Extensive lysis around certain colonies of bacteria growing on blood agar plates. See Alpha hemolysis and Gamma reaction.

Bdellovibrio bacteriovorus (del-low-VIB-bree-oh back-ter-ee-oh-VOR-us) Small curved or spiral, gram negative, polarly flagellated, aerobic bacteria.

Beggiatoa (bej-gee-ah-TOE-ah) Gram negative, filamentous, gliding, aerobic to microaerophilic, hydrogen sulfide–oxidizing bacteria with numerous sulfur granules within their cytoplasms. These bacteria require organic compounds as their carbon source.

Behring, Emil von (BEH-rink, EH-meel von)

Beijerinck, Martinus (BY-jer-rink, mar-TEE-nus)

Beijerinckia (by-jer-RINK-kia) Gram negative, oval or rod-shaped, peritrichously flagellated or nonmotile, aerobic bacteria, Molybdenum required for nitrogen fixation.

Bergey's Manual A book in four volumes that characterizes and classifies bacteria.

Bifidobacterium (by-fid-dough-back-TER-ee-um) Gram positive, highly variable, rod-shaped, nonmotile, anaerobic, catalase negative bacteria.

Bioluminescence The production and release of light by some microorganisms and insects.

Blue-green algae Photosynthetic bacteria that release oxygen as a consequence of their photosynthesis. Blue-green algae are also known as cyanobacteria.

BOD Biochemical oxygen demand. The amount of oxygen required to metabolize dissolved organic material to carbon dioxide and water is referred to as the biological oxygen demand.

Bordetella pertussis (bore-de-TELL-la pair-TUSS-sis) Gram negative, small coccobacilli; nonmotile, aerobic bacterium. Causes whopping cough.

Borrelia recurrentis (bore-REL-lee-ah ree-cur-REN-tis) Gram negative spirochete with axial filaments; anaerobic bacterium. Causes relapsing fever.

Branhamella catarrhalis (bran-ha-MEL-la ka-ta-RAH-lis) Gram negative, flattened, nonmotile, aerobic cocci.

Brucella (bru-SELL-la) Gram negative coccobacilli or short rod-shaped bacteria; nonmotile, aerobic, oxidase positive. Facultatively intracellular. Cause of brucellosis (undulant fever, Bang's disease).

Bubonic plague A bacterial infection of the lymphatic system which causes swelling of the lymph nodes. A swollen lymph node is known as a bubo.

2,3-Butanediol fermentation A fermentation in which the major waste products are 2,3-butanediol and carbon dioxide. Bacteria such as *Enterobacter aerogenes* carry out a 2,3-butanediol fermentation.

Butyric acid bacteria Bacteria that release butyric acid when they ferment. An example is *Clostridium.*

Butyric acid–butanol fermentation A fermentation in which the major waste products are butyric acid, or butanol, and carbon dioxide. Bacteria such as *Clostridium* carry out a butyric acid–butanol fermentation.

Calvin cycle A cyclic series of chemical reactions in which large amounts of carbon dioxide are fixed into organic molecules. The dark reactions of photosynthesis.

Campylobacter jejuni (kam-pill-low-BACK-ter je-JEW-nee) Gram negative, curved, rod-shaped, flagellated, microaerophilic, oxidase positive bacterium. Causes dysentery and enterocolitis.

Candida albicans (KAN-de-dah al-bee-kanz) A yeast belonging to the class Deuteromycetes that causes thrush of the mouth and yeast infections of the vagina.

Capsid The protein coat of viruses, made up of capsomeres.

Capsule A thick layer of polysaccharide (infrequently protein) that surrounds some bacteria. See also Glycocalyx.

Carcinogen A physical or chemical agent that causes cancer.

Catabolism (kah-TAB-bow-liz-im) The breakdown of nutrients or cellular material. See also Anabolism.

Caulobacter (COW-low-back-ter) Gram negative, aerobic, flagellated, rod-shaped or nonmotile, prosthecate bacterium.

Cellular immunity Immunity mediated by thymus-derived lymphocytes (T-cells) and macrophages.

Cell wall A rigid layer of material outside the cytoplasmic membrane which protects the cell. The cell wall of most bacteria contains a layer of peptidoglycan. The gram positive bacteria have a wall that consists of peptidoglycan and teichoic acids, but most gram negative bacteria have a wall that consists of a peptidoglycan layer and an outer membrane containing proteins and lipopolysaccharide.

Gene The segment of DNA or RNA that codes for a protein, rRNA, or tRNA. See also Cistron.

Cephalosporium acremonium (sef-fa-low-SPORE-ee-um ah-kreh-MO-nee-um) A fungus that produces the antibiotic cephalosporin C.

Chemiosmotic hypothesis The theory that proton gradients are created by the movement of electrons and protons through an electron transport system, and that this proton gradient provides energy for doing work (concentrating nutrients and moving flagella) or synthesizing ATP from ADP and inorganic phosphate.

Chemoautotroph An organism that derives its energy from various chemicals (vs. sunlight) and its carbon from carbon dioxide.

Chemoheterotroph An organism that derives its energy from various chemicals and its carbon from organic molecules other than carbon dioxide. Usually, the same chemicals supply both the energy and the carbon.

Chemolithotroph An organism that derives its energy from inorganic chemicals, such as ferrous iron, hydrogen sulfide, or ammonium.

Chemotaxis Movement toward or away from a chemical.

Chitin (KAI-tin) A homopolymer of N-acetyl-D-glucosamine, which is the major structural molecule in the cell walls of most classes of fungi and in the exoskeletons of insects and crustaceans.

Chlamydia psittaci (klah-MID-dee-ah SIT-tah-see) Gram negative, spherical, nonmotile, ATP-requiring, intracellular bacterium. Causes lung infections in birds (ornithosis or psittacosis) and humans.

Chlamydia trachomatis (trah-ko-MA-tiss) Causes trachoma, inclusion conjunctivitis, infant pneumonia, nonspecific urethritis, and lymphogranuloma venereum.

Chromatophore Vesicular membranes found in the cytoplasm of the anaerobic photosynthetic bacteria.

Chromophore The colored portion of a dye.

Chromoplast A colored organelle (plastid), such as a chromosome or a chromoplastid.

Chromosome Chromosomes are DNA molecules that are coiled and condensed into organized structures, such as the metaphase chromosomes in mammals and the polytene chromosomes in certain insects. In eukaryotes, protein also makes up a major portion of the chromosomes. The bulk of the hereditary material in bacteria is found in a single circular molecule of DNA, sometimes referred to as the chromosome but more frequently called the genome. Small, autonomous pieces of hereditary material are called plasmids. See also genome.

Chromatin Fibers of DNA.

Cistron A gene or region of DNA that codes for a single polypeptide.

Citric acid cycle Also known as the Krebs cycle and the tricarboxylic acid cyclic. A cycle sequence of chemical reactions that produces carbon dioxide, reducing electrons, and ATP. The citric acid cycle also provides building blocks for the synthesis of cellular material.

Citrobacter freundii (sit-trow-BACK-ter FRO-een-dee) Gram negative, rod-shaped, flagellated, facultative anaerobic bacterium.

Claviceps purpurea (KLAH-vee-seps poor-POOR-ee-ah) A fungus that belongs to the class Ascomycetes. Causes ergot of rye.

Clostridia (kloss-TRID-dee-ah) Gram positive, rod-shaped, flagellated or nonmotile, endospore forming, anaerobic or microaerotolerant, generally catalase negative bacteria.

Clostridium botulinum (kloss-TRID-dee-um bot-chew-LIE-num) Causes botulism.

Clostridium perfringens (per-FRIN-jens) Causes intestinal infections, gas gangrene, and food poisoning.

Clostridium tetani (TET-ah-neye) Causes tetanus.

Coenocytic (seen-no-SIT-tic) The condition in which a cell contains many nuclei.

Coenzyme A small organic molecule that works with enzymes, carrying chemical groups to or away from chemical reactions. Coenzymes are required for enzyme activity. Many vitamins are used to make coenzymes.

Cofactor Generally, a metal ion that is required for enzyme activity.

Coccidioidomycosis (cock-sid-dee-OYE-doe-my-co-sis) Valley fever.

Coccidioides immitis (cock-sid-dee-OYE-dees IM-me-tis) The fungus that causes coccidioidomycosis.

Colicin A bacteriocin produced by coliform bacteria (and some enterics).

Complement A group of serum proteins involved in immune reactions which have cytolytic and chemotactic properties.

Complementation In genetics, a situation in which two defective DNA molecules, together, supply a missing function.

Complex medium A medium that contains an unknown mixture of many different nutrients. Generally, a rich medium that is used to grow fastidious (demanding) microorganisms. See Minimal medium.

Conjugation In bacteria, the unidirectional transfer of hereditary material from one bacterium (donor) to another (recipient). Transfer requires cell-to-cell contact.

Contact inhibition Cells of higher animals generally cease growing and proliferating when they touch one another.

Continuous cultures Cultures that are maintained in the exponential phase of growth for extended periods of time by continuously supplying the culture with fresh medium.

Corynebacteria Gram positive, rod-shaped, nonmotile, aerobic or facultative anaerobic bacteria.

Corynebacterium diphtheriae (core-in-nee-back-TER-ee-um dip-THER-ee-eh) Causes diphtheria.

Coxiella burnetii (kocks-ee-ELL-ah ber-NET-tee-eye) Gram negative, rod-shaped, nonmotile, spore forming, obligate intracellular bacterium. Causes Q-fever.

Cryptococcus neoformans (krip-toe-COCK-kus nee-oh-FOR-mans) A fungus in the class Deuteromycetes or Fungi Imperfecti. Causes cryptococcosis, a mild pneumonia, meningitis, or systemic infection. *Fillobasidiella neoformans* is the perfect state (a Basidiomycete) of *C. neoformans.*

***Culex* sp.** (Ku-lex) Mosquito vectors for encephalitis (EEE, WEE, St. Louis).

Cuture Population of microorganisms growing in an artificial medium.

Cyanobacteria (sigh-ann-no-back-TER-ee-ah) Photosynthetic bacteria that release molecular oxygen as a consequence of their photosynthesis. Also known as blue-green bacteria, blue-green algae.

Cyclic AMP (cAMP) 3′,5′-cyclic adenosine monophosphate. The phosphate group is connected to the 3′ and 5′ carbons of the ribose sugar, thus creating a cyclic structure.

Cytochrome Complex metal-containing porphyrin-protein molecules that are readily reduced and oxidized. They are important components of electron transport systems.

Cytopathic effect A visible change in cell cultures due to infection by viruses, bacteria, fungi, or protozoans.

Cytophaga (sigh-TOE-fay-ga) Gram negative, rod-shaped, gliding, facultative anaerobic or aerobic bacteria.

Cytoplasmic membrane See Plasma membrane.

D-value (decimal reduction time) The time that it takes for an antimicrobial agent to reduce a population to 10 percent of its original size at a specified temperature.

Denature To alter the normal shape or activity of a molecule.

Denitrifying bacteria Bacteria that convert nitrate to nitrite or to nitrogen gas by anaerobic respiration.

Dental caries Areas of tooth decay.

Dermatomycosis A fungal skin infection.

Dermatophite A fungus that grows in the skin and causes infections. Dermatophytes digest keratin.

Desulfotomaculum (dee-SULL-fo-toe-MA-cue-lum) Gram positive, rod-shaped, flagellated, endospore forming, anaerobic, sulfate- and sulfite-reducing bacteria that produce hydrogen sulfide.

Desulfovibrio (dee-SULL-fo-VIB-bree-oh) Gram negative, curved, flagellated, anaerobic, sulfate- and sulfite-reducing bacterium that produces hydrogen sulfide.

Dextran Polysaccharides in which most of the glucose is held together by alpha (1,6) linkages, whereas alpha (1,2), alpha (1,3), or alpha (1,4) linkages are responsible for the branching. The dextrans form viscous, slimy solutions.

Deuteromycetes (DUE-ter-oh-MY-seats) A class of fungi also known as the Fungi Imperfecti. These fungi have no known sexual stage; they produce asexual spores called conidia. Examples are *Penicillium* and *Aspergillus.*

Dextrin A branched polysaccharide of glucose which remains after limited amylase hydrolysis of starch or glycogen. Most of the glucose is held together by alpha (1,4) linkages, whereas alpha (1,6) linkages are responsible for the branching.

Dextrose Glucose.

Diacetyl A compound relased by certain bacteria and yeast which gives an off-flavor to beer and a buttery aroma and taste to butter.

Dialysis The removal of low molecular weight ions and molecules by diffusion across a membrane.

Diatomaceous earth Earth consisting of diatom cell walls.

Diauxic (di-OX-ick) The two-step growth curve that results when a population of organisms metabolizes one carbon source completely before beginning to metabolize another.

Dictyostelium (dick-tee-oh-STILL-ee-um) A slime mold that produces a pseudoplasmodium.

Didinium (die-DIN-ee-um) A large ciliated protozoan.

Diffusion The net movement of a solute from a region of high concentration to a region of low concentration. In passive diffusion across a membrane, the rate of diffusion is directly proportional to the concentration of the solute; in facilitated diffusion across a membrane, the rate of diffusion at low solute concentrations is dependent upon the concentration of the solute, but at high solute concentrations the rate of diffusion is dependent upon the mechanisms that are aiding the diffusion (e.g., concentration of carrier molecules on membrane).

Dikaryon A cell that contains two different types of nuclei.

Dimorphic fungus A fungus that has two different morphologies that change, depending upon the environmental conditions. For example, the formation of a mycelium at room temperature and yeastlike structures at 35°C.

Dissimilatory sulfate reduction The conversion of cellular sulfate to environmental hydrogen sulfide.

Double helix The double-stranded DNA molecule.

Doubling time The time that it takes a population to double its size. Same as generation time.

Dysentery Clinical syndrome resulting from an inflammation of the large intestine. It is characterized by diarrhea with blood and mucus, tenesmus, fever, and malaise.

Eclipse period The time from the initiation of a virus infection until new infectious virus particles appear within the host cell. See Latent period.

Elementary bodies Intracellular chlamydia cells with an atypical structure which are the infectious form of the bacterium. Elementary bodies are sporelike structures that do not multiply, do not transport ATP, and do not carry out protein synthesis. The number of ribosomes in the elementary bodies is greatly reduced as compared with the number in the reticular bodies (initial bodies). The outer membrane proteins of elementary bodies are linked by disulfide bonds, making the cells rigid. See Reticular bodies (initial bodies).

Embden-Meyerhof pathway A specific sequence of chemical reactions that converts glucose to pyruvic acid. Same as glycolysis.

Endergonic reaction A reaction that absorbs heat or energy from its environment. See Exergonic reaction.

Endonuclease An enzyme that cuts within a DNA or RNA molecule. See Exonuclease.

Endosymbiont An organism that lives within another in a symbiotic relationship.

Endotoxin The lipopolysaccharide fraction from the outer membrane in gram negative bacteria which is toxic to many animals.

Enrichment Cultural selection for an organism, so that its relative abundance increases.

Entamoeba coli (ent-ah-ME-ba Ko-lee) A nonpathogenic ameba that colonizes the intestines of about 30% of the human population.

Entamoeba histolytica (his-toe-LIT-tee-kah) An ameba that causes dysentery.

Enteric bacteria Bacteria that normally live in the intestines.

Enterobacter aerogenes (en-ter-oh-BACK-ter air-RAH-jen-eez) Gram negative, rod-shaped, flagellated, facultatively anaerobic bacterium. Sometimes involved in urinary tract infections.

Enterococcus Spherical bacteria that live in the intestines, such as *Streptococcus faecalis.*

Enterotoxin A toxin that causes gastrointestinal problems when ingested or produced in the intestines.

Enterovirus Viruses that live or are commonly found in the intestines, such as the polioviruses and hepatitis A virus.

Epidemiology The study of disease incidence, distribution, and modes of transmission.

Epidermophyton (eh-pee-der-MO-fy-ton) A fungus that causes skin infections, such as jock itch (ringworm of the groin) and athlete's foot (ringworm of the foot).

Episome A plasmid that is able to integrate into the host's chromosomes.

Epizootic (eh-pee-zow-OT-tick) Disease that affects a large number of animals.

Ergot A fungus that grows on wheat and rye. See *Claviceps purpurea.*

Erysipelas A serious infection of the skin caused by Group A *Streptococcus.*

Erwinia amylovora (err-WIN-nee-ah ah-me-LOV-or-ah) Gram negative, rod-shaped, flagellated, facultatively anaerobic, oxidase negative bacterium.

Etiology (et-tee-OL-lo-gee) The cause of a disease.

Escherichia coli (eh-share-REE-kee-ahKO-lee) Gram negative, rod-shaped, flagellated, facultatively anaerobic bacterium commonly found in the intestines.

Ethanol fermentation A fermentation in which the major waste products are carbon dioxide and ethanol.

Eubacteria Classical bacteria that have peptidoglycan in their cell walls, lipids with unbranched fatty acids connected by ester bonds in their cell membranes, and distinctive rRNAs, tRNAs, and RNA polymerases. The mycoplasmas, which lack a cell wall, and the chlamydia, which lack peptidoglycan in their cell walls, are considered eubacteria but the cyanobacteria are not. The archaebacteria lack peptidoglycan in their cell walls, have lipids with branched hydrocarbon chains connected by ether bonds, and have distinctive rRNAs, tRNAs, and RNA polymerases are not considered to be eubacteria either.

Eukaryote A cell that has one or more nuclei and generally contains organelles and structures such as mitochondria, chloroplasts, Golgi bodies, endoplasmic reticulum, and vacuoles.

Eutrophication (you-trow-fi-KAY-shun) A massive growth of algae in lakes and other bodies of water, which leads to a decrease in dissolved oxygen and a decrease in animal life in the water. Eutrophication is due to an excess of nutrients from erosion or sewage.

Exergonic reaction A reaction that gives off heat or energy. See Endergonic reaction.

Exoenzyme Enzyme that functions outside the cell.

Exonuclease An enzyme that removes nucleotides from the ends of DNA or RNA molecules. See Endonuclease.

Exotoxin Large molecules, usually proteins, that are toxic to animals and plants.

Exponential phase A period of growth during which a population is growing at a constant rate. Same as log phase. See also Lag, Stationary, and Death phases.

F-factor (fertility factor) A plasmid that makes a bacterium able to transfer hereditary material to another bacterium during conjugation. In a conjugation, F^- cells receive herditary information; F^+ cells donate plasmid genes; F' cells donate plasmid genes and some genes also found on the main chromosome; and Hfr cells donate genes on the main chromosome and some genes on the integrated plasmid.

Facilitated diffusion Diffusion of molecules through a membrane by means of special pores or proteins that penetrate the membrane. See Diffusion.

Facultative anaerobe Organism that usually respires aerobically, but occasionally proliferates under anaerobic conditions by fermenting or anaerobically respiring.

Fermentation The production of energy exclusively by substrate level phosphorylation and the use of organic molecules as electron acceptors. Fermentations generally result in the release of organic and inorganic molecules, such as carbon dioxide and molecular hydrogen (gases); lactic, formic, acetic, succinic, butyric, and propionic acids; and ethanol, butanol, and propanol (alcohols).

Fimbria [pl. fimbriae] (FIM-bree-ah, FIM-bree-eh) Various types of pili that extend from the surface of bacteria.

Fix (1) To attach bacteria to a slide. (2) To alter material such as carbon dioxide and molecular nitrogen, and incorporate the carbon and nitrogen into cellular material. See Nitrogen fixation and Photosynthesis.

Flavobacterium (flay-vo-back-TER-ee-um) Gram negative, spherical or rod-shaped, flagellated or nonmotile, aerobic bacteria.

Flavoprotein A protein that contains a covalently bonded flavin group. A component of electron transport systems.

Fomite (FOE-might) An inanimate object or sputum that may be the source of microorganisms.

Franciscella tularensis (fran-sis-SELL-lah too-lar-REN-sis) Gram negative, elipsoid, rod-shaped, nonmotile, aerobic bacteria. Causes tularemia.

Free energy Energy that can do useful work.

Fungi imperfecti (FUN-jeye im-per-FEK-teye) See Deuteromycetes.

Fungus [pl. fungi] (FUN-gus; FUN-jee or FUN-jeye) Eukaryotic, nonphotosynthetic, micro- and macroscopic organisms that form tubular cells with a rigid cell wall. The cells are interconnected, forming hyphae. Fungi grow typically by extention of hyphal tips.

Furuncle (fir-UNK-kul) An abscess in the skin, similar to a boil, which sometimes results from an infection of a hair follicle.

Fusobacteria Gram negative, cigar-shaped, nonmotile or flagellated, anaerobic, catalase negative bacteria.

Gaffkya (GAFF-key-ah) Gram positive, spherical, nonmotile, facultative anaerobic bacterium, usually found in groups of four.

Galactoside Molecule that has the sugar galactose attached to it. Examples are lactose, isopropyl thiogalactoside (IPTG), and thiomethyl galactoside (TMG).

Gallionella (gal-yawn-NELL-ah) Gram negative, kidney-shaped, flagellated or nonmotile, microaerophilic, ferrous oxidizing bacteria that form a fibrous (noncellular) stalk.

Gardnerella vaginalis (gard-ner-RELL-ah vah-jin-NAL-iss) Gram negative, rod-shaped, nonmotile, facultatively anaerobic, catalase negative bacteria with many growth requirements. May be involved in nonspecific vaginosis.

Gamma globulins A group of serum proteins, some of which function as antibodies.

Gamma hemolysis The absence of hemolysis around bacterial colonies growing on blood agar plates.

Generation time The time it takes for a population to double its size. Same as doubling time.

Genetic code The representation of amino acids by specific sequences of three nucleotides in DNA or RNA. The specified amino acids are incorporated during protein synthesis.

Genetic map An ordered sequence of mutations or genes in a chromosome.

Genome A complete set of an organism's hereditary material. In polyploid organisms, a monoploid set of the chromosomes. In bacteria, the circular chromosome and any plasmids. In viruses, the DNA or RNA molecules.

Geotrichum candidum (gee-oh-TRICK-um KAN-dee-dum) A filamentous fungus in the class Deuteromycetes which forms arthrospores.

Giardia lamblia (gee-ARE-dee-ah lamb-lee-AH) A flagellated protozoan. The cause of beaver fever and traveler's diarrhea.

Gliding bacteria A large group of diverse bacteria that move slowly and smoothly over moist surfaces. They lack axial filaments, flagella, or other protrusions that could be involved in their movement.

Glycocalyx (gly-ko-KAY-licks) Stringy polysaccharides that emanate from the surfaces of cells and help the cells attach to one another and to inanimate objects. In bacteria, the glycocalyx can become so thick that it is indistinguishable from a capsule.

Glycolysis Same as Embden-Meyerhof pathway.

Granulocytes White blood cells that contain numerous large granules in their cytoplasm: neutrophils, basophils, and acidophils.

Growth curve A curve tracing the growth of a population

in a broth medium. It is divided into phases: lag, exponential (log), stationary, and death.

Growth factors Various organic molecules such as amino acids, nucleosides, and vitamins which an organism is unable to synthesize and with which it must be supplied.

Growth rate The increase in the number of individuals in a population as a function of time.

Growth yield The mass of cells that can be harvested.

Haemophilus (he-MAW-fill-us) Gram negative, spherical to rod-shaped, nonmotile, aerobic or facultatively anaerobic, intracellular bacteria that require growth factors.

Haemophilus ducreyi (due-KRAY-ee) Requires hemin (X-factor). Causes soft chancre known as chancroid, a venereal disease.

Haemophilus influenzae (in-flew-ENZ-eh) Requires hemin (X-factor) and NAD^+ (V-factor). Causes meningitis, sinusitis, and pneumonia.

Hafnia (HAF-nee-ah) Gram negative, rod-shaped, peritrichously flagellated, facultatively anaerobic bacteria that resemble *Enterobacter.*

Halobacterium (hal-low-back-TER-ee-um) Gram negative, rod-shaped, polarly flagellated or nonmotile, sodium chloride-requiring (more than 2 M), aerobic, oxidase positive bacteria.

Halococcus (hal-low-KOCK-kus) Gram negative, spherical, nonmotile, sodium chloride-requiring (more than 2.5 M), aerobic bacteria.

Halophiles Organisms that grow best at (or require) high concentrations of salt.

Hapten A molecule that is specifically bound by antibodies, but is unable to induce the production and release of antibodies by plasma cells. A hapten may induce a specific immune response if it is attached to a large protein (carrier molecule).

Hemicelluloses Polymers of D-xylose held together by beta (1,4) linkages with side chains of arabinose and other sugars.

Heterokaryosis A condition in which a cell contains more than one type of nucleus.

Heterofermentation A fermentation in which there are a number of waste products. Used in connection with lactic acid bacteria that have major waste products other than lactic acid.

Histoplasma capsulatum (hiss-toe-PLAZ-ma cap-sue-LA-tum) A fungus in the class Deuteromycetes which causes histoplasmosis, usually a mild coldlike disease.

Homofermentation A fermentation in which there is only one major waste product. Used in connection with lactic acid bacteria that produce lactic acid and very minor amounts of other acids.

Humoral immunity Immunity mediated by antibodies that circulate through the lymphatic system and the blood system. "Humor" refers to the body fluids.

Hydrogen bacteria Bacteria that derive their energy by oxidizing molecular hydrogen. Examples include some species of *Alcaligenes, Xanthobacter, Nocardia,* and *Paracoccus.*

Hydrophilic forces Forces that arise from chemical groups or molecules that arrange themselves in an aqueous environment in such a way as to be surrounded by water.

Hydrophobic forces Forces that arise from chemical groups or molecules that arrange themselves in an aqueous environment in such a way as to exclude water.

Hypertonic A solution with a relatively high concentration of solutes or dissolved materials. See Hypotonic and Isotonic.

Hypha [pl. hyphae] (HI-fah; plural should be HI-feh, but pronounced HI-fee) A filament of fungal growth.

Hyphomicrobium (HI-fo-my-CROW-bee-um) Rod-shaped, hyphae forming, nonmotile or motile, aerobic bacteria that reproduce by a budding process.

Hypotonic A solution with a relatively low concentration of solutes or dissolved materials as compared to the cellular environment. See Hypertonic and Isotonic.

Icosahedron A closed three-dimensional structure with 20 triangular faces (surfaces) and 12 corners.

IgA Antibodies of the A class, found in secretions.

IgD Antibodies of the D class, found attached to the surfaces of B-lymphocytes.

IgE Antibodies of the E class, found in the lymph and blood, which are involved in allergic reactions.

IgG Antibodies of the G class, found in the lymph and blood, which bind complement. IgG are the most abundant antibodies.

IgM Antibodies of the M class, found attached to the surfaces of B-lymphocytes and in the lymph and blood, which bind complement. They are the first antibodies to be produced by B-lymphocytes.

Immediate-type hypersensitivity An antibody-mediated allergic reaction that occurs within minutes after presentation of antigen to which the individual is sensitized.

Immunodeficiency The lack of part or all of the immune system.

Immunofluorescence Fluorescence from a dye that is attached (conjugated) to an antibody. This type of fluorescing antibody is used to visualize the presence of specific antigens.

Immunoglobulin Same as antibody.

IMViC A group of four biochemical tests (indole, methyl red, Voges-Proskauer, and citrate) used to differentiate enteric bacteria.

Inducer A physical or chemical agent that causes the development of lysogenic viruses or turns on gene systems (operons). For example, ultraviolet light is an inducer of the lambda prophage, whereas lactose if the inducer of the lactose operon. See Effector.

Initial bodies Same as reticular bodies: intracellular chlamydia cells that grow and divide. They transport ATP and carry out protein synthesis, but they are not infectious because they are extremely sensitive to mechanical and osmotic shock. In electron micrographs, they have the appearance of typical prokaryotes except that they appear to lack a peptidoglycan layer. Initial bodies are also known as reticulate bodies (RBs). See Elementary bodies.

Insertion sequence Regions of DNA that are able to recombine readily with different regions within and between chromosomes. Insertion sequences in plasmids are believed to be involved in the plasmid's ability to insert into the cell's chromosomes.

Interferon A class of proteins, released by virus-infected cells, which stimulate uninfected cells to produce proteins that inhibit the proliferation of viruses.

Intracytoplasmic membranes Membranes found in the cytoplasm of some bacteria. Many of these membranes are invaginations of the cytoplasmic membrane. See Chromatophore, Photosynthetic membrane, Thylakoid, and Chlorobium vesicle.

Isotonic A solution with the same osmotic pressure (usually the same concentration of solutes) as another solution. See Hypertonic and Hypotonic.

Iron bacteria Bacteria that derive energy by oxidizing ferrous iron (Fe^{+2}) to ferric iron (Fe^{+3}). An example is *Thiobacillus ferrooidans*.

In vitro Within an artificial container.

In vivo Within a living organism.

Isoantibodies Antibodies against foreign isoantigens. For instance, antibodies against A-type blood in persons with B- or O-type blood.

Isoantigens Antigens that are present in some individuals of a population but not in other individuals. For example, people with A-type blood have antigens that people with B- and O-type blood do not have.

Isotonic solution A solution with the same concentration of solute (atoms, small molecules, and ions) as another solution.

Ixodes persulcatus (icks-OH-deeze pair-sull-CAT-us) A tick that transmits the virus responsible for Russian spring-summer encephalitis (RSSE).

K-antigen The capsule material from enteric bacteria.

Killer T-Cells One class of T-lymphocytes involved in cell-mediated immunity.

Kinins Small peptides, released into the blood and lymph upon trauma, which are involved in blood clotting and inflammatory reactions.

Klebsiella pneumoniae (kleb-see-EL-lah new-MO-nee-eh) Gram negative, rod-shaped, nonmotile, encapsuled, facultatively anaerobic bacterium. A cause of pneumonia.

Koch's postulates (KAWK or KAWH; breathe out on the H) A set of procedures for proving that an organism is responsible for a disease.

Krebs cycle Same as the citric acid cycle and the tricarboxylic acid cycle.

Kuru A progressive degenerative brain disease caused by a prion.

L-forms Common bacteria that do not produce a normal peptidoglycan layer or have lost the ability to manufacture the peptidoglycan layer of their cell walls. See Protoplasts and Spheroplasts.

Lactic acid bacteria A group of gram positive, catalase negative, exclusively fermentative, spherical and rod-shaped bacteria that carry out a lactic acid fermentation. Examples are *Streptococcus* and *Lactobacillus*.

Lactic acid fermentation A fermentation in which the major waste product is lactic acid.

Lactobacilli (should be lack-toe-ba-SILL-ee, but always pronounced lack-toe-ba-SILL-eye) Lactic acid bacteria. Gram positive, rod-shaped, nonmotile, aerotolerant, anaerobic, catalase negative bacteria.

Lactobacillus (lack-toe-ba-SILL-us)

Lag phase The period of time during which a population is adapting to new environmental conditions before it reproduces exponentially. During the first part of a lag phase, a population may not increase in number. In the latter part of the lag phase, the population multiplies at an ever increasing rate. See Exponential, Stationary, and Death phases.

Latent period The time from the initiation of a virus infection until virus progeny are released.

Leghemoglobin A hemoglobin-like molecule coded for by both the endosymbiotic bacteria in root nodules of leguminous plants and the host plant cells. Leghemoglobin binds molecular oxygen and protects the bacterial nitrogenase, which is oxygen sensitive.

Legionella pneumophila (lee-jon-NELL-ah new-mow-FILL-ah) Gram negative, rod-shaped, nonmotile, aerobic, weakly oxidase positive bacteria with numerous growth requirements. Causes Legionnaires' disease, a pneumonia.

Legionnaires' disease A pneumonia. See *Legionella*.

Leishmania donovani (liesh-main-NEE-ah don-oh-VAN-ee) A flagellated protozoan that reproduces in phagocytes and causes a serious leishmaniasis, called kala-azar.

Leptospira iterrogans (lep-toe-SPY-rah in-TER-row-gans) Gram negative, helical, motile by axial filament, microaerophilic, aerobic bacterium. Causes leptospirosis.

Leptothrix (lep-tow-THRIX) Gram negative, rod-shaped, nonmotile, anaerobic, catalase negative, carbon dioxide-requiring bacteria.

Leucocytes White blood cells, such as neutrophils, basophils, and eosinophils.

Leuconostoc mesenteroides (lou-ko-NOS-stock mez-zen-ter-OYE-deez) A lactic acid bacterium that is gram positive, spherical, nonmotile, aerotolerant, anaerobic, and catalase negative

Leucothrix (lou-kow-THRIX) Gram negative, oval, filament-forming, nonmotile, aerobic, sulfide-oxidizing bacteria that do not form sulfur granules.

Levans A homopolymer of D-fructose connected by beta (1,2) linkages.

Lichen (LIE-ken) A symbiotic association between eukaryotic algae or cyanobacteria and fungi which creates an organism distinct from either of the mutuals.

Lignin A polymer of aromatic alcohols that makes up 25 percent of wood.

Limit of resolution The minimum distance that must separate two points so that their image can be resolved. See Resolving limit.

Lipopolysaccharide A molecule found in the outer membrane of gram negative bacteria, responsible for allergic reactions in some animals. An endotoxin from gram negative bacteria.

Lipoteichoic acid Teichoic acids attach to lipids in the cytoplasmic membrane rather than to peptidoglycan. Lipoteichoic acids help some bacteria to attach to inanimate objects and to other organisms.

Log phase Same as exponential phase.

LPS Lipopolysaccharides found in the outer membrane of gram negative bacteria.

Lymphocytes Cells derived from the lymphatic system which are involved in the immune response, specifically B-cells and T-cells.

Lyngbya (LING-by-ah) A filamentous, nonmotile cyanobacterium.

Lysogen A cell that carries a provirus.

M-protein A protein on the surface of the bacterium *Streptococcus pyogenes* which helps it attach to host tissue.

Macrophage A large ameboid-like cell generally found in the reticulo-endothelial system. Involved in phagocytosis and in the immune response.

Malassezia furfur (mal-es-SEZ-zia FER-fer) A fungus belonging to the class Deuteromycetes which is responsible for tinea versicolor, a pigmented fungal infection of the skin on the torso and upper legs.

Memory cell B-lymphocytes ot T-lymphocytes that have been sensitized to a specific antigen and which can mount an immune response more rapidly than unsensitized lymphocytes involved in a primary immune response. See Anamnestic response.

Meninges [sing. meninx] (meh-NIN-gees, MEN-inks) The three membranes that cover the brain and spinal cord.

Meningitis An inflammation of the membranes that surround the brain and spinal cord.

Methane-oxidizing bacteria Bacteria that utilize methane as a source of carbon and energy. Examples are *Methylobacter* and *Methylococcus.* See Methylotrophs.

Methane-producing bacteria Bacteria that release methane. These bacteria belong to the family Methanobacteriaceae. Examples are *Methanobacterium, Methanosarcina,* and *Methanococcus.* See Methanogenic bacteria.

Methanogenic bacteria Bacteria that generate methane during their metabolism. Same as methane-producing bacteria.

Methylotrophs Bacteria that obtain their energy by metabolizing methane or methanol. These bacteria belong to the family Methylomonadaceae.

Microaerophiles Organisms that require reduced concentrations of oxygen, and elevated concentrations of carbon dioxide, in order to grow.

Micrococcus luteus (my-crow-KOCK-kus LU-teh-us) Gram positive, spherical, nonmotile, aerobic, oxidase positive bacterium.

Microorganism Any microscopic single-celled organism or microscopic multicellular animal or mold: bacteria, protozoans, algae, fungi, and metazoans. May also be used when referring to viruses. "Infectious agent" is used when referring to viruses, viroids, and prions.

Microsporum (my-krow-SPORE-um) A fungus belonging to the Deuteromycetes which is responsible for ringworm of the face, arms, and body.

MIC Minimum inhibitory concentration of a chemical that inhibits the growth of an organism.

Minimal medium A medium in which all the constituents and their concentrations are known. Also known as a defined medium. A minimal medium generally consists of salts that supply phosphate, nitrate (or ammonium), sulfate, magnesium, and trace elements, and a carbon source such as glucose. See Complex medium.

Mixed acid fermentation A fermentation in which a mixture of acids, alcohols, and gases is released. Carried out by bacteria such as *Escherichia coli.* Tested for by the methyl red test.

Monocyte Mononuclear phagocytic leukocyte.

Morbidity The number of diseased individuals existing during a given period or during a particular incident.

Mortality The number of deaths occurring during a given period or during a particular incident.

Murein See Peptidoglycan.

Mucopeptide See Peptidoglycan.

Mutagen Physical or chemical agents that induce mutations. See also Carcinogen.

Mutualism A symbiotic association between two organisms in which both benefit.

Myco- Having to do with fungi. Many bacteria showing filamentous and/or branching growth originally were thought to be related to fungi, or were mistaken for fungi, and consequently contain the prefix referring to fungi in their genus designation.

Mycelium [pl. mycelia] (my-SEE-lee-um, my-SEE-lee-ah) The mass of intertwined, branched, filamentous growth associated with many fungi.

Mycobacteria One group of gram positive, rod-shaped, nonmotile, aerobic, acid-fast, slow-growing bacteria.

Mycobacterium leprae (my-ko-back-TER-ee-um LEP-preh) Causes leprosy.

Mycobacterium tuberculosis (too-ber-cue-LOW-sis) Causes tuberculosis.

Mycology The study of fungi.

Mycoplasmas (my-ko-PLAZ-mahz) A group of bacteria that lack cell walls.

Mycoplasma pneumoniae (my-ko-PLAZ-ma new-MO-nee-eh) Gram negative, pleiomorphic (spherical, filamentous, and branched), nonmotile, anaerobic, sterol-requiring, bacterium that lacks a cell wall. It causes primary atypical pneumonia.

Mycorrhiza (my-ko-RYE-zah) A mutualistic association between a fungus and plant roots.

Mycosis A fungal infection.

Mycotoxin A fungal toxin.

Myxobacteria (mix-oh-back-TER-ee-ah) Gliding bacteria that congregate upon nutrient deprivation and form fruiting bodies. The fruiting bodies contain microcysts, which are more tolerant of high heat, desiccation, and radiation than the vegetative cells. An example of a myxobacterium is *Chondromyces crocatus.*

Myxomycetes (mix-oh-my-SEATS) A class of fungi known as the plasmodial slime molds. During part of their life cycle, the Myxomycetes exist as flat, macroscopic cells, as much as 5 cm in diameter, with numerous diploid nuclei. This slimelike cell is called a plasmodium. Upon nutrient deprivation and/or exposure to light, the plasmodium develops into numerous fruiting bodies, which contain spores with a single nucleus. The spores germinate into a vegetative cell that develops into a plasmodium. The plasmodium is amebalike in that it is able to phagocytize bacteria and other small single-celled organisms.

Necrotic tissue Dead tissue.

Nematodes Wormlike animals, often known as roundworms.

Neisseria (nye-SEAR-ee-ah) Gram negative, spherical, nonmotile, aerobic, oxidase positive bacteria.

Neisseria gonorrhoeae (gon-nah-REE-eh) Causes the venereal disease gonorrhea.

Neisseria meningitidis (meh-nin-JEYE-tid-dis) Causes an inflammation of the meninges, known as meningitis.

Neutrophil White blood cells involved in phagocytosis, in which the granules stain pooly with basic and acidic dyes.

Nitrification The conversion of ammonium to nitrate by soil microorganisms.

Nitrate-reducing bacteria Bacteria that convert nitrate to nitrite and nitrogen gas. Examples are *Pseudomonas* and *Thiobacillus.* See Denitrifying bacteria.

Nitrobacter (neye-trow-BACK-ter) Gram negative, rod-shaped, nonmotile, aerobic, ammonium-oxidizing, carbon dioxide requiring bacterium. Chemolithotroph. Nitrifying bacterium.

Nitrogen cycle A cyclic series of chemical reactions carried out by various microorganisms in which molecular nitrogen is reduced to ammonium, the ammonium is converted to nitrite, the nitrite is further oxidized to nitrate, and the nitrate is reduced back to molecular nitrogen.

Nitrogen fixation The reduction of atmospheric molecular nitrogen to ammonium by enzyme nitrogenase, and the incorporation of the ammonium nitrogen into organic compounds.

Nitrogen-fixing bacteria Bacteria that are able to reduce molecular nitrogen to ammonium and then assimilate (fix) the nitrogen into organic compounds. Examples include *Azotobacter, Klebsiella, Rhizobium, Anabaena,* and *Nostoc.*

Nitrosococcus (Neye-trow-so-KOCK-kus) Gram negative, spherical, nonmotile, aerobic, ammonium-oxidizing, carbon dioxide–requiring bacteria. Chemolithotroph. Nitrifying bacteria.

Nitrosomonas (neye-trow-so-MO-nas) Gram negative, rod-shaped, nonmotile, aerobic ammonium-oxidizing, carbon dioxide–requiring, bacterium. Chemolithotroph. Nitrifying bacterium.

Nocardia asteroides (no-CAR-dee-ah ass-ter-OYE-dees) Gram positive, acid-fast, rod-shaped, mycelium-forming, nonmotile, aerobic bacterium. Causes pulmonary, cutaneous, and subcutaneous infections. May become systemic and infect various organs.

Nodule A small mass of cells found on the roots of legumes in which nitrogen-fixing bacteria reside.

Nosocomial infection An infection obtained while a patient or worker in a hospital.

Nostoc A cyanobacterium.

Nucleus (pl. nuclei) (NEW-klee-us; pl. should be NEW-kleh-ee, but always pronounced NEW-klee-eye). Membrane-enclosed organelle of eukaryotic cells containing the cell's genetic information.

O-antigen The outer portion of the polysaccharide that extends from the lipopolysaccharide found in the outer membrane of gram negative bacteria.

Oligotroph (OH-lee-go-trowf) An organism that can grow at very low concentrations of nutrients.

Oligotrophic waters (oh-lee-go-TROW-fik) Bodies of water that are low in nutrients.

Oomycetes (oh-oh-my-SEAT) A class of unicellular or filamentous fungi that produce flagellated gametes. The walls contain cellulose rather than chitin. Filamentous forms show nonseptated hyphae.

Oscillatoria (ah-sill-la-TOR-ee-ah) A filamentous cyanobacterium.

Onchocerca volvulus (on-ko-SIR-kah VOL-view-lus) A parasitic roundworm (nematode) that causes onchocerciasis and is responsible for the disease known as river blindness.

Oncogenic viruses Viruses that can cause cancer.

One-gene-one-enzyme hypothesis The idea that a region

of DNA, called a gene, contains the information for making a functional enzyme. It was originally proposed by Beadle and Tatum.

One-step growth curve The growth curve shown by viruses that lyse their host cells.

Operon One or more structural genes controlled by one or more controlling sites. The smallest operon consists of a gene that codes for a polypeptide and a promoter (RNA polymerase binding site is a controlling site).

Osmosis The diffusion of solvent (water) across a membrane from a region of high concentration to a region of low concentration.

Osmotic pressure The pressure required to prevent the influx of water into a volume that is hypertonic. The internal water pressure on the cell membrane and wall due to the influx of water into the cell because of a hypotonic environment.

OxidationThe removal of electrons

Oxidative phosphorylation Occurs when electrons and protons from oxidized substrates are used to create a proton gradient across a membrane, and this proton gradient is used to power the synthesis of ATP from ADP and inorganic phosphate. See Substrate phosphorylation and Photophosphorylation.

Paramecium (par-ah-MEE-see-um) A ciliated protozoan.

Passive diffusion See Diffusion.

Pasteurization (pass-tur-rye-ZAY-shun) The heating of material such as wine and milk to temperatures below the boiling point, in order to kill spoilage organisms or disease-causing organisms. Thermoduric organisms survive pasteurization. See Sterilization and Appertization.

Pectin A polymer of D-galacturonate.

Pediculus humanus (peh-DICK-ku-lus hue-MAN-us) Body louse. Vector of the bacterium that causes relapsing fever, *Borrelia recurrentis.*

Pediococcus (peh-dee-oh-KOCK-kus) Gram positive, spherical, nonmotile, microaerotolerant, anaerobic bacterium with complex growth requirements, often arranged in tetrads.

Pellicle (PEL-lah-kul) (1) A film of microorganisms on the top of a broth. (2) An envelope that surrounds a microorganism. (3) A convoluted and cytoskeleton-strengthened plasma membrane found in some protozoans.

Penicillium (pen-neye-SILL-lee-um) A fungus belonging to the class Deuteromycetes which forms conidia from annellides. Some species produce the antibiotic penicillin.

Peptidoglycan A polysaccharide layer in the walls of most eubacteria, constructed from alternating units of N-acetylglucosamine and N-acetyl-muramic acid. Peptidoglycan is also known as murein or mucopeptide.

Periplasmic space In gram negative bacteria, the region between the cytoplasmic membrane and the outer membrane of the wall. In gram positive bacteria, the region between the cytoplasmic membrane and the peptidoglycan layer.

Permease A protein involved in the transport of materials into or out of a cell.

Phage (FAGE) A bacterial virus.

Photophosphorylation The synthesis of ATP using energy derived from light. See Oxidative phosphorylation.

Photosynthesis The process that uses light to generate energy (ATP) and reducing power (NADPH), and that uses ATP and NADPH to fix carbon dioxide.

Phytophthora infestans (fie-TOFF-thor-ah in-FESS-tans) A fungus belonging to the class Oomycetes which causes late blight of potato.

Plaque (1) A clearing in a bacterial lawn due to a virus infection. (2) A virus colony. (3) The bacteria, glycocalyx, food particles, and minerals that form a mat on teeth.

Plasma cells Cells that release antibodies and that develop when a B-lymphocyte is stimulated by antigen.

Plasma membrane The membrane that contains the cytoplasm and defines a cell.

Plasmid Small, autonomous piece of hereditary material distinct from the main genome. In bacteria, small circular chromosomes 5,000–100,000 base pairs long (5–100 genes).

Plasmodium See Myxomycetes and Acrasiomycetes.

Plasmodium (plaz-MO-dee-um) Flagellated protozoans.

Plasmodium falciparum (fall-SIP-are-um) Causes malaria.

Plasmodium malariae (mah-LAIR-ee-eh) Causes malaria.

Plasmodium vivax (VEE-vahz) Causes malaria.

Plastids Organelles found in eukaryotic cells, such as chromosomes and chromoplasts.

Primary structure In proteins and nucleic acids, the linear sequence of amino acids and nucleotides, respectively.

Prion (PREE-on) An infectious agent that lacks nucleic acids and appears to consist entirely of protein. See Virus and Viroid.

Prophage (PRO-fage) The phage DNA that is integrated into the host's hereditary material.

Propionic acid bacteria Bacteria that carry out a propionic acid fermentation.

Propionic acid fermentation A fermentation in which the major waste products are propionic acid and carbon dioxide. Bacteria such as *Propionibacterium* carry out this type of fermentation.

Propionibacterium (pro-pee-on-nee-back-TER-ee-um) Gram positive, rod- to club-shaped diphtheroid, nonmotile, aerotolerant and anaerobic bacteria.

Prostheca [pl. prosthecae] (praws-THEE-ka, praws-THEE-keh) Celular extensions of the cell, such as the stalk in *Caulobacter* and the hypha in *Rhodomicrobium.*

Proteus (PRO-tee-us) Gram negative, rod-shaped, peritrichously flagellated, facultatively anaerobic bacterium. Causes urinary tract infections.

Proton-motive force (pmf) An energized state of a membrane due to electrical potentials and proton gradients (pH differences). The potential created across a membrane when protons (hydrogen ions) are concentrated on one side of the membrane is: p = U − 2.3RT pH/F, where U = membrane potential, R = gas constant, T = absolute temperature, pH = pH difference across the membrane, and F = faraday.

Protoplast Gram positive cell that lacks its peptidoglycan cell wall because of mutation or treatment with chemicals. In an isotonic solution, protoplasts become spherical. Lysozyme treatment of gram positive cells results in protoplasts. See Spheroplast.

Prototroph Naturally occurring strain of microorganisms (wild type).

Providencia (pro-vee-DEN-see-ah) Gram negative, rod-shaped, peritrichously flagellated or nonmotile, facultatively anaerobic bacterium.

Provirus The virus DNA that is integrated into the host's hereditary material.

Pseudomonads (sue-doe-MO-nads) Gram negative, rod-shaped, polarly flagellated, aerobic bacteria.

Pseudomonas aeruginosa (sue-doe-MO-nas ah-rue-jin-NO-sah) Gram negative, rod-shaped, polarly flagellated, aerobic bacterium. Produces a green pigment. An opportunistic pathogen.

Pseudoplasmodium See Acrasiomycetes; Slime mold.

Psychrophile (1) An organism that has its maximum growth rate between 0°C and 20°C. (2) An organism that grows at temperatures below 5°C.

Puerperal sepsis (PWEAR-per-ul SEP-sis) A bacterial invasion of the blood, acquired during childbirth. Also known as childbed fever.

Purple nonsulfur bacteria Generally anaerobic, photosynthetic bacteria that use light to generate their energy (ATP) by cyclic photophosphorylation only. They may use simple organic acids such as acetate to generate their reducing electrons to fix carbon dioxide.

Purple sulfur bacteria Anaerobic, photosynthetic bacteria that use light to generate their energy (ATP) by cyclic photophosphorylation and hydrogen sulfide as a source of reducing electrons to fix carbon dioxide. They release sulfate rather than molecular oxygen as a result of their metabolism. The purple sulfur bacteria may use organic compounds as a source of reducing electrons and carbon, as do the purple nonsulfur bacteria.

Putrefaction The decomposition of proteins by microorganisms, with resulting production of foul-smelling compounds from the amino acids.

Q-fever A disease caused by the bacterium *Coxiella burnetii* sometimes transmitted to humans through milk from diseased cattle. The bacterium is highly resistant to heat because of the sporelike structures it forms.

Quarternary structure The structure that results when two or more folded proteins from a complex molecule, such as an antibody molecule or the hemoglobin molecule.

R-factor A plasmid that confers fertility on the host cell and that carries genes that make the host cell resistant to antibiotics and drugs.

Reduction The addition of electrons to a molecule. See Oxidation.

Refractive index The ratio of the speed of light in a vacuum to the speed of light in a material. The index of refraction of a lens is approximately 1.4.

Resolving limit The resolving limit (R) is the minimum distance between two points such that the images of the two points can be distinguished from each other. The resolving limit (R) of a lens system depends upon the wavelength of light (λ), the index of refraction (n) of the material between the object and the objective lens, and the diameter and working distance of the lens, given by sin θ : $R = \lambda/n \sin \theta$. Same as the limit of resolution.

Resolving power The ability to distinguish between two closely spaced points. See Resolving limit.

Respiration Involves the oxidation of inorganic or organic molecules, the generation of energy (ATP) by running electrons (and hydrogen ions) through an electron transport system, and the donation of electrons to an inorganic electron acceptor. Aerobic respiration occurs when the electron acceptor is molecular oxygen, and anaerobic respiration occurs when the electron acceptor is an inorganic molecule other than oxygen, such as sulfate or nitrate.

Restriction endonucleases Bacterial enzymes that cut DNA within the DNA molecule. The most useful endonucleases are those that cut at specific sites. These enzymes are used in genetic engineering to splice genes. See Endonuclease and Exonuclease.

Reticular bodies The dividing, noninfectious chlamydia found in infected cells. Same as initial bodies. See Elementary bodies

Rhizobium (rye-ZOH-bee-um) Gram negative, rod-shaped, flagellated, aerobic bacteria able to grow within certain plant cells. When growing within root nodules, the cells are pleomorphic. Carries out symbiotic nitrogen fixation.

Rhizopus (RYE-zoh-puss) A fungus in the class Zygomycetes which forms asexual and sexual sporangiospores and sexual zygospores. The mycelium is coenocytic and nonseptated.

Rickettsia (rick-KET-see-ah) Gram negative, oval to rod-shaped, nonmotile, obligately intracellular parasitic bacteria. See also *Coxiella.*

Rickettsia prowazekii (pro-wah-ZE-key-eye) Causes epidemic typhus.

Rickettsia typhi (TIE-fee) Causes rat and mouse (murine) typhus.

Rickettsia rickettsii (rick-KET-see-eye) Causes Rocky Mountain spotted fever.

Saccharomyces cerevisiae (sak-kah-row-MY-sees ser-reh-VIS-ee-ee) A yeast belonging to the class Ascomycetes which forms asci and sexual ascospores and carries out alcoholic fermentation. It is used in making bread, wine, and beer.

Salmonella typhi (sal-mo-NEL-lah TIE-fee) Gram negative, rod-shaped, peritrichously flagellated, facultatively anaerobic bacteria. Causes typhoid fever.

Salmonella typhimurium (tie-fee-MUR-ee-um) Causes paratyphoid fever: gastroenteritis and bacteremia.

San Joaquin Valley Fever (SAN hwah-KEEN VA-lee FEE-ver) A pulmonary infection caused by the fungus *Coccidioides.* See Valley fever and Coccidioidomycosis.

Sarcina (sar-SIN-na) Gram positive, spherical, nonmotile, anaerobic, catalase negative bacteria that are found in packets of four, eight, or more cells. They have complex growth factors. The cell wall of this species has a thick outer layer of cellulose. This is the only bacterium known to have a wall with cellulose. Only one other bacterium, *Acetobacter xylinum,* is known to synthesize cellulose.

Sarcoptes scabei (sar-COP-tees SKAY-bee-ee) The mite (arachnid) that burrows under the outer layer of skin and causes scabies.

Schistosoma (shis-toe-SOW-mah) A flatworm (fluke) that is responsible for schistosomiasis, an infection of the intestinal veins and liver which leads to liver destruction. Causes schistosomiasis.

Secondary structure In proteins, the structure formed from the folding of the primary structure due to disulfide bonds, and the alpha helix or pleated sheets formed from the hydrogen bonding. In nucleic acids, the structure formed from the folding of the primary structure due to the hydrogen bonding between complementary bases. See Tertiary structure.

Sepsis Generally, infected tissue. See Asepsis.

Septa [pl. septae] (SEP-tah, SEP-teh) Generally used to describe the walls that segment fungal hyphae and filamentous and branching bacteria.

Septicemia A bacterial infection of the blood. See Viremia.

Serratia marcescens (ser-RAY-she-ah mar-SES-sens) Gram negative, rod-shaped, peritrichously flagellated, facultatively anaerobic bacterium. Colonies form a red pigment at 18°C but not at 37°C.

Sex pilus Pilus that is necessary if a bacterium is to be able to conjugate and donate genetic information to a recipient cell.

Shigella dysenteriae (she-GELL-lah dis-sen-TER-ee-eh) Gram negative, rod-shaped, nonmotile, facultatively anaerobic bacterium. Causes dysentery.

Shigella flexneri (flex-NER-ee) Causes dysentery.

Slime molds Two classes of fungi: Acrasiomycetes (which form pseudoplasmodia) and Myxomycetes (which form plasmodia). The Acrasiomycetes exist as amebae or as pseudoplasmodia (multicellular structures) during one part of their life cycle and form fruiting bodies, similar to molds, during another part. The Myxomycetes exist as plasmodia (single, macroscopic cells with multiple nuclei) during one part of their life cycle and form fruiting bodies when deprived of nutrients.

Solutes The dissolved ions and small molecules in a solution.

Solution A liquid mixture of compounds in which one material, the solvent, uniformly distributes and dissolves other compounds (solutes). Examples are salt in water and ethanol in water.

Solvent A material, usually water, that uniformly distributes and dissolves another material.

Sphaerotilus (sfer-AH-til-us) Gram negative, rod-shaped, chain-forming, sheathed, aerobic bacterium. Individual cells may have subpolar flagella.

Spheroplast Gram negative cell that has lost its peptidoglycan layer. In an isotonic solution, spheroplasts become spherical. Speroplasts can be formed by treating gram negative cells with lysozyme.

Spirillum (spear-RILL-um) Gram negative, spiral, polarly flagellated, aerobic, oxidase positive bacterium.

Spirochete (SPY-row-keet) A group of helical bacteria that have axial filaments rather than flagella. Examples are *Treponema* and *Leptospira.*

Spirogyra (spy-row-JI-rah) A filamentous eukaryotic green alga with spiral-shaped chloroplasts.

Spirulina (spee-rue-LIE-na) A photosynthetic prokaryotic cyanobacterium that forms trichomes. Molecular oxygen is evolved from photosynthesis.

Sporolactobacillus (spore-row-LAK-toe-bah-SILL-us) Gram positive, rod-shaped, peritrichously flagellated, endospore-forming, microaerophilic, catalase negative bacteria that lack cytochromes and ferments by the lactic acid route.

Sporosarcina (spore-row-sar-SIN-ah) Gram positive, spherical, tetrad-forming, endospore forming, nonmotile, aerobic bacterium.

Staphylococcus (staff-fee-low-KOCK-kus) Gram positive, spherical, cluster-forming, nonmotile, facultatively anaerobic bacteria.

Staphylococcus aureus (AH-ree-us) Commonly found in the nose. Causes skin infections and food poisoning.

Staphylococcus epidermidis (eh-pee-der-MID-dis) Commonly found on the skin. A cause of endocarditis.

Sterilization The killing of all microorganisms and infectious agents, such as prions and viroids.

Streptococcus (strep-toe-KOCK-cus) Grampositive,spherical, chain-forming, nonmotile, aerotolerant anaerobic, catalase negative bacteria that lack cytochromes.

Streptococci, Viridans group: *S. salivarius, S. sanguis,* and *S. mutans.* Alpha-hemolytic on blood agar plates. Normally found in the oral cavity. May cause subacute bacterial endocarditis.

Streptococcus agalactiae (a-gah-LACK-tee-eh) Group B. Causes bovine mastitis and infections in newborn infants.

Streptococcus bovis (BOW-vis) Group D nonenterococcus. Causes subacute bacterial endocarditis.

Streptococcus cremoris (kre-MOR-is) Group N. Used in the fermentation of dairy products.

Streptococcus dysgalactiae (des-gah-LAK-tee-eh) Group C. Causes bovine mastitis.

Streptococcus equinus (eh-QUI-nus) Group D nonenterococcus. Causes subacute bacterial endocarditis.

Streptococcus faecalis (fee-KAH-lis) Group D enterococcus. May cause endocarditis.

Streptococcus mitis (MY-tis) Viridans group. May cause subacute endocarditis.

Streptococcus mutans (MU-tans) Viridans group. Responsible for tooth decay.

Streptococcus pneumoniae (new-MOAN-nee-eh) Pneumococcus group. A major cause of pneumonia.

Streptococcus pyogenes (pie-AH-jen-nees) Group A. Causes strep throat, scarlet fever, rheumatic fever, impetigo, skin infections, and puerperal sepsis.

Streptococcus thermophilus (ther-MAH-fill-us) Participates in the fermentation of milk to produce yogurt.

Streptomyces (strep-toe-MY-seez) Gram positive, rod-shaped, mycelia-forming, aerobic bacteria. Many species produce antibiotics.

Substrate level phosphorylation Occurs when a phosphorylated substrate donates the phosphate in the synthesis of ATP from ADP. See Oxidative phosphorylation and fermentation.

Sulfate-reducing bacteria Bacteria that use sulfate as an electron acceptor during anaerobic respiration. Examples include *Desulfovibrio* and *Desulfotomaculum*.

Sulfur-oxidizing bacteria Bacteria that use reduced forms of sulfur, such as hydrogen sulfide, elemental sulfur, and thiosulfate, as a source of reducing electrons or energy. Examples include the photosynthetic bacteria, such as *Chromatium* and *Chlorobium*, and the aerobic *Beggiatoa* and *Thiothrix*.

Systemic infection An infection that has spread throughout the body.

T-cells (T-lymphocytes) Thymus-derived lymphocytes protect animals from intracellular infectious agents, such as viruses and some bacteria and protozoans. T-lymphocytes also modulate the antibody response to infectious agents.

Taenia (TEH-nee-ah) Tapeworms. Cause taeniasis, an infection of the intestines, heart, spinal cord, and brain. The symptoms are diarrhea, abdominal pain and weight loss.

Teichoic acids Large polymers that are attached to the peptidoglycan layer in gram positive bacteria and may contribute as much as 50 percent to the weight of the cell wall. Teichoic acids are polymers of phosphate and molecules such as glycerol or ribitol. When teichoic acids are attached to membrane lipids, they are referred to as lipoteichoic acids.

Tertiary structure In a protein, the final three-dimensional structure that forms from the folding of the polypeptide. The final folding is determined by the disulfide bonds, the hydrogen bonds, and the hydrophobic and hydrophilic interactions of the amino acids with water and with themselves.

Thermoduric Highly resistant to heat. Vegetative cells and many types of spores are thermoduric.

Thiobacillus (thi-oh-ba-SILL-us) Gram negative, rod-shaped, polarly flagellated or nonmotile, aerobic, carbon dioxide–fixing bacteria that oxidize hydrogen sulfide, sulfur, or thiosulfate to sulfate for energy and for reducing electrons.

Thiothrix Gram negative, oval to rod-shaped, filament-forming, aerobic, carbon dioxide–fixing, hydrogen sulfide- and sulfur-oxidizing bacteria that accumulate sulfur granules within the cell.

Tinea Fungal infections of the skin.

Toxoplasma gondii (talks-oh-PLAZ-mah GONE-dee-eye) The protozoan responsible for the disease toxoplasmosis. The protozoan is an intracellular parasite, often infecting various organs of the body. It is often spread to humans from cats.

Trace element Minerals required in very small amounts by living organisms. Examples are zinc, copper, cobalt, and molybdenum.

Transduction The transfer of hereditary material from one cellular organism to another by a virus, with subsequent recombination of the hereditary material with the recipient's genome and the transformation of the recipient.

Tranformation The alteration of an organism's hereditary material. In microbiology, the uptake of naked DNA by an organism and the subsequent recombination of the hereditary material with the organism's genome and alteration of the organism's genetics and physiology.

Treponema carateum (treh-po-NEE-mah kah-rah-TEH-um) Gram negative, long helical, axially filamented (motile), anaerobic, catalase negative bacterium. Causes pinta, a skin disease that resembles syphilis. Generally found in tropical America.

Treponema pallidium (PAL-lee-dum) The bacterium responsible for the venereal disease syphilis.

Treponema pertenue (per-TEH-new) Causes yaws, a skin disease that resembles syphilis. Generally found in tropical Africa.

Tricarboxylic acid cycle See Citric acid cycle and Krebs cycle.

Trichome A filament of cells.

Trichomonas vaginalis (trick-ka-MO-nas vah-jin-NAH-lis) A flagellated protozoan that infects the urinary and genital tract and is responsible for the venereal disease trichomoniasis.

Trichophyton (trick-KO-fee-ton) A fungus that cuases ringworm (tinea).

Trophic level Feeding level. Relative position of a population in a food chain.

Trophozoites An infectious vegetative (growing and reproducing) form of some protozoans.

Trypanosoma cruzi (try-pan-oh-SO-mah CREW-zee) The flagellated protozoan responsible for Chagas' disease.

Trypanosoma brucei gambiense (bru-SAY-ee gam-bee-EN-seh) A flagellated protozoan responsible for African sleeping sickness.

Trypanosoma brucei rhodesiense (row-DEE-zee-EN-seh) A flagellated protozoan responsible for African sleeping sickness.

Tubercle A mass of cells (granuloma) that develops in the lungs. The mass may contain the tubercle bacillus (*Mycobacterium tuberculosis*).

Ureaplasma urealyticum (you-ree-ah-PLAS-mah you-ree-oah-LEET-eit-cum) A mycoplasma. Gram negative, spherical, nonmotile, wall-less, anaerobic, catalase negative bacterium. Causes urethritis.

Vaccination The inoculation of an animal or human with microorganisms or material from microorganisms in order to induce an immune response that protects the animal or human from infections by a specific microorganism.

Variolation Vaccination against smallpox, using the smallpox virus.

Vector (1) The organism that carries and transmits a disease-causing organism. (2) The DNA (plasmid or virus) that is used to carry and clone specific pieces of DNA.

Vegetative cell A cell that grows and divides.

Veillonella (veye-lon-NEL-ah) Gram negative, spherical, nonmotile, anaerobic, carbon dioxide-requiring bacteria, unable to ferment carbohydrates or polymeric alcohols, with complex growth requirements.

Vibrio cholerae (VEE-bree-oh KAW-ler-eh) Gram negative, curved or rod-shaped, polarly flagellated, facultatively anaerobic, oxidase positive bacterium. Responsible for cholera.

Vibrio parahaemolyticus (pah-rah-heh-mo-LIT-tee-kus) A cause of gastroenteritis usually acquired from seafoods.

Viremia (vy-REE-me-ah) A virus infection of the blood. See Septicemia.

Virion A virus.

Viroids Naked, double-stranded RNA molecules that infect many different types of plant cells.

Virus Protein-covered nucleic acids that infect all types of cells.

Voges-Proskauer test (VOGE-PROS-cow-er) A chemical test for acetoin (acetylmethylcarbinol), which indicates whether or not a bacterium carries out the 2,3-butanediol fermentation. One of the IMViC tests for differentiating among enterics.

Vorticella (vor-tee-SELL-ah) A ciliated protozoan common in pond water.

Wild-type The type found in nature. The organism from which mutants (which have growth requirements, or cannot utilize a carbohydrate) have been derived in the laboratory.

Wuchereria bancrofti (wu-cher-AIR-ee-ah, ban-KROF-tee) A nematode (roundworm), transmitted by mosquito, that infects the lymphatics and connective tissue. In chronic cases, it may cause elephantiasis.

Yeast Fungi that usually exist as single-celled organisms. They may form short hyphae, but a mycelium never develops.

Yersinia pestis (i-er-SIN-nee-ah PES-tis) Gram negative, rod-shaped, nonmotile, facultatively anaerobic bacterium. Causes the plague.

Zygomycetes (zy-go-my-SEE-tees) A class of fungi that produces sexual spores called zygospores and asexual spores called basidiospores. An example is *Rhizopus*.

INDEX

PHOTO CREDITS

Chapter 1

Chapter 1 opener: Culver Pictures, Inc. 1.1a left: Z. Skobe/BPS. 1.1a center: Z. Skobe/BPS. 1.1a right: P.W. Johnson & J. McN. Sieburth, Univ. of Rhode Island/BPS. 1.1b left: J.N.A. Lott, McMaster Univ./ BPS. 1.1b left center: P.W. Johnson & J. McN. Sieburth, Univ. of Rhode Island/BPS. 1.1b right center: L.E. Roth, The Univ. of Tennessee/BPS. 1.1b right: J.N.A. Lott, McMaster Univ./BPS. 1.1c: Dr. Dennis E. Feely, Univ. Nebraska Medical Center and Dr. Stanley Erlandsen, U of MN. 1.1d center: G.T. Cole, Univ. of Texas-Austin/ BPS. 1.1d right: G.T. Cole, Univ. of Texas-Austin/BPS. 1.1e left and center: A.N. Broers, B.J. Panessa, and J.F. Gennaro, Jr. 1.4 and 1.5a: Culver Pictures, Inc. 1.5b and c: BPS. 1.8: Culver Pictures, Inc. 1.10: National Library of Medicine. 1.12a: Culver Pictures, Inc. 1.12b, 1.13a, and 1.13b: National Library of Medicine.

Chapter 2

Chapter 2 opener: American Soc. for Microbiology. 2.3a, b, c, and d: J.R. Waaland, Univ. of Washington/BPS. 2.4: Centers for Disease Control, Atlanta. 2.6a: S.C. Holt, Univ. of Texas Health Science Center, San Antonio/BPS. 2.6b: Z. Skobe/BPS. 2.7a: M.E. Bayer, Inst. for Cancer Research, Philadelphia. 2.7b: M.W. Steer and E.H. Newcomb, Univ. of Wisconsin-Madison/BPS. 2.8: C.L. Sanders/BPS. 2.11a right: S.C. Holt, Univ. of Texas Health Science Center, San Antonio/BPS. 2.11b right: S.W. Watson, Woods Hole Oceanographic Inst. 2.11c right: S.C. Holt, Univ. of Texas Health Science Center, San Antonio/BPS. 2.13 and 2.14: T.J. Beveridge, Univ. of Guelph/ BPS. 2.15: R.G.Mulder and M.H. Deinema, in *The Prokaryotes* (Starr, Stolp, Balows, and Schegel, eds.), Springer-Verlag, 1981. 2.16b and c: American Soc. for Microbiology. 2.18a: J.N.A. Lott, McMaster University./BPS. 2.18b: H.S. Pankratz, Michgian State Univ./ BPS. 2.19: S. Abraham and E.H. Beachey, V.A. Medical Center, Memphis, TN. 2.20: R. Kavenoff, Designergenes Posters, Ltd./BPS. 2.21: R. Welch, Univ. of Wisconsin Medical School/BPS. 2.23a: T.J. Beveridge, Univ. of Guelph/BPS. 2.23b: S.C. Holt, Univ. of Texas Health Science Center, San Antonio/BPS. 2.25: R. Blakemore and N. Blakemore. 2.26: Courtesy of Dr. D.T. John, Department of Microbiology and Immunology, Oral Roberts University School of Medicine. 2.27a: Courtesy of Dr. José F. Fahrni, Department of Animal Biology, University of Geneva. 2.27b and c: W.L. Dentler, The Univ. of Kansas/BPS. 2.29b: P.W. Johnson & J. McN. Sieburth, Univ. of Rhode Island/BPS. 2.30a: E.H. Newcomb, Univ. of Wisconsin-Madison/BPS. 2.32: Osborn, Webster, and Weber, *J. Cell Biol.* 77:R27-R34 (1978).

Chapter 3

Chapter 3 opener: P.W. Johnson & J. McN. Sieburth, Univ. of Rhode Island/BPS.

Chapter 4

Chapter 4 opener: Centers for Disease Control, Atlanta. 4.6: R. Cano. 4.7a: Centers for Disease Control, Atlanta. 4.7b and d: E.H. Runyon, V.A. Hospital, Salt Lake City, UT. 4.7c: R.H. Hawley, Holy Cross Hospital of Silver Spring, MD. 4.11: New Brunswick Scientific Co., Inc.

Chapter 5

Chapter 5 opener: H.W. Jannasch, Woods Hole Oceanographic Inst. 5.4: R. Humbert/BPS. 5.9b: H.W. Jannasch, Woods Hole Oceanographic Inst.

Chapter 6

Chapter 6 opener: R. Kavenoff, Designergenes Posters, Ltd./BPS. 6.17: J. Carnahan and C. Brinton, Univ. of Pittsburgh.

Chapter 7

Chapter 7 opener: R. Cano. 7.4a, b, c, and d: Wards Natural Science Establishment, Inc. 7.5: Centers for Disease Control, Atlanta. 7.6: T. Huber/BPS. 7.10a, b, c, d, and e: Vitek Systems, McDonnell Douglas Health Systems Co.

Chapter 8

Chapter 8 opener: K. Stephens, Stanford Univ./BPS. 8.1a: P.W. Johnson & J. McN. Sieburth, Univ. of Rhode Island/BPS. 8.1b: S.C. Holt, Univ. of Texas Health Science Center, San Antonio/BPS. 8.1c: P.W.

Johnson & J. McN. Sieburth, Univ. of Rhode Island/BPS. 8.2a: R. Cano. 8.2b: L. Thomashow, Washington State Univ./BPS. 8.2c: T.J. Beveridge, Univ. of Guelph/BPS. 8.3: J.J. Cardamone, Jr., Univ. of Pittsburgh/BPS. 8.4: J.G. Hancock & M.N. Schroth, *Science,* 216: 1378 © 1982 by the AAAS. 8.5a & b: Centers for Disease Control, Atlanta. 8.6: W. Burgdorfer, Rocky Mountain Lab., Hamilton, MT. 8.7a: Courtesy of Dr. J. Aznar. 8.7b: R.C. Cutlip, National Animal Disease Center. 8.8a: M.G. Gabridge, W. Alton Jones Cell Sci. Ctr., Lake Placid, NY. 8.8b: Courtesy of Prof. Karl Maramorosch, Rutgers Univ. 8.9a: Z. Skobe/BPS. 8.9b: D. Selinger and W.P. Reed, V.A. Medical Center, Albuquerque, NM. 8.10a, b, c: Centers for Disease Control, Atlanta. 8.11: Centers for Disease Control, Atlanta. 8.12a: Courtesy of Microbiology Department, University Hospital, Seville, Spain. 8.12b: R. Cano. 8.13a: Centers for Disease Control, Atlanta. 8.13b: Courtesy of Microbiology Department, University Hospital, Seville, Spain. 8.13c: J.J. Duda and J.M. Slack. *J. Gen. Microbiol.* 71:63-68 (1972). 8.14a, b, and c: J.R. Waaland, Univ. of Washington/BPS. 8.15: N.J. Lang, Univ. of California, Davis/BPS. 8.16a: P.W. Johnson & J. McN. Sieburth, Univ. of Rhode Island/BPS. 8.16b and c: S.C. Holt, Univ. of Texas Health Science Center, San Antonio/BPS. 8.16d: Peter Hirsch, Michigan State Univ. 8.17a: P.W. Johnson & J. McN. Sieburth, Univ. of Rhode Island/BPS. 8.17b: K. Stephens, Stanford Univ./BPS. 8.18: R.G. Mulder & M.H. Deinema, in *The Prokaryotes* (Starr, Stolp, Truper, Balows, & Schegel, eds.) Springer-Verlag 1981. 8.19a: T.J. Beveridge, Univ. of Guelph/BPS. 8.19b: R.L. Moore, BioTechniques Labs./BPS.

Chapter 9

Chapter 9 opener: G.T. Cole, Univ. of Texas-Austin/BPS. 9.1: G.T. Cole, Univ. of Texas-Austin/BPS. 9.2a, b, c, and d: G.T. Cole, Univ. of Texas-Austin/BPS. 9.3a: Centers for Disease Control, Atlanta. 9.3b, c, and d: R. Cano. 9.7a: G.T. Cole, Univ. of Texas-Austin/BPS. 9.7b: R. Cano. 9.8a, b, and c: Shirley R. Sparling, CA Polytechnic State U. 9.12a: J.R. Waaland, Univ. of Washington/BPS. 9.12b: C. Robinow, Univ. of Western Ontario. 9.12c: G.T. Cole, Univ. of Texas-Austin/BPS. 9.13: R. Cano. 9.15: Reprinted by permission from *Mycologia* 55: 35-38 © 1963, C.E. Bracker & E.E. Butler & The N.Y. Botanical Garden. 9.16a: J.R. Waaland, Univ. of Washington/BPS. 9.17: Carolina Biological Supply Co.

Chapter 10

Chapter 10 opener: J.J. Duda and J.M. Slack, *J.Gen. Microbiol.* 71:63-68 (1972). 10.3: R.N. Band and H.S. Pankratz, Michigan State Univ./BPS. 10.4b: R. Cano. 10.5a: Centers for Disease Control, Atlanta. 10.5b: Lawrence Ash. 10.5c: I. Armstrong, Washington Hospital Center, Washington, DC. 10.7: From Vetterling, J. Protozoology. *18*:248-60, 1971. 10.8a: M.L. Chiappino, B.A. Nichols, and G.R. O'Connor. J. Protozoology. *31*:288-292, 1984. 10.8b and c: R. Cano. 10.9a: Centers for Disease Control, Atlanta. 10.9b and c: Lawrence Ash. 10.10: P.W. Johnson & J. McN. Sieburth, Univ. of Rhode Island/BPS. 10.11: R. Cano. 10.12b and c: Centers for Disease Control, Atlanta. 10.14a: R.K. Burnard/BPS. 10.14b: Carolina Biological Supply Co. 10.14c: Wards Natural Science Establishment, Inc. 10.17a: Wards Natural Science Establishment, Inc. 10.17b: Reprinted with permission of ASCP Press, a division of the American Society of Clinical Pathologists. 10.18a and b: Wards Natural Science Establishment, Inc. 10.19: Carolina Biological Supply Co. 10.20a and b: Reprinted with permission of ASCP Press, a division of the American Society of Clinical Pathologists. 10.22a and b: R. Cano. 10.23a: Courtesy of Professor E.J. Perea. 10.23b: Courtesy of Professor F. Camacho, Department of Dermatology and Venereology, University Hospital, Seville, Spain. 10.24a: Centers for Disease Control, Atlanta. 10.24b and c: R. Cano.

Chapter 11

Chapter 11 opener: G.H. Smith, National Cancer Institute. 11.2a: R. Humbert/BPS. 11.2b: R. Cano. 11.2c: Centers for Disease Control, Atlanta. 11.3a: L. Caro and R. Curtiss. 11.3b: L. Simon, Rutgers U. 11.3c: J.D. Griffith, Cancer Research Center, U of NC. 11.4: L. Simon, Rutgers U. 11.10b: G.H. Smith, National Cancer Institute. 11.11b top and bottom: G. Wertz, Univ. of North Carolina School of Medicine/BPS. 11.12: T.O. Diener, U.S. Dept. of Agriculture. 11.13: S.B. Prusiner, et al. *Cell* 35:353 © by MIT Press, 1983.

Chapter 12

Chapter 12 opener: G.T. Cole, Univ. of Texas-Austin/BPS. 12.1: G.T. Cole, Univ. of Texas-Austin/BPS. 12.2: J.P. Rippon, *Medical Mycology,* 2e, W.B. Saunders Co., 1982. 12.4: J.W. Costerton, Univ. of Calgary. 12.5a: Runk/Schoenberger, Grant Heilman Photography. 12.5b: Luvenia Miller, Camera MD Studios. 12.6a: T.J. Beveridge, Univ. of Guelph/BPS. 12.8a and b: R. Rodewald, Univ. of Virginia/BPS. 12.8c: J.J. Cardamone, Jr., Univ. of Pittsburgh/BPS. 12.9b: R. Cano. 12.10a, b, and c: C.L. Sanders/BPS.

Chapter 13

Chapter 13 opener: R. Rodewald, Univ. of Virginia/BPS. 13.3a, b, and c: R. Rodewald, Univ. of Virginia/BPS. 13.14a right: L.J. Le Beau, Univ. of Illinois Hospital/BPS. 13.14b right: J.R. Barrett, *Textbook of Immunology,* 4e, C.V. Mosby, 1983. 13.16a: BPS.

Chapter 14

Chapter 14 opener: L. Winograd, Stanford Univ./BPS.14.1: Baylor College of Medicine, Houston, TX. 14.4a and b: Cedric S. Raine, Albert Einstein College of Medicine, Bronx, NY. 14.5a: Armed Forces Inst. of Pathology 57-15160-20. 14.5b: Armed Forces Inst. of Pathology 54-1548-3. 14.6a: L. Winograd, Stanford Univ./BPS. 14.7: Courtesy of Professor F. Camacho, Department of Dermatology and Venereology, University Hospital, Seville, Spain.

Chapter 15

Chapter 15 opener: Courtesy of Dr. J. L. Corral Arias. 15.4a and b: © 1985 Chris Grajczyk. 15.11a, b, and d: Courtesy of Pediatrics Department, Division of Infectious Diseases, University Hospital, Seville, Spain. 15.11c and 15.12: Courtesy of Dr. J.L. Corral Arias.

Chapter 16

Chapter 16 opener: Centers for Disease Control, Atlanta. 16.2a and c: Centers for Disease Control, Atlanta. 16.2b: Courtesy of Professor E.J. Perea. 16.3a and b: Courtesy of Professor F. Camacho, Department of Dermatology and Venereology, University Hospital, Seville, Spain. 16.3c: Centers for Disease Control, Atlanta. 16.4a: L. Winograd, Stanford Univ./BPS. 16.4b and c: Courtesy of Professor F. Camacho, Department of Dermatology and Venereology, University Hospital, Seville, Spain. 16.4d: Armed Forces Inst. of Pathology DL-4191. 16.4e: Charles Stoer, Camera MD Studios. 16.5a and b: California Polytechnic State U. 16.7a: L.M. Pope and D.R. Grote, Univ. of Texas, Austin/BPS. 16.7b and c: Courtesy of Microbiology Department, University Hospital, Seville, Spain. 16.8a: Courtesy of Professor E.J. Perea. 16.8b: Courtesy of Dr. J. Aznar. 16.9a and b: Centers for Disease Control, Atlanta. 16.9c and d: Courtesy of Professor F. Camacho, Department of Dermatology and Venereology, University Hospital, Seville, Spain. 16.10: G.T. Cole,

Univ. of Texas-Austin/BPS. 16.11a: Courtesy of Professor F. Camacho, Department of Dermatology and Venereology, University Hospital, Seville, Spain. 16.11b: J.W. Rippon. 16.12: Centers for Disease Control, Atlanta. 16.13: Courtesy of Professor F. Camacho, Department of Dermatology and Venereology, University Hospital, Seville, Spain. 16.14 a and b: Courtesy of Pediatrics Department, Division of Infectious Diseases, University Hospital, Seville, Spain. 16.15a and b: Courtesy of Professor F. Camacho, Department of Dermatology and Venereology, University Hospital, Seville, Spain. 16.17: Courtesy of Pediatrics Department, Division of Infectious Diseases, University Hospital, Seville, Spain.

Chapter 17

Chapter 17 opener: K.E. Muse/BPS. 17.3: J.W. Rippon. 17.5b: Courtesy of Dr. L.L. Corral Arias. 17.5c: Courtesy of Microbiology Department, University Hospital, Seville, Spain. 17.6: Courtesy of Dr. L.L. Corral Arias. 17.7a: K.E. Muse/BPS. 17.9a: R.B. Morrison, M.D., Austin, TX. 17.9b: R. Cano. 17.11a: J.W. Rippon. 17.11b and c: Centers for Disease Control, Atlanta. 17.12: Courtesy of the Biological Sciences Dept., CA Polytechnic State University, San Luis Obispo.

Chapter 18

Chapter 18 opener: D.C. Savage, Univ. of Illinois. 18.2b: D.C. Savage, Univ. of Illinois, Urbana. 18.3d and e: Z. Skobe/BPS. 18.4b: Camera M.D. Studios. 18.7a and b: G.T. Cole, Univ. of Texas-Austin/BPS. 18.11a: Armed Forces Inst. of Pathology 75-11243. 18.12: From R.L. Owen, et al. *Gastroenterology,* 76:757-759 (1979).

Chapter 19

Chapter 19 opener: Centers for Disease Control, Atlanta. 19.3: Courtesy of Professor F. Camacho, Department of Dermatology and Venereology, University Hospital, Seville, Spain. 19.4: R. Cano. 19.7a: Courtesy of Professor F. Camacho, Department of Dermatology and Venereology, University Hospital, Seville, Spain. 19.7b: Centers for Disease Control, Atlanta. 19.9: Martin M. Rotker/Taurus Photos. 19.11a-c: Courtesy of Professor F. Camacho, Department of Dermatology and Venereology, University Hospital, Seville, Spain. 19.12a: Armed Forces Inst. of Pathology 54-2488. 19.12b: L. Winograd, Stanford Univ./BPS. 19.14a: Courtesy of Professor F. Camacho, Department of Dermatology and Venereology, University Hospital, Seville, Spain. 19.14b: Courtesy of Professor E.J. Perea. 19.15: Courtesy of Dr. J. Aznar. 19.16: Courtesy of Professor F. Camacho, Department of Dermatology and Venereology, University Hospital, Seville, Spain. 19.17a: Carroll H. Weiss, Camera M.D. Studios. 19.17b: Harvey Blank, Camera M.D. Studios. 19.17c: Armed Forces Inst. of Pathology D-45421-1.

Chapter 20

Chapter 20 opener: Armed Forces Inst. of Pathology 57-17776. 20.2: Courtesy of Pediatrics Dept., Division of Infectious Diseases, University Hospital, Seville, Spain. 20.5: The Bettmann Archive. 20.9: The Bettmann Archive. 20.12a: Armed Forces Inst. of Pathology 55-8225. 20.12b: Armed Forces Inst. of Pathology 57-17776. 20.13: Centers for Disease Control, Atlanta.

Chapter 21

Chapter 21 opener: R.K. Burnard/BPS. 21.3A: Courtesy of Microbiology Department, University Hospital, Seville, Spain. 21.3b: L.J. Le Beau, Univ. of Illinois Hospital/BPS. 21.5: R. Cano. 21.6: Armed Forces Institute of Pathology. 21.11a: Centers for Disease Control, Atlanta, 21.11b: I. Armstrong, Washington Hospital Center, Washington, DC. 21.13b: M.L. Chiappino, B.A. Nichols, and G.R. O'Connor. J.Protozoology. *31*:288-292, 1984. 21.14: P.J. Fenton, *Arch. Ophthal.* 76:867, © 1966, Am. Medical Association. 21.15a: Stephen Lerner, Univ. of Chicago School of Medicine. 21.15b: World Health Organization. 21.16: R.K. Burnard/BPS.

Chapter 22

Chapter 22 opener: R. Cano.

Chapter 23

Chapter 23 opener: H.E. Evans, Oregon State Univ. 23.3a: H.E. Evans, Oregon State Univ. 23.3b: E.H. Newcomb, Univ. of Wisconsin-Madison/BPS. 23.6: H.W. Jannasch, Woods Hole Oceanographic Inst. 23.7a and b: Walter Reed Army Inst. of Research, Washington D.C. 23.11: M.J. Vinkesteyn, National Museum of Antiquities, Leyden, Holland.

Chapter 24

Chapter 24 opener: Pfizer, Inc. 24.2: Pfizer, Inc.

Contents photos

Chapters 1, 11, 12, 13, 17, 19: Sandra Silvers, Assistant Director of the EM Center, BioScience Dept., Florida State University; color by Kidd & Company. Chapter 6: Courtesy of Oscar Miller (from Miller and Beatty, Science 164: 955-957, 1969) and D.W. Fawcett. Color by Kidd & Company. Chapter 14: Courtesy of Emma Shelton, Jan Orenstein, and D.W. Fawcett. Color by Kidd & Company. Chapter 18: Courtesy of Sanford Palay and D.W. Fawcett. Color by Kidd & Company. Chapter 21: From D.W. Fawcett, *The Cell: An Atlas of Fine Structure*, 2d ed., Philadelphia, W.B. Saunders Co., 1981. Reprinted by permission. Color by Kidd & Company. Chapter 24: Courtesy of J.R. Paulson, U.K. Laemmli, and D.W. Fawcett. Color by Kidd & Company. Chapter 2: Tore Johnson, Woodfin Camp & Associates.